Tim L. McFall
1002 S. 8th

MICROBIOLOGY

An Introduction

The Benjamin/Cummings Series in the Life Sciences

F. J. Ayala
Population and Evolutionary Genetics: A Primer (1982)

F. J. Ayala and J. A. Kiger, Jr.
Modern Genetics (1980)

F. J. Ayala and J. W. Valentine
Evolving: The Theory and Processes of Organic Evolution (1979)

M. G. Barbour, J. H. Burk, and W. D. Pitts
Terrestrial Plant Ecology (1980)

L. L. Cavalli-Sforza
Elements of Human Genetics, second edition (1977)

R. E. Dickerson and I. Geis
Hemoglobin (1982)

P. B. Hackett, J. A. Fuchs, and J. Messing
Recombinant DNA Techniques (1983)

L. E. Hood, I. L. Weissman, W. B. Wood, and J. H. Wilson
Immunology, second edition (1983)

L. E. Hood, J. H. Wilson, and W. B. Wood
Molecular Biology of Eucaryotic Cells (1975)

J. B. Jenkins
Introduction to Human Genetics (1983)

T. Johnson and C. L. Case
Microbiology Lab Experiments (1983)

A. L. Lehninger
Bioenergetics: The Molecular Basis of Biological Energy Transformations, second edition (1971)

S. E. Luria, S. J. Gould, and S. Singer
A View of Life (1981)

E. N. Marieb
Human Anatomy and Physiology Laboratory Manual: Cat and Fetal Pig editions (1981)

E. B. Mason
Basic Human Physiology (1983)

D. Rayle, K. Johnson, and H. Wedberg
Introductory Biology (1983)

A. P. Spence
Basic Human Anatomy (1982)

A. P. Spence and E. B. Mason
Human Anatomy and Physiology, second edition (1983)

G. J. Tortora, B. R. Funke, and C. L. Case
Microbiology: An Introduction (1982)

J. D. Watson
Molecular Biology of the Gene, third edition (1976)

I. L. Weissman, L. E. Hood, and W. B. Wood
Essential Concepts in Immunology (1978)

N. K. Wessells
Tissue Interactions and Development (1977)

W. B. Wood, J. H. Wilson, R. M. Benbow, and L. E. Hood
Biochemistry: A Problems Approach, second edition (1981)

MICROBIOLOGY

An Introduction

Gerard J. Tortora
Bergen Community College

Berdell R. Funke
North Dakota State University

Christine L. Case
Skyline College

The Benjamin/Cummings Publishing Company, Inc.
Menlo Park, California • Reading, Massachusetts
London • Amsterdam • Don Mills, Ontario • Sydney

Sponsoring Editor: James W. Behnke
Developmental Editor: Jane R. Gillen
Production Editor: Patricia S. Burner
Copy Editor: Janet Greenblatt
Book and Cover Designer: Michael A. Rogondino
Artist: Barbara Haynes
Illustrator: Michael Fornalski
Photo Researcher: Carl W. May

About the cover: A technician is isolating plasmids, which are tiny circles of DNA found in bacteria. The plasmids are dissolved in a dye solution that fluoresces pink under ultraviolet light. Genetic engineering using plasmids is revolutionizing the biological sciences and industry (see pages 226–229 and 704–707).

Figure acknowledgments begin on page 749.

Library of Congress Cataloging in Publication Data

Tortora, Gerard J.
Microbiology: an introduction.

Bibliography: p.
Includes index.
1. Microbiology. I. Funke, Berdell R.
II. Case, Christine L., 1948–
III. Title.
QR41.2.T67 576 81-21712
ISBN 0-8053-9310-2 AACR2

cdefghij-DO-89876543

The Benjamin/Cummings Publishing Company, Inc.
2727 Sand Hill Road
Menlo Park, California 94025

Preface

Microbiology: An Introduction is a comprehensive text for students in a wide variety of programs, including allied-health sciences, biological sciences, environmental studies, animal science, forestry, agriculture, home economics, and liberal arts. It is a beginning text, assuming no previous study of biology or chemistry.

OUR APPROACH: SUPPORTING PRINCIPLES WITH APPLICATIONS

In writing this text, we have tried to achieve an appropriate balance between microbiological fundamentals and applications, and between medical applications and other applied areas. We have provided the solid grounding in fundamental facts and principles necessary to understand and adapt to the rapid developments in microbiology. At the same time, we have integrated applications throughout the text because we know that beginning students benefit from seeing the relevance of microbiology to their respective programs. Since many students who will use this book are in allied-health fields, we have made the text especially comprehensive in medically important areas of microbiology. For example, the principles and applications of immunology have been particularly emphasized. We believe this emphasis is deserved in the light of the tremendous importance of immunology, both as a basic science and as the source of many new valuable tools and techniques for microbiology and the health sciences.

We have also provided an overview of microorganisms in nature and of applied microbiology outside the health sciences. Thus, we believe the text will be useful for students with quite varied interests. We hope that all students who use this text will gain an appreciation of the fascinating diversity of microbial life, the central roles of microorganisms in nature, and the importance of microorganisms in our daily lives.

We have made a special effort to prevent both fundamentals and applications from being as dull as these ponderous words sometimes imply. We have tried to present the material in a logical, readable, and enjoyable manner, giving special attention to the art and photo programs. The artwork has been carefully developed to complement the text exposition. There are more than 250 photographs, including a large number that have never before appeared in a text. We have chosen both state-of-the-art micrographs that dramatically show microbial structure and good examples of conventional micrographs that more closely resemble what is usually seen in a microbiology laboratory. Many of the photos were hand-picked from the Centers for Disease Control's photo files.

Many topics of contemporary concern, such as viruses and cancer, interferon, Legionnaires' disease, drug resistance, and recombinant DNA, are included in the body of the text. Specialized aspects of various timely topics are described in **special topic boxes** throughout the book. There are primarily two types of boxes: *Microbiology in the News* items and reprints from the Centers for Disease Control's *Morbidity and Mortality Weekly Report (MMWR)*. These boxes motivate students by relating microbiology to real-life issues and applications.

SCOPE AND SEQUENCE

The book is divided into five parts. Part I, *Fundamentals of Microbiology*, emphasizes bacteriology. It opens with an introductory chapter giving an historical perspective of microbiology and an overview of the microbial world. It is followed by a chapter providing the chemistry background necessary for understanding microbial activities. In the remainder of this part, the principles of microbial cell structure, metabolism, growth, and genetics are presented, as well as the more applied topics of microscopy, laboratory methods for growing microorganisms, the control of unwanted microbial growth (outside the human body), and genetic engineering.

Part II, *Survey of the Microbial World*, expands upon the distinctive characteristics of the important groups of bacteria, viruses, fungi, protozoans, and helminths. A concise but thorough treatment of medically important helminths is included because these organisms, although not strictly microorganisms, are important agents of human disease but often will not be encountered in another course. The chapter on fungi, protozoans, and helminths may easily be omitted or covered selectively.

The first two chapters of Part III, *Interaction of Microbe and Host*, are devoted to the relationships between microorganisms and humans and how such interactions can lead to disease. Host defenses are then discussed in the next three chapters. Reflecting the increasing importance of immunology today, considerable attention is given to kinds of immunity, characteristics of antigens and antibodies, antibody formation, serology, hypersensitivity, and immune responses to cancer. The last chapter in this part discusses another aspect of defense against harmful microbial invasion, antimicrobial drugs.

Part IV, *Microorganisms and Human Disease*, describes a number of commonly encountered microbial diseases. The various diseases are organized into chapters according to host organ system affected. Each chapter opens with a brief discussion of the functional anatomy of the human organ system and its normal flora, proceeds with separate sections covering bacterial, viral, fungal, protozoan, and helminthic diseases, and concludes with a summarizing table.

Part V, *Microbiology, the Environment, and Human Affairs*, discusses topics that are nonmedical but nonetheless vital to human health and well-being. The chapter on soil and water microbiology discusses the microorganisms that inhabit soil and water and their contributions to biogeochemical cycles, water pollution, and sewage treatment. The chapter on food and industrial microbiology covers food preservation, the traditional roles of microorganisms in food and chemical production, and the new industrial importance of microorganisms resulting from recombinant DNA research.

COURSE SEQUENCES

We have organized the book in what we feel is a useful fashion, but we recognize that there are a number of alternative sequences in which the material might be effectively presented. For those who wish to follow another sequence, we have made each chapter as independent as possible and have included numerous cross-references. Thus, the survey of the microbial world, Part II, could be studied at the beginning of a course, immediately after Chapter 1. Or environmental and industrial microbiology, Part V, could follow Parts I and II. Since Chapters 7 and 18 both deal with the control of microbial growth, they could be covered together. The material on microbes and human disease, Part IV, readily lends itself to rearrangement or selective coverage. The Instructor's Guide provides detailed guidelines for organizing the disease material in several alternative ways.

IN-TEXT LEARNING AIDS

A major goal of writing this text was to create a book that would be an effective tool for learning. Therefore, we have included many student aids in each chapter.

- **Objectives** provide students with guidelines for what they should know after studying the chapter.
- **Tables** summarize, organize, and complement the text discussions.

- **Study Outlines** at the end of each chapter aid review.
- **Study Questions** test recall of information presented in the chapter *(Review Questions)* and provide an opportunity to apply knowledge in problem-solving and interpretation *(Challenge Questions)*.
- **Further Reading** suggestions give sources for further investigation of the topics in the chapter.

Several **appendices** at the end of the book heighten its usefulness. The first appendix is the *Classification of Bacteria According to Bergey's Manual of Determinative Bacteriology, 8th edition*. This is followed by a guide to *Pronunciation of Scientific Names*, which provides basic rules of pronunciation and phonetic pronunciations of genus and species names used in the text. Also included is a guide to *Word Roots Used in Microbiology*, a *Most Probable Numbers Table*, and a brief description of *Methods for Taking Clinical Samples*. A **Glossary** provides definitions of all important terms used in the text.

SUPPLEMENTARY MATERIALS

- A **Study Guide**, by Berdell Funke, is available to help students master and review major concepts and facts from the text. Each chapter of the study guide begins with a chapter summary organized under the text headings. Important terms are printed in boldface and defined, and important figures and tables from the text are included. Following the summary is an extensive self-testing section containing matching questions, fill-in questions, and an answer key.
- An **Instructor's Guide**, by Christine Case, includes many practical suggestions for using the text in a course. *Suggested course outlines* are provided. For presentation of microbial diseases by microbial agent (taxonomic group), mode of transmission, or portal of entry, sequences of topics and pertinent pages are listed. Also included are the *answers to Study Questions*, in a format that can be used for grading, or reproduced and distributed to students for self-study. The final section of the Instructor's Guide is devoted to *test items;* each chapter has two objective tests, each containing 15 questions. The tests can be reproduced and used directly from the Guide.
- **Acetate overhead transparencies** of 50 two-color line drawings from the text are available from the Publisher upon written request.

ACKNOWLEDGMENTS

In the preparation of this textbook, we have benefited from the guidance and advice of a number of microbiology instructors across the country. Six contributors provided early drafts of chapters or portions of chapters. Twenty-eight reviewers offered constructive criticism and suggestions at various stages of manuscript preparation. We gratefully acknowledge our debt to these individuals.

Gerard J. Tortora
Berdell R. Funke
Christine L. Case

Contributors:

Richard Bernstein, San Francisco State University
Frank Binder, Marshall University
Cynthia Moffet, San Francisco
Roger Nichols, Weber State College
J. Dennis O'Malley, University of Arkansas
Leleng To, Hardin-Simmons University

Reviewers:

Barry Batzing, SUNY Cortland
Lois Beishir, Antelope Valley College
Harold Bendigkeit, De Anza College
L. J. Berry, University of Texas
Frank Binder, Marshall University
John Boyd, Napa College
Richard Davis, West Valley College
Norman Epps, University of Guelph
Cindy Erwin, San Francisco City College

Roger Furbee, Middlesex County College
James Garner, C. W. Post College
Ronald Hochede, City College of San Francisco
Robert Janssen, University of Arizona
Diana Kaftan, Diablo Valley College
Alan Konopka, Purdue University
Walter Koostra, University of Montana
John Lammert, Miami University of Ohio
John Lewis, San Bernardino Valley College
Peter Ludovici, University of Arizona
William Matthai, Tarrant County College
Roger Nichols, Weber State College
Elinor O'Brien, Boston College
J. Dennis O'Malley, University of Arkansas
Dennis Opheim, Texas A & M University
Violet Schirone, Suffolk County College
Louis Shainberg, Mount San Antonio College
Bernice Stewart, Prince Georges Community College
John Trela, University of Cincinnati

Brief Table of Contents

Detailed Table of Contents

MICROBIOLOGY

An Introduction

PART

I

FUNDAMENTALS OF MICROBIOLOGY

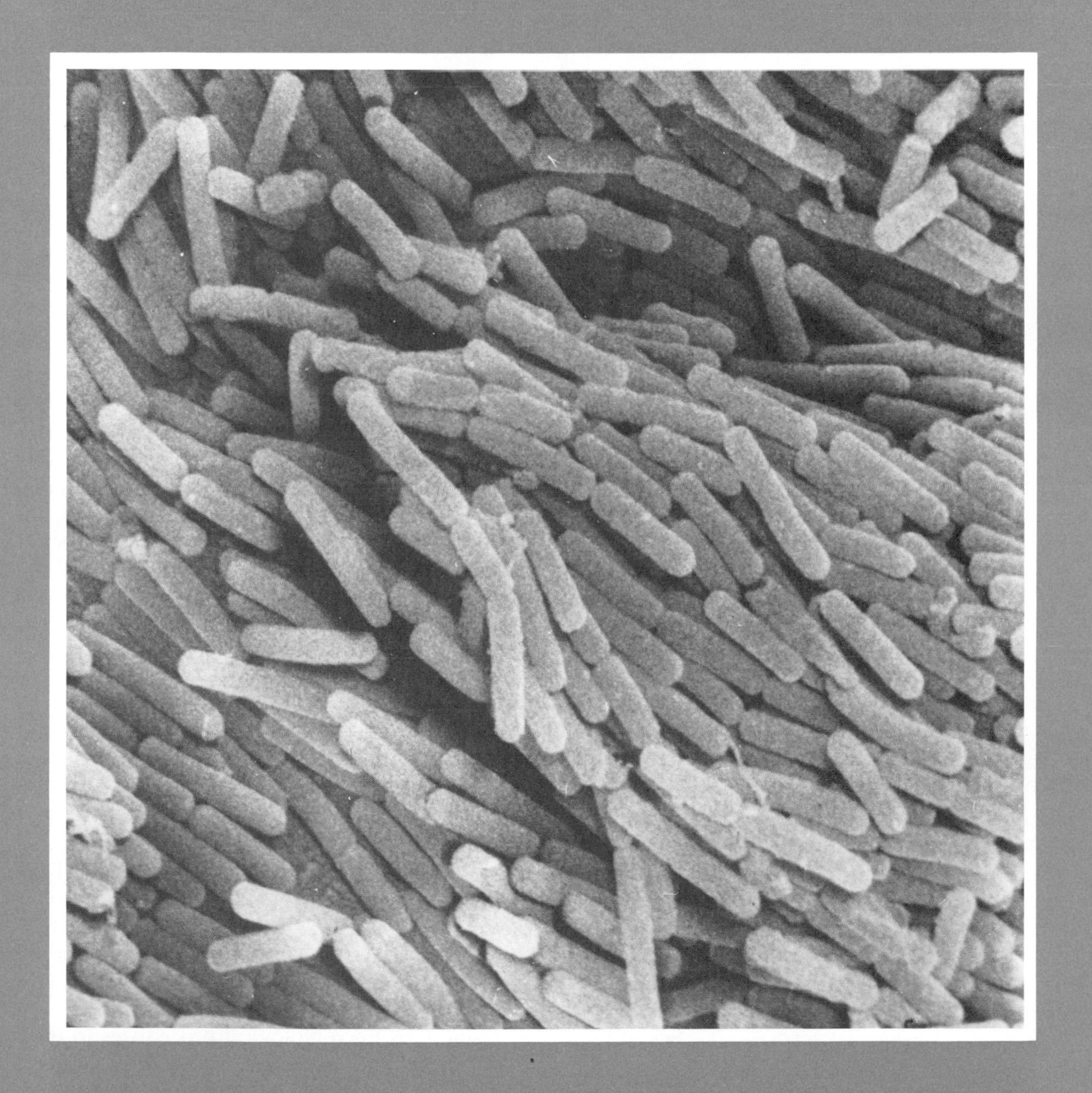

1

The Microbial World and You

Objectives

After completing this chapter you should be able to

- Identify the contributions to microbiology made by:
  Anton van Leeuwenhoek
  Robert Hooke
  Louis Pasteur
  Robert Koch
  Joseph Lister
  Paul Ehrlich
  Alexander Fleming
  Edward Jenner
- Compare the theories of spontaneous generation and biogenesis.
- Provide the rationale for scientific names.
- List the major groups of organisms studied in microbiology.
- List at least four beneficial activities of microorganisms.
- Define normal flora.

One hot summer, 182 guests at an elegant Philadelphia hotel became mysteriously ill. Within a few weeks, 29 had died. All the victims suffered fever, coughing, and pneumonia, and yet the hospital tests could not point to a microorganism or poison as the culprit. Thus, **Legionnaires' disease,** named after the 1976 American Legion Convention held at the hotel during the outbreak, became the subject of an intense investigation. Nationwide, scientists teamed up to find the source of the disease and how it was spread.

Modern microbiological techniques eventually helped reveal that the disease was caused by an airborne microorganism—a tiny bacterium not detectable by standard laboratory tests (Figure 1–1 and Color Plate IIIB). Researchers now believe that guests exposed to the hotel air conditioning system inhaled the bacterium, which had penetrated the system's ventilation filters. The bacterium was not new, only unfamiliar to microbiologists. It had, in fact, caused several earlier unexplained epidemics of pneumonia.

After careful documentation, the bacterium of Legionnaires' disease was assigned the scientific name *Legionella pneumophila,* and the disease itself was later renamed **legionellosis.** About 10% of pneumonias previously classified as "viral" are actually caused by the bacterium *Legionella.* Although it is still not clear which antibiotics are most effective against legionellosis, erythromycin and rifampin have been used successfully to treat a limited number of patients.

Scientists followed standard microbiological procedures in their effort to identify *L. pneumophila* and learn how to control it. You are about to begin your own study of microorganisms like *L. pneumophila.*

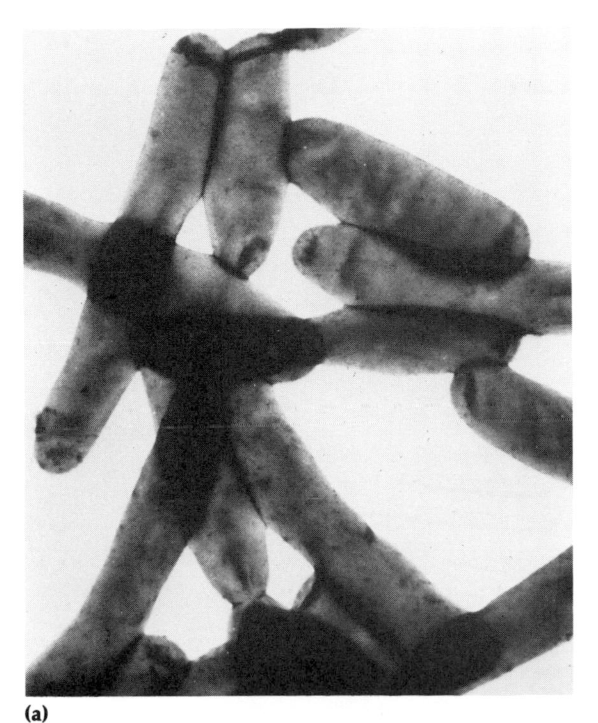

(a) (b)

Figure 1–1 **(a)** *Legionella pneumophila*, the bacterium that causes Legionnaires' disease, as seen under the electron microscope. **(b)** Joseph E. McDade and Charles C. Shepard, the microbiologists who conducted the first successful effort to isolate *Legionella pneumophila* at the Centers for Disease Control in Atlanta, Georgia.

Soon you will become familiar with microscopy techniques and culturing methods used to isolate and grow such microorganisms. And you will learn about the diseases some microbes cause, how the body responds to microbial infection, and how certain drugs combat microbial diseases.

MICROBES IN OUR LIVES

Most of us have had plenty of contact with the subject of this text—and not just from our last cold or bout with influenza. Any cheese or yogurt we eat has already been processed by the controlled activity of microorganisms.

For many people, words like *germ, microbe,* or *microorganism* bring to mind a group of tiny creatures that do not quite answer that old question, "Is it animal, vegetable, or mineral?" Microorganisms are minute living things, individually too small to be seen with the naked eye. They include bacteria, fungi (yeasts and molds), protozoans, and microscopic algae. They also include viruses, those noncellular entities sometimes regarded as being at the border of life and nonlife.

We tend to associate these small organisms with uncomfortable infections and diseases like legionellosis. The truth, however, is that the majority of microorganisms make a vital contribution to the welfare of the world's inhabitants. Many act as miniature benefactors, helping maintain the balance of living organisms and chemicals in our global environment. Marine and freshwater microorganisms are the basis of the food chain in oceans, lakes, and rivers. Soil microbes help break down wastes and incorporate nitrogen gas from the air into organic compounds, thus recycling chemical elements in the land and air. Certain bacteria and microscopic algae play important roles in carrying out the process of photosynthesis, a food- and oxygen-generating process vital to our life on earth. Such microbes are essential for the survival of life on the planet.

Practical knowledge of microbiology is necessary in the fields of medicine and related health sciences. For example, hospital workers must be

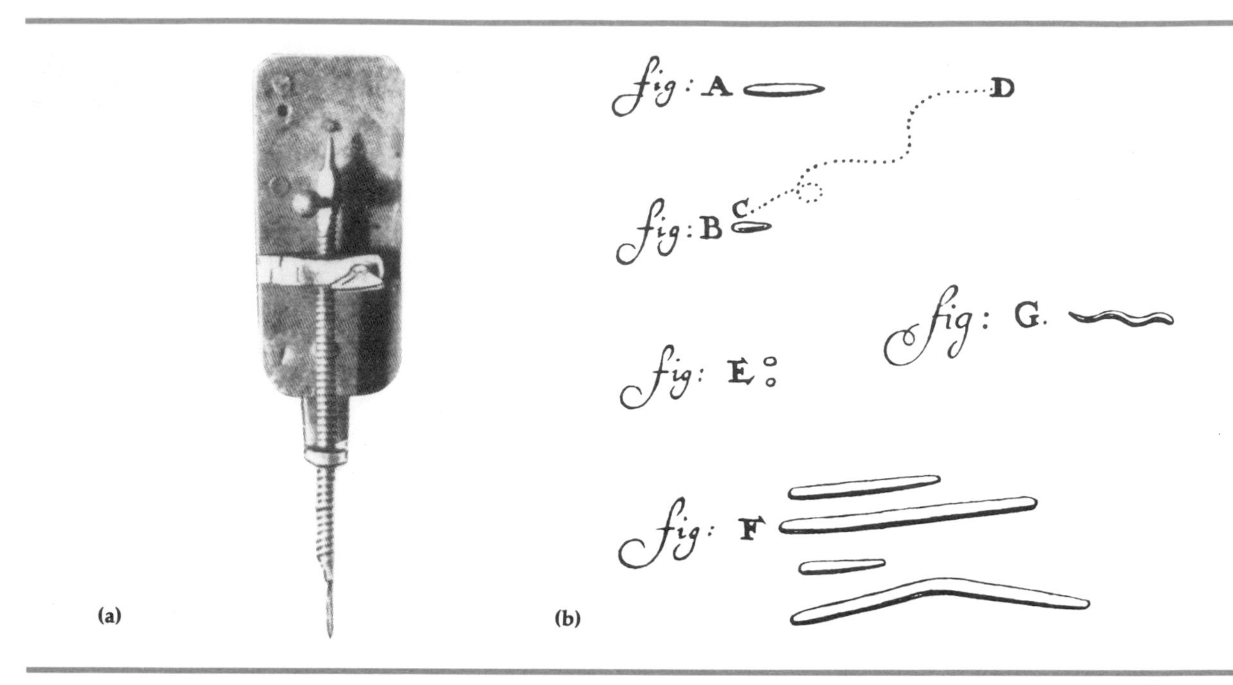

Figure 1–2 **(a)** A simple microscope made by Anton van Leeuwenhoek to observe living organisms too small to be seen with the naked eye. The specimen was placed on the tip of the adjustable point and viewed from the other side through the tiny round lens. The highest magnification possible with his lenses was about 300× (times). **(b)** Some of Leeuwenhoek's drawings of bacteria, made in 1683. He was the first to see the microorganisms we now call bacteria and protozoans.

able to protect patients from bacteria that normally flourish around our homes and work places, but which are more dangerous to the sick and injured. Yet, an antiseptic world would be undesirable—even uncomfortable. We humans (and many other animals) depend on the bacteria in our intestines for the synthesis of a number of vitamins that the body requires.

Microorganisms are not only natural assets for health; they provide commercial benefits as well. Microbiological techniques are used every day by industrial chemists and technicians. It is often easier and cheaper for manufacturers to culture microorganisms for the synthesis of their products than to have chemists synthesize them. Acetone, glycerin, organic acids, enzymes, alcohols, and many drugs are produced this way. And our food industry frequently relies upon microbes for crucial steps in food processing. Microorganisms help produce vinegar, sauerkraut, pickles, alcoholic beverages, buttermilk, cheeses, yogurt, and bread.

Today, we casually accept the fact that microorganisms are found practically everywhere. Yet there was a time, not long ago, when microbes could not be seen because there were no microscopes, when cells were unknown to scientists, and when people believed that life sprang up spontaneously from nonliving matter. In the not-so-distant past, food spoilage often could not be controlled, and entire families died because vaccinations and antibiotics were not available to fight an infection.

Perhaps we can get an idea of how our present concepts of microbiology developed if we look at a few of the breakthroughs that have changed our lives.

A BRIEF HISTORY OF MICROBIOLOGY

The First Observations

Astronomers had been using telescopes to look at the stars for centuries before anyone thought to turn an optic lens upon a drop of water. The Dutch merchant and amateur scientist Anton van Leeuwenhoek was one of the first to observe microorganisms with magnifying lenses. Starting in 1673, he wrote a series of letters to the Royal Society of London describing the "animalcules" he saw through his simple, single-lens microscope. Leeuwenhoek's detailed drawings of "animalcules" in rain water, in peppercorn infusions, and in material taken from teeth scrapings have since been identified as representing bacteria and protozoans (Figure 1–2).

While Leeuwenhoek was observing the microbial world, the Englishman Robert Hooke was using a microscope to investigate thin slices of cork—which is composed of dead plant cell walls. He called the pores between the walls "little boxes" or "cells." His discovery of the structure of cork, reported in 1665, marked the beginning of *a cell theory.*

By 1838–39, two German biologists, botanist Matthias Schleiden and zoologist Theodor Schwann, had consolidated the growing number of observations about cells in living matter. Their work clearly stated the theory that *all living things are composed of cells*. Subsequent investigations into the structure and functions of cells have been based on this cell theory, which is one of the most important generalizations in modern biology. We shall discuss the structure of cells in Chapter 4.

The Debate over Spontaneous Generation

After Leeuwenhoek discovered the "invisible" world of microorganisms, the scientific community became interested in the origin of these tiny, living things. Until the second half of the nineteenth century, it was generally believed that life could arise spontaneously from nonliving matter, a process known as **spontaneous generation.** People thought that toads, snakes, and mice could be born of moist soil, that flies could emerge from manure, and that maggots could arise from decaying corpses.

Experiments pro and con

As a strong opponent of spontaneous generation, the Italian physician Francesco Redi set out in 1668 to demonstrate that maggots, the larvae of flies, do not arise spontaneously from decaying meat, as was commonly believed. Redi filled three jars with decaying meat and sealed them tightly. Then he arranged three similar jars and left them open. Maggots appeared in the open vessels after flies were allowed to enter the jars and lay their eggs. But the sealed containers showed no forms of life (Figure 1–3a). Still, the antagonists were not con-

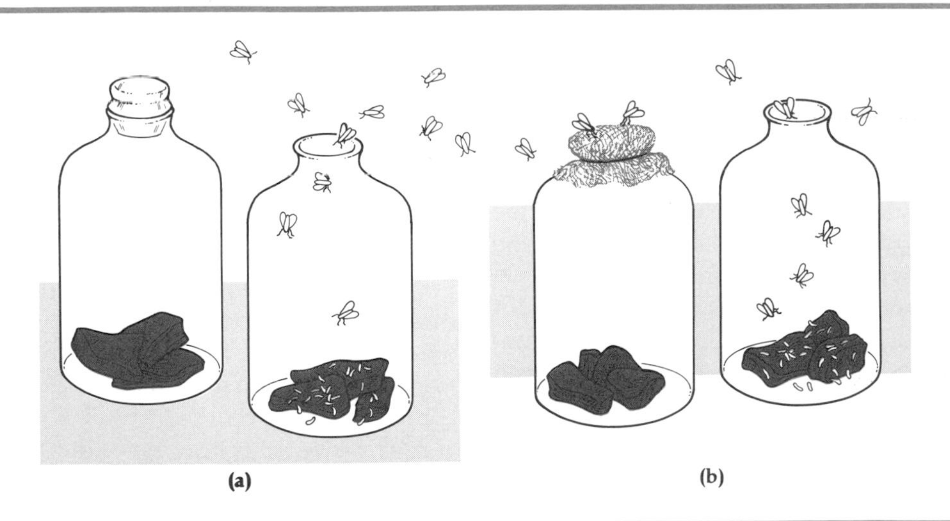

Figure 1–3 Redi's experiments to demonstrate that maggots do not arise spontaneously from decaying meat. **(a)** In his first experiment, Redi showed that maggots (fly larvae) would appear on uncovered meat but not on the meat in a sealed jar. **(b)** In his second experiment, Redi used gauze to cover half the jars, in order to allow the free entrance of air (but not flies). Again, the meat in the covered jars showed no maggots.

vinced, claiming that fresh air was needed for spontaneous generation. So Redi set up a second experiment in which three jars were covered with a fine net instead of being sealed. No worms appeared in the gauze-covered jars, even though air was present. Maggots appeared only if flies were allowed to leave their eggs on the meat (Figure 1–3b).

Redi's results were a serious blow to the long-held concept that larger forms of life could arise from nonlife. However, many scientists still believed that small organisms such as Leeuwenhoek's "animalcules" were simple enough to be generated by nonliving materials.

The case for spontaneous generation of microorganisms was strengthened in 1745 when John Needham found that even after he heated nutrient fluids (chicken broth and corn infusions) before pouring them into covered flasks, the cooled solutions were soon teeming with microorganisms. Needham claimed that the microbes developed spontaneously from the fluids. Twenty years later, Lazzaro Spallanzani suggested that microorganisms from the air probably had entered Needham's solutions after they were boiled, and he showed that nutrient fluids heated *after* being sealed in a flask did not develop microbial growth. Needham responded that the seals on Spallanzani's flasks kept out the "vital force" necessary for spontaneous generation, which, he claimed, had been destroyed by heating.

The theory of biogenesis

With the question still unresolved, the German scientist Rudolf Virchow challenged those who favored spontaneous generation when he introduced the concept of **biogenesis** in 1858. An extension of the cell theory, biogenesis is the claim that living cells can arise only from preexisting living cells. But arguments about spontaneous generation continued a few years more before the issue was resolved experimentally.

Pasteur settles the argument

In 1861, the French chemist Louis Pasteur designed a series of ingenious and persuasive experiments that finally ended the debate. He demonstrated that microorganisms are indeed present in the air and that they can contaminate seemingly sterile solutions, but that air itself does not give rise to microbial life.

Pasteur began by filling several short-necked flasks with beef broth and boiling them. Some were then left open and allowed to cool. In a few days, these flasks were found to be contaminated with microbes. Other flasks, sealed after boiling, remained free of microorganisms. From these test results, Pasteur reasoned that microbes in the air were the agents responsible for contaminating nonliving matter such as the broths in Needham's flasks.

Pasteur's next step was to place the broth in long-necked flasks, after which he bent the necks into S-shaped curves (Figure 1–4). The contents of these flasks were then boiled and cooled. The broth in the flasks did not decay, and there were no signs of life after days, weeks, and even months. (Some of these original vessels are on display at the Pasteur Institute in Paris. They still show no sign of contamination, more than 100 years later.) Pasteur's unique flask design allowed air to pass into the flask while the neck trapped any airborne microorganisms that might contaminate the broth.

Pasteur showed that microorganisms may be present in nonliving matter—on solids, in liquids, and in the air. Further, he demonstrated conclusively that microbial life can be destroyed by heat and that methods can be devised to block the access of airborne microorganisms to nutrient environments. These discoveries form the basis of the **aseptic techniques** (techniques to prevent contamination by unwanted microorganisms) now standard practice in the laboratory and in many medical procedures. Modern aseptic techniques are among the first and most important things that a beginning microbiologist learns.

Pasteur's work provided evidence that microorganisms cannot originate from mystical forces present in nonliving materials. Rather, any appearance of "spontaneous" life in nonliving solutions can be attributed to microorganisms already present in the air or in the fluids themselves. (Keep in mind that while a form of spontaneous generation probably did occur on primitive earth when life first began, scientists now agree that it does not happen under our present environmental conditions.)

Fermentation

After debunking the popular notion of spontaneous generation, Pasteur acquired a reputation as a microbiological problem solver. It thus came

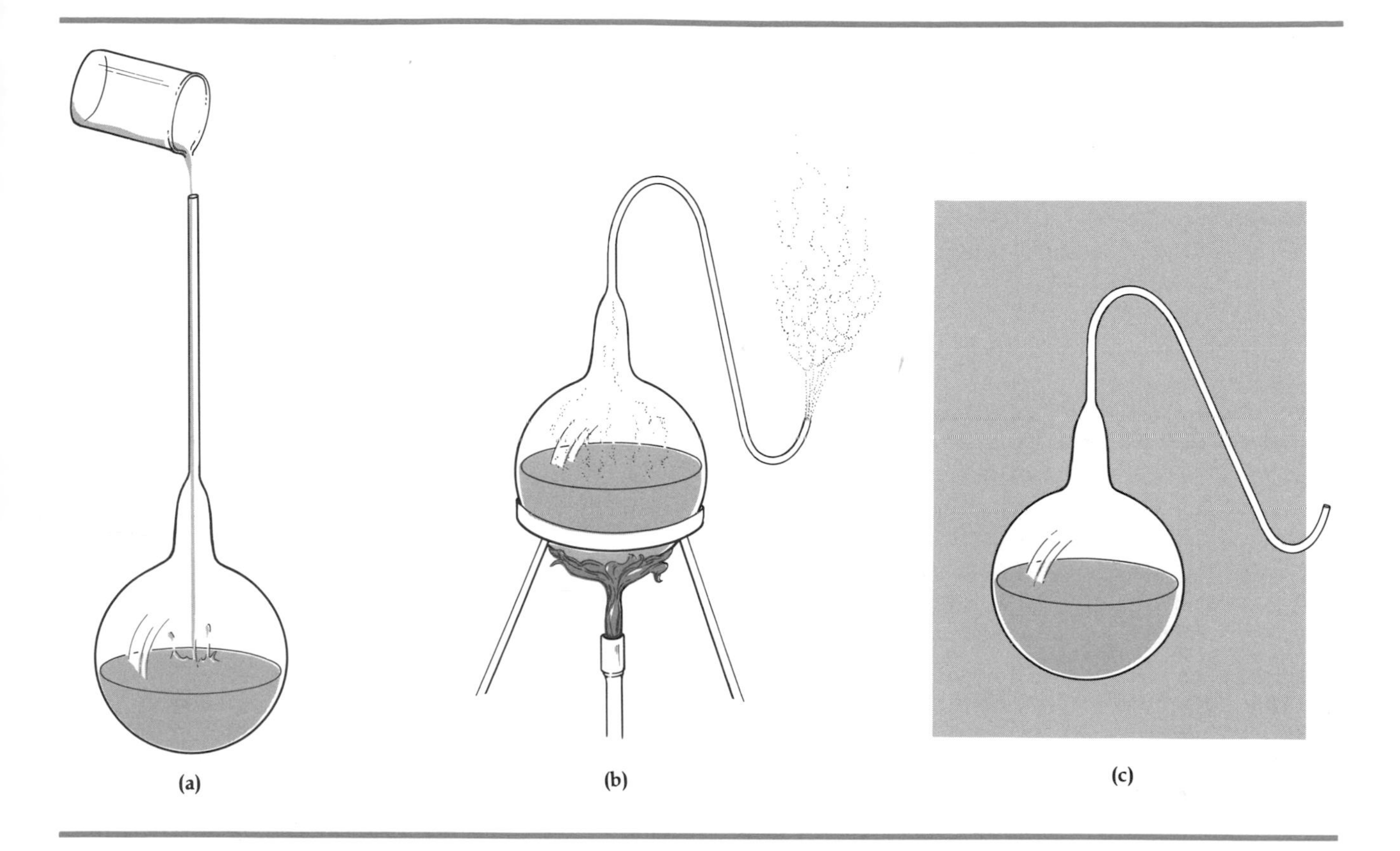

Figure 1–4 Pasteur's experiment that disproved the theory of spontaneous generation. **(a)** Pasteur first poured nutrient broth into a long-necked flask. **(b)** Next he heated the neck of the flask and bent it into an S-shaped curve; then he boiled the solution for several minutes. **(c)** Microorganisms did not appear in the cooled broth, even after long periods of time.

about that a group of French merchants asked him to find out why wine and beer soured. They hoped to develop a method that would prevent such beverages from spoiling when they shipped them long distances. At the time, many scientists believed that air acts on the sugars in these fluids to convert them into alcohol. Instead, Pasteur found that microorganisms called yeasts convert the sugars to alcohol in the absence of air. This process is called **fermentation** and is used to make wine or beer. Souring and spoilage occur later, caused by a different group of microorganisms, called bacteria. In the presence of air, bacteria change the alcoholic beverage into a sour waste product known as acetic acid (vinegar).

Pasteur's solution was to heat the alcohol just enough to kill most of the bacteria, a process that does not greatly affect the flavor of wine or beer. Today we call this process **pasteurization,** and it is used to kill bacteria in milk as well as in some alcoholic drinks.

The Germ Theory of Disease

The realization that yeasts play a crucial role in fermentation was the first concept to link a microorganism's activity to physical and chemical changes in organic materials. From this example, scientists were alerted to the possibility that microorganisms may have similar relationships with plants and animals—specifically that they may be able to cause disease. This idea was called the **germ theory of disease.**

The germ theory was a difficult concept for many to accept because for centuries people had believed that disease was punishment for an individual's crimes or misdeeds. In cases where the inhabitants of an entire village became ill, foul odors from sewage or poisonous vapors from swamps were often blamed for the sickness. Most people born in Pasteur's time found it inconceivable that "invisible" microbes could travel through the air to infect plants and animals, or remain on clothing and bedding to be transmitted from one person to another. But gradually, scientists accumulated the information to support the unorthodox germ theory.

In 1834, Agostino Bassi established the first association between a microorganism and a disease when he proved that a silkworm disease was caused by a fungus. Then, in 1865, Pasteur was called upon to help stop another silkworm disease, one that was ruining the silk industry in France and all of Europe. Pasteur found that the second infection was caused by a protozoan, and he developed a method for recognizing afflicted silkworm moths.

In the 1860s, Joseph Lister, an English surgeon, applied the germ theory to medicine. Disinfectants were unknown at the time, but Lister had heard of Pasteur's work connecting some microbes to animal diseases. Lister knew that carbolic acid (now called phenol) kills bacteria, so he began soaking surgical dressings in a mild solution of it. The practice so reduced infections and deaths that other surgeons quickly adopted it. Lister's technique was one of the earliest medical attempts to control infections caused by microorganisms.

The first proof that bacteria actually transmit disease was given by Robert Koch in 1876, just a century ago. Koch, a brilliant young German physician, was Pasteur's rival in the race to discover the cause of anthrax, a disease plaguing cattle and sheep in Europe.

Koch discovered rod-shaped bacteria in the blood of cattle that had died of anthrax. He cultured (grew) the bacteria on artificial media and then injected samples of the culture into healthy animals. When this second group of animals grew sick and died, Koch isolated the bacteria he found in their blood and compared them to the bacteria originally isolated. Koch found that the two sets of blood cultures contained the same bacteria (Figure 1–5).

Koch thus established a sequence of experimental steps for showing a direct relationship between a specific microbe and a specific disease. These steps are actually a set of criteria, known today as **Koch's Postulates.** Over the past hundred years, these same criteria have been invaluable for proving that specific microorganisms cause many diseases. Koch's Postulates, their limitations, and their application to Legionnaires' disease will be discussed in Chapter 13.

Vaccination

Often a treatment or preventative procedure is developed before scientists know why it works. The smallpox vaccine is an example of how preventative medicine can jump ahead of research. In 1798, almost seventy years before Koch established that a specific microorganism causes anthrax, Edward Jenner, a daring young British physician, embarked on an experiment to protect his patients from smallpox.

Smallpox epidemics were greatly feared. It was a disease that periodically swept through Europe killing thousands, and it wiped out 90% of the American Indians on the East Coast when European settlers first brought the infection to the New World. The disease usually starts with small red spots all over the body. The pox become hard pimples that break down into blisters filled with pus.

When a young milkgirl told Jenner that she couldn't get smallpox because she already had been sick from cowpox—a much milder disease—he decided to put the girl's story to a test. First Jenner collected scrapings from cowpox blisters. Then he inoculated people with cowpox material by scratching the patient's arm with a pox-infected needle. The scratch would turn into a raised bump, and in a few days the patient would become mildly sick, recover, and never contract either cowpox or smallpox again. The process was called **vaccination** from the Latin word "vaca" for cow. The protection from disease provided by vaccination (or by recovery from the disease itself) is called **immunity.** (We will discuss the mechanisms of immunity in Chapter 16.)

We now know that cowpox and smallpox are caused by viruses. Thus, Jenner's experiment was the first time in Western culture that a living viral agent—the cowpox virus—was used to produce immunity. Previously, ancient Chinese and Indians had rubbed material from smallpox victims into their skin, or breathed it into their nostrils, attempting to contract a mild form of the disease to protect themselves. But this technique would sometimes

Figure 1–5 Koch's procedure for demonstrating that anthrax is caused by a particular bacterium.

backfire, initiating an epidemic. The Royal Society of London had warned Jenner against risking his reputation with his experiment, but fortunately, Jenner's use of cowpox virus produced favorable results. Because the cowpox virus closely resembled the smallpox virus, it could induce immunity without giving humans a deadly infection.

Years later, around 1880, Pasteur discovered why vaccinations work. He found that the bacterium that causes a disease called chicken cholera lost its ability to cause disease (lost its *virulence*) after it was grown for long periods in the laboratory. He discovered that microorganisms with decreased virulence still retained their ability to induce immunity against subsequent infections by their virulent counterparts. This phenomenon provided a clue to Jenner's successful use of cowpox virus. Even though it is not a laboratory-produced derivative of smallpox virus, cowpox virus can induce immunity to both viruses because it is so closely related to the smallpox virus. Pasteur called cultures of avirulent microorganisms used for preventative inoculation **vaccines.**

Vaccines are produced today from avirulent microbial strains that still stimulate immunity to the related virulent strain. They are also made from killed virulent cells or from isolated components of virulent microorganisms. Vaccines are now available to protect us against such diseases as polio, whooping cough, measles, mumps, and rubella. In the future, vaccines may be manufactured by recombinant DNA techniques, to be discussed later.

The Birth of Modern Chemotherapy: Dreams of a "Magic Bullet"

After the relationship between microorganisms and disease was established, the next major focus for medical microbiologists was to search for substances that could destroy **pathogenic** (disease-causing) microorganisms without damaging the infected animal or human. The treatment of disease with chemical substances is called **chemotherapy.** Chemotherapeutic agents prepared from chemicals in the laboratory are called **synthetic drugs.** Agents produced naturally by bacteria and fungi are called **antibiotics.** Chemotherapy is based on the fact that some chemicals are more poisonous to microorganisms than to their hosts.

The first synthetic drugs

Paul Ehrlich, a German physician, is acknowledged as the imaginative thinker who fired the first shot in the chemotherapy revolution. As a medical student, Ehrlich speculated about a "magic bullet" that could hunt down and destroy a pathogen without harming the infected host. Ehrlich launched a search for such a bullet, and in 1910 he found a chemotherapeutic agent called *salvarsan,* an arsenic derivative effective against syphilis. Prior to this discovery, the only other known chemical in Europe's medical arsenal was an extract from South American tree bark, *quinine,* used by Spanish conquistadors to treat malaria.

By the late 1930s, researchers had developed several other synthetic drugs that could kill microorganisms. Most of them were derivatives of dyes. The *sulfa drugs* were one important group of drugs discovered at that time.

A fortunate accident—antibiotics

In contrast to the sulfa drugs, which were deliberately developed from a series of industrial chemicals, the first antibiotic was discovered by accident. Alexander Fleming almost tossed out some culture plates that had been contaminated by mold while he was culturing bacteria. Fortunately, he took a second look at the curious growth pattern on the contaminated plates. There was a clear area around the mold, where the bacterial culture had stopped growing (Figure 1–6). Fleming, a Scottish physician and bacteriologist, was looking at a mold that could inhibit the growth of a bacterium. The mold was later identified as *Penicillium notatum,* and in 1928, Fleming named its active inhibitor *penicillin.*

Penicillin is an antibiotic produced by a fungus. The enormous usefulness of penicillin was not apparent until the 1940s, when it was finally mass-produced and tested clinically. Since then, many other antibiotics have been found.

Unfortunately, antibiotics and other antimicrobial drugs are not without problems. Many antimicrobial chemicals are too toxic to humans for practical use; they kill the pathogenic microbe, but also have damaging side effects on the infected host. Toxicity for humans is a particular problem in the development of drugs for treating viral diseases, for reasons we shall discuss later.

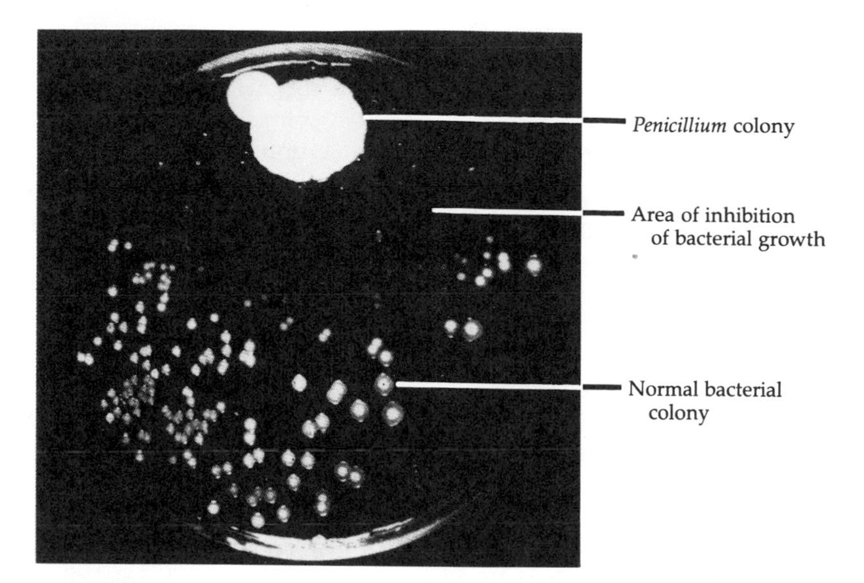

Figure 1–6 Inhibition of bacterial growth around the mold *Penicillium notatum*. In this figure published by Fleming in 1929, it is obvious that colonies of bacteria do not achieve normal growth in the vicinity of the contaminating *Penicillium* colony near the top of the photo.

The quest for better targeting

Antimicrobial drugs are sent to a diseased body by a variety of delivery systems: pills, salves, syrups, and syringes. A major drawback of current chemotherapy is that most such packages are general deliveries. Rarely can special messengers be directed to a particular, infected site; the exceptions are messengers to some localized infections.

Thus, the search for a better-targeted magic bullet continues—perhaps with more success soon to come. Advances in cell biology and immunology have provided researchers with information about three potential instruments that eventually may turn chemotherapy into a more potent medical weapon. Future bullets may include the "ghosts," or membranes, of red blood cells; a group of highly specific molecules called *monoclonal antibodies;* and artificial fatty globules known as *liposomes*.

Microbiologists, pharmacologists, and clinicians are now working on a variety of techniques to target antimicrobial drugs. The idea is to attach a drug molecule to a so-called bullet, which then travels like a homing device to its target, a specific site of infected cells. The bullet then releases the drug, which "attacks" without hurting nearby normal cells. Such a method would permit the use of drugs now regarded as too toxic for general use.

Inadequate targeting of drugs and the lack of antiviral drugs are only two limitations of modern chemotherapy. An equally important problem is the emergence and spread of new varieties of microorganisms that are resistant to antibiotics. The quest for more specific drugs and the means by which to use them will require sophisticated research techniques and correlated studies never dreamed of in the days of Koch and Pasteur. But before we continue our discussion of techniques to control microbes, we need to know more about the microbes themselves. How do microorganisms fit into the scheme of life? The next section is a short introduction to this important aspect of microbiology.

NAMING AND CLASSIFYING MICROORGANISMS

We name the things around us to communicate more easily. The common names we assign are often limited to a geographical area. One type of lumber tree, for example, has over twenty common names, depending on where you live in the United States. Sometimes, one common name refers to two different things. For example, "crown-of-thorns" is the name of both a plant and an animal. To avoid such confusion, scientists have developed a standardized system of scientific names.

The system of naming (nomenclature) we use was established in 1735 by Carolus Linnaeus, whose name is the latinized form of Carl von Linné. Scientific names are latinized because Latin was the traditional language used by scholars. Scientific nomenclature assigns each organism two names: the **genus** is the first name and it is always capitalized; the **species** name follows and is not capitalized. Genus and species names are underlined or italicized.

Scientific names can, among other things, describe the organism, honor a researcher, or identify the habitat of the species. For example, a bacterium commonly found on the skin of humans is *Staphylococcus aureus*. "Staphylo-" describes the clustered arrangement of the cells; "coccus" indicates that they are shaped like spheres. The species name, "aureus," is Latin for "golden," the color of this bacterium. As you recall, *Legionella pneumophila* was named after the famous epidemic that first brought it to the attention of physicians and researchers. The bacterium *Escherichia coli* has a genus named for a scientist, whereas its species name, "coli," reminds us that *E. coli* live in the colon, or large intestine. Notice that after a scientific name has been mentioned once, it can be abbreviated with the initial of the genus followed by the species name.

Similar species are included in the same genus; similar genera are placed in a **family;** related families are put in an **order;** related orders are in a **class;** related classes are part of a **phylum;** and related phyla constitute a **kingdom.**

Traditionally, all organisms were grouped into either the animal kingdom or the plant kingdom. But the discovery of numerous microscopic organisms with mixed characteristics presented classification problems. In 1969, H. R. Whittaker of Cornell University devised a five-kingdom classification, now widely accepted. It groups all organisms into Monera, Protista, Fungi, Plantae, and Animalia, based on cellular organization and nutritional patterns of organisms. This classification scheme will be discussed in more detail in Part II. For now, however, let us simply review the major groups of microorganisms covered in this text. These descriptions are only a brief overview; details and explanations will follow in later chapters.

THE DIVERSITY OF MICROORGANISMS

Bacteria

Bacteria are very small, relatively simple, one-celled organisms whose genetic material is not enclosed in a special nuclear membrane (Figure 1–7a). For this reason, bacteria are called **procaryotes,** from the Greek meaning "pre-nucleus." Bacteria make up the kingdom Whittaker calls Monera; others call it Procaryotae.

Bacterial cells generally have one of three shapes: **bacillus** (rodlike), **coccus** (spherical or ovoid), and **spirillum** (spiral or corkscrew) (see Figure 4–1). (There is also a genus *Bacillus*, but it includes only some of the rod-shaped bacteria.) Individual bacteria may form pairs, chains, clusters, or other groupings, and such formations are usually the same within a particular species.

Bacteria are enclosed in cell walls largely composed of a substance called *peptidoglycan*. (Cellulose is the main substance of plant cell walls.) Because of differences in the components of their cell walls, bacteria show characteristic reactions to a variety of stains (dyes). Bacteria generally reproduce by dividing into two equal daughter cells, a process called *binary fission*. For nutrition, most bacteria use organic material, which in nature may be derived from other, dead organisms or from a living host. Some bacteria can carry out photosynthesis, and some can derive nutrition from inorganic substances. Many bacteria can "swim" by means of whiplike appendages called *flagella*.

Fungi

Fungi are **eucaryotes**—organisms whose cells have a distinct nucleus containing the genetic material. Fungi may be unicellular or multicellular. Large, multicellular fungi, such as mushrooms, may look somewhat like plants, but they cannot carry out photosynthesis, as most plants can. True fungi have cell walls composed primarily of a substance called *chitin*. The unicellular forms of fungi, **yeasts,** are ovoid microorganisms larger than bacteria. The most typical fungi are **molds.** Molds form *mycelia*, which are long filaments that branch and intertwine (Figure 1–7b). The cottony growths sometimes found on bread and fruit are mold mycelia.

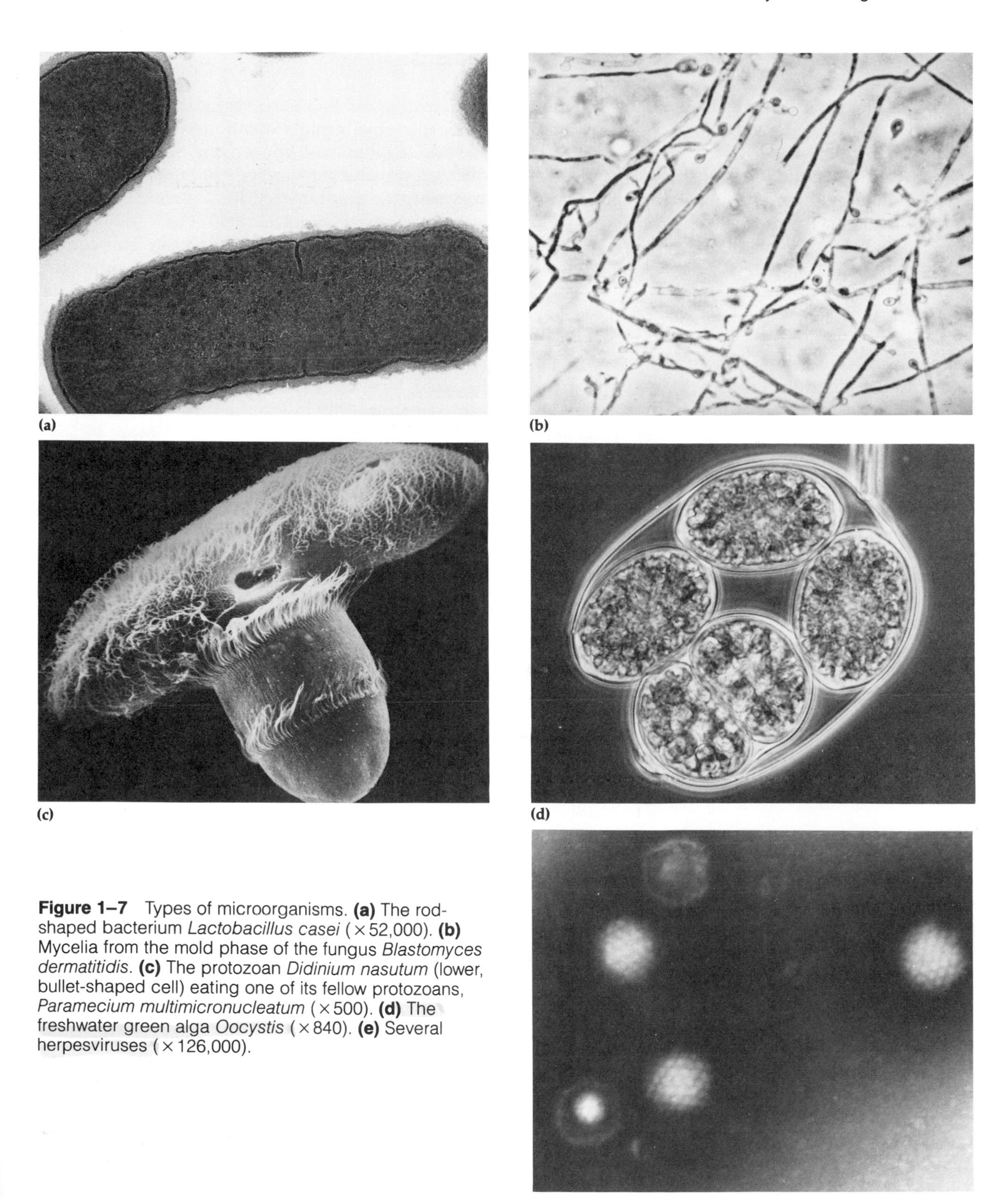

Figure 1–7 Types of microorganisms. **(a)** The rod-shaped bacterium *Lactobacillus casei* (×52,000). **(b)** Mycelia from the mold phase of the fungus *Blastomyces dermatitidis*. **(c)** The protozoan *Didinium nasutum* (lower, bullet-shaped cell) eating one of its fellow protozoans, *Paramecium multimicronucleatum* (×500). **(d)** The freshwater green alga *Oocystis* (×840). **(e)** Several herpesviruses (×126,000).

The crosswalls separating the nuclei in the mycelia of most molds are not complete, making these fungi exceptions to the strict version of the cell theory; in a sense, such a mycelium is one giant, multinucleate cell.

Fungi can reproduce sexually or asexually. They obtain nourishment by absorbing solutions of organic material from their environment—soil, water, or an animal or plant host.

Protozoans

Protozoans are unicellular, eucaryotic microbes that belong to the kingdom Protista (Figure 1–7c). Protozoans are classified according to their means of locomotion. Amoebas move using extensions of their cytoplasm called *pseudopods* ("false feet"). Other protozoans have flagella or numerous, shorter appendages called *cilia*. Protozoans have a flexible outer covering called a *pellicle*, rather than a rigid cell wall. They have a variety of shapes, and live as free entities or parasites, absorbing or ingesting organic compounds from their environment. Often they have a number of tiny, specialized, subcellular structures that perform separate functions analogous to animal and plant organs. *Paramecium* is an example of a protozoan with cilia, separate nuclei for growth regulation and sexual reproduction, an *oral groove* for the ingestion of food, and an *anal pore* for the excretion of wastes.

Algae

Algae are photosynthetic eucaryotes with a wide variety of shapes and both sexual and asexual reproductive forms (Figure 1–7d). The algae of interest to microbiologists are members of the kingdom Protista and are usually unicellular. Algae can be classified by the biochemical characteristics of their pigments, storage products, and cell walls, and by the kind of flagella they have. They are abundant in fresh and salt water, soil, plants, and other habitats. As photosynthesizers, algae need light and air for growth, but do not generally require organic compounds from the environment. Algae play an important role in the balance of nature by producing oxygen, which is utilized by other organisms, including animals.

Viruses

Viruses (Figure 1–7e) are very different from the other microbial groups mentioned here. They are so small that most can be seen only with an electron microscope, and they are not cellular. Structurally very simple, the virus particle contains a core made of only one type of nucleic acid, DNA (deoxyribonucleic acid) or RNA (ribonucleic acid). This core is surrounded by a protein coat. Sometimes the coat is encased by an additional layer, a lipid membrane called an envelope. All living cells have RNA *and* DNA, and they can carry out chemical reactions and reproduce as self-sufficient units. Viruses, however, have no "machinery" for metabolism and can reproduce only inside the cells of other organisms. Thus, viruses are parasites of other biological groups.

Multicellular Animal Parasites

Although not strictly microorganisms, multicellular parasites of medical importance will be discussed in this text. The two major parasitic groups are the flatworms and roundworms, collectively called **helminths.** During some stages of their life cycle, the helminths are microscopic in size, and laboratory identification of these organisms employs many techniques also used for the more traditional microorganisms.

MICROBES AND HUMAN WELFARE

As we mentioned earlier, pathogenic (disease-producing) microorganisms are in the minority. The vast majority of microbes are beneficial to humans in numerous ways.

Recycling Vital Elements

The chemical elements nitrogen, carbon, oxygen, sulfur, and phosphorus are essential for life and are available only in limited, though large, amounts. Pools of nitrogen, oxygen, and carbon (as carbon dioxide) exist as gases in the atmosphere. Sulfur and phosphorus are stored in the earth's crust. For the most part, it is the activity of microorganisms

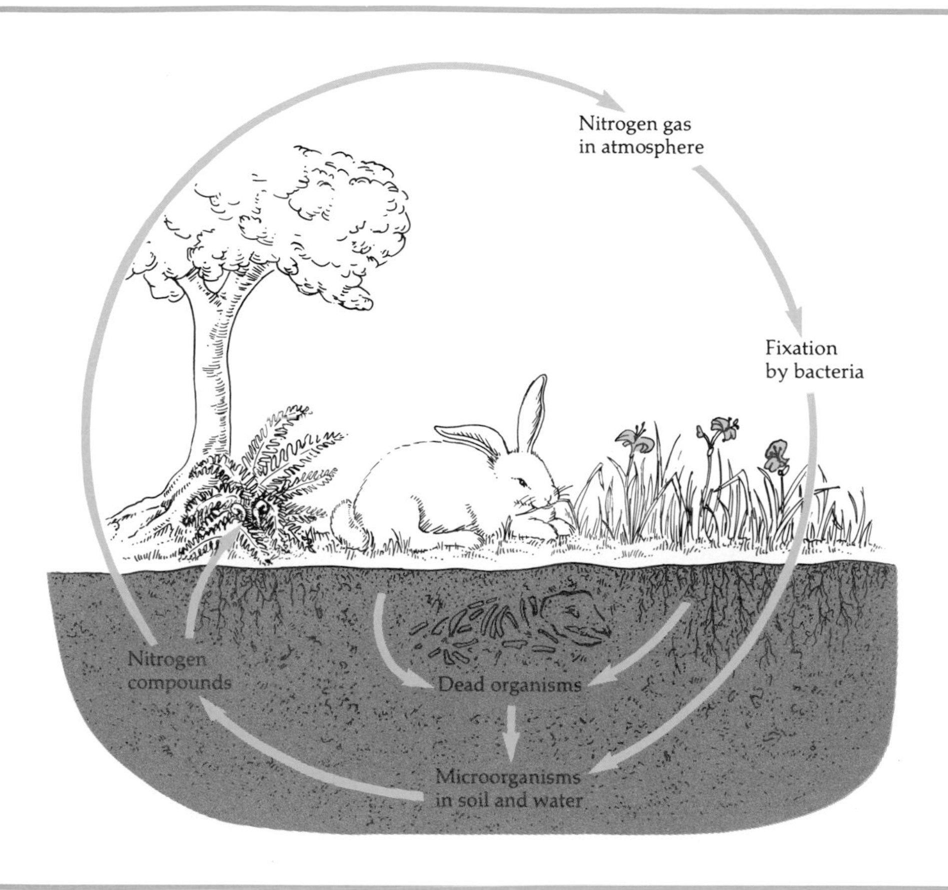

Figure 1–8 The nitrogen cycle. Microorganisms in the soil and water play key roles in recycling chemical elements essential to life, as shown here and in Figure 1–9.

that converts these elements into forms that can be used or stored by plants and animals.

The **nitrogen cycle** is one example of how microorganisms recycle an element (Figure 1–8). Nitrogen is a major constituent of all living cells. It is present in proteins and nucleic acids—DNA and RNA—which contain the genetic code for the function and reproduction of living cells. Although 79% of the atmosphere is nitrogen, it exists in a gaseous form that animals and plants cannot use. Certain soil bacteria and **cyanobacteria** (Monera formerly called blue-green algae) in the oceans combine nitrogen with other elements (fix it) so that it can be incorporated into living cells and enter the food chain. Other bacteria function to help decompose dead plants and animals, releasing nitrogen-containing chemicals into the soil, whereupon additional microbes convert the soil nitrogen back into nitrogen gas.

In the **carbon cycle,** green plants and algae remove carbon dioxide from the air and use sunlight to convert it into food by the process of photosynthesis (Figure 1–9). This food is consumed by other organisms, including humans. When they consume food, living organisms produce carbon dioxide by the process called respiration. Microorganisms, primarily bacteria, play a key role in the carbon cycle by returning carbon dioxide to the atmosphere when they decompose organic wastes and dead plants and animals. Algae, higher plants, and some bacteria also participate in the oxygen cycle by recycling oxygen to the air during photosynthesis. The nitrogen, carbon, and oxygen cycles, as well as those involving sulfur and phosphorus,

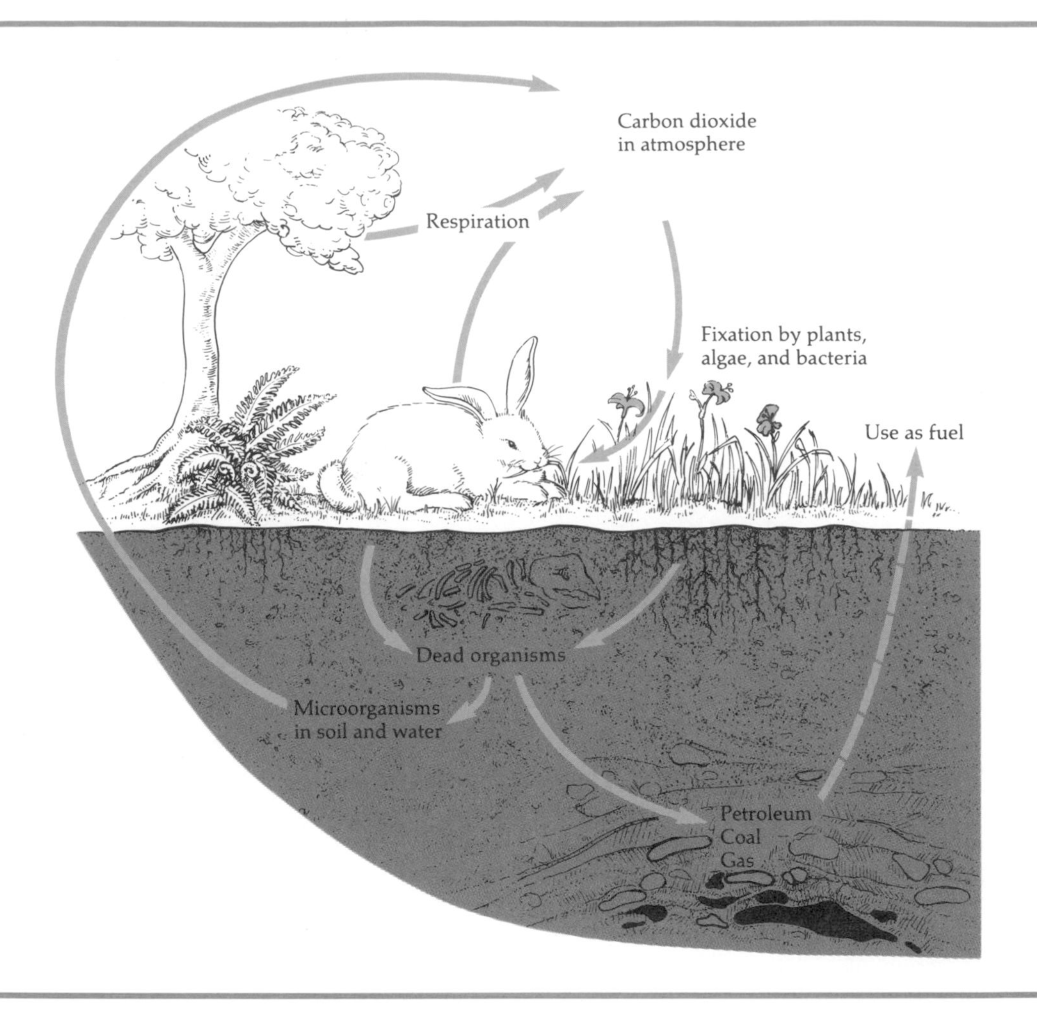

Figure 1–9 The carbon cycle.

will be discussed in greater detail in Chapters 5 and 26.

Sewage Treatment: Using Microbes to Recycle Water

With our growing awareness of the need to preserve the environment, we are more conscious of our responsibility to recycle precious water and prevent rivers and oceans from becoming polluted. For sanitary sewage disposal and safe drinking water, we rely on the activity of certain microorganisms used by sewage treatment facilities. **Sewage** consists of human and animal excrement, wash water, industrial wastes, and ground, surface, and rain water. Sewage is about 99.9% water with a few hundredths of a percent suspended solids. The remainder is a variety of dissolved materials. Sewage plant treatments aim at removing undesirable materials and harmful microorganisms. They combine various physical and chemical processes with treatment by beneficial microbes.

In a modern sewage treatment plant, large solids like paper, wood, glass, gravel, and plastic are first removed from sewage, leaving behind liquid and organic materials that bacteria convert into byproducts such as carbon dioxide, nitrates, phosphates, sulfates, ammonia, hydrogen sulfide, and methane. We shall discuss sewage treatment in detail in Chapter 26.

Insect Control by Microorganisms

Insect control is important for agricultural reasons, as well as for the prevention of human disease. The fact is, each year devastating crop damage is caused by insects. Although more research is needed into the use of microorganisms to control insects, several types of bacteria are already being used to reduce insect infestations.

The bacterium *Bacillus thuringiensis* has been used extensively to control pests in the United States, including alfalfa caterpillars, bollworms, corn borers, cabbageworms, tobacco budworms, and fruit tree leaf rollers. This particular species of bacteria produces protein crystals that are toxic to the digestive systems of the insects. Enormous quantities of the bacteria are grown, dried, and incorporated into a dusting powder applied to the crops upon which the insects feed.

Microbial insect control can have important implications for the environment. Many chemical insecticides, like DDT, are not easily degraded by soil organisms. They remain in the soil and air as toxic pollutants. After rain showers, they may be carried by erosion to streams and rivers, until they are eventually incorporated into the food chain. It has been shown that some insecticides ingested by fish and small animals are poisonous to them. Also, some of these chemicals may be responsible for birth defects in larger animals.

On the other hand, bacteria and other microorganisms used for insect control do not permanently disturb the ecology of the farm, forestland, or public water supply. And so far, insects have not developed a resistance to microbial insecticides. In contrast, many chemicals used to control insects have been needed in larger and larger concentrations, because new generations of resistant insects are no longer killed by the original levels of poison sprayed on crops.

Modern Industrial Microbiology and Genetic Engineering

Earlier in this chapter, we touched on the use of microorganisms to produce several common foods and chemicals commercially. More recently, two extensions of industrial microbiology have been attracting considerable attention: **single-cell protein (SCP),** a microbe-made food substitute, and **genetic engineering,** a new technology that greatly expands the potential use of bacteria as miniature biochemical factories.

The demand for food increases with the growing world population and the continuous consumption of decreasing energy supplies. In response to escalating food needs, scientists are investigating several new food sources. One answer is SCP—the protein (and other nutrients) produced by microorganisms cultivated on industrial wastes.

The advantage of SCP is that microorganisms grow rapidly and can produce a high yield, estimated to be 15 times greater than soybeans and 50 times greater than corn. For example, a half ton of yeast can produce 50 tons of protein daily. However, SCP is not very tasty and in some cases may need to be supplemented with essential amino acids to provide a balanced protein source. At present, SCP is used only in animal feed.

Besides growing food in tremendous quantities, microorganisms may soon be used to manufacture large amounts of human hormones and other badly needed medical substances. With the help of new techniques developed by modern biochemists, it is now possible to take fragments of human or animal DNA that code for important proteins and attach them to bacterial DNA. The tiny hybrid DNA is called **recombinant DNA.** When inserted into bacteria, recombinant DNA enables the bacteria to make large quantities of the desired protein.

Recombinant DNA techniques have been used thus far to produce a number of important proteins, including human insulin, human growth hormone, the brain chemical beta-endorphin, and some types of interferon. Insulin is a hormone required for normal blood sugar levels and is used to treat diabetes. Human growth hormone aids in the normal growth of all body tissues, especially bone and muscle. Beta-endorphin is a hormone in the brain that acts as a natural pain reliever. And interferon is an antiviral substance that may prove useful in combating viral infections and perhaps even cancer.

Eventually, recombinant DNA techniques may also be used in the production of vaccines. For example, an effective vaccine for hepatitis B was announced in 1980, but the manufacture of the vaccine requires a special protein that must first be extracted from a limited supply of infected human donors. It is hoped that recombinant DNA will soon be instructing bacteria to reproduce large amounts of the precious hepatitis B protein so that the vaccine may be manufactured in unlimited quantities.

It is possible that within a few decades certain genetic diseases like diabetes will be alleviated by the insertion of correct DNA sequences directly into unhealthy human cells. In the meantime, recombinant DNA techniques provide scientists with the means to cultivate large amounts of natural proteins and other substances that are otherwise very expensive and difficult to isolate and purify.

MICROBES AND HUMAN DISEASE

We all live in a microbial world, from birth until death. And we all have a normal group of microorganisms, or **flora,** on and inside our bodies (Figure 1–10). Although these normal flora do not usually disturb us, and often benefit us, they may under some circumstances cause us illness or be a source of infection to other people we contact.

When is a microbe a welcome part of the healthy human, and when is it a harbinger of disease? The difference between health and disease depends in large part on a balance between the natural defenses of the body and the disease-producing properties of the microorganism. Microorganisms may enter our bodies whenever we breathe, swallow, cut our skin, or have close contact with other people and animals. Usually, large numbers of microorganisms are needed to cause disease. And some microorganisms cause disease by invading body tissues, while others release poisons called toxins.

Whether our bodies overcome the offensive tactics of a particular microbe depends on our **resistance.** Important natural resistance factors include skin, mucous membranes, cilia, stomach acid, antimicrobial chemicals (such as interferon), destruction of microbes by white blood cells, inflammation, fever, and our powerful immune systems. Sometimes our natural defenses have to be supplemented by antibiotics or other drugs.

Certain people may be particularly susceptible to infection because their natural defenses are weakened by age, diet, or prior illnesses. This was true in the case of Legionnaires' disease. Not everyone at the convention developed the disease. And of those who contracted legionellosis, elderly persons and those with prior respiratory ailments were more likely to become seriously ill and suffer from complications. Researchers still do not completely understand why some victims of legionellosis had a mild case and others became seriously ill. Further information about the bacterium's natural habitat and growth requirements may provide clues.

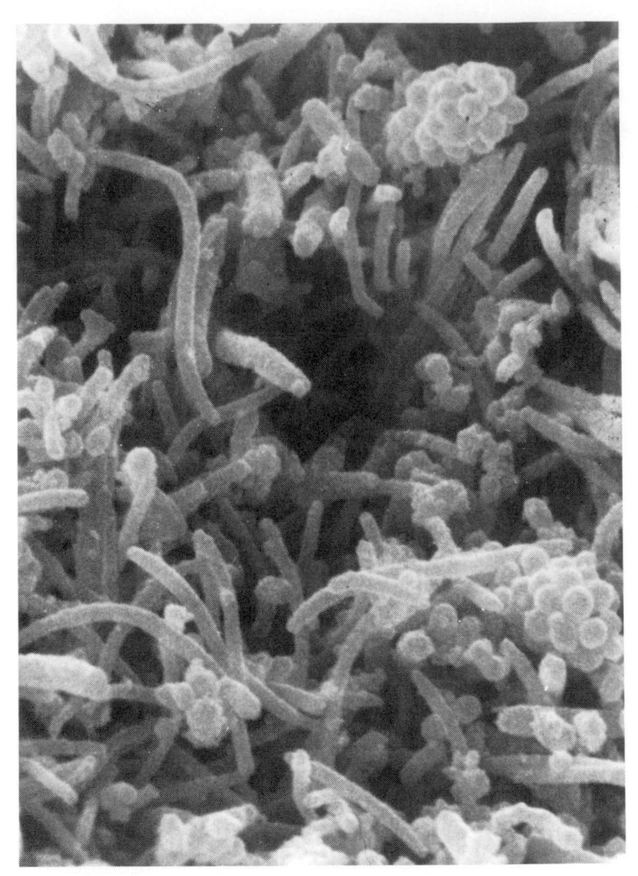

Figure 1–10 A variety of bacteria found as normal flora on human teeth at the edge of the gums ($\times 3000$). These bacteria are not necessarily associated with tooth decay or any other oral disease.

One of the ongoing tasks of microbiologists is to explore such questions as why an exposure to pathogenic microbes does not always lead to an infection. In the chapters that follow, you will become more acquainted with the principles and tools microbiologists use to understand the many interactions between microbes and their hosts.

STUDY OUTLINE

MICROBES IN OUR LIVES (pp. 3–4)

1. Living things too small to be seen with the naked eye are called microorganisms.
2. Microorganisms play important roles in maintaining an ecological balance on earth.
3. Some microorganisms live in humans and other animals and are essential to the health of the animal.
4. Some microorganisms are used to produce foods and useful chemicals.
5. Some microorganisms cause disease.

A BRIEF HISTORY OF MICROBIOLOGY (pp. 4–11)

The First Observations (pp. 4–5)

1. Anton van Leeuwenhoek, using a simple microscope, was the first to observe microorganisms (1673).
2. Robert Hooke observed that plant material was composed of "little boxes"; he introduced the term cell (1665).
3. Matthias Schleiden and Theodor Schwann introduced the concept that all living things are composed of cells (1838–39).

The Debate over Spontaneous Generation (pp. 5–6)

1. Until the mid-1800s, people believed in spontaneous generation, the idea that living organisms could arise from nonliving matter.
2. Francesco Redi demonstrated that maggots appear on decaying meat only when flies are able to lay eggs on the meat (1668).
3. John Needham claimed that microorganisms could arise spontaneously from heated nutrient broth (1745).
4. Lazzaro Spallanzani repeated Needham's experiments and suggested that Needham's results were due to microorganisms in the air entering his broth (1765).
5. Rudolf Virchow introduced the concept of biogenesis: living cells can arise only from preexisting cells (1858).
6. Louis Pasteur demonstrated that microorganisms are in the air everywhere and offered proof of biogenesis (1861).

Fermentation (pp. 6–7)

1. Pasteur found that yeast and bacteria are responsible for the conversion of sugars in fruits into alcohol and then acids.

2. Bacteria in some alcoholic beverages and milk are killed by pasteurization.

The Germ Theory of Disease (pp. 7–8)

1. Agostino Bassi (1834) and Pasteur (1865) showed a relationship between microorganisms and disease.
2. Joseph Lister introduced the use of a disinfectant to clean surgical dressings in order to control infections in humans (1860s).
3. Robert Koch proved that microorganisms cause disease. He used a sequence of procedures called Koch's Postulates (1876), which are used today to prove that a particular microorganism causes a particular disease.

Vaccination (pp. 8–10)

1. In 1798, Edward Jenner demonstrated that inoculation with cowpox material would provide humans with immunity from smallpox.
2. About 1880, Pasteur developed a vaccine for chicken cholera using avirulent bacteria; he coined the word *vaccine.*
3. Modern vaccines are prepared from living avirulent microorganisms, killed pathogens, or isolated components of pathogens.

The Birth of Modern Chemotherapy: Dreams of a "Magic Bullet" (pp. 10–11)

1. Chemotherapy is the process of treating a disease with chemicals.
2. Paul Ehrlich introduced an arsenic-containing chemical called salvarsan to treat syphilis (1910). This was the first chemotherapeutic agent.
3. Alexander Fleming observed that the mold *Penicillium* inhibited the growth of a bacterial culture. He named the active ingredient penicillin (1928).
4. Penicillin is an antibiotic (a substance produced by one microorganism that inhibits the growth of other microorganisms).
5. Researchers are continually looking for new antimicrobial drugs that harm pathogens without harming the host.

NAMING AND CLASSIFYING MICROORGANISMS (pp. 11–12)

1. In a nomenclature system designed by Carolus Linnaeus (1735), each living organism is assigned two names.
2. The two names consist of a genus and species, which must be underlined or italicized in writing.
3. All organisms may be classified into five kingdoms: Monera (or Procaryotae), Protista, Fungi, Plantae, and Animalia.

THE DIVERSITY OF MICROORGANISMS (pp. 12–14)

Bacteria (pp. 12–13)

1. Bacteria are one-celled organisms without a nucleus. The cells are described as procaryotic.
2. Bacteria can use a wide range of substances for their nutrition.

Fungi (pp. 12–14)

1. Fungi (mushrooms, molds, and yeasts) have eucaryotic cells (with a true nucleus). Most fungi are multicellular.
2. Fungi obtain nutrients by absorbing organic material from their environment.

Protozoans (pp. 13–14)

1. Protozoans are unicellular eucaryotes, classified according to their means of locomotion.
2. Protozoans obtain nourishment by absorption or ingestion through specialized structures.

Algae (pp. 13–14)

1. Algae are unicellular or multicellular eucaryotes that obtain nourishment by photosynthesis.
2. Most algae are of no medical importance.

Viruses (pp. 13–14)

1. Viruses are noncellular entities that are parasites of cells.
2. Viruses consist of a nucleic acid core (DNA or RNA) surrounded by a protein coat. An envelope may surround the coat.

Multicellular Animal Parasites (p. 14)

1. Although not microorganisms, multicellular animal parasites are medically important.
2. The principal groups of multicellular animal parasites are flatworms and roundworms.

MICROBES AND HUMAN WELFARE (pp. 14–18)

1. Microorganisms are responsible for degrading dead plants and animals and recycling chemical elements for use by living plants and animals.

2. Bacteria are used to decompose organic matter in sewage.
3. Bacteria that cause diseases in insects are being used as biological controls of insect pests. Biological controls are specific for the pest and do not harm the environment.
4. Microorganisms can be used to help produce foods. They are also food sources (single-cell protein) themselves.
5. Using recombinant DNA, bacteria can produce important proteins such as insulin.

MICROBES AND HUMAN DISEASE (p. 18)

1. Everyone has normal flora in and on the body.
2. Numbers of microbes, resistance, and general health are important factors that determine whether a person will contract a disease.

STUDY QUESTIONS

REVIEW

1. Match the following people to their contribution toward the advancement of microbiology.

____ Ehrlich	(a) First to observe bacteria
____ Fleming	(b) First to observe cells in plant material and name them
____ Hooke	(c) Disproved spontaneous generation
____ Koch	(d) Proved that microorganisms can cause disease
____ Lister	(e) Discovered penicillin
____ Pasteur	(f) Used the first synthetic chemotherapeutic agent
____ van Leeuwenhoek	(g) First to employ disinfectants in surgical procedures

2. How did the idea of spontaneous generation come about?
3. Some proponents of spontaneous generation believed that air is necessary for life. They felt that Spallanzani did not really disprove spontaneous generation because he hermetically sealed his flasks to keep air out. How did Pasteur's experiments address the air question without allowing the microbes in the air to ruin his experiment?
4. Briefly explain why scientific names are more useful than common names.
5. Match the following microorganisms to their descriptions.

____ Algae	(a) Not composed of cells
____ Bacteria	(b) Cell walls made of chitin
____ Fungi	(c) Cell walls made of peptidoglycan
____ Protozoans	(d) Cell walls made of cellulose; photosynthetic
____ Viruses	(e) Complex cell structure lacking cell walls

6. Briefly state the role played by microorganisms in each of the following.
   (a) Biological control of pests
   (b) Recycling of elements
   (c) Normal flora
   (d) Sewage treatment
   (e) Insulin production

CHALLENGE

1. How did the theory of biogenesis lead the way for the germ theory of disease?
2. The genus name of a bacterium is "erwinia" and the species name is "chrysanthemi." Write the scientific name of this organism correctly. Explain how scientific names are chosen, using this name as an example.
3. Find at least three products in the supermarket made by microorganisms. (*Hint:* The label will state the scientific name of the organism or include the word *culture.*)

FURTHER READING

De Kruif, P. 1953. *Microbe Hunters.* New York: Harcourt, Brace, and World. The stories of Leeuwenhoek, Koch, Pasteur, and others are presented in an interesting narrative.

Dixon, B. 1976. *Magnificent Microbes.* New York: Atheneum. A best-selling account of our dependence on microbes.

Fraser, D. W. and J. E. McDade. October 1979. Legionellosis. *Scientific American* 241:82–99. A summary of the mysterious epidemic that led to the discovery of a new bacterium.

Rosebury, T. 1969. *Life on Man.* New York: Viking Press. A humorous yet scientific account of the role of microbes on the human body.

Whittaker, R. H. and L. Margulis. 1978. Protist classification and the kingdoms of organisms. *BioSystems* 10:3–8. A discussion of the classification of living organisms.

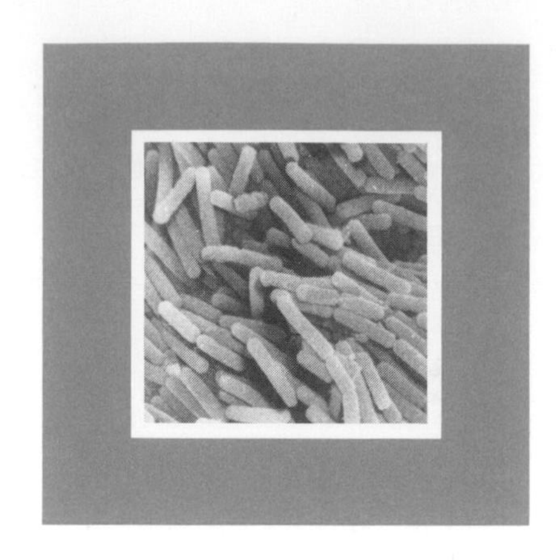

2
Chemical Principles

Objectives

After completing this chapter you should be able to

- Discuss the structure of an atom and its relation to the chemical properties of elements.
- Define the following: ionic bond; covalent bond; hydrogen bond; molecular weight; mole.
- Diagram three basic types of chemical reactions.
- Identify the role of enzymes in chemical reactions.
- List several properties of water that are important to living systems.
- Distinguish between inorganic and organic molecules.
- Define the following: acid; base; salt; pH.
- Name the building blocks of the following macromolecules: carbohydrates; simple lipids; proteins; nucleic acids.
- Identify the role of ATP in cellular activities.

Why study chemistry in a microbiology course? Microorganisms are so small that any examination of their functions quickly leads to the level of molecules. Most of the activities of microbes are, in fact, nothing more than a series of chemical reactions. Although we can see a tree rot and smell milk when it sours, we may not realize what is happening on a microscopic level. In both cases, microbes are conducting chemical operations. The tree rots because bacteria are inducing the hydrolysis (breakdown by water) of wood. And the milk proteins leave an odor as they are degraded by bacteria in a series of reactions that change sugar into lactic acid.

Like all organisms, microorganisms must be able to use nutrients to make chemical building blocks for growth and for carrying out all functions essential to life. For most microorganisms, synthesizing these essential organic building blocks requires the breaking apart of nutrient substances and the use of the energy released and the molecular fragments produced to assemble new substances. This complicated biochemistry takes place daily in countless microbial "laboratories"—any of which could fit on the head of a pin.

All matter—whether air, rock, or living organism—is made up of small units called **atoms.** Atoms interact with each other and in certain combinations to form **molecules.** Living cells are made up of molecules, some of which are very complex. The interaction of atoms and molecules is called **chemistry.**

The chemistry of microbes is the microbiologist's concern. To understand the changes that go on in microorganisms, and the changes they make on the world around us, you will need to know how molecules are formed and how they interact.

STRUCTURE OF ATOMS

Atoms are the smallest units of matter that can enter into chemical reactions. How big are atoms? It has been estimated that the largest atoms in nature are less than 0.00000005 centimeters (about 1/50,000,000 of an inch) in diameter, while the smallest, the atoms of hydrogen, are less than 0.00000001 cm in diameter. In other words, a line of 50 million of the largest atoms placed end to end would be, at most, 2.5 cm (1 inch) long.

It was once believed that atoms were the smallest particles of matter, but physicists have since demonstrated that with huge amounts of energy, atoms can be split into even smaller particles. However, for chemical reactions in our everyday world, atoms are still the smallest units.

All atoms have a centrally located **nucleus** and particles called **electrons** moving around the nucleus in arrangements called electronic configurations (Figure 2–1). The nuclei of most atoms are stable—that is, they do not change spontaneously—and nuclei do not participate in chemical reactions. The nucleus is made up of positively (+) charged particles called **protons** and uncharged (neutral) particles called **neutrons.** The nucleus, therefore, bears a net positive charge. Both neutrons and protons have approximately the same weight, which is about 1840 times heavier than an electron. The charge on electrons is negative (−), and in all atoms the number of protons is equal to the number of electrons. Since the total positive charge on the nucleus equals the total negative charge of the electrons, each atom is an electrically neutral unit.

The number of protons in an atomic nucleus ranges from one in a hydrogen atom to more than 100 in some of the largest known atoms. Different kinds of atoms are often listed by **atomic number,** the number of protons in the nucleus (see Table 2–1). The total number of protons and neutrons in an atom is its **atomic weight.**

Chemical Elements

All atoms with the same atomic number behave the same way chemically and are classified as the same **chemical element.** Each element has its own name and a one- or two-letter symbol, usually derived from the English or Latin name for the element. For example, the symbol for the element hydrogen is H, and the symbol for carbon is C. The symbol for sodium is Na—the first two letters of its Latin name *natrium*—to distinguish it from nitrogen, N, and sulfur, S. There are 92 naturally occurring elements. However, only about 23 elements are commonly found in living things; these are listed in Table 2–1. The most abundant elements in living matter—hydrogen, carbon, nitrogen, and oxygen—are shown in color in the table. The next most abundant elements appear in gray.

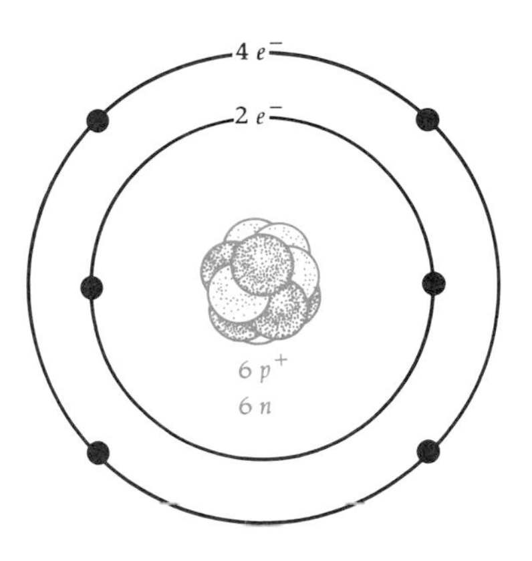

Figure 2–1 Atomic structure. Illustrated here is a very simplified version of a carbon atom (C). It contains six protons (p^+, dark color) and six neutrons (n, light color) in its centrally located nucleus, and six electrons (e^-, black) orbiting the nucleus.

Most elements have several **isotopes,** forms of the element in which the atoms differ in the number of neutrons in their nuclei. All isotopes of an element have the same number of protons in their nuclei, but different weights due to the difference in the number of neutrons present. For example, in a natural sample of oxygen, all the atoms will contain eight protons. However, 99.76% of the oxygen atoms will have eight neutrons, 0.04% will contain nine neutrons, and the remaining 0.2% will contain ten neutrons. Therefore, the three isotopes composing a natural sample of oxygen will have atomic weights of 16, 17, and 18, although all will have the atomic number 8. Atomic numbers are written as a subscript to the left of the chemical symbol of an element. Atomic weights are written as a superscript above the atomic number. Thus, natural oxygen isotopes are designated as $^{16}_{8}O$, $^{17}_{8}O$, and $^{18}_{8}O$. Rare isotopes of certain elements are extremely useful in biological research, medical diagnosis, and treatment of some disorders.

Electronic Configurations

In the atom, electrons are arranged in **electron shells,** which are regions corresponding to different **energy levels.** The arrangement is called an **electronic configuration.** Extending outward from the nucleus, each shell can hold a characteristic maximum number of electrons: two electrons in the innermost shell (lowest energy level), eight electrons in the second shell, and eight electrons in the third shell, if it is the atom's outermost (valence) shell. The fourth, fifth, and sixth electron shells can each accommodate 18 electrons, although there are some exceptions to this generalization. Table 2–2 shows the electronic configurations for atoms of some elements found in living organisms.

There is a tendency for an atom to fill its outermost shell with the maximum number of electrons. An atom may give up, accept, or share electrons with other atoms to complete this shell.

The chemical properties of atoms are largely a function of the number of electrons in the outermost electron shell. When the outer shell of an atom is filled, the element is very stable, or inert; it does not tend to react with other atoms. Helium (atomic number 2) and neon (atomic number 10) are examples of atoms that have filled outer shells. Helium has two electrons in the first and only shell, and neon has two electrons in the first shell and eight electrons in the second, its outermost, shell.

When an atom's outer electron shell is only partially filled, the atom is unstable. Such atoms react with other atoms to become more stable, and how they react depends, in part, on the degree to which the outer energy levels are filled. Note the number of electrons in the outer energy levels for the atoms represented in Table 2–2.

Table 2–1 The Elements of Life

Element	Symbol	Approximate Atomic Weight	Atomic Number
Hydrogen	H	1	1
Carbon	C	12	6
Nitrogen	N	14	7
Oxygen	O	16	8
Fluorine	F	19	9
Sodium	Na	23	11
Magnesium	Mg	24	12
Silicon	Si	28	14
Phosphorus	P	31	15
Sulfur	S	32	16
Chlorine	Cl	35	17
Potassium	K	39	19
Calcium	Ca	40	20
Vanadium	V	51	23
Chromium	Cr	52	24
Manganese	Mn	55	25
Iron	Fe	56	26
Cobalt	Co	59	27
Copper	Cu	64	29
Zinc	Zn	65	30
Selenium	Se	79	34
Tin	Sn	119	50
Iodine	I	127	53

Table 2–2 Electronic Configurations for the Atoms of Some Elements Found in Living Systems

Element	Electronic Configuration: First Electron Shell	Electronic Configuration: Second Electron Shell	Electronic Configuration: Third Electron Shell	Diagram	Number of Valence (Outermost) Shell Electrons
Hydrogen	1			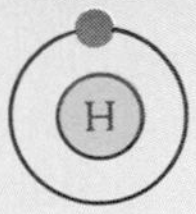 H	1
Carbon	2	4		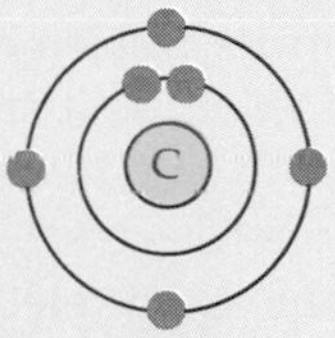 C	4
Nitrogen	2	5		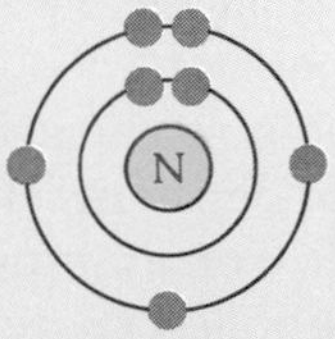 N	5
Oxygen	2	6		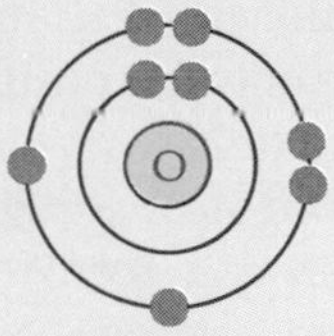 O	6
Magnesium	2	8	2	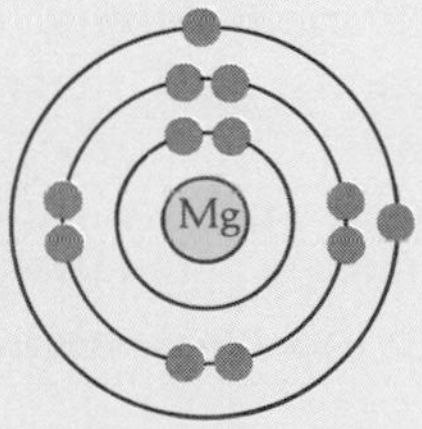 Mg	2
Phosphorus	2	8	5	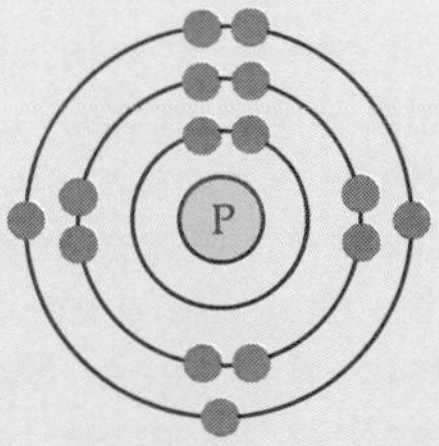 P	5
Sulfur	2	8	6	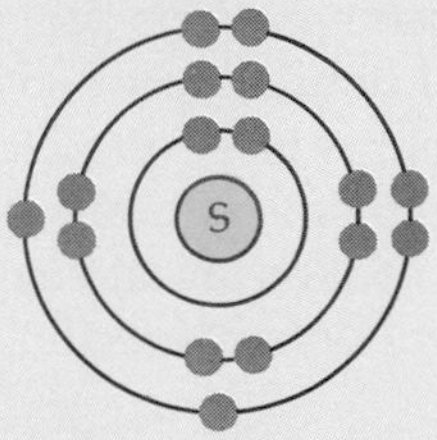 S	6

HOW ATOMS FORM MOLECULES: CHEMICAL BONDS

When the outermost energy level of an atom is not completely filled by electrons, it may be thought of as having either unfilled spaces or "extra" electrons in that energy level. For example, an atom of oxygen, with two electrons in the first energy level and six in the second, has two unfilled spaces in the second electron shell, while an atom of potassium has one extra electron in its outermost shell. For these two atoms to attain the most stable electron configuration, oxygen has to gain two electrons and potassium has to lose one electron. Atoms tend to combine such that the extra electrons in the outermost shell of one atom fill the spaces of the outermost shell of another atom.

The **valence,** or combining capacity, of an atom is the number of extra, or deficient, electrons in its outermost electron shell. For example, hydrogen has a valence of 1 (one unfilled space or one extra electron), oxygen a valence of 2 (two unfilled spaces), carbon a valence of 4 (four unfilled spaces), and magnesium a valence of 2 (two extra electrons). Valence may also be viewed as the bonding capacity of an element: hydrogen can form one chemical bond with another atom, oxygen can form two chemical bonds with various atoms, carbon can form four chemical bonds, and magnesium can form two chemical bonds.

Basically, atoms gain stability by completing the full complement of electrons in their outermost energy shells. They do this by combining to form molecules.

Two or more atoms may combine chemically to form a molecule, which can be made up of atoms of one or more elements. A molecule that contains at least two different kinds of atoms, such as H_2O, the water molecule, is called a **compound.** In H_2O, the subscript 2 indicates that there are two atoms of hydrogen to one atom of oxygen. Molecules of a compound hold together because the valence electrons of the combining atoms form attractive forces, called **chemical bonds,** between the atomic nuclei. And because energy is required for chemical bond formation, each chemical bond possesses a certain amount of potential chemical energy. Atoms of a given element form only a specific number of chemical bonds.

In general, atoms may form bonds in two ways: (1) by gaining or losing electrons from their outer electron shell or (2) by sharing outer electrons. When atoms gain or lose outer electrons, the chemical bond is called an ionic bond. By contrast, when outer electrons are shared, the bond formed is called a covalent bond. Although we will discuss ionic and covalent bonds separately, the kinds of bonds actually found in molecules do not entirely belong to either category. Molecules range from the highly ionic to the highly covalent.

Ionic Bonds

Atoms are electrically neutral because the number of positively charged protons equals the number of negatively charged electrons. But when an isolated atom gains or loses electrons, this balance is upset. If the atom gains electrons, it acquires an overall negative charge; if the atom loses electrons, it acquires an overall positive charge. Such a negatively or positively charged atom or group of atoms is called an **ion.**

Consider the following example. Sodium (Na) has 11 protons and 11 electrons, with one electron in its outer electron shell. Sodium tends to lose the single electron in this shell; it is an *electron donor.* When sodium donates an electron, it is left with 11 protons and only 10 electrons and so has an overall charge of +1. This positively charged sodium atom is called a sodium ion and is written as Na^+ (Figure 2–2a). Chlorine (Cl) has a total of 17 electrons, 7 of them in the outer electron shell. Since this outer shell can hold 8 electrons, chlorine tends to pick up an electron that has been lost by another atom; it is an *electron acceptor.* By accepting an electron, chlorine acquires a total of 18 electrons. However, it still has only 17 protons in its nucleus. The chloride ion therefore has a charge of −1 and is written as Cl^- (Figure 2–2b).

The opposite charges of the sodium ion (Na^+) and chloride ion (Cl^-) attract each other. The attraction, an ionic bond, holds the two atoms together, and a molecule is formed (Figure 2–2c). The formation of this molecule, called sodium chloride (NaCl) or table salt, is a common example of ionic bonding. Thus, an **ionic bond** is an attraction between atoms in which one atom loses electrons and another atom gains electrons. Strong ionic bonds, such as those that hold Na^+ and Cl^- ions together in salt crystals, have only limited importance in living cells. But the weaker ionic bonds

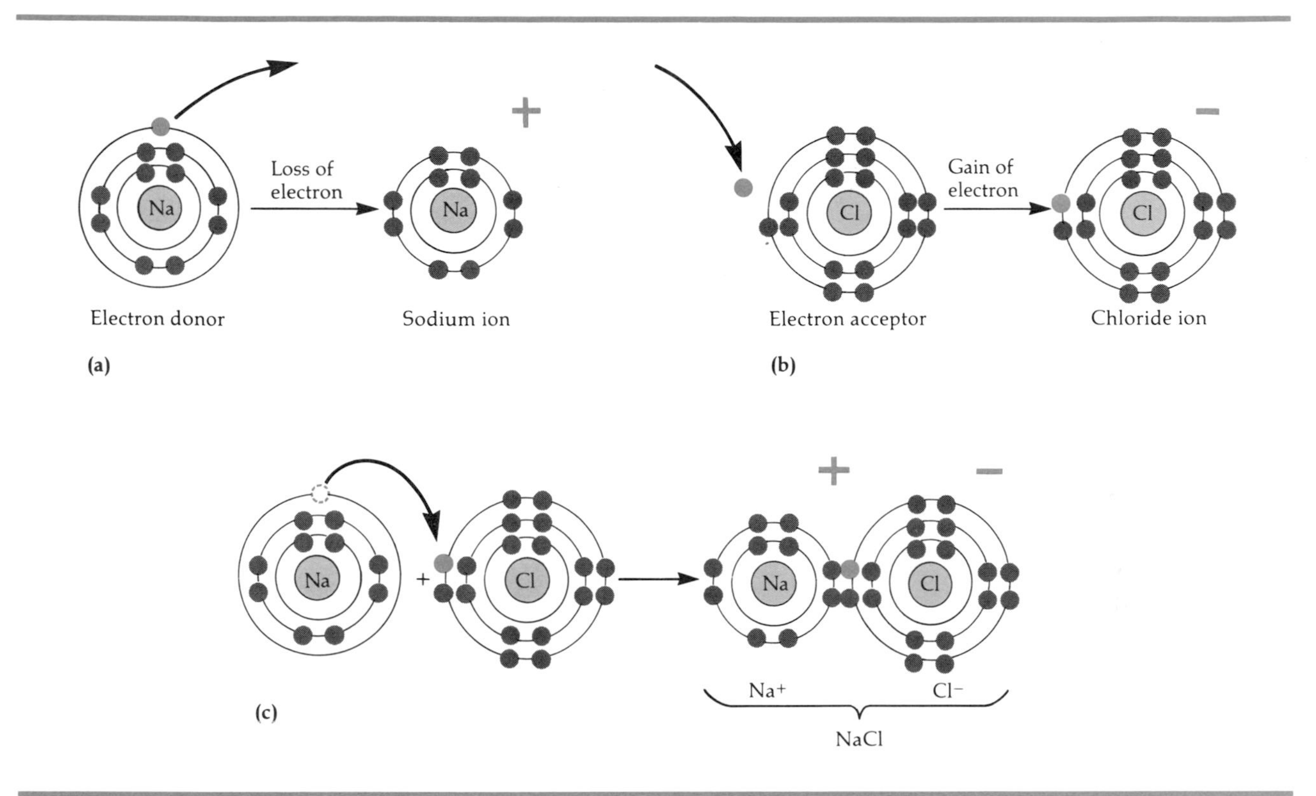

Figure 2–2 Ionic bond formation. **(a)** A sodium atom (Na) loses one electron to an electron acceptor and forms a sodium ion (Na^+). **(b)** A chlorine atom (Cl) accepts one electron from an electron donor to become a chloride ion (Cl^-). **(c)** The sodium and chloride ions are held together by an ionic bond, resulting in the formation of a molecule of sodium chloride (NaCl).

formed in aqueous solutions are important in biochemical reactions of microbes and other organisms.

In general, atoms whose outer electron shell is less than half filled lose electrons and form positively charged ions, called **cations.** Examples of cations are the potassium ion (K^+), calcium ion (Ca^{2+}), and sodium ion (Na^+). But when an atom's outer electron shell is more than half filled, it tends to gain electrons and form negatively charged ions, called **anions.** Examples are the iodide ion (I^-), chloride ion (Cl^-), and sulfur ion (S^{2-}). Notice that the symbol for an ion is written as the chemical abbreviation followed by the number of positive (+) or negative (−) charges the ion has.

Hydrogen is an example of an atom whose outer level is exactly half filled. The first energy level of an atom can hold two electrons, but the first energy level of a hydrogen atom contains only one electron. Hydrogen can lose its electron and become a positive ion (H^+), which is precisely what happens when a hydrogen ion combines with a chloride ion to form hydrochloric acid (HCl). Hydrogen can also gain an electron and become a negative ion (H^-), which can form ionic bonds with positive ions, but H^- ions are not very important in living systems. More importantly, hydrogen atoms can form an altogether different kind of bond: a covalent bond.

Covalent Bonds

A **covalent bond** is a chemical bond formed by the sharing of one or more pairs of electrons between two atoms. Covalent bonds are stronger and far more common in organisms than true ionic bonds. In the hydrogen molecule, H_2, two hydrogen atoms share a pair of electrons. Each hydrogen atom has its own electron plus one electron from the other

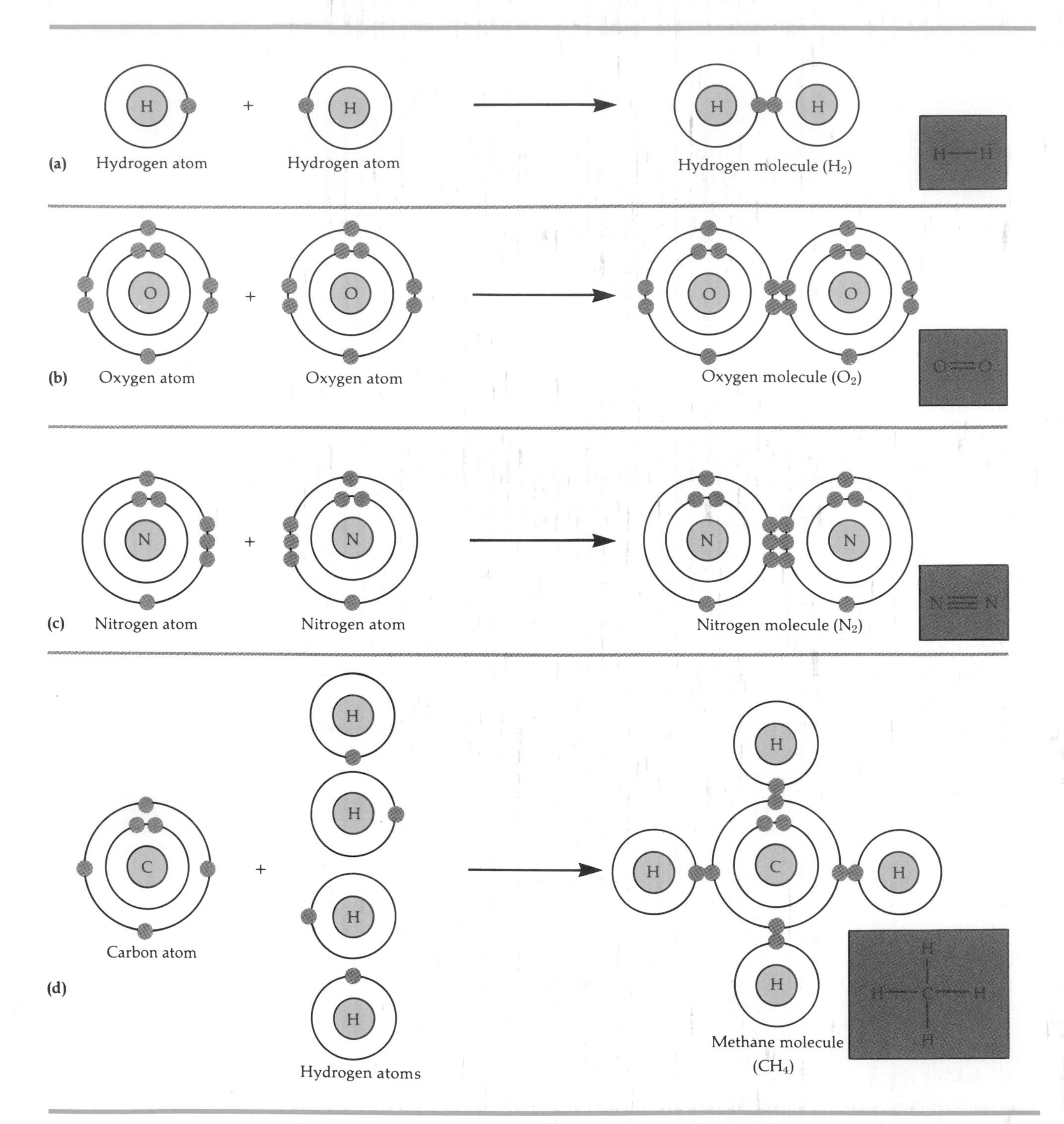

Figure 2–3 Covalent bond formation. **(a)** Single covalent bond between two hydrogen atoms. **(b)** Double covalent bond between two oxygen atoms. **(c)** Triple covalent bond between two nitrogen atoms. **(d)** Single covalent bonds between four hydrogen atoms and a carbon atom.

atom (Figure 2–3a). The shared pair of electrons actually orbits the nuclei of both atoms. Therefore, the outer electron shells of both atoms are filled. When only one pair of electrons is shared between atoms, a *single covalent bond* is formed. For simplicity, a single covalent bond is expressed as a single line between the atoms (H—H). When two pairs of electrons are shared between atoms, a *double covalent bond* is formed, expressed as two single lines (=) (Figure 2–3b). A *triple covalent bond,* expressed as three single lines (≡), occurs when three pairs of electrons are shared (Figure 2–3c).

The same principles of covalent bonding that apply to atoms of the same element apply to atoms of different elements. Methane (CH_4) is an example of covalent bonding between atoms of different elements (Figure 2–3d). The outer electron shell of the carbon atom can hold eight electrons but has only four of its own. Each hydrogen atom can hold two electrons but has only one of its own. Consequently, in the methane molecule, the carbon atom gains four hydrogen electrons to complete its outer shell requirement, while each hydrogen atom completes its pair by sharing one electron from the carbon atom. Each outer carbon electron orbits both the carbon nucleus and a hydrogen nucleus. Each hydrogen electron orbits its own nucleus and the carbon nucleus.

Elements whose outer electron shells are half filled, such as hydrogen and carbon, form covalent bonds quite easily. In fact, in living organisms, carbon always forms covalent bonds; it never becomes an ion. However, many atoms whose outer electron shells are almost full will also form covalent bonds. An example is oxygen. We will not discuss the reasons why some atoms tend to form covalent bonds rather than ionic bonds. But it is important to understand the basic principles of bond formation, because chemical reactions are nothing more than the making or breaking of bonds between atoms. *Remember:* Covalent bonds are formed by the sharing of electrons between atoms. Ionic bonds are formed by attractions between atoms that have lost or gained electrons.

Hydrogen Bonds

Another chemical bond of special importance to all organisms is the **hydrogen bond,** which consists of a hydrogen atom covalently bonded to one oxygen or nitrogen atom, but attracted to another oxygen or nitrogen atom. Such bonds are weak and do not bind atoms into molecules. However, they do serve as bridges between different molecules or between various portions of the same molecule.

When hydrogen combines with atoms of oxygen or nitrogen, the relatively large positive nucleus of these larger atoms attracts the electron of hydrogen more strongly than does the small hydrogen nucleus. Thus, in a molecule of water (H_2O), the electrons are all closer to the oxygen nucleus than to the hydrogen nuclei. As a result, the oxygen portion of the molecule has a slight negative charge, and the hydrogen portion of the molecule has a slight positive charge (Figure 2–4a). The positively charged end of one molecule can be attracted to the

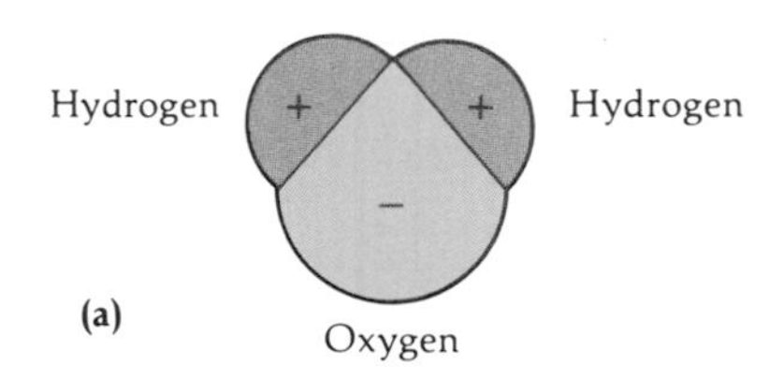

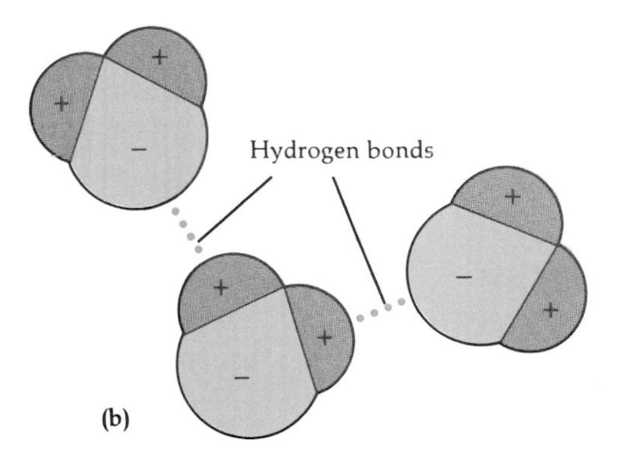

Figure 2–4 Hydrogen bond formation in water. **(a)** In a water molecule, the electrons of the hydrogen atoms are attracted to the oxygen atom. Therefore, the part of the water molecule containing the oxygen atom has a slight negative charge and the part containing hydrogen atoms has a slight positive charge. **(b)** In hydrogen bonding between water molecules, the hydrogen of one water molecule is attracted to the oxygen atom of another water molecule by means of a hydrogen bond (colored dots). Many water molecules may be attracted to each other as a result of hydrogen bonds.

negative end of another molecule, forming a hydrogen bond (Figure 2–4b). In very large molecules, the attraction may occur between hydrogen and other atoms within the same molecule. Because nitrogen and oxygen have unshared pairs of electrons, they are the elements most frequently involved in hydrogen bonding.

Hydrogen bonds are considerably weaker than either ionic or covalent bonds; they are only about 5% as strong as covalent bonds. Consequently, hydrogen bonds may be formed and broken relatively easily. It is this property that accounts for the temporary bonding between certain atoms within large complex molecules such as proteins and nucleic acids. Even though hydrogen bonds are relatively weak, such large molecules may contain several hundred of these bonds, resulting in considerable strength and stability.

Molecular Weight and Moles

The **molecular weight** of a molecule is the sum of the atomic weights of all its atoms. To go from the molecular level to the laboratory level, we use a unit called the mole. A **mole** of a substance is the number of grams equal to the molecular weight of the substance. For example, one mole of water weighs 18 grams, since the molecular weight of H_2O is 18 ($2 \times 1 + 16$). The word mole can also be applied to atoms, or even ions. Thus, one mole of hydrogen ions is equal to 1 gram.

CHEMICAL REACTIONS

As we said earlier, **chemical reactions** are the making or breaking of bonds between atoms. The total number of atoms remains the same, but the rearrangement of atoms after a chemical reaction results in different molecules with new properties. In this section, we will look at three basic types of chemical reactions common to all living cells. By becoming familiar with these reactions, you will be able to understand the specific chemical reactions we will discuss later.

Synthesis Reactions

When two or more atoms, ions, or molecules combine to form new and larger molecules, the reaction is called a **synthesis reaction.** To synthesize means "to put together," and a synthesis reaction *forms new bonds.* Synthesis reactions can be expressed in the following way:

A	+	B	⟶	AB
Atom, ion, or molecule A		Atom, ion, or molecule B	Combine to form	New molecule AB

The combining substances, A and B, are called the *reactants;* the substance formed by the combination is the *product.* The arrow indicates the direction in which the reaction proceeds.

All synthesis reactions in living organisms are collectively called anabolic reactions, or simply **anabolism.** The combining of sugar molecules to form starch, and amino acids to form protein, are two examples of anabolism.

Decomposition Reactions

The reverse of a synthesis reaction is a **decomposition reaction.** To decompose means "to break down into smaller parts," and in a decomposition reaction, *bonds are broken.* Typically, decomposition reactions split large molecules into smaller molecules, ions, or atoms. A decomposition reaction occurs in this way:

AB	⟶	A	+	B
Molecule AB	Breaks down into	Atom, ion, or molecule A		Atom, ion, or molecule B

All the decomposition reactions that occur in living organisms are collectively called catabolic reactions, or simply **catabolism.** The breakdown of food molecules in digestion is an example of catabolism.

Exchange Reactions

All chemical reactions are based on synthesis and decomposition. Many reactions, such as **exchange reactions,** are actually part synthesis and part decomposition. An exchange reaction works like this:

$$AB + CD \xrightarrow[\text{to form}]{\text{Recombines}} AD + BC$$

The bonds between A and B and between C and D are broken in a decomposition process. New bonds are then formed between A and D and between B and C in a synthesis process.

The Reversibility of Chemical Reactions

All chemical reactions are, in theory, reversible; that is, they can proceed in both directions. In practice, however, some chemical reactions do this more easily than others. When chemical reactions are readily reversible and the end product can revert to the original molecules, the **reversible reaction** is indicated by two arrows, as shown here:

$$A + B \underset{\text{Breaks down into}}{\overset{\text{Combines to form}}{\rightleftharpoons}} AB$$

Some reversible reactions occur because neither the reactants nor the end products are very stable. Other reactions reverse themselves only under special conditions:

$$A + B \underset{\text{Water}}{\overset{\text{Heat}}{\rightleftharpoons}} AB$$

Whatever is written above or below the arrows indicates the special condition under which the reaction occurs. In this case, A and B react to produce AB only when heat is applied, and AB breaks down into A and B only in the presence of water.

How Chemical Reactions Occur

The **collision theory** explains how chemical reactions occur and how certain factors affect the rates of those reactions. The basis of the collision theory is that all atoms, ions, and molecules are constantly moving and thus constantly colliding with one another. The energy gained by the particles in the collision may disrupt their electron structure enough to break down chemical bonds or form new ones.

Factors affecting chemical reactions

Several factors determine whether a collision will cause a chemical reaction: the velocity of the colliding particles, their energy, and their specific chemical configurations. Up to a point, the more rapidly the particles travel, the greater the probability that they will react. Also, different chemical reactions have specific minimum energy requirements that must be met before they can proceed. But even if colliding particles possess the minimum energy for reaction, no reaction will take place unless the particles are properly oriented toward each other.

Let us assume that molecules of substance X (the reactant) are to be converted to molecules of substance Y (the product). In a given population of molecules of substance X, at a specific temperature, there will be some molecules that possess relatively little energy; a large number of molecules, indeed the major portion of the population, will possess average amounts of energy; and a small portion of the population will have a high level of energy. Only the energy-rich X molecules are able to react and be converted to Y molecules. Therefore, only a relatively few molecules at any one time possess enough energy to react in a collision. The minimum collision energy required for a chemical reaction to occur is its **activation energy,** which is the amount of energy needed to disrupt the stable electronic configuration of a specific molecule so that a new arrangement of electrons can occur.

The *reaction rate,* the frequency of collisions of sufficient energy, depends on how many reacting molecules are at the activation energy level. One way to increase the reaction rate is to raise the temperature. This causes the molecules to move faster, thereby increasing the frequency of collisions and the number of molecules that attain activation energy. Increasing the pressure and concentrations of reactants reduces the distance between molecules and thereby also increases the number of collisions between reacting molecules.

Enzymes and chemical reactions

In living systems, there are large protein molecules called **enzymes** that accelerate chemical reactions. The three-dimensional enzyme molecule has an active site tailored to interact with a specific particle, or **substrate** (see Figure 5–2). The enzyme orients the substrate into a fixed position that increases the probability of a reaction. The appropriate enzyme will lower the activation energy of the reaction (Figure 2–5). The enzyme does this by combining with molecules that have a broad range of energies, not just those that are energy rich. The **enzyme–substrate complex** formed by the temporary binding together of enzyme and reactants enables the collisions to be more effective by lowering the activation energy required for the reaction. In this way, many more molecules of X can participate in the reaction than if the enzyme were absent. The presence of an enzyme, therefore, speeds up the reaction, since more of the X mole-

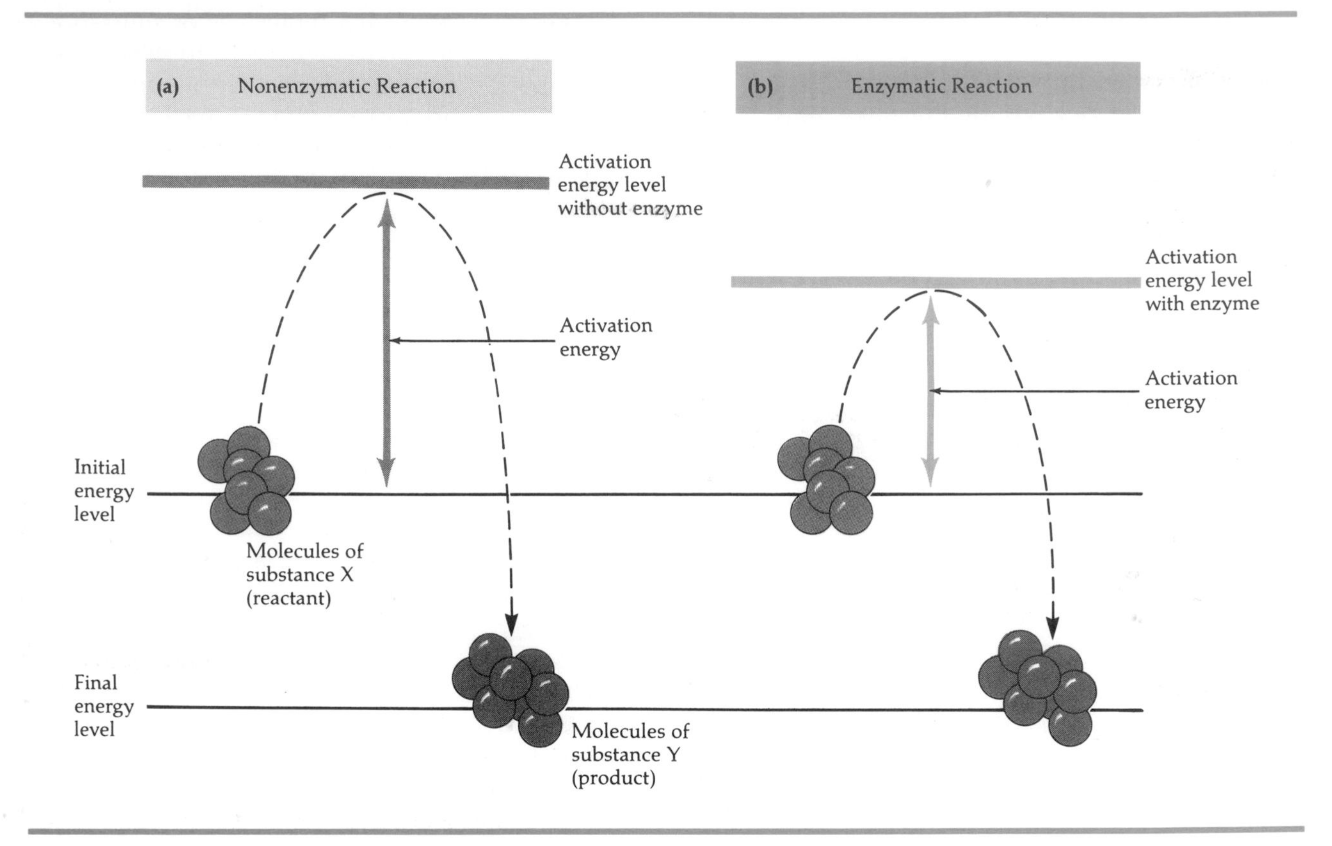

Figure 2–5 Energy requirements of a chemical reaction **(a)** without an enzyme and **(b)** with an enzyme. In **(b)**, the enzyme lowers the activation energy level. Thus, more molecules of X are converted to Y since more molecules of substance X possess the activation energy needed for the reaction to occur.

cules attain sufficient activation energy. (Heat, high pressure, and high concentration of reactants are all nonenzymatic agents of acceleration.)

An enzyme is able to accelerate a reaction without an increase in temperature. This is crucial for living systems, since a significant temperature increase would destroy cellular proteins. The essential function of enzymes, therefore, is to speed up biochemical reactions at a temperature that is compatible with the normal functioning of the cell.

Energy of Chemical Reactions

Some change of energy occurs whenever bonds between atoms in molecules are formed or broken during chemical reactions. When a chemical bond is formed, energy is required. When a bond is broken, energy is released. Thus, synthesis reactions require energy, whereas decomposition reactions give off energy. The building processes of organisms occur basically through synthesis reactions. The breakdown of foods, on the other hand, occurs through decomposition reactions. When foods are decomposed, they release energy that can be used by organisms for their building processes.

Forms of Energy

The ability to use energy and change it into different forms is basic to the survival of any living cell. Besides **chemical energy,** the energy of chemical reactions, there are three other common forms of energy that are fundamental in biological systems: mechanical, radiant, and electrical energy. These forms may exist as **potential energy** (which is stored or inactive) or **kinetic energy** (the energy of motion).

Mechanical energy is required for cell movement, reorganization of structures inside the cell, and changes in cell shape. **Radiant energy** comes from sources of heat or light. Heat will agitate the molecules in any system, promoting chemical activity. Light is the primary energy source for green plants, algae, and bacteria that use photosynthesis to make organic compounds for food. **Electrical energy** is produced whenever electrons move from one place to another. Although electrons do not usually flow through living cells, cells do have electrical currents associated with ionic flows where charged particles or molecules cross cell membranes.

IMPORTANT BIOLOGICAL MOLECULES

Biologists and chemists divide compounds into two principal classes: inorganic and organic. **Inorganic compounds** are usually small molecules in which ionic bonds may play an important role. Inorganic compounds include many salts, acids, bases, and water.

Organic compounds always contain carbon, a unique element in the chemistry of life. Carbon has four electrons in its outer shell, allowing it to combine with a variety of atoms, including other carbon atoms, to form straight or branched chains and rings. Organic compounds are held together mostly or entirely by covalent bonds. Carbon chains are the backbone for many substances of living cells. Organic compounds include sugars, amino acids, and vitamins. Some organic molecules, such as polysaccharides, proteins, and nucleic acids, are very large, usually containing more than 1000 atoms. Such molecules are referred to as *macromolecules*.

INORGANIC COMPOUNDS

Water

All living organisms require a wide variety of inorganic compounds for growth, repair, maintenance, and reproduction. **Water** is one of the most important, as well as one of the most abundant, of these compounds, and it is particularly vital to the needs of microorganisms. Outside the cell, nutrients are dissolved in water, enabling them to pass through cell membranes. And inside the cell, water is the medium for most chemical reactions. In fact, water is by far the most abundant component in almost all types of living cells, present in amounts ranging from 5 to 95% or more, the average being 65 to 75%. Simply stated, no organism can survive without water.

Water has structural and chemical properties that make it particularly suitable for its role in living cells. As we discussed, although the total charge on the water molecule is neutral, the oxygen region of the molecule has a slight negative charge and the hydrogen region has a slight positive charge (see Figure 2–4a). Any such molecule in which there is unequal distribution of charges is called a *polar molecule*. The polar nature of water gives it four characteristics that make it a useful medium for living cells.

First, every water molecule is capable of forming two hydrogen bonds with nearby water molecules. This property results in a strong attraction between water molecules. With this strong attraction, a great deal of heat is required to separate water molecules from each other and form water vapor; thus, water has a relatively high boiling point (100°C). With such a high boiling point, water exists in the liquid state on most of the earth's surface. Furthermore, the hydrogen bonding between water molecules gives ice a crystalline structure that is less dense than liquid water. Hence, ice floats and can serve as an insulating layer on the surface of lakes and streams that harbor living organisms.

Second, the polarity of water makes it an excellent dissolving medium, or *solvent*. Many polar substances dissociate (dissolve) into individual molecules in water because the negative part of the water molecule is attracted to the positive part of the molecules in the *solute* (dissolved substance), while the positive part of the water molecule is attracted to the negative part of solute molecules. Substances such as salts, composed of atoms (or groups of atoms) held together by ionic bonds, tend to dissociate into separate cations and anions in water. Thus, the polarity of water allows molecules of many different substances to separate and become surrounded by water molecules (Figure 2–6).

Third, the polarity of the water molecule accounts for its characteristic role as a reactant or

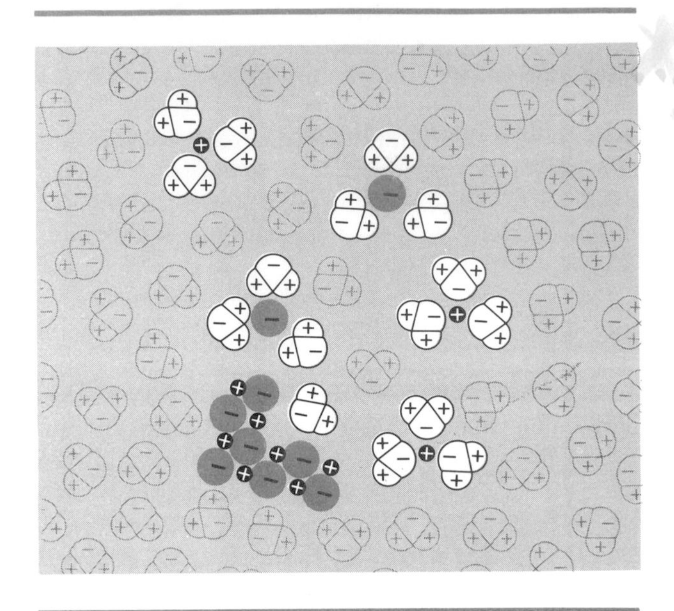

Figure 2–6 How water acts as a solvent for NaCl. The negative part of the water molecule is attracted to the positive sodium ion (Na^+) (small black circles), while the positive part of the water molecule is attracted to the negative chloride ion (Cl^-) (large colored circles).

product in many chemical reactions. Its polarity facilitates the splitting and rejoining of hydrogen (H^+) and hydroxyl (OH^-) groups. Water is a key reactant in the digestive processes of organisms—processes in which larger molecules are broken down into smaller ones. Water molecules are also involved in synthetic reactions. Water is a source of the hydrogen and oxygen that are incorporated into numerous organic compounds in living cells.

Finally, the relatively strong hydrogen bonding between water molecules (see Figure 2–4b) makes water an excellent temperature buffer. A given quantity of water, compared with the same quantity of many other substances, requires a much greater amount of heat to increase its temperature and a greater loss of heat to decrease its temperature. Normally, heat absorption by molecules increases their kinetic energy and thus their rate of motion, resulting in greater reactivity. In the case of water, however, heat absorption first results in the breaking of hydrogen bonds, rather than an increase in molecular motion. Therefore, much more heat must be applied to raise the temperature of water than to raise the temperature of a nonhydrogen-bonded liquid. The reverse is true as water cools. Thus, water maintains a more constant temperature than do other solvents and tends to protect a cell from fluctuations in environmental temperatures.

Acids, Bases, and Salts

As we saw in Figure 2–6, when inorganic salts such as sodium chloride (NaCl) are dissolved in water, they undergo **dissociation,** or **ionization.** That is, they break apart into ions. Substances called acids and bases show similar behavior.

An **acid** may be defined as a substance that dissociates into one or more hydrogen ions (H^+) and one or more negative ions (anions). Thus, an acid may also be defined as a proton (H^+) donor. A **base,** on the other hand, dissociates into one or more positive ions (cations) plus one or more negatively charged chemical groups that can accept, or combine with, protons. Among the most important proton acceptors is the hydroxyl ion (OH^-). Hydroxyl ions have a strong attraction for protons. Thus, sodium hydroxide (NaOH) is a base because it dissociates to release hydroxyl ions (OH^-). And finally, **salt** is a substance that dissociates in water into cations and anions, neither of which are H^+ or OH^-. Figure 2–7 shows common examples of each of these three types of compounds and how they dissociate in water.

Acid–Base Balance

An organism must maintain a fairly constant balance of acids and bases for it to remain healthy. In the aqueous environment within organisms, acids dissociate into hydrogen ions (H^+) and anions. Bases, on the other hand, dissociate into hydroxyl ions (OH^-) and cations. The more free hydrogen ions in a solution, the more acid the solution. Conversely, the more free hydroxyl ions in a solution, the more basic, or alkaline, it is.

Biochemical reactions, that is, reactions in living systems, are extremely sensitive to even small changes in the acidity or alkalinity of the environments in which they occur. In fact, H^+ and OH^- ions are involved in practically all biochemical processes, and the functions of cells are modified greatly by any deviation from narrow limits of nor-

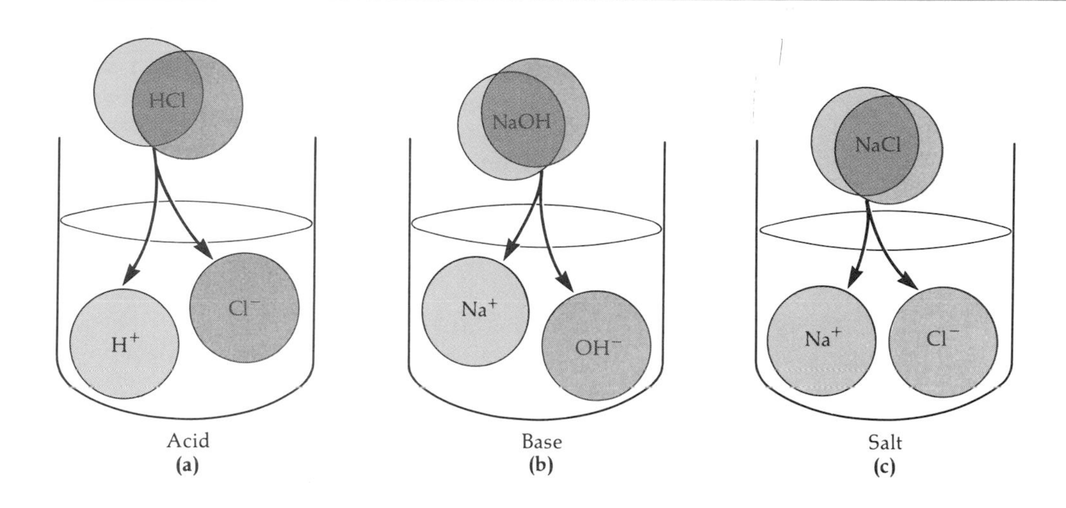

Figure 2–7 Acids, bases, and salts. **(a)** In water, hydrochloric acid (HCl) dissociates into H^+ ions and Cl^- ions. **(b)** Sodium hydroxide (NaOH), a base, dissociates into OH^- ions and Na^+ ions in water. **(c)** In water, table salt (NaCl) dissociates into positive ions (Na^+) and negative ions (Cl^-), neither of which are H^+ or OH^- ions.

mal H^+ and OH^- concentrations. For this reason, the acids and bases that are constantly formed in an organism must be kept in balance.

It is convenient to express the amount of H^+ in a solution by a logarithmic pH scale, which ranges from 0 to 14 (Figure 2–8). The **pH** of a solution is the negative logarithm to the base 10 of the hydrogen ion concentration in moles per liter $[H^+]$, or $-\log_{10} [H^+]$.

For example, if the H^+ concentration of a solution is 1.0×10^{-4} moles/liter, or 10^{-4}, its pH equals $-\log_{10}10^{-4} = -(-4) = 4$; this is about the pH of wine. You may wish to refer to the box on Exponents, Exponential Notation, and Logarithms to help you understand this calculation. In the laboratory, however, you will usually measure the pH of a solution with a pH meter or with chemical test papers, eliminating the need for calculations.

Acidic solutions contain more H^+ ions than OH^- ions and have a pH lower than 7. If a solution has more OH^- ions than H^+ ions, it is **basic,** or **alkaline.** In pure water, a small proportion of the molecules are dissociated into H^+ and OH^- ions, giving it a pH of 7. Since the concentrations of H^+ and OH^- are equal, this pH is said to be **neutral.**

Since the pH scale is logarithmic, a change of one whole number represents a tenfold change

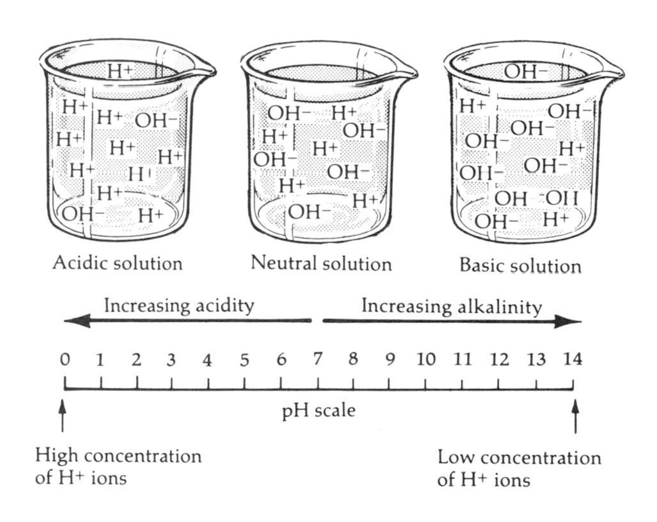

Figure 2–8 The pH scale. The concentrations of H^+ and OH^- ions are equal at pH 7, which is the neutral point. If the pH value of a solution is below 7, the solution is acidic; if the pH is above 7, the solution is basic (alkaline). As the pH values decrease from 14 to 0, the H^+ concentration increases. Thus, the lower the pH is, the more acidic the solution; the higher the pH is, the more basic the solution.

from the previous concentration. Thus, a solution with a pH 1 has 100 times more H^+ ions than a solution with a pH 3, and 10 times more H^+ ions than one with a pH 2.

Keep in mind that the pH of a solution can be changed. We can increase its acidity by adding substances that will increase the concentration of hydrogen ions. As living organisms take up nutrients, carry out chemical reactions, and excrete wastes, the balance of acids and bases tends to change, and the pH fluctuates. Fortunately, organisms possess natural **pH buffers,** compounds that help keep the pH from changing drastically. But the pH in our environment's water and soil can be

Exponents, Exponential Notation, and Logarithms

Very large and very small numbers—such as 4,650,000,000 and 0.00000032—are cumbersome to work with. It is more convenient to express such numbers in exponential notation, that is, as a power of 10. For example, 4.65×10^9 is written in **standard exponential notation,** or **scientific notation.** 4.65 is the **coefficient,** and 9 is the power or **exponent.** In standard exponential notation, the coefficient is a number between 1 and 10 and the exponent is a positive or negative number.

To change a number into exponential notation, follow two steps. First, determine the coefficient by moving the decimal point so you leave only one nonzero digit to the left of it. For example,

$$0.0000003.2$$

The coefficient is 3.2. Second, determine the exponent by counting the number of places you moved the decimal point. If you moved it to the left, the exponent is a positive number. If you moved it to the right, the exponent is negative. In the example, you moved the decimal point 7 places to the right, so the exponent is -7. Thus

$$0.00000032 = 3.2 \times 10^{-7}$$

Now suppose we are working with a large number instead of a very small number. The same rules apply, but our exponential value will be positive rather than negative. For example,

$$4,650,000,000. = 4.65 \times 10^{+9} = 4.65 \times 10^9$$

To multiply numbers written in exponential notation, multiply the coefficients and *add* the exponents. For example,

$$(3 \times 10^4) \times (2 \times 10^3) = (3 \times 2) \times 10^{4+3} = 6 \times 10^7$$

To divide, divide the coefficients and *subtract* the exponents. For example,

$$\frac{3 \times 10^4}{2 \times 10^3} = \frac{3}{2} \times 10^{4-3} = 1.5 \times 10^1$$

Microbiologists use exponential notation in many kinds of situations. For instance, exponential notation is used to describe the number of microorganisms in a population. Such numbers are often very large (Chapter 6). Another application of exponential notation is to express concentrations of chemicals in a solution—chemicals such as media components (Chapter 6), disinfectants (Chapter 7), or antibiotics (Chapter 18). Such numbers are often very small. Converting from one unit of measurement to another in the metric system requires multiplying or dividing by a power of 10, which is easiest to carry out in exponential notation.

A **logarithm** is the power to which a base number is raised to produce a given number. Usually we work with logarithms to the base 10, abbreviated $\mathbf{log_{10}}$. The first step in finding the $\log_{10}$ of a number is to write the number in standard exponential notation. If the coefficient is exactly 1, the $\log_{10}$ is simply equal to the exponent. For example,

$$\log_{10}0.00001 = \log_{10}(1 \times 10^{-5}) = -5$$

If the coefficient is not 1, as is often the case, a logarithm table or calculator must be used to determine the logarithm.

Microbiologists use logs for pH calculations and for graphing the growth of microbial populations in culture (Chapter 6).

altered by waste products from organisms, pollutants from industry, or fertilizers used in agriculture or gardening.

Different microbes function best within different pH ranges, but most organisms grow best between pH 6.5 and 8.5. The microbes most able to tolerate acid conditions are fungi, whereas the procaryotes called cyanobacteria (blue-green algae) tend to do well in alkaline habitats. *Propionibacterium acnes*, a bacterium that contributes to acne, has its natural environment on human skin, which tends to be slightly acidic, with pH around 4.

ORGANIC COMPOUNDS

Inorganic compounds, excluding water, constitute about 1 to 1.5% of living cells. These components, whose molecules have only a few atoms, cannot provide the chemical specificity that cells need to perform complicated biological functions. Such specificity is possible with organic molecules, whose carbon atoms can combine in an enormous variety of ways with other carbon atoms and atoms of other elements.

In the formation of organic molecules, carbon's four outer electrons can participate in up to four covalent bonds, and carbon atoms can bond to each other, forming straight-chain, branched-chain, or ring structures (Figure 2–9).

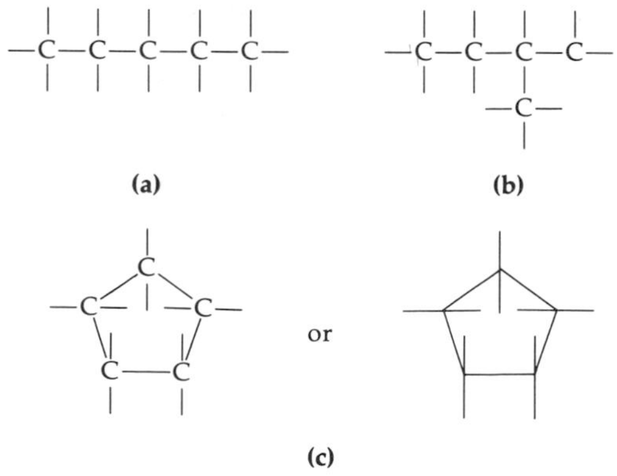

(a) (b) (c) or

Figure 2–9 Various bonding patterns of carbon. **(a)** Straight-chain structures. **(b)** Branched-chain structures. **(c)** Ring. In ring structures, the carbon atoms at the corners are often not shown.

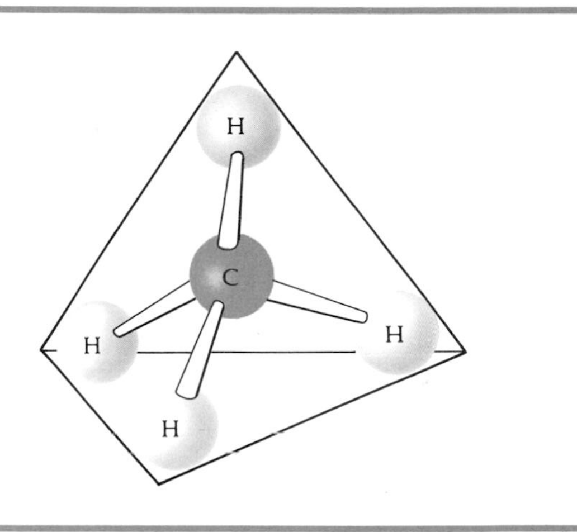

Figure 2–10 A tetrahedral carbon compound. Organic molecules containing carbon atoms with four single bonds have this distinctive shape due to the directions of the four bonds. Here the carbon atom is bonded to four hydrogen atoms to form the tetrahedral methane molecule (CH_4).

When a carbon atom is bonded to four other atoms, those four atoms form the skeleton of a three-dimensional shape called a **tetrahedron** (Figure 2–10). Carbon atoms can also bond with double or triple bonds, in which case the molecular shape is not tetrahedral.

In addition to carbon, the most frequently found elements in organic compounds are hydrogen (which can form one bond), oxygen (two bonds), and nitrogen (three bonds). Sulfur (two bonds) and phosphorus (five bonds) appear less often. Other elements are found, but only in a relatively few organic compounds. If you look back at Table 2–1, you will see that the elements most abundant in organic compounds are the same as those most abundant in living cells.

Functional Groups

The basic chain of carbon atoms in a molecule is called the **carbon skeleton.** Most of the carbons are bonded to hydrogen atoms. The bonding of other elements with carbon and hydrogen forms characteristic **functional groups** (Table 2–3). Functional groups are responsible for most of the characteristic chemical properties and many of the physical properties of a given organic compound. Since there is such a huge number of combinations possible for carbon skeletons, functional groups help classify organic compounds. One method of classification is based on the type of functional group. For example, the —OH group is present in each of the following molecules.

Methanol

Ethanol

Isopropyl alcohol

Toluol

Since the characteristic reactivity of the molecules is based on the —OH group, they are all classified as alcohols. The —OH group is called the *hydroxy group* and is not to be confused with the *hydroxyl ion* (OH^-) of bases. The hydroxy group of alcohols does not ionize; it is covalently bonded to a carbon atom.

To designate a class of compounds characterized by a certain functional group, the letter R is usually substituted for the remainder of the molecule. For example, alcohols in general may be written R—OH.

Frequently, more than one functional group is found in a single molecule. For example,

Amino group

Carboxyl group

If a molecule contains both amino and carboxyl groups, it is called an amino acid. Most of the organic compounds found in living organisms are quite complex, with a large number of carbon atoms forming the skeleton and many functional groups attached. In considering representations of organic and biochemical molecules, it is important to note that each of the four bonds of carbon is satisfied (attached to another atom) and each of the attaching atoms has its characteristic number of bonds satisfied.

Small organic molecules can be combined into the very large molecules called macromolecules. Macromolecules are usually **polymers,** that is, large molecules formed by covalent bonds between many repeating small molecules called **subunits** or **monomers.** When two monomers join together, the reaction usually involves the elimination of a hydrogen atom from one monomer and a hydroxy group from the other, producing a molecule of water:

$$X{-}OH + H{-}Y \longrightarrow X{-}Y + H_2O$$

This type of reaction is called **condensation** or **dehydration** because a molecule of water is released.

The synthesis of macromolecules such as carbohydrates, lipids, proteins, and nucleic acids is a dominant activity of cell chemistry.

Carbohydrates

A large and diverse group of organic compounds are the **carbohydrates,** which include sugars and macromolecules such as starches. The carbohydrates perform a number of major functions in living systems. For instance, one type of sugar (deoxyribose) is a building block of DNA, the molecule that carries hereditary information. Other sugars help to form the cell walls of bacterial cells. Macromolecular carbohydrates, meanwhile, function as food reserves. Simple carbohydrates may be utilized in the synthesis of amino acids and fats or fatlike substances, which are used to build structures and provide an emergency source of energy. The principal function of carbohydrates, however, is to fuel cell activities with a ready source of energy.

Carbohydrates are made up of the elements carbon, hydrogen, and oxygen. The ratio of hydrogen to oxygen atoms is always 2 to 1 in carbohydrates. This ratio can be seen in the formulas for

Table 2–3 Representative Functional Groups and the Compounds in Which They Are Found

Functional Group	Name of Group	Class of Compounds
R—O—H	Hydroxy	Alcohol
R—C(=O)—H	Carbonyl (terminal)	Aldehyde
R—C(=O)—R	Carbonyl (internal)	Ketone
R—CH$_2$—NH$_2$	Amino	Amine
R—C(=O)—O—R′	Ester	Ester
R—CH$_2$—O—CH$_2$—R′	Ether	Ether
R—CH$_2$—SH	Sulfhydryl	Sulfhydryl
R—C(=O)—OH	Carboxyl	Organic acid

the carbohydrates ribose ($C_5H_{10}O_5$), glucose ($C_6H_{12}O_6$), and sucrose ($C_{12}H_{22}O_{11}$). Although there are exceptions, the general formula for carbohydrates is $(CH_2O)_n$, where n symbolizes three or more CH_2O units. Carbohydrates can be divided into three major groups on the basis of size: monosaccharides, disaccharides, and polysaccharides.

Monosaccharides

Simple sugars are called **monosaccharides;** each molecule contains three to seven carbon atoms. Simple sugars with three carbons in the molecule are called trioses. (The number of carbon atoms in the molecule is indicated by the prefix.) There are also tetroses (four-carbon sugars), pentoses (five-carbon sugars), hexoses (six-carbon sugars), and heptoses (seven-carbon sugars). Pentoses and hexoses are extremely important to living organisms. Deoxyribose is a pentose found in DNA, the genetic material of the cell. And glucose, a common hexose, is the main energy-supplying molecule of living cells.

Disaccharides

Disaccharides are formed when two monosaccharides bond in a dehydration synthesis (condensation) reaction. For example, molecules of two monosaccharides, glucose and fructose, combine to form a molecule of the disaccharide sucrose (cane sugar):

$$\underset{\text{(Monosaccharide)}}{\underset{\text{Glucose}}{C_6H_{12}O_6}} + \underset{\text{(Monosaccharide)}}{\underset{\text{Fructose}}{C_6H_{12}O_6}} \longrightarrow \underset{\text{(Disaccharide)}}{\underset{\text{Sucrose}}{C_{12}H_{22}O_{11}}} + \underset{\text{Water}}{H_2O}$$

The formula for sucrose is $C_{12}H_{22}O_{11}$ and not $C_{12}H_{24}O_{12}$ because a molecule of water (H_2O) is lost in the process of disaccharide formation (Figure 2–11a). Similarly, the dehydration synthesis of the monosaccharides glucose and galactose forms the disaccharide lactose (milk sugar).

It may seem odd that glucose and fructose should have the same chemical formula. Actually, they are different monosaccharides. The positions of the oxygens and carbons vary in the two different molecules (Figure 2–11) and consequently, the molecules have different physical and chemical properties. Two molecules with the same chemical formula but different structures are called **isomers.**

Disaccharides can also be broken down into smaller, simpler molecules by adding water. This chemical reaction, the reverse of dehydration synthesis, is called **digestion** or **hydrolysis,** which means "to split by using water." A molecule of sucrose, for example, may be digested into its components of glucose and fructose by the addition of water. The mechanism of this reaction is represented in Figure 2–11b.

Polysaccharides

Carbohydrates in the third major group, the **polysaccharides,** consist of eight or more monosaccharides (glucose is a common sugar unit) joined together through dehydration synthesis, often with side chains branching off the main structure. Polysaccharides are macromolecules. Like disaccharides, polysaccharides can be split apart into their constituent sugars through hydrolysis. Unlike monosaccharides and disaccharides, however, they usually lack the characteristic sweetness of sugars like fructose and sucrose and usually are not soluble in water.

One important polysaccharide is *glycogen,* which is composed of glucose subunits and is synthesized as a storage material by animals and some bacteria. *Cellulose,* another important glucose polymer, is the main component of the cell walls of plants and most algae. The polysaccharide *dextran*

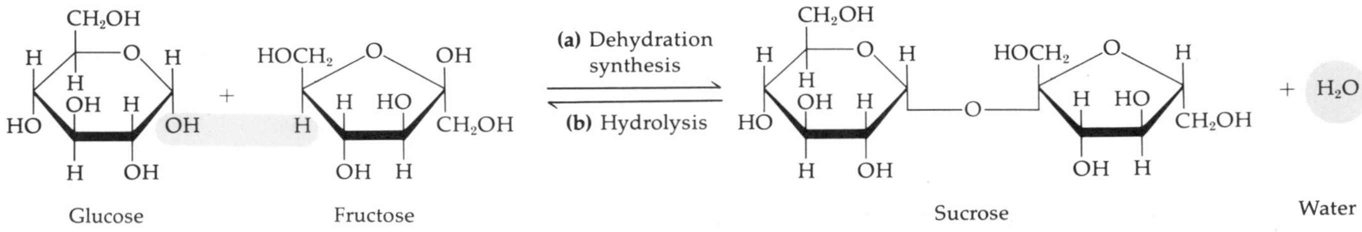

Glucose + Fructose (a) Dehydration synthesis ⇌ (b) Hydrolysis Sucrose + H_2O Water

Figure 2–11 (a) In dehydration synthesis (left to right), the monosaccharides glucose and fructose combine to form a molecule of the disaccharide sucrose. A molecule of water is lost in the reaction. **(b)** In hydrolysis (right to left), the sucrose molecule breaks down into smaller molecules, glucose and fructose. For the hydrolysis reaction to proceed, water must be added to the sucrose.

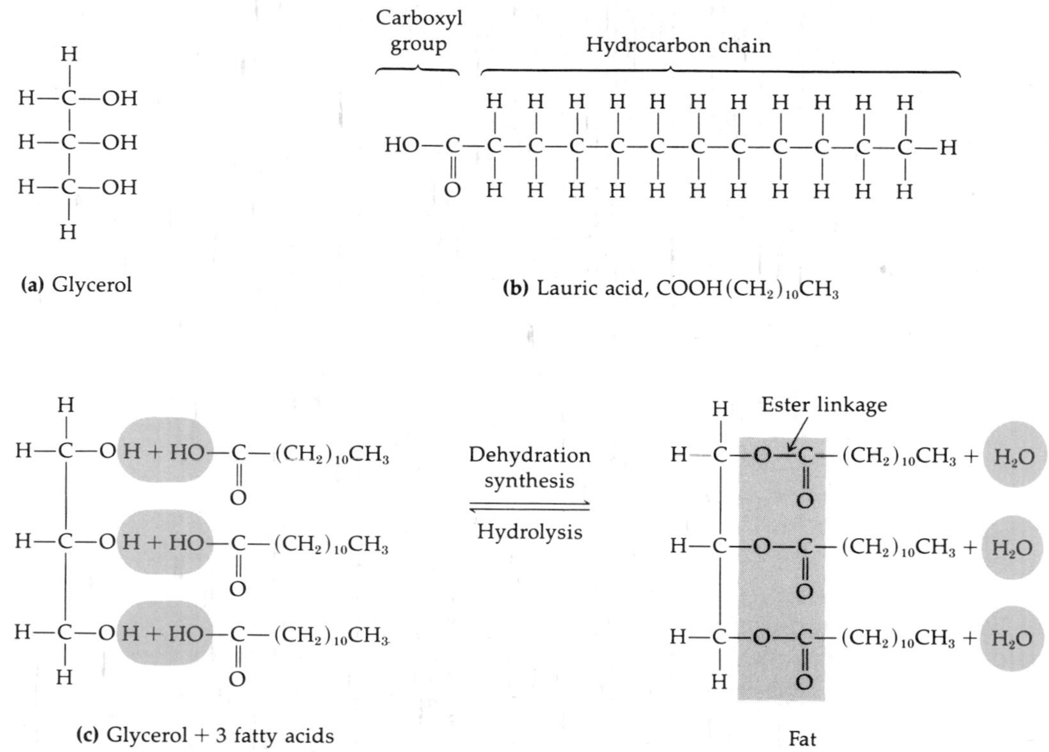

Carboxyl group

Hydrocarbon chain

(a) Glycerol

(b) Lauric acid, $COOH(CH_2)_{10}CH_3$

Dehydration synthesis

Hydrolysis

Ester linkage

(c) Glycerol + 3 fatty acids

Fat

Figure 2–12 Structural formulas for **(a)** glycerol and **(b)** lauric acid, a fatty acid. In the condensed formula for lauric acid, $COOH(CH_2)_{10}CH_3$, the subscript 10 means that the molecule has ten CH_2 units. **(c)** The chemical combination of a molecule of glycerol with three fatty acid molecules forms one molecule of fat and three molecules of water in a dehydration synthesis reaction. The addition of three water molecules to a fat forms glycerol and three fatty acid molecules in a hydrolysis reaction.

is produced as a sugary slime by certain bacteria. Dextran is used medically in a blood plasma substitute.

Lipids

Lipids constitute a second major group of organic compounds found in living matter. Like carbohydrates, they are composed of atoms of carbon, hydrogen, and oxygen, but lack the 2 to 1 ratio between hydrogen and oxygen atoms. Lipids are a very diverse group of compounds. Most are insoluble in water, but dissolve readily in nonpolar solvents such as ether, chloroform, and alcohol.

If lipids were suddenly to disappear from the earth, all living cells would collapse in a pool of fluid, because lipids are essential to the structure and function of membranes that separate living cells from their environment. In this role, lipids are indispensable to a cell's survival. They also function as fuel reserves within living organisms: examples are fats in animals and poly-β-hydroxybutyric acid in bacteria.

Simple lipids, or *fats,* contain an alcohol called *glycerol* and a group of compounds known as *fatty acids.* Glycerol molecules have three carbon atoms to which are attached three hydroxy (—OH) groups. The structural formula for glycerol is shown in Figure 2–12a. Fatty acids consist of long hydrocarbon chains (composed only of carbon and hydrogen atoms) ending in a carboxyl (—COOH) (organic acid) group (Figure 2–12b). Most common fatty acids contain an even number of carbon atoms.

A fat molecule is formed when a molecule of glycerol combines with one to three fatty acid molecules to form a monoglyceride, diglyceride, or triglyceride (Figure 2–12c). In the reaction, one to three molecules of water are formed (dehydration), depending on the number of fatty acid molecules reacting. The chemical bond formed where the water molecule is removed is an *ester linkage*. In the reverse reaction, hydrolysis, a fat molecule is broken down into its component fatty acid and glycerol molecules.

Since the fatty acids that form lipids have different structures, there is a wide variety of lipids. For example, three molecules of fatty acid A may combine with a glycerol molecule; or one molecule each of fatty acids A, B, and C may unite with a molecule of glycerol. Simple lipids serve as storage materials in higher organisms, but are not found in bacteria.

Complex lipids contain elements like phosphorus, nitrogen, and sulfur, in addition to the carbon, hydrogen, and oxygen found in simple lipids. The complex lipids called *phospholipids* are made up of glycerol, two fatty acids, and, in place of a third fatty acid, a phosphate group bonded to one of several organic groups (Figure 2–13). Phospholipids are the lipids that build membranes, and are thus essential to all procaryotic and eucaryotic cells. The structure of membranes will be discussed in detail in Chapter 4.

Structurally very different from the lipids described previously are the lipids called *steroids*. Figure 2–14 shows the structure of the steroid cholesterol. The —OH group present in cholesterol makes it a *sterol*. Sterols are important constituents

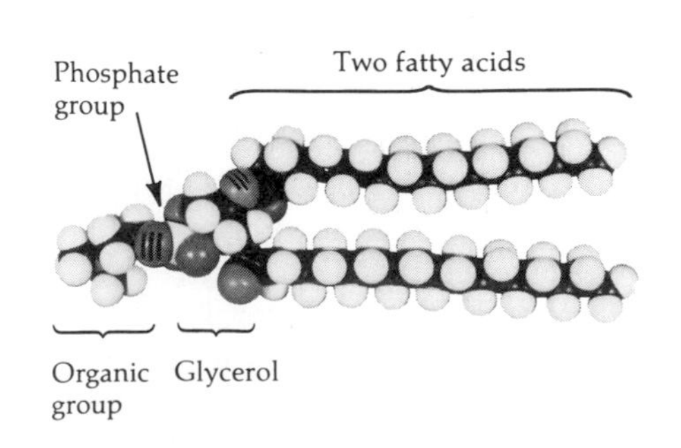

Figure 2–13 A model of a phospholipid showing the sizes and locations of the atoms in three dimensions.

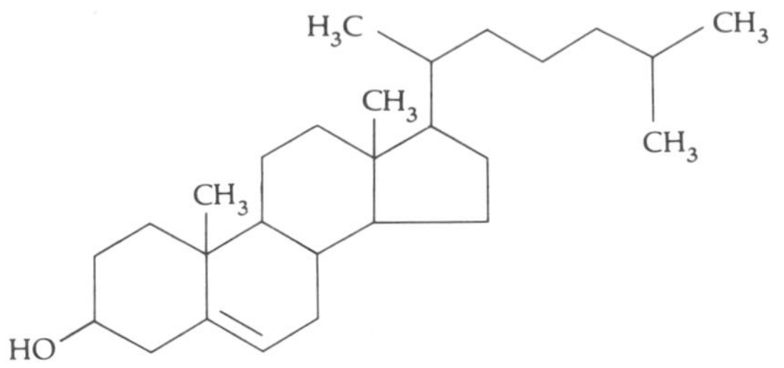

Figure 2–14 Cholesterol, a steroid. Note the four "fused" carbon rings, which are characteristic of steroid molecules. The hydrogen atoms attached to the carbons at the corners of the rings have been omitted.

of the plasma membranes of animal cells and of one group of bacteria (mycoplasmas), and they are also found in fungi and plants. Animals synthesize steroid hormones and vitamin D from sterols.

Proteins

If you were to separate out and weigh all the groups of organic compounds in a living cell, the proteins would tip the scale. Hundreds of different proteins can be found in any single cell, and together they make up 50% or more of the cell's dry weight. They are essential ingredients in all aspects of cell structure and function. We have already mentioned enzymes, the proteins that catalyze biochemical reactions. But proteins have other vital functions as well. Carrier proteins help to transport certain chemicals into and out of cells. Other proteins, such as the bacteriocins produced by many bacteria, serve to kill other bacteria. Toxins produced by certain disease-causing microorganisms also fall into the protein category.

Some proteins play a key role in cell contraction and locomotion. Others, such as the hormones of eucaryotic organisms, have regulatory functions. Then there are proteins that play a key role in vertebrate immune systems: antibodies, for instance, are proteins. And some proteins are simply structural, like the collagen of animal skin and bones.

Amino acids

Proteins are organic molecules that contain carbon, hydrogen, oxygen, and nitrogen. In addition, some

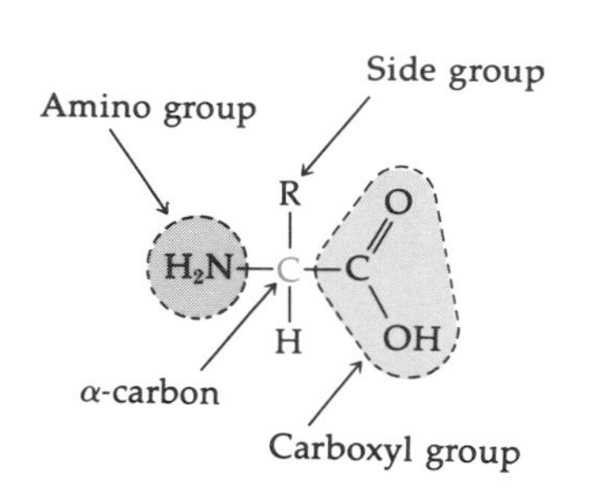

Figure 2–15 General structural formula for an amino acid. The alpha-carbon is shown in color. The letter R stands for any of a number of groups of atoms. Different amino acids have different R groups.

contain sulfur. Just as the monosaccharides are the building blocks of larger carbohydrate molecules, and fatty acids and glycerol are the building blocks of fats, **amino acids** are the building blocks of proteins. Amino acids contain at least one carboxyl (—COOH) group and one amino (—NH_2) group attached to the same carbon atom, called an alpha-carbon (α-carbon) (Figure 2–15). Such amino acids are called *alpha-amino acids.* Also attached to the alpha-carbon is a side group (R group), which distinguishes the various amino acids. The side group may be a hydrogen atom, an unbranched or branched chain, or a ring structure that may be cyclic (all carbon) or heterocyclic (an atom other than carbon is included in the ring). The side group may contain functional groups such as the sulfhydryl group (—SH), the hydroxy group (—OH), or additional carboxyl or amino groups. These side groups and the carboxyl and alpha-amino groups affect the total structure of a protein, as will be described later.

Amino acids can exist in either of two configurations called *stereoisomers,* designated by D and L. These two configurations are mirror images, corresponding to "right-handed" and "left-handed" three-dimensional shapes (Figure 2–16). The amino acids found in proteins are always the L-isomers (except for glycine, the simplest amino acid, which does not have stereoisomers). However, D-amino acids occasionally occur in nature, for example, in certain bacterial cell walls and antibiotics. (Many other kinds of organic molecules also can exist in D and L forms. One example is the sugar glucose, which occurs in nature as D-glucose.)

Although only about 20 different kinds of amino acids occur naturally in proteins, the total number of amino acids in a single protein molecule may vary from 50 to hundreds or more, and they may be arranged in an infinite number of ways. Thus, proteins may have different lengths, different relative quantities of the various amino acid units, and different specific sequences in which the amino acids are bonded. The number of proteins is practically endless, and every living cell produces many different proteins.

Peptide bonds

Amino acids bond together between the carboxyl (—COOH) group of one amino acid and the amino (—NH_2) group of another. For every bond between two amino acids, one water molecule is released

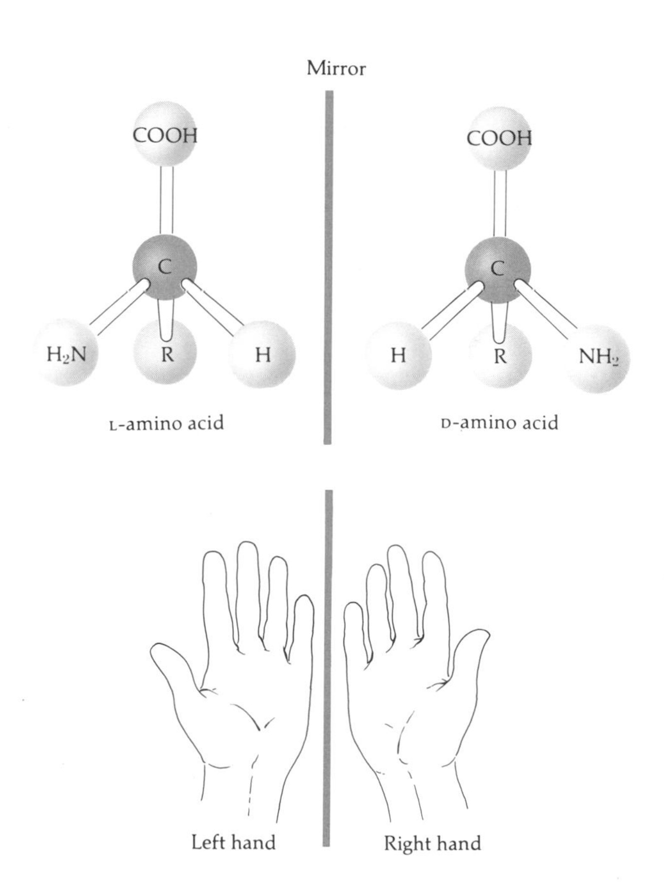

Figure 2–16 The L- and D-isomers of an amino acid, shown with ball-and-stick models. The two isomers, like left and right hands, are mirror images of each other and cannot be superimposed on one another (try it!).

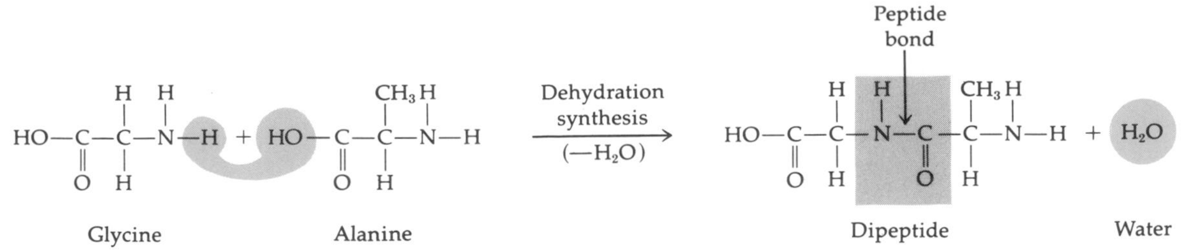

Peptide bond

$$\underset{\text{Glycine}}{HO-\overset{\overset{}{}}{\underset{\underset{O}{\|}}{C}}-\underset{\underset{H}{|}}{\overset{\overset{H}{|}}{C}}-\overset{\overset{H}{|}}{N}-H} + \underset{\text{Alanine}}{HO-\underset{\underset{O}{\|}}{C}-\underset{\underset{H}{|}}{\overset{\overset{CH_3}{|}}{C}}-\overset{\overset{H}{|}}{N}-H} \xrightarrow[(-H_2O)]{\text{Dehydration synthesis}} \underset{\text{Dipeptide}}{HO-\underset{\underset{O}{\|}}{C}-\underset{\underset{H}{|}}{\overset{\overset{H}{|}}{C}}-\overset{\overset{H}{|}}{N}-\underset{\underset{O}{\|}}{C}-\underset{\underset{H}{|}}{\overset{\overset{CH_3}{|}}{C}}-\overset{\overset{H}{|}}{N}-H} + \underset{\text{Water}}{H_2O}$$

Figure 2–17 Peptide bond formation. The amino acids glycine and alanine combine to form a dipeptide. The newly formed bond between the nitrogen atom of glycine and the carbon atom of alanine is called a peptide bond.

(dehydration). The bonds between amino acids are called **peptide bonds.**

In the example in Figure 2–17, a peptide bond is formed between two amino acids by dehydration synthesis. The carboxyl group of one amino acid supplies an OH^- and the amino group of the other releases an H^+ for the formation of water. The peptide bond is between the carbon atom of the carboxyl group of one amino acid and the nitrogen atom of the amino group of another amino acid. The resulting compound, because it consists of two amino acids joined via a peptide bond, is referred to as a *dipeptide*. The addition of another amino acid to a dipeptide would form a *tripeptide.* Further addition of amino acids would result in the formation of a long, chainlike molecule called a *polypeptide,* a large protein molecule.

Levels of protein structure

Proteins have four levels of organization: primary, secondary, tertiary, and quaternary. The *primary structure* is the order (the sequence) in which the amino acids are linked together (Figure 2–18a). Alterations in sequence can have profound metabolic effects. For example, a single substitution of an incorrect amino acid in a blood protein can produce a deformed hemoglobin molecule resulting in sickle-cell anemia.

Proteins are not simple linear chains of amino acids. A protein's *secondary structure* is any regular coiling or zigzag arrangement along one dimension. This aspect of a protein's shape generally comes from hydrogen bonds joining the atoms of peptide bonds at different locations along the polypeptide chain. Common secondary protein structures are clockwise spirals called helixes and pleated sheets (Figure 2–18b) that form from roughly parallel portions of the chain. Both types of structures are held together by hydrogen bonds.

Tertiary structure refers to how the protein is bent or folded into a three-dimensional shape (Figure 2–18c). This nonregular looping and twisting is the result of interactions between various functional groups or branches of specific amino acids. Hydrogen bonds and other relatively weak interactions play an important role in tertiary structure. In addition, sulfhydryl groups (—SH) on two amino acid subunits can form a covalent, disulfide link (—S—S—) by the removal of the hydrogen atoms (oxidation).

Some proteins have a *quaternary structure,* which results from an aggregation of two or more individual polypeptide units (Figure 2–18d). For example, an enzyme called phosphorylase *a* consists of four identical polypeptide subunits. Each of these units alone does not exhibit enzymatic activity, but becomes active when associated in groups of four. Certain other proteins with quaternary structure have two or more different kinds of polypeptide subunits.

The proteins we have been discussing are *simple proteins,* containing only amino acids. *Conjugated proteins* are those formed with a combination of amino acids and other organic or inorganic components. Conjugated proteins are named by their non-amino acid component. Thus, glycoproteins contain sugars, nucleoproteins contain nucleic acids, metalloproteins contain metal atoms, lipoproteins contain lipids, and phosphoproteins contain phosphate groups. An example of a microbial phosphoprotein is phospholipase C. Manufactured by the

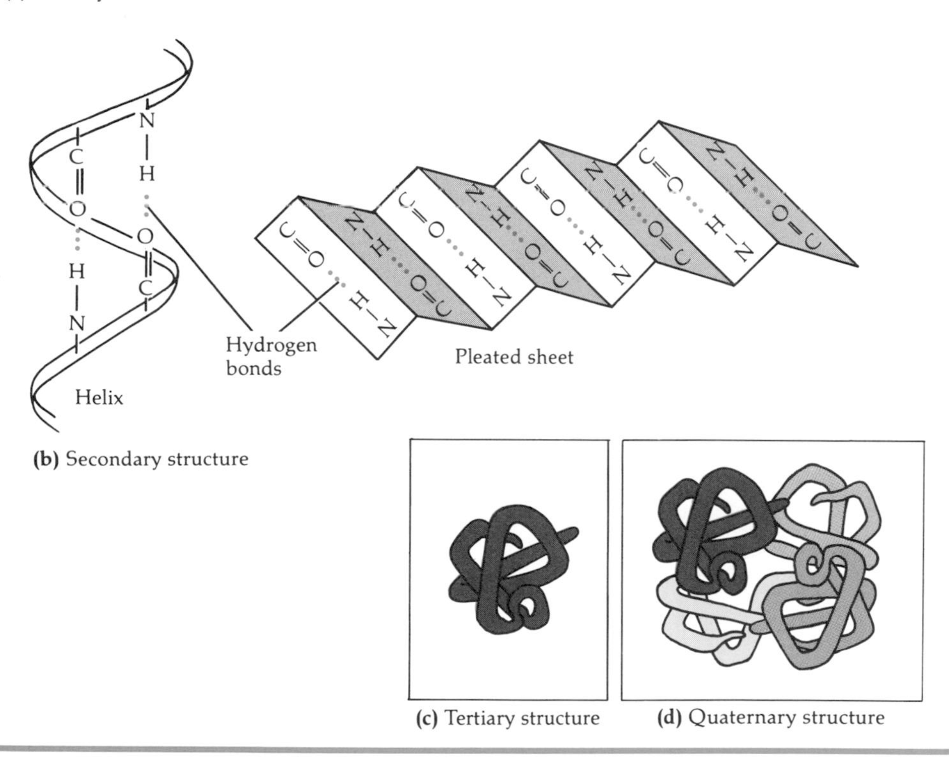

Figure 2–18 Protein structure. **(a)** Primary structure, the sequence of amino acids. Each amino acid is designated by a three-letter abbreviation. **(b)** Secondary structure, a helix and a pleated sheet. **(c)** Tertiary structure, the folding of a polypeptide chain. **(d)** Quaternary structure, the relationship between several polypeptide chains that make up a protein. Shown here is the quaternary structure of a protein with four chains, two of one type and two of another.

bacterium *Clostridium perfringens*, this enzyme breaks down red blood cells, inducing some of the symptoms of gas gangrene.

Nucleic Acids

First discovered in the nuclei of cells, **nucleic acids** are exceedingly large organic molecules containing carbon, hydrogen, oxygen, nitrogen, and phosphorus. Whereas the basic structural units of proteins are amino acids, the basic units of nucleic acids are *nucleotides*. Nucleic acids are of two principal kinds: *deoxyribonucleic acid (DNA)* and *ribonucleic acid (RNA)*.

DNA

A molecule of DNA is a chain composed of many nucleotide units. Each nucleotide of DNA has three basic parts: a nitrogen-containing base, a pentose (five-carbon sugar), and a phosphoric acid molecule (phosphate group) (Figure 2–19a). The nitrogen-containing base is either adenine, guanine, cytosine, or thymine. All the nitrogen-containing groups are ring structures containing atoms of carbon, hydrogen, oxygen, and nitrogen. Adenine and guanine are double-ring structures, collectively referred to as *purines*. Thymine and cytosine are smaller, single-ring structures, called *pyrimidines*. The pentose in DNA is *deoxyribose*.

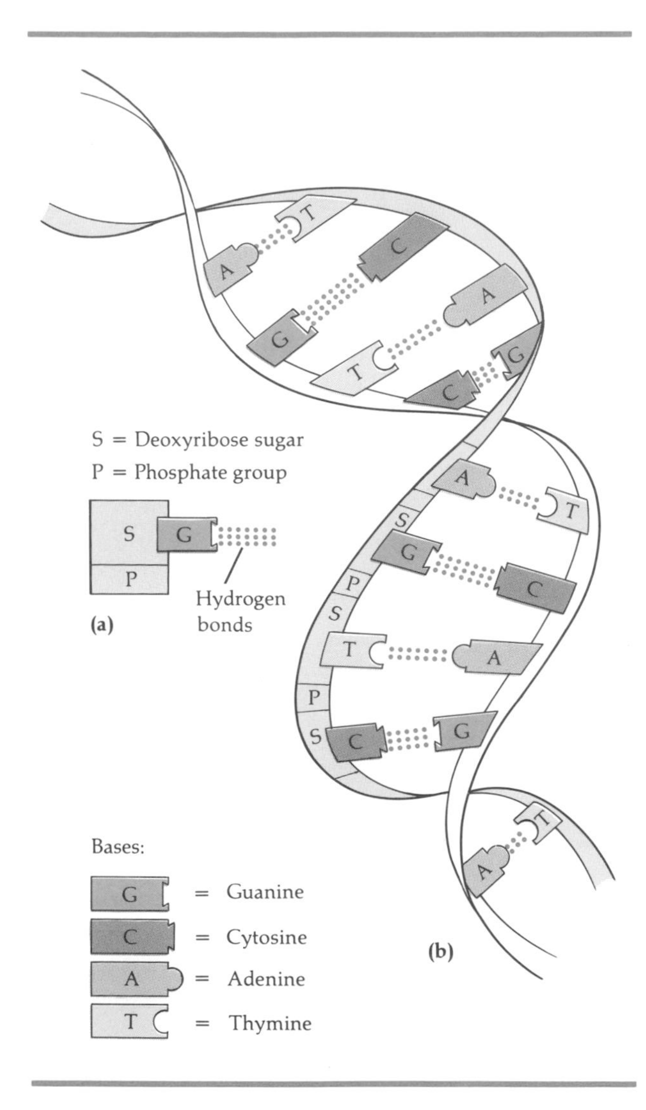

Figure 2–19 Structure of DNA. **(a)** Diagram of a guanine nucleotide. **(b)** Part of an assembled DNA molecule. The colored dots represent hydrogen bonds between nitrogenous bases.

Nucleotides are named according to the nitrogenous base that is present. Thus, a nucleotide containing thymine is called a *thymine nucleotide*. One containing adenine is called an *adenine nucleotide*, and so on.

Although the chemical composition of the DNA molecule was known before 1900, it was not until 1953 that a model of the organization of the chemical subunits was constructed. J. D. Watson and F. H. C. Crick proposed this model on the basis of data from many investigations. Figure 2–19b shows the following structural characteristics of the DNA molecule:

1. The molecule consists of two strands with crossbars. The strands twist around each other in the form of a *double helix*, so that the shape resembles a twisted ladder.
2. The uprights of the DNA ladder, referred to as the backbone, consist of alternating phosphate groups and the deoxyribose (sugar) portions of the nucleotides.
3. The rungs of the ladder contain nitrogenous bases in pairs joined by hydrogen bonds. As shown, adenine always pairs with thymine, and cytosine always pairs with guanine.

Cells contain hereditary material called genes, each of which is a segment of a DNA molecule. Genes determine all hereditary traits, and they control all the activities that take place within cells. When a cell divides, its hereditary information is passed on to the next generation of cells. This transfer of information is possible because of DNA's unique structure, and will be discussed further in Chapter 8.

RNA

RNA, the second principal kind of nucleic acid, differs from DNA in several respects. Whereas DNA is double-stranded, RNA is usually single-stranded. The five-carbon sugar in the RNA nucleotide is ribose, which has one more oxygen atom than deoxyribose. And one of RNA's bases is uracil instead of thymine. At least three different kinds of RNA have been identified in cells. As we shall see in Chapter 8, each type of RNA has a specific role in protein synthesis.

Adenosine Triphosphate (ATP)

A molecule that is indispensable to the life of the cell is **adenosine triphosphate (ATP).** ATP is the principal energy-carrying molecule of all cells. It stores chemical energy released by some chemical reactions, and it provides energy for energy-requiring reactions. Structurally, ATP consists of an adenosine unit composed of adenine and ribose joined

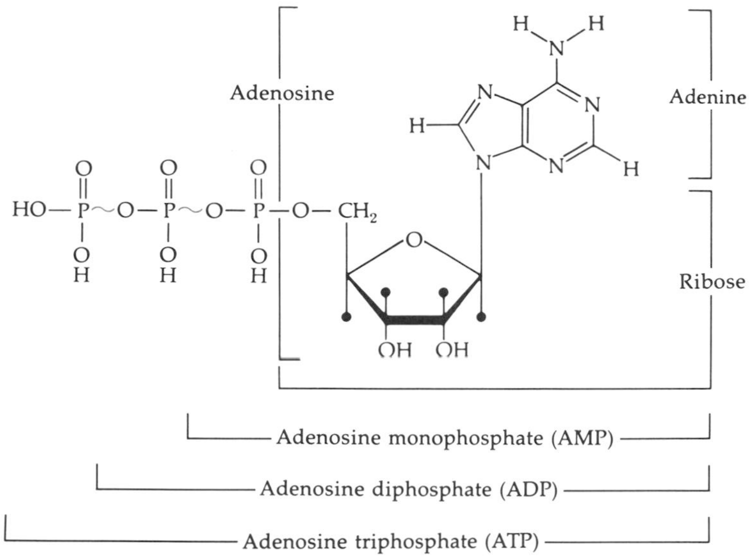

Figure 2–20 Structure of ATP. High-energy phosphate bonds are shown in color and are wavy. When ATP breaks down to ADP and inorganic phosphate, a large amount of chemical energy is released that can be used in other chemical reactions.

to three phosphate groups (Figure 2–20). In other words, it is an adenine nucleotide with two extra phosphate groups. ATP is regarded as a high-energy molecule because of the total amount of usable energy it releases when it is converted to adenosine diphosphate (ADP) by the addition of a water molecule and the loss of a terminal phosphate group (—PO_3 or Ⓟ). This reaction may be represented as follows:

ATP	⇌	ADP	+	Ⓟ	+	E
Adenosine triphosphate		Adenosine diphosphate		Phosphate		Energy

Since the supply of ATP at any given time is limited, a mechanism exists to replenish it: a phosphate group is added to ADP to manufacture more ATP. Since we can assume that energy is required to manufacture ATP, the reaction may be represented as follows:

ADP	+	Ⓟ	+	E	⇌	ATP
Adenosine diphosphate		Phosphate		Energy		Adenosine triphosphate

The energy required to attach a phosphate group to ADP is supplied by various decomposition reactions taking place in the cell, particularly by the decomposition of glucose. ATP can be stored in every cell, where it provides potential energy that is not released until needed. Energy released by catabolic reactions is stored in ATP and can be used to drive anabolic reactions.

STUDY OUTLINE

WHY STUDY CHEMISTRY IN A MICROBIOLOGY COURSE? (p. 24)

1. The interaction of atoms and molecules is called chemistry.
2. The metabolic activities of microorganisms involve complex chemical reactions.

3. Nutrients are broken down by microbes to obtain energy and molecules that are used to make new cells.

STRUCTURE OF ATOMS (pp. 25–27)

1. Atoms are the smallest units of chemical elements that enter into chemical reactions.
2. Atoms consist of a nucleus that contains protons and neutrons, and electrons that orbit around the nucleus.
3. Atomic number refers to the number of protons in the nucleus; the number of protons and neutrons is the atomic weight.

Chemical Elements (p. 25)

1. Atoms with the same atomic number and same chemical behavior are classified as the same chemical element.
2. Chemical elements are designated by letter abbreviations called chemical symbols.
3. There are about 23 elements commonly found in living cells.
4. Atoms of the same elements that have the same atomic number and different atomic weights are called isotopes.

Electronic Configurations (pp. 26–27)

1. In an atom, electrons are arranged around the nucleus in electron shells.
2. Each shell can hold a characteristic number of electrons.

HOW ATOMS FORM MOLECULES: CHEMICAL BONDS (pp. 28–32)

1. Atoms form molecules in order to fill their outermost electron shells.
2. Attractive forces that bind atomic nuclei together are called chemical bonds.
3. The combining capacity of an atom is its valence, the number of chemical bonds the atom can form with other atoms.
4. Any positively or negatively charged atom or group of atoms is called an ion.

Ionic Bonds (pp. 28–29)

1. A chemical attraction between ions of opposite charge is called an ionic bond.
2. Ionic bonds have only limited importance in living cells.

Covalent Bonds (pp. 29–31)

1. In covalent bond formation, atoms share pairs of electrons.
2. Covalent bonds are stronger and far more common in organisms than ionic bonds.

Hydrogen Bonds (pp. 31–32)

1. A hydrogen bond consists of a hydrogen atom covalently bonded to one oxygen or nitrogen atom but attracted to another.
2. Hydrogen bonds form weak links between different molecules or between parts of the same large molecule.

Molecular Weight and Moles (p. 32)

1. Molecular weight is the sum of the atomic weights of all the atoms in a molecule.
2. A mole of an atom, ion, or molecule is equal to the weight times the molecular weight.

CHEMICAL REACTIONS (pp. 32–35)

1. Chemical reactions are the result of the making or breaking of chemical bonds between atoms.
2. Three basic types of chemical reactions are synthesis, decomposition, and exchange reactions. In theory, all chemical reactions are reversible.

How Chemical Reactions Occur (pp. 33–34)

1. For a chemical reaction to take place, the reactants must collide with each other (collision theory).
2. The minimum energy of collision for a chemical reaction is called activation energy.
3. Specialized proteins called enzymes accelerate chemical reactions in living systems by lowering the activation energy.

Energy of Chemical Reactions (p. 34)

1. When a chemical bond is formed, energy is required.
2. When a bond is broken, energy is released.

Forms of Energy (pp. 34–35)

1. Chemical, mechanical, radiant, and electrical energy all are important to living cells.
2. These forms of energy may exist as either potential or kinetic energy.

IMPORTANT BIOLOGICAL MOLECULES (pp. 35–49)

INORGANIC COMPOUNDS (pp. 35–39)

1. Inorganic compounds are usually small, ionically bonded molecules.
2. Water, acids, bases, and salts are examples of inorganic compounds.

Water (pp. 35–36)

1. Water is the most abundant substance in cells.
2. Water is a polar molecule, which makes it an excellent solvent.
3. Water is an excellent temperature buffer.

Acids, Bases, and Salts (p. 36)

1. An acid dissociates into H^+ ions and anions.
2. A base dissociates into OH^- ions and cations.
3. A salt dissociates into negative and positive ions, neither of which are H^+ or OH^-.

Acid–Base Balance (pp. 36–39)

1. The term pH refers to the concentration of H^+ in a solution.
2. A solution with a pH of 7 is neutral; a pH below 7 indicates acidity; a pH above 7 indicates alkalinity.

ORGANIC COMPOUNDS (pp. 39–49)

1. Carbon is the characteristic element of organic compounds.
2. Carbon atoms form four bonds with carbon and other atoms.
3. Organic compounds are mostly or entirely covalently bonded, and many of them are large molecules.
4. Small organic molecules may combine into very large molecules called macromolecules.

Carbohydrates (pp. 40–43)

1. Carbohydrates consist of atoms of carbon, hydrogen, and oxygen, with hydrogen and oxygen in a 2:1 ratio.
2. Carbohydrates include sugars and starches.
3. Carbohydrates are divided into three groups: monosaccharides, disaccharides, and polysaccharides.

4. Monosaccharides may form disaccharides and polysaccharides by dehydration synthesis.
5. Polysaccharides and disaccharides may be broken down by hydrolysis.

Lipids (pp. 43–44)

1. Lipids are a diverse group of compounds distinguished by their insolubility in water.
2. Simple lipids (fats) consist of a molecule of glycerol and three molecules of fatty acids.
3. Complex lipids, such as phospholipids, may contain atoms of phosphorus, nitrogen, or sulfur in addition to carbon, hydrogen, and oxygen.

Proteins (pp. 44–47)

1. Amino acids are the building blocks of proteins.
2. Amino acids consist of carbon, hydrogen, oxygen, nitrogen, and sometimes sulfur.
3. Amino acids link together by peptide bonds to form polypeptide chains.
4. Twenty amino acids occur naturally.
5. Proteins are organized into four levels of structure: primary (linear chain of amino acids); secondary (regular coils or pleats); tertiary (irregular folds); and quaternary (two or more polypeptide chains).

Nucleic Acids (pp. 47–48)

1. Nucleic acids—DNA and RNA—are macromolecules consisting of repeating nucleotides.
2. A nucleotide is composed of a pentose, phosphoric acid, and a nitrogenous base.
3. A DNA nucleotide consists of deoxyribose (pentose) and one of these nitrogenous bases: cytosine, guanine, adenine, or thymine.
4. DNA consists of two strands of nucleotides wound in a double helix. The strands are held together by hydrogen bonds between nitrogenous bases.
5. An RNA nucleotide consists of ribose (pentose) and one of these nitrogenous bases: cytosine, guanine, adenine, or uracil.

Adenosine Triphosphate (ATP) (pp. 48–49)

1. ATP stores chemical energy for various cellular activities.
2. When ATP's terminal phosphate group is broken, energy is released.
3. ATP is regenerated from ADP using the energy from decomposition reactions.

STUDY QUESTIONS

REVIEW

1. What is a chemical element?
2. Diagram the electronic configuration of a carbon atom.
3. How does $^{14}_{6}C$ differ from $^{12}_{6}C$?
4. What type of bonds will hold the following atoms together?
   (a) Li^+ and Cl^- ions in LiCl
   (b) Carbon and oxygen atoms in CO_2
   (c) Oxygen atoms in O_2
   (d) A hydrogen atom from glutamic acid and a nitrogen atom from lysine in

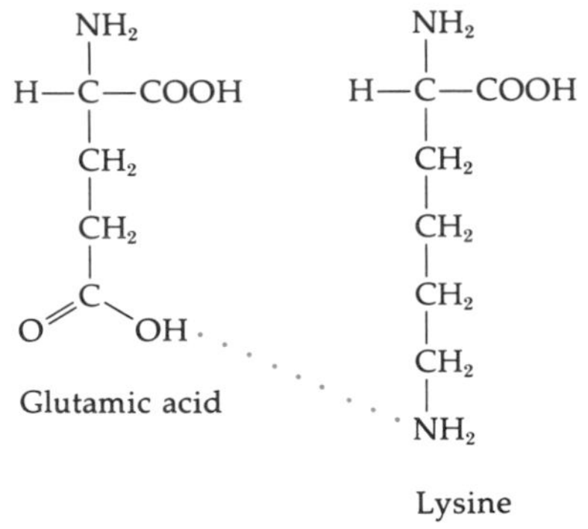

Glutamic acid

Lysine

   (e) A hydrogen atom of one nucleotide to nitrogen or oxygen atoms of another nucleotide in

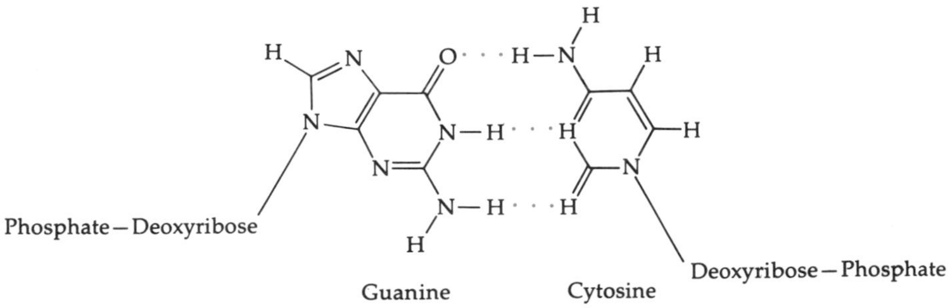

Phosphate—Deoxyribose

Deoxyribose—Phosphate

Guanine

Cytosine

5. What type of bonding exists between water molecules in a beaker of water?
6. Classify the following inorganic molecules as an acid, base, or salt. Their dissociation products are shown to help you.
   (a) $HNO_3 \rightarrow H^+ + NO_3^-$
   (b) $H_2SO_4 \rightarrow 2H^+ + SO_4^{2-}$
   (c) $NaOH \rightarrow Na^+ + OH^-$
   (d) $Mg_2SO_4 \rightarrow 2Mg^{2+} + SO_4^{2-}$
7. Vinegar, pH 3, is how many times more acidic than pure water?
8. Calculate the molecular weight of $C_6H_{12}O_6$.

9. How many moles are in 360 grams of glucose?

10. Classify the following types of chemical reactions.
    (a) Glucose + Fructose $\rightarrow$ Sucrose
    (b) Lactose $\rightarrow$ Glucose + Galactose
    (c) $NH_4Cl + H_2O \rightarrow NH_4OH + HCl$
    (d) ATP $\rightleftharpoons$ ADP + Ⓟ

11. The following reaction requires an enzyme. What purpose does the enzyme serve in this reaction?
    Lactose $\rightarrow$ Glucose + Galactose

12. Classify the following as subunits of either a carbohydrate, lipid, protein, or nucleic acid.
    (a) $CH_3—(CH_2)_7—CH=CH—(CH_2)_7—COOH$

    Oleic acid

    (b) 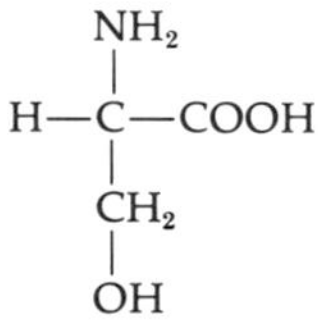
```
      NH2
      |
   H—C—COOH
      |
      CH2
      |
      OH
```
    Serine

    (c) $C_6H_{12}O_6$

    (d) Thymine nucleotide

13. Water plays an important role in these reactions:

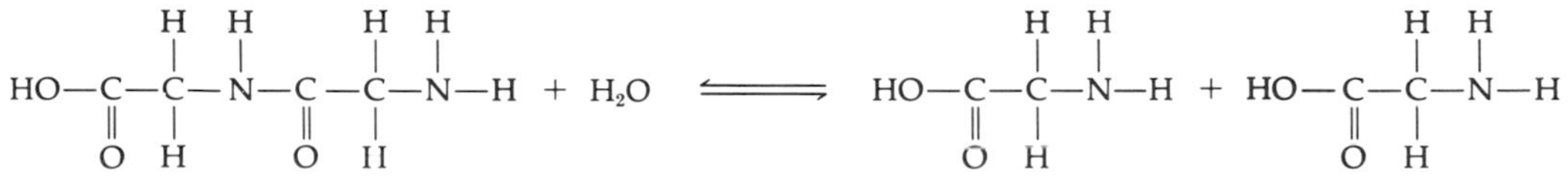

```
     H  H     H  H                          H  H              H  H
     |  |     |  |                          |  |              |  |
HO—C—C—N—C—C—N—H + H2O  ⇌  HO—C—C—N—H + HO—C—C—N—H
    ||  |    ||  |                  ||  |            ||  |
    O  H     O  II                  O  H             O  H
```

    (a) What direction is the hydrolysis reaction (left to right or right to left)?
    (b) What direction is the dehydration synthesis reaction?
    (c) Circle the atoms involved in the formation of water.
    (d) Identify the peptide bond.

14. This diagram shows a protein molecule. Indicate the regions of primary, secondary, and tertiary structure. Does this protein have quaternary structure?

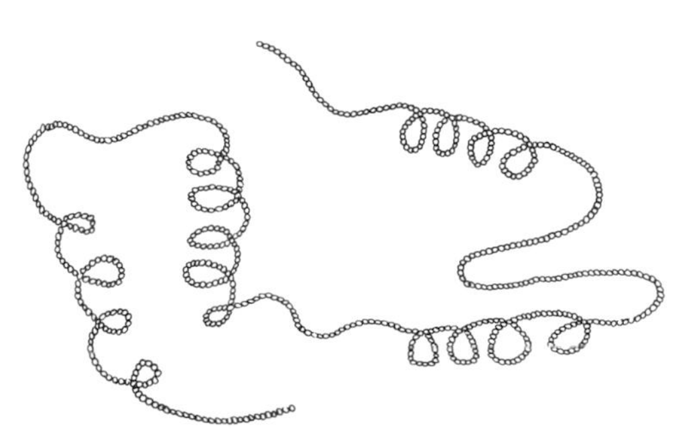

15. ATP is an energy storage compound. Where does it get this energy from?

CHALLENGE

1. When you blow bubbles into a glass of water, the following reactions take place:

$$H_2O + CO_2 \xrightarrow{A} H_2CO_3 \xrightarrow{B} H^+ + HCO_3^-$$

   (a) What type of reaction is *A*?
   (b) What does reaction *B* tell you about the type of molecule H_2CO_3 is?

2. The energy-carrying property of the ATP molecule is due to energy dynamics favoring breaking bonds between ________. These bonds are (ionic/covalent/hydrogen) bonds.

3. What are the common structural characteristics of ATP and DNA molecules?

FURTHER READING

Asimov, I. 1962. *The World of Carbon.* New York: Collier Books. An introduction to organic chemistry designed for the nonchemist.

Dickerson, R. E. and I. Geis. 1976. *Chemistry, Matter and the Universe.* Menlo Park, CA: Benjamin/Cummings. An introductory college chemistry textbook with more organic chemistry than most texts.

Luria, S. E., S. J. Gould, and S. Singer. 1981. *A View of Life.* Menlo Park, CA: Benjamin/Cummings. Chapter 2 provides an introduction to chemistry for the biology student.

Stryer, L. 1981. *Biochemistry,* 2nd ed. San Francisco, CA: W. H. Freeman. A technical survey of biochemistry, with a cellular approach.

Watson, J. D. *The Double Helix,* ed. Gunther S. Stent. 1980. New York: W. W. Norton. Original papers and Watson's account of the struggle to decipher the structure of DNA give insight to creative scientific processes. Includes commentaries and reviews by other scientists.

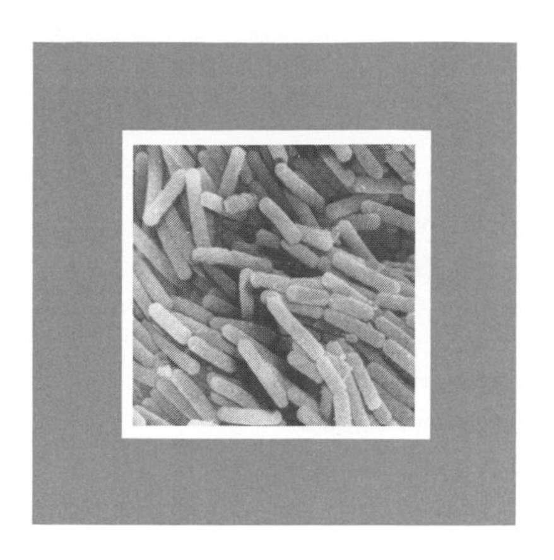

3

Observing Microorganisms Through a Microscope

Objectives

After completing this chapter you should be able to

- List the units of measurement used to measure microorganisms, and know their equivalents.
- Diagram the path of light through a compound microscope.
- Define resolution and total magnification.
- Cite an advantage of each of the following types of microscopy and compare each to brightfield illumination: (a) phase-contrast, (b) darkfield, (c) differential interference contrast, (d) fluorescence.
- Explain how electron microscopy differs from light microscopy.
- Differentiate between an acid dye and a basic dye.
- Compare simple, differential, and special stains.
- List the steps in a Gram stain and describe the appearance of a gram-positive and gram-negative cell after each step.

Since microorganisms are too small to be seen with the naked eye, they must be observed with the aid of a microscope (*micro* means "small," *skopein* means "to see"). In Leeuwenhoek's day, looking through a microscope meant looking at a specimen through rainbow rings and shadows and multiple images. Modern microbiologists, however, have access to microscopes that magnify, with great clarity, anywhere from 10 to thousands of times more than Leeuwenhoek's simple single lens. A major part of this chapter will deal with how different microscopes function and what kind of "seeing" they do best.

Some microbes are more visible than others. Many have to be escorted through several staining procedures before cell walls, membranes, and structures lose their opaque or colorless natural state. The last part of this chapter will explain the methods of preparing specimens for light microscopy.

With the prospect of microbial images dancing around in your head, you may be wondering just how we are going to sort out, measure, and count the specimens we will be studying. Therefore, we open this chapter with a short discussion of how to use the metric system for microbial measurements.

UNITS OF MEASUREMENT

Because microorganisms and their component parts are so very small, they are measured in units that are unfamiliar to many of us in everyday life. When measuring microorganisms, we use the metric system.

The standard unit of length in the metric system is the **meter (m).** (One meter is equal to 39.37 inches in the English system.) A major advantage of the metric system is that the units are related to each other by factors of 10. Thus, 1 m is the same as 10 decimeters (dm) or 100 centimeters (cm) or 1000 millimeters (mm). (With the English system of length, we would have to say that 1 yard is the same as 3 feet or 36 inches.)

Microorganisms and their structural components are measured in even smaller units, such as micrometers, nanometers, and angstroms. A **micrometer (μm),** formerly known as a micron (μ), is equal to 0.000001 m. The prefix *micro* indicates that the unit following it should be divided by one million (10^6) (see Box on Exponential Notation in Chapter 2). A **nanometer (nm),** formerly known as a millimicron (mμ), is equal to 0.000000001 m. *Nano* tells us that the unit after it should be divided by one billion (10^9). An **angstrom (Å)** is equal to 0.0000000001 m, or 10^{-10} m. Table 3–1 presents the basic metric units of length and some of their English equivalents. Here you can compare the microscopic units of measurement with other commonly known macroscopic units of measurement such as centimeters, meters, and kilometers.

MICROSCOPY: THE INSTRUMENTS

The microscopes used by Leeuwenhoek and Hooke were simple microscopes consisting of only one lens, similar to a magnifying glass. A combination of lenses was later used by Joseph Jackson Lister (the father of Joseph Lister) in 1830. This led to the modern compound microscope, the kind used today in a microbiology laboratory.

Compound Light Microscopy

A modern **compound light microscope** has two sets of lenses, objective and ocular, and uses visible light as its source of illumination (Figure 3–1a). With the compound light microscope, we can examine very small specimens, as well as their fine detail or *ultrastructure*. A series of finely ground lenses are arranged so that a clearly focused image is formed that is many times larger than the actual specimen itself. To achieve this magnification, light rays from an *illuminator,* the light source, are passed through a *condenser,* which directs the light rays through the specimen. From here, light rays pass into the *objective lens,* the lens closest to the specimen being examined. The image of the specimen then forms on a mirror and is magnified again by the *ocular lens,* or *eyepiece* (Figure 3–1b).

The total magnification of a specimen is calculated by multiplying the objective lens magnification (power) by the ocular lens magnification (power). Most microscopes used in microbiology have three objective lenses: 10 times (low power), 40 times (high power), and 100 times (oil immersion). Most oculars magnify specimens by 10 times. If we multiply the magnification of a specific objective lens with that of the ocular, the total magnifi-

Table 3-1 Metric Units of Length and English Equivalents

Metric Unit	Meaning of Prefix	Metric Equivalent	English Equivalent
1 kilometer (km)	*kilo* = 1000	1000 m = 10^3 m	3280.84 ft or 0.62 mi; 1 mi = 1.61 km
1 meter (m)		Standard unit of length	39.37 inches or 3.28 ft or 1.09 yd
1 decimeter (dm)	*deci* = 1/10	0.1 m = 10^{-1} m	3.94 inches
1 centimeter (cm)	*centi* = 1/100	0.01 m = 10^{-2} m	0.394 inch; 1 inch = 2.54 cm
1 millimeter (mm)	*milli* = 1/1000	0.001 m = 10^{-3} m	
1 micrometer (μm) (formerly micron (μ))	*micro* = 1/1,000,000	0.000001 m = 10^{-6} m	
1 nanometer (nm) (formerly millimicron (mμ))	*nano* = 1/1,000,000,000	0.000000001 m = 10^{-9} m	
1 angstrom (Å)		0.0000000001 m = 10^{-10} m	

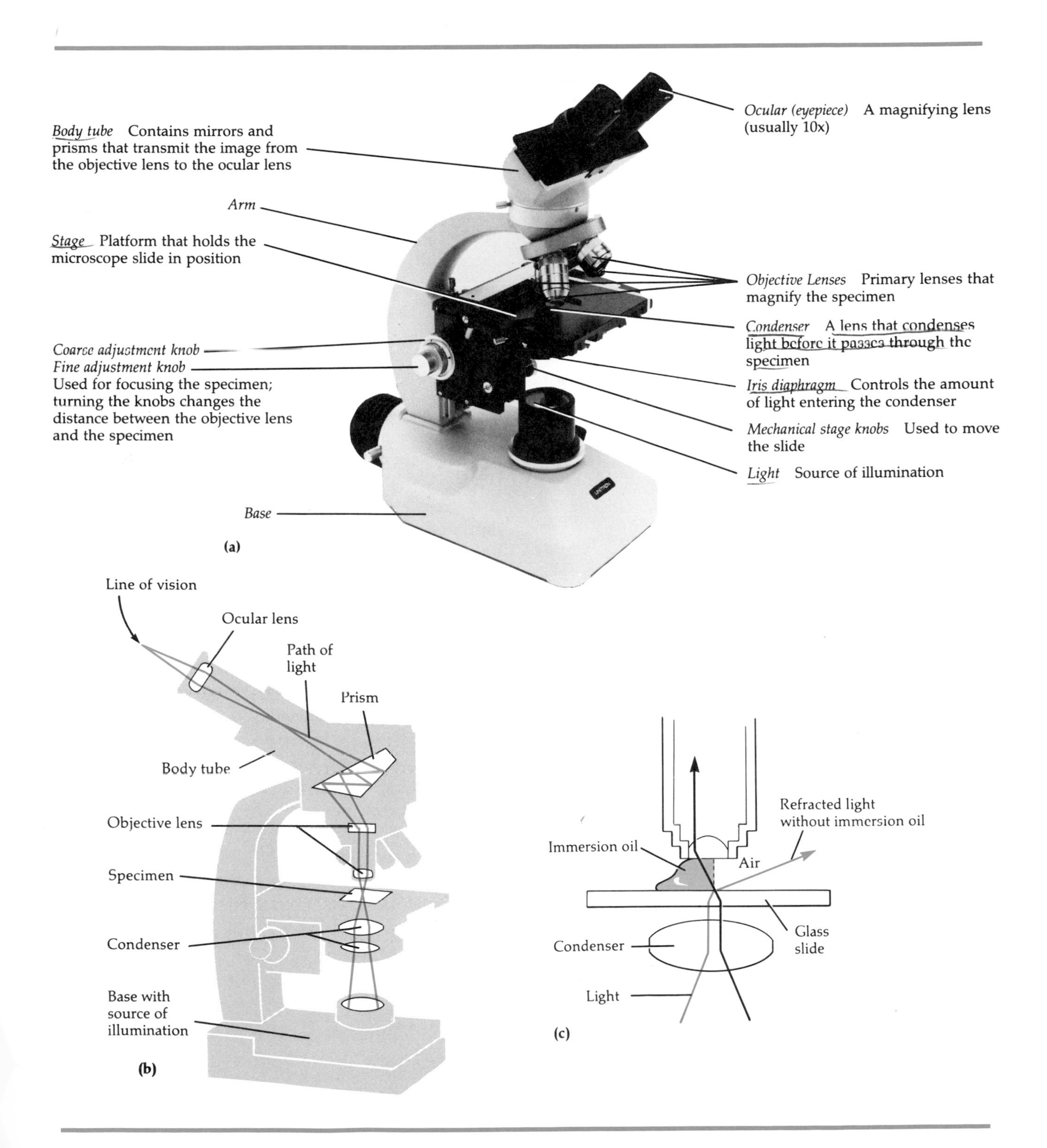

Figure 3–1 The compound light microscope. **(a)** Principal parts and their functions. **(b)** The path of light. **(c)** Refractive index. Since the refractive indexes of the glass microscope slide and immersion oil are the same, the oil keeps the light rays from refracting.

cations would be 100 times for low power, 400 times for high power, and 1000 times for oil immersion. Some compound light microscopes can achieve a magnification of 2000 times with oil immersion.

A general principle of microscopy is that the shorter the wavelength of light used in the instrument, the greater the resolution. **Resolution,** or **resolving power,** is the ability of the lenses of a microscope to distinguish fine detail and structure. Specifically, it refers to the ability of the lenses to distinguish between two points set at a specified distance apart. For example, if a microscope has a resolving power of 4 Å, it is capable of distinguishing two points as separate objects if they are at least 4 Å apart. The white light used in a compound light microscope has a relatively long wavelength and cannot resolve structures smaller than 0.3 μm. This fact and practical considerations limit the magnification achieved by even the best compound light microscope to about 2000 times.

It is usually necessary to stain a specimen being examined under a compound light microscope, because the refractive index of the specimen is nearly the same as the refractive index of the medium that surrounds it. The **refractive index** is the relative velocity at which light passes through a specimen. Light rays traveling in a single medium usually move in a straight line. But if the light rays pass through two materials with different refractive indexes, the rays change direction (refract) at the boundary between the materials, increasing the contrast between the specimen's image and the medium's.

Once the light rays pass through the stained specimen, it is important to preserve their direction, which is accomplished by using immersion oil between the glass slide and the objective lens. Since immersion oil has a refractive index similar to that of glass, the oil keeps the light rays from refracting again—which would result in a fuzzy image with poor resolution (Figure 3–1c).

Under usual operating conditions, the field of vision in a compound light microscope is brightly illuminated. The condenser focuses the light straight upward, producing a **brightfield** illumination (Figure 3–2a). An unstained cell has little contrast with its surroundings and is difficult to see. Unstained cells are more easily observed in the modified compound microscopes described in the next section.

Phase-Contrast Microscopy

Another way to observe microorganisms is with a **phase-contrast microscope.** This technique is especially useful because it permits the detailed examination of structures in *living* microorganisms. In this technique, a special condenser is used to take advantage of subtle differences in the refractive index of different parts of the living cell and its surrounding medium. The light is slowed down as it passes through the denser parts of the microorganism, and so certain structures stand out from other parts of the cell and the surrounding medium (Figure 3–2b). The advantage of this technique is that the microorganism can be studied in its natural state; it does not have to be killed or stained.

Darkfield Microscopy

A **darkfield microscope** is used for examining certain microorganisms that either are invisible in the living state in the ordinary light microscope, cannot be stained by standard methods, or are so distorted by staining that their characteristics cannot be identified. In a darkfield microscope, light rays do not pass directly through the microorganism. Instead, a special device is inserted into or below the condenser so that light hits only the sides of the specimen. Only light *scattered* by the specimen reaches the lens. The specimen appears white against a black background—the dark field (Figure 3–2c). Internal structures of the cell are often not visible. This technique is frequently used to examine unstained microorganisms suspended in liquid. One use for darkfield microscopy is the examination of very thin spirochetes, such as *Treponema pallidum,* the causative agent of syphilis.

Differential Interference Contrast (DIC) Microscopy

Differential interference contrast (DIC) microscopy is a relatively new technique for viewing transparent living microscopic entities. In DIC microscopy, beams of light are split and recombined by specially designed prisms. The effect is a spectacular three-dimensional view of the object (Figure 3–2d).

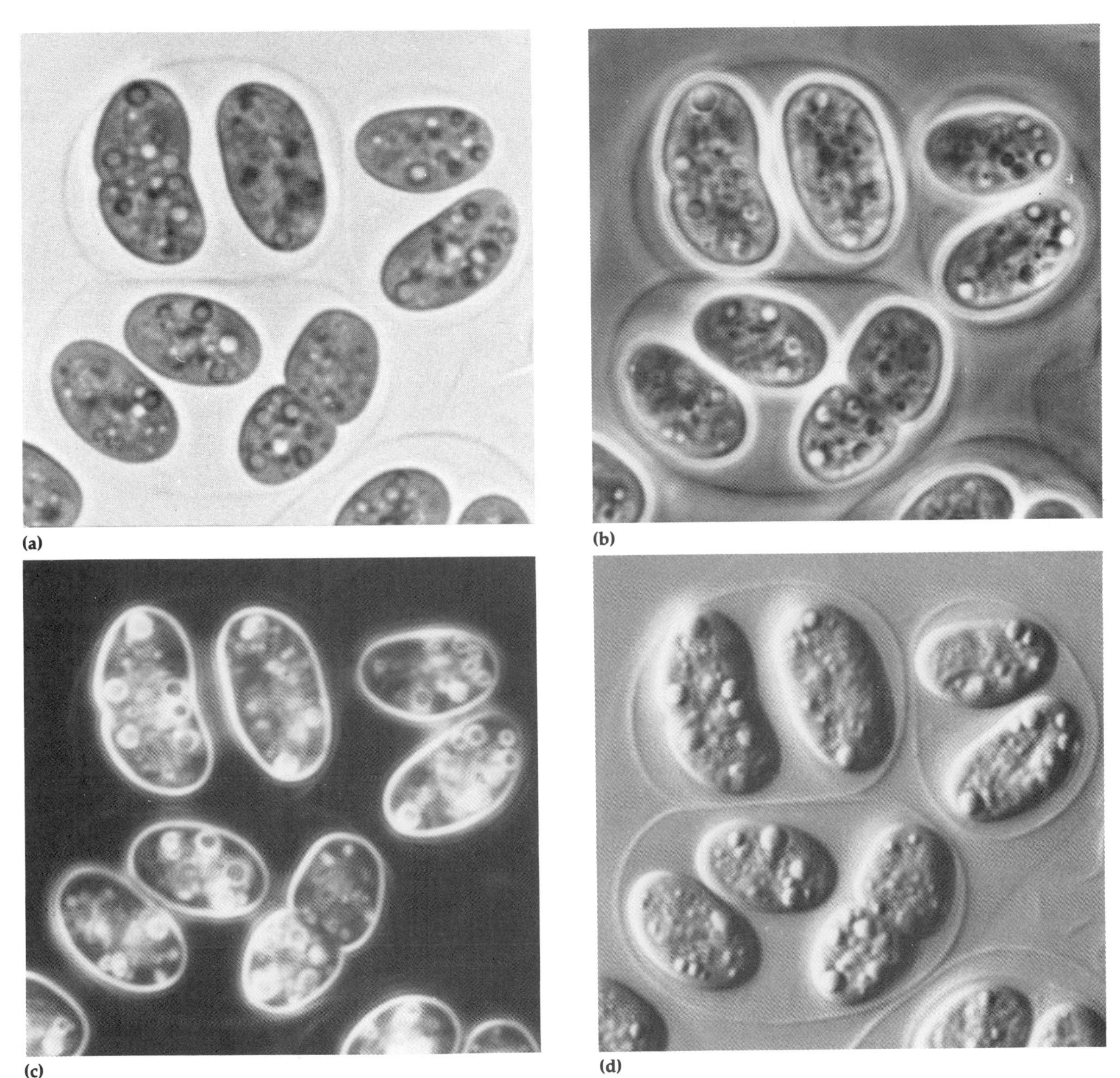

Figure 3–2 Types of images produced with a compound light microscope. Shown here are the same cells of a cyanobacterium (also known as a blue-green alga) in the genus *Gloeocapsa*, as seen using different microscopic techniques. All cells are printed at a magnification of $\times 2800$. **(a)** Brightfield illumination shows internal structures in the cells and the outline of the transparent sheath that often surrounds two or four recently divided cells. **(b)** Phase-contrast microscopy shows greater differentiation among the internal structures and also shows the sheath well. The wide light band around each cell is an artifact resulting from this type of microscopy—a "phase halo." **(c)** Against the dark background seen with darkfield microscopy, edges of the cell are bright, some internal structures seem to "sparkle," and the sheath is almost invisible. **(d)** Differential interference contrast microscopy results in a three-dimensional appearance for the cells and their inclusions. The surrounding sheaths are sharply defined.

Fluorescence Microscopy

Fluorescence microscopy takes advantage of the fluorescence of substances. Fluorescent substances give off light of one color when light of another color strikes them. Fluorescence microscopy uses an ultraviolet or near-ultraviolet light source. If the specimen to be viewed does not naturally fluoresce, it is stained with one of a group of fluorescent dyes called *fluorochromes*. Microorganisms stained with a fluorochrome and examined with a microscope using ultraviolet light appear as luminescent, bright objects against a dark background (Figure 3–3b).

Fluorochromes have special attractions for different microorganisms. For example, the fluorochrome auramine O, which glows yellow when exposed to ultraviolet light, is strongly absorbed by *Mycobacterium tuberculosis,* the bacterium that causes tuberculosis. When the dye is applied to a sample of material suspected of containing the bacterium, the presence of the bacterium is detected by the appearance of bright yellow organisms against a dark background. *Bacillus anthracis,* the causative agent of anthrax, appears apple-green in color when stained with another fluorochrome, fluorescein isothiocyanate (FITC).

The principal use of fluorescence microscopy is a diagnostic technique called the **fluorescent-antibody technique** or **immunofluorescence. Antibodies** are natural defense molecules produced by humans and many animals in reaction to a foreign substance, or **antigen.** Fluorescent antibodies for a particular antigen are obtained as follows: An animal is injected with a specific antigen, such as a bacterium, and it begins to produce specific antibodies against that antigen. After a sufficient amount of time, the antibodies are removed from the serum of the animal. Next, a fluorochrome is chemically combined with the antibodies. These fluorescent antibodies are then added to a microscope slide containing an unknown bacterium. If this unknown bacterium is the same bacterium that

Leeuwenhoek's Description of a Protozoan

Leeuwenhoek was an uneducated man, ignorant of the science of his day and of Latin, the language of scholars. But he was a superb lens grinder and a careful, patient observer of the world he saw through his microscope. The following is his description of a ciliated protozoan, translated from Leeuwenhoek's colloquial Dutch.*

In the year 1675, . . . I discovered living creatures in rain, which had stood but a few days in a new tub. . . . These little animals were, to my eye, more than ten thousand times smaller than the animalcules . . . called by the name of Water-flea, or Water-louse, which you can see alive and moving in water with the bare eye. . . . I . . . discovered a . . . sort of animalcules [sic], whose figure was an oval; and I imagined that their head was placed at the pointed end. . . . Their belly is flat, provided with divers incredibly thin little feet, or little legs, which were moved very nimbly, and which I was able to discover only after sundry great efforts, and wherewith they brought off incredibly quick motions. The upper part of their body was round, and furnished inside with 8, 10, or 12 globules: otherwise these animalcules were very clear. These little animals would change their body into a perfect round, but mostly when they came to lie high and dry. Their body was also very yielding: for if they so much as brushed against a tiny filament, their body bent in, which bend also presently sprang out again; just as if you stuck your finger into a bladder full of water, and then, on removing the finger, the inpitting went away.

*Translation by Clifford Dobell. Reproduced in H. A. Lechevalier and M. Solotorovsky, *Three Centuries of Microbiology,* New York: Dover, 1974.

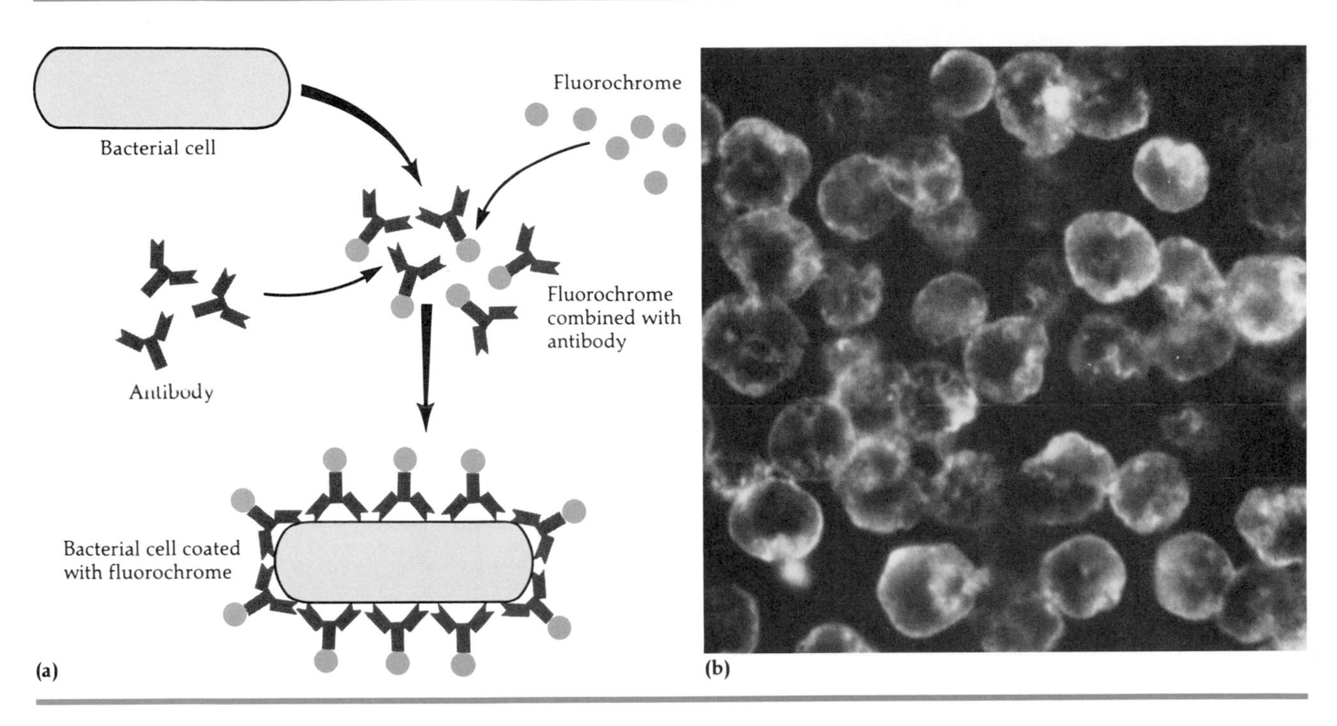

Figure 3–3 **(a)** The principle of immunofluorescence. A fluorochrome is combined with a specific bacterial antibody. The preparation is added to bacterial cells on a microscope slide, and the antibody attaches to the bacterial cells. As a result, the cells fluoresce. **(b)** With the fluorescent-antibody technique, the structures to which the specific antibodies attach show up as light regions against a darker background. Here amoebas of the species *Naegleria fowleri* (a protozoan) present a bright appearance in a fluorescent-antibody stain of brain tissue from a fatal case of amoebic meningoencephalitis ($\times$3000).

was injected into the animal, the fluorescent antibodies bind to the surface of the bacteria, causing them to fluoresce (Figure 3–3a). This technique can detect bacteria or other disease-producing microorganisms even within cells, tissues, or other clinical specimens (Figure 3–3b and Color Plate III). It is especially useful in diagnosing syphilis and rabies. We will say more about antigen-antibody reactions and immunofluorescence in Chapter 16.

Electron Microscopy

To examine objects smaller than 0.3 μm, such as viruses or the internal structures of cells, an **electron microscope** must be used. Its resolving power is far greater than the other microscopes mentioned. In electron microscopy, a beam of electrons is used instead of light. Free electrons travel in waves, just as light does. Instead of glass lenses, magnets are used in an electron microscope to focus a beam of electrons through a vacuum tube onto a specimen. Since the wavelengths of electrons are about 1/100,000 that of visible light, the resolving power of a very sophisticated electron microscope can come close to 2.5 Å, and therefore can resolve small molecules. Conventional electron microscopes have a resolving power of about 10 Å (1 nm) and can magnify objects up to 200,000 times.

Transmission electron microscopy

There are two types of electron microscopes: the transmission electron microscope and the scanning electron microscope. In the **transmission electron microscope,** a finely focused beam of electrons passes through specially prepared, ultrathin sections of the specimen. The beam is then refocused,

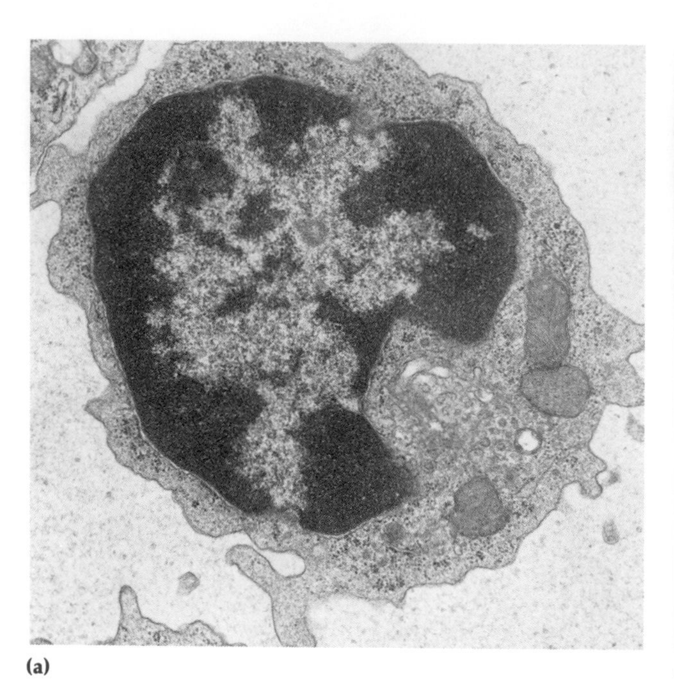

(a)

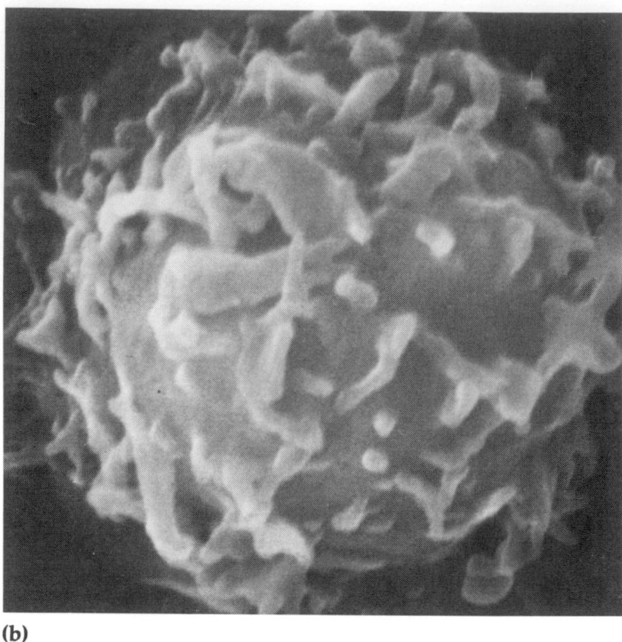

(b)

Figure 3–4 The same type of cell viewed through a transmission electron microscope and a scanning electron microscope. **(a)** This transmission electron micrograph shows a thin slice of a lymphocyte, a type of white blood cell. This type of microscopy allows one to see internal structures—at least those structures present in the slice. **(b)** In a scanning electron micrograph, surface structures can be seen, as demonstrated in this view of a lymphocyte. Note the three-dimensional appearance of this cell in contrast to the two-dimensional appearance of the cell in (a). (Both cells ×18,000).

and the image reflects what the specimen does to the transmitted electron beam. The image produced is called a **transmission electron micrograph** (Figure 3–4a). Since the density of most microscopic specimens is very low, the contrast between their ultrastructures and the background is weak. Contrast can be greatly enhanced by "staining" with salts of various heavy metals such as lead, tungsten, and uranium. These metals may be fixed on the specimen (positive staining) or used to increase the electron opacity of the surrounding field (negative staining). Negative staining is useful for studying the very smallest specimens, such as virus particles, bacterial flagella, and protein molecules.

Scanning electron microscopy

Until recently, a disadvantage of electron microscopes has been the necessity to section (cut into thin slices) the specimen before examination. Although transmission microscopy of thin sections is an extremely valuable technique for examining the inside of specimens at different layers, it does not provide a three-dimensional effect. Use of the **scanning electron microscope** overcomes this difficulty. This instrument provides striking three-dimensional views of specimens at magnifications of about 10,000 times (Figure 3–4b). In scanning electron microscopy, a finely focused beam of electrons is directed over the specimen and then *reflected from the surface* of the specimen onto a televisionlike screen or photographic plate. The scanning electron microscope is especially useful in studying the surface structure of intact cells and viruses.

PREPARATION OF SPECIMENS FOR LIGHT MICROSCOPY

Since most microorganisms appear almost colorless through a standard light microscope, it is often necessary to prepare them for observation by fixing and staining them. Occasionally, however, it is important to observe living microorganisms, which precludes staining. We therefore begin with a brief discussion of a hanging-drop preparation.

Hanging-Drop Preparation

In a **hanging-drop preparation,** a drop of a microbial suspension is placed on the center of a cover glass. Then the edges of the cover glass are coated with petroleum jelly, such as Vaseline (Figure 3–5a). Next, a special microscope slide with a concavity in its center (depression slide) is pressed against the petroleum jelly (Figure 3–5b) and quickly inverted. If this is done correctly, the drop of microorganisms hangs from the cover glass into the concavity of the slide. The slide is then examined with a microscope (Figure 3–5c). The petroleum jelly seals the cover glass to the slide and prevents excessive evaporation of the solution in which the microorganisms are suspended.

The primary reason for using a hanging-drop preparation is to observe microbial motility. **Motility** is the ability of an organism to move by itself. It should not be confused with **Brownian movement,** the movement of particles (and microorganisms) in a suspension due to the bombardment by moving molecules in the suspension. In Brownian movement, the particles or microbes all vibrate at about the same rate and maintain a fairly consistent spatial relationship with one another. In motility, movement is continuous in a given direction.

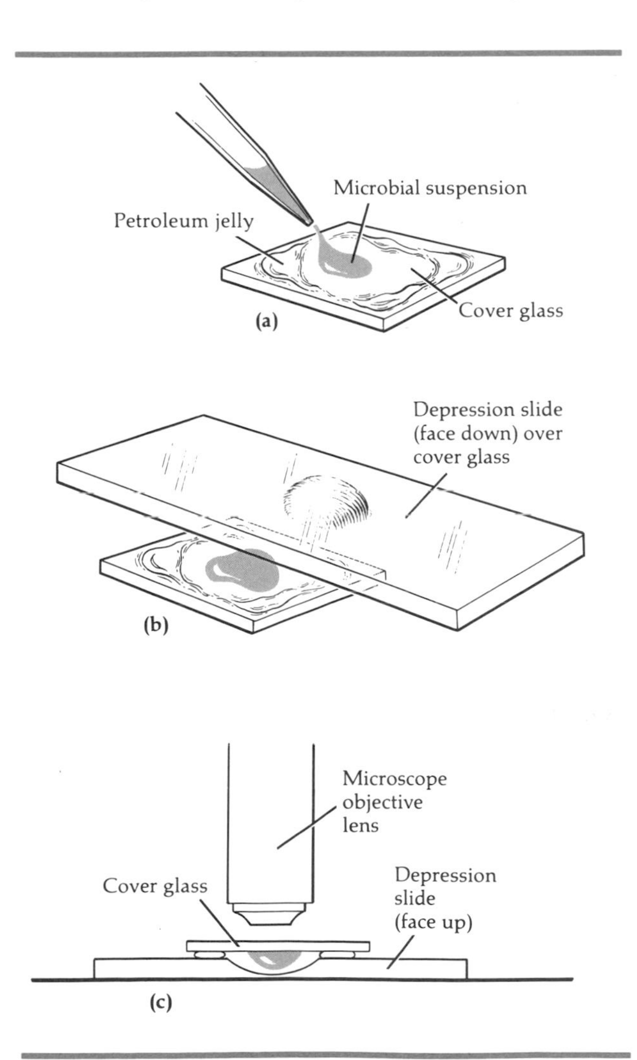

Figure 3–5 Procedure for making a hanging-drop preparation.

Staining and Smear Preparation

Most studies of the shapes and cellular arrangements of microorganisms are made from stained preparations. **Staining** simply means coloring the microorganisms with a dye to emphasize certain structures. Before they can be stained, however, the microorganisms must be attached, or **fixed,** to the microscope slide; otherwise, the stain may wash them from the slide.

To fix the specimen, a thin film of material containing the microorganisms is spread out over the surface of the slide (Figure 3–6a). This film is called a **smear.** After it is allowed to air dry (Figure 3–6b), the slide is slowly passed through the flame of a Bunsen burner several times, smear side up (Figure 3–6c). Flaming coagulates the microbial proteins and fixes the microorganisms to the slide. (The fixing procedure usually kills them.) Stain is applied (Figure 3–6d), then washed off with water, and the slide is blotted with absorbent paper. The stained microorganisms are now ready for microscopic examination.

Essentially, stains are salts composed of a positive and a negative ion, one of which is colored. With so-called **basic dyes,** the color is in the positive ion. With **acid dyes,** it is in the negative ion. Bacteria are slightly negatively charged. Thus, the colored positive ion in a basic dye is attracted to the negative bacterial cell. Examples of basic dyes are crystal violet, methylene blue, and safranin. Acid dyes are not attracted to most types of bacteria

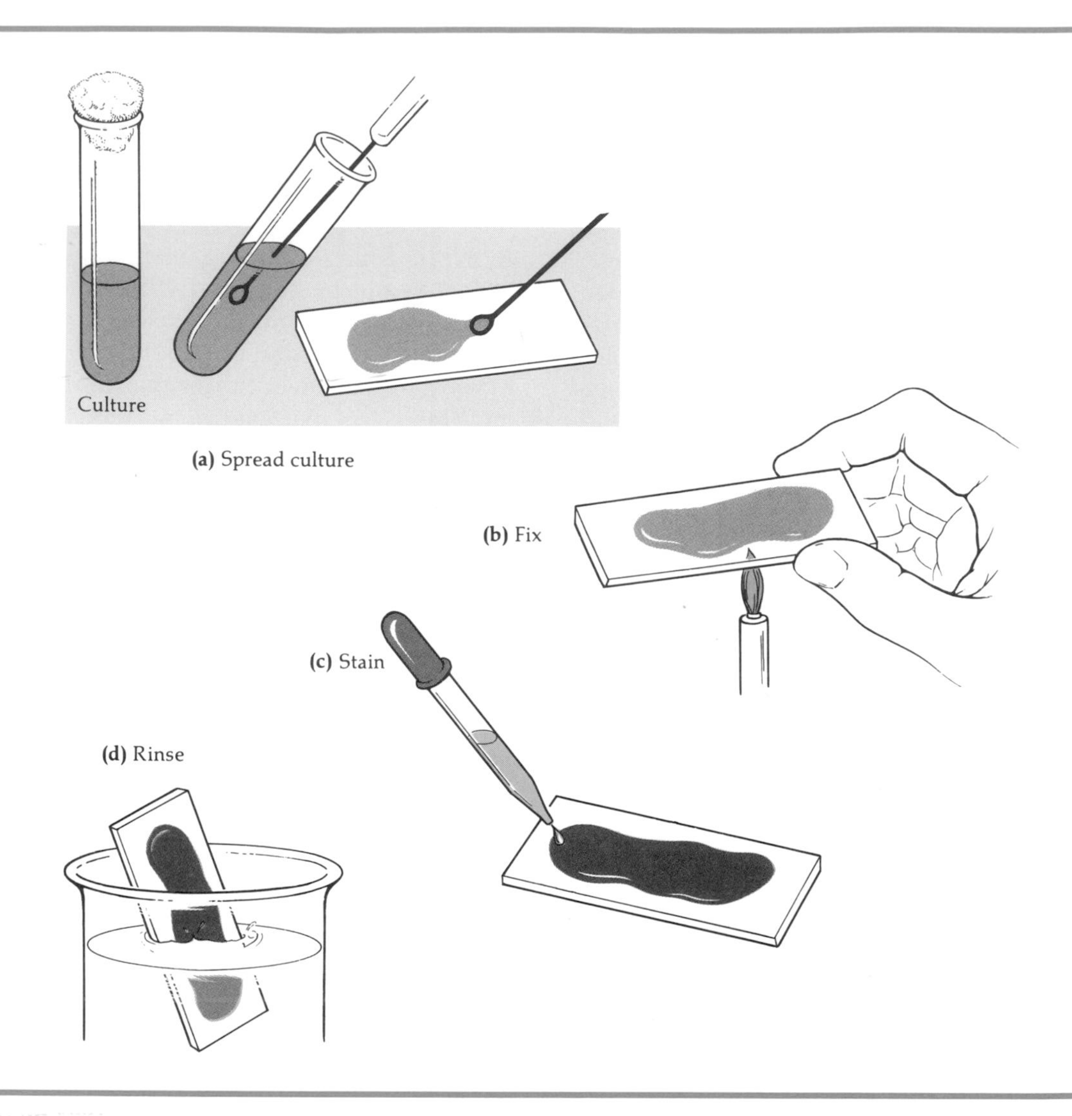

Figure 3–6 Smear preparation and staining procedures. **(a)** A suspension of microorganisms is spread in a thin film over a slide and air dried. **(b)** The smear is fixed by flaming. **(c)** The fixed smear is flooded with stain, then **(d)** rinsed. After drying, the slide is ready for microscopic examination.

because their negative ions are repelled by the bacterial surface. As a result, the stain is repelled by the bacteria and instead colors the background. This technique of preparing colorless bacteria against a colored background is called **negative staining.** It is valuable for observing overall cell shapes, sizes, and capsules, since the cells are made highly visible against a contrasting dark background. Distortions of cell size and shape are minimized because heat fixing is not necessary and the cells do not pick up the stain. Examples of acid dyes are eosin and nigrosin.

In general, microbiologists use three kinds of staining techniques: simple, differential, and special.

Simple Stains

A **simple stain** is an aqueous or alcohol solution of a single basic dye. It is used to stain the entire microorganism, primarily for visualization of cell shapes and arrangements. Although different dyes bind specifically to different parts of cells, the pri-

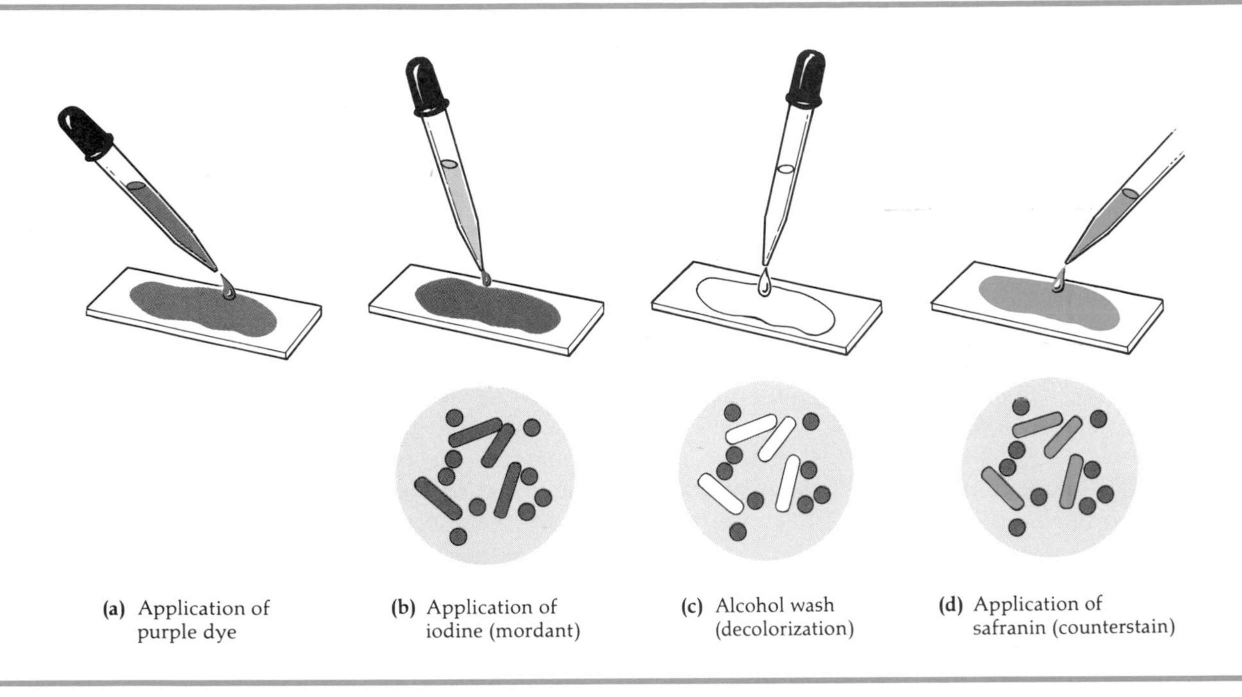

Figure 3–7 Gram-staining procedure. **(a)** A heat-fixed smear is covered with a basic purple dye such as gentian violet, and the dye is washed off. **(b)** Then the smear is covered with iodine (a mordant) and washed off. At this point in the procedure, both gram-positive and gram-negative bacteria appear purple. **(c)** The slide is next washed with ethyl alcohol or alcohol and acetone (a decolorizer) and then water. Now the gram-positive cells are purple and the gram-negative cells are colorless. **(d)** In the final step, safranin is added as a counterstain, and the slide is washed, dried, and examined microscopically. Gram-positive bacteria retain the purple dye, even through the alcohol wash. Gram-negative bacteria appear pink because they pick up the safranin counterstain.

mary purpose of a simple stain is to make the entire microorganism more visible. The stain is applied to the fixed smear for a given amount of time, then washed off, and the slide is dried and examined (Figure 3–6d). Occasionally, a chemical is added to the solution to make it stain more intensely. Such an additive is called a **mordant.** Some of the simple stains commonly used in the laboratory are methylene blue, carbolfuchsin, gentian violet, and safranin.

Differential Stains

Unlike simple stains, **differential stains** distinguish different kinds of bacteria on the basis of their reaction to staining. The most frequently used differential stains for bacteria are the Gram stain and the acid-fast stain.

Gram stain

The **Gram stain** was developed in 1884 by the Danish bacteriologist Hans Christian Gram. It is one of the most useful staining procedures available because it divides bacteria into two large groups: gram-positive (+) and gram-negative (−).

In this procedure, the heat-fixed smear is covered with a basic purple dye, usually crystal violet or gentian violet (Figure 3–7a). After a short time, the dye is washed off and the smear is covered with iodine (a mordant) (Figure 3–7b). When the iodine is washed off, both gram-positive and gram-negative bacteria appear dark violet or purple. Next, the slide is washed off with an ethyl alcohol or ethyl alcohol–acetone solution (Figure 3–7c). This solution acts as a decolorizing agent, removing the purple color from the cells of some species, but not from others. The alcohol is rinsed and the slide is stained with safranin, a basic red dye (Figure 3–7d). The slide is washed again, blotted dry, and examined microscopically.

The purple dye and the iodine combine with the bacterium and color it dark violet or purple. Bacteria that retain this color after attempted decolorization by the alcohol are classified as **gram-positive** (see Color Plate IA). Bacteria that lose the dark violet or purple color after decolorization are classified as **gram-negative** (see Color Plate IB). Since gram-negative bacteria are colorless after the alcohol wash, they are no longer visible. It is for this reason that the safranin is applied, causing the gram-negative bacteria to turn pink. Stains such as safranin that are used to give contrast in color are called **counterstains.** Since gram-positive bacteria retain the original purple stain, they are not affected by the safranin counterstain.

As you will see in the next chapter, bacteria react differently to the Gram stain because of chemical differences in their cell walls. Iodine forms a chemical complex with the dye in the cells. It is thought that the decolorizing agent causes some cell walls (gram-positive) to shrink, preventing the dye–iodine complex from leaving the cell. The cell walls of gram-negative bacteria are not affected in this way, and the dye–iodine complex is easily washed out.

The Gram method is one of the most important staining techniques in medical microbiology. But Gram staining results are not absolute, as some bacterial cells stain poorly or not at all. The Gram reaction is most consistent when used on young, growing bacteria. In many cases, the Gram reaction of a bacterium provides valuable information concerning the treatment of a disease. For example, it is known that gram-positive bacteria often tend to be killed easily by penicillin and sulfonamide drugs. Gram-negative bacteria resist these drugs, but are much more susceptible to drugs such as streptomycin, chloramphenicol, and tetracycline. Thus, Gram identification of a bacterium can help determine the most effective drug for treating a disease.

Acid-fast stain

Another important differential stain that divides bacteria into distinctive groups is the **acid-fast stain.** This stain is used specifically to identify members of bacteria belonging to the genera *Mycobacterium* and *Nocardia*. Although many bacteria in these groups are nonpathogenic, two are important disease producers, *Mycobacterium tuberculosis* and *Mycobacterium leprae* (see Color Plate IC).

In the acid-fast staining procedure, the red dye carbolfuchsin is applied to a fixed smear, and the slide is gently steamed over a flame for several minutes. Then the stain is washed with water. The bacteria now appear red. The slide is next treated with acid–alcohol, a decolorizer, which removes the red stain from bacteria that are not acid-fast. The acid-fast microorganisms retain the red color. The slide is then stained with a methylene blue counterstain. Since the acid-fast bacteria retain the carbolfuchsin and repel the counterstain, they remain red. But the non-acid-fast bacteria take up the counterstain and appear blue. The acid-fast bacteria probably stain as they do because of chemical components in the cell wall and cell membrane that have an attraction for carbolfuchsin. The acid-fast stain is invaluable in the identification of the tuberculosis microorganism.

Special Stains

Special stains are used to color and identify specific parts of microorganisms, such as endospores and flagella, and to reveal the presence of capsules.

Negative staining for capsules

Many microorganisms contain a gellike covering called a **capsule** or **slime layer,** which we will discuss in our examination of the procaryotic cell. In medical microbiology, it is often desirable to demonstrate the presence of a capsule as a means of determining the organism's identity and/or **pathogenicity,** that is, its ability to cause disease. Because of their chemical composition, capsules do not accept most biological dyes. To demonstrate the presence of capsules, a microbiologist can mix the bacteria in a solution containing a fine colloidal suspension of colored particles (usually India ink), stain the bacteria with a simple stain such as safranin, and observe the halo (capsule) surrounding each bacterial cell. This negative staining technique indicates the presence of a capsule. Negative stains do not penetrate the cell and thus provide a contrasting background between the cell and the surrounding medium (Figure 3–8a and Color Plate ID).

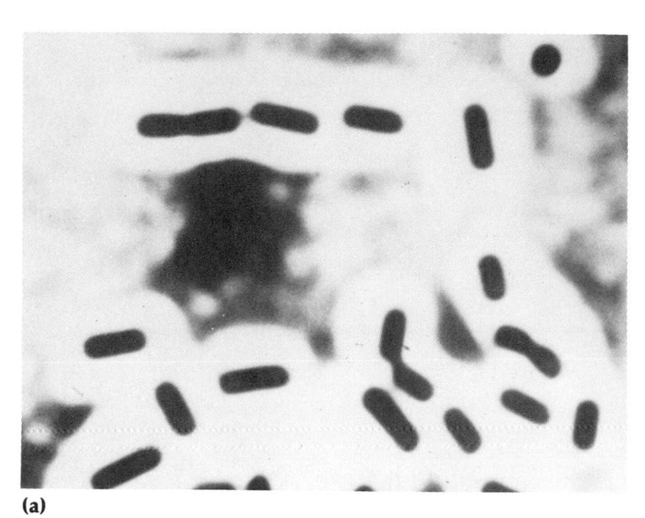

(a)

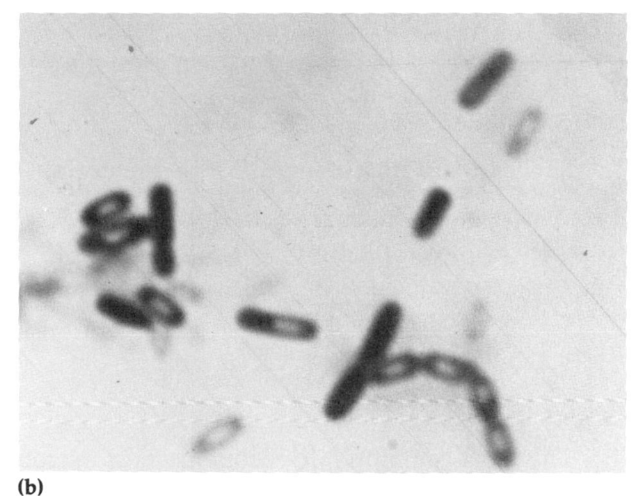

(b)

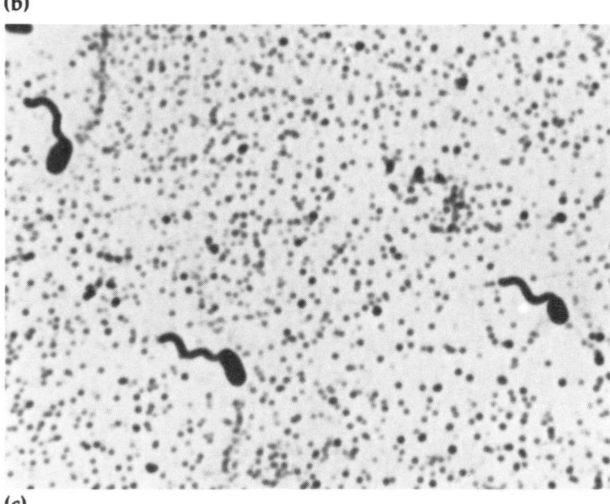

(c)

Figure 3–8 Negative staining of capsules and staining of endospores and flagella. **(a)** With India ink providing a dark background, the capsules of these bacteria *(Clostridium perfringens)* show up as light areas surrounding the stained cells. **(b)** Endospores are seen as light areas in the centers of these rod- to oval-shaped cells of the bacterium *Clostridium bifermentans.* **(c)** Flagella are shown as thick wavy extensions from one end of these cells of the bacterium *Vibrio cholerae.* The flagella are much thicker than normal in relation to the bodies of the cells because of buildup from the stain.

Endospore (spore) stain

Although relatively uncommon in bacterial cells, five genera of bacteria have the ability to form endospores. An **endospore** is a very resistant structure formed within a cell that protects the microorganism when environmental conditions are adverse. Endospores cannot be stained by ordinary methods such as simple staining and Gram staining because the dyes do not penetrate the wall of the endospore. But endospores are highly refractive and are readily detected under the light microscope (Figure 3–8b).

The most commonly employed endospore stain is the *Schaeffer–Fulton endospore stain.* Malachite green, the primary stain, is applied to a heat-fixed smear and heated to steaming for about 5 minutes. This helps the stain to penetrate the endospore wall. Then the preparation is washed for about 30 seconds with water to remove the malachite green stain from all cell parts except the endospores. Next, safranin, a counterstain, is applied to the smear to displace any residual malachite green in all cell parts except the endospores. In a properly prepared smear, the endospores appear green within red or pink cells.

Flagella staining

Bacterial **flagella** are structures of locomotion too small to be seen by light microscopes. In order to view them, a tedious and delicate staining procedure is used that builds up the diameter of the flagella with a mordant until they become visible with the light microscope. In medical microbiology, the number and arrangement of flagella are useful diagnostic aids. Figure 3–8c depicts a successful flagella stain.

STUDY OUTLINE

UNITS OF MEASUREMENT (pp. 57–58)

1. The standard unit of length is the meter (m).
2. Microorganisms are measured in micrometers (10^{-6} m), nanometers (10^{-9} m), and angstroms (10^{-10} m).

MICROSCOPY: THE INSTRUMENTS (pp. 58–64)

Compound Light Microscopy (pp. 58–61)

1. The most common microscope used in microbiology is the compound light microscope, which uses multiple lenses (ocular and objective).
2. Total magnification of an object is calculated by multiplying the magnification of the objective lens being used by the magnification of the ocular lens.
3. The compound light microscope uses visible light.
4. The maximum resolving power (ability to distinguish two points) is 0.3 μm; maximum magnification is 2000×.
5. Brightfield illumination is used for stained smears.
6. The following four types of microscopy use modified compound microscopes.

Phase-Contrast Microscopy (pp. 60–61)

1. The phase-contrast microscope uses a special condenser to enhance differences in refractive index between parts of the cell and its surroundings.
2. It allows detailed observation of living organisms.

Darkfield Microscopy (pp. 60–61)

1. The darkfield microscope shows a light silhouette of an organism against a dark background.
2. It is most useful to detect the presence of extremely small organisms.

Differential Interference Contrast (DIC) Microscopy (pp. 60–61)

1. The DIC microscope provides a three-dimensional image of the object being observed.
2. In the procedure, beams of light are split and recombined by prisms.

Fluorescence Microscopy (pp. 62–63)

1. In fluorescence microscopy, specimens are first stained with fluorochromes, then viewed through a compound microscope using an ultraviolet (or near-ultraviolet) light source.
2. The microorganisms appear as bright objects against a dark background.

Electron Microscopy (pp. 63–64)

1. A beam of electrons, instead of light, is used with an electron microscope.
2. Conventional electron microscopes have resolving power down to about 10 Å and achieve a magnification of about 200,000×.
3. Thin sections of organisms may be seen in an electron micrograph produced with a transmission electron microscope.
4. Three-dimensional views of the surfaces of whole microorganisms may be obtained with a scanning electron microscope.

PREPARATION OF SPECIMENS FOR LIGHT MICROSCOPY (pp. 64–69)

Hanging-Drop Preparation (p. 65)

1. This procedure is used to study microbial motility, the ability of an organism to move by itself.
2. In Brownian movement, particles or microorganisms are moved about at random because of bombardment by moving molecules in the suspension. Such particles maintain a fairly constant spatial relationship with one another.

Staining and Smear Preparation (pp. 65–66)

1. Staining means to color with a dye to make certain structures more visible.
2. Fixing refers to attaching microorganisms to a slide by heat.
3. A smear is a thin film of material for microscopic examination.
4. Bacteria are negatively charged, and the colored positive ion of a basic dye will stain bacterial cells.
5. The colored negative ion of an acid dye will stain the background of a smear, producing a negative stain.

Simple Stains (pp. 66–67)

1. A simple stain is an aqueous or alcohol solution of a single basic dye.
2. It is used to make cell shapes and arrangements visible.

Differential Stains (pp. 67–68)

1. Differential stains, such as the Gram stain and acid-fast stain, divide bacteria into groups based upon their reaction to the stains.
2. The Gram-stain procedure uses a purple stain, iodine as a mordant, and a red counterstain.
3. Gram-positive bacteria retain the purple stain after the decolorization step; gram-negative bacteria do not, and thus appear red from the counterstain.

Special Stains (pp. 68–69)

1. Stains such as the endospore stain and flagella stain color only certain parts of microbes.
2. Negative staining is used to make microbial capsules visible.

STUDY QUESTIONS

REVIEW

1. Fill in the following blanks.
   1 μm = ______ m
   1 ______ = 10^{-9} m
   1 Å = ______ m
2. Label the parts of the compound light microscope.

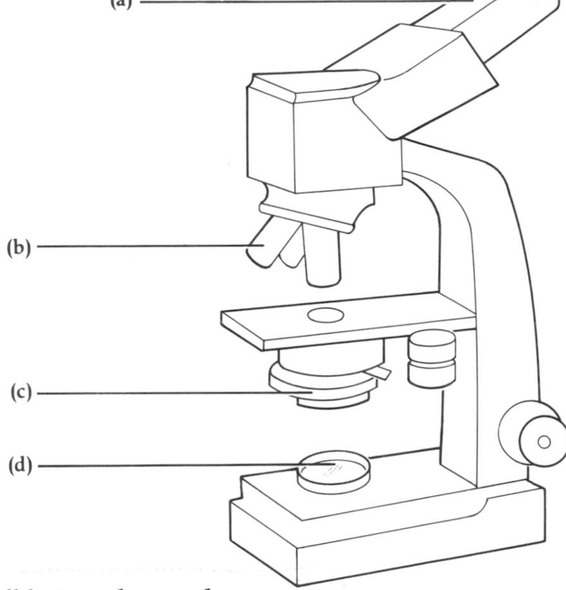

3. Calculate the total magnification of the nucleus of a cell being observed through a compound light microscope using a 10× ocular lens and the oil-immersion lens.
4. Which type of microscope would be best to use to observe the following?
   (a) A stained bacterial smear
   (b) Unstained bacterial cells where the cells are small and no detail is needed
   (c) Unstained live tissue, where it is desirable to see some intracellular detail
   (d) A sample that emits light when illuminated with ultraviolet light
   (e) Intracellular detail of a cell that is 1 μm long

5. An electron microscope differs from a light microscope because ________ is used instead of light and the image is not viewed through the ocular lenses but ________.
6. The maximum magnification of a compound microscope is ________, of an electron microscope, ________. The maximum resolution of a compound microscope is ________, of an electron microscope, ________.
7. One advantage of a scanning electron microscope over a transmission electron microscope is ________.
8. Acid dyes stain the (cells/background) in a smear and are used for (negative/simple) stains.
9. Basic dyes stain the (cells/background) in a smear and are used for (negative/simple) stains.
10. When is it most appropriate to use a simple stain? Differential stain? Negative stain? Flagella stain?
11. Why is a mordant used in the Gram stain? In the flagella stain?
12. What is the purpose of a counterstain in the acid-fast stain?
13. What is the purpose of a decolorizer in the Gram stain? In the acid-fast stain?
14. Fill in the following table regarding the Gram stain.

Steps	Appearance after this step of Gram (+) cells	Gram (−) cells
crystal violet		
iodine		
alcohol–acetone		
safranin		

CHALLENGE

1. In a Gram stain, one step could be omitted, still allowing you to differentiate between gram-positive and gram-negative cells. What is that one step?
2. Why do basic dyes stain bacterial cells? Why don't acid dyes stain bacterial cells?
3. 1 μm = ________ Å
4. Using a good compound light microscope with a resolving power of 0.3 μm, a 10× ocular lens, and a 100× oil-immersion lens, would you be able to discern two objects separated by 3 μm? 0.3 μm? 30,000 Å? 300 nm?

FURTHER READING

Barer, R. 1974. Microscopes, microscopy, and microbiology. *Annual Review of Microbiology* 28:371–389. An in-depth discussion of the basic principles of microscopy.

Branson, D. 1972. *Methods in Clinical Bacteriology.* Springfield, IL: C. C. Thomas. A good reference for stepwise directions for staining procedures and preparation of stains.

Gray, P., ed. 1973. *Encyclopedia of Microscopy and Microtechniques.* New York: Van Nostrand. An illustrated encyclopedia of terminology and techniques.

Lennette, E. H., E. H Spaulding, and J. P. Truant, eds. 1980. *Manual of Clinical Microbiology,* 3rd ed. Washington, D.C.: American Society for Microbiology. Contains a section on microscopy, specimen preparation, and staining.

Sieburth, J. M. 1975. *Microbial Seascapes.* Baltimore, MD: University Park Press. An atlas of electron micrographs of marine microbes. Introduction explains preparations for transmission electron microscopy and scanning electron microscopy.

4

Functional Anatomy of Procaryotic and Eucaryotic Cells

Objectives

After completing this chapter you should be able to

- Identify the three basic shapes of bacteria.
- Explain the differences between gram-positive and gram-negative cell walls.
- Describe the structure, chemistry, and functions of the procaryotic plasma membrane.
- Define the following: simple diffusion; osmosis; facilitated diffusion; active transport.
- Identify the functions of procaryotic cell structures.
- Compare and contrast procaryotes and eucaryotes with regard to the following: overall cell structure; flagella; nucleus.
- Distinguish between eucaryotic phagocytosis and pinocytosis.
- Explain what an organelle is.
- Describe the functions of the following: endoplasmic reticulum; Golgi complex; mitochondria; lysosomes; chloroplasts.

In the study of human anatomy, we must look carefully at the physical arrangement of organs and tissues in the body. Since most microorganisms are unicellular, microbial anatomy brings us to the inspection of a single cell's construction—how each cell component contributes to the structure of the cell as a whole.

Despite their complexity and variety, all living cells can be divided into two groups: procaryotic and eucaryotic cells. On the macroscopic level, plants and animals are entirely composed of eucaryotic cells. In the microbial world, bacteria and cyanobacteria (formerly known as blue-green algae) are procaryotes. Most other microbes are eucaryotes: fungi (yeasts and molds), protozoans, and true algae.

Viruses slip between this dichotomy of procaryote and eucaryote. As noncellular elements with some cell-like properties, viruses do not fit any scheme for living cells. They are genetic particles that reproduce but are unable to perform the usual cell chemistry of living things. Viral structure and activity will be discussed in Chapter 12. For now, we will concentrate on procaryotic and eucaryotic cells.

PROCARYOTIC AND EUCARYOTIC CELLS

Procaryotes and eucaryotes are chemically similar: they both contain nucleic acids, proteins, lipids, and carbohydrates. They use the same kinds of chemical reactions to metabolize food, build proteins, and store energy. It is primarily the *structure* of cell walls, membranes, and various internal compartments (organelles) that distinguish procaryotes from eucaryotes.

With so many chemical properties in common, procaryotes and eucaryotes are clearly related as to origin. Scientists think that procaryotic cells evolved first, and that the larger, more complex eucaryotic cells may have developed from a group of relatively large procaryotes that were parasitized by smaller bacteria. It is suggested that some of the organelles in eucaryotes are derived from these ancient remnants of bacteria.

The chief distinguishing characteristics of **procaryotic** (from the Greek for *pre-nucleus*) **cells** are as follows:

1. Their genetic material (DNA) is not enclosed within a membrane.
2. They lack other membrane-bounded organelles common to eucaryotic cells.
3. Their DNA is not associated with histone proteins.
4. They almost always have cell walls containing peptidoglycan, which is unique to procaryotes.

Eucaryotic (from the Greek for *true nucleus*) **cells** have their DNA organized into linear structures called chromosomes, which are found in the cell's nucleus, separated from the cytoplasm by a nuclear membrane. Eucaryotic chromosomes consist of DNA associated in a regular way with special chromosomal proteins, called histones and nonhistones. In addition, eucaryotes have a number of organelles, including a mitotic apparatus, mitochondria, endoplasmic reticulum, and, sometimes, chloroplasts. We will go over the particular characteristics and functions of these organelles later.

THE PROCARYOTIC CELL

The members of the procaryotic world, the bacteria and cyanobacteria, comprise a vast heterogenous group of unicellular organisms. The thousands of species of bacteria are differentiated by many factors, including morphology (shape), chemical composition (often detected by staining reactions), nutritional requirements, and biochemical activities.

Several general features of bacteria can be observed with the light microscope:

1. Almost all bacteria have a semirigid cell wall.
2. If bacteria are motile, their motility is usually by means of flagella.
3. Bacteria are fundamentally unicellular; although they are frequently found in characteristic groupings, each individual cell carries out all the functions of the organism.

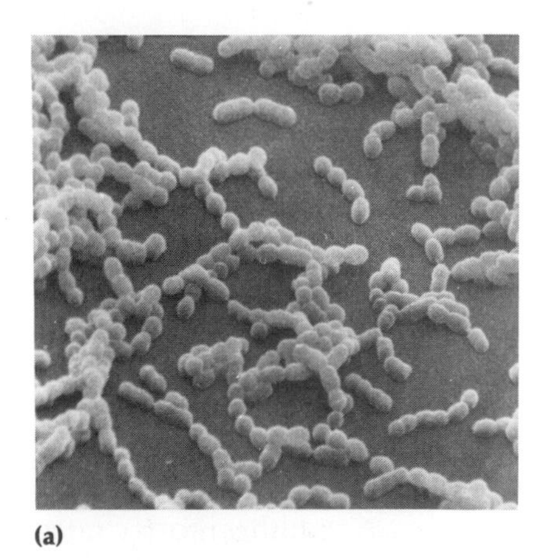

(a)

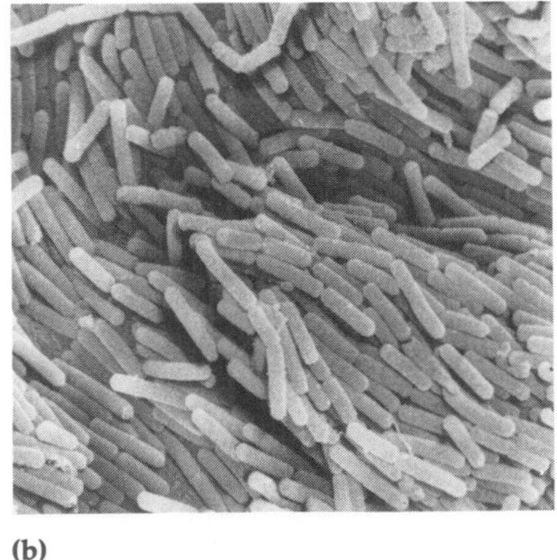

(b)

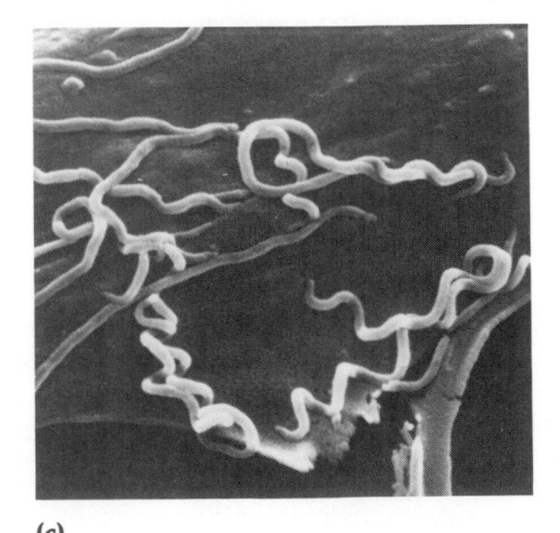

(c)

Figure 4–1 Basic shapes of bacteria as seen under the scanning electron microscope. **(a)** Spherical cocci of *Streptococcus mutans* are shown joined in short chains (×2500). **(b)** Rod-shaped bacilli of *Bacillus cereus* appear individually and joined end to end in this preparation (×1250). **(c)** The spirochete *Treponema pallidum* has spiral-shaped cells (×4000).

4. Most bacteria multiply by binary fission, a process by which a single cell divides into two identical daughter cells.

SIZE, SHAPE, AND ARRANGEMENT OF BACTERIAL CELLS

There are a great many sizes and shapes among bacteria. Most bacteria fall within a range of 0.20 to 2.0 μm in diameter, with the cells occurring in three basic shapes: the spherical **coccus** (plural *cocci,* meaning "berries"); the rod-shaped **bacillus** (plural *bacilli,* meaning "little staffs"); and the **spiral** shapes (Figure 4–1).

Cocci are usually round, but may also be oval, elongated, or flattened on one side. When cocci divide to reproduce, the cells may remain attached to one another; they do not separate. Cocci that divide and remain in pairs are called **diplococci;** those that divide and remain attached in chainlike patterns are referred to as **streptococci;** those that divide in two planes and remain attached in groups of four are known as **tetrads;** those that divide in three regular planes and remain attached in cubelike groups of eight are called **sarcinae;** and those that divide at random planes and form grapelike clusters or broad sheets are referred to as **staphylococci.** These groupings are frequently helpful in identifying certain cocci (Figure 4–2a–c and Color Plate IE).

Because bacilli divide only across their short axis, there are fewer groupings of bacilli than of cocci. **Diplobacilli** appear in pairs after division and **streptobacilli** occur in chains (Figure 4–2d and e and Color Plate IF). Some bacilli look like cigarettes. Others have tapered ends like cigars. Still others are oval and look so much like cocci that they are called **coccobacilli.** However, most bacilli appear as single rods.

Spiral bacteria have one or more twists; they are never straight. Some are curved rods that look like commas, and these are called **vibrios** (Figure 4–2f). Others, called **spirilla,** have a distinctive helical shape like a corkscrew, their cell bodies fairly rigid (Figure 4–2g). Yet another group of spirals are called **spirochetes** (Figure 4–2h). They, too, are spiral-shaped; but unlike the spirilla, which have flagella, they move by flexing and wiggling their bodies. As you will see later, some genera of bacteria have more complex shapes and arrangements.

The various shapes of bacteria are determined by heredity. However, a number of environmental conditions may alter their shape. If this happens, identification becomes even more difficult. Moreoever, some bacteria are genetically **pleomorphic,** which means they may have many shapes, not just one.

Figure 4–3 is an illustration of the structure of a typical procaryotic (bacterial) cell. We will discuss its components according to the following organization: structures external to the cell wall; the cell wall; and structures internal to the cell wall.

STRUCTURES EXTERNAL TO THE CELL WALL

Among the structures external to the cell wall are the capsule (slime layer), flagella, axial filaments, and pili (fimbriae).

Capsule, or Slime Layer

Many, but not all, bacterial cells have an outer viscous (sticky) covering called a **capsule,** or **slime layer** (see Figure 4–3). This loose, gellike structure can be demonstrated by negative staining, such as the India ink method mentioned in Chapter 3 (see Figure 3–8a and Color Plate ID). Chemically, the capsule is composed of a gelatinous polymer of polysaccharide, polypeptide, or both. Chemical complexity varies widely depending on the species of bacteria. The material of the capsule, for the most part, is made inside the cell and excreted to the cell surface where it adheres because of its high viscosity.

Capsules often protect pathogenic bacteria from phagocytosis by cells of the host, and therefore play an important role as a mechanism of bacterial virulence (the degree to which a pathogen causes disease). Phagocytosis, which is discussed in Chapter 15, is a process by which certain white blood cells engulf and destroy microbes. The identification of capsular material is often an important step in identifying certain pathogenic bacteria. *Streptococcus pneumoniae, Klebsiella pneumoniae,* and *Bacillus anthracis* are three important species of encapsulated bacteria that can be identified by the swelling of their capsules in the presence of specific chemicals (antibodies). This swelling of bacterial

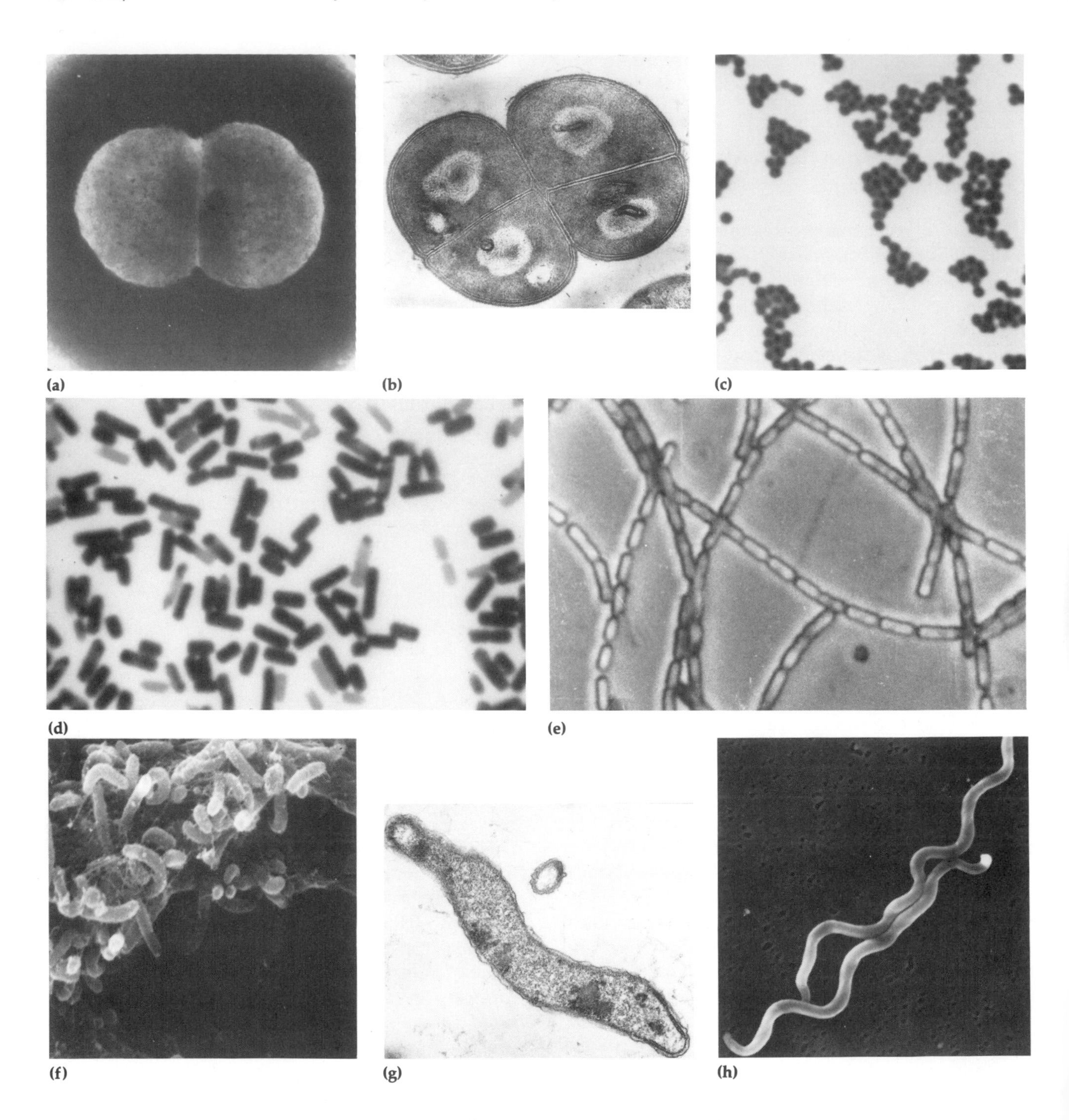

Figure 4–2 Basic bacterial shapes. **(a)–(c)** Cocci. **(a)** A diplococcus of *Neisseria gonorrhoeae*, Type III (×28,000). **(b)** Thin section through four cells that form a tetrad of *Sporosarcina ureae*. **(c)** Clusters of staphylococci of *Staphylococcus aureus* (×4700). **(d)–(e)** Bacilli. **(d)** Bacilli of *Clostridium perfringens* (×4700); the chains only two cells long serve as representatives of diplobacilli. **(e)** Streptobacilli of *Bacillus anthracis*. **(f)–(h)** Curved bacteria. **(f)** Several cells of *Vibrio cholerae* (×4800). **(g)** Thin section through a helical cell of the spirillum *Campylobacter fetus* (×42,000). **(h)** A scanning electron micrograph of two spirochetes (×18,000).

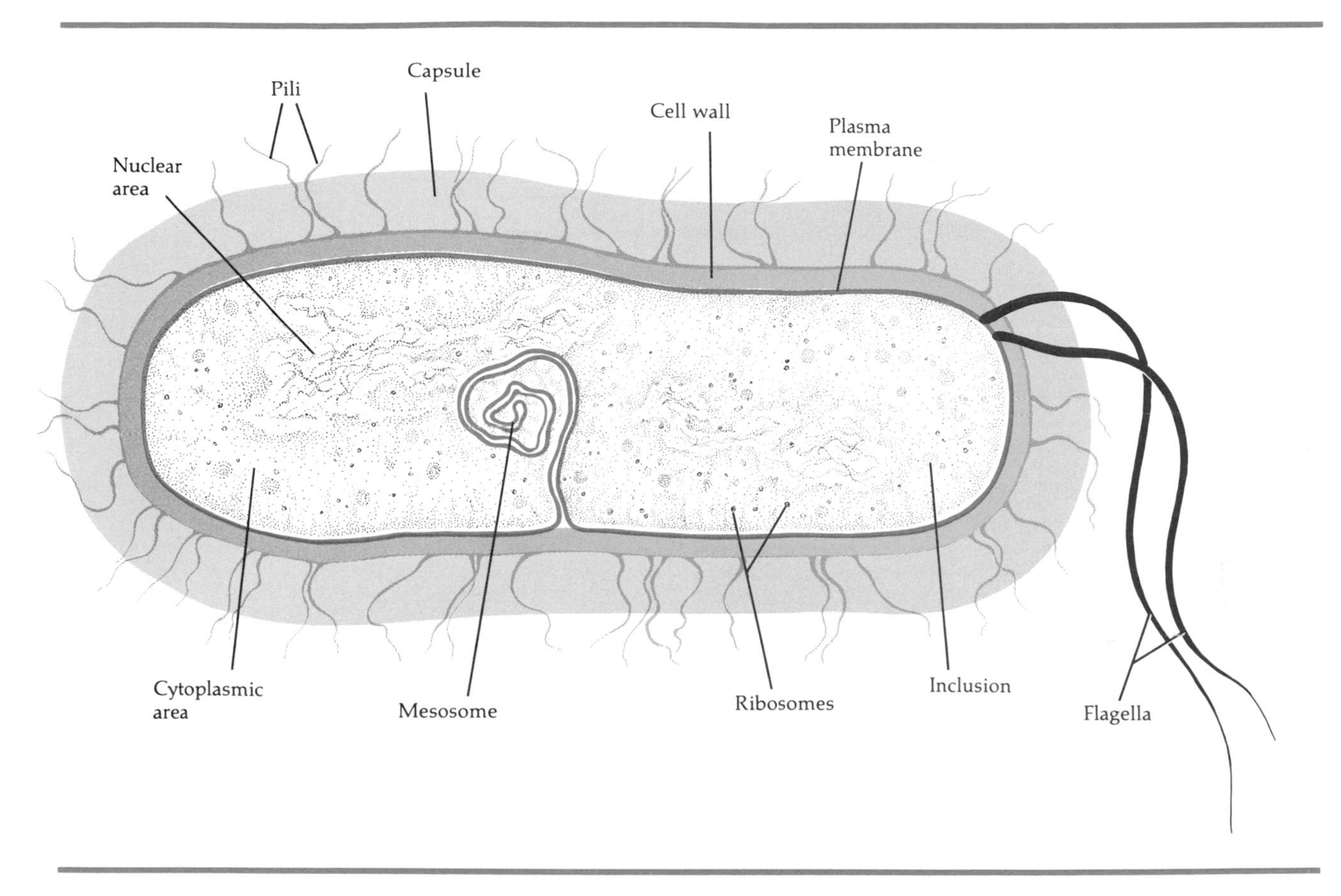

Figure 4–3 Structure of a typical procaryotic (bacterial) cell as seen in longitudinal section.

capsules is called the **quellung reaction.** Figure 4–4 shows the quellung reaction of *Streptococcus pneumoniae.* Many bacteria manufacture a polysaccharide capsule called a *glycocalyx* that helps them adhere to surfaces. The glycocalyx can help the bacteria survive in their natural environment, whether it is a stream or the human intestine.

Flagella

A second structural component of the procaryotic cell is flagella (see Figure 4–3). **Flagella** (singular, *flagellum,* meaning "whip") are long filamentous appendages that propel some bacteria. The flagellum has three basic parts (Figure 4–5a). The outermost region, the *filament,* is constant in diameter and is composed of the globular (roughly spherical) protein *flagellin* arranged in several chains forming

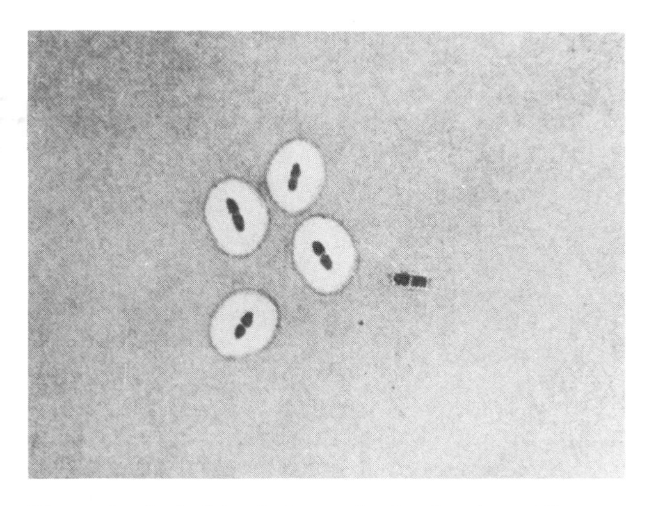

Figure 4–4 In this photograph of a drawing of the quellung reaction, the capsules around pairs of cells of *Streptococcus pneumoniae* appear swollen as a result of reaction with specific antibodies.

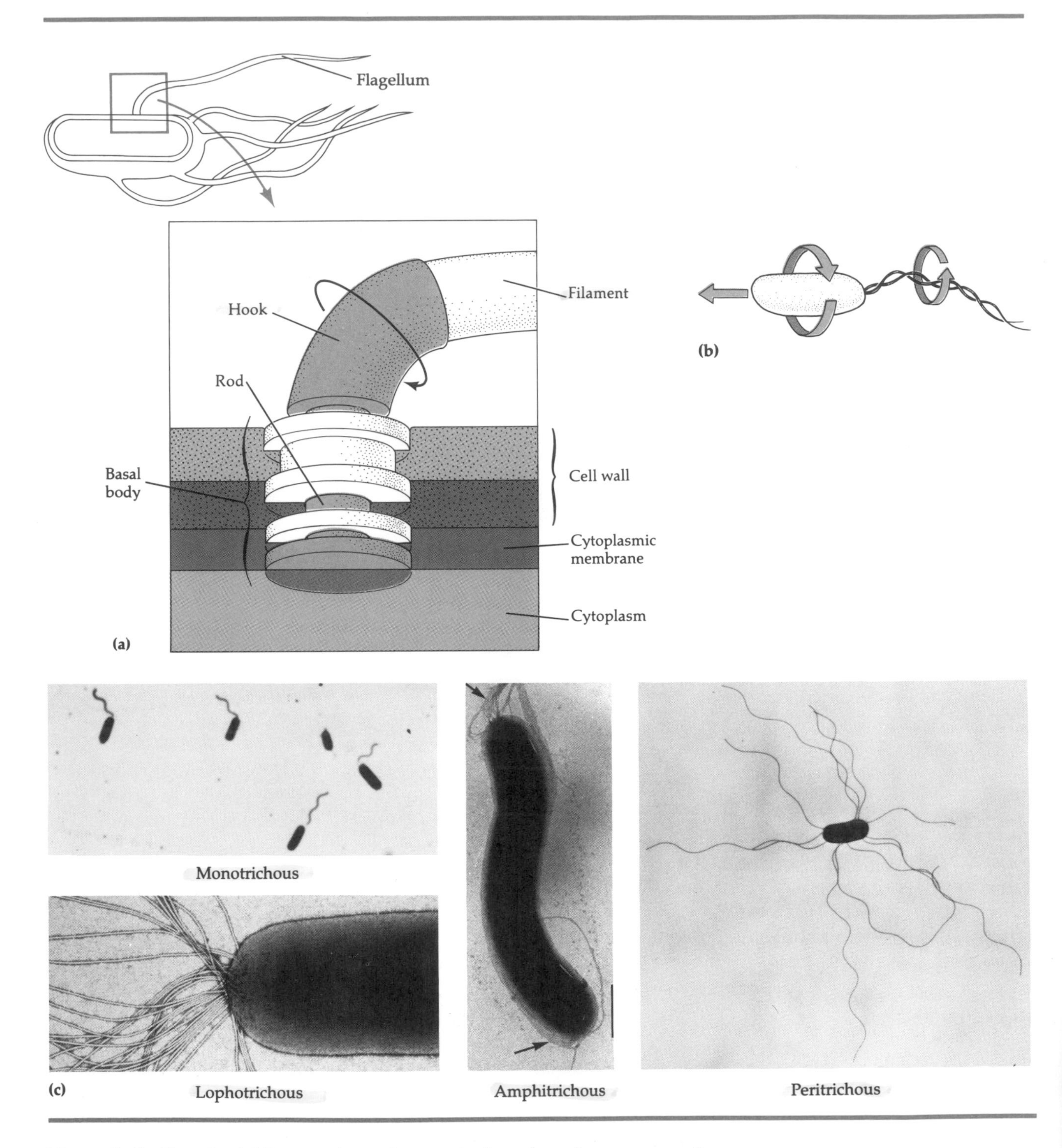

Figure 4–5 Flagella. **(a)** Parts and attachment of a flagellum of a gram-negative bacterium. **(b)** Mechanism of flagellar motion. **(c)** The four basic types of flagellar arrangements: monotrichous (*Pseudomonas aeruginosa*); lophotrichous (×32,000); amphitrichous (*Aquaspirillum bengal;* bar = 1 μm; arrows point to flagella); peritrichous (*Proteus vulgaris;* ×4200).

a helix around a hollow core. The filament is attached to a slightly wider *hook,* which consists of a different protein. The third portion of a flagellum is the *basal body,* which anchors the flagellum to the cell wall and cytoplasmic membrane.

The basal body is composed of a small, central rod inserted into a series of rings. Gram-negative bacteria contain two pairs of rings. The outer rings are anchored to the cell wall, while the inner rings are anchored to the plasma membrane. In gram-positive bacteria, only the inner pair is present.

Bacteria with flagella are motile. That is, they have the ability to actively move on their own. Motility can be seen by observing hanging-drop preparations microscopically (see Figure 3–5) or by growing bacteria on a semisolid "motility medium" and observing movement outward from the inoculum.

The mechanism by which bacteria are able to propel themselves by flagellar movement is known. Each flagellum is a semirigid, helical rotor that pushes the cell by spinning either clockwise or counterclockwise around its axis (Figure 4–5b). Although the exact mechno-chemical basis for this biological "motor" is not completely understood, we know that it depends on the continuous generation of energy by the cell.

The arrangements of flagella on the bacterial cell are of four basic types (Figure 4–5c): **monotrichous** (single polar flagellum); **lophotrichous** (two or more flagella at one pole of the cell); **amphitrichous** (tufts of flagella at both ends of the cell); and **peritrichous** (flagella distributed over the entire cell).

Flagellated bacteria can move in one direction, tumble, and reverse movement. Some species of bacteria endowed with large numbers of flagella (*Proteus,* for example) can "swarm" or show rapid, wavelike growth across a culture medium. Flagellar proteins serve to identify certain pathogenic bacteria in the laboratory.

Axial Filaments

Spirochetes are a group of bacteria that have unique structure and motility. One of the best known spirochetes is *Treponema pallidum,* the causative agent of syphilis. Spirochetes move by means of **axial filaments,** which arise at the poles of the cell within

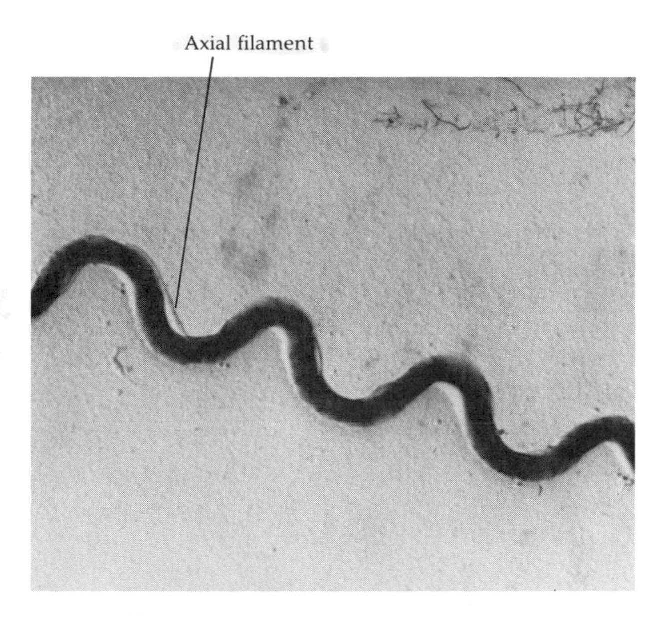

Figure 4–6 Axial filament of the spirochete *Treponema pallidum*, the cause of syphilis.

the cell wall and wrap around the cell in a spiral fashion (Figure 4–6). Axial filaments have a structure similar to that of flagella. Spirochetes move in a snakelike manner.

Pili

Pili, or **fimbriae,** are hairlike appendages attached to bacterial cells in a fashion similar to flagella, but which are considerably shorter and thinner than flagella (see Figure 4–3). Like flagella, pili consist of a protein (called *pilin*) arranged helically around a central core. Pili may occur at the poles of the bacterial cell, or they may be evenly distributed over the entire surface of the cell, with anywhere from a few to several hundred per cell (Figure 4–7). Many gram-negative bacteria have pili.

Pili have two primary functions. The first is adhesion to surfaces, including the surfaces of other cells. For example, pili associated with the bacterium *Neisseria gonorrhoeae,* the causative agent of gonorrhea, help the microbe to colonize mucous membranes. Once colonization occurs, the bacteria are capable of causing the disease. When pili are absent (as a result of genetic mutation), no disease occurs. The second function of pili is to aid in reproduction. Some bacteria have special pili that

How Bacteria Stick to Body Surfaces

To establish residence on or in the human body, bacteria must be able to adhere to external or internal body surfaces. Adherence to surfaces is important in colonization by both normal flora and pathogenic invaders. Without such attachment, bacteria could easily be washed away faster than they could multiply—particularly from surfaces such as those lining the mouth, upper intestinal tract, and urinary tract.

A variety of structures on bacterial cell surfaces aid adherence. These structures include capsules composed of polysaccharides and polypeptides, pili, and other, much finer filaments extending outward from the surface. In addition to general "stickiness," structures like these often have the ability to bind to specific receptors on the surfaces of host cells. Their inability to bind to cells of other species is part of the reason that a given species of bacterium will cause disease in some host species but not in others. Within a host, the presence of receptors on cells of some tissues and not others helps account for the selective colonization of certain parts of the body by certain bacteria. For example, *Streptococcus* adhere better to the lining of the mouth than to the intestinal lining, whereas *Escherichia coli* prefer the intestinal lining. The ability of *Neisseria gonorrhoeae* to cause disease (gonorrhea) correlates with the presence of pili that help attach the bacteria to the lining of the male urethra.

Mechanisms that block adherence of bacteria to surfaces can help the host resist colonization by invading pathogenic bacteria. If a host produces specific antibodies that bind to the bacterial surface or secretes other substances that nonspecifically cover host cell receptors, the bacteria may be unable to adhere. Thus, they may not be able to accumulate in large enough numbers to cause disease.

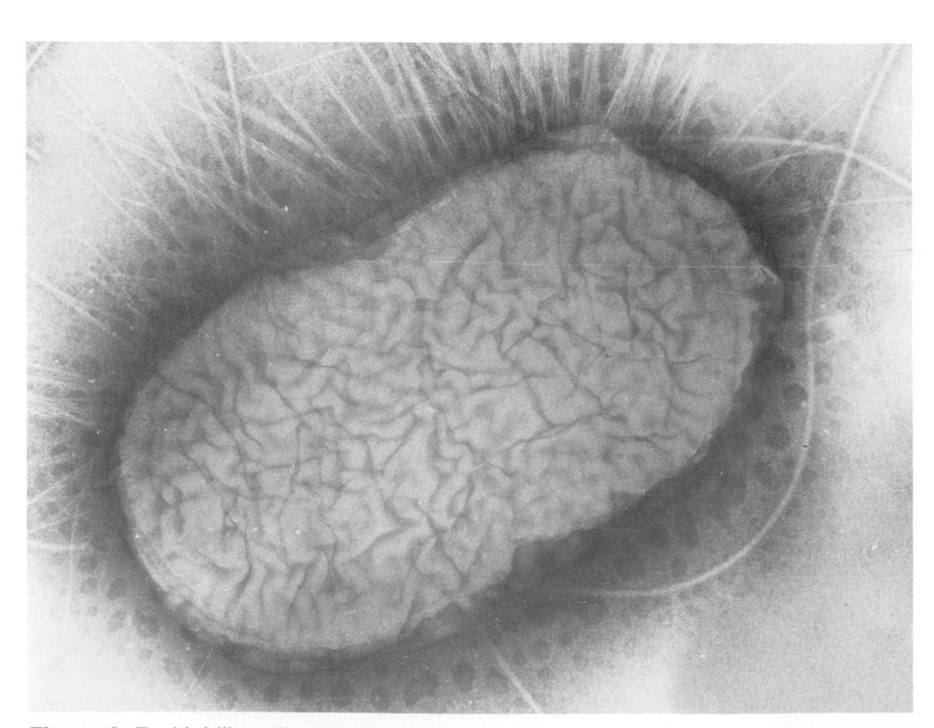

Figure 4–7 Hairlike pili seem to "bristle" from this cell of *Proteus vulgaris* (×48,000). Pili cover this cell, even though they are seen on only half of the cell. (Several longer, thicker flagella are also seen.)

assist in the transfer of genetic material between cells during bacterial mating (Chapter 8). These pili are called **sex pili.**

CELL WALL

The **cell wall** of the bacterial cell is a complex, semirigid structure that is responsible for the characteristic shape of the cell. It surrounds the underlying, fragile plasma (cytoplasmic) membrane, protecting it and the internal parts of the cell from adverse changes in the surrounding environment (see Figure 4–3).

Composition and Characteristics

Chemically, the bacterial cell wall is composed of a macromolecular network called *peptidoglycan (murein).* Peptidoglycan consists of two sugars related to glucose called N-acetylglucosamine and N-acetylmuramic acid (*murus,* meaning "wall") and chains of amino acids. The various components of peptidoglycan are assembled in the cell wall as follows. The N-acetylglucosamine and N-acetylmuramic acid alternate with each other in rows, each row forming a carbohydrate "backbone" (Figure 4–8). There are 10 to 65 sugars in each row. To each molecule of N-acetylmuramic acid is attached a *tetrapeptide side chain,* consisting of four amino acids (alternating D and L forms). Adjacent tetrapeptide side chains may be directly bonded to each other or linked by a *peptide cross bridge* consisting of one to five amino acids (Figure 4–8).

In Chapter 3 we discussed the classification of bacteria by a differential staining method called the Gram stain. In most gram-positive bacteria, the cell wall consists of several layers of peptidoglycan connected by peptide side chains and cross bridges,

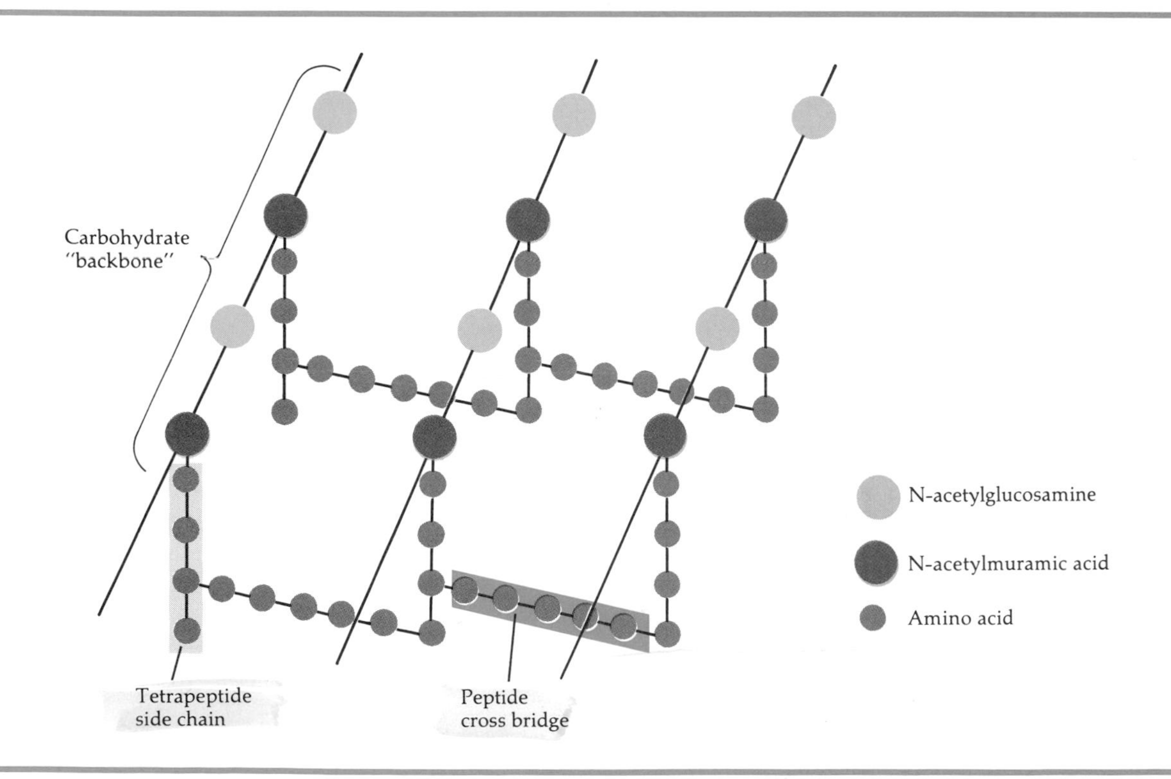

Figure 4–8 Chemical structure of peptidoglycan of the bacterial cell wall. The frequency of peptide cross bridges and the number of amino acids in these bridges varies with the species of bacterium. In addition to the bridges between chains in each peptidoglycan sheet, there are also bridges between different sheets. Thus, a peptidoglycan macromolecule may be quite thick.

the structure shown in Figure 4–8. This arrangement provides a very rigid framework. Also, many gram-positive cell walls contain polysaccharides called *teichoic acids* that make it possible to identify bacteria by immunological means, as will be described in Chapter 16. Teichoic acids may aid the transport of ions into and out of the gram-positive bacterial cell, and they may play a role in normal cell division.

Gram-negative bacteria also contain peptidoglycan, but in much smaller amounts, and they contain no teichoic acids at all. Because of the smaller amount of peptidoglycan, the cell walls of gram-negative bacteria are more susceptible to mechanical breakage. Gram-negative cells also have a lipopolysaccharide–phospholipid–lipoprotein layer (sometimes termed an *outer membrane*) surrounding the peptidoglycan layer (Figure 4–9). One important function of this outer layer is as a barrier to the passage of certain substances from the environment into the cell. These substances include antibiotics such as penicillins, certain dyes, and bile salts.

The lipoprotein portion of the surrounding layer is responsible for the gram-negative reaction of certain bacteria. The violet stain is initially retained by the lipoproteins, but is released when the decolorizer dissolves the lipoprotein. When the counterstain safranin is added, the red color is taken up by the bacterium. Gram-positive cells retain the violet stain after the decolorization process. Apparently, the violet stain complexed with iodine becomes trapped inside the thick peptidoglycan walls of these bacteria.

A primary function of the lipopolysaccharides of the outer layer of gram-negative cells is to prevent the rupture of the cell. The lipopolysaccharides also provide the bacteria with resistance to phagocytosis. The lipopolysaccharides of gram-negative cells are comparable to the teichoic acids of gram-positive cells in that they are useful for identifying specific bacterial species.

Another unique feature of gram-negative bacterial lipopolysaccharides is their toxic (poisonous) activity. Lipopolysaccharides are toxic when introduced into the host's bloodstream. They cause fever and intravascular hemolysis (disintegration of red blood cells). The nature and importance of these and other types of toxins will be discussed in Chapter 14.

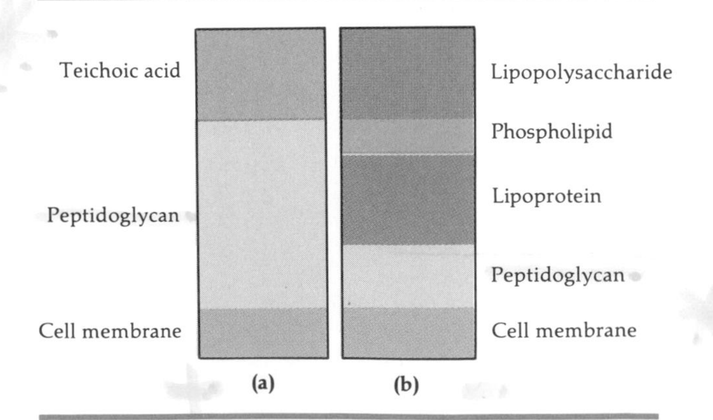

Figure 4–9 Comparison of the chemical components of the cell wall of **(a)** a gram-positive and **(b)** a gram-negative bacterium.

Damage to the Cell Wall

The cell wall is a good target for certain antimicrobial drugs, since the wall is made of chemicals unlike those in eucaryotic cells. Thus, chemicals that will damage bacterial cell walls, or interfere with their synthesis, often will not harm the cells of an animal host. One way that the cell wall can be damaged is by exposing it to the enzyme *lysozyme*. This enzyme occurs naturally in eucaryotic cells and is a constituent of tears, mucus, and saliva. In the presence of lysozyme the cell wall of gram-positive bacteria is seriously damaged, making it sensitive to rupture, or lysis. What happens is this: the lysozyme catalyzes the hydrolysis of the bonds between the sugars in the polysaccharide chain of peptidoglycan. This collapse is analogous to cutting the steel supports of a bridge with a cutting torch. The gram-positive cell wall is completely destroyed when treated with lysozyme. However, the remaining cellular contents surrounded by only the plasma membrane may remain intact; this structure is referred to as a **protoplast.** Typically, the protoplast is spherical and still capable of carrying on metabolism.

When the same procedure is carried out with gram-negative cells, usually the wall is not completely destroyed; much of the lipopolysaccharide-phospholipid-lipoprotein layer remains. In this case, the cellular contents, plasma membrane, and remaining outer wall layer are referred to as a **spheroplast,** also a spherical structure. Gram-negative cells that lose all of their wall material are referred to as protoplasts (Figure 4–10). Protoplasts and spheroplasts burst in pure water or very dilute salt or sugar solutions because of the rapid diffusion of water molecules from the surrounding fluid into the cell, which has a much lower concentration of water. This rupturing is called **osmotic lysis,** and it will be discussed in detail shortly.

Among procaryotes, there are cells that occur naturally without walls or that have very little wall material. These include members of the genus *Mycoplasma* and related organisms. Mycoplasmas are the smallest known bacteria that can grow and reproduce outside of living host cells. Because they have no cell walls, they pass through most bacterial filters and were first mistaken for viruses. They are unique among bacteria because they have lipids called *sterols* in their plasma membranes, which are thought to help protect them from osmotic lysis.

Other atypical bacterial cells are the **L forms** (named after the Lister Institute, where they were discovered). These are tiny mutant bacteria with defective cell walls. Certain chemicals and antibiotics like penicillin induce many bacteria to produce L forms. Although some L forms can revert to the original bacterial form, others are stable. L forms tend to contain just enough cell wall material to prevent lysis in dilute solutions. Antibiotics interfere with the synthesis of the peptide bridges of peptidoglycan, preventing the formation of a functional cell wall. This is how these antibiotics destroy bacteria.

In general, penicillins are more effective against gram-positive bacteria than against gram-negative. The additional lipopolysaccharide–phospholipid–lipoprotein layer of gram-negative bacteria seems to interfere with the entry of penicillin into the cells. Cell wall deficient organisms, such as mycoplasmas and L forms, are resistant to antibiotics that interfere with cell wall synthesis.

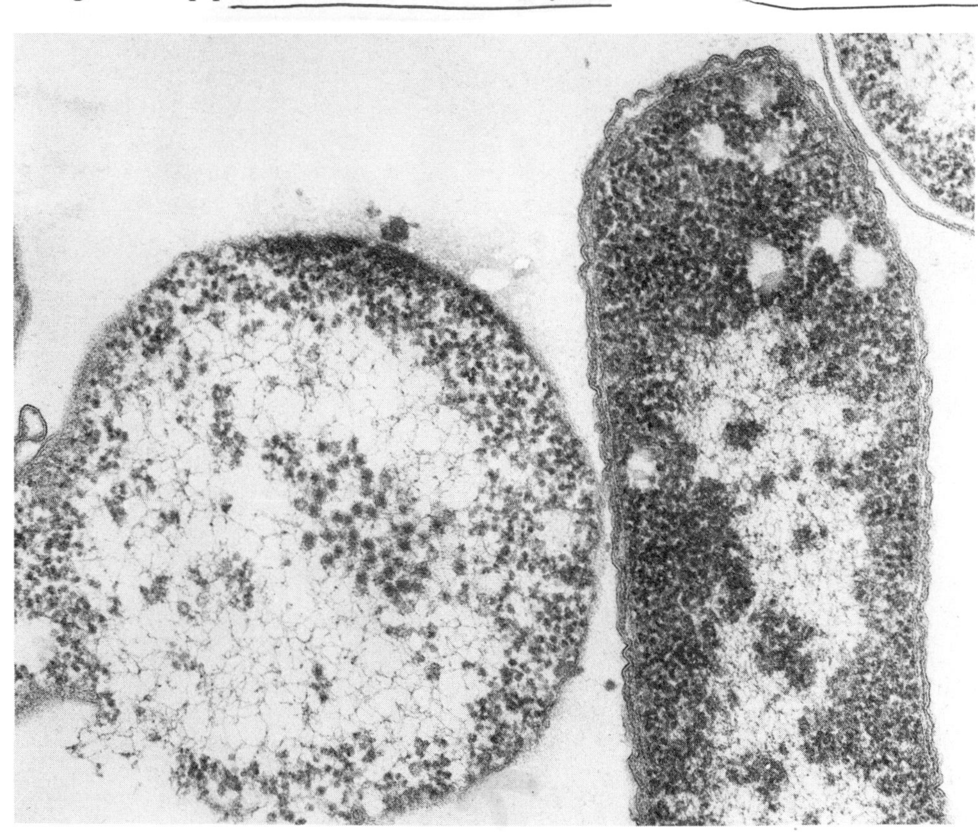

Figure 4–10 On the left is a rounded protoplast of the bacterium *Escherichia coli*; it is bounded only by the cytoplasmic membrane. To the right is one end of a normal, rod-shaped cell of *E. coli* (×61,000).

STRUCTURES INTERNAL TO THE CELL WALL

Thus far, we have discussed the procaryotic cell wall and structures external to it. We will now look inside the procaryotic cell and discuss the structure and function of the plasma membrane and other components within the cytoplasm of the cell.

Plasma (Cytoplasmic) Membrane

The **plasma (cytoplasmic) membrane** is a thin structure internal to the cell wall and enclosing the cytoplasm of the cell (see Figure 4–3). Chemically, the membrane is composed of about 60% protein and 40% lipid, most of which is phospholipid.

Structure

In electron micrographs, procaryotic and eucaryotic cell membranes look like a two-layered structure: there are two dark lines with a lighter space between the lines. The phospholipid molecules are arranged in two parallel rows, called a *phospholipid bilayer* (Figure 4–11a). Each phospholipid molecule contains a charged polar head (the phosphate end), which is soluble in water, and an uncharged nonpolar tail (the hydrocarbon end), which is insoluble in water. The phospholipids are arranged so that

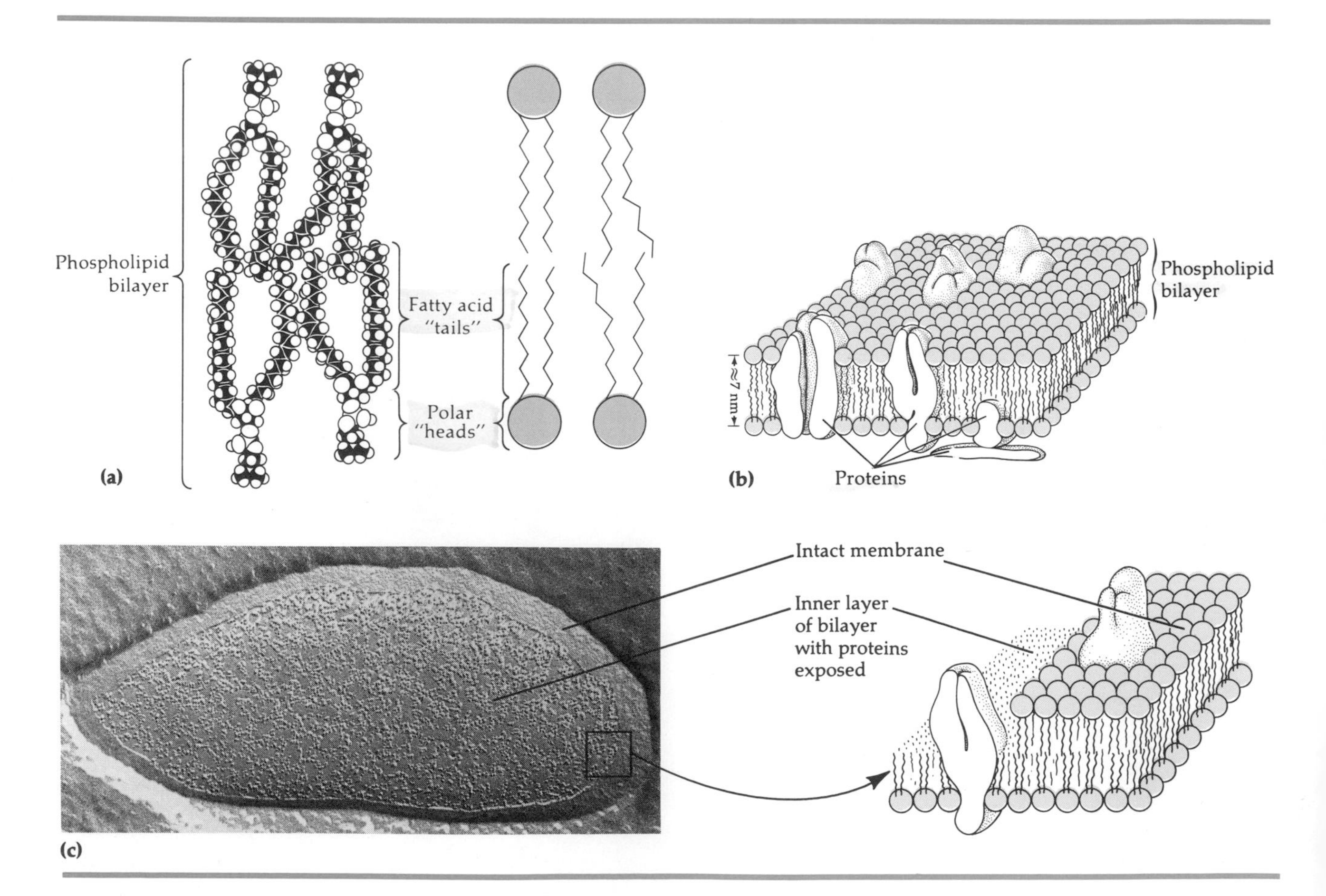

Figure 4–11 Plasma membrane. **(a)** Phospholipid bilayer. Space-filling models of several molecules next to a schematic drawing. **(b)** Drawing of membrane showing phospholipid bilayer and proteins. **(c)** An electron micrograph showing a freeze-etched membrane of a red blood cell ($\times 20{,}000$). The outer layer of the phospholipid bilayer has been chipped away, leaving proteins sticking up through the inner layer. The bilayer is still intact around the edge of the cell.

the polar ends are on the outside of the phospholipid bilayer and the nonpolar ends are on the inside.

The protein molecules are arranged in a variety of ways (Figure 4–11b). Some lie at or near the inner and outer surface of the membrane. Others penetrate the membrane partway, completely, singly, or in pairs. Recent physical and biological studies have demonstrated that the phospholipid and protein molecular arrangements in membranes are not static. The molecules seem to move quite freely within the membrane surface. This movement is most likely associated with the many functions performed by the plasma membrane. This dynamic arrangement of phospholipid and protein is referred to as the *fluid mosaic model*.

Functions

The most important function of the plasma membrane is to serve as a selective barrier through which materials enter and exit the cell. In this respect, plasma membranes are said to be **selectively permeable** (sometimes called **semipermeable**). This means that certain molecules and ions pass through the membrane, but others are restricted. The permeability of the membrane is related to several factors. While large molecules with molecular weights over several hundred (such as proteins) cannot pass through the membrane, smaller molecules (such as water, amino acids, and some simple sugars) usually pass through easily if they are uncharged. Substances that dissolve easily in lipids enter and exit more easily than other substances, since a large portion of the membrane consists of phospholipids. Ions penetrate only very slowly. The movement of materials across plasma membranes also depends on the presence of carrier molecules, which will be described shortly.

Another function of plasma membranes concerns the breakdown of foods and energy production. The plasma membranes of bacteria contain enzymes that are capable of catalyzing chemical reactions that break down nutrients and produce energy. In some bacteria, pigments and enzymes involved in photosynthesis (conversion of light energy into chemical energy) are found in infoldings of the plasma membrane that extend into the cytoplasm. These layers of membranes are called **chromatophores** (Figure 4–12).

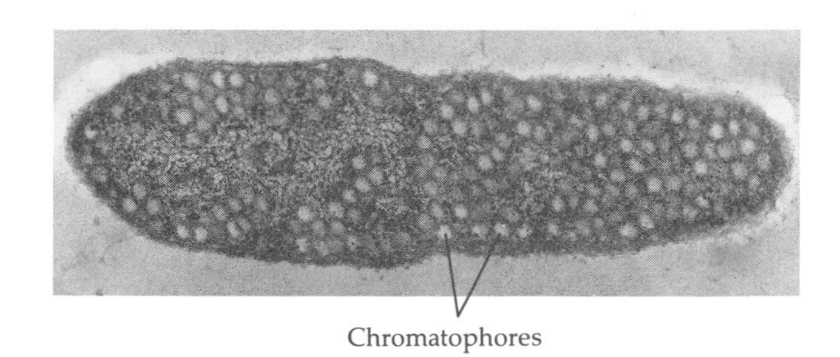

Figure 4–12 Chromatophores, membranous structures that are sites of photosynthesis, seen in a transmission electron micrograph of the photosynthetic bacterium *Rhodopseudomonas spheroides* (×43,000).

Mesosomes

Gram-positive bacteria often contain one or more large, irregular folds of the plasma membrane called **mesosomes** (see Figure 4–3). In gram-negative cells, mesosomes are considerably smaller. Generally, mesosomes appear as layered structures.

Mesosomes play a role in reproduction and metabolism. When a bacterial cell divides (binary fission), a cross wall called a *transverse septum* forms and partitions the genetic material of the parent cell into each of the two identical daughter cells (see Figure 6–10). Mesosomes begin the formation of the transverse septum and attach bacterial DNA to the plasma membrane (Figure 4–13). Mesosomes may also help separate DNA into each daughter cell following binary fission. And the ability of cells to concentrate nutrients is enhanced by the presence of mesosomes, because their folds increase the surface area of the plasma membrane.

Destruction of plasma membrane by antimicrobial agents

Because the plasma membrane is vital to the bacterial cell, it is not surprising that it is the site at which several antimicrobial agents exert their effect. In addition to the chemicals that damage the cell wall and hence indirectly expose the membrane to injury, many compounds specifically damage plasma membranes. These include certain alcohols and quaternary ammonium compounds, which are used as disinfectants. A group of antibiotics known

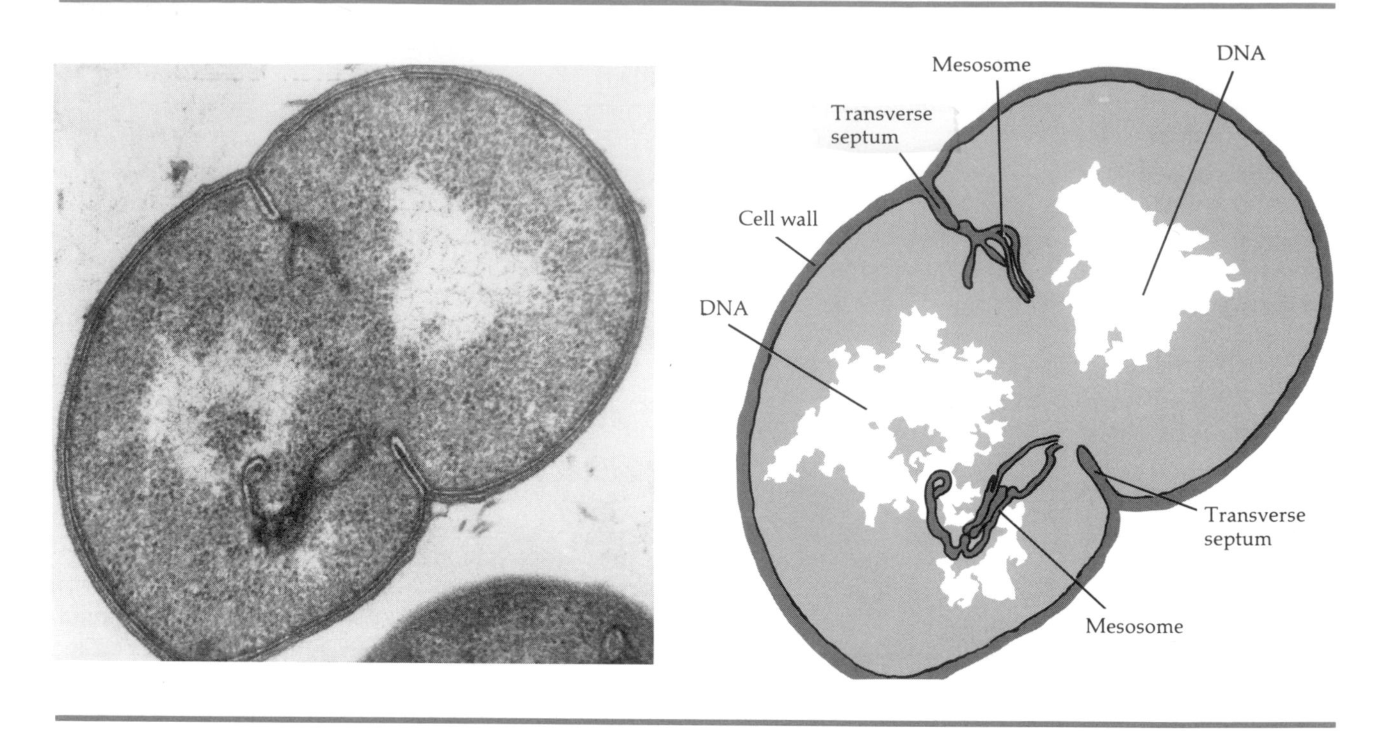

Figure 4–13 Transverse septum forming in a dividing cell of *Sporosarcina ureae* (×72,000). Two mesosomes can be seen extending into the cytoplasm from the innermost edge of the septum.

as the polymyxins disintegrate membrane phospholipids, causing leakage of intracellular contents and subsequent cell death. This mechanism is discussed in Chapter 18.

Movement of Materials Across Membranes

When a substance is more concentrated on one side of a membrane than another, it creates a **concentration gradient** (difference). If the substance can cross the membrane, it will move to the more dilute side until the concentrations are equal—or until other forces stop its movement.

Materials move across plasma membranes by two kinds of processes: passive and active. In **passive processes,** substances move across the membrane from an area of higher concentration to an area of lower concentration, without any expenditure of energy by the cell. Examples of passive processes are simple diffusion, osmosis, and facilitated diffusion. In **active processes,** the cell must use energy to move substances from areas of lower concentration to areas of higher concentration. An example of an active process is active transport.

Simple diffusion

Simple diffusion is the net (overall) movement of molecules or ions from an area of higher concentration to an area of lower concentration (Figure 4–14). The movement from higher to lower concentration continues until the molecules or ions are evenly distributed. The point of even distribution is called *equilibrium*. Bacterial cells rely on diffusion to transport certain small molecules such as oxygen and carbon dioxide across their cell membranes.

Osmosis

Osmosis is the net movement of solvent molecules across a selectively permeable membrane, from an area where the molecules are highly concentrated to an area of lower concentration. In living systems, the chief solvent is water.

Figure 4–14 Illustration of the principle of simple diffusion. The molecules of dye in the pellet are diffusing into the water, from an area of higher dye concentration to areas of lower dye concentration.

Osmosis may be demonstrated with the apparatus shown in Figure 4–15. A tube constructed from cellophane, a selectively permeable membrane, is filled with a colored, aqueous solution of 20% sucrose. The upper portion of the cellophane tube is plugged with a rubber stopper through which a glass tubing is fitted. The cellophane tube is placed into a beaker containing distilled water. Initially, the concentrations of water on either side of the membrane are different. There is a lower concentration of water inside the cellophane tube than outside. Because of this difference, water moves from the beaker into the cellophane tube. The force with which the water moves is called osmotic pressure.

Very simply, **osmotic pressure** is the force with which a solvent moves from a solution of lower solute concentration to a solution of higher solute concentration when the solutions are separated by a semipermeable membrane. There is no movement of sugar out of the cellophane tube into the beaker, however, since the cellophane is impermeable to molecules of sugar—the sugar molecules are too large to go through the pores of the membrane. As

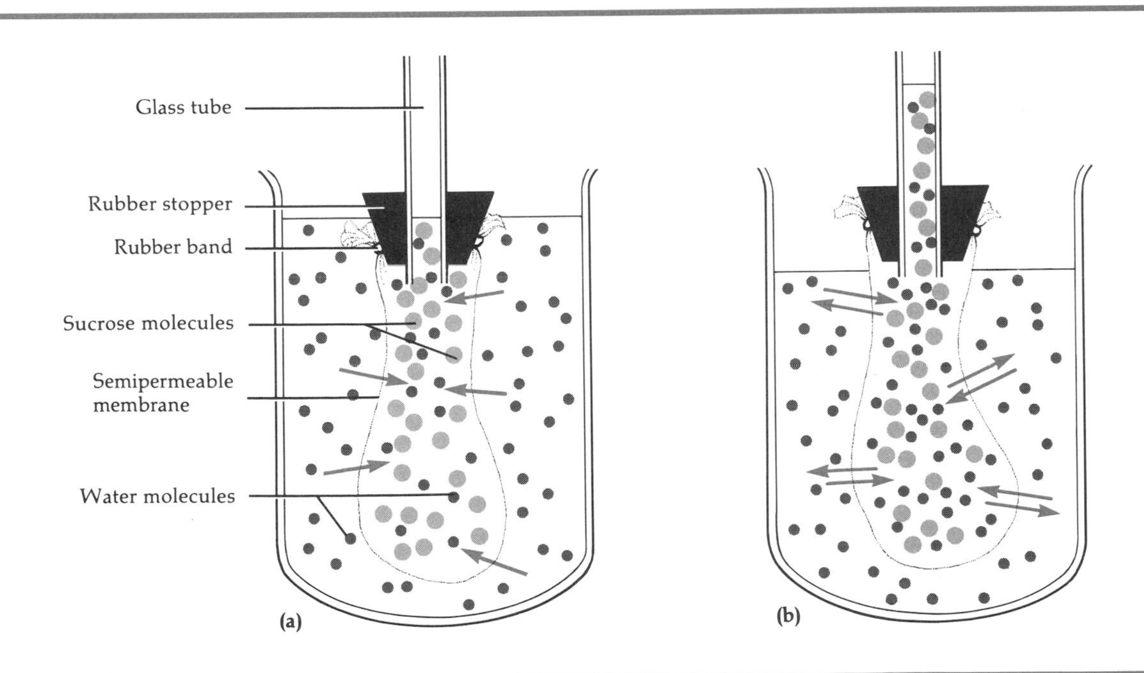

Figure 4–15 Illustration of the principle of osmosis. **(a)** Setup at the beginning of the experiment. **(b)** Setup at equilibrium. The final height of solution in the glass tube in **(b)** is a measure of the osmotic pressure.

water moves into the cellophane tube, the sugar solution becomes increasingly diluted and begins to move up the glass tubing. In time, the water that has accumulated in the cellophane tube and glass tubing exerts a downward pressure that forces water molecules out of the cellophane tube and back into the beaker. When water molecules leave the cellophane tube and enter the tube at the same rate, equilibrium is reached.

A bacterial cell may be immersed in three kinds of osmotic solutions: isotonic, hypotonic, or hypertonic. An **isotonic solution** (*iso* meaning "equal") is one where the overall concentrations of solutes are the same on both sides of the membrane. Neither water nor solutes tend to flow into or out of the cell; the cell contents are in equilibrium with the solution outside the cell wall.

Earlier we mentioned that lysozyme and certain antibiotics damage bacterial cell walls, causing the cells to rupture, or lyse. Such rupturing occurs because bacterial cytoplasm is usually so concentrated with solutes that when the wall is weakened or absent, water often enters the cell by osmosis. The damaged (or absent) cell wall cannot constrain the swelling of the cytoplasmic membrane, and the membrane bursts. This is an example of osmotic lysis caused by a hypotonic solution. A **hypotonic solution** (*hypo* meaning "under" or "less") is a medium outside the cell that has a lower concentration of solutes than the solution inside the cell. Most bacteria live in hypotonic solutions, where swelling is contained by the cell wall.

A **hypertonic solution** (*hyper* meaning "above") has a higher concentration of solutes outside the cell than inside. Most bacterial cells placed in a hypertonic solution tend to shrink and collapse because water leaves the cells by osmosis.

Keep in mind that the terms isotonic, hypotonic, and hypertonic describe the concentration of solutions outside the cell *relative to* the concentration inside the cell.

Facilitated diffusion

In **facilitated diffusion,** the substance to be transported combines with a *carrier protein* in the plasma membrane. Such carriers are sometimes called *permeases*. The carrier can transport a substance across the membrane from areas of higher concentration to those of lower concentration. Facilitated diffusion is similar to simple diffusion in that the substance moves from a higher to a lower concentration, forgoing any need for cellular expenditure of energy. The process differs from simple diffusion in that it is facilitated by carriers (Figure 4–16a).

In some cases, molecules required by bacterial cells are too large to be transported into the cells by the methods just described. Most bacteria, however, produce enzymes that can break down large molecules like proteins into smaller amino acids, polysaccharides into simple sugars. Such enzymes, when released by the bacteria into the surrounding medium, are appropriately called *extracellular enzymes*. Once the enzymes degrade the large molecules, the subunits are transported by permeases

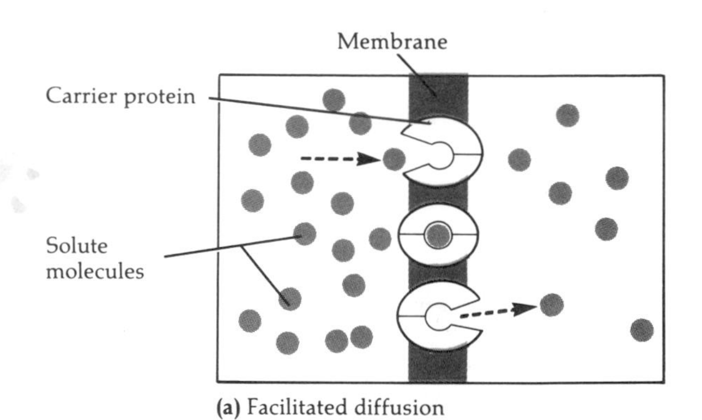

(a) Facilitated diffusion

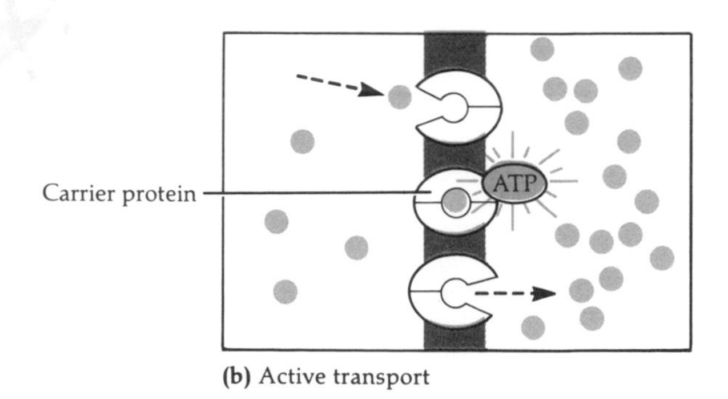

(b) Active transport

Figure 4–16 Facilitated diffusion compared with active transport. **(a)** In facilitated diffusion, carrier proteins in the membrane transport molecules across the membrane from an area of higher concentration to one of lower concentration (with the concentration gradient). This process does not require ATP. **(b)** In active transport, ATP supplies the energy needed for a carrier protein to transport a molecule across the membrane against a concentration gradient.

into the cell. For example, specific carriers retrieve DNA bases, like the purine guanine, from extracellular media, and bring them into the cell's cytoplasm.

Active transport

In **active transport,** the cell *uses energy* in the form of ATP to move substances across the plasma membrane. The movement is usually from outside to inside, even though the substance may be much more concentrated inside the cell. Like facilitated diffusion, active transport depends on the presence of carrier proteins in the plasma membrane (Figure 4–16b). There appears to be a different carrier for each transported substance or group of closely related transported substances.

Simple diffusion and facilitated diffusion are useful mechanisms for transporting substances into cells when their concentrations are greater outside the cell. However, when a bacterial cell is in an environment where nutrients are in low concentration, it must utilize active transport to accumulate the needed substances. For example, carrier proteins called galactoside permease enable some bacteria to concentrate certain sugars from growth media.

Cytoplasm

For a procaryotic cell, the term **cytoplasm** refers to everything inside the plasma membrane (see Figure 4–3). Cytoplasm is composed of about 80% water and contains nucleic acids, proteins, carbohydrates, lipids, inorganic ions, and many low-molecular-weight compounds. Functionally, cytoplasm is the substance in which the chemical reactions of the cell occur. The cytoplasm receives raw materials from the external environment, and enzymatic reactions within the cytoplasm degrade them to yield usable energy. The cytoplasm is also the place where new substances are synthesized for cellular use. It includes a fluid component (matrix), particles with various functions, and a nuclear region. The fluid component is composed of water in which are dissolved a vast array and high concentration of molecules. Inorganic ions are present in very high concentrations in cytoplasm with respect to most media. Physically, the fluid is thick, semitransparent, and elastic.

Inclusions

Within the cytoplasm are suspended accumulations of several kinds of reserve material, known as **inclusions.** Some inclusions are common to a wide variety of bacteria, whereas others are limited to a small number of species and therefore serve as a basis for identification. Among the more prominent inclusions are the following.

Metachromatic granules These inclusions stain red with certain blue dyes, such as methylene blue, and are collectively known as **volutin.** They represent a stored form of phosphate and are generally formed by cells that grow in phosphate-rich environments. Metachromatic granules are found in algae, fungi, and protozoans, as well as in bacteria. Since these granules are quite large and are characteristic of *Corynebacterium diphtheriae,* the causative agent of diphtheria, they do have diagnostic significance.

Polysaccharide granules These inclusions typically consist of glycogen and starch, and their presence may be demonstrated by applying iodine to the cells. In the presence of iodine, glycogen granules appear reddish brown, while starch granules appear blue.

Lipid inclusions Lipid inclusions appear in various species of *Mycobacterium, Bacillus, Azotobacter, Spirillum,* and other genera. A common lipid storage material unique to bacteria is the polymer *poly-β-hydroxybutyric acid.* Bacteria do not contain simple lipids (fats). Lipid inclusions are demonstrated with fat-soluble dyes, such as Sudan dyes.

Sulfur granules Certain bacteria known as the "sulfur bacteria," which belong to the genus *Thiobacillus,* derive energy by the oxidation of sulfur and sulfur-containing compounds. These bacteria may deposit sulfur granules in the cell, after which the stored granules serve as an energy reserve.

Besides inclusions, the cytoplasm contains thousands of very small structures called **ribosomes,** which give the cytoplasm of bacteria a granular appearance. These structures are composed of RNA and protein. As you will see in Chapter 8, ribosomes are protein factories that guide the synthesis of proteins needed for cellular activities. Several antibiotics, such as streptomycin, neomycin, and tetracyclines exert their antimicrobial effects by inhibiting protein synthesis.

Nuclear area

The **nuclear area,** or **nucleoid,** of bacterial cells contains a single, long, circular molecule of DNA, referred to as the **bacterial chromosome** (Figure 4–17). This is the cell's genetic information. Unlike the chromosomes of eucaryotic cells, bacterial chromosomes are not surrounded by a nuclear envelope. Nucleoids can be spherical, elongated, or dumbbell-shaped.

Bacteria often contain, in addition to the bacterial chromosome, small cyclic DNA molecules called **plasmids.** These are extrachromosomal genetic elements; that is, they are not connected to the main bacterial chromosome. Plasmids contain genes that are generally not essential for bacterial growth under the usual environmental conditions, and plasmids may be gained or lost without any harm to the cell. Under certain conditions, however, plasmids can provide an advantage to cells that contain them. We shall discuss plasmids further in Chapter 8.

Endospores

When essential nutrients are depleted, certain gram-positive bacteria are capable of forming specialized "resting" cells called **endospores.** Endospores are highly durable, dehydrated bodies with a thick wall. They are formed *inside* the bacterial cell wall. When released into the environment, they can survive extreme heat, lack of water, and exposure to many toxic chemicals.

The process of endospore formation within a vegetative (parent) cell is known as **sporulation** or **sporogenesis** (Figure 4–18a and b). It is not clear what biochemical events trigger this process. The first observable stage of sporogenesis is the ingrowth of the plasma membrane to form a *spore septum*. A newly replicated bacterial chromosome and a small portion of cytoplasm become surrounded by a double layer of membrane. This structure, entirely enclosed within the original cell, is called a *forespore*. Thick layers of peptidoglycan are laid down between the two membrane layers. Then a thick *spore coat* of protein forms around the outside membrane. It is this coat that is responsible for the resistance of endospores to many harsh chemicals.

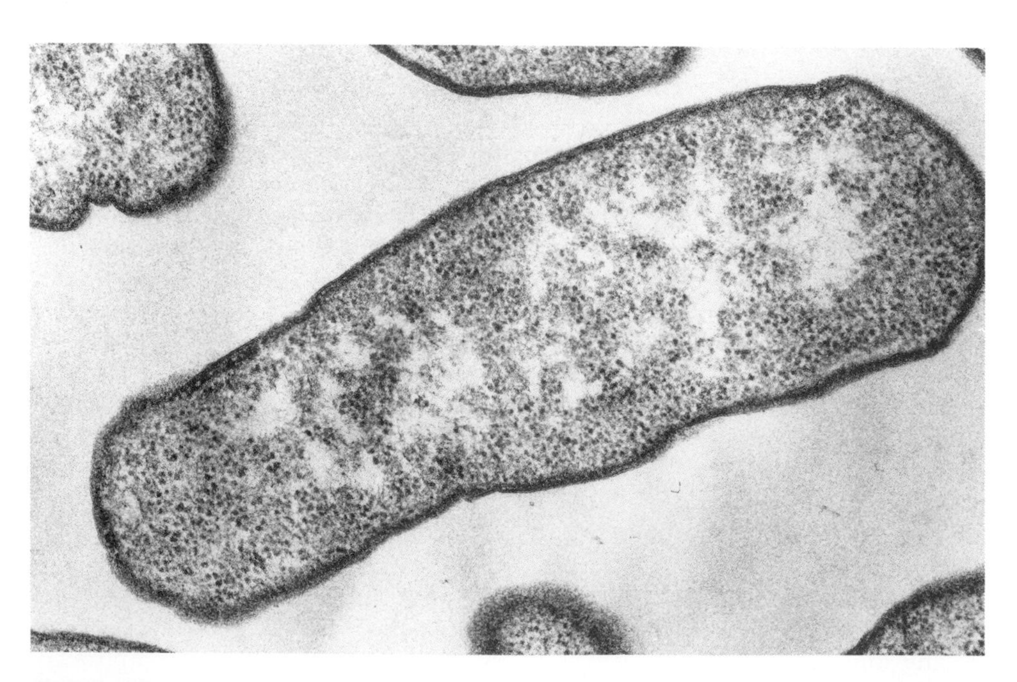

Figure 4–17 The light area throughout the center of this cell of *Escherichia coli* is the nucleoid, shown in a transmission electron micrograph (×55,000). The two light regions in the dividing cell of Figure 4–13 are also nucleoids.

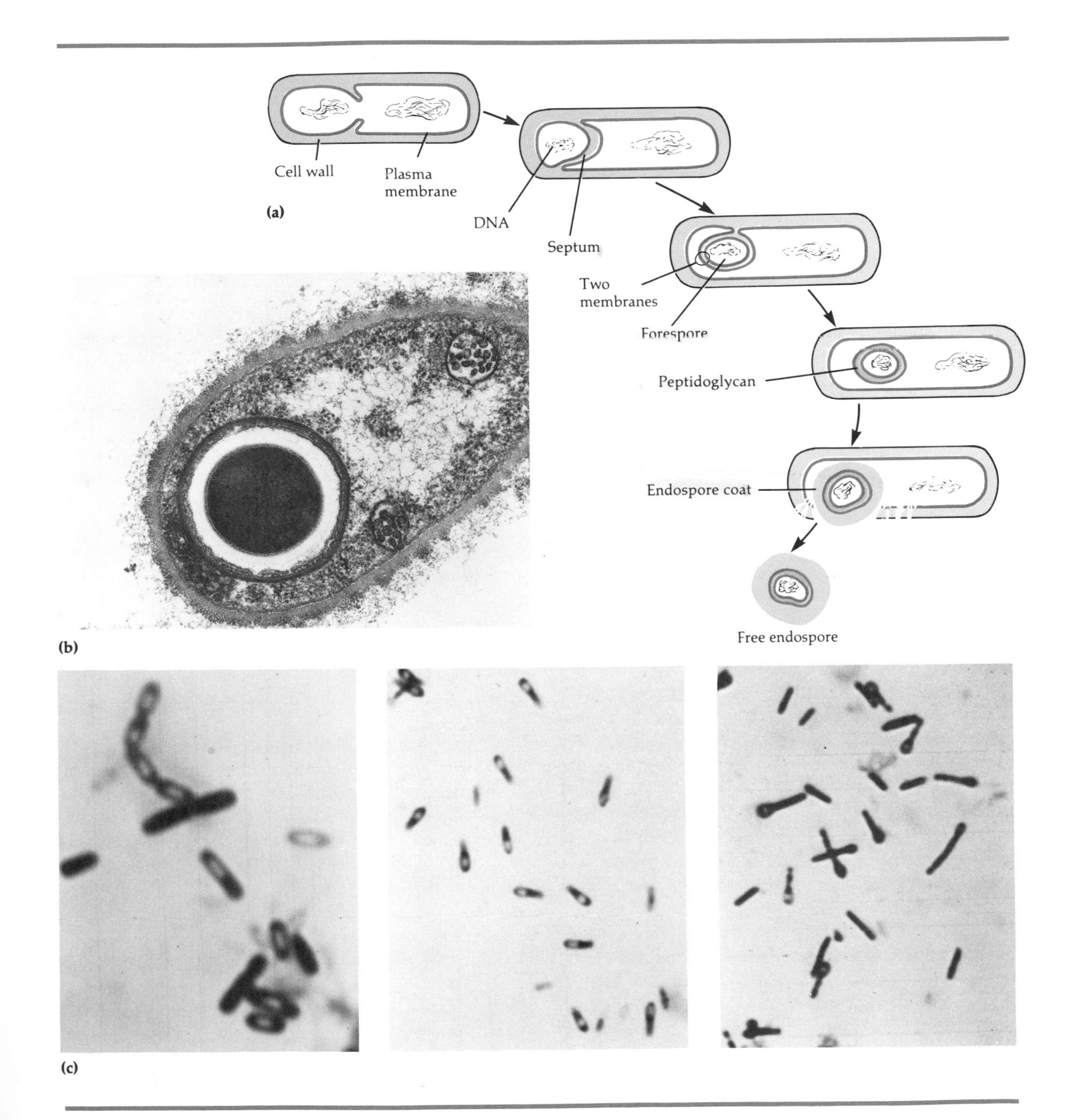

Figure 4–18 Endospores. **(a)** Sporogenesis, the process of endospore formation. **(b)** Electron micrograph of an endospore in *Bacillus sphaericus.* **(c)** Various locations of endospores. *Left:* central endospores in cells of *Clostridium bifermentans* (approx. ×4500); *center:* subterminal endospores in *C. subterminale* (approx. ×4500); *right:* round, terminal endospores at the ends of cells of *C. tetani* (approx. ×4500).

The diameter of the endospore may be the same as, smaller than, or larger than the diameter of the vegetative cell. Depending on the species, the endospore may be located *terminally* (at one end), *subterminally* (near one end), or *centrally* inside the vegetative cell (Figure 4–18c). When the endospore matures, the vegetative cell wall dissolves (lyses) and the endospore is freed.

Most of the water originally present in the forespore cytoplasm is eliminated by the time sporogenesis is complete, and endospores do not carry out metabolic reactions. The highly dehydrated endospore core contains only the cellular components that will be essential for resuming metabolism later: DNA, small amounts of RNA, ribosomes, enzymes, and a few important small molecules. The latter include a strikingly large amount of an organic acid called *dipicolinic acid,* which is accompanied by a large amount of calcium ion.

As noted in Chapter 3, endospores are difficult to stain. Thus, a specially prepared stain must be used, along with heat, to detect the presence of endospores. (The Schaeffer-Fulton endospore stain is commonly used.)

Endospores can remain dormant for long periods of time, even hundreds of years. However, an endospore may return to its vegetative state by a process called **germination.** Germination is triggered by physical or chemical damage to the endospore coat. Endospore enzymes break down the extra layers surrounding the endospore, water enters, and metabolism resumes. Since one vegetative cell forms a single endospore, which after germination remains one cell, sporogenesis in bacteria is *not* a means of reproduction. There is no increase in the number of cells.

Endospore formation is important from a clinical viewpoint, because endospores are quite resistant to processes that normally kill vegetative cells. Such processes include heating, freezing, desiccation, use of chemicals, and radiation. Whereas most vegetative cells are killed by temperatures above 70°C, endospores may survive in boiling water for an hour or more. Endospore-forming bacteria are a problem in the food industry, since some species produce toxins that result in food spoilage and disease. Special methods used to control organisms that produce endospores are discussed in Chapter 7.

Having examined the functional anatomy of the procaryotic cell, we will now look at the functional anatomy of the eucaryotic cell.

THE EUCARYOTIC CELL

As mentioned earlier, eucaryotic organisms include algae, protozoans (kingdom Protista), fungi, higher plants, and animals. The eucaryotic cell (Figure 4–19) is typically larger and structurally more complex than the procaryotic cell. By comparing the structure of the procaryotic cell in Figure 4–3 with that of the eucaryotic cell, we can see the differences between the two types of cells.

To review briefly, the genetic material (DNA) of procaryotic cells is not membrane-bounded or associated in a regular way with protein. The genetic material of eucaryotic cells *is* membrane-bounded, organized into chromosomes, and closely associated with histones and other proteins. In addition, eucaryotic cells contain membrane-bounded organelles; procaryotic cells do not. **Organelles** are specialized structures that perform specific functions. Although procaryotic cells carry on the same basic functions as eucaryotic cells, these functions are not localized in specific organelles.

The following discussion of eucaryotic cell anatomy will parallel our discussion of procaryotic cell anatomy by starting with structures that extend beyond the body of the cell. At the end of the discussion, the principal differences between procaryotic and eucaryotic cells will be summarized in Table 4–1.

FLAGELLA AND CILIA

Many types of eucaryotic cells have projections used for locomotion or for moving substances along the surface of the cell. These projections contain cytoplasm and are enclosed by the plasma membrane. If the projections are few and long in relation to the size of the cell, they are called **flagella.** If the projections are numerous and short, resembling hairs, they are referred to as **cilia.** Euglenoid protozoans use a flagellum for locomotion, whereas other protozoans, such as *Parame-*

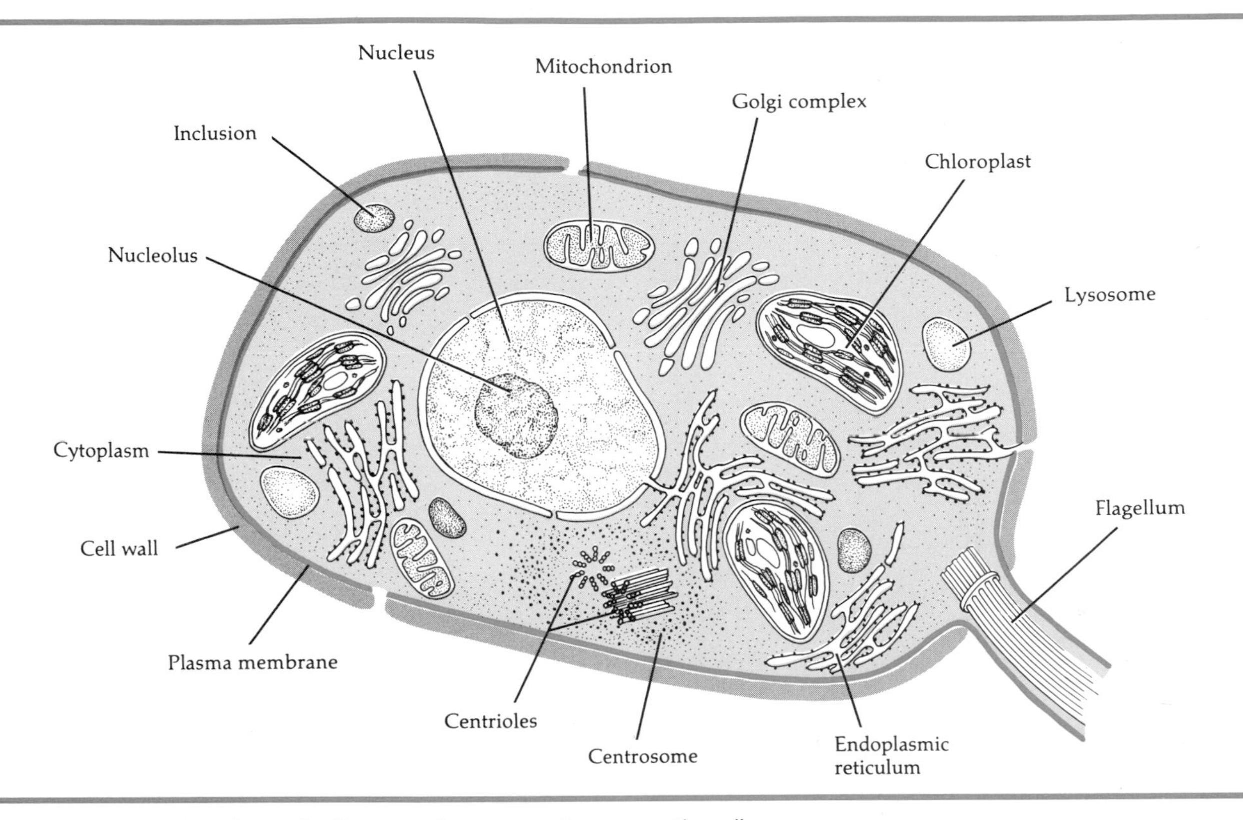

Figure 4–19 Highly schematic diagram of a composite eucaryotic cell.

cium, utilize cilia for locomotion (Figure 4–20a and b). In humans, the tail of a sperm cell is a flagellum and is used to propel the sperm cell through the male and female reproductive systems. Ciliated cells of the human respiratory system help to keep foreign material out of the lungs by moving the material along the surface of the cells in the bronchial tubes and trachea toward the throat and mouth (see Figure 15–3).

Structurally, eucaryotic flagella and cilia are composed of small tubules of protein called *microtubules* (Figure 4–20c). Both flagella and cilia consist of nine pairs of microtubules that form a ring around an inner central pair. The central microtubules arise from a plate near the surface of the cell, whereas the outer pairs arise from a structure called a centriole.

CELL WALL

In general, the eucaryotic **cell wall,** when present, is considerably simpler than that of the procaryotic cell. Most algae have cell walls consisting of the polysaccharide *cellulose*. Cell walls of some fungi also contain cellulose, but in most fungi, the principal structural component of the cell wall is the polysaccharide *chitin*, a polymer of N-acetylglucosamine units. (Chitin is also the main structural component of the exoskeleton of crustaceans and insects.) The cell wall of yeasts contains the polysaccharides *glucan* and *mannan*. Protozoans do not have a typical cell wall.

An important clinical consideration is that eucaryotic cells do not contain peptidoglycan, the framework of the procaryotic cell wall. This is important medically because human eucaryotic cells are unaffected by antibiotics, such as penicillins and cephalosporins, that act against peptidoglycan.

PLASMA (CYTOPLASMIC) MEMBRANE

In eucaryotic cells that lack a cell wall, the **plasma membrane** represents the external covering of the cell (see Figure 4–19). Functionally and structurally,

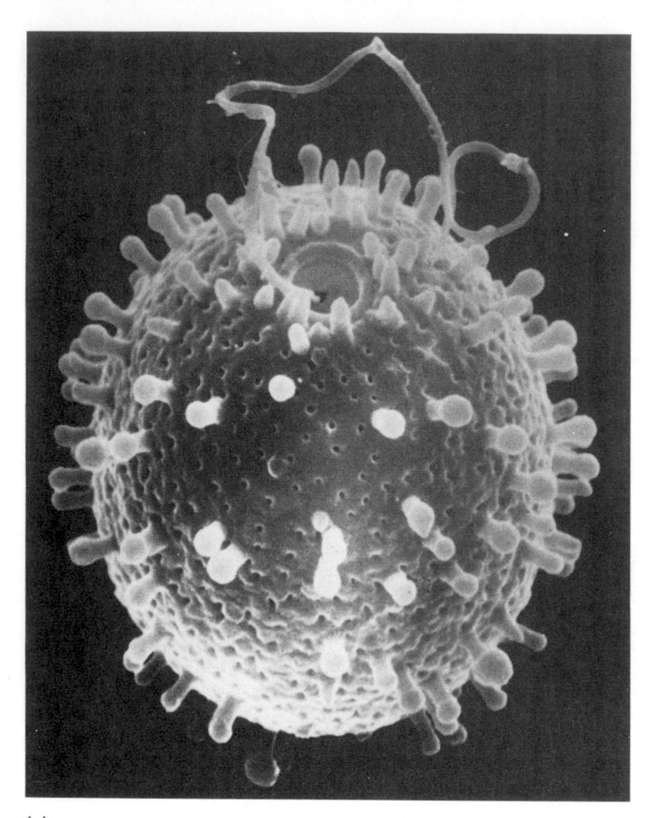
(a)

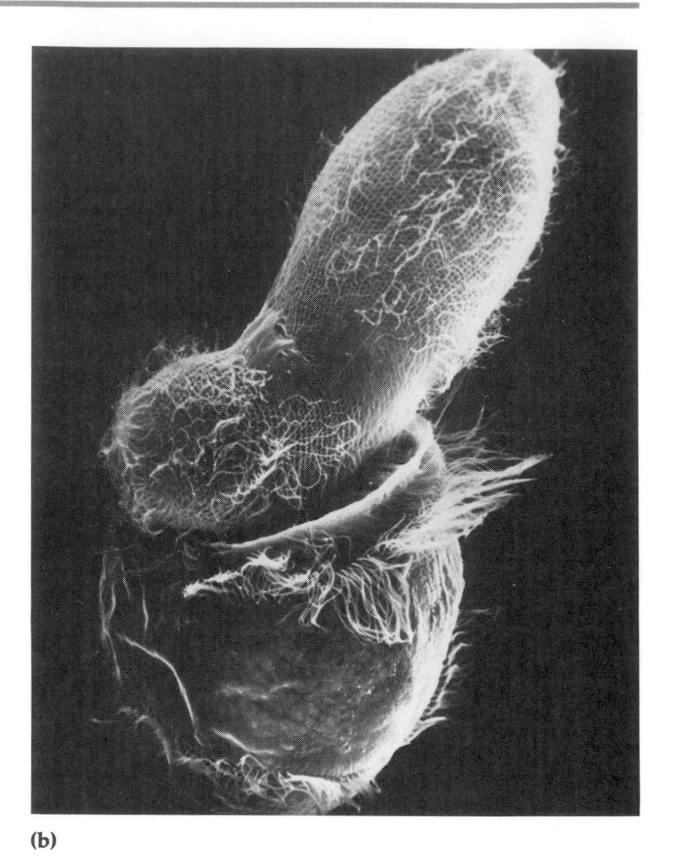
(b)

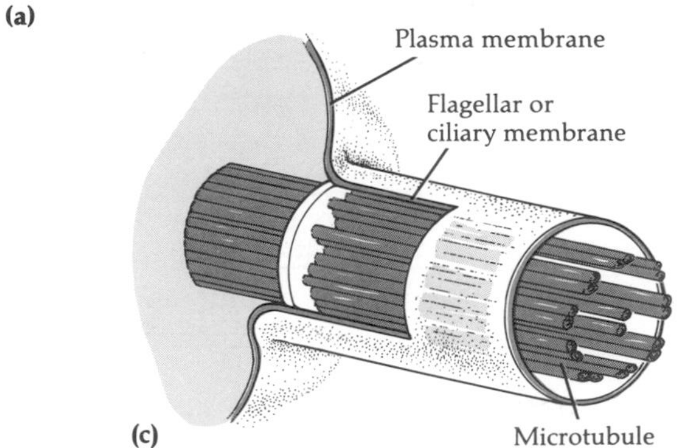

(c)

Figure 4–20 Eucaryotic flagella and cilia. **(a)** The freshwater euglenoid *Trachelomonas* sp. (×5000). Its long, whiplike flagellum is emerging from the cell body at the top of this scanning electron micrograph. **(b)** Ciliate eats ciliate. *Didinium nasutum*, the lower cell with two rows of cilia around its body, has begun to engulf a cilia-covered cell of *Paramecium multimicronucleatum* (×1500). **(c)** Structure of a flagellum or cilium.

the eucaryotic and procaryotic plasma membranes are very similar. There are, however, differences with respect to the proteins present and the carbohydrates attached to the proteins of some eucaryotic membranes. Another difference is that eucaryotic plasma membranes contain *sterols*, complex lipids not found in procaryotic plasma membranes (with the exception of the mycoplasmas). Sterols seem to be associated with the ability of the membranes to resist lysis due to increased osmotic pressure. (The chemical activity of sterols may be altered by a group of antimicrobial drugs called polyenes.)

Substances can cross eucaryotic plasma membranes by simple diffusion, osmosis, facilitated diffusion, or active transport, as in procaryotic cells. In addition, eucaryotic cells use two additional mechanisms called phagocytosis and pinocytosis. In **phagocytosis,** or "cell eating," cellular projections called *pseudopods* engulf solid particles exter-

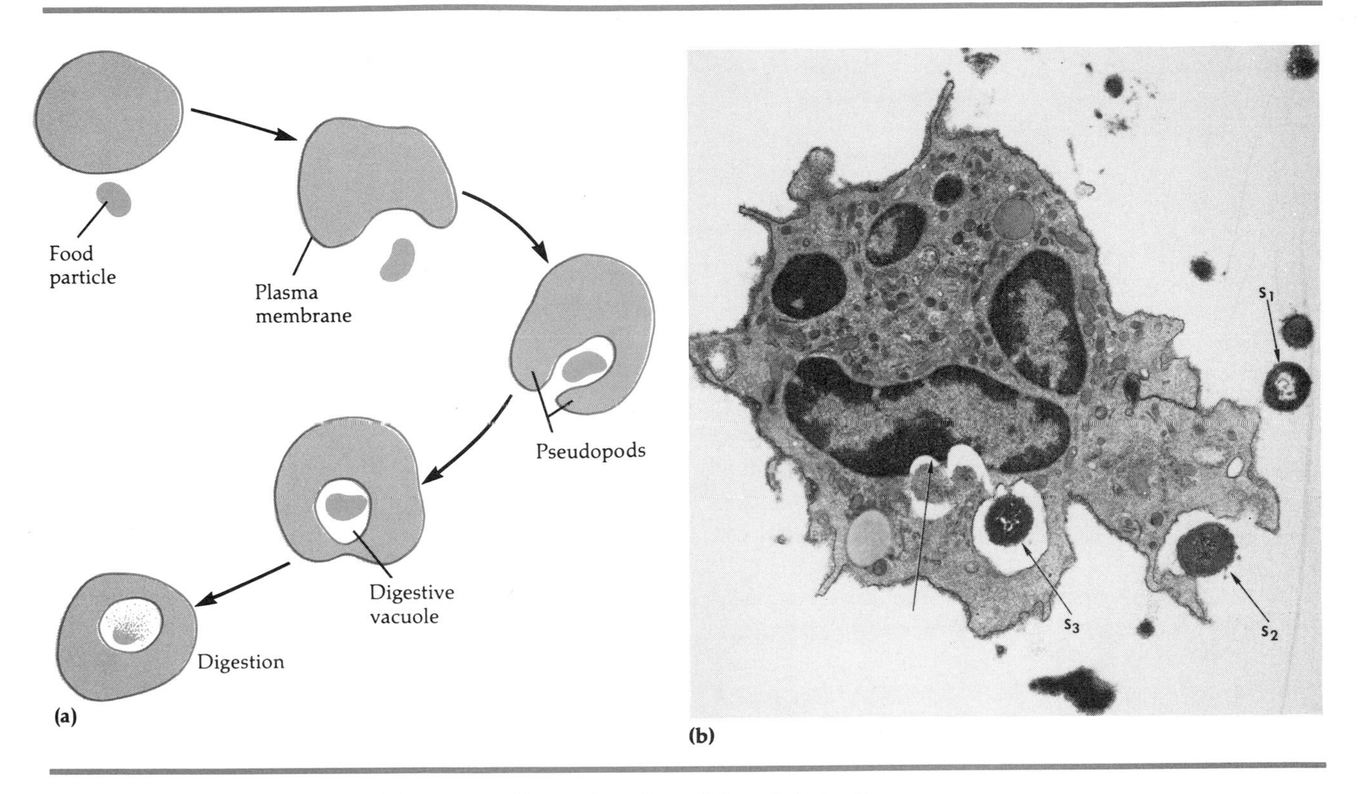

Figure 4–21 Phagocytosis. **(a)** Diagram of ingestion of a solid particle by the process of phagocytosis. **(b)** Phagocytosis of bacteria *(Streptococcus pyogenes)* by a white blood cell in the lung of a mouse (×23,500). S_1 is a free bacterium, S_2 is a bacterium that has been partially engulfed, and S_3 is a bacterium completely engulfed in a clear vacuole. (The unlabeled arrow points to an area of the nucleus that appears to have been digested away.)

nal to the cell. Once the particle is surrounded, the membrane folds inward, forming a membrane sac around the particle (Figure 4–21a). This newly formed sac, called a *digestive vacuole,* breaks off from the outer cell membrane, and the solid material inside the vacuole is digested. Indigestible particles and cell products are removed from the cell by a reverse phagocytosis. This process is important because molecules and particles of material that would normally be restricted from crossing the plasma membrane can be brought into or removed from the cell. Human phagocytic white blood cells constitute a vital defensive mechanism. Through phagocytosis, the white blood cells destroy bacteria and other foreign substances (Figure 4–21b).

In **pinocytosis,** or "cell drinking," the engulfed material is in liquid rather than solid form. Moreover, no cytoplasmic projections are formed. Instead, the membrane folds inward, surrounds the liquid, and detaches from the rest of the intact membrane. Few cells are capable of phagocytosis, but many cells can carry on pinocytosis.

When both phagocytosis and pinocytosis involve the inward movement of materials, they are together referred to as **endocytosis.** However, since phagocytosis and pinocytosis can also work in reverse, expelling materials from cells, they are also referred to as **exocytosis.**

CYTOPLASM

The **cytoplasm** of the eucaryotic cell encompasses everything inside the cell's membrane and external to the nucleus (see Figure 4–19). It is the matrix, or ground substance, in which various cellular components are found. Physically and chemically, the fluid component of the cytoplasm of eucaryotes is similar to that of procaryotes. A major difference, however, is that many of the important enzymes in the cytoplasmic fluid of procaryotes are sequestered in the organelles of eucaryotes.

ORGANELLES

Nucleus

The most characteristic eucaryotic organelle is the nucleus (see Figure 4–19). The **nucleus** is usually a spherical or oval organelle (Figure 4–22), and in addition to frequently being the largest structure in the cell, it also contains almost all of the cell's hereditary information (DNA). Some DNA is found in mitochondria and chloroplasts (in photosynthetic organisms).

The nucleus is separated from the cytoplasm by a double membrane called the *nuclear envelope.* Each of the two membranes resembles in structure the plasma membrane. Minute *pores* in the nuclear membrane allow the nucleus to communicate with the membranous network in the cytoplasm called the endoplasmic reticulum (see next section). Substances entering and exiting the nucleus are believed to pass through the tiny pores. Within the nuclear envelope is a gellike fluid called *nucleoplasm.* Spherical bodies called the *nucleoli* are also present. These structures serve as a center for the synthesis of ribosomal RNA, an essential constituent of ribosomes (see next section). Finally, there is the DNA, which is combined with a number of proteins, including several kinds of basic proteins called *histones* and *nonhistones.* When the cell is not reproducing, the DNA and its associated proteins appear as a threadlike mass called **chromatin.** Prior to cellular reproduction, the chromatin coils into shorter and thicker rodlike bodies called **chromosomes,** which have characteristic shapes in each species. Procaryotic "chromosomes" do not undergo this process and do not have histones.

Endoplasmic Reticulum and Ribosomes

Within the cytoplasm, there is a system consisting of pairs of parallel membranes enclosing narrow cavities of varying shapes. This system is known as the **endoplasmic reticulum,** or **ER** (Figure 4–23).

The ER is a network of canals running throughout the cytoplasm. These canals are continuous with both the plasma membrane and the nuclear membrane (see Figure 4–19). It is thought that the ER provides a surface area for chemical reactions, a pathway for transporting molecules within the cell, and a storage area for synthesized molecules. The ER plays a role in both lipid synthesis and protein synthesis.

Attached to the outer surface of some of the ER are exceedingly small, dense, spherical bodies called **ribosomes,** which are also found free in the cytoplasm. Ribosomes are the sites of protein synthesis in the cell.

(a)

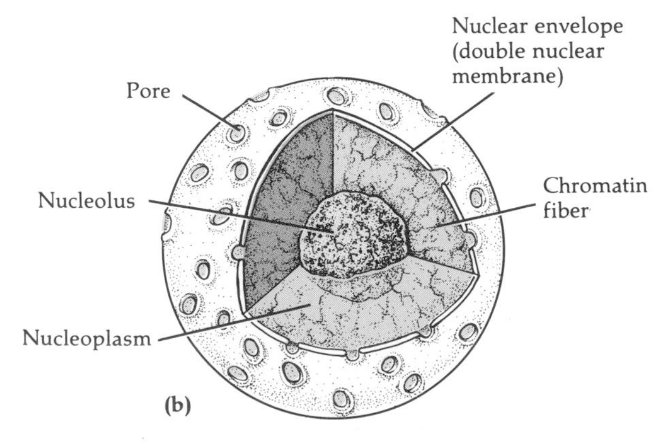

(b)

Figure 4–22 Nucleus. **(a)** Electron micrograph of a freeze-etched section of part of a nucleus in an onion root tip cell (×17,000). **(b)** Diagram of a nucleus.

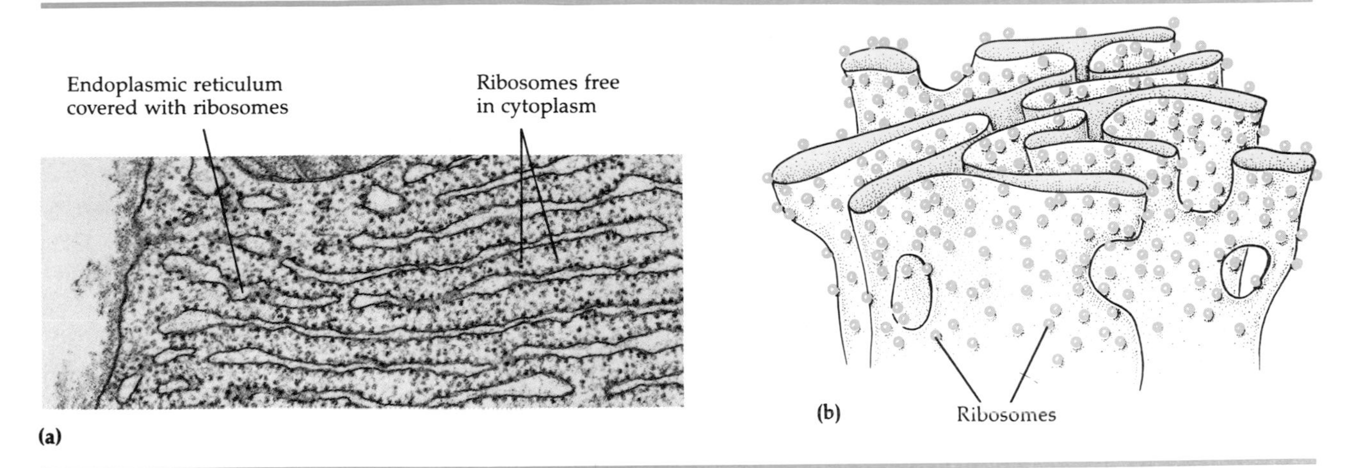

Figure 4–23 Endoplasmic reticulum and ribosomes. **(a)** Transmission electron micrograph of endoplasmic reticulum and ribosomes in cross section (×24,000). **(b)** Three-dimensional drawing of endoplasmic reticulum and ribosomes.

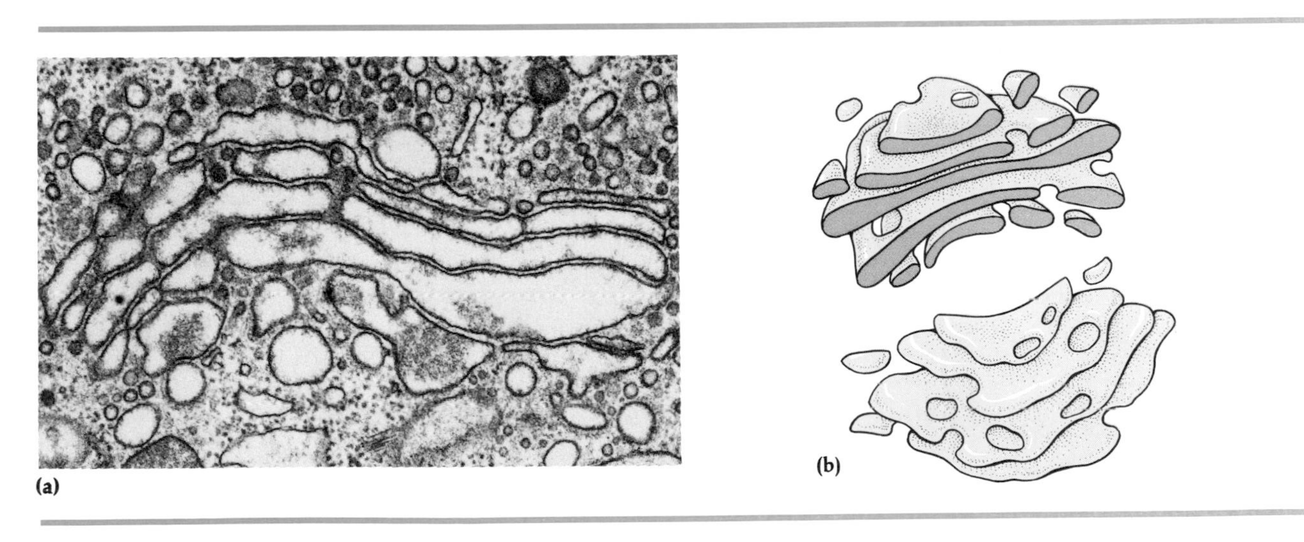

Figure 4–24 Golgi complex. **(a)** Transmission electron micrograph of a Golgi complex in cross section (×26,000). Note the vesicles forming at the top. **(b)** Three-dimensional drawing of a Golgi complex.

The ribosomes of eucaryotic cells are somewhat larger and more dense than those of procaryotic cells. Accordingly, eucaryotic ribosomes are called 80S ribosomes, and those of procaryotic cells are known as 70S ribosomes. The letter S refers to Svedberg units, which indicate the relative rate of sedimentation during ultra-high-speed centrifugation. The structure of ribosomes and their role in protein synthesis will be discussed in more detail in Chapter 8.

Golgi Complex

Another organelle found in the cytoplasm is the **Golgi complex.** This structure usually consists of four to eight flattened channels, stacked upon each other with expanded areas at their ends (Figure 4–24). The Golgi complex is sometimes connected to the ER (see Figure 4–19). Its function is the secretion (release from the cell) of certain proteins, lipids, and carbohydrates.

Lipids synthesized by the ER and proteins synthesized by the ribosomes associated with ER are transported from the ER tubules to the Golgi complex. As these substances accumulate in the Golgi complex, *vesicles* (membrane-bounded droplets) form and pinch off. Referred to as a *secretory granule,* the protein and its associated vesicle move toward the surface of the cell, where the contents of the vesicle are secreted.

The Golgi complex also functions in the synthesis of carbohydrates, which combine with proteins to form complexes called *glycoproteins*. Glycoproteins are secreted from the cell in vesicles.

Mitochondria

Spherical, rod-shaped, or filamentous organelles called **mitochondria** appear throughout the cytoplasm (see Figure 4–19). When sectioned and viewed under an electron microscope, each of these small organelles reveals an elaborate internal organization (Figure 4–25). A mitochondrion consists of a double membrane similar in structure to the plasma membrane. The outer mitochondrial membrane is smooth, but the inner membrane is arranged in a series of folds called *cristae*. The center of the mitochondrion is called the *matrix*. Because of the nature and arrangement of the cristae, the inner membrane provides an enormous surface area for chemical reactions. Enzymes involved in energy-releasing reactions that form ATP are located on the cristae. Mitochondria are frequently called the "powerhouses of the cell" because of their central role in the production of ATP (see Chapter 5).

Lysosomes

When viewed under the electron microscope, **lysosomes,** which are formed from Golgi complexes,

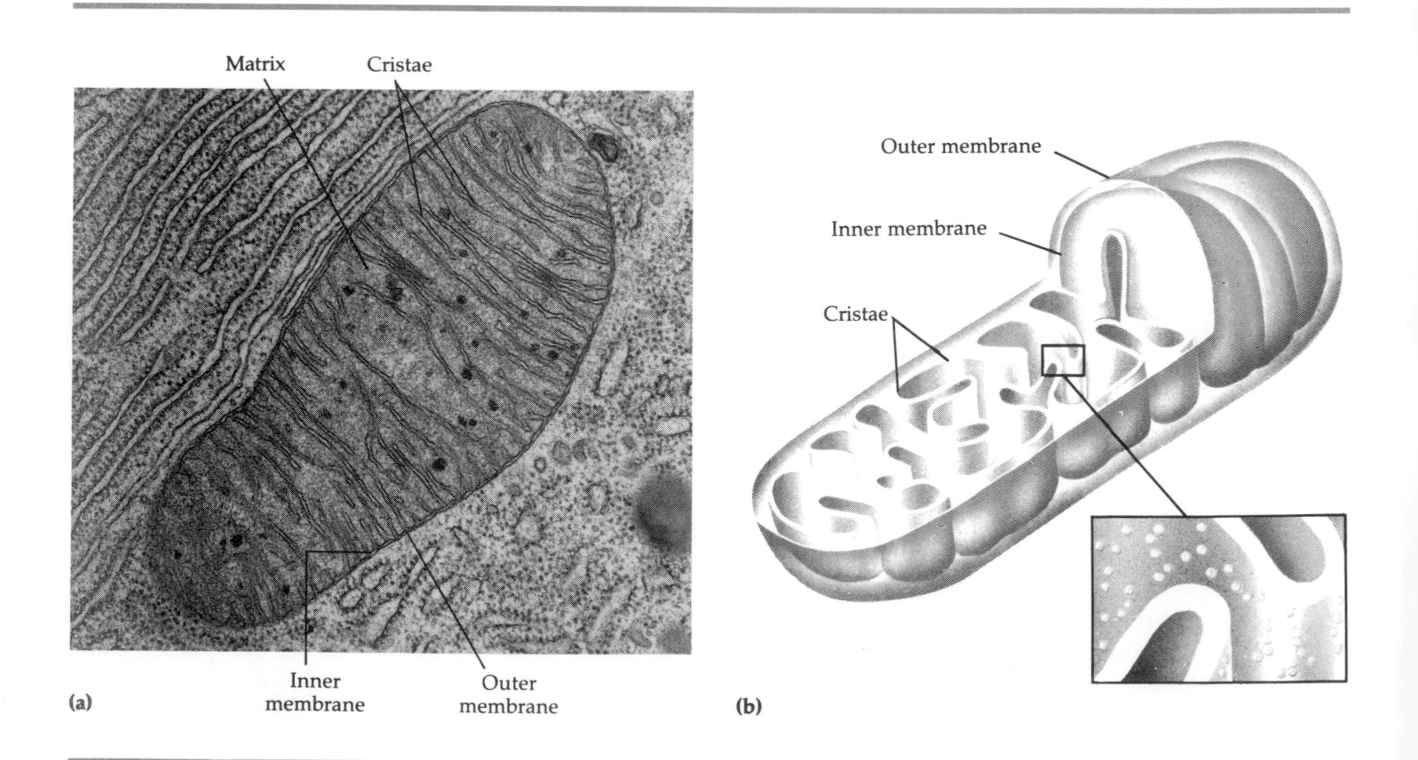

Figure 4–25 Mitochondria. **(a)** Transmission electron micrograph of a mitochondrion in longitudinal section from a bat pancreas cell ($\times 34{,}000$). **(b)** Three-dimensional drawing of a mitochondrion.

appear as membrane-enclosed spheres (Figure 4–26). Unlike mitochondria, lysosomes have only a single membrane and lack detailed structure (see Figure 4–19). But they contain powerful digestive enzymes capable of breaking down many kinds of molecules. Moreoever, these enzymes can also digest bacteria that enter the cell. White blood cells, which ingest bacteria by phagocytosis, contain large numbers of lysosomes. Scientists have wondered why these powerful enzymes do not also destroy their own cells. Perhaps the lysosome membrane in a healthy cell is impermeable to enzymes, preventing the enzymes from moving out into the cytoplasm. However, when a cell is injured, the lysosomes release their enzymes, which then promote reactions that break the cell down into its chemical constituents. The chemical remains are either reused by the body or excreted. Because of this function, lysosomes have been called "suicide packets."

Centrosome and Centrioles

There is a dense area of cytoplasm, generally spherical and located near the nucleus, that is called the **centrosome.** Within the centrosome is a pair of cylindrical structures, the **centrioles** (see Figure 4–19). Each centriole is composed of a ring of nine evenly spaced bundles of microtubules. Each bundle, in turn, consists of three microtubules. The two centrioles are situated so that the long axis of one is at right angles to the long axis of the other. Centrioles play a role in eucaryotic cell division.

Chloroplasts

Algae (and green plants) contain a unique organelle called a **chloroplast** (Figure 4–27). A chloroplast is a membrane-bounded structure that contains the pigment chlorophyll and enzymes required for the light-gathering functions involved in photosynthesis. The chlorophyll is contained in stacks of membranes called *thylakoids.* Like mitochondria, chloroplasts contain 70S ribosomes, DNA, and enzymes involved in protein synthesis. They are capable of multiplying by fission within the cell. Some of the needed genes are carried in the nuclear DNA.

CYTOPLASMIC INCLUSIONS

Eucaryotic cells contain cytoplasmic inclusions that have various influences on cellular function. The size and composition of these inclusions vary widely among the various species of Protista. *Zy-*

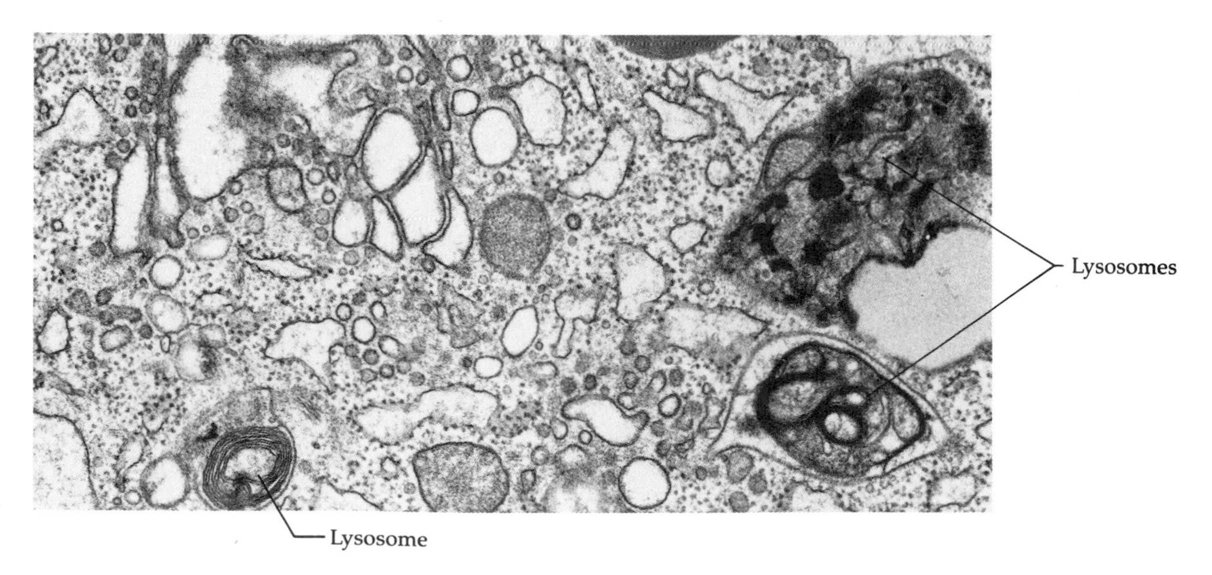

Figure 4–26 Lysosomes. Transmission electron micrograph of lysosomes in a rat pancreas cell (×26,000).

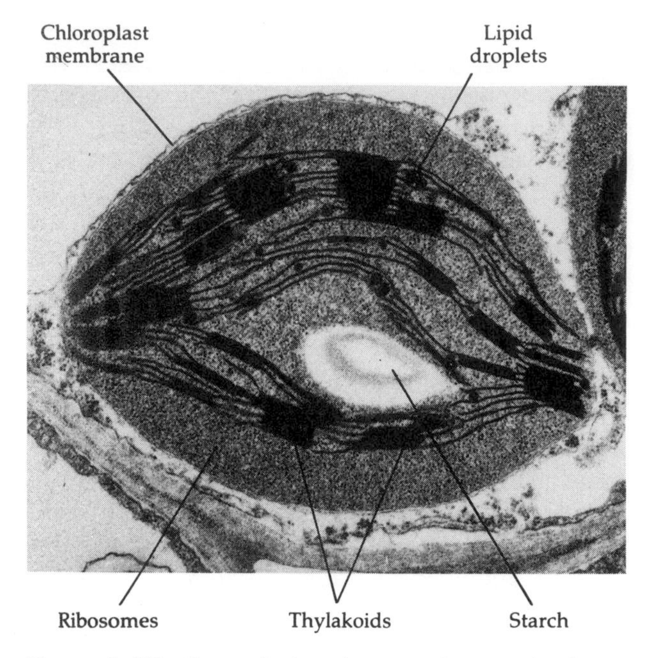

Figure 4–27 Transmission electron micrograph of a chloroplast from a plant cell (×19,000). The light-trapping pigments are located on the thylakoids.

mogen (an enzyme storage bank), *fat inclusions, vacuoles* (raw food forms), and *glycogen* (complex carbohydrates) are several examples of eucaryotic inclusions.

The principal differences between procaryotic and eucaryotic cells are presented in Table 4–1. Most of the differences have been discussed in this chapter; a few will be treated in greater detail in subsequent chapters.

Our next concern is to examine microbial metabolism. In Chapter 5 you will learn the importance of enzymes to microbes, and how microbes produce and utilize energy.

Table 4–1 Principal Differences Between Procaryotic and Eucaryotic Cells

Characteristic	Procaryotic	Eucaryotic
Nucleus	No nuclear membrane or nucleoli	True nucleus consisting of nuclear envelope and nucleoli
Membrane-bounded organelles	Absent	Present; examples include lysosomes, Golgi complex, endoplasmic reticulum, mitochondria, and chloroplasts
Flagella	Submicroscopic; simple	Microscopic; complex
Mitotic apparatus	Absent	Present during nuclear division
Chromosome (DNA) arrangement	Single circular chromosome; lacks histones	Several or many chromosomes with histones
Cell membrane	Generally lacks sterols	Sterols present
Ribosomes	Small size	Small and large sizes
Cell wall	When present, chemically complex	When present, chemically simple
Vacuoles	Atypical	Typical
Sexual reproduction	Fragmentary; no meiosis, only portions of chromosomes are reassorted	Regular; meiosis occurs, whole chromosome is reassorted

STUDY OUTLINE

PROCARYOTIC AND EUCARYOTIC CELLS (pp. 75–76)

1. Procaryotic and eucaryotic cells are similar in their chemical composition and chemical reactions.
2. Procaryotic cells lack membrane-bounded organelles (including a nucleus).
3. Peptidoglycan is found in procaryotic cell walls and not in eucaryotic cell walls.

THE PROCARYOTIC CELL (pp. 76–94)

1. Bacteria have a semirigid cell wall. If they are motile, it is by flagella. They are essentially unicellular, and most multiply by binary fission.
2. Bacteria are distinguished by morphology, chemical composition, nutritional requirements, and biochemical activities.

SIZE, SHAPE, AND ARRANGEMENT OF BACTERIAL CELLS (p. 77)

1. Most bacteria are between 0.20 and 2.0 μm in diameter.
2. The three basic bacteria shapes are coccus (spheres), bacillus (rods), and spiral.
3. Pleomorphic bacteria can assume several shapes.

STRUCTURES EXTERNAL TO THE CELL WALL (pp. 77–83)

1. The capsule, or slime layer, is a gelatinous polysaccharide and/or polypeptide covering. It may protect pathogens from phagocytosis or allow attachment to a surface.
2. Flagella are relatively long filamentous appendages consisting of a filament, hook, and basal body. They are used for motility.
3. Motility of spirochetes is by axial filaments, which wrap around the cell in a spiral fashion.
4. Pili (fimbriae) are appendages found on gram-negative bacteria. They help the cells attach to surfaces and transfer genetic material.

CELL WALL (pp. 83–85)

Composition and Characteristics (pp. 83–84)

1. The cell wall surrounds the plasma membrane and protects the cell from changes in osmotic pressure.

2. Chemically, the cell wall consists of peptidoglycan (sugars and amino acids).
3. Many gram-positive bacteria also contain teichoic acids.
4. Gram-negative bacteria have a lipopolysaccharide–phospholipid–lipoprotein layer surrounding the peptidoglycan. This layer is responsible for the gram-negative reaction.

Damage to the Cell Wall (pp. 84–85)

1. In the presence of lysozyme, the gram-positive cell wall is destroyed, and the remaining cellular contents are referred to as a protoplast.
2. In the presence of lysozyme, the gram-negative cell wall is not completely destroyed, and the remaining cellular contents are referred to as a spheroplast.
3. The mycoplasmas are bacteria that naturally lack cell walls. L forms are other bacteria that temporarily lack cell walls or have very little wall material.
4. Antibiotics like penicillin interfere with cell wall synthesis.

STRUCTURES INTERNAL TO THE CELL WALL (pp. 86–94)

Plasma (Cytoplasmic) Membrane (pp. 86–88)

1. The plasma membrane encloses the cytoplasm and is a phospholipid bilayer with protein (fluid mosaic).
2. Plasma membranes carry enzymes for metabolic reactions such as nutrient breakdown, energy production, and photosynthesis.
3. Plasma membranes can be destroyed by alcohols and polymyxins.
4. Some membrane proteins function as permeases.
5. Mesosomes are irregular infoldings of the plasma membrane that function in cell division and metabolism.

Movement of Materials Across Membranes (pp. 88–91)

1. Movement across the membrane may be by passive processes, where materials move from higher to lower concentration with no energy expenditure by the cell.
2. In simple diffusion, molecules and ions move until equilibrium is reached.
3. Osmosis is the movement of water from higher to lower concentration across a semipermeable membrane until equilibrium is reached.
4. In facilitated diffusion, substances are transported across membranes from higher to lower concentration by permeases.
5. In active transport, materials move from lower to higher concentrations and the cell must expend energy.

Cytoplasm (pp. 91–92)

1. Cytoplasm is the fluid component (cytoplasmic area) and nuclear area inside the plasma membrane.
2. The fluid component contains mostly water and high concentrations of molecules and inclusions (metachromatic granules, polysaccharide granules, lipid inclusions, and sulfur granules).
3. The cytoplasmic area contains numerous ribosomes.
4. The nuclear area contains the DNA of the main bacterial chromosome. Bacteria may also contain plasmids, which are extrachromosomal DNA circles.

Endospores (pp. 92–94)

1. Endospores are resting structures formed by some bacteria for survival during adverse environmental conditions.
2. The process of endospore formation is called sporulation (sporogenesis); the return of an endospore to its vegetative state is called germination.

THE EUCARYOTIC CELL (pp. 94–102)

FLAGELLA AND CILIA (pp. 94–95)

1. Whereas flagella are few and long in relation to cell size, cilia are numerous and short.
2. Flagella and cilia are used for motility, and cilia also move substances along the surface of the cells.
3. Both flagella and cilia consist of microtubules.

CELL WALL (p. 95)

1. The cell walls of most algae and some fungi consist of cellulose.
2. The main material of fungal cell walls is chitin.

PLASMA (CYTOPLASMIC) MEMBRANE (pp. 95–97)

1. Like the procaryotic plasma membrane, the eucaryotic plasma membrane is a phospholipid bilayer containing proteins.
2. Eucaryotic plasma membranes contain carbohydrates attached to the proteins and sterols not found in procaryotic cells.
3. Besides the methods used by procaryotic cells to move substances across the plasma membrane, eucaryotic cells use phagocytosis ("cell eating") and pinocytosis ("cell drinking").

CYTOPLASM (pp. 97–98)

1. The cytoplasm of eucaryotic cells includes everything inside the plasma membrane and external to the nucleus.
2. Physically and chemically the cytoplasm of eucaryotic cells resembles the cytoplasm of procaryotic cells.

ORGANELLES (pp. 98–101)

1. Organelles are specialized membrane-bounded structures in the cytoplasm.
2. They are characteristic of eucaryotic cells.
3. The nucleus contains DNA in the form of chromosomes. It is the most characteristic eucaryotic organelle.
4. The nuclear membrane is connected to a system of parallel membranes in the cytoplasm, the endoplasmic reticulum.
5. The endoplasmic reticulum provides a surface for chemical reactions, serves as a transporting network, and stores synthesized molecules.
6. Ribosomes may be found in the cytoplasm or attached to the endoplasmic reticulum.
7. Eucaryotic ribosomes are 80S, and procaryotic ribosomes are 70S.
8. The Golgi complex consists of stacked, flattened channels. It functions in secretion, carbohydrate synthesis, and glycoprotein formation.
9. Mitochondria are the primary sites of ATP production. They contain small amounts of ribosomes and DNA and multiply by fission.
10. Lysosomes are formed from Golgi complexes. They store powerful digestive enzymes and are referred to as "suicide packets."
11. The centrosome is a dense area of cytoplasm near the nucleus. It contains a pair of cylindrical structures called centrioles that are involved in cell division.
12. Chloroplasts contain chlorophyll and enzymes for photosynthesis. Like mitochondria, they contain some ribosomes and DNA and multiply by fission.

CYTOPLASMIC INCLUSIONS (pp. 101–102)

1. Cytoplasmic inclusions vary in size and composition.
2. Examples are zymogen, fat, vacuoles, and glycogen.

STUDY QUESTIONS

REVIEW

1. Draw each of the following bacterial shapes:
   (a) Spiral
   (b) Bacillus
   (c) Coccus
2. List three differences between procaryotic and eucaryotic cells.
3. Diagram each of the following flagellar arrangements:
   (a) Lophotrichous
   (b) Monotrichous
   (c) Peritrichous
4. Match the structures to their functions.

_____ Axial filament	(a) Motility
_____ Capsule	(b) Motility in spirochetes
_____ Cell wall	(c) Transfer of genetic material
_____ Endospore	(d) Protection from phagocytes
_____ Flagella	(e) Cell shape
_____ Mesosomes	(f) Selectively permeable
_____ Pili	(g) Resting
_____ Plasma membrane	(h) Cell wall formation
	(i) Attachment to surfaces

5. List three cytoplasmic inclusions in procaryotic cells.
6. Endospore formation is called _____________. It is initiated by _____________. Formation of a new cell from an endospore is called _____________. This process is triggered by _____________.
7. Why is an endospore called a resting structure? Of what advantage is an endospore to a bacterial cell?
8. Explain what would happen in the following experiments.
   (a) A suspension of bacteria is placed in distilled water.
   (b) A suspension of bacteria is placed in distilled water with lysozyme.
   (c) A suspension of bacteria is placed in an aqueous solution of lysozyme and 10% sucrose.
9. A cell requires an amino acid from the environment, where it is present in a higher concentration than within the cell. How could this amino acid be brought into the cell?
10. Describe the process called active transport.
11. Why are mycoplasmas resistant to antibiotics that interfere with cell wall synthesis?

12. Answer the following question using the diagrams shown, which illustrate cross sections of bacterial cell walls.

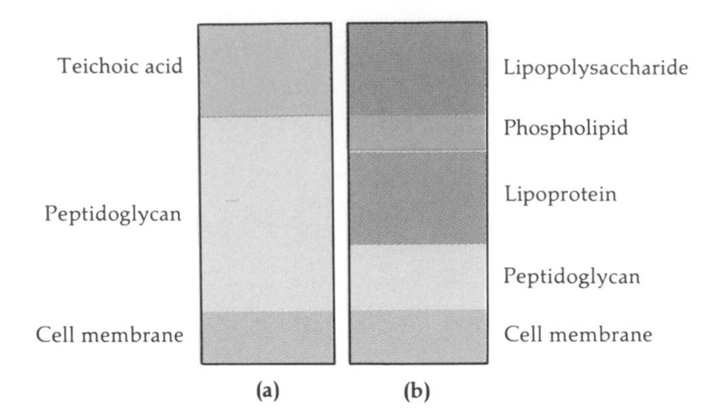

(a) Which diagram refers to a gram-positive bacterium? How can you tell?
(b) Explain how the Gram stain works to distinguish between these two types of cell walls.
(c) Why does penicillin have no effect on most gram-negative cells?

13. Match the following characteristics of eucaryotic cells with their functions.

____ Chloroplasts	(a) Intracellular transport
____ Endoplasmic reticulum	(b) Photosynthesis
____ Golgi complex	(c) ATP production
____ Lysosomes	(d) Digestive enzyme storage
____ Mitochondria	(e) Secretion

14. What process would a eucaryotic cell use to ingest a procaryotic cell?

CHALLENGE

1. Eucaryotic cells may have evolved from early procaryotic cells living in close association. What do you know about eucaryotic organelles that would support this theory?

FURTHER READING

Adler, J. April 1976. The sensing of chemicals by bacteria. *Scientific American* 235:40–47. Describes the chemotactic response of bacteria using molecules that detect the presence of chemicals in the environment.

Berg, H. C. August 1975. How bacteria swim. *Scientific American* 235:36–45. A detailed description of the rotational movement of procaryotic flagella.

Bittar, E. E., ed. 1980. *Membrane Structure and Function,* 3 vol. New York: John Wiley. A compilation of essays on the ultrastructure and function of the plasma membrane.

Costerton, J. W. 1979. The role of electron microscopy in the elucidation of bacterial structure and function. *Annual Review of Microbiology* 33:459–479. Uses scientific methodology to interpret micrographs; includes a discussion of structures.

Costerton, J. W., G. G. Geesey, and K. J. Cheny. January 1978. How bacteria stick. *Scientific American* 239:86–95. Describes how bacteria adhere to surfaces in nature.

Rosen, B. P. 1978. *Bacterial Transport*. New York: Marcel Dekker. A detailed discussion of the function of the plasma membrane.

Walsy, A. E. August 1977. The gas vacuoles of blue-green algae. *Scientific American* 237:90–97. Summarizes the structure of gas vacuoles and how they are used to regulate buoyancy in cyanobacteria.

5

Microbial Metabolism

Objectives

After completing this chapter you should be able to

- Define metabolism and contrast the fundamental differences between anabolism and catabolism.
- Describe the mechanism of enzyme action.
- List factors that influence enzyme activity.
- List and provide examples for three ways in which ATP is generated during metabolism.
- Explain what is meant by oxidation–reduction.
- Categorize the various nutritional patterns among organisms.
- Define fermentation and describe the chemical reactions of glycolysis.
- Compare aerobic and anaerobic respiration.
- Explain the products of the Krebs cycle.

Metabolism refers to the sum of all chemical reactions that occur within a living organism. Since chemical reactions either release or require energy, metabolism may be visualized as an energy-balancing act. Accordingly, metabolism may be divided into two general classes of chemical reactions: **anabolic (synthetic) reactions** and **catabolic (degradative) reactions.**

The chemical reactions in living cells that combine simpler substances into more complex molecules are collectively known as **anabolism.** Overall, any anabolic process requires energy. Examples of anabolic processes are the formation of proteins from amino acids, nucleic acids from mononucleotides, and polysaccharides from simple sugars. These biosynthetic reactions generate the materials for cell growth.

The chemical reactions that break down complex organic compounds into simpler ones are collectively known as **catabolism.** Catabolic reactions release chemical energy stored in organic molecules, energy that can then be used to drive anabolic reactions. However, only a small part of the energy released is actually available for cellular functions, since a major part of the energy is lost to the environment as heat. Thus, there is a continual need to provide new external sources of energy for the cell. The chemical composition of a living cell is constantly changing: some molecules are being broken down while others are being synthesized. This balanced flow of chemicals and energy maintains the life of a cell.

The energy liberated from catabolic processes is made available for anabolism in the *energy-rich bonds* of compounds produced by catabolic reactions. These compounds represent a high level of potential energy and therefore serve as energy carriers to drive energy-requiring reactions. The most common energy carrier in all biological systems is adenosine triphosphate (ATP); its structure can be reviewed in Figure 2–20. The role of ATP in the relationship between catabolic and anabolic processes is shown in Figure 5–1.

A little later in the chapter, we will examine some representative chemical reactions that deal with energy production (catabolic reactions) and energy utilization (anabolic reactions) in microorganisms. We will then look at how these various reactions are integrated within the cell. But first let us consider the principal properties of a group of proteins involved in almost all biologically important chemical reactions. These proteins, the enzymes, were described briefly in Chapter 2. Although it is beyond the scope of this text to name and discuss the actions of individual enzymes, you should be aware of the central role of enzymes in metabolic reactions. It is important to understand that a cell's metabolic pathways are determined by its enzymes, which are, in turn, determined by its genetic makeup.

ENZYMES

Many organic chemicals are so stable that they could remain unchanged in a cell for years. To activate these chemicals, living cells produce **enzymes,** proteins that act as catalysts in chemical reactions of importance to the cell. A *catalyst* is a substance that speeds up a reaction without being

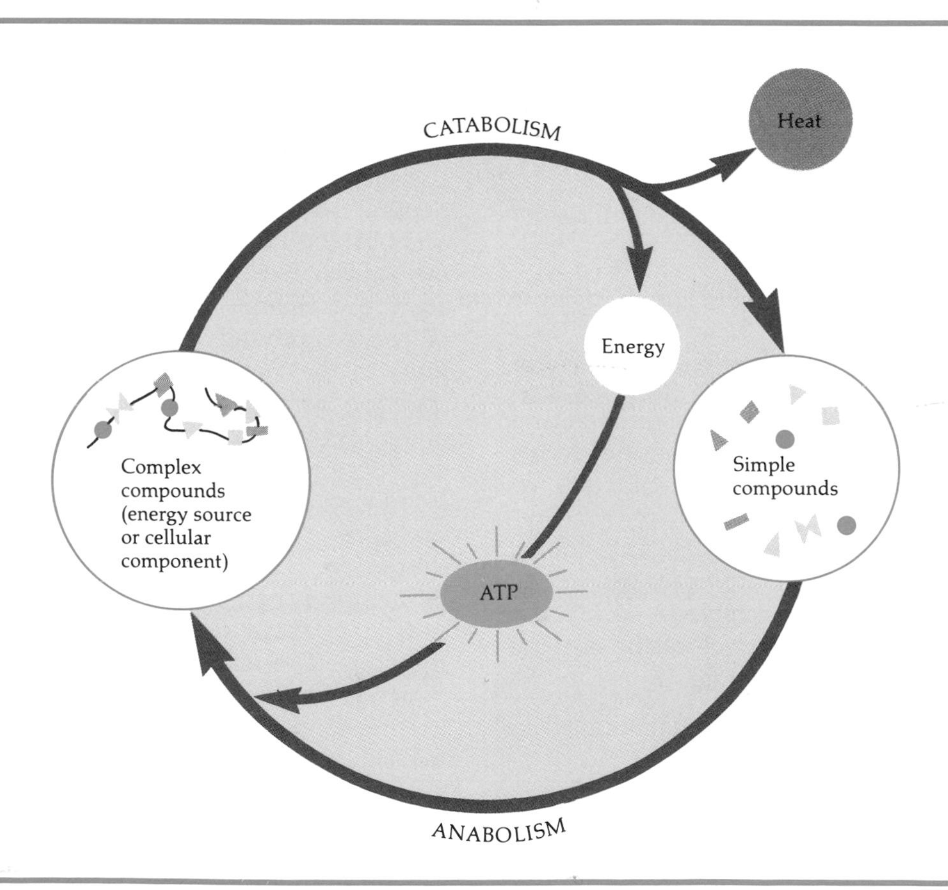

Figure 5–1 Relationship between anabolism and catabolism and the role of ATP. When simple compounds are combined to form complex compounds (anabolism), ATP provides the energy for synthesis. When large compounds are split apart (catabolism), heat energy is given off and some energy is trapped in ATP molecules.

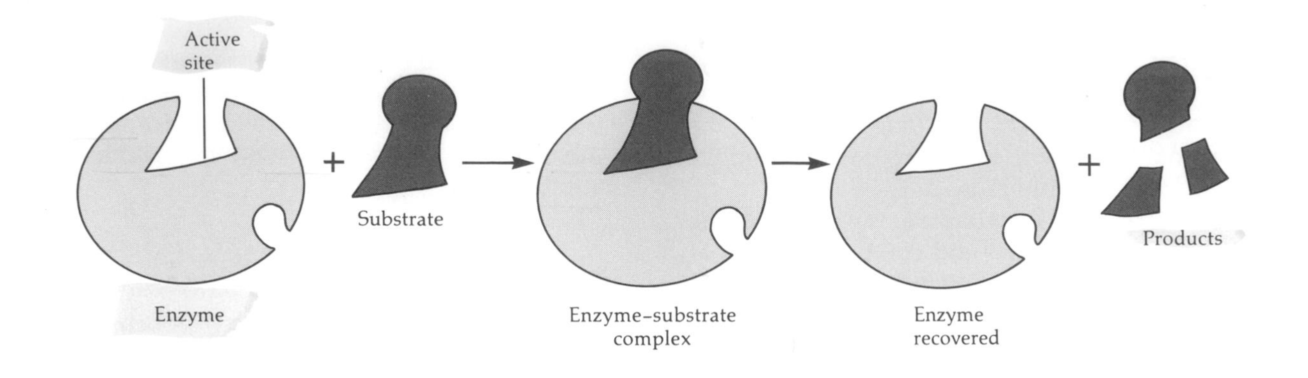

Figure 5–2 Mechanism of enzyme action. The surface of the substrate comes into contact with the active site on the surface of the enzyme to form an enzyme–substrate complex. The substrate is then transformed into products and the enzyme is recovered.

changed by it. Generally large globular proteins, enzymes range in molecular weight from about 10,000 to somewhere in the millions. Of the thousand or more known enzymes, each has a three-dimensional characteristic shape with a specific surface configuration due to its primary, secondary, and tertiary structures (see Figure 2–18).

Mechanism of Enzyme Action

As mentioned in Chapter 2, catalysts lower the *activation energy* required for a chemical reaction. Although scientists do not completely understand how an enzyme does this, the sequence of events is believed to be as follows (Figure 5–2):

1. The surface of the *substrate*—that is, the molecule or molecules that are reactants in the chemical reaction to be catalyzed—contacts a specific region on the surface of the enzyme molecule, called the *active site.*
2. A temporary intermediate compound called an *enzyme–substrate complex* forms.
3. The substrate molecule is transformed (by rearrangement of existing atoms, a breakdown of the substrate molecule, or the combining of several substrate molecules).
4. The transformed substrate molecules, the products of the reaction, move away from the surface of the enzyme molecule.
5. The recovered enzyme, now freed, reacts with other substrate molecules.

Enzyme reaction is characterized by its extreme *specificity* for a particular substrate. For example, a specific enzyme may be capable of hydrolyzing a peptide bond only between two specific amino acids. And other enzymes are capable of hydrolyzing starch, but not cellulose; even though both starch and cellulose are polysaccharides composed of glucose subunits, the orientations of the subunits in the two polysaccharides differ. Enzyme specificity results from the three-dimensional shape of the active site, which fits the substrate somewhat like a lock with its key. In most instances, the substrate is much smaller than the enzyme, and relatively few of the enzyme's amino acids make up the active site.

A given compound can be a substrate for a number of different enzymes that catalyze different reactions. The fate of a given reactant (substrate) depends on the specific enzyme that reacts upon it. For example, glucose-6-phosphate, an important molecule in cell metabolism, may be acted upon by at least four different enzymes, each of which will give a different product.

Enzymes are exceedingly efficient. Under optimum conditions, they can catalyze reactions at rates that are 10^8 to 10^{10} times (up to 10 billion times) more rapid than those of comparable reactions without enzymes. The *turnover number* (number of substrate molecules metabolized per enzyme molecule per second) is generally between 1 and 10,000 and in some instances may be as high as 500,000.

Yet, as previously mentioned, enzymes are specific in the reactions they catalyze, as well as in the substrates they act upon. From among the large number of diverse molecules in the cell, an enzyme must "find" the correct substrate. Moreover, the enzyme-catalyzed reactions tend to take place in aqueous solutions and at relatively low temperatures—conditions that otherwise would not favor rapid movement of molecules or rapid chemical reactions.

Enzymes are subject to various cellular controls. Their rate of synthesis and their concentration at any given time are under the control of the cell's genes and are influenced by various other molecules present in the cell, as will be discussed in Chapter 8. Many enzymes are present in the cell in both active and inactive forms. The rate at which the inactive form becomes active or the active form becomes inactive is determined by the cellular environment.

Enzymes are generally named after the substrate they react with or the type of reaction they perform, using the suffix *-ase* to identify the enzyme. Cellul*ase* is an enzyme that breaks down its substrate cellulose.

Nonprotein Components of Enzymes

Some enzymes consist entirely of proteins. But many enzymes contain a protein called an **apoenzyme** that is inactive without a nonprotein component called the **cofactor.** Together, the apoenzyme and cofactor are an activated **holoenzyme,** or whole enzyme. If the cofactor is removed, the apoenzyme will not function (Figure 5–3). The cofactor may be a metal ion or a complex organic molecule called a **coenzyme.**

Coenzymes assist the enzyme in transforming the substrate by acting as an acceptor of atoms being removed from the substrate or as a donor of atoms required by the substrate. Many coenzymes are derived from vitamins (Table 5–1; for the sake of completeness, vitamins not used by microorganisms are also included). Two of the most important coenzymes in cellular metabolism are *NAD* (nicotinamide adenine dinucleotide) and *NADP* (nicotinamide adenine dinucleotide phosphate). Both compounds contain derivatives of the B vitamin nicotinic acid (niacin) and function with their respective enzymes in the removal and transfer of hydrogen ions and electrons from substrate molecules. Enzymes that participate in reactions in which hydrogen atoms are removed (dehydrogenation) are called *dehydrogenases*.

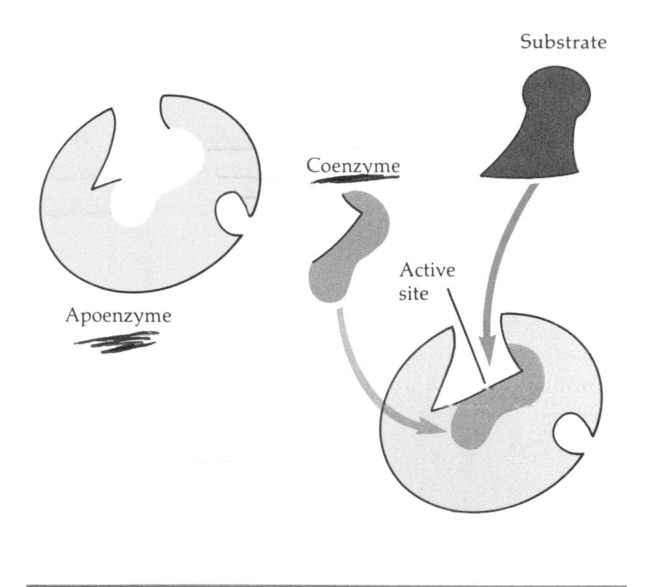

Figure 5–3 Components of a holoenzyme. Many enzymes require for their activity both an apoenzyme (protein portion) and a cofactor (nonprotein portion). The cofactor may be a metal ion or an organic molecule, called a coenzyme (as shown here). The apoenzyme and cofactor together make up the holoenzyme, or complete enzyme.

The flavin coenzymes, such as *FMN* (flavin mononucleotide) and *FAD* (flavin adenine dinucleotide), contain derivatives of the B vitamin riboflavin and are important in hydrogen transfer reactions as well as photosynthetic reactions. Like NAD and NADP, these coenzymes function in hydrogen transfer reactions with dehydrogenase enzymes.

Another important coenzyme, *coenzyme A (CoA)*, contains a derivative of pantothenic acid, another B vitamin. This coenzyme plays an important role in the synthesis and breakdown of fats and in a series of oxidizing reactions called the *Krebs cycle*. Coenzyme A is used in decarboxylation reactions (removal of CO_2) and is associated with a useful fragment of cellular catabolism, the acetyl group:

$$CH_3-\overset{\overset{\displaystyle O}{\|}}{C}-$$

Table 5–1 Vitamins and Their Functions

Vitamin	Function
Water-soluble vitamins:	
Vitamin B_1 (thiamine)	Part of coenzyme cocarboxylase; has many functions, including the metabolism of pyruvic acid
Vitamin B_2 (riboflavin)	Coenzyme in flavoproteins; active in electron transfers
Niacin	Part of NAD molecule; active in electron transfers
Vitamin B_6 (pyridoxine)	Coenzyme in amino acid metabolism
Vitamin B_{12} (cyanocobalamin)	Active in red blood cell formation, amino acid metabolism
Pantothenic acid	Part of coenzyme A molecule; involved in metabolism of pyruvic acid and lipids
Biotin	Involved in carbon dioxide fixation reactions, fatty acid synthesis
Folic acid	Coenzyme in synthesis of purines and pyrimidines
Vitamin C (ascorbic acid)	Involved in collagen deposition in connective tissue
Lipid-soluble vitamins:	
Vitamin A	Active in formation of visual pigment, bone and teeth growth
Vitamin D	Involved in absorption of calcium and phosphorus from intestine
Vitamin E	Needed for cellular and macromolecular syntheses
Vitamin K	Coenzyme in formation of blood-clotting proteins

When coenzymes are very tightly bonded to their apoenzymes, they are called **prosthetic groups.** One example is the heme (iron-containing) group of an enzyme called cytochrome *c*. **Cytochromes** are a group of enzymes that function as electron carriers in respiration and photosynthesis, as we shall see later in this chapter. These compounds contain pigments and metal ions and are structurally related to hemoglobin and chlorophyll.

As noted earlier, some cofactors are metal ions, including iron, copper, magnesium, manganese, zinc, calcium, and cobalt. It is believed that such cofactors serve as a bridge between the enzyme and the substrate, thus binding them together to facilitate substrate transformation. For example, magnesium (Mg^{2+}) is required by many phosphorylating enzymes that act together with ATP. The Mg^{2+} may act to form a bond between the enzyme and ATP. Most trace elements required by living cells are probably used to activate cellular enzymes.

Factors Influencing Enzyme Activity

Several factors influence the activity of an enzyme. Among the more important are temperature, pH, substrate concentration, and inhibitors.

Temperature

Most chemical reactions occur more rapidly as the temperature rises. In enzymatic reactions, however, once a certain temperature is attained, any further elevation results in a drastic decline in the reaction rate (Figure 5–4a). This decrease in reaction rate is due to denaturation of the enzyme, a phenomenon common to all proteins. Besides heat, enzymes may be denatured by concentrated acids, bases, heavy metal ions (such as copper, zinc, silver, arsenic, or mercury), alcohol, and ultraviolet radiation.

Denaturation (Figure 5–5) usually involves breakage of the hydrogen bonds and other weak bonds that hold the enzyme in its characteristic three-dimensional structure (tertiary configuration). A common example of denaturation is the transformation of uncooked egg white (albumin) to a hardened state after it is heated. As might be expected, this alteration in structure changes the arrangement of the amino acids in the active site with a loss in catalytic ability and of vital biological activity. In some cases, denaturation is partially or fully reversible. But if denaturation reaches a point where the enzyme loses its solubility and coagulates, denaturation becomes irreversible, because the enzyme cannot regain its original properties.

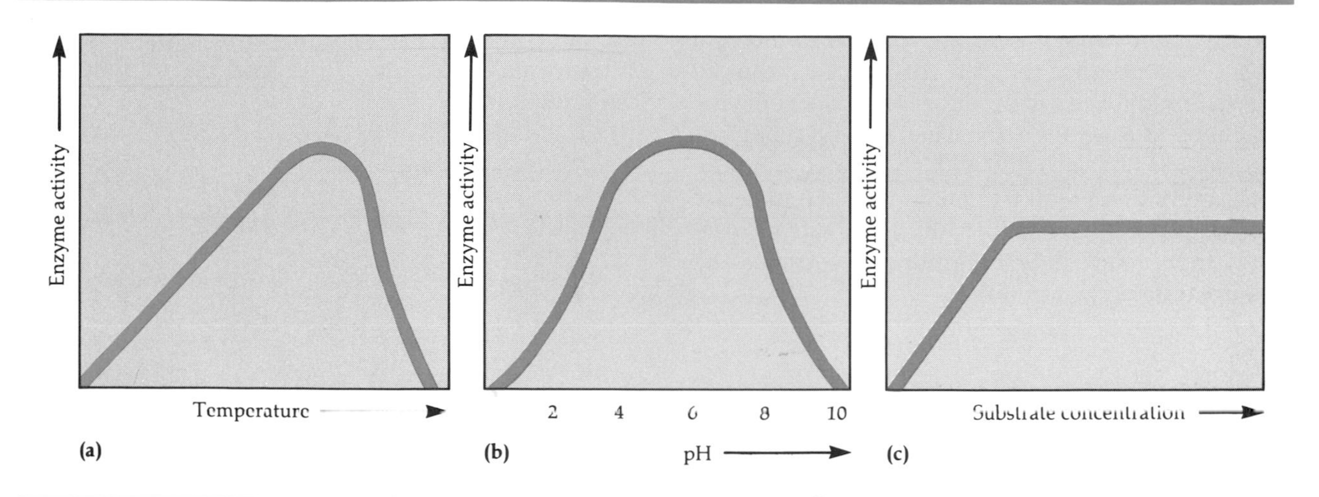

Figure 5–4 Factors that influence enzyme activity, plotted for a hypothetical enzyme. **(a)** Temperature. The enzyme activity (rate of reaction catalyzed by the enzyme) increases with increasing temperature until the enzyme, a protein, is denatured by heat and inactivated. At this point, the reaction rate falls suddenly. **(b)** pH. The pH at which this enzyme is most active is around pH 5. **(c)** Substrate concentration. With increasing concentration of substrate molecules, the rate of reaction increases until the active sites on all the enzyme molecules are filled, at which point the maximum rate of reaction is reached.

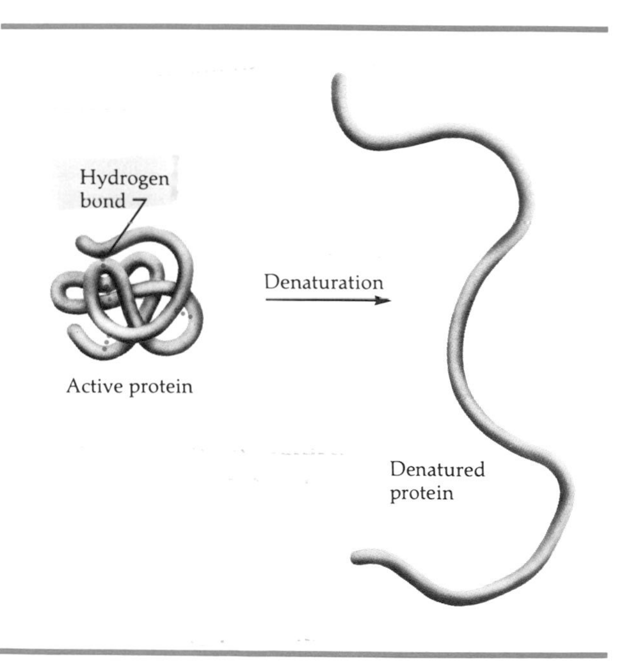

Figure 5–5 Denaturation of a protein comes about by the breakage of the weak bonds (such as hydrogen bonds) that hold the active protein in its three-dimensional shape. The denatured protein is no longer functional.

pH

Most enzymes have a characteristic pH at which their activity is maximal; this is referred to as the **pH optimum.** Above or below this pH value, enzyme activity, and therefore reaction rates, decline (Figure 5–4b). Changes in ion concentrations in the medium (pH) would, understandably, affect the many ionizable groups essential in stabilizing the enzymatic three-dimensional structure. As we have said, extreme changes in pH can cause denaturation.

Substrate concentration

There is a certain maximum rate at which a given amount of enzyme can catalyze a specific reaction. Only when the concentration of substrate(s) is extremely high can this maximum rate be attained. Under conditions of high substrate concentration, the enzyme is said to be **saturated,** that is, its active site is occupied by substrate or product molecules at all times. When this happens, further increase in substrate concentration will have no effect on the reaction rate because additional active sites are not available for reaction (Figure 5–4c). If a substrate concentration exceeds saturation levels in a cell, a further increase in the rate of reaction can be achieved only if the cell produces additional enzyme molecules. But under normal cell conditions, most enzymes are not saturated with substrate(s). At any given time, many of the enzyme molecules are inactive for lack of substrate; thus, the rate of reaction is more likely to be determined (restricted) by the substrate concentration.

Inhibitors

As you will see in later chapters, an effective way to control the growth of bacteria is to control their enzymes. Certain poisons such as cyanide, arsenic, mercury, and nerve gas combine with enzymes and prevent their functioning. This results in the inhibition or death of a cell. One group of enzyme inhibitors used in medicine are the antimicrobial drugs.

Based upon their mechanism of action, enzyme inhibitors are classified in two ways: competitive inhibitors and noncompetitive inhibitors. **Competitive inhibitors** compete with the normal substrate for the active site of the enzyme. The competitive inhibitor is able to do this because its shape and chemical structure are very similar to those of the normal substrate (Figure 5–6a and b). One good example of a competitive inhibitor is sulfanilamide (a sulfa drug). Para-aminobenzoic acid (PABA) is the normal substrate of the enzyme inhibited by sulfanilamide.

Sulfanilamide PABA

PABA is an essential nutrient used by many bacteria in the synthesis of folic acid, a vitamin that functions as a coenzyme. If sulfanilamide is administered to bacteria, the enzyme that normally converts PABA to folic acid combines with the sulfanilamide instead. This prevents the synthesis of folic acid, and the bacteria cannot grow. Since human cells do not use PABA to make their own folic acid, sulfanilamide selectively kills the bacteria but does not harm human cells.

Noncompetitive inhibitors do not compete with the substrate for the enzyme molecule's active site. Instead, they act on other parts of the enzyme

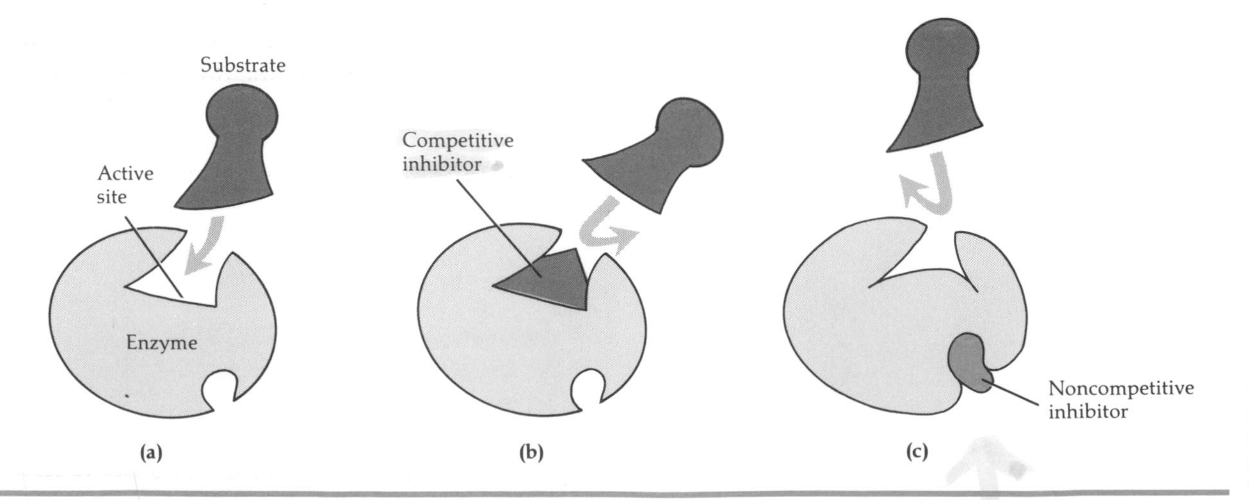

Figure 5–6 Enzyme inhibitors. **(a)** Uninhibited enzyme and its normal substrate. **(b)** Competitive inhibitor. **(c)** One type of noncompetitive inhibitor, causing allosteric inhibition.

and indirectly decrease the ability of the normal substrate to combine with the enzyme. In Figure 5–6c, we see an example of *allosteric inhibition*, in which the binding of the inhibitor to the enzyme at a site different from the active site changes the shape of the active site so that the substrate can no longer bind there (more on this in Chapter 8). Another type of noncompetitive inhibition can operate on enzymes that require metal ions for their activity. Certain chemicals have the ability to bind or tie up the metal ion activators and thus prevent an enzymatic reaction. Cyanide can bind the iron in iron-containing enzymes, and fluoride can bind the calcium or magnesium in enzymes that contain those ions.

Before we look at the actual chemical reactions by which organisms produce energy, let us examine a few basic concepts of energy production by cells.

ENERGY PRODUCTION CONCEPTS

Nutrient molecules, like all molecules, have energy stored in the bonds between their atoms. But this energy is often spread throughout the molecule, making it difficult for the cell to use. In a process that makes the energy readily available, cells use certain pathways to concentrate energy into the high-energy bonds of ATP. We will now consider how this is done, focusing on three important areas of energy production: oxidation–reduction, ATP generation, and nutritional patterns of organisms.

The Concept of Oxidation–Reduction

Oxidation is the addition of oxygen to a molecule, or, more generally, the removal of electrons (e^-) from a molecule. In many cell oxidations, two electrons and two hydrogen ions (H^+) are removed at the same time, the equivalent of removing two hydrogen atoms. For this reason, most biological oxidations are called **dehydrogenation** reactions; that is, they involve the loss of hydrogen atoms from a substance. Consider the oxidation of lactic acid to form pyruvic acid, in which the lactic acid loses two hydrogen atoms.

$$\underset{\text{Lactic acid}}{\begin{array}{c}COOH\\|\\CHOH\\|\\CH_3\end{array}} \xrightarrow{-2H\ (\text{oxidation})} \underset{\text{Pyruvic acid}}{\begin{array}{c}COOH\\|\\C{=}O\\|\\CH_3\end{array}}$$

When a compound picks up electrons or hydrogen atoms, it is said to be reduced. Thus, **reduction** is a gain of electrons, the opposite of oxidation. Consider the reduction of pyruvic acid to form lactic acid.

$$\underset{\text{Pyruvic acid}}{\begin{array}{c}COOH\\|\\C{=}O\\|\\CH_3\end{array}} \xrightarrow{+2H\ (\text{reduction})} \underset{\text{Lactic acid}}{\begin{array}{c}COOH\\|\\CHOH\\|\\CH_3\end{array}}$$

Within a cell, oxidations and reductions are always coupled. In other words, each time one substance is oxidized, another is almost simultaneously reduced. The pairing of these reactions is referred to as **oxidation–reduction.**

When a substance is oxidized in this way, the hydrogen atoms removed do not remain free in the cell, but are picked up immediately by another compound. Coenzymes transfer the hydrogen atoms from one substance to another. Two common coenzymes used by living cells to carry hydrogen atoms are derivatives of the vitamin niacin. They are nicotinamide adenine dinucleotide (NAD) and nicotinamide adenine dinucleotide phosphate (NADP). The oxidation and reduction states of NAD and NADP may be symbolized as follows.

$$NAD \underset{-\,2H}{\overset{+\,2H}{\rightleftharpoons}} NADH_2$$

$$\underset{\text{Oxidized}}{NADP} \underset{-\,2H}{\overset{+\,2H}{\rightleftharpoons}} \underset{\text{Reduced}}{NADPH_2}$$

Thus, when lactic acid is *oxidized* to form pyruvic acid, the two hydrogen atoms removed in the reaction are used to *reduce* NAD. This coupled, oxidation–reduction reaction may be written as follows.

NAD + 2H → $NADH_2$

COOH
|
CHOH
|
CH_3

Lactic acid

⟶

COOH
|
C=O
|
CH_3

Pyruvic acid

An important point to remember about oxidation–reduction reactions is that oxidation is usually an energy-producing reaction. Cells take foodstuffs and degrade them from highly reduced compounds (many hydrogen atoms) to highly oxidized compounds (many oxygen atoms or multiple bonds). For example, when a cell oxidizes a molecule of glucose ($C_6H_{12}O_6$), the energy in the glucose molecule is removed in a stepwise manner and ultimately trapped by ATP for storage. Compounds such as glucose that have many hydrogen atoms are highly reduced compounds, containing more potential energy than oxidized compounds.

Generation of ATP

The energy released during oxidation reactions is trapped within the cell to form ATP. As described in Chapter 2, this involves the addition of a phosphate group (symbolized Ⓟ) to ADP to form ATP.

Adenosine—Ⓟ ~ Ⓟ (ADP) + Energy + Ⓟ ⟶

Adenosine—Ⓟ ~ Ⓟ ~ Ⓟ (ATP)

The symbol ~ designates a high-energy bond, one that is readily broken to release usable energy. The high-energy bond that attaches the third phosphate contains the energy stored in the reaction shown. The addition of a phosphate to a chemical compound is called **phosphorylation.** Organisms use three mechanisms of phosphorylation to generate ATP.

In **oxidative phosphorylation,** electrons removed from organic compounds (usually by NAD) are passed to a series of electron acceptors and finally to molecules of oxygen (O_2) or other inorganic molecules. The series of electron acceptors is called the **electron transport chain.** During the transfer of electrons from one electron acceptor to the next, energy is released, a result of the oxidations that occur. This energy is used to generate ATP from ADP.

In **substrate-level phosphorylation,** no oxygen or other inorganic final electron acceptor is required. ATP is generated by the direct transfer of a high-energy phosphate group from an intermediate metabolic compound to ADP. The following example shows only the carbon skeleton and the phosphate group of the metabolic compound.

C—C—C~Ⓟ + ADP ⟶ C—C—C + ATP

The third mechanism of phosphorylation is **photophosphorylation,** and it occurs only in photosynthetic cells, which contain light-trapping pigments like the chlorophylls. In this case, light energy liberates an electron, which is passed along a series of electron acceptors. Each transfer releases energy for the conversion of ADP to ATP.

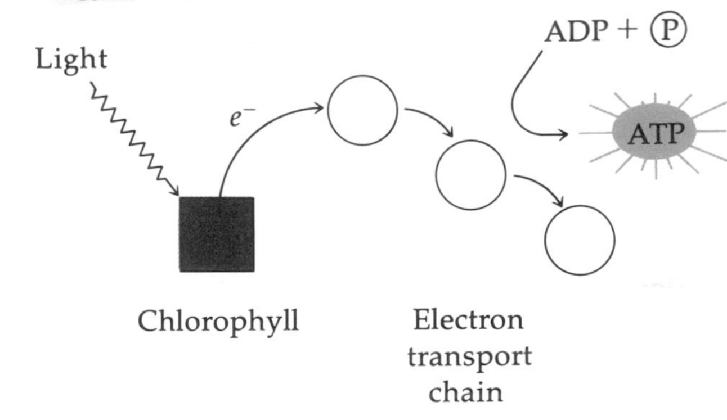

The significance of these three forms of phosphorylation will become clearer to you as we discuss in more detail their role in energy production later in this chapter.

Nutritional Patterns Among Organisms

In order to acquire the energy for various metabolic activities, microorganisms need raw materials—mainly a carbon source for making new organic compounds. Nutritional patterns among organisms may be distinguished on the basis of two criteria: energy source and principal source of carbon. If we first consider the energy source, we can generally classify organisms as phototrophs or chemotrophs. **Phototrophs** use light as their primary energy source, whereas **chemotrophs** depend on

Table 5–2 Nutritional Classification of Organisms

Nutritional Type	Energy Source	Carbon Source	Examples
Photoautotroph	Light	Carbon dioxide (CO_2)	Photosynthetic bacteria (green sulfur and purple sulfur bacteria), cyanobacteria, algae, plants
Photoheterotroph	Light	Organic compounds	Purple nonsulfur and green nonsulfur bacteria
Chemoautotroph	Inorganic compounds	Carbon dioxide (CO_2)	Hydrogen, sulfur, iron, and nitrifying bacteria
Chemoheterotroph	Organic compounds	Organic compounds	Most bacteria, fungi, protozoans, and all animals

oxidation–reduction reactions as their source of energy. With respect to a principal carbon source, **autotrophs** (self-feeders) use carbon dioxide and **heterotrophs** (feeders on others) require an organic carbon source.

If we combine the patterns of energy and carbon sources, we arrive at the following nutritional classification for organisms: photoautotrophs, photoheterotrophs, chemoautotrophs, and chemoheterotrophs. The nutritional classification of microbes is summarized in Table 5–2. Almost all the medically important microorganisms discussed in this book are chemoheterotrophs.

Photoautotrophs

Photoautotrophs use light as a source of energy, and carbon dioxide as their chief source of carbon. They include photosynthetic bacteria (green sulfur and purple sulfur bacteria), cyanobacteria, algal protists, and plants. The process by which photoautotrophs transform carbon dioxide and water into carbohydrates and oxygen gas is called **photosynthesis.** Essentially, photosynthesis is the conversion of light energy into chemical energy, using chlorophyll molecules to trap the light. The overall reaction for photosynthesis may be represented as follows:

$$\underbrace{\underset{\text{Carbon dioxide}}{6CO_2} + \underset{\text{Water}}{6H_2O}}_{\text{Raw materials}} \underbrace{\xrightarrow[\text{Chlorophyll Enzymes}]{\text{Light}}}_{\text{Necessary conditions}} \underbrace{\underset{\text{Sugar}}{C_6H_{12}O_6} + \underset{\text{Oxygen}}{6O_2}}_{\text{Products}}$$

The carbon (C) in the synthesized sugar comes from carbon dioxide (CO_2) and the hydrogen (H) in the synthesized sugar comes from water (H_2O). The source of oxygen (O) in the sugar is carbon dioxide (CO_2), and the oxygen (O) in water (H_2O) is eventually given off as oxygen gas (O_2). In the process of photosynthesis, the energy for the synthesis of sugar is derived from ATP. The ATP is generated by photophosphorylation.

The reactions of photosynthesis just described occur in the cyanobacteria (formerly called blue-green algae), some protists (algae), and plants. These organisms use the hydrogen atoms of water to reduce carbon dioxide, giving off oxygen gas in the process. The type of light-trapping pigment they use is called *chlorophyll a*.

In addition to the cyanobacteria, there are several other families of photosynthetic procaryotes, classified according to the way they reduce carbon dioxide. Two of the families are photoautotrophs: the green sulfur and purple sulfur bacteria. The **green sulfur** bacteria use sulfur, sulfur compounds, or hydrogen gas to reduce carbon dioxide and form organic compounds. Using the energy from light with chlorophyll and the appropriate enzymes, these bacteria oxidize sulfur to sulfuric acid, hydrogen sulfide to sulfur, or hydrogen gas to water. The **purple sulfur bacteria** also use sulfur compounds or hydrogen gas to reduce carbon dioxide. They are distinguished from the green sulfur bacteria on biochemical and morphological grounds.

Unlike cyanobacteria and photosynthetic eucaryotes, photosynthetic bacteria do not use water to reduce carbon dioxide. Consequently, they do not produce oxygen gas as a product of photosynthesis. In addition, the chlorophyll used by the photosynthetic bacteria is a group of pigments called *bacteriochlorophylls,* which absorbs longer wavelengths of light than chlorophyll *a.* Such bacteria must carry on photosynthesis in the complete absence of oxygen (anaerobic environment).

Photoheterotrophs

Photoheterotrophs use light as a source of energy but cannot convert carbon dioxide to sugar; rather, they use organic compounds as sources of carbon. The organic compounds used include alcohols, fatty acids, other organic acids, and carbohydrates. Among the photoheterotrophs are green nonsulfur and purple nonsulfur bacteria.

Chemoautotrophs

Chemoautotrophs use inorganic compounds as a source of energy and carbon dioxide as their principal source of carbon. Inorganic sources of energy include hydrogen sulfide, H_2S *(Beggiatoa)*; elemental sulfur, S *(Thiobacillus)*; ammonia, NH_3 *(Nitrosomonas)*; nitrites, NO_2^- *(Nitrobacter)*; hydrogen gas, H_2 *(Pseudomonas)*; and iron, Fe^{++} *(Thiobacillus ferrooxidans).* The energy derived from the oxidation of inorganic compounds is eventually stored in ATP. This involves oxidative phosphorylation.

Chemoheterotrophs

When we discuss photoautotrophs, photoheterotrophs, and chemoautotrophs, it is easy to categorize energy source and carbon source because they occur as separate entities. However, in the case of **chemoheterotrophs,** the distinction is not as clear, since both energy source and carbon source are usually the same organic compound—glucose, for example.

Based on the source of organic molecules, heterotrophs may be classified as **saprophytes,** which live on dead organic matter, or **parasites,** which derive nutrients from a living host. The vast majority of bacteria, and all fungi, protozoans, and animals, are chemoheterotrophs.

We will now consider the specific biochemical pathways organisms use to produce energy.

BIOCHEMICAL PATHWAYS OF ENERGY PRODUCTION

Organisms that produce and store energy from organic molecules do so in a series of controlled reactions rather than in a single burst. If the energy were released all at once, it would be released as a large amount of heat, which could not be readily used to drive chemical reactions and would, in fact, damage the cell. To extract energy from organic compounds and store it in cells, organisms pass electrons from one compound to another through a series of oxidation–reduction reactions.

A sequence of enzymatically catalyzed chemical reactions occurring in a cell is called a **biochemical pathway.** Pathways are usually written as follows.

$$A \xrightarrow{H_2O} B \xrightarrow{ATP \rightarrow ADP + (P)} C \xrightarrow{\rightarrow CO_2} D \rightleftharpoons E$$

Note that pathways shown in this way are not usually written as balanced chemical equations. This hypothetical pathway converts starting material *A* into the end product *E* in a series of four steps. (Of course, if this were an actual pathway, chemical structures or names would replace letters *A* through *E.*)

The curved arrow originating at H_2O indicates that water is a reactant in the reaction converting *A* to *B.* Similarly, the curved arrow pointing to CO_2 indicates that carbon dioxide is a product of the reaction converting *C* to *D.* To drive the reaction converting *B* to *C,* a molecule of ATP must be broken down to ADP and a phosphate in a coupled reaction; that is, the two reactions occur simultaneously, each depending on the other. The reaction converting *D* to *E* is readily reversible, as indicated by the double arrow. Keep in mind that virtually every reaction in a biochemical pathway is catalyzed by a specific enzyme; sometimes the name of the enzyme is printed near the arrow.

In most microorganisms, oxidation of carbohydrates provides most of the cell's energy. **Carbohydrate catabolism** (Figure 5–7), the break-

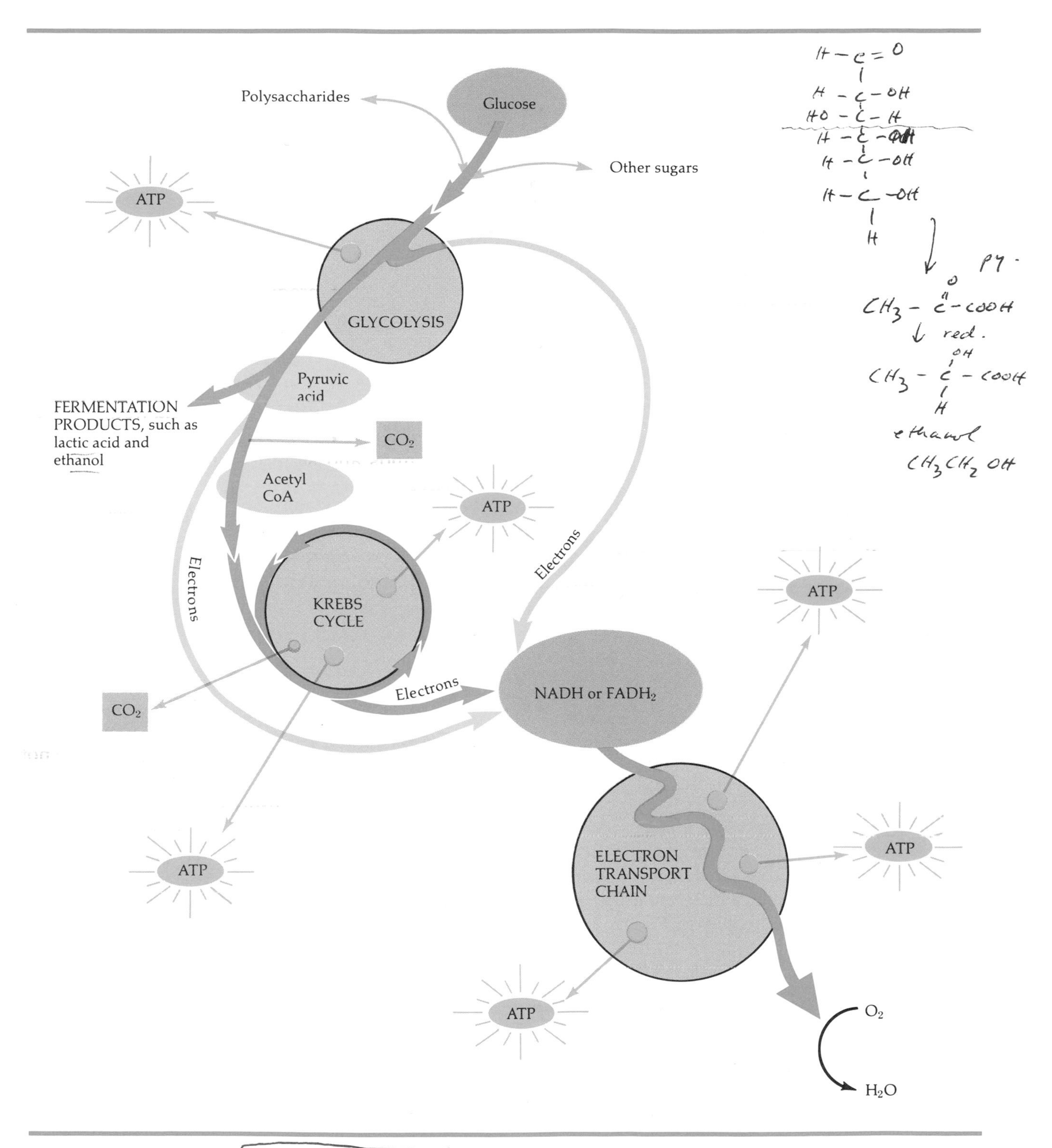

Figure 5–7 Overview of carbohydrate catabolism. Glucose is broken down completely to carbon dioxide and water and ATP is generated. This process has three major phases: glycolysis, the Krebs cycle, and the electron transport chain. The key event in the process is that electrons are removed from intermediates of glycolysis and the Krebs cycle and are carried by NAD or FAD to the electron transport chain. The electron transport chain produces most of the ATP by oxidative phosphorylation. Fermentation uses only the glycolysis portion of this sequence, generating a much smaller amount of ATP by substrate-level phosphorylation.

down of carbohydrate molecules to produce energy, is therefore of great importance in cell metabolism.

The six-carbon sugar glucose plays a central role in carbohydrate catabolism. Glucose is usually broken down to pyruvic acid by a process called **glycolysis.** Glycolysis is the first stage of both fermentation and respiration. Fermentation yields products such as lactic acid or alcohol and produces small amounts of ATP. Respiration yields much larger amounts of ATP by breaking glucose down completely. In respiration, pyruvic acid is split and a fragment is attached to a coenzyme molecule called coenzyme A or CoA, forming acetyl CoA. Acetyl CoA then enters the Krebs cycle, a series of reactions that releases electrons. Electrons released during glycolysis, the formation of acetyl CoA, and the Krebs cycle are passed to NAD or FAD molecules. These molecules carry the electrons to the electron transport chain, where ATP is formed. This summary of carbohydrate catabolism is diagrammed in Figure 5–7.

Let us now take a closer look at glycolysis, fermentation, and respiration.

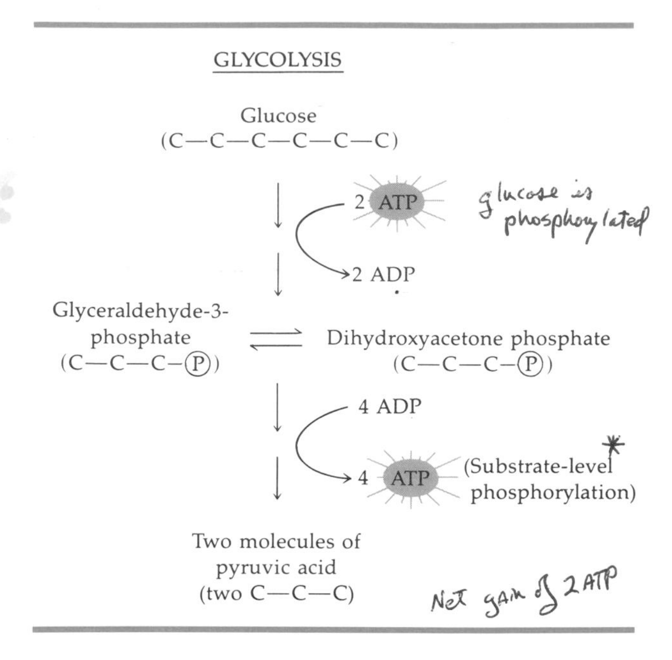

Figure 5–8 Outline of the principal reactions of glycolysis.

Glycolysis

As we said, glycolysis, usually the first step in fermentation and respiration, is the oxidation of glucose to pyruvic acid. Glycolysis is also called the **Embden–Meyerhof pathway.** Most microorganisms, as well as all higher organisms, use this pathway, which is a series of ten chemical reactions, each reaction catalyzed by a different enzyme. The principal reactions are shown in Figure 5–8 and summarized below:

1. First, using two molecules of ATP, a six-carbon glucose molecule is phosphorylated, rearranged, and split into two three-carbon compounds, glyceraldehyde-3-phosphate and dihydroxyacetone phosphate. These two compounds are interconvertible.
2. Next, the two three-carbon molecules are oxidized to two molecules of pyruvic acid. In these reactions, four molecules of ATP are formed by substrate-level phosphorylation.
3. Since two molecules of ATP are needed to get glycolysis started, and four molecules of ATP are generated when the process is completed, *there is a net gain of two molecules of ATP for each molecule of glucose that is oxidized.*

Pentose phosphate pathway

Many bacteria have, in addition to glycolysis, an alternate pathway for oxidizing glucose. This is called the **pentose phosphate pathway** (or **hexose monophosphate shunt),** and it operates simultaneously with glycolysis. The pentose phosphate pathway is a cyclic pathway that provides a means for the breakdown of five-carbon sugars (pentoses), as well as glucose. A key feature of this pathway is that it produces important intermediate pentoses that act as precursors in the synthesis of (1) nucleic acids, (2) glucose from carbon dioxide by photosynthesizing organisms, and (3) certain amino acids. The pathway is an important producer of $NADPH_2$ from NADP for various biosynthetic reactions in the cell. Unlike glycolysis, the pentose phosphate pathway produces only one ATP. Bacteria that utilize the pentose phosphate pathway include *Bacillus subtilis, Escherichia coli, Leuconostoc mesenteroides,* and *Streptococcus faecalis.*

Entner–Doudoroff pathway

The **Entner–Doudoroff pathway (EDP)** is still another pathway for oxidizing glucose to pyruvic acid. From each molecule of glucose oxidized, two molecules of $NADPH_2$ are produced for cellular biosynthetic reactions. Bacteria that have the enzymes for the EDP can metabolize without glycolysis and the pentose phosphate pathway. The only bacteria known to possess the EDP are some species of the genus *Rhizobium* and several of the genus *Pseudomonas*. Tests for the ability to oxidize glucose by this pathway are sometimes used to identify *Pseudomonas* in the clinical laboratory.

Fermentation of Carbohydrates

As we mentioned, after glucose is broken down to pyruvic acid, the pyruvic acid can undergo either further fermentation or respiration. Actually, fermentation can be defined several ways (see Box), but we will define it as a process that

1. Releases energy from sugars or other organic molecules such as amino acids, organic acids, purines, and pyrimidines.
2. Does not require oxygen (but sometimes can occur in its presence).
3. Does not require an electron transport chain.
4. Uses an organic molecule as the final electron acceptor.

In fermentation, pyruvic acid takes electrons (hydrogen) from $NADH_2$ and is turned into various end products, converting ADP to ATP by substrate-level phosphorylation. Various microorganisms are able to ferment various substrates; the end products depend on the particular microorganism, the substrate, and the enzymes that are present and active. Chemical analyses of these end products are useful in identifying microorganisms. Fermentation products produced from pyruvic acid by various microorganisms are shown in Figure 5–9.

Fermentation end products such as the ethanol and carbon dioxide produced by yeasts, are waste products for yeast cells but useful to humans. Ethanol is used to make alcoholic beverages by the brewing and distilling industries, and bakers need carbon dioxide to help bread dough rise. Table 5–3 lists some of the various microbial fermentations used by industry to convert inexpensive raw materials into useful end products.

What Is Fermentation?

To many people, fermentation simply means the production of alcohol. Grain and fruits ferment into beer and wine. If a food sours, you might say it was "off" or fermented. Here are some definitions of fermentation. They range from informal, general usage to more scientific definitions that you will need to know for this course.

Fermentation is:

1. Any process that produces alcoholic beverages (general use).
2. Any spoilage of food by microorganisms (general use).
3. Any large-scale microbial process, with or without air (common definition used in industry).
4. Any energy-releasing metabolic process that takes place only under anaerobic conditions (becoming more scientific).
5. The group of all metabolic processes that release energy from a sugar or other organic molecule, do not require oxygen or an electron transport system, and use an organic molecule as the final electron acceptor (the definition we will use in this text).

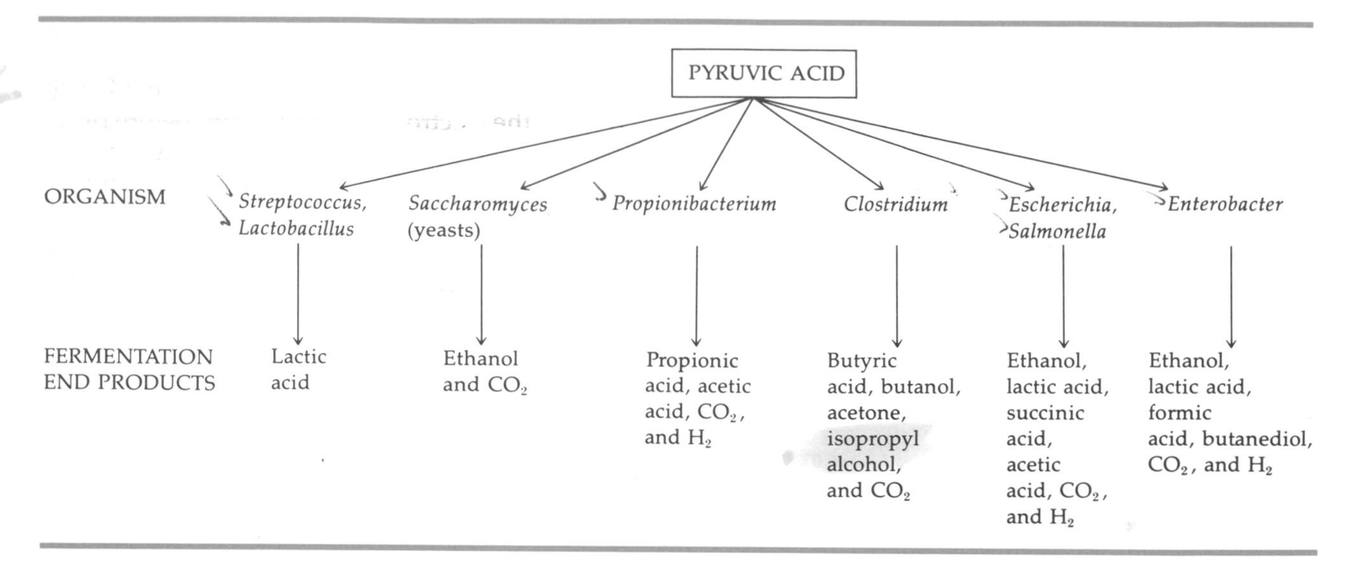

Figure 5–9 Fermentation end products of various microorganisms.

Table 5–3 Some Industrial Fermentations

Fermentation Product	Commercial Use	Starting Material	Microorganism
Ethanol	Beer	Malt extract	*Saccharomyces cerevisiae* (fungus)
	Wine	Grape or other fruit juices	*Saccharomyces ellipsoideus* (fungus)
Acetic acid	Vinegar	Alcohol	*Acetobacter* (bacterium)
Lactic acid	Cheese, yogurt	Milk	*Lactobacillus, Streptococcus* (bacteria)
	Rye bread	Grain, sugar	*Lactobacillus bulgaricus* (bacterium)
	Summer sausage	Meat	*Pediococcus* (bacterium)
Propionic acid and carbon dioxide	Swiss cheese	Milk	*Propionibacterium freudenreichii* (bacterium)
Acetone and butanol	Pharmaceutical, industrial uses	Molasses	*Clostridium acetobutylicum* (bacterium)
Glycerol	Pharmaceutical, industrial uses	Molasses	*Saccharomyces cerevisiae* (fungus)
Citric acid	Flavoring	Molasses	*Aspergillus* (fungus)
Methanol	Fuel	Agricultural wastes	*Clostridium* (bacterium)

Respiration

Respiration is an ATP-generating process in which chemical compounds are oxidized and the final electron acceptor is almost always an inorganic molecule. A major feature of respiratory processes is the operation of an electron transport chain, a group of chemicals that readily accept electrons from one compound and pass them to another. As electrons move through the chain, ATP is generated by oxidative phosphorylation. Any naturally occurring organic molecule can be degraded by some microbe during respiration. And, as you will see, the organic molecule is usually oxidized completely to carbon dioxide. Consequently, the yield of ATP per molecule of substrate is much greater in respiration than in fermentation.

In most respirations, the final hydrogen (electron) acceptor is free oxygen (O_2). Accordingly, this type of respiration is called **aerobic respiration.** A few bacteria carry on **anaerobic respiration,** in which the final hydrogen (electron) acceptor is an inorganic molecule other than free oxygen. Examples of such inorganic acceptors are nitrates (NO_3^-), sulfates (SO_4^{2-}), and carbonates (CO_3^{2-}). Anaerobic respiration can result in an energy yield almost as high as that of aerobic respiration.

We will first consider aerobic respiration.

Aerobic respiration

Only a small amount of the energy present in a glucose molecule is released when oxidized through fermentation, since much of the energy is retained in the chemical bonds of the organic end product (such as ethanol or lactic acid). However, if the final hydrogen acceptor is oxygen (O_2), the glucose molecule can be completely oxidized to carbon dioxide (CO_2) and water (H_2O) with a yield of 38 ATP molecules for each glucose molecule oxidized. In contrast, glycolysis alone yields only two ATP molecules for each glucose molecule, and the pentose phosphate pathway and EDP each yield only one ATP. From this you can see why brewing and wine making must be anaerobic. The microorganisms do not grow as efficiently, but they produce an end product desired by humans—ethanol. On the other hand, in the treatment of sewage we try to maximize aeration because with oxygen, microbes can break down waste matter all the way to carbon dioxide and water.

For pyruvic acid to be completely oxidized to carbon dioxide and water, two sets of reactions beyond glycolysis must occur. These are the Krebs cycle and the electron transport chain. Before pyruvic acid can enter the Krebs cycle, it must be converted into a two-carbon compound by the loss of carbon dioxide. The two-carbon compound, called an *acetyl group*, attaches to the coenzyme known as *coenzyme A (CoA)*, and the whole complex is known as *acetyl coenzyme A (acetyl CoA)* (Figure 5–10).

Krebs Cycle The **Krebs cycle,** also called the **tricarboxylic acid cycle (TCA)** or **citric acid cycle,** is a series of chemical reactions in which the large amount of potential chemical energy still stored in pyruvic acid is released step by step. In this cycle, a series of oxidations and reductions occurs so that the chemical energy stored in intermediate compounds derived from pyruvic acid is transferred, in the form of electrons, to a number of coenzymes. The pyruvate derivatives are oxidized; the coenzymes are reduced.

As acetyl CoA enters the Krebs cycle, CoA detaches from the acetyl group and then can pick up more acetyl groups for the next Krebs cycle. Meanwhile, the acetyl group combines with a substance called oxaloacetic acid to form citric acid. It is at this point that the Krebs cycle begins. The major chemical reactions of this cycle are outlined in Figure 5–11.

As the various acids move through the cycle, they undergo a number of changes, all controlled by specific enzymes. One of these changes is called *decarboxylation*, a process in which a chemical compound loses a molecule of carbon dioxide (CO_2). Isocitric acid, a six-carbon compound, is decarboxylated to form a five-carbon compound. By losing

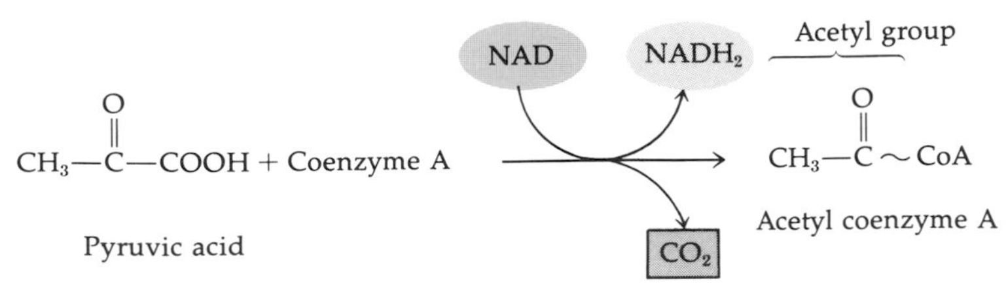

$$CH_3-\overset{\overset{O}{\|}}{C}-COOH + \text{Coenzyme A} \xrightarrow[CO_2]{NAD \rightarrow NADH_2} CH_3-\overset{\overset{O}{\|}}{C}\sim CoA$$

Pyruvic acid — Acetyl group — Acetyl coenzyme A

Figure 5–10 Formation of acetyl CoA from pyruvic acid.

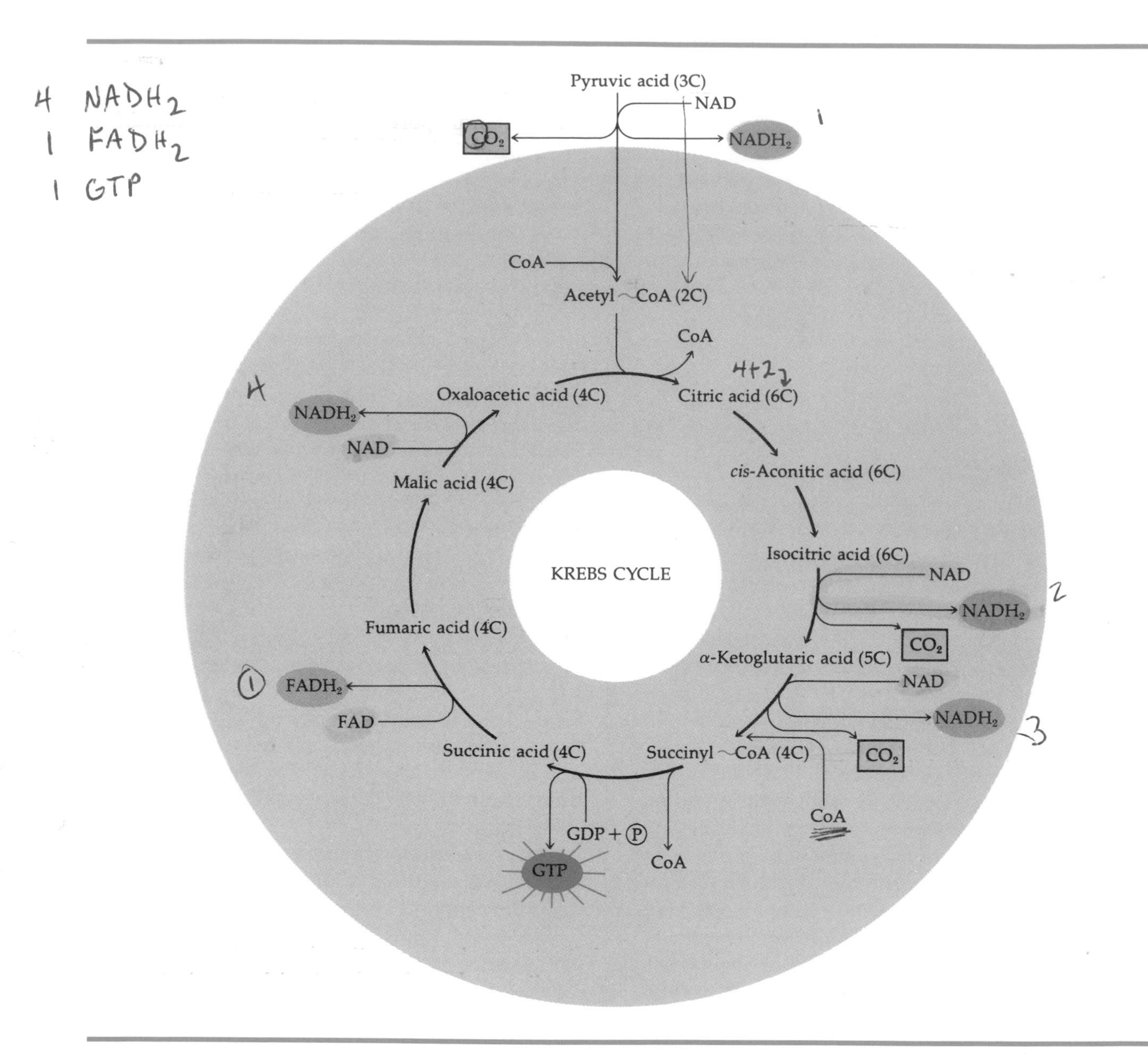

Figure 5–11 The Krebs cycle. Note the points where CO_2 is produced, $NADH_2$ and $FADH_2$ are formed, and GTP is generated. GTP is used directly in certain cell processes and can also be used to convert ADP to ATP.

a molecule of CO_2 and picking up a molecule of CoA, the five-carbon compound becomes a four-carbon compound. Another place where CO_2 is produced as a result of decarboxylation is between pyruvic acid (three-carbon) and acetyl CoA (two-carbon). Thus, one three-carbon molecule (pyruvic acid) is broken down to three one-carbon molecules (CO_2) with each full cycle.

As the acids go through the cycle, they are also involved in a series of oxidation–reduction reactions. In the conversion of the six-carbon isocitric acid to a five-carbon compound, two hydrogen atoms are lost. In other words, the six-carbon compound is oxidized. The hydrogen atoms released in the Krebs cycle are picked up by the coenzymes NAD and FAD. When NAD picks up hydrogen atoms, it is reduced and may be represented as $NADH_2$; similarly, FAD is reduced to $FADH_2$. Notice the points in the synthesis of acetyl CoA and in the Krebs cycle where coenzymes pick up hydrogen atoms (see Figure 5–11).

If we look at the Krebs cycle as a whole, we see that each time a molecule of pyruvic acid enters the cycle, three molecules of carbon dioxide are liberated by decarboxylation; four molecules of $NADH_2$ and one molecule of $FADH_2$ are produced by oxidation–reduction reactions; and one molecule of GTP (the equivalent of ATP) is generated. In addition, many of the intermediates in the Krebs cycle play a role in other pathways, especially in amino acid biosynthesis.

The carbon dioxide produced in the Krebs cycle is liberated into the atmosphere as a gaseous by-product of aerobic respiration. (In humans, the carbon dioxide produced from the Krebs cycle in each cell of the body is discharged by the lungs during exhalation.) The reduced coenzymes ($NADH_2$ and $FADH_2$) represent the most important outcome of the Krebs cycle, since they now contain the stored energy originally in glucose and then in pyruvic acid. The next phase of aerobic respiration involves a series of reductions that transfer the energy in the coenzymes to ATP. These reactions are collectively called the electron transport chain.

Electron Transport Chain The electron transport chain consists of a sequence of carrier molecules capable of oxidation and reduction. As electrons are passed through the chain, there is a stepwise release of energy, and ATP is generated. In aerobic respiration, the terminal electron acceptor is oxygen (O_2); in anaerobic respiration, the final acceptor is an inorganic molecule other than oxygen (or, rarely, an organic compound). The final oxidation in both respirations is irreversible. In eucaryotic cells, the electron transport chain is contained in membranes in mitochondria; in procaryotic cells, it is found in the plasma membrane.

The respiration of organic compounds involves three classes of carrier molecules in the electron transport chain: **flavoproteins,** proteins that contain a coenzyme derived from riboflavin (vitamin B_2), which are capable of alternating oxidations and reductions; **cytochromes,** proteins that contain an iron-containing group (heme) capable of alternating between a reduced form (Fe^{2+}) and an oxidized form (Fe^{3+}); and **quinones,** nonprotein carriers of low molecular weight.

Since considerably more is known about the electron transport chain in the mitochondria of eucaryotic cells, this is the one we will describe. The electron transport chains of bacterial systems are quite diverse, and the particular carrier molecules may differ markedly from each other and from those of eucaryotic mitochondrial systems. However, keep in mind that they achieve the same basic goal.

The first step in the electron transport chain shown in Figure 5–12 involves the transfer of hydrogen atoms from $NADH_2$ to FAD. In the process, $NADH_2$ is oxidized to NAD, and FAD is reduced to $FADH_2$. The importance of the hydrogen transfer from $NADH_2$ to FAD is that it releases energy. This energy is used to produce ATP from ADP and phosphoric acid (phosphate). The hydrogen atoms are then used to reduce a quinone. After that step, however, the hydrogen atoms do not stay intact. They ionize into hydrogen ions (H^+) and electrons (e^-) according to the following reaction.

$$\underset{\text{Hydrogen atom}}{H} \longrightarrow \underset{\text{Hydrogen ion}}{H^+} + \underset{\text{Electron}}{e^-}$$

In the next step of the electron transport chain, the electrons (e^-) from the hydrogen atoms are passed to cytochrome *b*. (At the same time, the H^+ ions are released into solution.) The electrons are successively passed from one cytochrome to another—from cytochrome *b* to cytochrome *c* to cytochrome *a*, and finally to cytochrome oxidase. Each cytochrome in the electron transport system is alternately reduced as it picks up electrons and oxidized as it gives up electrons. (The electron transport chains of some bacteria, such as *Escherichia coli,* lack cytochrome *c*. The presence of cytochrome *c* can be determined by the diagnostically useful **oxidase test.**)

In the process of electron transfer between cytochromes, more energy is liberated and then stored in ATP. At the end of the electron transport system, the electrons are passed to oxygen, which becomes negatively charged. The oxygen then combines with H^+ ions to form water. As you may recall, the process by which ATP is generated by the transfer of electrons from reduced coenzymes to oxygen is known as oxidative phosphorylation.

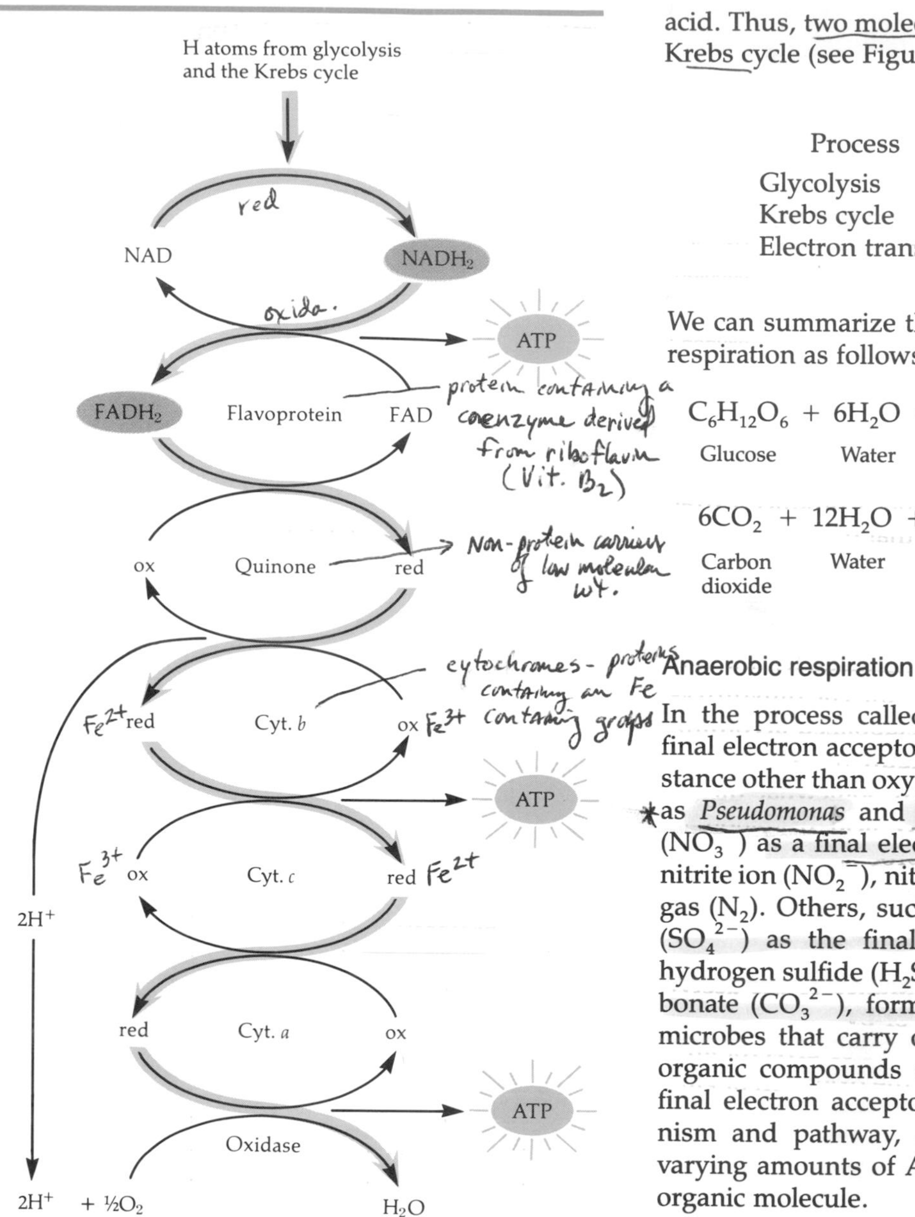

Figure 5–12 An electron transport chain. The heavy, colored arrows show the path of the electrons as they pass from one carrier to the next. ATP is produced at three points along the chain.

Reviewing the various phases of aerobic respiration, let us add up the number of ATP molecules produced during each phase. Keep in mind that each time a molecule of glucose is oxidized, it is split by glycolysis into two molecules of pyruvic acid. Thus, two molecules of pyruvic acid enter the Krebs cycle (see Figure 5–8).

Process	Net Gain of ATP (or GTP)
Glycolysis	2
Krebs cycle	2
Electron transport chain	34
	38

We can summarize the overall reaction for aerobic respiration as follows:

$$\underset{\text{Glucose}}{C_6H_{12}O_6} + \underset{\text{Water}}{6H_2O} + \underset{\text{Oxygen}}{6O_2} + 38\ ADP + 38\ Ⓟ \rightarrow$$

$$\underset{\text{Carbon dioxide}}{6CO_2} + \underset{\text{Water}}{12H_2O} + 38\ ATP$$

Anaerobic respiration

In the process called **anaerobic respiration,** the final electron acceptor is usually an inorganic substance other than oxygen (O_2). Some bacteria, such as *Pseudomonas* and *Bacillus,* can use nitrate ion (NO_3^-) as a final electron acceptor, reducing it to nitrite ion (NO_2^-), nitrous oxide (N_2O), or nitrogen gas (N_2). Others, such as *Desulfovibrio,* use sulfate (SO_4^{2-}) as the final electron acceptor, forming hydrogen sulfide (H_2S). Still other bacteria use carbonate (CO_3^{2-}), forming methane (CH_4). (A few microbes that carry on anaerobic respiration use organic compounds such as fumaric acid as the final electron acceptor.) Depending on the organism and pathway, anaerobic respiration yields varying amounts of ATP from the oxidation of an organic molecule.

A summary of fermentation, aerobic respiration, and anaerobic respiration is given in Table 5–4.

Lipid Catabolism

Thus far in our discussion of energy production, we have emphasized the oxidation of glucose, the principal energy-supplying carbohydrate. But microbes also oxidize lipids and proteins, and the oxidations of all these nutrients are related.

Table 5–4 Comparison of Fermentation, Aerobic Respiration, and Anaerobic Respiration

Energy-Producing Process	Growth Conditions	Final Hydrogen (Electron) Acceptor	Type of Phosphorylation Used to Build ATP	Molecules of ATP Produced
Fermentation	Aerobic or anaerobic	An organic molecule	Substrate-level	2
Aerobic respiration	Aerobic	Free oxygen (O_2)	Substrate-level and oxidative	38
Anaerobic respiration	Anaerobic	Usually an inorganic substance (such as NO_3^-, SO_4^{2-}, or CO_3^{2-}), but not free oxygen (O_2)	Oxidative	Variable

Fats, you may recall, are lipids consisting of fatty acids and glycerol. Microbes produce extracellular enzymes called *lipases* that break fats down into their fatty acid and glycerol components. Each component is then metabolized separately. Many microbes can convert glycerol into dihydroxyacetone phosphate, one of the intermediates formed during glycolysis (see Figure 5–8). The dihydroxyacetone phosphate is then metabolized according to the mechanism shown in Figure 5–13.

Fatty acids are catabolized somewhat differently. The mechanism of fatty acid oxidation is called **beta oxidation.** In this process, carbon fragments of a long chain of fatty acid are removed two at a time to form acetyl CoA. As the molecules of acetyl CoA are formed, they enter the Krebs cycle in a manner similar to the entrance of the acetyl CoA formed by the oxidation of pyruvic acid (see Figure 5–11). Thus, the oxidation of glycerol and fatty acids is directly linked to the oxidation of glucose by the Krebs cycle.

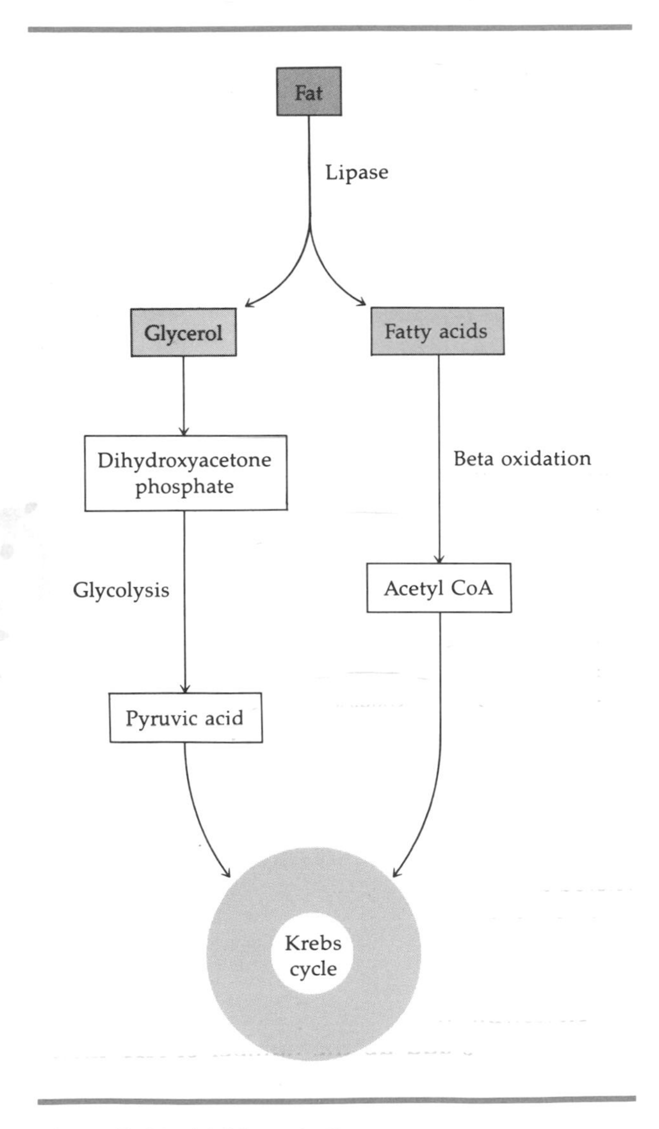

Figure 5–13 Lipid catabolism.

Protein Catabolism

Proteins are too large to pass through the plasma membranes of microbes. To get the proteins into the cells, microbes produce extracellular *proteases* and *peptidases,* which break down proteins into their component amino acids. However, before amino acids can be catabolized, they must be converted to various substances that can enter the Krebs cycle. In one such conversion, called **deamination,** the amino group of an amino acid is removed and converted to ammonium ion (NH_4^+), which can be excreted from the cell. The resulting organic acid can enter the Krebs cycle. Other conversions involve decarboxylation (removal of —COOH) and dehydrogenation. Although these conversions are complex, the important thing to remember is that amino acids can be prepared in various ways to enter the Krebs cycle.

BIOCHEMICAL PATHWAYS OF ENERGY UTILIZATION

Up to now we have been considering energy production. Through the oxidation of organic molecules, organisms produce energy by fermentation, aerobic respiration, and anaerobic respiration. Much of this energy is used up in heat production. In fact, although the complete metabolic oxidation of glucose to carbon dioxide and water is considered a very efficient process, about 60% of the energy of glucose is lost as heat. The energy that is trapped in molecules of ATP is used by cells in a variety of ways. Microbes use ATP to provide energy for the transport of substances across plasma membranes—a process called active transport, discussed in Chapter 4. Certain microbes also use some of their ATP for flagellar motion (also discussed in Chapter 4). The bulk of ATP generated, however, is used in the biosynthesis of new cell components. We will now consider the biosynthesis of a few representative classes of biological molecules: carbohydrates, lipids, amino acids, purines, and pyrimidines.

Biosynthesis of Polysaccharides

Nucleotides function as the carriers of sugars in the biosynthesis of more complex carbohydrates. Bacteria synthesize glycogen from *adenosine diphosphoglucose (ADPG)*, which is glucose activated by ATP. In the activation process, ADP is joined to a glucose molecule and a phosphate is released. Animals synthesize glycogen and many other carbohydrates from *uridine diphosphoglucose (UDPG)*. A compound related to UDPG called *UDP-N-acetyl glucosamine (UDPNAG)* is a key starting material in the biosynthesis of peptidoglycan, the substance that forms bacterial cell walls (Figure 5–14).

Biosynthesis of Lipids

Microbes synthesize lipids, such as fats, by uniting glycerol and fatty acids. The glycerol portion of the lipid is derived from dihydroxyacetone phosphate, an intermediate formed during glycolysis. Fatty acids, which are long-chained hydrocarbons (hydrogen linked to carbon), are built up from the

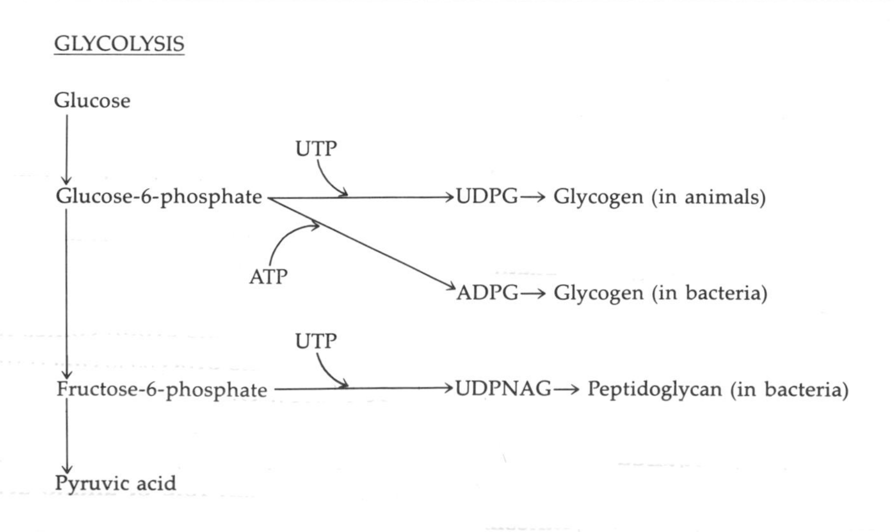

Figure 5–14 Biosynthesis of carbohydrates.

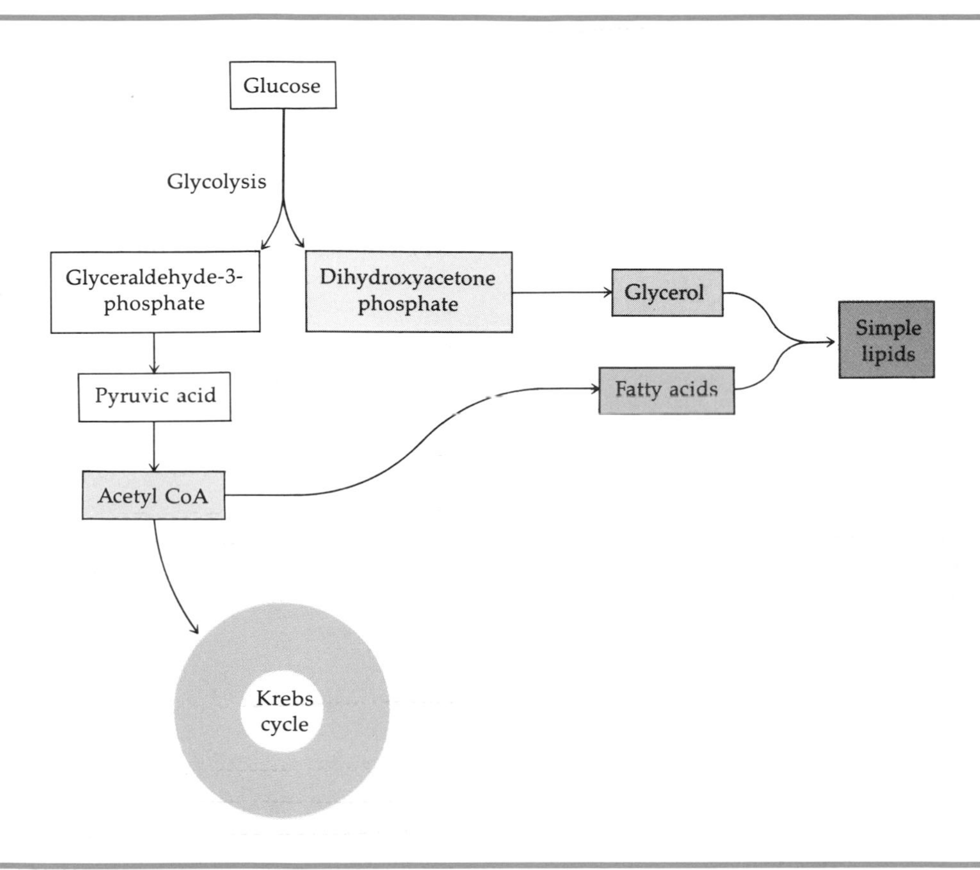

Figure 5–15 Biosynthesis of simple lipids.

successive addition of two carbon fragments of acetyl CoA (Figure 5–15). The enzymatic linkage of glycerol to fatty acids results in the biosynthesis of a variety of simple fats or lipids. As mentioned previously, the most important role of lipids is as structural components of plasma membranes. In addition, lipids help form the cell wall of gram-negative bacteria, and, like carbohydrates, they can also serve as storage forms of energy. Recall that the breakdown products after biological oxidation feed into the Krebs cycle.

Biosynthesis of Amino Acids

Amino acids are principally required for protein biosynthesis. Some microbes, such as *Escherichia coli,* contain the necessary enzymes to synthesize all the amino acids they need from starting materials such as glucose and inorganic salts. Other microbes require some preformed amino acids from the environment in order to metabolize. In organisms with the necessary enzymes, all amino acids can be synthesized directly or indirectly from intermediates of carbohydrate metabolism (Figure 5–16a).

The Krebs cycle, in particular, is an important source of precursors needed for amino acid synthesis. The addition of an amino group to pyruvic acid or an appropriate organic acid in the Krebs cycle *(amination)* converts the acid into an amino acid. If this amine group comes from a preexisting amino acid the process is called *transamination* (Figure 5–16b). Other sources of precursors for amino acid biosynthesis are derived from the pentose phosphate pathway and the Entner-Doudoroff pathway. The main role of amino acids is to serve as building blocks for protein synthesis. The mechanism of protein synthesis will be discussed in Chapter 8.

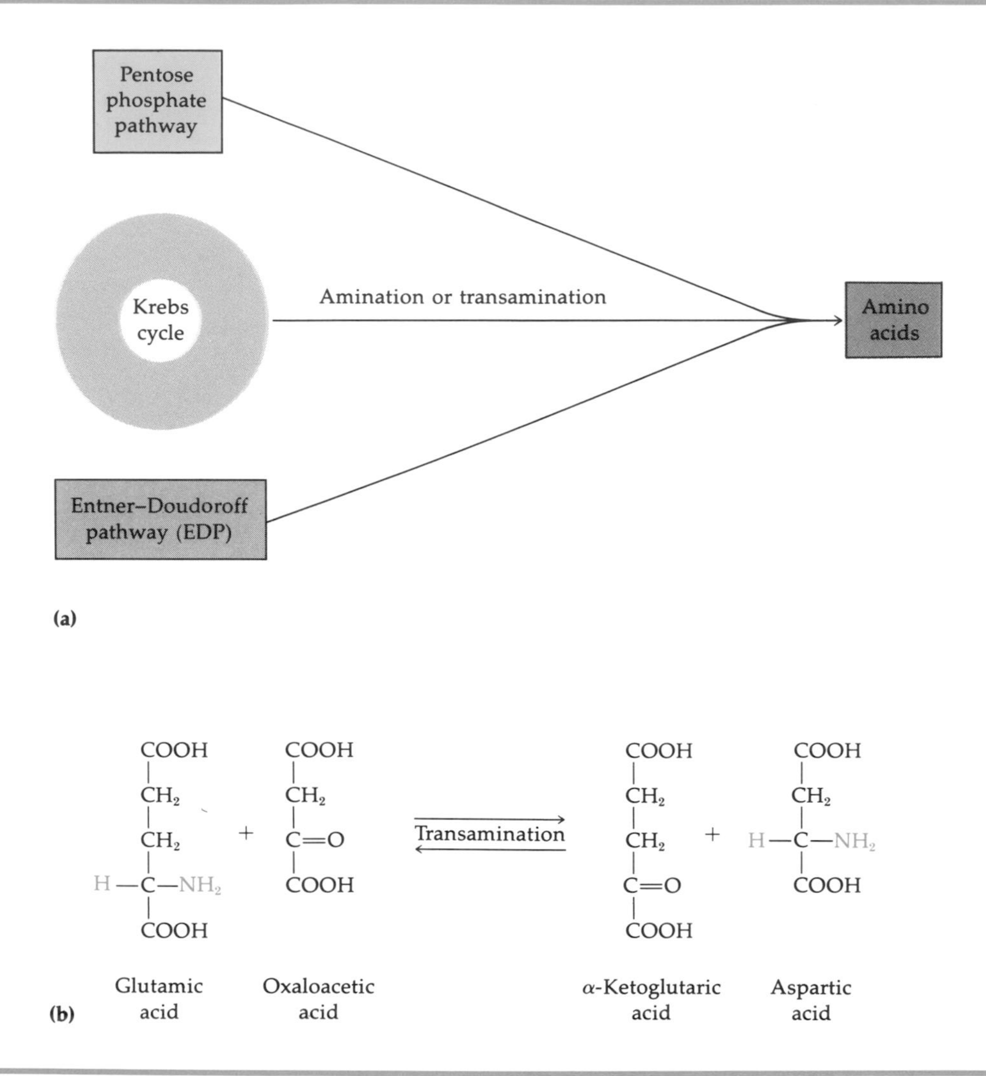

Figure 5–16 Biosynthesis of amino acids. **(a)** Pathways of amino acid biosynthesis. **(b)** Transamination, a process by which new amino acids are made using the amino groups from old amino acids. Glutamic acid and aspartic acid are both amino acids. The other two compounds are intermediates in the Krebs cycle.

Biosynthesis of Purines and Pyrimidines

As you may recall from Chapter 2, DNA and RNA, the informational molecules, consist of repeating units called *nucleotides*, which consist of a purine or pyrimidine, a five-carbon sugar, and a phosphate group. The sugars composing nucleotides are derived from either the pentose phosphate pathway or the Entner-Doudoroff pathway. Certain amino acids—aspartic acid, glycine, and glutamine—play an essential role in the biosynthesis of purines and pyrimidines (Figure 5–17). The carbon and nitrogen atoms derived from these amino acids

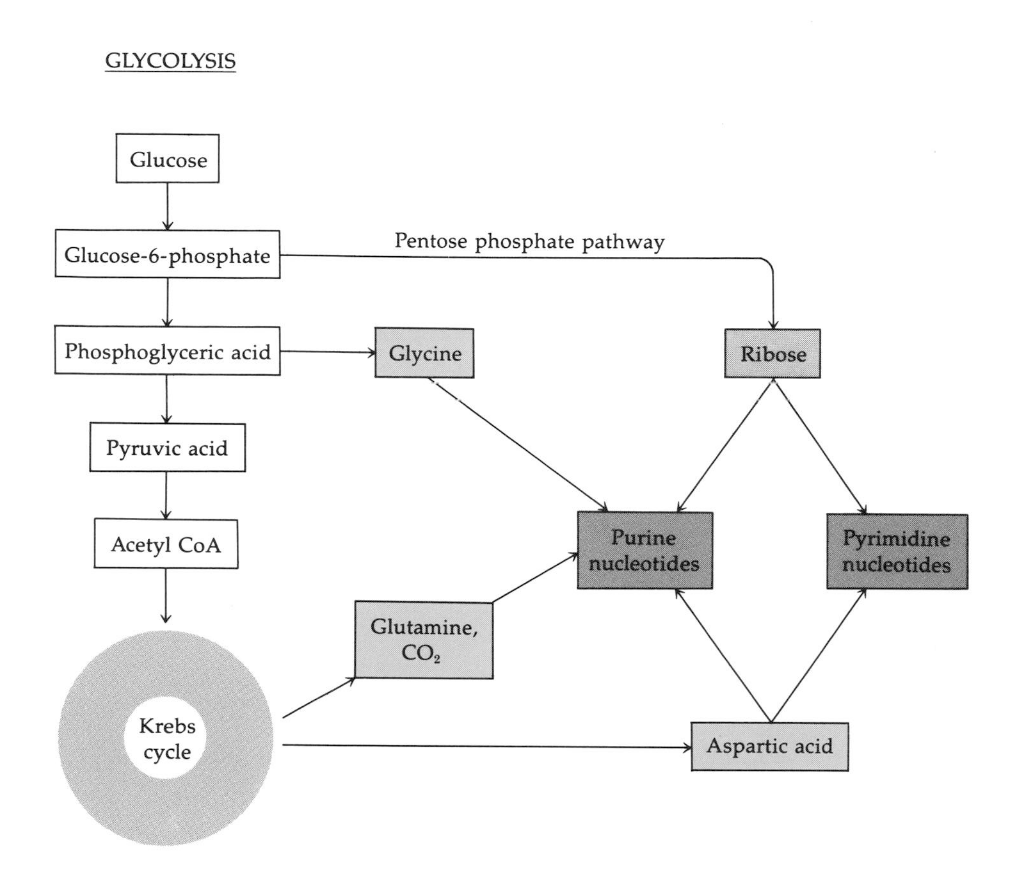

Figure 5–17 Biosynthesis of purine and pyrimidine nucleotides.

form the backbone of the purines and pyrimidines. The synthesis of DNA and RNA from nucleotides will be covered in Chapter 8.

INTEGRATION OF METABOLISM

We have seen thus far that microbes are involved in metabolic reactions that produce energy from light, inorganic compounds, and organic compounds. Reactions also occur in which energy is utilized for biosynthesis. With such an array of activity, it might appear that anabolic and catabolic reactions occur independently of each other in space and time. In actuality, anabolic and catabolic reactions are integrated through a group of common intermediates (Figure 5–18). Certain biochemical pathways, such as the Krebs cycle, operate in both anabolic and catabolic reactions. Reactions in the Krebs cycle not only form pyruvic acid and acetyl CoA as part of the oxidation of glucose, they also produce intermediates that lead to the synthesis of amino acids, fatty acids, and glycerol. Such pathways are called **amphibolic pathways.**

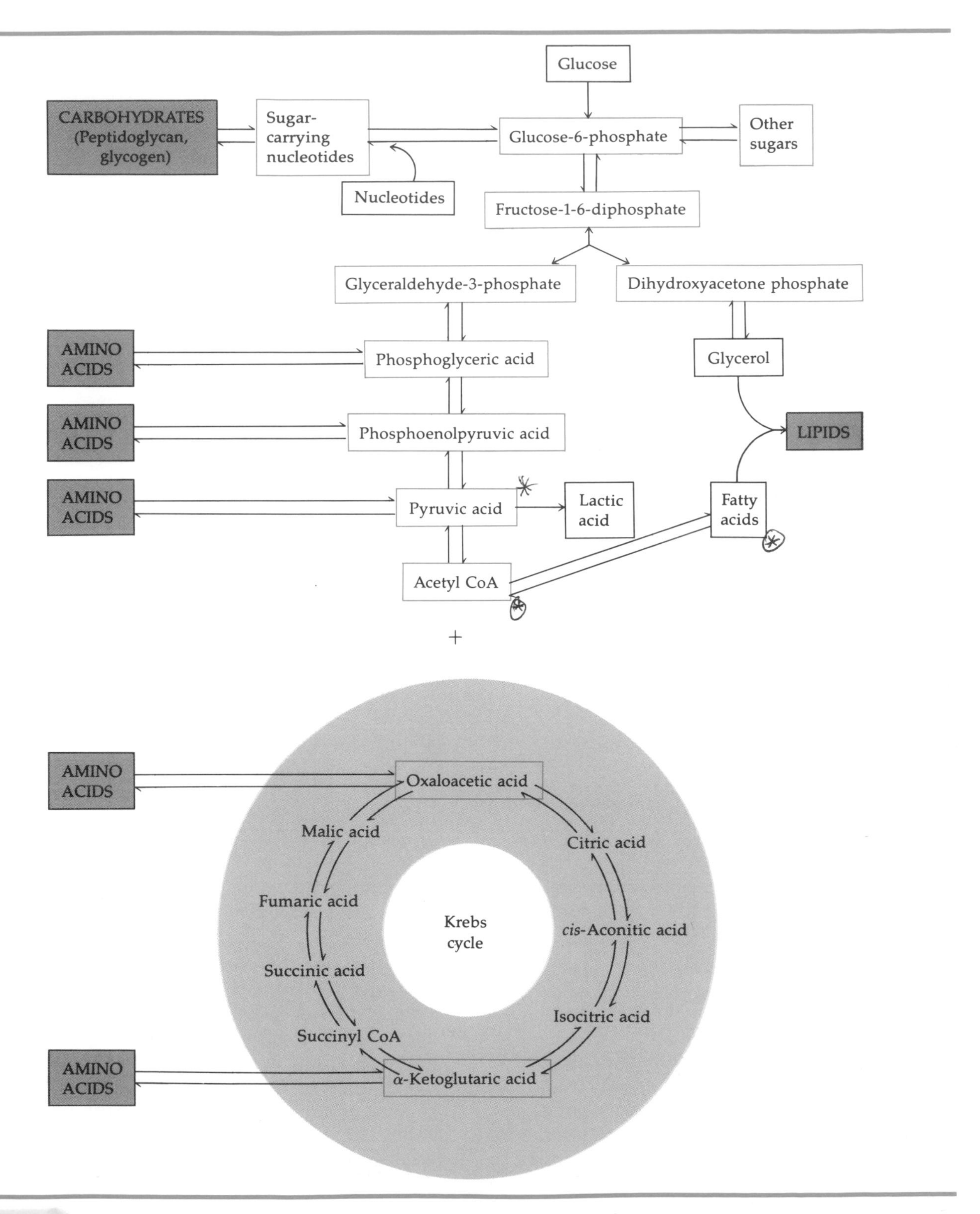

Figure 5–18 Integration of metabolism. Key intermediates are outlined in color. Although not indicated in the figure, amino acids and ribose are used in the synthesis of purine and pyrimidine nucleotides (see Figure 5–17).

STUDY OUTLINE

INTRODUCTION (pp. 110–111)

1. The sum of all chemical reactions that occur within a living organism is known as metabolism.
2. Anabolism refers to chemical reactions in which simpler substances are combined to form more complex molecules. Anabolic reactions usually require energy.
3. Catabolism refers to chemical reactions that result in the breakdown of more complex organic molecules into simpler substances. Catabolic reactions usually release energy.
4. The energy of catabolic reactions is used to drive anabolic reactions.
5. The energy for chemical reactions is stored in ATP.

ENZYMES (pp. 111–117)

1. Enzymes are proteins produced by living cells that catalyze chemical reactions.
2. Enzymes are generally globular proteins with characteristic three-dimensional shapes.

Mechanism of Enzyme Action (pp. 112–113)

1. Enzymes catalyze chemical reactions by lowering the activation energy.
2. When an enzyme and substrate combine, the substrate is transformed and the enzyme is recovered.
3. Enzymes are characterized by specificity, a function of their active sites.
4. Enzymes are efficient, able to operate at relatively low temperatures, and subject to various cellular controls.

Nonprotein Components of Enzymes (pp. 113–114)

1. Most enzymes are holoenzymes, consisting of a protein portion (apoenzyme) and a nonprotein portion (cofactor).
2. The cofactor may be a metal ion (iron, copper, magnesium, manganese, zinc, calcium, or cobalt) or a complex organic molecule known as a coenzyme (NAD, NADP, FMN, FAD, coenzyme A, and cytochromes).

Factors Influencing Enzyme Activity (pp. 114–117)

1. At high temperatures, enzymes undergo denaturation and lose their catalytic properties.
2. The pH at which enzymatic activity is maximal is known as the pH optimum.

3. Within limits, as substrate concentration increases, enzymatic activity increases.
4. Competitive inhibitors compete with the normal substrate for the active site of the enzyme. Noncompetitive inhibitors act on other parts of the enzyme and decrease the ability of the normal substrate to combine with the enzyme.

ENERGY PRODUCTION CONCEPTS (pp. 117–120)

The Concept of Oxidation–Reduction (pp. 117–118)

1. Biological oxidation means either the addition of oxygen or the removal of a pair of electrons from a substance (H^+ ions are usually removed with the electrons).
2. Reduction refers to the gain of a pair of electrons and H^+ by a substance.
3. Each time a substance is oxidized, another is simultaneously reduced.

Generation of ATP (p. 118)

1. Energy released during certain metabolic reactions may be trapped to form ATP from ADP and Ⓟ.
2. In oxidative phosphorylation, energy is released as electrons are passed to a series of electron acceptors and finally to oxygen or another inorganic compound.
3. In substrate-level phosphorylation, a high-energy phosphate group is added to ADP from an intermediate in catabolism.
4. In photophosphorylation, energy from light is trapped by chlorophyll, liberating an electron that passes to a series of electron acceptors.
5. Transfer of the electron releases energy for the synthesis of ATP.

Nutritional Patterns Among Organisms (pp. 118–120)

1. Photoautotrophs use light as an energy source and carbon dioxide as their principal carbon source.
2. Photoheterotrophs use light as an energy source and an organic compound for their carbon source.
3. Chemoautotrophs use inorganic compounds as their energy source and carbon dioxide as their carbon source.
4. Chemoheterotrophs use complex organic molecules as their carbon and energy sources.

BIOCHEMICAL PATHWAYS OF ENERGY PRODUCTION (pp. 120–130)

Glycolysis (pp. 122–123)

1. The most common pathway for the oxidation of glucose is glycolysis. Pyruvic acid is the end product.

2. Some organisms also use the pentose phosphate pathway or the Entner-Doudoroff pathway.

Fermentation of Carbohydrates (p. 123)

1. Fermentation releases energy from sugars or other organic molecules.
2. Oxygen is not required.
3. ATP is produced by substrate-level phosphorylation and the final electron acceptor is an organic molecule.

Respiration (pp. 124–128)

1. During respiration, organic molecules are oxidized completely to CO_2. Energy is generated during the electron transport chain.
2. The highest energy yield requires operation of the Krebs cycle, which produces CO_2, $NADH_2$, $FADH_2$, and GTP.
3. In aerobic respiration, the Krebs cycle and electron transport chain are used to oxidize a single molecule of glucose to carbon dioxide and water with a yield of 38 ATP molecules. Oxygen is the final electron acceptor.
4. In anaerobic respiration, the final electron acceptor may be nitrate, sulfate, or carbonate. Depending on the organism and other pathways employed, varying amounts of ATP are produced.

Lipid Catabolism (pp. 128–129)

1. Glycerol is catabolized by conversion to dihydroxyacetone phosphate, and fatty acids are catabolized by beta oxidation.
2. Catabolic products can be further broken down in glycolysis and the Krebs cycle.

Protein Catabolism (pp. 129–130)

1. Before amino acids can be catabolized, they must be converted to various substances that enter the Krebs cycle.
2. Transamination, decarboxylation, and dehydrogenation reactions convert the amino acids prior to catabolism.

BIOCHEMICAL PATHWAYS OF ENERGY UTILIZATION (pp. 130–133)

Biosynthesis of Polysaccharides (p. 130)

1. Glycogen is formed from ADPG.
2. UDPG is the starting material for the biosynthesis of peptidoglycan.

Biosynthesis of Lipids (pp. 130–131)

1. Lipids are synthesized from fatty acids and glycerol.

2. Glycerol is derived from dihydroxyacetone phosphate, and fatty acids are built from acetyl CoA.

Biosynthesis of Amino Acids (pp. 131–132)

1. Amino acids are required principally for protein biosynthesis.
2. All amino acids can be synthesized either directly or indirectly from intermediates of carbohydrate metabolism, especially the Krebs cycle.

Biosynthesis of Purines and Pyrimidines (pp. 132–133)

1. The sugars composing nucleotides are derived from either the pentose phosphate pathway or the Entner-Doudoroff pathway.
2. Carbon and nitrogen atoms from certain amino acids form the backbone of the purines and pyrimidines.

INTEGRATION OF METABOLISM (pp. 133–134)

1. Anabolic and catabolic reactions are integrated through a group of common intermediates.
2. Such integrated pathways are called amphibolic pathways.

STUDY QUESTIONS

REVIEW

1. Define metabolism.
2. Distinguish between catabolism and anabolism. How are these processes related?
3. Using the following diagrams show
   (a) Where the substrate will bind
   (b) Where the competitive inhibitor will bind
   (c) Where the noncompetitive inhibitor will bind.

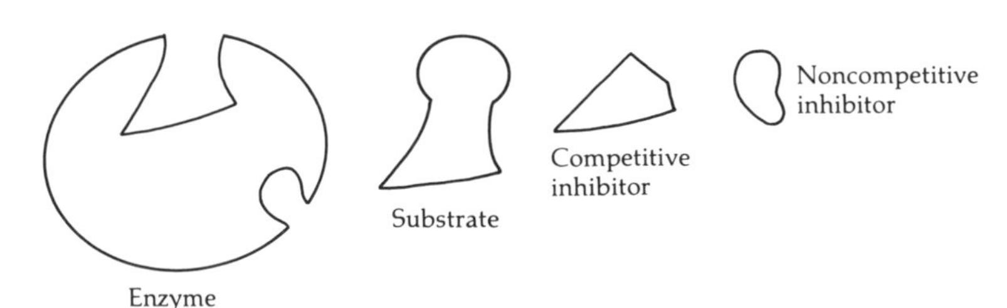

4. What will the effect of the reactions in question 3 be?
5. Why are most enzymes active at one particular temperature? Why are enzymes less active below this temperature? What happens above this temperature?

6. Which substances in each of these reactions are being oxidized? Which are being reduced?

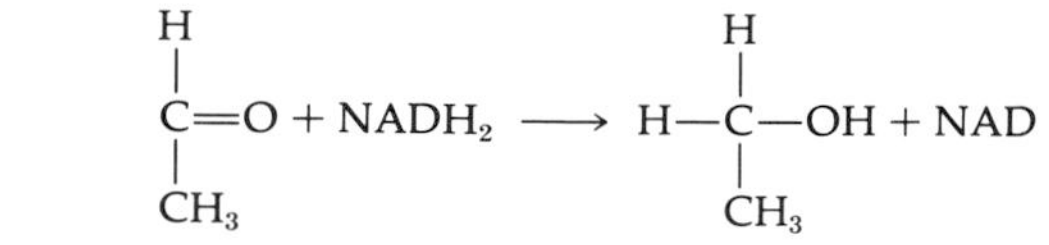

$$\begin{matrix} H \\ | \\ C{=}O \\ | \\ CH_3 \end{matrix} + NADH_2 \longrightarrow \begin{matrix} H \\ | \\ H{-}C{-}OH \\ | \\ CH_3 \end{matrix} + NAD$$

(a) Acetaldehyde → Ethanol

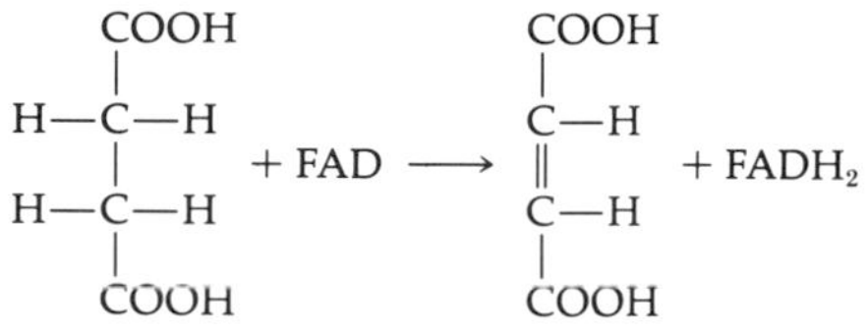

$$\begin{matrix} COOH \\ | \\ H{-}C{-}H \\ | \\ H{-}C{-}H \\ | \\ COOH \end{matrix} + FAD \longrightarrow \begin{matrix} COOH \\ | \\ C{-}H \\ \| \\ C{-}H \\ | \\ COOH \end{matrix} + FADH_2$$

(b) Succinic acid → Fumaric acid

7. There are three mechanisms for generating ATP. Write the name of the mechanism that describes each of the following reactions.

ATP generated by	Reaction
	An electron liberated from chlorophyll by light is passed down an electron transport chain.
	Cytochrome *c* passes two electrons to cytochrome *a*.
	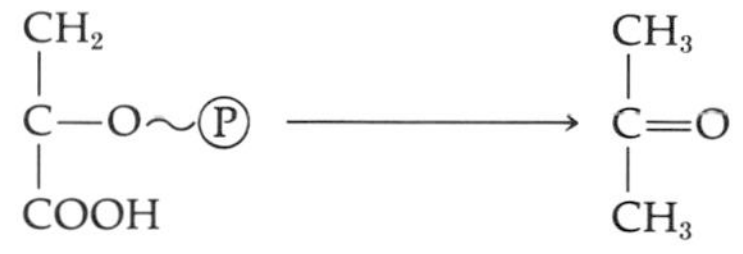 $\begin{matrix} CH_2 \\ \| \\ C{-}O{\sim}Ⓟ \\ \vert \\ COOH \end{matrix} \longrightarrow \begin{matrix} CH_3 \\ \vert \\ C{=}O \\ \vert \\ CH_3 \end{matrix}$ Phosphoenolpyruvic acid → Pyruvic acid

8. List four compounds that may be made from pyruvic acid by an organism that uses fermentation only.

9. What is the fate of pyruvic acid in an organism that uses aerobic respiration?

10. Fill in the table with the carbon and energy source of each type of organism.

Organism	Carbon Source	Energy Source
Photoautotroph		
Photoheterotroph		
Chemoautotroph		
Chemoheterotroph		

11. Define respiration and differentiate between aerobic and anaerobic respiration.
12. An enzyme and substrate are combined. The rate of reaction begins as shown in this graph. Complete the graph, showing the effect of increased substrate concentration on a constant enzyme concentration.

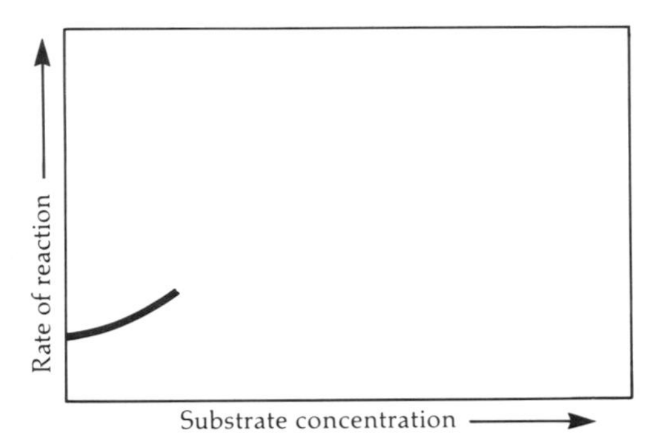

Use the following diagrams for questions 13–17:

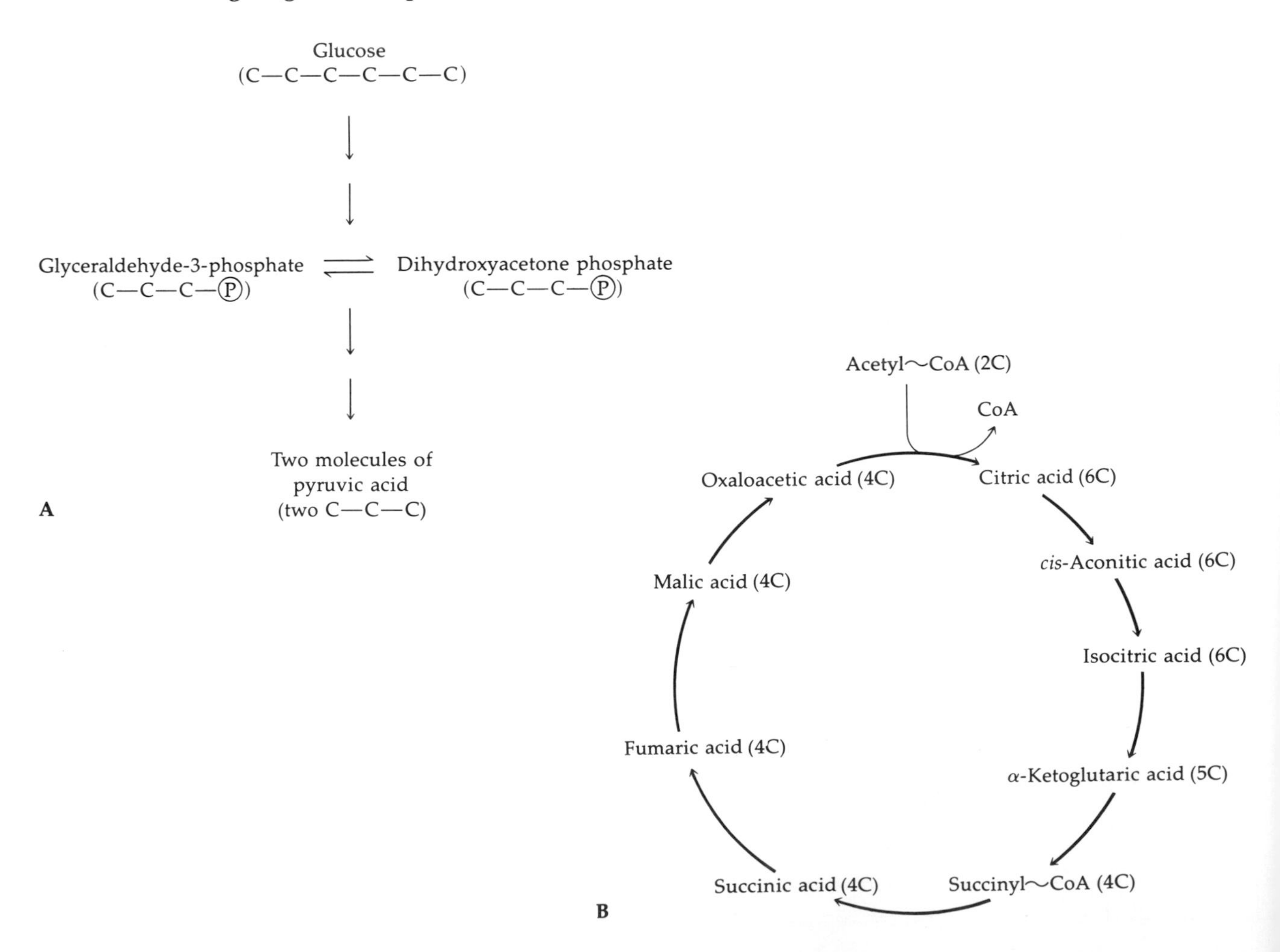

13. Name the pathway diagrammed in A. Name pathway B.
14. Glycerol is catabolized by pathway ____________; fatty acids are catabolized by pathway ____________.
15. Identify four places where anabolic and catabolic pathways become integrated.
16. Where is ATP required in pathways A and B?
17. Where is CO_2 released in pathways A and B?

CHALLENGE

1. The pentose phosphate pathway produces only one ATP. List four advantages of this pathway to the cell.
2. What will probably happen to the $FADH_2$ made in the reaction shown in review question 6b?
3. This graph shows the normal rate of reaction of an enzyme and its substrate (black), and the reaction when an excess of competitive inhibitor is present (color). Explain why the graph appears as it does.

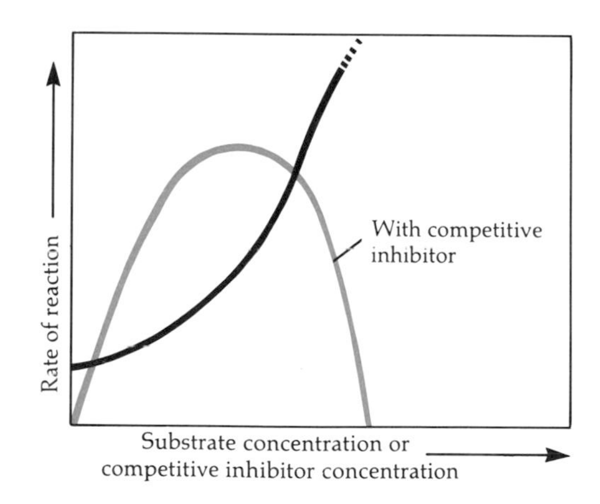

FURTHER READING

Doell, H. W., ed. 1974. *Microbial Metabolism.* Stroudsberg, PA: Dowden, Hutchinson, and Ross. A collection of papers on various aspects of metabolism including fermentation and oxidation of carbohydrates and the metabolism of inorganic compounds.

Glenn, A. R. 1976. Production of extracellular proteins by bacteria. *Annual Review of Microbiology* 30:41–62. A summary of how exoproteins are synthesized and excreted from cells and their medical and industrial importance.

Gottschalk, G. 1978. *Bacterial Metabolism.* New York: Springer-Verlag. A detailed description of metabolic pathways in bacteria.

Lascelles, J. 1973. *Microbial Photosynthesis.* Stroudsberg, PA: Dowden, Hutchinson, and Ross. Includes papers on the culturing and metabolism of photosynthetic organisms and comparisons of eucaryotic and procaryotic cells.

Lechtman, M. D., B. L. Roohk, and R. J. Egan. 1978. *The Games Cells Play: Basic Concepts of Cellular Metabolism.* Menlo Park, CA: Benjamin/Cummings. Innovative analyses and images make metabolism fun.

Lehninger, A. L. 1971. *Bioenergetics: The Molecular Basis of Biological Energy Transformations,* 2nd ed. Menlo Park, CA: Benjamin/Cummings. This authoritative reference provides clear explanations of bioenergetics.

Stoekenius, W. June 1976. The purple membrane of salt-loving bacteria. *Scientific American* 234:38–46. The discovery of a new photosynthetic mechanism in bacterial cells.

6

Microbial Growth

Objectives

After completing this chapter you should be able to

- Define bacterial growth, including transverse fission.
- Classify microbes into three principal groups on the basis of preferred temperature range.
- Explain the importance of pH and osmotic pressure to microbial growth.
- List the chemical requirements for growth in general terms.
- Explain how microbes are classified on the basis of oxygen requirements.
- Distinguish between chemically defined and complex media.
- Describe how pure cultures can be isolated.
- Compare the phases of microbial growth and their relation to generation time.
- Describe several direct and indirect measurements of microbial growth.

When we talk about microbial growth, we are really referring to the *number* of cells, not the *size* of the cells. Microbes that are "growing" are increasing in number, accumulating into clumps of hundreds, colonies of hundreds of thousands, or populations of billions. For the most part, we are not concerned with the size of an individual cell, because it does not vary much during the cell's lifetime. Thus, measurements of the total mass of a bacterial population are usually roughly proportional to measurements of the number of cells in the population; the rate of increase in cell mass is often a measure of the reproductive rate of a population.

By understanding the conditions necessary for microbial growth, we can predict how fast microorganisms will grow in various situations, and determine how to control their growth. Microbial populations can become very large in a very short time, and unchecked microbial growth can cause serious disease and food spoilage.

In this chapter we will examine the physical and chemical requirements for microbial growth, the various kinds of culture media used, bacterial division, the phases of microbial growth, and the methods of measuring microbial growth. Once you understand these concepts, you will have a better idea as to how microbial growth can be retarded or inhibited.

REQUIREMENTS FOR GROWTH

The requirements for microbial growth may be divided into two categories, physical and chemical. Physical aspects include temperature, pH, and osmotic pressure. Chemical requirements include water, sources of carbon and nitrogen, minerals, oxygen, and organic growth factors.

Physical Requirements

Temperature

Microorganisms tend to grow best at rather ordinary temperatures not much different from those favored by higher animals. However, certain bacteria are capable of growing in extreme cold or extreme heat, at temperatures that would certainly hinder the survival of most higher organisms.

Microorganisms are divided into three groups on the basis of their preferred range of temperature. These are the **psychrophiles** (cold-loving microbes), **mesophiles** (moderate-temperature-loving microbes), and **thermophiles** (heat-loving microbes). Most bacteria grow only within a temperature range of about 30°C, growing poorly at some temperatures within the range and better at others. Each bacterial species has a particular minimum, optimum, and maximum temperature at which it grows. The **minimum growth temperature** is the lowest temperature at which the species will grow. The **optimum growth temperature** is the temperature at which the species grows best (fastest). The **maximum growth temperature** is the highest temperature at which growth is possible. By graphing the growth response over a temperature range, we can see that the optimum growth temperature is usually near the top of the range, after which the rate of growth drops off rapidly (Figure 6–1). This is presumably because of the thermal inactivation of necessary enzyme systems of the cell.

The ranges and maximum growth temperatures that define bacteria as psychrophile, mesophile, or thermophile are not rigidly defined and vary from reference to reference. Psychrophiles, for example, were originally considered simply as organisms capable of growth at 0°C. However, there seem to be two fairly distinct groups of organisms capable of growth at that temperature. It has therefore been suggested that the category of psychrophile be reserved for organisms with an optimum growth temperature of about 15°C and a maximum growth temperature of about 20°C, requirements that would prevent their growth in most reasonably warm rooms (20°C is 68°F). Such organisms are mostly found in the ocean's depths or in certain arctic regions and are seldom of concern in human pursuits, such as preserving food. On the other hand, a common and often troublesome group of psychrophiles will grow at refrigerator temperatures, or even below 0°C, but can also grow above 20°C. However, few will grow well above 30°C, and their optimum growth temperatures tend to be relatively low (often in the upper teens or low twenties). It has been proposed that these organisms be called **psychrotrophs,** and many sources use this term to describe bacteria capable of growth at a common refrigerator temperature (4°C), while other sources retain the term psychrophile.

Refrigeration is the most common method of preserving household food supplies. It is based on the principle that microbial reproduction rates decrease at low temperatures. Microbes are usually not killed even at subfreezing temperatures, although they may become entirely dormant. However, they gradually decline in numbers, some species faster than others. Psychrophiles, or psychrotrophs, actually do not grow particularly well at low temperatures, except in comparison with other organisms, but given time they are able to cause slow

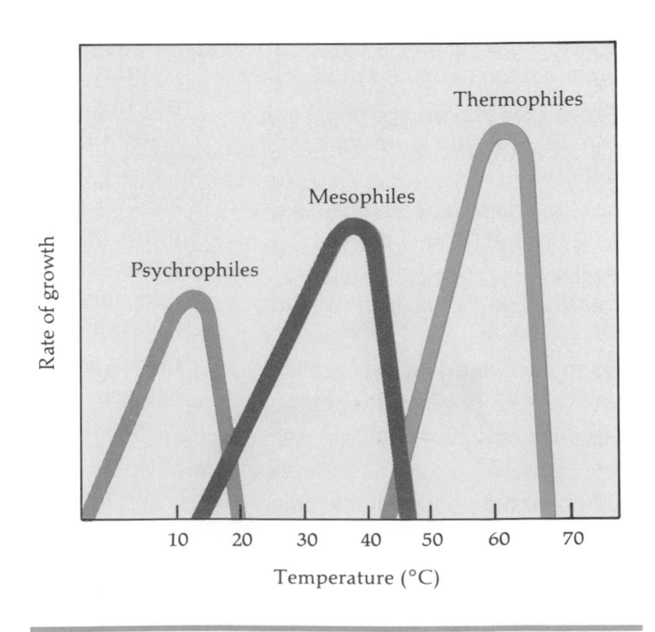

Figure 6–1 Typical growth responses to temperatures for psychrophiles, mesophiles, and thermophiles.

deterioration in food quality. This may take the form of mold mycelium, slime on food surfaces, or off tastes or colors in foods. A refrigerator set properly will prevent the growth of most spoilage organisms entirely and practically all of the pathogenic bacteria.

The mesophiles are the most common type of microbe, with an optimum growth temperature of between 25° and 40°C. Organisms that have adapted to life in the bodies of higher animals usually have an optimum temperature close to that of their host. The optimum temperature for many pathogenic bacteria is about 37°C, and incubators for clinical cultures are usually set at about this temperature. The mesophiles include most of our common spoilage and disease organisms.

A very interesting group of microorganisms are those capable of growth at high temperatures, the thermophiles. Many of these organisms have an optimum growth temperature of between 50° and 60°C. This is about the temperature of the water from a hot-water tap. Such temperatures can also be reached where sunlight falls on soil surfaces and in thermal waters, such as the hot springs in Yellowstone Park. Some of these organisms grow at temperatures well above 90°C, near the boiling point of water. Remarkably, many thermophiles are not capable of growth at all at ordinary temperatures below about 45°C. (Thermophilic fungi usually have an upper limit of 55° to 60°C, somewhat lower than most thermophilic bacteria.) Endospores formed by thermophilic bacteria tend to be relatively more heat resistant and are sometimes a problem in the preservation of foods by heat. Thermophiles are also important in the composting of organic matter, where the temperature can rise rapidly to the 50° to 60°C range.

pH

As you may recall from Chapter 2, pH refers to the relative acidity or alkalinity of a solution. Most bacteria grow best in a narrow range of pH near neutrality, between pH 6.5 and 7.5. Very few bacteria grow at an acid pH below about 4.0. For that reason, a number of foods, such as sauerkraut, pickles, and many cheeses, are preserved by the acids of bacterial fermentation. Nonetheless, some bacteria are remarkably tolerant of acidity. One autotrophic bacterium found in the drainage water from coal mines, and which oxidizes sulfur to form sulfuric acid, can survive at a pH of 1. Alkalinity also inhibits microbial growth, but is rarely used in food preservation.

When bacteria are cultured in the laboratory, acids they produce may interfere with desired bacterial growth. To neutralize the acids, chemicals called **buffers** are included in the growth medium. The peptones and amino acids in some media tend to act as buffers, and many media also contain phosphate salts. Phosphate salts have the advantage of exhibiting their buffering effect in the pH growth range of most bacteria. They are also nontoxic; in fact, they provide an essential nutrient element.

Osmotic pressure

Microbes require water for growth and are actually about 80 to 90% water themselves. When a microbial cell is in solution, and the solution contains a concentration of solutes higher than is present in the cell, the cellular water passes through the cytoplasmic membrane in the direction of the high salt concentration. (See the discussion of osmosis in Chapter 4.) This osmotic loss of water is called **plasmolysis** (Figure 6–2). What is so important about this phenomenon is that the growth of the cell is inhibited as the cytoplasmic membrane pulls away from the cell wall. Thus, the addition of salts (or other solutes) to a solution, and the resulting increase in osmotic pressure, can be used to preserve foods. Salted fish, honey, and sweetened condensed milk are largely preserved by this mechanism; the high salt or sugar concentrations draw water away from any microbial cells that are present, preventing their growth.

Some bacteria have become so adapted to an environment of high salt concentrations that they actually require them for growth. These are called **extreme halophiles.** Isolates from such saline waters as the Dead Sea often require nearly 30% salt to grow and must be transferred with an inoculating loop first dipped into a saturated salt solution. More common are **facultative halophiles,** which do not require high salt concentrations but are able to grow at salt concentrations up to 1 or 2%, a concentration that tends to inhibit the growth of many other bacteria. A few species of facultative halophiles can even tolerate 10 or 15% salt.

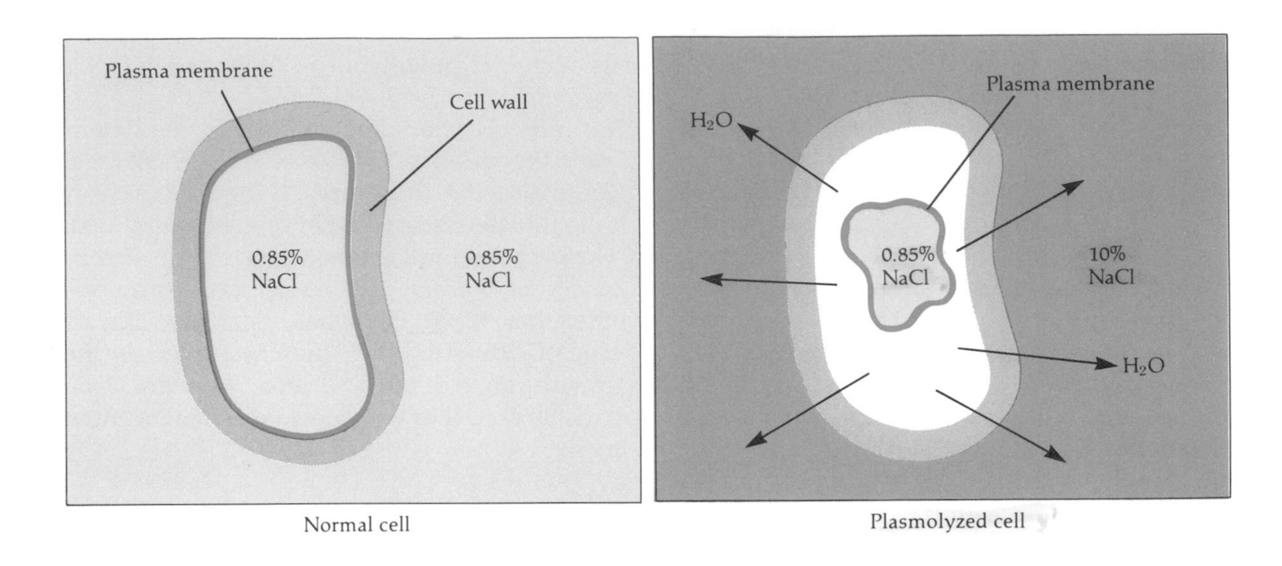

Figure 6–2 Plasmolysis. When the concentration of solutes in the medium exceeds the concentration within the cell, the cellular water leaves the cell by osmosis and the plasma membrane shrinks away from the cell wall.

The concentration of agar (a polysaccharide isolated from algae) used to solidify microbial growth media is usually about 1.5%. If markedly higher concentrations are used, the growth of some bacteria can be inhibited by the increased osmotic pressure. Because the effects of osmotic pressure are roughly related to the numbers of molecules in a given volume of solution, a low molecular weight compound such as sodium chloride has a greater antimicrobial effect than does, for example, sucrose, which has a somewhat higher molecular weight.

Chemical Requirements

Carbon

Besides water, one of the primary requirements for microbial growth is a source of the carbon needed for all the organic compounds that make up a living cell. Carbon is the structural backbone of living matter. As we saw in Chapter 2, its four valences allow it to be used in constructing extended and complicated organic molecules (see Figure 2–10). One-half of the dry weight of a bacterial cell is typically carbon.

Chemoheterotrophs get most of their carbon from the same source as their energy supply, namely, organic materials such as proteins, carbohydrates, and lipids. Chemoautotrophs derive some carbon from carbon dioxide. And photoautotrophs get their carbon entirely from carbon dioxide.

Since the utilization of carbon is such a basic trait of life on Earth, one way to test for life on another planet is to look for a sign that carbon is being metabolized. If carbon compounds are being either formed or broken down, then living organisms may be at work (see Box).

Nitrogen, sulfur, and phosphorus

In addition to carbon, there are other important elements needed by microbes for the synthesis of cellular material. For example, protein synthesis requires considerable amounts of nitrogen, as well as some sulfur. The synthesis of DNA and RNA requires nitrogen and phosphorus, as does the synthesis of ATP, the molecule so important for storage and transfer of chemical energy within the cell. Cell chemistry requires nitrogen, sulfur, and phosphorus in relatively large amounts. Together, these three elements constitute about 18% of the dry weight of the cell, of which 15% is nitrogen. Important sources of sulfur in nature include sulfate ion

Microbiology in the News

Is There Life on Other Planets?

If life has evolved on planets other than Earth, it probably includes microorganisms, the simplest forms of life we can imagine. In 1975, the U.S. National Aeronautics and Space Administration landed two spacecraft on the planet Mars. A major purpose of the Viking mission, as it was called, was to test for signs of life on Mars. Although the surface of Mars is cold and dry, Mars is the most likely planet in our solar system to have evolved living organisms.

For the Viking mission, scientists designed tests to detect the presence of microorganisms, the simplest forms of life we know of. As you know from Chapter 5, an essential feature of all living organisms is that they carry out chemical changes called metabolism. Living organisms take up molecules from their environment and metabolize them into other molecules. On Earth, the chemical compounds involved in metabolism are primarily organic (carbon) compounds, and some are relatively simple. Guessing that extraterrestrial life is likely to have the same basic features as life on Earth, the Viking scientists planned three experiments to be performed on samples of Martian soil (see diagram).

The *carbon assimilation experiment* incubated Martian soil with carbon dioxide labeled with ^{14}C, a radioactive form of carbon. The experiment looked for incorporation (fixation) of ^{14}C from CO_2 into other compounds. Since CO_2 fixation on Earth usually requires light (for photosynthesis), simulated Martian sunlight illuminated the test chamber. After five days the chamber atmosphere

(continued on next page)

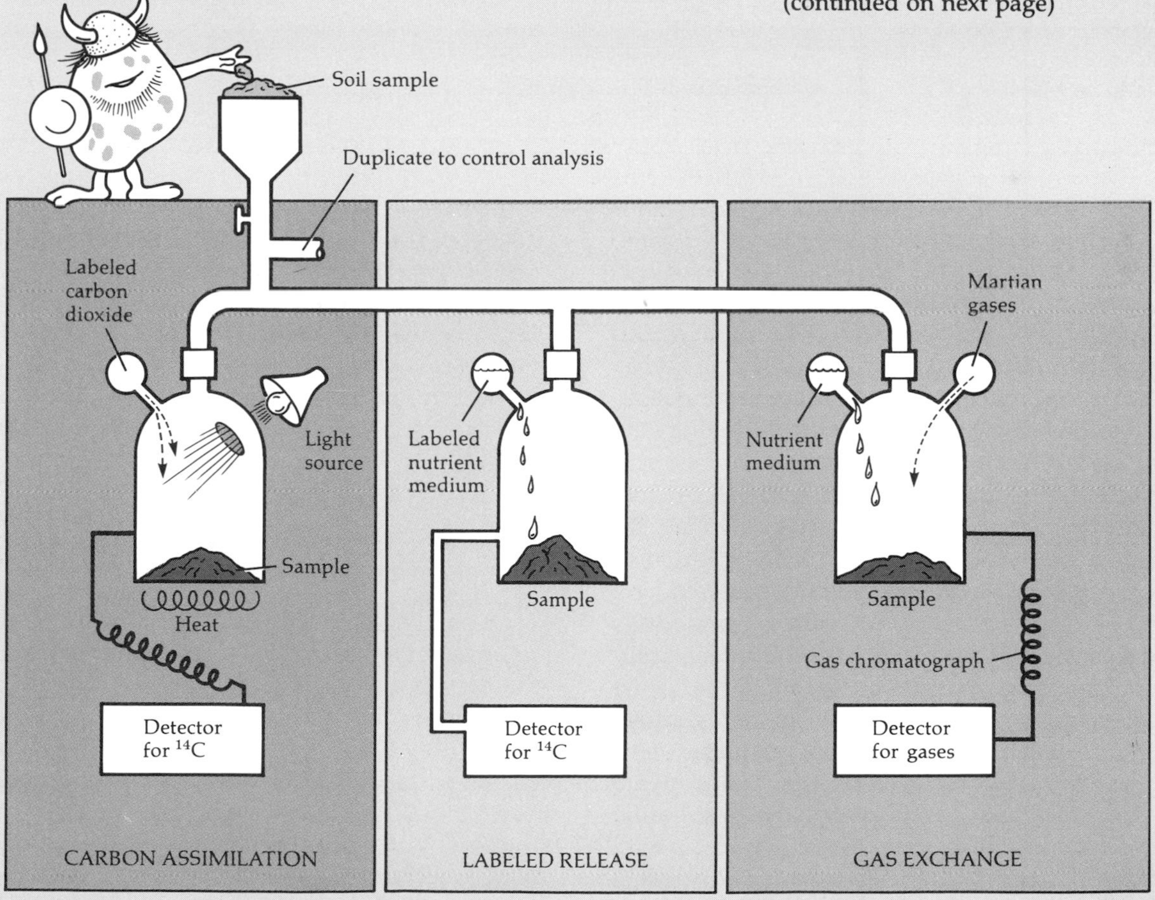

was flushed to remove any CO_2 that had not been utilized by microbes. The soil sample was then heated to 625°C to break down any cells and vaporize their organic material. The resulting gases would contain traces of any radioactive carbon fixed during incubation.

The other two experiments looked for evidence of metabolic reactions being carried on in a liquid nutrient medium. The addition of such a medium to a Martian soil sample might stimulate the metabolism of dormant microorganisms, perhaps remaining from a time when the Martian environment was less harsh. These experiments assumed that some of the metabolic waste-products released by the organisms would be gases.

In the *labeled release experiment*, a liquid nutrient medium containing organic compounds labeled with ^{14}C was incubated with Martian soil for 11 days. A radioactivity detector measured any gaseous end products labeled with ^{14}C (such as CO_2, CO, or CH_4).

In the *gas-exchange experiment*, the gases detected were not limited to carbon compounds. The soil sample was incubated with unlabeled nutrient medium for 12 days in an atmosphere similar to that of Mars (helium, krypton, and carbon dioxide). Any changes in the composition of the atmosphere were detected by an instrument that can identify a variety of simple gases, such as H_2, N_2, O_2, CH_4, and CO_2.

If any experiment showed a positive result, a control experiment was done. The soil samples used for the controls were heat-sterilized to insure that all living organisms were destroyed. Then, if the result of the control experiment was negative, scientists could be relatively sure that the original test result was due to the presence of living organisms.

To the disappointment of many people, the Viking mission did not detect any signs of life on Mars. But even if there is no life on Mars, it is still possible that life has evolved on some of the other billions of planets that orbit stars within our galaxy. Perhaps, someday, spacecraft will travel to other solar systems to search for living organisms there.

(SO_4^{2-}), hydrogen sulfide (H_2S), and the sulfur-containing amino acids. An important source of phosphorus is phosphate ion (PO_4^{3-}).

Nitrogen is used primarily to form the amino group of the amino acids of proteins. Many bacteria are able to meet this requirement by decomposing proteinaceous material and reincorporating the amino acids into newly synthesized protein and other nitrogen-containing compounds. Other bacteria can use ammonium ions (NH_4^+), which are already in the reduced form and usually found in the organic cellular material. Still other bacteria are able to derive nitrogen from nitrates (compounds that dissociate to give nitrate ion, NO_3^-, in solution). And some important bacteria and cyanobacteria (blue-green algae) are able to use gaseous nitrogen (N_2) directly from the atmosphere. This last process is called **nitrogen fixation.** Some organisms that can use this method are free living, mostly in the soil, but others live in symbiosis with the roots of certain plants. *Symbiosis* is a close association of two or more different kinds of organisms. The most important symbiotic nitrogen-fixing bacteria belong to the genus *Rhizobium* and are associated with legumes such as clover, soybeans, alfalfa, beans, and peas. The nitrogen fixed in the symbiosis is used for growth by both the plant and the bacterium. This reaction greatly increases soil fertility and is therefore very important in agriculture. Details of nitrogen fixation as a process in the nitrogen cycle are presented in Chapter 26.

Trace elements

Much smaller amounts of other mineral elements are required by microbes, such as iron, copper, and zinc, referred to as **trace elements.** Although these components are sometimes added to laboratory media, they are usually assumed to be naturally present in tap water or other media components. Even most distilled water contains adequate amounts, but tap water is sometimes specified for culture media to ensure the presence of these minerals.

Oxygen

We are accustomed to thinking of oxygen as a necessity of life, but it is actually a rather corrosive gas that was absent from the atmosphere during most of our planet's history. It is speculated that originally, life could not have arisen if oxygen were present. However, many forms of life have metabolic systems that require oxygen (for aerobic respiration). As we have seen, the oxygen eventually combines with hydrogen atoms stripped from organic compounds to form water. This process creates a high yield of energy, while neutralizing a potentially toxic gas—a very neat solution all in all. When oxygen is not available, the result is a much lower energy yield, because the organic material is only partially degraded to compounds such as acids or alcohols. These compounds still contain considerable amounts of unreleased chemical energy.

Microbes that use oxygen, called **aerobes,** produce more energy from nutrients than microbes that do not use oxygen. Organisms that require oxygen to live are called **obligate aerobes** (Figure 6–3a). Examples include *Mycobacterium tuberculosis* and some endospore-forming bacteria. However, obligate aerobes have a disadvantage in that oxygen is poorly soluble in water, and many parts of the environment are oxygen poor. Therefore, many of the aerobic bacteria have developed, or retained, the ability to continue growing in the absence of oxygen. Such organisms, including some that exist in your intestinal tract, are called **facultative anaerobes** (Figure 6–3b). In other words, facultative anaerobes can use oxygen when it is present, but are able to continue growth by fermentation or anaerobic respiration when oxygen is not available. However, facultative anaerobes do not grow as efficiently in the absence of oxygen. Examples of facultative anaerobes include the familiar *Escherichia coli* and yeasts.

Obligate anaerobes are bacteria that are totally unable to use oxygen for growth (Figure 6–3c). Often they lack cytochromes and have no means of reducing oxygen to water. In other cases, they have an anaerobic electron transport chain, which may indirectly cause their death in the presence of oxygen. Hydrogen atoms from the electron transport chain may be passed to oxygen (O_2), forming the toxic hydrogen peroxide (H_2O_2). Most organisms have the enzyme *catalase,* which breaks down hydrogen peroxide to water and oxygen before lethal concentrations accumulate. The genus *Clostridium* is probably the most familiar example of an obligate anaerobe that lacks catalase. (The ability to form endospores has probably aided the survival of these oxygen-sensitive bacteria.) A test for the presence of catalase is routinely done to differentiate certain gram-positive cocci. A drop of hydrogen peroxide is placed on a suspension of cells; if bubbles appear, the cells have catalase (catalase +).

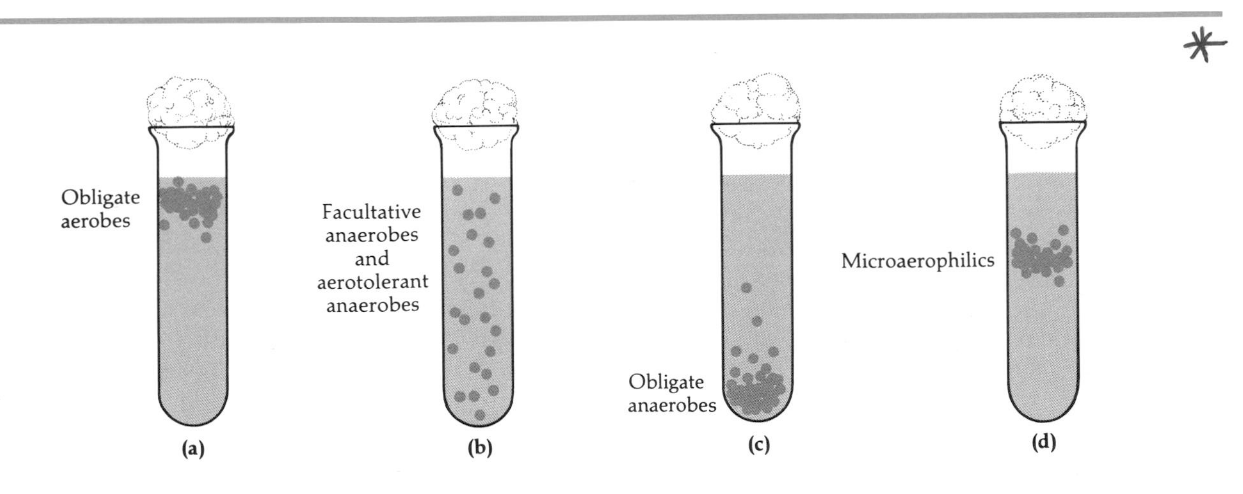

Figure 6–3 Effect of oxygen concentration on the growth of various types of bacteria in a tube of solid medium. Oxygen diffuses only a limited distance from the atmosphere into the solid medium.

Aerotolerant anaerobes cannot use oxygen for growth, but tolerate it fairly well. They will grow on the surface of a solid medium without the special techniques required for cultivation of less oxygen-tolerant anaerobes (Figure 6–3b). Many of these bacteria characteristically ferment carbohydrates to lactic acid. As lactic acid accumulates, it inhibits the growth of aerobic competitors, helping provide the lactic acid producers with an ecological niche. A common example of lactic acid-producing aerotolerant anaerobes are the lactobacilli used in making many acidic fermented foods, such as pickles and cheese. In the laboratory they are handled and grown much like any other bacterium, but they make no use of the oxygen in the air.

A few bacteria are **microaerophilic,** meaning they grow only in oxygen concentrations lower than those found in air. However, they are aerobic in the sense that they require oxygen. In a test tube of solid nutrient medium, their growth appears only at a depth where small amounts of oxygen have diffused into the medium, but they do not grow at the oxygen-rich surface (Figure 6–3d). This may reflect a requirement for high carbon dioxide concentrations, rather than a sensitivity to high concentrations of oxygen.

Organic growth factors

Organic growth factors are organic compounds needed for life that a given organism is unable to synthesize; they must be obtained from the environment. A group of organic growth factors for humans is vitamins. Most vitamins function as coenzymes, which are accessories required for certain enzymes to function. Many bacteria can synthesize all their own vitamins and do not need to have them supplied. However, some bacteria lack the enzymes needed for the synthesis of certain vitamins and therefore require those vitamins as organic growth factors. Other organic growth factors required by some bacteria are amino acids, purines, and pyrimidines.

CULTURE MEDIA

Any nutrient material prepared for the growth of microorganisms in a laboratory is called a **culture medium.** Some bacteria can grow well on just about any kind of culture medium, others require special media, and still others cannot grow on any nonliving medium yet developed. The microbes that grow and multiply in a container of culture medium are referred to as a **culture.**

Suppose we want to grow a culture of a particular microorganism, perhaps the microbes from a particular clinical specimen. What criteria must the culture medium meet? First of all, it must contain the right kinds of nutrients for the particular microorganisms we want to grow. It should also contain sufficient moisture and oxygen and a properly adjusted pH. To ensure that the culture will contain only the microorganisms we add to the medium (and their offspring), the medium must initially be **sterile,** that is, it must initially contain no living microorganisms. Finally, the growing culture should be incubated at the proper temperature.

There is a wide variety of media available for the growth of microorganisms in the laboratory. Most of these are available from commercial sources with all the components premixed and requiring only the addition of water and sterilization for use. For the isolation and identification of many bacteria of interest, in such areas as food, water, and clinical microbiology, there is a constant revision of old media formulas and introduction of new ones.

When it is desirable to grow bacteria on a solid medium, a solidifying agent such as agar is added to the medium. **Agar** is a polysaccharide derived from a marine alga, and it has had a long history as a thickener in foods such as jellies, soups, and ice cream. In fact, it is still used for that purpose.

Agar has some very important properties that make it valuable in microbiology, and no really satisfactory substitute has ever been found. Few microbes can degrade agar, so it remains a solid. Also important is the fact that it melts at about the boiling point of water but remains in a liquid state until the temperature drops to about 40°C. Agar is therefore held in water baths at about 50°C for laboratory use. At this temperature, it does not injure bacteria when it is poured over a bacterial inoculum in a Petri plate. When desired, bacterial suspensions can also be mixed with the melted agar in the water bath and poured to make a uniform suspension of bacteria, which can be tested for antibiotic susceptibility and other purposes. Once the agar has solidified, it can be incubated at temperatures approaching 100°C before liquefying, a useful property, particularly with thermophilic bacteria.

Chemically Defined Media

When a medium is being prepared for microbial growth, consideration must be given to providing an energy source, as well as a source of carbon, nitrogen, sulfur, phosphorus, and any necessary growth factors that the organism is unable to synthesize. A **chemically defined medium** is one in which the exact chemical composition is known. Table 6–1 shows a chemically defined medium for the growth of an organism capable of extracting energy from the oxidation of ammonium ions to nitrite ions.

Table 6–1 Chemically Defined Medium for the Growth of a Bacterium That Is Able to Use Ammonium Ions for Energy

Constituent	Concentration (Grams/Liter)
$(NH_4)_2SO_4$	0.5
$NaHCO_3$*	0.5
Na_2HPO_4	13.5
KH_2PO_4	0.7
$MgSO_4 \cdot 7H_2O$	0.1
$FeCl_3 \cdot 6H_2O$	0.014
$CaCl_2 \cdot 2H_2O$	0.18
Water	1000 ml (= 1 liter)

*The $NaHCO_3$ is a source of carbon dioxide (the carbon source) in solution.

Source: From D. Pramer and E. L. Schmidt, *Experimental Soil Microbiology*, Burgess Publishing Company, Minneapolis, MN, 1964.

As Table 6–2 demonstrates, many organic growth factors must be provided in the chemically defined medium used to cultivate species of *Lactobacillus*. Organisms that require many growth factors are described as *fastidious*.

A much simpler chemically defined medium is required for the growth of *Escherichia coli*, a less fastidious bacterium. The composition of this medium is shown in Table 6–3. It reflects the much greater synthesizing ability of *E. coli* as compared to the *Lactobacillus* species.

Table 6–2 Chemically Defined Medium for the Growth of a Fastidious Heterotrophic Bacterium, Such as *Lactobacillus* Species

Constituent	Concentration (Grams/Liter)
Carbon and Energy Sources:	
Glucose	10.0
Sodium acetate	6.0
Salts (added as two solutions of mixed salts):	
NH_4Cl	2.5
$(NH_4)_2SO_4$	2.5
KH_2PO_4	0.50
K_2HPO_4	0.50
NaCl, $FeSO_4$, and $MnSO_4$, each	0.005
Amino Acids:	
Arginine, cystine, glycine, histidine, hydroxyproline, proline, tryptophan, and tyrosine, each	0.20
Alanine, aspartic and glutamic acids, isoleucine, leucine, lysine, methionine, norleucine, phenylalanine, serine, threonine, and valine, each	0.10
Glutamine	0.025
Purines:	
Adenine, guanine, and uracil, each	0.010
Vitamins:	
Pyridoxamine-HCl	0.0004
Riboflavin, thiamine, niacin, and pantothenic acid, each	0.0002
Para-aminobenzoic acid	0.00004
Folic acid	0.00002
Biotin	0.0000002
Water	1000 ml (= 1 liter)

Source: From K. Thimann, *The Life of Bacteria*, 2nd ed., The Macmillan Company, New York, 1963.

Table 6–3 Chemically Defined Medium for the Growth of a Less Fastidious Heterotroph, Such as *Escherichia coli*

Constituent	Concentration (Grams/Liter)
Glucose	5.0
$NH_4H_2PO_4$	1.0
NaCl	5.0
$MgSO_4 \cdot 7H_2O$	0.2
K_2HPO_4	1.0
Water	1000 ml (= 1 liter)

Complex Media

Most heterotrophic bacteria and fungi are routinely grown on **complex media,** media for which the exact chemical composition varies slightly from batch to batch. These complex media are made up of nutrients such as extracts from yeasts, beef, or plants or digests of proteins from these and other sources. Table 6–4 gives one widely used recipe.

In complex media such as this one, the energy, carbon, nitrogen, and sulfur requirements of the growing microorganisms are largely met by protein. Protein itself is a large, relatively insoluble molecule that few microorganisms can utilize directly. But a partial digestion by acids or enzymes reduces the protein to shorter chains of amino acids, called *peptones*. These small, soluble fragments can be digested by the bacteria.

The vitamins and other organic growth factors are provided by meat extracts or yeast extracts. The soluble vitamins and minerals from the meats or yeasts go into solution in the extracting water, which is then evaporated to concentrate these factors. (Such extracts also supplement the organic nitrogen and carbon compounds.) Yeast extracts are particularly rich in the B vitamins. If this type of medium is in liquid form, it is called **nutrient broth**. When agar is added, it is called **nutrient agar**. (This terminology sometimes confuses students into thinking, wrongly, that agar itself is a nutrient—it is not.)

Table 6–4 Composition of Nutrient Agar, a Complex Medium for the Growth of Heterotrophic Bacteria

Constituent	Concentration (Grams/Liter)
Peptone	5.0
Beef extract	3.0
Sodium chloride	8.0
Agar	15.0
Water	1000 ml (= 1 liter)

Reducing Media

The cultivation of anaerobic bacteria poses a special problem. Because anaerobes may be killed by exposure to oxygen, special media called **reducing media** must be used. These media contain ingredients such as sodium thioglycollate that chemically combine with dissolved oxygen to deplete the oxygen content of the culture medium. For the routine growth and maintenance of pure cultures of obligate anaerobes, microbiologists use reducing media stored in ordinary, tightly capped test tubes.

When the culture must be grown in Petri plates for observation of individual colonies, special jars are used that hold several Petri plates in an oxygen-free atmosphere. To remove the oxygen, a packet of chemicals (sodium bicarbonate and sodium borohydride) in the jar is moistened with a few milliliters of water and then the jar is sealed. Hydrogen and carbon dioxide are produced by the reaction. A palladium catalyst in the jar combines the oxygen in the jar with the hydrogen produced by the chemical reaction, and water is formed. As

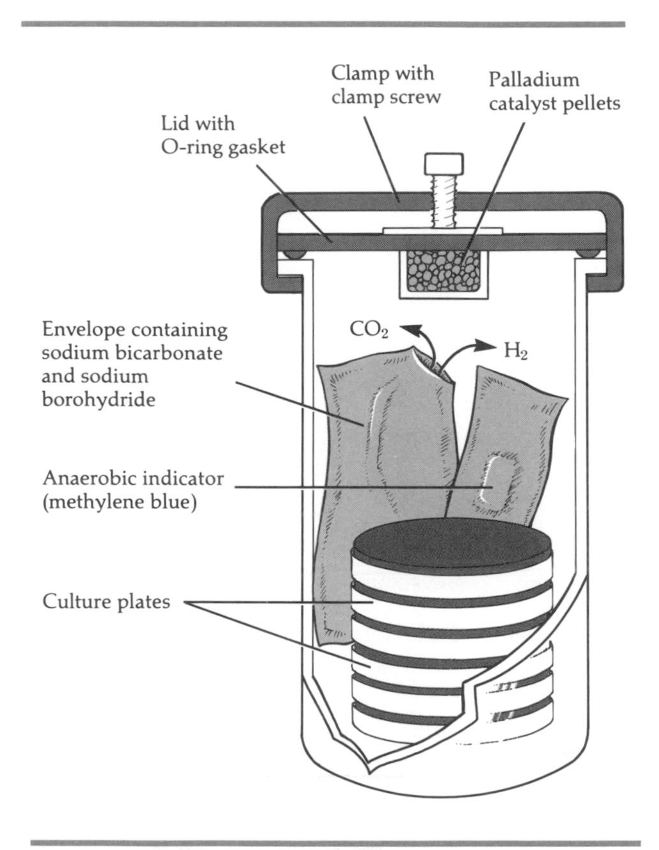

Figure 6–4 Anaerobic container for cultivation of anaerobic bacteria on Petri plates. When water is mixed with the chemical packet containing sodium bicarbonate and sodium borohydride, hydrogen and carbon dioxide are generated. The hydrogen reacts with the atmospheric oxygen in the jar on the surface of a palladium catalyst in a screened reaction chamber, forming water and removing oxygen. An anaerobic indicator is also in the container. It contains methylene blue, which is blue when oxidized, and which turns colorless as the oxygen is removed from the container.

a result, the oxygen disappears in a short time. Moreover, the carbon dioxide produced aids the growth of many anaerobic bacteria (Figure 6–4).

Laboratories working with anaerobes on a regular basis use more elaborate techniques. Anaerobic glove boxes, which are small, transparent chambers equipped with air locks and filled with inert gases, are sometimes used. The technicians are able to manipulate the equipment by inserting their hands in airtight rubber gloves fitted to the wall of the chamber.

It is also possible to grow colonies of anaerobes by substituting deep test tubes for Petri plates. The atmosphere in the test tube is replaced with an inert gas such as nitrogen. The bacterial inoculum is then mixed with melted nutrient medium in the tube, and the tube, tightly capped, is rolled on horizontal rollers. The nutrient medium solidifies against the interior walls of the test tube, and colonies that appear there can be counted and picked much as with a shallow layer of agar in a Petri plate.

Some bacteria have never been successfully grown on artificial laboratory media. Examples are the leprosy bacterium *(Mycobacterium leprae)*, usually grown in the laboratory on the foot pads of mice, and sometimes in armadillos, and the syphilis spirochete, although certain specialized strains of the latter have been grown on laboratory media. Obligate intracellular parasites such as the rickettsias and the chlamydia bacteria with few exceptions do not grow on artificial media. They, like viruses (which will be discussed later), can grow only on a living host cell.

Selective and Differential Media

In clinical and public health microbiology, it is frequently necessary to detect the presence of specific microorganisms associated with disease or poor sanitation. For this purpose, selective and differential media are used. **Selective media** are designed to suppress the growth of unwanted bacteria and encourage the growth of the desired microbes. For example, bismuth sulfite agar is used to isolate the typhoid bacterium from feces. Sabouraud's glucose agar, which has a pH of 5.6, is used to isolate fungi. Dyes such as crystal violet, methylene blue, and brilliant green inhibit gram-positive bacteria. **Differential media** make it easier to distinguish colonies of the desired organism from other colonies growing on the same plate. Blood agar (which contains red blood cells) is a dark, reddish brown medium used often by microbiologists to identify bacterial species that destroy blood cells. These species, such as *Streptococcus pyogenes*, the bacterium that causes strep throat, show a clear ring around their colonies where they have lysed the surrounding blood cells (Figure 6–5 and Color Plate IIA and B).

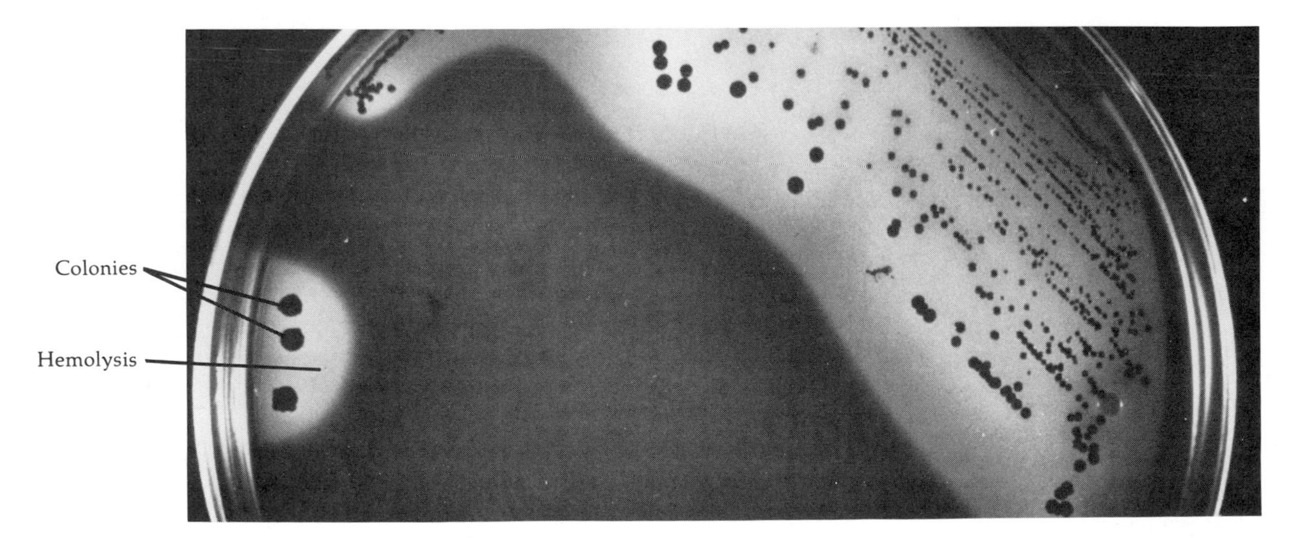

Figure 6–5 Differential medium containing red blood cells. The organism has lysed the red blood cells (hemolysis), forming a clear area around the colonies.

Sometimes selective and differential media are used together. Suppose we want to isolate the common bacterium *Staphylococcus aureus* from the nasal passages. One of the characteristics of this organism is its tolerance for high concentrations of sodium chloride, and another characteristic is its ability to ferment the carbohydrate mannitol to form acid. There is an isolation medium called mannitol salt (MS) agar medium that combines features of both selective and differential media. MS agar contains 7.5% sodium chloride, which will discourage the growth of competing organisms and thus *select* for (favor the growth of) *Staphylococcus*. This salty medium also contains a pH indicator that changes color if the mannitol in the medium is fermented to acid, thereby differentiating mannitol-fermenting colonies of *S. aureus* from colonies of bacteria that do not ferment mannitol. Bacteria that grow at the high salt concentration *and* ferment mannitol to acid can be readily identified. Such colonies will be surrounded by a zone in which the pH indicator has changed color, showing that the mannitol has been converted to acid. These are probably colonies of *S. aureus*, and their identification can be confirmed with a few additional tests.

Another selective and differential medium is MacConkey agar. This medium contains bile salts and crystal violet, which inhibit the growth of gram-positive bacteria. It also contains lactose. Gram-negative bacteria that can grow on this disaccharide can be differentiated from similar bacteria that cannot. The ability to distinguish between lactose fermenters and nonfermenters is useful in distinguishing the pathogenic *Salmonella* bacteria from other, related bacteria.

STOP

Inoculation

The **inoculating loop,** or a modification of it, the **inoculating needle,** is made of platinum or stainless steel wire and is used for inoculation and several other microbiological procedures. **Inoculation**

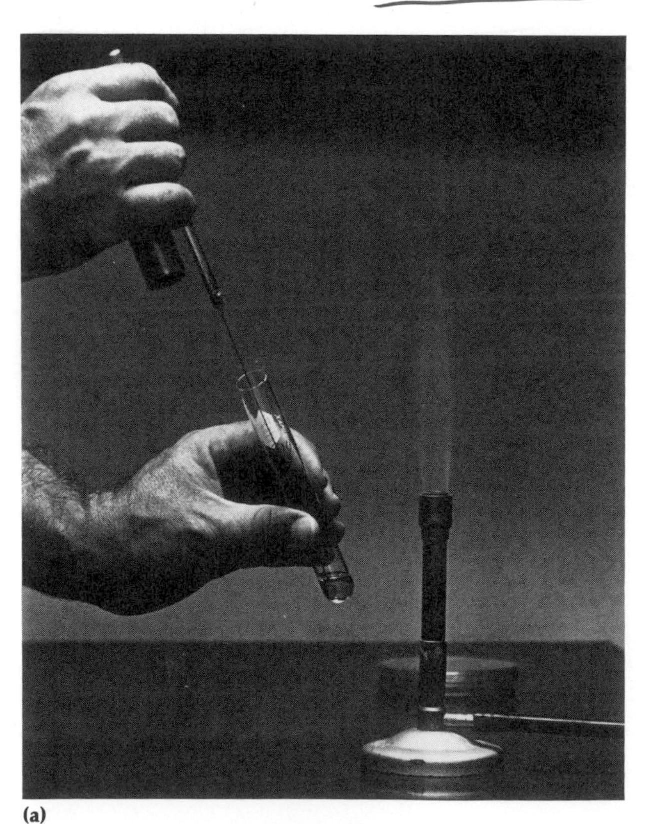

(a)

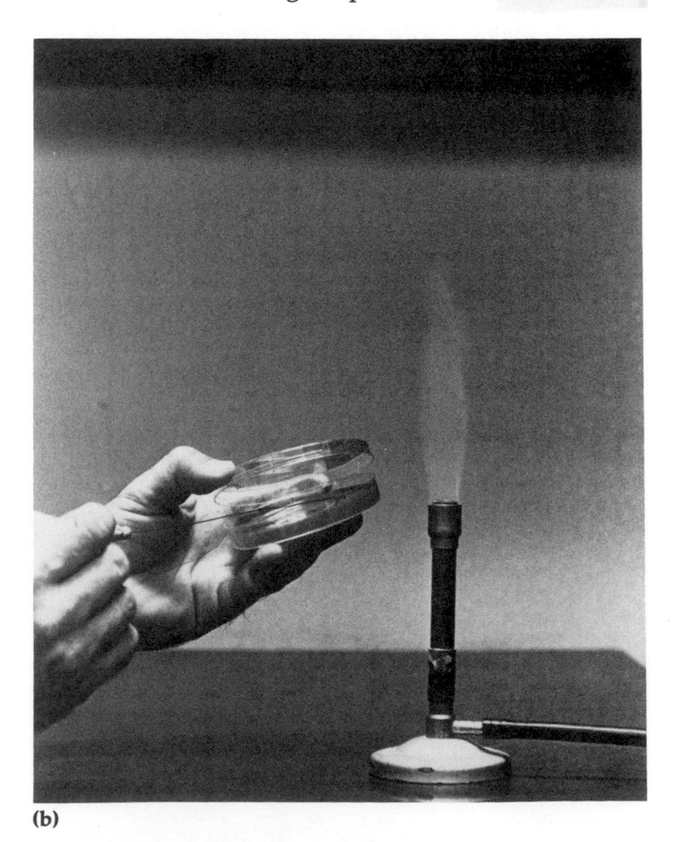

(b)

Figure 6–6 Inoculation of sterile media. **(a)** Inoculation of a liquid medium in a test tube. **(b)** Inoculation of a solid medium in a Petri plate. In both cases aseptic technique with an inoculation loop is employed.

is the transfer of microbes to a previously sterilized growth medium (Figure 6–6). After the medium is inoculated, it is placed in an incubator, usually for 24 to 48 hours. During this time, the bacteria actively grow and reproduce.

If a sufficiently dilute culture is inoculated onto a solid medium, the microorganisms that grow from each cell in the original sample form separate clumps, or masses, that are visible to the naked eye. These masses (each the progeny of a single cell) are called **colonies** (Figure 6–7 and Color Plate IIA, B, and D). Colonies have special characteristics such as texture, size, shape, color, and adherence to the medium. Such characteristics are fairly constant for each species and are useful in distinguishing one microbe from another.

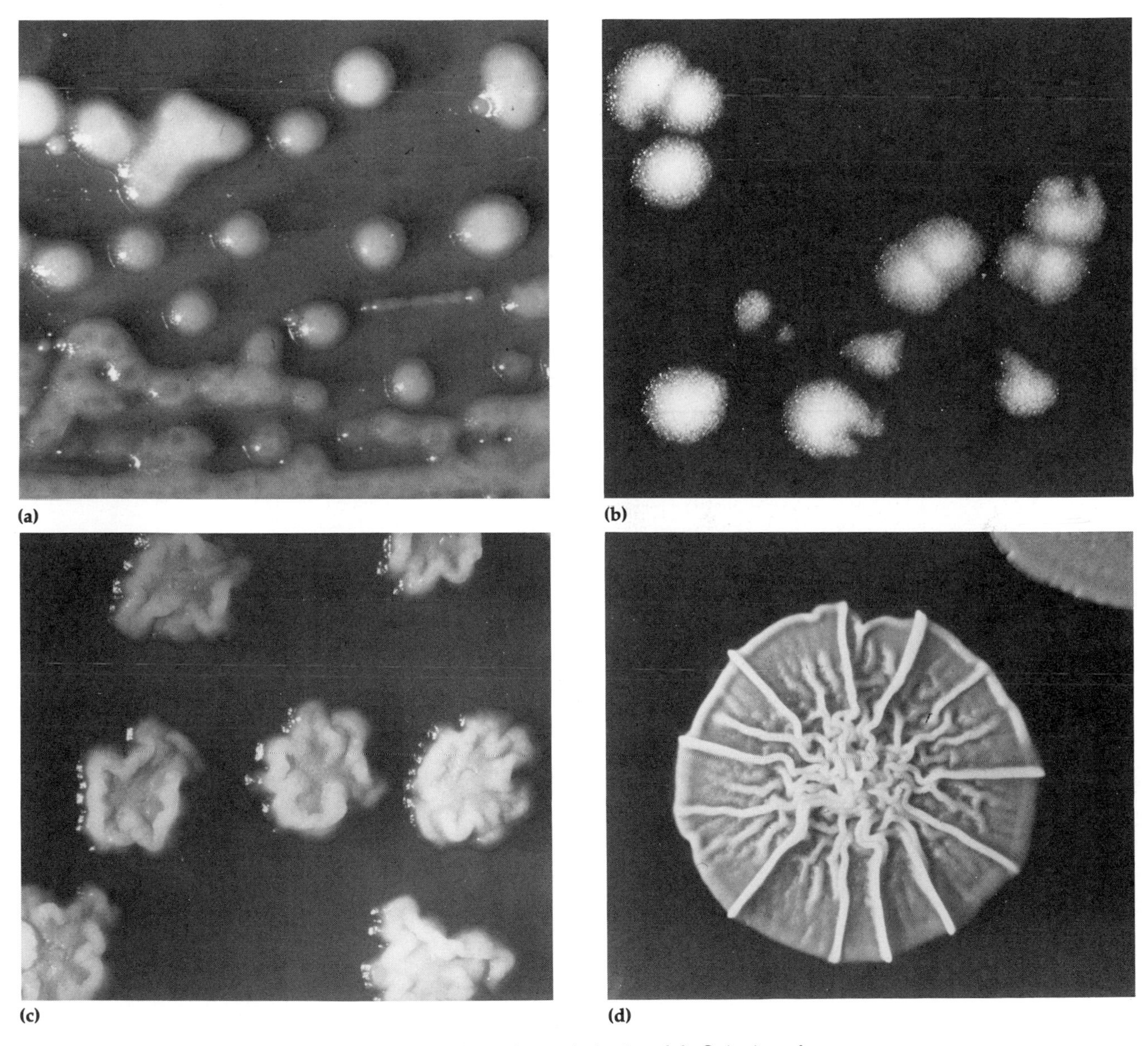

Figure 6–7 Bacterial colonies, showing a variety of morphologies. **(a)** Colonies of *Klebsiella pneumoniae* on MacConkey agar. **(b)** Colonies of *Bacillus anthracis* on artificial medium. **(c)** Rough colonies of *Mycobacterium smegmatis* on Penassay agar. **(d)** Colony of *Micrococcus luteus* on blood agar.

Mixed and Pure Cultures

Most infectious materials, like pus, sputum, and urine, contain several kinds of bacteria. A culture including more than one kind of microorganism is a **mixed culture,** and if grown on solid medium, it will often show several visibly different kinds of colonies. If we wish to study a particular microorganism from a mixed culture, it is necessary to take a sterile needle, touch it to the microbial colony we are interested in, and transfer sample cells from that colony to a separate, sterile medium. The microbial cells will multiply to form new colonies in a **pure culture,** that is, a culture where only one kind of microorganism is present.

Isolating Pure Cultures

There are several procedures for isolating and growing individual colonies of microorganisms.

Pour plate method

The use of solid media makes it a fairly simple procedure to isolate bacteria in pure culture. In the **pour plate method,** dilutions of bacteria suspended in liquid are mixed with melted nutrient agar. Then the agar is poured into a Petri plate, cooled to solidify, and incubated. On such a pour plate, individual bacteria will grow into colonies (Figure 6–8a). Isolated colonies can then be picked (touched) with a sterile inoculating needle and transferred to individual containers for growth in pure culture.

Spread plate method

An alternative method of isolating bacterial colonies is to dilute a bacterial suspension in liquid and then *spread* a measured volume of the diluted suspension on a nutrient agar medium. This is called a **spread plate method.** A sterilized bent glass rod is used to spread the bacteria (Figure 6–8b).

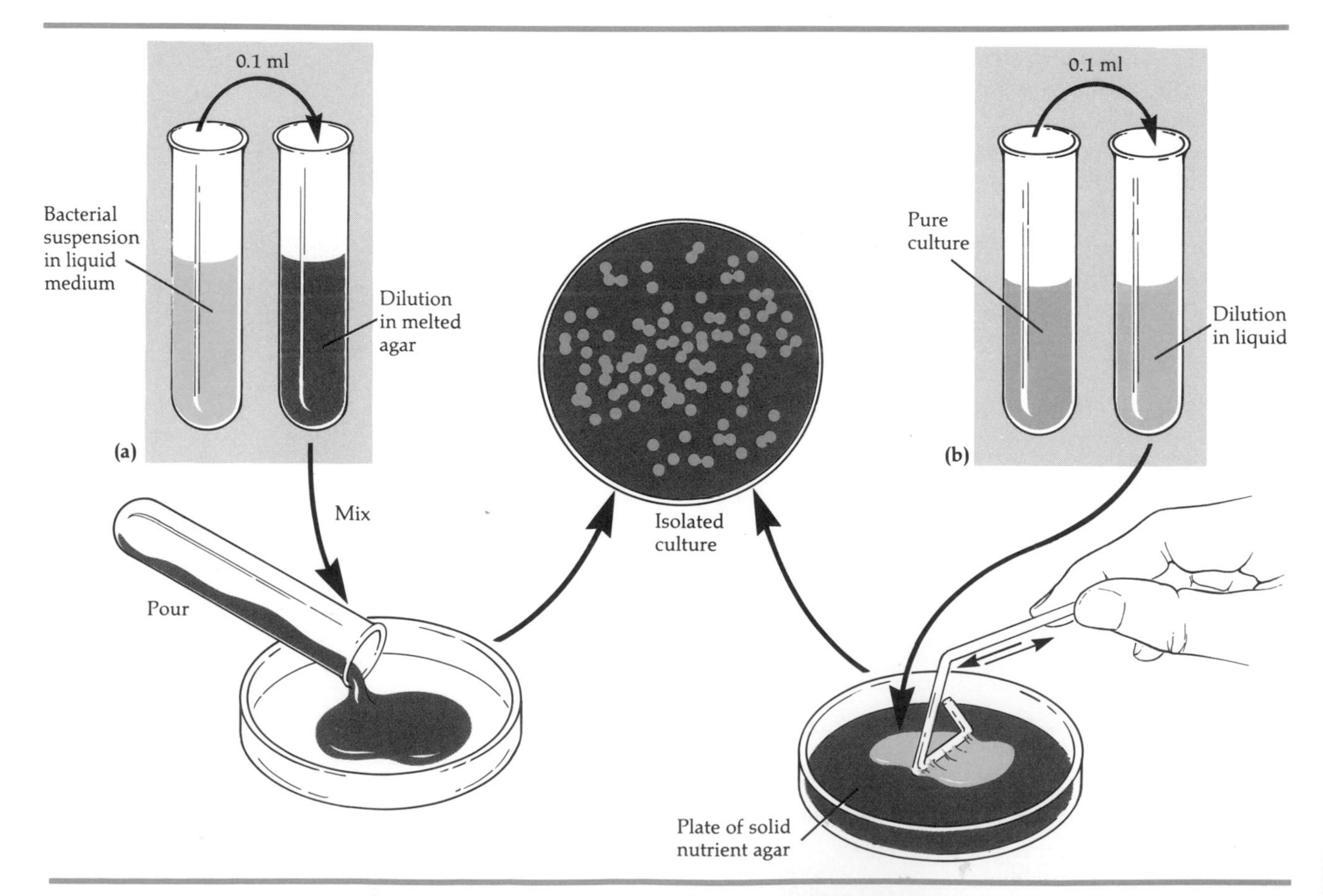

Figure 6–8 Isolating pure cultures. **(a)** Pour plate method. **(b)** Spread plate method.

Streak plate method

The isolation method most commonly used is the **streak plate method** (Figure 6–9). A sterile inoculating needle is dipped into a mixed culture and streaked in a pattern over the surface of the nutrient medium. As the pattern is traced, bacteria are rubbed off the needle onto the medium, putting down a path of fewer and fewer cells. The last cells that rub off are wide enough apart that they grow into isolated colonies.

Enrichment culture

Because bacteria present in small numbers can be missed by the previous three methods, and because the bacterium to be isolated may be an unusual physiological type, it is sometimes necessary to resort to an **enrichment culture.** Such a culture medium is usually liquid and provides nutrients and environmental conditions that favor the growth of a particular microbe but are not suitable for the growth of other types of microbes.

Let us assume that we want to isolate from a soil sample a microbe capable of growth on phenol, a standard disinfectant. If the soil sample is placed in an enrichment culture in which phenol is the only source of carbon and energy, microbes unable to metabolize phenol will not grow. The culture medium is allowed to incubate for a few days, and then a small amount of it is transferred into another culture medium of the same composition. A series of such transfers will eventually result in a population of bacteria capable of metabolizing phenol. The bacteria are given time to grow in the medium between transfers: this is the enrichment. Any nutrients in the original inoculum are rapidly diluted out with the successive transfers. After a time, the last dilution can be streaked out onto a solid medium of the same nutrient composition, and colonies of organisms capable of using phenol should grow.

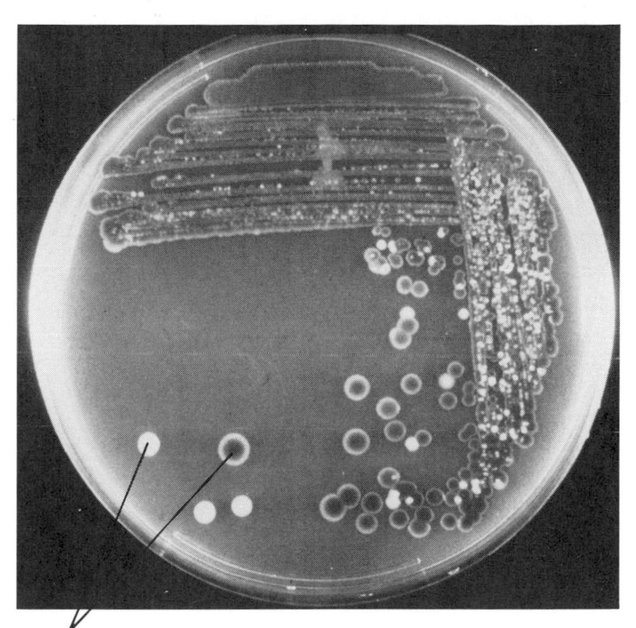

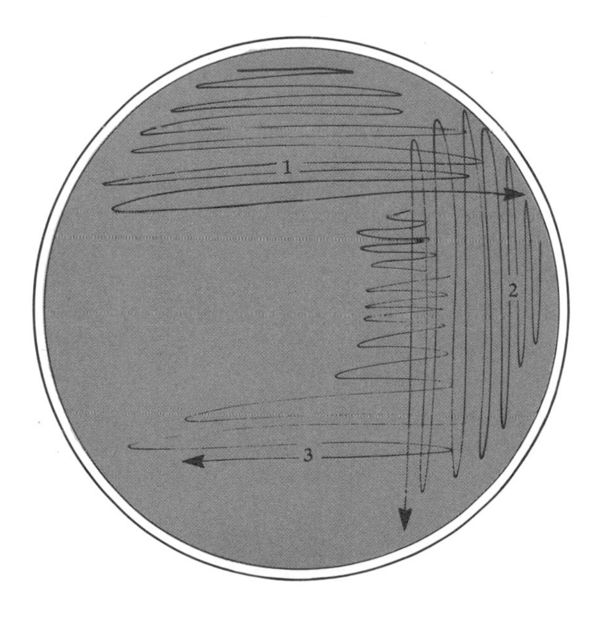

Figure 6–9 Streak plate procedure for pure culture isolation of bacteria. The direction of streaking is indicated by the arrows. Between each section, the needle is sterilized and reinoculated with a fraction of the bacteria by going back across the previous section. In this way, the bacteria are diluted out, and well-isolated colonies are obtained. Shown here are isolated colonies of two different bacteria.

PRESERVING BACTERIAL CULTURES

Two common methods of preserving microbial cultures for long periods are deep-freezing and lyophilization. In **deep-freezing,** a pure culture of microbes is placed in a suspending liquid and quick-frozen at temperatures ranging from −50° to −95°C. The culture can usually be thawed and used several years later.

In **lyophilization (freeze-drying)**, a suspension of microbes is quickly frozen at temperatures ranging from −54° to −72°C, and the water removed by a high vacuum. The remaining powder can be stored for years under vacuum at room temperature. The microbes can be revived at any time by hydrating them with suitable liquid nutrient media.

GROWTH OF BACTERIAL CULTURES

Bacterial Division

Bacteria normally reproduce by **transverse fission** (Figure 6–10). In preparation for fission, the cell elongates slightly and the plasma membrane near the center of the cell pinches inward. The cell wall thickens and grows inward at this same point. When the nuclear material is evenly distributed, a

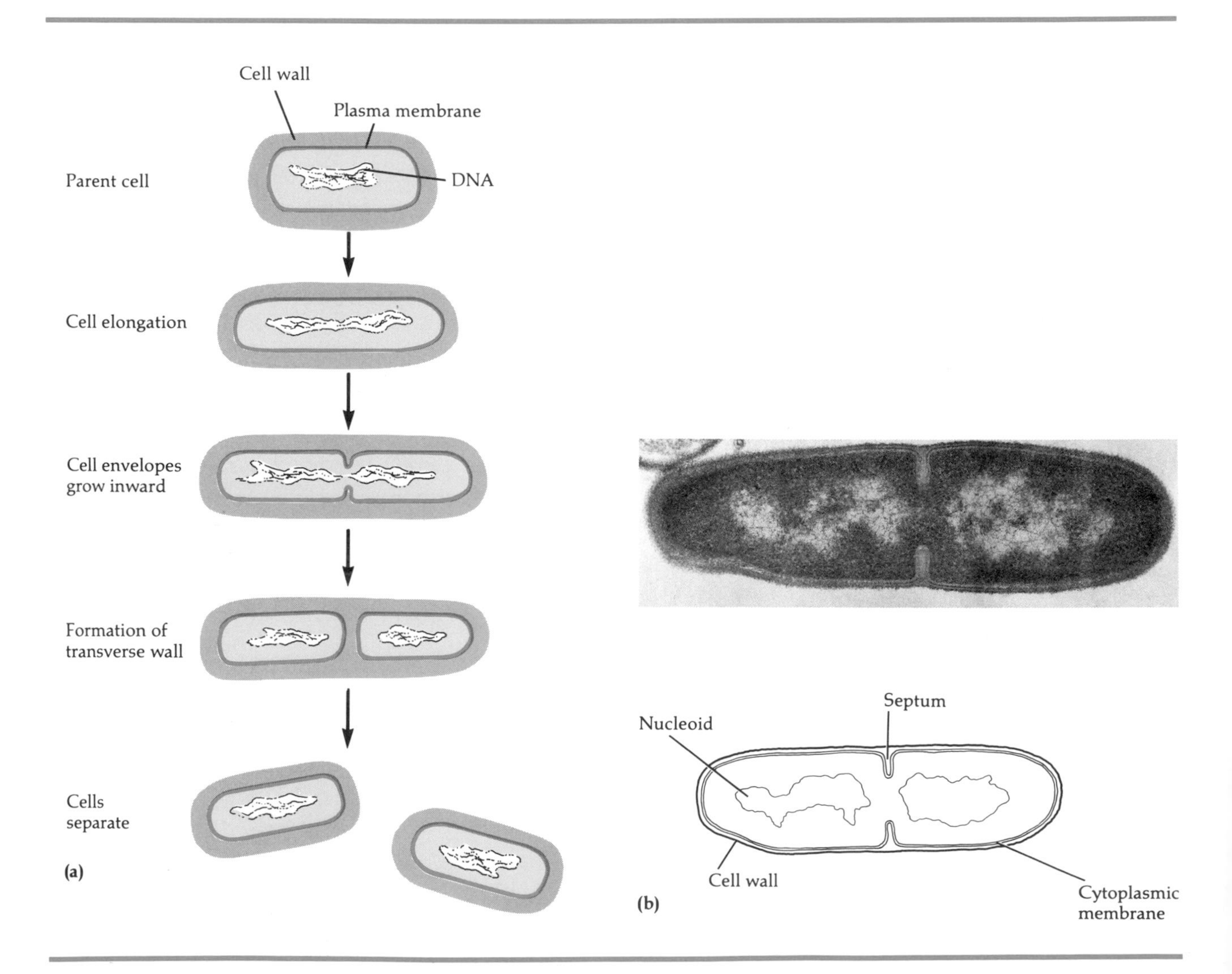

Figure 6–10 Transverse fission in bacteria. **(a)** Diagram of steps. **(b)** Electron micrograph of a thin section of a cell of *Bacillus licheniformis* starting to divide (×34,600), with art indicating the main structures.

transverse wall with plasma membranes is formed at the center of the cell.

There are minor variations on this process, and even gram-negative and gram-positive cells differ slightly. Also, a few bacterial species **bud;** that is, they form a small initial outgrowth that enlarges until it approaches the size of the parent cell, then separates. Then there are the filamentous bacteria (actinomycetes) that reproduce by producing spores from chains of cells. And a few filamentous species simply fragment, the fragments initiating the growth of new cells.

Generation Time

As a result of transverse fission, one cell divides and produces two cells, the two cells divide and produce four cells, and so on. Thus, if we start with a single bacterium, and transverse fission continues unchecked, an enormous number of cells will be produced in geometric progression (exponential growth):

$$\begin{array}{cccccc} 1 & 2 & 4 & 8 & 16 & 32 \ldots \\ & \rightarrow & \rightarrow & \rightarrow & \rightarrow & \rightarrow \\ 2^0 & 2^1 & 2^2 & 2^3 & 2^4 & 2^5 \ldots \end{array}$$

When the number of cells in each generation is expressed as a power of 2, the exponent tells the number of doublings that have occurred. (Before reading on, you may wish to review the box on exponential notation in Chapter 2.)

The time required for the cells to divide and the population to double is called the **generation (doubling) time.** It varies considerably among organisms. For example, although most bacteria have a generation time of 1 to 3 hours, others require over 24 hours per generation. Under favorable conditions, an *E. coli* cell will divide to form two cells every 20 minutes. After 4 hours (12 generations), 4096 cells will have been produced.

There are practical reasons to study the generation time of microbes. A urine sample may be collected and left in a laboratory, where it is tested for *E. coli* several hours later. If the sample initially contained a few *E. coli* cells (as it normally would), then during the time between collecting and testing, the *E. coli* may have reproduced in such quantities that test results indicate a serious infection.

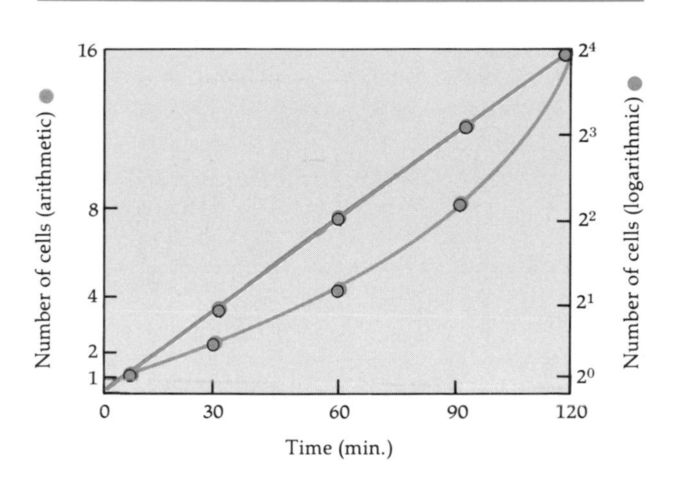

Figure 6–11 Growth curve for an exponentially increasing bacterial population, plotted logarithmically and arithmetically.

The easiest way of graphing immense increases in a microbial population is to use a logarithmic, rather than an arithmetic, scale (Figure 6–11). Note that if the size of an exponentially increasing microbial population is plotted logarithmically, the curve forms a straight line. The decline and death of microbial populations can be plotted the same way.

Phases of Growth

When a few bacteria are inoculated into a liquid growth medium, and the population is counted at intervals, it is possible to plot a typical **bacterial growth curve** that shows the growth of cells over time (Figure 6–12). There are four basic phases of growth.

Lag phase

For a while, there is little or no change in the number of cells because they do not begin to reproduce in a new medium right away. This period of little or no cell division is called the **lag phase,** and it may last for an hour or several days. But during this time, the cells are not dormant. There is intense metabolic activity in the microbial population, in

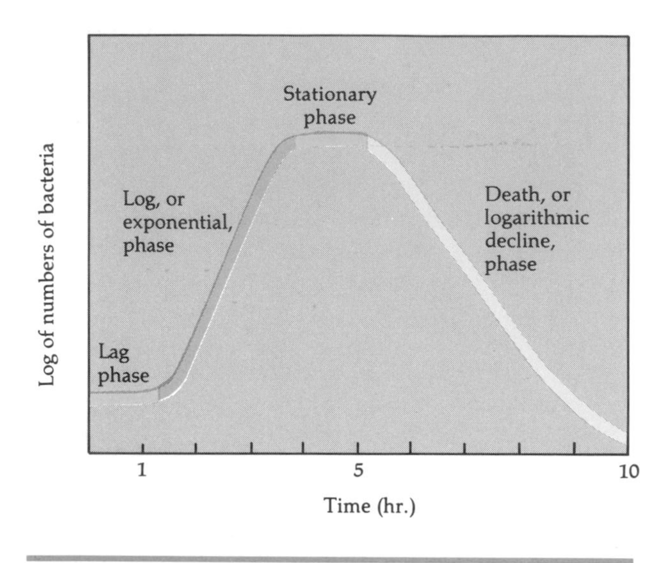

Figure 6–12 Bacterial growth curve, showing the four typical phases of growth.

particular, enzyme synthesis. The situation is analogous to equipping a factory to produce automobiles: there is considerable initial activity, but no increase in the automobile population. Near the end of the lag phase, some cells will double or triple in size, in preparation for reproduction.

Log phase

Eventually, the cells begin to divide and enter a period of growth, or logarithmic increase, called the **log phase** or **exponential growth phase.** During this period, the cells are reproducing most actively, and their generation time reaches a minimum and remains constant. There is apparently a characteristic minimum generation time—maximum rate of doubling—genetically determined for each species. Because the doubling time is constant, a logarithmic plot of growth during the log phase produces a straight line. During the log phase, cells show their visible characteristics: the shape, color, density, and groupings of their colonies. The log phase is also the time when cells are most active metabolically, and in industrial production, this is the period of peak activity and efficiency.

On the other hand, microorganisms are more sensitive than usual to adverse conditions during their log phase of growth. Radiation and many antimicrobial drugs—for example, the antibiotic penicillin—exert their effect by interfering with some important step in the growth process, and are therefore most harmful to cells during this phase.

Stationary phase

If exponential growth continues unchecked, some startling numbers of cells can arise. For example, a single bacterium dividing every 20 minutes for two days can theoretically produce an enormously large population of cells (2^{144}, a 44-digit number!). But this does not happen. Eventually, growth slows down, and sooner or later the number of microbial deaths balances the number of new cells, and the population stabilizes. This period of equilibrium is called the **stationary phase.**

The reason for cessation of exponential growth is not always clear. The accumulation of waste products toxic to cells and the exhaustion of certain required nutrients are usually involved, along with related changes in pH and temperature. In a specialized apparatus called a *chemostat,* it is possible to keep a population in the exponential growth phase indefinitely by continually draining off spent medium and adding fresh medium.

Death phase

Before long, the number of deaths begins to exceed the number of new cells formed, and the population enters the **death phase,** or **logarithmic decline phase.** This continues until the population is diminished to a tiny fraction of the more resistant cells, or it may die out entirely. Some species pass through the entire series of phases in only a few days, while others retain some surviving cells almost indefinitely. Microbial death will be discussed further in the next chapter.

Measurement of Microbial Growth

There are a number of ways bacterial growth can be measured. Some methods measure cell number; other methods measure the total mass of the population, which is often directly proportional to cell number. Population numbers are usually recorded as the number of cells in a milliliter of liquid or in a gram of solid material. Because bacterial populations are usually very large, most methods of counting them are based on direct or indirect counts of very small samples, followed by calculations to determine the size of the total population. Assume, for example, that a millionth of a milliliter of sour milk is found to contain 70 bacterial cells. Then there must be 70 times one million cells, or 70 million cells, per milliliter.

The problem in all of this is that it is not practical to measure out a millionth of a milliliter of liquid or a millionth of a gram of food. Therefore, the procedure is done indirectly, in a series of dilutions. For example, if we add one milliliter of milk to 99 milliliters of water, each milliliter of this dilution now has a hundredth as many bacteria as in each milliliter of the original sample. By making a series of such dilutions, we can readily estimate the number of bacteria in our sample. (This principle is illustrated in Figure 6–14.) To count the microbial population in foods (such as hamburger), a slurry of one part food to nine parts water is finely ground in a food blender.

As we now approach the various ways to measure bacterial growth, let us first examine two methods used to determine bacterial numbers in highly concentrated solutions: the direct microscopic count and the standard plate count.

Direct microscopic count

In the method known as the **direct microscopic count,** a measured volume of a bacterial suspension must be in a definite area on a microscope slide. For example, in the Breed count method for counting milk bacteria, a 0.01 ml sample is spread over a marked square centimeter of slide and stained. The area of the field of view of an oil immersion objective can be determined. It is then possible to inspect a number of microscope fields and average the number of bacteria seen. This number can be multiplied by a factor that will estimate the bacterial count in an entire milliliter of milk.

A specially designed slide called a *Petroff–Hausser counter* is also used in direct microscopic counts (Figure 6–13). A shallow well of known volume is indented into the surface of a microscope slide and covered with a thin glass inscribed with squares of known area. The well is filled with the microbial suspension. The bacterial numbers in a series of

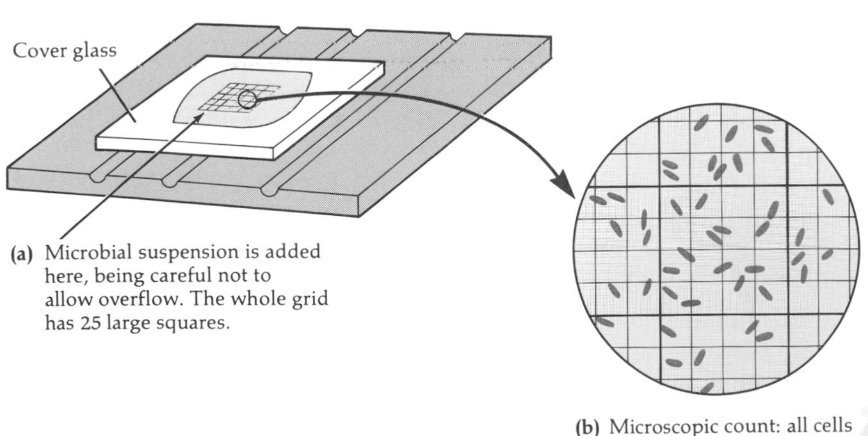

(a) Microbial suspension is added here, being careful not to allow overflow. The whole grid has 25 large squares.

(b) Microscopic count: all cells in several large squares are counted and the numbers averaged. The large square shown here has 14 bacterial cells.

(c) Number of cells per milliliter of bacterial suspension: 14 cells × 20,000,000 = 2.8×10^8 cells/ml.

Figure 6–13 Direct microscopic count of bacteria using a Petroff–Hausser cell counter. The average number of bacteria under a square is multiplied by a factor of 20 million.

these squares are averaged and multiplied by the necessary factor to arrive at the count per milliliter. Motile bacteria are difficult to count by this method, and as with other microscopic methods, dead cells are about as likely to be counted as live ones. In addition to these disadvantages, it also requires a rather high concentration of cells before it can be used, about 10 million bacteria per milliliter. The chief advantage of microscopic counts is that no incubation time is required, and their use is usually reserved for applications where this is the primary consideration. This advantage also holds for *electronic cell counters,* which automatically count the number of cells in a measured volume of liquid. These instruments are available in some research laboratories.

Standard plate count

A second method for measuring bacterial growth in highly concentrated solutions is the **standard plate count,** the most frequently used method for counting bacteria. A principal advantage of this technique over the direct microscopic count is that it reflects the number of viable (living) cells. Another advantage is that even small numbers of cells can be detected, since a relatively large volume of microbial suspension can be used. In one procedure, a known volume of 0.1 to 1.0 ml of inoculum is introduced into a Petri plate (Figure 6–14a). Then melted nutrient agar is poured into the plate and the two are thoroughly mixed by rotating the plate. (Either of the procedures shown in Figure 6–8 may also be used.) When the agar solidifies, the plate is incubated. The standard plate count is based on the assumption that each bacterium grows and divides to produce a single colony. It is also assumed that the original inoculum is homogeneous and that no aggregates of cells are present.

When a standard plate count is performed, it is important that only a limited number of colonies develop in the plate. When too many colonies are present, some cells are overcrowded and do not develop, resulting in inaccuracies in the count. Generally, only those plates with from 30 to 300 colonies are counted. To ensure colony counts in this range, the original inoculum is diluted several times in a process called **serial dilution** (Figure 6–14b).

The standard plate count has a number of shortcomings. As already noted, some bacteria are grouped tightly in aggregations, and unless these are broken up, which is not always possible, each colony on the plate will represent a group of cells rather than an individual cell in the original population. Technically, then, counts are sometimes referred to as **colony-forming units.** Another disadvantage is that the plate must be incubated long enough for a visible colony to form. This takes about 24 hours, sometimes longer. For perishable foods and in health-related bacteriology, this delay is often a problem.

Also, when foods or soils containing a wide variety of microorganisms are plated out, the nutrients and other growth conditions are not always suitable for the growth of all the organisms present. In the case of soil, as an extreme example, it is estimated that as few as 1% of the organisms present will grow on the usual test plate. In the case of foods, where the organisms are usually more uniform in their nutritional and growth requirements, results are much better. And when counting a pure culture of bacteria suspended in liquid, considerable accuracy can be achieved by the standard plate count.

Filtration

If bacterial numbers are very low, such as in lakes or relatively pure streams, bacteria can be counted by **filtration** methods. One hundred milliliters or more of water is passed through a thin membrane filter containing pores too small for bacteria to pass. The bacteria are sieved out and retained on the surface of the filter. This filter is then transferred to a pad of nutrient media, where colonies arise from the bacteria on the surface of the filter (Figure 6–15). This is essentially a method of concentrating the bacteria before doing a plate count.

Most probable number (MPN)

Another method for determining the number of bacteria in a population is the **most probable number (MPN)** method. It is a statistical estimating method based on the fact that the greater the number of bacteria, the more dilutions it will take to dilute them out entirely and leave a tube without

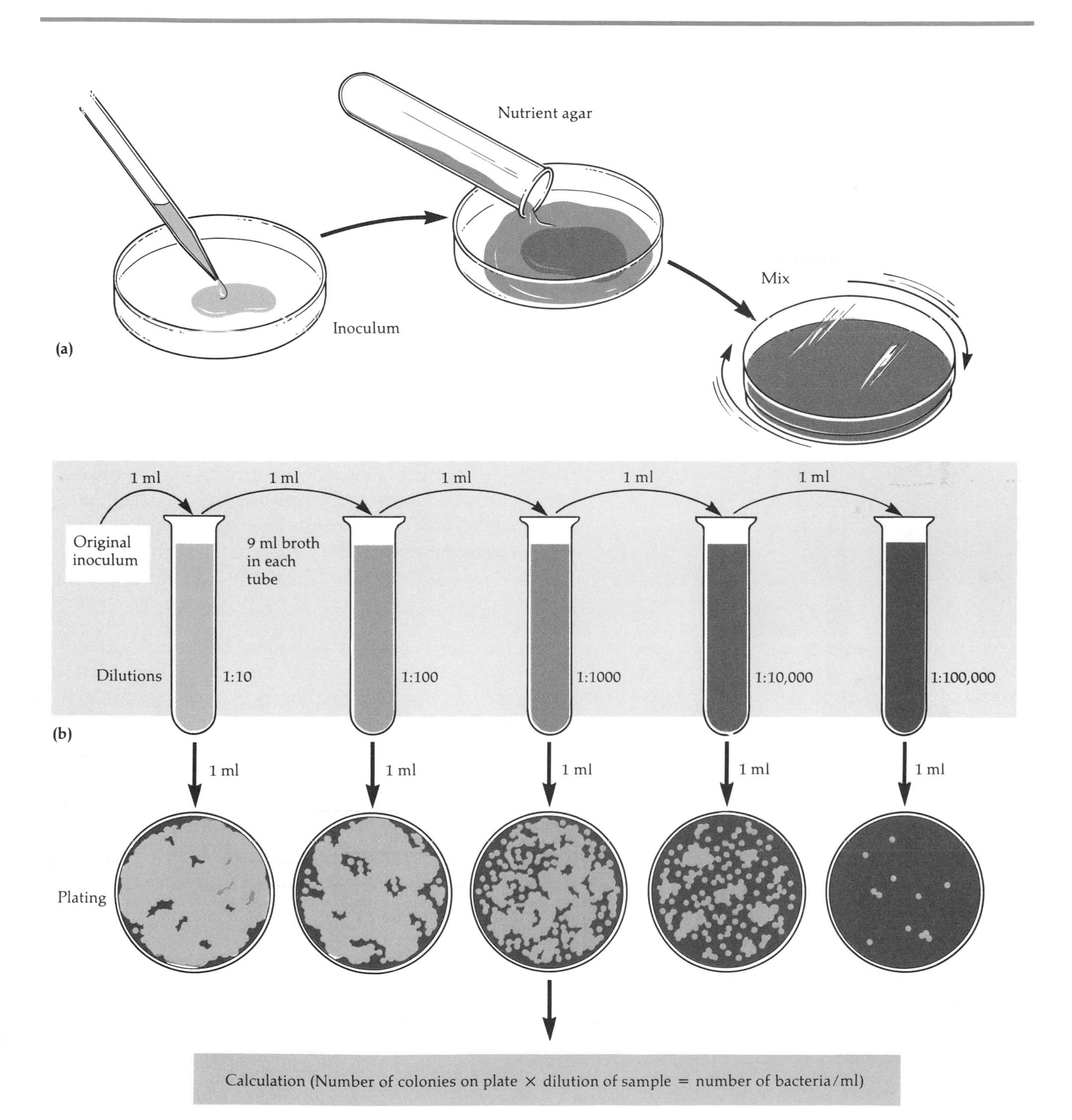

Figure 6–14 Standard plate count. **(a)** Pour plate. Inoculum is pipetted into a Petri plate, liquefied nutrient agar is added, and the agar is mixed with the inoculum by gentle swirling of the plate. **(b)** Serial dilutions. If the Petri plate at the 1:10,000 dilution has 50 colonies, there are 50 × 10,000, or 5×10^5, bacteria in the original inoculum.

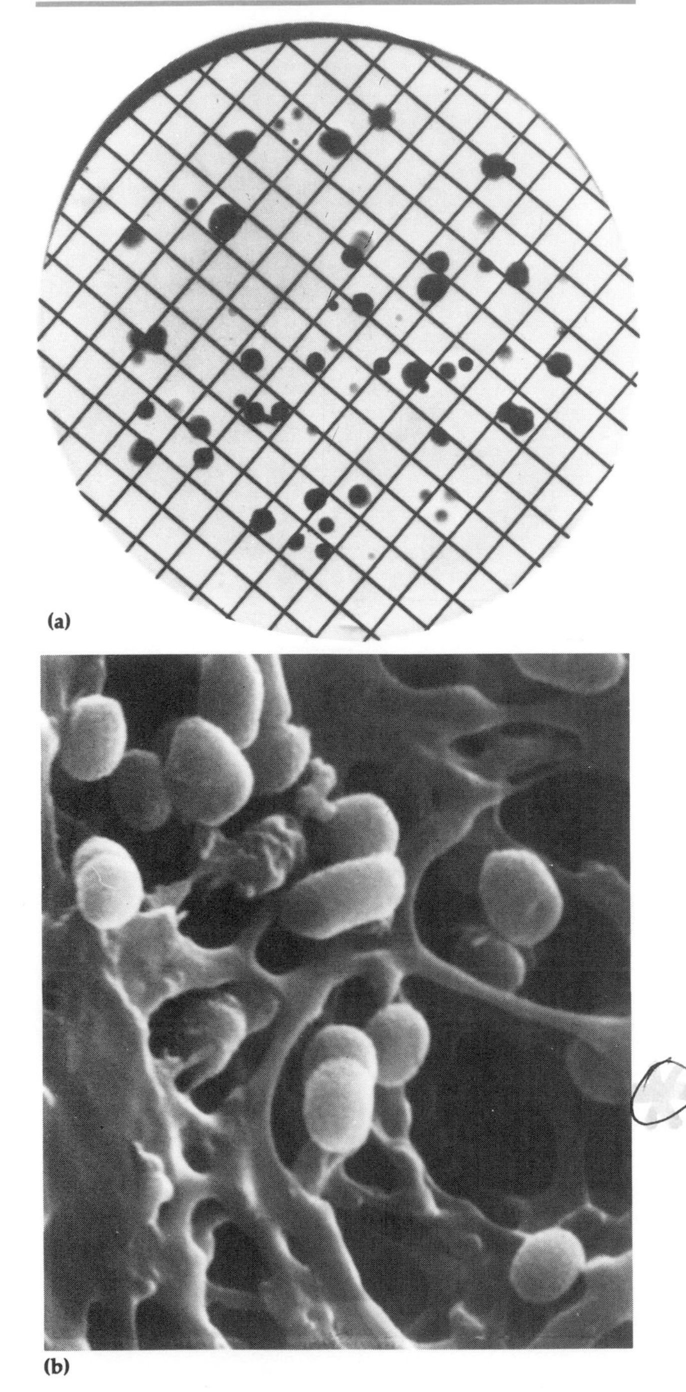

Figure 6–15 Bacteria on filters. **(a)** Fecal coliform bacteria were collected on this filter and grew into visible colonies. **(b)** Scanning electron micrograph showing cells of the bacterium *Pseudomonas diminuta* collected on the surface of a Millipore type HA filter. (Both courtesy of the Millipore Corporation.)

growth. In the MPN procedure, a sample is diluted until either zero cells or only one cell is present in a given portion of the dilution. Next, a number of tubes are inoculated with identical portions taken at the highest dilution. Some portions will contain a cell, others will not. A count is taken as to what fraction of tubes show growth, the figure is compared to statistical tables that have already been developed, and a cell count is estimated. The MPN technique is often used in food and water sanitation—for example, in determining water purity—and is used when counts of bacteria are lower than can be reliably counted by the standard plate count. The MPN is only a statement that there is a 95% chance that the bacterial population falls within a certain range, and that the MPN is statistically the most probable number. A 15-tube MPN table is provided in Appendix D.

We will now look at three ways to measure microbial growth by measuring total cell mass. These methods measure turbidity, metabolic activity, and dry weight.

Turbidity

For some types of experimental work, estimating **turbidity** is a practical way to follow bacterial growth. As bacterial numbers increase in a liquid medium, the medium becomes turbid (cloudy) with cells. Since something like 10 or 100 million cells per milliliter are needed to make a suspension turbid enough to be read, this method has its limitations. It is worth emphasizing here that over a million cells per milliliter must be present before the first traces of turbidity are visible to the eye. Therefore, visible turbidity is not a useful measure of contamination of liquids by relatively small numbers of bacteria.

To estimate turbidity, a beam of light is transmitted through a bacterial suspension to a photoelectric cell (Figure 6–16). As bacterial numbers increase, there will be less light reaching the photoelectric cell. This change of light will be registered on the scale of the instrument as the percentage of transmission. Also printed on the instrument scale is a logarithmic expression called the *absorbance* (sometimes *Optical Density* or *OD*), a value derived from the percentage of transmission. It is the absorbance that is used to plot bacterial growth. When the bacteria are in logarithmic growth or

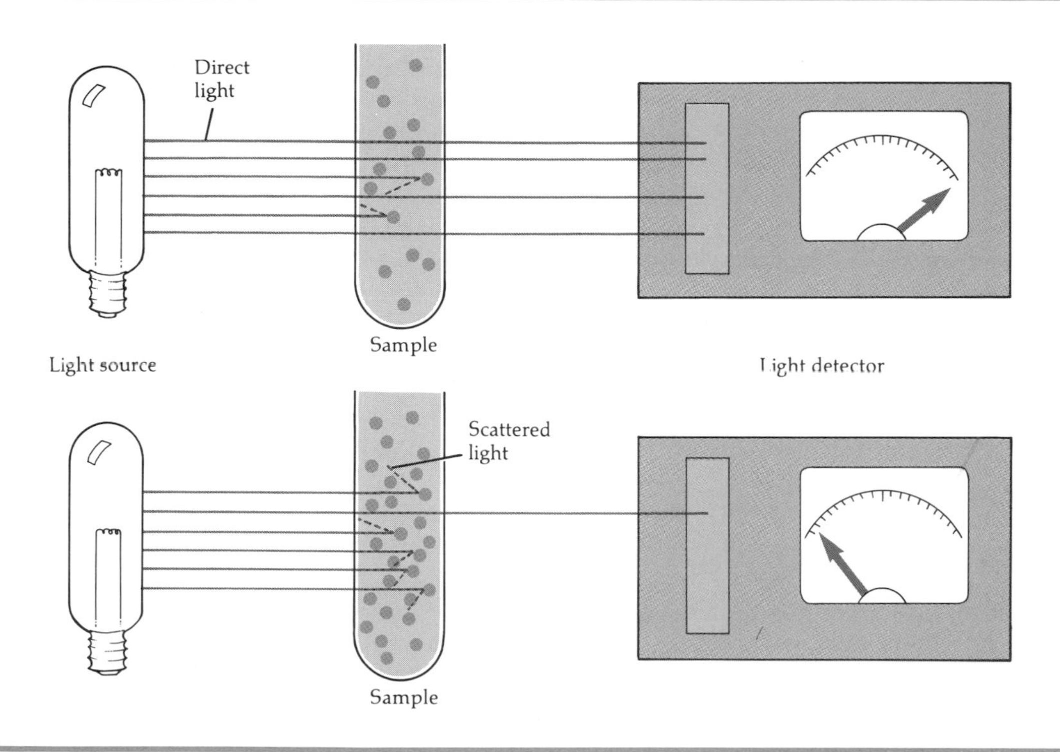

Figure 6–16 Estimation of bacterial numbers by turbidity. The amount of light picked up by the light detector is proportional to the number of bacteria. The less light transmitted, the more bacteria in the sample.

decline, the plot of the log of absorbance versus time will form an approximately straight line. If absorbance readings are matched with plate counts of the population, this correlation can be used in the future, assuming conditions are kept the same, to estimate bacterial numbers directly by turbidity. The instrument used for turbidity measurements is usually an electrically operated *spectrophotometer* or *colorimeter*.

Metabolic activity

Another indirect way of estimating bacterial numbers is by measuring the **metabolic activity** of the population. Rather than attempt to count bacterial colonies or estimate turbidity, we measure, say, the acid produced by the bacteria and assume it to be in direct proportion to the number of bacteria present.

Other ways to measure microbial numbers by metabolic activity are through the reduction tests. (The principle underlying reduction tests has already been discussed in relation to reducing media.) Reduction tests are sometimes used to estimate milk quality. A tube of milk has a dye added that changes color in the presence or absence of oxygen. Methylene blue, for example, is blue in the presence of oxygen and colorless in its absence. The filled tube of milk is tightly capped, and the bacteria present use the oxygen as they metabolize the milk. The faster the dye changes color, the faster the oxygen is being depleted and the more bacteria are presumed present in the milk.

Dry weight

For filamentous organisms, such as fungi, the usual measuring methods are less satisfactory. One of the better ways of measuring growth of filamentous organisms is by **dry weight.** In this procedure, the fungus is removed from the growth medium, filtered to remove extraneous material, placed in a weighing bottle, and dried in a desiccator. With bacteria, the same basic procedure is followed. It is customary to remove the bacteria from the culture medium by centrifugation.

You now have a basic understanding of the requirements for and measurements of microbial growth. In Chapter 7, we will proceed to look at how this growth is controlled in laboratories, clinics, and industry.

STUDY OUTLINE

REQUIREMENTS FOR GROWTH (pp. 143–150)

1. Growth is an increase in the number of cells in a population or an increase in mass.
2. Requirements for microbial growth are divided into physical and chemical needs.

Physical Requirements (pp. 144–146)

1. On the basis of preferred growth range of temperatures, microbes are classified as psychrophiles (cold-loving), mesophiles (moderate-temperature-loving), and thermophiles (heat-loving).
2. The minimum growth temperature is the lowest temperature at which a species will grow; the optimum growth temperature is the temperature at which it grows best; and the maximum growth temperature is the highest temperature at which growth is possible.
3. Most bacteria grow best at a pH between 6.5 and 7.5.
4. In a hypertonic solution, microbes undergo plasmolysis; halophiles can tolerate high salt concentrations.

Chemical Requirements (pp. 146–150)

1. All organisms require a carbon source; chemoheterotrophs use an organic molecule while chemoautotrophs usually use carbon dioxide.
2. Other chemicals required for microbial growth include nitrogen, sulfur, phosphorus, trace elements, and, for some microorganisms, oxygen and/or organic growth factors.
3. On the basis of oxygen requirements, organisms are classified as obligate aerobes, facultative anaerobes, obligate anaerobes, aerotolerant anaerobes, and microaerophiles.

CULTURE MEDIA (pp. 150–157)

1. Any material prepared for the growth of bacteria in a laboratory is referred to as a culture medium.
2. Microbes that grow and multiply in a culture medium are known as a culture.

3. Agar is a common solidifying agent for a culture medium.

Chemically Defined and Complex Media (pp. 151–152)

1. A chemically defined medium is one in which the exact chemical composition is known.
2. A complex medium is one in which the exact chemical composition is not known.

Reducing Media (pp. 152–153)

1. Reducing media remove molecular oxygen (O_2) that might interfere with the growth of anaerobes.
2. The procedure is carried out in an anaerobic container.

Selective and Differential Media (pp. 153–154)

1. Selective media allow growth of only the desired microbe by inhibiting unwanted organisms with the presence of salts, dyes, or other chemicals.
2. Differential media are used to distinguish between different organisms.

Inoculation (pp. 154–155)

1. The transfer of microbes to a previously sterilized culture medium is called inoculation.
2. It is performed with an inoculating loop or needle.

Mixed and Pure Cultures (p. 156)

1. Colonies are the progeny of a single cell that have grown in or on a culture medium.
2. A mixed culture contains several different species of bacteria. A pure culture contains only one species.

Isolating Pure Cultures (pp. 156–157)

1. Pure cultures may be isolated by any of several methods.
2. Included are the pour plate method, spread plate method, streak plate method, and enrichment culture.

PRESERVING BACTERIAL CULTURES (pp. 157–158)

1. Microbes may be preserved for long periods of time.
2. This may be accomplished by deep-freezing or lyophilization (freeze-drying).

GROWTH OF BACTERIAL CULTURES (pp. 158–166)

Bacterial Division (pp. 158–159)

1. The normal reproductive method of bacteria is transverse fission, in which a single cell divides and gives rise to two identical cells.
2. Some bacteria reproduce by budding, spore formation, or fragmentation.

Generation Time (p. 159)

1. The time required for a cell or population to divide is known as generation time.
2. Bacterial division occurs according to a logarithmic progression (2 cells, 4 cells, 8 cells, etc.).

Phases of Growth (pp. 159–160)

1. During the lag phase, there is little or no change in the numbers of cells, but metabolic activity is high.
2. During the log phase, the bacteria multiply at the fastest rate possible under the conditions provided.
3. During the stationary phase, there is an equilibrium between cell division and death.
4. During the death phase, the number of deaths exceeds the number of new cells formed.

Measurement of Microbial Growth (pp. 160–166)

1. In a direct microscopic count, the microbes in a measured volume of a bacterial suspension are counted using a specially designed slide.
2. A standard plate count reflects the number of viable microbes and assumes that each bacterium grows and divides into a single colony.
3. In filtration, bacteria are retained on the surface of a membrane filter and then transferred to culture medium for growth and subsequent enumeration.
4. The most probable number method can be used for microbes that will grow in a liquid medium; it is a statistical estimation.
5. Turbidity is used to determine bacterial growth by transmitting a light beam through a suspension of cells.
6. An indirect way of estimating bacterial numbers is by measuring the metabolic activity of the population, for example, acid production or oxygen consumption.
7. For filamentous organisms such as fungi, it is more convenient to measure growth by measuring dry weight.

STUDY QUESTIONS

REVIEW

1. Describe transverse fission.
2. What physical and chemical properties of agar make it useful in culture media?
3. Draw a typical bacterial growth curve. Label and define each of the four phases.
4. Which of these curves best depicts the log phase of
   (a) A thermophile incubated at room temperature?
   (b) *Listeria monocytogenes* growing in a human?
   (c) A psychrophile when incubated at 9°C?

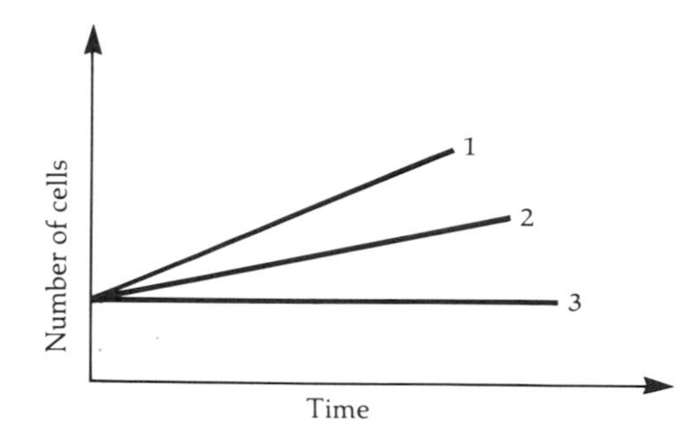

5. Macronutrients (needed in relatively large amounts) are often listed as CHONPS. What does each of these letters indicate, and why are they needed by the cell?
6. Most bacteria grow best at pH ____________.
7. Why can high concentrations of salt or sugar be used to preserve food?
8. Explain five categories of bacteria based on their requirements for oxygen.
9. Outline three procedures that might be used to culture obligate anaerobes.
10. Seven methods of measuring microbial growth were explained in this chapter. Categorize each as a direct or indirect method.
11. Flask *A* contains yeast in glucose-minimal salts broth, incubated at 30°C with aeration. Flask *B* contains yeast in glucose-minimal salts broth, incubated at 30°C in a Brewer Anaerobic Jar.
    (a) Which culture produced the most ATP?
    (b) Which culture produced the most alcohol?
    (c) Which culture had the shortest generation time?
12. A normal fecal sample has a high concentration of coliform bacteria. A patient with typhoid fever will excrete *Salmonella* along with normal bacteria. Feces from a suspected case of typhoid fever are inoculated on

an S-S agar plate. After proper incubation, coliform bacteria form red colonies and *Salmonella* form colorless colonies with a black center. Is S-S agar selective? Differential? How can you tell?

13. Differentiate between complex and chemically defined media.
14. Bacteria can be stored without harm to them for extended periods of time by deep-freezing. Why then does refrigeration and freezing work to preserve foods?

CHALLENGE

1. *E. coli* was incubated with aeration in a nutrient medium containing two carbon sources and the following growth curve was made from this culture.
   (a) Explain what happened at the time marked x.
   (b) Which substrate provided "better" growth conditions for the bacteria? How can you tell?

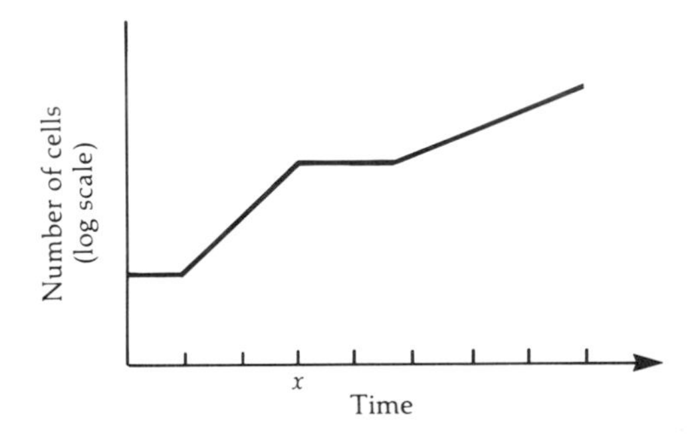

2. *Clostridium* and *Streptococcus* are both catalase-negative. *Streptococcus* grows by fermentation. Why is *Clostridium* killed by oxygen while *Streptococcus* is not?
3. Design an enrichment medium and procedure for an endospore-forming bacterium that fixes nitrogen and uses cellulose for its carbon source.
4. Most laboratory media contain a fermentable carbohydrate and peptone because the majority of bacteria require carbon, nitrogen, and energy sources in these forms. How are these three needs met by glucose-minimal salts medium? (See Table 6–3.)

FURTHER READING

Difco Manual of Dehydrated Culture Media and Reagents for Microbiological and Clinical Laboratory Procedures, 9th ed. 1971. Detroit, MI: Difco Laboratories. Formulae and applications of media from the manufacturer.

Hartman, P. A. 1968. *Miniaturized Microbiological Methods*. New York: Academic Press. This monograph on cultivation techniques is part of the series *Advances in Applied Microbiology*.

Kushner, D. J. 1978. *Microbial Life in Extreme Environments.* New York: Academic Press. Includes chapters on physiological adaptations of microbes exposed to extreme temperature, pressure, salt, pH, and radiation conditions in nature.

Laskin, A. I. and H. Lechevalier, eds. 1974. *Microbial Ecology.* Cleveland, OH: CRC Press. The effects of environmental factors, pesticides, and mixed cultures on microbial growth.

Payne, W. J. and W. J. Wiebe. 1978. Growth yield and efficiency in chemosynthetic microorganisms. *Annual Review of Microbiology* 32:155–183. A review of the effects of substrate and physical conditions on microbial growth and techniques to determine growth rate.

Shapton, D. A. and R. G. Board, eds. 1971. *Isolation of Anaerobes.* New York: Academic Press. Techniques for the isolation and cultivation of anaerobes from soil and mammals; culturing of photosynthetic and microaerophilic bacteria is included.

7

Control of Microbial Growth

Objectives

After completing this chapter you should be able to

- Define terms related to the destruction or suppression of microbial growth.
- Explain how microbial growth is affected by the type of microbe, the physiological state of the microbe, and the environmental conditions.
- Describe the actions of microbial control agents with respect to cell structures.
- Describe the pattern of microbial death as a result of treatment with a microbial control agent.
- Describe the physical methods of microbial control.
- Describe the factors related to effective disinfection and their effect on the evaluation of a disinfectant.
- Describe the method of action and preferred use for disinfectants.

The scientific control of microbial growth began only about one hundred years ago. Prior to this time, it was not uncommon to have massive epidemics that killed thousands of people. Surgical death rates were high because of an assortment of postoperative infections, and even childbirth carried a considerable risk. Today, however, the picture has changed.

Two individuals who introduced the concept of microbial control were Ignatz Semmelweis (1816–1865), a Hungarian physician working in Vienna, and Joseph Lister (1827–1912), an English physician. At the obstetrics ward in the Vienna General Hospital, Semmelweis required all personnel to wash their hands in chlorinated lime, a procedure that significantly lowered the infection rate. Lister, meanwhile, had read about Pasteur's work with microbes, and assumed that wound infections (sepsis) following surgery could be decreased through procedures that prevented the access of microbes to the wound. This system, known as **antiseptic surgery,** included the heat sterilization of surgical instruments and the application of phenol (carbolic acid) to wounds following surgery.

We have come a long way in controlling microbial growth since the time of Semmelweis and Lister. Today's procedures, far more sophisticated and effective, are used not only to control disease organisms, but also to curb microbial growth that results in food spoilage. In this chapter, we will consider how microbial growth can be controlled by physical methods and chemical agents. Physical methods include the use of heat, low temperatures, desiccation, osmotic pres-

5 ways

sure, and radiation. Chemical agents include several groups of substances that destroy or limit microbial growth on body surfaces or inanimate objects. Our present discussion will be limited to agents that prevent microbial infection. Later, in Chapter 18, we will consider methods for controlling microbes after infection has occurred. Chemotherapy—the use of antimicrobial drugs—is one such method.

Before we actually begin our survey of physical and chemical methods used to control microbial growth, it would be useful to first define the important terms related to control. Microbial control is needed to prevent the transmission of infection, contamination, and spoilage. But as you will see, it is not always necessary to kill the microbes; some situations require that they simply be inhibited or removed. A variety of agents and procedures are available, and each has its own practical application and range of controlling effects, depending on the situation. Table 7–1 contains a list of terms related to the destruction or suppression of microbial growth. (Note: In this chapter, we shall use the term *disinfection* to include antisepsis.)

Table 7–1 Terminology Related to the Control of Microbial Growth

Term	Definition
Terms Related to Destruction:	
Sterilization	The process of destroying all forms of microbial life on an object or in a material. This includes the destruction of endospores—the most resistant form of microbial life. Sterilization is absolute; there are no degrees of sterilization.
Disinfection	The process of destroying vegetative pathogens, but not necessarily endospores or viruses. Usually, a disinfectant is a chemical applied to an object or a material. Disinfectants tend to reduce or inhibit growth; they usually do not sterilize.
Antisepsis	Chemical disinfection of the skin, mucous membranes, or other living tissues.
Germicide (*cide* = killer)	A chemical agent that kills microbes rapidly, but not necessarily their endospores. A *bactericide* kills bacteria, a *sporicide* kills endospores, a *fungicide* kills fungi, a *virucide* kills viruses, and an *amoebicide* kills amoebas and other protozoans.
Terms Related to Suppression:	
Bacteriostasis (*stasis* = halt)	A condition in which bacterial growth and multiplication are inhibited, but the bacteria are not killed. If the bacteriostatic agent is removed, bacterial growth and multiplication may resume. *Fungistasis* refers to the inhibition of fungal growth, *virustasis* to the inhibition of viral growth.
Asepsis (*asepsis* = without infection)	The absence of pathogens from an object or area. Aseptic technique is designed to prevent the entry of pathogens into the body. Whereas *surgical asepsis* is designed to exclude all microbes, *medical asepsis* is designed to exclude microbes associated with communicable diseases.
Degermation	The removal of transient microbes from the skin by mechanical cleansing or by the use of an antiseptic.
Sanitization	The reduction of pathogens on objects to safe public health levels by mechanical cleansing or chemicals.
Deodorant	A chemical that destroys or masks offensive odors.

CONDITIONS INFLUENCING MICROBIAL CONTROL

Usually, if we want to sterilize an object, say, a laboratory flask, we do not bother to identify the species of microbes on it. Instead, we launch a general attack calculated to be strong enough to kill the most resistant microbial forms that could possibly be present—endospores. At any point in time, the flask could carry a multitude of microorganisms in different phases of their growth cycle, different metabolic states, even different microenvironments. Thus, it is quite possible that an antimicrobial agent will not kill all of the microbial population immediately. The time it takes to reduce or eliminate microbial growth on any object depends, in part, on the number and species of microbes dwelling on the object. It also depends on a number of other factors.

Temperature

Everyone is familiar with the use of ice or other means of refrigeration to control microbial growth. The biochemical reactions required for growth, like all chemical reactions, occur much more slowly at low temperatures. Persons who use chemical disinfectants should be aware that the action of many chemical agents is also inhibited by lower temperatures. Their activity is due to temperature-dependent chemical reactions, and they tend to work somewhat better in warm solutions. You may have noticed that label directions on disinfectant containers frequently specify using a warm solution.

Kind of Microbe

Gram-positive bacteria, as a group, tend to be more susceptible to action by many disinfectants and antiseptics than gram-negative bacteria, although such a differentiation is by no means as clear cut as it is with antibiotic activity. A certain group of gram-negative bacteria, the pseudomonads (genus *Pseudomonas*), is unusually resistant to chemical activity and will even grow actively *in* some disinfectants and antiseptics. These bacteria are able to maintain themselves in such unlikely substrates as simple saline (salt) solutions, and they are also resistant to many antibiotics. Such properties have made these common, and normally harmless, bacteria very troublesome in hospitals. Pseudomonads are examples of **opportunistic pathogens,** that is, microbes that do not ordinarily cause disease, but become pathogenic in the absence of normal competitive flora, such as when antibiotics suppress the growth of other microbes without adversely affecting the pseudomonad. *Mycobacterium tuberculosis*, the microbe that causes tuberculosis, is another non-endospore-forming bacterium that exhibits greater than normal chemical resistance. The endospores of bacteria and cysts of protozoans are highly resistant to chemical agents, as are some viruses.

Physiological State of the Microbe

Microorganisms that are actively growing tend to be more susceptible to chemical agents than older cells. This may be because an actively metabolizing and reproducing cell has more points of vulnerability than an older, possibly dormant cell. Moreover, if microorganisms have formed endospores, the endospores are generally more resistant than the vegetative cells. The endospores of *Clostridium botulinum*, for example, can withstand boiling for $5\frac{1}{2}$ hours.

Environment

Organic matter frequently interferes with the action of chemical agents. In hospitals, the presence of organic matter in vomit and feces influences the selection of a disinfectant, as some disinfectants are more effective than others under these conditions. Even bacteria found in food tend to be protected by the organic matter of the food. For example, bacteria in cream can survive more heat, protected by proteins and fats, than bacteria in skim milk. Also, heat is much more effective under acidic conditions than at a neutral pH. This is also true of a few chemicals.

Let us now examine the mechanisms of action of control agents, that is, the way various agents of control actually kill or inhibit microbes.

ACTIONS OF MICROBIAL CONTROL AGENTS

Alteration of Membrane Permeability

The plasma membrane, located just inside the cell wall, is the target of action of many control agents. This membrane actively controls the passage of nutrients into the cell and the elimination of wastes from the cell. Damage to the lipids or proteins of the plasma membrane by antimicrobial agents (phenol, synthetic detergents, soaps, and quaternary ammonium compounds) typically causes leakage of cellular contents into the surrounding medium and interferes with the growth of the cell. Several forms of chemical agents and antibiotics work, at least in part, in this manner.

Damage to Proteins and Nucleic Acids

You may recall that the functional properties of proteins are due to their three-dimensional shape. This shape is maintained by certain chemical bonds that hold together adjoining portions of the amino acid chain as it folds back and forth upon itself. Hydrogen bonds are susceptible to breakage by heat or certain chemicals, and the result is denaturation of the protein. Covalent bonds, which are stronger, are also subject to attack. For example, disulfide bridges, which play an important role in protein structure by joining amino acids with exposed sulfhydryl (—SH) groups, can be broken by certain chemicals or sufficient heat. In contrast, antimicrobial drugs that affect proteins and nucleic acids most often interfere with their synthesis.

Since DNA and RNA carry the genetic message, damage to them, by radiation, for example, is frequently lethal to the cell, preventing both replication and normal functioning. Chemicals also can prevent the proper replication of genetic material, as can a number of antibiotics (Chapter 18).

PATTERN OF MICROBIAL DEATH

When bacterial populations are heated or treated with antimicrobial chemicals, they tend to die off at a constant rate over a period of time. For example, let us start with a population of 1 million microbes, and let us assume that after one minute of treatment, 90% of the population dies. We are now left with 100,000 microbes. If the population is treated for another minute, 90% of *those* microbes die, leaving us with 10,000 survivors. In other words, for each minute that the treatment is applied, 90% of the population is killed (Table 7–2). If the death curve is plotted logarithmically, it can be seen that the death rate is constant (Figure 7–1). Obviously, the more microbes there are at the beginning, the longer it will take to kill them all.

Table 7–2 Pattern of Microbial Death

Time (in Minutes)	Deaths per Minute	Number of Survivors
0	0	1,000,000
1	900,000	100,000
2	90,000	10,000
3	9,000	1,000
4	900	100
5	90	10
6	9	1

Source: O. Rahn, *Physiology of Bacteria*, McGraw-Hill, New York, 1932.

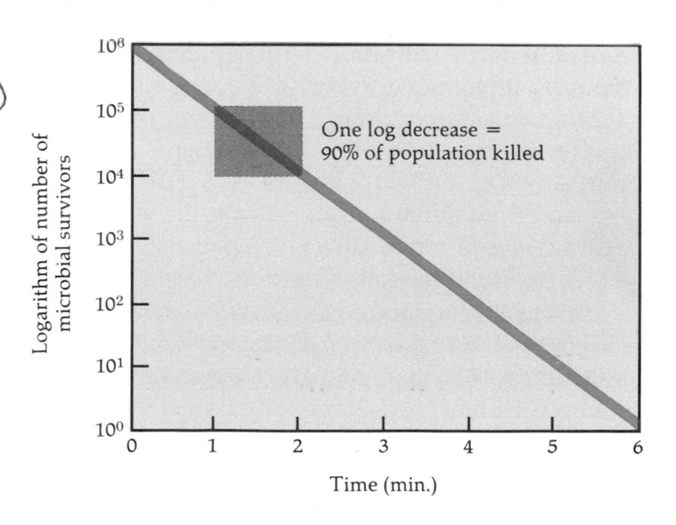

Figure 7–1 Microbial death curve plotted logarithmically.

Now that you have an understanding of the basic principles of microbial control, we will examine some representative physical and chemical methods.

PHYSICAL METHODS OF MICROBIAL CONTROL

Heat

Probably the most common method of eliminating microbes is by **heat.** A visit to any supermarket will demonstrate that canned goods, in which microorganisms have been killed by moist heat, easily outnumber the foods preserved by other methods. Also, laboratory equipment, such as media and glassware, and hospital instruments are usually sterilized by heat.

Heat is the most widely applicable and effective agent for sterilization. It is also the most economical and easily controlled. Heat appears to kill microbes by destroying their enzymes by denaturation. In sterilization by heat, the degree of heat resistance by bacteria must be considered. Heat resistance varies among different microbes, and a useful way of expressing these differences is through the concept of thermal death point. **Thermal death point (TDP)** is the lowest *temperature* required to kill a liquid culture of a certain species of bacteria in 10 minutes at a pH of 7. Another factor to be considered in sterilization is the length of time that the material must be treated to be rendered sterile. This is expressed as **thermal death time (TDT),** the length of *time* required to kill all bacteria in a liquid culture at a given temperature. Both TDP and TDT are useful guidelines that indicate the severity of treatment required to kill a given population of bacteria. **Decimal reduction time** is a third concept related to the degree of heat resistance of bacteria. It is the time, in minutes, required to kill 90% of the population of bacteria at a given temperature. Decimal reduction time is especially useful in the canning industry.

The heat used in sterilization may be applied in the form of moist heat or dry heat. Dry heat kills by oxidation effects. A simple analogy is the slow charring of paper in a heated oven, even though the temperature is below the ignition point of paper. Moist heat kills microorganisms because the presence of water hastens the breaking of hydrogen bonds that hold proteins in their three-dimensional structure.

Moist heat sterilization

One type of moist heat sterilization is **boiling.** Boiling (100°C) kills vegetative forms of bacterial pathogens, many viruses, and fungi within 10 minutes. Endospores and some viruses, however, are not destroyed as quickly. The hepatitis virus can survive up to 30 minutes of boiling, and some bacterial endospores have resisted boiling temperatures for up to 20 hours. The addition of certain chemicals to boiling water (such as 2% sodium carbonate) adds to the killing power of this method. Boiling is usually not recommended as a reliable sterilization procedure because of the varying resistances of microbes and their endospores.

Free-flowing (unpressurized) steam can also be used for sterilization. One method using unpressurized steam is called **fractional sterilization.** This method is used to sterilize culture media and certain foods in the canning process. In this procedure, the material to be sterilized is exposed to steam 30 minutes a day on three successive days. During each treatment, all vegetative cells are killed. Between steam treatments, the material remains at room temperature so that any endospores present will germinate. Thus, when the material is exposed to steam again on the second and third days, the vegetative cells arising from the endospores are killed. One problem with this procedure is that some endospores germinate after the third exposure to steam.

The most effective method of moist heat sterilization is **autoclaving,** the use of steam under pressure. It is carried out in a device called an **autoclave** (Figure 7–2). Autoclaving is the preferred method unless the material being autoclaved would be damaged by heat or moisture. The higher the pressure in the autoclave, the higher the temperature at which water vaporizes (boils), and the higher the temperature of the steam. For example, free-flowing steam has a temperature of 100°C. But if the steam is placed under a pressure of 1 atmosphere, that is, about 15 pounds per square inch, the temperature is increased to 121°C. An increase in pressure to 20 pounds per square inch raises the temperature to 126°C. The relationship between

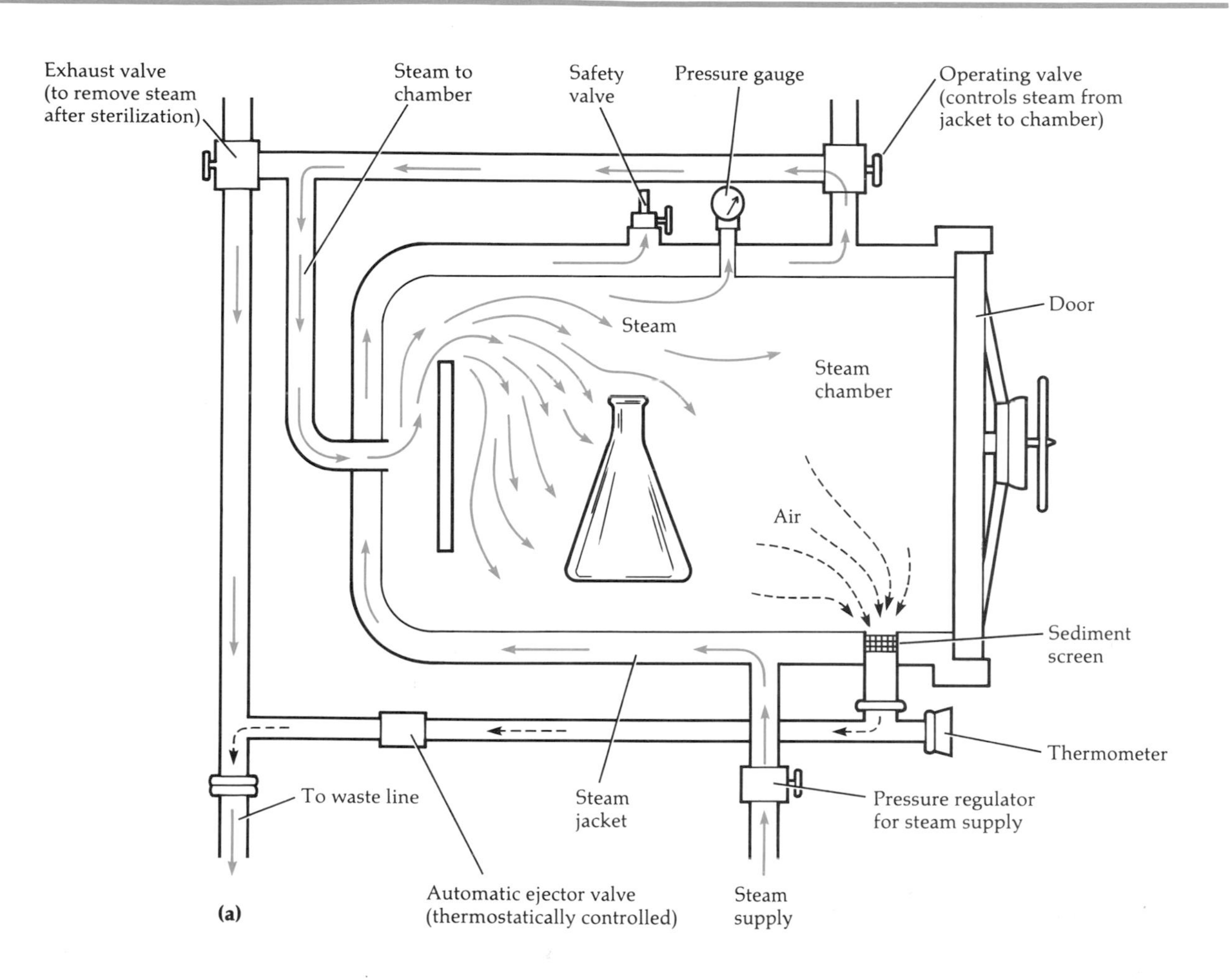

Figure 7–2 **(a)** Basic parts and path of steam flow (solid arrows) of an autoclave. The autoclave consists of a steam chamber capable of withstanding more than 1 atmosphere of pressure (15 pounds per square inch). As steam enters the chamber, air is forced out through a vent in the bottom of the chamber (broken arrows). The automatic ejector valve remains open as long as an air–steam mixture is passing out to the waste line. When all of the air has been ejected, the higher temperature of pure steam closes the valve. As steam continues to enter, the pressure in the chamber increases. **(b)** Front view of two modern autoclaves.

Table 7–3 Relationship Between Pressure and Temperature of Steam

Pressure in Pounds per Square Inch (psi) (In Excess of Atmospheric Pressure)	Temperature in °C
0 psi	100°C
5 psi	110°C
10 psi	116°C
15 psi	121°C
20 psi	126°C
30 psi	135°C

Table 7–4 Effect of Container Size on Sterilization Times for Liquid Solutions

Container Size	Sterilization Time in Minutes
Test tube: 18 x 150 mm	15
Erlenmeyer flask: 125 ml	15
Erlenmeyer flask: 2000 ml	30
Serum bottle: 9000 ml	70

temperature and pressure is shown in Table 7–3. Steam in an autoclave at a pressure of about 15 pounds per square inch (121°C) will kill *all* organisms and their endospores in about 15 minutes, or perhaps a bit longer, depending on the type and volume of material being sterilized.

Autoclaving is used to sterilize culture media, instruments, dressings, intravenous equipment, applicators, solutions, syringes, transfusion equipment, and numerous other items that can withstand high temperature and pressure. Large industrial autoclaves are called *retorts*, but the same principle applies to the common household pressure cooker, used for home canning of foods.

Heat requires more time to reach the center of solid materials such as canned meats, which depend on conduction of heat without the efficient convection currents in liquids. Also, more time is needed to heat larger containers. Table 7–4 shows the different time requirements for various container sizes. It is important to remember that steam has to actually contact surfaces to sterilize them. With liquids this is no problem, but with dry glassware, bandages, and the like, care must be taken to ensure that contact is made. For example, aluminum foil is impervious to steam and should not be used for wrapping; paper should be used instead. And air trapped in the bottom of a dry container is a situation that should be avoided, since this prevents its replacement by steam, which is lighter than air. The trapped air is the equivalent of a small hot-air oven, and we will see shortly that this requires a more vigorous combination of heat and time. Such containers should be tipped so that the air can be displaced by steam. Products such as mineral oil or petroleum jelly do not permit penetration by moisture and are not sterilized by the same treatments that would sterilize aqueous solutions.

Dry heat sterilization

One of the simplest methods of dry heat sterilization is **direct flaming.** You will use this procedure many times in the laboratory when you sterilize inoculating loops and needles before inoculation. In order to sterilize the inoculating loop or needle, all you have to do is heat the wire to a red glow. This method is 100% effective. A similar principle is used in **incineration.** This is an effective way to sterilize and dispose of contaminated paper cups, bags, and dressings.

Another form of dry heat sterilization is **hot-air sterilization.** In this procedure, items to be sterilized are placed in an ovenlike apparatus. Generally, a temperature of 170°C is maintained for nearly two hours to ensure sterilization. The longer period of time, relative to moist heat, is required because the heat of the water is very readily transferred to a cooler body. (Compare, if you wish, the effect on your hand of brief immersion in hot water versus holding it in a hot-air oven at the same temperature for the same time.)

In addition to moist and dry heat, another heating procedure, called pasteurization, is used to control microbes.

Pasteurization

You may recall from Chapter 1 that Louis Pasteur, in the early days of microbiology, found a practical method of preventing the "sickness" or spoilage of beer and wine. Pasteur used mild heating, which was sufficient to kill the organisms that caused the particular spoilage problem without seriously damaging the taste of the product. The same principle was later applied to milk to produce what we now call pasteurized milk. The pasteurization of milk was first undertaken for the elimination of the tuberculosis bacterium. Substantial numbers of relatively heat-resistant organisms survive pasteurization but are unlikely to cause disease, nor will they cause spoilage in a reasonable time in refrigerated milk.

The classic **pasteurization** treatment of milk consisted of exposing the milk to a temperature of about 63°C for 30 minutes. Most pasteurization today uses higher temperatures, at least 72°C for about 15 seconds. This treatment, known as **high-temperature, short-time (HTST) pasteurization,** is applied as the milk flows continuously past a heat exchanger. Higher temperatures and slightly longer times are common, and the aim today is to lower the bacterial counts so that the milk not only will be free of pathogens, but also will keep better.

We now come to the concept of **equivalent treatments.** For example, 115°C might require 70 minutes to destroy a suspension of very resistant endospores, whereas 125°C would accomplish this in only 7 minutes. Both treatments yield equivalent results. In other words, as the temperature is increased, it takes less time to kill the same number of microbes.

In addition to heat, microbial growth may also be controlled by filtration.

Filtration

Recall from Chapter 6 that filtration methods are sometimes used to concentrate microorganisms from a liquid source where the microbes are found in small numbers. **Filtration** is the passage of a liquid or gas through a screenlike material with pores small enough to retain microorganisms (see Figure 6–15). A vacuum is created in the receiving flask to aid gravity in pulling the liquid through the filter. As a sterilization method, filtration is used for materials that are sensitive to heat, such as some culture media, enzymes, vaccines, and antibiotic solutions. Materials used for filters in the past include diatomaceous earth (Berkefield filter), asbestos or compressed paper (Seitz filter), porcelain (Chamberland filter), and ground glass (sintered glass filter). Because of their thickness and absorption properties, however, such filters are rarely used today. They retain some of the filtrate, tend to have a slow rate of flow, and easily become clogged.

In recent years, **membrane filters,** composed of substances such as nitrocellulose and cellulose acetate, have become popular in industry and laboratories. These filters are only 0.1 mm thick, in contrast to the asbestos or ground glass filters, which are 5 mm thick. The pores of membrane filters are of uniform size, and because of the numerous pores in the filter, flow rate is very fast. The pore sizes of membrane filters range down to sizes that retain the smallest viruses, and even some larger protein molecules.

Low Temperature

The effectiveness of applying **low temperatures** to microbes depends on the particular microbe and the intensity of the application. For example, at ordinary refrigerator temperatures (0° to 7°C), the metabolic rate of most microbes is reduced to such a point that they do not reproduce or elaborate products. In other words, ordinary refrigeration has a bacteriostatic effect. Essentially, refrigerator temperatures preserve food and not the microbes. Yet, psychrophilic (cold-loving) species do grow slowly at refrigerator temperatures and will alter the appearance and taste of foods after a considerable time. Surprisingly, many bacteria can grow at temperatures several degrees below freezing, and although most microbes die after long periods in a storage freezer, this resistance varies among different species.

Desiccation

Microbes require water in order to grow and multiply. In the absence of water, a condition known as **desiccation,** microbes are not capable of growth or reproduction, but may remain viable for years. Then, when water is made available to them, they can resume their growth and division. This fact is used in the laboratory to preserve microbes by lyophilization, a process described in Chapter 6.

Resistance of vegetative cells to desiccation varies with the species and the organism's environment. For example, the gonorrhea organism can withstand drying for only about an hour or so, whereas the tuberculosis bacterium may remain viable for months. Also, a normally susceptible bacterium is much more resistant if embedded in mucus, pus, or feces. Viruses are generally resistant to desiccation, but not as resistant as endospores. In a hospital setting, it is important to remember the capacity of certain dried microbes and endospores to remain viable. Dust, clothing, bedding, and dressings may contain infectious microbes in dried mucus, urine, pus, and feces.

Nomadic peoples such as the native Americans made use of the principle of desiccation to preserve foods. If meat is sliced thin and sun dried, it can be preserved for long periods of time. Raisins and similar dried fruits are other examples.

Osmotic Pressure

The use of high concentrations of salts and sugars to preserve food is based on the effects of **osmotic pressure.** High concentrations of these substances create a hypertonic environment and cause water to leave the microbial cell (see Chapter 4). This process resembles preservation by desiccation in that both methods deny moisture needed for microbial growth. As water leaves the microbial cell, the plasma membrane shrinks away from the cell wall (plasmolysis), and the cell stops growing, although it may not be immediately killed. As mentioned in Chapter 4, the principle of osmotic pressure is used in the preservation of foods. For example, concentrated salt solutions are used to "cure" meats, and thick sugar solutions to preserve fruits.

As a general rule, molds and yeasts are much more capable of growth in materials with low mois-

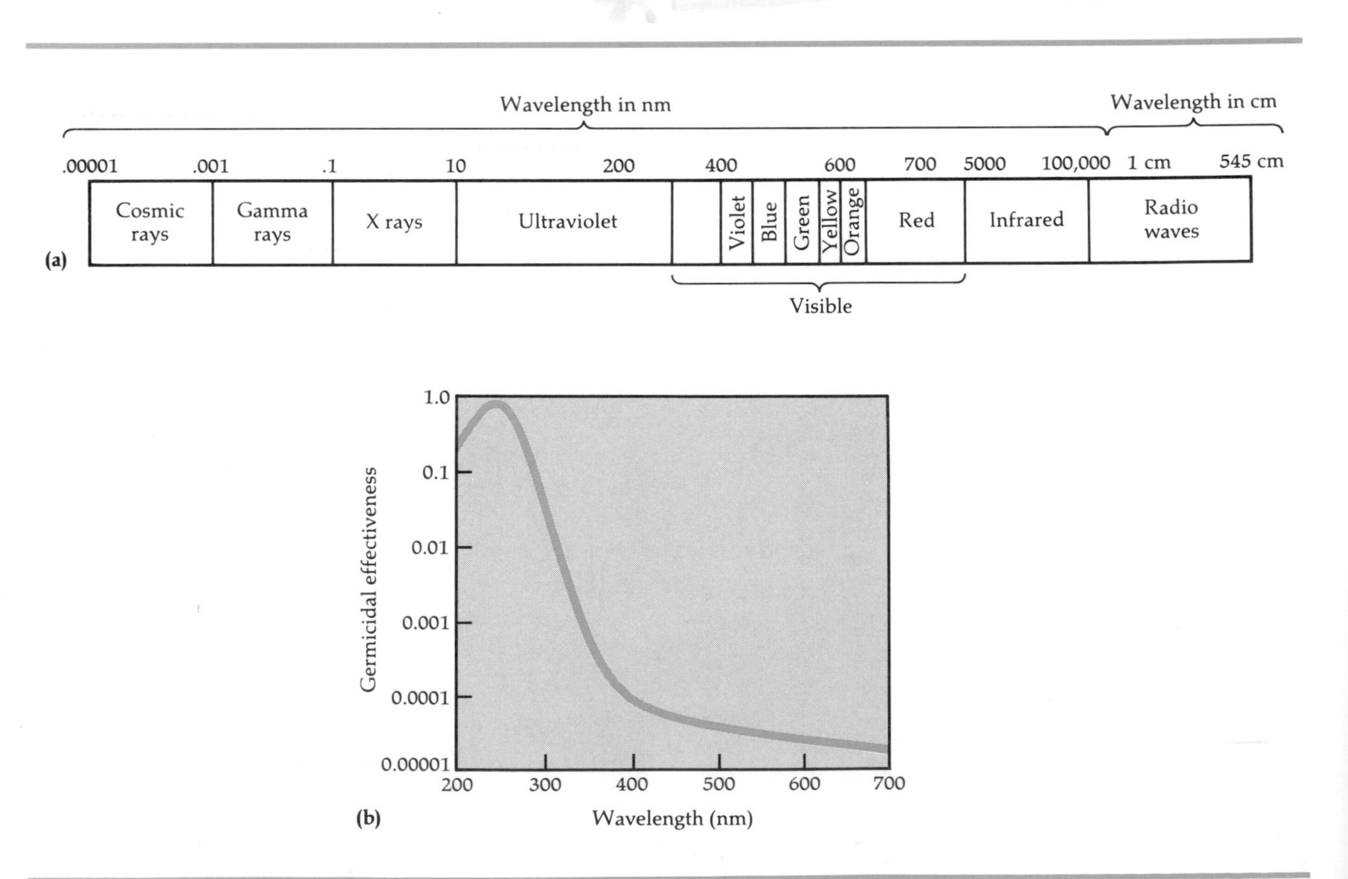

Figure 7–3 Radiant energy. **(a)** Radiant energy spectrum. **(b)** Germicidal effectiveness of radiant energy between 200 and 700 nanometers (nm) (from UV to visible red light).

ture or high osmotic pressures than are bacteria. This property, sometimes combined with a greater ability to grow under acidic conditions, is the reason fruits and grains are spoiled by molds rather than bacteria. It is also part of the reason why molds are able to form "mildew" growth on such unlikely places as a damp wall or a shower curtain.

Radiation

Radiation has varying effects on cells depending on its wavelength, intensity, and duration. There are two basic types of radiation: ionizing and nonionizing.

Ionizing radiation, such as **X rays** and **gamma rays,** has a shorter wavelength than nonionizing radiation, and therefore carries much more energy (Figure 7–3a). Ionizing radiation has a high degree of penetrating power, and when directed at microbes, it has damaging effects. Its principal effect is the ionization of water to form highly reactive free radicals (molecular fragments with unpaired electrons) that can break DNA strands. Unfortunately, ionizing radiation as a microbial control agent does not have widespread practical application in routine sterilization. Foods sterilized by X rays and gamma rays often undergo characteristic changes in color, odor, and flavor. Moreover, the use of ionizing radiation requires expensive equipment and special safety measures, since it is also damaging to human tissues. However, since ionizing radiation does not use heat or moisture and has great penetrability, it is useful for sterilizing pharmaceuticals and disposable dental and medical supplies, such as plastic syringes, surgical gloves, suturing materials, and catheters.

Nonionizing radiation has a longer wavelength than ionizing radiation and therefore carries less energy. A good example of nonionizing radiation is **ultraviolet (UV) light.** UV light damages the DNA of exposed cells. It causes bonds to form between adjacent thymines in DNA chains (see Figure 8–13). These *thymine dimers* inhibit correct replication of the DNA during reproduction of the cell. The UV wavelengths most effective for killing microorganisms are around 260 nm, wavelengths specifically absorbed by cellular DNA (Figure 7–3b). UV radiation is used to control microbes in the air. A UV or "germicidal" lamp is commonly found in hospital rooms, nurseries, operating rooms, and cafeterias. UV light is also used to sterilize vaccines, sera and toxins, and food in the food-packaging and -processing industries. A major drawback of UV light is that it has low penetrability, and this means that organisms must be directly exposed to the rays. Organisms protected by solids and coverings such as paper, glass, and textiles are not affected. Another drawback is that UV light can damage the eyes, and prolonged exposure to UV light can cause burns.

As you probably know, sunlight contains UV light and has some germicidal action. This action varies with the intensity of sunlight, time of day, and season of the year.

A summary of physical methods of microbial control is presented in Table 7–5.

We will now turn our attention to a discussion of several chemical methods of microbial control.

CHEMICAL METHODS OF MICROBIAL CONTROL

Chemical agents are used to control microbes on body surfaces and inanimate objects. Unfortunately, few chemical agents achieve sterility; most of them merely reduce microbial populations to safe levels or remove vegetative forms of pathogens from objects. A common problem in disinfection is to select an agent that will kill all organisms present in the shortest period of time without damaging the contaminated material. Just as there is no single physical method of microbial control that can be used in every situation, there is no one disinfectant that will be appropriate for all circumstances.

Qualities of an Effective Disinfectant

The more of the following qualities a disinfectant has, the more effective it is:

1. Acts rapidly.
2. Attacks all, or a wide range, of microbes.
3. Is able to penetrate thoroughly the material that is contaminated.

4. Readily mixes with water to form a stable solution or emulsion.
5. Is not hampered by organic matter on the substance that is to be disinfected.
6. Is not likely to decompose and thereby lose its activity after exposure to light, heat, or unfavorable weather.
7. Does not stain, corrode, or destroy the object being disinfected.
8. Is harmless to animals if it is to be used as an antiseptic and does not destroy body tissues or act as a toxin if inhaled or swallowed.

Also, the ideal disinfectant should have a pleasant odor and be economical to use and safe to transport.

Principles of Effective Disinfection

We can effectively promote disinfection by understanding the factors that influence the action of a particular disinfectant. For example, what are the properties of the disinfectant? In other words, what is it designed to do? Simply by reading the label, we can learn a great deal about a disinfectant's properties. We should also remember that the concentration of a disinfectant will affect its action. A disinfectant should always be diluted exactly as suggested by the manufacturer. Solutions that are too weak may be ineffective, or bacteriostatic instead of bactericidal. On the other hand, solutions that are too strong may be dangerous to humans who come in contact with them.

Also to be considered is the nature of the material being disinfected. Are there any organic materials present that might interfere with the action of the disinfectant? The presence of organic materials and the pH of the medium in which the microbes are present may determine whether a chemical control agent is only inhibitory or lethal to the microbes.

Another very important consideration is that the disinfectant make contact with the microbes. This may mean scrubbing and rinsing an area before applying the disinfectant. In general, disinfection is a gradual process. Thus, a disinfectant may have to be left on a surface for several hours to be effective.

A final factor to keep in mind is that the higher the temperature at which the disinfectant is applied, the more effective it is.

Evaluating a Disinfectant

Phenol coefficient

When considering the relative effectiveness of a disinfectant, some standard of comparison is needed. One such standard compares the disinfecting action of a chemical with that of phenol for the same length of time on the same organism under identical conditions. The test organisms used are *Staphylococcus aureus* (a gram-positive bacterium) and *Salmonella typhi* (a gram-negative organism). If the disinfectant being tested requires a greater concentration or a longer period of time than phenol, its efficiency is judged to be less than that of phenol. If the test disinfectant requires a lower concentration or shorter period of time than phenol, its efficiency is determined to be greater than that of phenol. The ratio of the activity of the test disinfectant to the activity of the phenol is called the **phenol coefficient (PC).** If the disinfectant has a phenol coefficient greater than 1, it is stronger than phenol; if the coefficient is less than 1, the disinfectant is weaker than phenol. You may learn how to determine phenol coefficients in your laboratory work.

Use-dilution test

The **use-dilution test,** a more elaborate test, is sometimes considered the official test for evaluating a disinfectant. In this test, standardized preparations of several test bacteria are added to a series of tubes containing increasing concentrations of the test disinfectant. The tubes are then incubated, and growth, or lack of it, is recorded. The more the chemical can be diluted and still be effective, the higher its evaluation rating.

Filter paper method

The **filter paper method** for evaluating a chemical agent consists of soaking a disk of filter paper with

Table 7–5 Summary of Physical Methods Used to Control Microbial Growth

Method	Mechanism of Action	Comment	Preferred Use
Heat:			
1. Moist heat sterilization			
a. Boiling	Denaturation	Kills vegetative bacterial and fungal pathogens and many viruses within 10 minutes; less effective for endospores	Dishes, basins, pitchers, various equipment
b. Fractional sterilization	Denaturation	Effectively kills vegetative cells but not all endospores	Microbiological media and certain foods that would be damaged by temperatures above boiling
c. Autoclaving	Denaturation	Very effective method of sterilization; at about 15 pounds of pressure (121°C), all vegetative cells and their endospores are killed in about 15 minutes	Microbiological media, solutions, linens, utensils, dressings, equipment, and other items that can withstand temperature and pressure
2. Dry heat sterilization			
a. Direct flaming	Burning to ashes	Very effective method of sterilization	Inoculating loops and needles
b. Incineration	Burning to ashes	Very effective method of sterilization	Paper cups, dressings, animal carcasses, bags, and wipes
c. Hot-air sterilization	Denaturation	Very effective method of sterilization, but requires about 2 hours at a temperature of 170°C	Empty glassware, instruments, needles, and syringes
3. Pasteurization	Denaturation	Heat treatment for milk (72°C for about 15 seconds) that kills all pathogens and some nonpathogens	Milk, cream, and certain alcoholic beverages (beer and wine)
Filtration	Separation of bacteria from suspending liquid	Passage of a liquid or gas through a screenlike material that traps microbes; most filters in use consist of cellulose acetate	Useful for sterilizing liquids (toxins, enzymes, vaccines) that are destroyed by heat
Low Temperatures:			
1. Refrigeration	Decreased chemical reactions and possible changes in proteins	Has a bacteriostatic effect	Food, drug, and culture preservation
2. Deep-freezing (see Chapter 6)	Decreased chemical reactions and possible changes in proteins	An effective method for *preserving* microbial cultures in which cultures are quick-frozen between −50° to −95°C	Food, drug, and culture preservation
3. Lyophilization (see Chapter 6)	Decreased chemical reactions and possible changes in proteins	Most effective method for long-term preservation of microbial cultures; water removed by high vacuum at low temperature	Food, drug, and culture preservation
Desiccation	Disruption of metabolism	Involves removing water from microbes; primarily bacteriostatic	Food preservation
Osmotic Pressure	Plasmolysis	Results in loss of water by microbial cells	Food preservation
Radiation:			
1. Ionizing	Destruction of DNA by X rays and gamma rays	Not widespread in routine sterilization	Used for sterilizing pharmaceuticals and medical and dental supplies
2. Nonionizing	Damage to DNA by UV light	Radiation not very penetrating	Practical application is the UV (germicidal) lamp

the chemical and placing it on the surface of agar plates that have been previously inoculated and incubated with the test organism. If the chemical agent is effective, a clear zone, representing inhibition, can be observed around the disk (Figure 7–4).

Kinds of Disinfectants

Phenol and phenolics

Phenol (carbolic acid), the substance first used by Lister in his operating room, is no longer used as an antiseptic and seldom as a disinfectant because it irritates the skin and has a disagreeable odor. However, as mentioned previously, it is still used as a standard for measuring the effectiveness of other disinfectants. The structure of a phenol molecule is shown in Figure 7–5a.

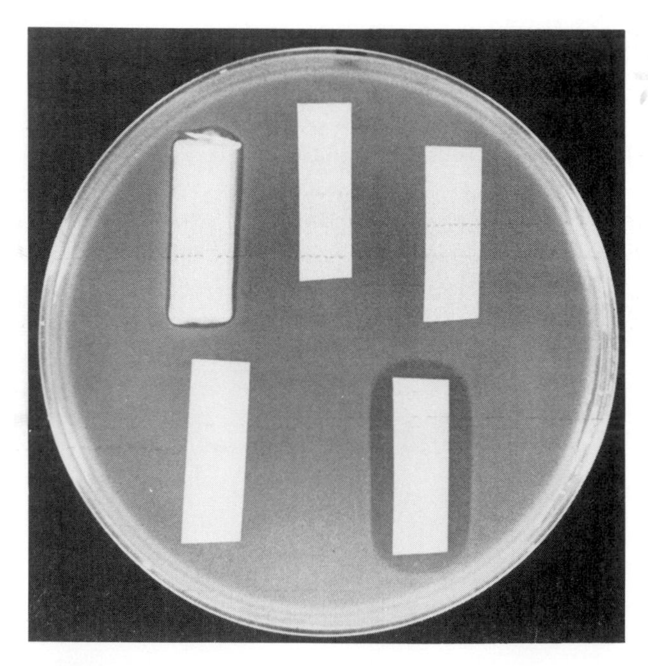

Figure 7–4 Evaluation of disinfectants by the filter paper method. *Top left:* narrow zone of inhibition around strip soaked in Lysol household disinfectant. *Top center and top right:* no obvious inhibition around strips soaked in Wescodyne (an iodine-containing disinfectant) and 50% ethanol, respectively. *Bottom left:* no inhibition around this control strip soaked in sterile water. *Bottom right:* wide area of inhibition around strip soaked in a 3% hydrogen peroxide solution. Wescodyne and 50% ethanol are known to kill bacteria (*Escherichia coli*, in this case), but were not as effective under the conditions used here.

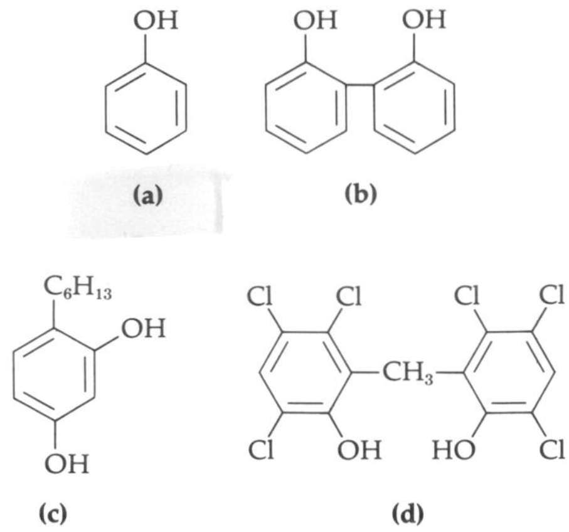

Figure 7–5 Structure of phenol and phenolics. **(a)** Phenol. **(b)** O-phenylphenol. **(c)** Hexylresorcinol. **(d)** Hexachlorophene.

Derivatives of phenol, called **phenolics,** contain a molecule of phenol chemically altered to reduce its irritating qualities or increase its antibacterial activity in combination with a soap or detergent. As a group, phenolics exert antimicrobial activity by injuring plasma membranes, inactivating enzymes, and denaturing proteins. Phenolics are frequently used as disinfectants because they remain active in the presence of organic compounds, they are stable, and they persist for long periods of time after application. For these reasons, phenolics are suitable agents for disinfecting pus, saliva, and feces.

One of the most frequently encountered phenolics is a group of chemicals derived from coal tar called *cresols.* A very important cresol is *O-phenylphenol* (Figure 7–5b), the main ingredient in most formulations of Lysol. Cresols make very good environmental disinfectants. Another familiar phenolic is *hexylresorcinol* (Figure 7–5c), which is used in throat lozenges and household antiseptics such as S.T. 37.

Another phenolic, used much more widely in the past than now, is *hexachlorophene,* consisting of two molecules of phenol joined together (Figure 7–5d). This phenolic was originally incorporated into soaps and lotions (pHisoHex) used for surgical scrubs and hospital microbial control procedures, cosmetic soaps, deodorants, feminine hygiene sprays, and even toothpaste. Hexachlorophene is an effective bacteriostatic agent, especially against staphylococcal and streptococcal bacteria, which can cause infections of the skin. Moreover, the hexachlorophene persists on the skin for long periods of time. For these reasons, it was once heavily used to control staphylococcal and streptococcal infections in hospital nurseries. However, in 1972, it was found that *excessive* use of hexachlorophene, such as bathing infants several times a day, could lead to neurological damage. Hexachlorophene is still used in hospitals to control nosocomial infections and as a "scrub" for hospital personnel. However, it is no longer in casual use by the general public. At present, a prescription is required for the purchase of a 3% or stronger solution of hexachlorophene.

Halogens

The **halogen** elements, particularly iodine and chlorine, are effective antimicrobial agents both alone (as I_2 or Cl_2 in solution) and as constituents of inorganic or organic compounds. *Iodine* (I_2) is not only one of the oldest antiseptics, but also one of the most effective. It is also used to disinfect water, air, and food utensils. Iodine is effective against all kinds of bacteria, many endospores, fungi, and some viruses. A proposed mechanism for the activity of iodine is that it combines with the amino acid tyrosine, a common component of many enzymes and other cellular proteins (Figure 7–6). As a result, microbial protein function is inhibited. Since iodine is also a powerful oxidizing agent, some of its antimicrobial activity may be due to this property.

Iodine is available as a *tincture*, that is, in solution in aqueous alcohol, and as an iodophor. An *iodophor* is a combination of iodine and an organic molecule, usually a detergent (for example, polyvinylpyrrolidone), in which the iodine is released slowly. Iodophors have the antimicrobial activity of iodine, but they do not stain and are less irritating than iodine. Commercially available iodophors are Wescodyne and Ioclide, which are used for environmental disinfection, and Betadine and Isodine, which are used as skin antiseptics.

Chlorine used as a gas or in combination with other chemicals is another widely used disinfectant. Its germicidal action is based on the formation of hypochlorous acid when added to water.

$$\underset{\text{Chlorine}}{Cl_2} + \underset{\text{Water}}{H_2O} \rightarrow \underset{\text{Hypochlorous acid}}{HClO} + \underset{\text{Hydrochloric acid}}{HCl}$$

Hypochlorous acid then dissociates into hydrochloric acid and a highly reactive form of oxygen that combines with and alters cellular components of microbes.

$$\underset{\text{Hypochlorous acid}}{HClO} \rightarrow \underset{\text{Hydrochloric acid}}{HCl} + \underset{\text{Oxygen}}{O}$$

The chlorine itself may also combine directly with cellular components to inactivate them.

Chlorine gas compressed into liquid form is used extensively for disinfecting municipal drinking water, swimming pools, and sewage. Several compounds of chlorine are also effective disinfectants. For example, solutions of *calcium hypochlorite* ($Ca{\cdot}(OCl)_2$) are used to disinfect dairies, barns, abattoirs, and eating utensils in restaurants. And another chlorine compound, *sodium hypochlorite* (NaOCL), is used as a household disinfectant and bleach (Clorox), as well as a disinfectant in dairies, food-processing establishments, and hemodialysis systems (see Box). When the quality of drinking water is in question, household bleach may be used

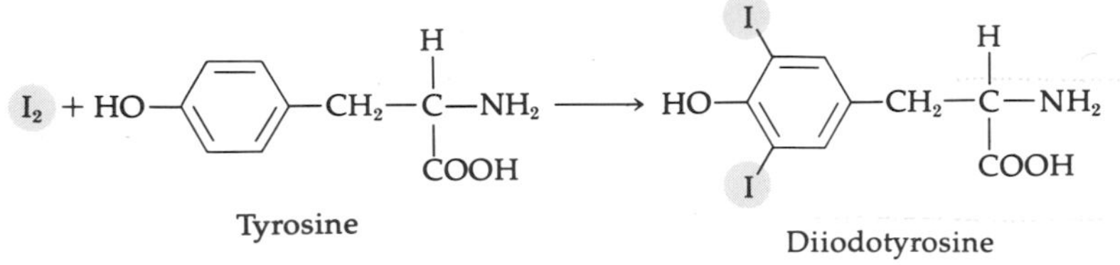

$$I_2 + \underset{\text{Tyrosine}}{HO{-}C_6H_4{-}CH_2{-}CH(NH_2){-}COOH} \longrightarrow \underset{\text{Diiodotyrosine}}{HO{-}C_6H_2I_2{-}CH_2{-}CH(NH_2){-}COOH}$$

Figure 7–6 Proposed mechanism for the activity of iodine.

to achieve a rough equivalent of drinking water chlorination. Two drops of bleach added to a liter of water (4 drops if the water is cloudy) and allowed to stand for 30 minutes is considered adequate to make water safe for drinking under emergency conditions.

Another group of chlorine compounds, the *chloramines,* consist of chlorine and ammonia. They are used as disinfectants, antiseptics, or sanitizing agents. Chloramines are very stable compounds that release chlorine effectively for long periods of time. They may be used to sanitize glassware and eating utensils and to treat dairy and food-manufacturing equipment. They may also be used for the emergency disinfection of small amounts of drinking water. Halazone is an example of a chloramine that may be used for this emergency purpose.

Alcohols

Alcohols are bacteriocidal and fungicidal but are not effective against endospores and nonenveloped viruses. For example, alcohols are not particularly effective against the hepatitis virus (which is not enveloped) spread by syringes and other skin-penetrating instruments. The mechanism of action of alcohols is protein denaturation, and because alcohol dissolves many lipids, the lipid component of enveloped viruses is susceptible. Alcohols are not effective against nonenveloped viruses because these viruses tend to have especially stable proteins. Alcohols have the advantage of rapid germicidal action, evaporating rapidly and leaving no residue. In a quick swabbing of the skin before an injection, most of the antiseptic action comes from a simple *wiping away* of dirt and microorganisms. An iodine-alcohol solution will kill even more microbes on skin. Alcohols are also used to disinfect clinical oral thermometers.

Two of the most commonly used alcohols are ethanol and isopropanol. *Ethanol,* or *ethyl alcohol* (CH_3CH_2OH), is one of the most widely used antiseptics and disinfectants. Its recommended effective optimum concentration is 70%, but concentrations between 60 and 95% seem to kill as fast (Table 7–6). Pure alcohol is less effective than aqueous solutions (alcohol mixed with water) because denaturation requires the presence of water. *Isopropanol,* or *isopropyl alcohol* ($(CH_3)_2CHOH$), better known as rubbing alcohol, is superior to ethanol as an antiseptic and disinfectant. Moreover, it is less volatile, less expensive, and more easily obtainable than ethanol.

Table 7–6 Germicidal Action of Various Concentrations of Ethanol Against *Streptococcus pyogenes* (*Note:* + = Growth; − = No Growth)

Percent Ethanol	Seconds				
	10	20	30	40	50
100	+	+	+	+	+
95	−	−	−	−	−
90	−	−	−	−	−
80	−	−	−	−	−
70	−	−	−	−	−
60	−	−	−	−	−
50	+	+	−	−	−
40	+	+	+	+	+

Source: H. E. Morton, *Ann. N.Y. Acad. Sci. 53:*191–196 (1950).

Alcohols are also used to enhance the effectiveness of other chemical agents. For example, an aqueous solution of Zephiran (described shortly) has been shown to kill about 40% of the population of a test organism in two minutes, whereas a tincture of Zephiran kills about 85% in the same period of time.

Heavy metals and their compounds

Several **heavy metals,** such as silver, mercury, and copper can be used as germicidals or antiseptics. The ability of very small amounts of heavy metals, especially silver and copper, to exert antimicrobial activity is referred to as **oligodynamic action.** This action can be seen when we place a clean piece of metal containing silver or copper, such as a coin, on an inoculated Petri plate. Extremely small amounts of metal diffuse from the coin and inhibit the growth of bacteria for some distance around the coin (Figure 7–7). This effect is explained by the known action of heavy metal ions on microbes. The metal ions combine with the —SH groups on cellular proteins, causing denaturation.

Silver is used as an antiseptic in a 1% *silver nitrate* solution. The solution is bactericidal for most organisms. Many states require that the eyes of newborns be treated with a few drops of silver nitrate as a precaution against a gonococcal infection of the eyes called gonorrheal ophthalmia neon-

Centers for Disease Control

MMWR

Morbidity and Mortality Weekly Report

Inadequate Disinfection Leads to an Outbreak of Bacteremia in a Hospital Dialysis Unit

On March 18, 1978, 3 patients undergoing hemodialysis at a university hospital in Pennsylvania became ill. All 3, who had previously been well, developed fever and chills during dialysis. One patient became hypotensive. Dialysis was discontinued, and all 3 were admitted for evaluation. *Pseudomonas aeruginosa* was isolated from blood cultures from 2 patients; *Klebsiella pneumoniae* was also isolated from one of these patients' blood. Blood cultures from the third patient were negative. On March 20, 2 more patients undergoing hemodialysis became ill with fever and chills and required hospitalization.

The outpatient hemodialysis facility is located in a separate building. Each unit contains a single-pass dialysis machine equipped with a disposable coil dialyzer. The dialyzers are not reused. An investigation revealed that a sodium hypochlorite pump attached to the dialysis mixing console had not been functioning for the previous 2 weeks because a replacement part had not arrived. Normally, a solution of sodium hypochlorite was pumped through the entire system between each shift. Alkaline glutaraldehyde solutions were also used on a variable schedule. During the 2 weeks before the outbreak, individual units were disinfected daily with sodium hypochlorite. However, without the pump the distribution system and pipes leading to the dialysis machine could not be disinfected.

Cultures taken in the unit grew *P. aeruginosa* in all samples of water and dialysate; heavy growth was present in the storage tank from which mixed dialysate was distributed to individual units and in the tubing leading to the dialysis machines. *K. pneumoniae* and *Enterobacter agglomerans* were also cultured from additional sites in the system. Bacteria were recovered from the inlet and outlet blood tubing as well as from the blood side of the dialyzer used for one of the ill patients.

Following the repair of the sodium hypochlorite pump and thorough disinfection of the unit, there have been no further cases of febrile illness associated with dialysis.

Editorial Note: Chronic hemodialysis as a treatment for patients suffering from end-stage renal disease has been used increasingly in the United States since the early 1960s. In 1978, the total number of patients undergoing chronic hemodialysis in private or hospital-based centers was 37,000—more than triple the 1973 figure. It is estimated that by 1980 there will be 45,000 such patients. Technological advances in hemodialysis systems have been significant in the past 10–15 years; they have allowed chronic hemodialysis to become a common procedure and accounted for a dramatic improvement in the clinical state of the art.

However, a significant number of microbiologic parameters were not taken into consideration in the design of many hemodialysis systems, and, as a result, there are many situations in which certain types of bacteria, notably gram-negative water bacteria, can persist and actively multiply in water associated with hemodialysis equipment. This can result in the production of massive levels of gram-negative bacteria in dialysis fluid—levels which have been associated with outbreaks of pyrogenic (fever-producing) reactions and septicemia among patients undergoing hemodialysis. These bacteria contain lipopolysaccharide or endotoxin, which can produce a pyrogenic response if introduced into the bloodstream. The Centers for Disease Control have investigated a number of outbreaks of pyrogenic reactions in the United States, some of which were complicated by septicemia, among patients in hemodialysis centers.

The most probable explanation of the outbreak reported here is that the gram-negative bacteria were able to grow within the water and dialysate distribution systems because of the cessation of routine sodium hypochlorite disinfection. Sodium hypochlorite is used as a disinfectant for hemodialysis systems by a large number of hemodialysis centers in the United States. However, it is corrosive, and consequently it is usually rinsed from the system after a minimum exposure time. Unfortunately, this practice commonly results in negation of the disinfection procedure because water used to rinse the disinfectant out of the system is not sterile and usually contains gram-negative bacteria. In recirculating dialysis machines—the most difficult type to disinfect—gram-negative bacteria present in the rinse water can multiply immediately and, if permitted to stand overnight, can significantly contaminate the system. This level of bacterial contamination will increase by 3–5 logs during dialysis.

The optimum strategy for the disinfection of dialysis systems may be to use aqueous formaldehyde (1.5–2.5%) and to allow it to remain in the system for prolonged periods of time when the system is not operational such as for a night or weekend. Although other high-level disinfectants such as alkaline glutaraldehyde could be used, the volume of these systems is so large that their use would be very expensive.

Abridged from *MMWR* 27:307 (8/25/78).

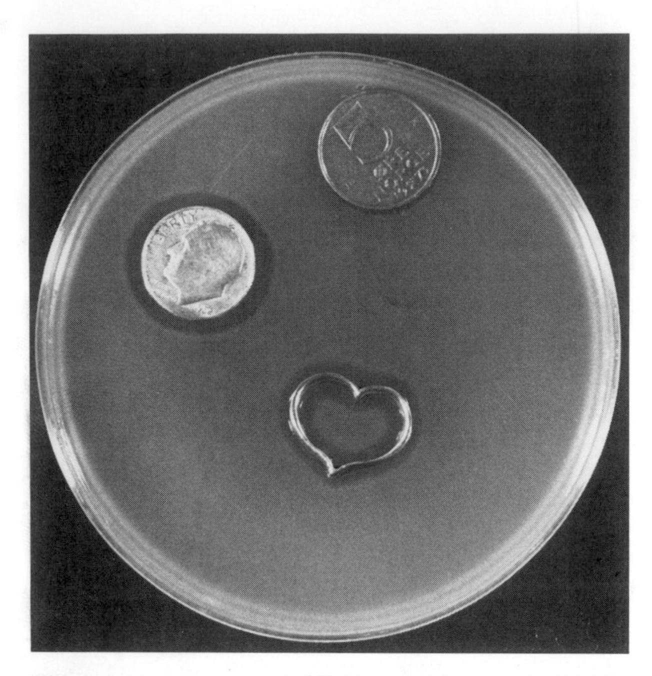

Figure 7–7 Oligodynamic action of silver. Clear zones where bacterial growth has been inhibited are seen around the silver coin and heart. The nonsilver coin has been pushed aside to show the growth beneath it.

atorum, which the infants might contract as they pass through the birth canal. In recent years, antibiotics have been replacing silver nitrate for this purpose. However, there is a trend back toward silver nitrate because of problems with antibiotic-resistant bacteria.

Inorganic mercury compounds, such as *mercuric chloride* and *mercuric oxide*, are bactericidal in solutions of 1:1000 (0.1%). However, their use is limited because of their high toxicity, corrosiveness, and ineffectiveness in the presence of organic matter. Organic mercury compounds, such as *Mercurochrome* and *Merthiolate* are less irritating and toxic than inorganic mercury compounds. Mercurochrome and Merthiolate are employed as skin and mucous membrane antiseptics. Unfortunately, their effects are reversed if removed (washed away) and they are ineffective in the presence of extraneous organic matter.

Copper in the form of *copper sulfate* is used chiefly to destroy green algae that grows in reservoirs, swimming pools, and fish tanks. If the water does not contain an excess of organic matter, copper sulfate is effective in concentrations of 1 part per million (ppm) of water. One other metal used as a mild antiseptic is zinc. *Zinc peroxide* is used to control infections caused by anaerobic and microaerophilic microbes in injuries such as deep puncture wounds and bites. Calamine lotion is a mixture of zinc oxide and ferric oxide.

Dyes

There are a number of **dyes** that have antimicrobial properties. For example, *brilliant green* is sometimes incorporated into nutrient microbiological media for selective diagnostic purposes. It suppresses the growth of unwanted organisms. *Crystal violet* and *gentian violet* are still occasionally used for treatment of superficial fungal infections, and *acridine dyes* are used experimentally for some herpesvirus infections. Dyes probably exert their action by interfering with cellular oxidation. The advent of antibiotics has displaced most antiseptic uses of dyes.

Surface-active agents

Surface-active agents, or **surfactants,** have the ability to decrease surface tension between molecules of a liquid. They include soaps and detergents.

Soaps are made from fats and lye (NaOH, a strong alkali). They have a limited bactericidal action on organisms such as streptococci, pneumococci, and gonococci, but no chemical effects on most skin microbes. However, soaps do have an important function in the mechanical removal of microbes through scrubbing. The skin normally contains dead cells, dust, dried sweat, microbes, and oily secretions from oil glands. Soap breaks up the oily film into tiny droplets, a process called **emulsification,** and the water and soap together lift up the emulsified oil and debris and float them away as the lather is washed off. In this sense, soaps are good degerming and emulsifying agents. Many cosmetic soaps, such as Dial, Safeguard, and Lifebuoy, contain disinfectants that increase their antimicrobial activity.

Detergents are synthetic surface-active agents primarily composed of hydrocarbons attached to other chemical groups. When detergents ionize in water, the ions attract the debris to be washed off. In general, detergents are regarded as superior to soaps because they do not form precipitates or deposits with water minerals as soaps do. Detergents are classified as anionic, cationic, and nonionic.

In *anionic detergents,* cleansing ability is related to the negatively charged portion (anion) of the molecule. Anionic detergents are incorporated into laundry powders. Generally, they are not disinfectants; they mechanically remove debris. *Cationic detergents* do have antimicrobial activity, a benefit of the positively charged portion (cation) of the molecule. The best known of the cationic detergents are the so-called quaternary ammonium compounds, which are the most effective detergents against microbes. They will be discussed next as a separate group of chemical control agents. *Nonionic detergents* do not ionize and have no antimicrobial activity by themselves. However, some are used as the organic molecules in iodophors.

Quaternary ammonium compounds

As just noted, **quaternary ammonium compounds** are cationic detergents. Their name comes from the fact that they are modifications of the four-valence ammonium ion (NH_4^+) (Figure 7–8). Quaternary ammonium compounds, or **quats,** as they are commonly called, have a very high bactericidal action against gram-positive bacteria, and somewhat less against gram-negative bacteria. They also have excellent bacteriostatic action. In addition, quats are fungicidal, amoebicidal, and virucidal against enveloped viruses. Quats probably exert their action by enzyme inhibition, denaturation, and disruption of plasma membranes.

Three popularly used quats are *Zephiran* (benzalkonium chloride), *Phemerol* (benzethonium chloride), and *Ceepryn* (cetylpyridinium chloride). Their popularity is due not only to their antimicrobial activity, but also to the fact that they are colorless, odorless, tasteless, nontoxic, stable, and easily diluted. Wearers of contact lenses will often find a quat in the storage solution. In fact, if your mouthwash foams heavily when shaken, it probably contains a quat. On the other hand, quats are not sporicidal, and they are readily absorbed by surfaces such as cotton and gauze. Moreover, organic matter interferes with their activity, and they are easily neutralized by soaps and anionic detergents.

Anyone associated with medical applications of quats should remember that certain bacteria, such as some species of *Pseudomonas*, not only survive in quaternary ammonium compounds but actively grow in them. This applies not only to the disinfectant solution, but to the moistened gauze and bandages as well, the fibers of which tend to neutralize the quats. These bacteria are also resistant to chlorines, which allow them to grow in swimming pools or whirlpool baths, where their presence can cause skin infections. This ability to metabolize unusual substrates for growth permits *Pseudomonas* to grow in such unlikely places as simple saline solutions, traces of soap residues, and cap-liner adhesives.

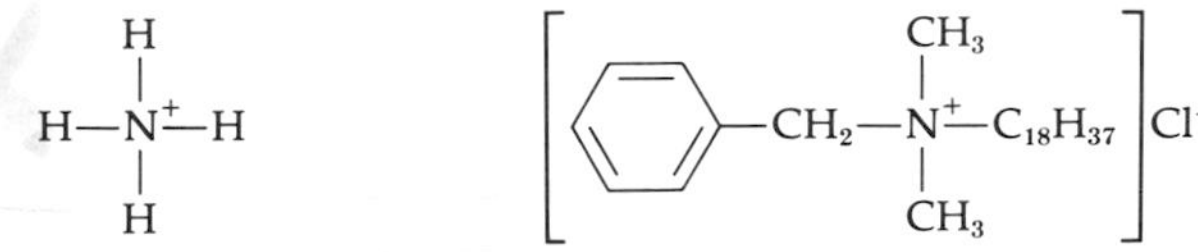

Before moving on to the next group of chemical agents of microbial control, refer to Figure 7–9, which compares the relative effectiveness of various substances discussed thus far.

H–N⁺(H)(H)–H — Ammonium ion

[C_6H_5–CH_2–N⁺(CH_3)(CH_3)–$C_{18}H_{37}$] Cl^- — Zephiran

Figure 7–8 The ammonium ion and a quaternary ammonium compound.

Acids and alkalies

Most bacteria grow in only a limited pH range near neutrality. This is why media for the growth of bacteria are adjusted to about pH 7. Molds, however, are an exception. They actually prefer acidic media and readily spoil foods if provided air for growth. The bactericidal effect of **mineral (inorganic) acids,** such as hydrochloric acid (HCl) and sulfuric acid (H_2SO_4), is due to hydrolysis and denaturation. However, since these acids are very destructive to tissues and corrosive to metals, they have little practical application. The bactericidal effect of **alkalies** is also due to hydrolysis and denaturation. And again, because they are destructive to tissues, alkalies have few uses in microbial control.

Organic acids have some practical use in controlling microbial growth. *Boric acid* is a mild antiseptic most often used as an eyewash. *Benzoic acid* and *salicylic acid* are used in combination in a mixture called Whitfield's ointment for treating fungus infections of the feet. *Undecylenic acid* is the active ingredient in Desenex, which is used to treat athlete's foot and other fungus infections of the skin.

There are a number of organic acids that are used as preservatives to control mold growth. Examples are *sorbic acid* and its salt, potassium sorbate; *benzoic acid* and its derivative *methylparaben;*

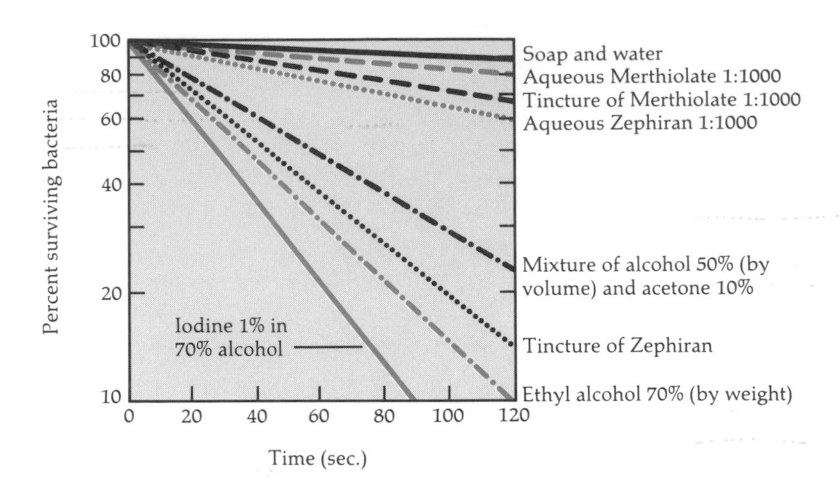

Figure 7–9 Comparison of the effectiveness of washing with various antiseptics. The steeper the curve, the more effective the antiseptic. For example, a 1% iodine in 70% alcohol is the most effective antiseptic, whereas soap and water is the least effective.

and *calcium propionate.* These compounds are listed on the labels of many foods and a few cosmetic preparations. By and large, the body metabolizes these organic acids readily, and so their use is considered quite safe.

Aldehydes

Aldehydes are among the most effective antimicrobials. Two examples are formaldehyde and glutaraldehyde. They combine with a number of organic functional groups on proteins (—NH_2, —OH, —COOH, and —SH), leading to protein inactivation. *Formaldehyde* gas is an excellent disinfectant. However, it is stable only in high concentrations and at high temperatures. The more commonly used form of formaldehyde is *formalin,* a 37% aqueous solution of formaldehyde gas. Formalin is used extensively to preserve biological specimens, embalm bodies in undertaking, and inactivate bacteria and viruses in vaccines. Formalin is sporocidal at high concentrations. Its sterilization and disinfecting applications are limited by its tissue-irritating qualities, poor penetration, slow action, unpleasant odor, and its property of leaving a white residue on treated materials.

Glutaraldehyde is a chemical relative of formaldehyde that is less irritating and more pleasant to handle than formaldehyde. When used in a 2% solution (Cidex), it is bactericidal, tuberculocidal, and virucidal in 10 minutes and sporocidal in 10 hours. It is probably the only chemical disinfectant that can be considered a possible sterilizing agent. The drastic sporocidal exposure time emphasizes the fact that chemical agents cannot be relied on to kill resistant endospores.

Gaseous chemosterilizers

Gaseous chemosterilizers are chemicals that sterilize in a closed chamber similar to an autoclave. A gas suitable for this purpose is *ethylene oxide.* Its activity is due to denaturation of proteins by a mechanism similar to that of the aldehydes discussed earlier. Ethylene oxide kills all microbes and endospores. Since it is toxic and explosive in pure form, it is usually mixed with an inert gas such as carbon dioxide or nitrogen. One of its principal advantages is that it is highly penetrative. Although materials to be sterilized with ethylene oxide require an exposure time of 4 to 18 hours, its remarkable penetrating power was a factor in choosing ethylene oxide to sterilize space craft sent to land on the moon and planets. Heat sterilization of electronic gear on these vehicles was not practical.

Because of their ability to sterilize without heat, these gases are also widely used on medical supplies and equipment. Examples include disposable sterile plasticware, such as syringes and Petri plates; textiles; sutures; lensed instruments; artificial heart valves; heart-lung machines; and mattresses. Many large hospitals have ethylene oxide chambers as part of their sterilizing equipment, some large enough to sterilize mattresses.

Oxidizing agents

Oxidizing agents exert antimicrobial activity by oxidizing cellular components of the treated microbes. Two examples of oxidizing agents are ozone and hydrogen peroxide. After a lightning storm, in the vicinity of electric sparking, or around an ultraviolet light, a rather fresh odor can be detected in the air. This is *ozone* (O_3), a highly reactive form of oxygen. There is presently much interest in the use of ozone as a replacement for chlorination in the decontamination of water. An advantage is that ozone leaves no chemical residue of any sort, and some concern is developing about this effect of chlorination. Ozone is usually generated by high-voltage electric discharges. At present, its use is not widespread, but there is reason to expect that this will change.

Hydrogen peroxide is an antiseptic found in many household medicine cabinets and hospital supply rooms. It is not a good antiseptic for open wounds since it deteriorates readily and is quickly broken down to hydrogen and oxygen. The bubbling of released oxygen as it is applied to a wound produces a mild cleansing effect. Hydrogen peroxide is used in the treatment of deep anaerobic wounds, where the release of peroxide tends to suppress the growth of anaerobic bacteria that lack the enzyme catalase (for example, *Clostridium*). The peroxide ion (O_2^{2-}) is a strong oxidizing agent and will react with many molecules in a cell. The bubbling of released oxygen (O_2) as hydrogen peroxide (H_2O_2) is applied to a wound results from the action of the enzymes catalase and peroxidase, which are present in human cells.

A summary of chemical agents that control microbial growth is presented in Table 7–7.

Table 7–7 Summary of Chemical Agents Used to Control Microbial Growth

Chemical Agent	Mechanism of Action	Comment	Preferred Use
Phenol and Phenolics:			
1. Phenol	Disruption of plasma membrane, denaturation, inactivation of enzymes	No longer used as a disinfectant or antiseptic because of its irritating qualities and disagreeable odor	Still used as a standard for measuring the effectiveness of other disinfectants (phenol coefficient); used to preserve specimens and embalm bodies
2. Phenolics	Disruption of plasma membrane, denaturation, inactivation of enzymes	Derivatives of phenol that are reactive even in the presence of organic material; examples include O-phenylphenol, hexylresorcinol, and hexachlorophene	Environmental surfaces, instruments, skin surfaces, and mucous membranes
Halogens	Iodine inhibits protein function and is a strong oxidizing agent; chlorine acts by forming the strong oxidizing agent hypochlorous acid, which alters cellular components	Iodine and chlorine may act alone or as components of inorganic and organic compounds	Effective antiseptic available as a tincture and an iodophor; chlorine gas is used to disinfect water; chlorine compounds are used to disinfect dairy equipment, eating utensils, household items, and glassware
Alcohols	Denaturation and lipid solvation	Bactericidal and fungicidal, but not effective against endospores and nonenveloped viruses; commonly used alcohols are ethanol (ethyl alcohol) and isopropanol (isopropyl alcohol)	Thermometers and other instruments; in a quick swabbing of the skin with alcohol before an injection, most of the disinfecting action probably comes from a simple wiping away (cleansing) of dirt and some microbes; usually another antiseptic is used in addition

(continued on next page)

Table 7–7, continued

Chemical Agent	Mechanism of Action	Comment	Preferred Use
Heavy Metals and Their Compounds	Denaturation of enzymes and other essential proteins	Heavy metals like silver and mercury are germicidal or antiseptic	Silver nitrate was used extensively to prevent gonococcal eye infections, was replaced by antibiotics, and is being used extensively again; Mercurochrome and Merthiolate are skin and mucous membrane disinfectants; copper sulfate is used to disinfect water supplies; zinc peroxide controls infections of deep puncture wounds and bites
Dyes	Interference with cellular oxidation	Antiseptic uses have been largely replaced by antibiotics	Dyes used extensively in the past are crystal or gentian violet for fungus infections and acridine dyes for burn and eye infections; acridine dyes being used experimentally against herpesviruses
Surface-Active Agents:		Have the ability to decrease surface tension	
1. Soaps	Mechanical removal of microbes through scrubbing	Many cosmetic soaps contain disinfectants that increase antimicrobial activity	Skin degerming and emulsification of debris
2. Detergents	Mechanical removal of microbes through scrubbing	Anionic detergents (laundry powders) mechanically remove microbes; cationic detergents include quaternary ammonium compounds; nonionic detergents are used as components of iodophors	Clothing, utensils, rubber goods
Quaternary Ammonium Compounds	Enzyme inhibition, protein denaturation, and disruption of plasma membranes	Cationic detergents that are bactericidal, bacteriostatic, fungicidal, amoebicidal, and virucidal against enveloped viruses; examples of quats are Zephiran, Phemerol, and Ceepryn	Skin antisepsis, instruments, utensils, rubber goods
Acids and Alkalies	Hydrolysis and denaturation	Mineral acids, such as hydrochloric acid and sulfuric acid, and alkalies are destructive to tissues; organic acids can control microbial growth	Mineral acids have little practical application; boric acid (eyewash), benzoic and salicylic acids (fungus infections), undecylenic acid (athlete's foot); sorbic acid, benzoic acid, and calcium propionate (control of molds in foods)
Aldehydes	Protein inactivation	Very effective antimicrobials	Formalin (37% aqueous solution) or formaldehyde is used to preserve specimens, embalm bodies, and inactivate microbes in vaccines; glutaraldehyde is less irritating than formaldehyde and is used in a 2% aqueous solution (Cidex)
Gaseous Chemosterilizers	Denaturation	Ethylene oxide is most commonly used	Excellent sterilizing agent, especially for objects that would be destroyed by heat
Oxidizing Agents	Oxidation	Ozone is gaining prominence as a disinfectant replacement for chlorination; hydrogen peroxide is a mild antiseptic that deteriorates rapidly	Contaminated waters and some deep wounds

STUDY OUTLINE

INTRODUCTION (pp. 172–173)

1. Ignatz Semmelweis introduced the concept of handwashing for hospital personnel to reduce infections.
2. Joseph Lister introduced antiseptic surgery, including heat sterilization of instruments and disinfection of wounds.
3. Microbial growth can be controlled by physical and chemical methods.
4. Control of microbial growth can prevent infections and food spoilage.
5. Sterilization is the process of destroying all microbial life on an object.
6. Disinfection is the process of reducing or inhibiting microbial growth.

CONDITIONS INFLUENCING MICROBIAL CONTROL (p. 174)

1. A general rule of disinfection is to try to kill the most resistant microbes that may be found on the object to be disinfected.
2. The number and species of microorganisms present will affect the rate of disinfection.

Temperature (p. 174)

1. Biochemical reactions occur more rapidly at warm temperatures.
2. Disinfectant activity is enhanced by warm temperatures.

Kind of Microbe (p. 174)

1. Gram-positive bacteria are generally more susceptible to disinfectants than gram-negative bacteria.
2. Pseudomonads can even grow in some disinfectants and antiseptics.

Physiological State of the Microbe (p. 174)

1. An actively growing microorganism tends to be less resistant to chemical agents than an older microorganism.
2. Endospores are more resistant to chemical agents.

Environment (p. 174)

1. Organic matter frequently interferes with the action of chemical control agents.
2. Examples of such organic matter include vomit, feces, pus, and food.

ACTIONS OF MICROBIAL CONTROL AGENTS (p. 175)

Alteration of Membrane Permeability (p. 175)

1. The susceptibility of the plasma membrane is due to its lipid and protein components and proximity to the cell surface.
2. Certain chemical control agents damage the plasma membrane by altering its permeability.

Damage to Proteins and Nucleic Acids (p. 175)

1. Some microbial control agents damage cellular proteins by breaking hydrogen and covalent bonds.
2. Others interfere with protein synthesis; still others interfere with DNA and RNA replication.

PATTERN OF MICROBIAL DEATH (p. 175)

1. Bacterial populations subjected to heat or antimicrobial chemicals die off at a constant rate over a period of time.
2. Such a death curve can be plotted logarithmically.

PHYSICAL METHODS OF MICROBIAL CONTROL (pp. 176–181)

Heat (pp. 176–179)

1. Heat is frequently employed to eliminate microorganisms; it is economical and easily controlled.
2. The mechanism of action of heat is denaturation of enzymes.
3. Thermal death point is the temperature required to kill all the bacteria in a liquid culture in 10 minutes at pH 7.
4. Thermal death time is the length of time required to kill all bacteria in a liquid culture at a given temperature.
5. Decimal reduction time is the time required to kill 90% of a bacterial population at a given temperature.
6. Methods of moist heat sterilization include boiling, fractional sterilization, and autoclaving (steam under pressure).
7. Methods of dry heat sterilization include direct flaming, incineration, and hot-air sterilization.
8. Pasteurization employs a high temperature for a short time (72°C for 15 seconds) to destroy pathogens without altering the flavor of the food.
9. Equivalent treatment refers to producing the same effect (reduction in microbial growth) using different methods.

Filtration (p. 179)

1. Filtration is the passage of a liquid or gas through a filter with pores small enough to retain microbes.
2. Membrane filters composed of nitrocellulose or cellulose acetate are commonly used to filter out bacteria, viruses, and even large proteins.

Low Temperature (p. 179)

1. The effectiveness of low temperatures depends on the particular microorganism and the intensity of the application.
2. Most microorganisms do not reproduce at ordinary refrigerator temperatures (0° to 7°C).

Desiccation (pp. 179–180)

1. In the absence of water, microorganisms cannot grow, but may remain viable.
2. Viruses and endospores can resist desiccation.

Osmotic Pressure (pp. 180–181)

1. Microorganisms undergo plasmolysis in high concentrations of salts and sugars.
2. Molds and yeasts are more capable of growth in materials with low moisture or high osmotic pressure than are bacteria.

Radiation (p. 181)

1. The lethal effects of radiation depend on wavelength, intensity, and duration.
2. Ionizing radiation (X rays and gamma rays) has a high degree of penetration and exerts its effect primarily by breaking DNA strands.
3. Ultraviolet radiation, a form of nonionizing radiation, has a low degree of penetration and causes cell damage by making thymine dimers in DNA that interfere with the DNA replication; the most effective germicidal wavelength is 260 nm.

CHEMICAL METHODS OF MICROBIAL CONTROL (pp. 181–192)

1. Chemical agents are used as antiseptics and disinfectants.
2. Few chemical agents achieve sterility.

Qualities of an Effective Disinfectant (pp. 181–182)

1. Among the qualities of a good disinfectant are rapid action, wide range of action, good penetration, capability of being mixed with water, activity in organic matter, and resistance to decomposition.
2. An effective disinfectant also will be nonstaining and noncorrosive, exhibit low toxicity to body tissues, and be odorless.

Principles of Effective Disinfection (p. 182)

1. Careful attention should be paid to the properties and application concentration of the disinfectant to be used.
2. The presence of organic matter, contact with microorganisms, and temperature should also be considered.

Evaluating a Disinfectant (pp. 182–184)

1. The phenol coefficient is the disinfecting action of a chemical compared with that of phenol for the same length of time on the same organism under identical conditions.
2. The use-dilution test involves a series of tubes containing increasing concentrations of disinfectant; the more the chemical can be diluted and still be effective, the higher its evaluation rating.
3. The filter paper method consists of soaking a disk of filter paper with a chemical and placing it on an inoculated agar plate; a clear zone of inhibition indicates effectiveness.

Kinds of Disinfectants (pp. 184–192)

Phenol and phenolics (pp. 184–185)

1. Phenolics exert their action by injuring plasma membranes, inactivating enzymes, and denaturing proteins.
2. Common phenolics are cresols, hexylresorcinol, and hexachlorophene.

Halogens (pp. 185–186)

1. Halogens include iodine and chlorine alone or as components of inorganic or organic molecules.
2. Iodine inactivates enzymes and other cellular proteins by combining with the amino acid tyrosine.
3. Iodine is available as a tincture (in solution with alcohol) or as an iodophor (combined with an organic molecule).
4. The germicidal action of chlorine is based on the formation of hypochlorous acid when chlorine is added to water.
5. Chlorine is used as a disinfectant in gaseous form (Cl_2), or in the form of a compound, such as calcium hypochlorite, sodium hypochlorite, and chloramines.

Alcohols (p. 186)

1. Alcohols exert their action by denaturation and lipid solvation.
2. In tinctures, they enhance the effectiveness of other antimicrobial chemicals.
3. Ethanol and isopropanol are used as disinfectants.

Heavy metals and their compounds (pp. 186–188)

1. Silver, mercury, copper, and zinc are used as germicidals or antiseptics.
2. They exert their antimicrobial action through oligodynamic action, which results in denaturation.

Dyes (p. 188)

1. Although dyes are no longer commonly used as antimicrobial agents, several have antimicrobial properties.
2. Their antimicrobial action is interference with cellular metabolism.

Surface-active agents (pp. 188–189)

1. Surface-active agents decrease the tension between molecules that lie on the surface of a liquid; included are soaps and detergents.
2. Soaps are made from fats and lye (NaOH); they have limited germicidal action but assist in the removal of microorganisms through scrubbing.
3. Detergents are hydrocarbons classified as anionic, cationic, and nonionic, depending on their dissociation in water.

Quaternary ammonium compounds (p. 189)

1. Quats are cationic detergents attached to NH_4^+.
2. They inhibit enzymes, cause denaturation, and disrupt plasma membranes.
3. They are most effective against gram-positive bacteria.

Acids and alkalies (pp. 189–190)

1. Mineral (inorganic) acids and alkalies bring about hydrolysis and denaturation but have little practical application because they destroy tissues and corrode metals.
2. Organic acids are used to control microbial growth on skin and in foods; they exert their action by hydrolysis and denaturation.

Aldehydes (p. 190)

1. Aldehydes such as formaldehyde and glutaraldehyde exert their antimicrobial effect by inactivating proteins.

2. They are among the most effective antimicrobials.

Gaseous chemosterilizers (pp. 190–191)

1. Ethylene oxide is the gas most frequently used for sterilization.
2. It penetrates most materials and kills all microorganisms by denaturation.

Oxidizing agents (p. 191)

1. Ozone and hydrogen peroxide are used as antimicrobial agents.
2. They exert their effect by the oxidization of molecules inside cells.

STUDY QUESTIONS

REVIEW

1. Name the cause of cell death resulting from damage to the following:
   (a) Cell wall
   (b) Plasma membrane
   (c) Proteins
   (d) Nucleic acids
2. A bacterial culture was in log phase. At time x, a bactericidal compound was added to the culture medium. Explain why the viable count does not immediately drop to zero.

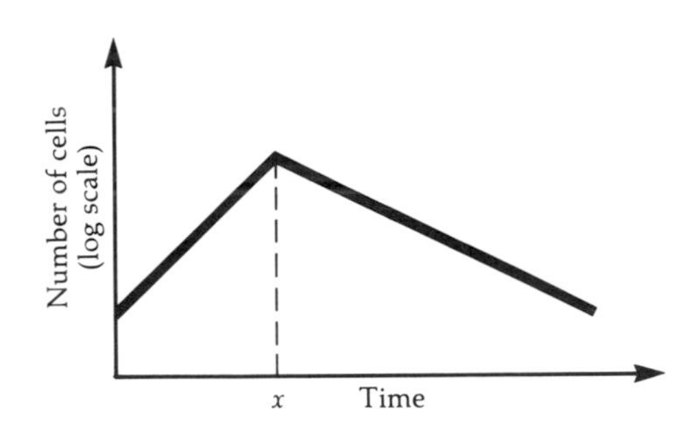

3. The antimicrobial activity of heat is due to ______________.
4. Fill in the following table.

Method of Sterilization	Temperature	Heat	Preferred Use
Fractional sterilization			
Autoclaving			
Dry heat			

5. (a) Pasteurization involves employing heat for how long at what temperature?
   (b) If pasteurization does not achieve sterilization, why is food treated by pasteurization?
6. Heat-labile solutions such as glucose-minimal salts broth are best sterilized by ______________.
7. Thermal death point is not considered an accurate measure of the effectiveness of heat sterilization. List three factors that may alter the thermal death point.
8. (a) The antimicrobial effect of gamma radiation is due to ______________.
   (b) The antimicrobial effect of ultraviolet radiation is due to ______________.
9. List ten factors to consider before selecting a disinfectant.
10. Give the method of action and at least one standard use of each of the following types of disinfectants:
    (a) Phenolics
    (b) Iodine
    (c) Chlorine
    (d) Alcohol
    (e) Heavy metals
    (f) Aldehydes
    (g) Ethylene oxide
    (h) Oxidizing agents
11. The phenol coefficients against *S. aureus* for ethanol and isopropanol are 0.039 and 0.054, respectively. Which is the more effective antimicrobial agent? What is a phenol coefficient?
12. Contrast soaps and detergents with respect to their chemical composition and antimicrobial activity.
13. The use-dilution values for two disinfectants tested under the same conditions are given below. Assuming that both disinfectants are designed for the same purpose, which would you select? Disinfectant A—1:2; Disinfectant B—1:10,000.
14. A large hospital washes burn patients in a stainless steel tub. After each patient, the tub is cleaned with a quat. It was noticed that 14 out of 20 burn patients acquired *Pseudomonas* infections after being bathed. Provide an explanation for this high rate of infection.

CHALLENGE

1. The filter paper method was used to evaluate three disinfectants. The results were as follows:

Disinfectant	Zone of Inhibition
X	0 mm
Y	5 mm
Z	10 mm

   (a) Which disinfectant was the most effective against the organism used?
   (b) Can you determine whether compound Y was bactericidal or bacteriostatic?

2. The thermal death time for a suspension of *B. subtilis* endospores in dry heat is 30 minutes, and in an autoclave, less than 10 minutes. Which type of heat is more effective? Why?

3. A bacterial culture was in log phase. At time x, a bacteriostatic compound was added to the culture. Explain why this graph of the number of viable organisms does not look like the graph in review question 2.

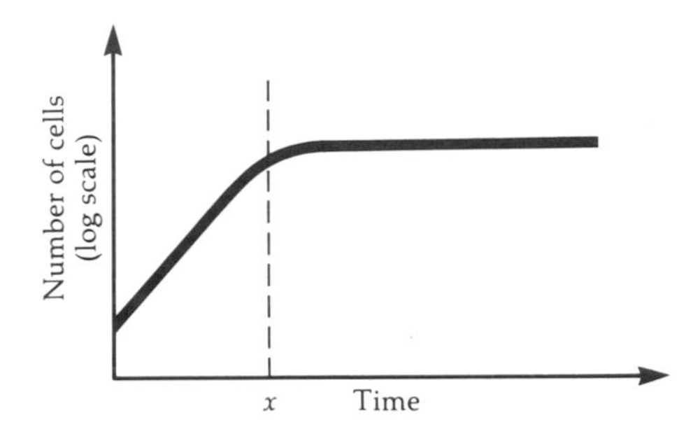

FURTHER READING

Block, S.S., ed. 1977. *Disinfection, Sterilization and Preservation*, 2nd ed. Philadelphia, PA: Lea and Febiger. A comprehensive reference on principles and procedures for disinfecting and sterilizing.

Board of Education and Training. 1976. *Infection Control in the Hospital Environment*. Washington, D.C.: American Society for Microbiology (Proceedings). Lectures and articles on disinfection and environmental monitoring in the hospital.

Brewer, J.H., ed. 1973. *Lectures on Sterilization*. Durham, NC: Duke University Press. Lectures addressed to hospital personnel who use supplies and are responsible for maintaining sterility and asepsis.

Castle, M. 1980. *Hospital Infection Control*. New York: John Wiley. An excellent reference for practical surveillance and maintenance of asepsis.

McInnes, B. 1977. *Controlling the Spread of Infection*, 2nd ed. St. Louis, MO: Mosby. A self-study book designed to help the student learn the definition of asepsis and procedures for control of infections.

Phillips, G.B. and W.S. Miller, eds. 1973. *Industrial Sterilization*. Durham, NC: Duke University Press. Methods and applications for the use of gas, formaldehyde, and radiation sterilization.

8

Microbial Genetics

Objectives

After completing this chapter you should be able to

- Define the following: genetics; chromosome; gene; genetic code; mutagen; genetic recombination.
- Describe DNA replication.
- Describe protein synthesis.
- Classify mutations by type.
- Describe the actions of mutagens on DNA.
- Outline how mutations are identified by direct and indirect selection.
- Compare mechanisms of genetic transfer in bacteria.
- Define plasmids and discuss their functions.
- Outline how recombinant DNA is produced in vitro and list its potential uses.
- Explain the regulation of gene expression through feedback inhibition, induction, and repression.

All the characteristics of microorganisms, including their morphology, metabolism, behavior, and pathogenicity, are inherited characteristics. Individual organisms transmit these characteristics to their offspring through genes, which are units of heredity that contain the information for determining these characteristics. **Genetics,** the science of heredity, is the study of what genes are, how they carry information, how they are replicated and passed to further generations of cells or passed between organisms, and how their information is expressed within an organism to determine the particular characteristics of that organism.

STRUCTURE AND FUNCTION OF THE GENETIC MATERIAL

Genes consist of DNA, deoxyribonucleic acid. In Chapter 2, we saw that DNA is a macromolecule composed of repeating units called *nucleotides*. You may recall that each nucleotide consists of a nitrogenous base (adenine, thymine, cytosine, or guanine), deoxyribose (a pentose sugar), and a phosphate group. How the nucleotides are arranged in a DNA molecule may be reviewed in Figure 2–19.

The DNA within a cell exists as two long strands of nucleotides twisted together to form a double helix. Each strand is a string of alternating sugar and phosphate groups, and to each sugar is attached a nitrogenous base. The two strands are held together by hydrogen bonds between their nitrogenous bases. The bases are paired in a specific, complementary way. Adenine (A) pairs with thymine (T), and cytosine (C) pairs with guanine (G). The thing to remember here is the notion of hydrogen bonding between the nitrogenous bases of the two strands leading to the formation of specific base pairs, A–T and C–G.

Later in this chapter, we will describe several important experiments that clearly identified DNA as the genetic material of cells. By saying that DNA is the genetic material, we mean that this molecule carries the cell's history as well as its future. The information locked within DNA sequences determines the characteristics of the cell and transfers these characteristics to subsequent generations of cells. But DNA does not work alone. A gene is not "expressed" (the chemical it codes for, usually a protein, is not actively produced) until RNA reads the code and directs the proper assembly of amino acids to form the protein prescribed by the gene. Proteins are so important that much of cellular machinery is concerned with translating the DNA of genes into specific proteins. The genetic information in a region of DNA is transcribed (copied) to produce a specific molecule of RNA. The information encoded in RNA is then translated into a specific sequence of amino acids that form a newly produced protein.

A **gene** may be defined as a segment of DNA (or a sequence of nucleotides in DNA) that codes for a functional product. The final product can be a molecule of ribosomal RNA, for example. Usually, however, the final product is a protein. Thus, DNA is transcribed to produce a specific molecule of RNA, and the information encoded in RNA is translated into protein. This can be symbolized as follows:

$$\text{DNA} \xrightarrow{\text{Transcription}} \text{RNA} \xrightarrow{\text{Translation}} \text{Protein}$$

This flow of genetic information within a cell and between generations of cells is summarized in Figure 8–1.

DNA is a particularly suitable molecule to act as the cell's genetic library. Looking at the outside of the molecule, we tend to focus on the repeating alteration of sugars and phosphates. But inside, purine and pyrimidine bases are arranged in specific sequences that form strings of information—the genes that make up the books in this library. These books use a genetic alphabet with only four letters: the bases A, T, G, and C. But a gene with 1000 bases can be arranged in 4^{1000} different ways. This is how the genes can carry so much information; they code it within the multitude of base sequences available to the DNA molecule.

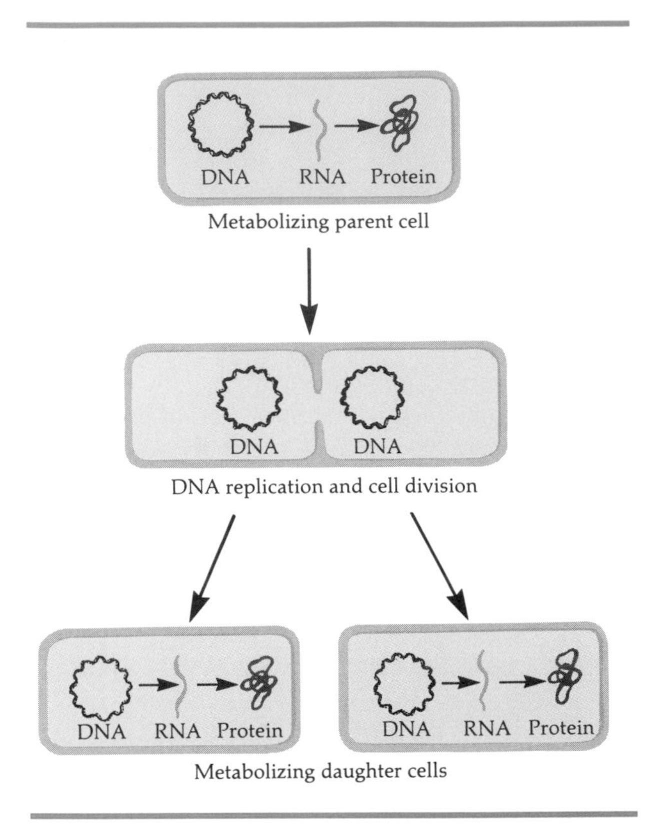

Figure 8–1 Flow of genetic information within a cell and between generations of cells.

Besides being able to store information, DNA can also carry this information from cell to cell and from generation to generation, because the specific sequence of nucleotides contained in the DNA usually duplicates accurately each time the cell divides. We shall see in the next section how this amazing feat is made possible by the distinctive structure of DNA, with its double helix formed by two complementary nucleotide chains.

Because the two DNA strands can open up and rejoin easily, genes are accessible to the rest of the cell. But the genes are not removed from the DNA to make proteins. Instead, copies of the genes are transcribed into RNA messengers, which carry these instructions to the ribosomes to make proteins as needed.

Finally, DNA can change, or mutate. Although some mutations can kill the cell, others permit the offspring a better chance of survival should they be subjected to new environments or stressful conditions. Thus, DNA's ability to mutate is in the long run an advantage that contributes to the evolution of the organism.

Before we discuss the details of how DNA replicates, functions, and changes, let us briefly discuss the way the DNA is arranged within the cells—the structure of the chromosomes.

DNA AND CHROMOSOMES

Although the structure of DNA is fairly well understood, the packaging of DNA in chromosomes is not. Evidence available so far suggests that the DNA in each chromosome—even in the large, complex chromosomes of eucaryotes—exists as one long double helix associated with various kinds of protein that regulate genetic activity.

The DNA of *Escherichia coli,* the common bacterium of the human large intestine, is contained in a typical procaryotic chromosome: a single long molecule of DNA with no nuclear membrane to enclose it (Figure 8–2a). The chromosome is attached at one or several points to the plasma membrane. *E. coli* DNA has about 4 million base pairs and is about 1 mm long—1000 times longer than the entire cell. However, DNA is also very thin and tightly packed inside the cell as a twisted, coiled macromolecule that takes up about 10% of the cell's volume. Bacterial DNA is circular—a closed loop with no free ends. Some proteins (including RNA polymerase and others) are associated with bacterial DNA. Because it is a very long molecule, it can contain large amounts of information. Essentially, it contains the blueprint for the entire cell.

Eucaryotic chromosomes (Figure 8–2b) contain DNA that is even more highly coiled (condensed) than procaryotic DNA, and they contain much more protein than procaryotic chromosomes do. Eucaryotic cells have histone proteins that are arranged in a very regular way with their DNA.

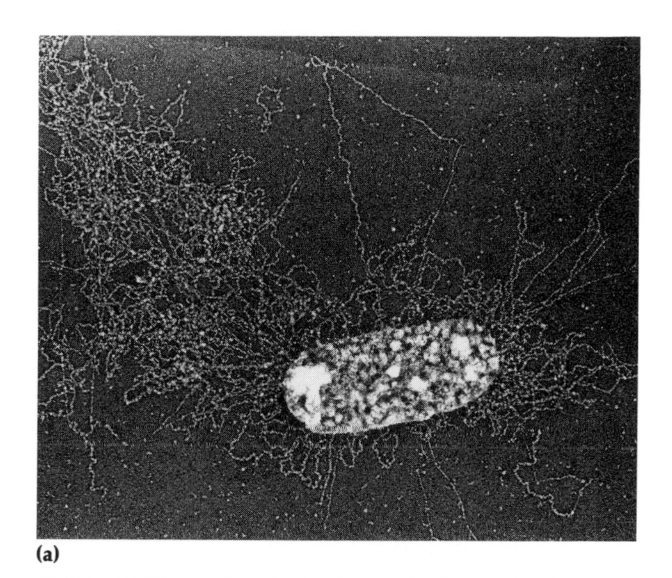

(a)

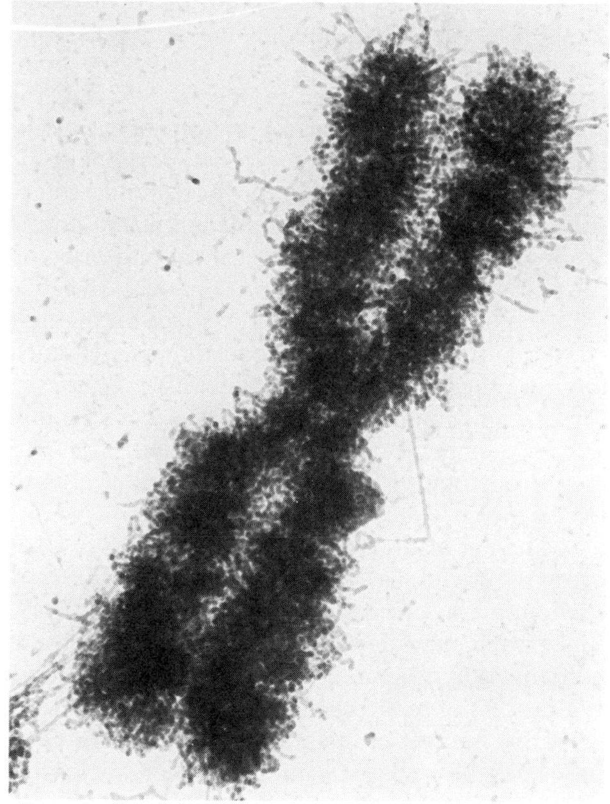

(b)

Figure 8–2 (a) Electron micrograph of a procaryotic chromosome. The tangled mass and looping strands of DNA emerging from this disrupted cell of *Escherichia coli* are part of its single chromosome (×6000). **(b)** Electron micrograph of a eucaryotic chromosome. This human chromosome is but one of 46 chromosomes found in a normal human cell. Note the individual strands of protein–DNA fibers (×28,000).

The detailed structure of the eucaryotic chromosome, and the precise arrangement of DNA with proteins, is still under investigation. Researchers feel that an understanding of the physical and chemical arrangement of the protein-wrapped DNA is likely to reveal how genes are turned on and off to produce crucial proteins when needed.

We will now examine how DNA functions as the genetic material through replication and protein synthesis.

DNA REPLICATION

The complementarity of the nitrogenous base sequences in the two strands of the double helix provides the key to understanding DNA replication. In the process of replication, the two strands of the original DNA double helix separate from each other (Figure 8–3). The separation point where new strands will be synthesized is called the *replication fork.* Following separation, new nucleotides are matched up to the exposed portions of the original strands according to the complementary rules of base pairing. If thymine is present on the original strand, only adenine can fit into place on the new strand; if guanine is present on the original strand, only cytosine can fit into place on the new strand; and so on. The new nucleotides are joined to each other by enzymes called *DNA polymerases.* Initially, only short fragments of DNA are synthesized. An enzyme called *DNA ligase* joins the fragments together, producing two complete strands of DNA.

The net result is that two new strands of DNA are synthesized, each having a base sequence complementary to one of the original strands of DNA. Since each new double-stranded DNA molecule contains one original strand and one new strand, the process of replication is referred to as **semiconservative replication.** Along the length of the DNA molecules, there are specific sites where replication is initiated; in bacterial DNA, there is usually one initiation site. In some cases, replication moves in one direction; other times it moves in two. Bacteria have circular molecules of DNA and in some bacteria, such as *E. coli,* replication is bidirectional. Two replication forks move in opposite directions around the DNA molecule and eventually meet to complete replication (Figure 8–4).

RNA AND PROTEIN SYNTHESIS

Although the previous section explains how DNA stores genetic information in nucleotide sequences and replicates itself by a process involving base pairing, it does not explain how the information in DNA is translated into the proteins that control cell activities. In a process called transcription, much of the genetic information in DNA is copied, or encoded, by a complementary base sequence of RNA. The RNA message is then used by the cell to synthesize specific proteins through the process of translation. We will now take a closer look at these two processes.

Transcription

In the process of **transcription,** a strand of RNA called **messenger RNA (mRNA)** is synthesized by using a specific portion of the cell's DNA as a template (Figure 8–5). In other words, the genetic information stored in the sequence of nitrogenous bases of DNA is rewritten so that the same information appears in the nitrogenous base sequence of RNA. As in DNA replication, a G in the DNA template dictates a C in the mRNA being made; a C in the DNA template dictates a G in the mRNA; and a T in the DNA template dictates an A in the mRNA. However, an A in the DNA template dictates a uracil (U) in the mRNA, since RNA does not contain T; it contains U instead. If, for example, the template portion of DNA has the base sequence ATGCAT, then the newly synthesized mRNA strand would have the complementary base sequence UACGUA. This may be represented as follows:

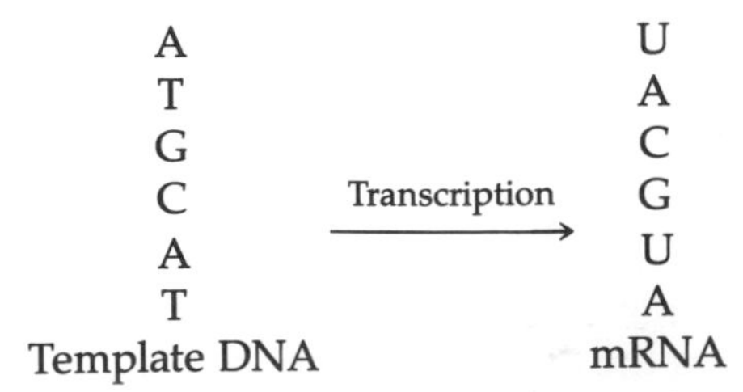

The process of transcription requires an enzyme called *RNA polymerase* and a supply of RNA nucleotides. Only one of the two DNA strands serves as the template for RNA synthesis. This is called the *sense strand.* The other strand, the *antisense strand,*

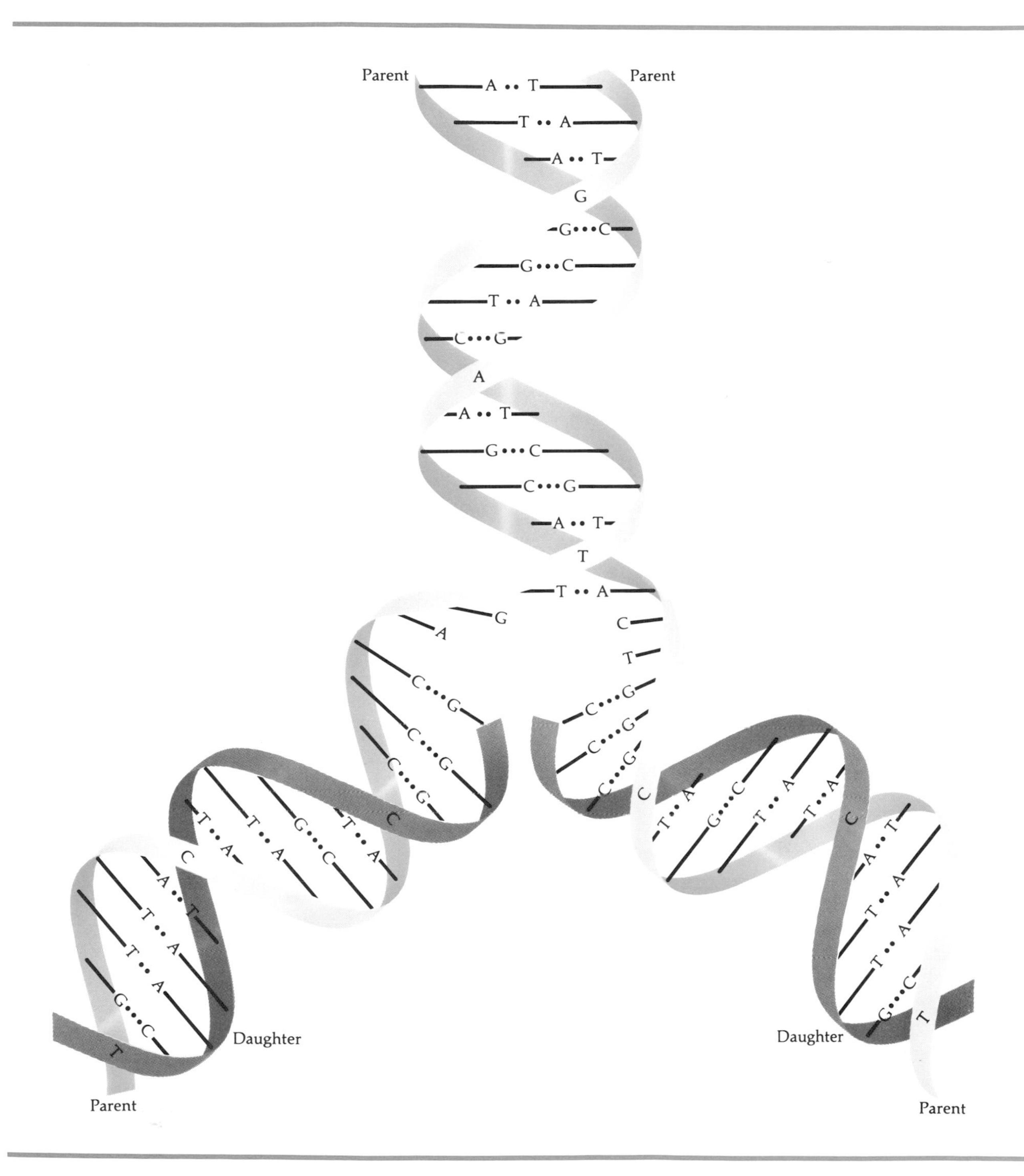

Figure 8–3 DNA replication. After the double helix separates, weak hydrogen bonds between nucleotides of the unspiraled strands break. Next, new complementary nucleotides are attached at the proper sites. Then hydrogen bonds are formed between complementary nucleotides and the process of replication is completed. Note that each daughter strand (gray) is a replica of one of the original (parent) strands (color).

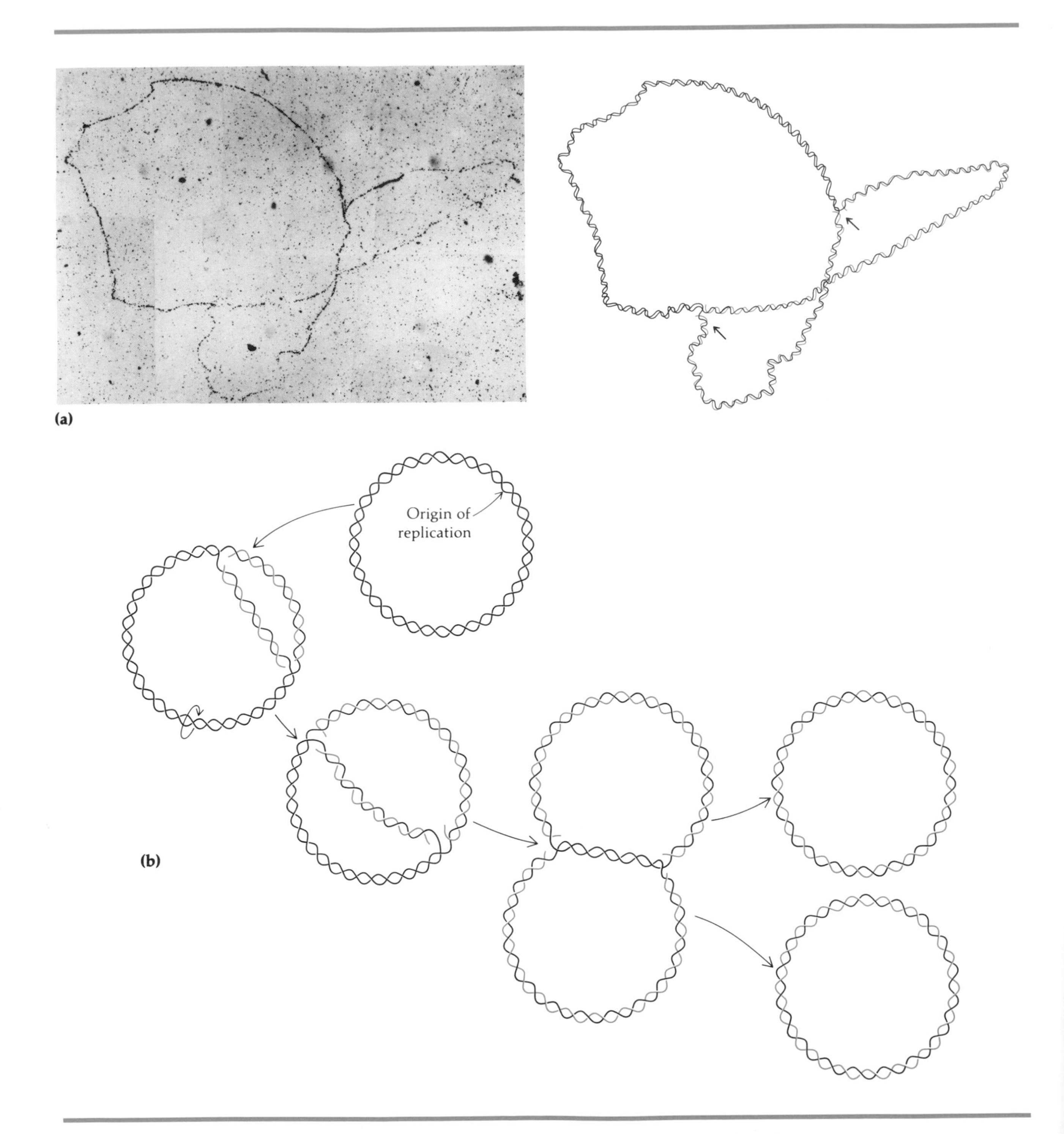

Figure 8–4 Replication of bacterial DNA. **(a)** Autoradiograph of the replication of an *E. coli* chromosome (corresponding diagram in the upper right). The arrows point to the two replication forks. The chromosome is about one-third replicated. Note that one of the new helices is crossed over the other one. **(b)** Diagrammatic representation of the bidirectional replication of a circular DNA molecule. The new strand is shown in color.

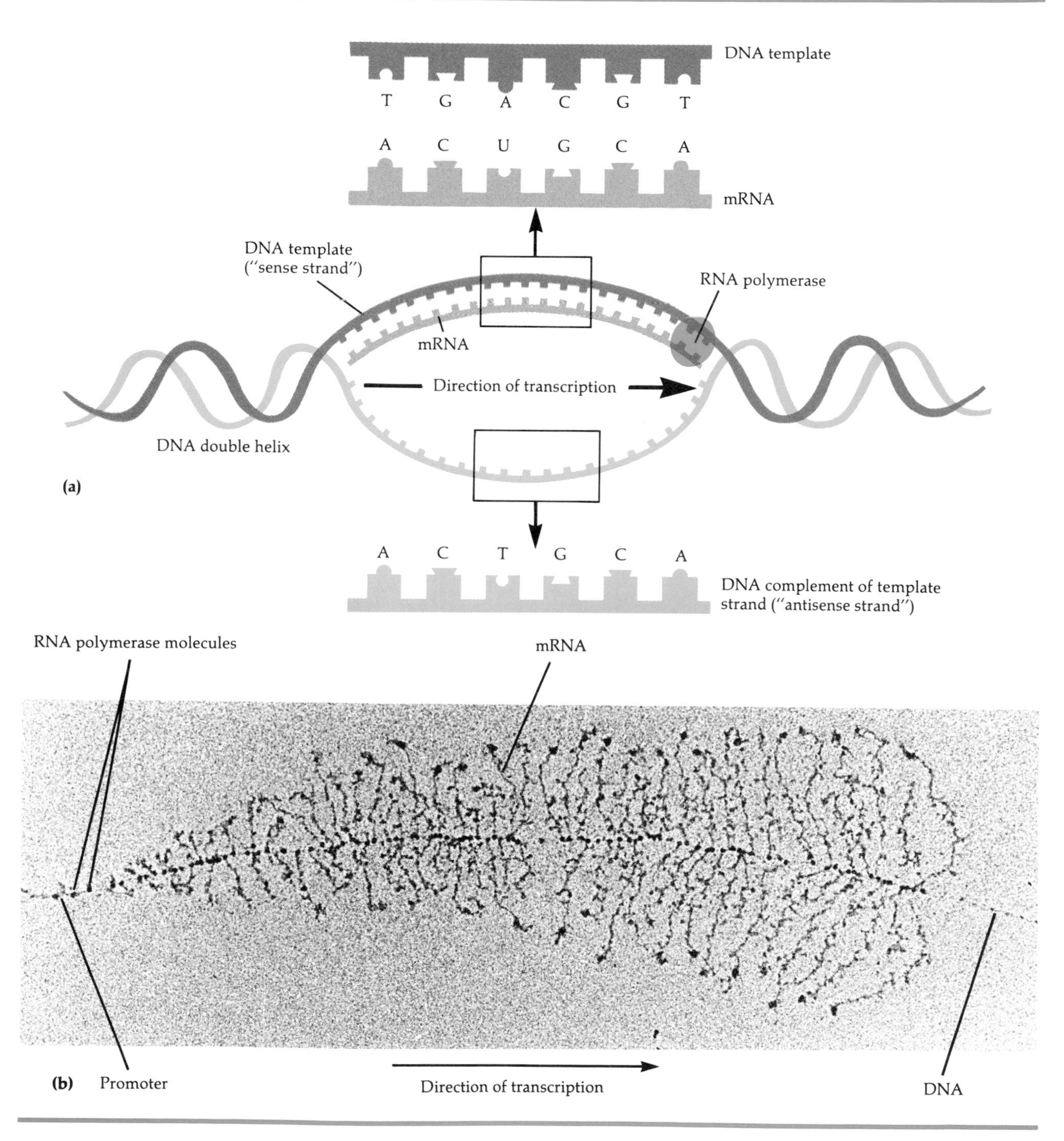

Figure 8–5 Transcription. **(a)** Diagram of a partly uncoiled DNA double helix. Transcription occurs in the unwound region of DNA. The upper box shows transcription, the lower box the base sequence of the antisense DNA strand. **(b)** Electron micrograph of transcription from a single gene (×51,000). Many molecules of mRNA are being synthesized simultaneously, starting at the promoter site. The longest mRNA molecules are the oldest. Note the many RNA polymerase molecules working all along the DNA.

is the complement of the sense strand. The region where RNA polymerase binds tightly to DNA is known as the **promoter** site; this is where transcription begins. The region of DNA that acts as the end point for transcription is referred to as the **terminator** site; at this site the RNA polymerase and newly formed mRNA are released from the DNA. The DNA double helix then reforms.

In summary, then, the genetic information stored in DNA for protein synthesis is passed to mRNA during transcription.

Translation

The process of using the information in the nitrogenous base sequence of mRNA to dictate the amino acid sequence of a protein is called **translation.** The events involved in translation are described in this section and shown in Figure 8–6.

First, one end of the mRNA molecule becomes associated with a ribosome, the site of protein synthesis (Figure 8–6a). Ribosomes (see Chapter 4) consist of a special type of RNA, called **ribosomal RNA** or **rRNA,** and protein. Each ribosome consists of two subunits.

In solution in the cytoplasm are 20 different amino acids that may participate in protein synthesis. These 20 amino acids and their abbreviations are listed in Table 8–1. These amino acids are synthesized by the cell or taken up from the external medium. However, before the appropriate amino acids can be joined together to form a protein, they must be *activated*. Another type of RNA, called **transfer RNA** or **tRNA,** participates in this step (Figure 8–7a). For each different amino acid, there is a different type of tRNA. In the process of amino acid activation, a specific amino acid is attached to its specific type of tRNA. This attachment is accomplished by an enzyme plus energy from ATP (Figure 8–7b).

One part of the tRNA molecule has a sequence of three nitrogenous bases that matches up with its complementary triplet on an mRNA strand. Each set of three nitrogenous bases on mRNA is called a **codon.** Their complement, the three nucleotides in tRNA, is the **anticodon** (Figure 8–7c). During translation, the anticodon of a molecule of tRNA attaches to its complementary codon on mRNA. For example, a tRNA with anticodon UAC pairs with the mRNA codon AUG (Figure 8–6b). The tRNA molecule is also bringing along its amino acid. The pairing of codon and anticodon occurs only where mRNA is attached to a ribosome.

Table 8–1 The Twenty Amino Acids Found in Proteins and Their Abbreviations

Name	Three-Letter Abbreviation
Alanine	Ala
Arginine	Arg
Asparagine	Asn
Aspartic acid	Asp
Cysteine	Cys
Glutamic acid	Glu
Glutamine	Gln
Glycine	Gly
Histidine	His
Isoleucine	Ile
Leucine	Leu
Lysine	Lys
Methionine	Met
Phenylalanine	Phe
Proline	Pro
Serine	Ser
Threonine	Thr
Tryptophan	Trp
Tyrosine	Tyr
Valine	Val

After the first tRNA, with its amino acid, attaches to mRNA, the ribosome moves along the mRNA strand and the second tRNA molecule, with its amino acid, moves into position (Figure 8–6b). The two amino acids are joined by a peptide bond, and the first tRNA molecule detaches itself from the mRNA strand (Figure 8–6c). It can now pick up another amino acid, for each tRNA can be used over and over again. As the proper amino acids are brought into line one by one, peptide bonds form between the amino acids and a polypeptide chain is formed (Figure 8–6d).

A special termination codon in the mRNA (called a **nonsense codon** because it does not specify any amino acid) signals the end of a polypeptide chain and its release from the ribosome (Figure 8–6e). The ribosome then comes apart into its two subunits.

As the ribosome moves along the mRNA, and before it completes translation of that gene, another ribosome may attach and begin translation of the

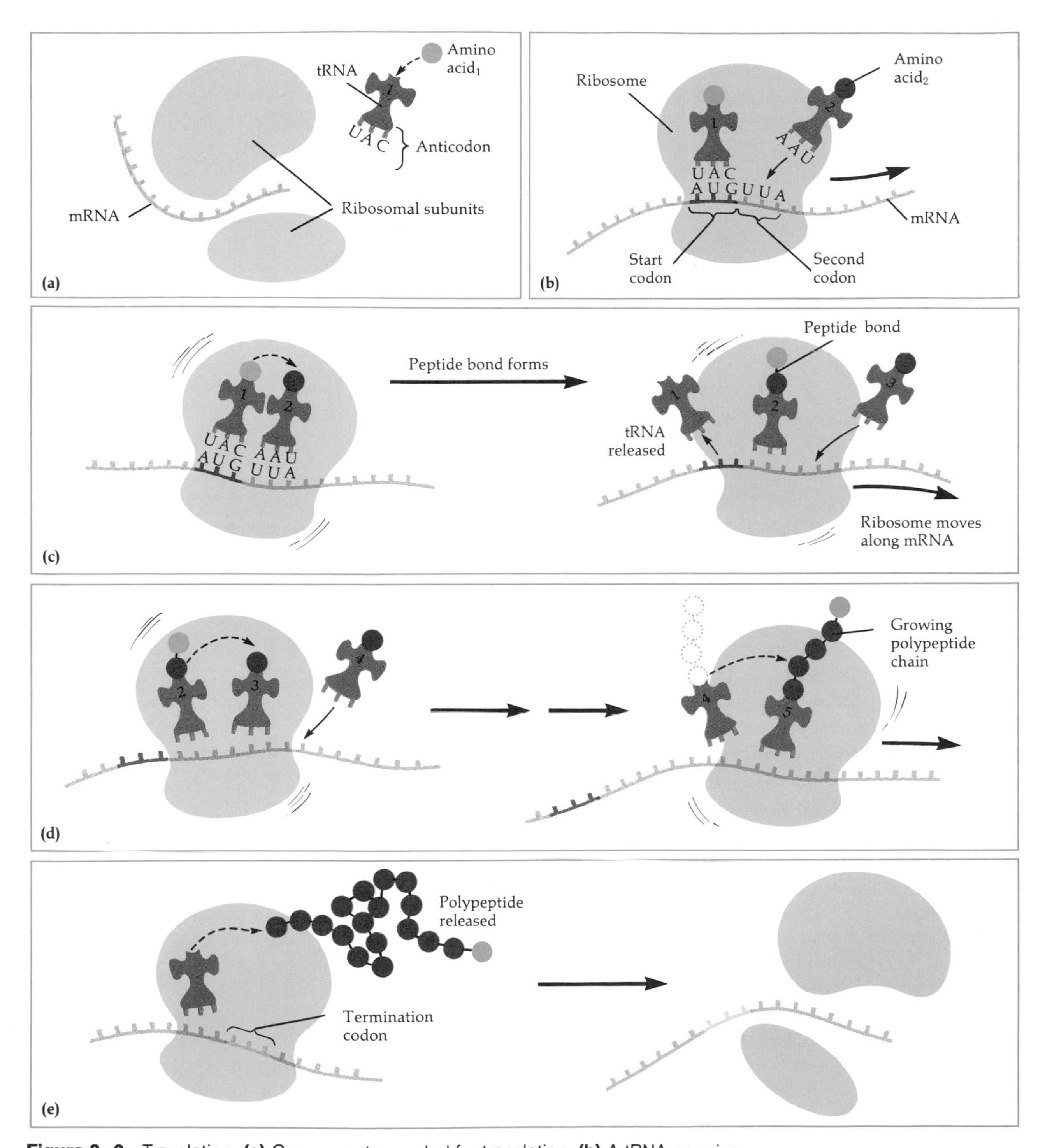

Figure 8–6 Translation. **(a)** Components needed for translation. **(b)** A tRNA carrying the first amino acid pairs with the start codon on the mRNA. **(c)** The second codon of the mRNA pairs with the tRNA carrying the second amino acid. The first amino acid is joined to the second by a peptide bond, and the first tRNA is released. **(d)** The ribosome moves along the mRNA to the next codon, and the process continues. (Nucleotide bases are shown only for the first two codons.) The chain of amino acids (polypeptide) grows. **(e)** When the ribosome reaches the termination codon, the polypeptide is released, the last tRNA is released, and the ribosome comes apart.

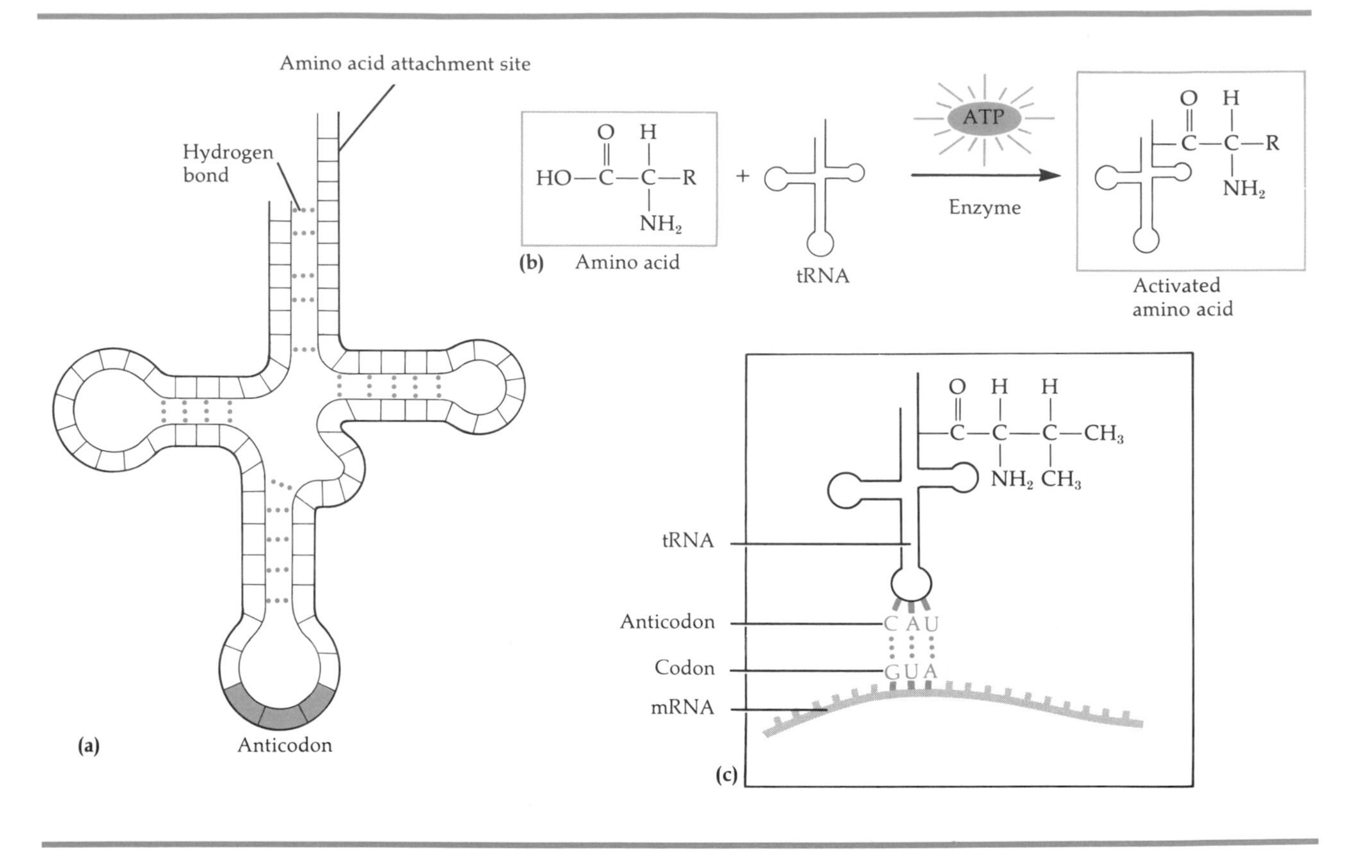

Figure 8–7 Transfer RNA. **(a)** Structure of tRNA. Each "box" represents a nucleotide. Note the regions of hydrogen bonding between base pairs, and the loops of unpaired bases. **(b)** Amino acid activation for attachment of an amino acid to tRNA. **(c)** The anticodon of tRNA pairs with its complementary codon on an mRNA strand.

same mRNA molecule, and then another, until there are a number of ribosomes attached at different positions to the same mRNA molecule. In this way, a single mRNA strand can be translated simultaneously into several identical protein molecules. An mRNA strand with several ribosomes attached is called a **polyribosome.**

In summary, the synthesis of a protein requires that the genetic information in DNA first be transferred to a molecule of mRNA by a process called transcription. Then, in a process called translation, the mRNA attaches to a ribosome, tRNA activates amino acids, and the amino acids are delivered to mRNA by tRNA according to complementary base pairings of codons and anticodons. The amino acids are then linked to form polypeptides. Proteins may be composed of one or more polypeptide chains, each of which may range in length from 50 to several hundred amino acids.

Since a typical polypeptide has about 300 amino acids, the typical mRNA must have about 300 codons and thus be about 900 bases long. As explained earlier, DNA is internally marked off in regions called genes, each region coding for a different kind of mRNA. A gene, therefore, can also be defined as a sequence of nitrogenous bases of DNA that codes for a functional product—and that product is almost always a protein, the exceptions being rRNA and tRNA. The DNA of *E. coli* is long enough to contain about 4000 genes and can therefore specify about 4000 different kinds of mRNA. In fact, *E. coli* has about 4000 different proteins. There are only a few copies of some proteins, but there are several thousand copies of other proteins.

THE GENETIC CODE

The **genetic code** specifically refers to the relationship between the nitrogenous base sequence of DNA, the corresponding codons of mRNA, and the amino acids that the codons stand for (Figure 8–8). Note that there are 64 possible codons but only 20 amino acids. This means that most amino acids are signaled by several codons, a situation referred to as the **degeneracy** of the code. For example, leucine has 6 codons and alanine has 4 codons.

Of the 64 codons, 61 are sense codons and 3 are nonsense codons. **Sense codons** code for amino acids, while **nonsense codons** do not. Rather, they signal the end of synthesis for the protein molecule. These nonsense codons are UAA, UAG, and UGA, and they are the terminator codons referred to earlier. The initiator codon that starts the synthesis of the protein molecule is AUG, which is also the sense codon for methionine. The base sequence that specifies the promoter site on DNA is not translated.

Second Position

First Position	U		C		A		G		Third Position
	U		**C**		**A**		**G**		
U	UUU	Phe	UCU	Ser	UAU	Tyr	UGU	Cys	U
U	UUC	Phe	UCC	Ser	UAC	Tyr	UGC	Cys	C
U	UUA	Leu	UCA	Ser	UAA	End	UGA	End	A
U	UUG	Leu	UCG	Ser	UAG	End	UGG	Trp	G
C	CUU	Leu	CCU	Pro	CAU	His	CGU	Arg	U
C	CUC	Leu	CCC	Pro	CAC	His	CGC	Arg	C
C	CUA	Leu	CCA	Pro	CAA	Gln	CGA	Arg	A
C	CUG	Leu	CCG	Pro	CAG	Gln	CGG	Arg	G
A	AUU	Ile	ACU	Thr	AAU	Asn	AGU	Ser	U
A	AUC	Ile	ACC	Thr	AAC	Asn	AGC	Ser	C
A	AUA	Ile	ACA	Thr	AAA	Lys	AGA	Arg	A
A	AUG	Met	ACG	Thr	AAG	Lys	AGG	Arg	G
G	GUU	Val	GCU	Ala	GAU	Asp	GGU	Gly	U
G	GUC	Val	GCC	Ala	GAC	Asp	GGC	Gly	C
G	GUA	Val	GCA	Ala	GAA	Glu	GGA	Gly	A
G	GUG	Val	GCG	Ala	GAG	Glu	GGG	Gly	G

Figure 8–8 The genetic code. The three nucleotides in an mRNA codon are designated, respectively, as the first position, second position, and third position of the codon. Each set of three nucleotides specifies a particular amino acid, represented by a three-letter abbreviation (see Table 8–1). The codon AUG (which specifies the amino acid methionine) is the start signal for protein synthesis. The word *End* stands for the nonsense codons that serve as signals to terminate protein synthesis.

GENOTYPE AND PHENOTYPE

The **genotype** of an organism is its genetic makeup, the information that codes for all the particular characteristics of the organism. The genotype represents the *potential* properties, but not the properties themselves. **Phenotype** refers to the *actual, expressed* properties, such as the ability to perform a particular chemical reaction. The phenotype, then, is the external manifestation of the genotype.

In molecular terms, an organism's genotype is its collection of genes, its entire DNA, its blueprint for determining its unique characteristics. What, then, constitutes the organism's phenotype in molecular terms? To put it simply, an organism's phenotype is its collection of proteins. Most of a cell's properties derive from the structure and function of its proteins. In microorganisms, most proteins are either enzymatic, catalyzing particular reactions, or structural, participating in large functional complexes like membranes or ribosomes. Even those phenotypes that depend on structural macromolecules other than protein rely indirectly on proteins. For instance, the structure of complex lipid or polysaccharide molecules is derived from the catalytic activities of enzymes that synthesize, process, and degrade them. Although it is not completely accurate to say that phenotypes are due only to proteins, it is a useful simplification.

MUTATION: CHANGE IN THE GENETIC MATERIAL

A **mutation** is a change in the base sequence of DNA. We can reasonably expect that a change in the base sequence of a gene will cause a change in the gene product coded by that gene. When a gene mutates, the enzyme coded by the gene may become inactive or less active because its amino acid sequence has changed. Such a change in genotype may be disadvantageous or even lethal, because the cell loses a phenotypic trait it needs. Yet, a mutation may also be beneficial, for instance, if the altered enzyme coded by the mutant gene has a new enzymatic activity that benefits the cell.

Many simple mutations are neutral, the change in DNA base sequence causing no change in the activity of the product coded by the gene. A common example of a neutral mutation is the substitution of one nucleotide for another in the DNA. Because of the degeneracy of the genetic code, the resulting new codon may still code for the same amino acid. Even if the amino acid is changed, there may be no change in protein function if it is in a nonvital portion of the protein.

TYPES OF MUTATIONS

The most common type of mutation involving single base pairs is **base substitution,** where at one point in the DNA a single base is replaced with a different one. Then, when the DNA replicates, the result is a substituted base pair (Figure 8–9). For example, A–T may become substituted for G–C or C–G for G–C. If a base substitution occurs in a portion of the DNA molecule that codes for a protein, then the mRNA transcribed from the gene will carry an incorrect base at some position. When the mRNA is translated into protein, the incorrect base may cause the insertion of an incorrect amino acid in the protein. Thus, the base substitution in DNA can result in an amino acid substitution in the synthesized protein. This is known as **missense mutation** (Figure 8–10a and b).

Some base substitutions effectively prevent the synthesis of a functional protein by creating a terminator (nonsense) codon in mRNA before the protein is synthesized. Thus, only a fragment of the protein is synthesized. A base substitution resulting in a nonsense codon is called a **nonsense mutation** (Figure 8–10c).

Besides base pair mutations, there are also changes in DNA called **frameshift mutations.** Here, one or a few base pairs are deleted or added to DNA (Figure 8–10d). This can shift the "translational reading frame," that is, the three-by-three grouping of nucleotides recognized by the tRNAs during translation. For example, inserting one base pair in the middle of a gene causes many amino acids downstream from the site of the original mutation to change. Frameshift mutations almost always result in an inactive protein product for the mutated gene. They lead to a long stretch of missense, and in most cases a nonsense codon will eventually be generated, terminating translation.

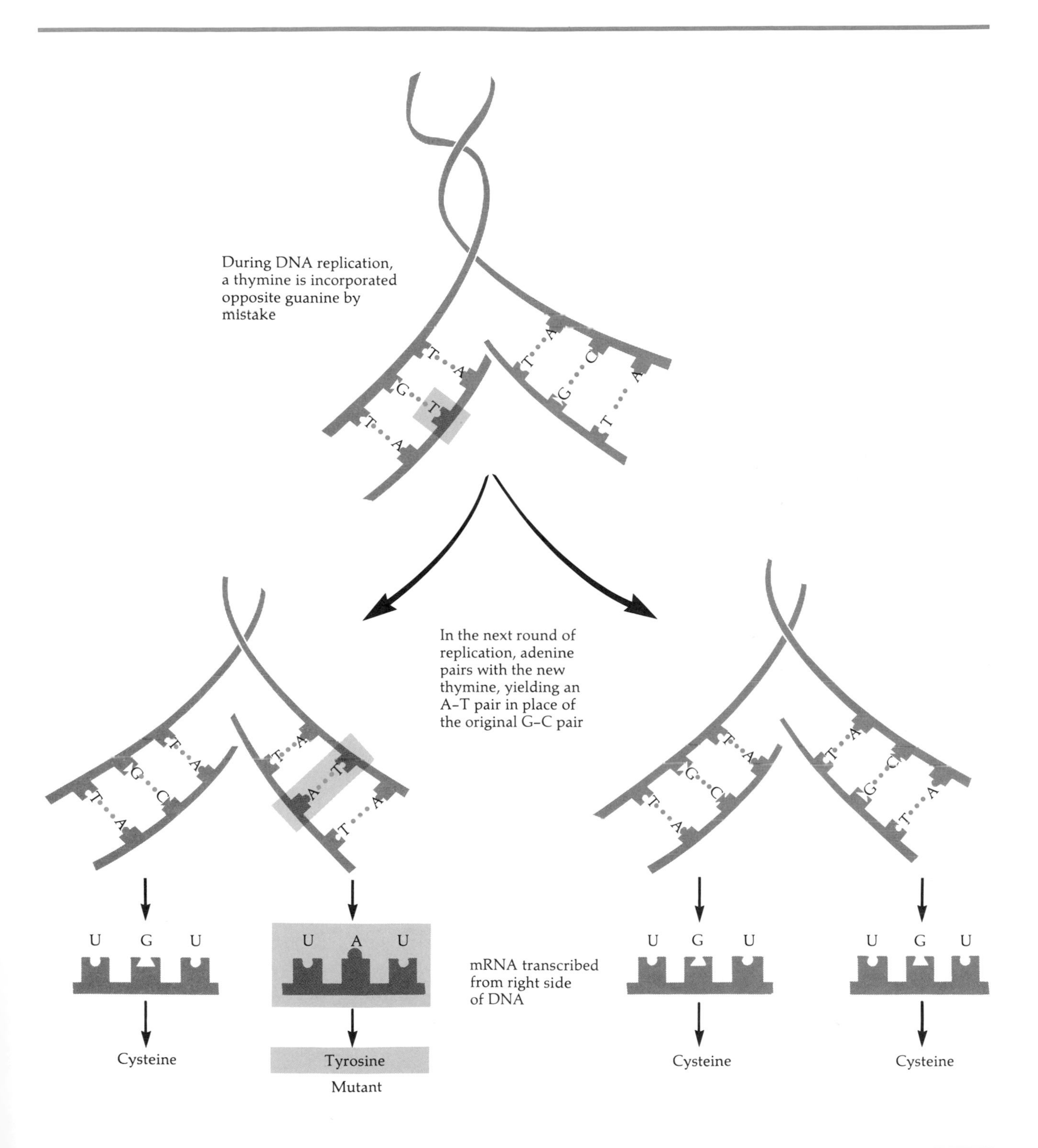

Figure 8–9 Base substitution, leading to base pair substitution.

(a) Normal:

DNA CTAGCATGTATAGGG
GATCGTACATATCCC

mRNA transcribed from bottom strand of DNA CUAGCAUGUAUAGGG

Amino acid sequence Leu - Ala - Cys - Ile - Gly

(b) Missense mutation:

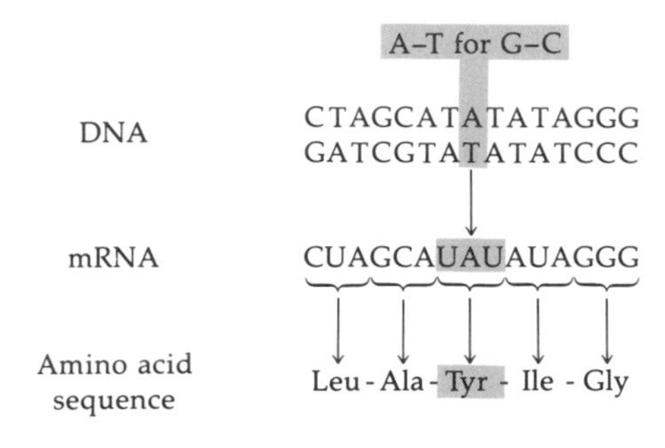

(c) Nonsense mutation:

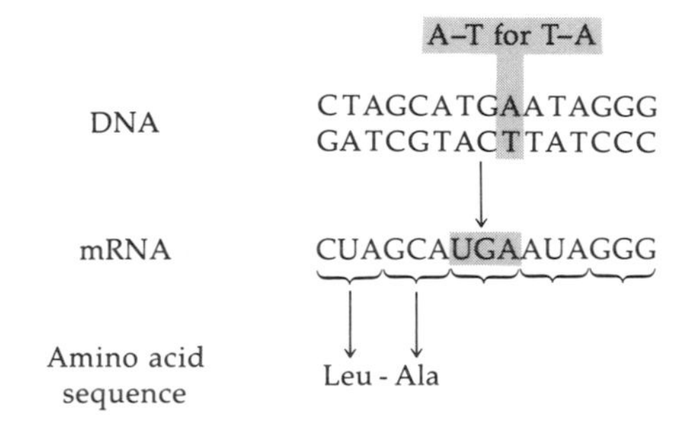

(d) Frameshift mutation:

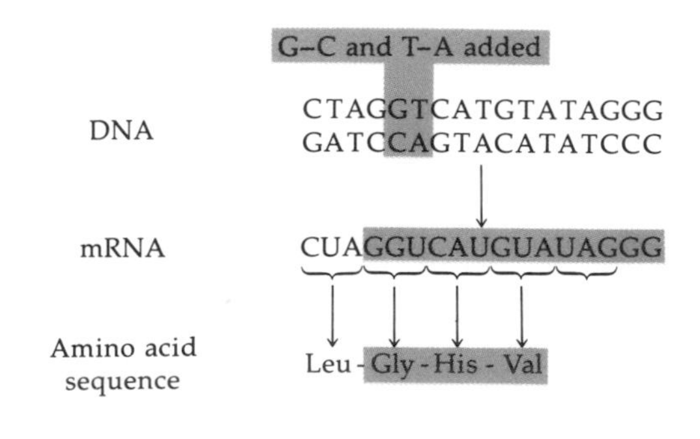

Figure 8–10 Types of mutations. **(a)** Normal DNA molecule. **(b)** Missense mutation. **(c)** Nonsense mutation. **(d)** Frameshift mutation.

MUTAGENESIS

Base substitutions and frameshift mutations may occur spontaneously because of occasional mistakes made during DNA replication. **Spontaneous mutations** are mutations that occur without known intervention of mutation-causing agents. Agents in the environment, such as certain chemicals and radiation, that directly or indirectly bring about additional mutations are called **mutagens.**

Chemical Mutagens

Among the many chemicals known to be mutagenic is *nitrous acid.* Figure 8–11 shows how exposure of DNA to nitrous acid can convert the base adenine (A) to a form that no longer base pairs with thymine (T) but with cytosine (C). When DNA containing such modified adenines replicates, one daughter DNA molecule will have a different base pair sequence than the parent DNA. Eventually, some A–T base pairs of the parent will have been changed to G–C base pairs in daughter cells. Nitrous acid is thus an effective **base pair mutagen,** making a specific kind of mutational change in DNA. Like all mutagens, it does not select which gene it will mutate, but alters DNA at random locations.

Other chemical mutagens are **base analogs,** like *2-aminopurine* and *5-bromouracil.* These molecules are structurally similar to normal nitrogenous bases, but have slightly altered base-pairing properties. The 2-aminopurine (Figure 8–12a) is incorporated into DNA in place of adenine and can pair with thymine. The 5-bromouracil (Figure 8–12b) is incorporated into DNA in place of thymine and often pairs with guanine. When base analogs are given to growing cells, the analogs are metabolized and incorporated into cellular DNA in place of the normal bases at random places. Then, during DNA replication, they cause mistakes in base pairing. The wrongly paired bases may be faithfully copied during further DNA replication, and a base pair substitution results. Some antiviral drugs are base analogs.

Still other mutagens can cause small deletions or insertions instead of substitutions. For instance, *benzpyrene,* which is present in smoke and soot, is

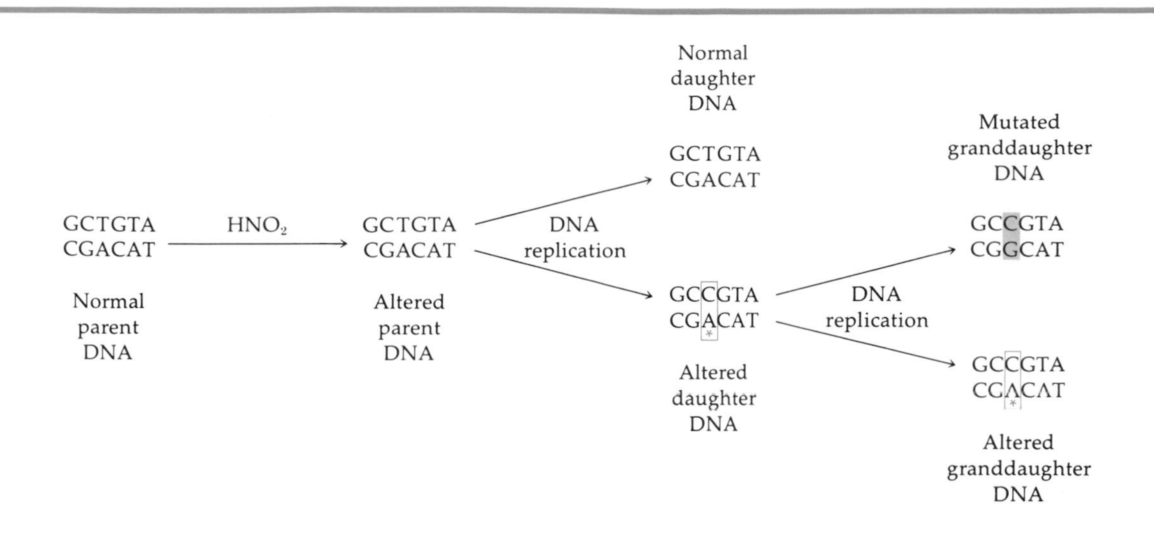

Figure 8–11 Mutagenesis by nitrous acid (HNO_2). The nitrous acid alters an adenine (asterisk) with the result that it pairs with cytosine instead of thymine. In the next round of replication, the new C will pair with G, introducing a new C–G pair into the mutated DNA.

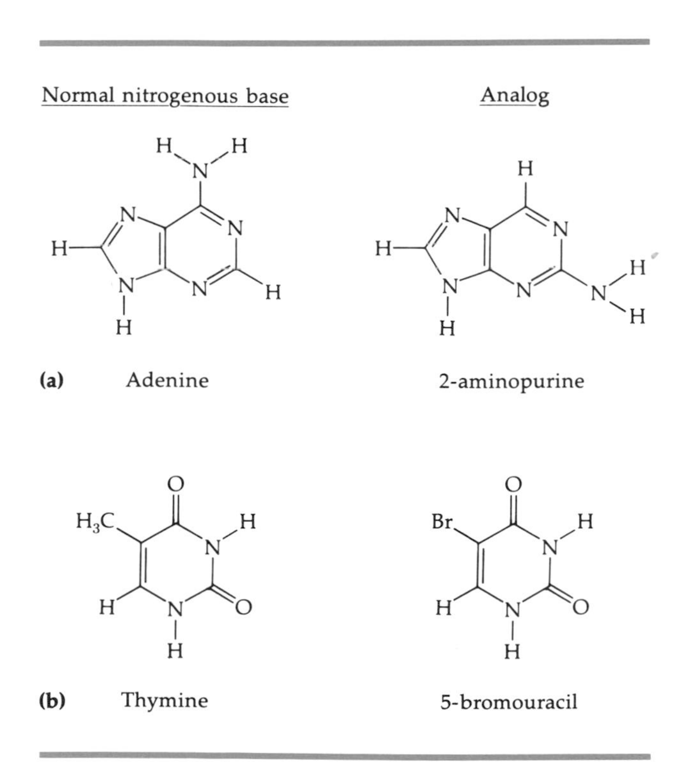

Figure 8–12 Base analogs and the nitrogenous bases they replace. **(a)** Adenine and 2-aminopurine. **(b)** Thymine and 5-bromouracil.

an effective **frameshift mutagen** under certain conditions. Likewise, *aflatoxin,* produced by a mold that grows on peanuts and grain, is a frameshift mutagen, as are the *acridine dyes,* used experimentally against herpesvirus infections. Frameshift mutagens usually have the right size and chemical properties to slip between the stacked base pairs of the DNA double helix. Though direct proof is lacking, it is believed that a frameshift mutagen works by offsetting the two strands of DNA slightly, leaving a gap in one strand or a bulge in the other. When the offset or staggered DNA strands are copied during DNA synthesis, the result is a deletion or insertion of one or more base pairs in the new double-stranded DNA. Interestingly, frameshift mutagens are often potent carcinogens.

Radiation

X rays, gamma rays, and other forms of ionizing radiation are potent mutagens. The penetrating rays of ionizing radiation cause electrons to pop out of their usual electron shell locations. The affected electrons bombard other molecules, causing more damage, and many of the resulting ions and free radicals (molecular fragments with unpaired electrons) are very reactive. Some of these ions may combine with bases in the DNA, resulting in mistakes in base pairing during replication and leading ultimately to base substitutions. An even more serious outcome is the breaking of covalent bonds in the sugar–phosphate backbone of DNA, causing physical breaks in chromosomes.

Another form of mutagenic radiation is **ultraviolet (UV) light,** a component of ordinary sunlight that is not very penetrating. The most important effect of UV light on DNA is the formation of chemical cross-links between adjacent thymines or cytosines in a DNA strand (Figure 8–13a). Cross-linked bases cause serious damage or death to the cell, since it cannot properly transcribe or replicate such DNA.

Many bacteria and other organisms can repair UV damage by the action of certain enzymes. In the presence of visible light, in a process called **photoreactivation,** an enzyme breaks the cross-links between adjacent thymines or cytosines. Other enzymes that repair UV damage operate in the absence of light. These enzymes snip out the cross-linked pair, widen the gap, and then fill in with nucleotides complementary to those in the undamaged strand, to restore the original base sequence (Figure 8–13b). Occasionally, this repair system

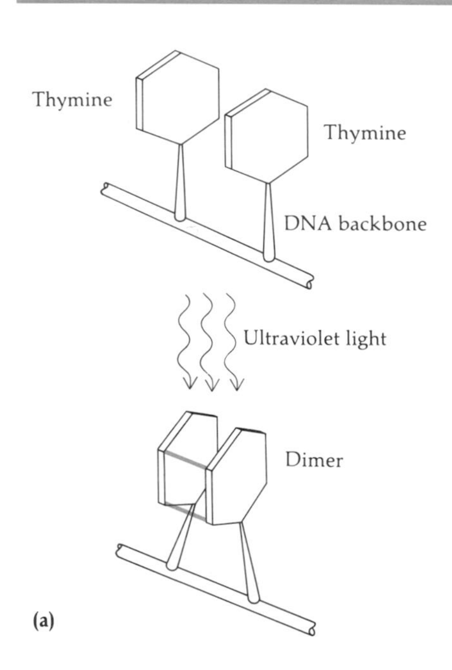

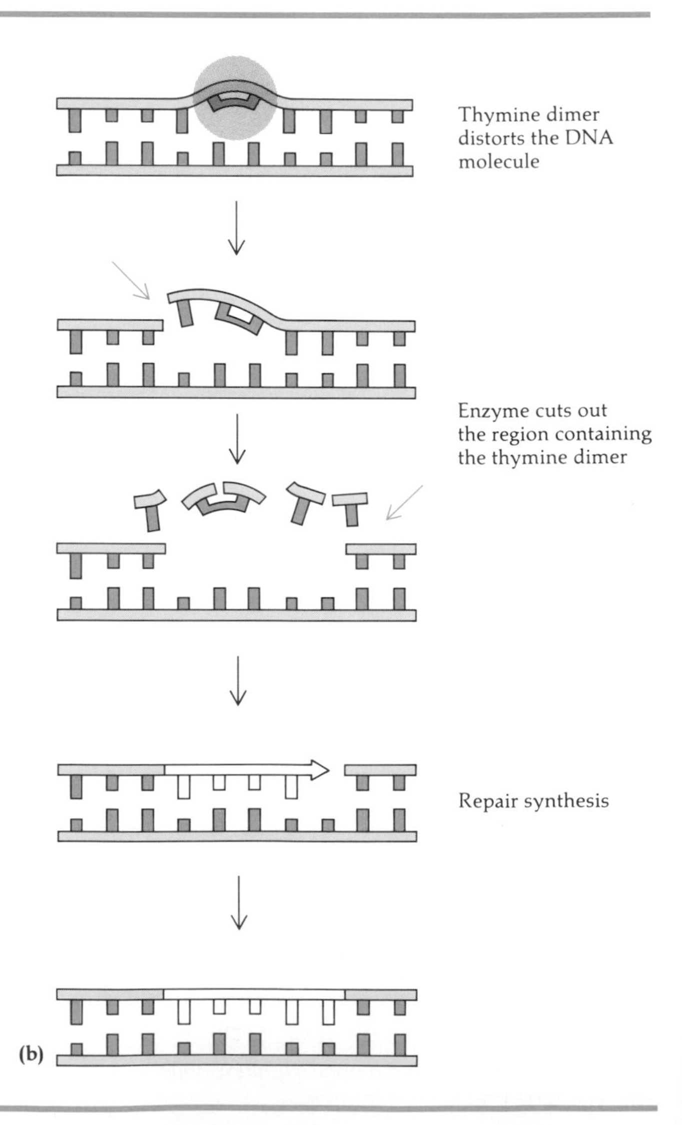

Figure 8–13 Mutagenesis by ultraviolet light. **(a)** Effect of ultraviolet light on adjacent thymines. **(b)** Mechanism of repair of cross-linked thymines. This mechanism does not require light.

makes errors and the original base sequence is not properly restored. The result is then a mutation.

MUTATION FREQUENCY

The **mutation rate** is the probability of a gene mutating each time a cell divides. The rate is usually stated as a power of 10, and because mutations are very rare, the exponent is always a negative number. For example, if there is one chance in 10,000 that a gene will mutate when the cell divides, the mutation rate is 1/10,000, which is expressed as 10^{-4}. Mistakes in DNA replication occur spontaneously at a very low rate because of the remarkably faithful DNA replication machinery. Perhaps only once in 10^9 replicated base pairs does an error occur. Since the average gene has about 10^3 base pairs, the spontaneous rate produces a mutation about once in 10^6 (a million) replicated genes. Mutations usually occur more or less randomly along a chromosome. No particular gene is singled out, and within a mutated gene, no particular A–T or G–C base pair is more likely to be mutated. Having random mistakes at low frequency is an essential aspect in the design of living organisms, for evolution requires that genetic diversity be generated at random and at a low rate. Typically, a mutagen will increase the spontaneous rate from 10 to 1000 times; that is, a mutagen will produce a mutation rate of 10^{-5} to 10^{-3}.

IDENTIFYING MUTATIONS

Mutations can be detected by selecting or testing for an altered phenotype. Whether or not a mutagen is used, mutant cells with specific mutations are always rare compared with other cells in the population. The problem is one of detecting a rare event.

Experiments are more easily done with bacteria than with other organisms, because bacteria reproduce rapidly and large numbers of organisms (more than 10^9 per milliliter of broth or per Petri plate) can easily be used. Furthermore, since bacteria generally have only one version of each gene per cell (they are haploid), the effects of a mutated gene are not masked by the presence of a normal version of the gene.

Positive, or **direct, selection** involves picking out the mutant cells but rejecting the unmutated parent cells. For example, suppose we were trying to find mutant bacteria that are resistant to penicillin. By plating the bacterial cells on a medium containing penicillin, the mutant can be identified directly. The few cells in the population that are resistant (mutants) will grow and form colonies.

For identifying mutations in many other kinds of genes, it may be necessary to use **negative,** or **indirect, selection.** This is made possible by the **replica plating technique.** In replica plating, a number of bacteria are inoculated onto an agar plate by some method that will separate individual cells. This plate, called the master plate, contains a rich medium on which all cells will grow. After several hours of incubation, each cell grows into a colony. Then a pad of sterile velvet is pressed over the master plate (Figure 8–14), causing some of the cells from each colony to adhere to the velvet. Next, the velvet is pressed down onto two (or more) sterile plates. One contains a rich medium and one contains a minimal medium on which the original, nonmutant bacteria can grow (for example, glucose and inorganic salts). A mutant colony growing on the rich medium of the master plate that has a new requirement for a growth factor will not grow on the minimal medium, which lacks that growth factor (Figure 8–14). Cells taken from the mutant colony on the master plate can then be tested on minimal media supplemented with various growth factors to determine precisely which factor is required.

The replica plating technique is very effective for isolating mutants that require one or more new growth factors. Any mutant microorganism with a nutritional requirement (such as an amino acid) not possessed by the parent type is known as an **auxotroph.** By using the replica plating technique, it is possible to identify auxotrophs because they will not grow in plates containing minimal medium.

THE AMES TEST FOR IDENTIFYING CHEMICAL CARCINOGENS

As we mentioned earlier, many known mutagens have also been found to be carcinogens. Very simply, a **carcinogen** is any cancer-producing substance. In recent years, a number of chemicals in

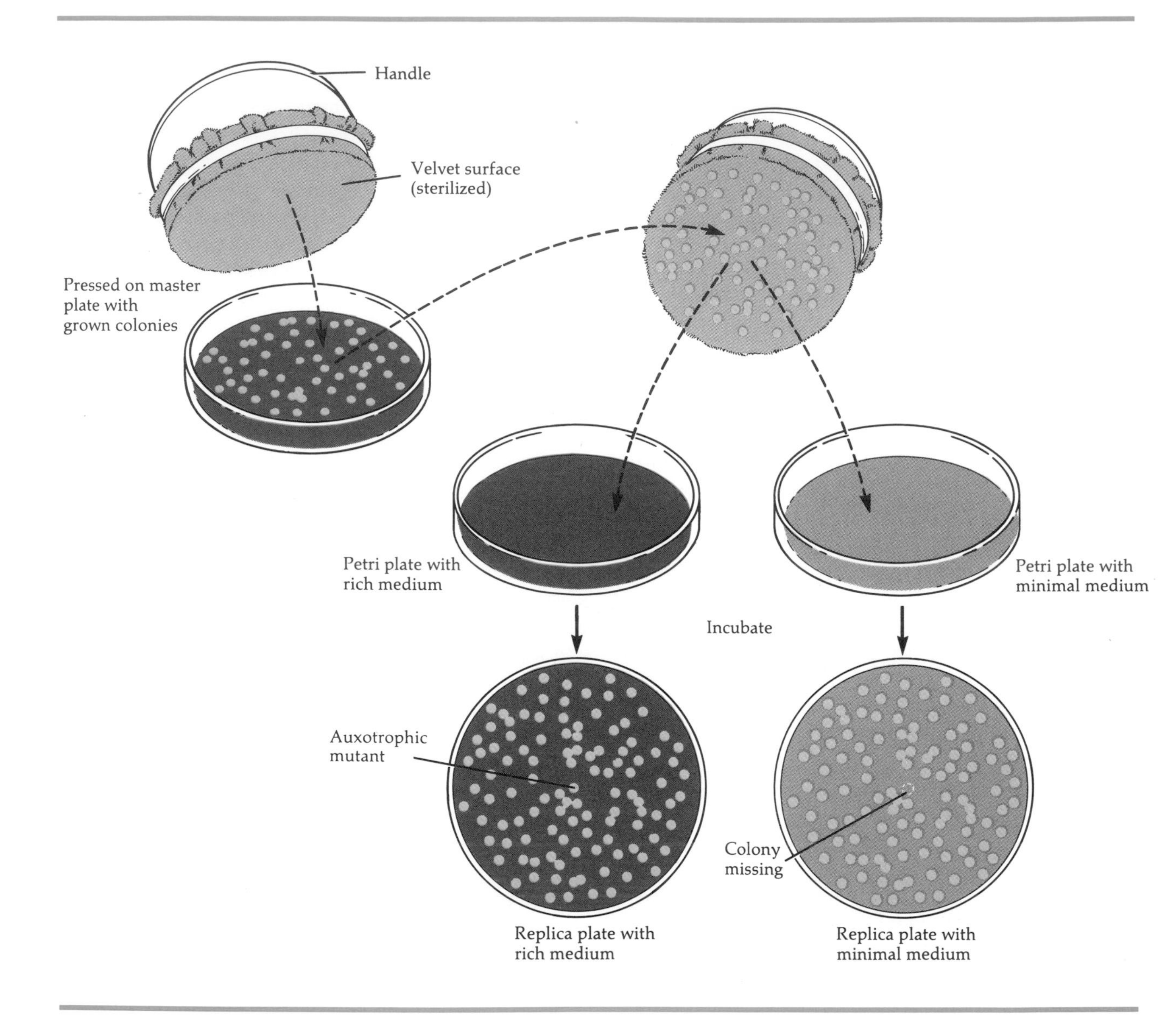

Figure 8–14 Replica plating. This technique transfers bacteria quickly and easily from colonies on a master plate (Petri plate) to a different medium in another plate. The procedure permits the identification of auxotrophic mutants, which form colonies on the supplemented medium of the master plate but are unable to grow on the minimal medium of the replica plate.

the environment have been linked to many types of cancers. It is hypothesized that carcinogens act by inducing mutations that in turn cause cancer. At one time, animals alone were used to determine potential carcinogens, but the testing procedures proved to be both time consuming and costly. Now there is a new procedure for the preliminary screening of potential carcinogens. Called the **Ames test,** this procedure is far less time consuming and more economical, employing bacteria instead of animals.

The Ames test is based on the fact that a mutated cell can often revert to a cell that resembles

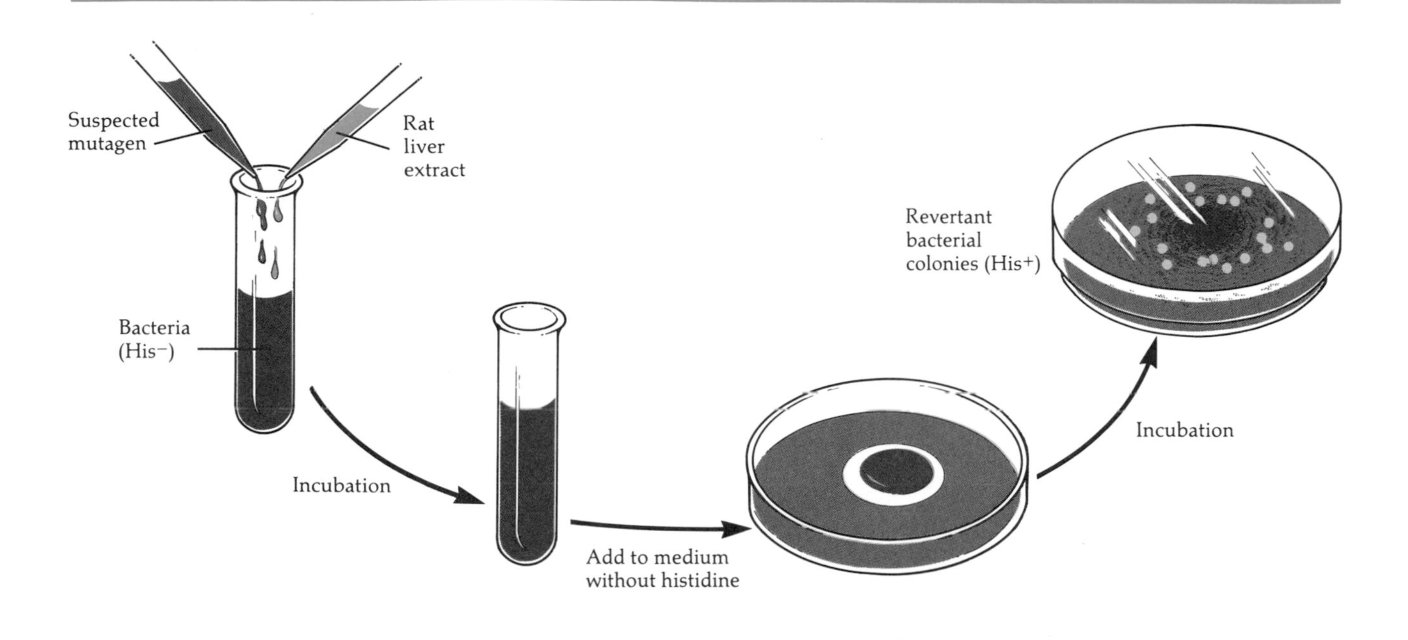

Figure 8–15 Ames test. Mutant *Salmonella* bacteria unable to synthesize histidine (His^-) are mixed with a suspected mutagen and rat liver extract and inoculated onto a medium lacking histidine. Only bacteria that have further mutated (reverted) to His^+ (able to synthesize histidine) will grow into colonies. The liver extract "activates" the suspected mutagen.

the original, nonmutant cell by undergoing *another* mutation. The test measures the reversion of histidine auxotrophs of *Salmonella*, mutants that have lost the ability to synthesize the amino acid histidine, to nonmutant cells in both the presence and absence of the chemical being tested (Figure 8–15). One further "trick" is required in that many chemicals must be activated by animal enzymes for mutagenic or carcinogenic activity to be exhibited. The chemical to be tested and the mutant bacteria are incubated together with rat liver extract, a rich source of activation enzymes. If the chemical substance being tested is mutagenic, it will cause a reversion rate higher than the spontaneous reversion rate. The number of observed revertants provides an indication of the degree to which a substance is mutagenic. About 90% of the substances found by the Ames test to be mutagenic have been shown also to be carcinogenic in animals, with the more mutagenic substances generally more carcinogenic as well.

GENETIC TRANSFER AND RECOMBINATION

Genetic recombination is the rearrangement of genes to form new combinations. Usually, when we use the term, we are referring to rearrangement between groups of genes from two different individuals of a species. Exceptions to this generalization will be discussed later.

Figure 8–16 shows one type of recombinational event occurring between two pieces of DNA, which we shall regard as chromosomes for the sake of simplicity. We have called one chromosome *A* and the other *B*. If these two chromosomes break and rejoin as shown, a process called **crossing over,** the result is a reshuffling of some of the genes carried by these chromosomes. Thus, each original chromosome has been *recombined* so that it now carries a portion of the other chromosome's genes.

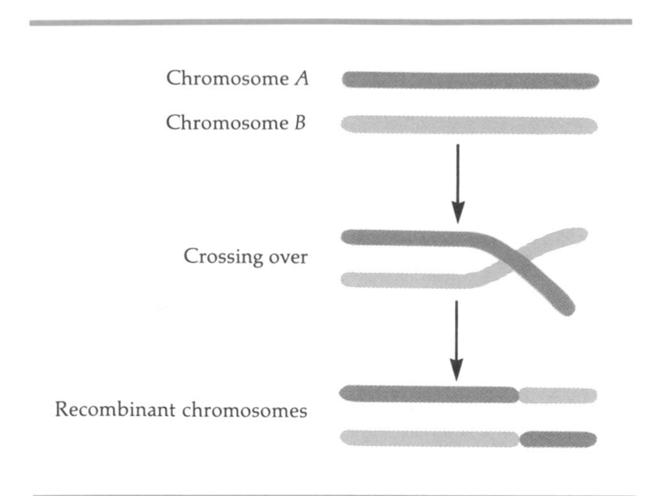

Figure 8–16 Simplified version of one type of genetic recombination between two chromosomes, *A* and *B*. The chromosomes cross over by breaking and rejoining. The result is two recombinant chromosomes, each of which carries genes originally present on two different parent chromosomes.

If *A* and *B* represent DNA from different individuals, how are they brought close enough to recombine? Genes are recombined in higher organisms during meiosis, when chromosomes originating from the two parents pair (a process called synapsis). Thus, the gametes resulting from meiosis contain recombined DNA. For eucaryotes, genetic recombination is an ordered process that usually occurs as part of the sexual cycle of the organism. In bacteria, genetic recombination may happen in a number of ways, which we are about to discuss.

Like mutation, genetic recombination contributes to a population's genetic diversity, which is the source of variation in evolution. Recombination is more likely to be beneficial than mutation, because it brings together different gene groups that may enable the organism to carry out a new function.

Genetic material can be transferred between bacteria in several ways to result in recombination. In all of the mechanisms, the transfer involves a **donor cell** that gives only a portion of its total DNA to a different **recipient cell.** Once transferred, part of the donor DNA is usually incorporated into the recipient's DNA; the remainder is degraded by cellular enzymes. In this process, there is no net change in the length of intact DNA in the **recombinant,** the recipient cell that has DNA from the donor added to its own DNA. The transfer of genetic material between bacteria is by no means a regular event, occurring perhaps in only 1% or less of an entire population. Let us now examine the specific types of genetic transfer in detail.

TRANSFORMATION IN BACTERIA

In the process of **transformation,** genes are transferred from one bacterium to another as "naked" DNA in solution. This process was first described more than 50 years ago, although it was not understood at that time. Not only did transformation demonstrate that genetic material could be transferred from one bacterial cell to another, but study of this event eventually led to the conclusion that DNA is the genetic material. The initial experiment on transformation was performed by F. Griffith in England in 1928 while he was working with two strains of *Streptococcus pneumoniae.* One, a virulent strain, has a polysaccharide capsule and causes pneumonia; the second, an avirulent strain, lacks the capsule and does not cause disease.

Griffith was interested in determining whether injections of heat-killed (60°C) encapsulated pneumococci could be used to vaccinate mice against pneumonia. He found that the injected mice did not become ill from the heat-killed strain. However, when these dead cells were mixed with live avirulent cells and injected into the mice, the mice frequently died. In the blood of the dead mice Griffith found living, encapsulated, virulent bacteria. Presumably, hereditary material from the dead bacteria had entered the live cells and changed them genetically so that their offspring were transformed into encapsulated, pathogenic forms (Figure 8–17).

Subsequent investigations based on Griffith's research revealed that bacterial transformation could be duplicated using standard microbiological culturing techniques in broth and on agar in Petri plates. Dead bacteria with capsules were added to a broth (liquid culture medium) that was inoculated with live bacteria without capsules, and then incubated for some hours. A drop of the culture was smeared over agar (solid culture medium) in Petri plates. After a period of incubation, it was found that some of the bacterial colonies that arose had the smooth appearance typical of encapsulated bacteria—that is, they had been transformed. The bacteria had acquired a new hereditary trait.

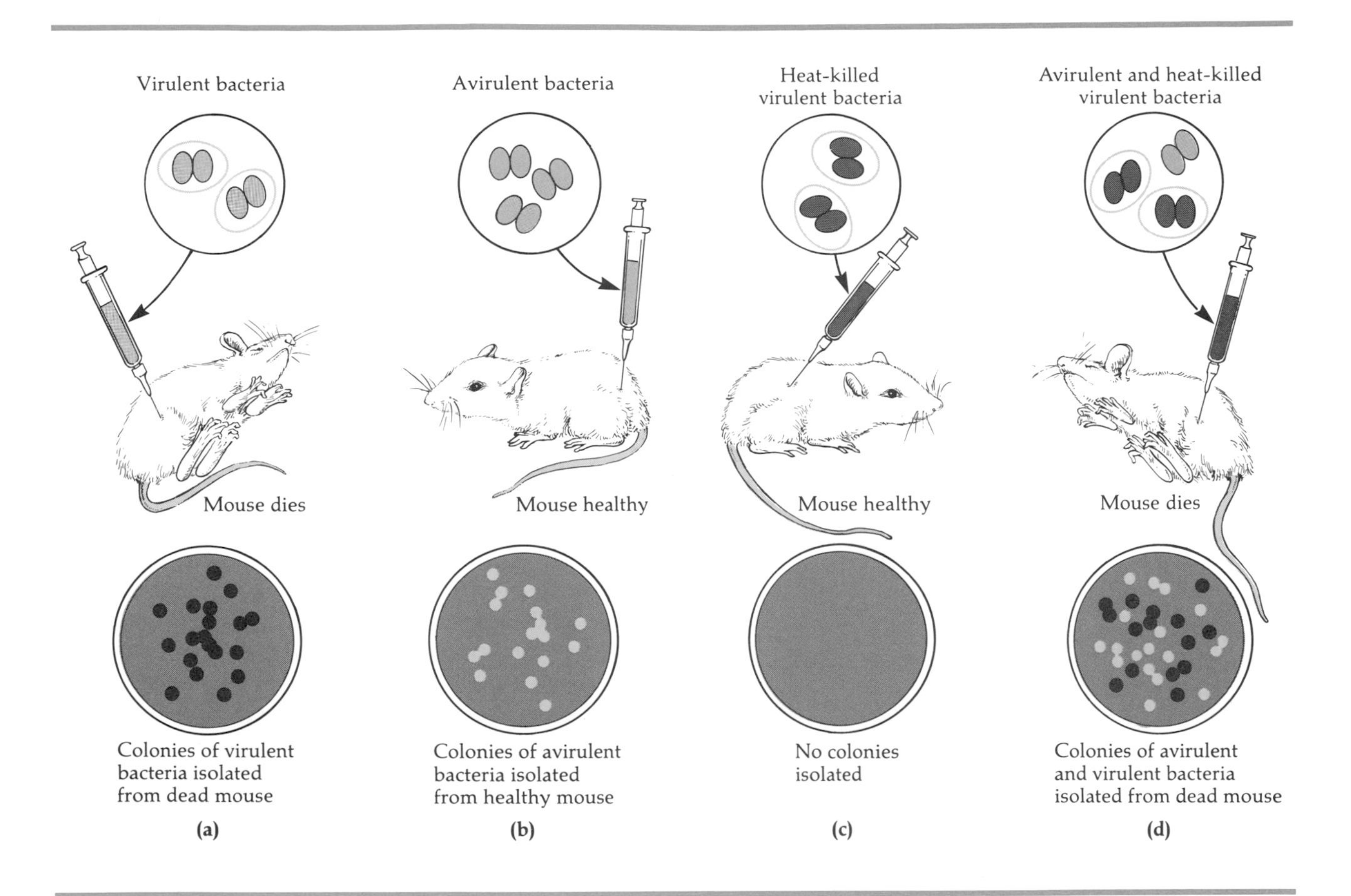

Figure 8–17 Griffith's experiment demonstrating genetic transformation. Some material from the heat-killed virulent bacteria transformed the living avirulent bacteria into virulent bacteria, which killed the mouse.

The next logical step was to extract various chemical components from the dead cells and determine whether any of these chemicals cause transformation. These crucial experiments were performed by O. T. Avery and his associates at the Rockefeller Institute in the United States. After some years of research, they announced in 1944 that the chemical substance responsible for transforming harmless pneumococci into virulent strains was DNA.

Since Griffith's experiment, considerable information has been gathered about transformation. In nature, some bacteria, perhaps after death and cell lysis, release their DNA into the environment. Other bacteria may then encounter the DNA, and depending on the particular species and growth conditions, may take up fragments of DNA into the cytoplasm and recombine the DNA into their own chromosomes. New combinations of genes can result in such recipient cells, resulting in a kind of hybrid—or recombinant—cell (Figure 8–18). Since it is the genes that have been changed, all the descendants of such a recombinant cell will be identical to it. Transformation occurs naturally among very few genera of bacteria. These include *Bacillus, Hemophilus, Neisseria, Escherichia, Rhizobium,* and certain strains of *Streptococcus* and *Staphylococcus.*

Transformation works best when the donor and recipient cells are very closely related and when the recipient cells are in the late log phase of growth. Even though only a small portion of a cell's DNA is transferred to the recipient, it is still a very large piece of DNA that must pass through the recipient cell wall and membrane. When a recipient cell is in a physiological state in which it can take

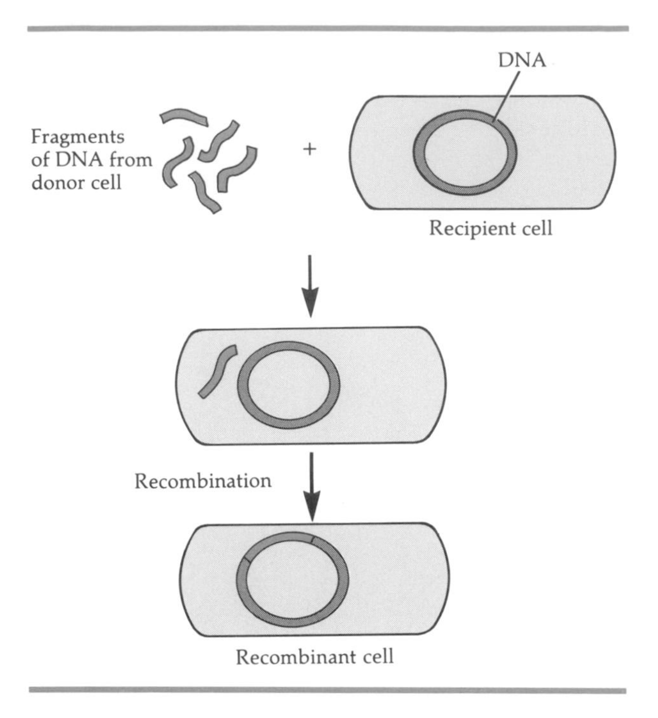

Figure 8–18 Mechanism of transformation.

up and incorporate the donor DNA, it is said to be **competent.** Competence may be related to alterations in the cell wall that make it permeable to the large DNA molecule, or it may be related to the synthesis of specific receptor sites on the bacterial cell surface. Among the traits whose transformation has been studied are capsule formation (already noted), nutritional needs, pathogenicity, pigment formation, and drug resistance.

CONJUGATION IN BACTERIA

Conjugation is another mechanism for the transfer of genetic material from one bacterium to another. It differs from transformation in that contact between living cells is required. This can be demonstrated by the following experiment. Two different auxotrophs are placed in a U tube separated by a perforated glass disk that is permeable to macromolecules like DNA, but not to entire cells (Figure 8–19). One auxotroph might require the amino acids phenylalanine and cysteine for growth,

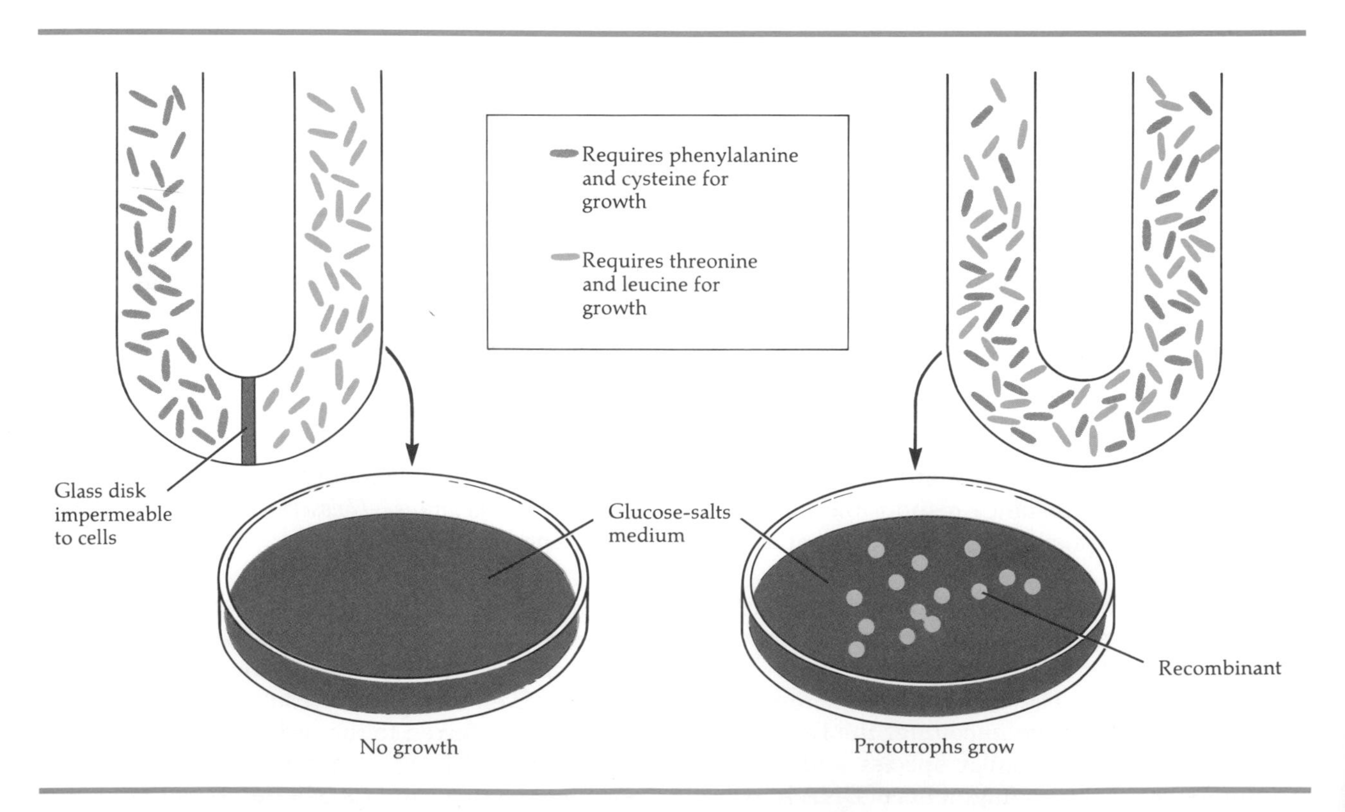

Figure 8–19 For conjugation to occur, contact between living cells is required.

while the other might require the amino acids threonine and leucine. Thus, if either auxotroph culture is plated on a minimal medium lacking amino acids, no growth occurs. However, if the glass disk is removed and cell-to-cell contact permitted, recombination occurs by conjugation. In this process, DNA is transferred from one auxotroph to another, and some of the recombinants, called **prototrophs,** are able to grow on the minimal medium, because they can synthesize all amino acids.

Since much experimental work on conjugation has been done with *E. coli,* we will describe the process in this organism. Conjugation is the closest that bacteria come to the sexual reproduction found among higher organisms. It not only requires direct cell-to-cell contact, but the cells must also be of opposite mating types, male and female. One type of male, or donor, cell is called an **F^+ cell.** Female, or recipient, cells are called **F^- cells.** F^+ cells have extra pieces of DNA in their cytoplasm called **F factors.** They are double-stranded pieces of DNA arranged in a closed circle and are not part of the bacterial chromosome. As you may recall from Chapter 4, F factors are one type of **plasmid**—a free genetic element in the cell that is not necessarily attached to, or part of, the bacterium's main chromosome. We shall describe some other types of plasmids later in this chapter.

When F^+ and F^- cells are mixed together, the F^+ cells attach to the F^- cells by means of sex pili. (F factors contain genes coding for the synthesis of sex pili.) During conjugation, the F factor is duplicated and the new copy is transferred across an intercellular bridge from the F^+ to the F^- cell. This converts the F^- cell into an F^+ cell. Once an F^+, it is capable of transferring its F^+ factor to another F^- cell. Remember, the F factor is a plasmid, separate from the bacterial chromosome. When an F factor is transferred, the bacterial chromosome of the F^+ cell is not passed to the F^- cell. Thus, in this event, no recombinants are produced (Figure 8–20a).

Some male cells have the F factor integrated into their chromosome (Figure 8–20b); they are called **Hfr (high frequency of recombination) cells.** During conjugation between an Hfr and F^- cell, the chromosome of the Hfr cell, with its integrated F factor, replicates, and the new copy of the chromosome is transferred to the recipient cell (Figure 8–20c). Only Hfr cells can transfer their chromosome. Replication of the Hfr chromosome begins within the F factor, and a small piece of the F factor leads the chromosomal genes into the F^- cell. Most of the integrated F factor enters the recipient cell last, if at all. Usually, the chromosome breaks before it is completely transferred. Once within the recipient cell, some donor DNA may recombine with the recipient's DNA, as was shown for transformation. Donor DNA that is not integrated is degraded.

TRANSDUCTION IN BACTERIA

A third mechanism of genetic transfer in bacteria is **transduction.** In this case, the DNA is not passed naked, as in transformation, or during cell-to-cell contact, as in conjugation. Rather, it is passed inside a bacterial virus, called a **bacteriophage** (or just simply **phage**) from the donor to the recipient cell. (Phages will be discussed further in Chapter 12.)

To understand how transduction works, let us consider the life cycle of one type of transducing phage of *E. coli,* a phage that carries out a process called **generalized transduction.** In the process of infection, the phage attaches to the bacterial cell wall and injects its DNA into the bacterium (Figure 8–21). The viral DNA acts as a template for the synthesis of new viral DNA and also directs the synthesis of viral protein coats to envelop new viral particles. During a viral infection of a bacterial cell, fragments of the bacterial chromosome may occasionally be accidentally incorporated into the viral genome. These particles now carry bacterial DNA in addition to viral DNA, and when a bacterial cell releases the newly synthesized viral particles, some bacterial DNA enclosed in viral coats is also released. If the viral-coated bacterial DNA infects a new population of bacteria, the new population will thus receive several bacterial genes. Transduction of the viral-coated bacterial DNA usually leads to recombination between this DNA and the DNA of the new host cell. This is one way that bacteria acquire new genotypes.

In generalized transduction, all genes are equally likely to be picked up in a phage coat and transferred. There is another type of transduction, called **specialized (restricted) transduction,** in which only certain bacterial genes are transferred. Specialized transduction will be discussed in Chapter 12.

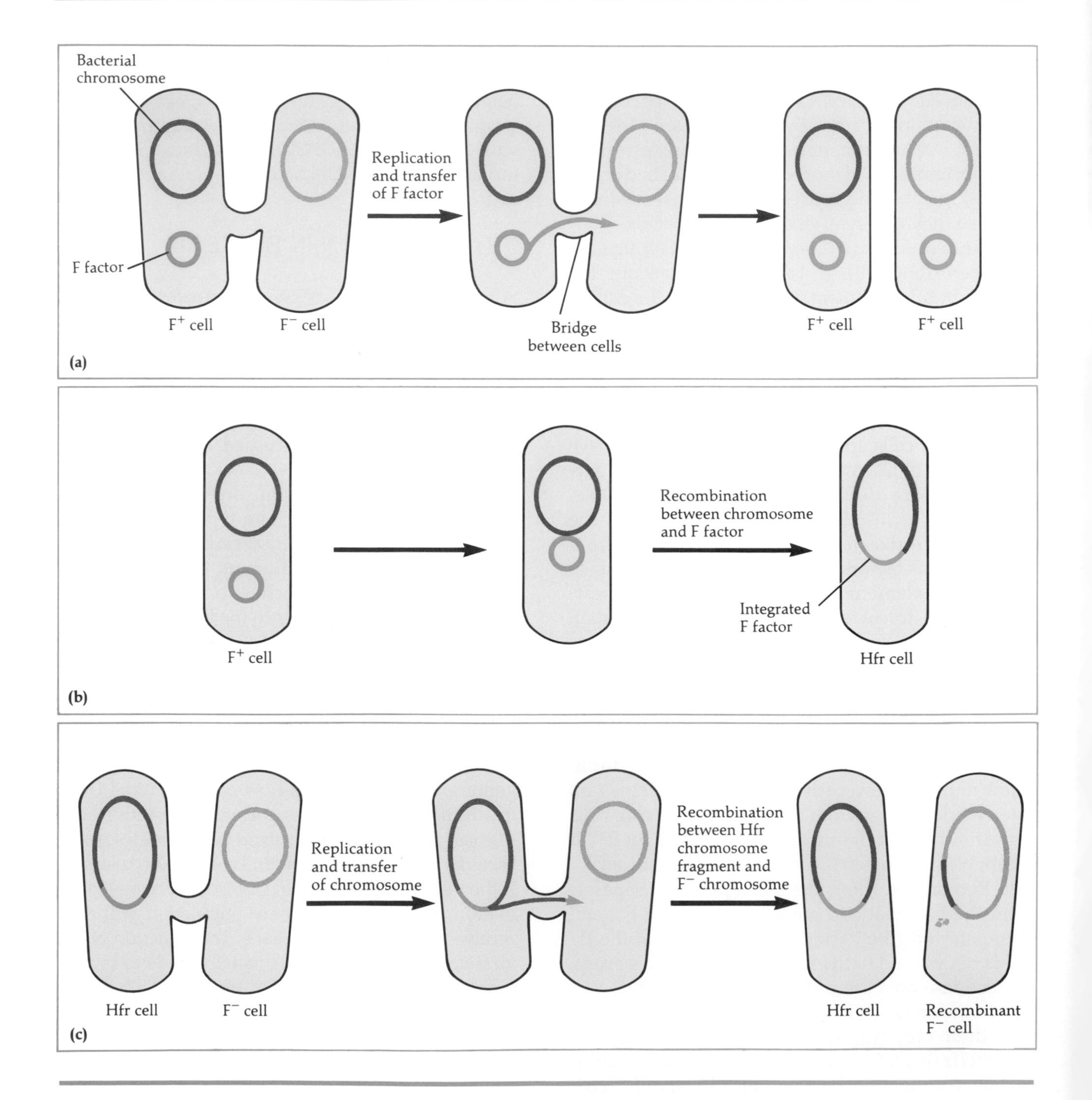

Figure 8–20 Mechanism for conjugation. **(a)** An F factor is transferred from donor (F$^+$) to recipient (F$^-$), converting the F$^-$ cell into an F$^+$ cell. **(b)** Integration of an F factor into the bacterial chromosome, making the cell an Hfr. **(c)** An Hfr donor passes a portion of its chromosome into an F$^-$ cell. A recombinant F$^-$ cell results.

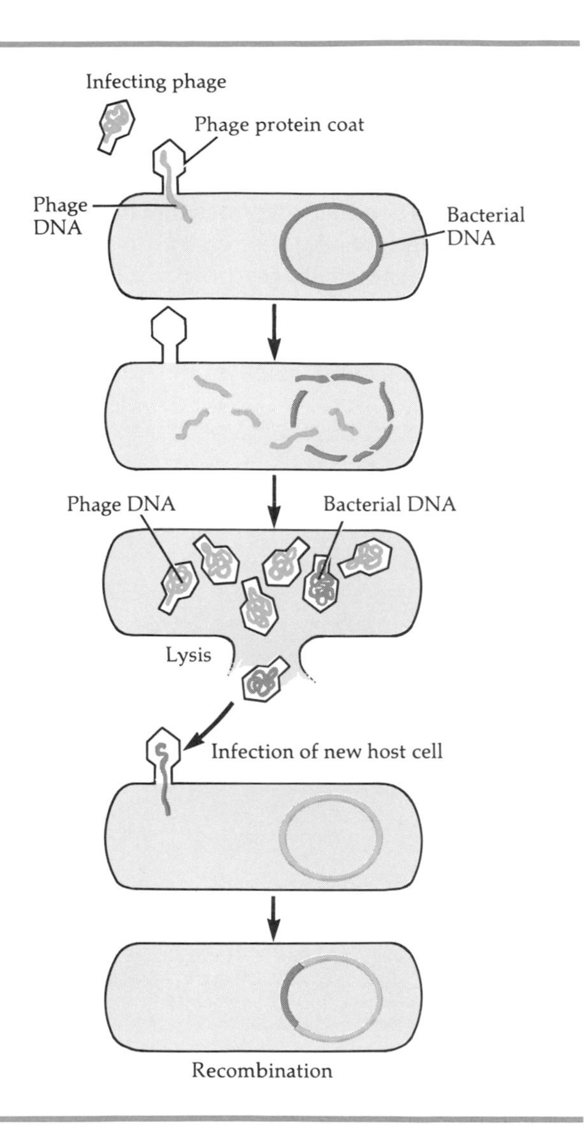

Figure 8–21 Transduction by a bacteriophage. Shown here is generalized transduction, in which any bacterial gene can be transferred.

RECOMBINATION IN EUCARYOTES

Transformation, conjugation, and transduction all result in genetic recombination and involve only the DNA of the donor cell entering the recipient cell. In eucaryotes, genetic recombination is the result of a sexual reproductive process. In contrast to the mechanisms in bacteria, sexual reproduction is a regular event for most eucaryotes and is necessary for the survival of many species. Sexual reproduction in eucaryotes involves the fusion of the haploid nuclei of two parent cells (Figure 8–22). A **haploid** cell is one that contains only one of each type of chromosome; the number of chromosomes in such a cell is called the **haploid number,** symbolized n. The haploid cells that fuse are called **gametes.** (In animals, the male gametes are sperm and the female gametes are ova.) Fusion of the gametes produces a fertilized ovum, or **zygote,** which receives half its chromosomes from one parent and half from the other.

Each species of eucaryotic organism has a characteristic **chromosome number,** that is, the number of chromosomes in each cell that is not a gamete. The human chromosome number is 46 (23 pairs). Each parent contributes one set of 23 chromosomes, which carries genes for all the activities of human cells. The other set of 23 chromosomes is more or less a duplicate set; it contains the same or a different version of each gene. Another term for chromosome number is **diploid number,** symbolized $2n$ for the two sets of chromosomes. The paired chromosomes are known as **homologous chromosomes.**

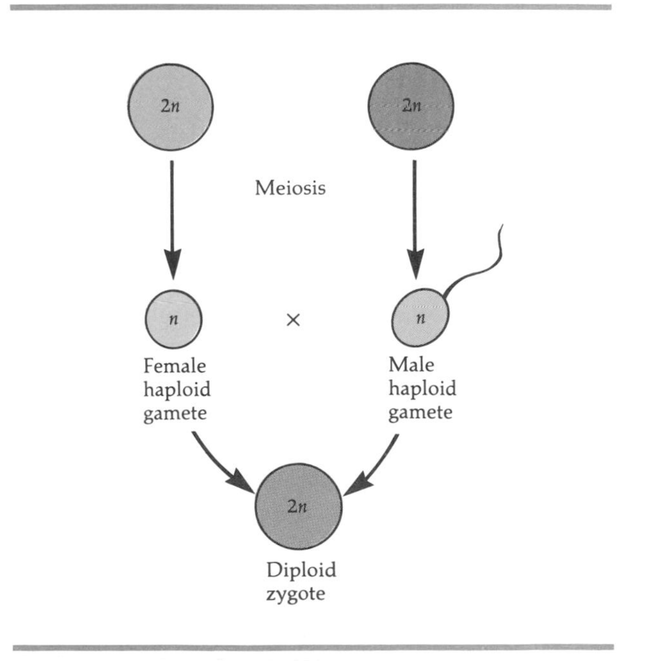

Figure 8–22 Sexual reproduction in eucaryotes.

For eucaryotes to reproduce sexually by fusing their gametes to form zygotes, the chromosome number must be reduced from diploid to haploid in the gametes. This is accomplished by a process called **meiosis,** in which certain cells undergo division to yield haploid gametes. In the process of meiosis, portions of chromosomes may be exchanged with one another by **crossing over** (see Figure 8–16). Crossing over in eucaryotic cells can often be observed microscopically; it is in fact the process of genetic recombination. Once haploid gametes are produced by meiosis, their nuclei fuse to form a diploid zygote. Subsequent growth and development of a zygote produces a mature organism again capable of sexual reproduction.

PLASMIDS

Plasmids are small, self-replicating circles of DNA found in many bacteria. They carry genes that the cell needs only under special conditions, and certain plasmids, like the F plasmid, can be integrated into the main chromosome. But all plasmids are dispensable in that the cell can survive under most conditions without the use of the proteins coded by plasmid genes. The F factor, for instance, is useful in gene transfer, which is usually not essential for survival. Many plasmids are **conjugative plasmids;** that is, they have genes for carrying out their transfer to another cell (and for transferring the bacterial chromosome, if the plasmid is integrated).

Plasmids often carry genes that have a more direct use to the cell than simple gene transfer. **Dissimilation plasmids** have genes coding for enzymes that catalyze the catabolism of certain unusual sugars and hydrocarbons. Some species of *Pseudomonas* can actually use such substances as toluene, camphor, and the high molecular weight hydrocarbons of petroleum as primary carbon and energy sources.

Bacteriocinogenic plasmids contain genes for the synthesis of *bacteriocins,* toxic proteins that kill other bacteria. These plasmids have been found in many genera, and they are useful markers for identifying certain bacteria in clinical laboratories.

Resistance factors (R factors) are plasmids of great medical importance. R factors carry genes that determine the resistance of their host cell to antibiotics, heavy metals, or cellular toxins. Many R factors are conjugative plasmids that contain two groups of genes. One group has the **resistance transfer factor (RTF)** and includes genes for replication and conjugation. The other group, the **R genes,** codes for resistance and contains the genetic instructions to produce enzymes that inactivate certain drugs (Figure 8–23). Different R factors, when present in the same cell, readily recombine to produce R factors with new combinations of R genes.

R factors present very serious problems in the treatment of infectious diseases by antibiotics. The

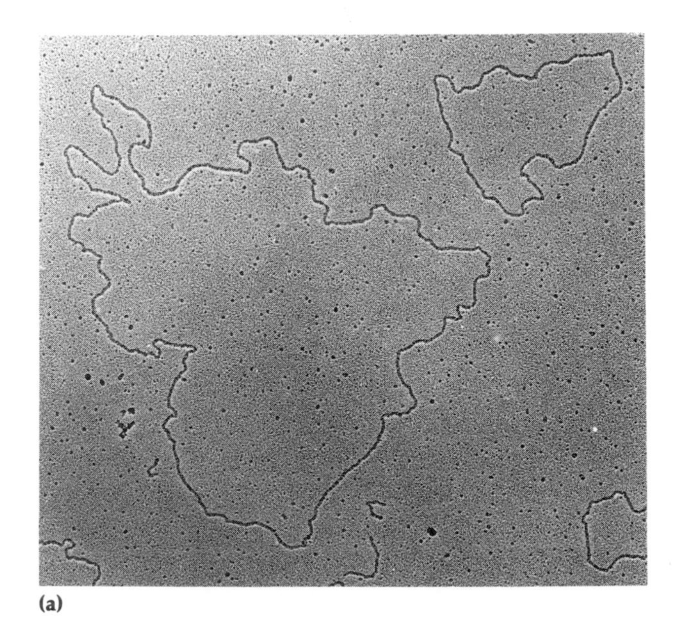

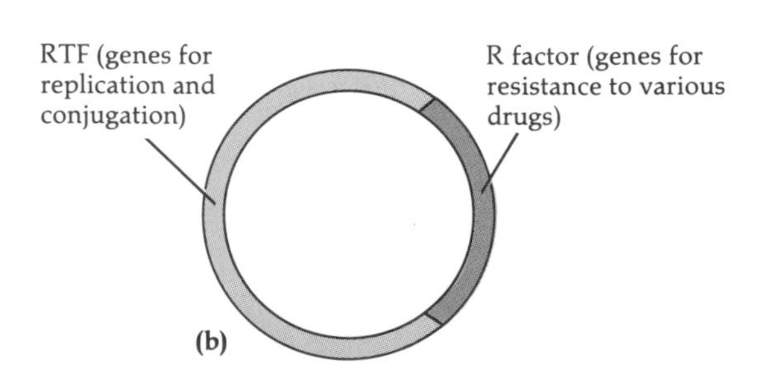

Figure 8–23 R factors, a type of plasmid. **(a)** Electron micrograph showing two different resistance (R) factors ($\times$ 15,000). The larger circle of DNA is a plasmid isolated from the bacterium *Bacteroides fragilis* and encodes resistance to the antibiotic clindamycin. The smaller loop of DNA is a plasmid from *Escherichia coli* and provides resistance to tetracycline. **(b)** Diagram of an R factor.

widespread use of antibiotics in medicine and agriculture (many animal feeds contain antibiotics) has led to the preferential survival (selection) of bacteria that have R factors. The transfer of resistance between bacterial cells of a population, and even between different genera, also contributes to the problem.

GENETIC ENGINEERING: RECOMBINANT DNA

Recombination, as discussed above, is a natural biological process in most organisms. In recent years, however, the term *recombination* has taken on a special, somewhat different meaning when applied to experiments involving genetic engineering. Our detailed knowledge of the plasmids and enzymes involved in DNA metabolism has made it possible to isolate genes from various organisms and incorporate them into bacterial cells. This procedure of manufacturing and manipulating genetic material is referred to as **genetic engineering. Recombinant DNA** results from the joining together of DNA from different sources, often from different species, in vitro. (*In vitro* means "in glass," that is, not within a living organism.) When the recombinant DNA is introduced into an appropriate bacterium, the bacterium can synthesize the products of whatever new genes it has acquired.

In the process of making recombinant DNA, special bacterial enzymes called **restriction enzymes** are used to cut open bacterial plasmids (Figure 8–24). These same enzymes are used to cut segments of DNA from some other organism (frog, rat, human), and the foreign DNA fragment is then inserted into the opened plasmid. Restriction enzymes usually cut uneven ends on DNA molecules, which allows base pairs to be matched as two different pieces of DNA are combined. The DNA fragment is inserted into the bacterial plasmid and the junctions are covalently sealed by DNA ligase.

The recombinant plasmid is then introduced into bacterial cells by transformation. Chemical treatment of the cell wall of the recipient bacteria removes structural barriers and increases the permeability to DNA during transformation. The resulting cells are hybrids with genotypes that never before existed in nature. Each time the hybrid bacteria divide, so do its plasmids carrying the recombinant DNA. Such rapidly dividing synthetic bacteria can produce an enormous amount of whatever substance is programmed by their new genotype.

Although one important goal of recombinant DNA research is to increase our understanding of how genes are regulated, the technique also has practical applications. First, animal genes can be inserted into bacterial DNA so that bacteria produce human substances of therapeutic importance. For example, bacteria have already been programmed to synthesize *somatostatin*, a brain hormone related to growth; *insulin*, a hormone used by diabetics; *human growth hormone*, a hormone that promotes growth in childhood; and *interferon*, an important antiviral (and possible anticancer) substance that will be discussed in Chapter 15. A second application of recombinant DNA research is to improve existing fermentation processes for making microbial products such as antibiotics. And a third application is its use to the chemical and energy industries. For example, ways of programming bacteria to convert ethylene to ethylene glycol or to ferment plants into ethanol for use as fuel are now being rigorously pursued.

Genetic engineering is not without dangers. There is always the possibility of creating a lethal organism or a highly resistant one that might cause widespread disease. In view of this possibility, scientists have agreed that any potentially dangerous genetic engineering experiments should only be performed using genetically weakened bacterial strains, and in laboratories from which the newly created recombinants cannot escape and contaminate the general public.

REGULATION OF GENE EXPRESSION IN BACTERIA

In Chapter 5 we learned that the bacterial cell is involved in an enormously large number of metabolic reactions. Some of these reactions are concerned with biosynthesis and are referred to as anabolic reactions. Others are concerned with degradation and are known as catabolic reactions. The

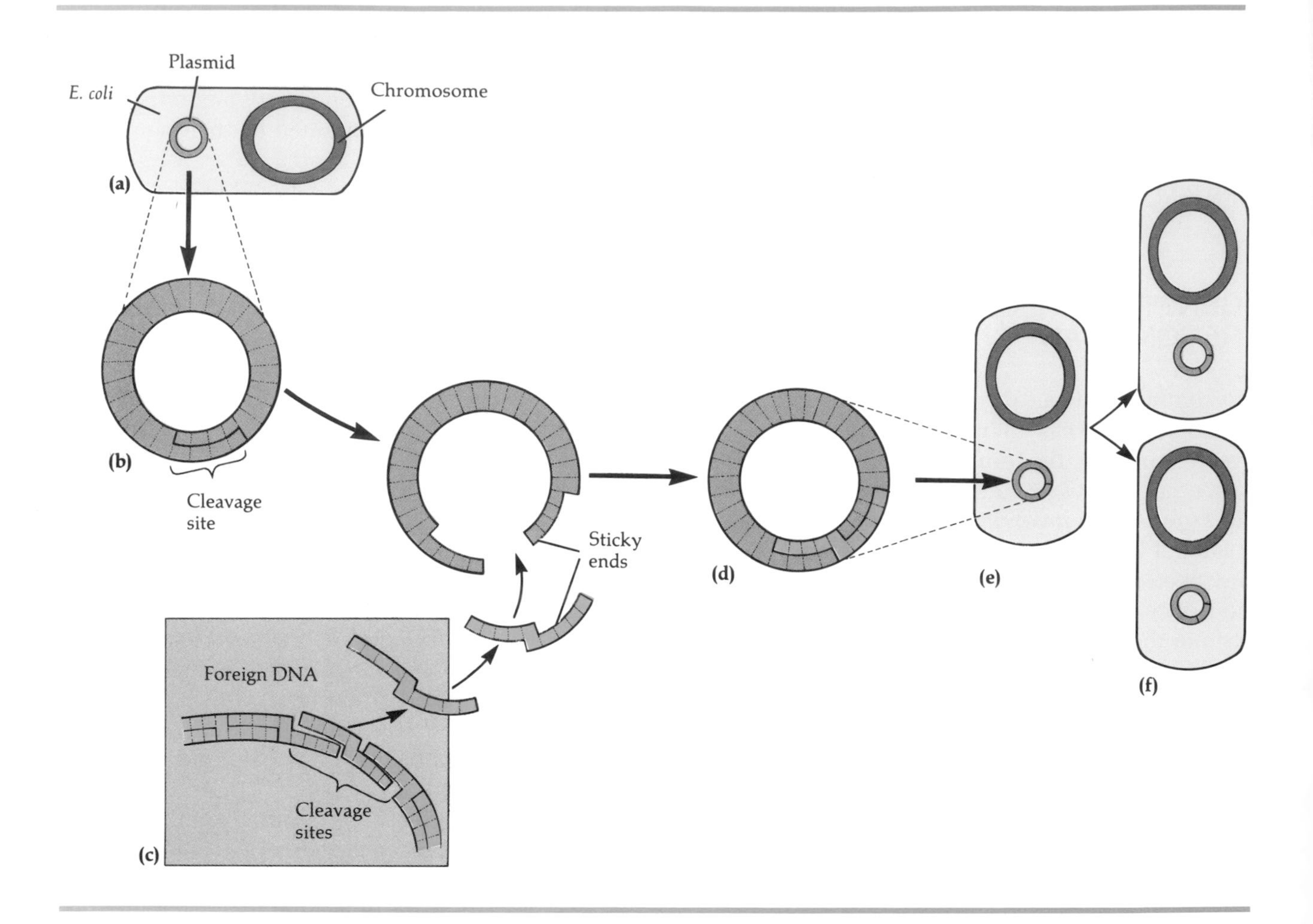

Figure 8–24 Method for in vitro construction of recombinant DNA. **(a)** A plasmid is first isolated from *E. coli.* **(b)** Restriction enzymes split the plasmid at a known cleavage site. **(c)** A segment of foreign DNA from another organism is split by restriction enzymes. **(d)** The foreign DNA is inserted into the plasmid and joined covalently by the enzyme DNA ligase. **(e)** The recombinant plasmid is introduced into bacterial cells by transformation. **(f)** Successive generations of the hybrid bacteria maintain the recombinant plasmid.

common feature of all metabolic reactions is that they are regulated by enzymes. In this chapter we have seen that genes, through transcription and translation, direct the synthesis of proteins, many of which serve as enzymes. These are the very enzymes used for cellular metabolism. Therefore, the genetic machinery and metabolic machinery of a cell are integrated and interdependent. We will now take a look at how gene expression, and therefore metabolism, is regulated. You should be aware of the fact that protein (enzyme) synthesis requires a tremendous expenditure of energy (ATP) by the cell. Therefore, the regulation of protein synthesis is important to the cell's energy economy, since the cells will synthesize only the proteins needed at a particular time. This is an excellent way to conserve energy.

In general, there are two types of mechanisms that control enzymatic activity. With one type, enzyme *activity* is affected, but not its synthesis. An important example of this type is feedback inhibition. With the other mechanism, enzyme *synthesis* is affected, and an important example is end-product repression.

Microbiology in the News

1980 Nobel Prize Split Among Three Gene-Splicers

STOCKHOLM (Oct. 14, 1980)—Sweden's Royal Academy of Sciences today awarded the Nobel Prize in Chemistry to an Englishman and two Americans whose studies of DNA, the molecule of life, have provided laboratories around the world with the foundations of a new technology known as gene-splicing or genetic engineering.

The distinguished scientists are Dr. Paul Berg, 54, professor of biochemistry at Stanford University; Dr. Walter Gilbert, 48, professor of molecular biology at Harvard University; and Dr. Frederick Sanger, 62, professor of molecular biology at Cambridge University in England. In its announcement, the Academy praised the men for increasing the world's knowledge of how DNA, as the carrier of all inherited traits, rules the chemical machinery of the cell. DNA, a nucleic acid with long, twisted strands, is the active ingredient of the genes in all living things. DNA carries a universal genetic code that tells every living cell how to make the proteins required for survival, and how to reproduce.

Dr. Berg was the first scientist to construct a recombinant DNA molecule—a molecule that contains parts of DNA from different species. He was studying the way proteins are put together in simple organisms, and in the process of trying to learn more about protein construction, Dr. Berg dreamed up an experiment to slice the circular DNA from a monkey tumor virus called SV-40, and insert it into a piece of DNA from the bacterial virus known as *lambda*. The next step was to place the circle of recombined DNA into *Escherichia coli*, a bacterium flourishing in the human gut that has been extensively studied by molecular biologists and microbiologists. The *E. coli* rapidly divide and reproduce the recombinant genes in multiple copies. This is called *cloning* of the recombinant genes. In this way, large quantities of identical genes can be produced for study, and theoretically, the protein products of some of these genes could be "farmed" from the bacteria. Dr. Berg's research was the first to demonstrate that DNA could be spliced and that spliced genes would actually continue to function.

Doctors Sanger and Gilbert independently devised a method called "rapid DNA sequencing" which determines the precise line-up of bases in large segments of DNA. These bases are four subunits of the DNA molecule that are repeated in various combinations thousands of times in order to spell out the messages that we call genes. Rapid DNA sequencing has made it possible for researchers to artificially manufacture the part of a gene that would code for a specific, desired protein. This gene could be manipulated by recombinant DNA techniques to produce large quantities of protein.

Frederick Sanger has made history by winning the Nobel Prize in Chemistry twice. His earlier prize was for determining the precise structure of insulin, a protein that helps regulate the body's use of sugar. In his past 40 years of research, Sanger has uncovered details on protein structure that led to such laboratory results as the chemical synthesis of insulin, and later, the bacterial cloning of insulin via recombinant DNA techniques.

In addition to his academic post at Harvard, Walter Gilbert is a part owner and founder of a Swiss-based company, Biogen, that has successfully manufactured human interferon using bacteria. Interferon is a substance that combats viral infections.

The work of these Nobel gene-splicers has made it possible to insert human and animal genes into bacteria, where these genes can be reproduced in large quantities. This has radically changed our access to substances like insulin, interferon, and other proteins that are ordinarily made only in the bodies of humans and animals, and are costly and time-consuming to produce by chemical reconstruction in the laboratory. Thus, the efforts of Doctors Berg, Sanger, and Gilbert and their colleagues have brought us yet another industrial revolution: where humans harness microbes rather than machines.

FEEDBACK INHIBITION

In some metabolic reactions, several steps are required for the synthesis of a particular chemical compound, the end product. The cell makes this end product by converting one substance to another in the presence of specific enzymes (Figure 8–25a). In many biosynthetic pathways, once the final product of the reaction accumulates, it can inhibit one of the earlier enzymes in the pathway. This phenomenon is called **feedback inhibition** (Figure 8–25b). This mechanism of regulation prevents the cell from making excess end product. As the end product already present is used up or diluted out by cell growth, the enzyme inhibition will decrease, and the pathway will become active again.

Feedback inhibition can be demonstrated in *E. coli* for the metabolic pathway leading to the synthesis of the amino acid isoleucine. In this metabolic pathway, the amino acid threonine is enzymatically converted to isoleucine in five separate steps (Figure 8–26). If isoleucine is added to the *E. coli* culture medium, the bacteria no longer synthesize isoleucine. This condition is maintained until the supply of isoleucine is depleted. The isoleucine control system of *E. coli* is only one example

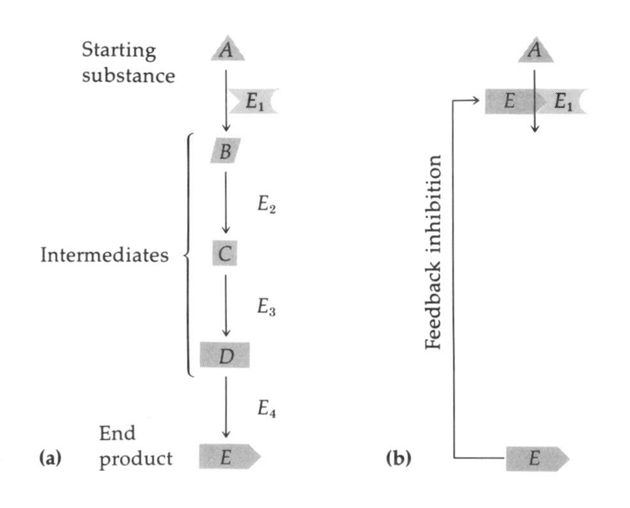

Figure 8–25 Feedback inhibition. **(a)** Synthesis of the end product. **(b)** Inhibition of an early step in the pathway by the end product. E_1 through E_4 represent different enzymes.

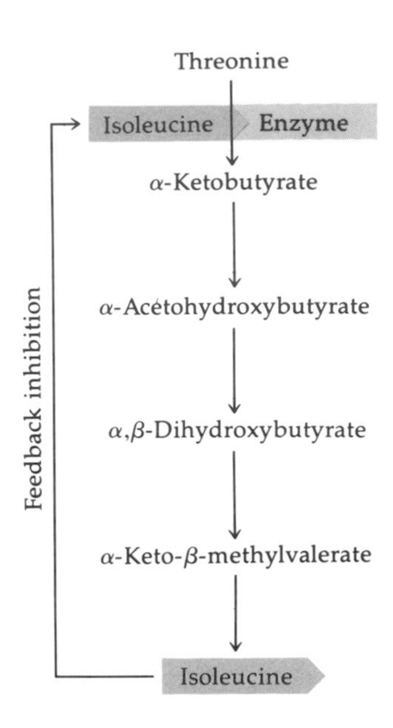

Figure 8–26 Feedback inhibition in *E. coli.* Shown here is a sequence of reactions in which threonine is converted to isoleucine. If isoleucine is added to the culture medium, *E. coli* no longer synthesizes isoleucine, since the isoleucine combines with the enzyme that catalyzes the first step of the synthesis. Thus, isoleucine inhibits the synthesis of more isoleucine.

of a feedback inhibition system. Feedback inhibition is also involved in regulating the cell's production of other amino acids, vitamins, purines, and pyrimidines.

Feedback inhibition is characterized by its action on enzymes that have already been synthesized; it does not affect the synthesis of these enzymes. A single enzyme, usually the first one in a metabolic pathway, is affected by feedback inhibition.

How can an enzyme acting on a compound in the early stages of a pathway be inhibited by the pathway's end product? The end product can inhibit such an enzyme because the enzyme is **allosteric** ("other-space"); that is, it has two unique binding sites: an **active site** specific for the substrate molecule, and a second, **allosteric site** that is specific for the inhibiting end product. The binding of the inhibitor somehow results in a change in the enzyme's shape so that it can no longer bind with the substrate. **Allosteric transition** is the process whereby an enzyme's activity is changed because of selective binding at a second site on the enzyme—a site that does not overlap with the substrate binding site (Figure 8–27). In some cases, allosteric mechanisms can activate an enzyme, rather than inhibit it.

When the enzyme is no longer capable of interacting with its substrate, the product of the first enzymatic reaction in the pathway is not synthesized. Since the unsynthesized product is also the substrate for the next enzyme in the pathway, the reaction stops immediately. Thus, even though only the first enzyme in the pathway is inhibited, the entire pathway shuts down and no end product is formed.

Feedback inhibition is an immediate type of control mechanism. It acts on preexisting enzymes and results in an immediate inhibition of the synthesis of an end product. As such, it is a precise regulator of gene expression and cell metabolism.

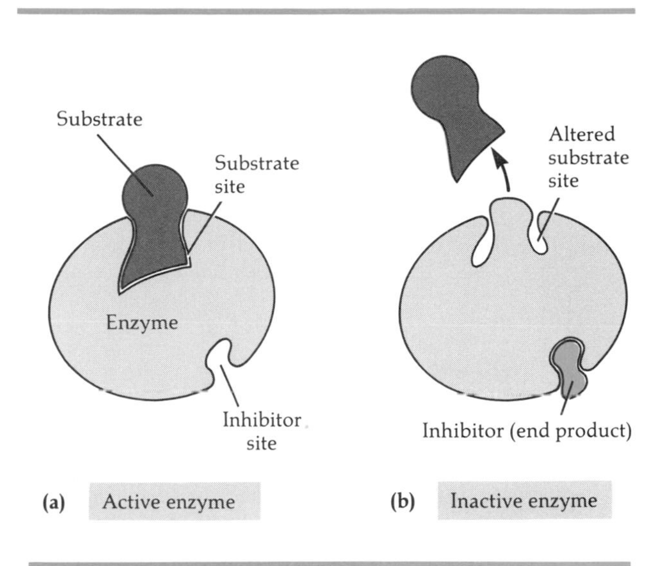

Figure 8–27 Allosteric transition. **(a)** In a normal biochemical reaction, the substrate interacts with the enzyme, and as a result, the substrate is transformed. **(b)** If an inhibitor (end product) combines with its site on the enzyme, however, there is an alteration of the substrate site on the enzyme preventing the normal substrate from combining with the enzyme. This alteration causes enzyme inhibition.

INDUCTION AND REPRESSION

In the control mechanisms called induction and repression, the synthesis of enzymes in a metabolic pathway is regulated. In end-product repression, for example, preexisting enzymes are not acted upon. Instead, the end product inhibits *synthesis* of new enzymes. Thus, control by end-product repression is less precise and immediate than control by feedback inhibition. Induction and repression involve control over enzymatic activity at the gene level; that is, genes are turned on or off so that the synthesis of specific enzymes is increased or decreased.

Induction

The genes required for lactose metabolism in *E. coli* are a well-known example of an inducible system. One of these genes codes for the enzyme β-galactosidase, which splits the substrate lactose into two simple sugars, glucose and galactose. If *E. coli* is placed in a medium in which no lactose is present, the organism is found to contain practically no β-galactosidase enzyme. However, when lactose is added to the medium, the bacterial cells make a large quantity of enzyme. In other words, the presence of lactose has induced the cells to synthesize more enzyme. A substance such as lactose, which brings about an increased amount of an enzyme, is called an **inducer,** and enzymes that are synthesized in the presence of inducers are referred to as **inducible enzymes.** This response, which is under genetic control, is termed **enzyme induction.**

Repression

The type of regulation that decreases the synthesis of enzymes is called **enzyme repression.** In this case, if cells are exposed to a particular end product of a metabolic pathway, there is a decrease in the rate of synthesis of all the enzymes leading to the formation of that end product.

For example, cells of *E. coli* grown in a medium lacking any amino acids contain the enzymes necessary for the synthesis of all the amino acids contained in the protein molecules of the bacteria. But the introduction of a particular amino acid to the culture medium greatly decreases the synthesis of the enzymes necessary for the production of that amino acid. Enzymes that are reduced in amount by the presence of the end product of a metabolic pathway are called **repressible enzymes,** and the molecule (end product) that brings about repression is termed the **corepressor.**

The control of gene function on the basis of induction and repression is explained according to a concept called the operon model.

Operon Model

In 1961, François Jacob and Jacques Monod, as a result of their studies on induction and repression, formulated a general model to account for the regulation of enzyme synthesis in bacteria. This model, called the **operon model,** was based on studies of enzyme systems in *E. coli*. Let us examine the components of the operon model and their role in regulating enzyme synthesis.

The cells of *E. coli* contain all the genetic information required for metabolic reactions, growth, and reproduction. Some of the enzymes needed to conduct these reactions are present at all times. Others, however, are produced only when they are needed by the cell; that is, they are **induced.** Included in this second group are certain enzymes needed for uptake and catabolism of the disaccharide lactose. These are β-galactosidase, which breaks lactose into glucose and galactose; β-galactoside permease, which is involved in the transport of lactose into the cell; and thiogalactoside transacetylase, which is involved in lactose utilization.

Genes that code for enzymes are often called **structural genes** because they specify the amino acid sequence responsible for the structure of particular enzymes. The structural genes for enzymes involved in lactose uptake and utilization turn out to be next to each other on the bacterial chromosome (Figure 8–28a) and produce their respective enzymes rapidly and simultaneously when lactose is introduced into the culture medium. We will now see how this occurs.

A region of DNA called the **regulator gene** codes for a protein called a **repressor.** Another region of DNA called the **operator** is located adjacent to the structural genes. It controls the transcription of the structural genes. And the **promoter site** is the region of DNA where RNA polymerase binds, prior to the initiation of transcription. When lactose is absent (Figure 8–28b), the repressor protein binds tightly to the operator site. This prevents the RNA polymerase enzyme from transcribing the adjacent structural genes, and consequently no mRNA is made and no enzymes are synthesized. This is why in the absence of lactose, no enzymes are produced. But when lactose, the inducer, is present (Figure 8–28c), it binds to the repressor and alters it so that it cannot bind to the operator site. In the absence of bound repressor, the RNA polymerase enzyme can transcribe the structural genes into mRNA, and the enzymes coded in the nucleic acid are synthesized. This is why in the presence of lactose, enzymes are produced. Lactose is said to **induce** enzyme synthesis.

The operator site and the structural genes it controls are referred to as an **operon.** This is the basis for the term **operon model.**

In other systems, the repressor requires the presence of a small molecule called a corepressor in order to bind to the operator gene and promoter site. In the absence of the corepressor, the repressor is unable to bind. The enzymes involved in the synthesis of many amino acids are regulated in this manner. In such cases, the corepressor is an amino acid produced by a set of enzymes. The cell stops synthesis of these enzymes when the amino acid is already present in the medium.

It should be remembered that many genes in bacteria are not subject to repression. Such genes produce enzymes at certain necessary levels, regardless of how much nutrient is present in the medium. Enzymes so produced are called **constitutive enzymes,** and genes responsible for their production are known as **constitutive genes.** These genes usually code for enzymes that the cell always needs in fairly constant amounts to carry out its major life processes. The enzymes of glycolysis are an example.

Both induction and repression of enzymes represent types of differential gene activity in which the synthesis of gene products is sensitive to a given set of environmental conditions. In this regard, induction and repression are adaptive mechanisms that appear to be of survival value to the organism. When these mechanisms are in operation, cells do not expend large amounts of energy in synthesizing enzymes not immediately required; rather, such mechanisms allow for the synthesis of a limited number of enzymes in response to conditions in the environment.

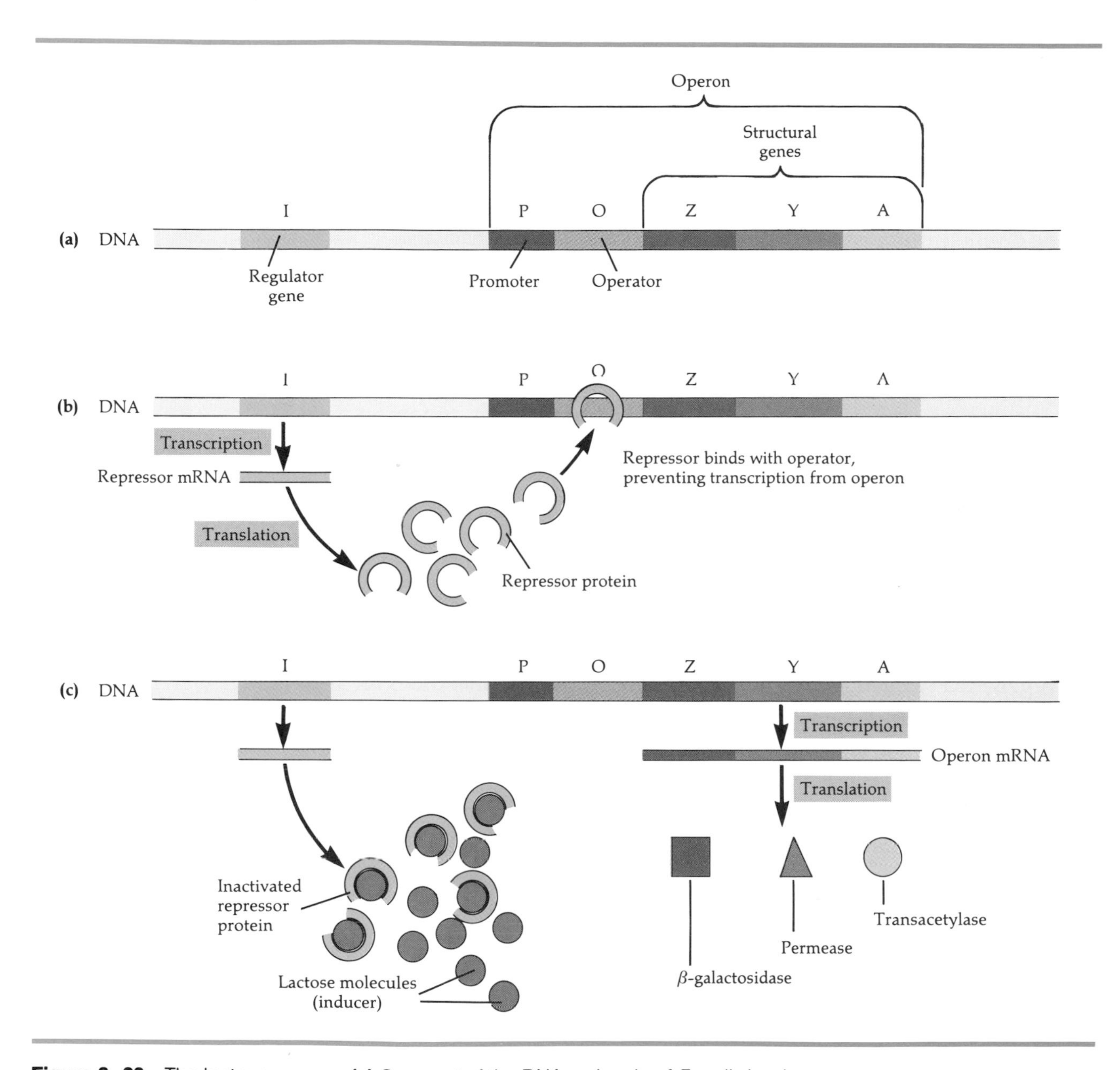

Figure 8–28 The lactose operon. **(a)** Segment of the DNA molecule of *E. coli* showing the genes involved in lactose catabolism. **(b)** Control of the operon in the absence of lactose (repression). **(c)** Control of the operon in the presence of lactose (induction).

Keep in mind that enzyme repression is similar to feedback inhibition because in both cases, the end product acts to prevent its own synthesis. However, repression differs from feedback inhibition in that the end product of enzyme repression regulates enzyme *synthesis*, whereas the end product of feedback inhibition regulates enzyme *activity*.

Although it is fairly common in bacteria for groups of genes to be close together on a chromosome when they govern a sequence of related steps, this is not typical of eucaryotes. In eucaryotes, genes with similar functions tend to be scattered throughout various chromosomes. Scientists do not yet have much information on the regulatory mechanisms that control genes in higher organisms, but it is apparent that natural selection favors those organisms that can regulate their gene expression.

GENES AND EVOLUTION

We have now seen how genes can be controlled by the cell's internal regulatory mechanisms and altered by mutation and recombination. All these processes provide diversity for the descendants of cells, and diversity is the driving force of evolution. Without genetic diversity, there can be no natural selection by environmental factors. The different kinds of microorganisms that exist today, all with relatively stable properties, are the result of a long history of evolution. During that process, microorganisms have continually changed by altering their genetic properties, adapting, as they did so, to many different habitats.

STUDY OUTLINE

STRUCTURE AND FUNCTION OF THE GENETIC MATERIAL (pp. 201–212)

1. Genetics is the study of what genes are, how they carry information, how their information is expressed, and how they are replicated and passed to subsequent generations or other organisms.
2. DNA in cells exists as a double-stranded helix; the two strands are held together by hydrogen bonds between specific nitrogenous base pairs: A–T and C–G.
3. A gene is a segment of DNA or sequence of nucleotides that codes for a functional product, usually a protein.
4. In order for DNA to serve as the genetic material, it works with RNA and proteins.
5. The genetic information in a region of DNA is transcribed to produce RNA, which is translated into proteins.

DNA AND CHROMOSOMES (pp. 203–204)

1. The DNA in a chromosome exists as one long double helix associated with various proteins that regulate genetic activity.
2. Bacterial DNA is circular; the chromosome of *E. coli* contains about 4 million base pairs and is approximately 1000 times longer than the cell.
3. The DNA of eucaryotic chromosomes is more condensed and is associated with much protein, including histones.

DNA REPLICATION (p. 204)

1. In DNA replication, the two strands of the double helix separate at the replication fork, and each strand is used as a template by DNA polymerase to synthesize two new strands of DNA according to the rules of nitrogenous base pairing.

2. The result of DNA replication is two new strands of DNA, each having a base sequence complementary to one of the original strands.
3. Since each double-stranded DNA molecule contains one original and one new strand, the replication process is called semiconservative.

RNA AND PROTEIN SYNTHESIS (pp. 204–210)

Transcription (pp. 204–208)

1. In transcription, a strand of mRNA is synthesized by using a specific enzyme of DNA as a template.
2. The process requires RNA polymerase. The starting point for transcription where RNA polymerase binds to DNA is the promoter site; the region of DNA that is the end point of transcription is the terminator site.

Translation (pp. 208–210)

1. The process of using the information in the nucleotide base sequence of mRNA to dictate the amino acid sequence of a protein is called translation.
2. The mRNA associates with ribosomes, which consist of rRNA and protein.
3. Specific amino acids are attached to molecules of tRNA. Another portion of the tRNA has a triplet of bases called an anticodon.
4. A codon is a segment of three bases of mRNA.
5. The tRNA delivers a specific amino acid to the codon.
6. The ribosome moves along an mRNA strand as amino acids are joined to form a growing polypeptide.
7. A polyribosome is an mRNA with several ribosomes attached to it.

THE GENETIC CODE (p. 211)

1. The genetic code refers to the relationship between the nucleotide base sequence of DNA, the corresponding codons of mRNA, and the amino acids for which the codons code.
2. Of the 64 codons, 61 are sense codons (which code for amino acids) and 3 are nonsense codons (which do not code for amino acids).
3. The genetic code is degenerate; that is, most amino acids are coded for by more than one codon.

GENOTYPE AND PHENOTYPE (p. 212)

1. Genotype is the genetic composition of an organism—its entire DNA.
2. Phenotype is the expression of the genes—the proteins of the cell and their ability to perform particular chemical reactions.

MUTATION: CHANGE IN THE GENETIC MATERIAL (pp. 212–219)

1. A mutation is a change in the nitrogenous base sequence of DNA, which causes a change in the gene product coded for by the mutated gene.
2. Many mutations are neutral, some are disadvantageous, and some are beneficial.

TYPES OF MUTATIONS (pp. 212–214)

1. In base substitutions, one base in DNA becomes replaced with a different base.
2. In a frameshift mutation, one or a few base pairs are deleted or added to DNA.
3. Alterations in DNA may result in missense mutations (amino acid substitutions) or nonsense mutations (creating terminator codons).

MUTAGENESIS (pp. 214–217)

1. Mutagens are agents in the environment that, directly or indirectly, cause mutations.
2. Some mutations occur without the presence of a mutagen; these are called spontaneous mutations.

Chemical Mutagens (pp. 214–215)

1. Chemical mutagens include base pair mutagens (e.g., nitrous acid), base analogs (e.g., 2-aminopurine and 5-bromouracil), and frameshift mutagens (e.g., benzpyrene).

Radiation (pp. 215–217)

1. Ionizing radiation causes the formation of ions and free radicals that react with DNA, resulting in base substitutions or breakage of the sugar–phosphate backbone.
2. Ultraviolet radiation is nonionizing; it causes bonding between adjacent thymines or cytosines.
3. UV damage may be repaired in the presence of light by photoreactivation, or in the dark by enzymes that cut out and replace the damaged portion of DNA.

MUTATION FREQUENCY (p. 217)

1. Mutation rate is the probability of a gene mutating each time a cell divides; the rate is expressed as 10 to a negative power.
2. Mutations usually occur randomly along a chromosome.

IDENTIFYING MUTATIONS (p. 217)

1. Mutations can be detected by selecting or testing for an altered phenotype.
2. In positive selection, mutant cells are selected and unmutated cells are rejected.
3. Replica plating is used for negative selection, detecting, for example, auxotrophs that have nutritional requirements not possessed by the parent (nonmutated) cell.

THE AMES TEST FOR IDENTIFYING CHEMICAL CARCINOGENS (pp. 217–219)

1. This is a relatively inexpensive and rapid test for identifying possible chemical carcinogens.
2. The test assumes that a mutant cell can revert to a normal cell in the presence of a mutagen, and that many mutagens are carcinogens.
3. Histidine auxotrophs of *Salmonella* are exposed to an enzymatically treated potential carcinogen and reversions to the nonmutant (prototroph) are selected.

GENETIC TRANSFER AND RECOMBINATION (pp. 219–227)

1. Genetic recombination refers to the rearrangement of genes from separate groups of genes, usually originating from different organisms; it contributes to genetic diversity.
2. In crossing over, genes from two chromosomes are recombined into one chromosome containing some genes from each original chromosome.
3. Mechanisms for recombination in bacteria involve a portion of the cell's DNA being transferred from donor to recipient.
4. When some of the donor's DNA has been integrated into the recipient's DNA, the resultant cell is called a recombinant.

TRANSFORMATION IN BACTERIA (pp. 220–222)

1. In this process, genes are transferred from one bacterium to another as "naked" DNA in solution.
2. This process was first demonstrated in *Streptococcus pneumoniae.*
3. It occurs naturally among a few genera of bacteria.

CONJUGATION IN BACTERIA (pp. 222–223)

1. This process requires contact between living cells.
2. One type of genetic donor cell is an F^+; recipient cells are F^-.

3. F^+ cells contain cytoplasmic DNA (plasmids) called F factors; these are transferred to the F^- cell during conjugation.
4. When the plasmid becomes incorporated into the chromosome, the cell is called an Hfr (high-frequency recombinant).
5. During conjugation, an Hfr can transfer its entire chromosome to an F^-. Usually, the Hfr chromosome breaks before it is fully transferred.

TRANSDUCTION IN BACTERIA (p. 223)

1. In this process, DNA is passed from one bacterium to another in a bacteriophage.
2. The new bacterial genes may be incorporated into the recipient cell.

RECOMBINATION IN EUCARYOTES (pp. 225–226)

1. In eucaryotes, genetic recombination is associated with sexual reproduction.
2. Sexual reproduction involves the fusion of haploid gametes to form a diploid zygote.
3. Haploid gametes are produced through meiosis, at which time portions of chromosomes may be exchanged, the process of genetic recombination.

PLASMIDS (pp. 226–227)

1. Plasmids are self-replicating circular strands of DNA found in the cytoplasm of bacteria.
2. They are not essential for the survival of the cell.
3. There are conjugative, dissimilation, bacteriocinogenic, and resistance factor plasmids.

GENETIC ENGINEERING: RECOMBINANT DNA (p. 227)

1. The procedure of manufacturing or manipulating genetic material in the laboratory is called genetic engineering.
2. Recombinant DNA results from the joining together of DNA from different sources, often different species; it requires restriction enzymes and ligase.
3. In the procedure, foreign DNA is inserted into a bacterial plasmid and the plasmid is introduced into bacterial cells by transformation.
4. Through genetic engineering, bacteria can produce human substances of therapeutic importance, carry out improved fermentation processes, and synthesize chemicals for industry.

REGULATION OF GENE EXPRESSION IN BACTERIA (pp. 227–234)

1. Gene regulation of protein synthesis is energy conserving, since synthesis occurs only when proteins (enzymes) are needed.
2. In enzymatic control mechanisms, the control is aimed at controlling either enzyme activity or enzyme synthesis.

FEEDBACK INHIBITION (pp. 230–231)

1. In feedback inhibition, if excess end product is produced, an early enzyme in the pathway is inhibited by the end product; this prevents a cell from making excess end product.
2. In addition to an active site, the enzyme has a binding site for the end product. When end product is bound, activity of the enzyme is altered (allosteric transition).

INDUCTION AND REPRESSION (pp. 231–233)

Induction (p. 231)

1. In the presence of certain chemicals (inducers), cells are induced to synthesize more enzymes, a process called induction.
2. An example is the production of β-galactosidase in the presence of lactose in *E. coli*.

Repression (pp. 231–232)

1. In repression, there is an alteration in the synthesis of one or several (repressible) enzymes.
2. When cells are exposed to a particular end product, enzyme synthesis decreases.
3. The end product that causes repression is called a corepressor.

Operon Model (pp. 232–233)

1. The formation of enzymes is controlled by structural genes. These genes and the operator site that controls their transcription is called an operon.
2. In the operon model, a regulator gene codes for repressor protein.
3. When the inducer is absent, the repressor binds to the operator and no enzymes for lactose degradation are produced.
4. When the inducer is present, it binds to the repressor so that it cannot bind to the operator; thus, enzyme synthesis is induced.

5. In some systems, the repressor requires a corepressor in order to bind to the operator and promoter; the corepressor stops enzyme synthesis.
6. Some genes, called constitutive genes, produce enzymes regardless of how much nutrient is present. Examples are genes for the enzymes of glycolysis.

GENES AND EVOLUTION (p. 234)

1. Diversity is the driving force of evolution.
2. Genetic mutation and recombination provide a diversity of organisms so that the process of natural selection will allow the growth of those best adapted for a given environment.

STUDY QUESTIONS

REVIEW

1. Briefly describe the components of DNA and explain its functional relationship to RNA and protein.
2. Draw a diagram showing a portion of a chromosome undergoing replication.
   (a) Identify the replication fork.
   (b) What is the role of DNA polymerase?
   (c) What is the role of DNA ligase?
   (d) How does this represent semiconservative replication?
3. Given below is a strand of DNA.

   T A T A A T G G A T T T C G C A T C A A T G T C T
   1 2 3 4 5 6 7 8 9 10 11 12 13 14 15 16 17 18 19 20 21 22 23 24 25
   TATAATG = Promoter sequence

   (a) Show the complementary strand of DNA.
   (b) Using the genetic code provided in Figure 8–8, identify the sequence of amino acids coded for by this strand of DNA (the complementary strand is the antisense strand).
   (c) What would the effect be if C was substituted for G at base 22?
   (d) What would the effect be if A was substituted for C at base 13?
   (e) What would the effect be if A was substituted for T at base 10?
   (f) What would the effect be if C was inserted between bases 12 and 13?
   (g) What would be the effect of ultraviolet radiation on this strand of DNA?
   (h) Identify a nonsense codon in this strand of DNA.
4. Describe translation, and be sure to include the following terms: ribosome, rRNA, amino acid activation, tRNA, anticodon, and codon.

5. Contrast the structure of a bacterial chromosome and a eucaryotic chromosome.
6. Match the following examples of mutagens.

___ A mutagen that is incorporated into DNA in place of a normal base	(a) Frameshift mutagen
___ A mutagen that causes the formation of highly reactive ions	(b) Base analog
___ A mutagen that alters adenine so that it base-pairs with cytosine	(c) Base pair mutagen
___ A mutagen that causes insertions	(d) Ionizing radiation
	(e) Nonionizing radiation

7. Explain the selection of an antibiotic-resistant mutant by direct selection, and the selection of an antibiotic-sensitive mutant by indirect selection.
8. Describe the principle of the Ames test for identifying chemical carcinogens.
9. Differentiate between transformation and transduction.
10. Define plasmids and explain the relationship between F plasmids and conjugation.
11. Outline the production of recombinant DNA in vitro.
12. List three practical applications of genetic engineering.
13. Use this metabolic pathway to answer the questions below:

$$\text{Substrate } A \xrightarrow{\text{enzyme } a} \text{Intermediate } B \xrightarrow{\text{enzyme } b} \text{End product } C$$

   (a) Explain what happens when an excess of *C* acts on enzyme *b* by allosteric transition.
   (b) If enzyme *a* is inducible and is not being synthesized at present, a ______________ protein must be bound tightly to the ______________ site. When the inducer is present, it will bind to the ______________ so that ______________ can occur.
   (c) If enzyme *a* is constitutive, what effect, if any, will the presence of *A* or *C* have on it?
14. Define the following terms:
   (a) Genotype
   (b) Phenotype
   (c) Recombination
15. Why are mutation and recombination important in the process of natural selection and the evolution of organisms?

CHALLENGE

1. Base analogs and ionizing radiation are used in the treatment of cancer. Since these are mutagens, how do you suppose they are used beneficially to treat diseases?

2. Why are semiconservative replication and degeneracy of the genetic code advantageous to survival?

3. You are provided with cultures with the following characteristics.

   Culture 1: F^+, genotype $A^+B^+C^+$
   Culture 2: F^-, genotype $A^-B^-C^-$

   (a) Indicate the possible genotypes of a recombinant cell resulting from the conjugation of cultures 1 and 2.
   (b) Indicate the possible genotypes of a recombinant cell resulting from conjugation of the two cultures after the F^+ has become an Hfr.

FURTHER READING

Ayala, F. J. and J. A. Kiger, Jr. 1980. *Modern Genetics.* Menlo Park, CA: Benjamin/Cummings. A recent genetics textbook with excellent coverage of bacterial and viral genetics.

Daven, C. I. (Introduction). 1981. *Genetics: Readings from Scientific American.* San Francisco, CA: W. H. Freeman. A collection of papers covering the chemical structure of DNA, the genetic code, and applied genetics.

Devoret, R. August 1979. Bacterial tests for potential carcinogens. *Scientific American* 241:40–49. Short tests using bacteria instead of animals to detect carcinogenic chemicals and the effects of carcinogens on DNA.

Freifelder, D. (Introduction). 1978. *Recombinant DNA: Readings from Scientific American.* San Francisco, CA: W. H. Freeman. *Scientific American* offprints on molecular biology of the gene and gene manipulation.

Gilbert, W. and L. Villa-Komaroff. April 1980. Useful proteins from recombinant bacteria. *Scientific American* 242:74–94. How genes are spliced into bacteria so the bacteria can produce nonbacterial proteins such as insulin and interferon.

Watson, J. D. 1976. *Molecular Biology of the Gene,* 3rd ed. Menlo Park, CA: Benjamin/Cummings. Fundamental principles of molecular biology using work done primarily on microorganisms.

PART

SURVEY OF THE MICROBIAL WORLD

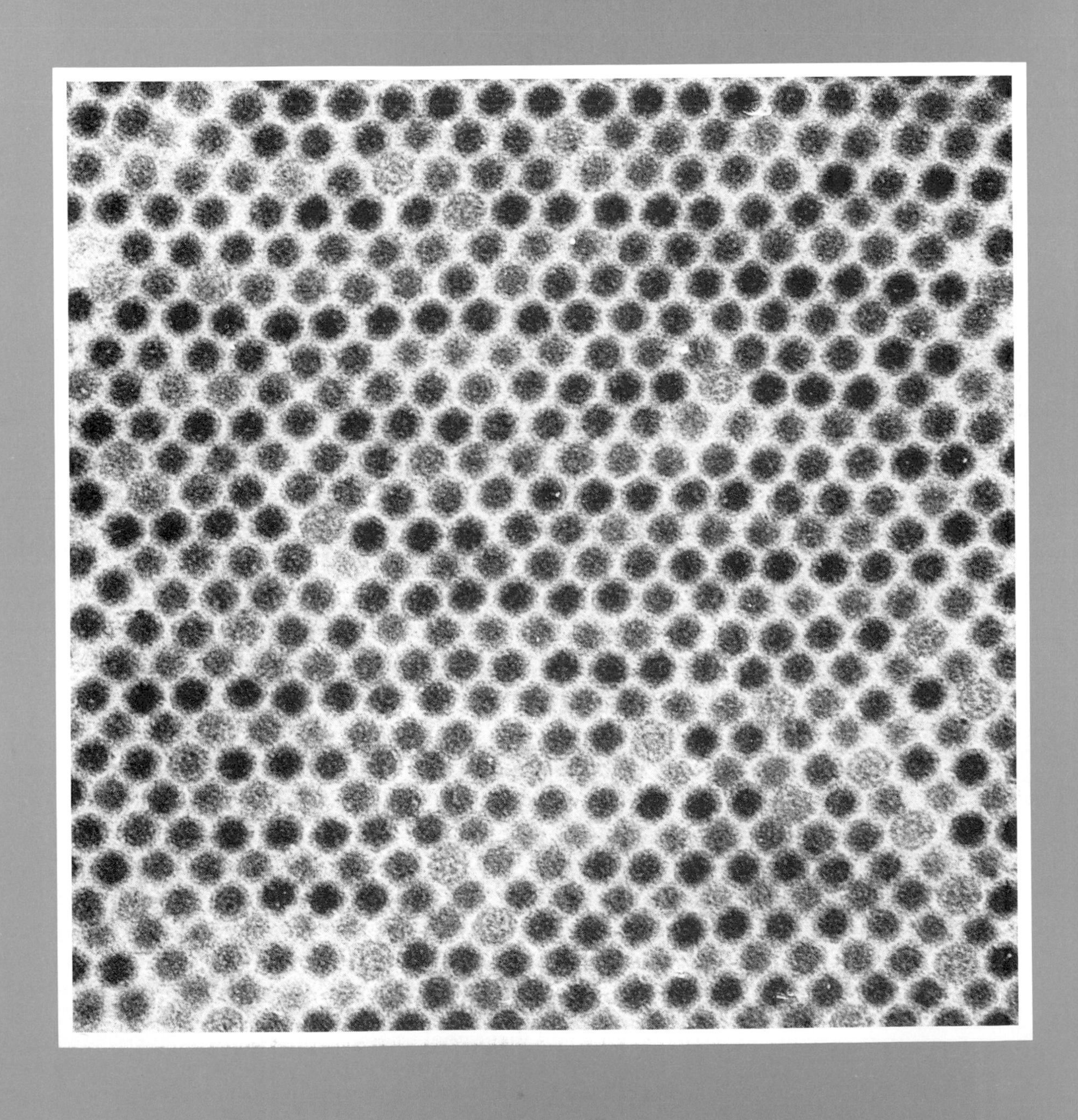

9

Classification of Microorganisms

Objectives

After completing this chapter you should be able to

- Define taxonomy.
- Name the five kingdoms of living organisms.
- Identify the characteristics of the kingdom Monera that differentiate it from other kingdoms.
- Define binomial nomenclature and discuss its purpose.
- Describe eight methods of identifying and classifying microorganisms.

The science of classification, especially of living forms, is called **taxonomy.** The object of taxonomy is to establish the relationship between one group of organisms and another, and to be able to differentiate between them. Until the end of the nineteenth century, all organisms were divided into two kingdoms: plant and animal. Then, when microscopes were developed and biologists came to understand the structure and physiological characteristics of microorganisms, it became apparent that microorganisms did not really belong to either the plant or animal kingdom, although many microbes have both plant and animal characteristics. In 1866, a third kingdom, the Protista, was proposed by Ernst Haeckel to include all microorganisms—bacteria, fungi, protozoans, and algae. But subsequent research indicated that among the Protista, two different basic cell types could be distinguished: procaryotic and eucaryotic. The morphological differences between these two cell types were described in detail in Chapter 4.

FIVE-KINGDOM SYSTEM OF CLASSIFICATION

In 1969, H. R. Whittaker proposed the **five-kingdom system** of biological classification, which has been widely accepted. Recall from Chapter 1 that in this system, there is a basic division between organisms with procaryotic cells and those with eucaryotic cells (Figure 9–1). The bacteria and the cyanobacteria (formerly called blue-green algae) are the procaryotic organisms, called **Monera** by Whittaker and **Procaryotae** by others, including *Bergey's Manual of Determinative Bac-*

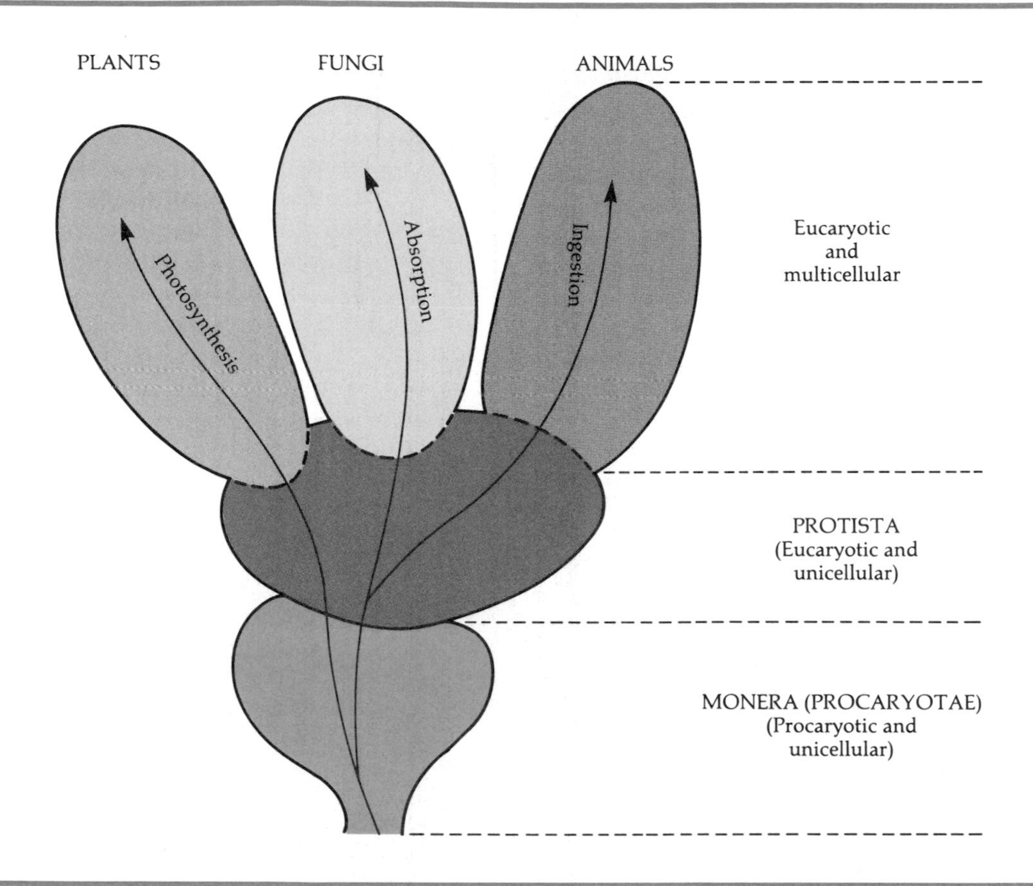

Figure 9–1 Five-kingdom system of classification.

teriology. As discussed in Chapter 4, procaryotes do not have a nucleus separated from the cytoplasm by a membrane. Instead, they have a nuclear region in which the genetic material is more or less localized. Fossil evidence for procaryotic organisms indicates their presence on earth over 3.5 billion years ago.

Higher organisms, structurally much more complicated, are based on the eucaryotic cell, in which a membrane separates the nucleus from the cytoplasm. This type of cell apparently evolved more recently, a little more than a billion years ago. Simpler eucaryotic organisms, mostly unicellular, are grouped as the **Protista** and include funguslike slime molds, animallike protozoans, and some algae. Some algae are more complex multicellular types. Algae are assigned to kingdoms based on evolutionary relationships. The more primitive algae are placed in the Protista, and other algae are included with the plant kingdom.

Plants, animals, and fungi comprise three kingdoms of more complex eucaryotic organisms, most of which are multicellular. The main distinguishing factor among these three kingdoms is the nutritional mode; another important difference is cellular differentiation.

Fungi are eucaryotic organisms that include the unicellular yeasts, multicellular molds, and macroscopic varieties such as mushrooms. To obtain raw materials for vital functions, a fungus absorbs dissolved organic matter through the membranes of cells in rootlike structures called hyphae.

Plantae are multicellular eucaryotes, and they include some algae and all mosses, ferns, conifers, and flowering plants. To obtain energy, a plant photosynthesizes, a process involving the conversion of carbon dioxide from the air to organic molecules for use by the cell.

Animalia are multicellular animals with eucaryotic cells. These include sponges, various worms, insects, and animals with backbones (vertebrates). Animals obtain carbon and energy by ingesting organic matter through a mouth of some kind.

In 1978, Whittaker and L. Margulis proposed a revised classification scheme, organizing the five kingdoms according to procaryote or eucaryote. Essentially, this new scheme is as follows:

Superkingdom Procaryotae
- Kingdom Monera

Superkingdom Eucaryotae
- Kingdom Protista
  - Branch Protophyta, plantlike protists
  - Branch Protomycota, funguslike protists
  - Branch Protozoa, animallike protists
- Kingdom Fungi
- Kingdom Plantae
- Kingdom Animalia

The Archaebacteria—Neither Procaryotes nor Eucaryotes?

The division of all living organisms into procaryote and eucaryote has been challenged recently by a number of biologists who claim that there is a third basic category. The organisms belonging to this new category are bacteria, that is, procaryotes. Now it is becoming clear that these bacteria are no more closely related to other bacteria (called *eubacteria* or *true bacteria* by these biologists) than they are to eucaryotic organisms such as animals and plants.

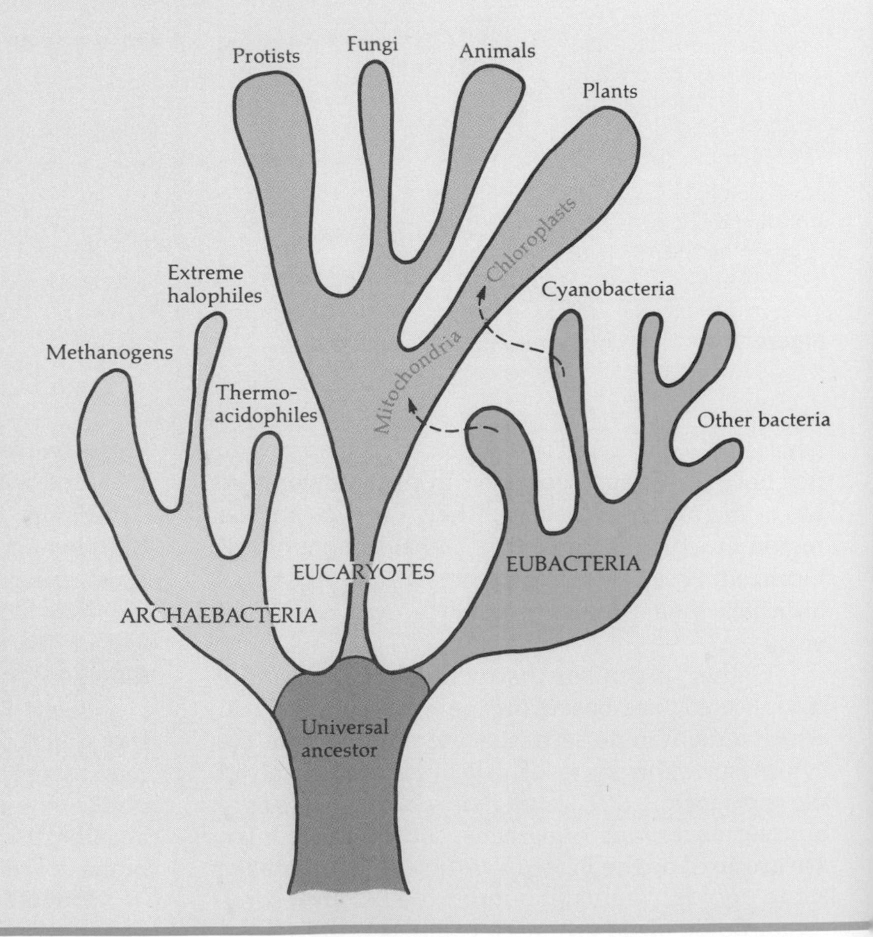

Archaebacteria

The bacteria belonging to this new category look like typical bacteria and lack a nucleus and other membrane-bounded organelles. For years, however, it has been known that they are unusual in a number of ways. For example, their cell walls never contain peptidoglycan, they live in extreme

NAMING ORGANISMS

In a world inhabited by millions of living organisms, biologists must be sure they know exactly which organism is being discussed. We cannot use common names because the same name is often used for many different organisms, based on regional occurrence. For example, there are two different plants with the common name Spanish Moss, and neither one is a moss. And three different animals are referred to as a gopher. Since common names are rarely specific and can often be misleading, a system of naming things, referred to as **scientific nomenclature,** was developed in the eighteenth century by Carolus Linnaeus.

All organisms have two names, usually derived from either Latin or Greek. These are the **genus** name and **species** name, and both names appear underlined or italicized. The genus name is always capitalized and is always a noun. The species name is written in lowercase and is usually an adjective. This system of two names for an organism is called

environments, and they carry out unusual metabolic processes. They include three groups: (1) the methanogens, strict anaerobes that produce methane (CH_4) from carbon dioxide and hydrogen; (2) extreme halophiles, which require high concentrations of salt for survival; and (3) thermoacidophiles, which normally grow in hot, acid environments. Because they are thought to have descended from very ancient organisms, these microbes have been dubbed *archaebacteria* (*archae*-meaning primitive).

Evidence

What is the evidence that the archaebacteria constitute a group separate from the true bacteria and the eucaryotes? The evidence is biochemical, and most of it comes from the sequencing of ribosomal RNA (rRNA) molecules from many kinds of organisms. Ribosomes are essential to all forms of life on Earth, and the sequence of nucleotides in rRNA reflects the evolutionary history of the cell in which the rRNA is found. On the basis of similarities and differences among the rRNAs of many different species, the archaebacteria form a group distinct from the other bacteria and the eucaryotes. Data on membrane lipids, transfer RNA molecules, and sensitivity to antibiotics also support this distinction.

Evolutionary Relationships

From the information gathered so far, researchers have concluded that the archaebacteria split off the evolutionary tree to form a separate branch at the same time that the eubacterial and eucaryotic branches diverged (see diagram). The physiology of the methanogenic bacteria certainly seems to be suited to the kind of atmosphere thought to have existed billions of years ago on Earth—one containing carbon dioxide and some hydrogen but little or no oxygen.

The research on archaebacteria suggests some changes in the current theory to explain the origin of eucaryotic cells. It has been commonly believed that mitochondria and chloroplasts are descended from bacteria that were engulfed by a larger procaryotic cell, and it has been assumed that the host cell was an ordinary bacterium. Now, it appears that the original host cell may have been a microorganism that evolved along a line separate from the true bacteria. This concept would help account for many of the basic molecular differences between the true bacteria, the eucaryotes, and the archaebacteria.

binomial nomenclature. Let us consider a few examples. Your own genus and species names are *Homo sapiens;* one bacterium that causes pneumonia is called *Klebsiella pneumoniae;* and a mold that contaminates bread is called *Rhizopus nigricans*. The genus and species names are spelled out the first time they are used *(Klebsiella pneumoniae)*, but later, the genus name may be abbreviated *(K. pneumoniae)*.

CLASSIFICATION OF EUCARYOTIC ORGANISMS

For eucaryotic organisms, a **species** is a group of closely related organisms that have a limited geographical distribution and that breed among themselves (interbreed), but not with other species. Consequently, members of one species have certain morphological characteristics that distinguish them from members of another species. A **genus** (plural **genera**) consists of a number of species different from each other in certain ways, but related to each other by descent. For example, *Quercus,* the genus name for oak, consists of all types of oak trees (white oak, red oak, bur oak, velvet oak, and so on). Even though each species of oak differs from every other species, they are all related genetically. And just as a number of species make up a genus, related genera make up a **family.** A group of similar families constitutes an **order,** and a group of similar orders makes up a **class.** Related classes, in turn, comprise a **phylum** or **division.** Thus, a particular organism has a genus and species name and belongs to a family, order, class, and phylum or division. All phyla or divisions that are related to each other make up a **kingdom**.

Charles Darwin (1809–1882) explained the many similarities and differences among organisms on the basis of descent from common ancestors. The arrangement of organisms into taxonomic categories, collectively called **taxa,** reflects degrees of relatedness among organisms. The hierarchy of taxas shows evolutionary or **phylogenetic** (from a common ancestor) relationships.

CLASSIFICATION OF BACTERIA

The classification of eucaryotic organisms and the classification of bacteria differ in several ways. For example, the term *species* as applied to higher organisms is defined on the basis of geographical distribution and interbreeding that result in distinctive morphological characteristics. Among bacteria, however, morphological traits are limited, and sexuality involves unilateral transfer of genetic material rather than the fusion of two cells such as an egg and a sperm. A **bacterial species** is accordingly defined as a population of cells with similar characteristics. (The types of characteristics will be discussed later in this chapter.) The members of a bacterial species are essentially undistinguishable from each other, but can be distinguished from members of other species, usually on the basis of several features. In some cases, not all pure cultures of the same species are identical in all ways, and each such group in a species is called a strain. A **strain** is a group of cells all derived from a single cell. Strains are identified by numbers, letters, or names following the species name. It is therefore possible to define a bacterial species as a collection of closely related strains. As with the classification of eucaryotic organisms, several species of bacteria constitute a genus, several genera comprise a family, and families are grouped into orders. No classes are officially designated for bacteria at this time. Rather, the bacteria are grouped into 19 "parts," which will be discussed in Chapter 10.

Some of the information employed to determine evolutionary relationships in higher organisms comes from fossil records. Bones, shells, or stems that contain mineral matter or have left imprints in rock that was once mud are examples of fossils. Most microorganisms do not contain structures that are readily fossilized. One exception is the white cliffs of Dover, England, which are actually fossilized marine protists. Another exception is the fossilized cyanobacteria found in 3- to 3.5-billion-year-old rocks in Canada. These are the oldest known fossils. Some younger fossil procaryotes are shown in Figure 9–2.

The taxonomic classification scheme for bacteria may be found in a reference book called *Bergey's Manual of Determinative Bacteriology.* This book groups bacteria according to different shared properties, such as morphology, staining characteristics, nutrition, and metabolism. While continuing to use most of the familiar taxonomic terms, the eighth edition of *Bergey's Manual* explicitly postpones the assignment of definite evolutionary relationships, pending new information to be derived from the methods of modern molecular biology. Only recently, the techniques of analyzing nucleic acid base com-

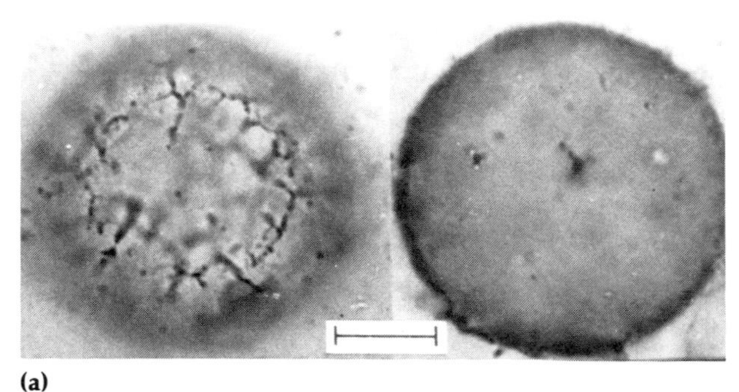
(a)

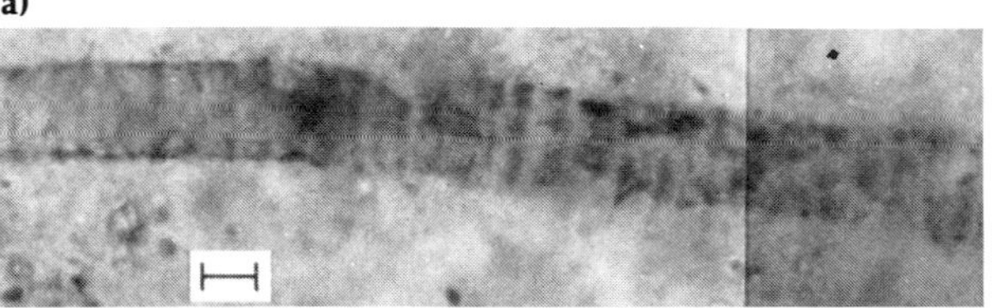
(b)

Figure 9–2 Fossil procaryotic cells. **(a)** Coccoid forms. **(b)** Part of a filament (scales = 10 μm). These microfossils, probably the remnants of cyanobacteria, were found in 650-million-year-old rocks from the Soviet Union.

position and protein sequences have provided means for determining evolutionary relationships among bacteria. Further information on the evolution of bacteria will tell us something about the origin of life.

BASES FOR CLASSIFYING AND IDENTIFYING MICROORGANISMS

Higher organisms are frequently classified according to observed anatomical detail. However, many microorganisms appear too similar for classification by structure. We will now discuss several criteria and methods for classifying microorganisms and, in some cases, for identifying them on a routine basis.

Morphological Characteristics

Attempting to separate microorganisms by evolutionary groupings—as one would separate whales from fish or opossums from rats—can be extremely frustrating. Through a microscope, there is a great deal of similarity between organisms that may otherwise be different with regard to metabolic or physiological properties. There are literally hundreds of bacterial species that are small rods or small cocci. However, morphological characteristics are still of some use in classifying and identifying bacteria. Sometimes the presence of structures such as endospores or flagella can be helpful.

One of the first steps in the identification of bacteria is differential staining. As we have seen from earlier chapters, just about all bacteria are either gram-positive or gram-negative. Other differential stains, such as the acid-fast stain, can be useful, but for a more limited group of microorganisms.

Biochemical Tests

Enzymatic activities are widely used to differentiate between bacteria. Even closely related bacteria can usually be separated into distinct species by testing their ability to ferment an assortment of selected carbohydrates or by subjecting them to other biochemical tests.

Enteric, gram-negative microbes are a large heterogenous group of bacteria whose natural habitat is the intestinal tract of humans and animals. Among the enteric bacteria are members of the genera *Escherichia, Enterobacter, Shigella,* and *Salmonella*. *Escherichia* and *Enterobacter* can be distinguished from *Salmonella* and *Shigella* on the basis of lactose fermentation; the former ferment lactose to produce acid and gas whereas the latter do not. Further biochemical testing as shown in Figure 9–3 can be used to separate the genera.

Microbial identification and classification often include the use of selective or differential media. Recall from Chapter 6 that selective media contain ingredients to suppress the growth of competing organisms and encourage the growth of desired

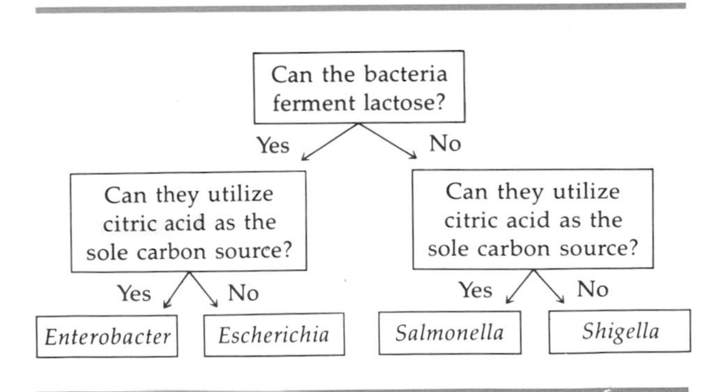

Figure 9–3 Identification of selected genera of enteric bacteria by metabolic characteristics.

ones, and differential media allow the desired organism to form a colony that is somehow distinctive.

Serology

Serology is the science that deals with the blood serum, and in particular, with immune responses evident in serum (see Chapter 16). Microorganisms are antigenic; that is, they are capable of stimulating antibody formation when injected into animals. For example, the immune system of a rabbit injected with killed typhoid bacteria will respond by producing antibodies against typhoid bacteria. The antibodies are proteins that circulate in the blood and combine in a highly specific way with the bacteria that caused their production. Solutions of such antibodies, called **antisera** (singular, **antiserum**), are commercially available for use in identifying many medically important bacteria. If an unknown bacterium is isolated from a patient, it can be tested against known antisera.

In a procedure called a **slide agglutination test,** the unknown bacterium is placed in a drop of salt solution on a slide and mixed with a drop of known antiserum. This test is done with different antisera on different slides. The bacteria will agglutinate or clump when mixed with antibodies produced against the same strain or species. A positive test is therefore determined by agglutination. A slide agglutination test is illustrated in Figure 9–4.

Not only can microbial species be identified with these serological techniques, but even strains within species can be differentiated.

Phage Typing

A similar concept is the basis for phage typing. Phages are bacterial viruses, and they are highly specialized in that they usually attack only members of a given species, or even strains within a species; strains of bacterial viruses are specific for strains of bacteria within a bacterial species. Both serology and phage typing are particularly useful in tracing the origin and course of disease outbreaks.

To see how phage typing can help trace the origin and course of a specific strain of a bacterium, let us consider the following example. Years ago there was a very high incidence of hospital-associated (nosocomial) staphylococcal infections. Moreover, the chain of transmission was unknown in many cases. Although such outbreaks still occur, they are not as frequent because their sources can be traced by phage typing. In this procedure, var-

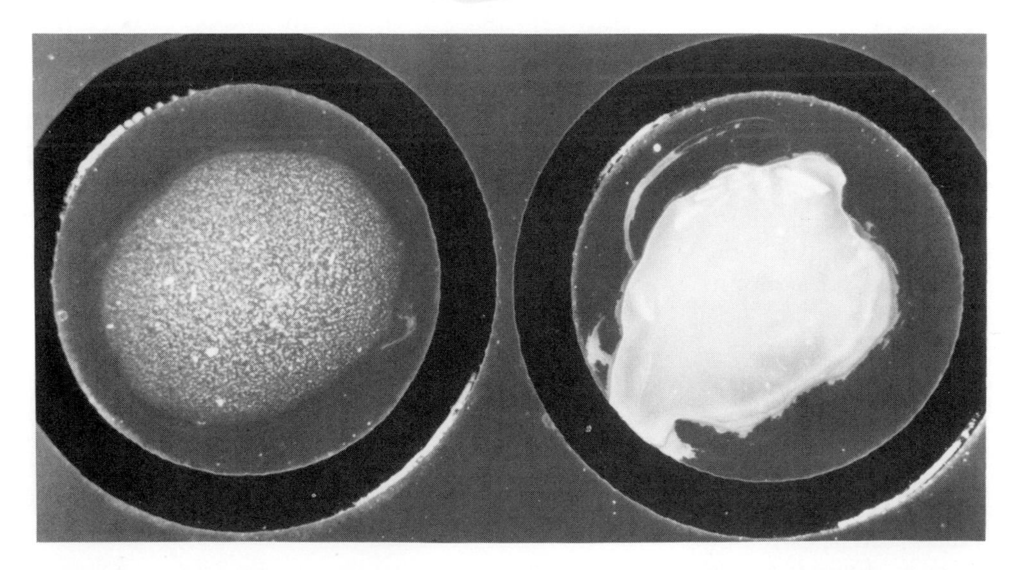

Figure 9–4 Slide agglutination test. On the left is a positive test; the grainy appearance is due to the clumping (agglutination) of the bacteria. Agglutination results when the bacteria are mixed with antibodies produced against the same strain. On the right is a negative test in which the bacteria are still evenly distributed in the liquid of the salt solution and antiserum. The bacterium being tested can be identified as strain *A* because agglutination occurred with antibodies against *A*.

ious phages are applied to staphylococci growing on an agar medium (Figure 9–5). The plate is marked off in small squares and a drop of each phage type is placed on the bacteria. In those squares where the phage causes lysis, plaques appear. Such a test might show that staphylococci isolated from a surgical wound have the same pattern as those isolated from the operating surgeon or surgical nurses.

Numerical Taxonomy

Numerical taxonomy (Figure 9–6) has been used to infer evolutionary relationships between organisms. Many microbial characteristics, such as the ability to use a certain carbohydrate, to form a pigment, or to produce a certain acid, are listed, and the presence or absence of each characteristic is scored for each organism. The characteristics of each organism are then matched against those of other organisms (a computer is helpful here). The greater the number of characteristics shared by two

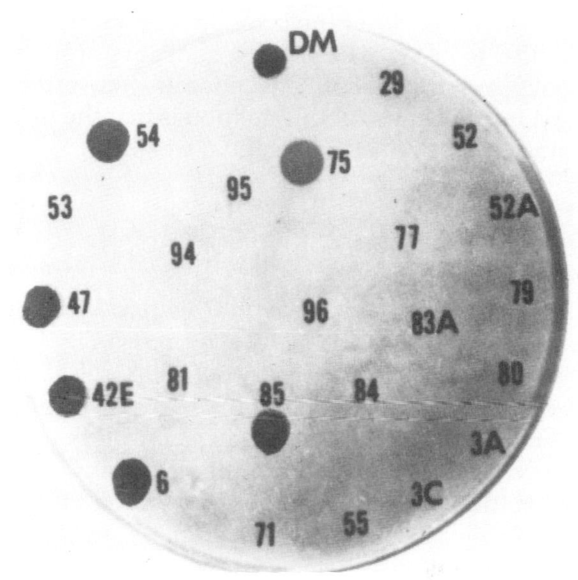

Figure 9–5 Phage typing of a strain of *Staphylococcus aureus*. The strain being tested was grown over the entire plate and was lysed by bacteriophages 6, 42E, 47, 54, 75, and 85. Accordingly, the strain is identified as phage type 6/42E/47/54/75/85. (DM is a dye marker.)

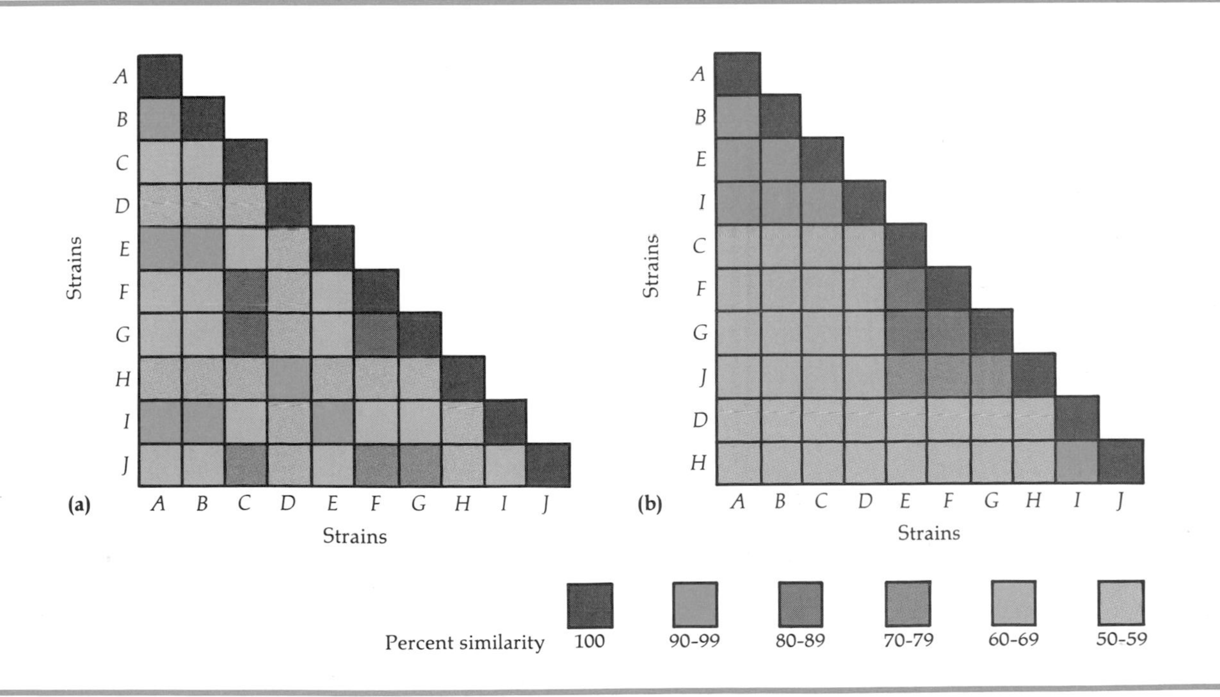

Figure 9–6 Numerical taxonomy. With the aid of a computer, a similarity matrix (array) can be constructed to show relationships between different organisms. As many as 100 different tests may be run on each organism. Each organism is then given a numerical score indicating the degree of its similarity with every other organism, and the score is shown on a matrix **(a).** In another matrix **(b),** organisms with similar test results are grouped together.

organisms, the closer the taxonomic relationship is assumed to be. A match of 90% similarity or higher usually indicates a single taxonomic unit, or species. One weakness of this method is that the listed characteristics are all given equal weight.

Base Composition of Nucleic Acids

A technique that has come into wide use among taxonomists because it has a good possibility of at least suggesting evolutionary relationships is the determination of the nitrogenous base composition of the DNA, usually expressed as percent guanine plus cytosine. The base composition of a single species is theoretically a fixed property and thus affords a measure of species relatedness. As we saw in the chapter on genetics, for each guanine (G) in DNA, there is a complementary cytosine (C). Similarly, for each adenine (A) in the DNA, there is a complementary thymine (T). Therefore, the percent G–C pairs also tells us the percent A–T pairs (subtract from 100%). Two organisms that are closely related and hence have many identical or similar genes will naturally have similar amounts of the various bases in their DNA. On the other hand, if there is a difference of more than 10% in their percent G–C pairs (for example, if one bacterium's DNA contains 40% G + C and another bacterium has 60% G + C) then these two organisms are probably not related. It should be remembered, however, that two organisms that have the same percent G + C need not be closely related; other supporting data are needed to draw conclusions about evolutionary relationships.

Actually determining the entire *sequence* of bases in an organism's DNA, while theoretically possible with modern biochemical methods, is impractical for all but the smallest microorganisms (viruses). However, the sequencing of bases in ribosomal RNA has provided useful information on evolutionary relationships among bacteria. Closely

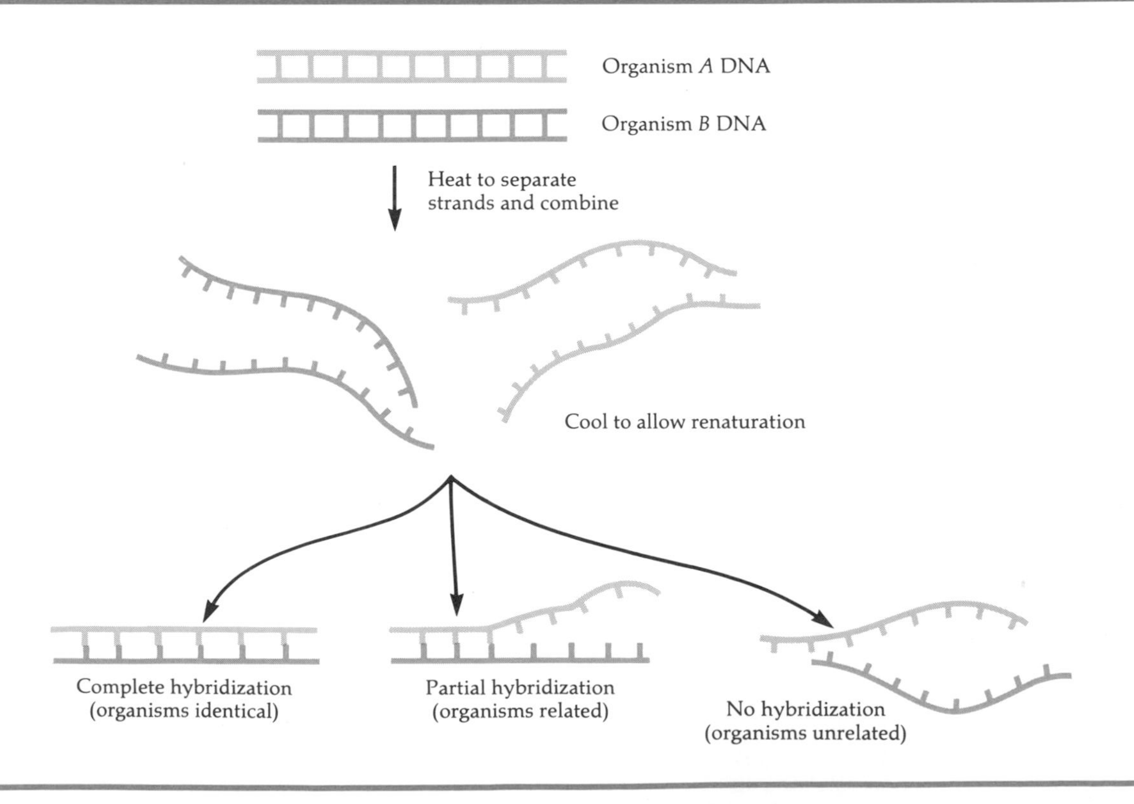

Figure 9–7 DNA hybridization. The greater the amount of pairing between DNA strands from different organisms (hybridization), the more closely related are the organisms.

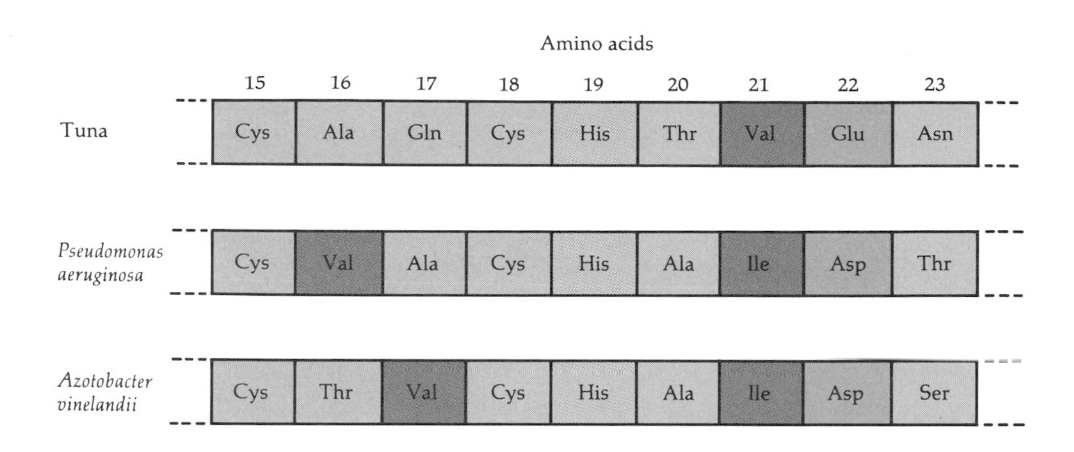

Figure 9–8 Amino acid sequences for a fragment of cytochrome *c* from tuna and two species of bacteria. Three-letter abbreviations indicate the amino acids, and amino acids of the same chemical type are shown in the same color (e.g., Glu and Asp are both "acidic amino acids," having an extra —COOH group). The similarity among the three sequences is striking, although clearly the two bacteria are more closely related to each other than to tuna.

related genera have a similar base sequence different from nonrelated genera. An indirect determination of the degree of similarity between DNA nucleotide sequences can be made by the method of nucleic acid hybridization.

Nucleic Acid Hybridization

If a double-stranded molecule of DNA is subjected to heat, the complementary strands will separate as the hydrogen bonds between nitrogenous bases break. If the complementary single strands are then incubated, they will unite to form a double-stranded molecule similar to the original double strand. (This union occurs because the single strands are complementary.) Using this technique on separated strands from two different organisms, it is possible to determine the extent of similarity of base sequence of the two organisms. This is known as **nucleic acid hybridization.** The procedure assumes that if two species are similar or related, a major portion of their nucleic acid sequences will also be similar. The procedure measures the ability of DNA strands from one organism to hybridize (bind through complementary base pairing) with the DNA strands of another organism (Figure 9–7). The degree of hybridization measures the degree of relatedness between two organisms. The greater the degree of hybridization, the greater the degree of relatedness.

Amino Acid Sequencing

A great deal of information can be derived from the study of the amino acid sequences of proteins, the products of gene translation. The amino acid sequence of a protein is a direct reflection of the base sequence of the encoding gene. Thus, by comparing the amino acid sequences of proteins from two different organisms, it is possible to determine relatedness. The more similar the proteins, the more closely related are the organisms. The phylogenetic tree derived for animals from traditional methods such as fossils and comparative anatomy is reinforced now by a comparison of protein sequences.

The sequencing of amino acids in several important proteins, such as cytochrome *c*, has been useful in the study of phylogenetic relationships among many different organisms (Figure 9–8). It is important to realize, however, that this method, as well as nucleic acid composition and hybridization, is too complex to use for routine identification in the clinical laboratory.

In the remaining chapters of this part, we will look at representative groups of microorganisms.

STUDY OUTLINE

FIVE-KINGDOM SYSTEM OF CLASSIFICATION (pp. 244–246)

1. Taxonomy is the science of classifying organisms with the goal of showing relationships between organisms.
2. Living organisms are classified into five kingdoms.
3. Procaryotic organisms include bacteria and cyanobacteria and are placed in the kingdom Monera or Procaryotae.
4. Eucaryotic organisms may be classified into the kingdoms Protista, Fungi, Plantae, or Animalia.

NAMING ORGANISMS (pp. 247–248)

1. According to scientific nomenclature, each organism is assigned two names (bionomal nomenclature), a genus and a species.

CLASSIFICATION OF EUCARYOTIC ORGANISMS (p. 248)

1. Organisms are grouped into taxa according to degrees of relatedness.
2. A eucaryotic species is a group of organisms that interbreeds but does not breed with individuals of another species.
3. Similar species are grouped into a genus; similar genera are grouped into a family; families into an order; orders into a class; classes into a phylum; and phyla into a kingdom.

CLASSIFICATION OF BACTERIA (pp. 248–249)

1. A group of bacteria derived from a single cell is called a strain.
2. Closely related strains constitute a species.
3. Related species are arranged into a genus; related genera into a family; and related families into an order.
4. *Bergey's Manual of Determinative Bacteriology* (8th edition) groups related orders of bacteria into "parts."

BASES FOR CLASSIFYING AND IDENTIFYING MICROORGANISMS (pp. 249–253)

1. Morphological characteristics, especially when aided by differential staining techniques, are useful for identifying microorganisms.
2. The possession of various enzymes as determined by biochemical tests is used to identify microorganisms.

3. Serological tests, involving the reactions of microorganisms with specific antibodies, are useful in determining the identify of strains and species as well as relationships between organisms.
4. Phage typing is the identification of bacterial species and strains by determining their susceptibility to various phages.
5. Grouping many different characteristics to show relatedness is called numerical taxonomy.
6. The percent G+C in the nucleic acid of cells may be used to classify organisms.
7. Single strands of DNA from related organisms will hydrogen-bond to form a double-stranded molecule; this is called DNA hybridization.
8. The sequence of amino acids in proteins of related organisms is similar.

STUDY QUESTIONS

REVIEW

1. What is taxonomy?
2. List the five kingdoms used in the five-kingdom system of classification.
3. What characterizes the kingdom Monera, and what organisms are placed in this kingdom?
4. What is binomial nomenclature?
5. Why is binomial nomenclature preferable to the use of common names?
6. Put the following terms in the correct sequence from the most general to the most specific: order; class; genus; kingdom; species; phylum; family.
7. Define species.
8. List the eight bases discussed in this chapter for classifying microorganisms. Separate your list into those tests used primarily for taxonomic classification and those used primarily for identification of microorganisms already classified.
9. Higher organisms are arranged into taxonomic groups on the basis of evolutionary relationships. Why is this type of classification not presently used for bacteria?

CHALLENGE

1. In *Bergey's Manual*, Part 14 (gram-positive cocci) includes the genera *Micrococcus* and *Staphylococcus*. The reported G+C content of *Micrococcus* is 66 to 75 moles %, and of *Staphylococcus*, 30 to 40 moles %. Based on this information, would you conclude that these two genera are closely related?

FURTHER READING

Barghoorn, E. S. May 1971. The oldest fossils. *Scientific American* 224:30–42. Evidence including photographs of bacteria and algae that lived more than three billion years ago.

Dickerson, R. E. March 1980. Cytochrome *c* and the evolution of energy metabolism. *Scientific American* 242:137–153. A practical application of amino acid sequencing.

Fox, G. E. et al. 1980. The phylogeny of prokaryotes. *Science* 209:457–463. A discussion of the use of RNA base sequencing to determine relatedness between microorganisms.

Gillies, R. R. and T. C. Dodds. 1976. *Bacteriology Illustrated,* 4th ed. New York: Churchill Livingstone. Biochemical tests used to identify medically important bacteria are explained and illustrated.

Isenberg, H. D. and J. D. MacLowry. 1976. Automated methods and data handling in bacteriology. *Annual Review of Microbiology* 30:483–505. An excellent summary of techniques for detection and identification of bacteria, including radiometric and gas chromatography methods; also includes a chapter on computerized data handling.

MacFaddin, J. F. 1976. *Biochemical Tests for Identification of Medical Bacteria.* Baltimore, MD: Williams and Wilkins. Provides useful methods for standard bacteriological tests and identification schemes.

Olds, R. J. 1975. *Color Atlas of Microbiology.* Chicago, IL: Year Yook Medical Publishers. Color photographs are used to illustrate laboratory procedures for biochemical, serological, and pathogenicity tests.

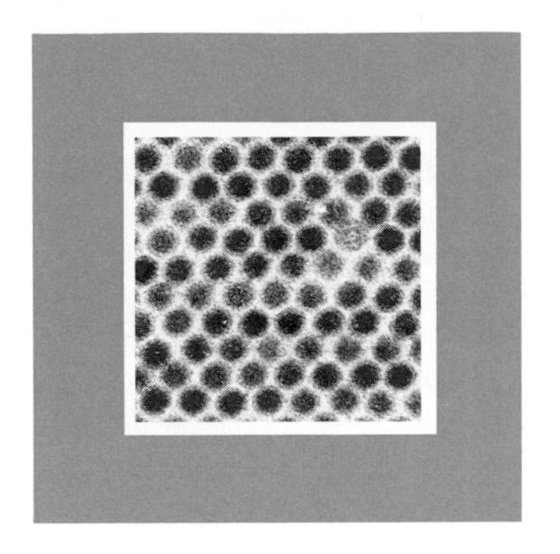

10

Bacteria

Objectives

After completing this chapter you should be able to

- List at least six characteristics used to classify and identify bacteria according to *Bergey's Manual of Determinative Bacteriology.*
- List one major characteristic of each of *Bergey's* 19 groups of bacteria.
- Identify the groups that contain species of medical importance.

Bacteria are the most diverse and most important of all microorganisms. Relatively few species of bacteria cause disease in humans or other organisms, and without the activities of the many other kinds of bacteria, all life on Earth would cease. Indeed, eucaryotic organisms probably evolved from bacterialike organisms. Because bacteria are simple in structure and many are readily cultured and controlled in the laboratory, microbiologists have devoted intensive study to their life processes. From such studies we have learned that the basic processes of life are the same for all organisms.

This chapter will introduce you to common bacteria, emphasizing those that cause disease in humans. In those cases where the bacterium also has an important nonmedical role, that role will also be noted. Many of the bacteria covered in this chapter will be discussed in more detail in the chapters on microbial disease, in Part IV.

BERGEY'S MANUAL: BACTERIAL TAXONOMY

There is no universally accepted evolutionary classification of bacteria as yet. Aside from the shared property of procaryotic cellular organization, bacteria exhibit great diversity, and it is difficult to correlate structural and functional characteristics to show evolutionary relationships. The most widely accepted taxonomic classification for bacteria is actually a classification of convenience rather than one of evolutionary relationships. It is *Bergey's Manual of Determinative Bacteriology.*

Bergey's Manual, first published in 1923, is the microbiologist's most important taxonomic reference. In the first seven editions, bacteria were classified in a hierarchical way according to kingdom, phylum, class, order, family, genus, and species—similar to schemes used for higher organisms. However, modern molecular evidence from base composition, nucleic acid hybridization, and amino acid sequencing studies began to challenge many aspects of the old hierarchy. Thus, in the recent eighth edition, the editors of *Bergey's Manual* have proposed that the old hierarchy be replaced by the kingdom Procaryotae, which has two divisions: Cyanobacteria (Figure 10–1) and Bacteria. The bacteria are grouped into 19 numbered **parts** according to characteristics such as Gram-stain reaction, cell shape, cell arrangements, oxygen requirements, motility, and nutritional and metabolic properties, among others. Accordingly, there are group names such as spirochetes, gram-negative aerobic rods and cocci, methane-producing bacteria, and endospore-forming rods and cocci. Each part consists of a number of genera. In some parts, genera are grouped into families and orders; in other parts, they are not (see Appendix A at the back of the book).

This new arrangement of bacteria in *Bergey's* eighth edition is essentially a holding pattern until enough evidence is available to construct a new, more valid hierarchy showing true evolutionary relationships. For the microbiologist, *Bergey's* represents a very comprehensive compilation of names and descriptions and a practical (though artificial) key to the genera.

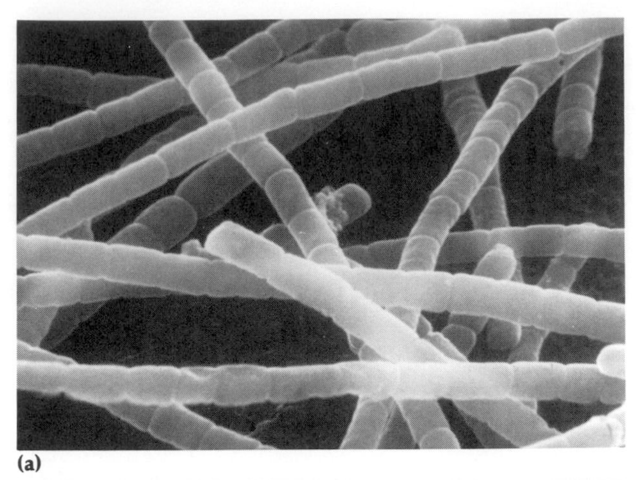

(a)

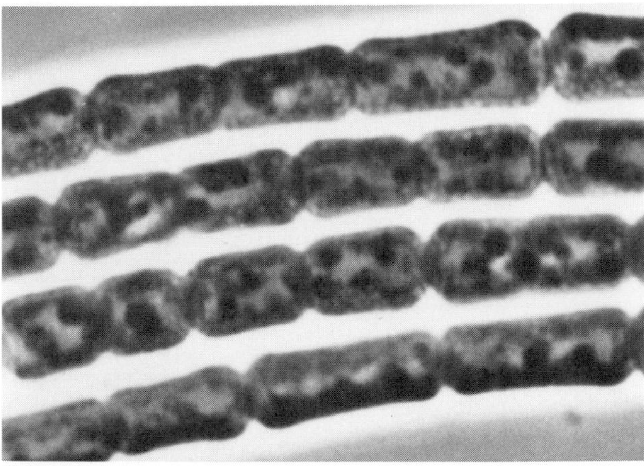

(b)

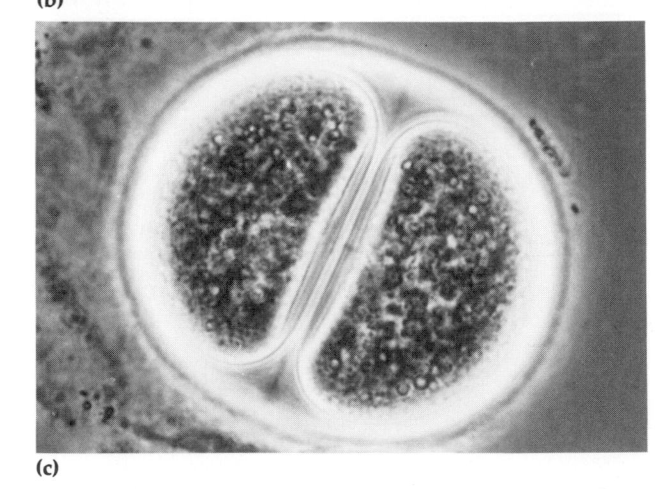

(c)

Figure 10–1 Cyanobacteria. **(a)** Scanning electron micrograph of *Phormidium luridum*, showing long chains of rod-shaped cells (×4400). **(b)** Phase-contrast micrograph of chains of cells typical of *Anabaena* (×2900). **(c)** *Chroococcus turgidus*, as seen with phase-contrast optics (×930); two or four of the large cells of this species often remain within a single surrounding sheath following cell division.

MEDICALLY IMPORTANT GROUPS OF BACTERIA

We will now discuss the groups of bacteria that contain genera of medical significance (Table 10–1). With a basic understanding of the characteristics of these groups, we can begin to identify bacteria.

Spirochetes

Spirochetes are found in contaminated water, sewage, soil, decaying organic matter, and within the bodies of humans and animals. These bacteria are typically coiled, like a metal spring. Some are tightly coiled, while others resemble a spring that is stretched. All spirochetes are actively motile and achieve their motility by means of an axial filament (Figure 10–2a). The spiral bacteria vary in length from a few μm to approximately 500 μm. Spirochetes may be aerobic, facultatively anaerobic, or anaerobic, and they do not have flagella or endospores. All divide by transverse fission. Of medical importance is the sensitivity of most spirochetes to heavy metals (arsenic, antimony, and bismuth) and very small amounts of penicillin. Spirochetes are extremely thin and not readily visible with an ordinary light microscope. Conventional staining, such as the Gram stain, causes distortion of the cells. Spirochetes are best studied by darkfield microscopy of unstained cells (Figure 10–2b).

Three genera of spirochetes contain pathogens. These are *Treponema, Borrelia,* and *Leptospira.* The genus *Treponema* includes the causative organism of syphilis. This bacterium, *Treponema pallidum* (see Figures 10–2 and 4–1c), is usually transmitted by sexual contact. Its life on dry surfaces removed from body fluids is very short. The organism is also sensitive to heat and grows markedly better in cooler regions of the body, one reason why it is sometimes cultivated on the testes of living rabbits. It does not grow on artificial media. A number of other *Treponema* species are common inhabitants of the mouth. One of the first microorganisms from

Table 10–1 Summary of Selected Characteristics of Bacterial Groups That Include Pathogens, According to *Bergey's Manual*, 8th edition, 1974

Name of Group	*Bergey's* Part	Mode of Motility	Human Pathogens	Other Features/Tests Used to Identify Genera
Spirochetes	5	Axial filament	*Treponema, Borrelia, Leptospira*	Difficult to culture in vitro
Spiral and curved bacteria	6	Polar flagella	*Spirillum, Campylobacter*	Amino acids and Krebs cycle intermediates used as carbon and energy source
Gram-negative aerobic rods and cocci	7	If motile, polar flagella	*Pseudomonas, Brucella, Bordetella, Francisella, Legionella*	Pigments; nonfermenting; dye inhibition
Gram-negative facultatively anaerobic rods	8	If motile, peritrichous flagella	*Salmonella, Shigella, Klebsiella, Yersinia, Vibrio, Hemophilus, Pasteurella, Streptobacillus, Calymmatobacterium*	Lactose fermentation; IMViC; oxidase; X and V factors
Gram-negative anaerobic bacteria	9	If motile, peritrichous flagella	*Bacteroides, Fusobacterium*	Anaerobes
Gram-negative cocci and coccobacilli	10	Flagella	*Neisseria, Moraxella*	Oxidase; enriched medium with increased CO_2
Gram-negative anaerobic cocci	11	Nonmotile	*Veillonella*	Anaerobes; oxidase; catalase
Gram-positive cocci	14	Flagella	*Staphylococcus, Streptococcus*	Catalase; hemolysis; mannitol fermentation
Endospore-forming rods and cocci	15	If motile, peritrichous flagella	*Bacillus, Clostridium*	Oxygen tolerance; catalase
Gram-positive asporogenous rod-shaped bacteria	16	Nonmotile	*Lactobacillus, Listeria, Erysipelothrix*	Catalase; produce 50% or more lactic acid from carbohydrate fermentation
Actinomycetes and related organisms	17	Generally nonmotile	*Corynebacterium, Propionibacterium, Actinomyces, Mycobacterium, Nocardia*	Branching rods and filaments, some with conidiospores; acid fast
Rickettsias	18	Nonmotile	*Rickettsia, Coxiella, Chlamydia*	Intracellular parasites
Mycoplasmas	19	Nonmotile	*Mycoplasma*	Pleomorphic; lack cell wall

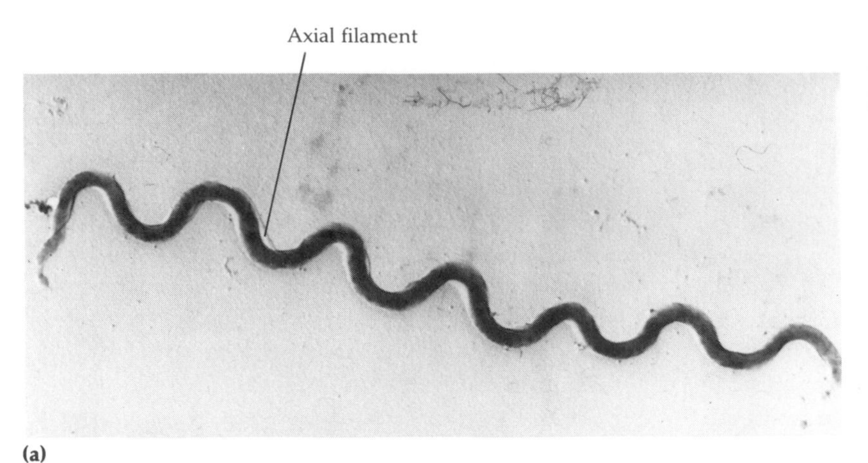

(a)

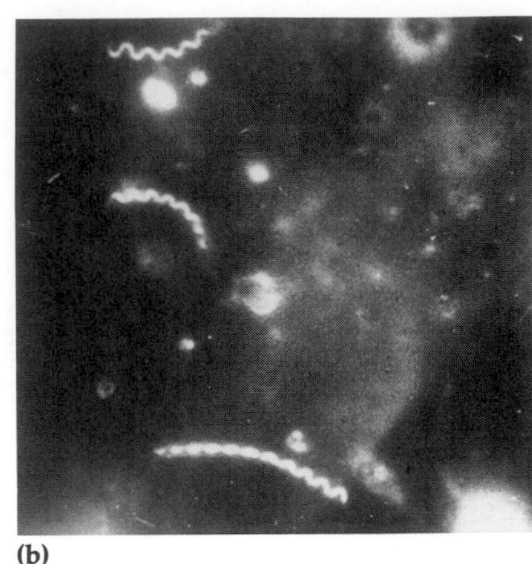

(b)

Figure 10–2 *Treponema pallidum*, a representative spirochete and the cause of syphilis. **(a)** Electron micrograph of a single corkscrew-shaped cell; the axial filament is seen as a thread alongside portions of the cell body. **(b)** Darkfield view of several cells.

saliva and tooth scrapings described by Leeuwenhoek in the 1600s was a large spirochete.

The genus *Borrelia* is the cause of relapsing fever, a serious disease that is characteristically transmitted by either ticks or lice. *Borrelia* species closely resemble *Treponema* species, and differentiation may require electron microscopy.

Leptospirosis is a disease usually spread to humans from waters contaminated by *Leptospira* species excreted in the urine of animals such as dogs, rats, and swine. The pathogen enters the body through minute breaks or abrasions in the skin, and the most common form of the disease involves a kidney infection. It is one of the diseases domestic dogs are routinely immunized against.

Spiral and Curved Bacteria

One would think that possession of a distinctly helical (spiral) shape would serve as a universal criterion for classifying a bacterium as a spirochete. However, spiral and curved bacteria are not included with the spirochetes because they lack an axial filament, having, instead, flagella for motility. There may be a single flagellum at one or both poles or sometimes tufts of flagella in these locations (Figure 10–3). Spiral bacteria, unlike spirochetes, are rigid helices or curved rods. Although most spiral bacteria are harmless aquatic organisms, one species of this group, *Campylobacter fetus* (see Figure 4–2h), can cause septicemia and food-borne illness, and bacteria of the genus *Spirillum* cause "rat-bite fever."

Gram-Negative Aerobic Rods and Cocci

For many reasons, the organisms of greatest interest in the group of gram-negative aerobic rods and cocci are those of the genus *Pseudomonas* (Figure 10–4). The common name for these bacteria is *pseudomonads*. These organisms have polar flagella, and many species excrete extracellular, water-soluble pigments that diffuse into the medium. One species, *Pseudomonas aeruginosa*, the organism of blue pus, produces a characteristic soluble, blue-green pigmentation. Under the right conditions, particularly in weakened hosts, this organism can cause urinary tract infections, burn and wound infections, septicemia, abscesses, and meningitis. Other pseudomonads produce soluble pigments that are fluorescent, glowing when illuminated by ultraviolet light.

Pseudomonads are very common in soil and other environments and are generally no threat to a healthy individual. They do not utilize many of the more common nutrients as efficiently as some other heterotrophic bacteria but have compensating characteristics. For example, many pseudomonads are psychrophilic (or psychrotrophic) and grow at refrigerator temperatures, imparting off tastes and colors to foods. They are also capable of synthe-

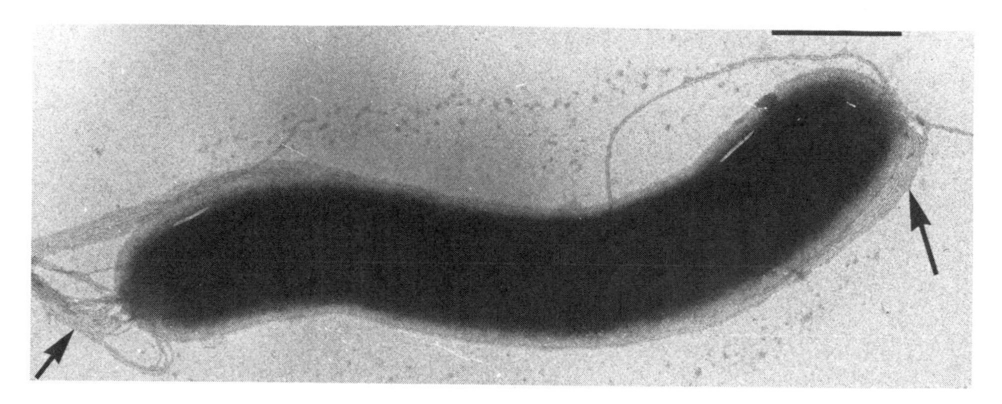

Figure 10–3 Spiral cell of *Aquaspirillum bengal;* arrows point to the tufts of flagella at each end of the cell (bar = 1 μm).

sizing an unusually large number of enzymes, and they probably contribute significantly to the decomposition of chemicals added to soil, such as pesticides. In hospital or pharmaceutical situations, the ability of pseudomonads to grow on minute traces of unusual carbon sources has led to their growth in antiseptic solutions, whirlpool baths, and other unexpected places where they have been troublesome. And their resistance to most antibiotics has also been a source of medical interest. Some of these considerations will be discussed in more detail later. Many pseudomonad genes for unusual metabolic characteristics, including antibiotic resistance, are carried on plasmids (see Chapter 8).

While pseudomonads are classified as aerobic, some of them in the soil will also grow anaerobically. In a suitable environment, they may use nitrate as a terminal electron acceptor for anaerobic respiration. Nitrate is the form of fertilizer nitrogen most easily used by plants. Under anaerobic conditions, such as in waterlogged soil, nitrate utilization by pseudomonads can cause important losses in valuable fertilizer and soil nitrogen.

Three other genera included in gram-negative aerobic rods and cocci are of medical importance: *Brucella, Bordetella,* and *Francisella.*

Brucella is a nonmotile coccobacillus about 1 μm in length. All species of *Brucella* are obligate parasites of mammals. They are transmitted to humans through the ingestion of unpasteurized dairy products. Bacteria of this genus cause undulant fever.

Bordetella is a nonmotile rod. Virulent forms have capsules. Found only in humans, *Bordetella pertussis* is the primary cause of whooping cough.

Francisella is a genus of small, pleomorphic bacteria that grow only on complex media enriched with blood or tissue extracts. *Francisella tularensis* causes rabbit fever (tularemia). This disease is most often transmitted to humans by direct contact with infected rabbits, although it may also be transmitted by the bite of a deer fly.

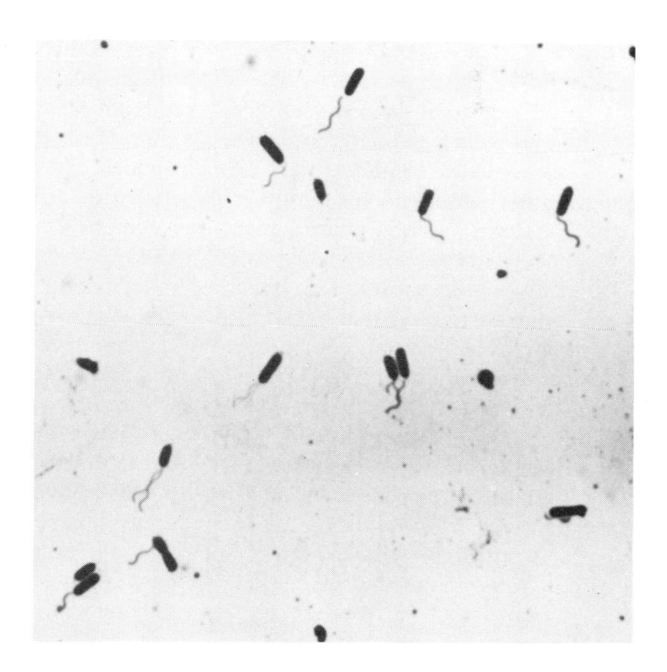

Figure 10–4 Flagella stain of *Pseudomonas aeruginosa.* The polar flagella are clearly visible.

Gram-Negative Facultatively Anaerobic Rods

From a medical viewpoint, gram-negative facultatively anaerobic rods represent a very important group of bacteria. Included here are three subgroups: Enterobacteriaceae; Vibrionaceae; and a number of genera not included in these families.

Enterobacteriaceae (enterics)

The family Enterobacteriaceae, or **enterics,** as they are commonly called, are a group of bacteria that inhabit the intestinal tract of humans and other animals. Some species are permanent residents, others are found in only a fraction of the population, and still others are present only under conditions of disease. Enterics vary in length from 1 to 8 μm. Most enterics are active fermenters of glucose and other carbohydrates. In fact, certain pathogenic enterics, such as *Salmonella* and *Shigella,* can be distinguished from nonpathogenic enterics, such as *Escherichia* and *Enterobacter,* on the basis of the inability of the pathogens to ferment lactose (see Figure 9–3).

Because of the diversity and importance of the enterics, a great many biochemical tests are used to differentiate them (see Color Plate IIC and E). A

Microbiology in the News

Microbial Fabrics—The Cotton of the Future?

Fashion designers may have a new line of synthetic fabrics if microbiologists can gather the threads spun by certain bacteria.

Even in the days of Louis Pasteur, scientists knew that bacteria called *Acetobacter xylinum,* gram-negative aerobes, could weave a film of fibers over fermented wine and rotten fruit. These fibers are pure cellulose, a polysaccharide that also forms the skeleton of plant cell walls. In nature, *A. xylinum* make narrow, twisted bands of cellulose (Figure a), which may serve to help the bacteria bind to their host. Researchers now find that when certain chemical dyes are added to the bacteria, they begin to produce broad bands of cellulose that accumulate in culture (Figure b). These bands may be used as a new thread for future textiles. Some microbiologists expect the tough, intertwined cellulose strands to be useful in developing paper products as well.

Cellulose ribbons are made when sugar units are added to simple chains in the outer envelope of the bacterial cell. Pores in the envelope extrude the chains, which assemble into microfibrils. Bundles of microfibrils then twist into ribbons. Each bacterium can produce 3 to 4 millimeters of ribbon in a day, and ribbon-making ability is passed on every time the bacterium divides. So a colony of *A. xylinum* might discharge large quantities of cellulose thread within hours.

There is one sticky point left to resolve. *A. xylinum* makes cellulose from glucose—an expensive raw material. Cotton plants that produce their own glucose using carbon dioxide from the air and energy from the sun are still cheaper to harvest than any known cellulose-making microbe.

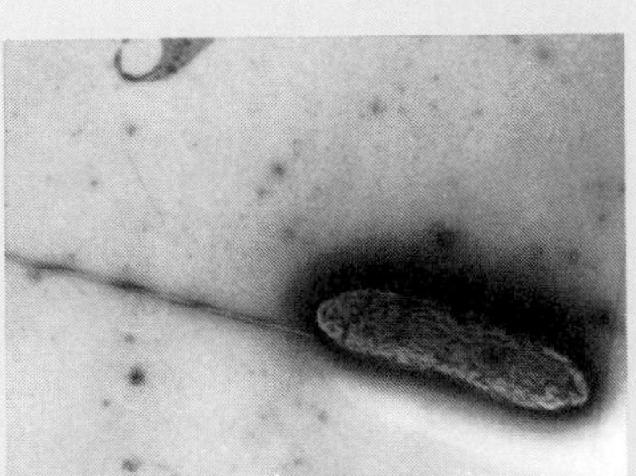

(a) × 10,400

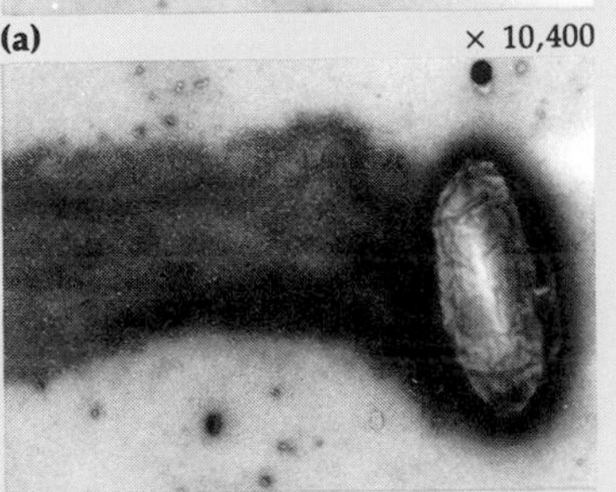

(b) × 13,600

The solution may arrive when genetic engineers learn to transfer the cellulose-making ability into a photosynthetic microorganism. Gene splicing techniques may allow scientists to insert the genes required for cellulose production into cyanobacteria, or some other appropriate microbe.

After that happens, it may not be long before designer jeans are made from designer microbial genes.

standard combination of four of the most useful tests is known as **IMViC.** Each capital letter in IMViC represents a specific test; the "i" is added for easier pronunciation. *I* stands for indole, a compound produced from tryptophan by some bacteria. *M* stands for methyl red, a pH indicator that turns red when substantial amounts of acid are produced from glucose. *V* stands for Voges-Proskauer, a test for the production of acetoin from glucose. And *C* stands for citric acid, in a test to determine whether an organism can grow using only citric acid as a carbon source. An identification scheme for some enteric genera is given in Table 10–2, using the IMViC reactions. Results are recorded in terms of + and −; a + indicates that the organism produced the product tested for (or grew in the case of the citric acid test), and a − indicates the organism did not produce the desired product. Each test employs an indicator color change to determine the results. Enterics may also be distinguished from each other on the basis of antigens present on their surfaces. The cell wall antigen, or *O antigen,* is a lipoprotein-polysaccharide; the flagellar antigen, or *H antigen,* is a protein; and the capsular antigen, or *K antigen,* is a polysaccharide. Since the chemistry of these antigens varies from species to species, and even from strain to strain within a species, they are useful in identifying a species or strain of enteric bacterium.

Table 10–2 The IMViC Reactions Used to Differentiate Between Genera of Enterics

Genus	Indole	Methyl Red	Voges–Proskauer	Citric Acid
Escherichia	+	+	−	−
Enterobacter	−	−	+	+
Salmonella	−	+	−	+
Shigella	d*	+	−	−

*d = different reactions given by different strains

Enterics include motile as well as nonmotile species, and those that are motile have peritrichous flagellation. Many enterics are also characterized by the presence of pili, which are used in conjugation (Figure 10–5). Within the intestinal tract, many enterics can produce toxins called **bacteriocins.** These substances are toxic to related species of bacteria and not to the species of bacteria producing them. Bacteriocins may help to maintain the ecological balance of various enterics in the small intestine.

Among the genera included as enterics are *Escherichia, Salmonella, Shigella, Klebsiella, Serratia, Proteus, Yersinia, Erwinia,* and *Enterobacter.*

Escherichia *Escherichia coli* is one of the most common inhabitants of the intestinal tract and probably

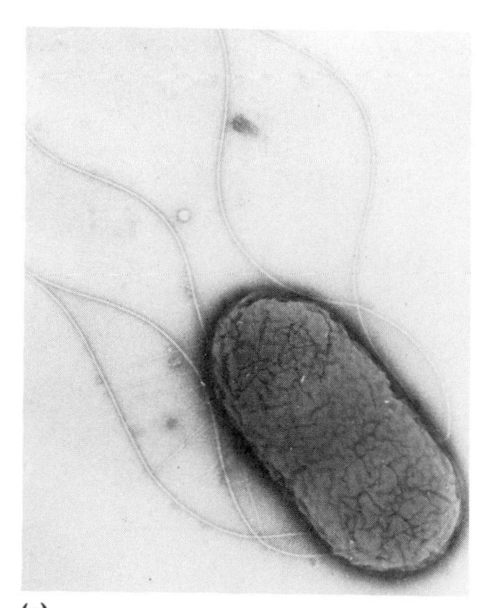
(a)

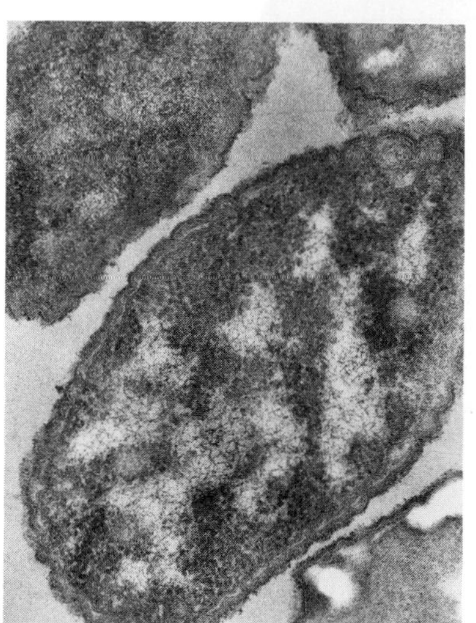
(b)

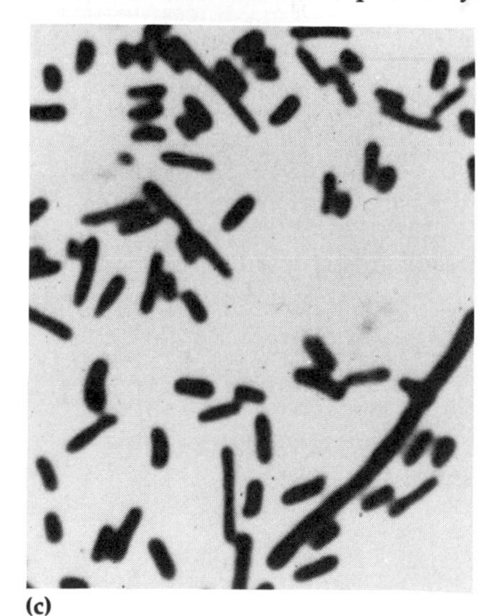
(c)

Figure 10–5 Enteric bacteria. **(a)** Negative stain of *Proteus vulgaris*, showing several long flagella and short, hairlike pili extending from the cell (×16,000). **(b)** Thin section through *Salmonella typhimurium* (×32,900); the gram-negative cell wall is evident. **(c)** Light micrograph showing *Shigella dysenteriae* from a laboratory culture.

the most familiar organism used in microbiology (see Color Plate IB). As you may remember from earlier chapters, a great deal is known about the biochemistry and genetics of *E. coli*, and it continues to be an important tool for basic biological research. Its presence in water or foods is an indication of fecal contamination. Although *E. coli* is not usually pathogenic, under exceptional circumstances it can produce urinary tract infections and diarrhea.

Salmonella Almost all members of this group are pathogenic, at least to some degree. In humans, the most common diseases caused by salmonellae are typhoid fever *(Salmonella typhi)*, characterized by a high fever and rose spots on the abdomen and chest, and gastroenteritis or salmonellosis *(Salmonella typhimurium)*, characterized by headache, chills, vomiting, diarrhea, and fever.

Shigella Species of *Shigella* are responsible for a disease called bacillary dysentery or shigellosis. Symptoms include mild diarrhea, abdominal pains, and the presence of pus or blood in the feces.

Klebsiella The organism *Klebsiella pneumoniae* (see Color Plate ID) is a major cause of septicemia in pediatric wards and also a cause of one form of pneumonia, especially among individuals with upper respiratory tract infections and chronic alcoholism.

Serratia *Serratia marcescens* is distinguished by its production of red pigment and has been used as an experimental organism to test air dispersal of bacteria for biological warfare. *S. marcescens* has become increasingly important in recent years in connection with nosocomial infections, that is, infections acquired in hospitals. The organism has been found on catheters, in saline irrigation solutions, and in other supposedly sterile solutions, and it causes urinary and respiratory tract infections.

Proteus *Proteus* bacteria are very actively motile and are implicated in urinary tract and wound infections and in infant diarrhea.

Yersinia *Yersinia pestis* is the causative agent of bubonic plague ("black death") (see Color Plate IIIC). This bacterium was formerly known as *Pasteurella pestis*. When introduced into the body, the bacteria infiltrate lymph nodes, causing their enlargement to form "buboes." Thus the name bubonic plague. Later, they may pass into other parts of the body.

Erwinia *Erwinia* species are primarily plant pathogens. One species, however, *Erwinia herbicola*, can cause systemic infection in humans when introduced in contaminated intravenous solutions.

Enterobacter Two *Enterobacter* species, *Enterobacter cloacae* and *Enterobacter aerogenes*, play a part in urinary tract infections.

Vibrionaceae

It was noted earlier that the gram-negative facultative anaerobic rod group includes the family Vibrionaceae. Among the genera in this family are *Vibrio* and *Aeromonas*.

Vibrio Members of the genus *Vibrio* are comma-shaped rods (Figure 10–6) and most are nonpathogenic. One important pathogen is *Vibrio cholerae*, the causative agent of cholera. The disease is characterized by a profuse and watery diarrhea. *V. parahaemolyticus* causes gastroenteritis and is transmitted to humans primarily in shellfish. The normal habitat of *V. parahaemolyticus* is the ocean.

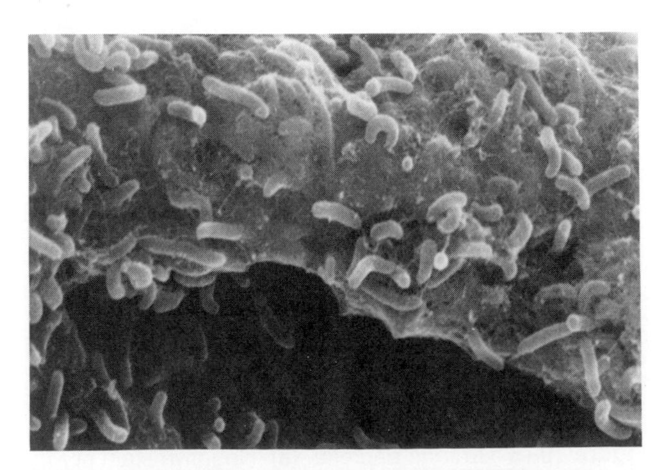

Figure 10–6 Numerous cells of *Vibrio cholerae* are visible here on the surface of a ball of mucus from the small intestine of a mouse (×4000).

Aeromonas Members of this genus are associated with septicemia, pneumonia, and intestinal disorders.

Hemophilus

The third medically important group among the gram-negative facultative anaerobic rods includes *Hemophilus* (also spelled *Haemophilus*). *H. influenzae* is the most common cause of meningitis in young children. *Hemophilus* is cultured on complex media enriched with hemoglobin. Species of *Hemophilus* can be identified partially by their requirements of X and V factors (Table 10–3). **X factor** is a precursor needed to synthesize respiratory enzymes. **V factor** is NAD or NADP. Figure 10–7 illustrates the **satellite phenomenon,** which results from the requirement for V factor. In laboratory culture, *Hemophilus* is heavily inoculated over an entire sterile blood agar plate, and a spot of staphylococci is placed on the plate. *X* factor is supplied by the hemin in the blood and is not destroyed by autoclaving. *Hemophilus* requiring V factor, which is destroyed by autoclaving, will grow only around the colonies of *Staphylococcus*, since the staphylococci supply V factor.

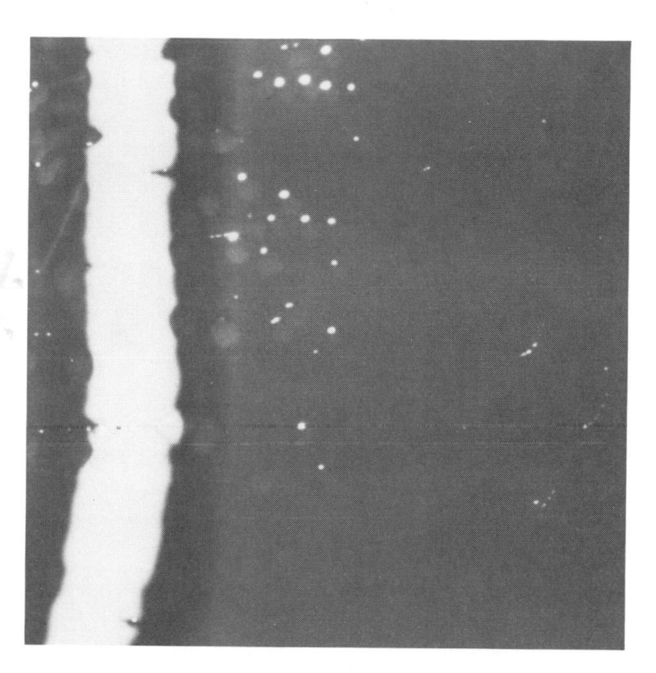

Figure 10–7 The satellite phenomenon. The small colonies of *Hemophilus influenzae* grow only in the vicinity of the streak of *Staphylococcus* on autoclaved blood agar. The staphylococci provide V factor, which was originally present in the blood agar but was destroyed by autoclaving.

Gram-Negative Anaerobic Bacteria

Among the gram-negative anaerobic bacteria, the genus *Bacteroides* represents a large group of microbes in the human intestinal tract (Figure 10–8a). Some also reside in the oral cavity, genital tract, and upper respiratory tract. *Bacteroides* are non-endospore-forming and nonmotile. Infections due to *Bacteroides* often occur as a result of puncture wounds or surgery and are a frequent cause of peritonitis.

Table 10–3 Differentiation of Selected Species of *Hemophilus* Based on Requirements for X and V Factors

Species	Requirement for X factor	Requirement for V factor
H. influenzae	+	+
H. parainfluenzae	–	+
H. haemoglobinophilus	+	–

Another genus of gram-negative anaerobic bacteria is *Fusobacterium*. These microbes are long and slender with pointed rather than blunt ends (Figure 10–8b). They are found most often in the gums and are responsible for dental abscesses.

Gram-Negative Cocci and Coccobacilli

Neisseria and *Moraxella* are two medically important genera in the group of gram-negative cocci and coccobacilli. *Neisseria* is particularly important. Members of the genus *Neisseria* are non-endospore-forming diplococci that are aerobic or facultatively anaerobic, parasitic on human mucous membranes, and able to grow well only near body temperature. Pathogenic *Neisseria* include the gonococcus bacterium *Neisseria gonorrhoeae*, the causative agent of gonorrhea (see Color Plate IE), and the meningococcus bacterium *Neisseria meningitidis*, the agent of meningococcal meningitis, also called epidemic cerebrospinal fever (Figure 10–9a). The *Neisseria* possess the enzyme cytochrome oxidase; they are "oxidase +," an important biochemical characteristic.

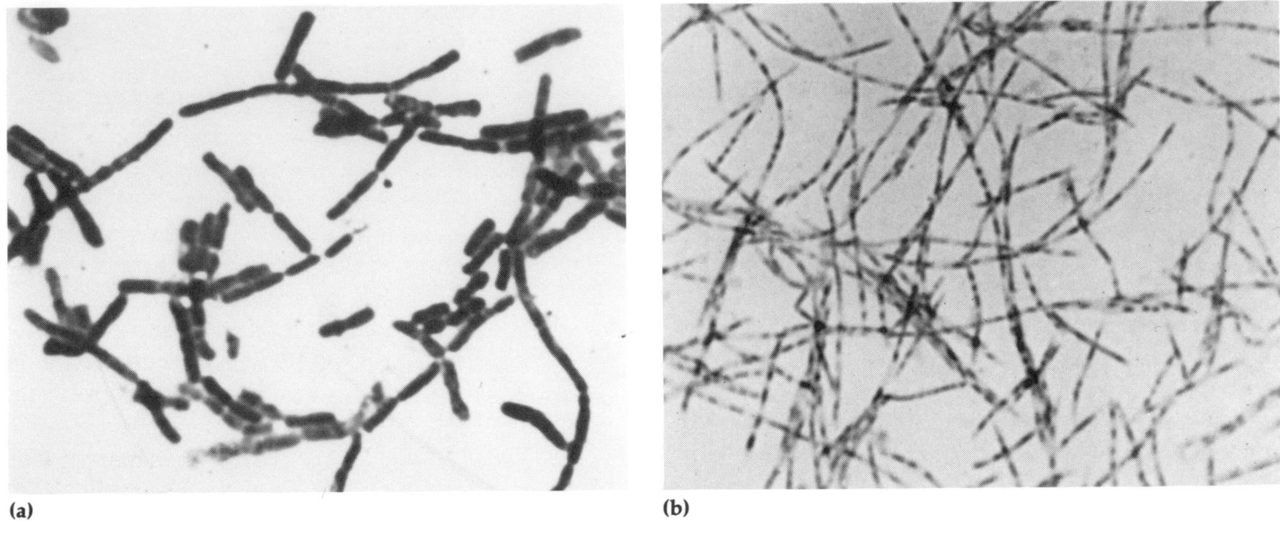

Figure 10–8 Gram-negative anaerobes. **(a)** Chains of *Bacteroides hypermegas*. **(b)** Chains of slender cells of *Fusobacterium*.

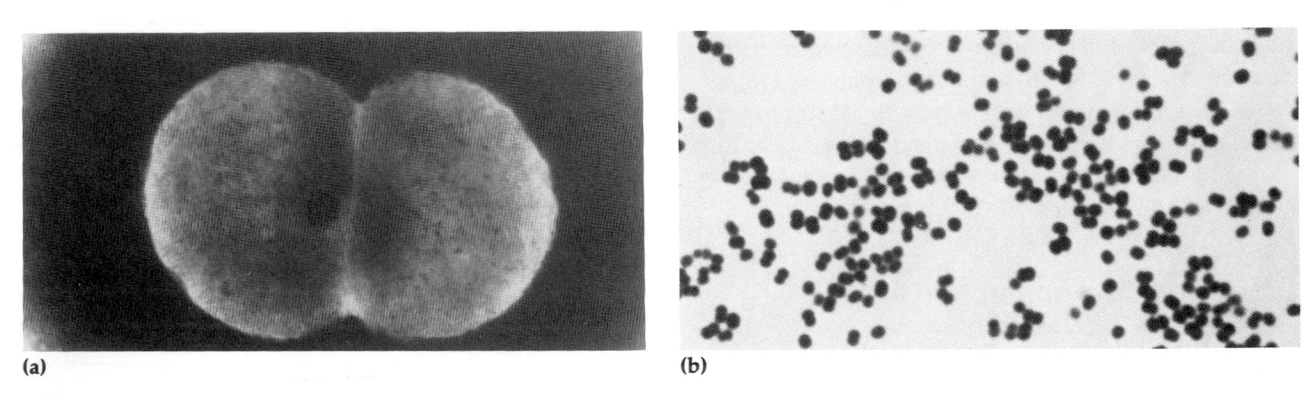

Figure 10–9 Gram-negative cocci and coccobacilli. **(a)** *Neisseria gonorrhoeae*, Type III (approx. ×35,000). **(b)** *Moraxella lacunata*.

Members of the genus *Moraxella* are strictly aerobic coccobacilli. Coccobacilli are egg-shaped, intermediate in structure between cocci and rods. *Moraxella lacunata* is implicated in conjunctivitis, an inflammation of the conjunctiva (Figure 10–9b).

Gram-Negative Anaerobic Cocci

The cells of gram-negative anaerobic cocci typically occur in pairs, but also may occur singly, in clusters, and in chains. They are nonmotile and nonendospore-forming. Bacteria of the genus *Veillonella* are found as part of the normal flora of the mouth and are components of dental plaque (Figure 10–10).

Gram-Positive Cocci

Gram-positive cocci of medical importance are members of the genera *Staphylococcus* and *Streptococcus*. Staphylococci typically occur in grapelike clusters (Figure 10–11a and Color Plate IA). They are nonmotile and do not form endospores. All members of the *Staphylococcus* genus produce the enzyme **catalase,** which breaks down hydrogen peroxide (H_2O_2) to oxygen gas and water. Members of this genus are aerobes or facultative anaerobes. When grown on blood agar, staphylococci produce characteristic pigments ranging in color from deep gold to lemon yellow to white. Staphylococci are quite resistant to heat, desiccation, and chemicals. Many are also resistant to penicillin. As a group,

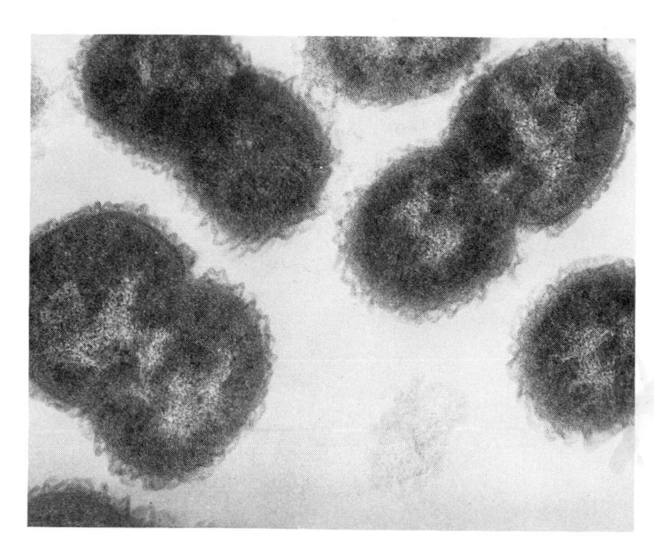

Figure 10–10 Pairs of cocci of *Veillonella* sp., a gram-negative anaerobic coccus from a human mouth (×48,000).

they produce a number of toxic substances that destroy red blood cells *(staphylolysins)*, destroy white blood cells *(leucocidins)*, cause tissue necrosis *(necrotizing exotoxin)*, and cause gastrointestinal disturbances *(staphylococcal enterotoxin)*. Some staphylococci also produce the enzyme **coagulase,** which causes blood plasma to clot (form fibrin). The coagulase causes the host to surround the bacterial cells of infected tissue with fibrin, which protects the bacteria against phagocytic attack. The ability of staphylococci to produce coagulase is detected by a laboratory test called a **coagulase test.** A positive test result is evidence that the microbe is probably pathogenic. Coagulase production and pathogenicity have a 95% correlation. The production of coagulase is correlated with the synthesis of other toxins.

Several staphylococcal infections are caused by *S. aureus.* This microbe is associated with skin infections (boils, carbuncles, furuncles, and impetigo), pneumonia, endocarditis, meningitis, brain abscesses, and cystitis. The characteristic feature of staphylococcal infections is the presence of an abscess, a localized collection of pus formed by tissue breakdown and surrounded by inflammation.

Members of the genus *Streptococcus* are probably responsible for more illness and cause a greater variety of disease than any other group of bacteria. Streptococci typically appear in chains ranging from as few as 4 to 6 cocci to as many as 50 or more (Figure 10–11b). They are nonmotile and do not form endospores, but a few species form capsules. Streptococci do not produce catalase; thus, the catalase test is useful in distinguishing this genus from *Staphylococcus.* Because they are unable to use oxygen for growth and lack catalase, streptococci grow best in the absence of oxygen. Most are able to grow in its presence (aerotolerance), and a few species are strict anaerobes. Streptococci grow best at body temperature, but some may grow between 15° and 45°C.

One method of classifying streptococci is on the basis of their action on blood agar. **Alpha-hemolytic** (greening) species produce a substance called alpha hemolysin that lyses red blood cells and brings about a narrow, greenish zone of hemolysis on blood agar (see Color Plate IIA). Alpha hemolysins reduce the hemoglobin in red blood cells to methemoglobin. **Beta-hemolytic** species produce a hemolysin that forms a clear zone of

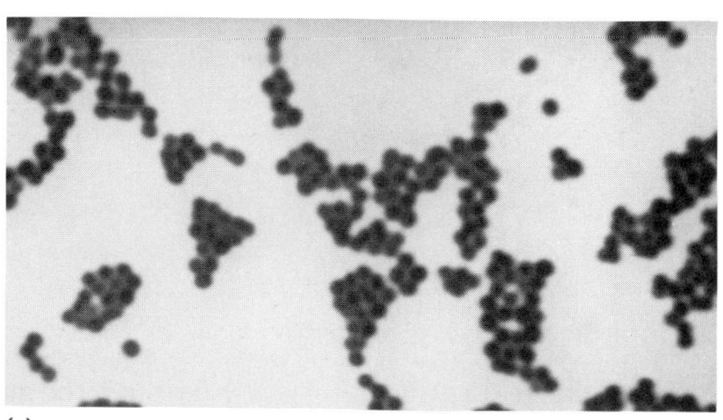

(a)

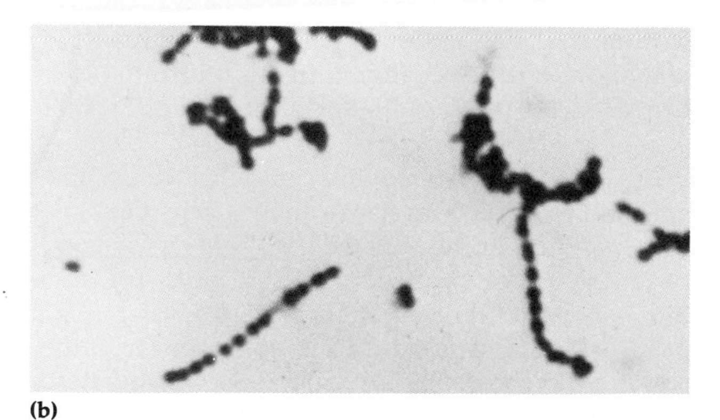

(b)

Figure 10–11 Gram-positive cocci. **(a)** Gram stain showing clusters of *Staphylococcus aureus* (approx. ×4700). **(b)** Chains of streptococci in a Gram stain of an unidentified species of *Streptococcus*.

hemolysis on blood agar (see Color Plate IIB). Species that do not hemolyze red blood cells are called **gamma-hemolytic.** Streptococci are also classified on the basis of the specific antigens they contain. Furthermore, like the staphylococci, streptococci produce a number of extracellular substances that contribute to disease and are useful in identification. These include substances that produce erythema (redness) of the skin *(erythrogenic toxin)* and lyse blood cells *(streptolysins),* as well as enzymes that help spread the infection to other tissues *(hyaluronidase, proteinase, streptokinase, and nucleases).*

Over 90% of streptococcal infections are caused by a group of beta-hemolytic streptococci called *group A.* (The various groups are determined by specific antigens in the cell walls of these microbes.) Among the diseases associated with **group A beta-hemolytic streptococci** are puerperal fever, erysipelas, scarlet fever, strep throat, rheumatic fever, and acute glomerulonephritis. Alpha-hemolytic *Streptococcus mutans* is a cause of tooth decay.

Endospore-Forming Rods and Cocci

The formation of endospores by bacteria is important both medically and in the food industry, because of their resistance to heat and many chemicals. The majority of endospore-forming rods and cocci are gram-positive. With respect to oxygen requirements, they may be strict aerobes, facultative anaerobes, obligate anaerobes, or microaerophiles. All species produce endospores. The two important genera are *Bacillus* and *Clostridium.*

Bacillus anthracis causes anthrax, a disease of cattle, sheep, and horses that may be transmitted to humans. The anthrax bacillus is a nonmotile, facultative anaerobe ranging in length from 4 to 8 μm (Figure 10–12a and b and Color Plate IF). It is one of the largest bacterial pathogens. The endospores of *Bacillus anthracis* are centrally located, and smears prepared from tissue show a capsule.

Members of the genus *Clostridium* are obligate anaerobes. They vary in length from 3 to 8 μm, and in most species the sporangia appear swollen (Figure 10–12c). Some clostridial endospores can withstand temperatures of 120°C for 15 to 20 minutes. Diseases associated with clostridia include tetanus, or lockjaw *(Clostridium tetani),* botulism *(Clostridium botulinum),* and gas gangrene *(Clostridium perfringens).*

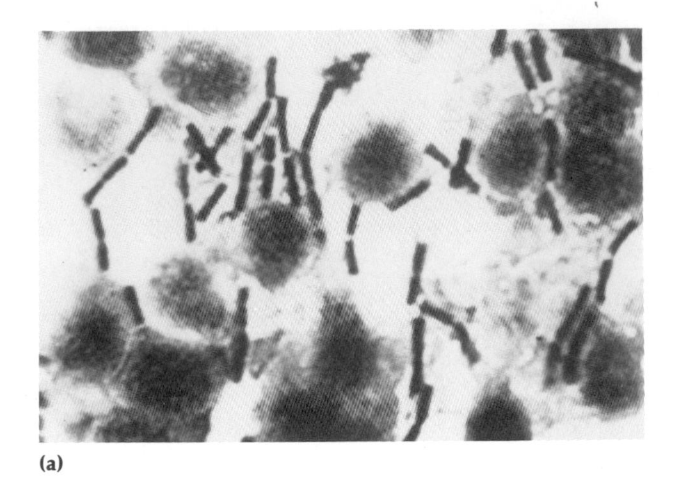

(a)

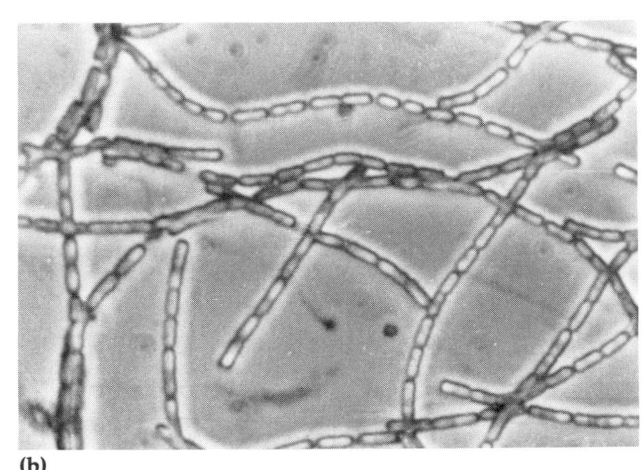

(b)

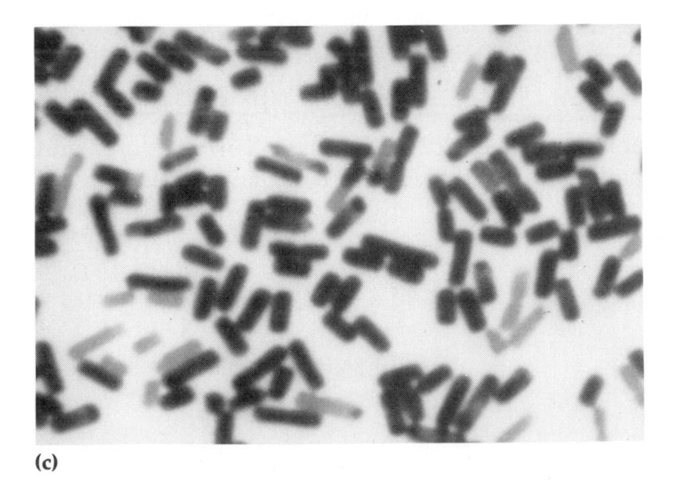

(c)

Figure 10–12 Endospore-forming rods. **(a)** Thick, blunt bacilli of *Bacillus anthracis* as they appear in infected tissue. **(b)** *Bacillus anthracis* after growth on an artificial medium. The light areas in the cells are endospores. **(c)** Gram stain of *Clostridium perfringens* after growth on blood agar, showing endospores (approx. ×4700). For examples of endospores in the genus *Clostridium*, see Figure 4–18.

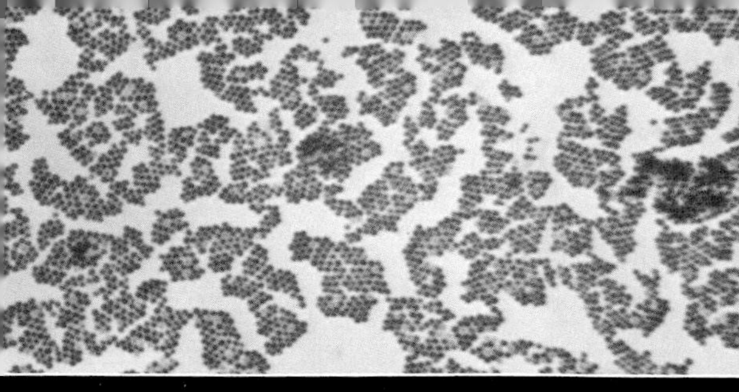

A. Clusters of cocci of *Staphylococcus aureus* demonstrate a gram-positive stain.

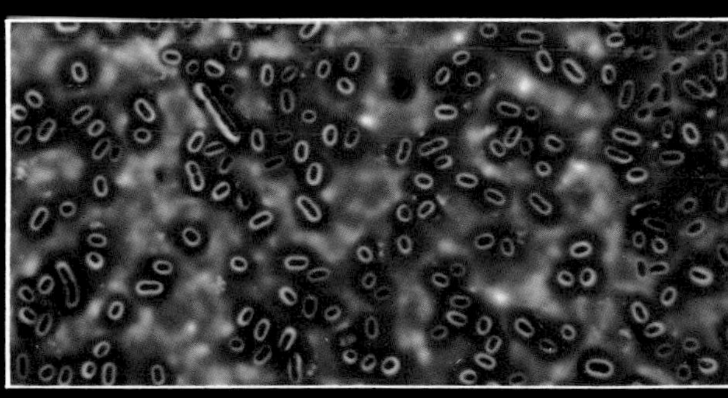

B. Gram-negative rods of the intestinal bacterium *Escherichia coli.*

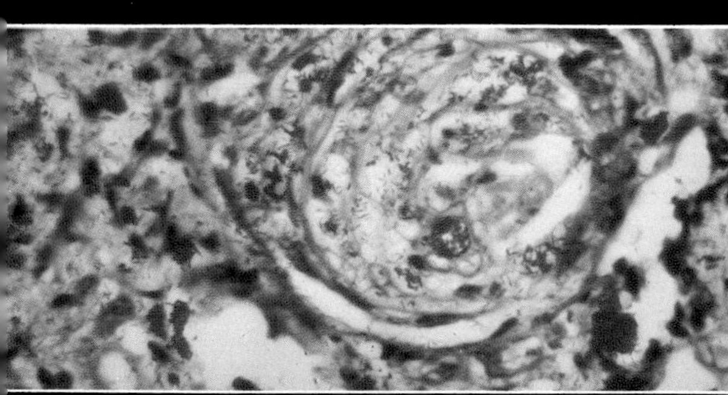

C. Section of a cutaneous nerve showing red, acid-fast bacteria of the leprosy-causing species *Mycobacterium leprae.*

D. A capsule stain of *Klebsiella pneumoniae*, clearly showing the light, unstained capsules around the dark cell bodies.

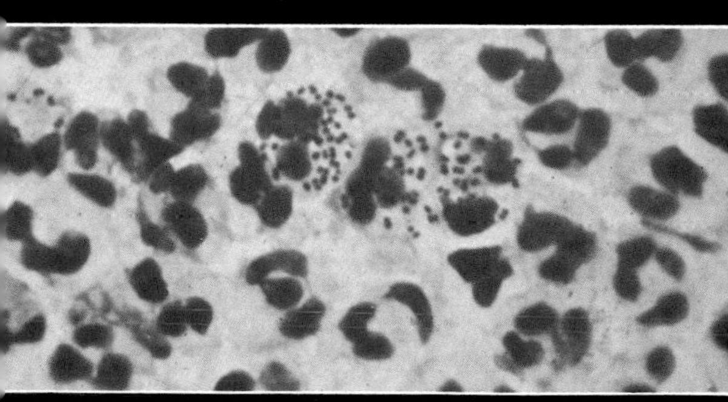

E. Numerous intracellular diplococci of *Neisseria gonorrhoeae* are seen in this smear of urethral discharge from a case of gonorrhea. (The large, irregular red bodies are the nuclei of the white blood cells.)

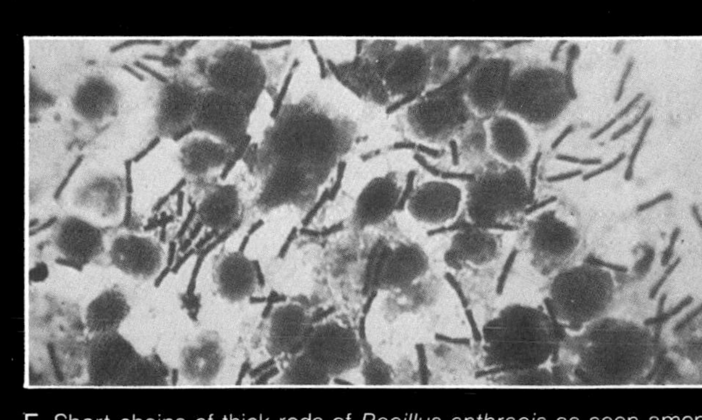

F. Short chains of thick rods of *Bacillus anthracis* as seen among tissue cells from a case of anthrax.

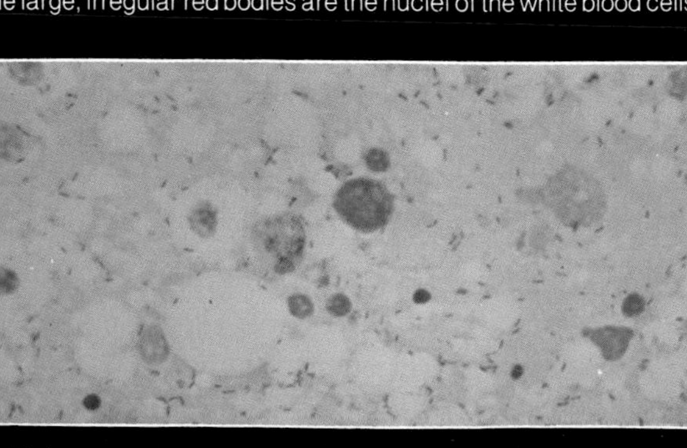

G. Rickettsias of the species *Rickettsia rickettsii* show up as tiny, red, intracellular forms in this Giménez stain of a yolk sac smear.

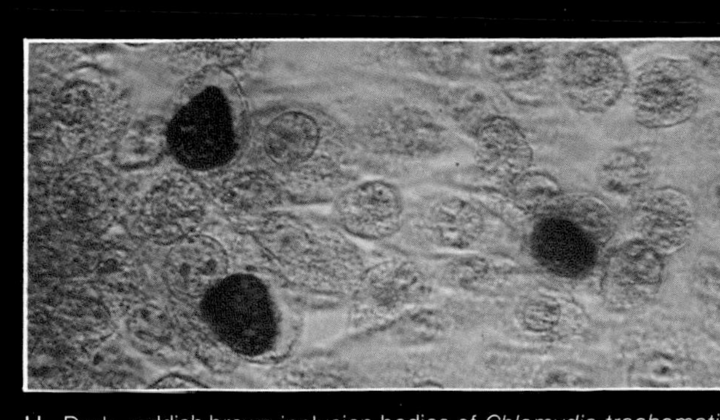

H. Dark, reddish-brown inclusion bodies of *Chlamydia trachomatis* are seen in several cells of a cultured monolayer of McCoy cells.

PLATE II Bacterial Cultures and Tests

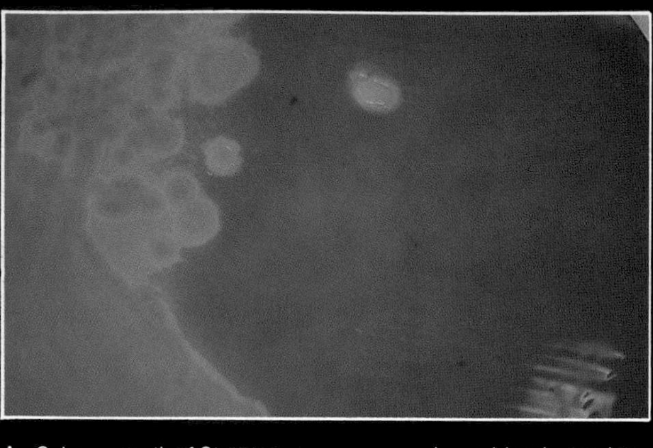

A. Colony growth of *Streptococcus pneumoniae* on blood agar demonstrates alpha-hemolysis; note the lack of clear areas around the colonies and the green hue of the bacterial growth.

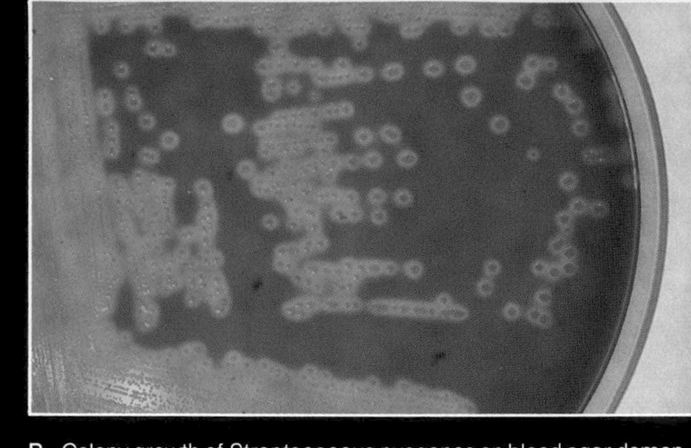

B. Colony growth of *Streptococcus pyogenes* on blood agar demonstrates beta-hemolysis. Extensive lysis of the red blood cells produces the very light areas around the colonies.

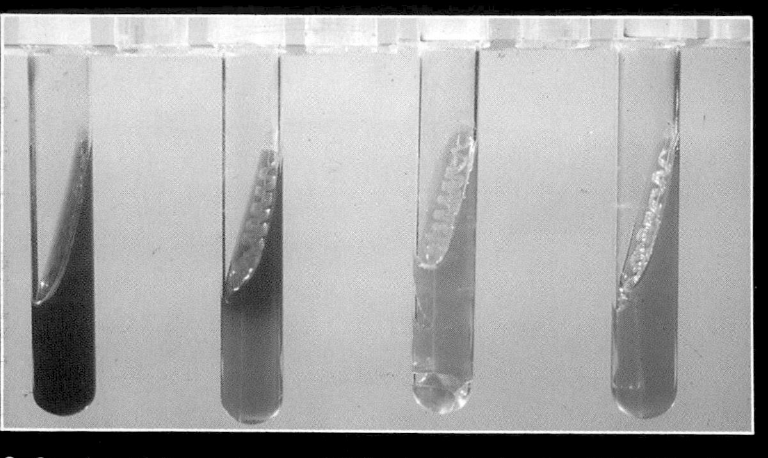

C. Growth on triple sugar iron agar (TSIA) to test for hydrogen sulfide (H_2S) and gas production by members of the Enterobacteriaceae. Tube 1 shows blackening of the medium due to H_2S production by a strain of *Proteus*; tube 2 shows growth of *Proteus* without H_2S production; tube 3 shows gas production by *Escherichia coli* (note disrupted agar near the bottom of the tube); and tube 4 shows growth of *E. coli* on TSIA without gas production.

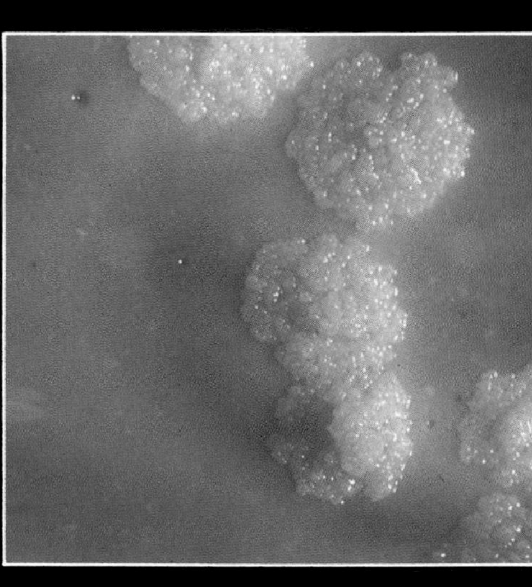

D. Colonies of *Mycobacterium tuberculosis*; note the rough colony surface characteristic of this species.

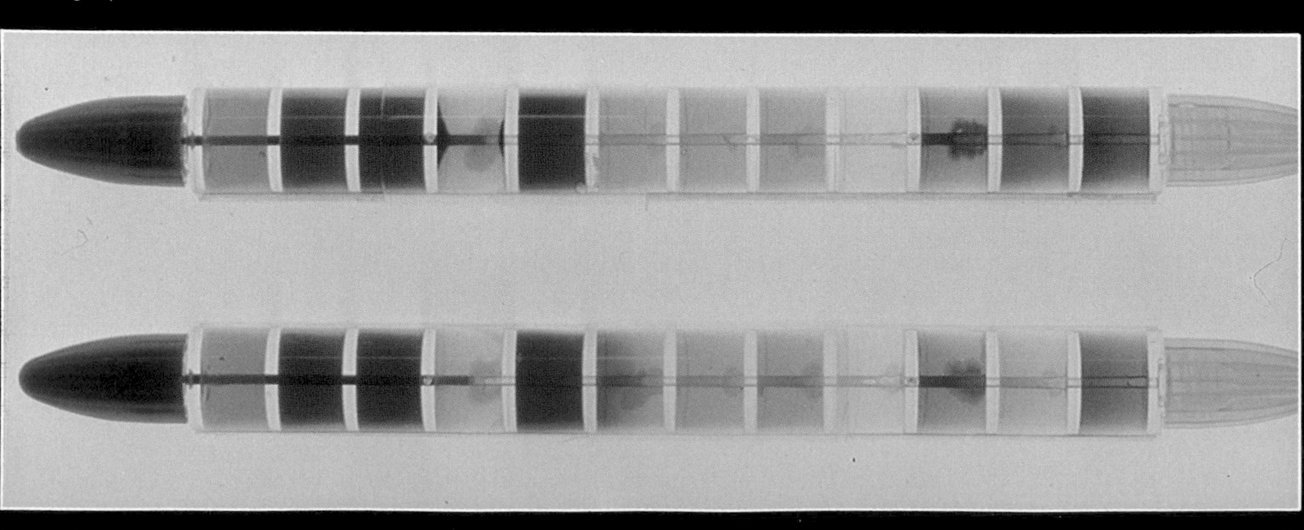

E. The Enterotube apparatus is one modern method for identifying members of the Enterobacteriaceae. A different chemical test is conducted in each of the small compartments along the tube, and the results following growth are matched against a color key.

PLATE III Fluorescent Antibody Staining

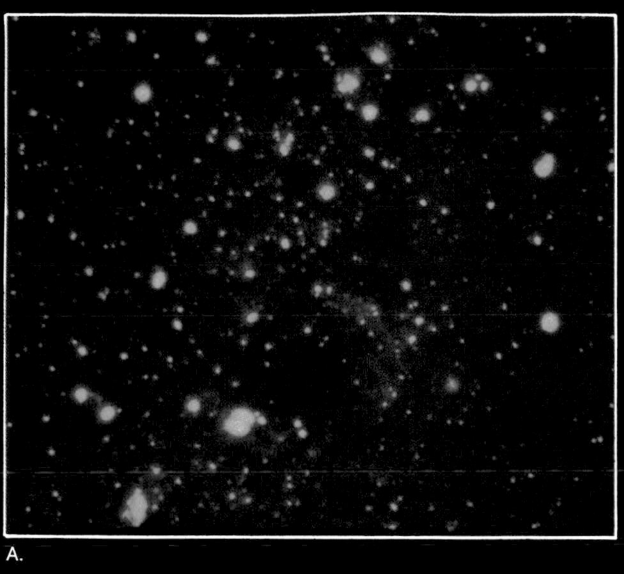

A.

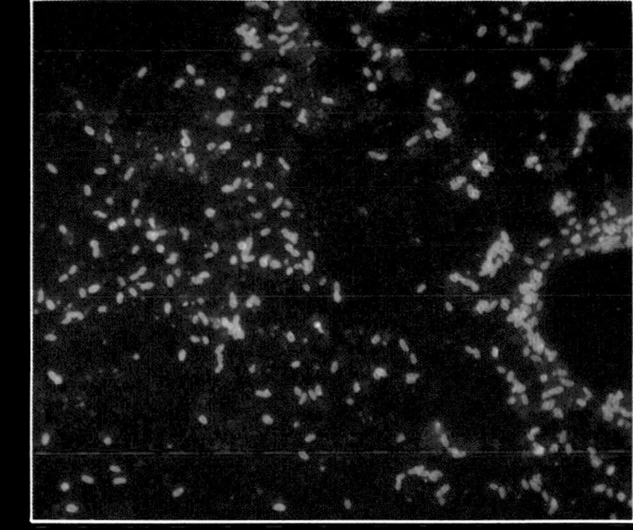

B.

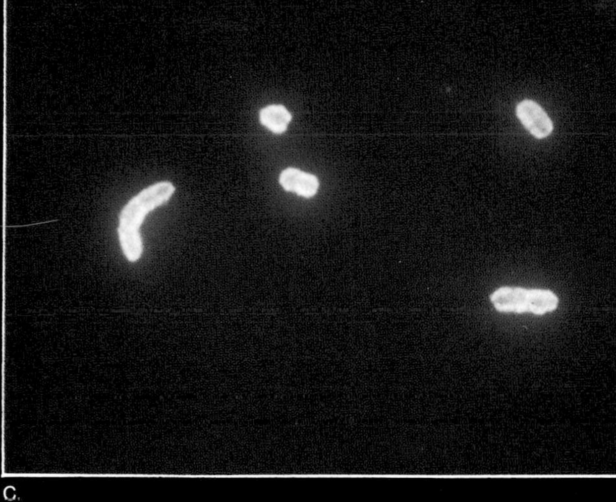

C.

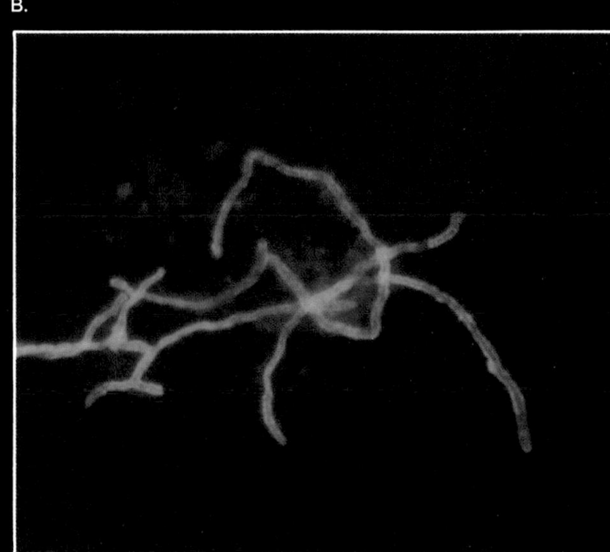

D.

Fluorescent antibody staining has become a powerful technique for locating and identifying a wide variety of pathogenic microorganisms.

A. Rabies virus antigen in a preparation of brain tissue from a naturally-infected skunk. Only the largest few fluorescing bodies would have appeared as Negri bodies using older techniques.

B. *Legionella pneumophila* in lung tissue from a fatal case of Legionnaires' disease.

C. Cells of the plague-causing bacterium *Yersinia pestis.*

D. Chains of cells of *Actinomyces israelii* in material from a brain abscess.

E. Amoebas of the species *Naegleria fowleri* are the green, fluorescing objects in this smear of brain tissue from a case of amoebic meningoencephalitis.

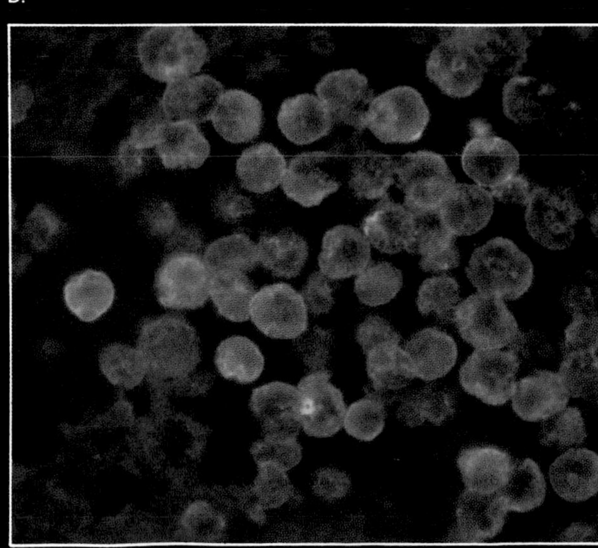

E.

A.

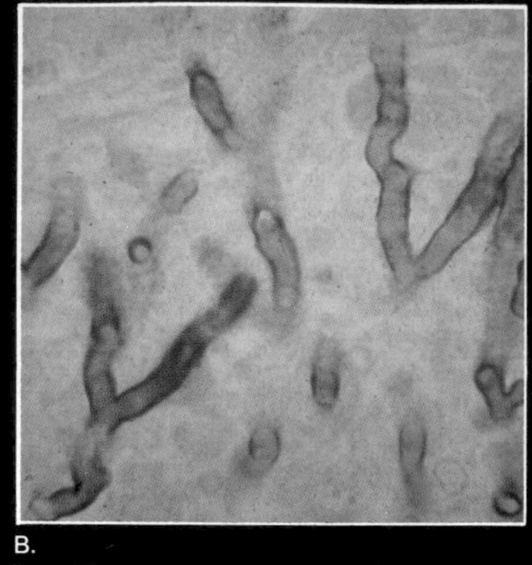

B.

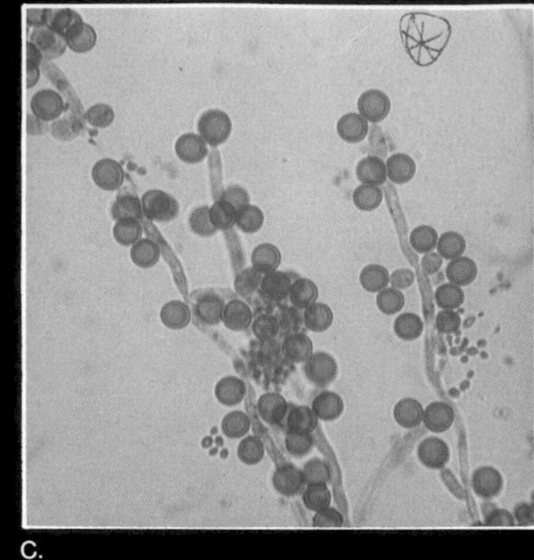

C.

Fungi A. Macroconidia of the fungal dermatophyte *Microsporum gypseum*. B. Hyphae of *Aspergillus fumigatus* in lung tissue from a case of aspergillosis. C. Appearance of *Candida albicans* grown on corn meal agar; the large spherical cells are chlamydospores, the much smaller rounded cells are blastoconidia, and the elongated strands of cells are pseudohyphae.

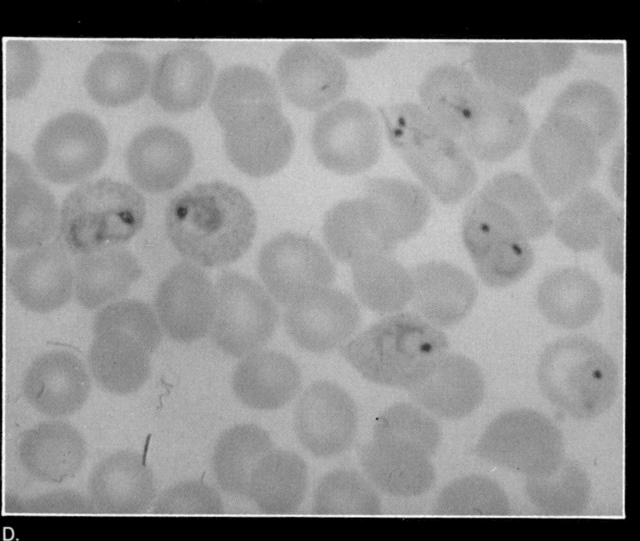

D.

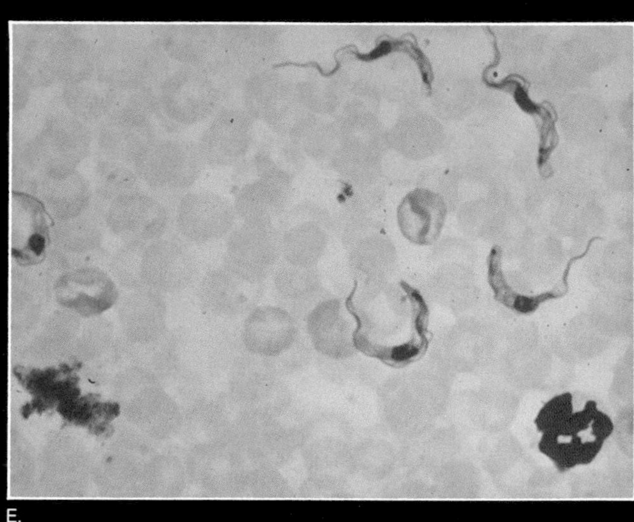

E.

Protozoans D. Blood smear showing intracellular rings and trophozoites of the malarial parasite *Plasmodium vivax*. E. Trypomastigotes of *Trypanosoma brucei* in a blood smear from an experimentally infected animal; note the undulating membrane along the body of each trypanosome.

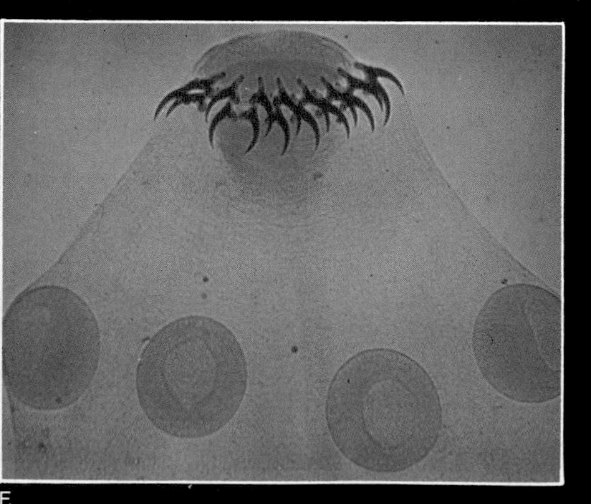

F.

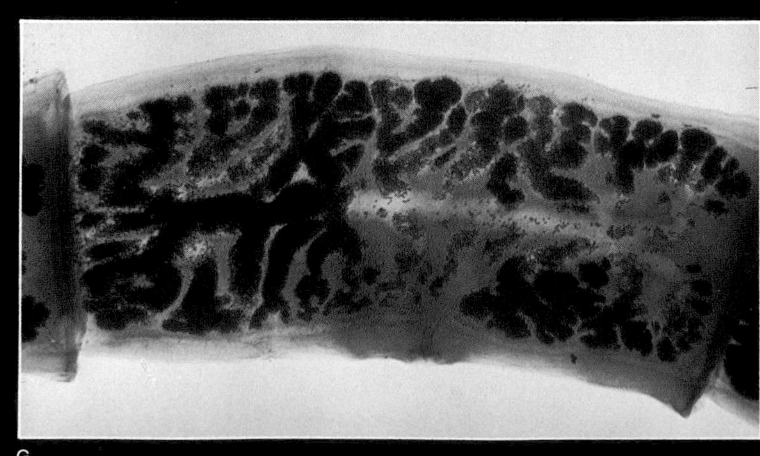

G.

Helminth F. Scolex of the pork tapeworm *Taenia solium*; note the sucking cups and hooks used to attach to the intestinal wall. G. A gravid proglottid of *Taenia solium*.

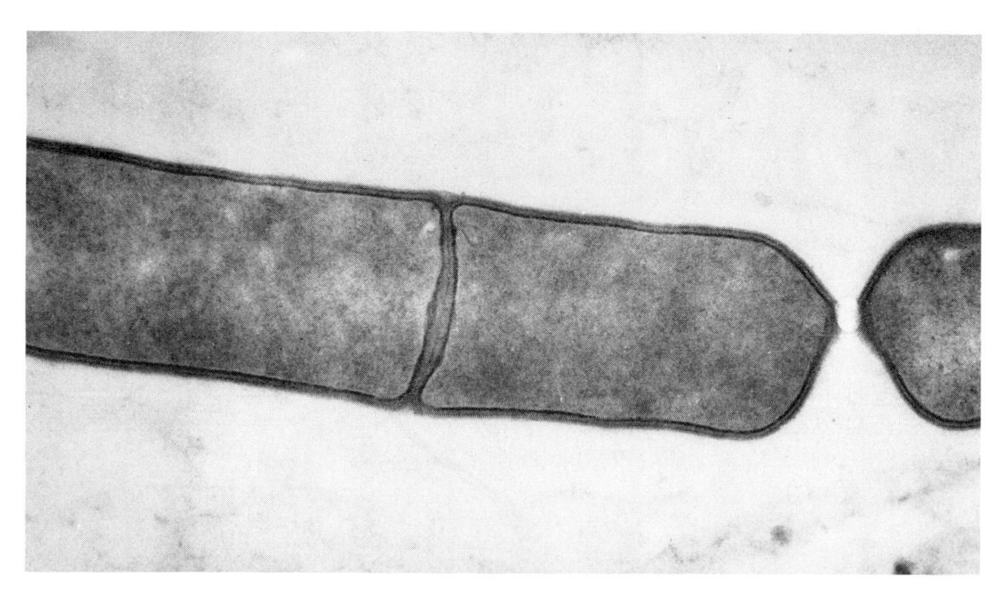

Figure 10–13 Electron micrograph of a few cells of *Lactobacillus acidophilus*, a gram-positive, non-endospore-forming rod (×46,600).

Gram-Positive Asporogenous (Non-Endospore-Forming) Rod-Shaped Bacteria

A chief representative of gram-positive asporogenous rods is the genus *Lactobacillus* (Figure 10–13). Lactobacilli are aerotolerant, anaerobic rods that produce lactic acid from simple carbohydrates and grow well in acidic environments (pH 5). Lactobacilli are located in the vagina, intestinal tract, and oral cavity. In the mouth, they may play a role in tooth decay. A common industrial use of lactobacilli is in the production of sauerkraut, pickles, buttermilk, and yogurt. A pathogen in this group is *Listeria monocytogenes*, which is associated with abscess formation, encephalitis, and endocarditis.

Actinomycetes and Related Organisms

Many of the actinomycetes and related organisms are important from both a medical and commercial point of view. Among the important genera are *Corynebacterium*, *Mycobacterium*, *Nocardia*, *Actinomyces*, and *Streptomyces*. All genera except *Corynebacterium* exhibit some branching.

The genus *Corynebacterium* contains gram-positive, aerobic, nonmotile, non-endospore-forming rods. Corynebacteria have swollen ends, making the rods club shaped. (*Koryne* means "club shaped," hence the origin of the name.) One related species, *Propionibacterium acnes*, is commonly found on the human skin and has been implicated in acne. *Propionibacterium* produces propionic acid from the fermentation of carbohydrates. The best known and most widely studied species is *Corynebacterium diphtheriae*, the causative agent of diphtheria (Figure 10–14a).

Mycobacteria are aerobic, non-endospore-forming, nonmotile rods (Figure 10–14b). Most pathogenic species have the distinguishing characteristic of being acid-fast, while many saprophytic species are gram-positive. Although a number of species of *Mycobacterium* are found in the soil, the most important pathogens of humans are *Mycobacterium tuberculosis*, the causative agent of tuberculosis, and *Mycobacterium leprae*, the causative agent of leprosy (see Color Plate IC and IID).

Members of the genus *Nocardia* are aerobic, acid-fast or gram-positive microbes that cause two types of infections in humans. *Nocardia asteroides* produces a condition called pulmonary nocardiosis, a condition that simulates tuberculosis. The organism is also one of the causative agents of mycetoma, a localized destructive infection usually involving the feet and hands. *Nocardia* species are characterized by their long, branching filaments (Figure 10–14c).

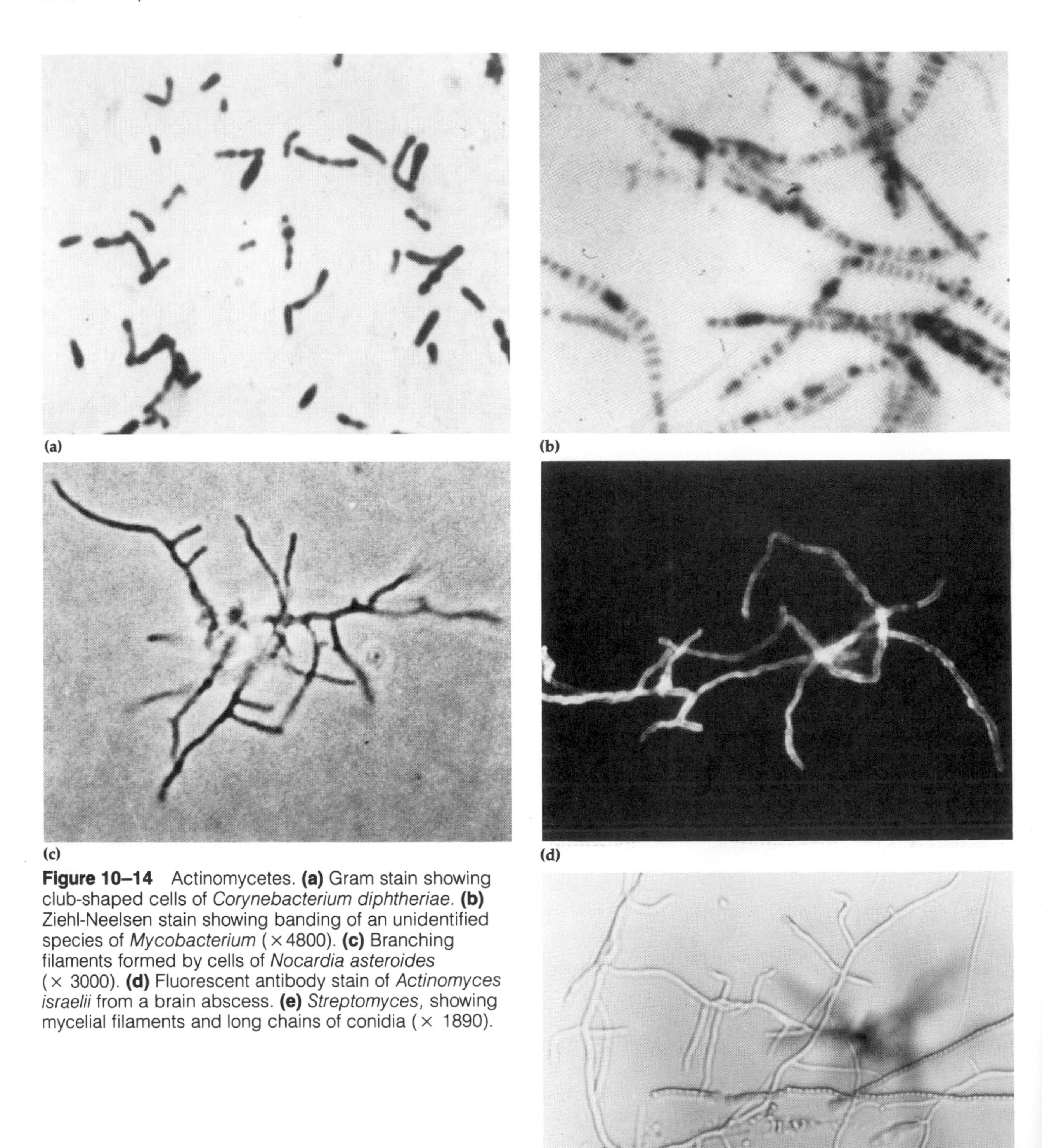

Figure 10–14 Actinomycetes. **(a)** Gram stain showing club-shaped cells of *Corynebacterium diphtheriae*. **(b)** Ziehl-Neelsen stain showing banding of an unidentified species of *Mycobacterium* (×4800). **(c)** Branching filaments formed by cells of *Nocardia asteroides* (× 3000). **(d)** Fluorescent antibody stain of *Actinomyces israelii* from a brain abscess. **(e)** *Streptomyces*, showing mycelial filaments and long chains of conidia (× 1890).

Species of *Actinomyces* are similar in appearance to *Nocardia*. However, unlike *Nocardia*, this genus consists of obligate anaerobes. As a group, *Actinomyces* is widely distributed in nature and plays a vital role in decomposing organic matter in the soil. One species, *Actinomyces israelii* (Figure 10–14d and Color Plate IIID), causes actinomycosis, an infectious disease characterized by lesions of the face, lungs, or abdomen.

Streptomyces species also form long filaments that branch (Figure 10–14e); masses of these filaments are called **mycelia** (singular, mycelium). *Streptomyces* is identified partially on the basis of the arrangement of **conidia**, which are spores formed from the aerial filaments. These spores are quite different from endospores. Thick cross walls form, separating the cells of the filaments. These cells separate into the conidia, which then become detached, free to spread in the air. Each conidium is capable of germinating into a new colony if it lands on a suitable substrate.

Streptomyces is commonly found in the soil, where it produces a chemical called *geosmin* that gives fresh soil its typical odor. Today, most of our commercial antibiotics are produced by various species of *Streptomyces*.

Rickettsias

Rickettsias include both the rickettsias and chlamydias. They are obligate, intracellular parasites, which means that they can reproduce only within a host cell. In this respect, they are similar to viruses. In fact, they are not much larger than some of the largest viruses. However, morphologically and biochemically they resemble bacteria and are therefore classified as such. A comparison of rickettsias, chlamydias, and viruses is in Table 10–4.

Rickettsias (Figure 10–15a) are rod-shaped bacteria or coccobacilli that have a high degree of pleomorphism. They are gram-negative and nonmotile and divide by transverse binary fission. They range in length from 1 to 2 μm. One of the distinguishing features of most rickettsias is that they are transmitted to humans by insects and ticks. The one exception is *Coxiella burnetii*, which causes Q fever and can be transmitted by airborne or food-borne routes, as well as by insects. Examples of diseases caused by rickettsias include epidemic typhus (*Rickettsia prowazekii*), transmitted by lice; endemic murine typhus (*Rickettsia typhi*), transmitted by rat fleas; and Rocky Mountain spotted fever (*Rickettsia rickettsii*), transmitted by ticks (see Color Plate IG).

Chlamydias (Figure 10–15b) are coccoid bacteria that range in size from 0.2 to 1.5 μm. They are gram-negative and nonmotile and, unlike most rickettsias, do not require insects for transmission. They are transmitted by person-to-person contact or by airborne respiratory routes. The developmental cycle of chlamydias is perhaps their most distinguishing characteristic. First the infectious form of the microbe, called the **elementary body**, attaches itself to a host cell. The host cell phagocytizes the elementary body and houses it in a vacuole. Within the host cell cytoplasm, the elementary body reorganizes and develops into a larger, less infective **initial body**. This then divides successively and condenses to produce smaller infectious

Table 10–4 Comparison of Biochemical Properties of Rickettsias, Chlamydias, and Viruses

Property	Rickettsias	Chlamydias	Viruses
Nucleic acid	RNA and DNA	RNA and DNA	RNA or DNA
Ribosomes	Present	Present	Absent
Structural integrity maintained during multiplication	Yes	Yes	No
Macromolecular synthesis	Carried out	Carried out	Only with use of host cell
ATP-generating system	Present	Absent	Absent
Sensitivity to antibacterial antibiotics	Sensitive	Sensitive	Resistant

Source: Thomas D. Brock, *Biology of Microorganisms*, 3rd ed., © 1979, p. 720. Reprinted by permission of Prentice-Hall, Inc., Englewood Cliffs, NJ.

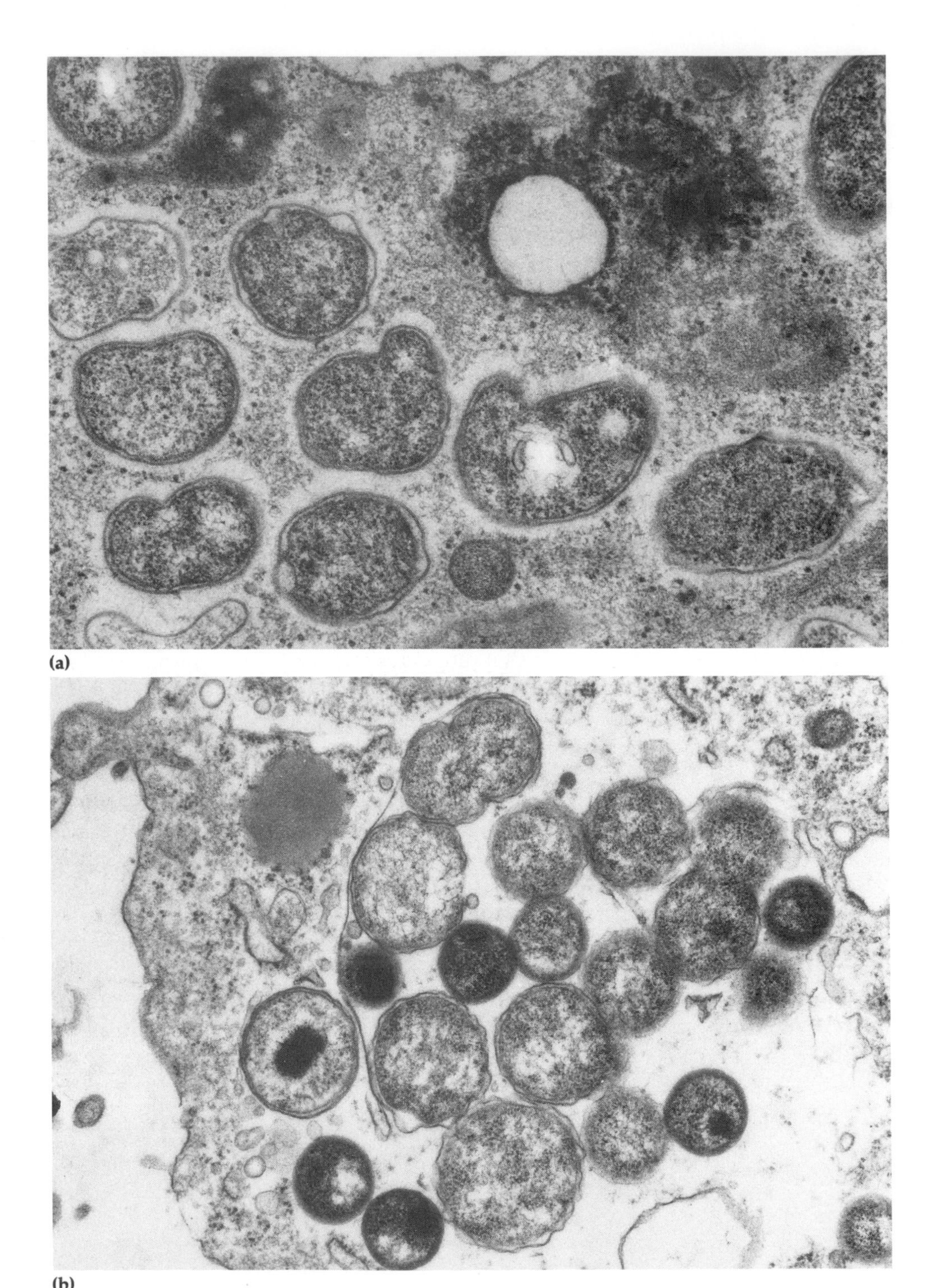

Figure 10–15 Rickettsias and chlamydias. **(a)** Electron micrograph showing *Rickettsia prowazekii* in experimentally-infected tick tissue (× 42,500). The inner and outer membranes are clearly visible around the gram-negative cells. **(b)** Inclusion body of *Chlamydia psittaci* in the cytoplasm of a McCoy cell (×25,000). Several small, dense, infectious elementary bodies are shown, including two at the lower left. The larger members of the inclusion are the less dense initial bodies, and one of these (at the top of the photo) is in the process of dividing.

elementary bodies. Finally, these are released from the host cell, causing infection in surrounding host cells.

There are only two species of chlamydias. *Chlamydia trachomatis* is the causative agent of trachoma, the most common cause of blindness in humans (see Color Plate IH). *Chlamydia trachomatis* seems to be the primary causative agent of nongonoccocal urethritis (NGU), which may now be the most common venereal disease in the United States. *Chlamydia psittaci* is the causative agent of psittacosis (ornithosis). And lymphogranuloma venereum, another venereal disease, is also caused by a chlamydia.

Mycoplasmas

Mycoplasmas are bacteria that do not form cell walls. They should not be confused with L forms, which are mutants of wall-forming bacteria that fail to form normal cell walls. Mycoplasmas cannot revert to normal cell-wall-containing types, as L forms can. Moreover, mycoplasmas do not require high concentrations of salt to maintain cellular integrity, whereas L forms do. And the plasma membranes of mycoplasmas have a high sterol content, while those of L forms do not. Finally, the growth of mycoplasmas is not inhibited by penicillin. These differences between mycoplasmas and L forms are summarized in Table 10–5.

The majority of *Mycoplasma* species are aerobes or facultative anaerobes, and because of the absence of cell walls, they are highly pleomorphic (Figure 10–16). They can produce filaments that resemble fungi, hence their name (*Myco* means "fungus").

Table 10–5 Comparison Between Mycoplasmas and L Forms

Property	Mycoplasmas	L Forms
Stability	Cannot revert to cell-wall-containing type	May revert to cell-wall-containing type
Medium	Do not require high salt medium to maintain cellular integrity	Require high salt medium to maintain cellular integrity
Plasma membrane	High sterol content	No sterols
Reaction to penicillin	No adverse effect	Reproduction is inhibited

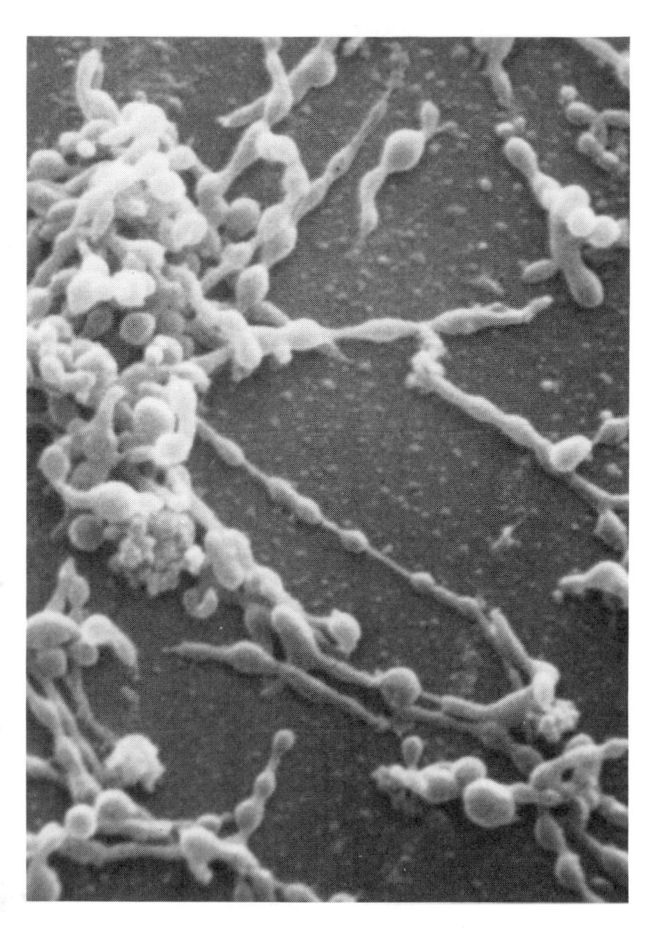

Figure 10–16 Scanning electron micrograph showing pleomorphic cells of *Mycoplasma pneumoniae* (×10,000).

Mycoplasma cells are very small, ranging in size from 100 to 250 nm. Most mycoplasmas grow best at temperatures between 30° and 36°C in alkaline environments. The most significant human pathogen among mycoplasmas is *Mycoplasma pneumoniae*, the causative agent of primary atypical pneumonia. This disease typically involves the upper respiratory system and rarely involves the lungs as other pneumonias do.

OTHER GROUPS OF BACTERIA

All the bacteria included in the medically important groups are heterotrophs. *Bergey's Manual* contains six additional parts. The characteristics of the bacteria included in these parts are summarized in Table 10–6, and several of the bacteria are shown in Figure 10–17. These bacteria are not medically important, but they do play a significant role in the biogeochemical cycles of elements (see Chapter 26). Some of these bacteria are photoautotrophic, others are chemoautotrophic, and still others are heterotrophic.

In the next chapter, we will turn our attention to fungi, protozoans, and helminths.

Table 10–6 Summary of Selected Characteristics of Nonpathogenic Bacterial Groups According to *Bergey's Manual*, 8th edition, 1974

Name of Group	*Bergey's* Part	Gram Reaction	Mode of Motility	Morphology	Habitat	Special Features
Phototrophic bacteria	1	–	If motile, polar flagella	Rods, cocci, spirilla	Water	Photoautotrophs using CO_2 and light and H_2S as an electron donor, includes green sulfur bacteria and purple sulfur bacteria (see Chapter 5); facultative photoautotrophs (purple nonsulfur bacteria) that grow by fermentation in the dark.
Gliding bacteria	2	–	Gliding	Rods, spirilla, filaments	Compost, manure, water, mouth	Related to the gliding cyanobacteria, but incapable of photosynthesis. Some form brightly colored fruiting bodies, the cells of which form resting bodies, called microcysts.
Sheathed bacteria	3	–	Polar flagella	Rods, filaments	Fresh water, sewage	Form filaments of cells enclosed in a thin sheath consisting of lipoprotein–polysaccharide derived from cell walls. *Sphaerotilus* forms tangled masses that interfere with operation of activated sludge sewage treatment systems.
Budding and/or appendaged bacteria	4	–	Polar flagella	Rods	Water	Form prosthecae, projections from the main body, for attachment to surfaces. Prosthecae contain cytoplasm and are enclosed by a cell wall. *Hyphomicrobium* buds a new cell from the tip of the prostheca.
Gram-negative chemolithotrophic bacteria	12	–	Polar flagella	Rods, cocci, spirilla	Soil, water	Oxidize a wide variety of inorganic compounds. The nitrifying bacteria, such as *Nitrobacter* and *Nitrosomonas,* are important in the nitrogen cycle. *Thiobacillus* and other sulfur bacteria are part of the sulfur cycle (see Chapter 26).
Methane-producing bacteria	13	–, +	Flagella	Rods, cocci, spirilla	Sewage, gastrointestinal tract of humans and other animals	Obtain energy by the oxidation of hydrogen and formic acid and reduction of CO_2 to form methane.

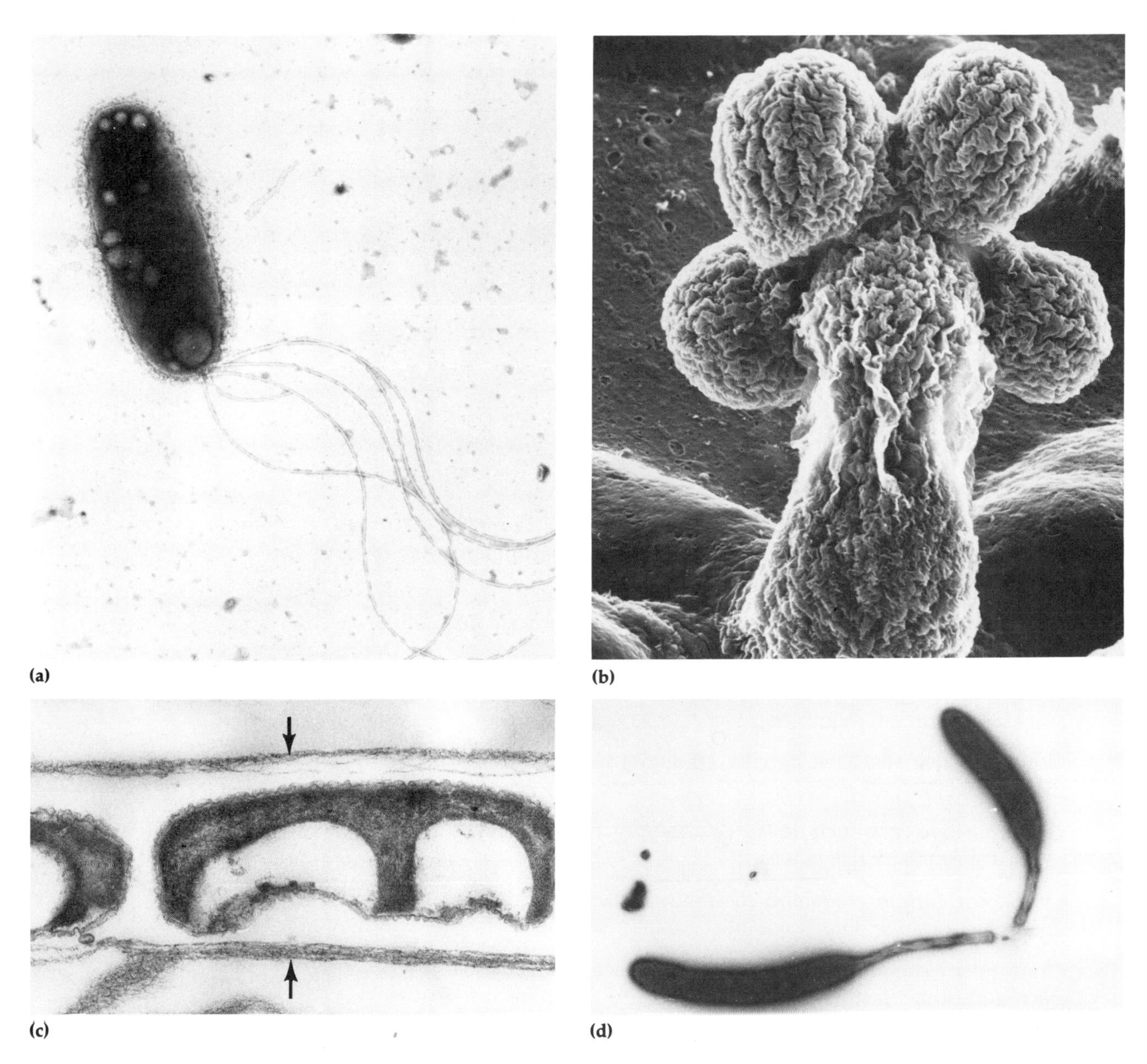

Figure 10–17 Several nonpathogenic bacteria. **(a)** The purple sulfur bacterium *Ectothiorhodospira mobilis* (×10,750). **(b)** Scanning electron micrograph of the fruiting body of the gliding myxobacterium *Stigmatella aurantiaca;* such fruiting bodies are formed by the aggregation of numerous individual bacterial cells (×350). **(c)** Thin section through the sheathed bacterium *Sphaerotilus natans*, showing the sheath (arrows) and portions of two cells within it (×40,000). **(d)** Two cells of the stalked bacterium *Caulobacter bacteroides* (×9000).

STUDY OUTLINE

BERGEY'S MANUAL: BACTERIAL TAXONOMY (pp. 257–258)

1. *Bergey's Manual* divides bacteria into 19 parts based on Gram-stain reaction, cellular morphology, oxygen requirements, and nutritional properties.
2. In some cases, the parts include families and orders, and some bacteria are included as genera of uncertain affiliation.

MEDICALLY IMPORTANT GROUPS OF BACTERIA (pp. 258–273)

1. Spirochetes are long, thin, helical cells that move by means of an axial filament.
2. Spiral and curved bacteria move by means of one or more polar flagella.
3. Gram-negative aerobic rods and cocci have polar flagella, if flagellated, and can utilize a wide variety of organic compounds.
4. Gram-negative facultatively anaerobic rods have peritrichous flagella and include the enterics and the genus *Vibrio*.
5. Gram-negative anaerobic bacteria are long, slender rods found in humans.
6. Gram-negative cocci and coccobacilli are aerobic or facultatively anaerobic; some are normal human flora.
7. Gram-negative anaerobic cocci occur as normal flora of the human mouth.
8. Gram-positive cocci include the catalase-positive *Staphylococcus* and catalase-negative *Streptococcus*.
9. Endospore-forming rods and cocci may be aerobic, facultatively anaerobic, or anaerobic.
10. Gram-positive asporogenous rod-shaped bacteria include aerobic, microaerophilic, facultatively anaerobic, and anaerobic organisms.
11. Actinomycetes and related organisms are gram-positive pleomorphic rods; they include some acid-fast genera.
12. Rickettsias and chlamydias are obligate intracellular parasites.
13. Mycoplasmas are bacteria that lack cell walls.

OTHER GROUPS OF BACTERIA (pp. 273–275)

1. *Bergey's Manual* contains six parts that do not contain bacteria of medical importance.
2. They are the phototrophic bacteria, the gliding bacteria, the sheathed bacteria, the budding and/or appendaged bacteria, the gram-negative chemolithotrophic bacteria, and the methane-producing bacteria.
3. These bacteria play an important role in the cycles of elements in the environment.

STUDY QUESTIONS

REVIEW

1. The following is a key that can be used to identify the medically important groups of bacteria. Fill in the name of the group indicated by the key.

	Name of Group
I. Cells helical or curved	
A. Axial filament	______________
B. Polar flagella	______________
II. Cells rod or coccus	
A. Gram-negative	
1. Aerobic, nonfermenting	______________
2. Facultative anaerobes	
a. Rods	______________
b. Cocci	______________
3. Anaerobic	
a. Rods	______________
b. Cocci	______________
B. Gram-positive	
1. Endospores	______________
2. No endospores	
a. Rods	______________
b. Cocci	______________
c. Acid-fast or conidia-producing	______________
III. Pleomorphic, gram-negative	
A. Intracellular parasites	______________
B. Lacking cell wall	______________

2. *Bergey's Manual* divides the kingdom procaryotae (Monera) into two divisions (phyla). What are they?

3. List six criteria used to classify bacteria according to *Bergey's Manual.*

4. Matching:

___ Form fruiting bodies; some resemble cyanobacteria	(a) Photosynthetic bacteria
___ Form projections from the cell called prosthecae	(b) Gliding bacteria
___ Long filaments, found in sewage	(c) Budding and/or appendaged bacteria
___ Photosynthetic	(d) Gram-negative chemolithotrophic bacteria
___ Reduce CO_2 to CH_4	(e) Methane-producing bacteria
___ Oxidize inorganics such as NO_2	(f) Sheathed bacteria

CHALLENGE

1. Does *Bergey's Manual* provide any phylogenetic information regarding the relationships between bacteria?
2. Should *Bergey's Manual* be considered a classification or identification tool? Briefly explain your answer.

FURTHER READING

Buchanan, R. E. and N. E. Gibbons, eds. 1974. *Bergey's Manual of Determinative Bacteriology,* 8th ed. Baltimore, MD: Williams and Wilkins. The standard reference of identification and classification of bacteria.

Edwards, P. R. and W. H. Ewing. 1972. *Identification of Enterobacteriaceae,* 3rd ed. Minneapolis, MN: Burgess. Biochemical and serological reactions of the genera of enterics are listed and discussed in this laboratory reference.

Finegold, S. M., W. J. Martin, and E. G. Scott. 1978. *Diagnostic Microbiology,* 5th ed. St. Louis, MO: C. V. Mosby. A good reference for procedures used in the clinical microbiology laboratory.

Kloos, W. E. 1980. Natural populations of the genus *Staphylococcus. Annual Review of Microbiology* 34:559–592. Illustrates the use of modern technology to determine a species; includes a discussion of staphylococci that live on birds and mammals.

Konenman, E. W., S. D. Allen, V. R. Dowell, and H. M. Sommers. 1979. *Color Atlas and Textbook of Diagnostic Microbiology.* Philadelphia, PA: J. B. Lippincott. An excellent reference to laboratory tests used to identify medically important microorganisms; chapters on important genera, test kits, and eucaryotic parasites.

Slack, J. A. and M. A. Gerencser. 1975. *Actinomyces, Filamentous Bacteria.* Minneapolis, MN: Burgess. Discussion of each genus as well as identification methods for actinomycetes.

11

Fungi, Protozoans, and Multicellular Parasites

Objectives

After completing this chapter you should be able to

- Differentiate between asexual and sexual reproduction, using a fungal example and a protozoan example.
- List the defining characteristics of the four phyla of fungi described in this chapter.
- Distinguish between systemic, subcutaneous, cutaneous, superficial, and opportunistic mycoses on the basis of mode of entry and tissues affected.
- Describe the outstanding characteristics of the four phyla of protozoans described in this chapter.
- List the characteristics of the three classes of parasitic helminths and give an example of each class.
- Define intermediate host, definitive host, and arthropod vector.

In this chapter, we will examine the eucaryotic microorganisms that cause human disease, including fungi, protozoans, and parasitic helminths, or worms. Although most adult helminths are not microscopic, they do have microscopic stages in their development and are therefore studied by microbiologists.

FUNGI

As noted in Chapter 9, fungi constitute a separate kingdom of eucaryotes. All fungi are heterotrophs, requiring organic compounds for energy and carbon. Fungi are aerobic or facultatively anaerobic; no strictly anaerobic fungi are known. The majority of fungi are saprophytes in soil and water, where they primarily decompose plant material. Like bacteria, fungi are important in the decomposition and recycling of elements. Table 11–1 lists the basic differences between fungi and bacteria. Of the more than 100,000 species of fungi, only about 100 are pathogenic for humans and other animals. The study of fungi is called **mycology,** and the study of parasitic fungi is **medical mycology.**

Fungi include yeasts, molds, and fleshy fungi. Yeasts are unicellular organisms. Molds are multicellular, filamentous organisms such as mildews, rusts, and smuts. And fleshy fungi include multicellular mushrooms, puffballs, and coral fungi.

Table 11–1 Comparison of Selected Features of Fungi and Bacteria

Property	Fungi	Bacteria
Cell type	Eucaryotic with well-defined nuclear membrane	Procaryotic
Cell membrane	Sterols present	Sterols absent except in *Mycoplasma*
Cell wall	Glucans; mannans; chitin	Muramic acid peptides; teichoic acid
Spores	Produce a wide variety of sexual and asexual reproductive spores	Endospores (not for reproduction); conidia produced by Actinomycetales
Metabolism	Heterotrophic; aerobic, facultatively anaerobic	Heterotrophic, chemoautotrophic, photoautotrophic; aerobic, facultatively anaerobic, anaerobic
Sensitivity to chemotherapy	Sensitive to polyenes and griseofulvin; not sensitive to penicillin, tetracyclines, chloramphenicol, streptomycin	Sensitive to penicillin, tetracyclines, chloramphenicol, streptomycin; not sensitive to polyenes or griseofulvin

Source: Modified from B. D. Davis, et al., *Microbiology*, 3rd ed., Harper & Row, New York, 1980, p. 829.

VEGETATIVE STRUCTURES

A **thallus** (colony) of a mold or fleshy fungus consists of long filaments of cells joined together, called **hyphae** (singular, **hypha**). In most molds, the hyphae contain crosswalls, termed **septa** (singular, **septum**), that divide the hyphae into distinct, uninucleate cell-like units. These are called **septate hyphae** (Figure 11–1a). In a few classes of fungi, however, the hyphae contain no septa and appear as continuous, long cells with many nuclei. These hyphae are referred to as **coenocytic hyphae** (Figure 11–1b). Even in fungi with septate hyphae, there are usually openings in the septa that make the cytoplasm of adjacent "cells" continuous; so these fungi are actually coenocytic organisms too. The hyphae of a thallus grow by elongating at the tips (Figure 11–1c). Each part of a hypha is capable of growth, and when a minute fragment breaks off, it can elongate to form a new hypha.

When environmental conditions are suitable, the hyphae grow, intertwine, and form a mass called a **mycelium,** which is visible to the naked eye. The portion of the mycelium concerned with obtaining nutrients is called the **vegetative mycelium;** the portion concerned with reproduction is the **reproductive** or **aerial mycelium,** so called because it projects above the surface of the medium on which the fungus is growing, as shown in Figure 11–2. The aerial mycelium often bears reproductive spores, which we will discuss later.

Yeasts

Yeasts are nonfilamentous, unicellular fungi that are typically spherical or oval in shape. Like molds, yeasts are widely distributed in nature, frequently found on fruits and leaves, often appearing as a white powdery coating. Yeasts usually grow—that is, increase in number—by **budding.** In budding, a protuberance (bud) is formed on the outer surface of the parent cell. As the bud elongates, the parent cell nucleus divides, and one nucleus migrates into the bud. Cell wall material is then laid down between the bud and parent cell, and the bud eventually breaks away (Figure 11–3). One yeast cell may produce up to 24 daughter cells by budding.

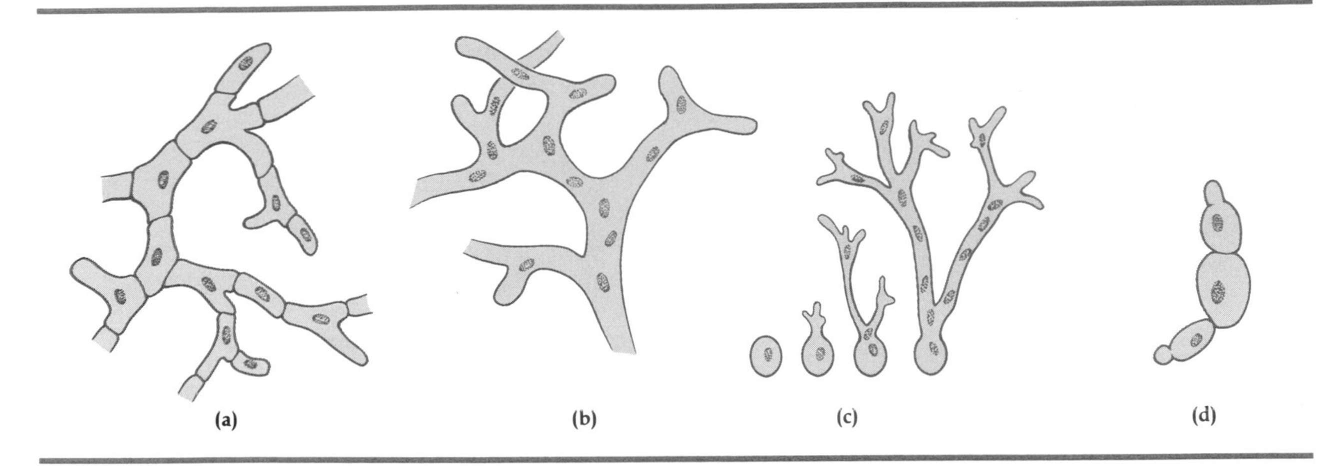

Figure 11–1 Vegetative structures of fungi. **(a)** Septate hyphae have crosswalls dividing the hyphae into cell-like units. **(b)** Coenocytic hyphae lack crosswalls. **(c)** Hyphae grow by elongating at the tips. **(d)** Pseudohyphae are short chains of cells formed by some yeasts.

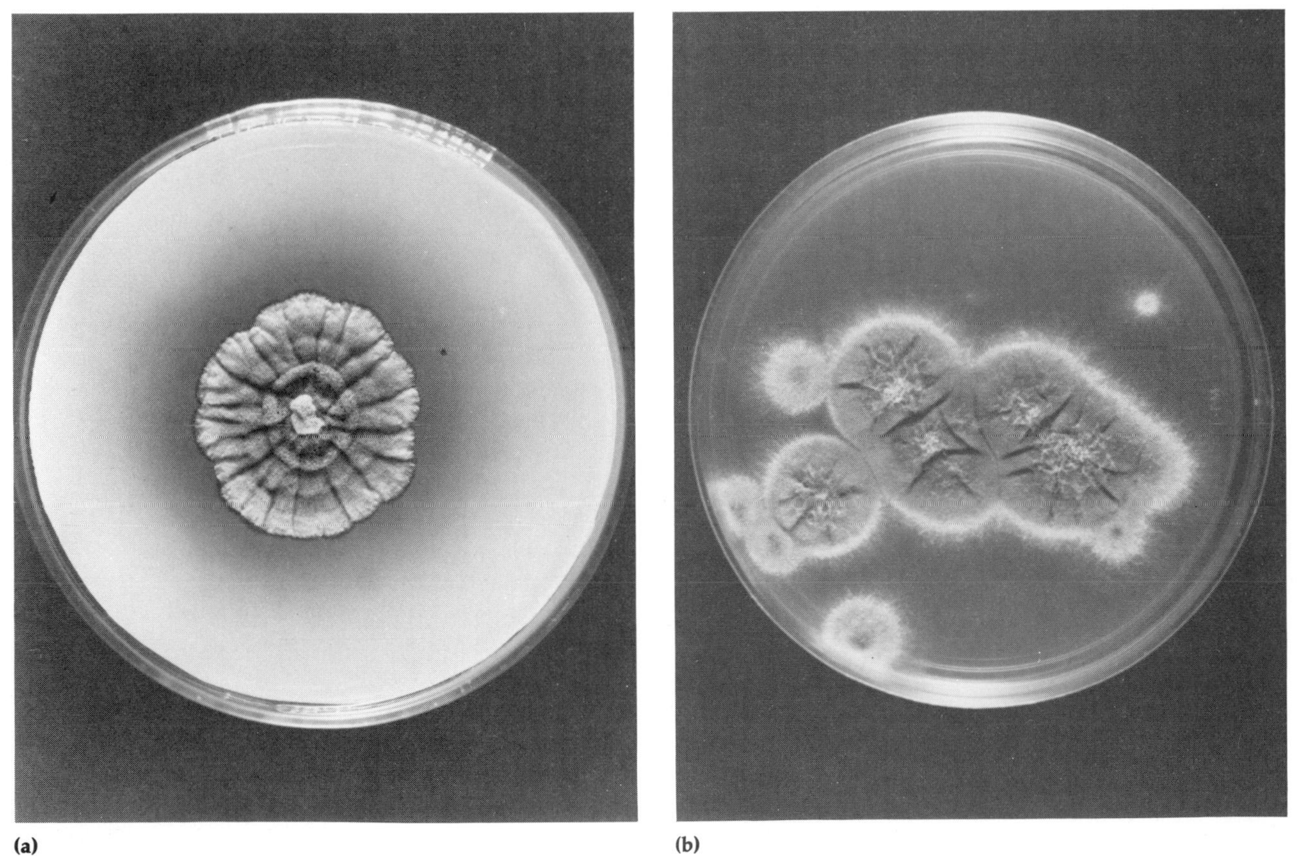

Figure 11–2 Aerial mycelia of fungi grown on agar. **(a)** *Penicillium marneffei*. **(b)** *Aspergillus fumigatus*.

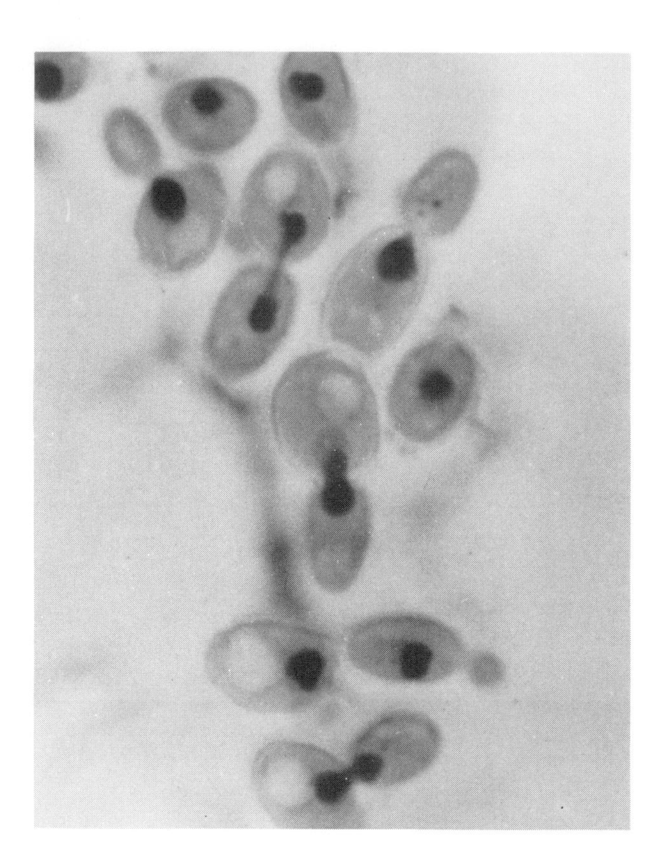

Figure 11–3 Bakers' yeast (*Saccharomyces* sp.) showing various stages of budding (×1570). Note the distribution of the darkly-stained nuclear material from some of the parent cells into the buds.

In some species of yeasts, buds are produced that fail to detach themselves, resulting in a short chain of cells called a **pseudohypha** (Figure 11–1d). In a few cases, yeast cells grow by fission, whereby the parent cell elongates, the nucleus divides, and two daughter cells are produced. Increases in the number of yeast cells on a solid medium produce a colony similar to a bacterial colony.

Yeasts are capable of facultative anaerobic growth. As you may recall from Chapter 5, this means that yeasts can use oxygen or an organic compound as the final electron acceptor, a valuable attribute. If given access to oxygen, yeasts metabolize carbohydrates to carbon dioxide and water by aerobic respiration. Denied oxygen, they ferment carbohydrates and produce ethanol and carbon dioxide. This is the basis of the brewing, winemaking, and baking industries. Species of *Saccharomyces* are used to produce ethanol for brewing beverages and carbon dioxide for raising breads.

Dimorphic Fungi

Some fungi, most notably the pathogenic species, exhibit **dimorphism,** that is, two growth forms. Such fungi can grow either as a mold or as a yeast. Frequently, dimorphism is temperature dependent: at 37°C the fungus is yeastlike, and at 25°C it is moldlike.

REPRODUCTIVE STRUCTURES

Viewed with a microscope, the hyphae of almost all fungi look alike. To identify fungi, the reproductive structures, or **spores,** must be examined.

Reproduction in fungi is by spore formation. These spores, however, are quite different from

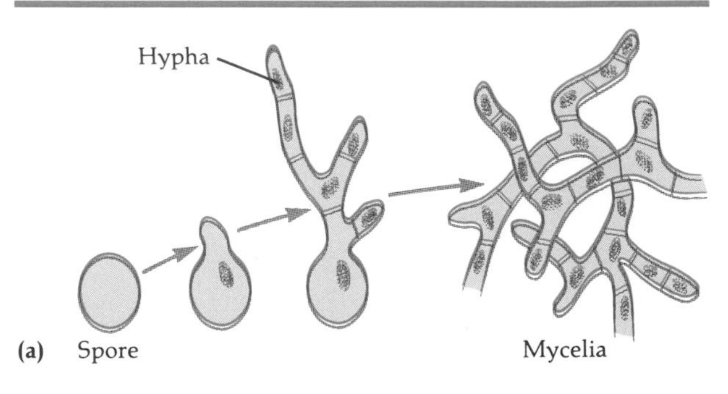

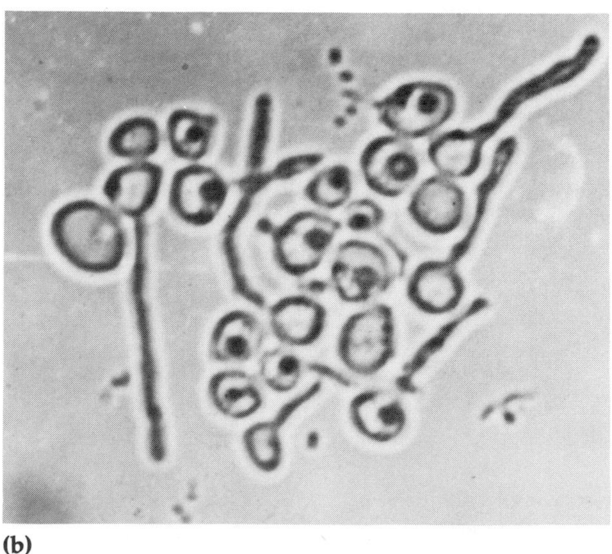

(b)

Figure 11–4 Spore germination. **(a)** Mycelia develop as outgrowths of a single spore. **(b)** Germinating spores of *Candida albicans.*

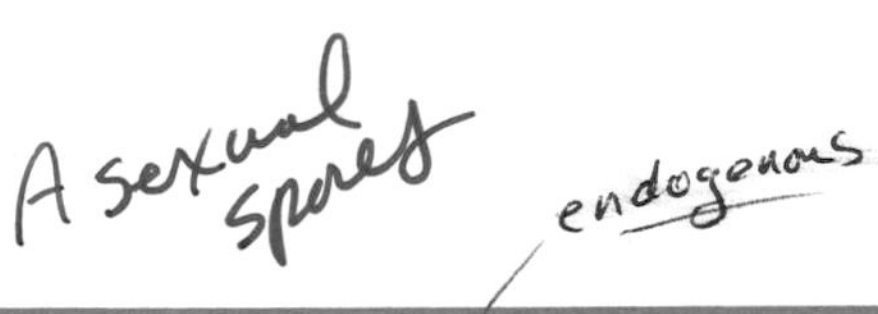

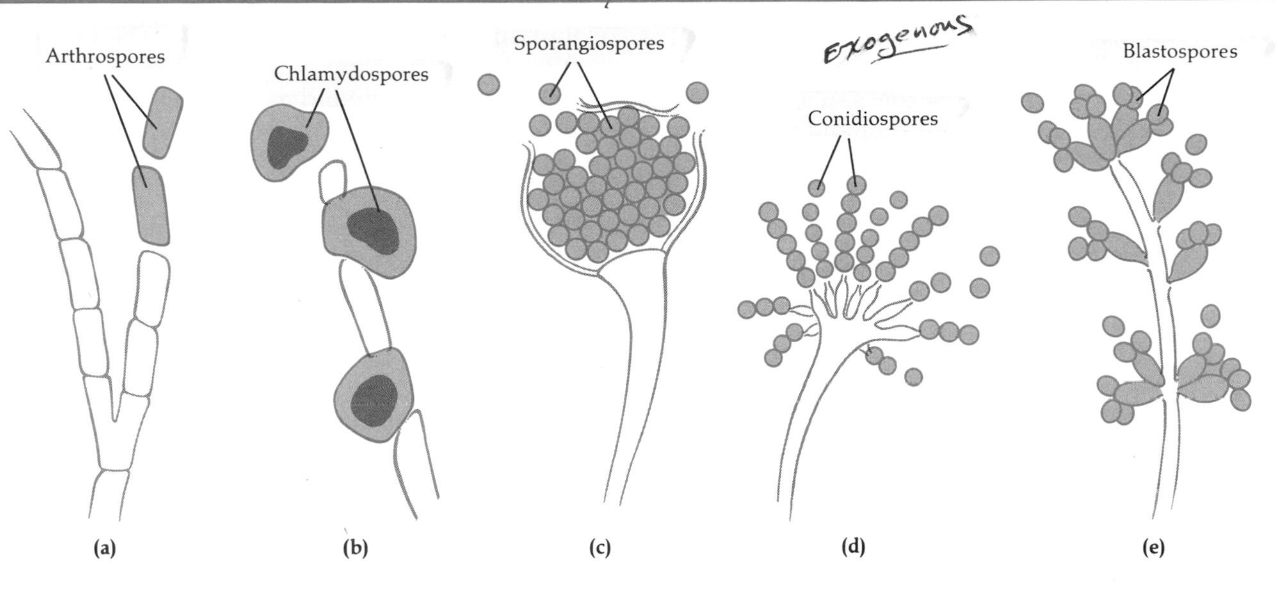

Figure 11–5 Representative asexual spores. **(a)** Fragmentation of hyphae results in formation of these arthrospores. **(b)** Chlamydospores are thick-walled cells within the hyphae. **(c)** Sporangiospores are formed within a sporangium (spore sac). **(d)** Conidiospores are arranged in chains at the end of a conidiophore. **(e)** Blastospores are formed from buds of the parent cell.

bacterial endospores. Bacterial endospores are formed for survival under adverse environmental conditions. A single vegetative bacterial cell forms an endospore, which eventually germinates to produce a single vegetative bacterial cell. It is not reproduction in the sense that the total number of bacterial cells increases. But when a mold forms a spore, the spore detaches from the parent, and upon germination, a new mold is produced (Figure 11–4). Unlike the bacterial endospore, this is a true reproductive spore, with one cell giving rise to an entire multicellular organism. Spores are formed in a great variety of ways from the aerial mycelium, depending on the species.

Fungal spores can be asexual or sexual. **Asexual spores** are formed by the aerial mycelium of one organism. When the spores germinate, they result in organisms that are genetically identical to the parent. **Sexual spores** result from the fusion of nuclei from two opposite mating strains of the same species of fungus. Organisms that grow from sexual spores will have genetic characteristics of both parent strains. Since spores are of key importance in the identification of fungi, we will now look at some of the various types of asexual and sexual spores that are formed.

Asexual Spores

Asexual spores are produced by an individual fungus; there is no fusion of the nuclei of cells. Several types of asexual spores are produced by fungi. One type, an **arthrospore,** is formed by the fragmentation of a septate hypha into single, slightly thickened cells (Figure 11–5a). One species that produces such spores is *Coccidioides immitis*. Another type of asexual spore is referred to as a **chlamydospore,** a thick-walled spore formed as segments within a hypha (Figure 11–5b). A fungus that produces chlamydospores is *Candida albicans* (see Color Plate IVC). A **sporangiospore** is an asexual spore formed within a sac **(sporangium)** at the end of an aerial hypha called a **sporangiophore.** The sporangium may contain hundreds of sporangiospores (Figure 11–5c). Such spores are produced by *Rhizopus*. A fourth principal type of asexual spore is a **conidiospore,** which is a unicellular or multicellular spore that is not enclosed in a sac (Figure 11–5d). Conidiospores are produced in a chain at the end of a **conidiophore.** Such spores are produced by *Penicillium*. A final type of asexual spore, known as a **blastospore,** consists of a bud coming off the parent cell (Figure 11–5e). Such spores are found in yeasts.

Sexual Spores

Among fungi, a sexual spore results from sexual reproduction consisting of three phases: (1) A haploid nucleus of a donor cell (+) penetrates the cytoplasm of a recipient cell (−). (2) The (+) and (−) nuclei fuse to form a diploid, zygote nucleus. (3) By meiosis, the diploid nucleus gives rise to haploid nuclei (sexual spores), some of which may be genetic recombinants. Among fungi, sexual spores are produced less frequently than asexual spores. Often, the sexual spores are produced only under special circumstances.

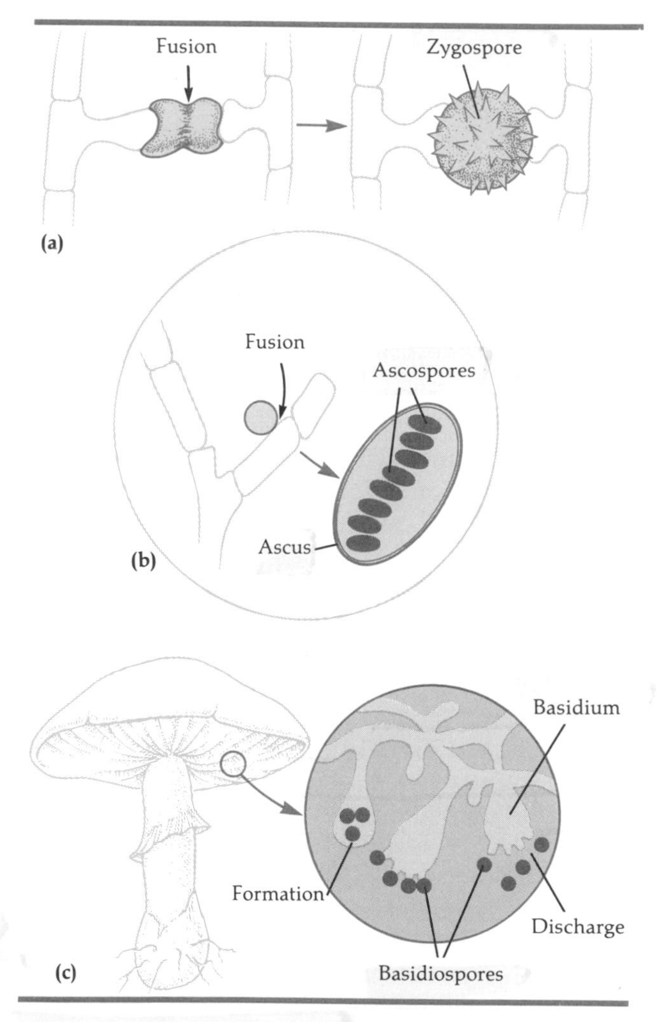

Figure 11–6 Sexual spores. **(a)** Zygospores, characteristic of the phylum Zygomycota, are produced from the fusion of two cells that are morphologically alike. **(b)** Ascospores, produced by the Ascomycota, are formed within an ascus. **(c)** Basidiospores, produced only by Basidiomycota, are formed on the tip of a pedestal called a basidium.

One kind of sexual spore is a **zygospore,** a large spore enclosed in a thick wall (Figure 11–6a), which results from the fusion of the nuclei of two cells that are morphologically similar to each other. Such spores are produced by fungi of the phylum Zygomycota. A second type of sexual spore is an **ascospore,** a spore resulting from the fusion of the nuclei of two cells that may be morphologically similar or dissimilar. These spores are produced in a saclike structure called an **ascus** (Figure 11–6b). There are usually two to eight ascospores in an ascus. Such spores are produced by the phylum Ascomycota. A final type of sexual spore is the **basidiospore,** a spore formed externally on a base pedestal called a **basidium** (Figure 11–6c). There are usually four basidiospores per basidium. Basidiospores are produced only by the phylum Basidiomycota.

The sexual spores produced by fungi serve as a basis for grouping the fungi into several phyla. We will now examine the phyla that contain genera of medical importance: Zygomycota, Ascomycota, Basidiomycota, and Deuteromycota.

MEDICALLY IMPORTANT PHYLA OF FUNGI

Zygomycota

The Zygomycota, or conjugation fungi, are saprophytic molds that have coenocytic hyphae. A common example is *Rhizopus nigricans,* the common black bread mold (Figure 11–7). The asexual spores of *Rhizopus* are sporangiospores. The dark sporangiospores inside the sporangium give *Rhizopus* its descriptive common name. When the sporangium breaks open, the sporangiospores are dispersed, and if they fall on a suitable medium, they will germinate into a new mold thallus. The sexual spores are zygospores.

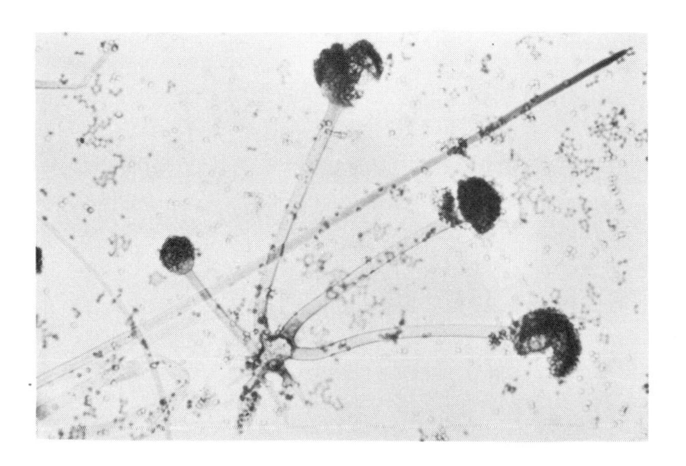

Figure 11–7 The mold *Rhizopus arrhizus*. At the top of each of the four sporangiophores shown here is a sporangium. Several sporangia have been ruptured, releasing numerous sporangiospores.

Ascomycota

The Ascomycota, or sac fungi, include molds with septate hyphae and yeasts. They are called sac fungi because their sexual spores are ascospores produced in an ascus. The asexual spores are usually conidiospores produced in long chains from the conidiophore. The arrangements of conidiospores in *Penicillium* and *Aspergillus* are shown in Figure 11–8. Conidia means "dust," and these spores freely detach from the chain at the slightest disturbance and float in the air like dust.

Basidiomycota

The Basidiomycota, or club fungi, also possess septate hyphae. The common name club fungi is derived from the shape of the basidium that bears the sexual basidiospores. Asexually, some of the basidiomycota produce conidiospores. Representative basidiomycetes are shown in Figure 11–9.

Deuteromycota

The Deuteromycota are also known as the Fungi Imperfecti. These fungi are "imperfect" because no sexual spores have been demonstrated as yet. Members of this phylum produce the asexual chlamydospores, arthrospores, and conidiospores, and budding also occurs. Deuteromycota have septate hyphae.

Most of the pathogenic fungi are, or once were, classified as Deuteromycota. This phylum might be described as a "holding category" until sexual spores are observed and the particular fungus can be properly classified. Table 11–2 provides a list of some well-known Fungi Imperfecti that have only recently been classified. Note that the generic names are changed with reclassification. When different species within a deuteromycete genus are observed to have morphologically different sexual spores, they may be reclassified into two or more different genera, as with *Penicillium*.

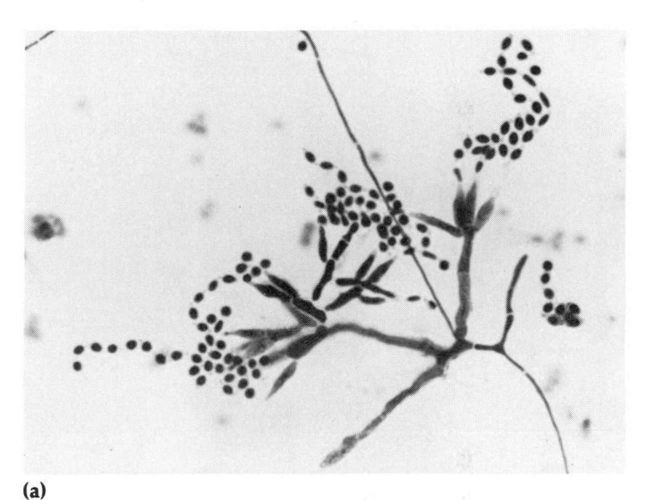

(a)

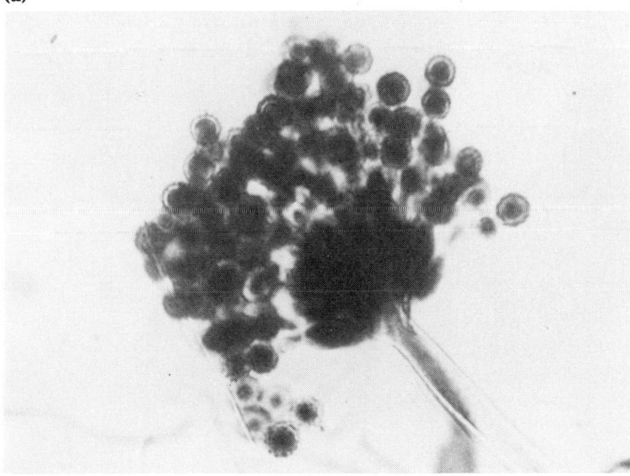

(b)

Figure 11–8 Conidiospores. The arrangement of conidiospores is useful in identification of fungi. **(a)** *Penicillium marneffei* produces conidiospores from a branched conidiophore. **(b)** Conidiospores of *Aspergillus tamarii* are produced from the enlarged terminal end (vesicle) of the conidiophore.

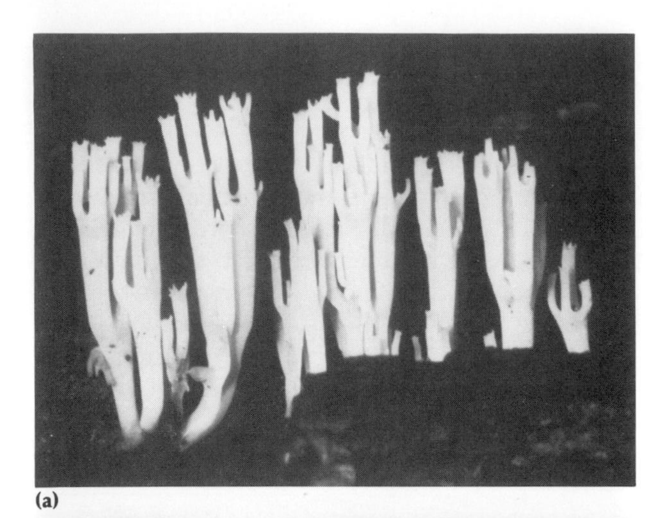
(a)

(b)

(c)

Figure 11–9 Some representative basidiomycetes. **(a)** *Calvulina cristata*, one kind of coral fungus. **(b)** *Agaricus augustus* is a popular produce item. **(c)** Puffballs of the species *Lycoperdon perlatum*.

Table 11–2 Several Genera of Reclassified Imperfect Fungi

Imperfect Name	Reclassified to Phylum	Perfect Name
Aspergillus	Ascomycota	*Sartorya, Eurotium, Emericella*
Blastomyces	Ascomycota	*Ajellomyces*
Cryptococcus	Basidiomycota	*Filobasidiella*
Histoplasma	Ascomycota	*Emmonsiella*
Microsporum	Ascomycota	*Nannizia*
Penicillium	Ascomycota	*Talaromyces, Carpenteles*
Petriellidium	Ascomycota	*Allescheria*
Trichophyton	Ascomycota	*Arthroderma*

FUNGAL DISEASES

A summary of the medically important fungi is given in Table 11–3. Any fungus infection is referred to as a **mycosis.** Mycoses may be divided into five groups, depending on the level of infected tissue and mode of entry into the host. Mycoses are classified as systemic, subcutaneous, cutaneous, superficial, or opportunistic.

Systemic Mycoses

Systemic mycoses are fungal infections deep within the patient. They are not restricted to any particular region of the body and may affect a number of tissues and organs. Systemic, or deep, mycoses are usually caused by saprophytic fungi that live in the soil. Inhalation of spores is the route of transmission, and these infections typically begin in the lungs and then spread to other body tissues. They are not contagious from other animals to humans or from human to human.

Subcutaneous Mycoses

Subcutaneous mycoses are fungal infections beneath the skin, and they are caused by saprophytic fungi that live in soil and on vegetation. Infection occurs by direct implantation of spores or mycelial fragments into a puncture wound in the skin. The wounds are usually initiated by splinters, thorns, or rocks. Such diseases progress slowly and are characterized by localized subcutaneous abscesses. As the abscesses spread, they often break through the skin surface and form chronic, ulcerated, crusted lesions. Although dissemination may occur to lymph nodes, spread to the viscera is rare. Subcutaneous mycoses are extremely disfiguring and may persist for years.

Cutaneous Mycoses

Fungi that infect only the epidermis, hair, and nails are known as **dermatophytes,** and their infections are called **dermatomycoses** or **cutaneous mycoses.** Dermatophytes secrete *keratinase,* an enzyme that degrades keratin. Keratin is a protein found in hair, skin, and nails. The lesions produced by dermatophytes are roughly circular and tend to expand equally in all directions. *Microsporum* (see Color Plate IVA), *Trichophyton,* and *Epidermophyton* are representative dermatophytes, causing diseases known clinically as **tineas** (ringworms). Infection is transmitted from person to person or animal to person by direct contact or by contact with infected hairs and epidermal cells (as from barber shop clippers or shower room floors).

Superficial Mycoses

The fungi that cause **superficial mycoses** are localized along hair shafts and in superficial (surface) epidermal cells. One example of a superficial mycosis is *tinea nigra,* a disease characterized by dark lesions, mostly confined to the palms. Infection is prevalent in the tropics. Another example is *white piedra.* This disease is characterized by soft, pale nodules that appear on the scalp and beard (Figure 11–10).

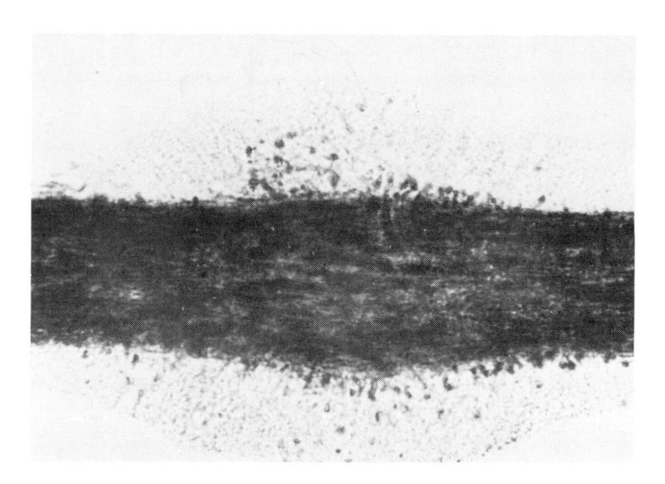

Figure 11–10 The growth of *Trichosporon* on hair shafts produces the pale, hard nodules that are characteristic of the superficial mycosis, white piedra. This micrograph shows a single nodule of white piedra on a dark hair shaft.

Opportunistic Mycoses

An **opportunistic** pathogen is an organism that is generally harmless in its normal habitat but which may become pathogenic in a host who is seriously debilitated or traumatized or who is under treatment with broad-spectrum antibiotics or immunosuppressive drugs. A person's normal flora can become opportunistic pathogens under these conditions.

Mucormycosis is an opportunistic mycosis caused by *Rhizopus* and *Mucor* primarily in patients with ketoacidosis resulting from diabetes mellitus, leukemia, or treatment with immunosuppressive drugs. **Aspergillosis** is also an opportunistic mycosis, and it is caused by *Aspergillus* (see Color Plate IVB). This disease occurs in individuals with debilitating lung diseases or cancer who have inhaled *Aspergillus* spores. The mycosis **Candidiasis** is most frequently caused by *Candida albicans* and may occur during pregnancy in the form of vulvovaginal candidiasis. **Thrush,** a mucocutaneous candidiasis, is an inflammation of the mouth and throat that frequently occurs in newborns.

ECONOMIC DISADVANTAGES OF FUNGI

Some of the beneficial industrial uses of fungi are discussed in Chapter 27. But for now let us discuss the less desirable economic effects.

Table 11–3 Characteristics of Some Parasitic Fungi

Phylum	Growth Characteristics	Asexual Spore Types	Human Pathogens
Zygomycota	Nonseptate hyphae	Sporangiospores	*Rhizopus*, *Mucor*
Ascomycota	Septate hyphae	Conidiospores	*Allescheria boydii*
			*Aspergillus**
	Dimorphic	Conidiospores	*Blastomyces dermatitidis**
			*Histoplasma capsulatum**
	Septate hyphae, strong affinity for keratin	Conidiospores, arthrospores	*Microsporum** *Trichophyton**
Basidiomycota	Septate hyphae includes the fleshy fungi (mushrooms), rust and smuts, and plant pathogens; yeastlike encapsulated cells	Conidiospores	*Cryptococcus neoformans**
Deuteromycota	Septate hyphae	Conidiospores	*Epidermophyton*
		Chlamydospores	*Cladosporium werneckii*
	Dimorphic	Conidiospores	*Sporothrix schenkii*
		Arthrospores	*Coccidioides immitis*
	Yeastlike, pseudohyphae	Chlamydospores	*Candida albicans*
	Dimorphic	Chlamydospores, arthrospores, conidiospores	*Paracoccidioides brasiliensis*
	Septate hyphae, produce melaninlike pigments	Conidiospores	*Phialophora* *Fonsecaea*
	Yeastlike pseudohyphae	Arthrospores	*Trichosporon*

*Imperfect name. Refer to Table 11–2 for current classification and names.

Fungi have a number of nutritional and physiological characteristics that contribute to their economic relations to humans:

1. Fungi usually prefer an acidic pH (5.0), a pH that is too acid for the growth of most common bacteria.
2. Most molds are aerobic.
3. Most fungi are more resistant to osmotic pressures than bacteria and are therefore able to grow in higher sugar or salt concentrations.
4. Fungi are capable of growing on substances with a very low moisture content, generally too low to support the growth of bacteria.
5. Fungi require somewhat less nitrogen for growth than do bacteria, and they are nutritionally very adaptable and efficient. This is an important factor for growth on such unlikely substrates as painted walls or shoe leather.

As most of us have observed, mold spoilage of fruits, grains, and vegetables is relatively common, while bacterial spoilage of such foods is not. There is little moisture on the unbroken surfaces of such foods, and the interior of fruits is too acidic for the growth of many bacteria. Jams and jellies also tend to be acidic, and what's more, they have a high osmotic pressure from the sugars they contain. These factors all discourage bacterial growth but readily support the growth of mold. A paraffin layer on top of a jar of homemade jellies will help deter mold growth because molds are aerobic and the paraffin layer keeps out the oxygen. On the other hand, foods like fresh meats are such good

Table 11–3, continued

Clinical Notes	Type of Mycosis	Habitat	Refer to Chapter
Opportunistic pathogen	Systemic	Ubiquitous	23
Opportunistic pathogen	Systemic	Ubiquitous	23
Primary cause of maduromycosis	Subcutaneous	Soil	19
Opportunistic pathogen	Systemic	Ubiquitous	23
Inhalation	Systemic	Unknown	23
Inhalation	Systemic	Soil	23
Tinea capitis	Cutaneous	Soil, animals	19
Tinea pedis ✓	Cutaneous	Soil, animals	19
Inhalation	Systemic	Soil, bird feces	21
Tinea cruris, tinea unguium	Cutaneous	Soil, humans	19
Tinea nigra	Superficial	Ubiquitous	19
Puncture wound	Subcutaneous	Soil	—
Inhalation	Systemic	Soil	23
Opportunistic pathogen	Cutaneous, systemic, mucocutaneous	Human normal flora	19, 25
Inhalation	Systemic	Presumably soil	—
Chromoblastomycosis	Subcutaneous	Soil, plant debris	19
Chromoblastomycosis	Subcutaneous	Soil, plant debris	19
White piedra	Superficial	Soil (?), humans	—

substrates for bacterial growth that bacteria not only outgrow molds, but will actively suppress mold growth in such foods.

The ability of fungi to grow at low moisture levels is of particular importance in their role as plant pathogens. Bacterial plant pathogens are far less common. In fact, the fungus that caused the great potato blight in Ireland during the early 1800s (*Phytophthora infestans)* was one of the first microorganisms to be associated with a disease.

The human disease called **ergotism,** which was prevalent in Europe during the Middle Ages, is caused by a toxin produced by an ascomycete plant pathogen *(Claviceps purpurea)* that grows on grains. The toxin is contained in **sclerotia** (Figure 11–11), which are highly resistant mycelia. The toxin itself, **ergot,** is an alkaloid that can cause hallucinations resembling those of LSD; in fact, ergot is a natural

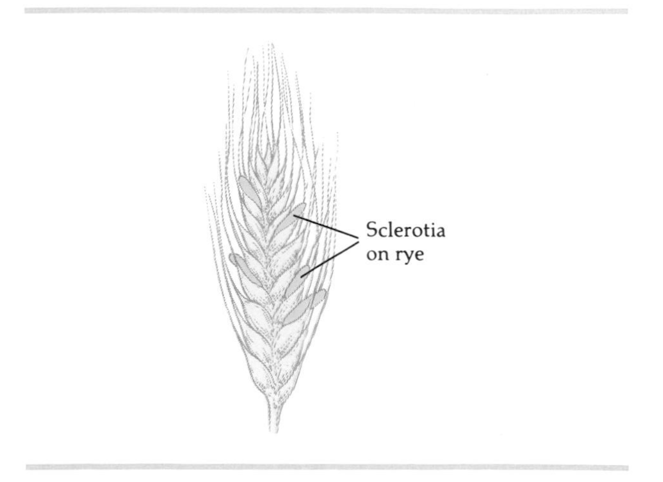

Figure 11–11 The sclerotia of *Claviceps purpurea* (wheat rust) are visible on this rye flower. *Claviceps* is the source of the drug LSD.

source of LSD. Ergot also constricts capillaries and can cause, through prevention of proper blood circulation, gangrene of the extremities. Although *Claviceps* still occasionally occurs on grains, modern milling usually removes the sclerotia.

A number of other toxins are produced by fungi that grow on grains. For example, peanut butter is occasionally recalled because of excessive amounts of **aflatoxin,** a toxin that has been found to cause cancer of the liver in laboratory animals. No link to human cancer has been found as yet, but it is believed that aflatoxin may be altered in the animal's body to a mutagenic compound. Aflatoxin is produced by the growth of the mold *Aspergillus flavus,* from which the name of the toxin is abridged.

A few mushrooms produce toxins that exert their effects after ingestion of the mushroom. Perhaps the most notable mycotoxins are those produced by *Amanita phalloides,* commonly known as the death angel. **Phalloidin** and **amanitin** are potent neurotoxins that affect nerve transmission. Ingestion of the mushroom may result in death.

The spreading chestnut tree, of which the poet wrote, no longer grows in this country except in a few widely isolated sites. An important fungal blight killed essentially all the trees. First seen in the United States in 1904, the ascomycete *Endothia parasitica* was introduced from China. The tree roots live and put forth shoots regularly, but the shoots are just as regularly killed by the fungus. Another devastating fungal plant disease that was also imported to this country is the Dutch elm disease. It is carried from tree to tree by a bark beetle. The fungus blocks the tree's circulation.

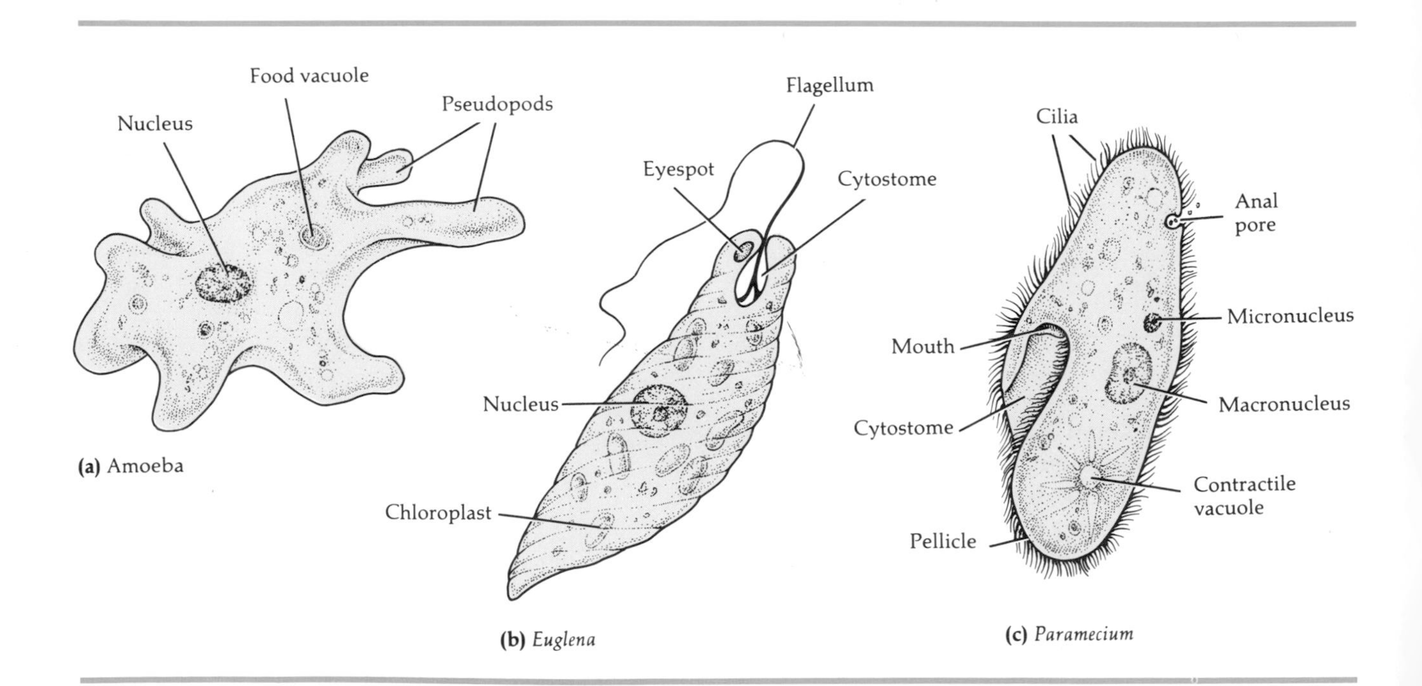

Figure 11–12 Examples of protozoans. **(a)** Amoeba. Amoebas approach and engulf food by extending those portions of the cytoplasm called pseudopods. Once surrounded by the pseudopods, the food is in a food vacuole. **(b)** *Euglena*, a flagellate. A flagellum is used to pull a flagellate along. *Euglena* is not a typical protozoan in that it is a photoautotroph. *Euglena* has a red "eye spot" that is used to detect light. **(c)** *Paramecium*, a ciliate. Rows of cilia cover the cells of ciliates. The cilia are moved in unison for locomotion and to bring food particles to the protozoan. *Paramecium* has specialized structures for ingestion (mouth), elimination of wastes (anal pore), and regulation of osmotic pressure (contractile vacuoles). The macronucleus is involved with protein synthesis and other ongoing cellular activities. The micronucleus functions in sexual reproduction.

PROTOZOANS

Protozoans are one-celled eucaryotic organisms that belong to the kingdom Protista. Recall what you have learned about eucaryotic cell structure in Chapter 4. Among the protozoans there are many variations on this cell structure, as we shall see. Protozoans inhabit water and soil, feeding upon bacteria and small particulate nutrients. And some protozoans are part of the normal flora of animals. For example, termites and cows have protozoans in their gastrointestinal systems that help them digest cellulose. Of the nearly 20,000 species of protozoans, relatively few cause disease.

Protozoans may be classified into phyla on the basis of their means of motility. We will restrict our study to the four phyla of protozoans that include disease-causing species. One such phylum consists of the **Sarcodina,** or amoebas (Figure 11–12a). The amoebas move by extending usually blunt, lobelike projections of the cytoplasm called **pseudopods.** Any number of pseudopods may flow from one side of the amoeba cell, and the rest of the cell will flow toward the pseudopods. Another phylum, the **Mastigophora,** or flagellates, possess flagella. Flagella are capable of whiplike movements that pull the cell through the medium (Figure 11–12b). Most flagellates have one or two flagella, but some of the parasitic species have as many as eight. Species in the phylum **Ciliata** have projections similar to flagella only shorter, called cilia. The cilia are in a precise arrangement over the cell (Figure 11–12c) and are moved to propel the cell through its medium. The species in the fourth phylum we will study, the **Sporozoa,** are incapable of movement.

PROTOZOAN BIOLOGY

Nutrition

Most protozoans are aerobic heterotrophs. A few contain chlorophyll and are photoautotrophs. One chlorophyll-containing flagellate, *Euglena,* can also be grown in the dark in the laboratory as a heterotroph.

All protozoans live in areas where there is a large supply of water. Some protozoans take in food by absorption through the cell membrane. However, some have a protective covering called the **pellicle** and require specialized structures to take in food. The ciliates take in food by waving their cilia toward a mouthlike opening called the **cytostome.** The amoebas engulf food by surrounding it with pseudopods and phagocytizing it. In the protozoans, digestion takes place in membrane-bounded **vacuoles,** and waste may be eliminated through the cell membrane or through a specialized **anal pore.**

Reproduction

Protozoans reproduce asexually by fission, budding, or schizogony. **Schizogony** is *multiple* fission. The nucleus undergoes multiple divisions before the cell divides. After many nuclei are formed, a small portion of cytoplasm concentrates around each nucleus, and separation of daughter cells follows.

Sexual reproduction has been observed in some protozoans. The ciliates reproduce sexually by **conjugation** (Figure 11–13), which is very different from the bacterial process of the same name. During protozoan conjugation, two cells fuse and a haploid nucleus from each migrates to the other cell. The haploid nucleus fuses with another haploid nucleus within the cell. The parent cells separate, each now a fertilized cell. The cells will later

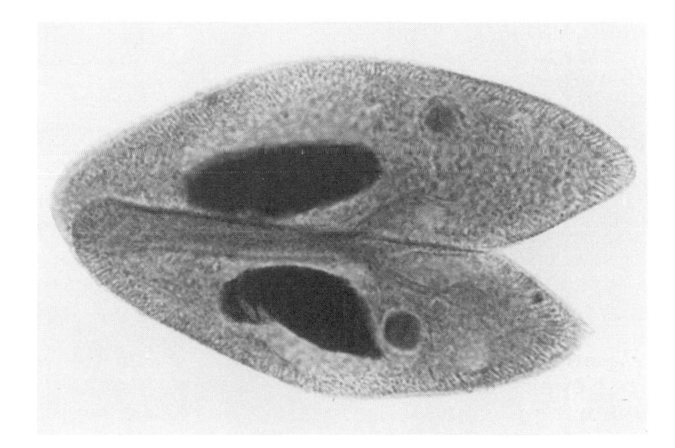

Figure 11–13 *Paramecium* conjugation. Sexual reproduction in ciliates is by conjugation. Each cell has two nuclei, one of which is specialized (haploid) for conjugation. One haploid nucleus from each cell will migrate to the other cell during conjugation. (Courtesy of Carolina Biological Supply Company.)

divide, producing daughter cells with recombinant DNA. Some protozoans produce **gametes,** haploid sex cells. Two gametes fuse to form a diploid zygote.

Encystment

Under certain adverse conditions, some protozoans are capable of producing a protective capsule called a **cyst.** A cyst permits the organism to survive when food, moisture, or oxygen are lacking, when temperatures are not suitable, and when toxic chemicals are present. A cyst also enables a parasitic species to survive outside a host.

Parasitic protozoans may have life cycles that are quite complex, involving more than one host. We will now examine representatives of the medically important phyla of protozoans.

MEDICALLY IMPORTANT PHYLA OF PROTOZOANS

Sarcodina

A well-known parasitic amoeba is *Entamoeba histolytica* (Figure 11–14), the causative agent of amoebic dysentery. *E. histolytica* is transmitted from human to human through cysts passed out with feces and ingested by the next host.

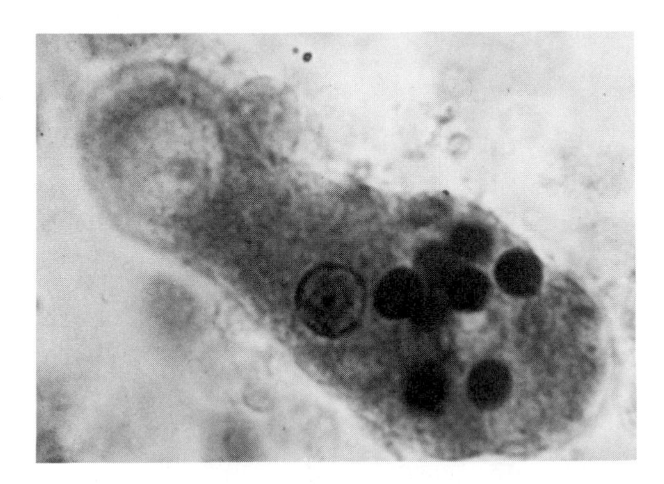

Figure 11–14 *Entamoeba histolytica* in the trophozoite form. The dark circles inside the amoeba are engulfed red blood cells from its human host.

Mastigophora

As previously mentioned, members of the phylum Mastigophora move by means of flagella. Their outer membrane is a tough, flexible pellicle. Flagellates are typically spindle shaped with flagella projecting from the front end. Food is ingested through a cytostome. Some flagellates have an **undulating membrane,** which appears to consist of highly modified flagella (see Figure 11–16).

An example of a flagellate that is a human parasite is *Giardia lamblia.* Figures 11–15 and 24–12 show the vegetative form, or **trophozoite,** of *G. lamblia,* as well as the cyst stage. The parasite is found in the small intestine of humans and other mammals. It is passed out of the intestine and survives in a cyst before being ingested by the next host. Diagnosis of giardiasis, the disease caused by *G. lamblia,* is based on identification of cysts in feces.

Another parasitic flagellate is *Trichomonas vaginalis* (Figure 11–16). *T. vaginalis* does not have

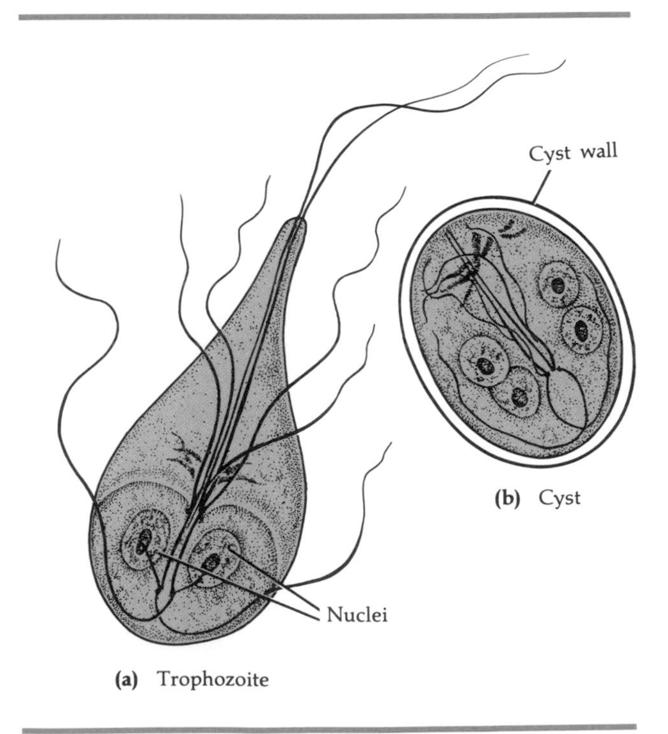

Figure 11–15 *Giardia lamblia.* **(a)** The trophozoite of this intestinal parasite has eight flagella and two prominent nuclei, giving it a distinctive appearance. **(b)** The cyst provides protection in the environment before being ingested by a new host.

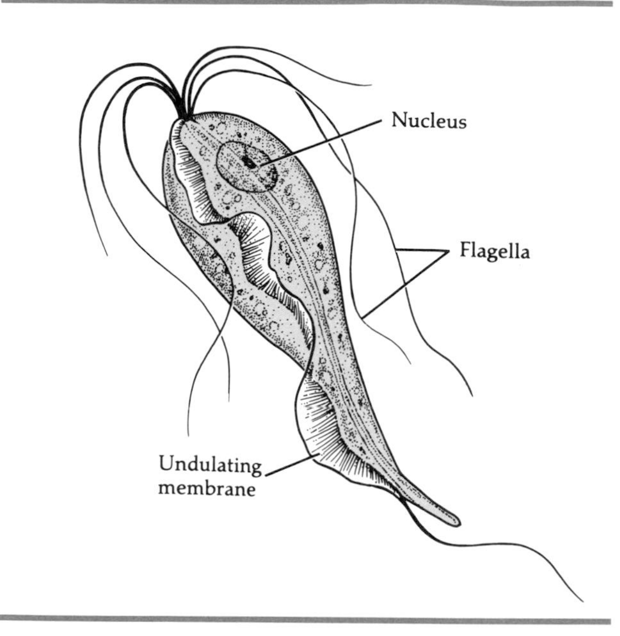

Figure 11–16 *Trichomonas vaginalis.* This flagellate is the cause of urinary and genital tract infections. It has a small undulating membrane. This flagellate does not have a cyst stage.

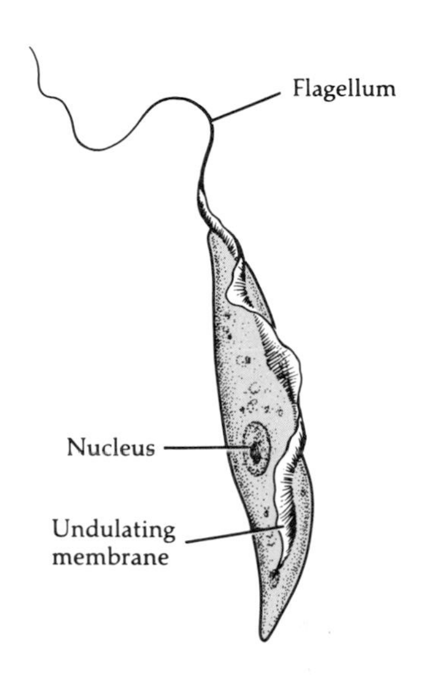

Figure 11–17 *Trypanosoma cruzi.* This elongated flagellate causes Chagas' disease. Trypanosomes have one flagellum that appears as an extension of the undulating membrane.

a cyst stage and must be transferred from host to host quickly to avoid desiccation. *T. vaginalis* is found in the vagina and male urinary tract. It is usually transmitted by sexual intercourse but may be transmitted by toilet facilities or towels.

The hemoflagellates are transmitted by the bite of bloodsucking insects and are found in the circulatory system of the host. To survive in this viscous fluid, hemoflagellates have long, slender bodies and an undulating membrane. The genus *Trypanosoma* includes the species that cause African sleeping sickness (*T. brucei;* see Color Plate IVE) and Chagas' disease. *T. cruzi* (Figure 11–17), the causative agent of Chagas' disease, is transmitted by the "kissing bug." When the trypanosome enters the insect, it rapidly multiplies by fission. Then, when the insect bites a human, it may defecate and release trypanosomes that can contaminate the bite wound.

Ciliata

The only ciliate that is a human parasite is *Balantidium coli,* the causative agent of a severe though rare type of dysentery. When cysts are ingested by the host, they enter the colon, where the trophozoites are released. The trophozoites feed on bacteria and fecal debris as they multiply, and cysts are passed out with feces (see Figure 24–13).

Sporozoa

Sporozoans are not motile in their mature forms and are obligate intracellular parasites. They have complex life cycles that ensure their survival and transmission from host to host. An example of a sporozoan is *Plasmodium,* the causative agent of malaria. A vaccine against malaria is now being investigated (see box).

Plasmodium grows by schizogony in human red blood cells (Figure 11–18), where it completes its asexual cycle. The young trophozoite looks like a ring in which the nucleus and cytoplasm are visible. This is called a **ring stage.** The ring stage enlarges and divides, and the red blood cells eventually rupture and release the microorganisms, now called **merozoites.** Upon release of the merozoites, their waste products, which cause the fever and chills, are also released. Most of the merozoites infect new red blood cells and perpetuate their cycle of asexual reproduction. However, some develop into male and female sexual forms called **gametocytes.** Even though the gametocytes themselves cause no further damage, they can be picked up by the bite of an *Anopheles* mosquito, at which point they enter the mosquito's intestine and begin their sexual cycle. Here the male and female gametocytes unite to form a zygote. The zygote forms a cyst in which cell division occurs, and asexual **sporozoites** are formed. When the cyst ruptures, the sporozoites migrate to the salivary glands of the mosquito. They can then be injected into a new human host by the biting mosquito.

When the sporozoites enter a new human host, they are carried by the blood to the liver. Here, the sporozoites undergo schizogony and produce thousands of merozoites. These cells enter the bloodstream and infect red blood cells—the start of another cycle.

Malaria is diagnosed in the laboratory by

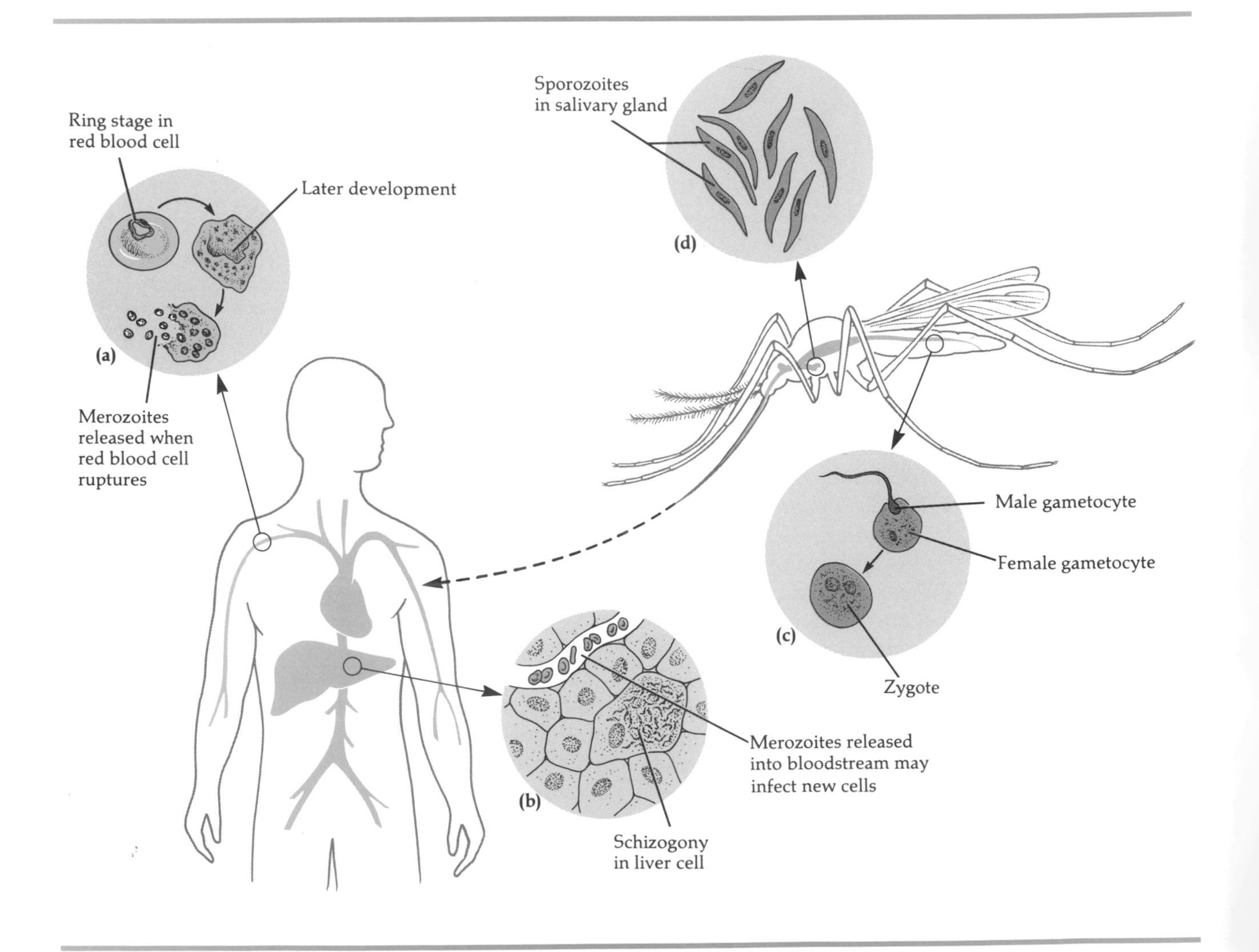

Figure 11–18 Life cycle of *Plasmodium vivax.* Asexual reproduction, schizogony, of the merozoites takes place **(a)** in the red blood cells or **(b)** in the liver of a human host. **(c)** Sexual reproduction of the parasite occurs in the intestine of an *Anopheles* mosquito after ingestion of gametocytes by the mosquito. **(d)** Sporozoites resulting from sexual reproduction migrate to the salivary gland to be injected into the mosquito's next host.

observing thick blood smears microscopically for the presence of *Plasmodium* (see Figure 20–4 and Color Plate IVD).

Toxoplasma gondii is another intracellular parasite of humans. The life cycle of this sporozoan is not known but appears to involve domestic cats. The trophozoites reproduce sexually and asexually in an infected cat, and cysts are passed out with feces. If the cysts are ingested by humans or other animals, the trophozoites may reproduce in the tissues of the new host (see Figure 22–9). *T. gondii* is dangerous to pregnant women, as it can cause congenital infections in utero. Tissue examination and observation of *T. gondii* is used for diagnosis.

Table 11–4 lists some typical parasitic protozoans and the diseases they cause.

HELMINTHS

There are a number of parasitic animals that spend part or all of their lives in humans. Most of these animals belong to two phyla: **Platyhelminthes** (flatworms) and **Aschelminthes** (roundworms). These worms are commonly called **helminths.** There are also free-living species in these phyla, but we will limit our discussion to the parasitic species.

HELMINTH BIOLOGY

Helminths are multicellular eucaryotic animals possessing digestive, circulatory, nervous, excre-

Microbiology in the News

Malaria Vaccine Via Gene-Splicing Proposed

The hundreds of millions of people inhabiting areas of the world where malaria is endemic may soon be able to avoid the disease by means of an inexpensive vaccine.

Genentech, a California genetic engineering firm, has begun research on a malaria vaccine using modern techniques of immunology and bacterial genetics.

Most successful vaccines for other microbial disease consist of dead or weakened microbes, or surface components of microbes. When the vaccine is injected into a human, the human's immune system makes antibodies against the vaccine. These antibodies protect the person against the disease upon later exposure to the infectious microbe.

A vaccine against *Plasmodium*, the protozoan that causes malaria, has proven extremely difficult to develop. *Plasmodium* assumes a number of different forms during its complex life cycle in mosquito and mammal, and each form of the parasite was thought to have a complex array of different antigens on its surface. Antigens are the large molecules that trigger antibody production.

Researchers have concentrated their attention on the sporozoite—the form of the protozoan that the mosquito injects into the human host and that initiates infection by penetrating the liver wall. Using monoclonal (very pure) antibodies (see Box in Chapter 16) as a research tool to help isolate surface antigens, researchers have recently made the surprising discovery that the entire surface of the sporozoite is covered by a single antigen, a protein. The researchers are encouraged by the fact that the antibody itself, when injected into mice and monkeys, temporarily protects these animals against malaria.

Scientists at Genentech plan to isolate the gene that codes for the sporozoite surface antigen and, using genetic engineering techniques, insert the gene into the genes of the bacterium *Escherichia coli*. The bacterial cells should then be able to turn out vast quantities of the malarial parasite's antigenic protein. When purified and injected as a vaccine into humans, this protein should mobilize antibody production and thus produce long-lasting protection against the live parasite.

Table 11–4 Some Representative Parasitic Protozoans

Phylum	Human Pathogens	Distinguishing Features	Disease	Source of Human Infections	See Figure	Refer to Chapter
Sarcodina (amoebas)	*Entamoeba histolytica*	Pseudopods	Amoebic dysentery	Fecal contamination of drinking water	11–14	24
Mastigophora (flagellates)	*Giardia lamblia*	Two nuclei, eight flagella	Giardial dysentery	Fecal contamination of drinking water	11–15	24
	Trichomonas vaginalis	No encysting stage	Urethritis; vaginitis	Contact with vaginal/ urethral discharge	11–16	25
	Trypanosoma cruzi	Undulating membrane	Chagas' disease	Bite of *Triatoma* (kissing bug)	11–17	22
Ciliata	*Balantidium coli*	Only parasitic ciliate of humans	Balantidial dysentery	Fecal contamination of drinking water	—	24
Sporozoa	*Plasmodium*	Complex life cycles may require more than one host	Malaria	Bite of *Anopheles* mosquito	11–18	20
	Toxoplasma gondii		Toxoplas- mosis	Cats, other animals; congenital	—	22

tory, and reproductive systems. Parasitic helminths are highly specialized to live inside their hosts. When compared to their free-living relatives, the following generalizations can be made about parasitic helminths:

1. Their digestive system is reduced. They can absorb nutrients from the host's food, body fluids, and tissues.
2. Their nervous system is reduced. They do not need an extensive nervous system because they do not have to search for food or adapt to their environment. The environment within a host is fairly constant.
3. Their means of locomotion is either reduced or completely lacking. They are transferred from host to host and do not have to search for a suitable habitat on their own.
4. Their reproductive system is often more complex, producing a greater number of fertilized eggs by which to infect a suitable host.

Reproduction

Adult helminths may be **dioecious,** with male reproductive organs in one individual and female reproductive organs in another. In those species, reproduction depends on two adults of opposite sex being in the same host.

Adult helminths may also be **hermaphroditic,** where one animal has both male and female reproductive organs. Two hermaphrodites may copulate and simultaneously fertilize each other. A few hermaphrodites fertilize themselves.

Life Cycle

The life cycle of parasitic helminths may be extremely complex, involving a succession of hosts. The term **definitive host** is given to the organism that harbors the adult, sexually mature helminth. One or more **intermediate hosts** may be necessary for each **larval,** or developmental, stage of the parasite.

PLATYHELMINTHES

Members of the phylum Platyhelminthes, flatworms, are dorsoventrally flattened. They have what is called an *incomplete* digestive system. This type of digestive system has only one opening (mouth) through which food enters and wastes leave. The classes of parasitic flatworms include the trematodes and cestodes.

Trematodes

Trematodes, or flukes, have flat, leaf-shaped bodies with a ventral sucker and an oral sucker (Figure

11–19). The suckers are used to hold the organism in place and to suck fluids from the host. Flukes can also obtain food by absorption through the nonliving outer covering called the cuticle. Flukes are named according to the tissue of the definitive host in which the adults live (for example, lung fluke, liver fluke, blood fluke).

To exemplify a fluke's life cycle, let us look at the lung fluke, *Paragonimus westermanni* (Figure 11–20). The adult lung fluke lives in the bronchioles of humans and other mammals. The adults are approximately 6 mm wide by 12 mm long. The hermaphroditic adults liberate eggs into the bronchi. The eggs are usually excreted from the host in the feces, as sputum is frequently swallowed.

To continue the life cycle, the eggs must be excreted into a body of water. Inside the egg a **miracidial larva,** or **miracidium,** then develops. When

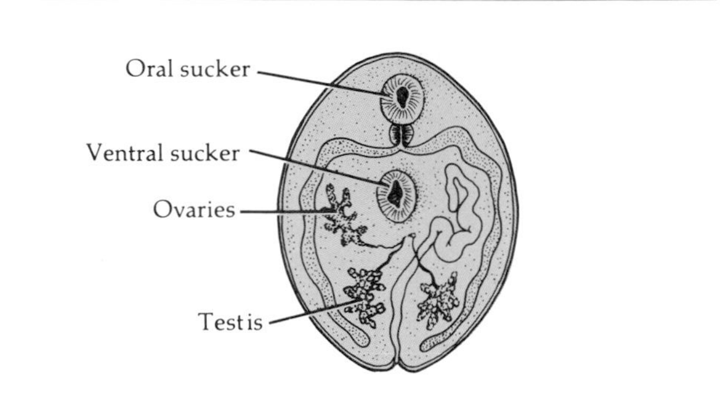

Figure 11–19 This generalized diagram of the anatomy of an adult fluke shows the oral and ventral suckers. The suckers are used for attachment to the host. The mouth is located in the center of the oral sucker. Flukes are hermaphroditic with testes and ovaries in each animal.

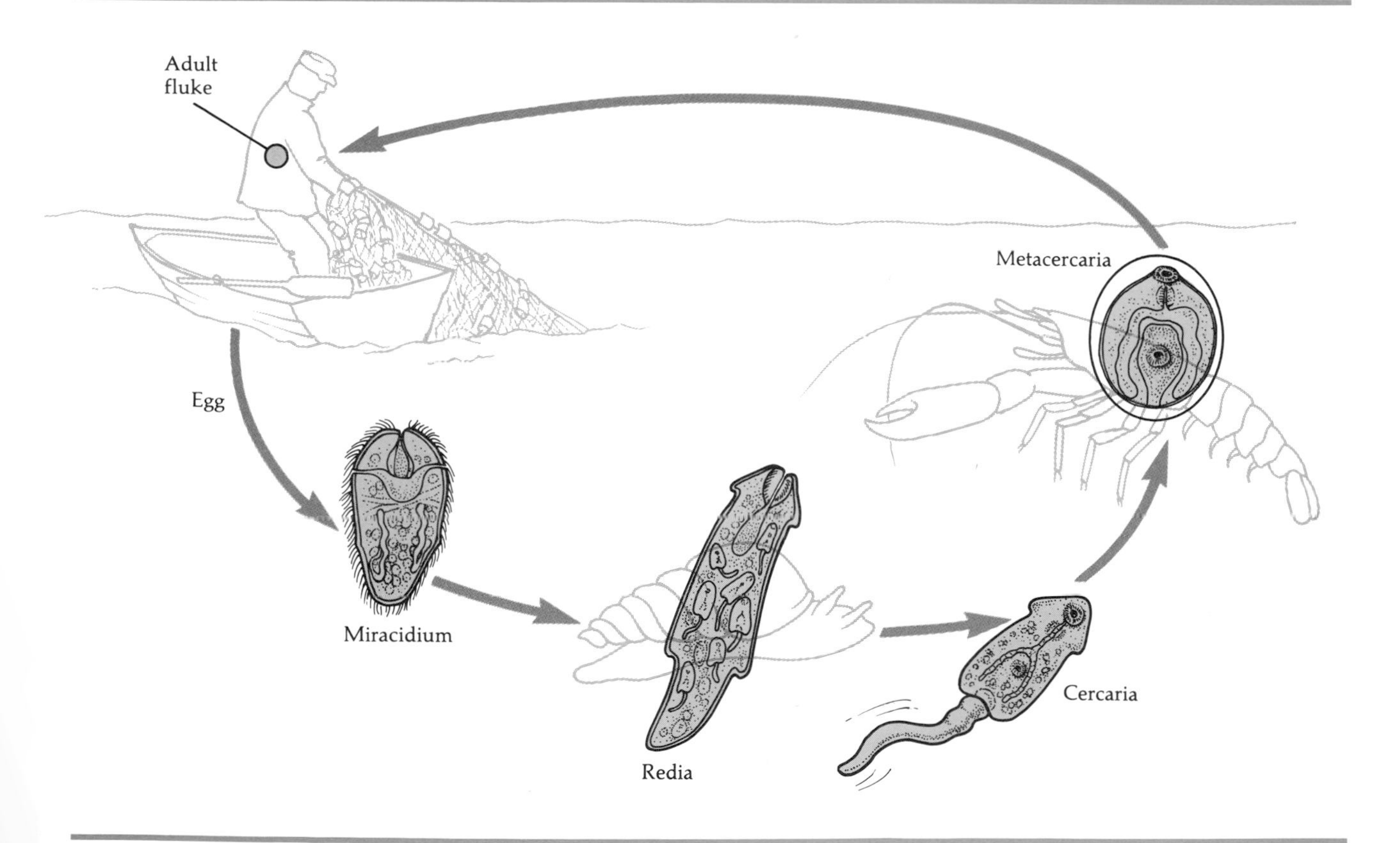

Figure 11–20 Life cycle of *Paragonimus westermanni.* The free-swimming miracidium invades the first intermediate host, a snail. Two generations of rediae develop within the snail, and the rediae give rise to cercariae. The cercariae leave the snail and penetrate the second intermediate host, a crayfish. Here they encyst as metacercariae. When a human eats raw or undercooked crayfish, the metacercariae are released to develop into adults.

the egg hatches, the larva enters a suitable snail. Only certain species of aquatic snails can be the intermediate host. Inside the snail, the lung fluke undergoes asexual reproduction, producing **rediae.** Each redia develops into a **cercaria** that bores out of the snail and penetrates the cuticle of a crayfish. The parasite encysts as a **metacercaria** in the muscles and other tissues of the crayfish. When the crayfish is eaten by a human, the metacercaria is freed in the small intestine. It bores out and wanders around until it penetrates the lungs, enters the bronchioles, and develops into an adult lung fluke.

Laboratory diagnosis is made on the basis of microscopic examination of sputum and feces for eggs. Infection results from eating undercooked crayfish, and the disease could be prevented by thoroughly cooking crayfish.

The cercariae of the blood fluke *Schistosoma* are not ingested. Instead, they burrow through the skin of the human host and enter the circulatory system. The adults are found in mesenteric and pelvic veins. The disease schistosomiasis is a major world health problem and will be discussed further in Chapter 22.

Cestodes

Cestodes, or tapeworms, are intestinal parasites. Their structure is shown in Figure 11–21. The head, or **scolex,** has suckers and may have small hooks for attaching to the intestinal mucosa of the host (see Color Plate IVF). Tapeworms do not ingest host tissues; they obtain nutrients by absorbing food from the small intestine through their cuticle. The body consists of segments called **proglottids** (see Color Plate IVG). The immature proglottids are produced by the neck region of the scolex. Each proglottid contains both male and female reproductive organs. The proglottids farthest away from the scolex are the mature ones containing fertilized eggs. Proglottids are continually being formed as long as the scolex is attached and alive.

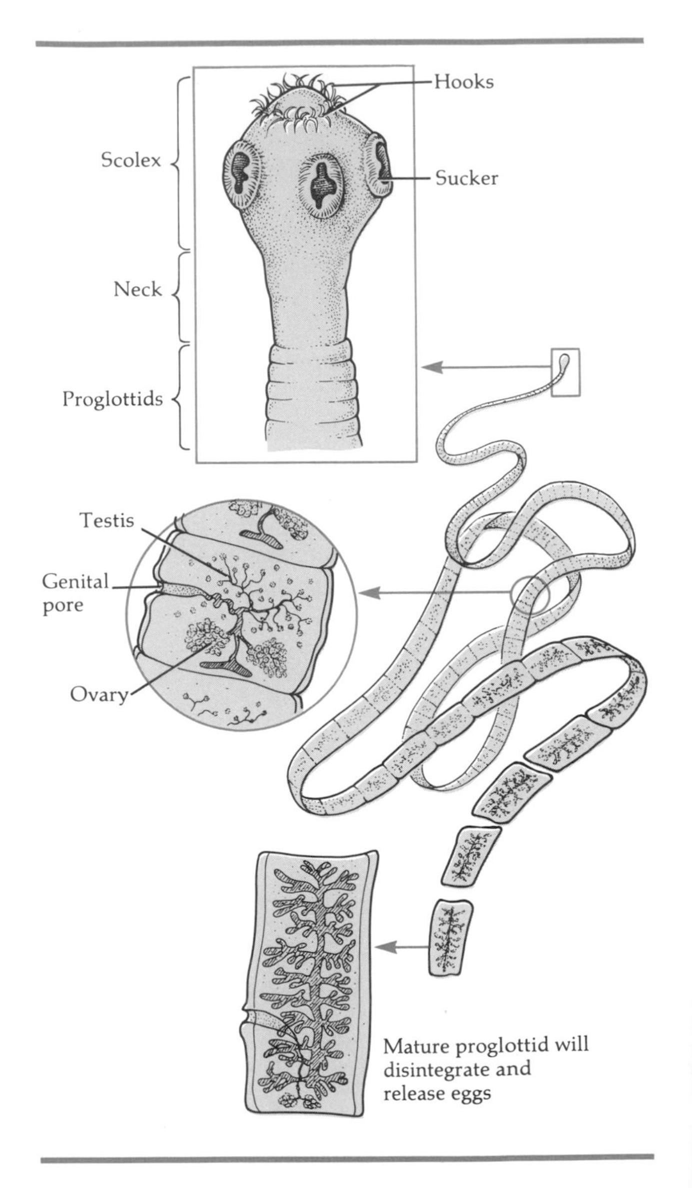

Figure 11–21 The general anatomy of an adult tapeworm. The scolex consists of suckers and hooks that attach to host tissues. The body lengthens as new proglottids form at the neck. Each proglottid contains both testes and ovaries.

Humans as definitive hosts

Taenia saginata (beef tapeworm) adults live in humans and can reach a length of 6 meters. The scolex is about 2 mm in length followed by a thousand or more proglottids. The life cycle of the beef tapeworm is illustrated in Figure 24–15. An infected human expels mature proglottids with feces, each proglottid containing thousands of eggs. The proglottids wiggle away from the fecal material, increasing their chances of being ingested by an animal that is grazing. Upon ingestion by cattle, the larvae hatch from the eggs and bore through the intestine. The larvae migrate to muscle (meat), where they encyst as **cysticerci.** When the cysticerci are ingested by humans, all but the scolex is digested. The scolex anchors itself in the small intestine and begins producing proglottids.

Diagnosis is made on the presence of mature proglottids and eggs in feces. Cysticerci can be seen macroscopically in meat, their presence is referred to as "measly beef." Beef for human consumption is inspected for "measly" appearance, and such inspection is one way to prevent infections by beef tapeworm. Another method of prevention is to avoid the use of untreated human sewage as fertilizer in grazing pastures.

Humans as intermediate hosts

Humans are the intermediate hosts for *Echinococcus granulosus* (Figure 11–22). Dogs and cats, both wild

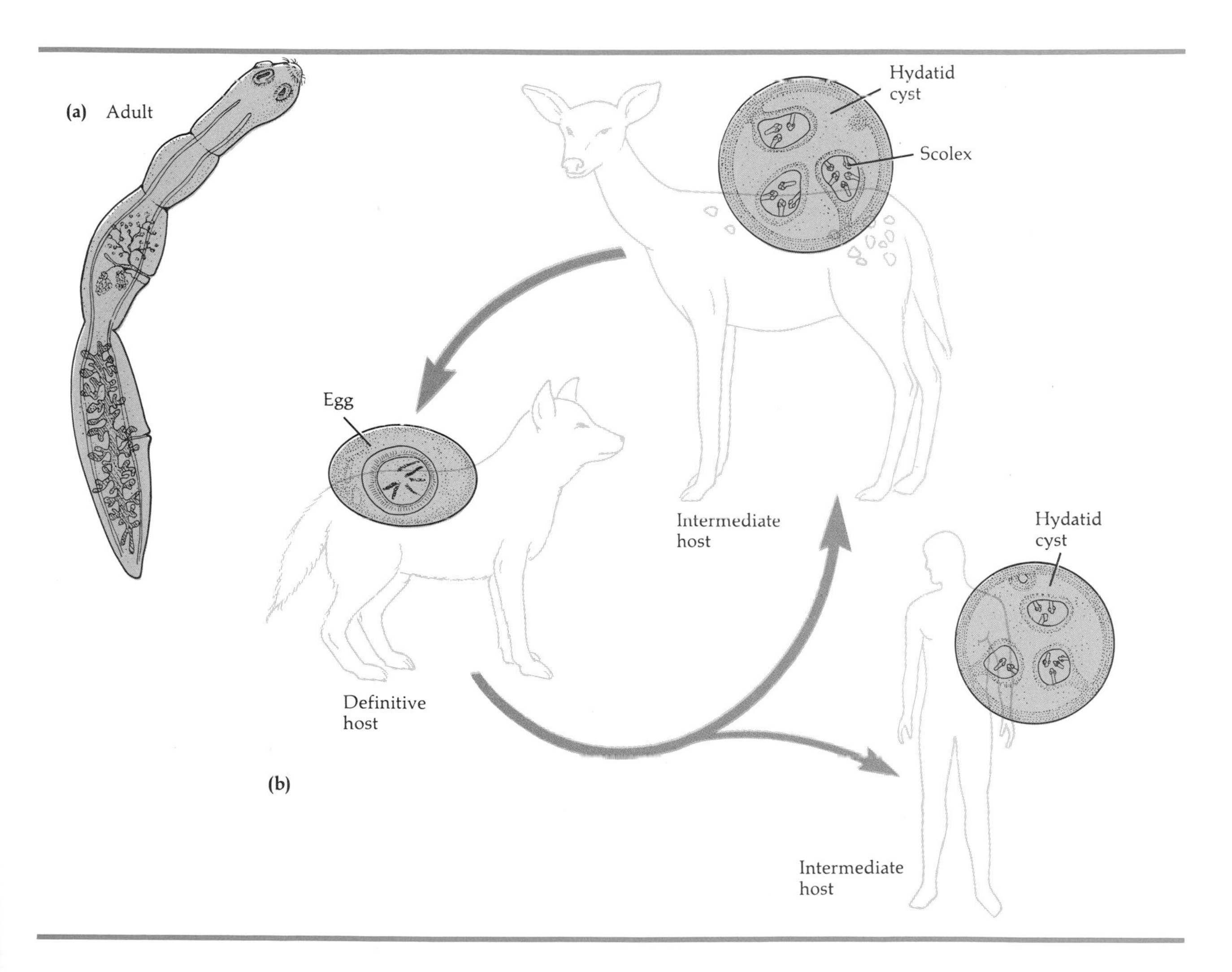

Figure 11–22 *Echinococcus granulosus.* This tiny tapeworm is found in the intestines of dogs, cats, wolves, and foxes. **(a)** Adult. **(b)** Life cycle. Eggs are excreted from the definitive host and ingested by an intermediate host, such as a deer. In the intermediate host, the eggs hatch and the larvae form hydatid cysts in host tissues. A hydatid cyst is a fluid-filled sac containing many scolexes. For the cycle to be complete, the cysts must be ingested by a definitive host eating the intermediate host. The cycle involving humans as the intermediate host is a dead end for the parasite unless the human is eaten by an animal.

and domestic, are the definitive hosts for this minute (2 to 8 mm) tapeworm. Eggs are excreted with feces and ingested by deer or humans. Humans can contaminate their hands with dog feces or can be contaminated by a dog's tongue. In the small intestine, the eggs hatch and the larvae migrate to the liver or lungs. The host forms a cyst around the larvae in these organs. This cyst, called a **hydatid cyst,** is a "brood capsule" in which thousands of scoleces may be produced. In the wild, the cysts might be in a deer, which might then be eaten by a wolf. The scoleces would be able to attach themselves in the wolf's intestine and produce proglottids.

Diagnosis of hydatid cysts is frequently only made on autopsy, although X rays can detect the cysts in lungs.

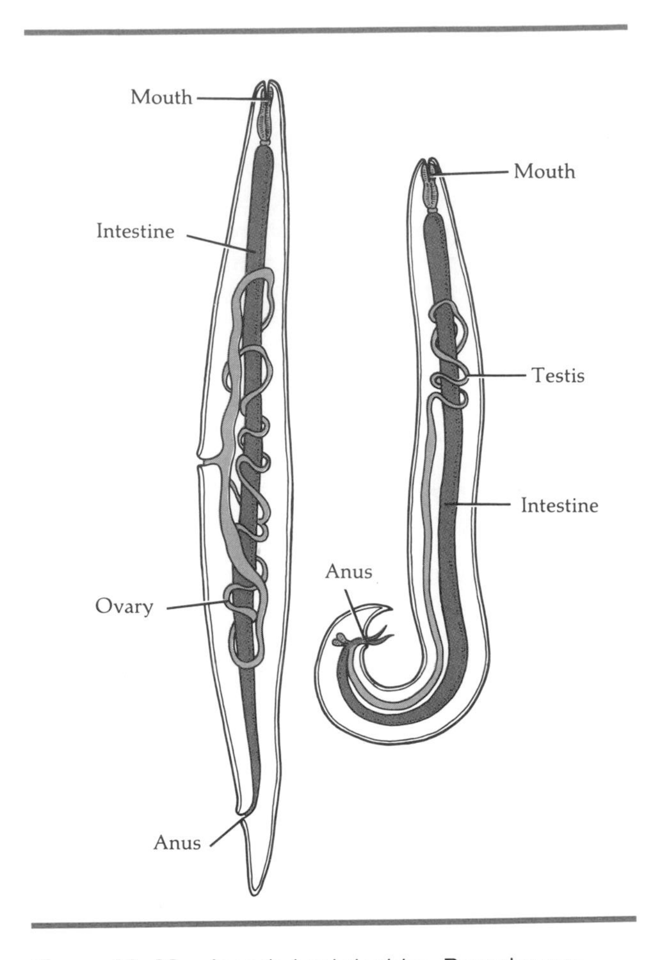

Figure 11–23 *Ascaris lumbricoides.* Roundworms have a complete digestive system with a mouth, intestine, and anus. Most roundworms are dioecious, and the female *(left)* is often distinctly larger than the male *(right).*

ASCHELMINTHES

Members of the phylum Aschelminthes, the roundworms, are cylindrical in shape and tapered at each end. Roundworms have a *complete* digestive system, consisting of a mouth, an intestine, and an anus. Most species are dioecious. Human parasites are found only in the class consisting of nematodes.

Nematodes

Some species of nematodes are free-living in soil and water, while others are parasites on plants and animals. Parasitic nematodes do not have the succession of larval stages exhibited by flatworms. In some nematodes, the eggs are produced and mature into adults in a single host.

Nematode infections can be divided into two categories: those in which the egg is infective for humans and those in which the larva is infective.

Eggs infective for humans

Ascaris lumbricoides (Figure 11–23) is a large nematode (30 cm in length). It is dioecious with **sexual dimorphism;** that is, the male and female worms look distinctly different, the male being smaller with a curled tail. Adult *Ascaris* live in the small intestine of humans and domestic animals (such as pigs and horses), where it feeds primarily on semi-digested food. Eggs, excreted with feces, can survive for long periods in the soil until accidentally ingested by another host. The eggs hatch in the intestine and mature into adults in one host.

Diagnosis is frequently made when the adult worms are excreted with feces. Prevention of the disease in humans is managed by proper sanitary habits. The life cycle of *Ascaris* in pigs can be interrupted by keeping pigs in areas free of fecal material.

The pinworm *Enterobius vermicularis* (Figure 11–24) spends its entire life in a human host. Adult pinworms are found in the large intestine. The

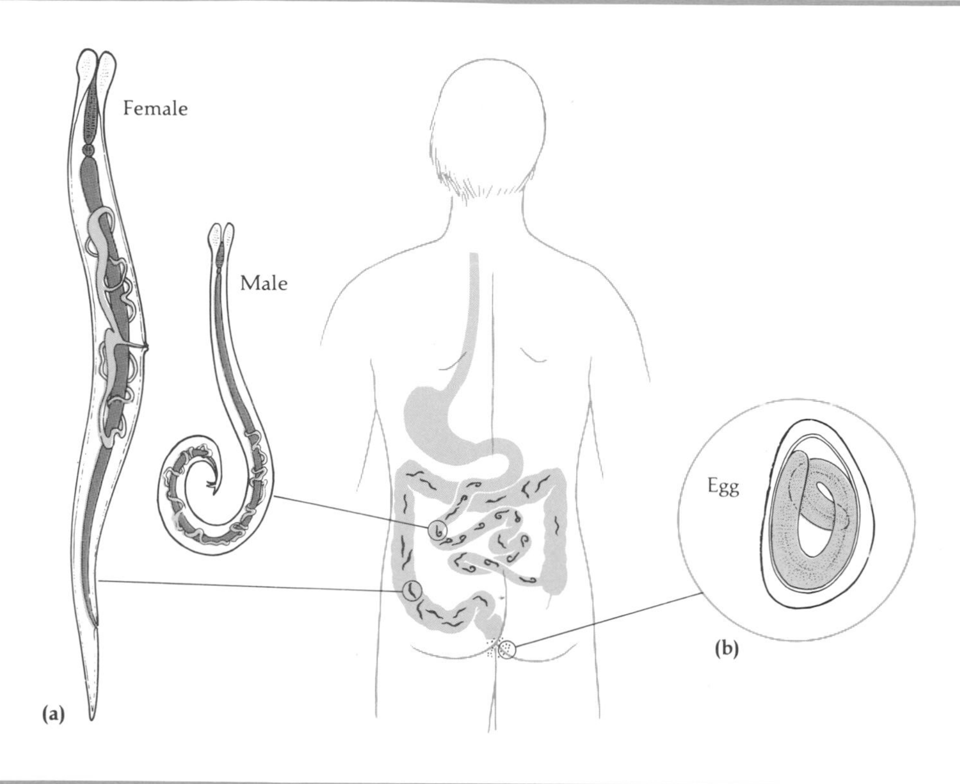

Figure 11–24 *Enterobius vermicularis.* **(a)** Adult pinworms, which live in the intestines of humans. **(b)** Eggs, which are deposited on the perianal skin at night by the female.

female pinworm migrates to the anus to deposit her eggs in the perianal skin. The eggs can be ingested by the same person or another person through contaminated clothing. Pinworm infections are diagnosed by the Graham sticky tape method. A piece of tape is placed on the perianal skin in such a way that the sticky side is exposed. The tape is microscopically examined for the presence of eggs adhering to it.

Larvae infective for humans

Adult hookworms, *Necator americanus* (Figure 11–25), live in the small intestine of humans; the eggs are excreted with feces. The larvae hatch in the soil, where they feed on bacteria. A larva enters its next host by penetrating through the skin. The larva then enters a blood or lymph vessel, is carried to the lungs, and is swallowed with sputum to finally reach the small intestine.

Diagnosis is based on the presence of eggs in feces. Hookworm infections can be prevented by wearing shoes while walking in soil.

Trichinella spiralis infections, called *trichinosis,* are usually acquired by eating encysted larvae in pork or bear meat. The larvae are freed from the cysts in the human digestive tract. They mature into adults in the intestine and sexually reproduce. Eggs develop in the female, and she gives birth to live nematodes. The larvae enter lymph and blood vessels and migrate throughout the body. They encyst in muscles and other tissues, where they remain until ingested by another host (Figure 11–26).

Diagnosis of trichinosis is made by microscopic examination for larvae in a muscle biopsy. Trichinosis can be prevented by thoroughly cooking meat prior to consumption.

Table 11–5 lists representative parasitic helminths of each phylum and class, and the diseases they cause.

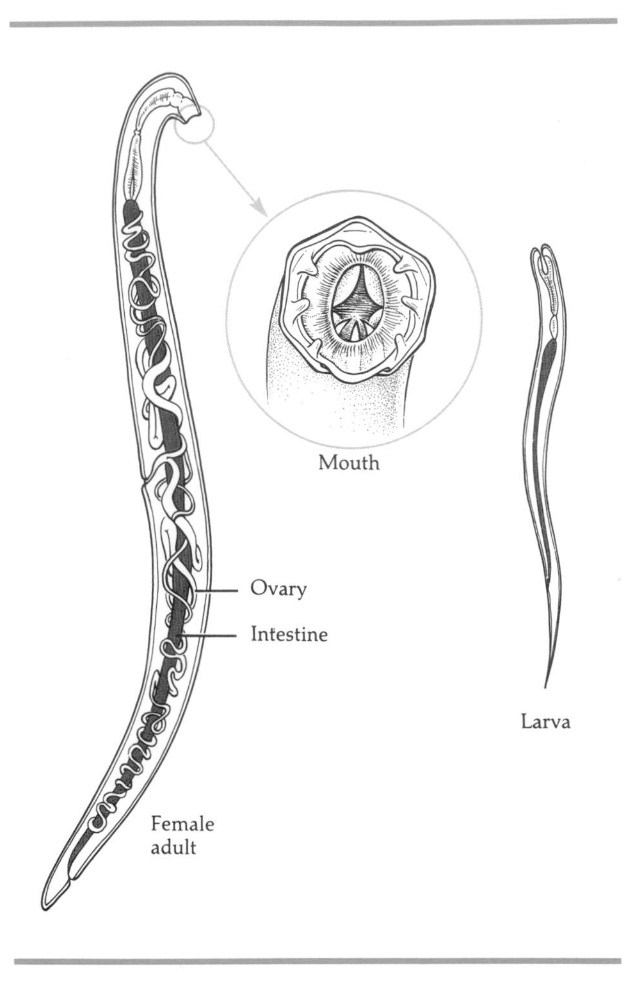

Figure 11–25 The hookworm *Necator americanus*, which is actually a roundworm. Adults are found in the intestine of humans. The hooks located around the mouth are used for attachment and feeding on the host tissue. The free-living larvae are found in the soil and penetrate the skin of the definitive host.

Table 11–5 Representative Parasitic Helminths

Phylum	Class	Human Parasites
Platyhelminthes	Trematodes	*Paragonimus westermanni*
		Schistosoma
	Cestodes	*Taenia saginata*
		Echinococcus granulosus
Aschelminthes	Nematodes	*Ascaris lumbricoides*
		Enterobius vermicularis
		Necator americanus
		Trichinella spiralis

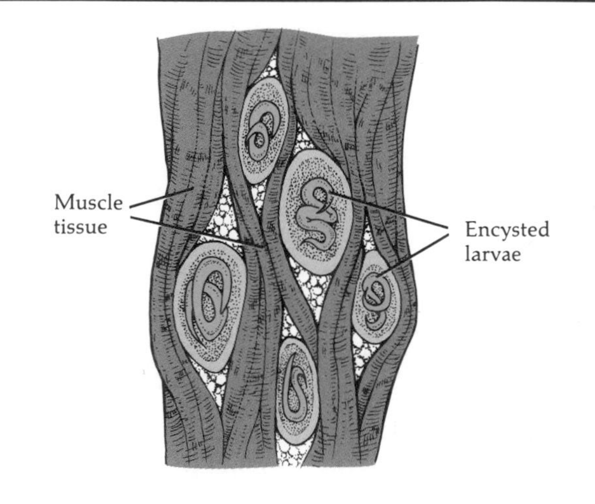

Figure 11–26 *Trichinella spiralis* encysted in muscle. The larvae of *T. spiralis* migrate from the intestines of the definitive host to tissues to encyst. The encysted larvae must be ingested by the next host.

ARTHROPODS

Arthropods are jointed-legged animals. This is the largest phylum in the animal kingdom, with nearly 1 million species. We will briefly describe arthropods here because a few suck the blood of humans and other animals, and while doing so can transmit microbial diseases. Arthropods that carry disease-causing microorganisms are called **vectors.**

Representative classes of arthropods include the following:

1. Arachnida (eight legs): spiders, mites, ticks
2. Crustacea (four antennae): crabs, crayfish
3. Chilopoda (two legs per segment): centipedes
4. Diplopoda (four legs per segment): millipedes
5. Insecta (six legs): bees, flies

Table 11–5, continued

Intermediate Host	Definitive Host	Stage Passed to Humans	Disease	Location in Humans	See Figure
Freshwater snails and crayfish	Humans, lungs	Metacercaria in crayfish	Paragonimiasis (lung fluke)	Lungs	11–20
Freshwater snails	Humans	Cercariae through skin	Schistosomiasis	Veins	—
Cattle	Humans, small intestine	Cysticercus in beef	Tapeworm	Small intestine	11–21
Humans	Dogs and other animals, intestines	Eggs from other animals	Hydatid cyst	Lungs	11–22
—	Humans, small intestine	Ingestion of eggs	Ascariasis	Small intestine	11–23
—	Humans, large intestine	Ingestion of eggs	Pinworm	Large intestine	11–24
—	Humans, small intestine	Penetration of larvae through skin	Hookworm	Small intestine	11–25
—	Humans, swine, and other mammals, small intestine	Ingestion of larvae in meat, especially pork	Trichinosis	Muscles	11–26

Table 11–6 lists those arthropods that are important as vectors, and Figures 11–27 and 11–28 illustrate some of them. These insects and ticks are only found on an animal when they are feeding. An exception to this is the louse, which spends its entire life on its host and cannot survive for long away from a host. Among insect vectors only the adult female flies (including mosquitoes) bite.

Some vectors are just a mechanical means of transport for a pathogen. For example, houseflies are attracted to decaying organic matter in which to lay their eggs. A housefly may pick up a pathogen on its feet or body from feces and transport the pathogen to food.

Some parasites multiply in their vectors. When this happens, the parasites may accumulate in the vector's feces. Large numbers of parasites may be deposited on the skin of the host while the vector is feeding there.

Plasmodium is an example of a parasite that requires that its vector also be its host. *Plasmodium* can sexually reproduce in the gut of an *Anopheles* mosquito only. *Plasmodium* is introduced into a human host with the mosquito's saliva. The mosquito's saliva serves as an anticoagulant to keep blood flowing.

Elimination of vector-borne diseases is concentrated in efforts to eradicate the vectors.

Table 11–6 Important Arthropod Vectors of Human Diseases

Class	Order	Vector	See Figure	Disease	Refer to Chapter
Arachnida	Mites and ticks	*Dermacentor* (tick)	11–28a	Rocky Mountain spotted fever	20
Insecta	Sucking lice	*Pedicularis* (human louse)	11–28b	Epidemic typhus	20
	Fleas	*Xenopsylla* (rat flea)	11–28c	Endemic murine typhus, plague	20
	True flies	*Chrysops* (deer fly)	11–28d	Tularemia	22
		Aedes (mosquito)	11–27a	Dengue fever, yellow fever	20
		Anopheles (mosquito)	11–27b	Malaria	20
		Culex (mosquito)	—	Arboviral encephalitis	20
	True bugs	*Triatoma* (kissing bug)	11–28e	Chagas' disease	22

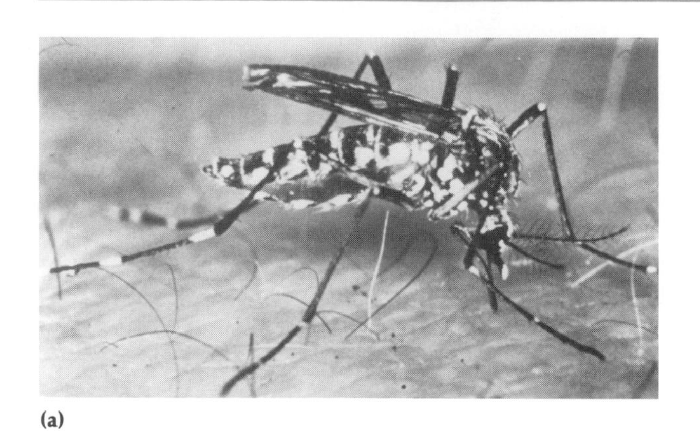

(a)

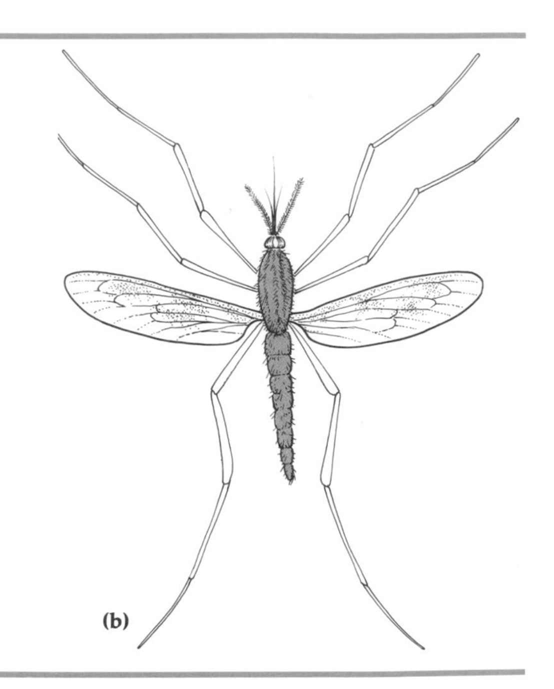

Figure 11–27 Mosquitoes. **(a)** *Aedes aegypti* (female), sucking blood from human skin. This insect vector transmits several diseases from person to person, including the viral diseases yellow fever and dengue fever. **(b)** *Anopheles*.

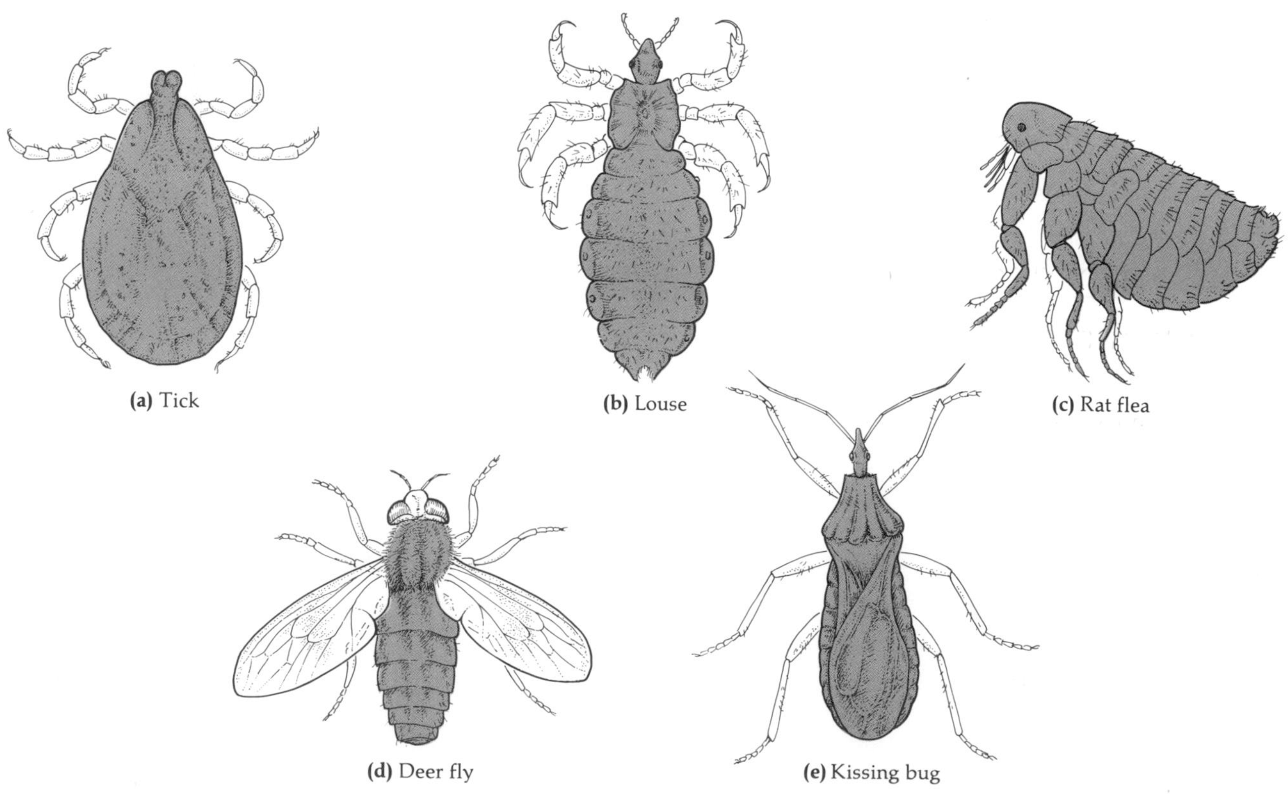

Figure 11–28 Arthropod vectors. **(a)** The tick, *Dermacentor.* **(b)** The human louse, *Pedicularis.* **(c)** The rat flea, *Xenopsylla.* **(d)** The deer fly, *Chrysops.* **(e)** The kissing bug, *Triatoma.* See Table 11–6.

STUDY OUTLINE

FUNGI (pp. 279–290)

1. The kingdom Fungi includes yeasts, molds, and fleshy fungi (mushrooms).
2. Fungi are aerobic or facultatively anaerobic heterotrophs.
3. Most fungi are decomposers, and a few are parasites of plants and animals.

VEGETATIVE STRUCTURES (pp. 280–282)

1. A colony of molds and fleshy fungi consists of filaments of cells called hyphae; many hyphae are called a mycelium.
2. Yeasts are unicellular fungi that grow by budding and occasionally by fission.
3. Buds that do not separate form pseudohyphae.
4. Dimorphic fungi are yeastlike at 37°C and moldlike at 25°C.

REPRODUCTIVE STRUCTURES (pp. 282–284)

1. The following spores may be produced asexually: arthrospores, chlamydospores, sporangiospores, conidiospores, and blastospores.
2. Fungi are classified on the basis of the type of sexual spore that is formed.
3. Sexual spores include zygospores, ascospores, and basidiospores.

MEDICALLY IMPORTANT PHYLA OF FUNGI (pp. 284–286)

1. The Zygomycota have coenocytic hyphae and produce sporangiospores and zygospores.
2. The Ascomycota have septate hyphae and produce ascospores and frequently conidiospores.
3. Basidiomycetes produce basidiospores and may produce conidiospores.
4. Deuteromycota reproduce asexually by one or more of the following: chlamydospores, arthrospores, conidiospores, or budding.
5. Deuteromycota are called the Fungi Imperfecti because sexual spores have not been seen as yet.

FUNGAL DISEASES (pp. 286–287)

1. Systemic mycoses are fungal infections deep within the body, affecting many tissues and organs.
2. Subcutaneous mycoses are fungal infections beneath the skin.

3. Cutaneous mycoses affect keratin-containing tissues such as hair, nails, and skin.
4. Superficial mycoses are localized on hair shafts and superficial skin cells.
5. Opportunistic mycoses are caused by normal flora or fungi that are not normally considered pathogenic.
6. Opportunistic mycoses include mucormycosis, caused by some Zygomycetes; aspergillosis, caused by *Aspergillus;* and candidiasis, caused by *Candida*.
7. Opportunistic mycoses may infect any tissues. However, they are usually systemic.

ECONOMIC DISADVANTAGES OF FUNGI (pp. 287–290)

1. Mold spoilage of fruits, grains, and vegetables is more common than bacterial spoilage of these products.
2. *Claviceps purpurea* and *Aspergillus flavus* grow on grains and can produce toxins that may affect humans.
3. Mushrooms in the genus *Amanita* produce neurotoxins that may be fatal to humans.

PROTOZOANS (pp. 291–295)

1. Protozoans are unicellular eucaryotes in the kingdom Protista.
2. Protozoans may be found in soil, water, and as normal flora in animals.
3. Protozoans are classified by their means of locomotion: members of the phylum Sarcodina move by amoeboid motion; the Mastigophora (flagellates) use flagella for motility; the Ciliata possess cilia; and the Sporozoa lack a means of locomotion and are obligate parasites.

Sarcodina
Mastigophora
Ciliata
Sporozoa

PROTOZOAN BIOLOGY (pp. 291–292)

1. Most protozoans are heterotrophs and feed on bacteria and particulate organic material.
2. Protozoans have complex cells that may include a pellicle, a cytostome, and an anal pore.
3. Asexual reproduction is by fission, budding, or schizogony.
4. Sexual reproduction is by conjugation.
5. Protozoan conjugation involves the fusion of two haploid nuclei to produce a zygote.
6. Some protozoans can produce a cyst for protection during adverse environmental conditions.

MEDICALLY IMPORTANT PHYLA OF PROTOZOANS (pp. 292–295)

1. Parasitic sarcodinae that cause amoebic dysentery are found in the genus *Entamoeba*.
2. Parasitic flagellates (Mastigophora) include the folllowing: *Giardia lamblia*, causing an intestinal infection called giardiasis; *Trichomonas vaginalis*, causing genitourinary infections that may be transmitted by coitus; and *Trypanosoma cruzi*, which is found in the blood of humans with Chagas' disease.
3. The only ciliate that is a parasite of humans is *Balantidium coli*, the cause of one form of dysentery.
4. *Plasmodium* is the sporozoan that causes malaria.
5. Asexual reproduction of *Plasmodium* occurs in red blood cells and the liver of humans.
6. Sexual reproduction of *Plasmodium* takes place in the intestine of the female *Anopheles* mosquito.
7. *Toxoplasma gondii* is a sporozoan that causes infections in humans and may be transmitted to a fetus in utero.

HELMINTHS (pp. 295–302)

1. Some parasitic helminths (worms) belong to the phylum Platyhelminthes.
2. Parasitic roundworms belong to the phylum Aschelminthes.

HELMINTH BIOLOGY (pp. 295–296)

1. Helminths are multicellular animals, a few of which are parasites of humans.
2. The anatomy and life cycles of parasitic helminths are modified for parasitism.
3. Helminths may be hermaphroditic or dioecious.
4. The adult stage of a parasitic helminth is found in the definitive host.
5. Larval stages of parasitic helminths are found in one or more intermediate hosts.

PLATYHELMINTHES (pp. 296–300)

1. Flatworms are dorsoventrally flattened animals without a complete digestive system.
2. Adult trematodes or flukes have an oral and ventral sucker to attach to and feed on host tissue.

3. Eggs of trematodes hatch into free-swimming miracidia that enter the first intermediate host; two generations of rediae develop in the first intermediate host; the rediae become cercariae that bore out of the first intermediate host and penetrate the second intermediate host; cercariae encyst as metacercariae in the second intermediate host; after ingestion by the definitive host, the metacercariae develop into adults.
4. A cestode, or tapeworm, consists of a scolex (head) and proglottids.
5. Humans serve as the definitive host for the beef tapeworm and cattle are the intermediate host.
6. Humans serve as the intermediate host for *Echinococcus granulosus;* the definitive hosts are dogs, wolves, and foxes.

ASCHELMINTHES (pp. 300–302)

1. Members of the phylum Aschelminthes are roundworms with a complete digestive system.
2. Parasitic members of Aschelminthes are in the class consisting of nematodes.
3. Members of Aschelminthes in which eggs are infective for humans are *Ascaris lumbricoides* and *Enterobius vermicularis* (pinworm).
4. Members of Aschelminthes in which the larvae are infective for humans are *Necator americanus* (hookworm) and *Trichinella spiralis.*

ARTHROPODS (pp. 302–304)

1. Jointed-legged animals including ticks and insects belong to the phylum Arthropoda.
2. Arthropods that carry diseases are called vectors.
3. Elimination of vector-borne diseases is best done by controlling or eradicating the vectors.

STUDY QUESTIONS

REVIEW

1. Contrast the mechanism of conidiospore and ascospore formation by *Penicillium.*
2. Fill in the following table.

Phylum	Spore Type(s)	
	Sexual	Asexual
Zygomycota		
Ascomycota		
Basidiomycota		
Deuteromycota		

3. Fungi are classified into phyla on the basis of ____________.
4. The following is a list of fungi, their method of entry into the body, and sites of infections they cause. Categorize each type of mycosis (systemic, subcutaneous, cutaneous, superficial).

Genus	Method of Entry	Site of Infection	Mycosis
Blastomyces	Inhalation	Lungs	
Sporotrix	Puncture	Ulcerative lesions	
Microsporum	Contact	Fingernails	
Trichosporon	Contact	Hair shafts	

5. Why is it significant that *Trichomonas* does not have a cyst stage? Name a protozoan parasite that does have a cyst stage.
6. Protozoans are classified on the basis of ____________.
7. Recall the life cycle of *Plasmodium*. Where does asexual reproduction occur? Where does sexual reproduction occur? Identify the definitive host. Identify the vector.
8. Transmission of helminthic parasites to humans is usually by ________.
9. List two characteristics of the beef tapeworm that put it in the phylum Platyhelminthes.
10. To what class does this animal belong? Name the body parts. What is the name of the encysted larva of this animal?

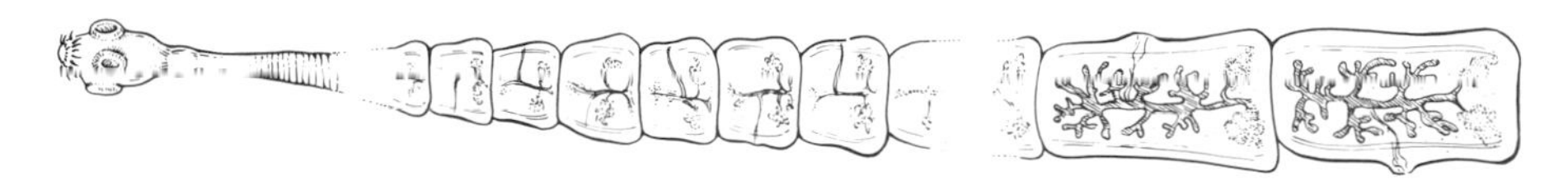

11. Most nematodes are dioecious. What does this mean? To what phylum do nematodes belong?
12. Vectors can be divided into three major types based on the role they play for the parasite. List the three types of vectors.

CHALLENGE

1. A generalized life cycle of the liver fluke *(Clonorchis sinensis)* is shown on the next page. Identify the intermediate host(s). Identify the definitive host(s). To what phylum and class does this animal belong?

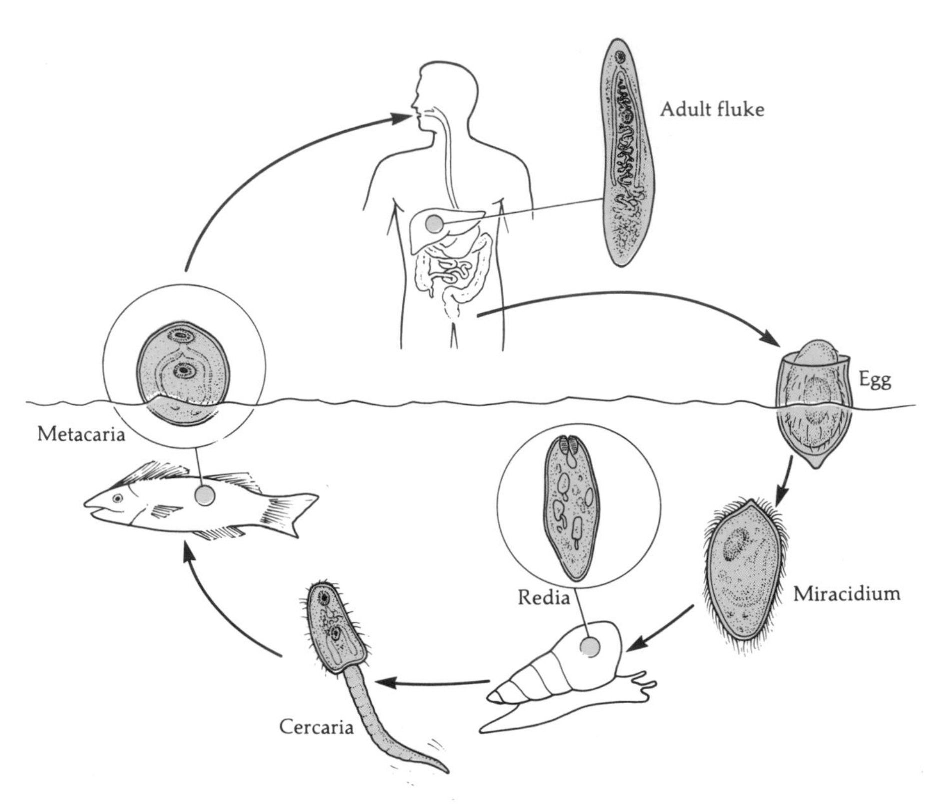

2. *T. gambiense* is the causative agent of African sleeping sickness. To what phylum and class does it belong? (Look at part (a).) Part (b) shows a generalized life cycle for *T. gambiense*. Identify the host and vector of this parasite.

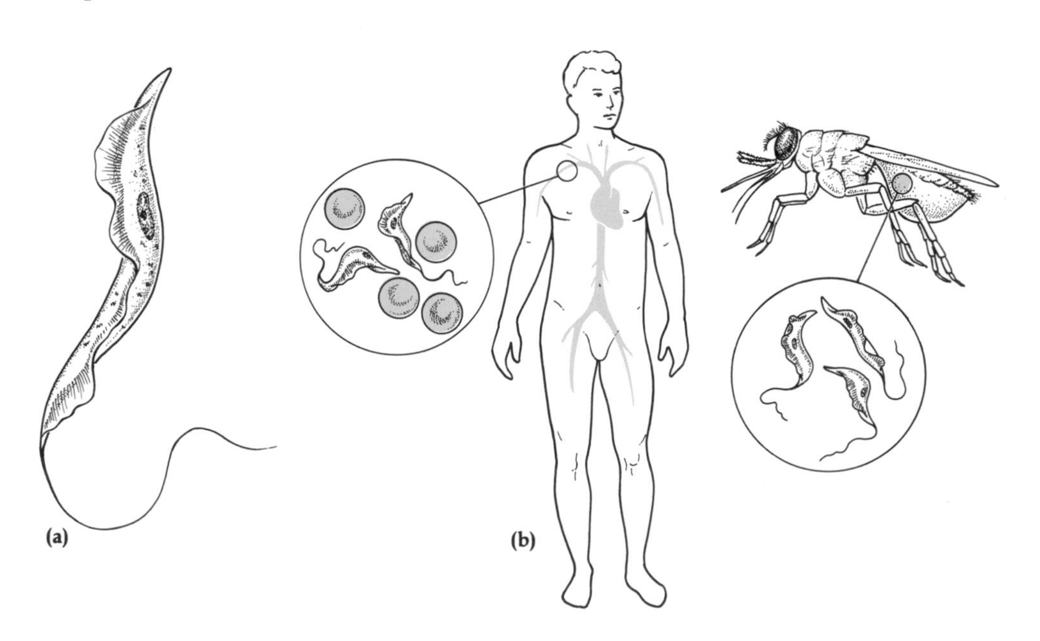

Further Reading

Alexopoulos, C. J. and C. W. Mims. 1975. *Introductory Mycology,* 3rd ed. New York: Wiley. The authoritative reference on the biology of fungi.

Beck, J. W. and J. E. Davies. 1976. *Medical Parasitology.* St. Louis, MO: C. V. Mosby. A concise textbook on protozoans and helminths that cause diseases in humans.

Board of Education and Training. 1978. *Identification of Saprophytic Fungi Commonly Encountered in a Clinical Environment.* Washington, D.C.: American Society for Microbiology. Lectures on macroscopic and microscopic identification of saprophytic fungi.

Chandler, A. C. and C. P. Read. 1961. *Introduction to Parasitology,* 10th ed. The authors' anecdotes add special interest to this textbook on the biology and epidemiology of parasitic protozoans and helminths.

Emmons, C. W., C. H. Binford, J. P. Utz, and K. J. Kwon-Chung. 1977. *Medical Mycology,* 3rd ed. Philadelphia, PA: Lea and Febiger. Discussions of fungal diseases that affect humans.

Faust, E. C., P. C. Beaver, and R. C. Jung. 1975. *Animal Agents and Vectors of Human Disease,* 4th ed. Phildelphia, PA: Lea and Febiger. Includes a discussion of arthropod vectors and culture techniques for protozoans and helminths.

Rose, A. H. and J. S. Harrison, eds. 1969. *The Yeasts.* New York: Academic Press. A good reference on the biology of yeasts as well as their occurrence in nature and as pathogens.

12

Viruses

Objectives

After completing this chapter you should be able to

- Define the following terms: virus; viroid; transformed cell.
- Describe the chemical composition of a typical virus.
- Classify viruses on the basis of morphology and list other criteria used to classify viruses.
- Explain how viruses are cultured.
- Describe the lytic cycle of T-even bacteriophages.
- Compare and contrast the multiplication cycle of DNA- and RNA-containing animal viruses.
- Differentiate between slow viral infections and latent viral infections.
- List the effects of animal viral infections on host cells.
- Explain what a tumor is and distinguish between malignant and benign tumors.
- Discuss the relationship of DNA- and RNA-containing viruses to cancer.

Up until the latter part of the nineteenth century, the term **filterable virus** was used to designate all infectious agents capable of passing through filters that retained all known bacteria, fungi, and protozoans. Shortly thereafter, the term *filterable* was dropped and the word *virus* (meaning poison or venom) was specifically used to refer to submicroscopic, filterable, infectious agents, that is, infectious agents too small to be viewed with a light microscope. During the first decades of the twentieth century, most scientists believed that viruses were a distinct group of infectious agents differing from others only in size. However, it was soon discovered that viruses had their own method of reproduction and a distinctive chemical makeup. With the advent of the electron microscope and advanced analytical procedures, exciting strides were made regarding the structural and functional characteristics of viruses.

Viruses are of particular interest to microbiologists for several reasons. As you will see, viruses differ fundamentally from the microorganisms discussed in previous chapters, in both structure and life cycle. With so many drugs available to combat bacterial infections, but so few antiviral drugs, viruses have become the most threatening agents of infectious disease in developed countries such as the United States. In addition, viruses may be related to certain types of cancer in humans.

GENERAL CHARACTERISTICS OF VIRUSES

As our introduction to viruses, let us first consider their definition, host range, and size.

Definition

Before we attempt to define a virus, we should consider whether or not viruses are living organisms. One definition of life is that it is a complex set of processes resulting from the action of instructions specified by nucleic acids. The nucleic acids of living cells are in action all the time. Since viruses are inert outside of living host cells, in this sense they are not considered living organisms. However, once viruses enter a host cell, viral nucleic acids become active, resulting in viral multiplication. In this sense, viruses are alive when they multiply in host cells they infect. From a clinical point of view, viruses may also be considered alive, because they cause infection and disease just like pathogenic bacteria, fungi, and protozoans do. Depending on one's viewpoint, a virus may be regarded as an exceptionally complex aggregation of nonliving chemicals or an exceptionally simple living microorganism.

How, then, do we define a virus? Viruses were originally distinguished from other infectious agents because they are especially small (filterable viruses) and because they are **obligatory intracellular parasites**—that is, they absolutely require living host cells in order to multiply. However, both of these properties are also shared by certain small bacteria, such as some rickettsias. The truly distinctive features of viruses are now known to lie in their simple structural organization and composition and their mechanism of multiplication. Accordingly, **viruses** are entities that (1) contain a single type of nucleic acid, either DNA or RNA; (2) contain a protein coat (sometimes enclosed by an envelope composed of lipids, proteins, and carbohydrates) that surrounds the nucleic acid; (3) multiply inside living cells using the synthetic machinery of the cell; and (4) cause the synthesis of specialized elements that can transfer the viral nucleic acid to other cells.

Since viruses have few or no enzymes of their own for metabolism—such as those for synthesizing proteins and nucleic acids and generating ATP—they must take over the metabolic machinery of the host cell in order to multiply. This fact has considerable medical significance for the development of antiviral drugs, because most drugs that would interfere with viral multiplication would also interfere with host cell function and hence are too toxic for clinical use. However, the presence of lipids in the coverings of some viruses makes these viruses sensitive to disinfection or damage by lipid solvents, such as ether, and emulsifying agents, such as bile salts and detergents.

When we describe the structure and multiplication of viruses in more detail, reference will be made to the term virion. A **virion** is a complete, fully developed viral particle composed of nucleic acid surrounded by a coat that protects it from the environment and serves as a vehicle of transmission from one host cell to another.

Host Range

The **host range** of a virus refers to the spectrum of host cells the virus can infect. Viruses multiply only in cells of particular species and thus are divided into three main classes: **animal viruses, bacterial viruses (bacteriophages),** and **plant viruses.** (Protists and fungi may also be hosts for viruses, but they will not concern us here.) Within each class, each virus is usually able to infect only certain species of cells.

The particular host range of a virus is determined by the requirements for the specific attachment of the virus to the host cell and the availability of host cellular factors required for viral multiplication. In order for the virus to infect the host cell, the outer surface of the virus must interact with specific receptor sites on the surface of the cell. The interaction between the virus and receptor site is a chemical one, in which the two complementary components are held together by weak bonds, such as hydrogen bonds. For some bacteriophages, the receptor site is a chemical constituent of the cell wall of the host; in other cases, it is a constituent of pili or flagella. For animal viruses, the receptor sites are on the plasma membranes of the host cells.

Size

The size of viruses was first estimated by filtration through membranes of known pore diameter. Viral sizes are also determined today by ultracentrifugation and electron microscopy. This last technique seems to offer the most accurate results. Although viruses vary considerably in size, most are quite a bit smaller than bacteria, although some of the larger viruses (such as the smallpox virus) are about the same size as some very small bacteria (such as the mycoplasmas, rickettsias, and chlamydias).

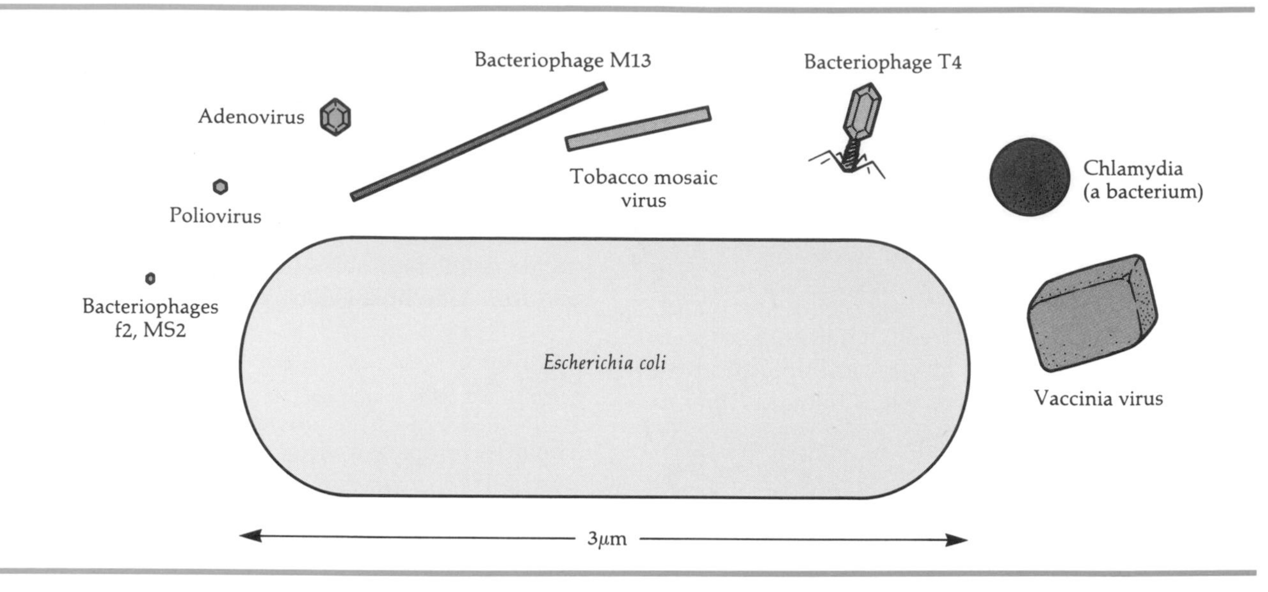

Figure 12–1 Comparative sizes of several viruses and bacteria.

Viruses fall into the range of 20 to 300 nm in diameter. The comparative sizes of several viruses and bacteria are shown in Figure 12–1.

VIRAL STRUCTURE

Nucleic Acid

As noted earlier, a virus contains a core of a single kind of nucleic acid, either DNA or RNA, which is the genetic material (Figure 12–2). The proportion of nucleic acid in relation to protein ranges from about 1% for the influenza virus to about 50% for certain bacteriophages. The total amount of nucleic acid varies from a few thousand nucleotides (or pairs) to as many as 250,000 nucleotides.

In contrast to procaryotic and eucaryotic cells, which always have DNA as the primary genetic material (with RNA playing an auxiliary role), viruses may have either nucleic acid, but never both in the same virus. Also, the nucleic acid may be single- or double-stranded. Thus, there are viruses with the familiar double-stranded DNA, with single-stranded DNA, with double-stranded RNA, and with single-stranded RNA. Depending on the virus, the nucleic acid may be linear or circular, and in some viruses (such as the influenza virus), the nucleic acid is in several separate molecules.

Capsid and Envelope

The nucleic acid of a virus is surrounded by a protein coat called the **capsid** (Figure 12–2). The capsid, whose architecture is ultimately determined by the viral nucleic acid, accounts for most of the mass of a virus, especially of small ones. Each capsid is composed of protein subunits referred to as **capsomeres.** In some viruses, the proteins composing the capsomeres are of a single type; in other viruses, several types of protein may be present. Individual capsomeres are often visible in electron micrographs (see Figure 12–4b, for example). The capsomeres can be arranged in several characteristic ways, as we shall see.

In some viruses, the capsid is covered by an **envelope** (Figure 12–2b), which usually consists of some combination of lipids, proteins, and carbohydrates. The molecular organization of these envelopes is generally unknown. Some animal viruses are released from the host cell by an extrusion process that coats the virus with a layer of the plasma membrane of the host cell, which becomes the envelope. In many cases the envelope contains proteins determined by viral nucleic acid as well as materials derived from the normal cell components. Depending on the virus, envelopes may or may not be covered by **spikes,** which are carbohydrate–protein complexes that project from the

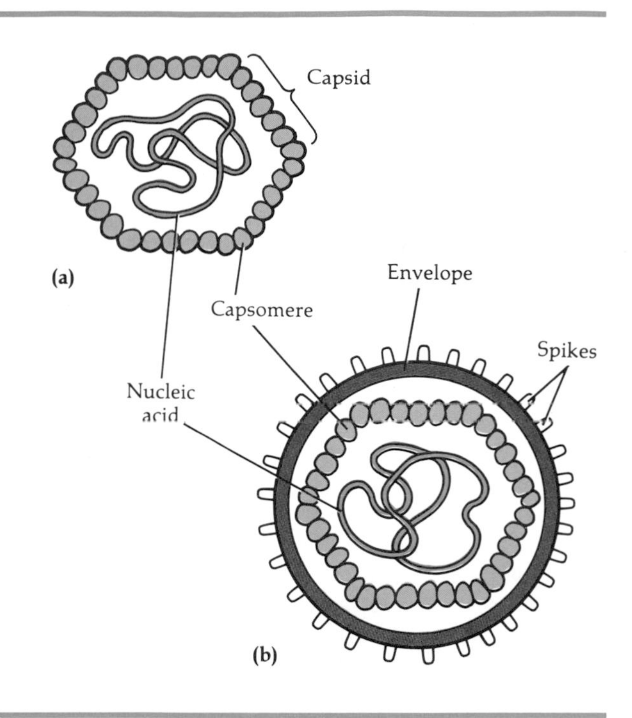

Figure 12–2 General structure of two types of viruses. **(a)** Naked virus. **(b)** Enveloped virus with spikes.

surface of the envelope. Spikes are so characteristic of some viruses that they may be used as a means of identification. The ability of certain viruses, such as the influenza virus, to clump red blood cells is associated with spikes. Such viruses bind to red blood cells and form bridges between them. The resulting clumping is called **hemagglutination** and is the basis for several useful laboratory tests.

Viruses whose capsid is not covered by an envelope are known as **naked viruses** (Figure 12–2a). In naked viruses, it is the capsid that protects the nucleic acid from nuclease enzymes in biological fluids and that promotes attachment to susceptible host cells.

General Morphology

Based on the architecture of capsids as revealed by electron microscopy, viruses may be classified into several morphological types.

Helical Viruses Helical viruses resemble long rods that may be rigid or flexible. Their capsid is a hollow cylinder with a helical structure, which surrounds the nucleic acid. An example of a helical virus that is a rigid rod is the tobacco mosaic virus (Figure 12–3). Another is bacteriophage M13.

Polyhedral Viruses Many animal, plant, and bacterial viruses are polyhedral viruses; that is, they are many-sided. In most polyhedral viruses, the capsid is in the shape of an *icosahedron*, a regular polyhedron with 20 triangular faces and 12 corners (Figure 12–4a). The icosahedrons form an equilateral triangle. An example of a polyhedral virus in the shape of an icosahedron is the adenovirus (Figure 12–4b). Another polyhedral virus is the poliovirus.

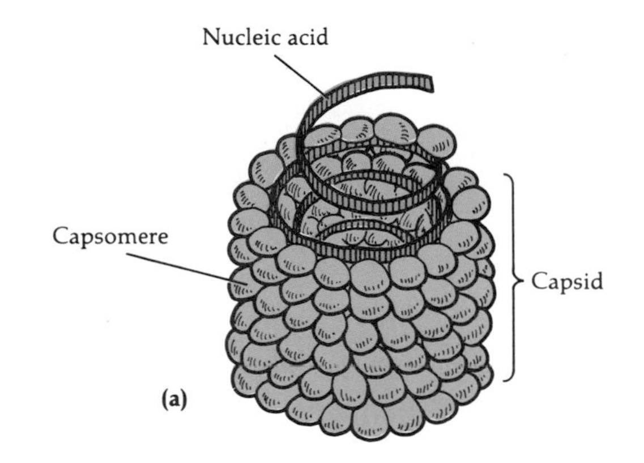

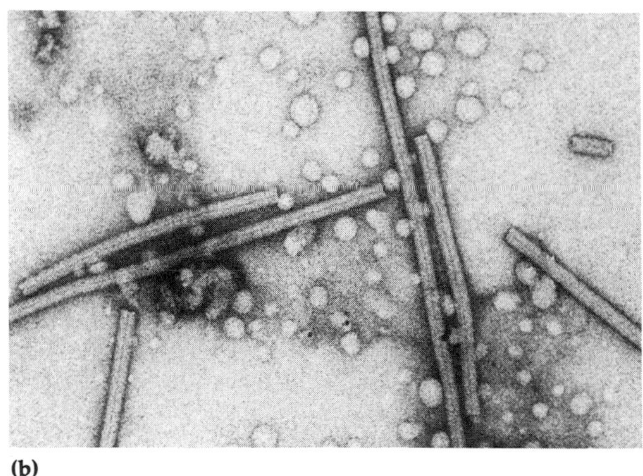

Figure 12–3 Morphology of a helical virus. **(a)** Diagram of a portion of a tobacco mosaic virus. Several rows of capsomeres have been removed to reveal the nucleic acid. **(b)** Electron micrograph of a tobacco mosaic virus showing helical rods (× 147,000).

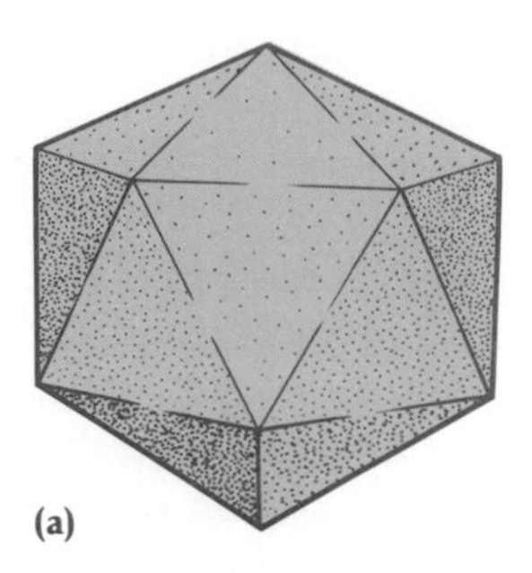

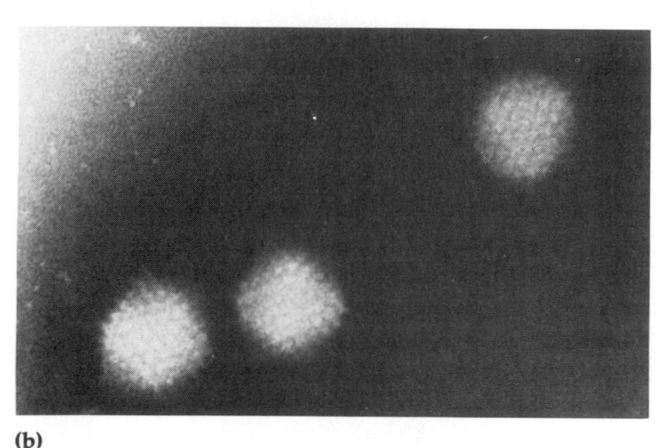

Figure 12–4 Morphology of a polyhedral virus in the shape of an icosahedron. **(a)** Diagram of an icosahedron. **(b)** Electron micrograph of several icosahedral particles of adenovirus ($\times$267,000). Individual capsomeres in the protein coat are visible.

Enveloped Viruses As noted earlier, the capsid of some viruses is covered by an envelope, and these viruses are referred to as enveloped viruses. Enveloped viruses are roughly spherical but highly pleomorphic (variable in shape) because the envelope is not rigid. When helical or polyhedral viruses are enclosed by envelopes, they are referred to as **enveloped helical** and **enveloped polyhedral** viruses. An example of an enveloped helical virus is the influenza virus (Figure 12–5). An example of an enveloped polyhedral (icosahedral) virus is the herpes simplex virus (Figure 12–6).

Complex Viruses Some viruses have a very complicated structure, especially bacterial viruses, and are referred to as complex viruses. Examples of complex viruses are poxviruses, which do not contain clearly identifiable capsids, but have several coats around the nucleic acid (Figure 12–7a), and certain bacteriophages that have a capsid to which additional structures are attached (Figure 12–7b and c). If you take a close look at the bacteriophage shown in Figure 12–7b and c, you will note that the capsid (head) is polyhedral and the tail is helical. The head contains the nucleic acid. Later in the chapter we will discuss the functions of the additional structures such as the tail sheath, tail fibers, plate, and pin.

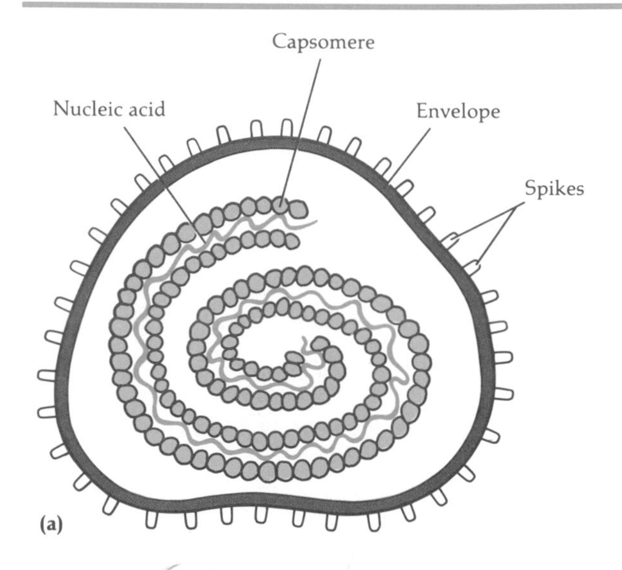

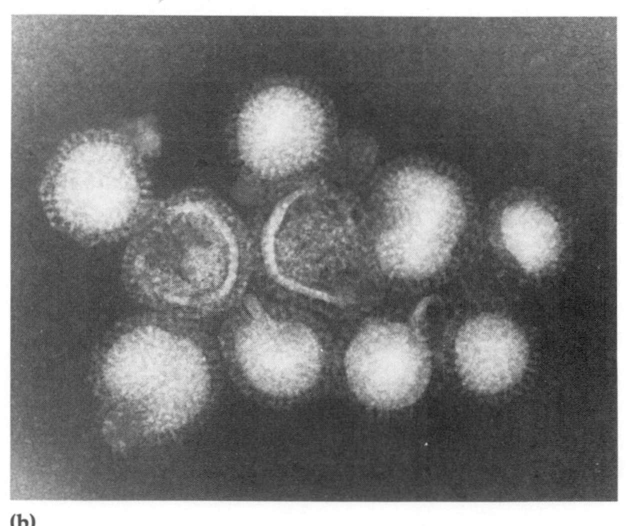

Figure 12–5 Morphology of an enveloped helical virus. **(a)** Diagram of an enveloped helical virus. **(b)** Electron micrograph of influenza viruses ($\times$ 119,000). Note the halo of spikes projecting from the outer surface of each envelope (see Chapter 23).

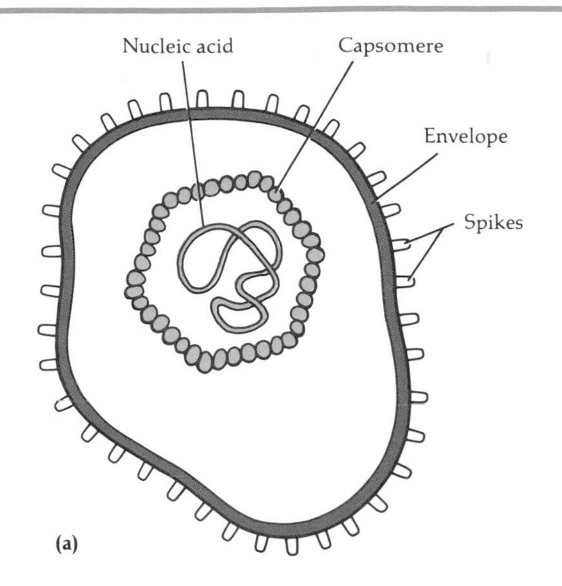

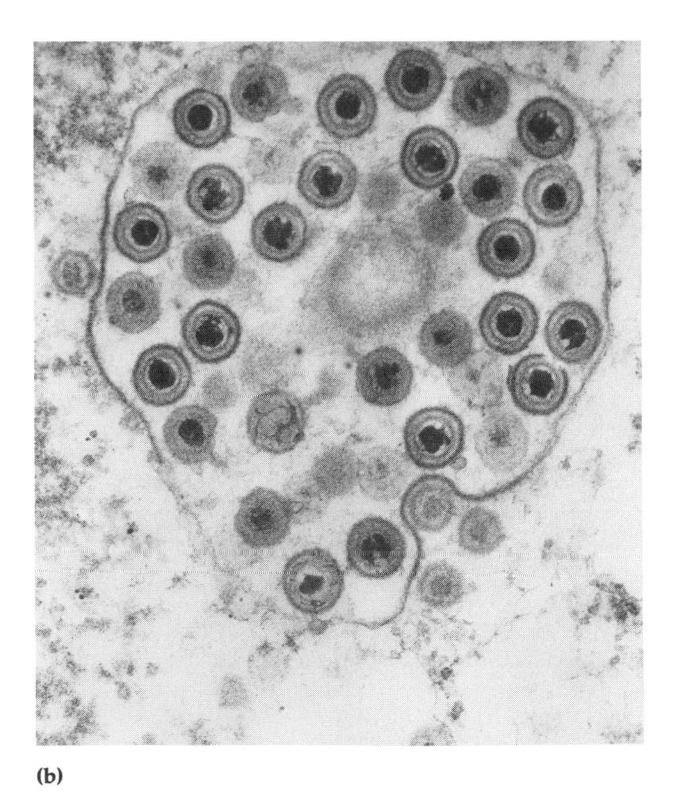
(b)

Figure 12–6 Morphology of an enveloped polyhedral (icosahedral) virus. **(a)** Diagram of an enveloped polyhedral virus. **(b)** Electron micrograph of a group of herpes simplex viruses (×48,000). To the lower right, a virus particle is acquiring its envelope as it buds out through a nuclear membrane of the host cell.

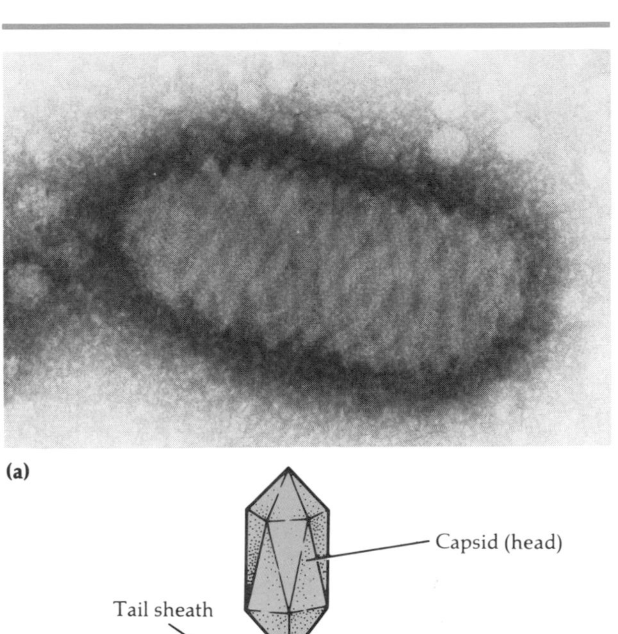
(a)

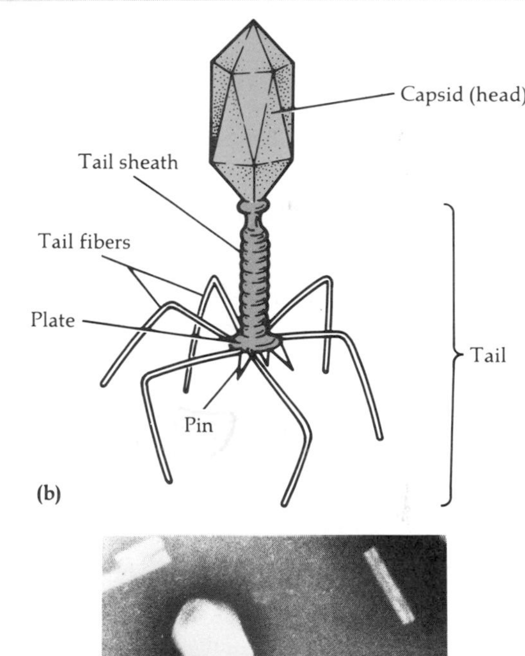

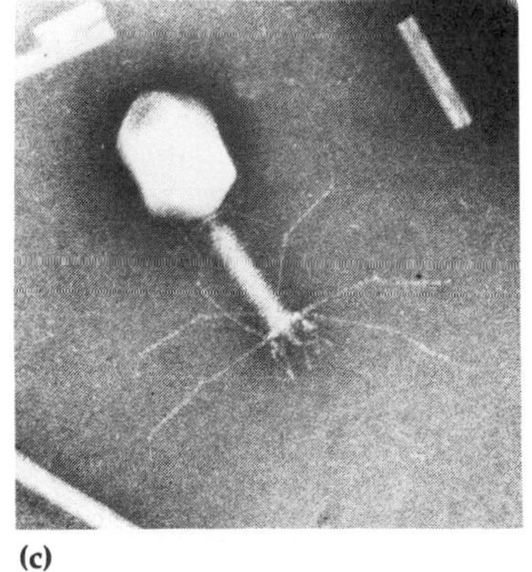
(c)

Figure 12–7 Morphology of complex viruses. **(a)** Electron micrograph of milker's nodule virus, a poxvirus that may cause lesions on human hands after being picked up from the teats of infected cows (×212,000). **(b)** Diagram of a T-even bacteriophage, showing its component parts. **(c)** A T4 bacteriophage (×110,000).

CLASSIFICATION OF VIRUSES

It was mentioned earlier that for practical purposes, viruses are classified as animal viruses, bacterial viruses, or plant viruses, depending on host range. This classification is convenient, but not scientifically acceptable.

The oldest classification of animal viruses is based on the organs affected by the diseases produced. This is a classification by symptomatology (Table 12–1). Since the same virus may cause more than one disease, depending on the organ attacked, this scheme is not satisfactory for the microbiologist, but it does offer certain conveniences for the physician.

In the past few decades, hundreds of viruses have been isolated from plants, animals, and humans. As the list grows, the problem of viral classification becomes more and more complex. Present classification systems are based on factors such as morphological class, type of nucleic acid present, size of capsid, and number of capsomeres. A summary of the classification of animal viruses based on these morphological, chemical, and physical properties is presented in Table 12–2. Other classification schemes also take into account sus-

Table 12–1 Classification of Viruses by Symptomatology

Classification	Diseases
Generalized Diseases: Diseases in which a virus spreads throughout the body via the blood or lymph, affecting several organs. In some cases, skin rash occurs.	Smallpox, cowpox, measles, German measles, chickenpox, yellow fever, and dengue
Diseases Primarily Affecting Specific Organs:	
Nervous System	Poliomyelitis, aseptic meningitis, rabies, encephalitis
Respiratory system	Influenza, common cold, pharyngitis, respiratory syncytial pneumonia, bronchitis
Skin and mucous membranes	Cold sores, warts, shingles, molluscum contagiosum
Eye	Various types of conjunctivitis
Liver	Infectious hepatitis, serum hepatitis, yellow fever
Salivary glands	Mumps and cytomegalovirus
Gastrointestinal tract	Gastroenteritis A virus and gastroenteritis B virus

Table 12–2 Classification of Animal Viruses According to Morphological, Chemical, and Physical Properties

Viral Group and Specific Examples	Morphological Class	Nucleic Acid*	Dimensions of Capsid (Diameter) (nm)	Clinical or Special Features
Parvoviruses (adenosatellite)	Naked polyhedral	SS DNA	18–26	Very small viruses, most of which depend on coinfection with adenoviruses for growth; probably infect only rats, mice, and hamsters
Papovaviruses (papilloma, polyoma, simian)	Naked polyhedral	DS circular DNA	40–57	Small viruses that induce tumors; the human wart virus (papilloma) and certain viruses that produce cancer in animals (polyoma and simian) belong to this family
Adenoviruses	Naked polyhedral	DS DNA	70–80	Medium-sized viruses that cause various respiratory infections in humans; some cause tumors in animals

Table 12–2, continued

Viral Group and Specific Examples	Morphological Class	Nucleic Acid*	Dimensions of Capsid (Diameter) (nm)	Clinical or Special Features
Herpesviruses (herpes simplex, herpes zoster)	Enveloped polyhedral	DS DNA	150–250	Medium-sized viruses that cause various diseases in humans, such as fever blisters, chickenpox, shingles, and infectious mononucleosis; implicated in a type of human cancer called Burkitt's lymphoma
Poxviruses (variola, cowpox, vaccinia)	Enveloped complex	DS DNA	200–350	Very large, complex, brick-shaped viruses that cause diseases such as smallpox (variola), molluscum contagiosum (wartlike skin lesion), cowpox, and vaccinia; vaccinia virus gives immunity to smallpox
Picornaviruses (poliovirus, rhinovirus)	Naked polyhedral	SS RNA	18–38	Smallest RNA-containing viruses; at least 70 human enteroviruses are known, including the polio-, coxsackie-, and echoviruses; more than 100 rhinoviruses exist and are the most common cause of colds
Togaviruses (alpha virus, flavivirus)	Enveloped polyhedral	SS RNA	40–60	Included are many viruses transmitted by arthropods; diseases include eastern equine encephalitis (EEE), Venezuelan equine encephalitis (VEE), St. Louis encephalitis (SLE), yellow fever, and dengue
Orthomyxoviruses (influenza A, B, C)	Enveloped helical	SS segmented RNA	80–200	Medium-sized viruses with a spiked envelope; have the ability to agglutinate red blood cells; cause influenza
Paramyxoviruses (measles, mumps)	Enveloped helical	SS segmented RNA	150–300	Morphologically similar to myxoviruses, but generally larger; cause parainfluenza, measles, mumps
Coronaviruses	Enveloped helical	SS RNA	80–130	Associated with upper respiratory tract infections and the common cold
Retroviruses	Enveloped helical	SS segmented RNA	100–120	Includes all of the RNA tumor viruses; cause leukemia and tumors in animals; some members produce "slow" viral infections
Rhabdoviruses (rabies)	Enveloped helical	SS RNA	70–180	Bullet-shaped viruses with a spiked envelope; cause rabies and Newcastle disease of chickens
Arenaviruses (lassa)	Enveloped helical	SS segmented RNA	50–300	Viruses contain RNA-containing granules; some members produce "slow" viral infections
Reoviruses	Naked polyhedral	DS segmented RNA	60–80	Relation to human disease not clear; may be involved in mild respiratory infections and infantile gastroenteritis

*SS = Single-stranded, DS = Double-stranded.

ceptibility to physical and chemical agents, immunologic properties, site of multiplication (nucleus or cytoplasm), and natural methods of transmission as criteria.

Figure 12–8 shows electron micrographs of several DNA-containing animal viruses, and Figure 12–9 shows electron micrographs of several RNA-containing animal viruses.

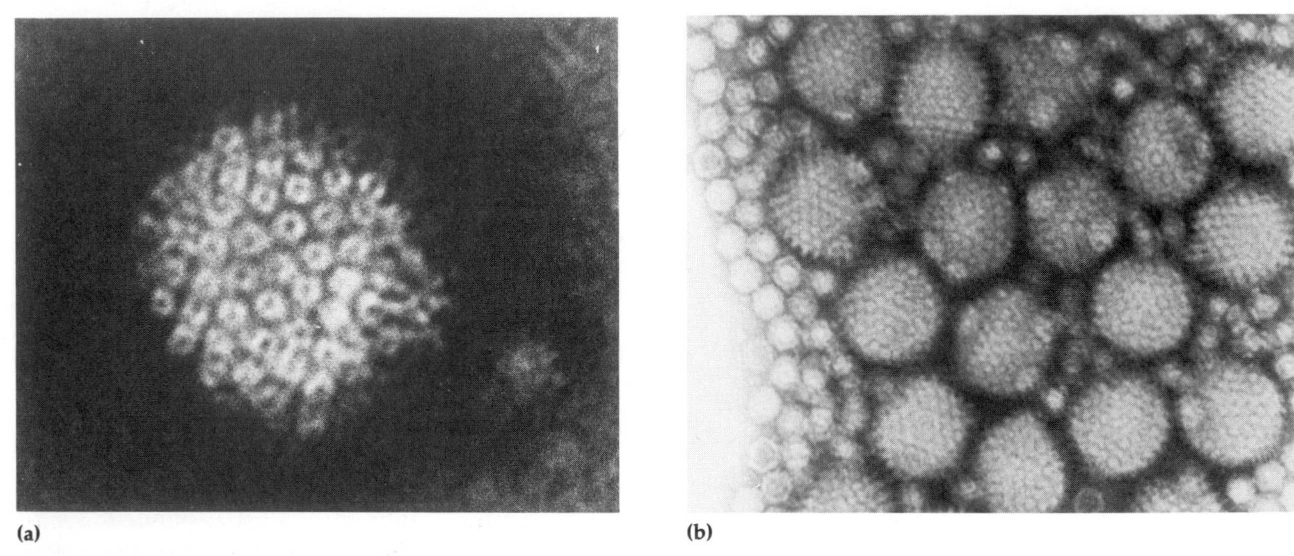

Figure 12–8 DNA-containing viruses. **(a)** Negatively stained capsid of a herpesvirus. The individual capsomeres are clearly visible. **(b)** Negatively stained viruses that have been concentrated in a centrifuge gradient (× 198,000). The larger capsids are an adenovirus and the smaller "adeno-associated" particles are a parvovirus.

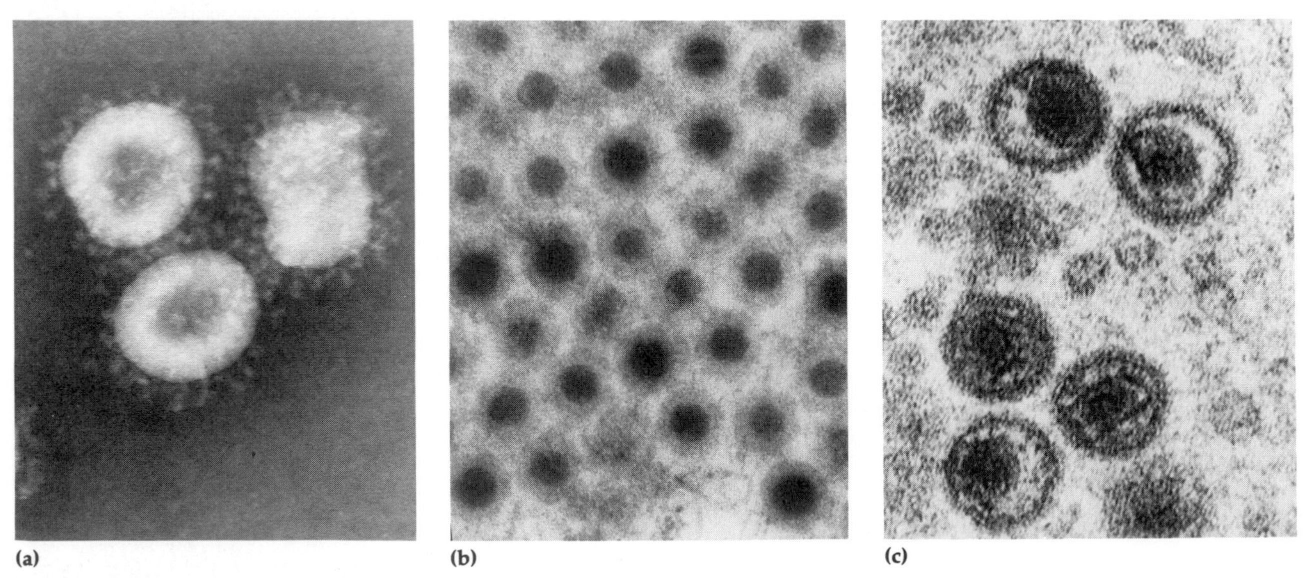

Figure 12–9 RNA-containing viruses. **(a)** Capsids of a coronavirus (× 172,500). The "corona" consists of a halo of spikes projecting from the capsid. **(b)** Reovirus particles in an aggregation from a cultured cell (× 95,000). **(c)** Mouse mammary tumor virus, a B-type retrovirus (× 160,000).

ISOLATION, CULTIVATION, AND IDENTIFICATION OF VIRUSES

The requirement of a living host cell for the multiplication of viruses complicates their detection, enumeration, and identification. Instead of a fairly simple chemical medium, it is necessary to provide viruses with living cells. Living plants and animals are difficult and expensive to maintain, and disease-causing viruses that grow only in higher primates and human hosts cause additional complications. On the other hand, viruses that use bacterial cells (bacteriophages) as a host are rather easily grown on bacterial cultures. This is one reason why so much of our understanding of viral multiplication has come from this latter type of virus.

Growth of Bacteriophages in the Laboratory

Bacteriophages can be grown either in suspensions of bacteria in liquid medium or in bacterial cultures on solid medium. The use of solid medium makes possible the **plaque method** for detecting and counting viruses, where only simple materials and equipment are needed. In this procedure, first developed for bacteriophages, a sample of bacteriophage is mixed with host bacteria and melted agar. The agar is then poured into a Petri plate containing a hardened layer of agar growth medium. The virus–bacteria mixture solidifies into a thin top layer that contains a layer of bacteria which is approximately one cell thick. Each virus infects a bacterium, multiplies, and releases several hundred new viruses. These newly produced viruses infect other bacteria in the immediate vicinity, and more new viruses are produced. Following several viral multiplication cycles, all the bacteria in the area surrounding the original virus are destroyed. This produces on the surface of the agar a number of clearings, or **plaques,** visible against a "lawn" of bacterial growth (Figure 12–10). At the same time that plaques form, uninfected bacteria elsewhere in the Petri plate multiply rapidly by fission, producing a turbid background.

Each plaque theoretically corresponds to a single virus in the initial suspension. Because some plaques may arise from more than one virion and

Figure 12–10 In the Petri plate on the left, clear viral plaques of varying sizes have been formed by the bacteriophage lambda on a lawn of *E. coli*. For comparison, the Petri plate on the right contains a bacterial culture without phage.

some virions may not be infectious, the concentrations of viral suspensions measured by counting plaques are usually given in terms of **plaque-forming units (pfu).**

Growth of Animal Viruses in the Laboratory

In living animals

Some animal viruses can only be cultured in **living animals,** such as mice, rabbits, and guinea pigs. Thus, most experiments to study the immune response to viral infections must also be performed in virally infected whole animals. Generally, animal inoculation is used as a diagnostic procedure for identifying a virus from a clinical specimen. Following inoculation of the animal with viruses in the specimen, the animal is observed for signs of disease or is killed so that infective tissues can be examined and evaluated.

In embryonated eggs

For many animal viruses, the **embryonated egg** offers a fairly convenient and inexpensive form of animal host if the virus will grow in it (Figure 12–11). A hole is drilled in the shell of the embryonated egg, and a viral suspension or suspected virus-containing tissue is injected into the fluid of the egg. There are several membranes in the egg on which the virus can be made to grow, and the virus is injected into the proper location in the egg. Viral growth is detected by the death of the embryo, by embryo cell damage, or by the formation on the membranes of the egg of typical pocks or lesions that result from viral growth. This method was once the most widely used for viral isolation and growth and is still used to grow viruses for some vaccines. You may be asked if you are allergic to eggs before receiving a vaccination, since egg proteins may be present in the viral vaccine preparations. (Allergy is discussed in Chapter 17.)

In cell culture

More recently, **cell culture** (sometimes called **tissue culture,** although that is not the best term) has replaced embryonated eggs for the growth of many viruses. Cell cultures are animal cells grown in culture medium in the laboratory. Since they are generally rather homogeneous collections of cells that can be propagated and handled much like bacterial cultures, they are more convenient to work with than whole animals or embryonated eggs.

Cell culture lines are readily started by treating a slice of animal tissue with enzymes that separate the individual cells. These cells are suspended in a solution that provides the proper osmotic pressure, nutrients, and growth factors required by the cell to grow. The cells tend to adhere to the glass or plastic container and reproduce to form a monolayer. Viruses infecting such a monolayer sometimes cause deterioration of the cells of the monolayer as they multiply. This tissue deterioration is called the **cytopathic effect (CPE)** and is illustrated in Figure 12–12. The CPE can be detected and counted in much the same way as plaques caused by bacteriophages on a lawn of bacteria.

Primary cell lines, derived from tissue slices, tend to die out after only a few generations. Certain cell lines from human embryos can be maintained for about a hundred generations and are widely used for diagnostic work in human diseases. For routine laboratory growth of viruses, **continuous cell lines** are used. These are "transformed" cells that can be maintained through an indefinite number of generations, and they are sometimes called "immortal" cell lines (see the discussion of transformation at the end of this chapter). One of these, the HeLa cell line, was isolated from the cancer of a woman who died in 1951. After years of laboratory cultivation, many such cell lines have lost almost all the original characteristics of the cell as it was when first isolated, but these changes have not interfered with the use of the cells for viral propagation. In spite of the success of cell culture in viral isolation and growth, there are still some viruses that have never been successfully cultivated in cell culture.

A major problem with cell culture is that the cell lines must be kept free of microbial contamination. In fact, the idea of cell culture is not really new, dating back to the end of the last century. However, it was not a practical laboratory technique until the development of antibiotics in the years following World War II. The maintenance of cell

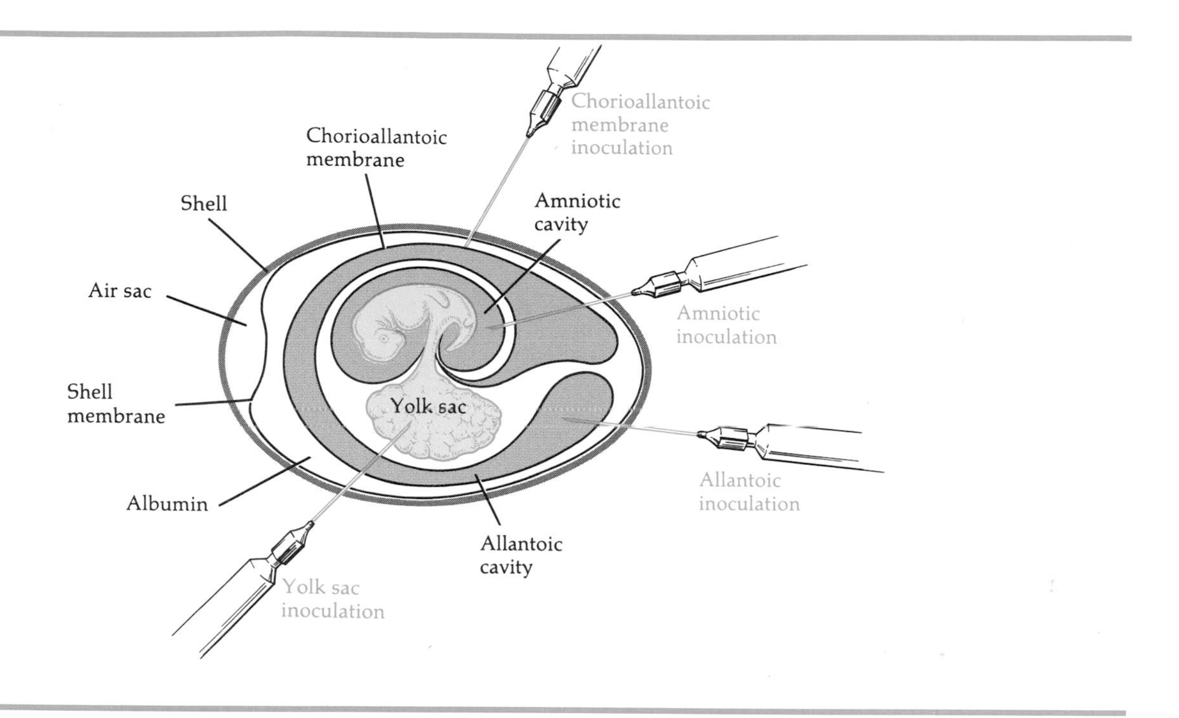

Figure 12–11 Inoculation of embryonated eggs. The injection site determines the membrane on which the viruses will grow.

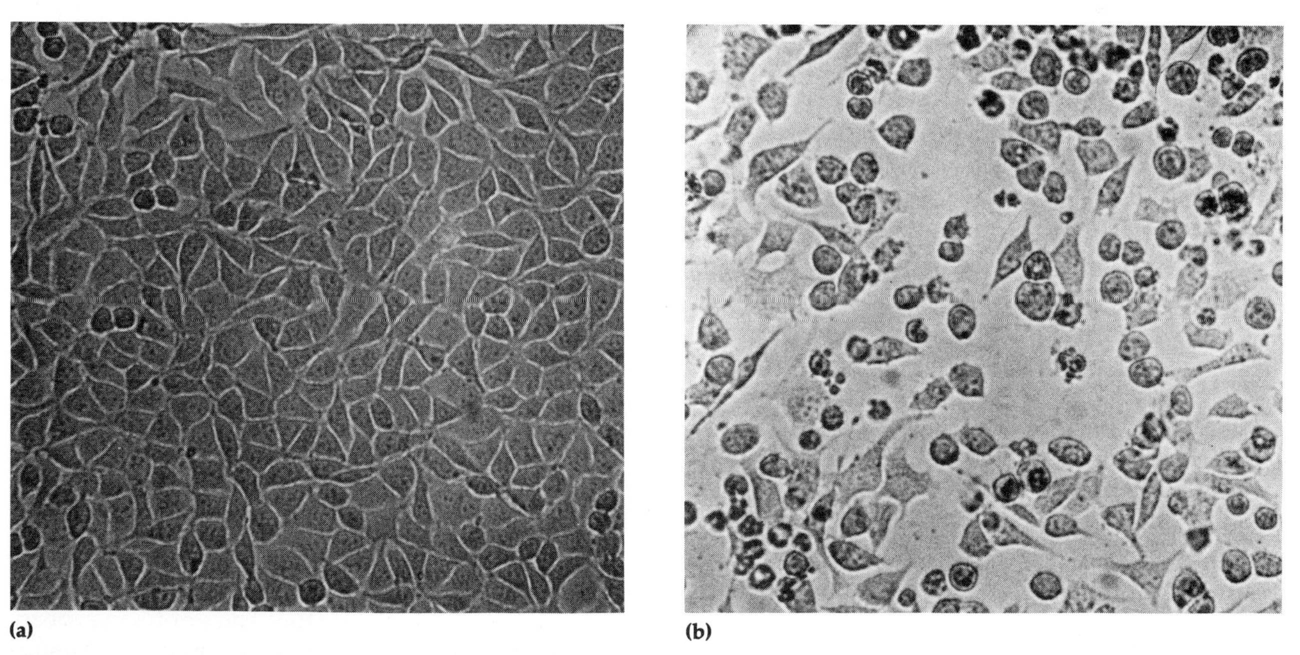

Figure 12–12 Cytopathic effect of viruses on a cell culture. **(a)** A monolayer of uninfected mouse L cells. **(b)** The same cells 24 hours after infection with vesicular stomatitis virus. Note the separation and "rounding up" of the cells.

culture lines still requires considerable experience, with trained technicians working on a full-time basis. For this reason, most hospital laboratories, and many state health laboratories, do not make isolations and identifications of viruses in clinical work. Instead, the tissue or serum samples are sent to central laboratories that specialize in this work.

The identification of viral isolates is not an easy task. For one thing, viruses cannot be seen at all without the use of an electron microscope. Immunological methods are the most commonly used means of identification. In these tests, the virus is detected and identified by its reaction with antibodies. Antibodies, which are proteins produced by animals in response to exposure to a virus, are highly specific for the virus that caused their formation. We shall discuss antibodies in detail in Chapter 16, along with a number of immunological tests for identifying viruses. Later in this chapter we will describe how viruses can be identified by their effects on host cells.

VIRAL MULTIPLICATION

The nucleic acid in virions contains only a few of the genes needed for the synthesis of new viruses. These include genes for structural components of the virion, such as the capsid proteins, and genes for a few of the enzymes used in the viral life cycle. Most **viral enzymes** (or **virus-specific enzymes**)—the enzymes encoded in viral nucleic acid—are not part of the virion, but rather are synthesized and function only within the host cell. The function of viral enzymes is almost entirely concerned with replicating or processing viral nucleic acid, and almost never with the machinery of protein synthesis or energy production. Although the smallest naked virions do not contain any preformed enzymes, the larger virions may contain one or a few enzymes, which usually function in helping the virus penetrate the host cell or replicate its nucleic acid.

Thus, for a virus to multiply, it must invade a host cell, taking over its metabolic machinery. In the process of viral multiplication, a single virus may give rise to several, or even thousands of, similar viruses in a single host cell. This results in drastic changes in the host cell, often ending in the death of the host cell.

Multiplication of Bacteriophages

Although the means of entering and exiting from a host cell may vary, the basic mechanism of viral multiplication is similar for animal viruses, plant viruses, and bacteriophages. The best-understood viral life cycles are those of the **bacteriophages,** or just simply **phages.** Since the so-called *T-even bacteriophages* (T2, T4, and T6) have been studied most extensively, we will first describe the multiplication of T-even bacteriophages in its host, *E. coli.*

T-even bacteriophages

The T-even bacteriophages have large, complex, naked virions, whose characteristic head-and-tail structure was shown in Figure 12–7b and c. These bacteriophages contain double-stranded DNA equal in length to about 6% of *E. coli* DNA, enough for over 100 genes. The multiplication cycle of these phages, like those of all viruses, can be divided into several distinct stages: adsorption, penetration, biosynthesis of viral components, maturation, and release.

Adsorption of Phage to Host Cell Following a chance collision between phage particles and bacteria, **adsorption** occurs. In this process, an adsorption site on the virus attaches to a complementary receptor site on the bacterial cell. T-even bacteriophages use fibers at the end of the tail as adsorption sites. The complementary receptor sites are on the cell wall (Figure 12–13a). (Other phages may adsorb to flagella or pili.) Adsorption is based on chemical interactions in which weak bonds are formed between the adsorption and receptor sites.

Penetration Following adsorption, the T-even bacteriophage injects its DNA (nucleic acid) into the bacterium. To do this, the bacteriophage tail releases an enzyme, *phage lysozyme,* which breaks down a portion of the bacterial cell wall. In the process of **penetration,** the tail sheath of the phage contracts, driving the tail core through the cell wall. When the tip of the core reaches the plasma membrane, the DNA from the bacteriophage head passes through the tail core, through the plasma membrane, and enters the contents of the bacterial cell. The capsid remains outside the bacterial cell for most bacteriophages (Figure 12–13b).

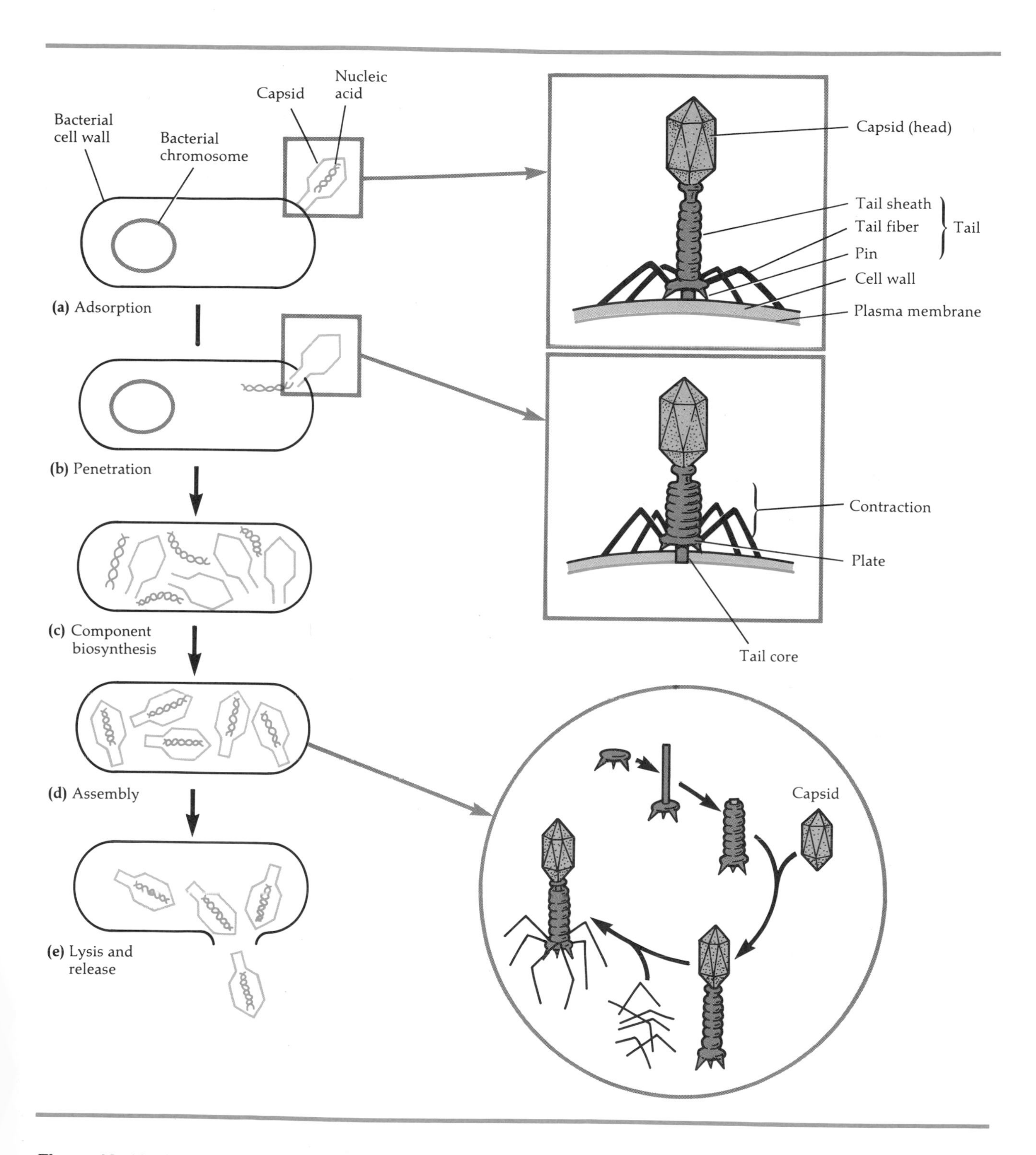

Figure 12–13 Multiplication cycle of a T-even bacteriophage. **(a)** Adsorption. **(b)** Penetration. **(c)** Biosynthesis of viral components (capsids and DNA are shown). **(d)** Maturation. Virions are assembled. **(e)** Lysis of the host cell and release of new virions.

Biosynthesis of Viral Components Once the bacteriophage DNA has reached the cytoplasm of the host cell, the **biosynthesis** of viral nucleic acid and protein occurs. In this process, the viral DNA takes over the metabolic machinery of the host cell. Transcription of RNA from the host chromosome stops, since the host DNA is degraded. Any RNA subsequently transcribed is mRNA transcribed from phage DNA, and it is through this mechanism that the phage commandeers the metabolic machinery of the host cell for its own biosynthesis. Since host enzymes continue to function, they produce energy for the biosynthesis of phage DNA and protein. Along with host enzymes, enzymes encoded in the phage DNA are synthesized and used by the phage.

Initially, the phage uses the host cell's nucleotides and several of its enzymes to synthesize many copies of phage DNA. Soon after, viral protein biosynthesis begins. The host cell's ribosomes, enzymes, and amino acids are used to synthesize viral proteins, including capsid proteins (Figure 12–13c).

Recall that during penetration, the protein coat (capsid) remains outside while the DNA is injected into the host cell. This means that the phage DNA provides the template for the multiplication of more phage DNA *and* for mRNA transcription for the biosynthesis of phage enzymes and capsid proteins.

For several minutes following infection, complete phages cannot be found in the host cell. Only separate components are detectable: DNA and proteins. The time during viral multiplication when complete, infective virions are not present is called the **eclipse period.**

Maturation In the next sequence of events, **maturation** occurs. In this process, bacteriophage DNA and capsids are assembled into complete virions. The assembly process is guided by the products of certain viral genes in a step-by-step sequence. The phage heads and tails are assembled separately from protein subunits, the head is packaged with phage DNA, and the tail is attached (Figure 12–13d). (For many simpler viruses, the nucleic acid and capsid proteins assemble spontaneously to form virions, without the intervention of other phage gene products.)

Release The general term for the final stage of viral multiplication is **release,** and it refers to the release of the new virions from the host cell. In the case of T-even phages, the term **lysis** is usually used for this stage. Lysozyme coded for by a phage gene is synthesized within the cell, causing a breakdown of the bacterial cell wall. The newly produced bacteriophages are thus released from the host cell (Figure 12–13e).

Lysozyme is not synthesized early in the multiplication cycle, since it would cause lysis of the host cell before complete phages could be assembled. Accordingly, there are controls regulating the regions of phage DNA that are transcribed into mRNA at various stages of the multiplication cycle. For example, there are early messages that are translated into early phage proteins, the enzymes of phage DNA synthesis. Also, there are late messages that are translated into late phage proteins for the synthesis of capsid proteins and lysozyme. Such a control mechanism is mediated by RNA polymerase. The released bacteriophages infect other susceptible cells in the area, and the viral multiplication cycle is repeated. The time required from phage adsorption to release is known as **burst time** and averages between 20 to 40 minutes. The number of newly synthesized phage particles released from a single cell is referred to as **burst size** and usually ranges from about 50 to 200.

Lysogeny

In the sequence of events just described for the multiplication of T-even bacteriophages, the release of the phages causes lysis and death of the host cell. Thus, such a sequence is referred to as a **lytic cycle.** However, some phages exist that can either proceed through a lytic cycle or have their DNA incorporated into the host cell's DNA, existing in a latent state without causing lysis of the host cell. Such a state is called **lysogeny,** and such phages are referred to as **lysogenic phages** or **temperate phages.** The participating bacterial host cells are known as **lysogenic cells.**

Let us go through the steps in lysogeny for the bacteriophage λ (lambda), a well-studied temperate phage (Figure 12–14). (Lambda is the name of a Greek letter.) Upon injection into an *E. coli* cell, the linear phage DNA forms a circle. This circle can multiply and be transcribed, leading to the production of new phage and lysis. Or, the circle can recombine with the circular bacterial DNA, becoming a part of it. The inserted phage DNA is now

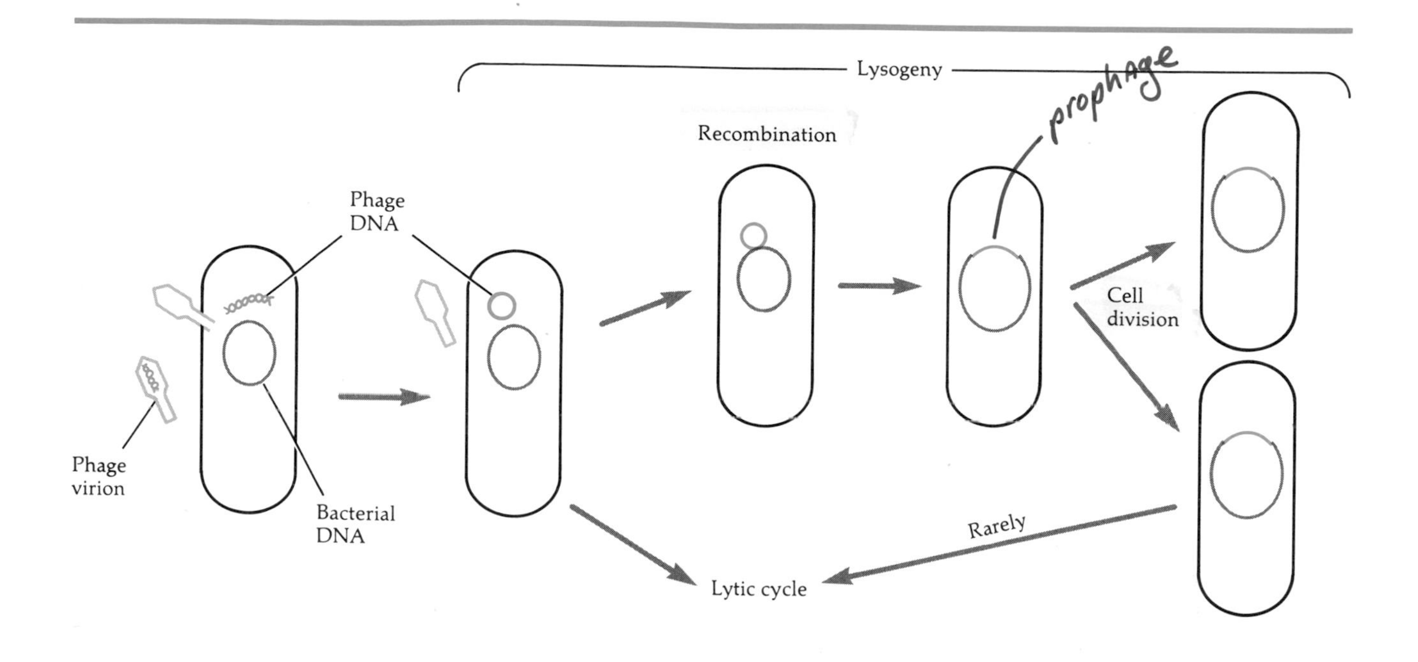

Figure 12–14 Lysogeny. After entering the host cell, the DNA from the temperate phage forms a circle, and by a single recombination event, becomes integrated into the host DNA. The host cell, now called a lysogenic cell, replicates the phage DNA (prophage) every time it divides. Infection by a temperate phage can also lead directly to a lytic cycle, and on rare occasions, a prophage excises from the bacterial chromosome and initiates a lytic cycle.

called a **prophage.** Most of the prophage genes are repressed by a repressor protein that is the product of one phage gene. Thus, the phage genes that would otherwise direct the synthesis and release of new virions are turned off in much the same way that the genes of the *E. coli* lac operon are turned off by the lac repressor (see Chapter 8).

Every time the cell machinery replicates the bacterial chromosome, it also replicates the prophage DNA. The prophage remains latent within the progeny cells. However, a rare spontaneous event or the action of ultraviolet light or certain chemicals can lead to the excision (popping-out) of the phage DNA—and initiation of the lytic cycle.

There are two important outcomes of lysogeny. First, the lysogenic cells become immune to reinfection by the same phage. (However, the host cell is not immune to infection by other phage types.) The second outcome of lysogeny is that the host cell may exhibit new properties. For example, the bacterium *Corynebacterium diphtheriae*, which causes diphtheria, is a pathogen whose disease-producing properties are related to the synthesis of a toxin. But the organism is capable of producing disease only when it carries a temperate phage, because the prophage carries the gene coding for the toxin. As another example, only streptococci carrying a temperate phage are capable of producing the toxin associated with scarlet fever. And the toxin produced by *Clostridium botulinum*, which causes botulism, is coded by a prophage gene.

Certain animal viruses may be able to undergo processes very similar to lysogeny. Animal viruses that can remain latent in cells for long periods without multiplying or causing disease may become inserted in a host chromosome or remain in a repressed state separate from host DNA (like some lysogenic phages). Cancer-causing viruses may also be lysogenic, as will be discussed later.

Multiplication of Animal Viruses

The multiplication of animal viruses follows the basic pattern of bacteriophage multiplication, but with several notable differences. Animal viruses differ from phages in their mechanisms of entering the host cell. And once inside, the synthesis and assembly of the new viral components are somewhat different, partly because of the differences between procaryotic cells and eucaryotic cells. Animal viruses may have certain types of enzymes not found in phages. Finally, the mechanisms of viral maturation and release, and the effects on the host cell, are different from those of phages.

In the following discussion of the multiplication of animal viruses, we will first consider the processes that are shared by both DNA- and RNA-containing animal viruses: adsorption, penetration, and uncoating. Then we will examine how DNA- and RNA-containing viruses differ with respect to biosynthesis and release from host cells.

Adsorption

Like bacteriophages, animal viruses attach by means of adsorption sites to complementary receptor sites on the host cell surface. In the case of animal cells, however, the cell surface is the plasma membrane (Figure 12–15a). Moreover, animal viruses do not possess appendages such as the tail fibers used by some bacteriophages. The adsorption sites of animal viruses vary from one group of viruses to another and are distributed over the surface of the virus. In adenoviruses, which are naked icosahedral viruses, the adsorption sites are small fibers at the corners of the icosahedron. In many of the enveloped viruses, such as the myxoviruses, the adsorption sites are spikes located on the surface of the envelope (see Figure 12–5a). The receptor sites on host cells also vary. Whereas they are mucoproteins for the myxoviruses, they are lipoproteins for the poliovirus.

Penetration

Following adsorption, penetration occurs. For enveloped animal viruses, penetration occurs by fusion of the envelope with the plasma membrane of the host cell (Figure 12–15b to c). For naked animal viruses, the complete virus penetrates the host cell plasma membrane and enters the cytoplasm. Complete viruses can also be taken into host cells by phagocytosis, a process termed **viropexis.**

Uncoating

Uncoating refers to the separation of the viral nucleic acid from its protein coat. It is a poorly understood process, and one that apparently varies with the type of virus. For some viruses, uncoating is accomplished by the action of lysosomal enzymes contained inside phagocytic vacuoles, which degrade the viral capsid proteins. For poxviruses, the

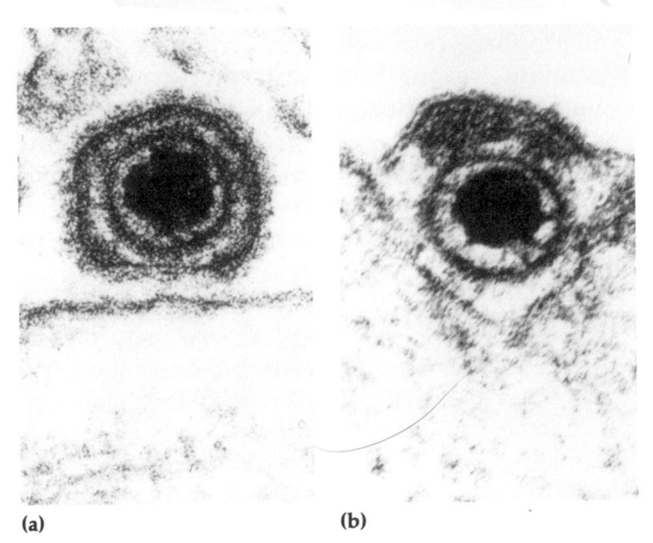

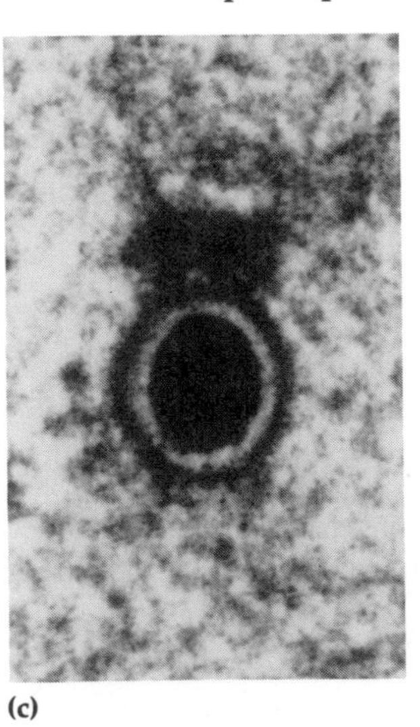

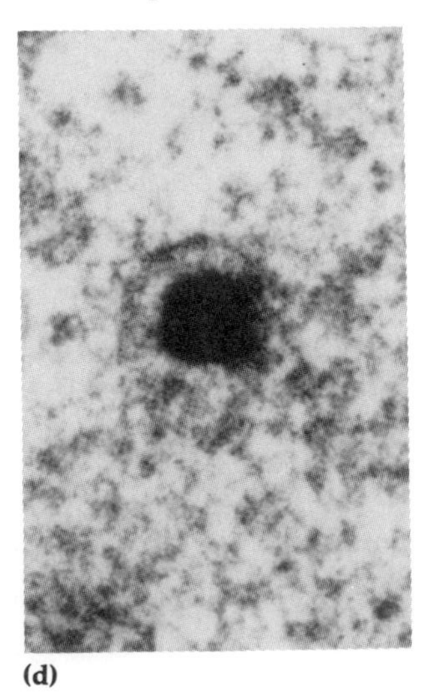

(a) (b) (c) (d)

Figure 12–15 Entry of herpes simplex virus into an animal cell. **(a)** Adsorption of the viral envelope to the plasma membrane (×180,000). **(b)** Penetration of the membrane, resulting in loss of the envelope (×180,000). **(c)** Unenveloped capsid in the cytoplasm of the cell (×180,000). **(d)** Digestion of the capsid, leaving only the nucleic acid core (×180,000).

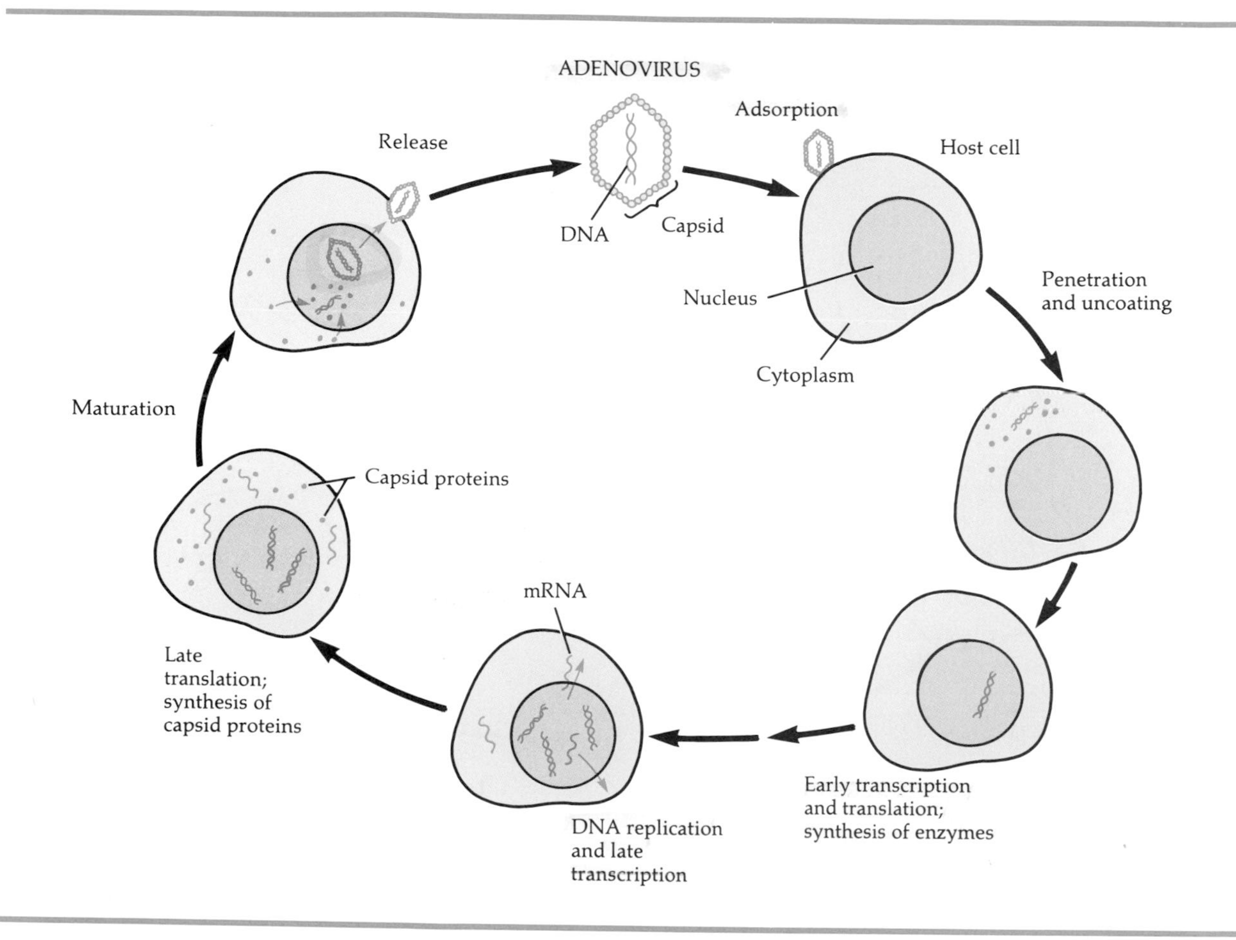

Figure 12–16 Multiplication of adenovirus, a DNA-containing virus.

uncoating is completed by a specific enzyme coded by the viral DNA and synthesized soon after infection. For other viruses, uncoating appears to be exclusively the result of enzymes in the host cell cytoplasm (Figure 12–15d). For at least one virus, the poliovirus, uncoating seems to begin while the virus is still attached to the host cell plasma membrane.

Biosynthesis for DNA viruses

In poxvirus multiplication, the biosynthesis of viral components and subsequent assembly occur in the cytoplasm of the host cell. The multiplication of other DNA-containing viruses (parvoviruses, papovaviruses, adenoviruses, and herpesviruses) differs in that viral DNA is replicated in the nucleus of the host cell, whereas viral protein is synthesized in the cytoplasm. This is followed by the migration of the proteins into the nucleus of the host cell for assembly into complete viruses. As an example of the multiplication of a DNA virus, we will follow the sequence of events as they occur in adenovirus (Figure 12–16).

Following adsorption, penetration, and uncoating, the viral DNA is released into the nucleus of the host cell. Next, the transcription of a portion of the viral DNA—the "early" genes—occurs, followed by translation. The products of these genes are enzymes required for the multiplication of viral DNA. In most DNA viruses, early transcription is carried out by means of host transcriptase (RNA polymerase); with poxviruses, however, the viruses contain their own transcriptase. Sometime after the initiation of DNA multiplication, transcription and translation of the

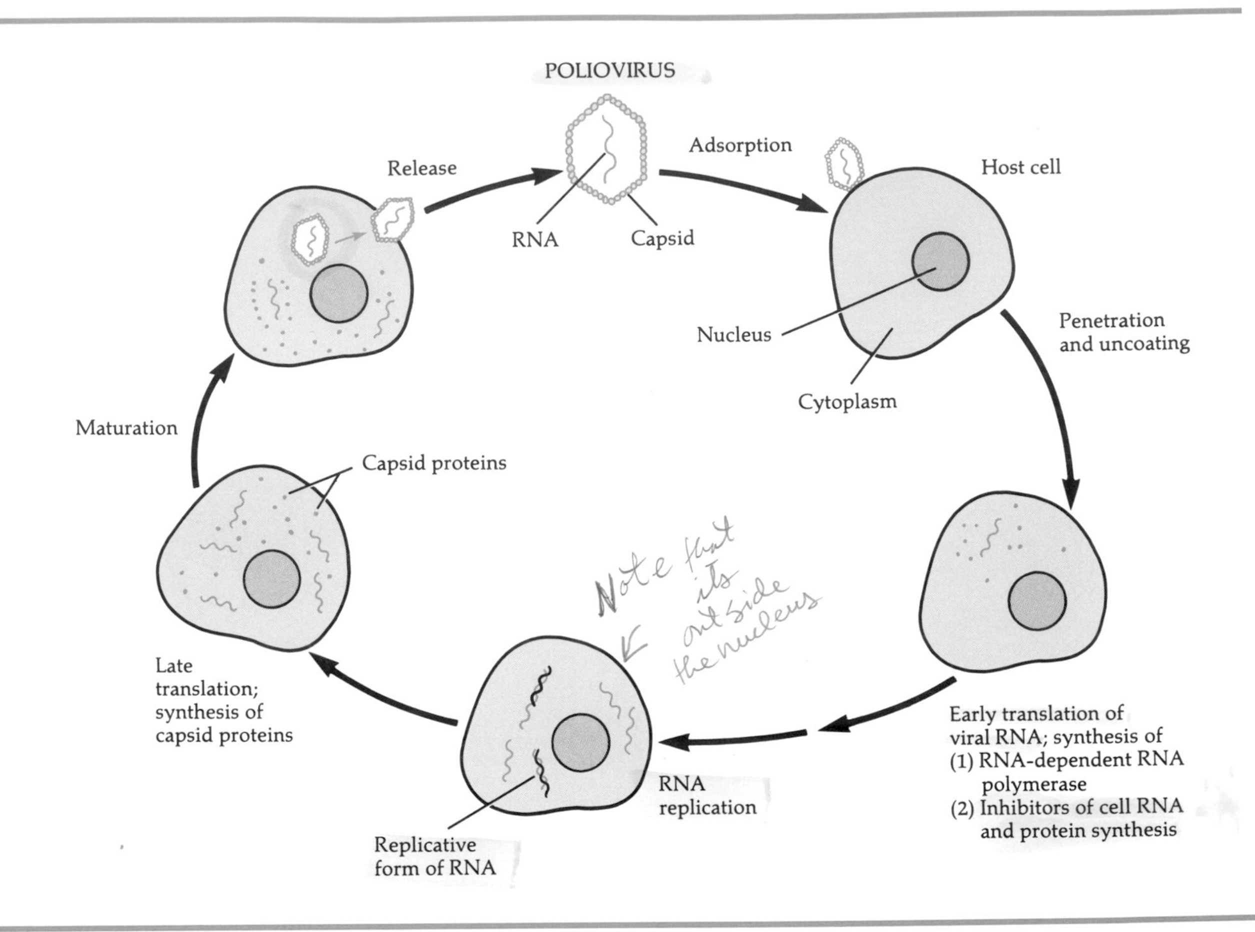

Figure 12–17 Multiplication of poliovirus, an RNA-containing virus.

remaining, "late," viral genes occur. This leads to the synthesis of capsid proteins, which occurs in the cytoplasm of the host cell. Following the migration of capsid proteins into the nucleus of the host cell, maturation occurs by the self-assembly of viral DNA and capsid proteins to form complete viruses.

Biosynthesis for RNA viruses *

The multiplication of RNA viruses is essentially the same as for DNA viruses, except that several different mechanisms of mRNA formation occur among different groups of RNA viruses. Although the details of these mechanisms are far beyond the scope of this text, for comparative purposes we will trace the multiplication cycle of the poliovirus, a picornavirus containing single-stranded RNA (Figure 12–17).

The multiplication of poliovirus takes place in the host cell cytoplasm. Following adsorption, penetration, and uncoating, the single-stranded viral RNA is translated into two principal proteins, which inhibit host cell RNA and protein synthesis and form an enzyme called *RNA-dependent RNA polymerase*. This enzyme catalyzes the synthesis of another strand of RNA, complementary in base sequence to the original infecting strand. This double-stranded RNA is called a **replicative form** and consists of one + and one − strand. The + strand may serve as mRNA for the translation of capsid protein, become incorporated into capsid protein to form a new virus, or serve as a template for continued RNA multiplication. Once viral RNA and viral protein are synthesized, maturation occurs.

Among the RNA-containing animal tumor viruses (retroviruses), the formation of mRNA and

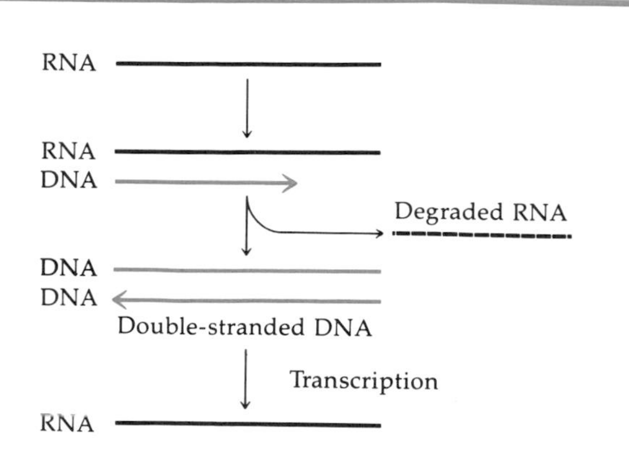

Figure 12–18 Reverse transcriptase. This enzyme catalyzes the synthesis of a strand of DNA on an RNA template. The RNA strand is then degraded, and another DNA strand, complementary to the first, is synthesized. The double-stranded DNA thus produced serves as a template for transcription to form both mRNA and RNA for new virions.

RNA for new virions is quite interesting. These viruses carry their own polymerase, which uses the RNA of the virus to synthesize a complementary strand of DNA, which in turn is replicated to form double-stranded DNA. The enzyme is an RNA-dependent DNA polymerase called *reverse transcriptase,* so called because it carries out a reaction (RNA → DNA) that is exactly the reverse of the familiar transcription of DNA → RNA (Figure 12–18). The formation of complete viruses requires that DNA be transcribed back into RNA to serve as mRNA for viral protein synthesis and to be incorporated into new virions. However, before transcription can take place, the viral DNA must be integrated into the DNA of a host cell chromosome. In this integrated state, it is called a **provirus.**

Maturation and release

The first step in viral maturation is the assembly of the protein capsid, usually a spontaneous process. The capsids of many RNA-containing animal viruses are enclosed by an envelope consisting of proteins, lipid, and carbohydrate, as noted earlier. Examples of such viruses include myxoviruses and paramyxoviruses. The envelope protein is synthesized by the virus and is incorporated into the plasma membrane of the host cell. The envelope lipid and carbohydrate are synthesized by host cell enzymes and are present in the plasma membrane. The envelope actually develops around the capsid by a process called **budding** (Figure 12–19). Following adsorption, penetration, uncoating, and biosyn-

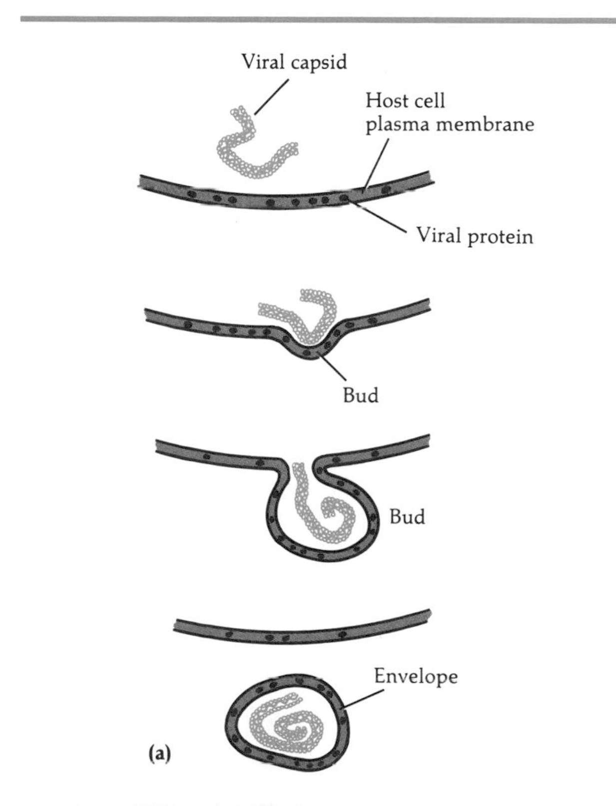

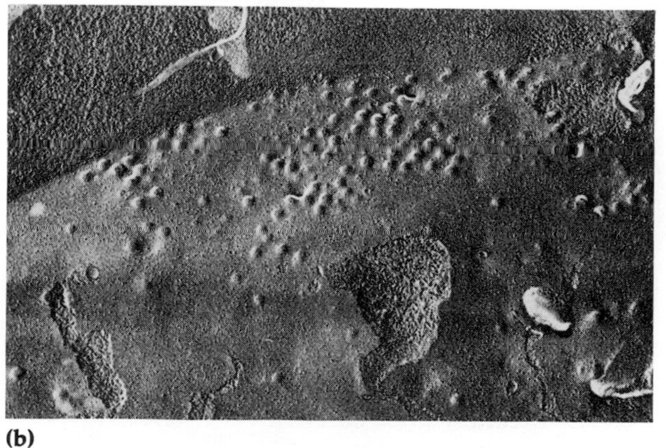

Figure 12–19 Budding of an enveloped virus. **(a)** Diagram of the budding process. **(b)** The small "bumps" seen on this freeze-fractured plasma membrane are Sindbis virus particles caught in the act of budding out from an infected cell (×38,000).

thesis of viral nucleic acid and protein, the assembled capsid-containing nucleic acid pushes through the plasma membrane. As a result, a portion of the plasma membrane, now the envelope, adheres to the virus. This extrusion of a virus from a host cell represents one method of release. Budding does not immediately kill the host cell, and in some cases the host cell survives indefinitely.

Naked viruses are released through ruptures in the host cell plasma membrane. As contrasted with budding, this type of release usually results in the death of the host cell.

LATENT VIRAL INFECTIONS

In some viral infections, the virus remains in equilibrium with the host without actually producing disease for a long period, often many years. The classic example of such a **latent viral infection** is the infection of the lip by herpes simplex virus, which produces cold sores. In this case, the virus can inhabit the host cells (nerve cells) without causing any damage until it is activated by stimuli such as fever or sunburn—hence the term *fever blister.*

It is interesting that in some individuals, viral production occurs but the symptoms never appear. Even though a large percentage of the human population carries the herpes simplex virus, most people never exhibit the disease. In some latent viral infections, the virus might exist in a lysogenic state within host cells. The varicella–zoster virus (also referred to as herpes zoster virus) is another example of a virus that can exist in a latent state. Chickenpox (varicella) is a childhood disease in which the viral agent is indistinguishable from the viral agent that causes shingles (herpes zoster), an adult neurological disorder. Shingles occurs in only a small fraction of the population who have had chickenpox. Many researchers believe that shingles may be caused by reactivation of chickenpox virus that has remained latent in certain nerve cells for years.

SLOW VIRAL INFECTIONS

The term **slow viral infection** refers to a disease process that occurs gradually over a long period of time. (It does not imply that viral multiplication is unusually slow.) Typically, slow viral infections are fatal. One example is a disease called *kuru,* first observed in a tribe of cannibals in New Guinea in 1957. It is a degenerative disease of the cerebellum that ends in death within one or two years after the onset of the first symptoms. Other examples of slow viral infections are listed in Table 12–3.

As Table 12–3 indicates, the causative agents of kuru and several other slow viral infections have not been identified, although there is evidence that

Table 12–3 Examples of Slow Viral Infections

Disease	Host	Organ Primarily Affected	Virus
Kuru	Human	Brain	Unknown (viroid?)
Subacute sclerosing panencephalitis	Human	Brain	Measles (Paramyxovirus)
Progressive encephalitis	Human	Brain	Rubella (togavirus)
Creutzfeldt-Jakob disease	Human	Brain	Unknown (viroid?)
Progressive multifocal leukoencephalopathy	Human	Brain	Papovavirus
Scrapie	Sheep	Brain	Unknown (viroid?)
Progressive pneumonia	Sheep	Lung	Retrovirus
Lymphocytic choriomeningitis	Mouse	Kidney, brain, liver	Arenavirus
Aleutian mink disease	Mink	Reticuloendothelial system	Parvovirus
Canine demyelinating encephalomyelitis	Dog	Brain, spinal cord	Distemper (paramyxovirus)

Source: From B. D. Davis, et al., *Microbiology,* 3rd ed., Harper & Row, New York, 1980, p. 1043.

the agents might be **viroids.** Viroids are very short pieces of naked RNA, only 300 to 400 nucleotides long, without a protein coat. There is much internal base pairing, which gives the molecule a closed, folded, three-dimensional structure and presumably helps protect it from attack by cellular enzymes. Thus far, they have only been conclusively identified as pathogens of plants. One of the best-studied viroids is the potato spindle tuber viroid, or PSTV (Figure 12–20). The PSTV seems capable of infection and multiplication without incorporation into a capsid. In some hosts, such as potatoes and tomatoes, disease results.

Common viruses are sometimes the source of slow viral infections. For example, the measles virus, several years after causing measles, may be responsible for a rare form of encephalitis called *subacute sclerosing panencephalitis.* The mechanism by which a common virus can also cause a slow viral infection is not known.

EFFECTS OF ANIMAL VIRAL INFECTION ON HOST CELLS

Infection of a host cell by an animal virus usually results in the death of the host cell. This can be caused by the accumulation of large numbers of multiplying viruses, by the effects of viral proteins on host cell plasma membrane permeability, or by inhibition of host DNA, RNA, or protein synthesis. The different types of abnormalities that can lead to damage or death of a host cell are known as **cytopathic effects (CPE).** These effects are frequently used as diagnostic tools for many viral infections.

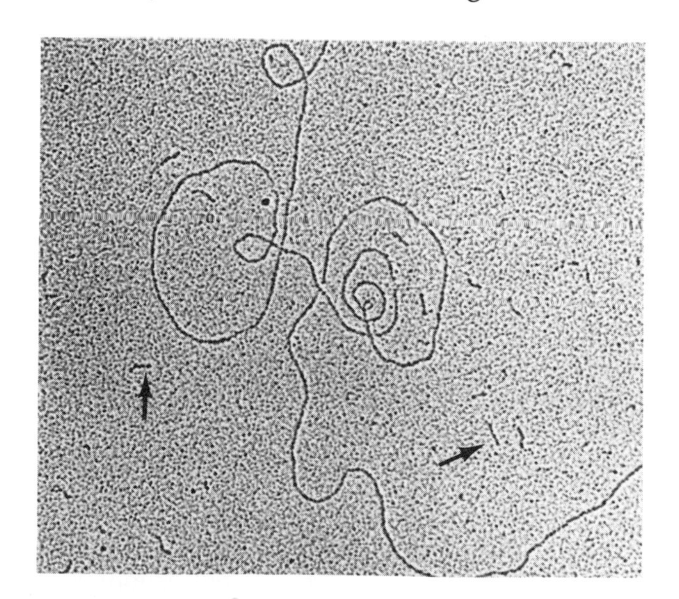

Figure 12–20 Electron micrograph of the potato spindle tuber viroid (PSTV) (×56,300). The arrows point to the viroids, which are adjacent to much longer viral nucleic acid from bacteriophage T7 for comparison.

One type of CPE is called an **inclusion body.** Some inclusion bodies arise as a result of the accumulation of assembled or unassembled viruses in the nucleus or cytoplasm of the host cell, or both. Other inclusion bodies arise at sites of earlier viral synthesis but do not contain assembled viruses or their components. Inclusion bodies are important because they can help to identify the particular virus causing an infection. For example, the rabies virus produces inclusion bodies (Negri bodies) in the cytoplasm of nerve cells, and their presence in the brain tissue of animals suspected of being rabid is still sometimes used today as a diagnostic tool for rabies. Diagnostic inclusion bodies are also associated with the measles virus, vaccinia virus, smallpox virus, herpesvirus, and adenoviruses (Figure 12–21).

As a result of viral infection, some host cells exhibit another type of CPE. At times, several adjacent infected cells fuse to form giant cells called **polykaryocytes.** These are produced from infections from myxoviruses. Other viral infections result in chromosomal damage to the host cell, most often chromosomal breakage.

One very significant response of some virus-infected cells is the production by infected cells of a substance called **interferon.** Although induced by viral infection, interferon is coded by the host cell DNA. Interferon protects neighboring uninfected cells from viral infection. More will be said about interferon in Chapter 15.

VIRUSES AND CANCER

When cells multiply in an uncontrolled way, the excess tissue that develops is called a **tumor.** A cancerous tumor is known as a **malignant tumor,** whereas a noncancerous tumor is referred to as a **benign tumor.** As a general rule, tumors are named by attaching the suffix *-oma* to the name of the tissue from which the tumor arises. Cancer is not a

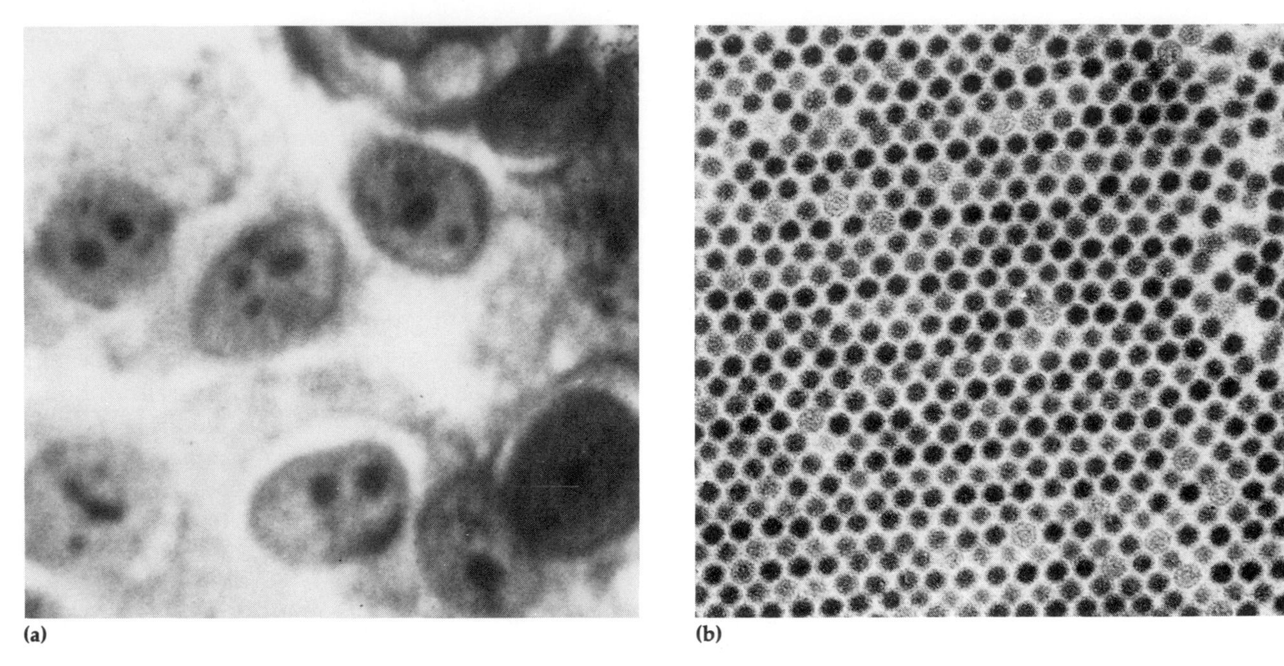

Figure 12–21 Viral inclusion bodies. **(a)** The dark spots in the gray, oval nuclei of the HeLa cells shown here are intranuclear inclusion bodies of adenovirus type 3 ($\times$5300). **(b)** Electron micrograph showing an intranuclear crystalline array of virions of adenovirus at a much higher magnification ($\times$44,100) than in part (a).

single disease, but many diseases. The human body contains over a hundred different kinds of cells, each of which can malfunction in its own distinctive way to cause cancer. Malignant cells multiply in an uncontrolled way, often rapidly. Most patients who die from cancer are not killed by the primary tumor that develops, but by **metastasis,** the spread of the cancer to other parts of the body. Metastatic tumors are harder to detect and eliminate than primary tumors.

Not all cancerous cells form solid tumors. When the malignant cells are those that give rise to white blood cells, the resulting cancer, characterized by an excess of white cells in the circulation, is called **leukemia.** The earliest relationship between cancers and viruses was in fact demonstrated in 1908 when it was shown that certain chicken leukemias could be transferred to healthy chickens by cell-free filtrates. Three years later it was found that a chicken **sarcoma** (cancer of connective tissue) can be similarly transmitted. Virus-induced **adenocarcinomas** (cancers of glandular tissue) in mice were discovered in 1936. At that time, it was clearly shown that mouse mammary tumors were transmitted from mother to offspring through the mother's milk.

The etiology of cancer can go unrecognized for several reasons. First, with some viruses, most of the viral particles infect cells without inducing cancer. Second, cancer may not develop until long after infection. And third, cancers do not seem to be contagious.

Transformation of Normal Cells into Tumor Cells

Almost anything that can alter the genetic material of a cell has the potential to make a normal cell cancerous. Examples include certain chemicals, high-energy radiation, and, as just noted, viruses. Viruses capable of producing tumors in animals are referred to as **oncogenic viruses.**

In experimental animals, both DNA- and RNA-containing viruses are capable of inducing tumors. When this occurs, the tumor cells are **transformed** in such a way that they acquire properties that are distinct from uninfected cells or infected cells in which tumors are not produced. An outstanding feature of all oncogenic viruses is that the viral genetic material becomes integrated with host cell DNA, replicating along with the host cell chromosome. This mechanism is similar to the phenomenon of lysogeny in bacteria and can result in an alteration of the characteristics of the host cell. In addition, transformed cells lose a property called contact inhibition. Normal animal cells in tissue culture move about randomly by amoeboid movement and divide repeatedly until they come into contact with each other. Then both movement and cell division stops—the phenomenon known as **contact inhibition.** But transformed cells in tissue culture do not exhibit contact inhibition, forming, instead, tumorlike cell masses. Transformed cells sometimes also produce tumors when injected into susceptible animals. Following transformation by viruses, many tumor cells contain a virus-specific antigen on their cell surface called **tumor-specific transplantation antigen (TSTA)** or an antigen in the nucleus called the **T antigen.** Finally, transformed cells tend to be rounder than normal cells and exhibit certain chromosomal abnormalities, such as unusual numbers of chromosomes and fragmented chromosomes.

DNA-Containing Oncogenic Viruses

Oncogenic viruses are found within several groups of DNA-containing viruses. These include adenoviruses, herpesviruses, poxviruses, and papovaviruses. Among the papovaviruses are the papilloma viruses that cause benign warts in humans and other animals, polyoma viruses that cause several kinds of tumors when injected into newborn mice, and simian virus 40 (SV40), which was originally isolated from cell cultures being used to cultivate polioviruses for vaccine production.

During polyoma and SV40 infection, there is an increase in host cell DNA synthesis. This is followed by the appearance of TSTA and T antigen. When the host cell is transformed, viral DNA is integrated into the host cell DNA. This mechanism, similar to lysogeny in bacteria, results in the distinctive properties of transformed cells noted earlier.

Among the herpesviruses that affect humans are the herpes simplex virus, the herpes zoster virus, and the Epstein–Barr (EB) virus. The EB virus has an attraction for lymphocytes (a type of white blood cell) and has the potential for transforming lymphocytes into highly proliferating cells. The EB virus, in addition to being the cause of infectious mononucleosis, has recently been implicated as the causative agent of two human cancers: Burkitt's lymphoma and nasopharyngeal carcinoma. Burkitt's lymphoma is a rare cancer of the lymphatic system primarily affecting children in certain areas of Africa. Nasopharyngeal carcinoma, a cancer of the nose and throat, is worldwide in distribution. Some researchers have also suggested that EB virus is involved in Hodgkin's disease, a cancer of the lymphatic system (see box).

It is estimated that about 80% of the U.S. population carries the EB virus in their lymphocytes in the latent stage, without any disease being produced. This is indicated by the presence of antibodies to EB virus in blood serum. Although the EB virus leads to no apparent symptoms in healthy individuals, it can cause infectious mononucleosis, mostly in teenagers. In specimens from patients with both Burkitt's lymphoma and nasopharyngeal cancer, TSTA and T antigen specific for the EB virus have been found. Moreover, EB virus DNA is always found in malignant cells from both cancers.

One type of herpes simplex virus (HSV 1) produces cold sores. Another type, HSV 2, is associated with more than 90% of genital herpes infections. Since women with cervical cancer have a higher level of HSV 2 antibodies than asymptomatic patients, it has been suggested that HSV 2 may be associated with cervical cancer.

RNA-Containing Oncogenic Viruses

Among the RNA-containing viruses, only the retroviruses seem to be oncogenic. These include the leukemia, lymphoma, and sarcoma viruses of cats, chickens, and mice and the mammary tumor viruses of mice. The ability of these viruses to induce tumors seems to be related to the production of a reverse transcriptase by the viruses, according to the mechanism described earlier. The double-stranded DNA molecule synthesized from the viral RNA, called a provirus, becomes integrated into the host cell's DNA.

Once the provirus is integrated into the host cell's DNA, several things can happen. In some cases, the provirus will simply remain in a lysogenic state, duplicating when the DNA of the host cell duplicates. In other cases, the provirus may become activated, leave the host cell DNA, and produce new viruses, which may infect adjacent cells. In still other cases, the provirus may convert the host cell into a tumor cell.

At present, there are three theories concerning the relationship of RNA-containing viruses to cancer and their transmission. According to the **provirus theory,** RNA tumor viruses are transmitted from individual to individual after the provirus excises and produces new viruses. In such a method of transmission, the viruses cause cancer at some later time. According to the **oncogenic theory,** a set of genes **(virogenes)** that leads to RNA virus production is part of the makeup of all vertebrate cells and was acquired early in evolution. One of these genes, the *oncogene,* has the potential to bring about malignant transformation. Thus, a host cell contains the potential information for producing tumors, but this information remains unexpressed until activated by agents such as radiation, chemicals, or hormones. A third theory, the **protovirus theory,** holds that certain normal genes (protoviruses) are duplicated by transcription into RNA and back to DNA. Then the duplicated segments are integrated at new sites, creating new gene sequences. The new sequences may result in the development of malignancies.

Microbiology in the News

Possible Viral Cause for Hodgkin's Disease

January 1981—A report by researchers at the Harvard School of Public Health suggests that Hodgkin's disease, a cancer of the lymphatic system, may be caused by a virus.

The recent evidence is epidemiological (see Chapter 13). The researchers compared the family backgrounds of 225 Hodgkin's patients with those of 447 healthy persons of the same ages. The Hodgkin's patients were more likely than the controls to come from small, well-educated, relatively high-income families and to have had relatively few playmates as children. Such backgrounds are similar to those common to victims of serious polio in the past. It is suggested that in both the polio and Hodgkin's cases, a relatively sanitized childhood prevented the development of immunity to viruses that later caused serious disease.

Other researchers have suggested previously that the Epstein–Barr (EB) virus is involved in Hodgkin's disease. They have found antibodies to EB virus and an unusually high rate of infectious mononucleosis in Hodgkin's patients. Mononucleosis is known to be caused by EB virus.

STUDY OUTLINE

GENERAL CHARACTERISTICS OF VIRUSES (pp. 312–314)

Definition (p. 313)

1. Depending on one's viewpoint, viruses may be regarded as exceptionally complex aggregations of nonliving chemicals or exceptionally simple living microbes.
2. Viruses contain a single type of nucleic acid (DNA or RNA) and a protein coat, sometimes enclosed by an envelope composed of lipids, proteins, and carbohydrates.

3. Viruses are obligate intracellular parasites. They multiply by using the synthetic machinery of the host cell to cause the synthesis of specialized elements that can transfer the viral nucleic acid to other cells.
4. A virion is a complete, fully developed viral particle composed of nucleic acid surrounded by a coat.

Host Range (p. 313)

1. Host range refers to the spectrum of host cells in which viruses can multiply.
2. According to host range, viruses are generally classified as animal viruses, bacterial viruses (bacteriophages), and plant viruses. A virus can only infect certain species within each class.
3. Host range is determined by the specific attachment site on the host cell surface and the availability of host cellular factors.

Size (pp. 313–314)

1. Viral size is determined by filtration through membrane filters, ultracentrifugation, and electron microscopy.
2. Viruses range from 20 to 300 nm in diameter.

VIRAL STRUCTURE (pp. 314–317)

Nucleic Acid (p. 314)

1. The proportion of nucleic acid in relation to protein in viruses ranges from about 1% to about 50%.
2. Viruses contain either DNA or RNA, never both, and the nucleic acid may be single- or double-stranded, linear or circular, or divided into several separate molecules.

Capsid and Envelope (pp. 314–315)

1. The protein coat surrounding the nucleic acid of a virus is called the capsid.
2. The capsid is composed of subunits, the capsomeres, which may be of a single type of protein or several types.
3. In some viruses, the capsid is enclosed by an envelope consisting of lipids, proteins, and carbohydrates.
4. Some envelopes are covered with carbohydrate–protein complexes called spikes.
5. Viruses without envelopes are called naked viruses.

General Morphology (pp. 315–317)

1. Helical viruses (for example, tobacco mosaic virus) resemble long rods, and their capsids are hollow cylinders surrounding the nucleic acid.

2. Polyhedral viruses (for example, adenovirus) are many-sided. Usually the capsid is an icosahedron.
3. Enveloped viruses are covered by an envelope and are roughly spherical but highly pleomorphic. There are also enveloped helical viruses (for example, influenza) and enveloped polyhedral viruses (for example, herpes simplex).
4. Complex viruses have complex structures. For example, many bacteriophages have a polyhedral capsid with a helical tail attached.

CLASSIFICATION OF VIRUSES (pp. 318–320)

1. Classification of viruses is based on morphological class, type of nucleic acid, size of capsid, and number of capsomeres.
2. Some other classification schemes account for susceptibility to microbial control agents, immunological properties, site of multiplication, and method of transmission.

ISOLATION, CULTIVATION, AND IDENTIFICATION OF VIRUSES (pp. 321–324)

1. Viruses must be grown in living cells.
2. The easiest viruses to grow are bacteriophages.

Growth of Bacteriophages in the Laboratory (pp. 321–322)

1. In the plaque method, bacteriophages are mixed with host bacteria and nutrient agar.
2. After several viral multiplication cycles, the bacteria in the area surrounding the original virus are destroyed, and the area of lysis is called a plaque.
3. Each plaque may originate with a single viral particle or more than one; the number of viral particles required to initiate a plaque is termed a plaque-forming unit.

Growth of Animal Viruses in the Laboratory (pp. 322–324)

1. Some animal viruses require whole animals for cultivation.
2. Some animal viruses can be cultivated in embryonated eggs.
3. Cell cultures are animal cells that will grow in culture media in the laboratory.
4. Continuous cell lines can be maintained in vitro indefinitely.
5. Viral growth may cause cytopathic effects in the cell culture.

VIRAL MULTIPLICATION

1. Viruses do not contain enzymes for energy production or protein synthesis.

2. For a virus to multiply, it must invade a host cell and direct the host's metabolic machinery to produce viral enzymes and components.

Multiplication of T-even Bacteriophages (pp. 324–326)

1. In adsorption, sites on the phage tail fibers attach to complementary receptor sites on the bacterial cell.
2. In penetration, phage lysozyme opens a portion of the bacterial cell wall, the tail sheath contracts to force the tail core through the cell wall, and phage DNA enters the bacterial cell. The capsid remains outside.
3. In biosynthesis, the host DNA is degraded and transcription of phage DNA produces mRNA coding for proteins necessary for phage multiplication. Phage DNA is replicated, followed by the production of capsid proteins. During the eclipse period, separate phage DNA and protein can be found.
4. During maturation, phage DNA and capsids are assembled into complete viruses.
5. During release, phage lysozyme breaks down the bacterial cell wall, and the multiplied phages are released.
6. The time required from phage adsorption to release is called burst time (20 to 40 minutes). Burst size, the number of newly synthesized phages produced from a single infected cell, ranges from 50 to 200.

Lysogeny (pp. 326–327)

1. In a lytic cycle, a phage causes the lysis and death of a host cell.
2. Some viruses can either cause lysis or have their DNA incorporated into the DNA of the host cell as a prophage. The latter situation is called lysogeny.
3. Prophage genes are regulated by a repressor coded for by the prophage. The prophage is replicated each time the cell divides.
4. Exposure to certain mutagens can lead to excision of the prophage and initiation of the lytic cycle.
5. As a result of lysogeny, lysogenic cells become immune to reinfection with the same phage, and the host cell may exhibit new properties.

Multiplication of Animal Viruses (pp. 327–332)

1. Animal viruses adsorb to the plasma membrane of the host cell.
2. For enveloped viruses, penetration occurs by fusion of the envelope with the plasma membrane. For naked viruses, the complete virus enters the cell by viropexis.
3. Animal viruses are uncoated by viral or host cell enzymes.

4. For most DNA viruses, viral DNA is released into the nucleus of the host cell. Transcription of viral DNA and translation produce viral DNA and, later, capsid protein. Capsid protein is synthesized in the cytoplasm of the host cell.
5. Multiplication of RNA viruses occurs in the cytoplasm of the host cell. RNA-dependent RNA polymerase synthesizes a double-stranded RNA called a replicative form. Viral RNA acts as mRNA for translation.
6. RNA-containing tumor viruses carry reverse transcriptase (RNA-dependent DNA polymerase), which transcribes DNA from RNA.
7. After maturation, viruses are released. One method of release (and envelope formation) is budding. Naked viruses are released through ruptures in the host cell membrane.

LATENT VIRAL INFECTIONS (p. 332)

1. In latent viral infections, the virus remains in the host cell for long periods of time without producing an infection.
2. Examples are cold sores and shingles.

SLOW VIRAL INFECTIONS (pp. 332–333)

1. Slow viral infections are disease processes that occur over a long period of time and are generally fatal.
2. They may be caused by viroids (infectious pieces of naked RNA).

EFFECTS OF ANIMAL VIRAL INFECTION ON HOST CELLS (p. 333)

1. Cytopathic effects (CPE) are abnormalities that lead to damage or death of a host cell.
2. Cytopathic effects include inclusion bodies and polykaryocytes.
3. Interferon is produced by virus-infected cells and protects neighboring cells from viral infection.

VIRUSES AND CANCER (pp. 333–336)

1. An excess of tissue due to unusually rapid cell multiplication is called a tumor. Tumors may be malignant (cancerous) or benign (noncancerous). Metastasis refers to the spread of cancer to other parts of the body.
2. Tumors are usually named by attaching the suffix *-oma* to the name of the tissue from which the tumor arises.
3. The earliest relationship between cancer and viruses was demonstrated in the early 1900s when chicken leukemia and chicken sarcoma were transferred to healthy animals by cell-free filtrates.

Transformation of Normal Cells into Tumor Cells (pp. 334–335)

1. Viruses capable of producing tumors are called oncogenic viruses.
2. The genetic material of oncogenic viruses becomes integrated into the host cell DNA.
3. Transformed cells lose contact inhibition, contain virus-specific antigens (TSTA and T antigen), exhibit chromosomal abnormalities, and may produce tumors when injected into susceptible animals.

DNA-Containing Oncogenic Viruses (p. 335)

1. Oncogenic viruses are found among adenoviruses, herpesviruses, poxviruses, and papovaviruses.
2. The EB virus, a herpesvirus, causes infectious mononucleosis and has been implicated in Burkitt's lymphoma and nasopharyngeal carcinoma. One type of herpes simples (HSV 2), associated with over 90% of genital herpes infections, may be implicated in cervical cancer.

RNA-Containing Oncogenic Viruses (pp. 335–336)

1. Among the RNA viruses, only retroviruses seem to be oncogenic.
2. The ability to produce tumors is related to the production of reverse transcriptase. The DNA synthesized from the viral RNA becomes incorporated into the host cell DNA as a provirus.
3. There are three major theories that attempt to explain the relationship of RNA-containing viruses to cancer: the provirus theory, the oncogene theory, and the protovirus theory.

STUDY QUESTIONS

REVIEW

1. Viruses were first detected because they are filterable. What do we mean by the term *filterable* and how could this property have helped their detection before invention of the electron microscope?
2. Why do we classify viruses as obligate intracellular parasites?
3. List the four properties that define a virus. What is a virion?
4. Describe the four morphological classes of viruses, then diagram and give an example of each.
5. Describe how bacteriophages are detected and enumerated by the plaque method.
6. Explain how animal viruses are cultured in each of the following:
   (a) Whole animals
   (b) Embryonated eggs
   (c) Cell cultures

7. Why are continuous cell lines of more practical use than primary cell lines for culturing viruses? What is unique about continuous cell lines?
8. Describe the multiplication of a T-even bacteriophage. Be sure to include the essential features of adsorption, penetration, biosynthesis, maturation, and release.
9. *Streptococcus pyogenes* produces erythrogenic toxin and is capable of causing scarlet fever only when it is lysogenic. What does this mean?
10. Describe the principal events of adsorption, penetration, uncoating, biosynthesis, maturation, and release of an enveloped DNA-containing virus.
11. Assume this strand of RNA is the nucleic acid for an RNA-containing animal virus: UAGUCAAGGU.
    (a) Describe the steps of RNA replication for a virus that contains RNA-dependent RNA polymerase.
    (b) Describe the steps of RNA replication for a virus that contains reverse transcriptase.
12. Provide an explanation for the chronic recurrence of cold sores in some people. (*Note:* The cold sores almost always return at the same site.)
13. Slow viral infections such as ______________ might be caused by ________ that are ______________.
14. What are cytopathic effects?
15. Recall from Chapter 1 that Koch's postulates are used to determine the etiology of a disease. Why is it difficult to determine the etiology of
    (a) a viral infection like influenza?
    (b) cancer?
16. The DNA of DNA-containing oncogenic viruses can become integrated into the host DNA. When integrated, it is called a ______________. This results in transformation of the cell. Describe the changes of transformation. How can an RNA-containing virus be oncogenic?

CHALLENGE

1. Discuss the arguments for and against viruses as living organisms.
2. In some viruses, capsomeres function as enzymes as well as structural supports. Of what advantage is this to the virus?
3. Prophages and proviruses have been described as being similar to bacterial plasmids. What similar properties do they exhibit? How are they different?

FURTHER READING

Champe, S. P., ed. 1974. *Phage.* Stroudsburg, PA: Dowden, Hutchinson, and Ross. Papers on the morphology and genetics of phage; chapters on lysogeny and phages of *E. coli.*

Diener, T. O. 1979. *Viroids and Viroid Diseases.* New York: Wiley. A comprehensive treatise on the identification of viroids, viroids pathogenic to plants, and a chapter on possible animal viroids.

Fenner, F. and D. O. White. 1976. *Medical Virology,* 2nd ed. New York: Academic Press. A textbook on the clinical aspects of human viral diseases.

Holland, J. J. February 1974. Slow, inapparent and recurrent viruses. *Scientific American* 230:32–40. Evidence is provided for the involvement of viruses in certain chronic degenerative diseases.

Horne, R. W. 1978. *The Structure and Function of Viruses.* London: Edward Arnold. A short book on the basic biology of viruses.

Hughes, S. S. 1977. *The Virus.* New York: Science History Publications. An "easy reading" history of viruses from the early theories of "contagion" to molecular biology.

Jonathan, P., G. Butler, and A. King. November 1978. The assembly of a virus. *Scientific American* 239:62–69. How the tobacco mosaic virus assembles itself inside a host cell.

Luria, S. E., J. E. Darnell, D. Baltimore and A. Campbell. 1978. *General Virology,* 3rd ed. New York: Wiley. A general textbook on the biology of viruses with chapters on insect, plant, and mammalian viruses.

PART

INTERACTION OF MICROBE AND HOST

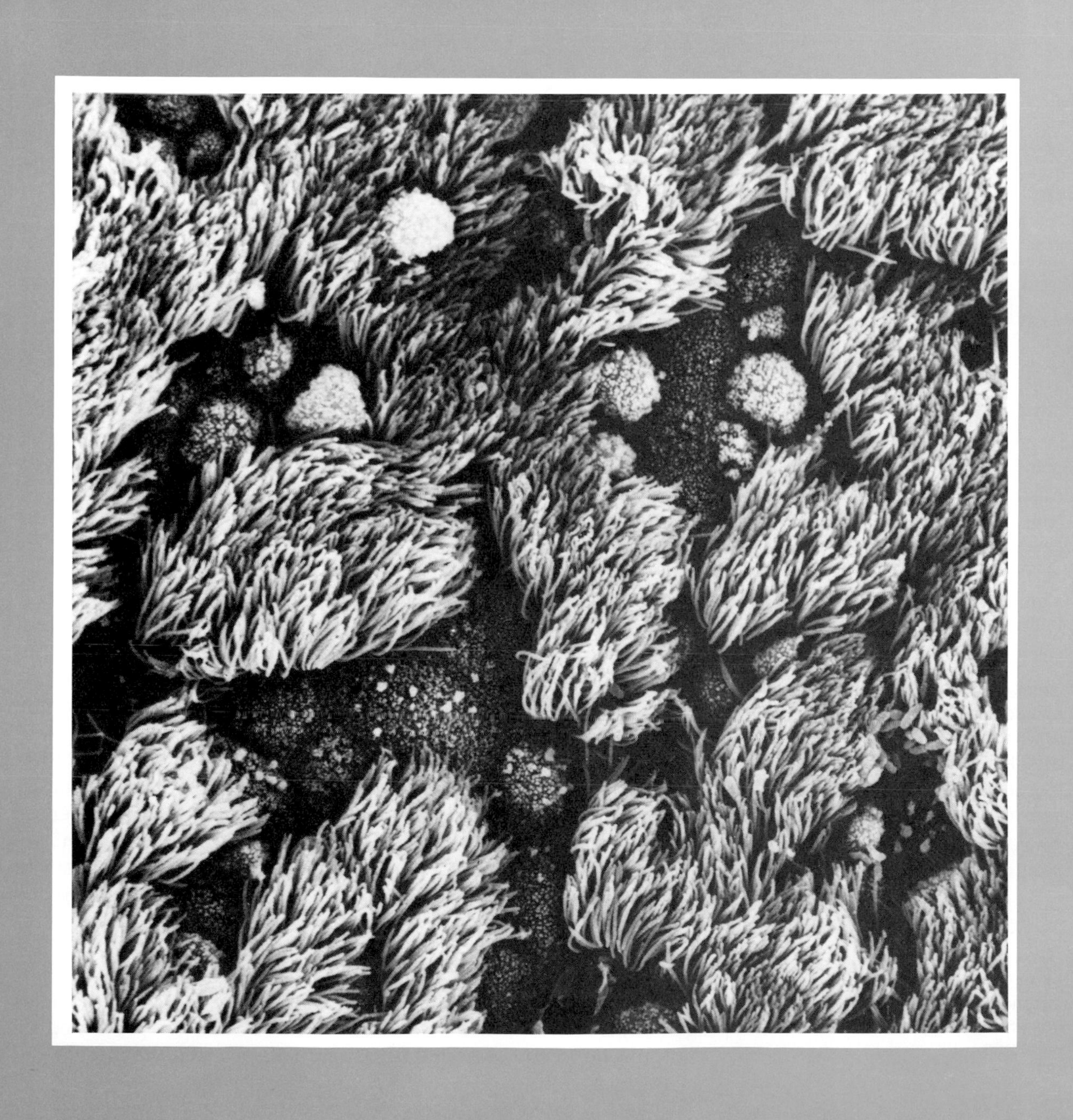

13

Principles of Disease and Epidemiology

Objectives

After completing this chapter you should be able to

- Define normal flora and compare commensalism, mutualism, and parasitism, providing one example of each.
- Identify four predisposing factors to disease.
- List Koch's postulates.
- Define a reservoir of infection and contrast human, animal, and nonliving reservoirs, using one example of each.
- Explain three methods of transmission of disease.
- Explain how to categorize diseases according to incidence.
- Define epidemiology and explain its importance.
- Define each of the following terms: pathogen; etiology; infection; host; disease; nosocomial; acute; chronic; subacute.

Now that you have a basic understanding of the structural and functional aspects of microorganisms, and of the variety of microorganisms that exist, we can consider the relationship between the human body and microorganisms from the viewpoint of health and disease.

We all have certain defense mechanisms that are always operating to keep us healthy. For instance, our unbroken skin is an effective barrier against microbial invasion. And under certain conditions, we can produce proteins called antibodies, which combine with specific microorganisms and contribute to their destruction. However, microorganisms also have their own special properties that make them **pathogenic,** that is, capable of causing disease. Some bacteria can invade our tissues and resist our defenses by producing capsules or certain enzymes. Other bacteria are capable of releasing poisonous substances, called toxins, that can seriously affect our health. A rather delicate balance exists between our defenses and the disease-producing mechanisms of microorganisms. Essentially, when our defenses resist the disease-producing capabilities of the microorganism, we maintain our health. But when the ability of the microorganism to cause disease overcomes our defenses, disease results. Following disease, an individual may recover completely, suffer permanent damage, or die, depending on many factors.

In Part III, we shall examine some of the principles of infection and disease, the mechanisms of pathogenicity, body defenses against disease, and the way microbial disease can be prevented by immunization and controlled by drugs. This chapter will discuss the general principles related to disease, starting with a discussion of the meaning and scope of pathology.

PATHOLOGY, INFECTION, AND DISEASE

Pathology is the science that deals with the study of disease (*pathos,* meaning "suffering"; *logos,* meaning "science"). First, pathology is involved with the cause, or **etiology,** of disease. Second, it deals with the manner in which a disease develops, called **pathogenesis.** And third, pathology is concerned with the structural and functional changes brought about by disease and the final effects on the body as a result of the disease.

Although the terms *infection* and *disease* are sometimes used interchangeably, they do differ somewhat in meaning. **Infection** is the invasion or colonization of the body by potentially pathogenic microorganisms. In this relationship, the body is referred to as the **host,** an organism that shelters and supports the growth of a microorganism. We are in contact with microorganisms throughout our lives, for microbial life is all around us. Moreover, microorganisms are spread from one person to another in the course of everyday living by bodily contact, sneezing, coughing, contact with the same objects, and even speaking and breathing. For the most part, the balance between health and disease favors the health of the host. But when the scale is tipped—such as when our defenses are penetrated by a pathogen—disease results. **Disease** is any change from a state of health. It is an abnormal state in which part or all of the body is not properly adjusted or is not capable of carrying on its normal functions.

The mere presence of microorganisms does not necessarily mean that a disease is in progress. For example, most throat cultures taken at random will contain large numbers of certain kinds of streptococci. Since these bacteria are normally present there, they do not necessarily indicate disease. However, if the same bacteria are repeatedly cultured from an individual's blood, which is normally sterile, then it can be concluded that disease is present. Similarly, although *Escherichia coli* is normally present in large numbers in stools, its presence in the urinary tract is usually an indication of disease.

The presence of microorganisms may even benefit the host. Let us consider this concept by examining the relationship of normal flora to the human body.

NORMAL FLORA

Animals, including humans, are germ-free in utero. Upon birth, however, normal flora begins establishing itself. Just prior to giving birth, lactobacilli in the human mother's vagina multiply rapidly. The newborn's first contact with microorganisms is usually with these lactobacilli, and they are the predominant organisms in the newborn's intestine. After birth, *E. coli* and other bacteria acquired from foods and contact with other humans begin to inhabit the intestine. "Germfree" animals can be reared in the laboratory, however (see box).

Many other microorganisms grow abundantly inside the normal adult body and on its surface. The microorganisms that establish more or less permanent residence (colonize) without producing disease are known as **normal flora** or **normal microbiota** (Figure 13–1). Others may be present for several days, weeks, or months and then disappear.

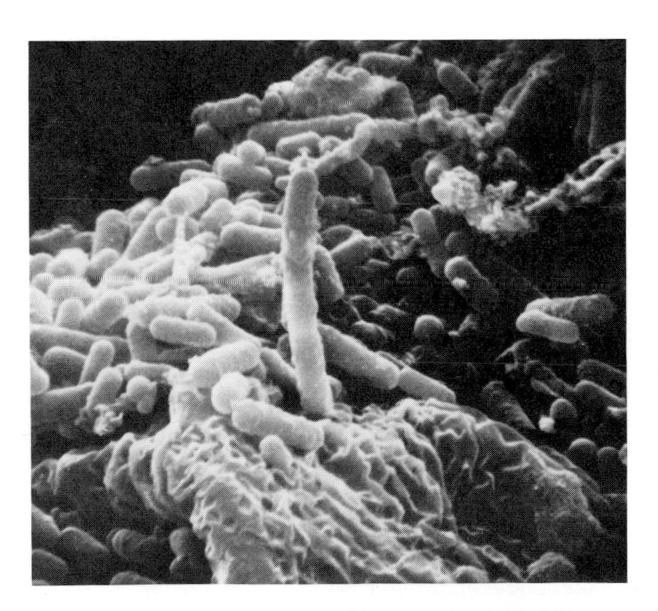

Figure 13–1 Normal flora. This scanning electron micrograph shows lactic acid bacteria on the surface of stomach epithelium (×3900).

These are called **transient flora.** Microorganisms are not found throughout the entire human body, but are localized in certain organs, as shown in Figure 13–2. (We will discuss some of the reasons for this localization in Chapter 15.)

Most microorganisms in the external environment do not colonize the human body because they cannot adjust to the temperature, pH, and available nutrients of certain body areas. Those that do adjust colonize the host. Once established, the normal flora can actually benefit the host by preventing the overgrowth of harmful microorganisms. Only when the balance is upset does disease result. For example, the lactobacilli of the normal human adult vagina maintain a pH of 4.0 to 4.5, which inhibits overpopulation by the yeast *Candida albicans* (see Color Plate IVC). Elimination of lactobacilli with antibiotics, excessive douching, or deodorants allows *Candida* to flourish and become the dominant microorganism in the vagina, which may lead to diseases of the mucous membranes and skin.

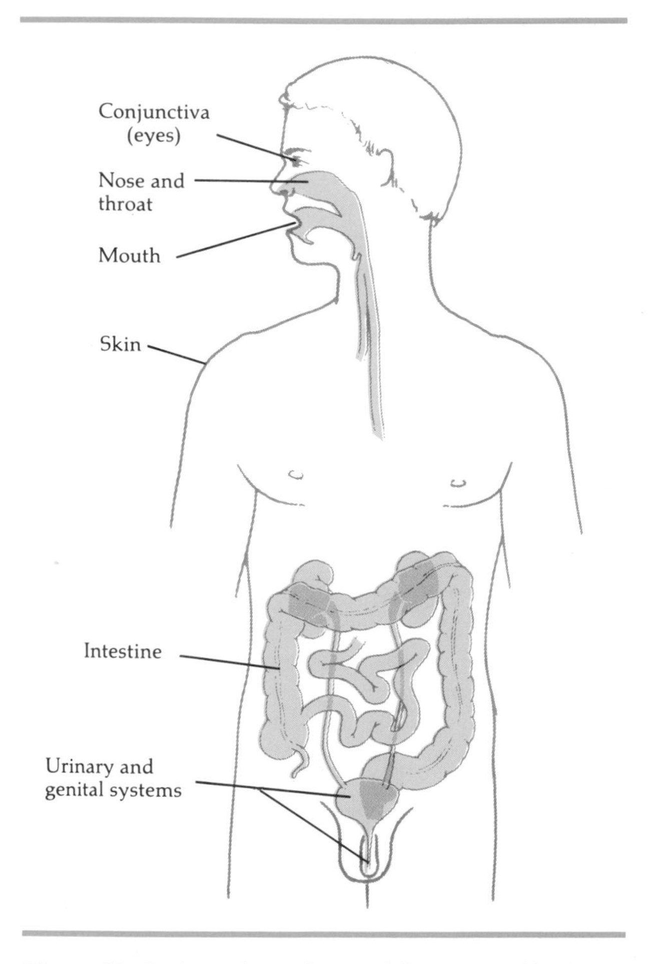

Figure 13–2 Locations of normal flora on and in the human body (shown in color). Various species of bacteria are found in all these regions, as will be described in Part IV of the text. In addition, the mouth, intestines, and vagina are commonly inhabited by the yeast *Candida* and several species of the protozoan *Trichomonas.*

The relationship between the normal flora of a healthy person and that person is called **symbiosis** (literally "living together"). In one type of symbiotic relationship known as **commensalism**, one of the organisms is benefited and the other is unaffected. Many of the microorganisms that make up our normal flora are commensals, including the corynebacteria that inhabit the surface of the eye and certain saprophytic mycobacteria that inhabit the ear and external genitals. These bacteria live on secretions and sloughed-off cells, and they bring about no apparent benefit or harm to the host. In another kind of symbiosis, both organisms are benefited; this is referred to as **mutualism.** For example, bacteria in the large intestine, such as *E. coli*, synthesize vitamin K and certain B vitamins, which are absorbed by the blood and distributed for use by body cells. In exchange, the large intestine provides nutrients for the bacteria so that they can carry on their metabolism. In still another kind of symbiosis, one organism is benefited at the expense of the other, a relationship called **parasitism.** Many disease-causing bacteria are parasites.

Although it is convenient to categorize symbiotic relationships by type, we must keep in mind that under certain conditions the relationship may change. For example, given the proper circumstances, a mutualistic organism such as *E. coli* may become harmful. Opportunists are potentially pathogenic organisms that ordinarily do not cause disease in their normal habitat in a healthy person. But they can cause disease—if, for example, another disease is in progress or if the skin and mucous membranes are broken accidentally or surgically. *E. coli*, as mentioned previously, is generally harmless as long as it remains in the large intestine. But if it gains access to other body sites, such as the urinary bladder, lungs, spinal cord, or wounds, it may cause urinary tract infections, pulmonary infections, meningitis, or abscesses.

In addition to the usual symbionts, there are present in or on the bodies of many people other microorganisms that are generally regarded as pathogenic but which may not actually cause disease in those people. Among the pathogens that are frequently carried without causing disease are enteric cytopathogenic human orphan (ECHO) viruses, which can cause intestinal diseases; adenoviruses, which can cause respiratory diseases; *Neisseria meningitidis*, which often resides peacefully in the respiratory tract but can cause a disease that inflames the coverings of the brain and spinal cord; and *Streptococcus pneumoniae*, which can cause a type of pneumonia.

CAUSES OF DISEASE

Some diseases have a well-understood etiology, some have an etiology that is not completely understood, and still others have no known etiology. As you know, not all diseases are caused by microorganisms. To put microbial diseases in perspective, we will now outline the main etiologies of disease and examine several important predisposing factors. Note that a given disease may fall into more than one etiological category—for example, it may be both inherited and degenerative.

Main Categories of Disease

Infectious Disease Disease-producing microorganisms (pathogens) probably account for half of all human diseases. Examples of infectious diseases are gonorrhea, measles, and meningitis. In Part 4 of this book, we will describe some of the more important pathogens that affect the human body.

Germfree Animals

In nature, all animals beyond the embryo stage are hosts for microorganisms, their normal flora. Impressed by this fact, Pasteur speculated that microbes were essential for animal life. We now know, however, that animals with no microbial flora whatsoever can be reared in the laboratory. Such germfree animals are very useful in studying the functions of microorganisms in the body, as well as the interactions among microorganisms in the body. Germfree animals that are infected with known microorganisms are called **gnotobiotic** animals.

Germfree animals can be obtained by surgically removing the fetus from the mother's uterus under aseptic conditions and maintaining it in a sterile environment. It is also possible to breed animals in a sterile environment, and today most germfree animals used in research are obtained in that way.

Early researchers working with germfree animals were handicapped by technical problems. Introducing air and food and manipulating the animals in any way frequently led to contamination. The first really successful apparatuses for raising germfree animals were stainless steel drums with sterile chambers for introducing materials, glass viewing ports, and other ports sealed to long rubber gloves for handling animals and materials. More modern apparatuses are made of plastic. Containers with walls of transparent, flexible vinyl have proved particularly useful; these can easily be constructed in almost any size and shape.

Research with germfree animals has shown that microbial flora are not absolutely essential to animal life. On the other hand, this research has also shown that germfree animals are at a number of disadvantages compared with normal animals. Germfree animals have undeveloped immune systems: they have little lymphoid tissue, and the levels of immune proteins in their body fluids are very low. When they do come in contact with microorganisms, these animals are unusually susceptible to infection and serious disease. In addition, germfree animals have higher requirements for certain nutrients than other animals, in particular for vitamins normally produced by intestinal bacteria. Clearly, normal flora benefit the host both directly and indirectly.

Nutritional Deficiency Diseases These diseases are caused by a lack of nutrients, such as carbohydrates, lipids, proteins, vitamins, and minerals, which are vital to the metabolism of cells. A few examples of nutritional deficiency diseases are rickets (vitamin D deficiency), scurvy (vitamin C deficiency), and simple goiter (iodine deficiency). On a worldwide basis, infectious diseases and malnutrition are the two most important health problems.

Congenital Diseases A congenital disease is one that is present at birth as a result of some condition that occurs in the uterus. It can result from mutation, maternal infection, drugs, or physical agents. Examples of congenital diseases include cleft palate, heart defects, and clubfoot.

Inherited Diseases Inherited diseases are those that are passed on to the child from the parents through their reproductive cells (sperm and ovum). A few examples are hemophilia, Down's syndrome, and Tay-Sachs disease. Diseases resulting from congenital and inherited factors are sometimes grouped together as **birth defects.** Some birth defects can be diagnosed by a procedure called *amniocentesis,* in which a sample of amniotic fluid containing cells sloughed off the fetus is withdrawn from the uterus; the cells are grown in culture and then analyzed for abnormalities.

Metabolic Diseases Metabolic diseases are those that result from abnormalities in the biochemistry of bodily functions. Some examples are phenylketonuria (PKU), cretinism, and diabetes mellitus. Many metabolic diseases are congenital or inherited.

Degenerative Diseases Degeneration is the wearing down of parts of the body, resulting in a loss of function. Some degeneration is due to the aging process, and both degeneration and aging are linked to excessive caloric intake, radiation, errors in gene function, and loss of immunities, as well as other factors. Osteoarthritis, cirrhosis, and emphysema are examples of degenerative diseases.

Neoplastic Diseases Neoplasms (*neo,* meaning "new"; *plasia,* meaning "molding") are new growths of cells or tissues. They are also called **tumors.** Some are *benign* (noncancerous) and some are *malignant* (cancerous). Cancer has been linked to chemicals, viruses, ultraviolet light in sunlight, X-irradiation, and chronic irritation.

Immunologic Diseases Our immunologic defenses protect us from disease. But there are instances when these same immunologic defenses may attack our own bodies (*autoimmunity*) or overreact (*hypersensitivity*). Autoimmune diseases include systemic lupus erythematosus (LE) and rheumatoid arthritis. Hypersensitive reactions are exemplified by asthma and allergies.

Iatrogenic Diseases These are diseases caused by a physician or health professional during administration of health care. Such diseases may be caused by certain procedures that result in infection (contaminated equipment or a surgical error), drugs (women whose mothers were given diethylstilbesterol to prevent abortion have developed cancer of the vagina 20 to 25 years later), or misuse of radiation. Diseases acquired in a hospital are called **nosocomial diseases.**

Psychogenic Diseases Psychogenic, or emotional, factors are linked to certain diseases, and such factors have an important bearing on the course taken by a given disease. Disorders such as ulcers and colitis are either caused by or associated with psychic stress. Other diseases, such as asthma and dermatitis, are known to have psychological components.

Idiopathic Diseases The term *idiopathic* means undetermined cause. Of the many examples of idiopathic disease, hypertension is perhaps the most common. For 90% of persons with hypertension, there is no known cause.

Predisposing Factors

In addition to the foregoing etiologies of disease, certain predisposing factors must also be considered. A **predisposing factor** is one that makes the body more susceptible to a disease and may alter the course of the disease. Gender may sometimes be a predisposing factor. For example, females show a higher incidence of scarlet and typhoid fevers than males. Males, on the other hand, show increased rates of pneumonia and meningitis. Other aspects of genetic background may play a role, too. Individuals with sickle-cell anemia, for instance, are actually more resistant to malaria than others.

Climate and weather seem to have some effect on the incidence of infectious disease. The incidence of respiratory diseases increases during the winter in temperate regions. This correlates with people staying indoors and having closer contact with each other, facilitating the spread of respiratory pathogens. Other predisposing factors include inadequate nutrition, fatigue, age, unhealthy environment, habits, life style or occupation, preexisting illness, chemotherapy, and emotional disturbances. It is often difficult to know what the exact relative importance is of different predisposing factors, such as gender and life style.

Koch's Postulates

In the historical overview of microbiology presented in Chapter 1, we briefly discussed Robert Koch, one of the founders of microbiology, and his famous postulates. Koch, you may recall, was a German physician who played a major role in establishing that specific microorganisms cause specific diseases. We will now discuss Koch's work in greater detail to see how microbiologists determine the etiology of an infectious disease.

In 1877, Koch published some early papers on anthrax, a disease of cattle that can also occur in humans. Koch demonstrated that certain bacteria (today known as *Bacillus anthracis*) were always present in the blood of animals that had the disease and were not present in healthy animals. Since he knew that the mere presence of the bacteria did not prove that they caused the disease, but may instead have been there as a result of the disease, he experimented further. He took a sample of blood from a diseased animal and injected it into another, healthy animal. The second animal became diseased and died. He repeated this procedure many times, always with the same results. (A key criterion in any scientific proof is that experimental results be repeatable.) Koch also cultivated the microorganism in fluids outside the animal's body, and even after many culture transfers, demonstrated that the bacterium would still cause anthrax. In short, Koch showed that a specific infectious disease (anthrax) is caused by a specific microorganism *(Bacillus anthracis)* that can be isolated and cultured on artificial media.

Koch later used the same methods to show that the bacterium *Mycobacterium tuberculosis* was the causative agent of tuberculosis. As part of this work, he also developed a specific stain for microscopic examination of the bacteria. Koch's research provides a framework for the study of the etiology of any infectious disease. Today, we refer to Koch's experimental requirements as **Koch's postulates.** They are summarized as follows:

1. The same pathogen must be present in every case of the disease.
2. The pathogen must be isolated from the diseased host and grown in pure culture.
3. The pathogen from the pure culture must cause the disease when inoculated into a healthy, susceptible laboratory animal.
4. The pathogen must again be isolated from the inoculated animal and must be shown to be the same pathogen as the original organism.

It should be noted that although Koch's postulates are useful in determining the causative agent of most bacterial diseases, there are some exceptions. For instance, it is known that the bacterium *Treponema pallidum* is the causative agent for syphilis, but virulent strains have never been cultured on artificial media. Moreover, many rickettsial and all viral pathogens cannot be cultured on artificial media because they multiply only within cells.

The discovery of microorganisms that cannot grow on artificial media has made it necessary to modify Koch's criteria and to use alternative methods of culturing and detecting the microbes. For example, when investigators searching for the microbial cause of Legionnaires' disease were unable to isolate the microbe directly from a victim, they took the alternative step of first inoculating lung tissue from a victim into guinea pigs. These guinea pigs developed the disease's coldlike symptoms, whereas guinea pigs inoculated with tissue from an unafflicted person did not. Then tissue samples from the diseased guinea pigs were cultured in yolk sacs of chick embryos (a method that reveals the growth of extremely small bacteria). After incubation, electron microscopy uncovered rod-shaped bacteria in the chick embryos. Finally, modern immunologic techniques (which will be discussed in Chapter 16) were used to show that the bacteria in the chick embryos were the same as the bacteria in the guinea pigs and in afflicted humans (see Color Plate IIIB).

There are a number of situations in which a human host exhibits certain signs and symptoms that can be distinguished from all others and which are associated only with a certain pathogen and its disease. Good examples are pathogens responsible for diphtheria and tetanus. These microorganisms always give rise to their own distinguishing signs and symptoms that can be produced by no other microbe. They are unequivocally the only organisms that produce their respective diseases. But some infectious diseases are not quite as clear-cut. For example, with nephritis (inflammation of the kidneys), several different pathogens might be involved, all of which give rise to the same signs and symptoms of nephritis. Thus, it is often difficult to know which particular microorganism is causing a specific disease. Other infectious diseases that have similar, poorly defined etiologies are pneumonia, peritonitis, and meningitis.

There also exist certain pathogens that can cause several pathologies. *Mycobacterium tuberculosis*, for example, is implicated in diseases of the lungs, skin, bones, and internal organs. *Streptococcus pyogenes* can cause sore throat, scarlet fever, skin infections (erysipelas), puerperal fever, and osteomyelitis, among other diseases. Such infections can usually be recognized and distinguished from infections of the same organs caused by other pathogens by laboratory methods and clinical signs and symptoms.

We will now turn our attention to how infections are spread among a population, taking into account the sources of pathogenic microorganisms and the different ways these microbes are transmitted.

SPREAD OF INFECTION

In considering how diseases are spread throughout a population, we will first take a look at the various places microorganisms inhabit prior to their transmission.

Reservoirs

For a disease to perpetuate itself, there must be a continual source of the infection. This source must be either a living organism or a nonliving substance that provides a pathogen with adequate conditions for survival and multiplication and an opportunity for transmission. Such a source is called a **reservoir of infection.** These reservoirs may be classified into human reservoirs, animal reservoirs, and nonliving reservoirs.

Human reservoirs

The principal living reservoir of microorganisms that cause human disease is the human body itself. Many people harbor pathogens and transmit them to others, directly or indirectly. People who exhibit signs and symptoms of a disease are obvious transmitters of the disease. However, some people can harbor pathogens and transmit them to others without themselves exhibiting any sign of illness. These people, called **carriers,** represent an important living reservoir of infection. Some carriers have infections for which no signs or symptoms are ever exhibited. Other people are carriers during symptom-free stages of an illness: during the incubation period (before symptoms appear) or during the convalescent period (recovery). Human carriers play an important role in the spread of diseases such as diphtheria, salmonellosis, epidemic meningitis, amoebic dysentery, pneumonia, and streptococcal infections.

Animal reservoirs

Animals other than humans represent another group of living reservoirs that can transmit diseases to humans. This group includes both sylvatic (wild) and domestic animals. Diseases that occur primarily in wild and domestic animals but which can be transmitted to humans are called **zoonoses.** About 150 zoonoses are known. The transmission of zoonoses to humans can occur by direct contact with the infected animals; by contamination of food and water; by contact with contaminated hides, fur, or feathers; by consumption of infected animal products; or by insect vectors (insects that transmit pathogens). A few representative zoonoses are presented in Table 13–1.

Nonliving reservoirs

The two major nonliving reservoirs of infectious disease are soil and water. The soil environment harbors pathogens such as fungi that cause mycoses and *Clostridium botulinum*, the agent of botulism. Water that has been contaminated by the feces of humans and other animals is a reservoir for several pathogens, most notably the gastrointestinal pathogens.

Table 13–1 Selected Zoonoses That Can Be Transmitted to Humans

Disease	Etiology	Reservoir	Method of Transmission	Refer to Chapter
Viral:				
Influenza (some strains)	Myxovirus	Swine	Direct contact	23
Rabies	Rhabdovirus	Bats, skunks, foxes	Direct contact (bite)	21
Western equine encephalitis	Arbovirus	Horses, birds	*Culex* mosquito bite	20
Yellow fever	Arbovirus	Monkeys	*Aedes* mosquito bite	20
Bacterial:				
Anthrax	*Bacillus anthracis*	Domestic livestock	Direct contact with contaminated hides or animals; air; food	22
Brucellosis	*Brucella*	Domestic livestock	Direct contact with contaminated milk, meat, or animals	22
Bubonic plague	*Yersinia pestis*	Rodents	Flea bites	20
Cat-scratch fever	*Chlamydia*	Domestic cats	Direct contact	20
Leptospirosis	*Leptospira*	Wild mammals, domestic dogs and cats	Direct contact with urine, soil, water	25
Pneumonic plague	*Yersinia pestis*	Rodents	Direct contact	20
Psittacosis and Ornithosis	*Chlamydia*	Birds, especially parrots	Direct contact	23
Q fever	*Coxiella burnetii*	Domestic livestock	Inhalation, tick bites	23
Rocky Mountain spotted fever	*Rickettsia rickettsii*	Rodents	Tick bites	20
Salmonellosis	*Salmonella* spp.	Poultry, rats, turtles	Ingestion of contaminated food, water	24
Tularemia	*Francisella tularensis*	Wild and domestic mammals, especially wild rabbits	Direct contact with infected animals; deer-fly bites	22
Typhus fever	*Rickettsia typhi*	Rodents	Flea bites	20
Fungal:				
Ringworms	*Trichophyton* *Microsporum* *Epidermophyton*	Domestic mammals	Direct contact; fomites	19
Protozoan:				
Chagas' disease	*Trypanosoma cruzi*	Wild mammals	"Kissing bug" bite	21
Malaria	*Plasmodium*	Monkeys	*Anopheles* mosquito bite	20
Toxoplasmosis	*Toxoplasma gondii*	Cats and other mammals	Direct contact with infected tissues or fecal material	22
Helminthic:				
Hydatid cyst	*Echinococcus granulosus*	Dogs	Direct contact with fecal material	11
Tapeworm (beef)	*Taenia saginata*	Cattle	Ingestion of contaminated beef	24
Trichinosis	*Trichinella spiralis*	Pigs, bear	Ingestion of contaminated meat	24

Nosocomial (hospital-acquired) infections

The hospital is a unique environment. Bedding, utensils, and instruments may become contaminated with microorganisms, and the hospital personnel are human reservoirs. Meanwhile, hospital patients are usually in a weakened condition from disease or surgery and may be predisposed to infections. A **nosocomial infection** is one that develops during a hospital stay—that is, it was not present when the patient was admitted. One-third of all infections in hospitalized patients are nosocomial, and most of these are not caused by microorganisms usually regarded as pathogenic, but by opportunistic human commensals and saprophytes. The lowered resistance of the patient increases susceptibility to infections by bacteria introduced by inanimate objects or hospital staff and visitors.

Transmission of Disease

The causative agents of disease may be transmitted from the source of the infection to the host by direct contact, indirect contact, or arthropod vectors (Figure 13–3).

Direct contact

With **direct contact,** the infection is spread more or less directly from one host to another through some kind of close association between the hosts. In some cases, actual body contact is involved, as in the transmission of venereal disease during sexual intercourse. Other forms of body contact that serve as vehicles of transmission include handshaking and kissing. In other cases of direct contact, the hosts are in close association but do not really make direct physical contact. In **droplet infection,** for example, the microorganisms are carried in small liquid drops, a result of the discharge of microorganisms in a fine spray from the mouth and nose during coughing, sneezing, laughing, and even talking (Figure 13–4). Diseases spread by direct contact include tuberculosis, measles, diphtheria, the common cold, scarlet fever, and smallpox. Some diseases, such as rabies, are transmitted from animals to humans by direct contact.

Indirect contact

Indirect contact refers to the transmission of pathogens by food, water, and objects contaminated with infectious materials from a diseased host (sputum, feces, pus). Nonliving objects that spread infection are called **fomites** and include such familiar examples as handkerchiefs, towels, bedding, diapers, and drinking cups. Transmission by indirect contact does not involve immediate contact with another host. Diseases spread by indirect contact are those in which the infectious material usually enters the body by the mouth, either in air or in food.

Although microorganisms do not grow while suspended in air, they will often survive. Thus, air is a medium that provides for the transfer of many diseases of the respiratory tract, frequently by droplets carried by air currents.

Gastrointestinal pathogens may be transmitted in foods that become contaminated during processing or in foods prepared from contaminated animals. Poultry products may be contaminated with *Salmonella* found in the intestines of the birds, and human carriers of *Salmonella* or *Staphylococcus aureus* may further contaminate foods during preparation. Milk and dairy products (such as cheese and yogurt) may be contaminated with pathogenic bacteria. Unpasteurized milk has been responsible for the transmission of tuberculosis, brucellosis, and salmonellosis. (See Chapter 7 for a discussion of pasteurization.)

Arthropod vectors

Arthropods are the most important group of disease **vectors,** animals that carry pathogens from one host to another. (Insects and other arthropod vectors were discussed in Chapter 11.) Arthropod vectors transmit disease by two general methods. In *mechanical transmission,* the insects carry the pathogens on their feet and other body parts. If the insect makes contact with food, the pathogens transferred to the food may later be swallowed by the host. Houseflies, for instance, are capable of transferring the pathogens of typhoid fever and bacillary dysentery (shigellosis) from the feces of patients to food.

In *biological transmission,* the situation is more complex. Here, the arthropod bites an infected per-

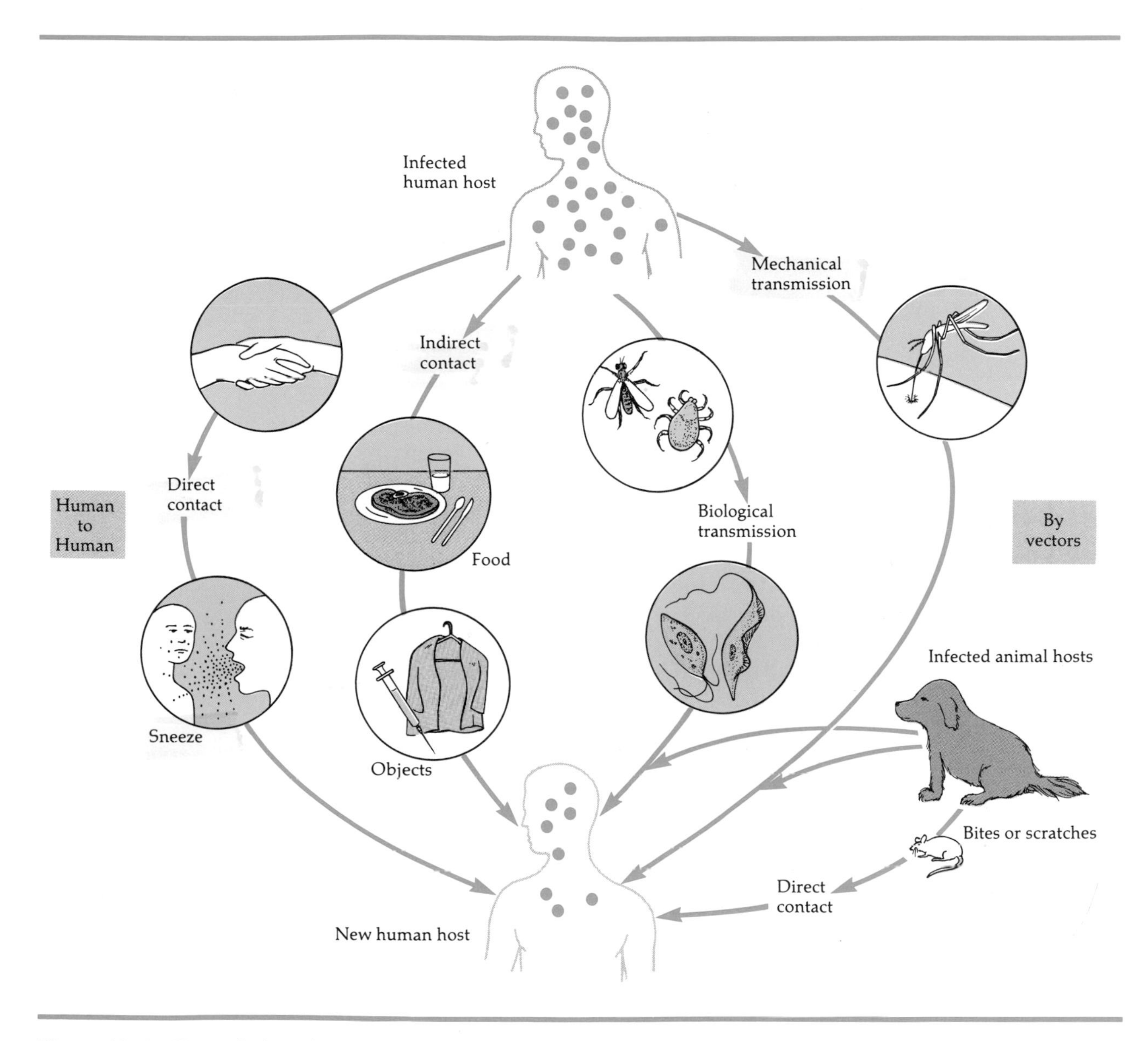

Figure 13–3 Transmission of microbial diseases.

son or animal and ingests some of the infected blood. The pathogens then reproduce in the vector, and the increased numbers of pathogens increase the possibility of being transmitted to another host. Some parasites reproduce in the gut of the arthropod and are passed with feces. If the arthropod defecates while biting, the parasite can enter the wound. Other parasites reproduce in the vector's gut and migrate to the salivary gland for direct injection into a bite. Some protozoan and helminthic parasites use the vector as a host for a developmental change in their life cycle.

Table 11–6 (Chapter 11, p. 303) lists a few important arthropod vectors and the diseases they transmit, including malaria, typhus fever, and bubonic plague. Figure 13–3 summarizes all the ways pathogenic microorganisms are transmitted from the source of infection to the host.

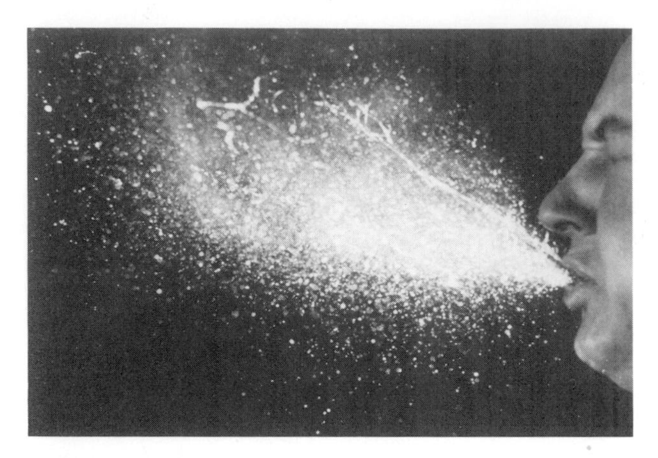

Figure 13–4 This high-speed photograph shows the spray of small droplets that comes from the mouth during a sneeze.

KINDS OF DISEASES

For purposes of discussing disease in this and later chapters, we will define certain key terms that relate to the nature and scope of disease. Any disease spread from one host to another, either directly or indirectly, is said to be a **communicable disease.** Typhoid fever and tuberculosis are examples. A **noncommunicable disease** is one whose microorganisms normally inhabit the body, only occasionally producing disease, or reside outside the body, producing disease only when introduced into the body. An example is tetanus. *Clostridium tetani* produces disease only when it is introduced into the body via abrasions or wounds. A **contagious disease** is one that is *easily* spread from one person to another.

To understand the scope of a disease, it is useful to know something about its occurrence. The **incidence** of a disease is the fraction of a population that contracts it during a particular length of time. The **prevalence** of a disease is the fraction of the population having the disease at a given time. These numbers give indications of the range of occurrence of a disease and its tendency to affect certain groups of people more than others.

One useful way of defining the scope of a disease is in terms of its severity or duration. Accordingly, an **acute disease** is one that develops rapidly but lasts only a short time. A good example is influenza. A **chronic disease** develops more slowly, and the body reactions are less severe, but it is likely to be continuous or recurrent for long periods of time. Tuberculosis, syphilis, and leprosy fall into this category. A disease that is intermediate between acute and chronic is described as a **subacute disease.**

Frequency of occurrence is another criterion that is used to classify diseases. If a particular disease occurs occasionally, it is called a **sporadic disease.** Polio might be considered such a disease. A disease constantly present in a certain population is called an **endemic disease,** and an example here is the common cold. If many people in a given area acquire a certain disease in a relatively short period of time, it is referred to as an **epidemic disease.** Influenza and measles are examples of diseases that often achieve epidemic status. Some authorities consider gonorrhea and certain other venereal diseases epidemic at this time. Finally, an epidemic disease that occurs worldwide is referred to as a **pandemic disease.** We experienced a pandemic of influenza in the late 1950s.

Infections may also be classified based on the extent to which the host's body is affected. A **local infection** is one in which the invading microorganisms are limited to a relatively small area of the body. Examples of local infections are boils and abscesses. In a **systemic,** or **generalized, infection,** microorganisms or their products are spread throughout the body by the blood or lymphatic system. Typhoid fever is an example. Very frequently, agents of a local infection may enter a blood or lymph vessel and spread to other parts of the body. In this case, we refer to the condition as a **focal infection.** Focal infections may arise from infections in the teeth, tonsils, or sinuses. The presence of bacteria in the blood is known as **bacteremia,** and if the bacteria actually multiply in the blood, the condition is called **septicemia.**

The health of the body also provides a basis for classifying infections. A **primary infection** is an acute infection that causes the initial illness. A **secondary infection** is one caused by an opportunist only after the primary infection has weakened the body's defenses. Secondary infections of the skin and respiratory tract are common and are sometimes more dangerous than the primary infections. Streptococcal bronchopneumonia following whooping cough, measles, and influenza is an example of a secondary infection. An **inapparent,** or **subclinical, infection** is one that does not cause any noticeable illness. Examples can be polio and infectious hepatitis, in which the individuals carry the viruses but show no illness because they are protected by antibodies.

When a particular disease affects the body, it may manifest itself in certain ways. In other words, it may signal the body that something is wrong. Let us now consider some of the signals of disease.

SIGNALS OF DISEASE

The specific disease affecting the body alters body structures and functions in certain ways, and these alterations are usually indicated by several kinds of evidence. For example, the patient may experience certain **symptoms,** changes in body function felt by the patient, such as pain and malaise. These are subjective changes and are not apparent to the observer. The patient may also exhibit **signs,** which are objective changes that the physician can observe and measure. Signs frequently evaluated include lesions (changes produced in tissues by disease), swelling, fever, and paralysis. Sometimes a specific group of symptoms or signs always accompanies a particular disease. Such a group is called a **syndrome.** The **diagnosis** of a disease, that is, its identification, is achieved by evaluation of the signs and symptoms, as well as the results of certain laboratory tests.

EPIDEMIOLOGY

To control and treat a disease effectively, it is desirable to identify the causative agent. It is also desirable to understand the mode of transmission and distribution of the disease. The science that deals with when and where diseases occur and how they are transmitted in the human population is called **epidemiology.** In today's crowded, overpopulated world, where frequent travel and mass production and distribution of foods and other goods are a way of life, epidemiology assumes an important and ever-growing role. A contaminated water supply or contaminated food can affect many thousands of people. The occurrence of Legionnaires' disease not so long ago (see Chapter 1) underscored the importance of an epidemiologic investigation in determining the source, means of transmission, and distribution of a disease.

An epidemiologist not only determines the etiology of a disease, but also identifies other factors that may be important, such as geographical distribution and nutrition, gender, and age of the persons affected. The epidemiologist also evaluates how effectively a disease is being controlled in a community—for example, by vaccination programs. By determining the frequency of a disease in a population and identifying the factors responsible for its transmission, an epidemiologist provides physicians with important information concerning the treatment and prognosis of a disease. (Figure 13–5 shows the graphs obtained from some hypothetical epidemiologic data, namely, the number of new cases of a disease reported each day within a certain time period. Such graphs provide information about whether disease outbreaks were sporadic or epidemic, and, if epidemic, how the disease was probably spread.) Finally, an epidemiologist can provide useful data for evaluating and planning overall health care for a community.

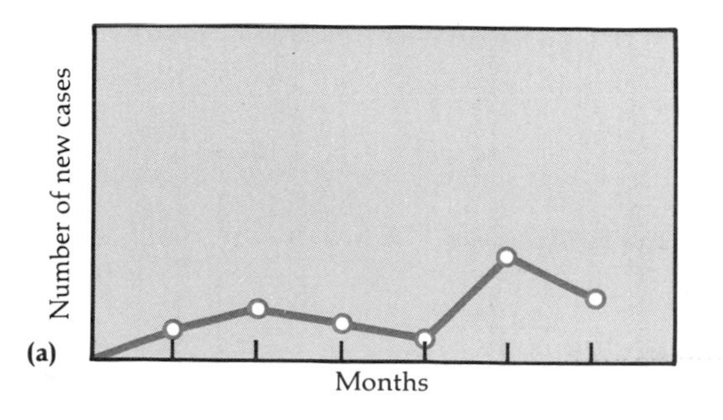

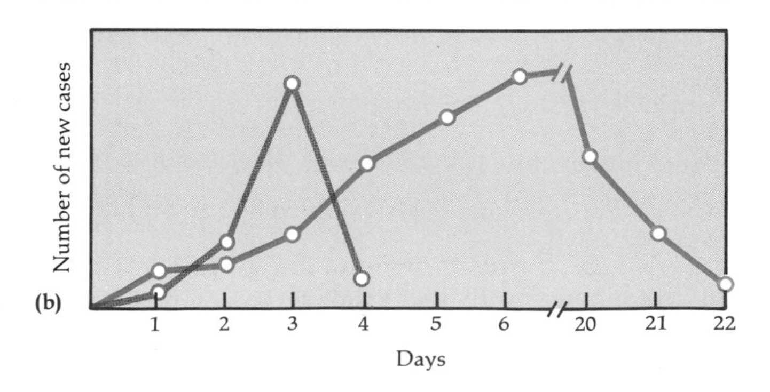

Figure 13–5 Patterns of disease spread. **(a)** Sporadic cases of a disease. **(b)** An epidemic spread from a common source, such as contaminated food (gray line), and an epidemic spread person-to-person (colored line).

Epidemiology is a major concern of the state and federal public health departments. The Centers for Disease Control (CDC), a branch of the U.S. Public Health Service located in Atlanta, Georgia, are a central source of epidemiologic information in the United States.

The CDC issues a publication called *Morbidity and Mortality Weekly Report (MMWR)*, which is read by microbiologists, physicians, and other hospital and public health personnel. The *MMWR* contains data on the incidence of specific notifiable diseases **(morbidity)** and the deaths from these diseases **(mortality),** usually organized by state. (**Notifiable diseases** are those for which physicians must report cases to the Public Health Service. Examples are measles, tetanus, and gonorrhea.) Publication articles include reports of disease outbreaks, case histories of special interest, and summaries of the status of particular diseases over a recent period. These articles often include recommendations for procedures for diagnosis, immunization, and treatment. A number of graphs and other data in this book are from *MMWR*, and some of the boxes are direct excerpts from this publication.

In the next chapter, we will consider the mechanisms of pathogenicity. There we will discuss in more detail how microorganisms enter the body, how they cause disease, the pattern of infection, the effects of disease on the body, and how pathogens leave the body.

STUDY OUTLINE

INTRODUCTION (pp. 346–347)

1. Pathogenic microorganisms have special properties that allow them to invade the human body or produce toxins.
2. When the microorganism overcomes the body's defenses, a state of disease results.

PATHOLOGY, INFECTION, AND DISEASE (p. 347)

1. Pathology is the science that deals with the study of disease.
2. Pathology is concerned with the etiology (cause) of disease, pathogenesis, and effects of disease.
3. Infection is the invasion and growth of pathogens in the body.
4. A host is an organism that shelters and supports the growth of pathogens.
5. Disease is an abnormal state in which part or all of the body is not properly adjusted or not capable of carrying on normal functions.

NORMAL FLORA (pp. 347–349)

1. Animals, including humans, are germ-free in utero.
2. Microorganisms begin colonization in and on the surface of the body soon after birth.
3. Microorganisms that establish permanent colonies inside or on the body without producing disease are called normal flora.
4. Transient flora are present for varying amounts of time and then disappear.
5. The normal flora and human exist in a symbiosis (living together).

6. The three types of symbiosis are commensalism (one organism benefits and the other is unaffected); mutualism (both organisms benefit); and parasitism (one organism benefits and one is harmed).
7. Opportunists (opportunistic pathogens) do not cause disease under normal conditions, but may cause infections under special conditions.

CAUSES OF DISEASE (pp. 349–352)

Main Categories of Disease (pp. 349–350)

1. Diseases that have a known etiology include infectious diseases (e.g., pneumonia), nutritional deficiency diseases (e.g., rickets), congenital diseases (e.g., cleft palate), inherited diseases (e.g., hemophilia), metabolic diseases (e.g., diabetes mellitus), degenerative diseases (e.g., emphysema), neoplastic diseases (e.g., cancer), immunologic diseases (e.g., rheumatoid arthritis), iatrogenic diseases (e.g., staphylococcus infections), and psychogenic diseases (e.g., ulcers).
2. Diseases that have no known etiology are referred to as idiopathic diseases.
3. Diseases acquired in the hospital are called nosocomial diseases.

Predisposing Factors (pp. 350–351)

1. A predisposing factor is one that makes the body more susceptible to disease or alters the course of a disease.
2. Examples include climate, age, fatigue, and inadequate nutrition.

Koch's Postulates (pp. 351–352)

1. Koch's postulates establish that specific microbes cause specific diseases.
2. Koch's postulates state the following: (1) the same pathogen must be present in every case of the disease; (2) the pathogen must be isolated in pure culture; (3) the pathogen isolated from pure culture must cause the same disease in a healthy, susceptible laboratory animal; and (4) the pathogen must be reisolated from the inoculated laboratory animal.

SPREAD OF INFECTION (pp. 352–356)

Reservoirs (pp. 352–354)

1. A continuous source of infection is called a reservoir of infection.
2. People who have a disease or are carriers of pathogenic microorganisms constitute human reservoirs of infection.
3. Zoonoses (diseases that occur in wild and domestic animals) can be transmitted to humans from animal reservoirs of infection.
4. Some pathogenic microorganisms grow in nonliving reservoirs such as soil and water.

5. Opportunists and saprophytes may cause nosocomial infections in hospitalized patients.

Transmission of Disease (pp. 354–356)

1. Transmission by direct contact includes close physical contact between two hosts and droplet infection resulting from sneezing or talking.
2. Transmission by fomites (inanimate objects), food, and water constitutes indirect contact.
3. Arthropod vectors carry pathogens from one host to another by both mechanical and biological transmission.

KINDS OF DISEASE (pp. 356–357)

1. Communicable diseases are transmitted directly or indirectly from one host to another.
2. Noncommunicable diseases are caused by microorganisms that normally grow outside the human body and are not transmitted from one host to another.
3. A contagious disease is one that is easily spread from one person to another.
4. Disease occurrence is reported by incidence (number of people with the disease) and prevalence (incidence at a particular time).
5. The scope of a disease may be defined as acute, chronic, or subacute.
6. Diseases are classified by frequency of occurrence: sporadic, endemic, epidemic, and pandemic.
7. A local infection affects a small area of the body; a systemic infection is spread throughout the body through the circulatory system.
8. A secondary infection may occur after the host is weakened from a primary infection.
9. An inapparent, or subclinical, infection does not cause any signs of disease in the host.

SIGNALS OF DISEASE (p. 357)

1. A patient may exhibit symptoms (subjective changes in body functions) and signs (measurable changes), which are used by a physician to make a diagnosis (identification of the disease).
2. A specific group of symptoms or signs that always accompanies a specific disease is called a syndrome.

EPIDEMIOLOGY (pp. 357–358)

1. The science of epidemiology is the study of the transmission, incidence, and frequency of disease.

2. The Centers for Disease Control are the main source of epidemiologic information in the United States.
3. The CDC publishes *Morbidity and Mortality Weekly Report* to provide information on morbidity (incidence) and deaths (mortality).

STUDY QUESTIONS

REVIEW

1. Differentiate between the following pairs of terms.
   (a) Etiology and pathogenesis
   (b) Infection and disease
   (c) Communicable disease and noncommunicable disease
2. What is meant by normal flora? How do they differ from transient flora?
3. Define symbiosis. Differentiate between commensalism, mutualism, and parasitism, and give an example of each.
4. List four predisposing factors to disease.
5. Describe how Koch's postulates establish the etiology of many infectious diseases.
6. What is a reservoir of infection? Match the following diseases with their reservoirs.

___ Influenza	(a) Nonliving
___ Rabies	(b) Human
___ Botulism	(c) Animal

7. Describe the various ways diseases can be transmitted in each of the following categories.
   (a) Transmission by direct contact
   (b) Transmission by indirect contact
   (c) Transmission by arthropod vectors
8. Indicate whether each of the conditions described is typical of subacute, chronic, or acute infections.
   (a) Patient experiences rapid onset of malaise; symptoms last five days.
   (b) Patient experiences cough and breathing difficulty for months.
   (c) Patient has no apparent symptoms and is a known carrier.
9. Of all the hospital patients with infections, one-third do not enter the hospital with an infection. How do they acquire these infections? What is the method of transmission of these infections? What is the reservoir of infection?
10. Differentiate between an endemic and epidemic state of infectious disease.
11. What is epidemiology? What is the role of the Centers for Disease Control (CDC)?
12. Distinguish between symptoms and signs as signals of disease.

13. How can a local infection become a systemic infection?

CHALLENGE

1. Why don't Koch's postulates apply to all infectious diseases?
2. Why are some organisms that constitute normal flora described as commensals while others are described as mutualistic?
3. Name the method of transmission of the following diseases.
   (a) Malaria (b) Influenza
   (c) Nosocomial infections (d) Salmonellosis
   (e) Syphilis
4. Mark the following graph to show when this disease occurred sporadically, endemically, and epidemically. What would have to be shown to indicate a pandemic of this disease?

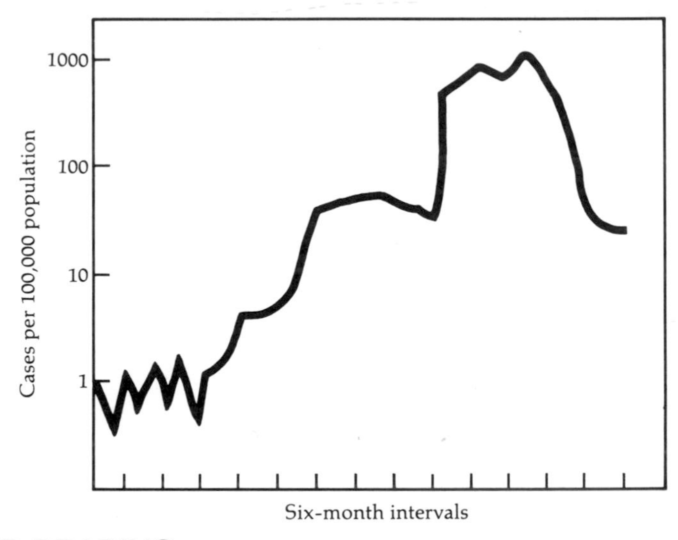

FURTHER READING

Austin, D. F. and S. B. Werner. 1974. *Epidemiology for the Health Sciences.* Springfield, IL: C. C. Thomas. A primer on epidemiological concepts.

Burnet, M. and D. O. White. 1972. *Natural History of Infectious Diseases*, 4th ed. Cambridge: Cambridge University Press. This short book has chapters on the evolution of infectious diseases, how diseases are acquired, and how epidemics occur.

Busvine, J. R. 1975. *Arthropod Vectors of Disease.* London: Edward Arnold. Describes vector-borne diseases and the different ways vectors can transmit diseases.

Mausner, J. S. and A. K. Bahn. 1974. *Epidemiology.* Philadelphia, PA: W. B. Saunders. A comprehensive reference on the collection and interpretation of epidemiological data.

Pike, R. M. 1979. Laboratory-associated infections: incidence, fatalities, causes, and prevention. *Annual Review of Microbiology* 33:41–66. A summary of the particular diseases acquired by laboratory personnel.

14

Mechanisms of Pathogenicity

Objectives

After completing this chapter you should be able to

- Define each of the following terms: portal of entry; pathogenicity; virulence; LD_{50}; portal of exit.
- Explain how capsules, cell wall components, and enzymes contribute to the invasiveness of pathogens.
- Compare the effects of hemolysins, leukocidins, coagulase, kinases, and hyaluronidase.
- Contrast the nature and effects of exotoxins and endotoxins.
- List seven cytopathic effects of viral infections.
- Discuss the causes of symptoms in fungal, protozoan, and helminthic diseases.
- Put the following terms in their proper sequence as they relate to the pattern of disease: period of decline; period of convalescence; period of illness; crisis; prodromal period; period of incubation.

Now that you have a basic understanding of the principles of disease, we will take a look at some of the specific aspects of microorganisms that contribute to **pathogenicity,** the ability to cause disease in a host. Our starting point will be a discussion of the routes taken by microorganisms to gain entrance into the human body.

PORTALS OF ENTRY

Pathogens invade the body by several avenues. We call the avenue by which a microbe gains access to the body a **portal of entry.**

Respiratory Tract

The respiratory tract offers the easiest and most frequently traveled access route for infectious microorganisms. Microbes are taken into the nose or mouth by the inhalation of drops of moisture and dust particles that contain them. Some diseases contracted via the respiratory tract are the common cold, pneumonia, tuberculosis, influenza, measles, and smallpox.

Gastrointestinal Tract

Another common portal of entry is the gastrointestinal tract. Microorganisms contracted from food, water, milk, and fingers enter the body this way, although most of these microbes are destroyed by hydrochloric acid and enzymes in the stomach and by bile and enzymes in the small intestine. Those that survive may cause diseases such as poliomyelitis, infectious hepatitis, typhoid fever, amoebic dysentery, bacillary dysentery (shigellosis), and cholera. The pathogens are eliminated with feces and may be transmitted via water, food, or fingers. Most pathogens enter through the gastrointestinal and respiratory tracts.

Skin and Mucous Membranes

Most microorganisms cannot penetrate unbroken skin. Only the hookworm *Necator americanus* has actually been demonstrated to bore through intact skin, and some fungi grow on the keratin in skin and infect the skin itself. Microorganisms that gain access to the body through the unbroken skin use existing openings such as hair follicles and sweat ducts.

Many bacteria and viruses can get access to the body by penetrating the mucous membranes lining the conjunctiva, nose, and mouth. Bacteria that penetrate mucous membranes may first adhere to the membrane surface (Figure 14–1). An important pathogen capable of penetrating mucous membranes of the genitourinary system is *Treponema pallidum*, the causative agent of syphilis.

Parenteral Route

Other microorganisms gain access to the body when they are deposited directly into the tissues beneath the skin and mucous membranes when these barriers are penetrated or injured (traumatized). This route is referred to as the **parenteral route.** The parenteral route is established by punctures, injections, bites, cuts, wounds, surgery, and splitting due to swelling or drying.

Even though microorganisms enter the body, they do not necessarily cause disease. Whether or not disease results depends on several related factors. One of these factors is the portal of entry. Many pathogens have a preferred portal of entry as a prerequisite to their being able to cause disease. If they gain access to the body by an alternate route, disease may not occur. For example, the bacterium of typhoid fever, *Salmonella typhi*, produces all the signs and symptoms of the disease when swallowed (preferred route). But if the same bacteria are rubbed on the skin (alternate route), no reaction, or only a slight inflammation, occurs. Streptococci that are inhaled (preferred route) may cause pneumonia. Those that are swallowed (alternate route) generally do not produce signs or symptoms. Some pathogens, like the microorganism that causes plague, may enter the body by more than one preferred route. The portals of entry for some common pathogens are given in Table 14–1.

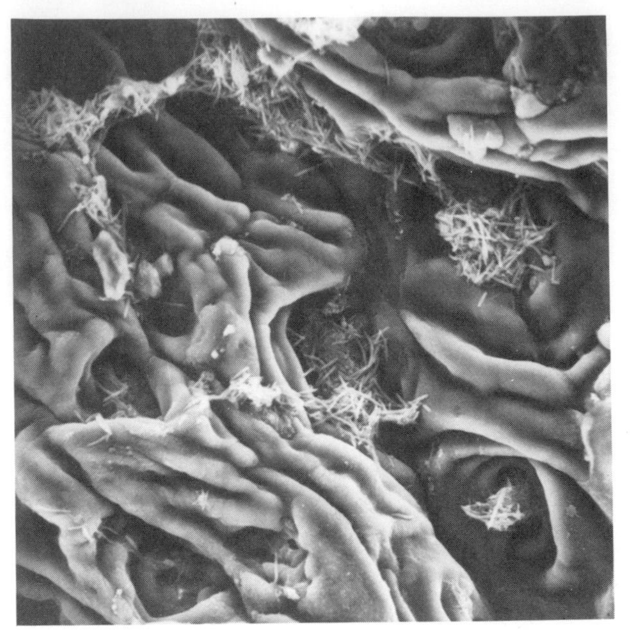

Figure 14–1 This scanning electron micrograph shows mucous membranes of the large intestine with bacteria in the crypts (×420).

PATHOGENIC PROPERTIES OF BACTERIA

The ability of a pathogen to produce a disease is called **pathogenicity,** and the degree of pathogenicity is referred to as **virulence.** Sometimes the two terms are used interchangeably. Our major concern here is to outline some of the factors that contribute to virulence. If these factors overpower the defenses of the host, disease results. But if the factors contributing to virulence are neutralized or overcome by the defenses of the host, health is maintained.

One important factor related to virulence is the *number of microorganisms* involved. If only a few microbes enter the body, it is likely that they will be overcome by the host's defenses. If, on the other hand, large numbers of microbes gain entry, then it is probable that the stage is set for disease. Thus, the likelihood of disease increases with greater numbers of pathogens and decreases with greater

Table 14–1 Causative Agents for Some Common Diseases Arranged by Portal of Entry

Portal of Entry	Causative Agent*	Disease	Incubation Period
Respiratory tract	*Corynebacterium diphtheriae*	Diphtheria	2–5 days
	Neisseria meningitidis	Bacterial meningitis	1–7 days
	Streptococcus pneumoniae	Pneumococcal pneumonia	Variable
	*Mycobacterium tuberculosis***	Tuberculosis	Variable
	Bordetella pertussis	Whooping cough (pertussis)	12–20 days
	Myxovirus	Influenza	18–36 hours
	Paramyxovirus	Measles (rubeola)	11–14 days
	Togavirus	German measles (rubella)	2–3 weeks
	Epstein–Barr virus (herpesvirus)	Infectious mononucleosis	2–6 weeks
	Zoster (herpesvirus)	Chickenpox (varicella)	14–16 days
	Poxvirus	Smallpox (variola)	12 days
	Coccidioides immitis (fungus)	Coccidioidomycosis (primary infection)	1–3 weeks
	Histoplasma capsulatum (fungus)	Histoplasmosis	5–18 days
Gastrointestinal tract	*Shigella* species	Bacillary dysentery (shigellosis)	1–2 days
	Brucella melitensis	Brucellosis (undulant fever)	6–14 days
	Vibrio cholerae	Cholera	1–3 days
	Salmonella enteritidis, Salmonella typhimurium, Salmonella cholerae-suis	Salmonellosis	7–22 hours
	Salmonella paratyphi	Paratyphoid fever	7–24 days
	Salmonella typhi	Typhoid fever	5–14 days
	Hepatitis A virus (picornavirus)	Infectious hepatitis	15–50 days
	Paramyxovirus	Mumps	2–3 weeks
	Picornavirus	Poliomyelitis	4–7 days
	Trichinella spiralis (helminth)	Trichinosis	2–28 days
Skin, mucous membrane, or parenteral route	*Clostridium perfringens*	Gas gangrene	1–5 days
	Clostridium tetani	Tetanus	3–21 days
	Neisseria gonorrhoeae	Gonorrhea	3–8 days
	Leptospira interrogans	Leptospirosis	2–20 days
	Yersinia pestis	Plague	2–6 days
	Rickettsia rickettsii	Rocky Mountain spotted fever	3–12 days
	Treponema pallidum	Syphilis	10–90 days
	Hepatitis B virus** (picornavirus)	Serum hepatitis	6 weeks–6 months
	Rhabdovirus	Rabies	10 days–1 year
	Togavirus	Yellow fever	3–6 days
	Plasmodium species (protozoan)	Malaria	2 weeks

*All causative agents are bacteria, unless indicated otherwise. For viruses, only the viral group is given, except where the virus has a name different from the disease it causes.
**These pathogens can also cause disease after entering the body via the gastrointestinal tract.

resistance of the host. The virulence of a microbial pathogen (or a toxin) is often expressed as the **LD_{50} (lethal dose),** the dose that will kill 50% of inoculated hosts within a given time. The dose required to produce a demonstrable infection in 50% of the hosts is called the **ID_{50} (infectious dose).**

Pathogenic bacteria cause disease by two basic mechanisms: invasiveness and damage to tissues or physiological processes by toxins. **Invasiveness** is the ability of microorganisms to establish residence in a host and, as a result of microbial metabolism and multiplication, cause structural damage to cells and interfere with metabolic reactions. Depending on the species, pathogens may colonize external or internal body surfaces or penetrate into deeper tissues. Invasiveness leads to significant changes in the host cells in the immediate vicinity of the invasion, and it is facilitated by the presence of bacterial capsules, components of the cell wall, and enzymes. Toxins, which are transported by the blood and lymph, cause their structural and functional effects on host cells far removed from the original lesion. Toxins are classified as exotoxins and endotoxins.

We will now look at how capsules, components of the cell wall, and enzymes contribute to invasiveness.

Invasiveness

Capsules

It was noted in Chapter 4 that some bacteria form capsules around their cell walls (Figure 14–2), a property that increases the virulence of these species. The capsule resists the host's defenses by impairing phagocytosis, a process by which certain cells of the body engulf and destroy microbes (discussed in Chapter 15). It appears that the chemical nature of the capsule does not allow the phagocytic cell to adhere to the bacterium. However, the human body is capable of producing antibodies against the capsule, and when these antibodies are present, the encapsulated bacteria are more easily destroyed by phagocytosis.

One bacterium that owes its virulence to the presence of a polysaccharide capsule is *Streptococcus pneumoniae,* the causative agent of lobar pneu-

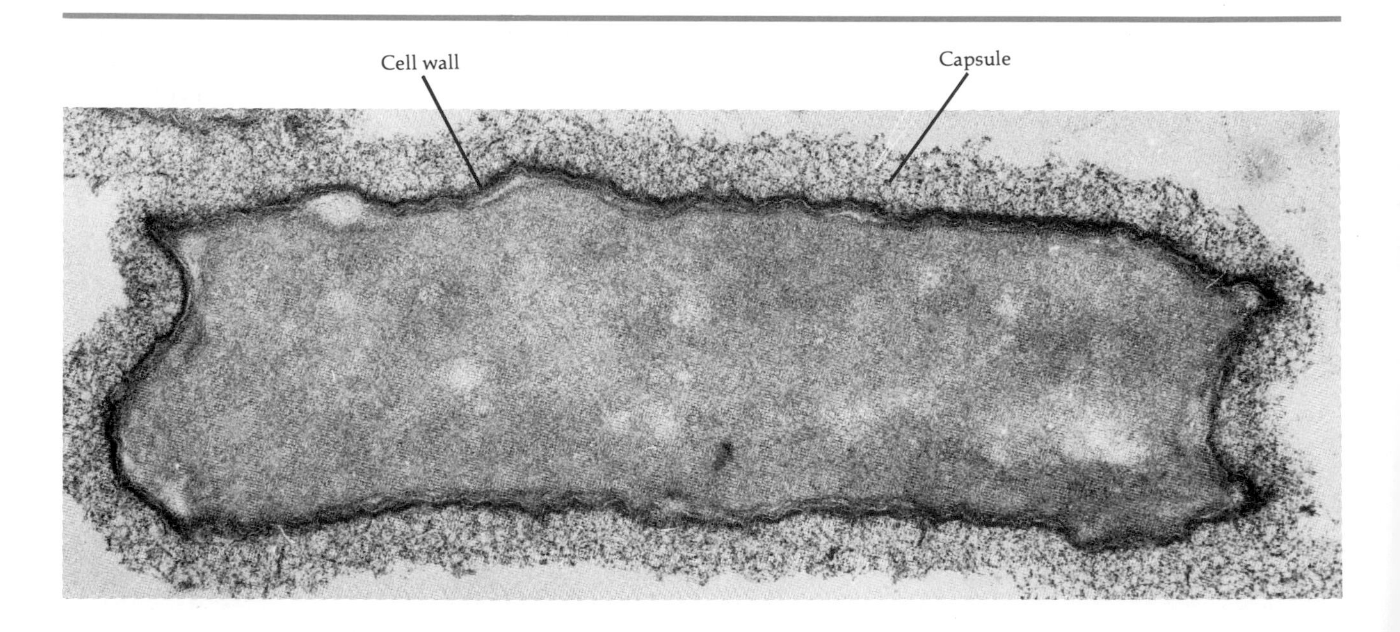

Figure 14–2 Electron micrograph of a thin section through a cell of an encapsulated strain of *Klebsiella pneumoniae* (×53,000). Here the capsule appears as a thick gray layer of material external to the dark, gram-negative cell wall around the cell. Capsules play an important role in the invasiveness of pathogenic bacteria.

monia (see Figure 4–4). This organism is virulent with its capsule but avirulent without it and easily susceptible to phagocytosis. Avirulent *S. pneumoniae* normally resides in the upper respiratory tract. In experiments where the capsule is removed by enzymes, the bacterium loses its pathogenicity. Other bacteria that produce capsules related to virulence are *Klebsiella pneumoniae,* a causative agent of bacterial pneumonia, and *Hemophilus influenzae,* a cause of pneumonia and meningitis in children. Keep in mind that many nonpathogenic bacteria produce capsules and that the virulence of certain other pathogens is not related to the presence of a capsule.

Components of the cell wall

The cell walls of certain bacteria contain chemical substances that contribute to invasiveness. For example, certain streptococcal bacteria, such as *Streptococcus pyogenes,* contain a heat- and acid-resistant cell surface protein called **M protein.** This protein helps the bacterium resist phagocytosis by white blood cells, thereby increasing the virulence of the microorganism. Immunity to *S. pyogenes* depends on the body's production of an antibody specific to M protein. And the cell walls of certain staphylococci, such as *Staphylococcus aureus,* contain another kind of protein that contributes to invasiveness, called **A protein.** This substance has antiphagocytic properties similar to M protein.

Enzymes

The invasiveness of some bacteria is thought to be aided by the production of extracellular enzymes (exoenzymes) and related substances. These chemicals have the ability to break cells open, dissolve materials between cells, form blood clots, and dissolve blood clots, among other functions. However, the importance of most of these enzymes to bacterial invasiveness has not been proved conclusively.

Leukocidins are substances produced by some bacteria that have the ability to destroy neutrophils, white blood cells (leukocytes) that are very active in phagocytosis. Leukocidins are also active against phagocytic cells (macrophages) present in tissue. Among the bacteria that secrete leukocidins are staphylococci and streptococci. Leukocidins produced by streptococci cause degradation of lysosomes within leukocytes, thus causing the death of the white blood cell. Hydrolytic enzymes released from the leukocytic lysosomes may damage other cellular structures and thus intensify streptococcal lesions. Damage to white blood cells decreases host resistance.

Hemolysins are another group of enzymes produced by bacteria that might contribute to virulence in some cases. Hemolysins cause the lysis (breakage) of red blood cells. A number of different hemolysins are produced by bacteria, and they differ from one another according to the kind of red blood cell they lyse (human, sheep, rabbit, and so on) and the type of lysis they cause. One "type" of lysis is the "hot and cold" lysis exhibited by staphylococcal beta toxin. Beta toxin enzymatically destroys the membrane surrounding a red blood cell. In laboratory culture, the effectiveness of beta toxin can be increased by incubating the culture first at 37°C and then incubating it in a colder environment (in a refrigerator, for example). Important producers of hemolysins are staphylococci, streptococci, and *Clostridium perfringens,* the causative agent of gas gangrene.

Coagulases are bacterial enzymes that coagulate (clot) the fibrinogen in the blood. (Fibrinogen is a plasma protein produced by the liver that is converted into fibrin, the threads that form a blood clot.) The resulting fibrin clot may protect the bacterium from phagocytosis and isolate it from other defenses of the host. Coagulase is produced by some members of the genus *Staphylococcus,* and it may, in fact, be involved in the walling-off process in boils produced by staphylococci. It should be noted, however, that some staphylococci do not produce coagulase and are still virulent. For these bacteria, a capsule, and not coagulase, may be the more important factor in invasiveness.

Bacterial kinases are another group of enzymes that may contribute to bacterial virulence. The kinases break down fibrin and thus dissolve clots formed by the body to isolate the infection. One of the better-known kinases is **streptokinase (fibrinolysin),** which is produced by streptococci. Another is **staphylokinase,** produced by staphylococci.

Hyaluronidase is yet another substance possibly related to microbial virulence. It is secreted by certain bacteria and dissolves hyaluronic acid, a mucopolysaccharide that holds together certain cells of the body, especially in connective tissue. As a result of this action, the microorganism is thought to be able to spread more easily from its initial site of infection. Hyaluronidase is produced by streptococci and some clostridia that cause gas gangrene. The spread of gas gangrene is facilitated by another enzyme, too. **Collagenase,** produced by several species of *Clostridium,* breaks down the protein collagen, which forms the framework of muscles.

Other bacterial substances believed to contribute to virulence are *necrotizing factors,* which cause the death of body cells, *hypothermic factors,* which affect body temperature, and *edema-producing factors.*

Toxins

We will now turn our attention to the second principal mechanism related to pathogenicity, the production of toxins. **Toxins** are poisonous substances that are produced by certain microorganisms, and which may be almost entirely responsible for the pathogenic properties of those microbes. The capacity of microorganisms to produce toxins is called **toxigenicity.** Toxins transported by the blood or lymph can cause serious, and sometimes fatal, effects. Some toxins produce fever, circulatory disturbances, diarrhea, and shock. And toxins may inhibit protein synthesis, destroy blood cells and blood vessels, and disrupt the nervous system by causing spasms. The term **toxemia** is applied to chemical blood poisoning by toxins. Toxins are of two types: exotoxins and endotoxins.

Exotoxins

Exotoxins are produced by some bacteria as part of their growth and metabolism and are released into the surrounding medium. Most bacteria that produce exotoxins are gram-positive. Exotoxins are proteins, and the genes for most of them (perhaps all) are carried on bacterial plasmids or phages. Because exotoxins are soluble in body fluids, they easily diffuse into the blood and are rapidly transported throughout the body. Actually, diseases caused by bacteria that produce exotoxins are often the result of the exotoxins, and not the bacteria themselves. It is the exotoxins that produce the signs and symptoms of the disease. Thus, exotoxins are disease specific.

Exotoxins work by destroying specific parts of the host's cells or by inhibiting certain metabolic functions. Exotoxins are among the most lethal substances known, and only 1 mg of botulism toxin is enough to kill 1 million guinea pigs. Fortunately, only a few bacteria produce exotoxins. We can briefly describe the most notable exotoxins as follows.

Botulinum Toxin Eight different types of botulinum toxin are produced by *Clostridium botulinum,* each toxin possessing a different potency. Botulinum toxin is not a typical exotoxin in that it is made within the clostridial cell and not released into the medium until cell death occurs. Botulinum toxin acts at the neuromuscular synapse (junction between nerve cell and muscle cell), preventing transmission of nerve impulses. The result is paralysis.

Tetanus Toxin *Clostridium tetani* produces tetanus toxin. This toxin causes excitation of the central nervous system, resulting in the convulsive symptoms of tetanus, or "lockjaw."

Diphtheria Toxin *Corynebacterium diphtheriae* produces diphtheria toxin only when it is infected by a lysogenic phage carrying the *tox* gene. Diphtheria toxin inhibits protein synthesis in eucaryotic cells.

Staphylococcal Enterotoxin *Staphylococcus aureus* produces a type of exotoxin called an **enterotoxin** because it affects the intestines. The enterotoxin induces vomiting and, by preventing absorption of water from the intestine, diarrhea.

***Vibrio* Enterotoxin** The enterotoxin produced by *Vibrio cholerae* growing in a host's intestines alters water and electrolyte balance, causing diarrhea.

In response to exotoxins, the body produces antibodies called **antitoxins** to provide immunity. When exotoxins are inactivated by heat, formaldehyde, iodine, or other chemicals, they no longer cause the disease but are still able to stimulate the body to produce antitoxins. Such altered exotoxins are called **toxoids.** As will be discussed in Chapter 16, toxoids are injected into the body to stimulate antitoxin production for immunity to diseases such as diphtheria and tetanus.

Endotoxins

Endotoxins differ from exotoxins in several ways. Unlike exotoxins, they are not released into the surrounding medium. Instead, endotoxins are actually part of the outer portion of the cell wall of most gram-negative bacteria. Moreover, endotoxins are not proteins, but are composed of lipid and carbohydrate (lipopolysaccharide). All endotoxins produce the same signs and symptoms, regardless of the microorganism. These include fever, weakness, generalized aches, and in some cases, shock. Endotoxins do not promote the formation of effective antitoxins. Antibodies are produced, but they tend not to counteract the effect of the toxin; sometimes, in fact, they actually enhance its effect.

Representative microorganisms that produce endotoxins are *Salmonella typhi* (the causative agent of typhoid fever), *Shigella species* (the causative agents of bacillary dysentery, or shigellosis), and *Neisseria meningitidis* (the causative agent of epidemic meningitis). A comparison of exotoxins and endotoxins appears in Table 14–2.

CYTOPATHIC EFFECTS OF VIRUSES

The production of substances coded for by the viruses may cause observable changes in infected cells. These observable changes are called **cytopathic effects (CPE)**. The point in the viral infection cycle when cytopathic effects occur varies with the virus. Some viral infections result in early changes in the host cell, while in other infections, changes are not seen until a much later stage. Moreover, some viruses cause **cytocidal** effects (changes resulting in cell death), while other viruses are **noncytocidal.** A virus may produce one or more of the following cytopathic effects:

1. At some stage in viral multiplication, cytocidal viruses cause the cellular macromolecular synthesis to stop. Some viruses, such as herpes simplex, irreversibly stop mitosis.
2. Cell death resulting from infection by a cytocidal virus may be due to the release of enzymes from lysosomes. These enzymes cause autolysis of the cell.
3. Inclusion bodies (Figure 14–3a) are granules found in the cytoplasm or nucleus of some infected cells. They are viral parts, nucleic acids or proteins, in the process of being assembled into virions. The granules may vary in size, shape, and staining properties, depending on the virus. Some inclusion bodies stain with an acidic stain (acidophilic), while others stain with a basic stain (basophilic). Table 14–3 lists some representative viruses that form inclusion bodies.
4. Some viral infections cause host cells to fuse, producing multinucleated "giant" cells (Figure 14–3b).

Table 14–2 Comparison of Exotoxins and Endotoxins

Property	Exotoxin	Endotoxin
Bacterial source	Mostly from gram-positive bacteria	Almost exclusively from gram-negative bacteria
Relation to microorganism	Metabolic product of growing cell	Present in cell wall and released only with destruction of cell
Chemistry	Protein or short peptide	Lipopolysaccharide complex
Heat stability	Unstable; can usually be destroyed at 60°–80°C (except staphylococcal enterotoxin)	Stable; can withstand autoclaving (120°C for one hour)
Toxicity (power to cause disease)	High	Low
Immunology (relation to antibodies)	Can be converted to toxoids and neutralized by antitoxin	Cannot be converted to toxoids and are not easily neutralized by antitoxin
Pharmacology (effect on body)	Specific for a particular cell structure or function in the host	General, such as fever, weakness, aches, and shock; all produce the same effects
Lethal dose	Small	Considerably larger
Representative diseases	Gas gangrene, tetanus, botulism, diphtheria, scarlet fever	Bacillary dysentery (shigellosis), epidemic meningitis, and tularemia

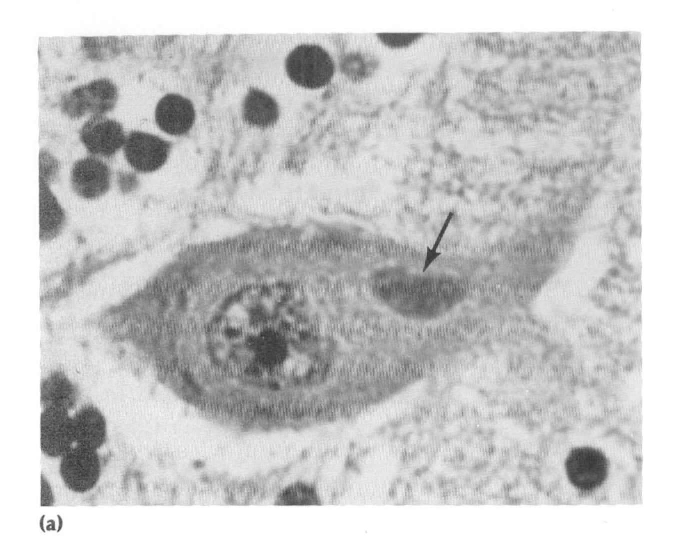
(a)

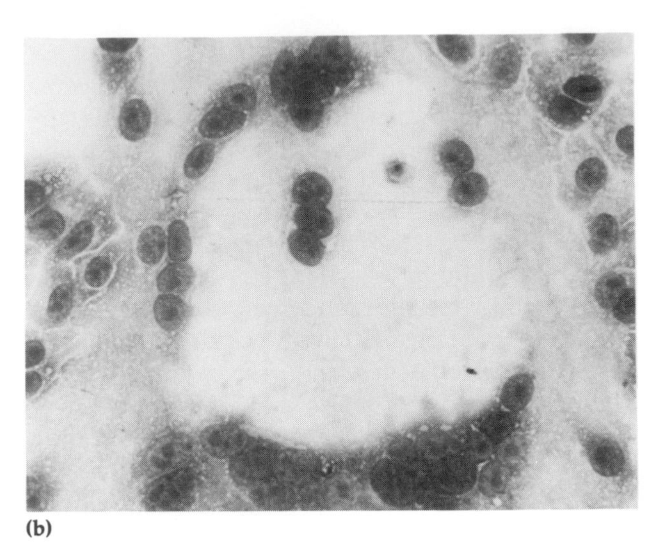
(b)

Figure 14–3 Effects of viruses on cells. **(a)** Cytoplasmic inclusion body (arrow) in a human brain cell from a fatal case of rabies. **(b)** The relatively clear area in the center of the micrograph is a giant cell formed in a culture of Vero cells infected with measles virus. The numerous dark, oval bodies around the cell's edges are its multiple nuclei.

Table 14–3 Cytopathic Effects (CPE) of Selected Viruses

Virus	CPE
Poliovirus	Cell death
Papovavirus	Acidophilic inclusion bodies in nucleus
Adenovirus	Basophilic inclusion bodies in nucleus
Rhabdovirus	Acidophilic inclusion bodies in cytoplasm
Measles, cytomegalovirus	Acidophilic inclusion bodies in nucleus and cytoplasm
Measles	Cell fusion
Polyoma	Transformation

5. Many viral infections induce antigenic changes on the surface of the infected cells. These antigens elicit a host antibody response against the infected cell, thus targeting the infected cell for destruction by the host immune system.
6. Some viruses induce chromosomal changes in the host cell.
7. Viruses capable of causing cancer *transform* host cells, as discussed in Chapter 12. Transformation results in an abnormal, spindle-shaped cell that does not recognize contact inhibition (Figure 14–4). Normal cells cease

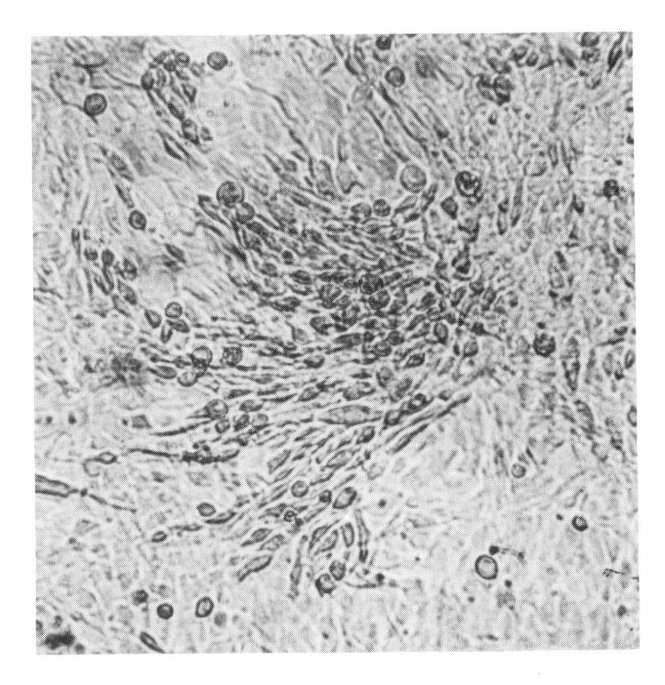

Figure 14–4 Transformed cells in culture. In the center of this micrograph is a "jumble" of chick embryo cells transformed by Rous sarcoma virus. Such a concentration of transformed cells in a cell culture is called a focus and results from multiplication of a single transforming virus that originally infected a cell at the site. Note how the transformed cells of the focus appear dark in contrast to the monolayer of light, flat, normal cells around them. This is due to their spindle shapes and uninhibited growth on top of one another; there is no contact inhibition.

growing when they come close to another cell, the phenomenon of *contact inhibition*. Loss of contact inhibition results in unregulated cell growth.

Some representative viruses that cause cytopathic effects are presented in Table 14–3. In subsequent chapters, the pathological properties of viruses will be discussed in more detail.

PATHOGENIC PROPERTIES OF OTHER MICROORGANISMS

Although fungi do cause disease, they do not have a well-defined set of virulence factors. *Cryptococcus neoformans* is a fungus that produces a capsule that helps this organism resist phagocytosis. Some products of fungal growth are toxic to human hosts. But toxins are an indirect cause of a fungal disease, as the fungus is already growing in or on the host. An allergic response may also result from a fungal infection, and the symptoms of these fungal infections are typical of allergic reactions.

The presence of protozoans and helminths often produces the symptoms of protozoan and helminthic diseases. In some cases, these organisms actually use host tissues for their own growth, with the resulting cell damage evoking symptoms. Waste products of the metabolism of these parasites may also contribute to the symptoms of a disease.

Specific diseases caused by fungi, protozoans, and helminths will be discussed in later chapters, along with more on the pathological properties of these organisms.

PATTERN OF DISEASE

A definite sequence of events usually occurs during infection and disease. As you already know, a reservoir of infection must be present as a source of pathogens before an infectious disease can occur. Next, there must be transmission of the pathogen either by direct contact, indirect contact, or vectors to a susceptible host. This is followed by invasion, which is the entrance of the microorganism into the host and an increase in the number of microorganisms. Entrance is usually through a preferred portal of entry. Following invasion, there is injury to the host by the microorganism through a process called pathogenicity. This injury is dependent on capsules, enzymes, and/or toxins produced by the microorganism. Given all these factors, whether or not disease actually occurs depends on the resistance of the host to the offensive weapons of the pathogen.

Assuming that the microorganism does overcome the defenses of the host, a certain pattern of disease evolves. This pattern usually follows the sequence described in the following sections.

Period of Incubation The period of incubation is the time interval between the actual infection and the first appearance of any signs or symptoms. The incubation period in some diseases is constant; in others it is quite variable. The time of incubation depends on the specific microorganism involved, the virulence of the microorganism, the number of infecting microorganisms, and the resistance of the host. Table 14–1 lists the incubation periods for a number of microbial diseases.

Prodromal Period The prodromal period is a relatively short period of time that sometimes follows the period of incubation. It is characterized by the first symptoms of disease, such as headache and malaise.

Period of Illness During the period of illness, the disease is most acute. The person exhibits overt signs and symptoms of disease, such as fever, chills, muscle pain (myalgia), sensitivity to light (photophobia), sore throat (pharyngitis), lymph node enlargement (lymphadenomegaly), and gastrointestinal disturbances such as diarrhea. It is during the period of illness that increases or decreases in the number of white blood cells occur.

Period of Decline During the period of decline, the signs and symptoms subside. The fever decreases and the feeling of malaise diminishes. If the period of decline is rapid, such as within 24 hours, it is said to occur by *crisis*. If, instead, it takes several days, with the fever decreasing a little each day until it returns to normal, then the period of decline is said to occur by *lysis*.

Period of Convalescence During the period of convalescence, the individual regains strength and the body returns to its prediseased state. Recovery has occurred.

We all know that during the period of illness, people serve as reservoirs of disease and can easily spread infections to other persons. However, you should also know that infections can be spread by persons during incubation and convalescence. This is especially true in cases where the convalescing person carries the pathogenic microorganism for months or even years.

Let us now conclude the chapter by considering how microbes exit the body.

PORTALS OF EXIT

To spread disease throughout a population, a pathogen must exit the body. Just as pathogens have preferred portals of entry, they also have definite routes of exit, called **portals of exit.** In general, portals of exit are related to the part of the body that has been infected.

The most common portals of exit for pathogens are the respiratory and gastrointestinal tracts. For example, many pathogens living in the respiratory tract exit in discharges from the mouth and nose as a result of coughing or sneezing. These microorganisms are found in droplets formed from mucus. Pathogens that cause tuberculosis, whooping cough, pneumonia, scarlet fever, epidemic meningitis, measles, mumps, smallpox, and influenza are discharged through the respiratory route. Other pathogens exit from the gastrointestinal tract in feces and saliva. Feces may be contaminated with path-

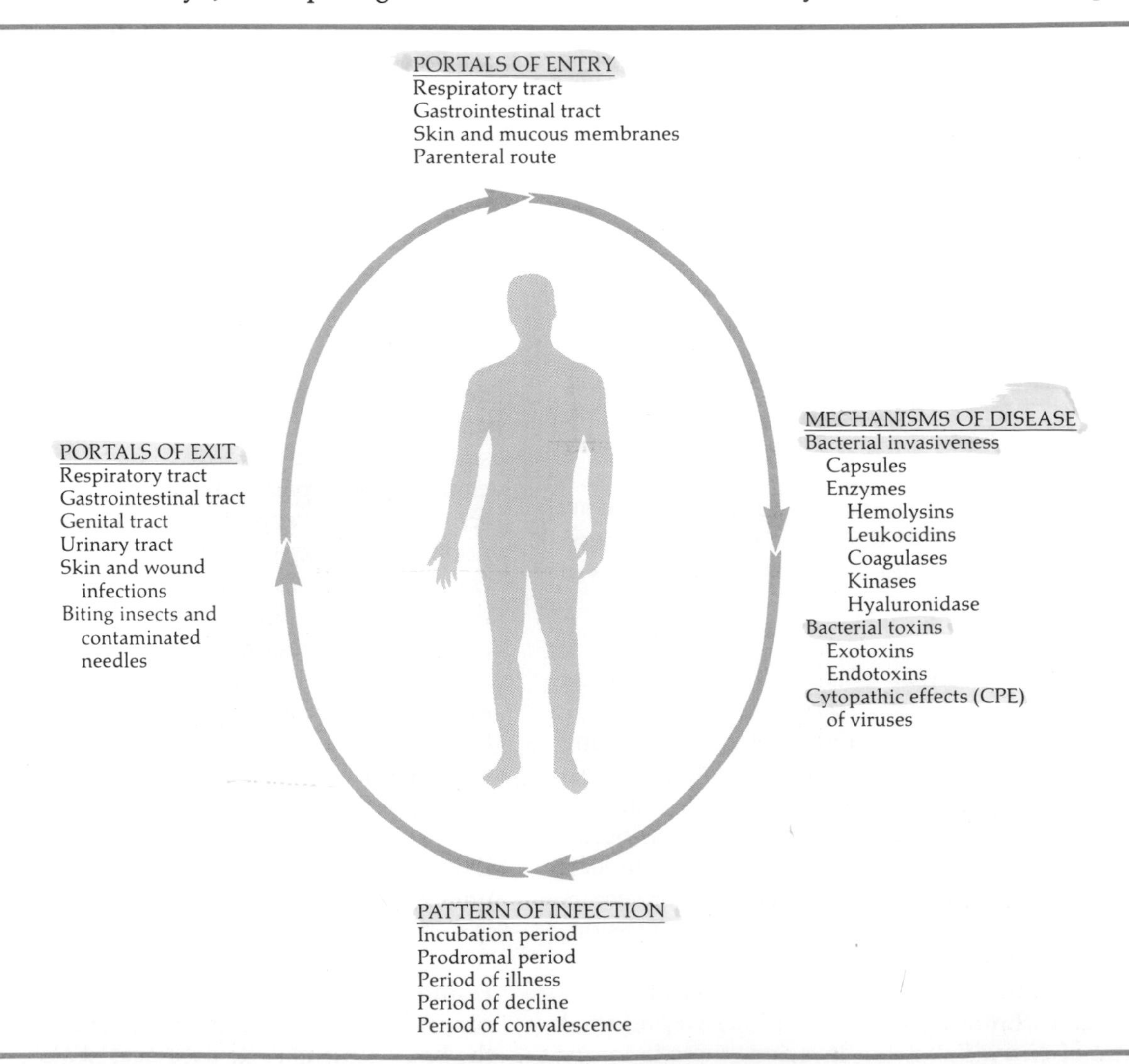

Figure 14–5 How microbes cause disease: a summary of the key concepts related to pathogenicity.

ogens associated with cholera, typhoid fever, paratyphoid fever, bacillary dysentery (shigellosis), amoebic dysentery, and poliomyelitis. And saliva can contain pathogens such as the rabies virus.

Another important route of exit is the genital tract. Microbes responsible for venereal diseases may be found in secretions from the penis or vagina. Urine may also contain pathogens responsible for typhoid fever, paratyphoid fever, and undulant fever, since these pathogens exit via the urinary tract. Skin or wound infections are other portals of exit in which drainage from these wounds may spread infections directly to another person or by contact with a contaminated fomite. Finally, infected blood removed and reinjected by biting insects or injections with contaminated needles and syringes may spread infections within the population. Examples of diseases so transmitted are yellow fever, typhus fever, Rocky Mountain spotted fever, tularemia, malaria, and serum hepatitis.

In the next chapter, we will examine a group of nonspecific defenses of the host against disease. But before proceeding, examine Figure 14–5 very carefully. It summarizes some of the key concepts we have discussed in this chapter related to the mechanisms of pathogenicity.

STUDY OUTLINE

PORTALS OF ENTRY (pp. 363–364)

1. The specific route taken by a particular pathogen to gain access to the body is called its portal of entry.
2. Many microorganisms can only cause infections when they gain access through their specific portal of entry.

Respiratory Tract (p. 363)

1. Microorganisms that are inhaled with droplets of moisture and dust particles gain access to the respiratory tract.
2. This is the most frequently used portal of entry.

Gastrointestinal Tract (pp. 363–364)

1. Microorganisms enter the gastrointestinal tract via food, water, and fingers.
2. Most microorganisms are destroyed by the stomach's hydrochloric acid.

Skin and Mucous Membranes (p. 364)

1. Most microorganisms cannot penetrate intact skin; they enter hair follicles and sweat ducts.
2. Some fungi infect the skin itself.
3. Many microorganisms can penetrate mucous membranes.

Parenteral Route (p. 364)

1. Some microorganisms can gain access to tissues by inoculation through the skin and mucous membranes in bites, injections, and other wounds.
2. This route of penetration is called the parenteral route.

PATHOGENIC PROPERTIES OF BACTERIA (pp. 364–369)

1. Pathogenicity is the ability of a pathogen to produce a disease by overcoming the defenses of the host.
2. Virulence is the degree of pathogenicity.
3. Virulence can be expressed as LD_{50} (lethal dose for 50% of the inoculated hosts) or ID_{50} (infectious dose for 50% of the inoculated hosts).

Invasiveness (pp. 366–368)

1. The ability of a microorganism to colonize a host (invasiveness) is dependent on several factors.
2. Some pathogens have capsules that prevent them from being phagocytized.
3. Proteins in the cell wall may prevent a pathogen from being phagocytized.
4. Some pathogens possess exoenzymes that lyse host cells and dissolve substances between cells.

Toxins (pp. 368–369)

1. Poisonous substances produced by microorganisms are called toxins.
2. The ability to produce toxins is called toxigenicity.
3. Exotoxins are produced by gram-positive bacteria and released into the surrounding medium.
4. Exotoxins, not the bacteria, produce the disease symptoms.
5. Antibodies produced against exotoxins are called antitoxins.
6. Endotoxins are a lipopolysaccharide component of the cell wall of gram-negative bacteria.
7. Endotoxins cause fever, weakness, generalized aches, and shock.

8. Antibodies may enhance the effects of the endotoxins.

CYTOPATHIC EFFECTS OF VIRUSES (pp. 369–371)

1. Signs of viral infections are called cytopathic effects (CPE).
2. Some viruses cause cytocidal effects (cell death), and others cause noncytocidal effects.
3. Cytopathic effects include stopping mitosis, lysis, formation of inclusion bodies, cell fusion, antigenic changes, chromosomal changes, and transformation.

PATHOGENIC PROPERTIES OF OTHER MICROORGANISMS (p. 371)

1. Symptoms of fungal infections may be due to capsules, toxins, and allergic responses.

2. Symptoms of protozoan and helminthic diseases may be due to damage to host tissue or metabolic waste products of the parasite.

PATTERN OF DISEASE (pp. 371–372)

1. The period of incubation is the time interval between the actual infection and the first appearance of signs and symptoms.
2. The prodromal period is characterized by the appearance of the first mild signs and symptoms.
3. During the period of illness, the disease is at its height, and all disease signs and symptoms are apparent.
4. During the period of decline, the signs and symptoms subside.
5. During the period of convalescence, the body returns to its prediseased state and health is restored.

PORTALS OF EXIT (pp. 372–373)

1. The preferred route by which a pathogen leaves the body is called the portal of exit.
2. The most common portals of exit are the respiratory and gastrointestinal tracts; the genitourinary tract and skin (from removal of blood with needles) or by vectors are also portals of exit.

STUDY QUESTIONS

REVIEW

1. List four portals of entry and describe how microorganisms gain access through each.
2. Compare pathogenicity with virulence.
3. How are capsules and cell wall components related to invasiveness? Give specific examples.
4. Describe how hemolysins, leukocidins, coagulase, kinases, and hyaluronidase might contribute to invasiveness.
5. Complete the following table comparing exotoxins with endotoxins.

	Exotoxin	Endotoxin
Bacterial source		
Chemistry		
Toxicity		
Pharmacology		
Example		

6. Define cytopathic effects and give five examples.

7. Write a one-sentence description of factors contributing to the pathogenicity of fungi. Of protozoans and helminths.
8. Put the following in the correct order to describe the pattern of disease: period of convalescence; crisis; prodromal period; period of decline; period of incubation; period of illness.
9. List four portals of exit and describe how microorganisms escape from the body by each route.

CHALLENGE

1. The LD_{50} for botulin toxin is 0.000025 μg. The LD_{50} for *Salmonella* toxin is 200 μg. Which of these is the more potent toxin? How can you tell from the LD_{50} values?
2. Food poisoning can be divided into two categories: food infection and food intoxication. Explain the difference between these two categories on the basis of toxin production by bacteria.
3. The following is a case history of a 49-year-old man. Identify each period in the pattern of disease that he experienced.

   On February 7, he handled a parakeet with a respiratory illness. On March 9, he experienced intense pain in the legs, followed by severe chills and headaches. On March 16, he had chest pains, cough, and diarrhea, and his temperature was 40°C. Appropriate antibiotics were administered on March 17, and his fever subsided within 12 hours. He continued taking antibiotics for 14 days. (*Note:* The disease is psittacosis. Can you find the etiology?)

FURTHER READING

Braude, A. I. March 1964. Bacterial endotoxins. *Scientific American* 210:36–45. The effects of endotoxins on the human host.

Hirschhorn, N. and W. B. Greenough. August 1971. Cholera. *Scientific American* 225:15–21. The scientific method is used to find out how cholera toxin causes loss of body fluids in the host.

Roueche, B. 1967. *Field Guide to Diseases.* Boston: Little, Brown and Company. A concise reference of specific diseases to show the patterns of diseases.

Smith, I. M. February 1968. Death from staphylococci. *Scientific American* 218:84–94. Effects of the growth of staphylococci in animal hosts.

Van Heyningen, W. E. April 1968. Tetanus. *Scientific American* 218:69–77. An overview of the effects of bacterial exotoxins and specific treatment of the incidence of tetanus.

von Graevenitz, A. 1977. The role of opportunistic bacteria in human disease. *Annual Review of Microbiology* 31:447–471. A good review of bacteria that cause infections in a compromised host.

15

Nonspecific Defenses of the Host

Objectives

After completing this chapter you should be able to

- Define resistance and susceptibility.
- Define nonspecific resistance.
- Describe the role of the skin and mucous membranes in nonspecific resistance, differentiating between mechanical and chemical factors.
- Define phagocytosis and include the stages of adherence and ingestion.
- Classify phagocytic cells, describing the role of granulocytes and monocytes.
- Describe the stages of inflammation and their relation to nonspecific resistance.
- Discuss the role of fever in nonspecific resistance.
- Discuss the role of interferon.
- Discuss the function of complement.

From what has been said thus far, you can see that pathogenic microorganisms are endowed with special properties that—given the right opportunity—enable them to cause disease. If microorganisms never encountered resistance from the host, we would be constantly ill and would eventually die of various diseases. But in most cases, our body defenses prevent this from happening. Some of our body defenses are designed to keep out the microorganisms altogether. Other defenses remove them if they do get in, and still others combat them if they remain inside. Our ability to ward off disease through our defenses is called **resistance.** Vulnerability or lack of resistance is known as **susceptibility.**

In discussing resistance, we will divide our body defenses into two general kinds: nonspecific and specific. **Nonspecific resistance** refers to all our defenses that tend to protect us from *any kind* of pathogen. **Specific resistance,** or **immunity,** is the defense that the body offers against a particular microorganism. Specific defenses are based on the production of specific proteins called antibodies and special cells of the immune system. Chapter 16 deals with specific resistance. In this chapter, we will consider nonspecific resistance.

Contributing to nonspecific resistance are the skin and mucous membranes, phagocytosis, inflammation, fever, and the production of antimicrobial substances other than antibodies. As we discuss these factors, keep in mind that nonspecific resistance protects us from invasion by pathogens in general; it is not a resistance to a specific microorganism.

SKIN AND MUCOUS MEMBRANES

The skin and mucous membranes of the body are commonly regarded as the first line of defense against disease-causing microorganisms. This function results from both mechanical and chemical factors.

Mechanical Factors

The **intact skin** consists of two distinct portions. The *dermis* is the inner portion and is composed of connective tissue. The *epidermis* is the outer portion, the part that is in direct contact with the external environment. It consists of several layers of tightly packed epithelial cells, with little or no material between the layers. The cells are arranged in continuous sheets. In addition, the top layer of epidermal cells contains a waterproofing protein called *keratin*.

If we consider the closely packed cells, continuous layering, and presence of keratin, we can see why the intact skin provides such a formidable physical barrier to the entrance of microorganisms (Figure 15–1). The intact surface of healthy epidermis is rarely, if ever, penetrated by bacteria. But when the epithelial surface is broken, a subcutaneous (below the skin) infection often develops. The bacteria most likely to cause such an infection are staphylococci that normally inhabit the hair follicles and sweat glands of the skin. Moreover, when the skin is moist, as in hot, humid climates, skin infections, especially fungus infections such as athlete's foot, are quite common.

Mucous membranes also consist of an epithelial layer and an underlying connective tissue layer. Mucous membranes line body cavities that open to the exterior. In fact, mucous membranes line the entire digestive, respiratory, urinary, and reproductive tracts. The epithelial layer of a mucous membrane secretes a fluid called *mucus*, which prevents the cavities from drying out. Some pathogens that can thrive on the moist secretions of a mucous membrane are able to penetrate the membrane if the microorganism is present in sufficient numbers. *Treponema pallidum*, *Mycobacterium tuberculosis*, and *Streptococcus pneumoniae* are examples of such pathogens. This penetration may be related to toxic products produced by the microorganism, prior

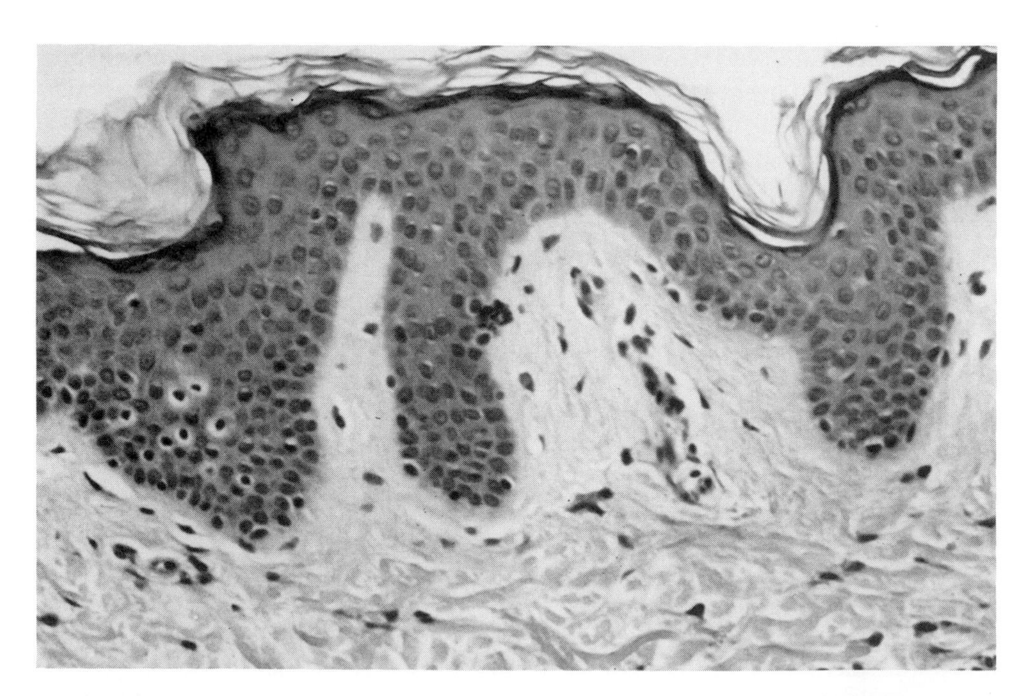

Figure 15–1 Section through human skin (approx. ×200). The thin layer at the top of the photo is keratin. This layer and the darker cells beneath it make up the epidermis. The lighter material in the bottom half of the photo is the dermis.

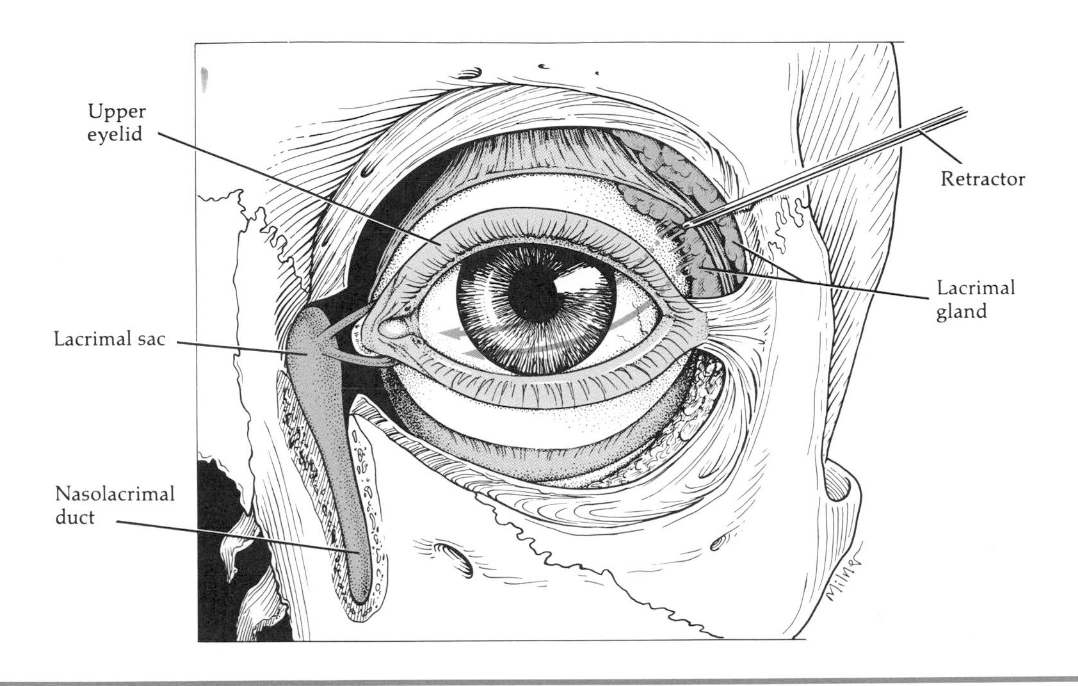

Figure 15–2 The lacrimal apparatus. The washing action of the tears, shown by the colored arrows, prevents microorganisms from settling on the surface of the eyeball.

injury by viral infection, or mucosal irritation. Although mucous membranes do inhibit the entrance of many microorganisms, they offer less protection than the skin.

While the skin and mucous membranes serve as physical barriers, there are several other mechanical factors that help protect certain epithelial surfaces. One such mechanism that protects the eyes is the **lacrimal apparatus** (Figure 15–2), which consists of a group of structures that manufactures and drains away tears. The lacrimal glands, located toward the upper and outer portions of each socket of the eye, produce the tears and pass them under the upper eyelid. From here, tears pass toward the corner of the eye near the nose into two small holes that lead to the nose. After being secreted by the lacrimal glands, the tears are spread over the surface of the eyeball by blinking. Normally, the tears evaporate or pass into the nose as fast as they are produced. This continual washing action of the tears helps to keep microorganisms from settling on the surface of the eye. If an irritating substance or large numbers of microorganisms come in contact with the eye, the lacrimal glands start to secrete heavily, and the tears then accumulate more rapidly than they can be carried away. This is also a protective mechanism, because the excess tears dilute and wash away the irritating substance or microorganisms.

In a cleansing action very similar to that of tears, **saliva,** produced by the salivary glands, washes microorganisms from the surface of the teeth and the mucous membrane of the mouth. This mechanism helps prevent colonization by microbes. Since mucus is slightly viscous (thick), it traps many of the microorganisms that enter the respiratory and digestive tracts. The mucous membrane of the nose also has mucus-coated hairs that trap and filter inspired air containing microorganisms, dust, and pollutants. And the cells of the mucous membrane of the lower respiratory tract contain **cilia,** microscopic, hairlike projections (Figure 15–3). These cilia move synchronously, propelling inhaled dust and microorganisms that have

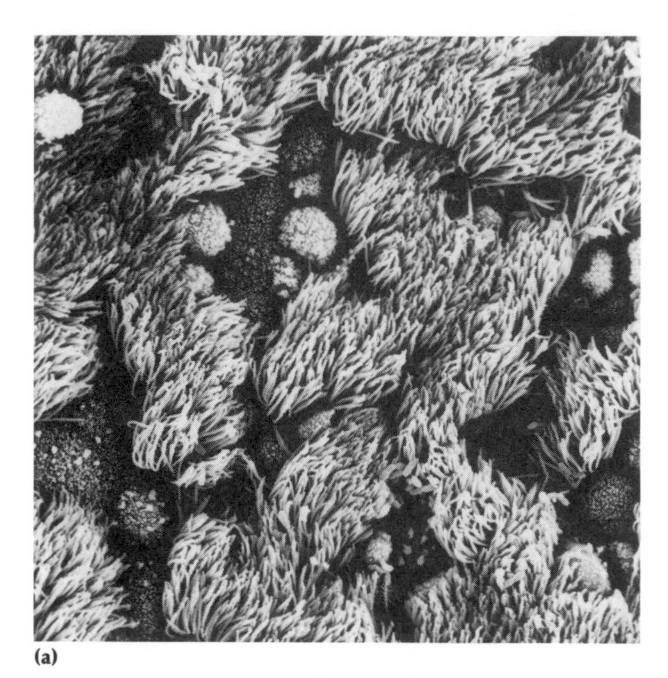

(a)

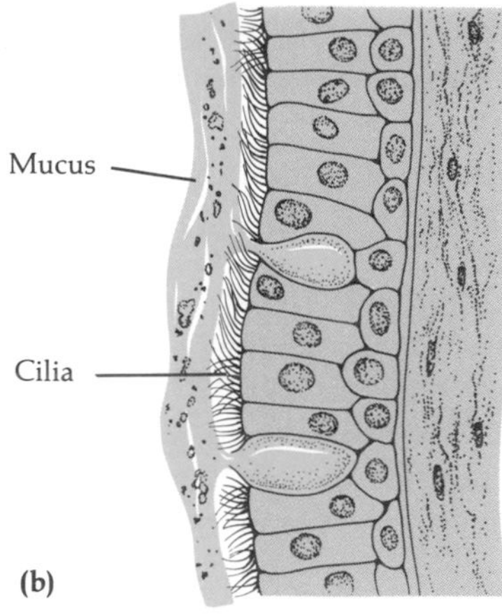

(b)

Figure 15–3 Mucous membrane of the trachea (windpipe). **(a)** Micrograph of cilia and mucus-secreting goblet cells, from which blobs of mucus are emerging (×1500). **(b)** Action of the ciliary escalator.

become trapped in mucus upward toward the throat. This so-called *ciliary escalator* keeps the *mucous blanket* moving toward the throat at a rate of 1 to 3 cm per hour, and coughing and sneezing speed up the escalator. Microorganisms are also prevented from entering the lower respiratory tract by a small lip of cartilage called the *epiglottis,* which covers the larynx (voice box) during swallowing.

The cleansing of the urethra by the flow of urine is a mechanical factor that prevents microbial colonization in the urinary system. Vaginal secretions likewise move microorganisms out of the female body.

Chemical Factors

Mechanical factors alone do not account for the high degree of resistance of skin and mucous membranes to microbial invasion. Certain chemical factors also play an important role.

Oil (sebaceous) glands of the skin produce an oily substance called **sebum** that prevents hair from drying and becoming brittle and also forms a protective film over the surface of the skin. One of the components of sebum is unsaturated fatty acids, which inhibit the growth of certain pathogenic bacteria. The low pH of the skin, between pH 3 and 5, is due in part to the secretion of fatty acids and lactic acid. Commensal bacteria on the skin decompose sloughed-off skin cells and these organic molecules, and the end products of this metabolism produce body odor. The skin's acidity probably discourages the growth of many other microorganisms.

The sweat (sudoriferous) glands of the skin produce perspiration, which helps to maintain body temperature, eliminate certain wastes, and flush microorganisms from the surface of the skin. Perspiration also contains **lysozyme,** an enzyme capable of breaking down cell walls of gram-positive bacteria and a few gram-negative bacteria under certain conditions (see Figure 4–8). Lysozyme is also found in tears, saliva, nasal secretions, and tissue fluids, where it exhibits its antimicrobial activity.

Gastric juice is a mixture of hydrochloric acid, enzymes, and mucus produced by the glands of the stomach. The very high acidity of gastric juice (pH 1.2 to 3.0) is sufficient to preserve the usual sterility of the stomach. This acidity destroys bacteria and most bacterial toxins except those of *Clostridium botulinum* and *Staphylococcus aureus.*

On the skin and on the mucous membranes that do carry normal flora, the presence of these microorganisms prevents colonization by other, potentially pathogenic microorganisms. The resident flora usually compete successfully for available nutrients and may produce metabolic end products that inhibit the growth of other microorganisms.

The remainder of this chapter is devoted to several aspects of nonspecific resistance commonly regarded as the body's second line of defense against infection. These defense mechanisms all operate within the body. We shall start the discussion with phagocytosis.

PHAGOCYTOSIS

Phagocytosis (from the Greek words for "eat" and "cell") is the ingestion of a microorganism or any particulate matter by a cell. We have previously mentioned phagocytosis as the method of nutrition of certain protozoans. In this chapter, we are concerned with phagocytosis as a means of countering infection in the human body. The human cells that perform this function, collectively called **phagocytes,** are all blood cells or derivatives of blood cells.

Formed Elements of the Blood

Blood consists of a fluid called *plasma* and *formed elements,* that is, cells and cell fragments (Table 15–1). (*Serum* is plasma from which all the clotting factors have been removed.) Of the cells listed in Table 15–1, those that concern us at present are the leukocytes, or white blood cells. Leukocytes are divided into two categories, granulocytes and agranulocytes. *Granulocytes* owe their name to the presence of granules in their cytoplasm. They are differentiated into three types on the basis of how the granules stain. The granules of *neutrophils* stain with a mixture of acidic and basic dyes; those of *basophils* stain with the basic dye methylene blue; and those of *eosinophils* stain with the acid dye eosin. Neutrophils are also commonly called *polymorphonuclear leukocytes* (or *PMNs*). *Agranulocytes* are so named because they lack granules in their cytoplasm. The two kinds of agranulocytes are *lymphocytes* and *monocytes.*

During many kinds of infections, there is an increase in the total number of white blood cells *(leukocytosis).* During the active stage of infection, the leukocyte count might double, triple, or quadruple, depending on the severity of the infection. Diseases that might cause such an elevation in the leukocyte count are meningitis, mononucleosis, appendicitis, pneumonia, and gonorrhea. Other diseases, such as typhoid fever, cause a decrease in the leukocyte count *(leukopenia).* The source of leukocyte increase or decrease can be detected by a *differential count,* calculating the percentage of each kind of white cell in a sample of white blood cells. In a normal differential count of leukocytes, the percentages would be as shown in parentheses in the first column in Table 15–1.

Table 15–1 Formed Elements in Blood

Type of Cell	Numbers per Cubic Millimeter (mm^3)	Function
Erythrocytes (Red Blood Cells)	4.8–5.4 million	Transport of O_2 and CO_2
Leukocytes (White Blood Cells)	5000–9000	
A. Granulocytes		
1. Neutrophils (PMNs) (60–70% of leukocytes)		Phagocytosis
2. Basophils (0.5–1%)		Production of heparin and histamines
3. Eosinophils (2–4%)		Production of histamines; some phagocytosis
B. Agranulocytes		
1. Lymphocytes (20–25%)		Antibody production
2. Monocytes (3–8%)		Phagocytosis
Thrombocytes (Platelets)	250,000–500,000	Blood clotting

Kinds of Phagocytic Cells

Phagocytic cells fall into two broad categories: granulocytes and macrophages, which include monocytes. Not all granulocytes exhibit the same phagocytic capabilities. Neutrophils (PMNs) have the most prominent phagocytic activity. They normally wander about in the blood and can pass through capillary walls to reach sites of trauma. Eosinophils are believed to have some phagocytic capability, but the role of basophils in phagocytosis is debatable. Granulocytic phagocytes are sometimes called **microphages.**

When an infection occurs, both granulocytes (especially neutrophils) and monocytes migrate to the infected area. During this process, monocytes enlarge and develop into actively phagocytic cells called **macrophages** (Figure 15–4). Since these cells leave the blood and migrate through tissue to infected areas, they are called *wandering macrophages*. Some macrophages, called *fixed macrophages* or *histiocytes*, enter certain tissues and organs of the body and remain there. Fixed macrophages are found in the liver (Kupffer cells), lungs (alveolar macrophages), spleen, lymph nodes, and bone marrow. Since they can be either reticular (forming a supporting network) or endothelial (lining the sinuses) in different parts of the body, the fixed macrophages are referred to as cells of the *reticuloendothelial system* or *RES*.

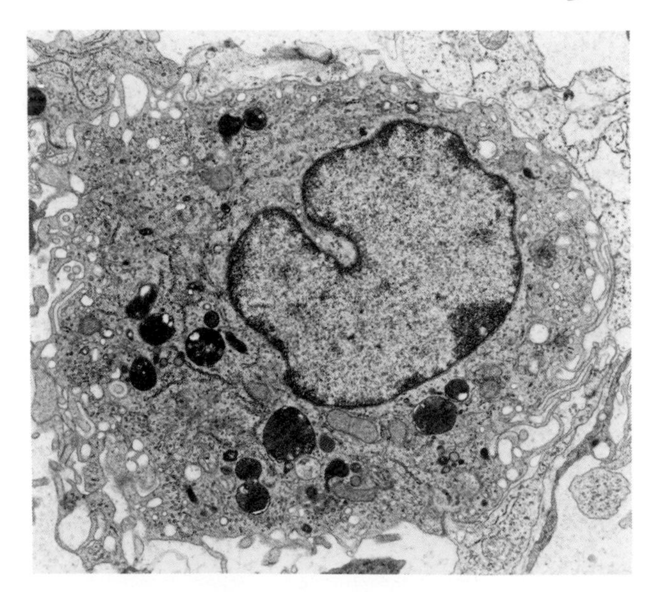

Figure 15–4 Electron micrograph of a macrophage from a rat's lymph node (×6000). Note the numerous fingerlike extensions of the plasma membrane.

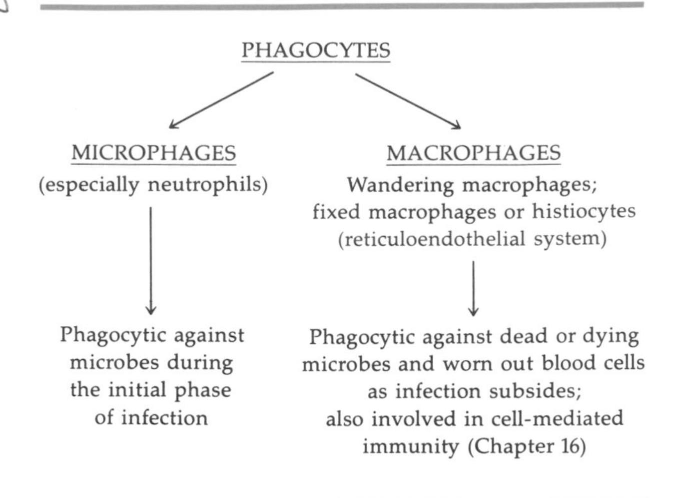

Figure 15–5 Classification of phagocytes.

During the course of an infection, there is a shift in the type of white blood cell that predominates. Granulocytes, especially neutrophils, predominate during the initial phase of infection, as indicated by their number in a differential count. But as the infection begins to subside, the monocytes predominate, acting as scavengers and phagocytizing dead or dying bacteria following phagocytosis by neutrophils. (This increased activity of monocytes can be determined by a differential count.) As blood and lymph containing microorganisms pass through organs with fixed macrophages, cells of the RES remove the microorganisms by phagocytosis. The RES also disposes of worn out blood cells (Figure 15–5).

Mechanism of Phagocytosis

How does phagocytosis occur? For convenience of study, we will divide phagocytosis into two phases: adherence and ingestion.

Adherence

Adherence refers to *attachment* between the cell membrane of the phagocyte and the surface of the microorganism (or other foreign material). Adherence is facilitated by **chemotaxis,** the attraction of

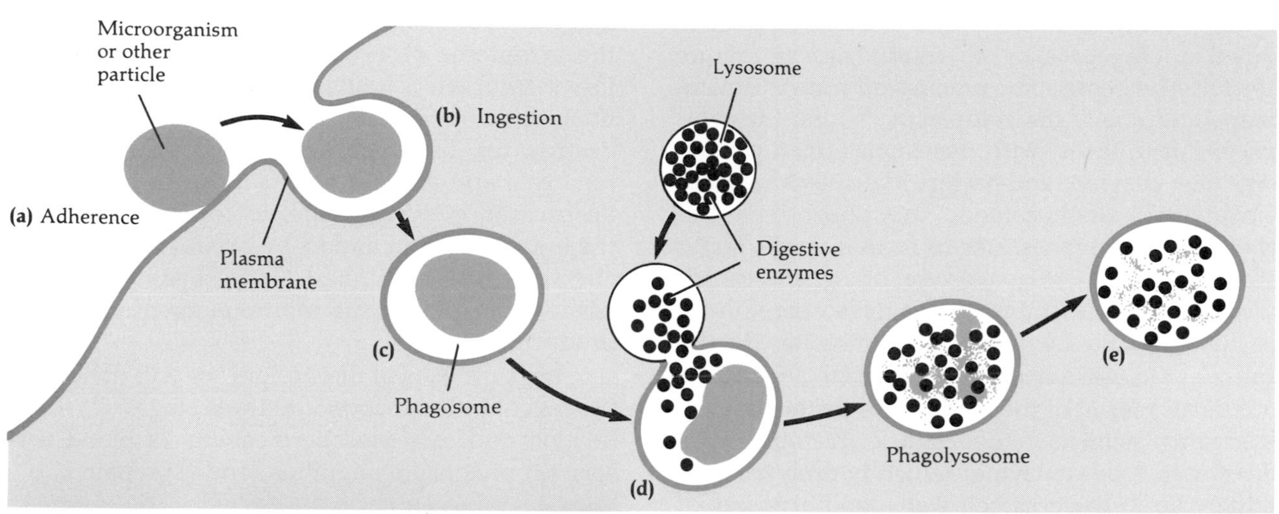

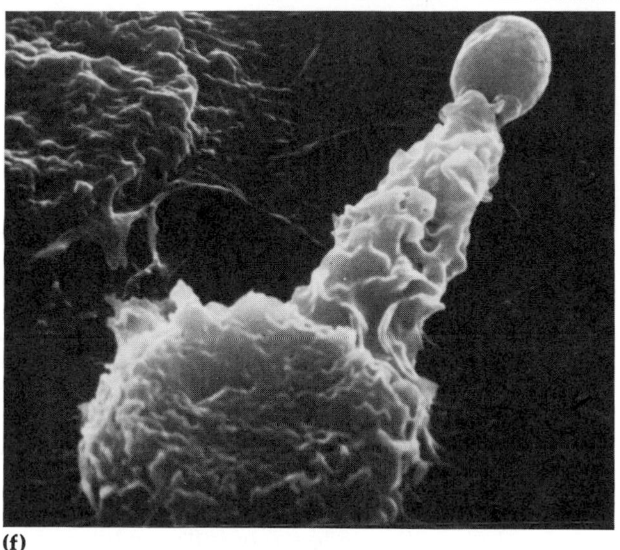
(f)

Figure 15–6 Phagocytosis. **(a)** Adherence. **(b)** Ingestion. **(c)** Formation of the phagosome. **(d)** Fusion of the phagosome with lysosome to form a phagolysosome. **(e)** Destruction of the ingested microorganism. **(f)** Early stage of phagocytosis by an alveolar macrophage (a type of macrophage found in the alveoli of the lungs). The "wrinkled" macrophage has contacted and adhered to a smooth, roughly-spherical yeast cell by means of a pseudopod (×3100).

phagocytes to microorganisms by certain chemicals. In some instances, adherence occurs easily and the microorganism is readily phagocytized (Figure 15–6a). One example is the adherence of *Streptococcus pyogenes* to phagocytes. In other cases, adherence is more difficult, as exemplified by *Streptococcus pneumoniae* and *Klebsiella pneumoniae,* both of which have large capsules. Such heavily encapsulated microorganisms can be phagocytized if the phagocyte traps the microorganism against a rough surface, like a blood vessel, blood clot, or connective tissue fibers, where the microorganism cannot slide away. This is sometimes called *nonimmune* or *surface phagocytosis.* Microorganisms can be more readily phagocytized if they are first coated with certain plasma proteins that promote the attachment of the microorganism to the phagocyte, a process called **opsonization.** The proteins that act as *opsonins* include, among others, some components of the complement system (described later in this chapter) and antibody molecules (Chapter 16).

Ingestion

Following adherence, ingestion occurs. In the process of **ingestion,** projections of the cell membrane of the phagocyte, called pseudopods, engulf the microorganism (Figure 15–6b). Once the microorganism is surrounded, the membrane folds inward, forming a sac around the microorganism called a *phagosome* or *phagocytic vacuole* (Figure 15–6c). The phagosome pinches off from the membrane and enters the cytoplasm. Within the cytoplasm, it collides with lysosomes that contain digestive enzymes and bactericidal substances (see Chapter 4). Upon contact, the phagosome and lysosome membranes fuse to form a single, larger structure called a *phagolysosome,* or *digestive vacuole* (Figure 15–6d). Within the phagolysosome, many bacteria are killed within 10 to 30 minutes (Figure 15–6e). It is assumed that microbial destruction occurs as a result of the contents of the lysosomes. Lysosomal enzymes that attack microbial cells directly include lysozyme, which hydrolyzes peptidoglycan in bacterial cell walls, and a variety of other enzymes, which hydrolyze other macromolecular components of microorganisms. The hydrolytic enzymes are most active at around pH 4, the usual pH in a phagolysosome due to production of lactic acid by the phagocyte. Another kind of lysosomal enzyme, myeloperoxidase, reacts with hydrogen peroxide and chloride ions to bond chlorine atoms to bacteria and viruses—resulting in death of the microorganisms.

Not all phagocytized microorganisms are killed by lysosomal enzymes. Some, such as toxin-producing staphylococci, may be ingested but not necessarily killed. In fact, their toxins can actually kill the phagocytes. Other microorganisms, such as *Mycobacterium tuberculosis,* may multiply within the phagolysosome and eventually destroy the phagocyte. And still others, such as the causative agents of tularemia and brucellosis, may remain dormant in phagocytes for months or years at a time.

Phagocytosis often occurs as part of another nonspecific resistance mechanism, inflammation.

INFLAMMATION

When tissues of the body are damaged, a host response called **inflammation** occurs. The damage may be caused by microbial infection, physical agents (such as heat, radiant energy, electricity, or sharp objects), or chemical agents (acids, bases, and gases). Inflammation is usually characterized by four fundamental symptoms: redness, pain, heat, and swelling. Sometimes, a fifth symptom, loss of function, is present, depending on the site and extent of damage. In apparent contradiction to the symptoms observed, however, the inflammatory response is actually a beneficial one. The functions served by inflammation are as follows: (1) to destroy the injurious agent, if possible, and to remove it and its by-products from the body; (2) if destruction is not possible, to confine or wall off the injurious agent and its by-products to limit the effects on the body; and (3) to repair or replace tissue damaged by the injurious agent or its by-products.

For purposes of discussion, we will divide the process of inflammation into three stages: (1) vasodilation and increased permeability of blood vessels, (2) phagocyte migration, and (3) repair.

Vasodilation and Increased Permeability of Blood Vessels

Immediately following tissue damage, there is vasodilation and increased permeability of blood vessels. **Vasodilation** is an increase in diameter of the blood vessels in the area of the injury. Vasodilation allows increased blood flow to the damaged area, and **increased permeability** permits defense substances normally retained in the blood to pass through the walls of the blood vessels and enter the injured area. Vasodilation is responsible for the redness and heat associated with inflammation. Increased permeability, which permits fluid to move from the blood into tissue spaces, is responsible for the swelling (edema) of inflammation. The pain of inflammation may be caused by nerve damage, irritation by toxins, or the pressure of edema.

What causes vasodilation and increased permeability of blood vessels? Damaged cells release certain chemicals in response to injury. One such substance is **histamine,** a chemical present in many tissues of the body, especially in mast cells in connective tissue, circulating basophils, and blood platelets. Histamine is released in direct response to any injury of cells that contain it, or by certain factors of the complement system (to be discussed later). Phagocytic granulocytes attracted to the site of injury can also produce chemicals that cause the release of histamine. **Kinins** are another group of substances that cause vasodilation and increased permeability of blood vessels. These chemicals are produced in blood plasma. Kinins also attract phagocytic granulocytes—chiefly PMNs—to the injured area. However, they also affect some nerve endings, accounting for much of the pain associated with inflammation. **Prostaglandins** are yet another group of chemicals that cause vasodilation.

Vasodilation and increased permeability of blood vessels also deliver clotting elements of the blood into the injured area. By forming blood clots around the site of activity, the microorganism (or its toxins) is prevented from spreading to other parts of the body. This focus of infection is called an **abscess.** Common examples include pimples and boils.

The next stage in inflammation involves phagocyte migration to the injured area.

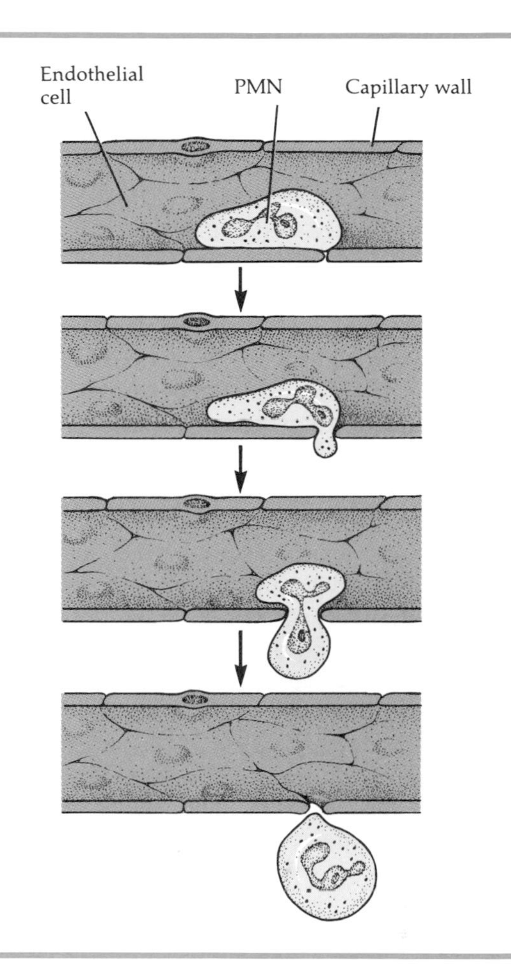

Figure 15–7 Diapedesis of a PMN through the wall of a capillary in inflammation. Monocytes can also squeeze through blood vessel walls in this way.

Phagocyte Migration

Generally within an hour after the process of inflammation is initiated, phagocytes appear on the scene. As the flow of blood decreases, phagocytes (both PMNs and monocytes) begin to stick to the inner surface of the endothelium (lining) of blood vessels. This is called **margination.** Then the phagocytes begin to squeeze through the wall of the blood vessel to reach the damaged area. This migration, which resembles amoeboid movement, is called **diapedesis,** and the whole process can take as little as two minutes (Figure 15–7). As mentioned earlier, PMNs are attracted to the site of injury by certain chemicals (chemotaxis). These include chemicals produced by microorganisms and even other PMNs, as well as kinins and certain components of the complement system. A steady stream of phagocytic granulocytes is ensured by the production and release of additional cells from bone marrow. This is brought about by a substance called **leukocytosis-promoting factor,** which is released from inflamed tissues.

As the inflammatory response continues, monocytes follow the granulocytes into the infected area. Once in the tissue, they become wandering macrophages. In addition, fixed macrophages in various tissues of the body reproduce, become motile, and also migrate to the infected area. Although the granulocytes predominate in the early stages of infection, they tend to die off rapidly. Macrophages enter the picture during a later stage of the infection. They are several times more phagocytic than granulocytes and large enough to phagocytize tissue that has been destroyed, granulocytes that have been destroyed, and invading microorganisms.

When both granulocytes and macrophages engulf large numbers of microorganisms and damaged tissue, they themselves eventually die. After a few days, a cavity forms in the inflamed tissue containing dead phagocytes and damaged tissue. This collection of dead cells and fluid is called **pus.** Pus formation usually continues until the infection subsides. At times the pus pushes to the surface of the body or into an internal cavity for dispersal. On other occasions, the pus remains, even after the infection is terminated. In this case, the pus is gradually destroyed over a period of days and is absorbed by the body.

Repair

The process by which tissues replace dead or damaged cells is called **tissue repair,** and is the final stage of inflammation. It begins during the active phase of inflammation, but cannot be completed until all harmful substances have been removed or neutralized at the site of injury. The ability of a tissue to repair depends, in part, on the tissue involved. For example, skin has a high capacity for regeneration, whereas nervous tissue in the brain and spinal cord does not regenerate at all.

Tissue repair occurs from the production of new cells by the stroma or parenchyma of a tissue. The *stroma* is the supporting connective tissue, whereas the *parenchyma* is the functioning part of the tissue. If only parenchymal cells are active in repair, a perfect or near-perfect reconstruction of the tissue occurs. A good example is a minor cut on the epidermis of the skin. However, if fibroblasts of the stroma are involved in repair, scar tissue is formed.

A summary of the process of inflammation is illustrated in Figure 15–8.

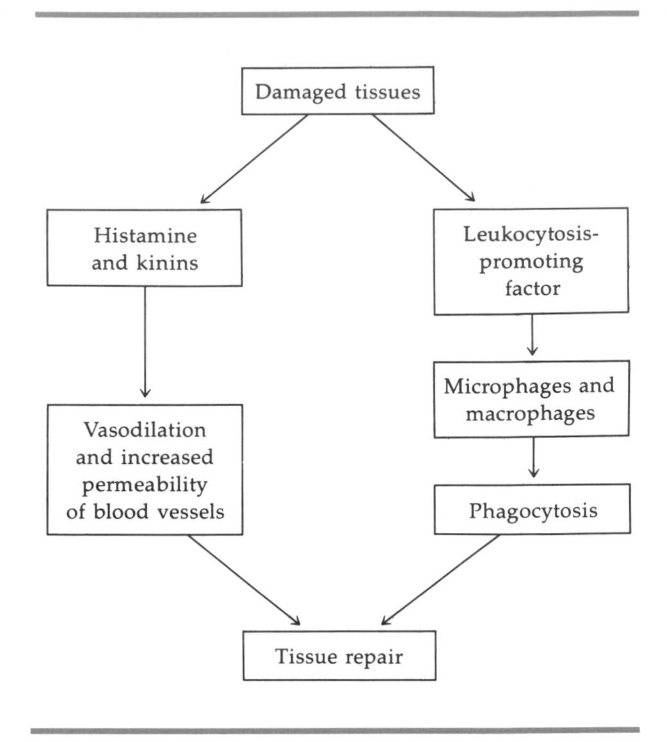

Figure 15–8 Summary of the process of inflammation.

FEVER

In the preceding section, we talked about inflammation, the local response of the body to microbial invasion. There are also systemic, or overall, responses, one of the most important being **fever,** an abnormally high body temperature. The most frequent cause of fever is infection from bacteria (and their toxins) and viruses.

Body temperature is controlled by a part of the brain called the hypothalamus. The hypothalamus is sometimes referred to as the body's thermostat, and it is normally set at 37°C (98.6°F). It is believed that antigens (foreign macromolecules) affect the hypothalamus by setting it at a higher temperature. As little as 1 mg of lipopolysaccharide from the bacterium that causes typhoid fever can set the thermostat as high as 43°C, and the body temperature will continue to be regulated at this temperature until the antigen is eliminated. Perhaps more important is the fact that during inflammation, phagocytic granulocytes, especially neutrophils, release a protein called *leukocytic pyrogen* that has the ability to raise the thermostat.

Assume that the body is invaded by pathogens and the thermostat setting is increased to 39°C. The body's response to adjust to the new thermostat setting is blood vessel constriction, increased rate of metabolism, and shivering, all of which raise body temperature. Thus, even though body temperature is climbing higher than normal, say 38°C, the skin remains cold and shivering occurs. This condition, called a **chill,** is a definite sign that body temperature is rising. When body temperature reaches the setting of the thermostat, the chills disappear. But the body will continue to regulate temperature at 39°C until the bacterial lipopolysaccharides or pyrogen is eliminated. When this occurs, the thermostat is reset at 37°C. As the infection subsides, heat-losing mechanisms such as vasodilation and sweating go into operation. (The skin becomes warm and the person begins to sweat.) This phase of the fever is called the **crisis** and indicates that body temperature is falling.

Up to a certain point, fever may be considered beneficial. The higher body temperature is believed to inhibit the growth of some microorganisms. And since the higher temperature speeds up body reactions, it may help body tissues to repair themselves more quickly. As a rule, however, death results if body temperature rises to about 45°C. But most body tissues can withstand marked cooling to less than 7°C, a fact that is used in some types of surgery.

ANTIMICROBIAL SUBSTANCES

In addition to the chemical barriers to infection mentioned earlier, the body produces certain antimicrobial substances. Among the most important of these are interferon and the proteins of the complement system.

Interferon

Because viruses depend on their host cells for many functions of viral multiplication, it is difficult to inhibit viral multiplication without at the same time affecting the host cell itself. One way the infected host counters viral infections is by the production of interferon. **Interferons** are a class of similar antiviral proteins produced by certain animal cells in response to viral infection, and their function is to interfere with viral multiplication. One of the most interesting features of interferons is that they are host-cell-specific but not virus-specific. This means that interferon produced by human cells protects human cells and has little antiviral activity for cells of other species, such as mice or chicks. However, the interferon of a species is active against a number of different viruses.

Not only do different animal species produce different interferons, but different types of cells in an animal produce different interferons. There are three types of interferon-producing cells: fibroblasts (cells of connective tissue), lymphocytes, and other leukocytes. The three types of interferon produced by these cells may have slightly different effects on the body.

All interferons are small proteins, with molecular weights between 15,000 and 30,000. They are quite stable at low pH and fairly resistant to heat.

Produced by host cells only in very small quantities, interferon is released from virus-infected cells and diffuses to uninfected neighboring cells, where it exerts its antiviral activity. (In this way, interferon protects these neighboring cells.) Interferon molecules bind to the surface of uninfected cells and somehow induce these cells to manufacture another antiviral protein. This newly synthesized protein remains in the cell and acts to prevent the synthesis of viral nucleic acid, in this way inhibiting viral multiplication (Figure 15–9).

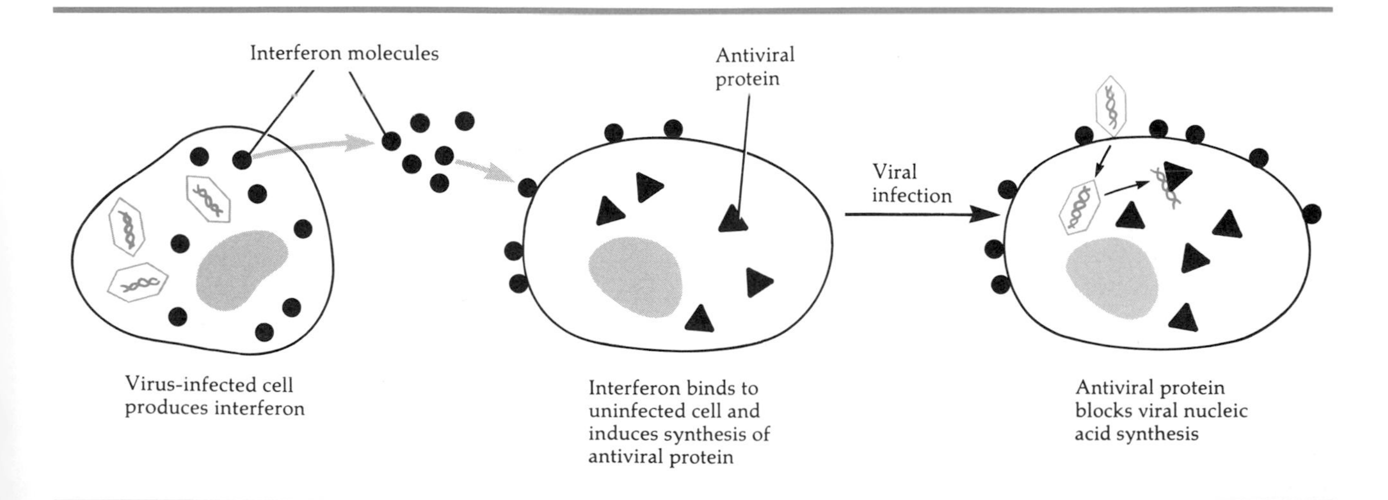

Figure 15–9 Action of interferon.

At low concentrations, interferon is nontoxic to uninfected cells; it inhibits viral multiplication only. Given all the beneficial properties, it would seem that interferon would be an ideal antiviral substance. But certain problems do exist. For one thing, interferon is effective for only short periods of time, typically playing a major role in infections that are acute and short-term, such as colds and influenza. Another problem is that it has no effect on viral multiplication in cells already infected. However, the importance of interferon in protecting the body against viruses, as well as its potential as an anticancer agent, has made its production in large quantities a top priority for medical scientists. At first the most promising method of interferon production was by cultured human cells. Recently, however, several groups of scientists have succeeded in "engineering" bacteria to produce the chemical, as described in Chapter 8.

Complement and Properdin

Complement is a group of proteins found in normal blood serum that is very important to both nonspecific and specific defenses against microbial infection. The overall function of the complement system is to attack and destroy invading microorganisms (Figure 15–10). Complement is so named because in its "classical" role, it complements (fills out or completes) certain immune reactions involving antibodies. For this reason, many textbooks do not discuss complement until after a detailed presentation of antigen–antibody reactions. We will discuss it here, however, because we regard the complement system as basically an area of *nonspecific* resistance. For one thing, complement has the same general role in all antigen–antibody reactions in which it participates. Also, complement itself is not specific for particular antigens and does not appear in greater amounts after immunization. Furthermore, complement participates in important defense reactions that do not involve antibody. (If you wish, however, you may delay this topic until after Chapter 16.)

Components of the complement and properdin system

The complement system, abbreviated **C** (formerly C′), is made up of eleven proteins. These proteins are designated C1 through C9, with C1 actually a complex of three proteins. The molecular weights of these proteins are high, ranging from about 80,000 to over 400,000. Three other serum proteins make up the related **properdin** system: properdin itself, factor B, and factor D. Together all these proteins make up over 10% of the proteins in serum.

The proteins of the complement and properdin system act in an ordered sequence. With the exception of C4, the order follows the numerical sequence. In a series of steps, the proteins *activate* one another, primarily by cleaving the next protein in the series. When a protein is cleaved, the resulting protein fragments have new enzymatic or physiological functions. For example, a whole protein cannot cleave the next protein in the series, but one of its fragments can.

Pathways of complement activation

C3 plays a central role in the complement system. As you can see in Figure 15–11, there are two pathways to the activation of C3. On the upper left is the *classical pathway,* which is initiated by interaction between an antigen-antibody complex (described in Chapter 16) and the C1 complex. On the upper right is the so-called *alternative pathway,* which is initiated by the interaction between certain polysaccharides and the proteins of the properdin system. The effective polysaccharides are chiefly those of certain bacterial and fungal cell walls (although they also include molecules on the surface of some foreign mammalian red blood cells). The properdin pathway is of particular importance in combating enteric gram-negative bacteria. The cell wall lipopolysaccharide (endotoxin) of these bacteria triggers the pathway. Note that this pathway does not use antibody, C1, C2, or C4.

Consequences of complement activation

How does complement contribute to microbial destruction? Both the classical and alternative pathways lead to the cleavage of C3 into two fragments, C3a and C3b. These fragments induce three kinds of consequences destructive to microorganisms:

1. C3b initiates a sequence of reactions of C5 through C9, leading to cell lysis caused by circular holes in the cell membrane (see Figure

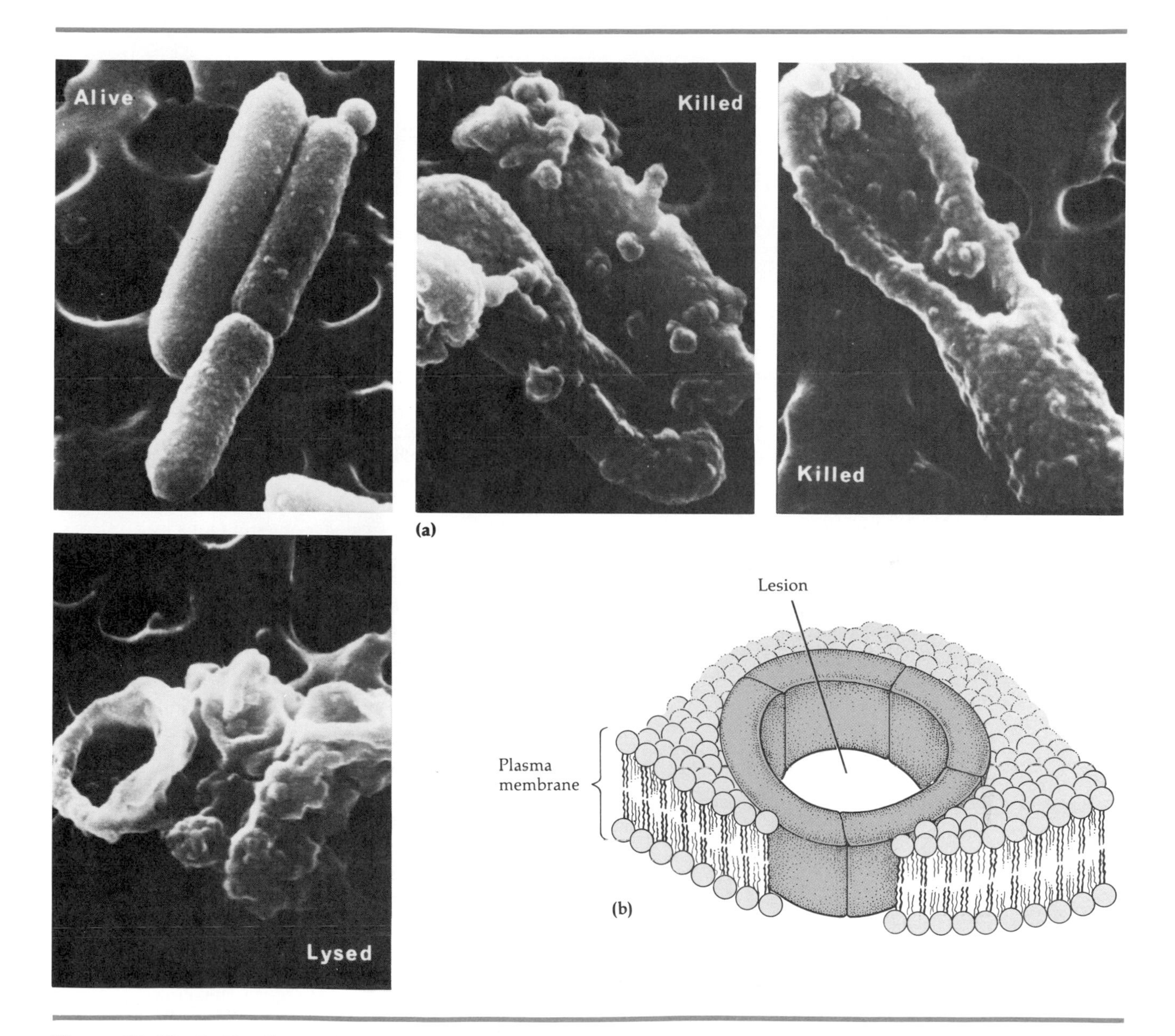

Figure 15–10 Destructive action of complement. **(a)** Scanning electron micrographs of *Escherichia coli* killed and then lysed by the action of complement. **(b)** Diagram of a complement lesion.

15–10). The utilization of complement components in this process is called **complement fixation** and is the basis of an important clinical laboratory test (see Chapter 16).

2. C3b, already bound to the surface of a microorganism, interacts with special receptors on phagocytes, promoting phagocytosis. This phenomenon is called **immune adherence** or opsonization.

3. C3a, as well as cleavage products from C5, C6, and C7, contribute to the development of acute inflammation. In particular, they increase blood vessel permeability and act as chemotactic agents, attracting large numbers of PMNs to the area. (The antimicrobial effects of inflammation were discussed earlier in this chapter.)

Once complement is activated, its destructive capabilities usually cease very quickly in order to

Microbiology in the News

How Sickle-Cell Genes Protect Against Malaria

The mysterious relationship between sickle-cell anemia and malaria has finally been explained by researchers working in the United States and Italy.

For many years it has been known that persons who carry a gene for the lethal genetic disease sickle-cell anemia are resistant to malaria, a protozoan disease which has been a major cause of death throughout human history. Fortunately, the resistance is most effective against the most lethal type of malaria, which is caused by *Plasmodium falciparum*. Apparently, if a sickle-cell gene carrier is infected by this protozoan, the infected cells sickle and lose potassium, which is needed for the parasite's survival.

Persons with sickle-cell anemia have an unusual form of hemoglobin, the oxygen-carrying molecule in red blood cells. This abnormal hemoglobin is called hemoglobin S, to differentiate it from normal hemoglobin, called hemoglobin A. When the oxygen concentration is low, molecules of hemoglobin S aggregate into long fibers, which push the red blood cell into a sickle shape. Sickled cells can block the local circulation of blood and interfere with the delivery of oxygen to the tissues, eventually causing tissue death. Without advanced medical treatment, most afflicted individuals die at a young age.

The genetic cause of sickle-cell anemia is the presence of two hemoglobin S genes in an individual, one inherited from each parent. Individuals with one hemoglobin S gene and one hemoglobin A gene are called "carriers" of the sickle-cell gene, and usually have no symptoms. Their red blood cells contain both types of hemoglobin and do not usually sickle in the short time they spend in regions of the body with low oxygen concentration. The children of two carriers have a 75% chance of having at least one hemoglobin S gene and a 25% chance of having two such genes.

Recent studies by William Trager of the United States, Lucio Luzzatto of Italy, and others have established a sequence of events that leads to the death of *Plasmodium* within the red blood cells of sickle-cell carriers. Infection of a red blood cell by *Plasmodium* lowers the pH within the cell. At lower pH, a red blood cell containing hemoglobin S has a greater tendency to sickle. When the cell sickles, potassium ions leak out through the cell membrane. The lowered potassium level kills the parasite (see figure). The death of even some of the infecting protozoans may give the host time to develop its immune response.

The recent breakthroughs were made possible by a technique developed by Trager for culturing *Plasmodium falciparum* outside an animal host. Red blood cells are grown in a culture dish or U tube as a thin layer of cells covered by an artificial nutrient medium. Infected cells from a malaria patient or from another culture are used to inoculate the culture.

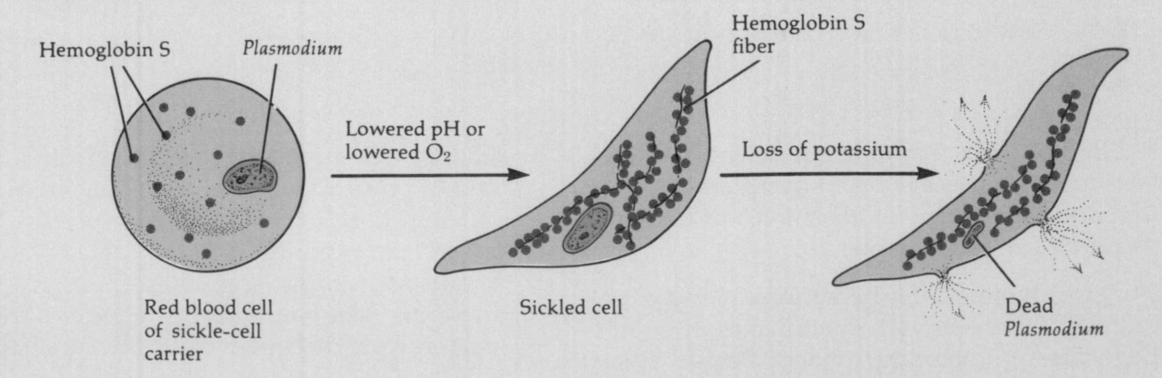

minimize destruction of the host's cells as well. This is accomplished by the spontaneous breakdown of activated complement and by interference from inhibitors and destructive enzymes in the host's body fluids.

There are other types of defense mechanisms besides the nonspecific defenses discussed in this chapter and immunity, which will be discussed in Chapter 16. For example, a genetic trait may provide a defense, as in the case of the relationship between sickle-cell anemia and malaria (see box).

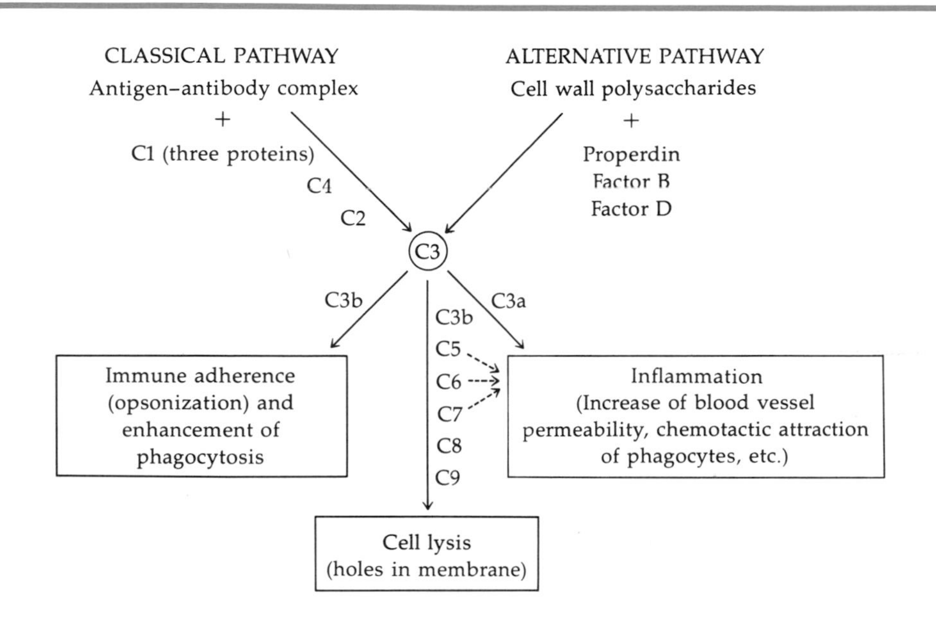

Figure 15–11 Activation pathways and functions of the complement and properdin systems.

STUDY OUTLINE

INTRODUCTION (p. 377)

1. The ability to ward off disease through body defenses is called resistance.
2. Lack of resistance is called susceptibility.
3. Nonspecific resistance refers to all body defenses that protect the body from any kind of pathogen.
4. Specific resistance refers to defenses (antibodies) against specific microorganisms.

SKIN AND MUCOUS MEMBRANES (pp. 378–381)

Mechanical Factors (pp. 378–380)

1. The structure of intact skin plus keratin provide resistance to microbial invasion.

2. The lacrimal apparatus protects the eyes from irritating substances and microorganisms.
3. Saliva washes microorganisms from teeth and gums.
4. Mucus traps many microorganisms that enter the respiratory and gastrointestinal tracts; in the lower respiratory tract the ciliary escalator moves mucus up and out.
5. The flow of urine moves microorganisms out of the urinary tract, and vaginal secretions move microorganisms out of the vagina.

Chemical Factors (pp. 380–381)

1. Sebum contains bacteriostatic fatty acids.
2. Perspiration washes microorganisms off the skin.
3. Lysozyme is found in tears, saliva, nasal secretions, and perspiration.
4. The high acidity (pH 1.2 to 3.0) of gastric juice prevents microbial growth in the stomach.

PHAGOCYTOSIS (pp. 381–384)

1. Phagocytosis is the ingestion of microorganisms or particulate matter by a cell.
2. Phagocytes are blood cells or derivatives of blood cells.

Formed Elements of the Blood (p. 381)

1. Blood consists of plasma (fluid) and formed elements (cells and cell fragments).
2. Leukocytes (white blood cells) are divided into two categories: granulocytes (neutrophils, basophils, and eosinophils) and agranulocytes (lymphocytes and monocytes).
3. During many infections, the number of leukocytes increase (leukocytosis); some infections are characterized by leukopenia (decrease in leukocytes).

Kinds of Phagocytic Cells (p. 382)

1. Among the granulocytes, neutrophils are the most important phagocytes.
2. Enlarged monocytes become wandering macrophages and fixed macrophages.
3. Granulocytes predominate during the early stages of infection, whereas monocytes predominate as the infection subsides.

Mechanism of Phagocytosis (pp. 382–384)

1. Phagocytes are attracted to microorganisms by chemotaxis.

2. The phagocyte then adheres to the microbial cell; adherence may be facilitated by opsonization.
3. Pseudopods engulf the microorganism, enclosing it in a phagosome to complete ingestion.
4. Many phagocytized microorganisms are killed by lysosomal enzymes.

INFLAMMATION (pp. 384–386)

1. Inflammation is a body response to cell damage and is characterized by redness, pain, heat, and swelling.
2. Sometimes loss of function results.

Vasodilation and Increased Permeability of Blood Vessels (pp. 384–385)

1. The release of histamine, kinins, and protaglandins cause vasodilation and increased permeability.
2. Blood clots may form around an abscess to prevent dissemination of the infection.

Phagocyte Migration (pp. 385–386)

1. Phagocytes have the ability to stick to the lining of the blood vessels (margination).
2. They also have the ability to squeeze through blood vessels (diapedesis).

Repair (p. 386)

1. Tissue repair occurs from the production of new cells by the stroma or parenchyma of a tissue.
2. Stromal repair by fibroblasts produces scar tissue.

FEVER (pp. 386 387)

1. Fever is an abnormally high body temperature produced in response to a bacterial or viral infection.
2. A chill indicates a rising body temperature; crisis indicates that body temperature is falling.

ANTIMICROBIAL SUBSTANCES (pp. 387–391)

Interferon (pp. 387–388)

1. Interferon is an antiviral substance produced in response to viral infection.
2. Its mode of action is to induce uninfected cells to produce an antiviral protein that prevents viral replication.

Complement and Properdin (pp. 388–391)

1. Complement and properdin make up about 10% of the serum proteins.
2. These proteins activate one another to destroy invading microorganisms.

STUDY QUESTIONS

REVIEW

1. Define the following terms.
   (a) Resistance
   (b) Susceptibility
   (c) Nonspecific resistance
2. Describe the mechanical factors of the skin and mucous membranes that assume a role in nonspecific resistance.
3. Describe the chemical factors of the skin and mucous membranes that assume a role in nonspecific resistance.
4. Define phagocytosis.
5. Compare the structure and function of granulocytes and monocytes in phagocytosis.
6. How do fixed and wandering macrophages differ?
7. Define inflammation, and list its characteristics.
8. Why are redness, heat, and swelling observed during inflammation?
9. Why is inflammation beneficial to the body?
10. How is fever related to nonspecific defense?
11. What is the importance of chill and crisis during fever?
12. What is interferon? Discuss its role in nonspecific resistance.
13. What is complement?
14. Summarize the major outcomes of complement activation.
15. What is properdin? What are its functions in nonspecific resistance?

CHALLENGE

1. Diagram the following processes that result in phagocytosis: margination, diapedesis, adherence, and ingestion.
2. A variety of drugs with the ability to reduce inflammation are available. Comment on the danger of misuse of these anti-inflammatory drugs.
3. A hematologist will often do a differential count on a blood sample. A differential count is used to determine the relative numbers of white blood cells. Why is this important? What do you think the hematologist would find in a differential count of a patient with mononucleosis?

FURTHER READING

Allison, A. November 1967. Lysosomes and disease. *Scientific American* 217:62–72. An investigation into the function of lysosomal enzymes and the possible role of lysosomes in malignancies.

Burke, D. C. April 1977. The status of interferon. *Scientific American* 236:42–50. Overview of the function of interferon and its effectiveness in treating infections by hepatitis B virus.

Collier, H. O. J. August 1962. Kinins. *Scientific American* 207:111–118. Roles of chemical mediators in inflammation.

Mayer, M. M. November 1973. The complement system. *Scientific American* 229:54–66. Detailed description of the functions of the complement enzymes.

Schlessinger, D. (ed.). 1980. Interferon: induction and action. In *Microbiology 1980*, pp. 199–216. Washington, D.C.: American Society for Microbiology (Proceedings). Six papers discussing production of interferon by host cells, inhibition of protein synthesis by interferon, and inhibition of murine leukemia virus by interferon.

Wood, W. B. February 1951. White blood cells versus bacteria. *Scientific American* 184:48–52. Electron micrographs illustrate the action of phagocytes.

16

Specific Defenses of the Host: Immunology

Objectives

After completing this chapter you should be able to

- Contrast the kinds of immunity.
- Name and define the two most important characteristics of an antigen.
- Explain what an antibody is and describe the structural and chemical characteristics of antibodies.
- Provide at least one function for each of the five classes of antibodies.
- Distinguish between humoral and cellular immunity.
- Explain how each of the following can be used in the diagnosis of a disease: precipitation; agglutination; complement fixation; neutralization.
- Define the following terms: immunity; serology; immunodiffusion; opsonization; radioimmunoassay; vaccine; attenuated; toxoid; anamnestic response.

In the last chapter, we looked at a number of nonspecific defenses of the host, including the skin and mucous membranes, phagocytosis, inflammation, fever, and relatively nonspecific antimicrobial substances. Despite their variety, nonspecific defenses have one thing in common: they will combat any microorganism that invades the body in numbers, rather than being specifically directed against a particular microbe. Now we are going to take a look at our specific defenses to disease. As mentioned earlier, one type of specific defense involves the production of specific proteins called antibodies. Unlike nonspecific defenses, antibodies are directed against specific microorganisms. If microorganism *A* invades the body, *A* antibodies are produced against it; if microorganism *B* invades the body, *B* antibodies are produced; and so on. Antibodies and other *specific* defenses against microbial invaders provide the body with a resistance to disease that we call **immunity.**

THE DUALITY OF THE IMMUNE SYSTEM

The immune system of humans, like that of all vertebrates, consists of two parts. One part is called the **humoral immune system** because it involves antibodies that are dissolved in the blood plasma and lymph—the body's "humors," or fluids. These antibodies are produced by lymphocytes called *B cells.* The other part of the immune system is the **cellular,** or **cell-mediated, immune system.** The cellular immune response depends on lymphocytes called *T cells,* which are located both in the blood and in the lymphoid tissues. T cells do not secrete antibodies into body fluids, but they do make antibodylike molecules that remain attached to the cell surface.

The humoral immune response primarily defends against bacteria and viruses present in body fluids. The cellular response is particularly effective against intracellular viruses, fungi, protozoans, helminths, transplanted tissue, and cancer cells. There is overlap, however, in the protection each of these two distinct types of responses provides.

KINDS OF IMMUNITY

Immunity is traditionally divided into native immunity and acquired immunity.

NATIVE "IMMUNITY"

The kind of immunity discussed in this chapter is sometimes called *acquired* immunity to distinguish it from *native*, or *innate*, immunity. The term **native immunity** refers to resistance to disease that is present at birth and which does not involve either humoral or cell-mediated antigen-specific immunity. Native immunity can be further divided into species immunity and individual immunity. *Species immunity* is the resistance of some animal species to certain diseases that seriously affect other species. For example, humans are resistant to certain infectious animal diseases such as canine distemper, cattle plague, and chicken and hog cholera. *Individual (native) immunity* is the resistance of certain individuals to diseases that affect other individuals of the same species. Among the factors that affect such resistance are sex, age, nutritional status, and the general health of organ systems (Chapter 13). Neither type of native immunity is truly "immunity" in the sense used in this chapter.

ACQUIRED IMMUNITY

Acquired immunity refers to the resistance to infection obtained during the life of the individual that results from the production and activity of antibodies against specific microbial components (antigens). These antibodies include the humoral antibodies that circulate freely in the blood and lymph and the antibodylike molecules on the surface of the lymphocytes involved in cell-mediated immunity. The individual may acquire antibodies either *actively* or *passively*. Actively acquired antibodies are those produced by the individual who carries them. Passively acquired antibodies are produced by another source (another human or a horse, for example) and then transferred to the individual who needs the antibodies. Acquired immunity is classified into four categories, based on the source of the antibodies and the means by which an individual acquires them.

Naturally Acquired Active Immunity

In *naturally acquired active immunity*, the immune system of an individual comes in contact with microbial antigens by natural processes such as infection. In response to an infectious disease, the individual actively produces antibodies against the pathogen. In some cases, the immunity may be lifelong, as with measles, chickenpox, smallpox, and yellow fever. In other cases, immunity against reinfection may last only a few years, as with diphtheria and tetanus. And there are also cases where immunity to reinfection appears to be minimal or nonexistent, as with pneumonia and influenza. A variety of agents may cause each of these diseases, and immunity to one does not necessarily confer immunity to the others. It should also be noted that naturally acquired active immunity is developed in instances of subclinical infections. Immunity to scarlet fever and polio can be acquired in this way.

Naturally Acquired Passive Immunity

Naturally acquired passive immunity is derived from the natural transfer of humoral antibodies from an immunized donor to a susceptible recipient. This immunity most commonly results from the passage of certain antibodies from the mother to the fetus across the placenta (placental transfer). An expectant mother with antibodies against diseases such as diphtheria, German measles, and polio will pass some of these antibodies to the fetus. Certain antibodies are also passed from mother to infant in

1st milk

human colostrum and milk during nursing. This immunity is natural because transfer of antibodies from mother to baby occurs under natural conditions, and it is passive because the baby does not synthesize the antibodies but picks them up from an outside source. Naturally acquired passive immunity generally lasts from a few weeks to a few months.

Artificially Acquired Active Immunity

In *artificially acquired active immunity,* a carefully chosen antigen is purposely introduced into the body. This kind of immunity involves the use of killed or inactivated microorganisms, inactivated toxins (toxoids), and living but attenuated (weakened) microorganisms. Such preparations are known as *vaccines* and are altered to the point where they are unable to produce the signs and symptoms of disease while still able to stimulate antibody formation against that specific disease. Vaccines are available against the microorganisms that produce pneumococcal pneumonia, typhoid fever, smallpox, yellow fever, polio, tetanus, diphtheria, tuberculosis, cholera, plague, whooping cough, German measles, mumps, rabies, influenza, and measles. Vaccines usually provide long-term immunity.

Artificially Acquired Passive Immunity

Artificially acquired passive immunity usually refers to the transfer of humoral antibodies formed by one individual to a susceptible individual by means of injection of antibody-containing serum. (*Serum* is blood plasma with clotting factors removed. The plural is *sera* or *serums*.) Passive immunization is particularly useful when an individual has been exposed to rattlesnake venom or other antigens such as those that cause botulism, diphtheria, whooping cough, rabies, and tetanus—diseases that must be treated quickly to avoid serious consequences. Such individuals must be immunized as soon as possible. Serum containing antibodies directed against either purified toxins (antitoxins) or whole microorganisms is called *immune serum* and may be prepared from the blood of humans, horses, and sometimes cows. Serum prepared against whole microorganisms is sometimes called *antimicrobial serum*. Usually human serum is obtained from donors who have acquired natural immunity by chance exposure to the disease-causing organism. Artificially acquired passive immunity is immediate but short-lived (two to three weeks).

The various kinds of immunity are summarized in Table 16–1.

ANTIGENS AND ANTIBODIES

Let us now take a closer look at what antigens and antibodies are, how they interact, and how they function in the body.

ANTIGENS

Definition

An **antigen** is a chemical substance that, when introduced into the body, causes the body to produce specific antibodies that the antigen can then

Table 16–1 Summary of the Kinds of Immunity

NATIVE (INNATE) IMMUNITY (Inherited; depends on nonspecific defenses of host)
1. Species: Some species are resistant to diseases that affect other species
2. Individual: Some individuals are resistant to certain diseases to which other members of the race are susceptible
ACQUIRED IMMUNITY (Obtained during life; depends on the production and activity of antibodies)
1. Naturally acquired active immunity: Antigens enter the body → Antibodies produced
2. Naturally acquired passive immunity (placental and milk transfer): Antibodies from an immunized mother → Fetus or baby receives antibodies
3. Artificially acquired active immunity: Prepared antigens in vaccines → Injected into susceptible individual → Antibodies produced
4. Artificially acquired passive immunity: Immune serum → Exposed individual receives antibodies

combine with. Based on this definition, antigens have two very important characteristics: *immunogenicity*, or the ability to stimulate the formation of specific antibodies, and *reactivity*, the ability to react specifically with antibodies. An antigen with both of these characteristics is called a **complete antigen.** But antigens have several other distinctive characteristics, as well.

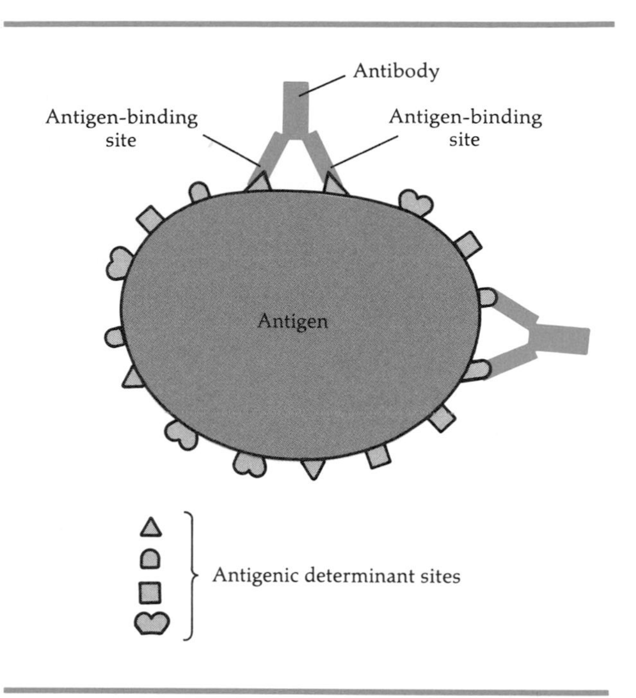

Figure 16–1 Relationship of an antigen to antibodies. Since most antigens contain more than one antigenic determinant site, they are referred to as multivalent. Antibodies have at least two sites that are complementary to the antigenic determinant sites for which they are specific. Most human antibodies are bivalent. Each antibody molecule has sites for one type of specific antigenic determinant site only.

Characteristics

Chemically, the vast majority of antigens are proteins, nucleoproteins (nucleic acid + protein), lipoproteins (lipid + protein), glycoproteins (carbohydrate + protein), and certain large polysaccharides. From what we already know about microorganisms, we can see how many microorganisms or their components may function as antigens. For example, an entire bacterium or virus may act as an antigen. And bacterial structures such as flagella, capsules, cell walls, and pili are also antigenic. Moreover, bacterial toxins are highly antigenic. Nonmicrobial examples of antigens include pollen, egg white, blood cells from other individuals or species, and transplanted tissues and organs. The myriad of antigens in the environment provides much stimulation for the production of antibodies by the body.

In general, antigens have molecular weights of 10,000 or greater. Moreover, antibodies are not formed against the whole antigen, but only against specific portions of the antigen. These specific regions on the surface of the antigen are called *antigenic determinant sites* (Figure 16–1), and it is at these sites that specific chemical groups of the antigen combine with the antibody. The nature of this combination depends on the size and shape of the determinant site and the way in which it corresponds to the chemical structure of the antibody. The antigen and antibody fit together like a lock and key, much the same way that enzyme and substrate do. The number of antigenic determinant sites on the surface of an antigen is called *valence.*

Most antigens are *multivalent*, that is, they have more than one antigenic determinant site. It is believed that in order to induce antibody formation, an antigen must have at least two determinant sites. If an antigen is chemically broken down, it is possible to separate a determinant site from the remainder of the molecule. The molecular weight of the determinant site is small compared to that of the antigen as a whole, perhaps as small as 200 to 1000. An isolated determinant site still has the ability to react with an antibody that has already formed in response to the original antigen (reactivity), but it does not have the ability to stimulate the production of antibodies (immunogenicity) when injected into an animal. The reason is that its lower molecular weight prevents it from serving as a complete antigen. A determinant site that has reactivity but not immunogenicity is called an **incomplete (partial) antigen** or **hapten** (*haptein*, "to grasp"). A hapten can be made into a complete antigen by combining it with a larger carrier molecule, such as a protein (Figure 16–2). As a matter of fact, some low-molecular-weight drugs like penicillin act as haptens. Penicillin may combine with high-molecular-weight proteins in the body, a combination that is antigenic. Antibodies formed against the complex are responsible for the allergic reactions to drugs and other chemicals.

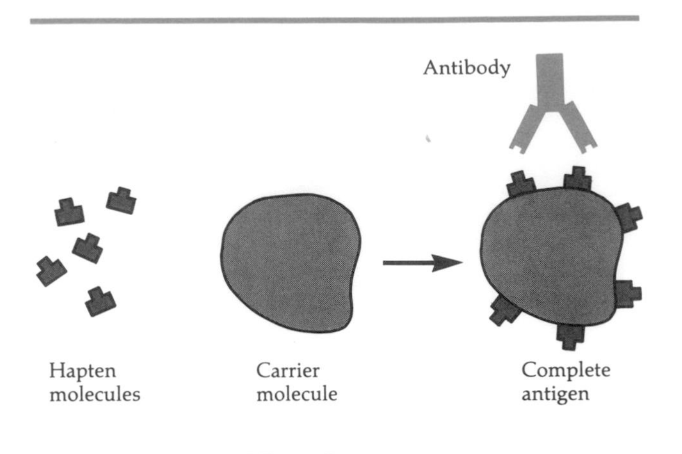

Figure 16–2 Haptens. A hapten (partial antigen) is a relatively small molecule that has reactivity but not immunogenicity. When combined with a larger molecule, such as a protein, the hapten and larger molecule together function as a complete antigen with both reactivity and immunogenicity.

Table 16–2 Classes of Antigens

Class	Description
Complete	Antigens that possess both reactivity and immunogenicity
Incomplete (haptens)	Low-molecular-weight compounds that have reactivity but not immunogenicity
Autoantigens	"Self" antigens against which the body produces antibodies called *autoantibodies* (see Chapter 17)
Isoantigens	Genetically determined antigens found in normal human red blood cells; responsible for the common A, B, AB, and O blood groups; the plasma of normal human blood also contains antibodies called *isoantibodies* (see Chapter 17)
Heterophile	Substances that stimulate the antibodies capable of reacting with antigens of a wide variety of unrelated species; for example, cardiolipin (present in beef heart extract) is a heterophile antigen used in the complement fixation test for syphilis; antibodies against *Treponema pallidum*, the pathogen of syphilis, react with cardiolipin

Our immune system recognizes components of our own body as "self" and not as foreign matter, which is why we do not usually produce antibodies against our own molecules. But as you will see later, there are certain circumstances under which a person can manufacture antibodies against his or her own molecules, a situation that results in what we call autoimmune disease.

Now that we have considered the basic nature of antigens, refer to Table 16–2, which lists the various classes of antigens. As you read the list, keep in mind that the classes are not mutually exclusive and that a particular antigen may represent more than one class (for example, autoantigens are also complete antigens).

ANTIBODIES

Definition

An **antibody** is a protein produced by the body in response to the presence of an antigen and capable of combining specifically with that antigen. This is essentially the complementary definition of an antigen. The specific fit of the antibody with the antigen depends not only on the size and shape of the antigenic determinant site, but also on the corresponding antibody site, much like the lock and key analogy (see Figure 16–1). An antibody, like an antigen, also has valence, the number of specific antigen–antibody combining sites. But whereas an antigen usually has several different combining sites, the combining sites of an antibody molecule are identical, as we shall see. The majority of human antibodies are bivalent. Since antibodies belong to a group of proteins called globulins (for their globular shape), they are also known as **immunoglobulins (Ig).**

Immunoglobulins are found in the blood serum. Fractions or parts of serum are identified by *electrophoresis*. When serum is subjected to an electric field, the negatively charged proteins move toward the anode (+ pole of the electric field). The small, highly charged serum albumin molecules move rapidly, while the globulins move more slowly, dividing into three groups: alpha, beta, and gamma. Because antibodies are found in the gamma globulin fraction, the term **gamma globulin** is used synonymously with immunoglobulin (Figure 16–3).

Structure

Antibodies are proteins, as just noted. Most antibody molecules consist of four polypeptide chains

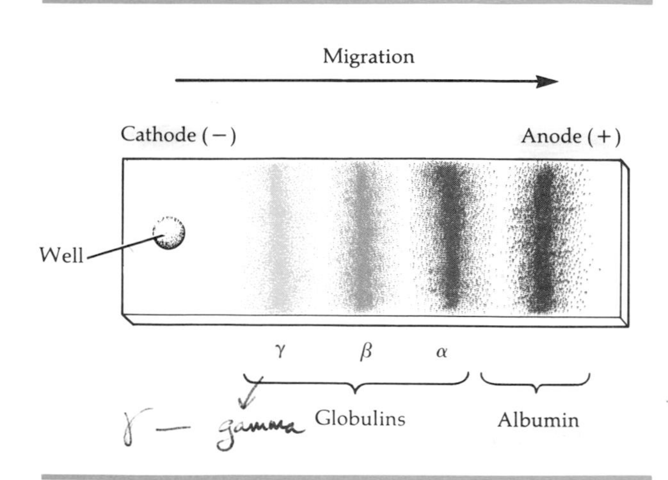

Figure 16–3 Electrophoresis. In this procedure, a slide is coated with a gel adjusted to an alkaline pH. Serum to be tested is placed in a well in the gel. An electric current is then passed through the gel and the negatively charged serum proteins migrate from the cathode (−) to the anode (+). Antibodies are gamma (γ) globulins.

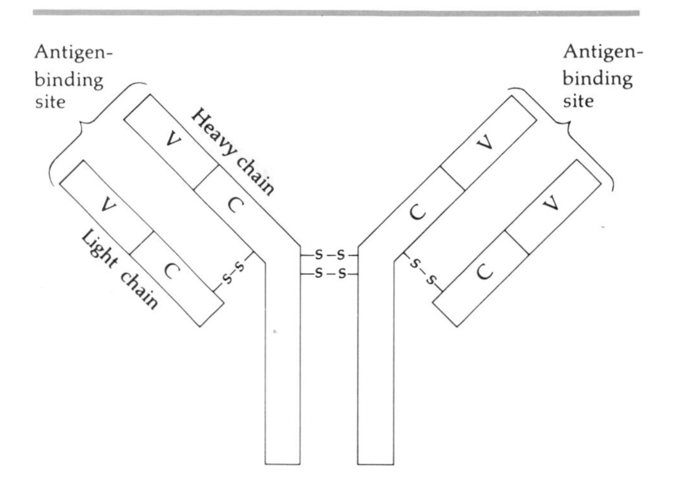

Figure 16–4 Approximate shape of an antibody molecule. Since this molecule has two antigen-binding sites, it is bivalent. V is the variable portion, C is the constant portion, and S—S indicates a disulfide bond.

(Figure 16–4). Two of the chains are identical to each other and are of high molecular weight. They are therefore called *heavy (H) chains*. Each consists of more than 400 amino acids. (There are also carbohydrate groups attached to the H chains, making immunoglobulins actually *glycoproteins*.) The other two chains, also identical to each other, are of lower molecular weight and are thus called *light (L) chains*. Each consists of approximately 200 amino acids. The antibody is structured so that it consists of identical halves, each half consisting of a heavy and a light chain held together by a disulfide bond (S—S) (see Chapter 2). The halves of the molecule are also held together by a disulfide bond. The amino (NH_2) ends of the polypeptide chains combine with antigens. The total antibody molecule is sometimes shaped like the letter Y, at other times like the letter T. These differences in shape will be explained shortly.

Within each H and L chain, there are two distinct regions. The *variable (V) portions* of the H and L chains, shown at the top of Figure 16–4, are where the antigen binding occurs. The variable portion is different for each kind of antibody, and it is this portion that allows the antibody to attach specifically to a particular type of antigen. Since most antibodies have two variable portions for attachment of antigens, they are bivalent. The remainder of each polypeptide chain is called the *constant (C) portion*, and it is this region that is responsible for the type of antigen–antibody reaction that occurs. The constant portion differs from one class of antibody to another, and the structure of the H chain constant portion serves as a basis for distinguishing the classes of antibodies.

In performing its role in immunity, the antibody is believed to behave like a switch. When antibodies are viewed in the electron microscope before combination with antigen, they resemble the letter T. But after combination with antigen, antibodies appear to be smaller and to resemble the letter Y. It is possible that the binding of antigen causes a rearrangement in the structure of the antibody. The region of the antibody molecule where the H chains are joined by disulfide bonds (Figure 16–4) is a hinge that provides the antibody molecule with flexibility. When antigen and antibody combine, the rearrangement of antibody seems to consist of a pivoting movement in which the H chain moves from a T shape (Figure 16–5) to a Y shape (Figure 16–6). This pivoting exposes the constant portions of the H chains so that they can function in further reactions, such as complement fixation.

Let us now take a look at the classes of antibodies.

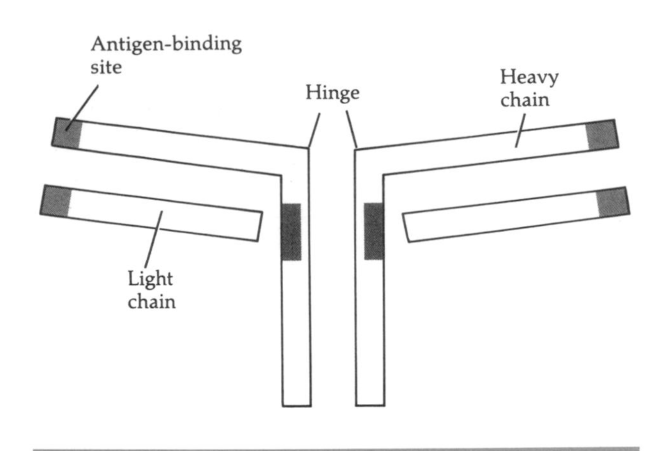

Figure 16–5 The T shape of an antibody before it combines with antigen.

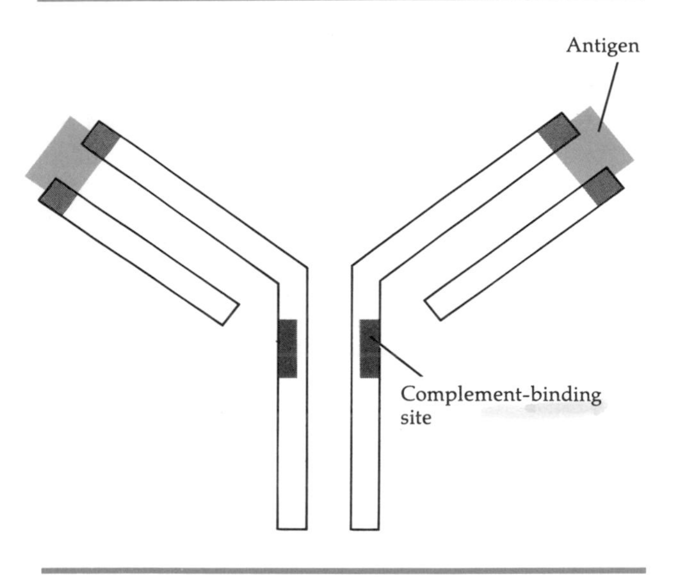

Figure 16–6 The Y shape of an antibody after it combines with antigen. Combination with an antigen may make the heavy chain pivot at the hinge, causing the antibody to assume a Y shape. This change in shape may expose parts of the antibody that function in further reactions, such as complement fixation.

Classes

The constant portions of the H chains of antibodies have distinctive amino acid sequences for each class of antibody. Using the abbreviation Ig for immunoglobulin, the five classes are designated as IgG,

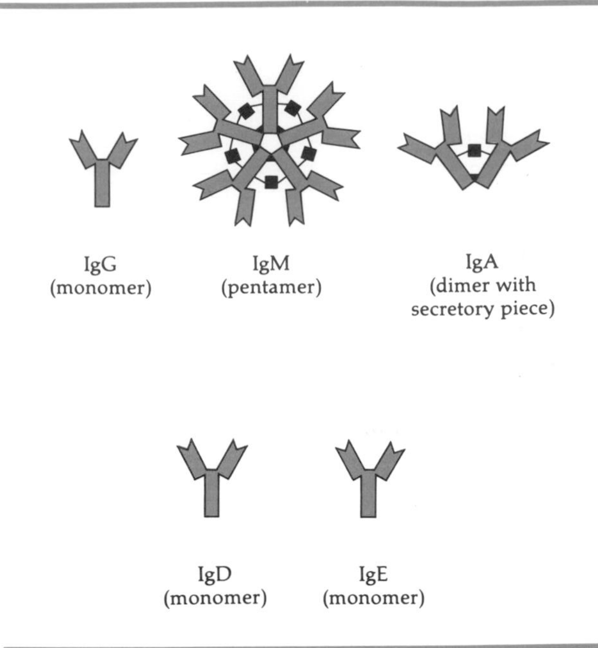

Figure 16–7 Structures of the five principal classes of human immunoglobulins.

IgM, IgA, IgD, and IgE. Molecules of IgG, IgD, and IgE resemble the structure shown in Figure 16–4; they are called monomers. Molecules of IgM and IgA are usually made up of several monomers joined together (Figure 16–7). The characteristics of the immunoglobulins are summarized in Table 16–3.

IgG

IgG amounts to about 80 to 85% of all antibody in the serum. The structure of IgG has been studied most extensively, and this is the antibody described earlier (see Figure 16–4). Antibodies of the IgG class have the ability to cross the placenta and fix complement. Among the functions of IgG antibodies are providing naturally acquired passive immunity to the newborn, neutralization of bacterial toxins, and attachment to phagocytes to enhance phagocytosis.

IgM

Antibodies of the IgM class comprise from 5 to 10% of the total antibody in serum. Structurally, IgM consists of a pentamer, five Y-shaped units similar to IgG. Although IgM has the ability to fix complement, it does not have the ability to cross the placenta. IgM is the first antibody to appear following an injection of antigen, and it is especially effective against invading microorganisms. When bound to microbial cells, IgM antibodies stimulate phagocytosis by macrophages and also cell destruction by complement fixation. Because they are multivalent, IgM antibodies agglutinate target cells very effectively.

IgA

Antibodies of the IgA class make up about 15% of the total antibody. IgA appears in two forms. One form is quite similar to IgG and is found in the blood. It is called *serum IgA*. The other form is twice as large as IgG and is called *secretory IgA*. It is unique in that its two immunoglobulin subunits are attached by a polypeptide called a *secretory piece*. The secretory piece is produced by epithelial cells and functions to help move the IgA through epithelial cells to mucosal surfaces and body secretions. The secretory piece also protects IgA from destruction by protein-digesting enzymes. Although IgA cannot fix complement or cross the placenta, it protects mucosal surfaces from invasion by pathogens, particularly viruses.

IgD

IgD antibodies comprise only about 0.2% of the total serum antibody. Structurally, they resemble IgG. IgD antibodies are found in blood and lymph and on the surfaces of antibody-producing cells called B cells. They do not have the ability to fix complement and cannot cross the placenta. Very little is now known about the functions of IgD antibodies. The interaction of IgD and a specific antigen somehow triggers B cells to produce certain specific antibodies.

IgE

Antibodies of the IgE class are a bit larger than IgG antibodies and they constitute 0.002% of total serum antibody. They cannot fix complement or cross the placenta. Apparently, IgE antibodies are involved in allergic reactions. When an antigen contacts the antibody, a reaction takes place on the surface of the cell to which the antibody is attached, whereby histamine and heparin are released from the cell. The histamine and heparin cause dilation of local blood vessels and contraction of smooth muscle, a major part of the allergic response. Further details of the allergic response are discussed in Chapter 17. IgE antibodies may also be involved in protecting the body from protozoan parasites.

Table 16–3 Summary of Immunoglobulins

Characteristics	IgG	IgM	IgA	IgD	IgE
% total serum antibody	80–85	5–10	15	0.2	0.002
Location	Blood, lymph	Blood, lymph	Secretions (tears, saliva, mucus, intestine, milk), blood, lymph	B cell surface; blood, lymph	Bound to mast cells throughout body; blood
Molecular weight	150,000 (monomer)	900,000 (pentamer)	170,000 (serum, monomer); 400,000 (secretory, dimer)	185,000 (monomer)	190,000 (monomer)
Half-life (days)* in serum	25	5	6	3	2
Complement fixation	+	+	–	–	–
Placental transfer	+	–	–	–	–
Functions	Enhances phagocytosis, neutralizes toxins, and protects fetus and newborn	Especially effective against microorganisms; agglutinates, promotes lysis, and enhances phagocytosis	Localized protection on mucosal surfaces	Stimulates B cells to produce antibodies (?)	Allergic reactions; possibly expulsion or lysis of protozoan parasites

*Time required for one-half the antibodies to disappear.

Our next concern is how antibodies are made by the body.

MECHANISM OF ANTIBODY FORMATION

It was noted earlier that the body has the ability to produce antibodies against invading agents such as bacteria, toxins, viruses, and foreign tissues. As we have said, this type of acquired immunity consists of humoral and cellular immunity. In humoral immunity, the body produces antibodies that circulate in the blood and other body fluids and are capable of attaching to a foreign antigen. In cellular immunity, specially sensitized lymphocytes are formed that have the capacity to attach to the foreign antigen and destroy it.

Humoral and cellular immunity are the products of the body's lymphoid tissue. The bulk of lymphoid tissue is located in the lymph nodes, but it is also found in the spleen, gastrointestinal tract, and bone marrow. With these locations, lymphoid tissue can intercept an invading agent before it can spread too extensively into the general circulation.

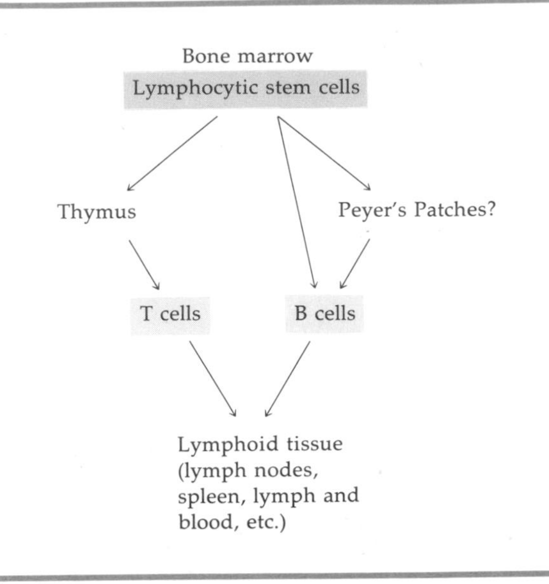

Figure 16–8 Differentiation of T cells and B cells. T cells originate in bone marrow, are processed in the thymus gland, and migrate to lymphoid tissue. B cells originate in bone marrow, are probably processed there and in the intestine (Peyer's patches), and migrate to lymphoid tissue.

B Cells and T Cells

Lymphoid tissue consists primarily of lymphocytes that are morphologically indistinguishable but which may be divided into two types of cells. **B cells** are the antibody-producing cells that provide humoral immunity, while **T cells** are responsible for cellular immunity. Both types of lymphocytes are derived from lymphocytic stem cells found in the liver in the fetus and in bone marrow after birth. Before migrating to their positions in lymphoid tissue, the descendents of the stem cells follow two distinct pathways (Figure 16–8). About half of them first migrate to the thymus gland, where they are processed to become thymus-dependent lymphocytes, or T cells. (They are called T cells because the processing occurs in the thymus gland.) They then leave the thymus gland and become embedded in lymphoid tissue. The thymus gland in some way confers what is called **immunologic competence** on the T cells. This means that the T cells develop the ability to differentiate into cells that perform specific immune reactions against specific antigens. We will look at this mechanism shortly. Immunologic competence is conferred by the thymus gland shortly before birth and for a few months after birth. Removal of the thymus gland before it has processed the T cells results in the failure of an animal to develop most cellular immunity. Removal after processing has little effect.

In mammals, the remaining stem cells are processed in other areas of the body to become B cells. The bone marrow itself is probably a major site of processing, and parts of the intestine (for example, Peyer's patches) may play a role. B cells were given this name because in birds they are processed in the bursa of Fabricius, a small pouch of lymphoid tissue attached to the intestine. Once processed, B cells migrate to lymphoid tissue throughout the body and take up their positions there. Although both T cells and B cells occupy the same lymphoid tissues, they localize in separate parts of the tissues.

The immune response also requires participation by macrophages. Although macrophages do not actually synthesize antibodies, they are believed to prepare and present antigen to T cells and B cells. One theory holds that macrophages act nonspecifically by phagocytizing antigens. In this way, antigenic materials become attached to the surface of the macrophage for presentation to T cells and B cells. Current research indicates that antibody production by cells developed from B cells is greatly increased if the B cells first interact with both macrophages and T cells.

B Cells and Humoral Immunity

The body contains literally thousands of different B cells, each capable of responding to a specific antigen. B cells develop into cells that produce specific antibodies that circulate in the lymph and blood to reach the antigen.

How does this happen? When an antigen—for example, an antigen on the surface of an invading microorganism—comes in contact with B cells in lymph nodes, spleen, or lymphoid tissue in the gastrointestinal tract, B cells specific for the antigen (because of special receptors on the cell surface) are activated. Some of them enlarge, divide, and differentiate into a clone of **plasma cells** (Figure 16–9). Plasma cells actively secrete antibody (Figure 16–10) at the phenomenal rate of about 2000 antibody molecules per second for each cell, for about 4 to 5 days until the plasma cell dies. The activated B cells that do not differentiate into plasma cells remain as **memory cells** ready to respond more rapidly and forcefully should the same antigen appear at a future time.

Different antigens stimulate different B cells to develop into plasma cells and their accompanying memory cells because the B cells of a particular clone are capable of secreting only one kind of antibody. Each specific antigen activates only those B cells already predetermined to secrete antibody specific to that antigen. The specific antigen stimulates a specific B cell because that particular B cell has on its surface the antibody molecules (IgD) that it is capable of producing. The surface antibodies serve as receptor sites with which the antigen can combine and thus provide the basis of "nonself" recognition by the immune system. This mechanism whereby a specific antigen stimulates a specific B cell to develop into a clone of plasma cells that produce specific antibodies is known as the **clonal selection** mechanism of antibody formation.

Once the antigen–antibody complex is formed, the antibody may activate complement for attack and fix the complement to the surface of the antigen. Complement, as you may recall from Chapter 15, can participate in attacking and destroying the antigen by inducing histamine release, chemotaxis, and opsonization.

Techniques have been developed to clone one type of B cell in the laboratory and thus produce one type of antibody—monoclonal antibodies. This breakthrough may revolutionize diagnosis and treatment of certain diseases, as well as prove useful in recombinant DNA technology (see box).

T Cells and Cellular Immunity

There are also thousands of different types of T cells, each capable of responding to a specific antigen. At any given time, the vast majority of T cells are inactive. As in the case of B cells, when an antigen enters the body, only the particular T cell specifically programmed to react with the antigen becomes activated. Such an activated T cell is said to be **sensitized,** and sensitization occurs when macrophages phagocytize the antigen and present it to the T cell. Antigen-binding receptors on T cells recognize the foreign antigen (see Figure 17–3). Sensitized T cells increase in size and divide, each cell giving rise to a *clone* or population of cells identical to itself (Figure 16–11).

The T cells do not secrete antibody. Instead, some of the T cells of the clone, called **killer T cells,** leave the lymphoid tissue and migrate to the site of invasion. Here, several things happen. The killer T cells attach to the invading cells and secrete *cytotoxic factors* that destroy microorganisms directly. Killer T cells can also release a protein substance called *transfer factor.* This substance reacts with nonsensitized lymphocytes at the site of invasion, causing them to take on the same characteristics as the sensitized T cells. In this way, additional lymphocytes are recruited and the effect of the sensitized T cells is intensified. Killer T cells also secrete a substance called the *macrophage chemotactic factor,* which attracts macrophages to the site of invasion, where they destroy the antigens by phagocytosis. Yet another substance released by killer T cells, *macrophage activating factor (MAF),* greatly increases the phagocytic activity of macrophages. All these substances

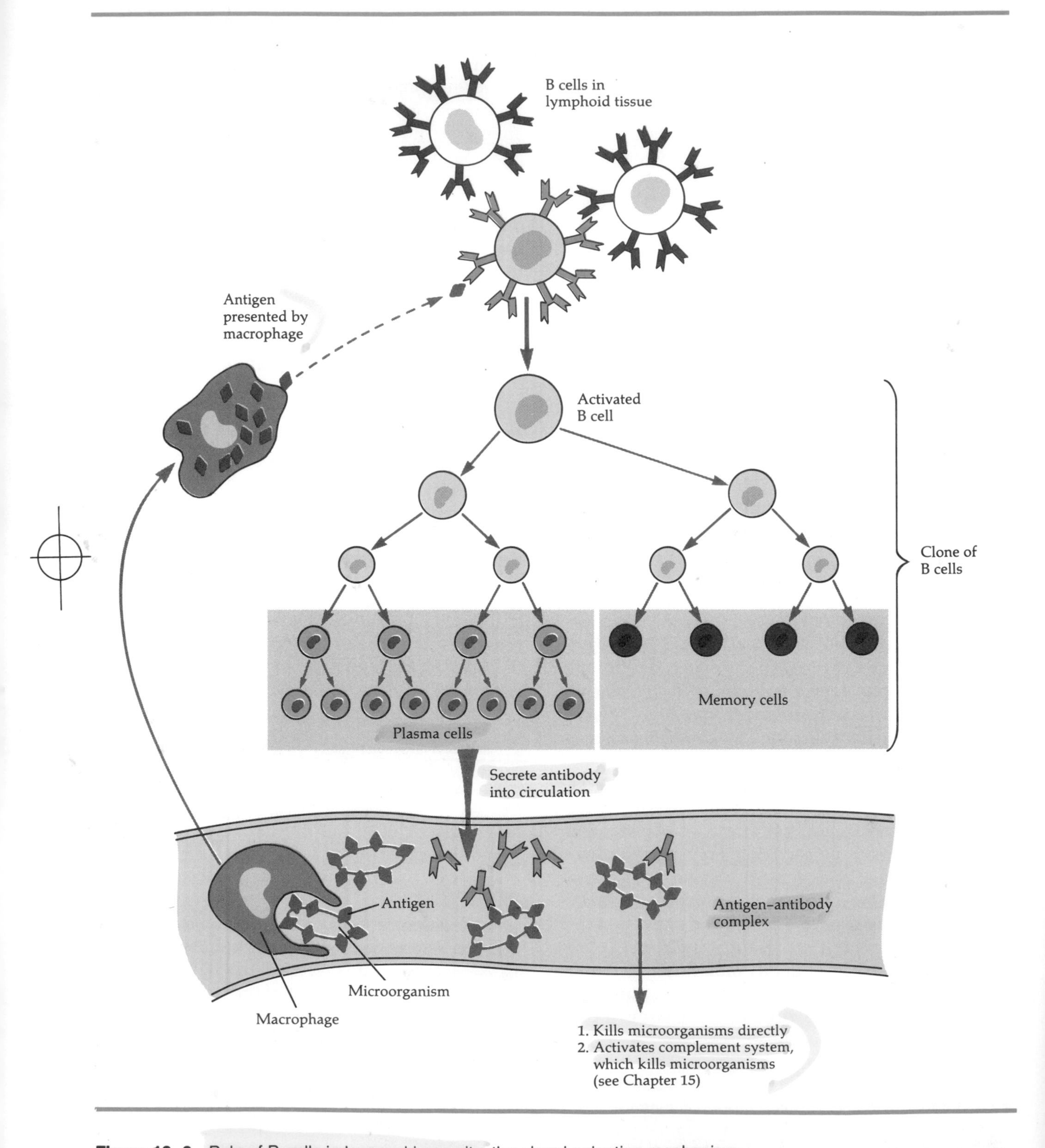

Figure 16–9 Role of B cells in humoral immunity; the clonal selection mechanism.

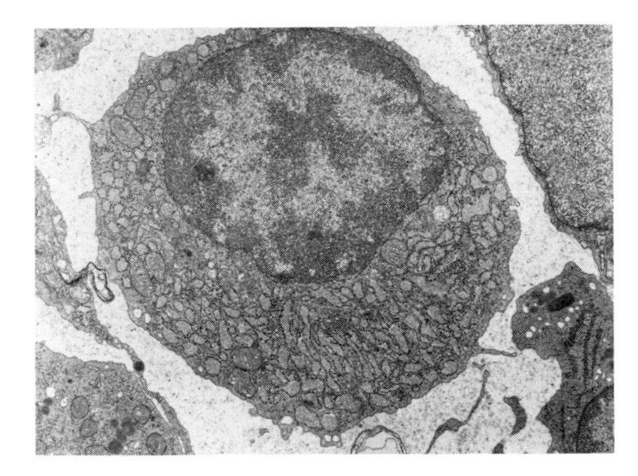

Figure 16–10 Electron micrograph of a plasma cell from a rat's lymph node (×16,700). The extensive areas of endoplasmic reticulum are engaged in manufacturing antibody molecules.

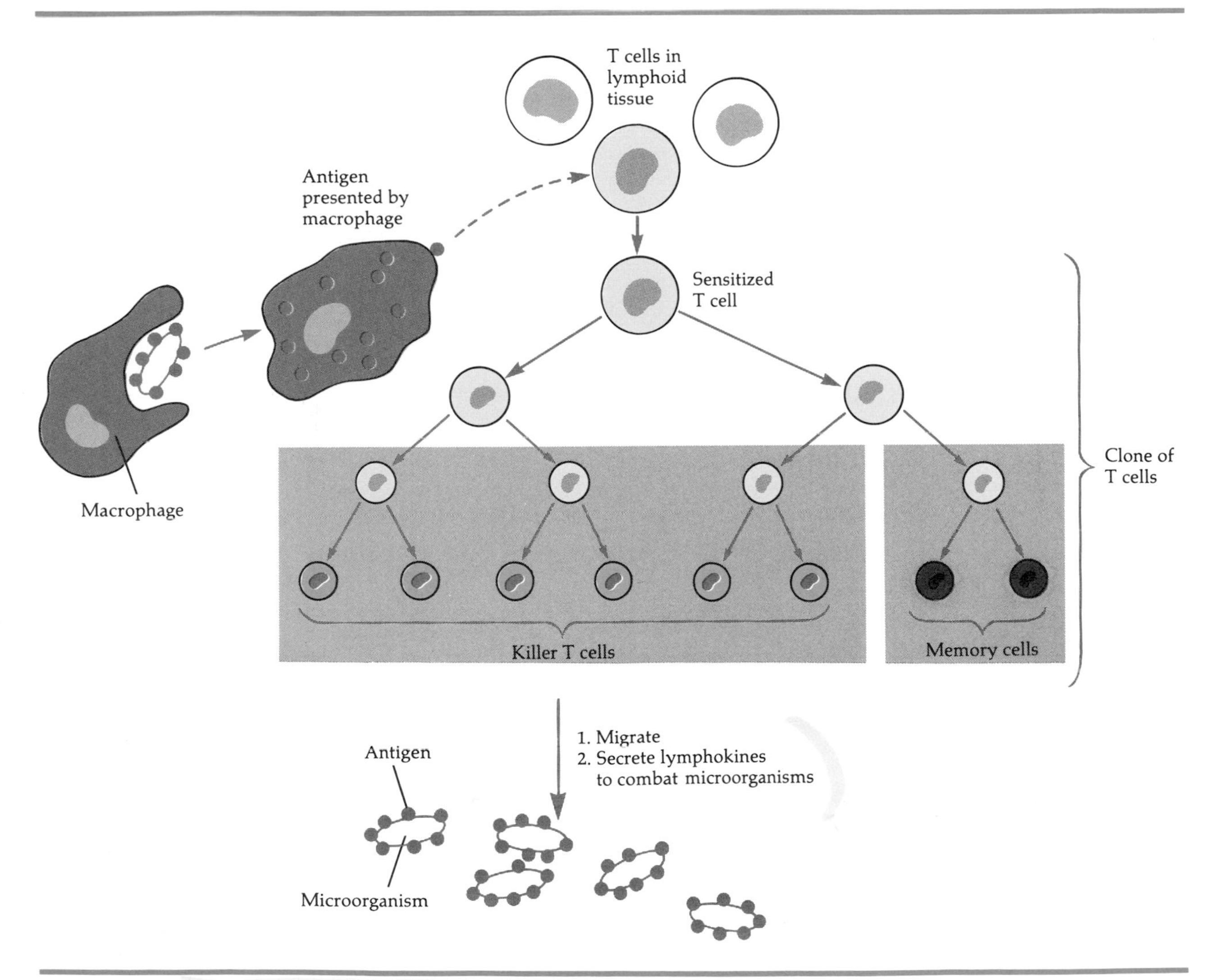

Figure 16–11 Role of T cells in cellular immunity. In this figure, a clonal selection mechanism similar to that for B cells is shown to be operating. Unlike B cells, however, T cells do not secrete antibodies, but migrate to the site where the antigens are and secrete other chemicals (lymphokines) to respond to the antigens.

Microbiology in the News

Monoclonal Antibodies and Hybridomas: New Immunologic Tools Are a Match For Any Disease

June 1981—Cloned armies of antibodies may one day be routinely dispatched on "search and destroy" missions to wipe out enemy viruses, bacteria, protozoans, and even cancers that have set up camp in the human body. These specially-produced antibodies, called monoclonal antibodies, are powerful for three reasons: they are uniform, they are highly specific, and they can readily be produced in large quantities. Like a well-trained battalion, monoclonal antibodies go after one and only one foreign substance, or antigen, in the body.

The Problem and the Solution

Scientists have long known how to stimulate laboratory animals (or humans) to produce antibodies to a particular antigen by injecting the antigen into the animal. In response to the antigen, special cells of the immune system, called B lymphocytes or B cells, release antibody molecules into the blood that will combine specifically with the antigen. Unfortunately, the antibodies produced in a living animal are a diverse lot. Produced by many different B cells, they vary chemically and physically (and may combine with different parts of the same antigen). For this reason, chemically pure antibodies are virtually impossible to isolate from an animal's blood.

Since a single B cell manufactures just one kind of antibody molecule, an obvious way to obtain pure antibodies is to grow antibody-producing B cells in laboratory culture. Although B cells isolated from the body do not survive long in culture, a major breakthrough was the discovery that they can be made "immortal" by fusing them with a cancer cell that is capable of proliferating endlessly in the laboratory. The resulting hybrid cell is called a *hybridoma* cell.

Grown in culture, the descendants or "clones" of a single hybridoma cell continue to produce only the type of antibody characteristic of the ancestral B cell—which is why these antibodies are called "monoclonal." Every antibody molecule produced is identical. Because the hybridoma grows rapidly like its cancerous parent, it blooms into an extremely rich source of antibody.

Hybrid Construction

The technique for making hybridomas was pioneered in 1975 by César Milstein and Georges Köhler of Britain's Medical Research Council. As shown in the diagram, antigen (for example, a virus or a human tumor cell) is injected into a mouse, whose B cells begin to produce antibodies against the antigen. Then the mouse's spleen, which contains many B cells, is removed and its cells teased apart. These cells are incubated with cancerous tumor cells derived from myeloma tumors from bone marrow. Under the right conditions, some of the B cells fuse with tumor cells to produce hybrid cells. Scientists screen the hybrids for the ones that produce the antibody they want. The hybridoma formed by the permanent union of a myeloma cell and an activated B cell takes the best features from both parents: it grows continuously and it keeps pumping out a never-ending supply of a single antibody.

Clinical Uses for Monoclonal Antibody

Initially, the greatest clinical use for monoclonal antibodies will be for diagnosis. In May 1981, the U.S. Food and Drug Administration approved an allergy diagnostic kit manufactured by Hybridtech, Inc., in La Jolla, California. Similar kits are under development by other companies for the diagnosis of infectious diseases such as hepatitis, rabies, and certain venereal diseases. Monoclonal antibodies are also a potential tool for detecting cancer at very early stages; this is possible because cancer cells have antigens on their surfaces that differ from those of normal cells.

Still more exciting than the diagnostic potential is the

prospect of using monoclonal antibodies to treat cancer and infectious diseases. Over 50 years ago, Dr. Paul Ehrlich imagined the possibility of a "magic bullet" that could hunt down and destroy a pathogen without damaging surrounding healthy tissue. Monoclonals used alone or chemically attached to powerful drugs appear to have qualities of such medically-intelligent bullets. The monoclonal antibody approach to cancer treatment is particularly attractive because until now cancer therapies have usually destroyed normal tissue along with cancerous tissue.

In a clinical study at Johns Hopkins, physicians actually used monoclonals to treat human patients with liver cancer. First, they prepared monoclonal antibodies to a liver cancer antigen (a protein) and attached radioactive iodine to the antibodies. When injected into the patients, the antibody delivered pinpoint radiation therapy directly to the cancerous tissue. Although results are only preliminary, researchers are encouraged by the remission of 9 of 14 patients for one year, three or four times longer than the usual remission time with more conventional therapy.

Sometimes an individual's immune system has a harmful effect, as when it rejects a transplanted organ in a transplant patient, or when it turns against the body's own tissues in an autoimmune disease. Monoclonal antibodies may someday be useful in these cases also.

Purification of Recombinant DNA Products

In a more indirect but extremely promising approach, recombinant DNA firms are already starting to use monoclonals for purification of their products. Vaccines to prevent malaria and hepatitis B and improved influenza vaccines are already under development by several firms, which are finding monoclonals invaluable for the purification of the vaccines. Using monoclonals as "handles," scientists can efficiently pluck the substances they want out of a mixture of molecules. Thus, monoclonals will be useful for purifying almost any molecule produced by recombinant DNA techniques.

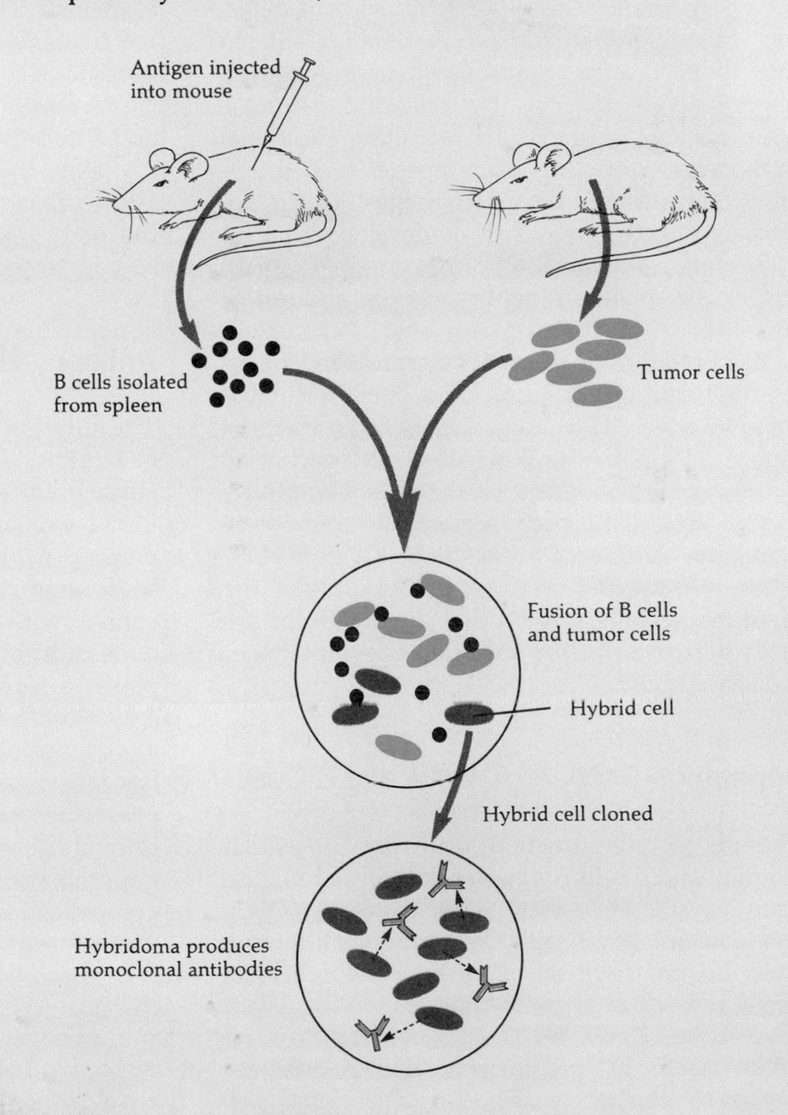

Table 16–4 Comparison of B Cells and T Cells

Cell	Type of Immunity	Source	Site of Processing	Functions
B cells	Humoral	Lymphocytic stem cells in bone marrow	Primarily bone marrow	Differentiate into plasma cells that secrete antibodies; activate and fix complement
T cells	Cellular	Lymphocytic stem cells in bone marrow	Thymus gland	Killer T cells secrete cytotoxic factors, transfer factor, macrophage chemotactic factor, macrophage activating factor, and other lymphokines

released by T cells are polypeptides and are collectively called **lymphokines** (see Table 17–3). In short, then, T cells can destroy antigen-carrying microorganisms directly by releasing cytotoxic substances and indirectly by recruiting additional lymphocytes, attracting macrophages, and intensifying phagocytosis by macrophages. T cells are especially effective against slow-developing bacterial diseases, such as tuberculosis and brucellosis, some viruses, fungi, transplanted cells, and cancer cells.

As pointed out earlier, not all sensitized T cells leave the lymphoid tissue as killer T cells to combat foreign invaders. Those that remain behind are the memory cells. Like the B memory cells, such cells are programmed to recognize the original invading antigen. Should the pathogen invade the body at a later date, the memory cells initiate a far swifter reaction than during the first invasion. In fact, the second response is so swift that the pathogens are usually destroyed before any signs or symptoms of the disease occur.

Cooperation Between T Cells and B Cells

Although T cells function primarily in cellular immunity and B cells function primarily in humoral immunity, there is some degree of cooperation between the T and B cells. For example, although T cells do not themselves secrete antibodies, in many instances antigens first interact with T cells before inducing antibody formation by the appropriate B cells. In this respect, such T cells are referred to as **helper T cells.** There are also instances when some T cells can inhibit immune responses. These are called **suppressor T cells.** The mechanisms of action of the helper cells and suppressor cells are not well understood at this time.

A summary of the differences between T cells and B cells is presented in Table 16–4.

Now that you understand the role of B cells and T cells in immunity, we will take a look at how the body responds to antigenic stimulation with respect to time.

Antibody Responses to Antigens

The immune response of the body, whether cellular or humoral, is much more intense after a second or subsequent exposure to an antigen than after the initial exposure. The humoral response can be demonstrated by measuring the *antibody titer,* that is, the amount of antibody in serum. After an initial contact with antigen, there is a period of several days during which no antibody is present in the patient's serum. Then there is a slow rise in the antibody titer, followed by a gradual decline. Such a response of the body to the first contact is called the **primary response** (Figure 16–12). During the primary response, in which the body is said to be primed or sensitized, there is proliferation of immunocompetent lymphocytes. When the antigen gains entrance into the body again, whether for the second time or two hundredth time, there is an immediate proliferation of immunocompetent lymphocytes, and the antibody titer is far greater, and remains high longer, than for the primary response. This accelerated, more intense response is called the **anamnestic** (*anamne,* meaning "recall") or **secondary response** (Figure 16–12). The reason

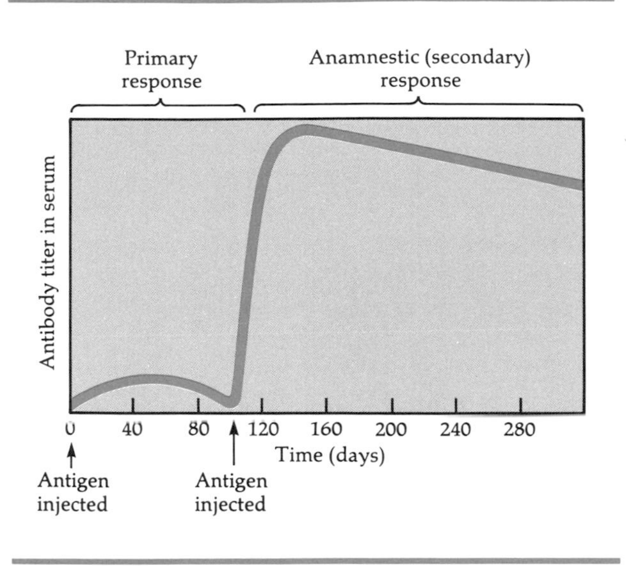

Figure 16–12 Primary and anamnestic responses to the injection of an antigen.

for the anamnestic response is that some of the immunocompetent lymphocytes formed during the primary response remain as memory cells, as we mentioned earlier. Not only do these memory cells add to the pool of cells that can respond to the antigen, but their response is more intense.

Primary and anamnestic responses may occur during microbial infection. When we recover from an infection without the help of antibiotics, it is usually because of the primary response. If, at a later time, we contact the same microorganism, the anamnestic response can be so swift that the microorganisms are quickly destroyed and we do not exhibit any signs or symptoms of the disease.

You have probably guessed by now that the anamnestic response provides the basis for immunization against certain diseases. When we receive the initial immunization, our body is sensitized. Should we encounter the antigenic stimulus again as an infecting microorganism or "booster dose" of vaccine, our body experiences the anamnestic response. A booster dose literally boosts the antibody titer to a higher level. As a rule, booster doses must be given periodically, with immunization schedules designed to maintain high antibody titers.

We will now examine how antigen–antibody reactions are used in the diagnosis of disease.

SEROLOGY

Thus far, we have discussed antigen–antibody reactions that occur within a host (in vivo). But such reactions can also occur outside the body (in vitro), a fact that has been used extensively in the laboratory in the diagnosis of disease. The branch of immunology concerned with the study of antigen–antibody reactions in vitro is called **serology.** Serological tests are extremely valuable because they can be performed quickly and offer a high degree of specificity and sensitivity. In this section, we will consider certain tests commonly employed in the diagnosis of disease, examining such immunologic reactions as precipitation, agglutination, complement fixation, neutralization, and opsonization. In addition, immunofluorescent and radioimmunoassay techniques will be discussed.

PRECIPITATION REACTIONS

Precipitation reactions occur when multivalent, low-molecular-weight, *soluble* antigens react with bivalent or multivalent antibodies (such as IgG or IgM) so that large, interlocking aggregates termed *lattices* develop that precipitate from solution. Antibodies that cause the formation of precipitates are called **precipitins.**

Precipitation reactions occur in two distinct stages. First there is the rapid interaction between antigen and antibody to form small antigen–antibody complexes. This interaction occurs within seconds and is followed by a slower reaction, from minutes to hours, in which the antigen–antibody complexes form lattices that precipitate from solution. Precipitation reactions are optimal only when antigen and antibody are mixed in the proper ratio, and an excess of either markedly decreases the amount of lattice formation and subsequent precipitation. Such effects of excess antigen or antibody on lattice formation are shown in Figure 16–13.

Precipitation tests have been used in several ways in the clinical laboratory. For example, in the **precipitin ring test,** antiserum containing specific antibody is added to a capillary tube, and soluble antigen is carefully layered onto the surface of the antiserum. A white precipitation ring develops at

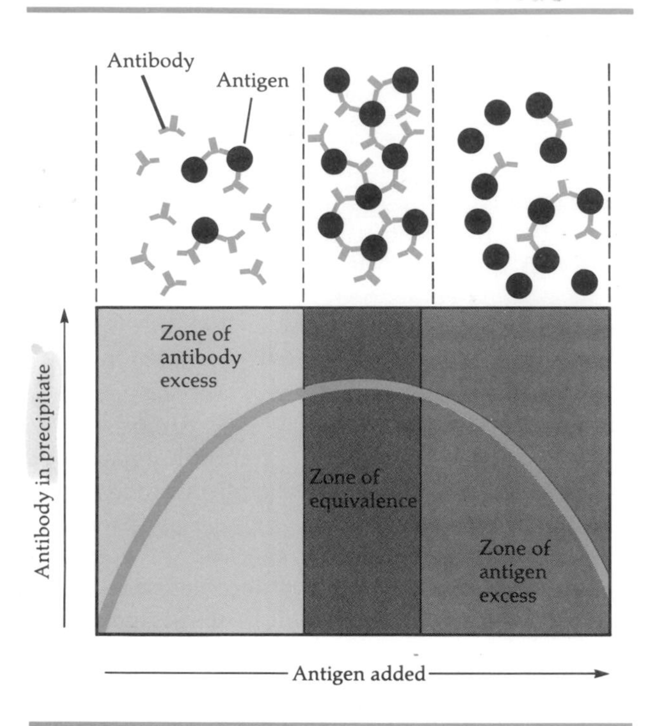

Figure 16–13 Precipitation curve based on the ratio of antigen to antibody. The maximum amount of precipitate forms in the zone of equivalence, where there is no free antigen or antibody.

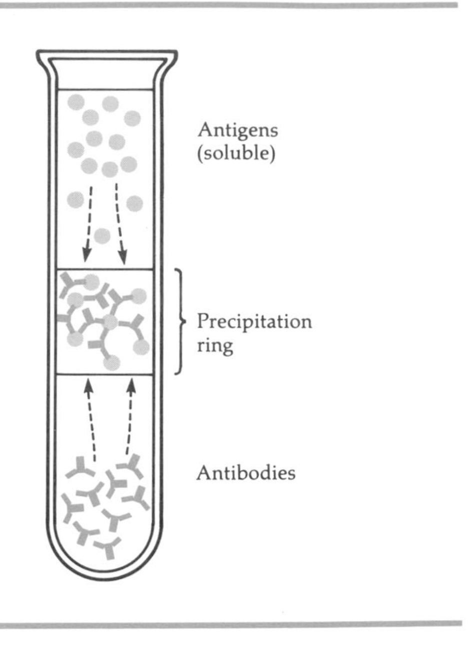

Figure 16–14 Diagrammatic representation of a precipitin ring test.

the antigen–antibody interface in a positive reaction (Figure 16–14). Ring precipitin tests have been used in the serological grouping of streptococci and in forensic medicine to determine if blood stains are of human origin.

Precipitation reactions may also be carried out in an agar gel medium, and such procedures are called **immunodiffusion tests.** One type of immunodiffusion test is the **Ouchterlony test,** which can be used to determine if multiple antigens and antibodies are present in a serum, as well as the relationship between two antigens. In the Ouchterlony test, a purified agar gel such as agarose is added to a Petri plate and wells are cut out of the agar. To a center well is added antiserum containing antibody, and an antigen preparation is added to each surrounding well. The reactants diffuse toward each other through the gel, and a visible precipitate develops where the ratio of antigen to antibody is optimal (equivalence point). If the antigenic preparation contains only a single antigen, one line of precipitation develops. If two or more antigens are present in the preparation, multiple precipitation lines are observed. If two antigen preparations contain identical antigens, precipitation lines come together to form a *line of identity.* But if such preparations contain different antigens, the lines of precipitation cross and form *lines of nonidentity.* If the antigenic preparations contain antigens that are not identical but closely related, *lines of partial identity* are formed (Figure 16–15).

Immunodiffusion reactions have been used to determine if certain isolates of *Corynebacterium diphtheriae* produce exotoxin. In the **Elek test,** filter paper saturated with antitoxin specific for *C. diphtheriae* exotoxin is embedded in an agar medium. A known toxigenic culture of *C. diphtheriae,* and the unknown isolates being tested, are streaked across the plate perpendicular to the filter paper. After 48 hours at 37°C, a positive result is shown by the development of lines of precipitation where optimal proportions of antitoxin and exotoxins occur in the agar (Figure 16–16). This test has replaced the older methods in which toxigenicity studies were carried out in laboratory animals.

A modification of the precipitation reaction combines the techniques of immunodiffusion with electrophoresis in a procedure called **immunoelectrophoresis.** This procedure is used in research in the separation of proteins in human serum, and

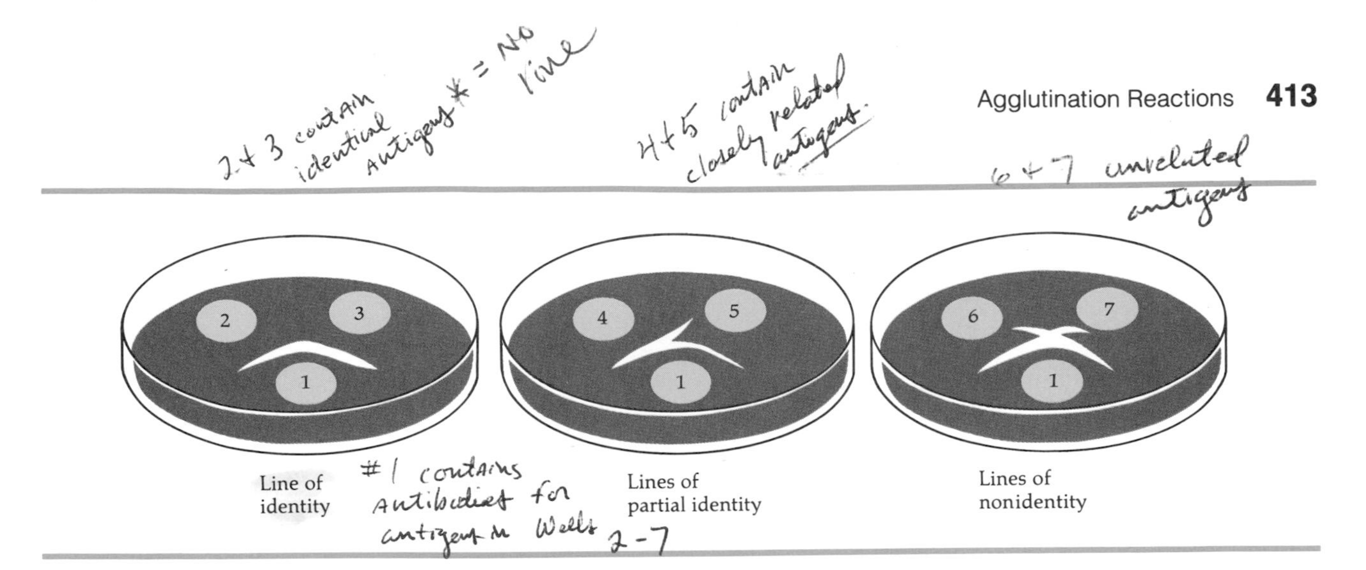

Figure 16–15 Immunodiffusion (Ouchterlony) precipitation test. Well 1 contains antibodies to antigens contained in wells 2 through 7. Wells 2 and 3 contain identical antigens; wells 4 and 5 contain closely related antigens; and wells 6 and 7 contain different, unrelated antigens.

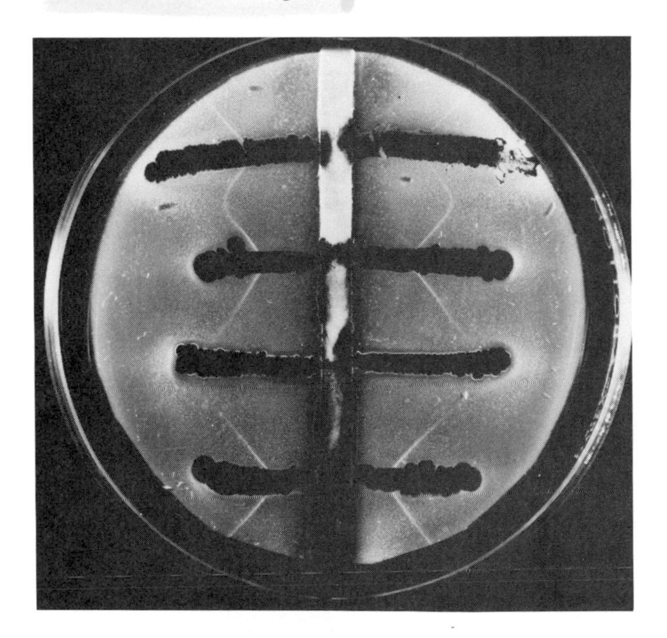

Figure 16–16 Elek test. Four strains of *Corynebacterium diphtheriae* have been streaked across an antitoxin-impregnated strip of filter paper. White lines of precipitate form where optimal concentrations of antitoxin and bacteria-produced exotoxin meet in the agar. Such lines can be seen angling off the first, second, and fourth streaks, showing that they produce exotoxin and are therefore toxigenic. The third strain down does not produce lines of precipitate and is, therefore, nontoxigenic.

today more than 30 such proteins have been detected. In this technique, a small amount of serum is placed in wells on a glass slide coated with an agar gel. The serum proteins are then exposed to an electric current, which causes them to migrate various distances in the electrical field because of differences in their surface charges, size, and other properties. To detect these proteins, antiserum containing antibodies specific for the proteins being investigated is added to a center trough between the wells and allowed to diffuse through the agar. Visible precipitates develop at the points of equivalence (Figure 16–17). In this procedure, some of the protein antigens are themselves antibodies; immunoglobulins, like all proteins, can act as antigens in foreign hosts.

AGGLUTINATION REACTIONS

Unlike precipitation reactions, **agglutination reactions** occur between high-molecular-weight, *particulate (cell-bound) antigens* and bivalent or multivalent antibodies. Since these antigens are not in solution but are cell bound, such reactions lead to the clumping of cells, a process termed **agglutination.** Antibodies that cause agglutination reactions are called **agglutinins.** Agglutination reactions are also based on lattice formation and are optimal only when antigen and antibody react in the proper proportion. Agglutination reactions have a wide variety of applications in the clinical laboratory and are useful in the diagnosis of many diseases. For example, in the **Widal test,** serum from a patient with typhoid fever contains antibodies that agglutinate cells of *Salmonella typhi*. Similarly, certain rickettsial infections cause the formation of a heterophilic antibody that agglutinates nonmotile strains of *Proteus*. A *heterophilic antibody* is one that reacts with

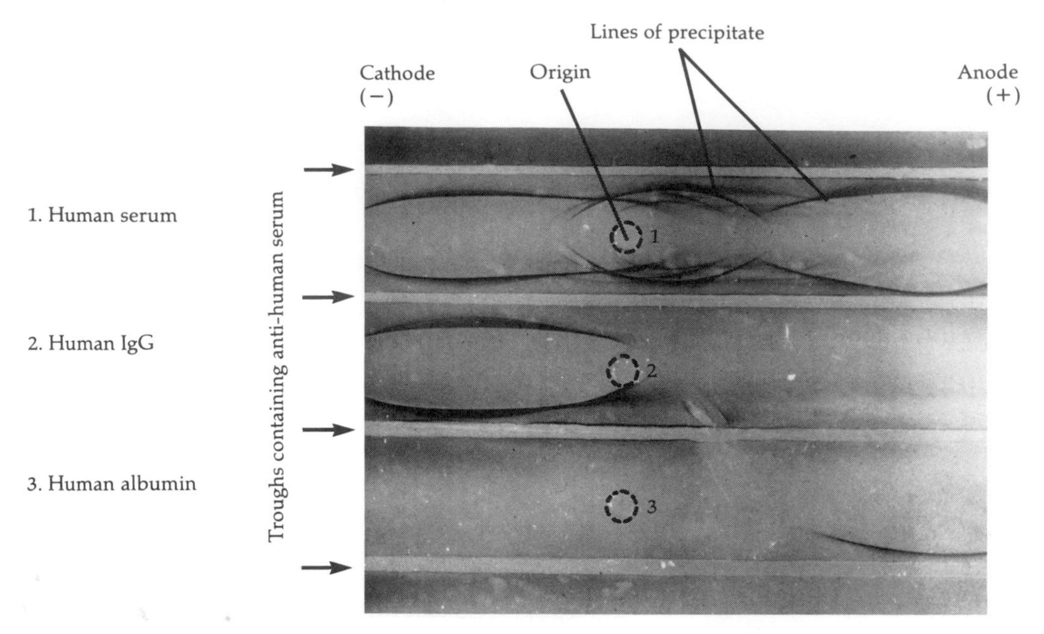

Figure 16–17 Immunoelectrophoresis of human serum. Human serum was placed in origin (well) 1, human IgG in origin 2, and human albumin in origin 3. The preparation was then subjected to an electric current, causing proteins from the wells to migrate through the agar gel. Lines of precipitate formed for each protein where it was met by anti-human serum added to troughs between the wells. From the lines of precipitate for the known human IgG and albumin, the lines for IgG and albumin in the whole human serum preparation can be identified.

distantly related antigens, and the response is called a *cross reaction*. The **Weil–Felix test,** which is based on the reaction between heterophilic antibody and strains of *Proteus*, is a diagnostic test for such rickettsial diseases as murine typhus and Rocky Mountain spotted fever.

Agglutination reactions are also used in estimating the concentration (titer) of antibody in serum. Such information is useful in the diagnosis of disease, since an increase in the titer of antibody to a particular microorganism during the course of a disease strongly suggests that this microorganism may be the causal agent. The **tube agglutination test,** shown in Figure 16–18, provides a simple means to determine an antibody titer. In this procedure, a suspension of cellular antigen, such as bacterial cells, is added to a series of tubes containing various dilutions of serum (1:5, 1:10, 1:20, 1:40, and so on). After a period of incubation, the greatest dilution of serum showing an agglutination reaction is determined. The reciprocal of this dilution is the titer. For example, if there is visible agglutination in the tubes through 1:160, but none in dilution 1:320, then the titer is 160.

When agglutination reactions involve red blood cells, the reaction is termed **hemagglutination.** Such reactions are used routinely in the typing of blood and in the diagnosis of certain diseases. In blood typing, the presence or absence of two very similar carbohydrate antigens (designated A and B) located on the surface of red blood cells is determined using specific antisera. One antiserum contains antibodies to the A antigen while a second serum contains antibodies to the B antigen. A person whose red cells are agglutinated by both antisera is in blood group AB, and if no agglutination occurs, the blood type is O (Figure 16–19).

Hemagglutination is a valuable diagnostic tool for some diseases. For example, certain infections cause the formation of antibodies that agglutinate red blood cells only at low temperatures (2°C), but not at 37°C. These unique antibodies are called *cold hemagglutinins* and are detected in protozoan infections such as trypanosomiases and in atypical primary pneumonia of mycoplasmal origin. Hemagglutination is also used in the diagnosis of infectious mononucleosis. This disease causes the production in humans of a heterophilic antibody

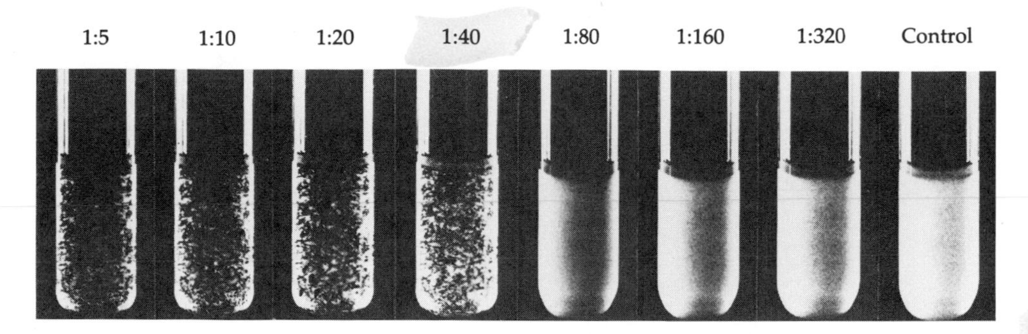

Figure 16–18 Tube agglutination test for determining antibody titer. Here the titer is 40, since there is no agglutination in the next tube in the dilution series (1:80). Nowadays, such tests are rarely performed in tubes, but rather on plates (microtitration plates) that have a large number of small wells.

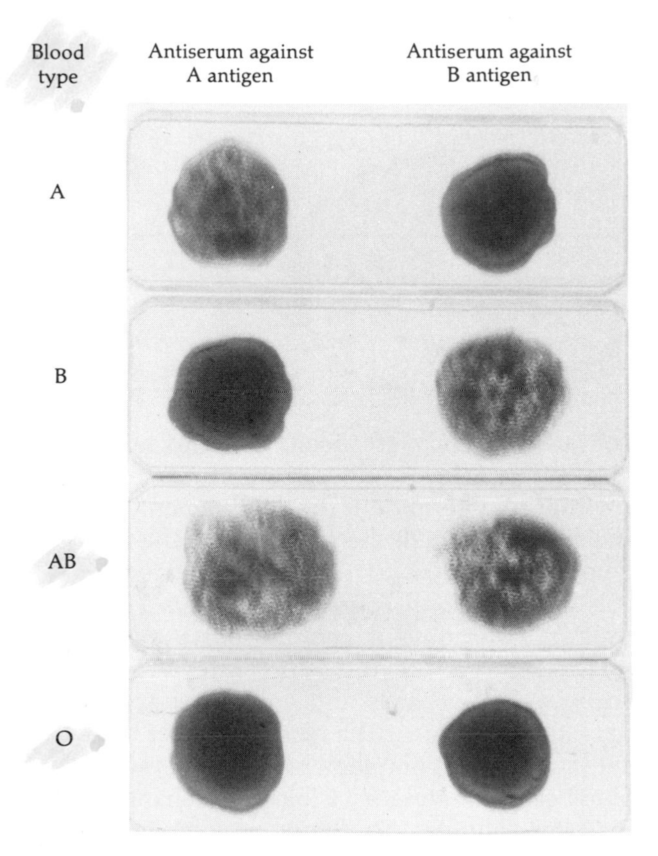

Figure 16–19 Determination of human ABO blood groups using specific antisera. With type A blood *(top)*, antiserum against A antigen causes the cells to agglutinate, whereas antiserum against B antigen has no effect; the opposite situation prevails with type B blood *(second row)*. Type AB blood agglutinates with antisera against both A and B antigen, and there is no agglutination with these antisera in the case of type O blood.

that agglutinates red blood cells of sheep. This specific diagnostic test for mononucleosis is known as the **Paul–Bunnell test.** Certain viruses, such as the causal agents of mumps, measles, influenza, and smallpox, possess the ability to agglutinate red blood cells, a process called *viral hemagglutination.* In response to infections caused by such agents, antibodies are produced that inhibit this process. The inhibition of viral hemagglutination forms the basis for the **hemagglutination-inhibition test,** a procedure useful for the detection of specific viral antibody in serum.

In recent years, the agglutination reaction has been modified to detect antibodies that react with soluble antigens. Such antigens undergo agglutination reactions if they are first attached to insoluble carrier particles such as latex spheres or red blood cells. This procedure, called the **passive agglutination test,** is used to detect the presence of antibodies that develop during certain mycotic infections or helminth diseases. Figure 16–20 illustrates both precipitation reactions and passive agglutination and compares their sensitivity in detecting antibody in serum.

COMPLEMENT FIXATION REACTIONS

In Chapter 15, we learned about a group of serum proteins collectively called *complement,* whose main function is to attack and destroy microorganisms (antigens). In most antigen–antibody reactions, the complement is used up, or "fixed." This process of

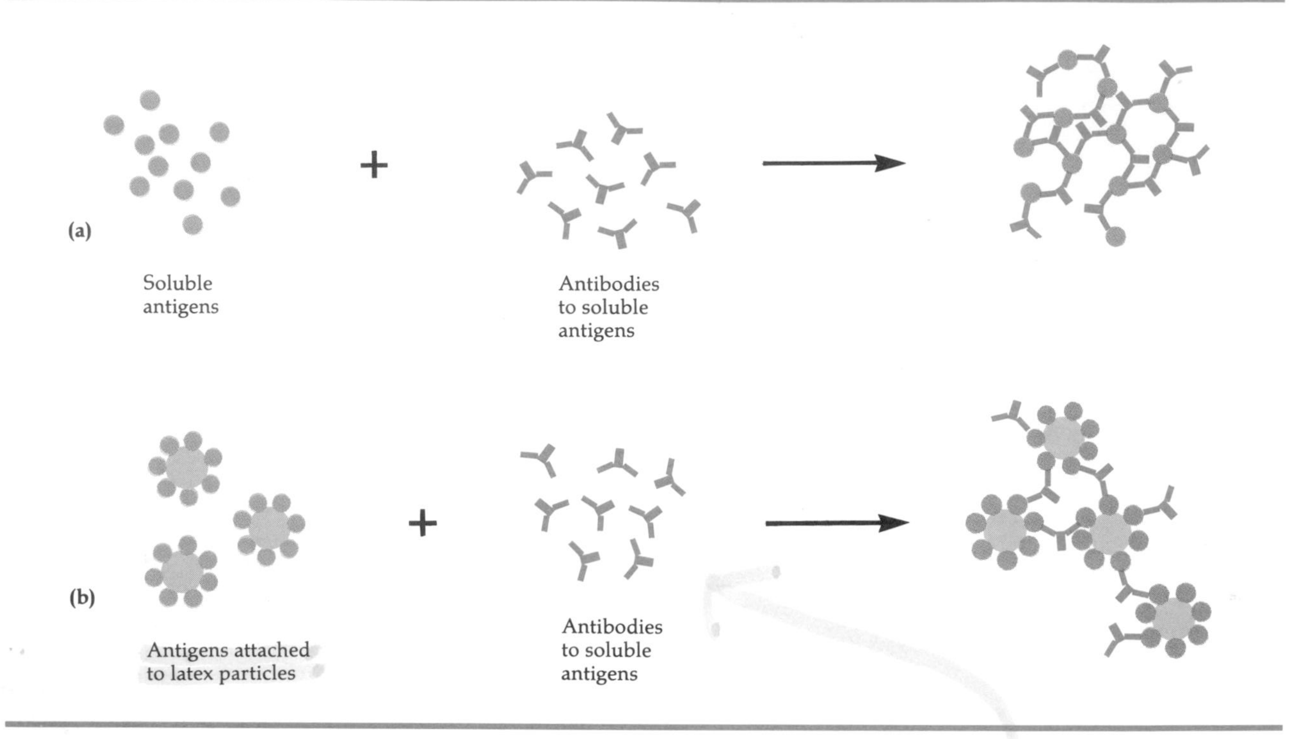

Figure 16–20 Comparison of **(a)** precipitation reactions and **(b)** passive agglutination for detecting antibody in serum. Fewer antibodies are required for a visible passive agglutination reaction than for a visible precipitation reaction.

complement fixation can be used to detect the presence of antibody, even in very small amounts. Some antibodies do not produce a visible reaction, such as agglutination or precipitation, after combining with antigen. But once complement is fixed in a reaction, the amount of antibody can be determined by determining the amount of complement that has been fixed. Complement fixation tests were once widely used in the diagnosis of syphilis **(Wasserman test)** and are still used in the diagnosis of certain viral, mycotic, and protozoan infections.

The complement fixation test is performed in two stages (Figure 16–21). In the first stage, serum obtained from an individual is heated at 56°C for 30 minutes. This inactivates any complement present in the serum but does not alter the antibody being detected. The treated serum is then diluted in test tubes and mixed with a specific amount of known antigen and fresh complement, usually from a guinea pig. The mixture is incubated for about 30 minutes. Since very dilute concentrations of antigen and antibody are used, it is not possible to visualize an antigen–antibody reaction at this point. In order to determine if the complement added to the mixture is fixed, the next stage in the test must be performed.

In the second stage, an indicator system is used to determine if complement is present or absent (fixed). The indicator system consists of sheep red blood cells and antibodies against the red blood cells, since the combination of these two substances with complement causes lysis of the red blood cells, creating a visible color change in the indicator solution. Therefore, if complement has been fixed during the first stage, it is not available to cause hemolysis of the blood cells in the second stage (Figure 16–21a). But if the complement has *not* been fixed during the first stage, then it *is* available to cause hemolysis of the red blood cells in the second stage (Figure 16–21b). The degree of hemolysis indicates the amount of complement fixed in the reaction and thus the amount of antibody present.

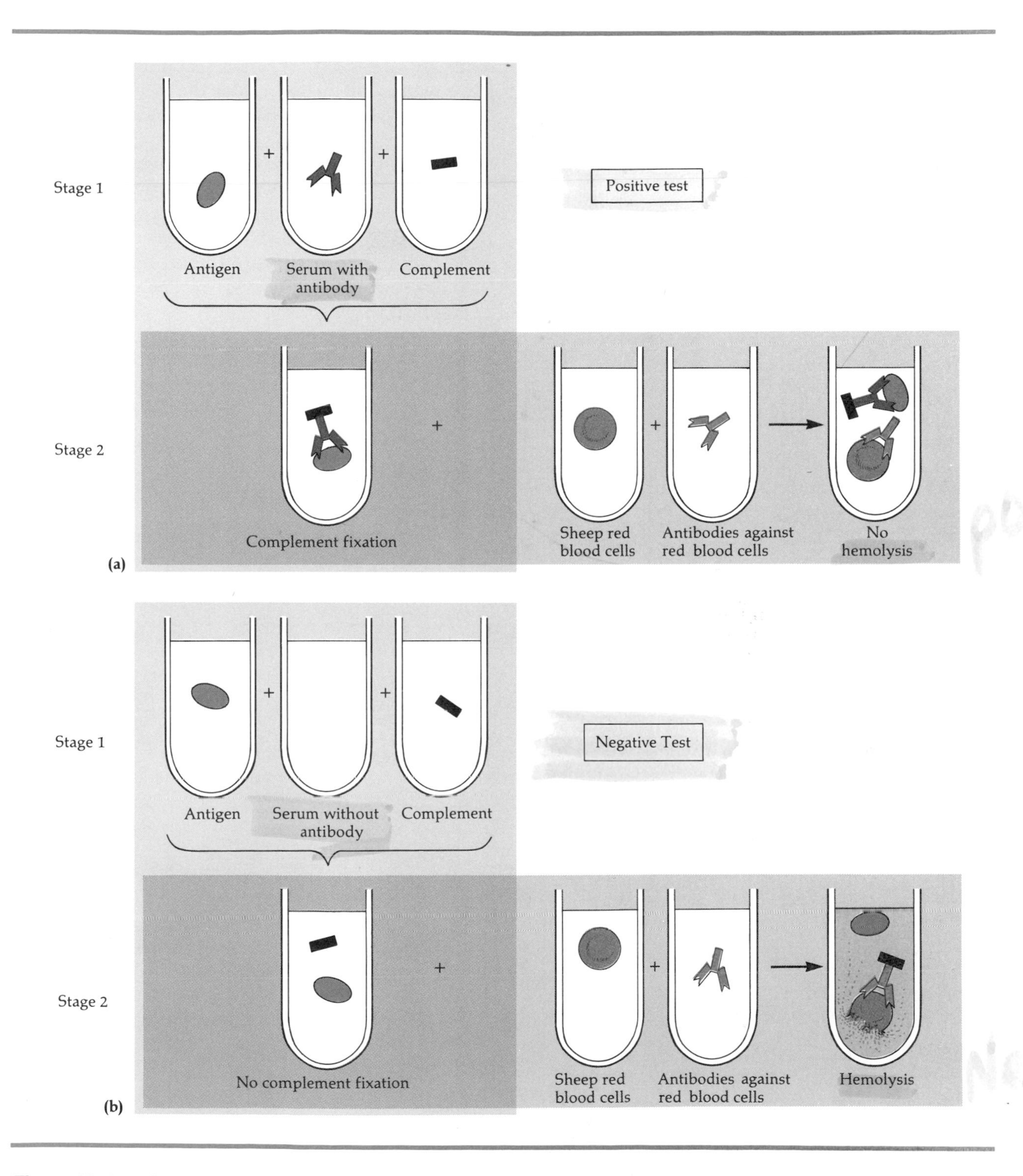

Figure 16–21 Complement fixation test. This test is based on the fact that if the complement is fixed (used up) by reacting with antigen and antibody, there is no hemolysis of red blood cells, even though specific antibody to the red blood cells is present. **(a)** Positive test. **(b)** Negative test.

NEUTRALIZATION REACTIONS

Neutralization reactions are antigen–antibody reactions in which the harmful effects of a bacterial exotoxin or virus are eliminated by a specific antibody. These reactions were first described in 1890 by Emil von Behring and Shibasaburo Kitazato when they noted the ability of immune serum to neutralize the toxic substances produced by the tetanus bacillus. They coined the term *antitoxin* for the neutralizing substance in serum. **Antitoxins** are specific antibodies produced by a host in response to a bacterial exotoxin or its corresponding toxoid. Antitoxins exert their action by combining with exotoxin in such a way that the antigenic determinants on the toxin responsible for cell damage are blocked (Figure 16–22a).

In vitro neutralization tests for bacterial exotoxins are not commonly done in the clinical laboratory today. More frequently, skin tests such as the **Dick test** for scarlet fever or the **Schick test** for diphtheria may be performed. The Schick test is used to determine the extent of a person's immunity to diphtheria. A small amount of diphtheria exotoxin is inoculated under the skin, and if sufficient antitoxin is present, no reaction is seen because of neutralization of the exotoxin. If, however, insufficient antitoxin is present, the exotoxin damages the tissues at the site of entry and causes a swollen, tender, reddish area that turns to brown in 4 to 5 days. This reaction indicates a lack of satisfactory immunity to diphtheria. The Dick test is similar, using scarlet fever exotoxin instead (see Chapter 23).

Neutralization tests are frequently used in the diagnosis of viral infections. In response to such infections, a host produces specific, protective, neutralizing antibodies that bind to receptor sites on the surface of the virus. The binding of these antibodies prevents viral attachment to a host cell, thus destroying the virus's infectivity (Figure 16–22b).

Viruses that exert their cytopathic effects (see Chapter 14) in tissue culture or embryonated eggs

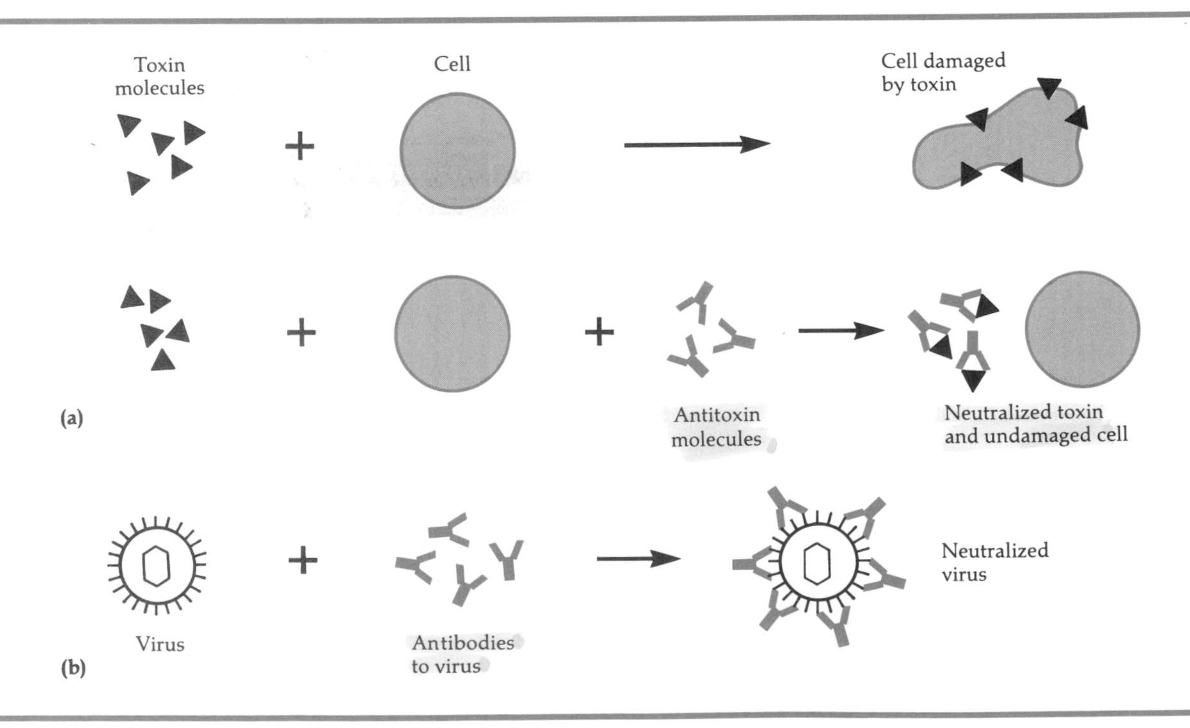

Figure 16–22 Neutralization reactions. **(a)** Effects of a toxin on a susceptible cell and neutralization of toxin by antitoxin. The antitoxin prevents the toxin from damaging the cell. **(b)** Neutralization of a virus by antibody to the virus. The antibody prevents the virus from infecting a cell.

can be used to detect the presence of neutralizing antibodies. Antibodies against a particular virus will prevent that virus from infecting cells, and cytopathic effects will not be seen. The inability of a specific virus to cause cytopathic effects in the presence of immune serum can thus be used to determine the identity of a virus as well as the viral titer.

OPSONIZATION REACTIONS

Phagocytic cells such as monocytes and macrophages possess membrane-bound receptors that can bind a component of complement (C3b) as well as antibody (IgG). When these serum factors are bound to phagocytic cells, the capacity of the cells for phagocytosis is markedly stimulated. Recall from Chapter 15 that this enhancement of phagocytosis is called **opsonization** and that the serum proteins that bring it about are sometimes called **opsonins.** Opsonization can be applied to the diagnosis of certain bacterial diseases such as brucellosis.

IMMUNOFLUORESCENCE (FLUORESCENT-ANTIBODY TECHNIQUES)

Fluorescent-antibody (F.A.) techniques (Figure 16–23 and Color Plate III) are used today to identify isolated microorganisms from clinical specimens

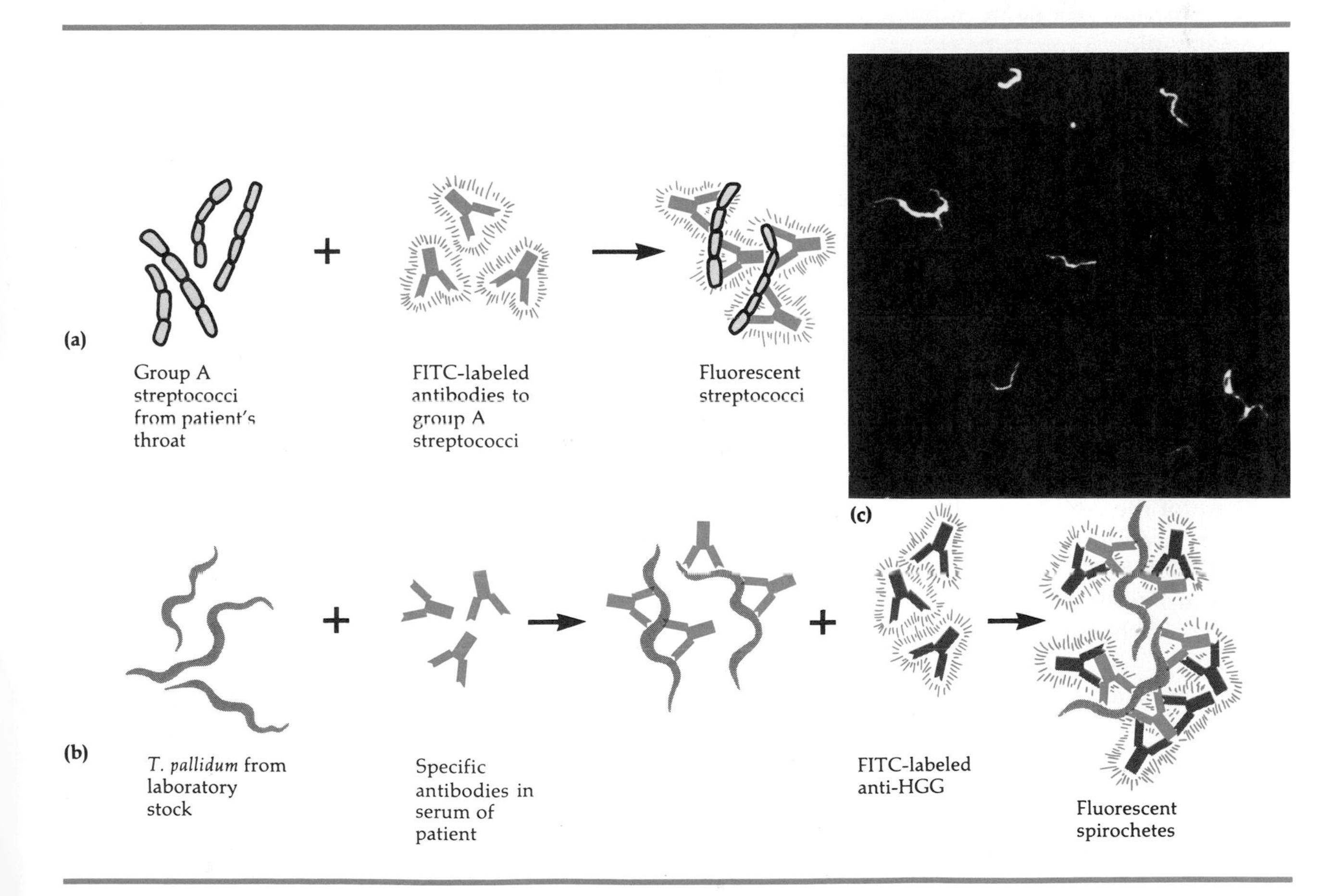

Figure 16–23 Fluorescent-antibody techniques. **(a)** Use of direct F.A. technique for the identification of group A streptococci. **(b)** Use of an indirect F.A. technique for the diagnosis of syphilis (FTA test). **(c)** Fluorescent cells of *Treponema pallidum* in a positive FTA test for syphilis.

and to detect the presence of specific antibody in serum following exposure to a microbial pathogen. These techniques utilize fluorescent dyes such as *fluoresceine isothiocyanate (FITC)* which can be combined with antibodies to make them fluorescent when exposed to ultraviolet light. These procedures are very valuable because they are quick, sensitive, and very specific.

Fluorescent-antibody tests are of two types: direct and indirect. **Direct F.A. tests** are usually used to determine the identity of a microorganism in a clinical specimen. In this procedure, the specimen containing the antigen to be identified is fixed onto a slide. Fluoresceine-labeled antibody is then added to the slide and incubated for a brief period. The slide is washed to remove any antibody not bound to antigen and then examined under the UV microscope for yellow-green fluorescence (Figure 16–23a). **Indirect F.A. tests** are used to detect the presence of specific antibody in serum following exposure to a microorganism. In this procedure, a known antigen (that is, microbial cells) is fixed onto a slide. Serum is then added, and if antibody specific to this microbe is present, it reacts with the antigen, forming a bound complex. To visualize this antigen–antibody complex, fluorescent-labeled *anti-human gamma globulin (anti-HGG)*, antibody that specifically reacts with human antibody, is added to the slide. After incubation and washing (to remove excess antibody), the slide is examined under a UV microscope. If the known antigen fixed to the slide appears fluorescent, the indication is that specific antibody to this antigen is present. A diagram of the reactions in the indirect F.A. technique as used in the **fluorescent treponemal antibody (FTA) test** for the diagnosis of syphilis is shown in Figure 16–23b. The results of an actual FTA test are shown in Figure 16–23c.

Table 16–5 lists some of the microorganisms that can be detected with F.A. techniques.

Table 16–5 Infectious Agents That Can Be Identified by Immunofluorescence (Fluorescent-Antibody Techniques) (D = direct; ID = indirect)

Type of Agent	Examples
Bacterial	*Neisseria gonorrhoeae* (D), *N. meningitidis* (D), *Hemophilus influenzae* (D, ID), group A streptococci (D), *Bacteroides fragilis* (D), *Treponema pallidum* (ID)
Viral	Herpesvirus (D, ID), rabiesvirus (D), influenza virus (D), parainfluenza virus (D), rubella virus (ID), Epstein–Barr virus (ID)
Mycotic	*Candida albicans* (D), *Cryptococcus neoformans* (D, ID), *Blastomyces dermatitidis* (D), *Histoplasma capsulatum* (D)
Protozoan	*Toxoplasma gondii* (ID), *Plasmodium* (ID), *Entamoeba histolytica* (ID)
Helminthic	*Schistosoma japonicum* (trematode) (ID)

RADIOIMMUNOASSAY

Radioimmunoassay (RIA) techniques have recently been developed to aid in the detection of a variety of compounds, including hormones, drugs, and immunoglobulins. Suppose, for example, we wish to determine the amount of insulin in a tissue sample. Radioactively labeled insulin is combined with antibodies against insulin and the tissue sample. The radioactive insulin will compete for antibody with the insulin in the tissue sample. Antigen–antibody complexes may be separated by electrophoresis or some other method and then analyzed for their radioactivity. An antigen–antibody complex with a high level of radioactive antigen indicates that the tissue sample did not have much antigen to combine with the antibody (Figure 16–24). Enzyme-linked immunosorbant assay (ELISA) is based on a similar principle. The competing antigen is labeled by joining it to an enzyme. Enzyme activity is assayed rather than radioactivity.

PUBLIC HEALTH VACCINATIONS

The study of immunity has had practical applications beyond the development of the diagnostic laboratory tests just described. The application with the greatest impact on human health has been the development of vaccines to protect people from contracting certain microbial diseases.

HISTORICAL ASPECTS

The concept that immunity to a disease can develop following recovery from the disease is over 2000

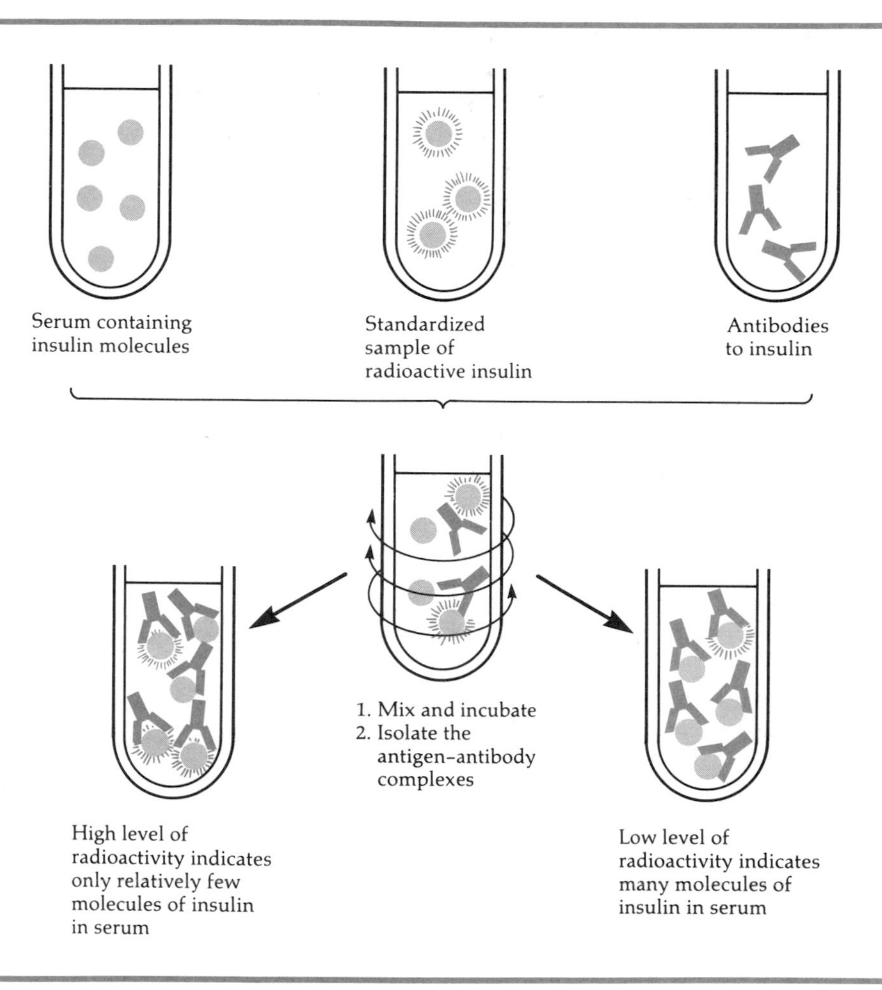

Figure 16–24 Radioimmunoassay, used to determine the concentration of material in a clinical sample. Here the procedure is used to detect insulin.

years old. Written reports from ancient Greece document the fact that persons who recovered from epidemics of plague or typhus often cared for the dead and dying in succeeding outbreaks of these diseases. It was observed that such individuals rarely became ill a second time or developed only a mild form of the illness. During the Middle Ages, this awareness of disease resistance led to deliberate attempts to induce a state of resistance to smallpox by inoculating individuals with scrapings taken from persons with a mild form of the disease. Although sometimes successful, this process, called *variolation,* often led to severe and sometimes fatal infections. It was not until the end of the eighteenth century that the concept of immunization was shown to be feasible. At that time, Edward Jenner demonstrated that inoculating humans with cowpox scrapings provided subsequent immunity to smallpox. As discussed in Chapter 1, Pasteur's success a century later in developing vaccines for chicken cholera, anthrax, and rabies demonstrated conclusively the valuable application of immunology to the prevention of disease. Today, we continue to see these beneficial aspects of immunology with the availability of vaccines that can induce a state of artificial active immunity to bacterial and viral diseases.

DEFINITION

A **vaccine** may be defined as a suspension of microorganisms (or some part of them) that will induce a state of immunity in a host. In most instances, vaccines are suspensions of attenuated microorganisms (living microbes with their virulence reduced or eliminated yet retaining their antigenicity) or microorganisms that have been killed by heat or chemical treatment. Exceptions are vaccines that utilize inactivated exotoxins *(toxoids)* and vaccines produced from cellular components of microorganisms (for example, capsules, pili, or ribosomes).

PREPARATION OF VACCINES

Vaccines are often grouped according to the microbial agent used in their preparation. Let us now take a brief look at the more common vaccines in use today.

Bacterial Vaccines

Most bacterial vaccines are suspensions of killed bacteria in physiological saline. Such vaccines are prepared by growing the bacteria, harvesting the cells, and then killing them with either chemical treatment or heat. Chemical agents such as phenol, acetone, formalin, or Merthiolate are usually used for this purpose. The use of heat to kill microorganisms for vaccines sometimes causes reduced antigenicity, as such treatment may destroy important antigenic determinants. Because of this, the lowest possible temperatures and shortest possible exposures that will lead to sterilization are used.

As we have already said, not all vaccines are made from killed bacteria. The vaccine for immunization against tuberculosis, referred to as the BCG vaccine (bacillus of Calmette and Guerin), is prepared from *Mycobacterium bovis* that has been attenuated by several hundred serial transfers on an unfavorable medium containing bile salts. This organism is antigenically indistinguishable from *M. tuberculosis* and has been used extensively in Europe as a vaccine to protect individuals in high-risk professions against tuberculosis. This vaccine is rarely used in the United States as a prophylactic measure to prevent tuberculosis, but because of its value in stimulating cell-mediated immunity, it has been used in cancer immunotherapy.

Table 16–6 Principal Vaccines Used in Prevention of Bacterial Diseases in Humans

Disease	Vaccine	Recommendation	Booster
Cholera	Crude fraction of *Vibrio cholerae*	For persons who work and live in endemic areas	Every 6 months as needed
Diphtheria	Purified diphtheria toxoid	See Table 16–8	Every 10 years for adults
Meningococcal meningitis	Purified polysaccharide from *Neisseria meningitidis*	For persons with substantial risk of infection	Duration of immunity by vaccine is unknown
Pertussis	Killed *Bordetella pertussis*	Children prior to school age; see Table 16–8	For high-risk adults
Plague	Crude fraction of *Yersinia pestis*	For persons who come in regular contact with wild rodents in endemic areas	Every 6–12 months as needed
Pneumococcal pneumonia	Purified polysaccharide from *Streptococcus pneumoniae*	For adults over 50 years with chronic systemic diseases	To date, studies show no booster effect with additional doses
Tetanus	Purified tetanus toxoid	See Table 16–8	Every 10 years for adults
Tuberculosis	*Mycobacterium bovis* BCG	For persons who are tuberculin negative and who are exposed for prolonged periods of time to tuberculosis	Every 3–4 years as needed
Typhoid fever and paratyphoid fever	Killed *Salmonella typhi, S. schottmulleri*, and *S. paratyphi*	For persons in endemic areas or areas having outbreak	Every 3 years as needed
Typhus fever	Killed *Rickettsia prowazekii*	For scientists and medical personnel in rural areas where typhus is endemic	Every 6–12 months as needed

Toxoid Vaccines

Exotoxins from some bacteria (for example, *Corynebacterium diphtheriae* and *Clostridium tetani*) may be rendered nontoxic by treatment with heat or formalin. Such inactivated toxins, called *toxoids*, retain their immunogenicity and induce the formation of specific antitoxin when inoculated into a host. Toxoid vaccines for tetanus and diphtheria are prepared by inactivating the toxin with formalin and then precipitating the antigen component with alum or absorbing it onto aluminum hydroxide or aluminum phosphate. The alum or aluminum compounds causes the slow release of the antigenic toxoid component into the system, which often enhances the immune response. Substances such as these that are used to prolong or intensify an antigenic stimulus are termed *adjuvants*. Following inactivation, precipitation, or absorption, toxoids are purified and then resuspended in sterile physiological saline for administration.

Microbial Components

In recent years, several bacterial vaccines have been investigated that use components of the bacterial cell. For example, it is hoped that purified pili can be used to stimulate the formation of antibodies that can bind to these cell components and thus prevent colonization of a host. Such vaccines are being developed for enteropathogenic *E. coli*, *Neisseria gonorrhoeae* and *Vibrio cholerae*. Ribosomal preparations are also being investigated for their immunogenicity for both bacterial and fungal diseases. And pneumococcal polysaccharide vaccines are being used in some instances to immunize against pneumococcal pneumonia in high-risk individuals. Table 16–6 presents information regarding bacterial vaccines presently in use.

Viral Vaccines

Viral vaccines consist of either killed suspensions of viruses (inactivated viral vaccines) or live, attenuated viral preparations.

Inactivated viral vaccines consist of virions that have been killed by treatment with formalin or UV radiation. The use of heat is avoided, since it may alter important surface components. Inactivated viral vaccines stimulate the formation of circulating antibodies against viral coat or receptor proteins. One disadvantage of inactivated viral vaccines is the problem of establishing long-term immunity. Antibody levels generally decrease about three months after immunization, and immunity is often lost within six months. Moreover, subcutaneous inoculation of inactivated virus produces only low levels of IgA antibodies in the respiratory tract, providing insufficient protection against some diseases. Often, repeated immunization is required, which in turn may produce hypersensitivity reactions (see Chapter 17). As new strains of viruses evolve with different antigenicity, new vaccines are needed. Some viruses, such as the causative agents of influenza, change especially rapidly, making those vaccines soon outdated.

Because these vaccines are usually administered by injection, localized resistance may not adequately develop at the natural point of entry for the virus. In addition, extreme care is needed to ensure that no residual live viruses are present in these vaccines since they could induce the disease, as happened when the Salk vaccine was first used against poliomyelitis. Examples of vaccines employing inactivated viruses include the formalin-inactivated poliomyelitis vaccine (IPV) developed by Salk and vaccines used to prevent influenza and rabies. A vaccine made from inactivated hepatitis B virus antigens is currently being developed. Other viral vaccines under investigation include ones for recurrent genital, ocular, and facial herpes; herpes zoster; and cytomegalovirus (CMV) disease.

In addition to problems of hypersensitivity, several other kinds of side effects have also limited the use and possible effectiveness of inactivated viral vaccines. In some people, especially infants and young children, fever is a common side effect. More threatening and poorly understood complications have also occasionally resulted from immunization with certain inactivated viral vaccines, such as those for rabies and influenza. Following immunization of over 35 million persons with inactivated swine influenza virus in 1976, the Guillain-Barré syndrome affected 354 recipients, causing 28 deaths. The syndrome consists of muscular pain and weakness and abolition of motor reflexes.

Attenuated viral vaccines are prepared by selecting naturally attenuated viral strains or by culturing the selected virus by serial passage through chick embryo or tissue culture cell lines. Serial passage of viruses under these conditions results in the loss of viral virulence without an alteration in antigenicity.

Live, attenuated viruses have the advantage of acting like a natural infection. They may be administered orally, such as with the Sabin poliomyelitis vaccine; through inhalation, as with the paramyxovirus vaccines; or by injection. Live viruses multiply in the host and tend to produce a longer-lasting immunity, since the immune system is constantly stimulated. However, there can be problems with the use of live viral vaccines. Occasionally, an attenuated viral strain mutates to a more virulent form. And sometimes other viruses in the cell lines used to grow the vaccine virus, or protein contaminants from the culture, can cause problems. Moreover, the viruses of some live viral preparations are unstable and may lose infectivity during storage.

Table 16–7 lists several important vaccines used in the prevention of viral diseases in humans.

Protozoan and Helminthic Vaccines

Protozoan and helminthic parasites are complex organisms that produce many different antigens on their cell surfaces and in the form of their metabolic products. Different antigens appear during different stages in the life cycles of these microorganisms. Perhaps because of these changes, no vaccines have yet been developed to combat infections caused by these organisms. Antibodies formed against these organisms are useful in serological tests, but there is little direct evidence with regard to their effectiveness in providing immunity. Cellular immunity does play a role in immunity to these infections.

Research toward a useful vaccine for malaria is intense at this time. Infants are protected against malaria by naturally acquired passive immunity. Recent laboratory studies have shown that mice produce antibodies against attenuated strains of *Plasmodium*. These antibodies inactivate *Plasmodium* in vitro. Sera from patients with *Trichomonas vaginalis* and *Trypanosoma cruzi* infections have provided passive immunity against these infections in laboratory mice. The most successful artificial active immunity has been achieved against cestodes. Vaccinations with attenuated tapeworm eggs and larvae have resulted in immunity, although not as good as that following infection.

IMMUNIZATION FOR CHILDHOOD DISEASES

Several childhood diseases, some with potentially serious complications, are no longer prevalent

Table 16–7 Principal Vaccines Used in Prevention of Viral Diseases in Humans

Disease	Vaccine	Recommendation	Booster
Influenza	Inactivated virus	For chronically ill persons, especially with respiratory diseases and over 65 years old	Annual
Measles	Attenuated virus	For infants 15–19 months old	*
Mumps	Attenuated virus	For infants 15–19 months old	*
Poliomyelitis	Attenuated or inactivated virus	For children, see Table 16–8; for adults as risk to exposure warrants (duration unknown)	Adults—as needed
Rabies	Inactivated virus	For field biologists in contact with wildlife in endemic areas	*
Rubella	Attenuated virus	For infants 15–19 months old; for females of childbearing age who are not pregnant	*
Smallpox	Attenuated virus	Personnel in laboratories culturing virus (duration to 20 years)	
Yellow fever	Attenuated virus	For persons traveling to endemic areas	Every 10 years

*Note: The duration of immunity is not known because these vaccines have been in use only a short time.

today because of active immunization programs using vaccines. Table 16–8 presents information about these vaccines as well as a recommended immunization schedule. Many of the vaccines used to prevent these diseases are **combined vaccines,** that is, they contain several immunizing agents. The DPT vaccine, for example, used to prevent diphtheria, pertussis, and tetanus, contains alum-precipitated toxoid preparations plus killed cells of *Bordetella pertussis*. The vaccines for measles, mumps, and rubella (German measles) contain live, attenuated viruses. A combined "MMR" vaccine is often given for these three diseases.

The present use of vaccines to prevent some of the common childhood diseases represents a significant application of immunology. Some of these diseases, particularly those of viral origin, although mild in children, can cause severe secondary complications. Thus, measles and poliomyelitis represent serious childhood diseases because of their potential involvement with the central nervous system. Mumps also represents a relatively mild childhood illness except for rare complications that affect the central nervous system or lead to sterility in males. Likewise, rubella (German measles) is not a serious disease in children, but childhood immunization helps keep pregnant women from being exposed to the rubella virus, which can cause birth defects (congenital rubella syndrome; see Chapter 19).

Table 16–8 Recommended Schedule for Active Immunization and Skin Testing of Children

Recommended Age	Immunizing Agent
2–3 months	DPT vaccine (diphtheria toxoid, pertussis vaccine, and tetanus toxoid) OPV (oral poliomyelitis vaccine), trivalent preparation
4–5 months	DPT vaccine OPV, trivalent preparation
6–7 months	DPT vaccine OPV, trivalent preparation
15–19 months	DPT vaccine OPV, trivalent preparation Mumps vaccine, Measles vaccine, Rubella vaccine } or combined MMR vaccine
4–6 years	DPT vaccine OPV, trivalent preparation Tuberculin skin test (see Chapter 17)
12–14 years	TD vaccine (tetanus and diphtheria toxoid) Tuberculin skin test

IMMUNIZATION FOR TRAVELERS

Certain vaccines are not routinely used for the entire U.S. population but are administered only when exposure to the disease may occur as a result of travel through an endemic area. Vaccines for diseases such as cholera, plague, typhus, typhoid fever, and yellow fever are available and are recommended for travelers to endemic areas where such diseases occur. Current recommendations are available from the U.S. Public Health Service.

STUDY OUTLINE

INTRODUCTION (p. 396)

1. Immunity is the ability of the body to specifically counteract foreign antigens.
2. One mechanism of immunity is the production of specific antibodies against pathogens.

THE DUALITY OF THE IMMUNE SYSTEM (p. 396–397)

1. Humoral immunity involves antibodies that are found in blood plasma and lymph.

2. Humoral antibodies are produced by B cells.
3. Cellular immunity depends on T cells and does not involve antibody production.
4. T lymphocytes are coated with antibodylike molecules.
5. Cellular immunity is primarily a response to intracellular viruses, multicellular parasites, transplanted tissue, and cancer cells.

KINDS OF IMMUNITY (pp. 397–398)

NATIVE "IMMUNITY" (p. 397)

1. Native immunity includes species immunity and individual immunity.
2. Individual immunity is affected by sex, age, nutritional status, and general health.

ACQUIRED IMMUNITY (pp. 397–398)

1. Acquired immunity is specific resistance to infection obtained during the life of the individual.
2. Acquired immunity results from the production of humoral antibodies and cellular immune response.
3. Immunity resulting from infection is called naturally acquired active immunity, an immunity that may be long lasting.
4. Humoral antibodies transferred from a mother to a fetus results in naturally acquired passive immunity in the newborn, an immunity that may last up to a few months.
5. Immunity resulting from vaccination is called artificially acquired active immunity, and may be long lasting.
6. Vaccines may be prepared from attenuated, inactivated, or killed microorganisms and toxoids.
7. Artificially acquired passive immunity refers to humoral antibodies acquired by injection, and may last for a few weeks.
8. Humoral antibodies made by a human or other mammal may be injected into a susceptible individual.

infection → active

ANTIGENS AND ANTIBODIES (pp. 398–411)

ANTIGENS (pp. 398–400)

Definition (pp. 398–399)

1. An antigen is a chemical substance that causes the body to produce specific antibodies that the antigen can then combine with.

2. Complete antigens possess immunogenicity (ability to stimulate antibody formation) and reactivity (ability to react with specific antibodies). 

Characteristics (pp. 399–400)

1. Most antigens are proteins, nucleoproteins, lipoproteins, glycoproteins, or large polysaccharides.
2. Generally, antigens have a molecular weight greater than 10,000.
3. Specific regions on the surface of an antigen that combine with antibodies are called antigenic determinant groups; the number of these groups is the valence.
4. Most antigens are multivalent.
5. A hapten (incomplete antigen) is a molecule that has reactivity (combines with antibody) but not immunogenicity.
6. As a rule, antigens are foreign substances; they are not part of the body's chemistry.

ANTIBODIES (pp. 400–403)

Definition (p. 400)

1. An antibody is a protein produced by the body in response to the presence of an antigen and is capable of combining specifically with the antigen.
2. Since antibodies are globulins, they are called immunoglobulins (Ig).
3. Immunoglobulins are found in blood serum, and when separated by electrophoresis, are found in the gamma fraction of serum; they are sometimes called gamma globulins.

Structure (pp. 400–401)

1. Most antibodies consist of four polypeptide chains. Two are heavy chains and two are light chains.
2. Within each chain is a variable portion (where antigen bonding occurs) and a constant portion (which serves as a basis for distinguishing the classes of antibodies).

Classes (pp. 402–403)

1. IgG antibodies are the most prevalent in serum; they provide naturally acquired passive immunity, neutralize bacterial toxins, and enhance phagocytosis.
2. IgM antibodies are involved in agglutination and complement fixation.
3. IgA antibodies protect mucosal surfaces from invasion by pathogens.
4. IgD antibodies appear to trigger B cells to produce specific antibodies.

5. IgE antibodies are involved in allergic reactions.

MECHANISM OF ANTIBODY FORMATION (pp. 404–411)

1. Humoral and cellular immunity are products of lymphoid tissue (lymph nodes, spleen, gastrointestinal tract, and bone marrow).
2. Lymphoid tissue is strategically located to intercept pathogens.

B Cells and T Cells (pp. 404–405)

1. Lymphocytic stem cells give rise to B cells and T cells.
2. T cells are processed in the thymus gland, where immunologic competence is conferred upon them. They then migrate to lymphoid tissue.
3. B cells are probably processed in bone marrow and Peyer's patches before migrating to lymphoid tissue.
4. Antibody production is enhanced by macrophages.

B Cells and Humoral Immunity (pp. 405–406)

1. A B cell becomes activated when an antigen reacts with receptor sites (IgD antibody) on its surface.
2. The activated B cell produces a clone of plasma cells and memory cells.
3. Plasma cells secrete antibody.
4. Memory cells recognize pathogens from previous encounters.

T Cells and Cellular Immunity (pp. 405–410)

1. When a particular antigen reacts with a particular T cell, the T cell becomes sensitized.
2. Sensitized T cells form clones, some of which become killer cells. Others remain as memory cells.
3. Killer T cells destroy antigens directly, secrete transfer factor to make other lymphocytes competent, and release macrophage chemotactic factor, which attracts macrophages.

Cooperation Between T Cells and B Cells (p. 410)

1. In many cases antigens first interact with helper T cells before inducing antibody formation by the appropriate B cells.
2. Suppressor T cells inhibit immune responses.

Antibody Responses to Antigens (pp. 410–411)

1. The amount of antibody in serum is called the antibody titer.

2. The response of the body to the first contact with an antigen is called primary response.
3. Subsequent contact with the same antigen results in very high antibody titers and is called the anamnestic response.
4. Anamnestic responses provide the basis for immunization against certain diseases.

SEROLOGY (pp. 411–420)

---

1. Serology is the study of antigen–antibody reactions in vitro and has many diagnostic applications.
2. Serological tests can be performed quickly and are quite specific and sensitive.

PRECIPITATION REACTIONS (pp. 411–413)

1. The interaction of soluble, multivalent antigens with multivalent antibodies (precipitins) leads to precipitation reactions.
2. Precipitation reactions depend on the formation of lattices and occur best when antigen and antibody are present in optimal proportions. Excesses of either component decrease lattice formation and subsequent precipitation.
3. The precipitin ring test is used in forensic medicine and in typing of streptococci.
4. Immunodiffusion procedures such as the Ouchterlony test involve precipitation reactions carried out in an agar gel medium.
5. Such tests detect the presence of multiple antigens and provide information about their relatedness.
6. The Elek test is an immunodiffusion test used to detect strains of *Corynebacterium diphtheriae* that produce exotoxin.
7. Immunoelectrophoresis combines electrophoresis with immunodiffusion for the analysis of serum proteins.

AGGLUTINATION REACTIONS (pp. 413–415)

1. The interaction of cell-bound (particulate) antigens with multivalent antibodies (agglutinins) leads to agglutination reactions.
2. Agglutination reactions depend on a lattice formation.
3. Diseases may be diagnosed by combining the patient's serum with a known antigen in tests such as the Widal test and Weil–Felix test.
4. Agglutination reactions can be used to determine antibody titer (tube agglutination test).
5. Hemagglutination reactions involve agglutination reactions using red blood cells.

6. Hemagglutination reactions are used in blood typing, diagnosis of certain diseases, and identification of viruses.

COMPLEMENT FIXATION REACTIONS (pp. 415–417)

1. Complement fixation reactions are serological tests based on the depletion of a fixed amount of complement in the presence of an antigen–antibody reaction.
2. Hemolysis or its absence is used as an indicator in evaluating complement fixation.

NEUTRALIZATION REACTIONS (pp. 418–419)

1. In these reactions, the harmful effects of a bacterial exotoxin or virus are eliminated by a specific antibody.
2. Neutralization reactions to determine the presence of antibodies are the Dick test (scarlet fever) and Schick test (diphtheria).

OPSONIZATION REACTIONS (p. 419)

1. The enhancement of phagocytosis by opsonins is called opsonization.

IMMUNOFLUORESCENCE (FLUORESCENT-ANTIBODY TECHNIQUES) (pp. 419–420)

1. Immunofluorescent techniques utilize antibodies labeled with fluorescent dyes.
2. Direct fluorescent-antibody tests are used to identify specific microorganisms.
3. Indirect fluorescent-antibody tests are used to demonstrate the presence of antibody in serum.

RADIOIMMUNOASSAY (p. 420)

1. Radioimmunoassay is used to detect the presence of a desired compound.
2. Antibodies against the compound are combined with a radioactively labeled antigen and a sample containing an unknown amount of the antigen.
3. Analysis of radioactivity in the resulting antigen–antibody complexes indicates the amount of antigen in the sample.

PUBLIC HEALTH VACCINATIONS (pp. 420–425)

HISTORICAL ASPECTS (pp. 420–421)

1. Variolation was an early attempt to immunize people against smallpox.
2. Edward Jenner performed the first successful vaccination against smallpox using cowpox virus.

DEFINITION (pp. 421–422)

1. A vaccine is a suspension of infectious agents (or some part of them) that will induce a state of immunity.
2. Most vaccines contain attenuated or killed microorganisms on their parts.

PREPARATION OF VACCINES (pp. 422–424)

1. Bacterial vaccines usually consist of suspensions of killed bacteria in physiological saline. Attenuated bacteria may also be used.
2. Toxoids (heat or chemically inactivated bacterial exotoxins) may be used in vaccines (for example, against diphtheria).
3. Vaccine employing capsular polysaccharide or pili are used in immunotherapy (for example, against gonorrhea).
4. Viral vaccines may consist of suspensions of inactivated viruses or attenuated viruses.
5. Attenuated cestode eggs and larvae may be used in vaccines.

IMMUNIZATION FOR CHILDHOOD DISEASES (pp. 424–425)

1. Immunization using vaccines has dramatically reduced the incidence of many childhood diseases (such as poliomyelitis, whooping cough, diphtheria, and measles).
2. Many of the vaccines used are combined vaccines.

IMMUNIZATION FOR TRAVELERS (p. 425)

1. Many vaccines are available when travel through an endemic area may result in possible exposure to certain diseases.
2. Current recommendations are supplied by the U.S. Public Health Service.

STUDY QUESTIONS

REVIEW

1. Define immunity.
2. Distinguish between the following sets of terms.
   (a) Nonspecific and specific resistance
   (b) Immunology and serology
   (c) Active and passive immunity
   (d) Acquired and native immunity
3. Classify the following examples of immunity as naturally acquired active immunity, naturally acquired passive immunity, artificially acquired active immunity, or artificially acquired passive immunity.
   (a) Immunity following injection of diphtheria toxoid
   (b) Immunity following an infection
   (c) Newborn's immunity to malaria
   (d) Immunity following an injection of antirabies serum

4. Explain what an antigen is. Distinguish between immunogenicity and reactivity.

5. Explain what an antibody is, describing the characteristics of antibodies. Diagram the structure of IgG antibodies.

6. How do humoral immunity and cellular immunity differ?

7. By means of a diagram, explain the role of T cells and B cells in immunity.

8. (a) At time *A*, the host was injected with tetanus toxoid. At time *B*, the host was given a booster dose. Explain the meaning of the areas of the curve marked *a* and *b*.
   (b) Indicate in the graph the antibody response of this same individual to exposure to a new antigen at time *B*.

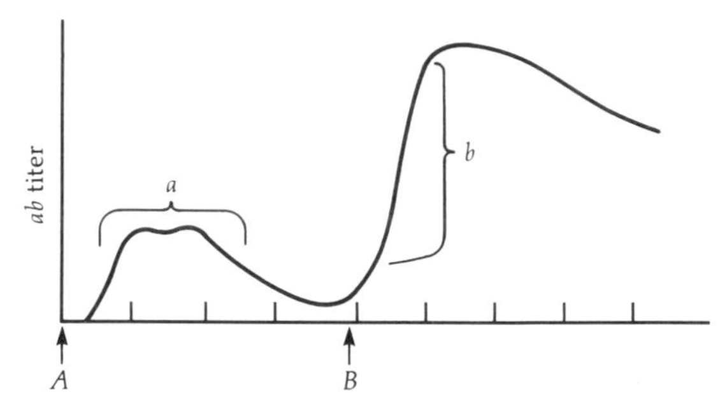

9. What is a precipitin? Explain the effects of excess antigen and antibody on the precipitation reaction. How is the precipitin ring test different from an immunodiffusion test?

10. How does the antigen in an agglutination reaction differ from that in a precipitation reaction?

11. Define the following terms and give an example of how each reaction is used diagnostically.
    (a) Viral hemagglutination
    (b) Passive hemagglutination
    (c) Hemagglutination inhibition

12. Explain the complement fixation test.

13. Match the following serological tests to the descriptions.

    ___ Precipitation
    ___ Immunoelectrophoresis
    ___ Agglutination
    ___ Complement fixation
    ___ Neutralization
    ___ Radioimmunoassay
    ___ Opsonization

    (a) Occurs with particulate antigens
    (b) Depends on enhancement of phagocytosis
    (c) Employs radioactive labels
    (d) Hemolysis is used as an indicator
    (e) Employs anti-human gamma globulin
    (f) Occurs with a soluble antigen
    (g) Used to determine the presence of antitoxin

14. What are the preparations used in the following immunizations for children: DPT vaccine, OPV, mumps vaccine, measles vaccine, and rubella vaccine.
15. Write a policy statement regarding recommendations for vaccination for travelers.

CHALLENGE

1. How would the addition of excess complement to a complement fixation test affect the results?
2. What problems are associated with the use of live vaccines?
3. Discuss the clonal selection mechanism.
4. Pooled human gamma globulin is sometimes administered to a patient after exposure to hepatitis. What is human gamma globulin? What type of immunity might this confer on the patient?
5. The World Health Organization has announced the complete eradication of smallpox. Why would vaccination be more likely to eradicate a viral disease than a bacterial disease?

FURTHER READING

Arnon, R. 1980. Chemically defined antiviral vaccines. *Annual Review of Microbiology* 34:593–618. Immunologic responses to viral vaccines.

Burnet, F. M. (Introduction). 1976. *Immunology: Readings from Scientific American*. San Francisco, CA: W. H. Freeman. Papers from *Scientific American* dealing with the structure and function of the immune system.

Capra, J. D. and A. B. Edmundson. January 1977. The antibody combining site. *Scientific American* 236:50–59. The details of the lock-and-key mechanism of an antibody–antigen reaction.

Hood, L. E., I. L. Weissman, and W. B. Wood. 1978. *Immunology*. Menlo Park, CA: Benjamin/Cummings. A concise textbook emphasizing the essential concepts of immunology.

Langer, W. L. January 1976. Immunization against smallpox before Jenner. *Scientific American* 234:112–117. Interesting history of the use of smallpox itself as a vaccination against smallpox.

Raff, M. C. May 1976. Cell-surface immunology. *Scientific American* 234:30–39. The action of antibodies can be used to find antigens on cell surfaces.

Sabin, A. B. 1981. Evaluation of some currently available and prospective vaccines. *Journal of the American Medical Association* 246:236–241. A summary of the currently available vaccines against bacterial pneumonia, poliomyelitis, measles, rubella, and influenza with a discussion of new vaccines under investigation.

Thaler, M. S., R. D. Klausner, and H. J. Cohen. 1977. *Medical Immunology*. Philadelphia, PA: J. B. Lippincott. A comprehensive immunology textbook.

17

Hypersensitivity and Cancer Immunology

Objectives

After completing this chapter you should be able to

- Explain the following terms: hypersensitivity; desensitization; histocompatibility antigens; HLA; immunosuppression; immunologic tolerance.
- Differentiate between immediate and delayed hypersensitivity and provide examples of each.
- Describe the mechanism of anaphylaxis.
- Describe the basis of human blood groupings and their relation to blood transfusions and hemolytic disease of the newborn.
- Explain how rejection of a transplant occurs.
- Define and give three possible explanations for autoimmunity.
- Discuss one immune-complex disease.
- Describe the problems and benefits of immunotherapy.

HYPERSENSITIVITY (ALLERGY)

Not all immune responses to an antigen induce a desirable state of immunity or resistance, and sometimes the reactions brought about by the immune system are even harmful. The immune state that results in altered or enhanced (hyper) reactions to an antigen, leading to pathological changes in tissue, is called **hypersensitivity** or **allergy** (meaning "altered reaction").

These pathological immune responses, or hypersensitivity reactions, only occur in people who have been previously "sensitized" to an antigen (in this context also called an **allergen**). These reactions are usually classified as either immediate or delayed. Although, as you would expect, the time it takes to develop the reaction in a sensitized person is usually much shorter in the "immediate" response, the primary difference between the two types is in the nature of the immune reaction to the antigen. Also, even though we will discuss immediate and delayed hypersensitivities separately, most hypersensitivity reactions observed clinically are actually a combination of both types. Let us now examine the details of these "altered" immune reactions to an antigen.

IMMEDIATE HYPERSENSITIVITY

Immediate hypersensitivities are reactions that involve *humoral* antibodies and that occur rapidly, often in minutes, when a sensitized person is exposed to the offending antigen. Because the antibodies involved are in solution in the blood, these responses to an antigen can be passively transferred with serum from an allergic person to a nonallergic person, causing temporary sensitivity. Immediate hypersensitivity reactions may be divided into three principal types as shown in Table 17–1. Type 1, *anaphylactic reactions,* are by far the most common of the immediate hypersensitivity reactions and will be discussed in detail next. Type 2, *cytotoxic reactions,* are those that relate to incompatible blood transfusions and will be discussed later in this chapter. Type 3, *immune-complex reactions,* are those that relate to serum sickness and certain autoimmune diseases.

Anaphylaxis

Anaphylaxis (*ana,* "against"; *phylaxis,* "protection") is defined as hypersensitivity that involves the interaction of humoral antibodies of the IgE class with mast cells and basophils, which are scattered throughout the body. Anaphylaxis-type responses may be divided into *systemic reactions,* which produce shock and asphyxia and are sometimes fatal, and *localized reactions,* which include common allergic conditions such as hayfever, asthma, and hives. To understand these immunologic responses to an antigen, we must first understand their basic mechanism.

Mechanism for Anaphylaxis

In response to certain antigens, some individuals produce IgE antibodies (Figure 17–1a). These antibodies are **cytotropic,** meaning they have the unique ability to bind to target cells such as basophils, which are found in the blood, and mast cells, which are especially prevalent in the skin, respiratory tract, and vascular endothelium. Once such antibodies have become fixed to these target cells, they may remain bound for weeks, causing sensitivity to the antigen. The binding of the IgE antibodies to the target cells occurs via special points of attachment on the antibodies different from the antigen-binding sites, which remain free to react with antigen when more of it later enters the system. The target cells possess numerous secretory granules in their cytoplasm that contain a variety of chemicals that are pharmacologically active mediators. The principal mediators and their activities are summarized in Table 17–2. They include *histamine, eosinophil chemotactic factor of anaphylaxis (ECF-A), heparin, platelet-activating factors (PAFs), slow-reacting substance of anaphylaxis (SRS-A),* and *serotonin.* It is believed that once an antigen has entered a sensitized host and reached the target cells, it binds to the cell-bound IgE antibodies, linking them together. This, in turn, stimulates the release of the mediators from the cells (Figure 17–1b). The release of these mediators into the tissue causes the observed symptoms for many allergic responses. Consider, for example, the pharmacological effects of the mediator histamine. Histamine causes increased permeability of the blood capillaries, which results in edema (swelling) and erythema (redness). It also causes smooth muscle contraction in the bronchi and intestines, resulting in breathing difficulty and intestinal discomfort. And finally, histamine increases mucous secretion and the gastric secretion of hydrochloric acid.

Table 17–1 Types of Immediate Hypersensitivities

Type	Antibody Involved	Complement Involved	Examples	Mediator Cells
Type 1: Anaphylactic reactions	IgE	No	Anaphylactic shock, cutaneous anaphylaxis, hives, asthma (some forms), hayfever, drug allergies	Mast cells and basophils
Type 2: Cytotoxic reactions	Any class, but primarily IgG	Yes	Transfusion reactions, Rh incompatibility, hemolytic anemia	None
Type 3: Immune-complex reactions	Any class, but primarily IgG or IgM	Yes	Serum sickness, certain autoimmune diseases (e.g., rheumatoid arthritis, systemic lupus erythematosus)	Primarily neutrophils

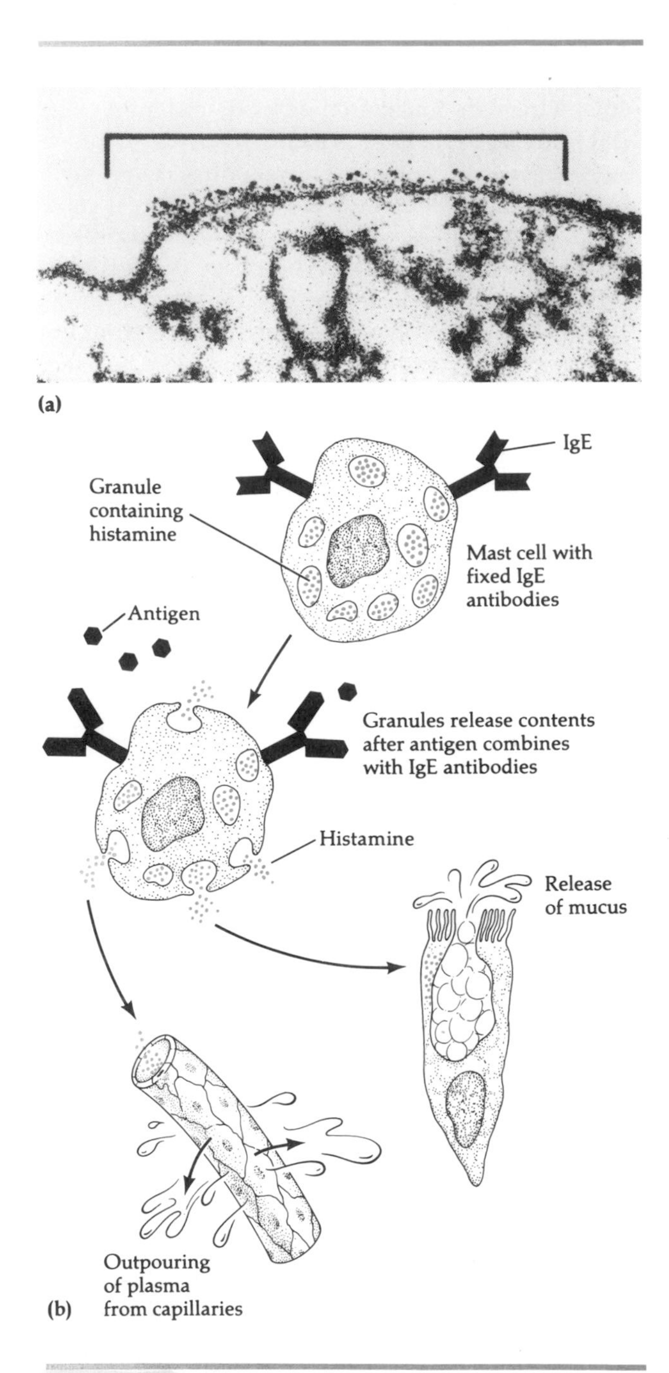

Figure 17–1 Mechanism of anaphylaxis. **(a)** Electron micrograph of IgE antibodies (bracket) on the surface of a basophil (×77,500). **(b)** In response to certain antigens, the immune system produces humoral IgE antibodies, which bind to mast cells (in connective tissue) and basophils (in the blood). Later, additional antigens bind to the cell-bound (fixed) IgE, triggering the release of mediators from the cell. Two of the effects of the mediator histamine are shown.

Table 17–2 Mediators of Anaphylaxis

Mediator	Comment and Function
Histamine	Found in mast cells. Increases blood capillary permeability and smooth muscle contraction of bronchial tubes and attracts eosinophils. Increases gastric secretions of hydrochloric acid.
Eosinophil chemotactic factor of anaphylaxis (ECF-A)	Found in mast cells and basophils. Attracts eosinophils and enhances complement reactions.
Heparin	Found in mast cells and basophils. Inhibits coagulation.
Platelet-activating factors (PAFs)	Found in mast cells and basophils. Aggregates and causes the lysis of platelets which release serotonin.
Slow-reacting substance of anaphylaxis (SRS-A)	Formed and secreted by mast cells and basophils after they are activated. Increases blood capillary permeability and smooth muscle contraction of bronchial tubes.
Serotonin	Found in blood platelets. Increases blood capillary permeability.

Systemic Anaphylaxis

Systemic anaphylaxis is a severe, sometimes fatal, reaction that develops rapidly after an antigen is injected into (or, occasionally, consumed by) a sensitized host. This condition can develop in persons sensitized to insect stings, certain drugs, and certain foods. This type of reaction is commonly seen in antibiotic therapy involving penicillin. In some individuals, the metabolic products of this drug function as haptens and bind to host serum proteins. These complexes stimulate the production of cytotropic IgE antibodies, which fix to mast cells, causing sensitization. If a large dose of penicillin is later taken by the person, a rapid reaction between the hapten and cell-bound IgE develops, triggering the release of large amounts of histamines into the circulation. Symptoms such as flushing of skin, wheezing, shortness of breath, circulatory collapse, and shock develop in minutes. When death results in humans, it is usually due to respiratory failure caused by the constriction of the smooth muscles in the bronchial tubes and laryngeal (voice box) edema. Treatment of this condition involves the prompt administration of epinephrine by intravenous or subcutaneous injection.

Localized Anaphylaxis

The spontaneous development of an allergic response is called *atopy* ("out of place"), and such reactions occur in about 10% of the U.S. population. Hypersensitivity reactions such as hayfever, asthma, and hives (elevated patches of reddened skin, often itchy) exemplify these responses and can be considered types of **localized anaphylaxis.** The antigens involved in these reactions are common environmental allergens such as plant pollens, fungal spores, dust, animal hair, and various foods. The symptoms that develop depend primarily on how the antigen gained entry into the body. With allergies involving the upper respiratory system, such as allergic rhinitis (hayfever), sensitization usually occurs by deposition of the antigen on the mucous membranes of the respiratory tract; IgE antibodies are produced and attach to mast (and basophil) cells. Subsequent exposure to the airborne antigen initiates the release of large amounts of histamine from mast cells in the affected tissues. This chemical, in turn, causes such symptoms as itchy, tearing eyes, congested nasal passages, cough, and sneezing. Antihistamine drugs are often useful in treating these symptoms, since this reaction depends on the release of histamine from tissue mast cells.

Asthma is an allergic reaction primarily affecting the lower respiratory system. It is characterized by wheezing and shortness of breath, which develop when a sensitized person comes in contact with the offending antigen. The symptoms result from spasms caused by constriction of smooth muscles in the bronchial tubes. Antihistamines are not effective in the treatment of this condition, since other mediators (such as SRS-A) are more important than histamine in this reaction. Treatment usually involves administration of epinephrine or aminophylline.

Allergens that enter the body through the digestive tract can also sensitize a host. Subsequent ingestion of the allergen can then produce symptoms such as vomiting, diarrhea, or hives, which result from the same kinds of mechanisms as those previously mentioned for histamines.

Prevention of Anaphylactic Reactions

Prevention of immediate hypersensitivity reactions is best accomplished by avoiding contact with the sensitizing antigen. Unfortunately, this is not always possible, since many persons are unaware of the identity of such allergens. To identify the allergen to which a person has been sensitized, skin tests can be performed (Figure 17–2). In this procedure, small amounts of antigen are inoculated under the skin. A rapid inflammatory reaction characterized by redness, swelling, and itching may develop, indicating sensitivity to the antigen. The small swollen area is called a *wheal*. Once an allergen has been identified, the sensitized person can either avoid contact with it or undertake desensitization to prevent the allergic response. **Desensitization** can be accomplished by carefully injecting small but repeated dosages of the allergen into the skin of the sensitized person. When the allergen is presented to the immune system in this fashion, it is believed to cause the production of IgG antibodies rather than those of the IgE class. It is hoped that the circulating IgG antibodies, called *blocking antibodies*, will intercept and neutralize the allergens

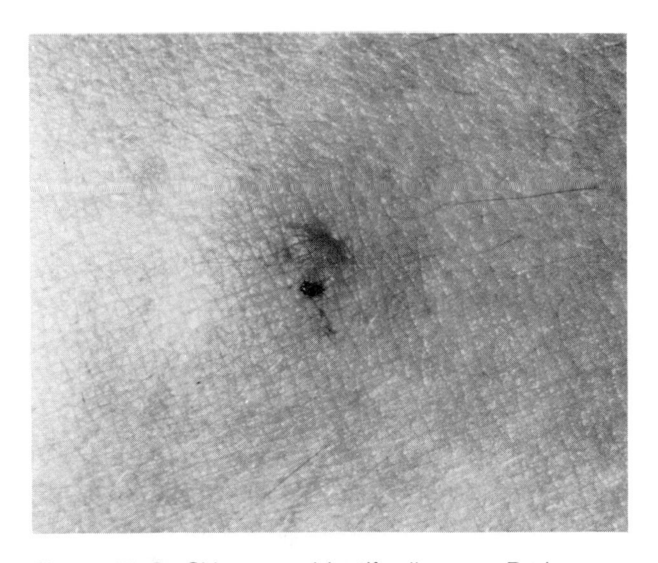

Figure 17–2 Skin test to identify allergens. Redness and a raised area, called a wheal, develop where the allergen responsible has been inoculated into the skin.

(such as pollen) before the allergens can react with cell-bound IgE. This prevents the allergic reaction. This usually works best for inhalation-induced allergies, where injections are used to cause formation of the blocking antibodies.

CELL-MEDIATED (DELAYED) HYPERSENSITIVITY

Another type of immunologic response that can cause intense, localized tissue damage is called **cell-mediated,** or **delayed, hypersensitivity.** These reactions are sometimes called type 4 hypersensitivity reactions. In cell-mediated hypersensitivity, recognition of an antigen by T cells leads to sensitization of a host. Subsequent contact between the sensitized T cells and their specific antigen induces a delayed tissue reaction often taking one to three days to develop. This reaction is characterized by the activities of lymphocytes, monocytes, and macrophages that initiate pathological changes in the tissue. Delayed hypersensitivity to an antigen can be passively transferred to an unsensitized host by using sensitized T cells or their extracts (transfer factor), but not by using serum (since serum antibodies are not involved). The tissue responses seen in the positive skin test for tuberculosis or with contact sensitivities such as poison ivy exemplify this type of immune reaction.

Mechanism

The basic mechanism of response in this type of hypersensitivity is similar to that of the cell-mediated immune response discussed in the previous chapter. It is believed that certain antigens, particularly those that bind to tissue cells, are phagocytized by macrophages and then presented to receptors on the surface of T cells (Figure 17–3). Contact between the antigen and the appropriate T cell stimulates the T cell to proliferate and become sensitized. When a person carrying such sensitized T cells is again exposed to antigen, a cell-mediated hypersensitivity may develop.

It is believed that the interaction of antigen with the sensitized T cells causes these cells to release a variety of mediators collectively called **lymphokines** (Table 17–3). These lymphokines then destroy antigens directly, recruit additional lymphocytes with characteristics of sensitized T cells, attract and aggregate macrophages, increase macrophage activity, induce lymphocytes to carry on DNA synthesis, and prevent viral multiplication on target cells. The activities of the lymphokines result in pathological tissue changes. For example, in the positive tuberculin skin test, the intracutaneous injection of tuberculin protein provokes a cell-mediated response involving lymphokines, which leads to redness, swelling, and induration (hardening). The redness and swelling tend to disappear rapidly, but the induration of the skin reaction may remain for days. Lymphocytes and macrophages accumulate in the area of tissue death.

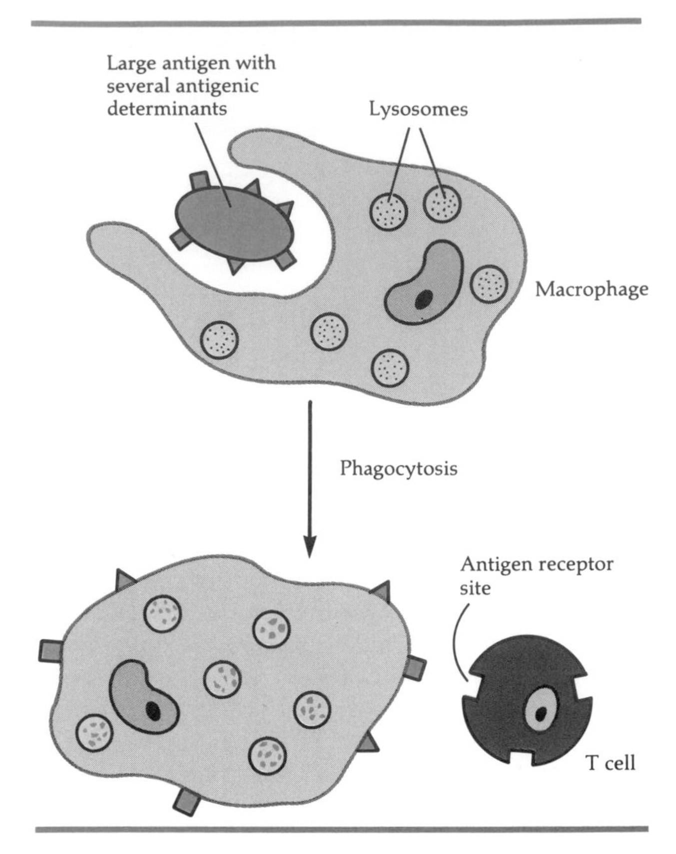

Figure 17–3 Suggested mechanism for the role of macrophages in delayed hypersensitivity and cell-mediated immunity. According to this mechanism, after phagocytosis, most of the antigen is degraded within the macrophage. But some antigen ends up on the surface of the macrophage for presentation to the appropriate T cells. A major role of the macrophage is to break down the large antigen into smaller components. It is not known exactly what alterations in the antigen the macrophage might cause or how such alterations might facilitate presentation to the T cells.

Table 17–3 Representative Lymphokines

Lymphokine	Comment
Cytotoxic factors	Destroy antigens directly
Transfer factor	Recruits additional lymphocytes, causing them to take on the same characteristics as sensitized T cells
Macrophage chemotactic factor	Attracts macrophages to destroy antigens by phagocytosis
Macrophage-activating factor	Increases phagocytic activity of macrophages
Macrophage aggregation factor (MAF)	Causes macrophages to aggregate
Lymphocyte blastogenic factor (BF)	Induces lymphocytes to carry on DNA synthesis
Interferon	Prevents viral replication in target cells

Cell-mediated hypersensitivities probably develop in response to many microbial infections and may play an important role in their pathogenesis. Delayed skin reactivity to an antigen has been useful in the diagnosis of certain bacterial infections such as tuberculosis and tularemia, chlamydial infections, systemic mycotic infections, and certain viral infections. A positive skin reaction must be interpreted carefully, since it only indicates *previous* exposure to the specific agent (or one closely related) sufficient to make the individual sensitive to the microbial proteins. Thus, a positive reaction to a *tuberculin skin test,* in which proteins from the bacterium that causes tuberculosis are injected into the skin, does not necessarily indicate that the individual has tuberculosis. Only the conversion of a negative to a positive test *during the illness* is indicative of an active infection. In addition, certain diseases (such as cancer and measles), as well as treatments using corticosteroids, can cause a marked reduction in skin reactivity to an antigen, a condition known as *anergy.*

Skin Contact Delayed Hypersensitivity

Some individuals exhibit *allergic contact dermatitis* (skin reactions) to drugs, plant products, or other chemicals, which can be attributed to a delayed hypersensitivity reaction. For example, when a previously sensitized person is later exposed to substances called *catechols* present on the leaves of poison ivy or poison oak, an intense local inflammatory reaction characterized by itching, swelling, and blistering develops in one to three days. Histologically, such a reaction is similar to that seen in the positive skin test for tuberculosis. Since many of the molecules causing contact hypersensitivities are of low molecular weight, they function as haptens but are not themselves antigenic until they form complexes with skin proteins. Skin lipids probably function as adjuvants (substances that improve antigen effectiveness) by helping retain the antigen in the tissues, a condition that favors delayed hypersensitivity reactions. Identity of skin allergies initiating these types of reactions can be determined by applying the allergen to the skin for 48 hours and observing the response *(Vollmer patch test).* Corticosteroids, which are known to depress cell-mediated immunity, are used to treat some of these reactions (such as for poison ivy). Figure 17–4 presents a schematic illustration showing the acquisition of allergic contact dermatitis to poison ivy.

TRANSFUSION REACTIONS AND Rh INCOMPATIBILITY

We have seen how the reaction of our immune system to an antigen sometimes leads to a harmful immunologic response (hypersensitivity). The ability of our immune system to react to surface antigens on foreign red blood cells plays an important role in blood transfusion reactions as well as Rh incompatibility, and such reactions may be regarded as immediate hypersensitivity reactions (primarily type 2).

More than 7 million transfusions are given each year in the United States, usually without complications. At one time, however, such transfusions were very risky, often killing the recipients. To understand the problems that develop in an incompatible blood transfusion, we must first examine the major blood group antigens of the ABO and Rh system.

The ABO Blood Group System

In the early 1900s, it was found that human blood could be grouped into four principal blood types, which were designated A, B, AB, and O. This has been called the **ABO blood group system.** Since then, at least fifteen other blood groups have been

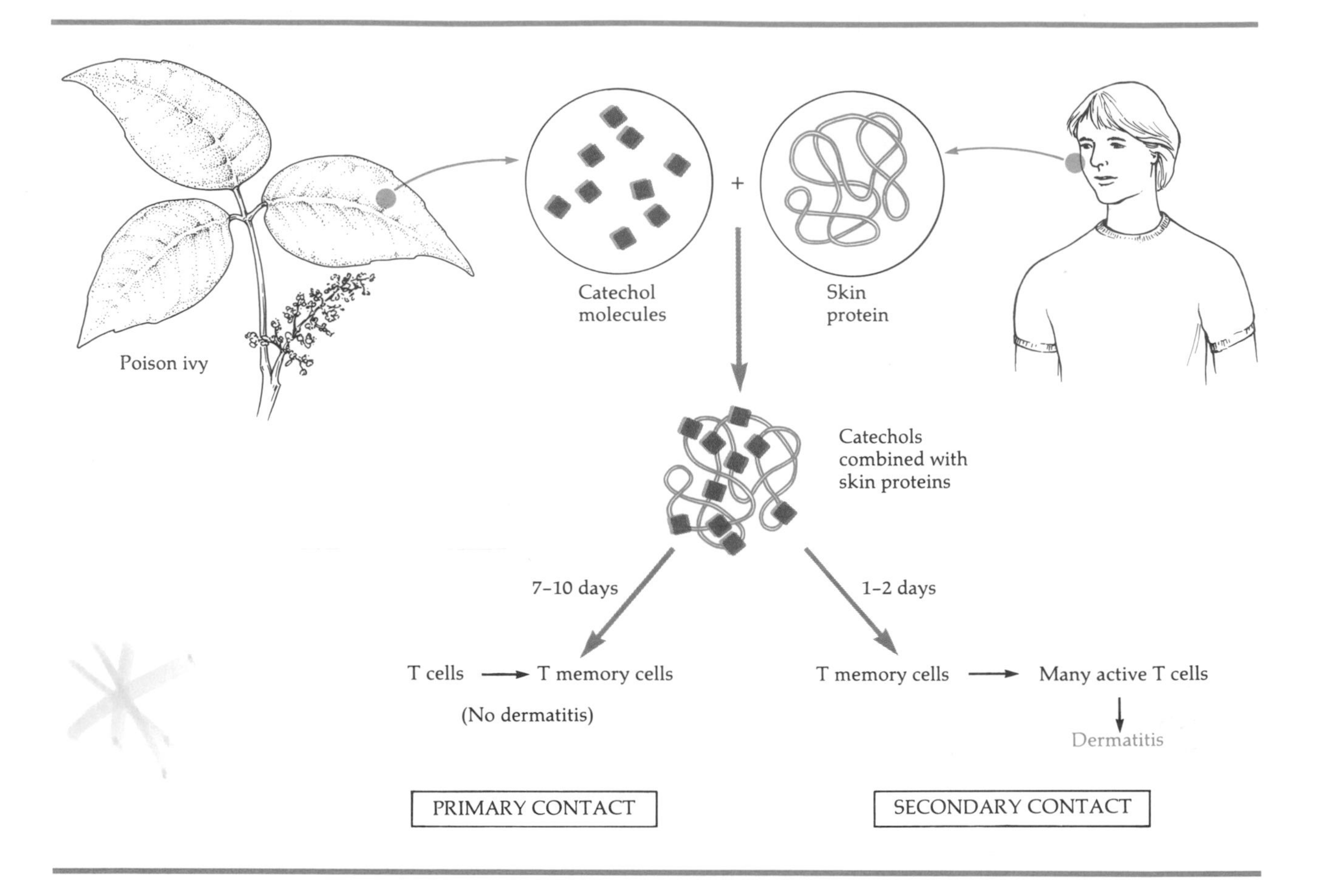

Figure 17–4 Development of allergic contact dermatitis, a delayed hypersensitivity reaction. Shown here is the development of allergy to catechols from poison ivy. No dermatitis results from the primary contact because the antigens (catechols) are gone before many sensitized T cells are produced. Upon secondary contact, T memory cells are quickly activated to secrete the substances that cause dermatitis.

Table 17–4 The ABO Blood Group System

Characteristic	Blood Type			
	A	B	AB	O
Isoantigen present on the red blood cells	A	B	Both A and B	Neither A nor B
Isoantibody normally present in the plasma	anti-B	anti-A	Neither anti-A nor anti-B	Both anti-A and anti-B
Plasma causes agglutination of red blood cells of these types	B, AB	A, AB	None	A, B, AB
Cells agglutinated by plasma of red blood cells of these types	B, O	A, O	A, B, O	None
Percent in a mixed Caucasian population	41	10	4	45
Percent in a mixed Black population	27	20	7	46

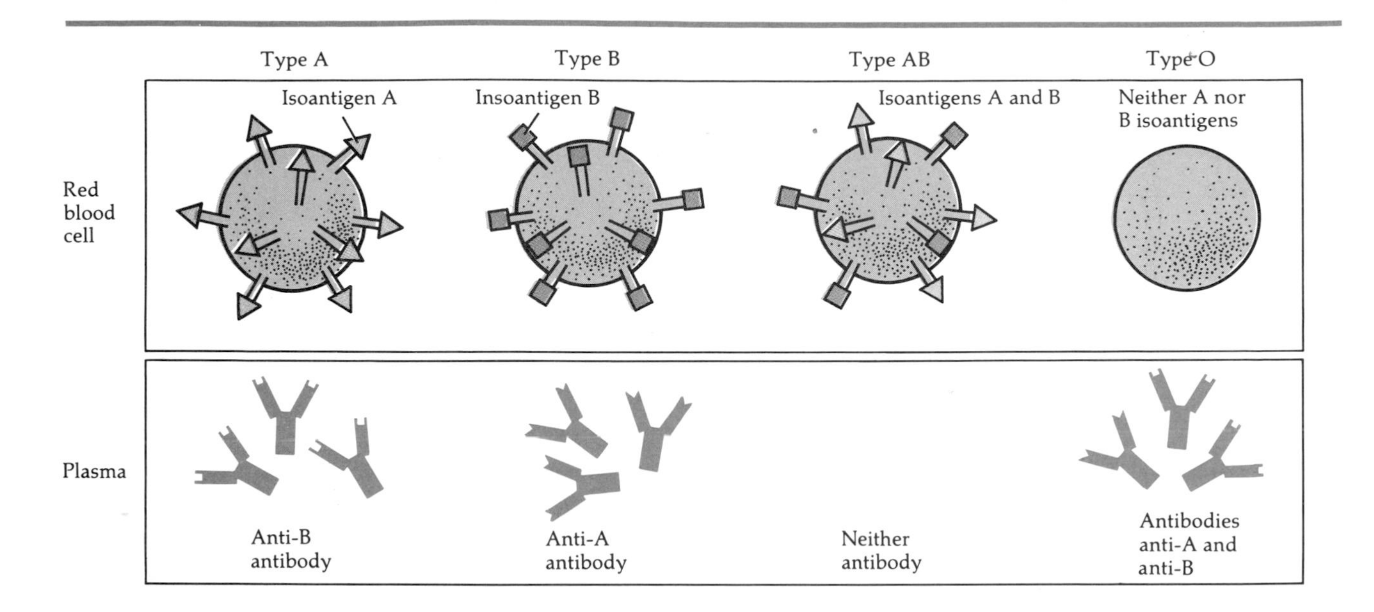

Figure 17–5 Relationship of isoantigens and isoantibodies involved in the ABO blood group system.

discovered. Our discussion, however, will be limited to the ABO and Rh blood group systems.

An individual's blood type is based on the presence or absence of two very similar carbohydrate antigens located on the cell membranes of the red blood cells, or erythrocytes (Figure 17–5). Individuals who are blood type A possess a mosaic of antigens designated A on their red blood cells, while persons of blood type B contain antigens designated B on their red cells. Persons with blood type AB possess both A and B antigens on their red cells, while persons with blood type O lack both A and B surface antigens. Since these blood group antigens are present in some individuals but absent in others of the same species, they are referred to as **isoantigens** (see Table 16–2). They are also called **agglutinogens,** which are found in serum. In addition to the surface red blood cell antigens, a person's blood type is also characterized by the presence or absence of IgM antibodies to these antigens, called **isoantibodies (agglutinins).** Individuals do not form isoantibodies to their own isoantigens, but uniquely possess *naturally* occurring isoantibodies to the *opposite* isoantigen. Thus, persons of blood type A will have serum containing isoantibodies to the B antigen. Likewise, individuals of blood type B will have isoantibodies against antigen A in their serum. Persons of blood type O have neither A nor B antigens but possess isoantibodies to both. And individuals of blood type AB have both A and B antigens but lack naturally occurring isoantibodies to either. These features of the ABO system are summarized in Table 17–4.

The mechanism that causes the production of naturally occurring blood group isoantibodies is unclear. However, such isoantibodies may arise from repeated stimulation of the immune system by common environmental antigens having a similar antigenic composition. The surface antigens present in some of the intestinal microorganisms, for example, have been shown to have an antigenic composition similar to that of the blood group isoantigens, and thus may contribute to this phenomenon.

The basic features of the ABO system, as outlined in Table 17–4, place restrictions on how blood may be transfused from one person to another. For example, individuals of blood type O, sometimes called *universal donors,* can donate blood to any of the other blood groups in the ABO system because their red blood cells lack both the A and B antigens and therefore are unreactive to the naturally occurring serum isoantibodies present in other blood types. The isoantibodies against the A and B antigens in the serum of type O individuals present only minimal problems during a transfusion since they are diluted by the total blood volume of the recipient. Similarly, individuals who are type AB are sometimes referred to as *universal recipients;* they can receive blood from any of the four blood types in the ABO system since their serum is lacking in isoantibodies to these antigens. If, however, a mistake is made in the typing of a blood sample and a person of blood type A is given blood from a donor of type B or AB, a severe hemolytic reaction will rapidly develop in the recipient. In such an *incompatible transfusion,* the B antigens of the donor red blood cells react with the isoantibodies that are naturally present in the serum of persons of blood type A. This antigen–antibody reaction activates complement, which in turn causes lysis of the donor's red blood cells as they enter the recipient's system. Clinical features of such incompatible transfusions include chills, lumbar pain, nausea, vomiting, fever, jaundice, and sometimes death. Because of the existence of blood antigens besides ABO, incompatibility reactions can occur even when the ABO systems are compatible. For this reason, the terms *universal donor* and *universal recipient* are misleading.

The Rh Blood Group System

In the 1930s, the presence of another surface antigen on human red blood cells was discovered. It was found that when rabbits were immunized with red blood cells from rhesus monkeys, their serum contained antibodies that would agglutinate 85% of all human red blood cells. Such results indicated that a common antigen was present on both types of red blood cells. This antigen was named the **Rh factor** (Rh for rhesus). We now know that the Rh factor is not a single antigen but a complex of over 30 different antigens, whose chemical nature is still obscure. For our purposes here, only the strongest antigen, referred to as Rh_o antigen or D antigen, will be considered as the Rh factor. This antigen is present on the surface of human red blood cells in approximately 85% of the population. Those persons possessing this antigen are referred to as Rh$^+$, while those lacking this antigen (15%) are Rh$^-$. Antibodies that react with the Rh antigen do not occur naturally in serum from Rh$^-$ individuals. But as a result of exposure to this antigen, such persons become sensitized and produce anti-Rh antibodies. Such an immune response must be considered in blood transfusions as well as in a naturally occurring Rh incompatibility, such as seen in hemolytic disease of the newborn.

Blood Transfusions and Rh Incompatibility

Individuals who are Rh$^+$ run little risk with respect to Rh blood group problems because their immune system recognizes the presence of the Rh antigen as "self" and does not produce anti-Rh antibodies. Such individuals may receive blood from either Rh$^+$ or Rh$^-$ donors without an adverse reaction, since the red blood cells from Rh$^-$ donors lack the surface antigens to stimulate an immune response. If, however, blood from an Rh$^+$ person is donated to an Rh$^-$ individual, the entry of the donor's red blood cells into the recipient's system will stimulate the production of anti-Rh antibodies. If the recipient receives Rh$^+$ red blood cells in a subsequent transfusion, a rapid hemolytic reaction will develop.

Hemolytic Disease of the Newborn

Hemolytic disease of the newborn is a relatively rare example of Rh incompatibility that develops under natural conditions (Figure 17–6). When an Rh$^-$ female and an Rh$^+$ male produce a child, the chances are at least 50% that the child will be Rh$^+$, since this antigen is inherited as a Mendelian dominant trait. If the developing fetus is Rh$^+$, the opportunity exists for the Rh$^-$ mother to become sensitized to this antigen during birth with the tearing of the placental membranes and subsequent entry of Rh$^+$ fetal red blood cells into maternal circulation. In later pregnancies involving an Rh$^+$ fetus, incomplete anti-Rh antibodies (univalent IgG) produced in response to the Rh antigen cross the placenta and react with fetal red blood cells, enhancing their destruction and clearance from the fetal bloodstream. The fetal blood-forming tissues respond to this loss of red blood cells by producing and releasing large numbers of immature red blood cells, or erythrocytes, called *erythroblasts*. Hence, the name *erythroblastosis fetalis* was once used to describe this condition. While in utero, the fetus is usually spared the severe effects of this condition, since maternal circulation removes most of the toxic

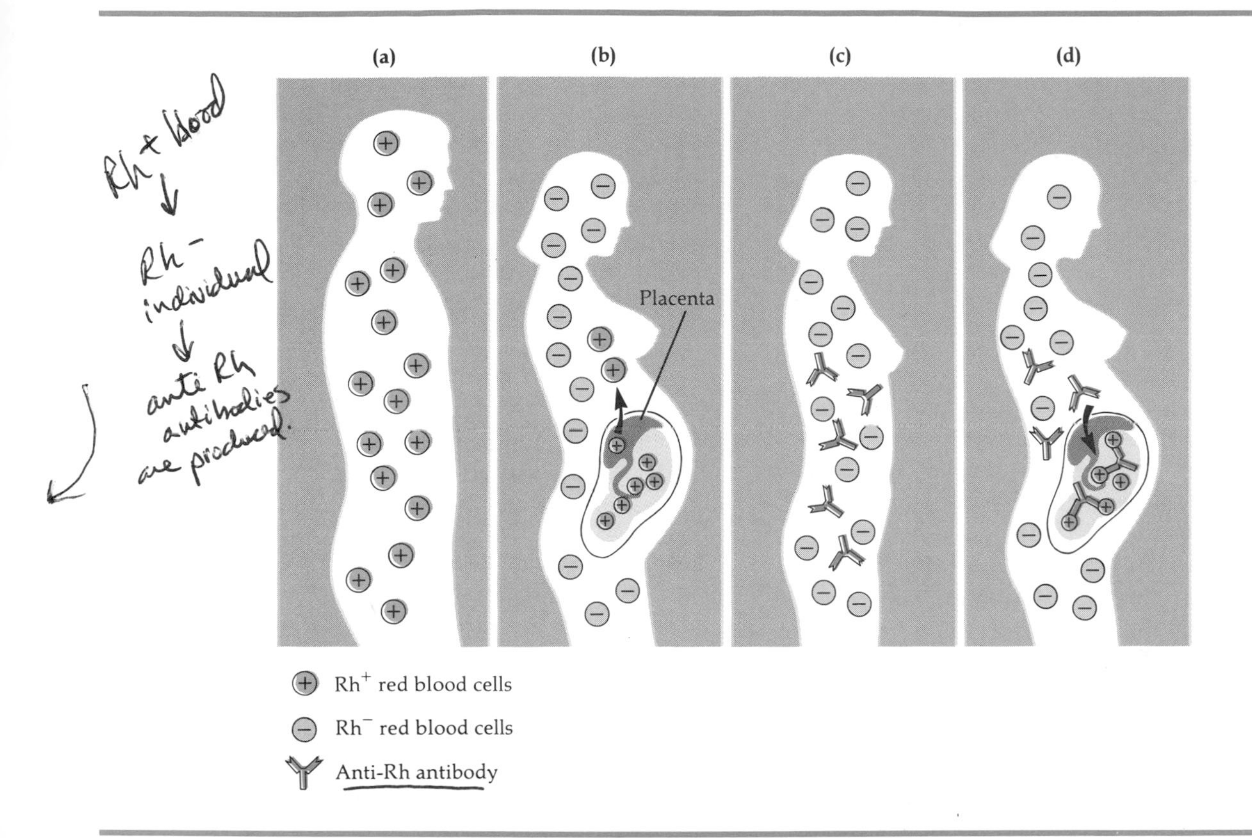

Figure 17–6 Hemolytic disease of the newborn. **(a)** Rh^+ father. **(b)** Rh^- mother carrying an Rh^+ fetus. If an Rh^- female is impregnated by an Rh^+ male and the developing fetus is Rh^+, then fetal Rh^+ antigens may enter the mother's blood across the placenta during delivery. **(c)** In response to exposure to the fetal Rh^+ antigens, the mother will produce anti-Rh antibodies. **(d)** If the female becomes pregnant again with an Rh^+ fetus, her anti-Rh antibodies will pass through the placenta into the blood of the fetus. The result is hemolytic disease of the newborn.

by-products of red cell destruction. At birth, however, with the disruption of maternal circulation, the fetal blood is no longer purified and the child rapidly develops jaundice, in addition to the severe anemia already present. Since these symptoms develop at birth or shortly after, this condition is often called *hemolytic disease of the newborn*. Treatment for this condition involves the removal of the fetal Rh^+ blood and transfusion with Rh^- blood. This eliminates the toxic by-products of red blood cell destruction and allows the introduction of red blood cells that are unreactive to any maternal anti-Rh antibodies in the serum. After several weeks, these transfused red blood cells are replaced by the normal Rh^+ cells of the child.

Hemolytic disease of the newborn is prevented today by passive immunization of the Rh^- mother with anti-Rh antibodies (commercially called Rhogam) shortly after birth. These anti-Rh antibodies combine with fetal Rh^+ red blood cells that may have entered maternal circulation and enhance the clearance of these cells, greatly reducing the sensitization of the mother's immune system to this antigen. Thus, passive immunization is used to prevent the acquisition of natural immunity.

TRANSPLANTATION

The replacement of injured or diseased tissues and organs with natural ones is called **transplantation.**

Human organ transplantation, although still an imprecise science, has added years of productive life to many individuals who would have otherwise succumbed to a variety of untreatable illnesses. Since the first kidney transplant in 1954, over 50,000 kidney transplants have been performed. Other transplants carried out include bone marrow, thymus, heart, liver, lung, cornea, and pancreas.

Types of Grafts

Tissues and organs for transplantation are taken either from living donors, as with many kidney grafts, or from recently deceased individuals (cadaveric donors) as in the case of essential organ grafts (heart grafts, for example).

Four types of transplants, or grafts, have been defined based on the genetic relationship between the person from whom the organ is taken (the *donor*) and the person who receives the grafted organ (the *recipient*). The four types of grafts are described in Table 17–5.

The major obstacle to successful transplant surgery is the immune system, which provides the basis for *rejection* of foreign tissues by the body, primarily by the cellular immunity system. The human body accepts only what is genetically identical to the part replaced. The ability of the body to recognize some substances as "self" and others as "nonself" becomes quite clear when we consider the success rates of various types of transplantation. It has been known for years that if a piece of skin is grafted from one individual to another, the recipient may soon reject the graft. Similar reactions occur when organs such as lungs, hearts, and kidneys are transplanted from one individual to another. If, however, a tissue or organ is transplanted from one identical twin to another, the transplant is accepted by the recipient; there is no rejection. Autografts and isografts are not rejected because there are no genetic differences between donor and recipient. Xenografts are almost always rejected because of the severe genetic differences between species and are seldom used in human transplantation. Allografts are the most frequently encountered type of transplant, and various procedures, described later, are undertaken to minimize the rejection process by the host.

Table 17–5 Types of Grafts

Graft	Description
Autograft (*autos*, "self")	A graft transplanted from one body site to another body site on the same individual (skin from the thigh grafted over a burn area on the arm)
Isograft (*isos*, "equal")	A graft between genetically identical individuals such as between identical twins
Allograft (*allos*, "other")	A graft between genetically different individuals of the same species
Xenograft (*xenos*, "strange")	A graft between different species (a transplant of a baboon liver to a human)

Major Histocompatibility Complex

One expression of the genetic differences between individuals are the antigens on the surfaces of cells, called **histocompatibility** antigens. There are four clusters of genes that code for the production of these antigens, and they are found on a portion of the sixth chromosome in humans. Collectively, the genes are known as the **major histocompatibility complex (MHC)** or the **human leukocyte antigen (HLA) complex** (see box). For each of these genes, there are a number of alternative alleles (versions) in the population, and there are known to be a total of at least 51 different possible antigens. Because each gene produces two antigens, cells may thus have as many as eight different MHC antigens on their surface—with the exception of human red blood cells, which lack these antigens. The antigens on a cell may be any of a variety of combinations of the 51 possible antigens, and the number of combinations is enormous. Two individuals chosen at random would not be expected to have many MHC antigens in common. But since each person inherits the genes that code for the four antigens from their parents, family members can be expected to share many of their MHC antigens and are usually the best source of organ donors for allograft transplantation.

To prevent rejection in allografting, the donor and recipient are *matched* as closely as possible with respect to their MHC antigens. Generally, the closer the MHC match (that is, the more MHC antigens in common), the more likely that the graft

Microbiology in the News

Scientific Sleuths Win Nobel Prize for Uncovering Cell "Fingerprints" Made by Genes

Thanks to careful detective work by three immunologists, cell "fingerprints" produced by genes may soon make tissue matching for a kidney transplant as easy as getting your blood type.

The 1980 Nobel Prize for medicine or physiology was awarded to George Snell, Jean Dausset, and Baruj Benacerraf, whose independent investigations led to the discovery of unique arrangements of proteins called histocompatibility antigens, found on the surfaces of all nucleated cells in humans and animals. These antigen groups are like biological fingerprints because they make it possible to distinguish people immunologically by their tissue type. The antigens are regulated by a set of genes now known as the major histocompatibility complex, or MHC.

Besides its role in producing tissue types, the MHC is relevant to microbiologists because it is intimately associated with the immune response of the body. MHC genes signal cells to recognize "self" and "nonself," and they also help control the production of antibodies against a number of foreign microbes. In addition, MHC genes may contain clues to explain why some people are more disposed than others to certain diseases like rheumatoid arthritis and multiple sclerosis. The MHC antigens have been linked to several disorders, including some forms of arthritis, bowel diseases, skin conditions, complications of infections, and thyroid problems.

Each of the Nobel researchers contributed to our present knowledge of the MHC genes. George Snell, 76, laid the groundwork with his painstaking studies of tumor cell transplants in mice. In the 1930s, Snell, a senior scientist from Jackson Labs in Bar Harbor, Maine, began to look for the genes in mice that control whether a transplant is accepted or rejected.

Snell performed the tedious task of breeding mice that were genetically alike except for a small set of genes. Then he tested the mice to see how well a transplant between two strains would "take." Snell found that the more antigens the mice had in common, the more likely it was that the graft would not be rejected. This is because the surface antigens on the cells of the donated tissue that are different from recipient surface antigens are recognized as foreign and destroyed by circulating T cells. Gradually, Snell worked out the details and found that the outcome of transplants was governed by genes at the H-2 locus on mice chromosomes. Snell's study of H-2 antigens in mice gave scientists the technology and theoretical background to recognize the MHC in humans.

It was the 1950s before Jean Dausset applied Snell's work on mice to humans. Dausset, 63, a physician and immunologist at the University of Paris, found a comparable set of antigens in humans. Dausset called them the Human Leukocyte Antigens, or HLA Complex, because they were first found while testing white blood cells. There are at least 51 of these antigens and they are now known to exist in all tissues, not just human leukocytes. HLAs define tissue types in the same way that the principal blood groups ABO distinguish blood types. HLAs are coded and controlled by MHC genes. (HLA typing is already becoming a useful tool in forensic (legal) medicine. HLA tests have been used to identify rapists via semen stains and resolve paternity cases.)

Baruj Benacerraf, 59, now at Harvard University, further explored the MHC in mice and guinea pigs. He identified a subset of genes in the MHC, called Ir genes (for "immune response"). These genes seem to control how well an animal can mount a counterattack against invading microbes. It is not certain how Ir genes work, but they appear to provide a protein on the surface of white blood cells that interacts with lymphocytes and macrophages, two immune system scouts that

(continued on next page)

roam the body hunting down foreign substances to attack.

One of the goals of MHC research is to learn to manipulate the body's immune response to histocompatibility antigens so as to increase the chances of successful transplant of a heart, kidney, or other organ. Understanding the MHC may also eventually make it possible to prevent a juvenile (insulin-dependent) form of diabetes and some thyroid diseases.

Doctors have known for years that there are strong links between certain diseases and histocompatibility antigens. For example, presence of the antigen called HLA-B27 increases by 120 times a person's chance of developing a type of arthritis called ankylosing spondylitis. This same antigen is also linked to an arthritis that develops as a consequence of bacterial infections caused by *Yersinia*, *Salmonella*, and *Neisseria gonorrhoeae*.

With the aid of genetic engineering and more knowledge of how the MHC genes work, it may be possible to tailor a vaccine to an individual's HLA type in order to bolster the person's immune response. This might protect individuals who, from statistical studies, appear to be biologically predisposed to certain microbial infections or diseases. Nobelists Snell, Dausset, and Benacerraf have provided us with the biological fingerprint file on MHC antigens and their genes. It is up to microbiologists and immunologists to develop the medical and scientific potential of MHC antigens.

will be successful. There are several criteria used to determine relatedness between donor and recipient for transplantation:

1. The donor and recipient must be of the same ABO blood type, since most tissues have ABO blood group antigens on the surface of their cells that can lead to rejection if not matched.
2. The recipient's serum is tested for the presence of antibodies to the donor's MHC antigens. These antibodies may have been produced as a result of exposure to other MHC antigens through blood transfusions, multiple pregnancies, or previous transplants.
3. The donor and recipient are tissue typed, and the MHC antigens are identified for each individual.

Even when the organ is well matched with the recipient, the graft is still sometimes rejected, indicating that we still have much to learn about the MHC antigens and the rejection process.

Immunosuppression

In an attempt to prevent or arrest rejection, the allograft recipient is usually treated to suppress an immune response against the graft. To bring about **immunosuppression,** that is, inhibition of the immune response, drugs (such as the steroid prednisone), antilymphocyte serum, or X-irradiation may be used. These treatments, unfortunately, suppress the immune response to *all* antigens, causing patients to become very susceptible to infectious diseases and cancer. Another severe consequence of immunosuppression is the development of **graft versus host (GVH) disease.** This occurs when the recipient cannot make an immune response to the graft, but the graft contains immunocompetent cells that can respond against the recipient (host). Essentially, the graft attempts to reject the recipient. Most commonly, GVH disease is a complication of bone marrow transplantation. Marrow grafts contain large numbers of B and T cells that can mount an immune response against the immunosuppressed host. Severe GVH reactions are often fatal.

Although human organ transplantation is a fairly common procedure, long-term survival and permanent acceptance of allografts are still difficult to achieve. Much of the current research in transplantation is aimed toward the specific blocking of the immune response to the graft antigens while leaving the immune system generally intact to deal with other environmental stresses.

IMMUNOLOGIC TOLERANCE AND AUTOIMMUNITY

Immunologic Tolerance

Just as immunity is a state of specific *reactivity* to an antigen resulting from prior exposure, **immunologic tolerance** is a state of specific *unresponsiveness* to an antigen following initial exposure to that antigen. This type of unresponsiveness occurs under natural conditions, as seen by the inability of our immune system to elicit a reaction against our own tissue antigens. Experimental evidence suggests that tolerance to "self" antigens comes about during fetal development. If an individual is exposed to an antigen before birth, further exposure to this antigen after birth *does not* stimulate the production of antibodies or sensitized T cells. Recognition of this tolerance led to the formulation of the clonal selection theory for antibody production (see Figure 16–9). The original explanation for self-tolerance was that during embryonic development, clones of lymphocytes, called *forbidden clones,* react with self-antigens and are destroyed, leaving only those lymphocytes that can react with foreign antigens to function in the mature immune system. Today, it is believed that such forbidden clones are not destroyed but are *suppressed* by specific classes of T lymphocytes.

Immunologic tolerance for an antigen is most easily established if the antigen is similar to the recipient's own molecules and if the recipient's immune system is immature, such as during fetal development or shortly after birth. However, mature animals may also be made tolerant to an antigen under certain experimental conditions. If highly purified polysaccharide or protein antigen is injected in large and repeated doses, the animal is unable to make antibody to the antigen. This form of tolerance, called **immunologic paralysis,** requires the constant exposure of the immune system to high concentrations of antigen and causes both T cells and B cells to become unresponsive. When the antigen concentration decreases, however, antibody formation resumes and tolerance is lost.

The constant exposure to low concentrations of antigen can also induce tolerance by causing T cells to become unresponsive. Thus, tolerance to microbial antigens sometimes develops very late in the course of certain diseases. Patients with advanced tuberculosis or certain systemic mycotic infections often lose the ability to exhibit a positive skin test when exposed to the appropriate antigen. This condition, anergy, demonstrates that T cells have become unresponsive to the microbial antigens.

The mechanism for immunologic tolerance is not completely understood at present. It is believed that certain populations of T cells *(suppressor T cells)* may function to prevent the humoral response to an antigen by repressing the differentiation of B cells into plasma cells or suppressing helper T cells. Suppressor T cells also probably function to inhibit T cell population growth and are believed to exert some control over cell-mediated immunity.

Autoimmunity

For reasons not well understood, the immune system occasionally loses its ability to discriminate between "self" and "nonself," and this loss of natural immunologic tolerance leads to **autoimmunity.** Autoimmunity (self-immunity) is really a misnomer, since this condition does not involve a state of protection, but rather represents an immunologic response mediated by antibodies or sensitized T cells against a person's own tissue antigens. Such conditions often induce a disease state in the host and might be better named **autoallergic responses.** An understanding of autoimmunity is complicated by the fact that certain diseases (for example, syphilis) lead to the formation of autoantibodies as a result of tissue destruction, although such antibodies are not involved in the pathogenesis of the disease. The association of an autoantibody with a given disease therefore does not always mean such antibodies are the cause of the disease.

Serum sickness is an immune-complex disease in which the host makes antibodies against foreign immunoglobulins. When a person is passively immunized with horse serum, for example, antibodies may be produced against the horse immunoglobulins. These new antibodies can cross-react with human immunoglobulins, causing serum sickness. Symptoms of serum sickness, which are caused by the formation of immune complexes and the activation of complement, include fever, rash, joint lesions, and kidney dysfunction.

There are many diseases in which autoantibodies or sensitized T cells reactive to self-antigens are thought to be a primary cause of the pathology. *Rheumatoid arthritis* is a fairly common inflammatory disease of the joints and connective tissue (Figure 17–7). Serum from patients with this disease contains autoantibodies of the IgM class (rheumatoid factor), which react with IgG antibody present in the synovial tissue of joints. Thus, these patients are making antibodies to some of their own antibodies! The immune complexes formed between the IgM and the altered IgG activate the complement system, attract polymorphonuclear leukocytes, and are phagocytized, causing the release of lysosomal enzymes that initiate the tissue damage seen in the disease. It is not known what triggers the production of the IgG found in the joints.

In *systemic lupus erythematosus (LE),* a disease characterized by collagen degeneration and tissue damage to kidneys, blood vessels, blood cells, and heart, antibodies reactive to host DNA are prevalent in the serum. In this condition, immune complexes develop, causing tissue damage where they become deposited. Localization of these complexes causes blood vessel damage, while deposition in the kidney leads to inflammation and often kidney failure, a major cause of death in the SLE patient.

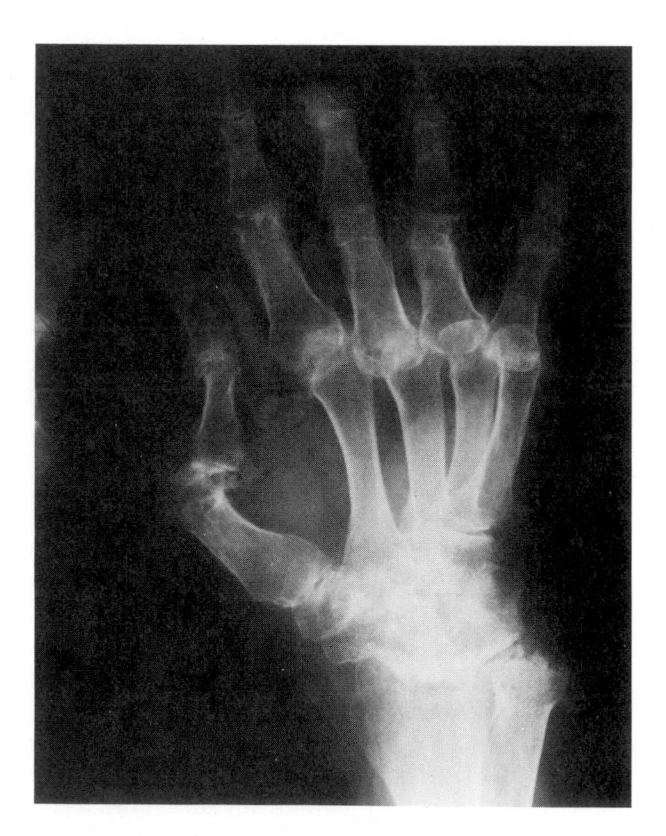

Figure 17–7 X ray of the hand of a rheumatoid arthritis patient with a 20-year history of joint destruction. Note the tissue damage and the severe misalignment of the joints.

Table 17–6 Representative Autoimmune Diseases Occurring in Humans

Disease	Autoantigen
Allergic encephalitis	Brain cell surface
Hashimoto's disease	Thyroid cell surface
Hemolytic anemia	Red blood cell surface
Myasthenia gravis	Receptor for acetylcholine
Pernicious anemia	Intrinsic factor (necessary for vitamin B_{12} absorption)
Rheumatoid arthritis	IgG in synovial membrane
Systemic lupus erythematosus (LE)	DNA of many cells
Thrombocytopenic purpura	Platelet surface

Source: James T. Barrett, *Textbook of Immunology,* 3rd ed., 1978, The C. V. Mosby Company, St. Louis, MO.

Hashimoto's disease, characterized by a deficiency in thyroid hormone and an enlarged thyroid gland heavily infiltrated with plasma cells and lymphocytes, is also considered an autoimmune disease. Serum from patients with this disorder contains antibodies (IgG and IgM) demonstrating anti-thyroid cell activity, while normal individuals lack these autoantibodies.

Examples of diseases that may result from autoimmune responses are given in Table 17–6.

Mechanism for Autoimmunity

Several theories have been proposed to explain how an autoimmune condition may arise. One theory suggests that not all self-antigens are exposed to immune recognition during fetal development, a time when lymphocytes reactive to self-antigens are believed to be eliminated or suppressed. Hidden or sequestered antigens such as those on sperm cells, the lens of the eye, or on tissues of the central nervous system (brain and spinal cord) may escape recognition of the immune system during this time and induce a specific immune reaction if they later enter circulation. For example, the development of sperm cells is not complete until sometime after birth when maturation antigens are formed, and if infection or injury occurs at a later date, exposure of sperm cells to the immune system can cause autosensitization, leading to sterility.

Another theory explaining autoimmunity suggests that in response to certain foreign antigen, antibodies are formed that **cross-react** with host tissue antigens. In cross-reactivity, an antibody that binds a given antigen will also bind other molecules with similar antigenic sites, although not as strongly. Thus, severe, recurrent infections caused by group A beta-hemolytic streptococci sometimes lead to the development of *rheumatic fever* (causing swollen joints, rashes, and heart damage) long after the streptococcal infection has subsided. Such observations suggest an autoimmune reaction. In support of this, it can be demonstrated that antibodies to cell membrane antigens of certain group A streptococci cross-react with heart tissue in humans. In fact, cross-reactivity between heart valve glycoproteins and streptococcal group-specific carbohydrates has been proved. Similarly, *glomerulonephritis,* another suspected autoimmune disease, may result from a cross-reactivity between shared kidney and streptococcal antigens. This disease is characterized by an inflammation of the kidney.

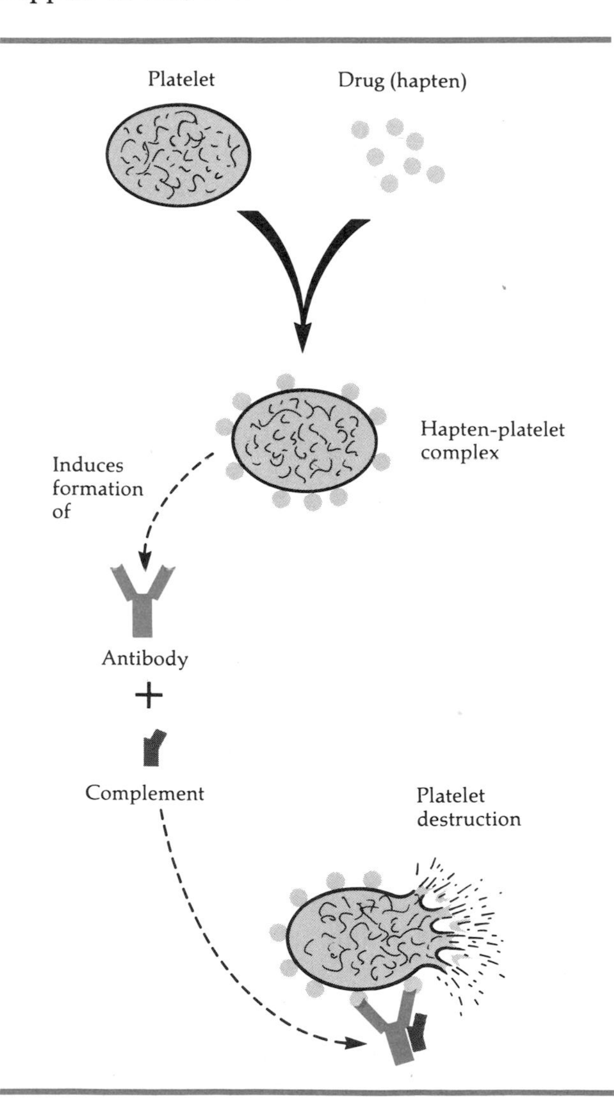

Figure 17–8 Drug-induced autoimmune thrombocytopenia.

Perhaps one of the best explanations for the development of autoimmunity suggests the formation of *new* or *altered* antigens on the surface of host cells. Tissue antigens may be altered in a number of ways by drugs or viruses. Certain drugs, for example, may bind to host cells or platelets and function as haptens to induce the formation of autoantibodies. When such antibodies bind to a hapten–platelet complex, the resulting activation of complement causes platelet destruction, a condition termed *autoimmune thrombocytopenia* (Figure 17–8).

Certain viruses, especially those such as measles virus, which bud from infected host cells, induce the formation of viral antigens on host cell surfaces. Such infected host cells are susceptible to destruction by the immune system. Thus, postinfectious *encephalomyelitis* (inflammation of the brain and spinal cord), which develops following immunization or infections caused by measles or mumps viruses, is believed to represent a cell-mediated response against host tissue possessing altered antigens. Autoimmune reactions to viral infections may also play an important role in neurological diseases such as multiple sclerosis, kuru, and lymphocytic choriomeningitis (LCM). (See Chapter 12.)

IMMUNE RESPONSE TO CANCER

The change of a normal cell into one with the properties of a cancer cell is called **cell transformation.** Normal cells can be transformed in the laboratory spontaneously or by a variety of agents, including certain chemicals, UV light, and several kinds of viruses. As described in Chapter 12, transformed cells, when grown in vitro, multiply rapidly and lose their ability to stop growing upon contact with adjacent cells, a property of normal cells called *contact inhibition.* In vivo, these transformed cells may

Table 17–7 Principal Types of Cancers

Type of Cancer	Description
Carcinoma	The most common form of cancer; arises from the cells forming the skin, the glands (such as the breast, uterus, and prostate), and the membranes of the respiratory and gastrointestinal tracts; metastasizes (spreads to other parts of the body) mainly through the lymph vessels
Leukemia	Cancer of the tissues that form blood; characterized by uncontrolled multiplication and accumulation of abnormal white blood cells
Lymphoma	Cancer of the lymph nodes
Sarcoma	Another form of malignant tumor; arises from connective tissues, such as muscle, bone, cartilage, and membranes covering muscles and fat; metastasizes through the blood vessels

Source: Courtesy of Barbara J. Combs, et al., *An Invitation to Health*, The Benjamin/Cummings Publishing Co., Menlo Park, CA, 1980.

grow into **tumors,** masses formed by uncontrolled cell proliferation. The principal types of cancers are summarized in Table 17–7.

Tumor-Specific Antigens

As a direct consequence of transformation, tumor cells express unique cell surface components known as **tumor-specific antigens** (Figure 17–9). The genes coding for these antigens may be mutant genes, oncogenic viral genes, or genes expressed during embryonic development but normally repressed in adult life. For example, an antigen called **carcinoembryonic antigen (CEA)** is present in a normal fetal gut, liver, and pancreas between two and six months gestation, but disappears during the last trimester of pregnancy. This antigen can be extracted from malignant colon tissue but is not found in nonmalignant colon tissue. Another cancer antigen, **alpha-feto protein (AFP),** is found during embryonic development but not in later life and is associated with liver and embryonal carcinomas.

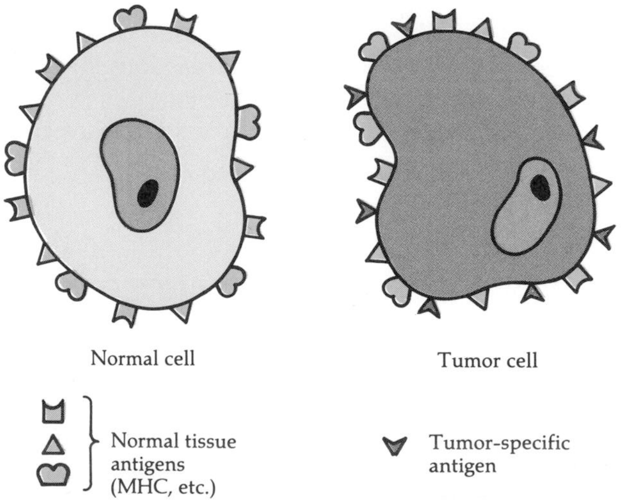

Assays for detecting tumor-specific antigens in the body before other signs of cancer appear offer hope of early treatment. But more important to human survival may be the role that such antigens play naturally in the body, for it is thought that the immune system usually recognizes tumor-specific antigens as nonself and destroys the malignant cells carrying them.

Figure 17–9 Tumor-specific antigens. Although these antigens are found on most tumor cells, they are not on equivalent normal cells.

Immunologic Surveillance

As we have said, the immune system is believed to play an important role in defending the body against cancer. Immune responses to cancer, presumably directed against tumor-specific antigens, are known as **immunologic surveillance.** Humoral antibody will bind to tumor-specific antigens, but even high concentrations appear to be insufficient to destroy rapidly proliferating tumor cells. Most investigators believe that cell-mediated immunity represents the major mechanism for tumor destruction in a host, and this idea is supported by the observation that persons having defects in this system have a higher incidence of certain types of cancer than normal individuals. It is thought that

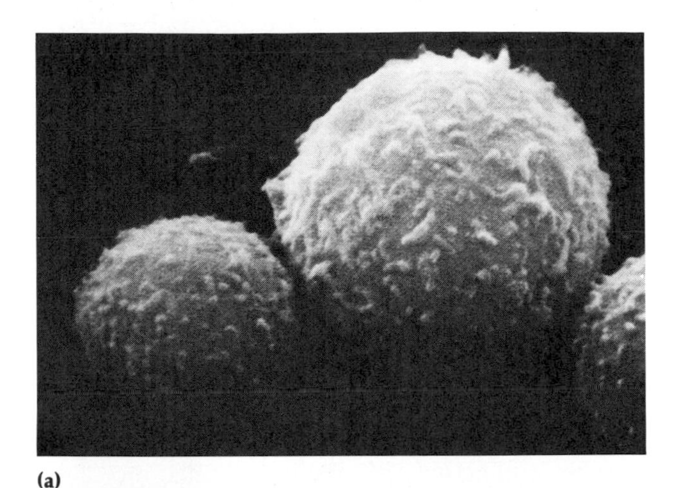

(a)

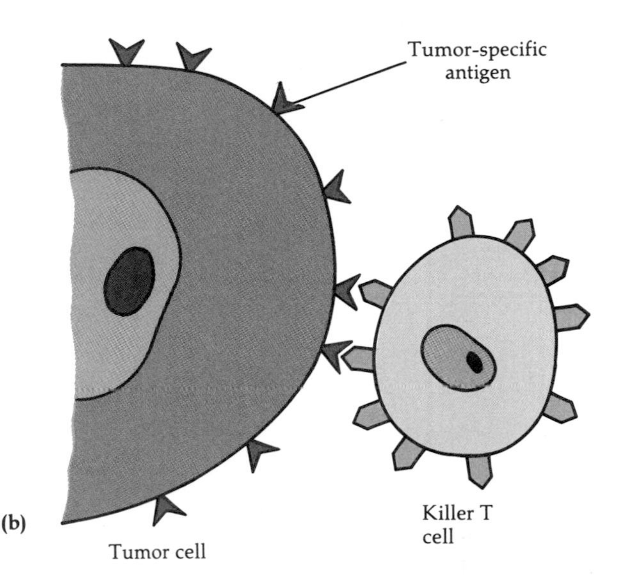

(b)

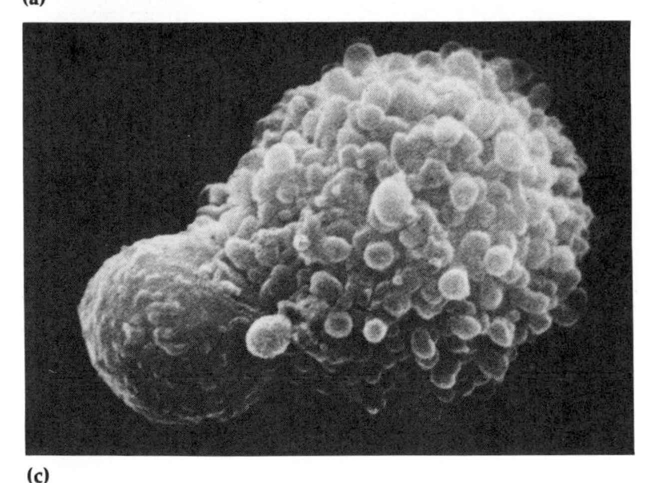

(c)

Figure 17–10 Interaction of a killer T cell and a tumor cell. **(a)** Killer T cell (smaller sphere at lower left) attaches to the tumor cell (×7250). **(b)** Diagram of a killer T cell binding to a tumor-specific antigen on the surface of a tumor cell. This is the mechanism thought to be responsible for the attachment shown in part (a). **(c)** Lysis of the tumor cell indicated by the deep folds in its surface membrane (×7250).

sensitized *killer T cells* react with tumor-specific antigens, initiating direct lysis of the tumor cells (Figure 17–10). More recently, sensitized macrophages bearing cytophilic antibodies have also been implicated in tumor destruction.

Immunologic Escape

In spite of the immune response mounted against tumor cells, some malignant cells possess the ability to escape destruction, a phenomenon called **immunologic escape.** A number of mechanisms have been proposed to explain this observation. In **antigen modulation,** tumor cells are believed to shed their specific antigens, thus evading recognition by humoral antibodies or sensitized killer T cells. In **immunologic enhancement,** the binding of humoral antibodies to tumor-specific antigens may prevent recognition by sensitized T cells, as shown in Figure 17–11.

Immunotherapy

The fact that it is possible to demonstrate an immune reaction to certain tumors forms the basis of a relatively new approach to cancer treatment called **immunotherapy.** In general, however, treating cancer by immunotherapy presents several problems. Injections of humoral antitumor antibodies are not practical, since it is cell-mediated

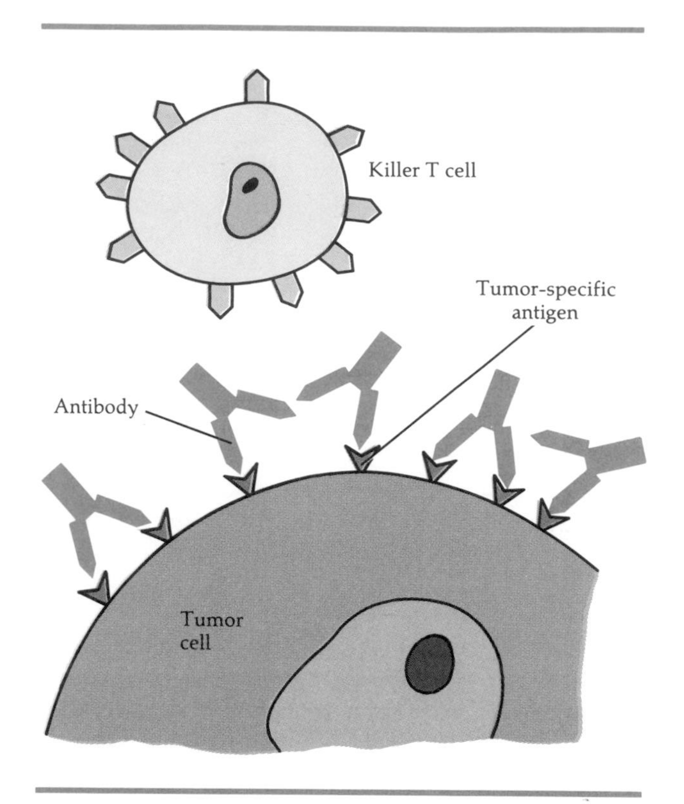

Figure 17–11 Immunologic enhancement. Humoral antibodies bound to tumor-specific antigens are thought to block binding by killer T cells.

immunity that apparently assumes the principal role against cancer cells. Moreover, injections of such antibodies into malignant tumors would probably trigger immunologic enhancement, making the situation even worse. Another approach considered is the transfer of live antitumor antigen lymphocytes from one patient to another. But such a transfer would activate the GVH response, affecting several body systems, and would probably be fatal. Other attempts to increase the number of cancer-fighting lymphocytes in a patient also would probably trigger immunologic escape mechanisms for the cancer cells.

One promising type of immunotherapy procedure employs the use of BCG vaccine to enhance the cell-mediated immune reactions to an antigen. BCG is a live, attenuated (weakened) strain of *Mycobacterium bovis*, which is the bacterium that causes bovine-transmitted tuberculosis. BCG vaccinations in animals act as an immunologic *adjuvant* and produce a generalized enhancement of immune responsiveness against a great variety of antigens. The effects of BCG include a great increase in the number and activity of phagocytic cells, increased humoral immunity, accelerated rejection of transplants, and increased resistance to infection. BCG has been found to cause tumor regression in animals and has been used, with limited success, in the treatment of certain human cancers accessible to injection (such as melanomas).

Immunology may also prove valuable in diagnosing and monitoring cancer. For example, many cancerous tumors release tumor-specific antigens into the blood. These antigens, given certain conditions, can be detected by radioimmunoassay (see Chapter 16), a procedure that can also be used to measure levels of serum antibody to a tumor-specific antigen. In this way, the progress of a particular cancer can be monitored.

It is entirely possible that at some future time, immunologic techniques will be used in the prevention of cancer. Once the etiologic agents of cancer have been identified, the development of vaccines against these agents will be a viable method of prevention.

STUDY OUTLINE

HYPERSENSITIVITY (ALLERGY) (p. 434)

1. Hypersensitivity reactions represent immunologic responses to an antigen (allergen), which lead to tissue damage rather than immunity.
2. Hypersensitivity reactions occur only when a person has been sensitized to an antigen.
3. Hypersensitivity reactions are divided into two groups: immediate and delayed.

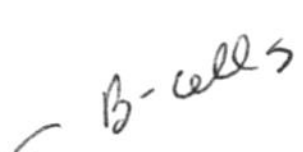

IMMEDIATE HYPERSENSITIVITY (pp. 434–438)

1. Immediate hypersensitivity is antibody mediated, develops rapidly in a sensitized host, and can be passively transferred in serum.
2. Types of immediate hypersensitivity reactions are anaphylactic, cytotoxic, and immune-complex reactions.
3. Anaphylaxis-type reactions involve the production of IgE antibodies that bind to target cells (basophils and mast cells) to sensitize the host.
4. Subsequent exposure to the antigen causes these target cells to release chemical mediators, such as histamine, which cause the observed allergic reactions.
5. Systemic anaphylaxis may develop in minutes after injection or ingestion of the antigen; this may result in respiratory failure.
6. Localized anaphylaxis is exemplified by hives, hayfever, and asthma.
7. Skin testing is useful in determining sensitivity to an antigen.
8. Desensitization to an antigen can be achieved by repeated injections of the antigen, which leads to the formation of blocking antibodies.

CELL-MEDIATED (DELAYED) HYPERSENSITIVITY (pp. 438–439)

1. Delayed hypersensitivity responses exemplified by the positive tuberculin skin test or contact sensitivities are cell-mediated reactions.
2. They develop slowly (one to three days) and can be passively transferred with sensitized lymphocytes.
3. Sensitized T cells secrete lymphokines in response to the appropriate antigen.
4. Lymphokines attract and activate lymphocytes and macrophages and initiate tissue damage.
5. Contact dermatitis is an allergic skin reaction to certain chemicals in the environment.
6. The environmental chemicals are probably haptens that combine with skin proteins; skin oils may act as adjuvants.
7. Identity of the allergens involved in contact dermatitis can be determined by the Vollmer patch test.

TRANSFUSION REACTIONS AND Rh INCOMPATIBILITY (pp. 439–443)

The ABO Blood Group System (pp. 439–442)

1. Human blood may be grouped into four principal types designated A, B, AB, and O.
2. The presence or absence of either one or two carbohydrate antigens (isoantigens) designated A and B on the surface of the red cell determines a person's blood type.

3. Naturally occurring IgM antibodies (isoantibodies) are present or absent in serum against the opposite isoantigen.
4. The transfusion of blood is most affected by the isoantigens on the donor's red cells and the isoantibodies in the recipient's serum.
5. Incompatible blood transfusions lead to the complement-mediated lysis of the donor red blood cells.

The Rh Blood Group System (p. 442)

1. Approximately 85% of the human population possesses another blood group antigen designated the Rh antigen.
2. The absence of this antigen in certain individuals (Rh^-) can lead to sensitization upon exposure to it.

Blood Transfusions and Rh Incompatibility (p. 442)

1. An Rh^+ person can receive Rh^+ or Rh^- blood transfusions.
2. When an Rh^- person receives Rh^+ blood, that person will produce anti-Rh antibodies.
3. Subsequent exposure to Rh^+ blood will result in a rapid hemolytic reaction.

Hemolytic Disease of the Newborn (pp. 442–443)

1. An Rh^- mother carrying an Rh^+ fetus will produce anti-Rh antibodies.
2. Subsequent pregnancies involving Rh incompatibility may result in hemolytic disease of the newborn.
3. The disease may be prevented by passive immunization of the mother with anti-Rh antibodies just after giving birth.

TRANSPLANTATION (pp. 443–446)

1. Transplantation is the replacement of injured or diseased tissues and organs with natural ones from living or cadaveric donors.
2. Four types of grafts (transplants) have been defined on the basis of genetic relationships between the donor and the recipient: autografts, isografts, allografts, and xenografts.
3. Rejection of transplanted tissue represents an immune response by the recipient to foreign tissue antigens on the transplanted tissue.
4. Histocompatibility antigens located on cell surfaces express genetic differences between individuals; these antigens are coded for the major histocompatibility complex (MHC), also called the human leukocyte antigen (HLA) gene complex.
5. To prevent tissue rejection in an allograft, MHC antigens and blood group antigens are matched as closely as possible.

6. Rejection may be suppressed by drugs, X-irradiation, or antilymphocyte serum.
7. Immunosuppression can lead to susceptibility to infectious disease, cancer, and graft versus host disease.

IMMUNOLOGIC TOLERANCE AND AUTOIMMUNITY (pp. 447–449)

Immunologic Tolerance (p. 447)

1. Immunologic tolerance represents a state of unresponsiveness to a specific antigen.
2. Tolerance to self-antigens occurs under natural conditions and develops during fetal development.
3. Tolerance to foreign (nonself) antigens is most easily achieved if antigenic exposure occurs when the immune system is immature (during fetal development or shortly after birth) and if such antigens are similar to the self-antigens.
4. Immunologic unresponsiveness (paralysis) may be induced in adults if repeated doses of antigen are presented to the immune system; constant exposure to the antigen is required to maintain tolerance.
5. Tolerance to certain microbial antigens develops late during the course of several chronic diseases, a condition termed anergy.
6. Suppressor T cells function in tolerance by repressing B cell differentiation or helper T cell function.

Autoimmunity (pp. 447–448)

1. Autoimmunity is a humoral or cell-mediated immune response against self-antigens.
2. Autoimmune responses frequently result in disease.

Mechanism for Autoimmunity (pp. 448–449)

1. The immune response may be against a formerly sequestered self-antigen.
2. Some antibodies formed against foreign antigens can cross-react with self-antigens (for example, rheumatic fever).
3. The immune response may be against a cell whose surface antigens have been altered by drugs or viruses.

IMMUNE RESPONSE TO CANCER (pp. 449–452)

1. Tumor cells are normal cells that have undergone transformation.
2. Transformation may occur spontaneously or from exposure to certain chemicals, UV radiation, or viruses.
3. Transformed cells fail to exhibit contact inhibition when grown in vitro and may produce tumors in vivo.

Tumor-Specific Antigens (p. 450)

1. Transformed cells possess tumor-specific antigens.
2. Examples include carcinoembryonic antigens (CEA) and alpha-feto protein (AFP) antigens.

Immunologic Surveillance (pp. 450–451)

1. The response of the immune system to cancer is called immunologic surveillance.
2. Tumor-specific antigens induce specific immunologic responses, with cell-mediated immunity the most important.

Immunologic Escape (p. 451)

1. The ability of tumor cells to escape immune responses mounted against them is called immunologic escape.
2. Antigenic modulation and immunologic enhancement represent two ways tumor cells escape detection and destruction by the immune system.

Immunotherapy (pp. 451–452)

1. Immunotherapy against cancer has serious drawbacks, triggering immunologic enhancement or GVH response.
2. BCG vaccination has been demonstrated to improve immunologic responsiveness and has been used to treat cancers accessible to injection.
3. Radioimmunoassay can be used to detect the presence of tumor-specific antigens.

STUDY QUESTIONS

REVIEW

1. Define hypersensitivity.
2. Compare and contrast the characteristics of immediate and delayed hypersensitivity reactions. Give an example of each type of hypersensitivity.
3. List five mediators released in anaphylactic hypersensitivities and explain their effects.
4. Contrast systemic and localized anaphylactic reactions. Which type is more serious? Cite an example of each.
5. Explain what happens when a person develops a contact sensitivity to poison oak.
   (a) What causes the observed symptoms?
   (b) How did the sensitivity develop?
   (c) How might this person be desensitized to poison oak?

6. Discuss the roles of isoantibodies and isoantigens in an incompatible blood transfusion.
7. What happens to the recipient of an incompatible blood type?
8. Explain how hemolytic disease of the newborn develops and how this disease might be prevented.
9. Which of the following blood transfusions are compatible? Explain your answers.

	Donor	Recipient
(a)	AB, Rh^-	AB, Rh^+
(b)	B, Rh^+	B, Rh^-
(c)	A, Rh^+	O, Rh^+

10. Which type of graft (autograft, isograft, allograft, or xenograft) is most compatible? Least compatible?
11. What three groups of antigens are tested when matching tissue for transplantation?
12. Define immunosuppression. Why is it used and what problems does it cause?
13. Define autoimmunity. Present three theories that have been formulated to explain autoimmune responses. Discuss one autoimmune disease in relation to one (or more) of these theories.
14. In what ways do tumor cells differ antigenically from normal cells? Explain how tumor cells may be destroyed by the immune system.
15. If tumor cells can be destroyed by the immune system, why isn't immunotherapy a simple, direct treatment for cancer?

CHALLENGE

1. When and how does our immune system discriminate between self and nonself antigens?
2. The first preparations used for artifically acquired passive immunity were horse serum antibodies. A complication that resulted from the therapeutic use of horse serum was serum sickness. Why did this occur?
3. After working in a mushroom farm for several months, a worker develops these symptoms: hives, edema, and swelling of lymph nodes.
   (a) What do these symptoms indicate?
   (b) What mediators cause these symptoms?
   (c) How may sensitivity to a particular antigen be determined?
   (d) Other employees do not appear to have any immunologic reactions. What could explain this?
   (*Note:* The allergen is conidiospores from molds growing in the mushroom farm.)

FURTHER READING

Bach, J. F. (ed.). 1978. *Immunology.* New York: Wiley. A collection of papers on immediate and delayed hypersensitivites, autoimmunity, and cancer.

Cunningham, B. A. October 1977. The structure and function of histocompatibility antigens. *Scientific American* 237:96–107. The role of cell-surface antigens in transplant rejection and protection against infections and cancer.

Notkins, A. L. and H. Koprowski. January 1973. How the immune response to a virus can cause cancer. *Scientific American* 228:22–31. Allergies, immune-complex diseases, and autoimmunity may be due to immune responses to viruses.

Old, L. J. May 1977. Cancer immunology. *Scientific American* 236:62–79. An explanation of why cancer cells can escape the immune system.

Parker, C. W. (ed.). 1980. *Clinical Immunology,* 2 vols. Philadelphia, PA: W. B. Saunders. A comprehensive textbook and reference for immunology emphasizing topics covered in this chapter.

Samter, M. (ed.). 1971. *Immunological Diseases,* 2nd ed., 2 vols. Boston, MA: Little, Brown and Company. Mechanisms and symptoms of hypersensitivities, atopic diseases, allergies, and other immunologic diseases.

18

Antimicrobial Drugs

Objectives

After completing this chapter you should be able to

- Define a chemotherapeutic agent and distinguish between a synthetic drug and an antibiotic.
- Identify the contributions of Ehrlich and Fleming in the field of chemotherapy.
- List the criteria used to evaluate antimicrobial agents.
- Identify five methods of action of antimicrobial agents.
- Describe the methods of action of each of the commonly used antimicrobial drugs.
- Explain the actions of currently used antiviral drugs.
- Describe two problems of chemotherapy for protozoan and helminthic infections.
- Describe three tests for microbial susceptibility to chemotherapeutic agents.
- Describe the mechanisms of drug resistance.

Sometimes the balance between microorganism and host tilts in the direction of disease to the host, and the normal defenses of the human body cannot prevent or overcome the disease. In such cases, we may turn to **chemotherapy,** the treatment of disease with chemicals (drugs) taken into the body. The term **chemotherapeutic agent** applies to any drug used for any disease—whether a simple headache, a strep throat, malaria, high blood pressure, or cancer. For example, aspirin and penicillin are both chemotherapeutic agents.

In this chapter, we limit our discussion to **antimicrobial drugs,** the class of chemotherapeutic agents used to treat infectious diseases. Like the disinfectants discussed in Chapter 7, these chemicals act by interfering with the growth of infectious microorganisms. Unlike disinfectants, they must act within the host. For this reason, the question of their effect on cells and tissues of the host is important; the ideal antimicrobial drug kills the harmful microorganism without damaging the host.

The drugs used in the chemotherapy of infectious disease fall into two groups. Some drugs have been synthesized by chemical procedures in the laboratory and are called **synthetic drugs.** Others are produced by bacteria and fungi and are called **antibiotics.**

HISTORICAL DEVELOPMENT

The birth of modern chemotherapy was largely due to the efforts of Dr. Paul Ehrlich, in Germany, during the early part of this century. While attempting to stain bacteria without staining the surrounding tissue, he speculated about some "magic bullet" that would selectively find and destroy pathogens but not harm the host. This idea provided the basis for *chemotherapy,* a term he coined. Ehrlich eventually found a chemotherapeutic agent called *salvarsan,* an arsenic derivative effective against syphilis (at least by the standards of the day). Prior to Ehrlich's discovery, there had really been only one effective chemotherapeutic agent in the medical arsenal—*quinine,* for the treatment of malaria.

In the late 1930s, the miracle sulfa drugs touched off a new interest in chemotherapy. The sulfa drugs had come out of a systematic survey of chemicals, which included many synthesized derivatives of aniline dyes. The chemical *prontosil,* which had actually been first synthesized as a dye many years before, was found to be an effective antimicrobial agent. Oddly enough, it worked only within living organisms, such as rabbits, and was completely ineffective in the test tube against the same bacteria. It turned out that the active ingredient was *sulfanilamide,* formed as the rabbit metabolized prontosil. This rapidly led to the development of a number of related drugs, the *sulfonamides,* or *sulfa drugs,* which are still used today. The sulfa drugs closely resemble a bacterial metabolite known as para-aminobenzoic acid (PABA, see Figure 18–5), which susceptible bacteria require to synthesize the vitamin folic acid. Sulfa drugs competitively inhibit the incorporation of PABA into folic acid in bacteria. Since humans meet their need for folic acid by ingesting it in food, they are not adversely affected by sulfonamides.

In London, in 1929, Dr. Alexander Fleming observed the inhibition of growth of the bacterium *Staphylococcus aureus* in the area surrounding the colony of a mold that had contaminated a Petri plate (Figure 18–1). The mold was identified as *Penicillium notatum,* and the active compound, which was isolated a short time later, was named *penicillin.* Similar inhibitions between colonies on solid media are commonly observed in microbiology and the mechanism is referred to as *antibiosis.* From this we get the term **antibiotic,** defined as a substance that is produced by microorganisms and that in small amounts will inhibit another microorganism. Technically, therefore, the wholly synthetic sulfa drugs are not antibiotics.

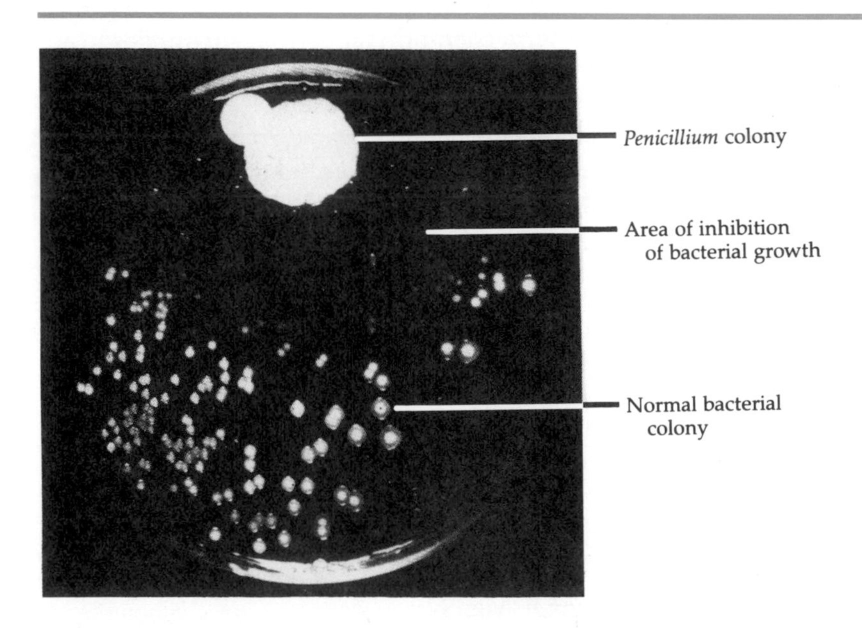

Figure 18–1 Discovery of penicillin. Near the contaminating colony of *Penicillium,* the growth of *Staphylococcus aureus* is inhibited, and cells are lysing. This is a photograph of a plate from Fleming's laboratory.

Since microorganisms capable of antibiosis are fairly common, it is surprising that antibiotics were not developed much earlier. Indeed, there is evidence in the scientific literature that penicillin had been seriously investigated in the late 1800s. But the researchers then had the same problem Fleming had later. Because the extract of the mold was unstable and production by the mold unreliable, it was simply not very useful clinically. In 1940, a group of scientists at Oxford University headed by H. Florey and E. Chain succeeded in the first clinical trials with penicillin. Intensive research in the United States then led to the isolation of especially productive *Penicillium* strains for use in mass production of the antibiotic. (The most famous of these high-producing strains was originally isolated from a cantaloupe bought at a market in Peoria, Illinois.) Penicillin is still one of our most effective antibiotics, and its enormous success led to the search for others, a search that revolutionized medicine.

Although many antibiotics have been developed in the last few decades (Table 18–1), relatively few are used in chemotherapy. This is because many antibiotics damage body cells in concentrations needed to kill microorganisms, while others are too limited in their action. As with disinfectants and antiseptics, there are certain criteria for selecting a chemotherapeutic agent.

Table 18–1 Representative Sources of Antibiotics from Microbes

Microorganism	Antibiotic (Trademarked Drug Names in Parentheses)
Gram-Positive Rods:	
Bacillus subtilis	Bacitracin
Bacillus polymyxa	Polymyxin
Bacillus brevis	Tyrothricin
Actinomycetes:	
Streptomyces parvullus	Actinomycin D (Dactinomysin)
Streptomyces nodosus	Amphotericin B (Fungizone)
Streptomyces venezuelae	Chloramphenicol (Chloromycetin)
Streptomyces aureofaciens	Chlortetracycline (Aureomycin), Demeclocycline (Declomycin), and Tetracycline
Streptomyces erythraeus	Erythromycin (Ilotycin)
Streptomyces kanamyceticus	Kanamycin (Kantrex)
Streptomyces fradiae	Neomycin
Streptomyces noursei	Nystatin (Mycostatin)
Streptomyces rimosus	Oxytetracycline (Terramycin)
Streptomyces griseus	Streptomycin
Micromonospora purpureae	Gentamicin
Nocardia lurida	Ristocetin (Spontin)
Fungi:	
Arachniotus aureus	Aranotin
Cephalosporium	Cephalothin (Keflin)
Penicillium griseofulvum	Griseofulvin (Fulvisin)
Penicillium notatum	Penicillin

CRITERIA FOR ANTIMICROBIAL DRUGS

The value of a chemotherapeutic agent can be measured against a number of criteria, including the following:

1. The drug should demonstrate *selective toxicity.* This means that it should be toxic for the microorganism but not for the host. With some exceptions (for example, penicillin), most chemotherapeutic agents are somewhat toxic to host cell metabolism, but usually not toxic enough to interfere with normal body functions.

2. The drug should *not produce hypersensitivity* in most hosts (see Chapter 17). A serious problem that arises with many chemotherapeutic agents—and penicillin is a major offender—is that individuals may develop a sensitivity to them. This sensitivity is usually characterized by symptoms typical of allergic reactions, such as skin rashes and fever. However, some reactions are more serious, even resulting in fatal anaphylactic shock.

3. A drug must have *solubility* in body fluids so that it can rapidly penetrate body tissues. The *rates at which the drug is broken down and/or excreted from the body* must be low enough that the drug remains in the infected body tissue long enough to exert its effects. Furthermore, a good chemotherapeutic agent has a *long shelf life* at normal refrigeration temperatures.

4. Use of the drug should not too readily lead to the development of *drug resistance,* where the

drug is no longer effective in suppressing the growth of the microorganism. Resistance may be due to any number of mechanisms and will be discussed later in this chapter.

5. The drug should *not eliminate the normal flora* of the host. When flora that ordinarily compete with and check the growth of potentially pathogenic flora are destroyed, the potential pathogens flourish and may cause a secondary infection called a *superinfection*. Examples of opportunistic pathogens that may cause superinfections are *Candida albicans* and *Pseudomonas*.

6. Because it is not often practical to identify the specific microorganism causing a disease before starting treatment, drugs with a broad spectrum of activity are extremely valuable. Such *broad-spectrum drugs* are active against many pathogenic microorganisms. For example, broad-spectrum antibacterial drugs are usually effective against a large number of gram-positive and gram-negative species of bacteria. One problem with broad-spectrum drugs, however, is that they are especially likely to destroy normal flora. Other drugs, called *narrow-spectrum drugs*, are effective against only a few species of microorganisms (see Table 18–2).

Table 18–2 Representative Broad-Spectrum and Narrow-Spectrum Antibiotics

Antibiotic	Common Brand Name
Broad-Spectrum:	
Chloramphenicol	Chloramycetin
Chlortetracycline	Aureomycin
Demeclocycline	Declomycin
Oxytetracycline	Terramycin
Tetracycline	—
Kanamycin	Kantrex
Ampicillin	—
Cephalothin	Keflin
Gentamicin	—
Narrow-Spectrum:	
Penicillin	—
Streptomycin	—
Erythromycin	Ilotycin
Polymyxin B	—
Nystatin	Mycostatin

Not all of these properties are displayed by every chemotherapeutic agent. However, the best possible combination is sought.

ACTION OF ANTIMICROBIAL DRUGS

Antimicrobial drugs interfere chemically with the synthesis or function of vital components of microorganisms. The effectiveness of a particular drug depends, in part, on whether the disease-causing microorganism and host cell share the same kinds of molecules affected by the drug, and how important the common molecules are to the functioning of the microbial and host cells. Fortunately, the cellular structures and functions of procaryotic cells such as bacteria differ significantly from the cellular structures and functions of the eucaryotic cells of the human body (Chapter 4). These differences provide the basis for selective toxicity, and we have a much wider array of chemotherapeutic agents against bacteria than we do against the eucaryotic fungi, protozoans, and multicellular parasites. Because of the similarities between the eucaryotic cells of these microorganisms and those of humans, eucaryotic microorganisms are more difficult to harm without also damaging the human host. The same can be said for viruses, which multiply within host cells. Antimicrobial drugs may either kill microorganisms outright or simply prevent them from growing. (Thus, like the chemical agents described in Chapter 7, drugs effective against bacteria are categorized as bactericidal or bacteriostatic.) In cases where the drug merely inhibits microbial growth, the host is given time to call on its own defenses, such as phagocytosis and antibody production, to destroy the microorganisms.

We will now examine the various ways in which antibiotics exhibit their antimicrobial activity.

Inhibition of Cell Wall Synthesis

You may recall from Chapter 4 that the cell wall of bacteria consists of a macromolecular network called peptidoglycan. This chemical is composed of two sugars (N-acetylglucosamine and N-acetylmuramic acid), peptide cross bridges, and tetrapeptide side chains (see Figure 4–8). The action of penicillin and certain other antibiotics prevents the formation of the peptide cross bridges in actively growing bacteria. As a result, the cell wall is greatly weakened and the cell lyses (Figure 18–2). Since only actively growing cells are synthesizing cell walls, only actively growing cells are affected by antibiotics like penicillin. And since human cells do not have peptidoglycan cell walls, penicillin has very little toxicity for the host cells, a fact that causes penicillin to rank high as a model antibiotic. You may also recall that gram-negative bacteria have additional layers around their cell walls. These layers prevent penicillin from reaching the site of action, which explains why many gram-negative bacteria are resistant to penicillin. Other antibiotics that inhibit cell wall synthesis in ways other than preventing the synthesis of peptide cross bridges are cephalosporins, cycloserine, bacitracin, vancomycin, novobiocin, and ristocetin.

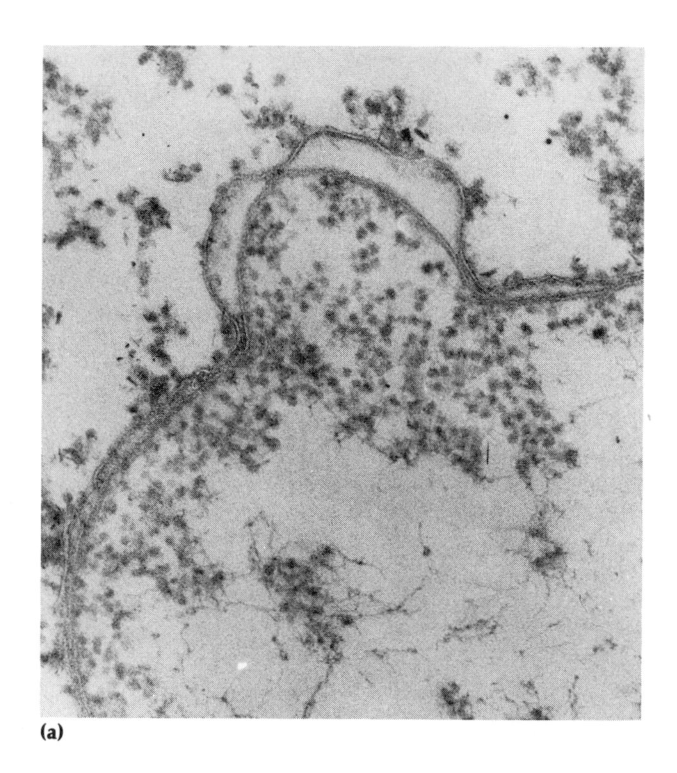
(a)

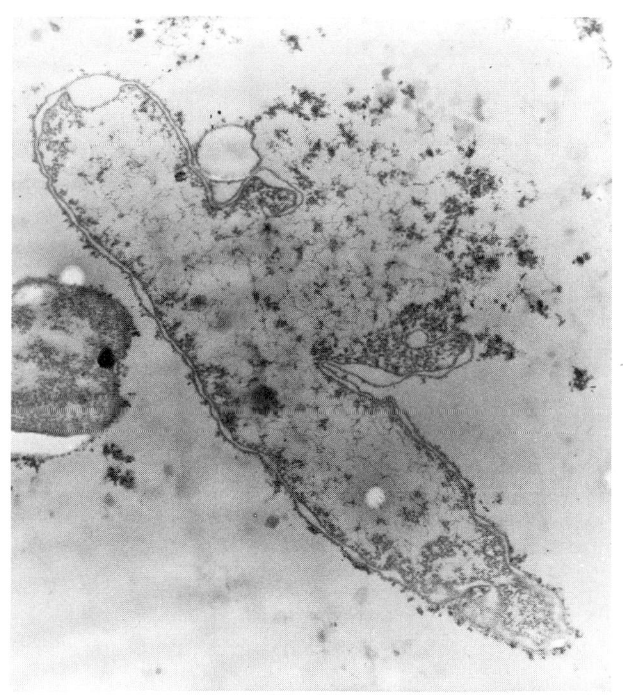
(b)

Figure 18–2 Effect of penicillin on a bacterial cell wall. **(a)** After treatment with penicillin for 20 minutes, a bulge has developed in the wall of this cell of *Escherichia coli* (×100,000). **(b)** Lysis of an *E. coli* cell through a portion of weakened cell wall (×16,500).

Inhibition of Protein Synthesis

Whereas procaryotic cells and eucaryotic cells can easily be distinguished by the presence of a cell wall, their differentiation based on protein synthesis is much more difficult, since protein synthesis is a common feature of all cells. One notable difference, however, is the structure of the ribosomes. As discussed in Chapter 4, eucaryotic cells have 80S ribosomes; procaryotic cells, with different ribosomal proteins and RNA, have 70S ribosomes. (Recall that the 70S ribosomes are made up of 50S and 30S units.) This difference is sufficient to account for the selective toxicity of antibiotics that affect protein synthesis. However, remember that some eucaryotic organelles, like mitochondria, also contain 70S ribosomes. Thus, the antibiotics specifically designed to inhibit procaryotic protein synthesis may also have adverse effects on host cells, causing certain side effects.

Among the antibiotics that interfere with protein synthesis are chloramphenicol, lincomycin, erythromycin, tetracyclines, and streptomycin (Figure 18–3).

Chloramphenicol and lincomycin These antibiotics react with the 50S portion of the 70S procaryotic ribosome, where they inhibit peptide bond formation in the growing polypeptide chain.

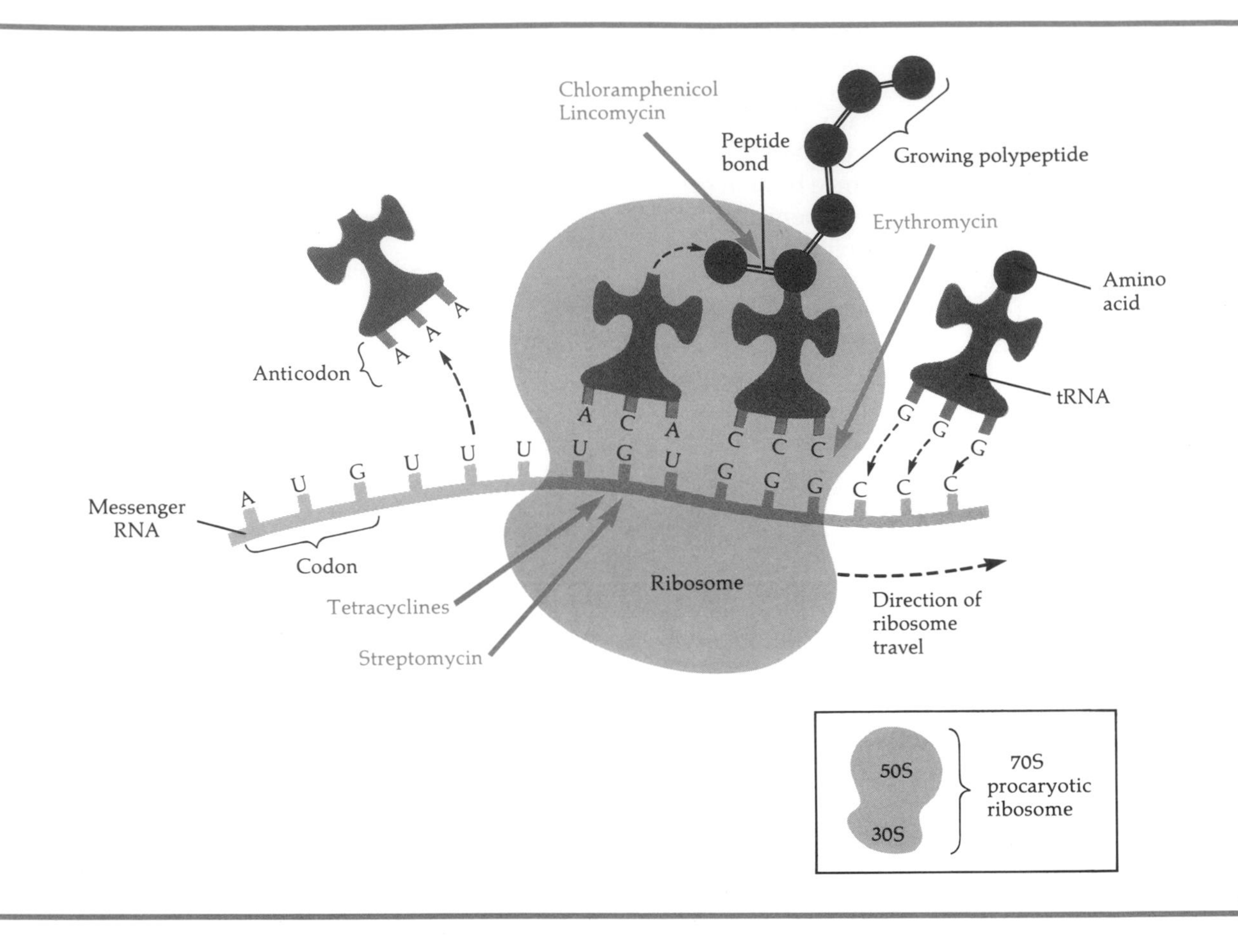

Figure 18–3 Inhibition of protein synthesis by antibiotics. The arrows indicate the specific points where chloramphenicol, lincomycin, erythromycin, tetracyclines, and streptomycin exert their activity.

Erythromycin This antibiotic reacts with the 50S portion of the 70S procaryotic ribosome, where it stops protein synthesis by freezing the ribosome so that it cannot move along the mRNA.

Tetracyclines These antibiotics react with the 30S portion of the 70S procaryotic ribosome by interfering with the attachment of the amino acid carrying tRNA to the ribosome, preventing the addition of amino acids to the growing polypeptide chain.

Streptomycin This antibiotic inhibits the initial steps of protein synthesis by changing the shape of the 30S portion of the 70S procaryotic ribosome, thereby causing the genetic code on the mRNA to be read incorrectly.

Injury to the Plasma (Cytoplasmic) Membrane

Certain antibiotics, especially polypeptide antibiotics, bring about changes in plasma membrane permeability that result in the loss of important metabolites from the microbial cell. One such antibiotic is polymyxin B, which attaches to the phospholipids of the plasma membrane, causing membrane disruption. Gramicidin acts in a similar way.

Two antifungal drugs, nystatin and amphotericin B, are effective against many systemic mycoses, combining with sterols in the fungal plasma membrane to cause disruption. Since bacterial plasma membranes generally lack sterols, these antibiotics do not act on bacteria. But since animal cell plasma membranes do contain sterols, it is not surprising that high doses of these drugs can produce toxic side effects, such as lysis of red blood cells (hemolytic anemia).

Inhibition of Nucleic Acid Synthesis

A number of antibiotics interfere with the nucleic acid metabolism of microorganisms. However, many of these drugs are also toxic for host cells because of the similarities between microbial and host cell DNA and RNA metabolism. Thus, most of these drugs have only a limited clinical application, their greatest use being in research related to nucleic acid metabolism. Most of these antibiotics inhibit DNA or RNA synthesis. Among the more clinically useful drugs are rifamycin, nalidixic acid, 5-fluorocytosine, and trimethoprim, which are discussed later in this chapter.

Inhibition of Enzymatic Activity

You may recall from Chapter 5 that an enzymatic activity of a microorganism can be competitively inhibited by a substance (antimetabolite) that closely resembles the normal substrate for the enzyme. An example already considered is the relationship between the antimetabolite sulfanilamide (a sulfa drug) and para-aminobenzoic acid (PABA). In many microorganisms PABA is the substrate for an enzymatic reaction leading to the synthesis of folic

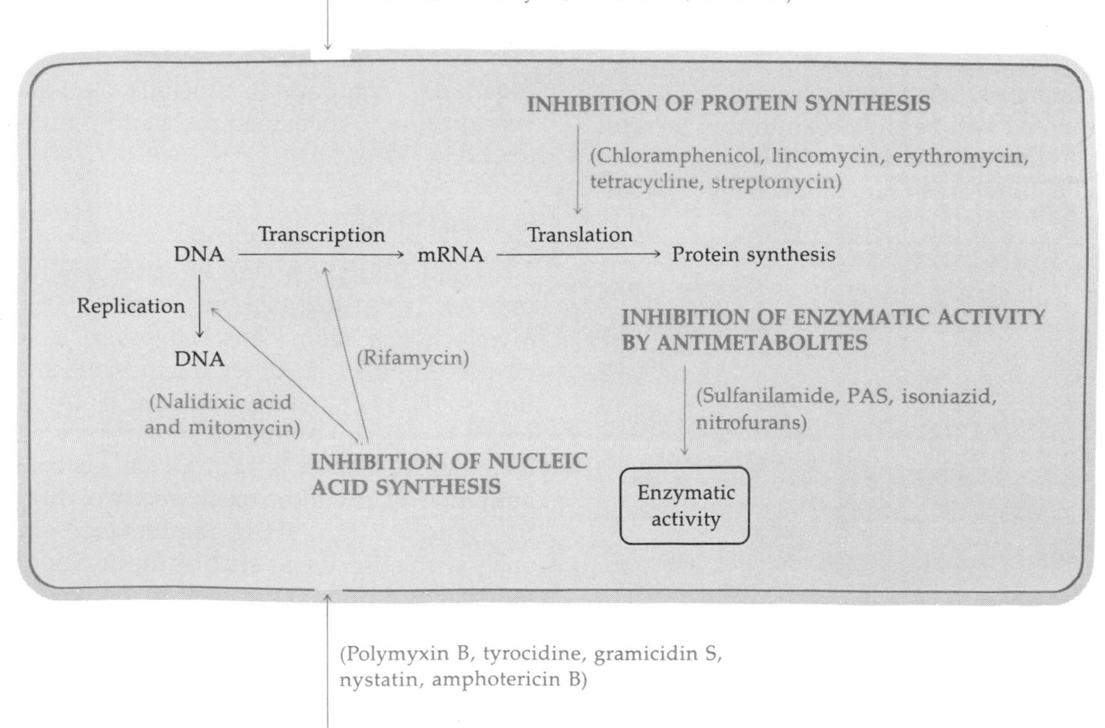

Figure 18–4 Summary of the actions of antimicrobial drugs in a highly diagrammatic, composite microbial cell. All the drugs shown act selectively on procaryotic (bacterial) cells, except for nystatin and amphotericin B, which act on eucaryotic (fungal) cells.

acid, a coenzyme required for the synthesis of purine and pyrimidine nitrogen bases and some amino acids. In the presence of sulfanilamide, the enzyme that normally converts PABA to folic acid combines with the drug instead of with PABA. This prevents folic acid synthesis and stops the growth of the microorganism. Since humans do not produce folic acid from PABA (they obtain it as a vitamin in ingested foods), sulfanilamide exhibits selective toxicity by affecting microorganisms that synthesize their own folic acid without harming the human host. Many bacteria, some fungi, and some protozoans must synthesize folic acid. Other chemotherapeutic agents that act as antimetabolites are para-aminosalicylic acid (PAS), isoniazid, and nitrofurans.

A summary of the actions of antimicrobial drugs is given in Figure 18–4 and Table 18–3.

SURVEY OF COMMONLY USED ANTIMICROBIAL DRUGS

We will now examine the properties and activities of some commonly used antimicrobial drugs. The drugs discussed will be divided into four groups: synthetic antimicrobial drugs, antibiotics, antiviral drugs, and antiprotozoan and antihelminthic drugs.

Table 18–3 Summary of Actions of Chemotherapeutic Agents

Chemotherapeutic Action	Example
Inhibition of cell wall synthesis	Penicillin, cephalosporins, cycloserine, bacitracin, vancomycin, novobiocin, ristocetin (Spontin)
Inhibition of protein synthesis	Chloramphenicol (Chloromycetin), lincomycin (Lincocin), erythromycin (Ilotycin), tetracycline, streptomycin
Injury to plasma membrane	Polymyxin B, tyrocidine, gramicidin S, nystatin (Mycostatin), amphotericin B
Inhibition of nucleic acid synthesis	Rifamycin, nalidixic acid, 5-fluorocytosine, trimethoprim, mitomycin, actinomycin D (Dactinomycin)
Inhibition of enzymatic activity	Sulfanilamide, para-aminosalicylic acid (PAS), isoniazid, nitrofurans

Synthetic Antimicrobial Drugs

Sulfonamides

As noted earlier, **sulfonamides (sulfa drugs)** were among the first synthetic antimicrobial drugs to be used to treat microbial disease. But due to the widespread use of these drugs, many pathogens have developed resistance to them. Moreover, sulfonamides cause allergic reactions in many people. For these reasons, antibiotics have generally replaced the sulfonamides as antimicrobial drugs.

A very important characteristic of sulfonamides is their rapid absorption (uptake), especially by the meninges, the covering around the brain and spinal cord. Thus, sulfonamides are still used to treat bacterial meningitis (inflammation of the membranes around the brain and spinal cord). They also continue to be used to treat certain urinary tract infections. Sulfonamides are bacteriostatic, and their action is due to their structural similarity to para-aminobenzoic acid (PABA). They prevent the synthesis of folic acid by competitive inhibition, thereby inhibiting the synthesis of certain amino acids and nucleotides. Important sulfonamide compounds include sulfanilamide, sulfadiazine, sulfasuxidine, and sulfisoxazole (Figure 18–5).

Isoniazid (INH)

Isoniazid (INH) is a very effective synthetic antimicrobial drug that exhibits bacteriostasis against *Mycobacterium tuberculosis*. However, it seems to have little effect on other microorganisms. Structurally, INH is similar to vitamin B_6 (pyridoxine), which serves as a coenzyme in metabolic reactions. Its mode of action is thought to be competitive inhibition of chemical reactions involving vitamin B_6 (Figure 18–6). INH is administered simultaneously with other drugs such as the antibiotics streptomycin and rifampin to treat tuberculosis.

Para-aminosalicylic acid (PAS)

Like the sulfonamides, **para-aminosalicylic acid (PAS)** exhibits its action as an enzyme inhibitor that competes with PABA (Figure 18–7). In the treatment of tuberculosis, PAS is administered simultaneously with streptomycin to reduce the development of bacterial strains resistant to the streptomycin.

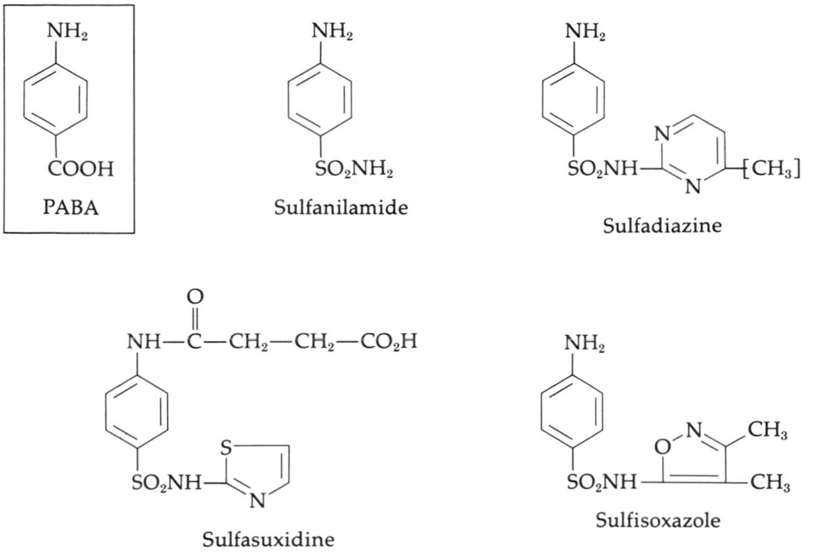

PABA

Sulfanilamide

Sulfadiazine

Sulfasuxidine

Sulfisoxazole

Figure 18–5 Structure of representative sulfonamides. These drugs are competitive inhibitors of an enzyme whose normal substrate is PABA, shown for comparison.

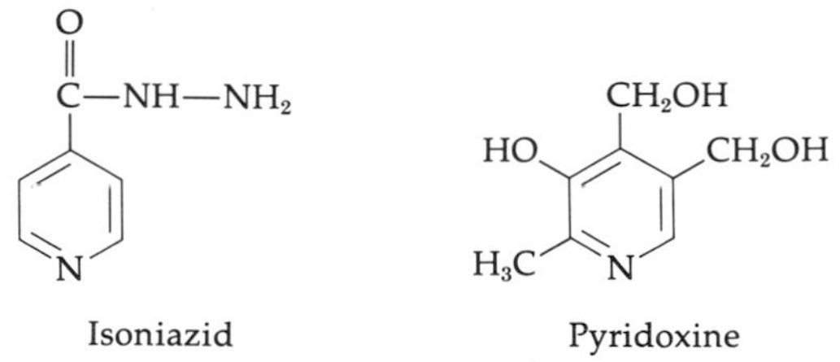

Isoniazid

Pyridoxine

Figure 18–6 A comparison of the structure of isoniazid (INH) and pyridoxine (Vitamin B_6). INH is a competitive inhibitor of enzymes that act on Vitamin B_6.

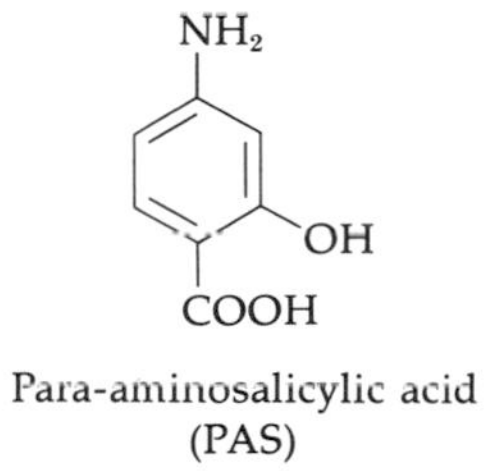

Para-aminosalicylic acid (PAS)

Figure 18–7 Structure of para-aminosalicylic acid (PAS). Compare this structure with that of PABA, shown in Figure 18–5.

Ethambutol

Ethambutol is also used to treat tuberculosis, often in combination with INH. It is believed that ethambutol acts as a competitive inhibitor that interferes with the synthesis of RNA. Like PAS, ethambutol decreases the development of resistant bacterial strains. And it has the added advantage of being less toxic than PAS when given in small doses.

Other synthetic antimicrobial drugs

Nitrofurans are active against a large number of gram-positive and gram-negative bacteria, protozoans, and fungi. Moreover, they exhibit little toxicity. Their mode of action is thought to be interference with acetyl coenzyme A. Nitrofurans are used to treat urinary tract infections, intestinal infections, burns, and vaginitis.

A number of synthetic antimicrobial drugs act by inhibiting DNA synthesis. **Nalidixic acid** stops bacterial DNA synthesis and is especially effective against gram-negative enterics. Because this drug accumulates in the urine, it is used to treat urinary tract infections in which *Escherichia*, *Proteus*, and *Klebsiella* species are implicated. Its precise mode of action is unknown but may involve damage to the plasma membrane, to which the machinery for bacterial DNA synthesis is attached.

The drug **5-fluorocytosine** is used exclusively to treat fungal infections. It probably functions as an antimetabolite of cytosine, the pyrimidine base, but it is not metabolized significantly by humans when given orally.

Trimethoprim is a broad-spectrum drug that interferes with nucleic acid synthesis by inhibiting an enzyme needed for purine and pyrimidine synthesis. It is frequently used in conjunction with sulfonamides, and the combination of two drugs affecting different metabolic steps has the advantage of reducing the development of resistant bacterial strains.

DNA synthesis is an essential activity for all organisms, and it is not surprising that many inhibitors of DNA synthesis harm both the microorganism and the human host. Several drugs have been found, however, that have selective toxicity for certain microorganisms, presumably because of permeability or metabolic differences between microbial and human cells.

Antibiotics

Penicillin

The term **penicillin** refers to a group of chemically related antibiotics. Penicillin extracted from cultures of *Penicillium* exists in six closely related forms. These are the so-called *natural penicillins*, designated as penicillin F, G, X, K, O, and V. Of these, penicillin G is the most stable and the most active against microorganisms. Penicillins act by inhibiting cell wall synthesis in actively growing bacteria.

Natural penicillins have several limitations. When taken orally, most are quickly destroyed by

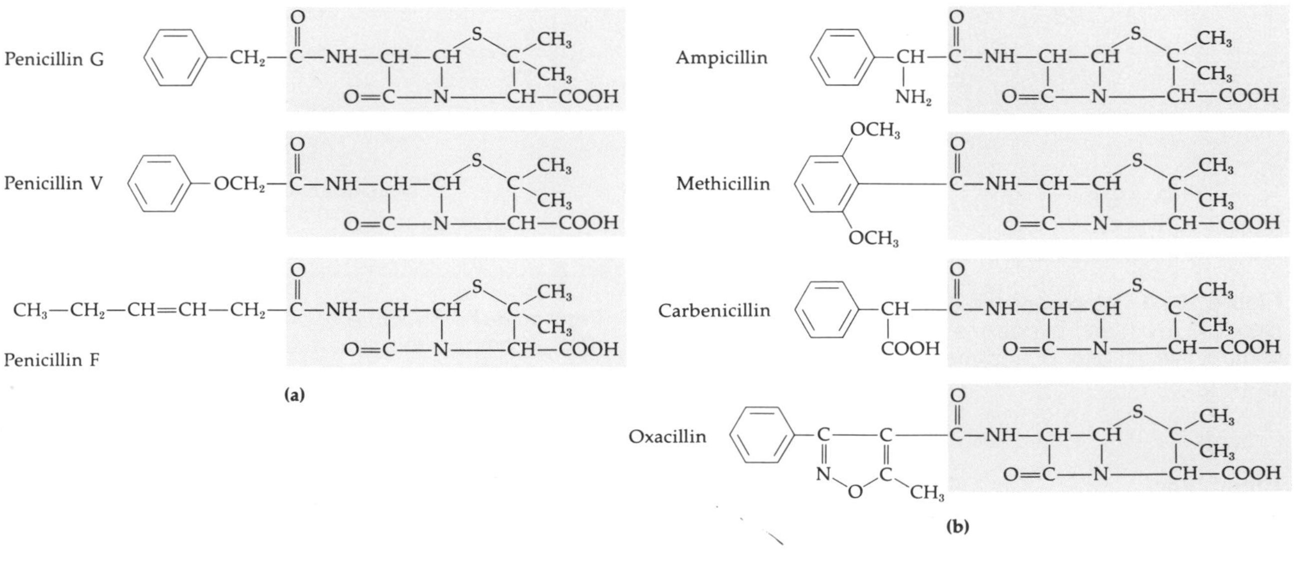

Penicillin G
Penicillin V
Penicillin F
(a)

Ampicillin
Methicillin
Carbenicillin
Oxacillin
(b)

Figure 18–8 Structure of penicillins. The shaded portions of the diagrams represent the common nucleus, which includes the beta-lactam ring. The unshaded portions represent the side chains that distinguish one penicillin from another. **(a)** Natural penicillins. **(b)** Semisynthetic penicillins.

stomach acid. Also, their effect is almost entirely limited to gram-positive bacteria. Moreover, natural penicillins are destroyed by *penicillinase,* an enzyme produced by many bacteria, especially *Staphylococcus.* On the other hand, natural penicillins have certain advantages over other antibiotics, one of these being the almost total absence of toxicity to human tissue, except in the small percentage of people who are allergic to penicillin.

In an attempt to overcome the disadvantages of natural penicillins while still retaining their desirable features, scientists developed the so-called *semisynthetic penicillins.* As noted earlier, natural penicillins are a group of chemically related substances, and what they all have in common is a nucleus called a beta-lactam ring. Depending on which chemicals (side chains) are attached to the common nucleus, different types of natural penicillins result (Figure 18–8a). In the development of semisynthetic penicillins, scientists perfected a technique whereby they could stop synthesis by *Penicillium* to obtain only the common penicillin nucleus, or remove the side chains from the completed natural molecules and then chemically add other side chains. Thus the term semisynthetic: part of the penicillin is produced by the mold and part is added synthetically.

Semisynthetic penicillins have overcome a number of the problems of natural penicillins. They may be acid-resistant (and therefore not readily destroyed by stomach acid) and resistant to penicillinase, and their spectrum of activity is greater than that of the natural penicillins. Commonly used semisynthetic penicillins are *ampicillin* (broad-spectrum and acid-resistant), *methicillin* (penicillinase resistant), *carbenicillin* (broad spectrum), and *oxacillin* (both penicillinase- and acid-resistant) (Figure 18–8b).

Aminoglycosides

Aminoglycosides are a group of antibiotics consisting of amino sugars and an aminocyclitol ring held together by glycoside bonds. Common examples are streptomycin (Figure 18–9a), neomycin, kanamycin, and gentamicin (Figure 18–9b). All are bactericidal, and their mode of action is the alteration of bacterial ribosome structures, causing inhibition of protein synthesis or misreading of the genetic code.

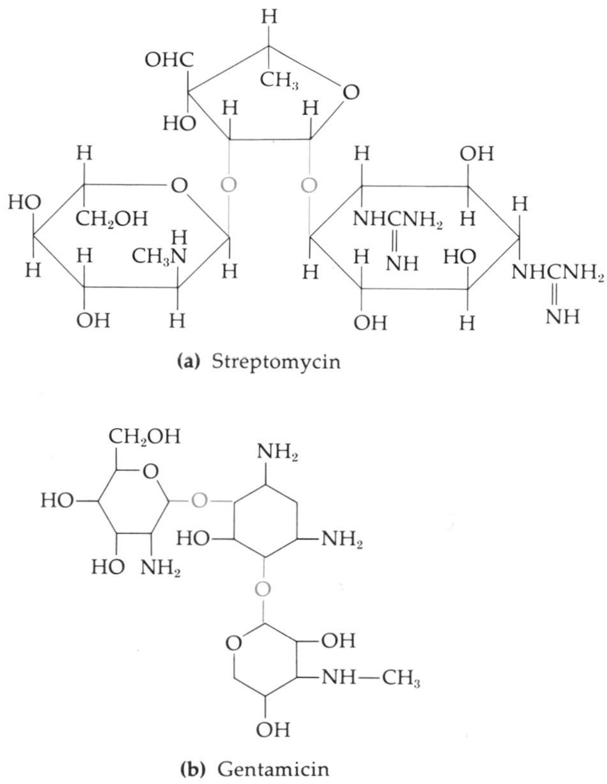

(a) Streptomycin

(b) Gentamicin

Figure 18–9 Structure of representative aminoglycosides. **(a)** Streptomycin. **(b)** Gentamicin. Glycosidic bonds (color) are bonds between sugars.

Streptomycin is primarily effective against gram-negative bacteria and *Mycobacterium tuberculosis.* It is also used in combination with penicillin against alpha-hemolytic streptococci to treat bacterial endocarditis. Unfortunately, practically all species of bacteria can become resistant to streptomycin. For this reason, it is usually administered with PAS and INH in treating tuberculosis. Still another problem with streptomycin is how poorly it is absorbed from the digestive tract, necessitating administration by injection. And finally, streptomycin can produce toxic effects, causing kidney damage and damage to the auditory portion of the vestibulocochlear nerve, resulting in deafness.

Neomycin is one of the few antibiotics that can be purchased without a prescription. It is used almost exclusively as a topical ointment for treating minor wounds, usually in combination with two other antibiotics, polymyxin B and bacitracin. *Kanamycin* is a broad-spectrum antibiotic effective against many strains of staphylococci. It is used to treat infantile gastroenteritis and as a suppressant of the intestinal flora prior to surgery. Its toxicity is similar to that of streptomycin—kidney damage and auditory nerve damage. *Gentamicin* is a newer aminoglycoside that is quite effective against most gram-negative bacteria. It is the drug of choice for treating coliform infections of the urinary tract and systemic infections caused by *Pseudomonas,* which is often highly resistant to antibiotics.

Cephalosporins

Structurally, the nucleus of **cephalosporins** resembles that of penicillin (Figure 18–10). Cephalosporins are similar in action to penicillins, inhibiting cell wall synthesis. In addition, they are effective against some gram-negative bacteria. Although cephalosporins are structurally similar to penicillin, they are sufficiently different so that they are resistant to penicillinase. For this reason, they can be used on strains of staphylococci that are resistant to penicillin and can be administered to people who are allergic to penicillin.

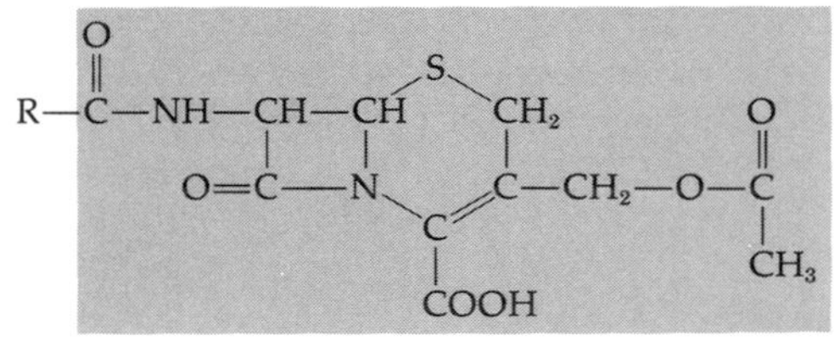

```
        O
        ‖          S
R—C—NH—CH—CH      CH2          O
            |     |     |            ‖
         O=C——N       C—CH2—O—C
                  C                  |
                  |                 CH3
                COOH
```

Cephalosporin nucleus

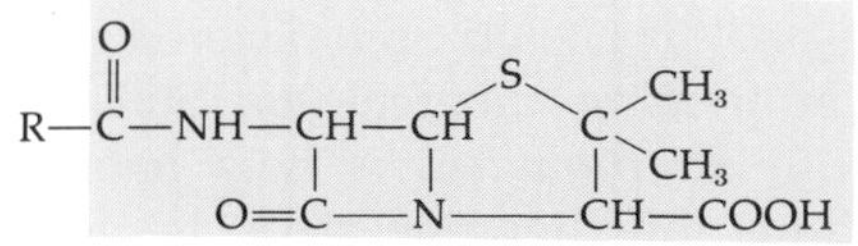

```
        O
        ‖          S       CH3
R—C—NH—CH—CH       C
            |     |     |  CH3
         O=C——N———CH—COOH
```

Penicillin nucleus

Figure 18–10 Comparison of the nucleus of cephalosporins with that of penicillin. R is an abbreviation for groups of atoms that make one compound different from another.

Tetracyclines

Tetracyclines are a group of closely related broad-spectrum antibiotics that act by inhibiting protein synthesis. They are bacteriostatic. They are not only effective against gram-positive and gram-negative bacteria, but also against rickettsias and chlamydias. Three of the more commonly encountered tetracyclines are *oxytetracycline (Terramycin), tetracycline,* and *chlortetracycline (Aureomycin)* (Figure 18–11).

Tetracyclines have the broadest spectrum of antibacterial activity of any known drug. Unfortunately, this has led to their indiscriminate use in the past, resulting in several problems. A number of bacterial species once susceptible to the drug are now resistant, including *Neisseria gonorrhoeae, Streptococcus pyogenes,* and species of *Bacteroides.* Another problem relates to suppression of the normal flora of the digestive and respiratory tracts. Excessive use of tetracyclines destroys much of the normal flora, resulting in the overgrowth of microorganisms such as *Proteus* and *Klebsiella* species, *Pseudomonas aeruginosa,* and *Staphylococcus aureus.* This, in turn, results in superinfection caused by one's own normal flora. Tetracyclines also produce toxic effects, such as diarrhea, discoloration of teeth in children, and liver and kidney damage.

Chloramphenicol

Chloramphenicol is a broad-spectrum bacteriostatic antibiotic that interferes with protein synthesis on 70S ribosomes (Figure 18–12). Due to its relatively simple structure, it is cheaper for the pharmaceutical industry to synthesize it chemically rather than isolate it from *Streptomyces.* Its relatively small molecular size promotes its diffusion into portions of the body that are normally inaccessible to many other drugs. Unfortunately, it has been found that high doses or prolonged use can interfere with the formation of red blood cells in the bone marrow, causing a serious disorder called *aplastic anemia.* While this reaction is relatively rare (about 1 in 40,000 persons, as opposed to the normal rate of about 1 in 500,000), responsible agencies do not recommend using the drug under most circumstances. However, it is still the drug of choice in the treatment of typhoid fever, where its high effectiveness against this disease seems to justify the risk.

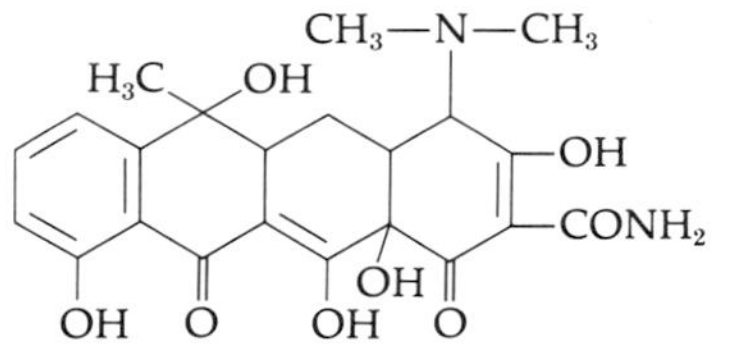

Tetracycline

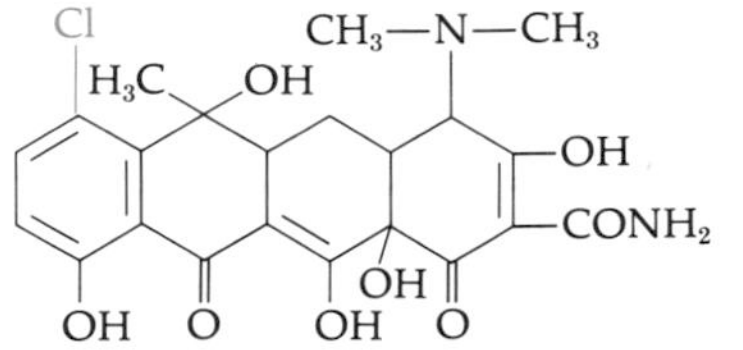

Chlortetracycline

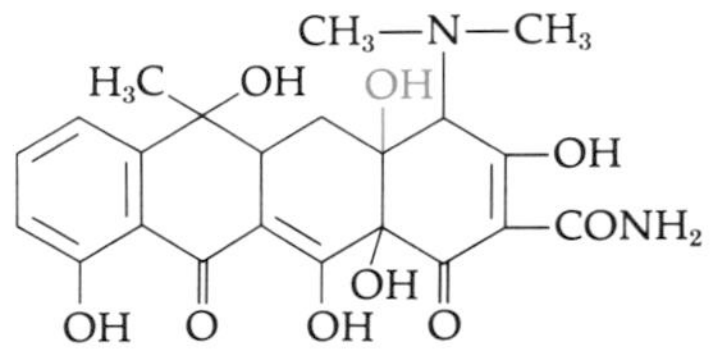

Oxytetracycline

Figure 18–11 Structure of representative tetracyclines. Chlortetracycline has the same structure as tetracycline with the addition of a chlorine (Cl) atom, shown in color. Oxytetracycline is like tetracycline with an additional hydroxyl group (—OH), shown in color.

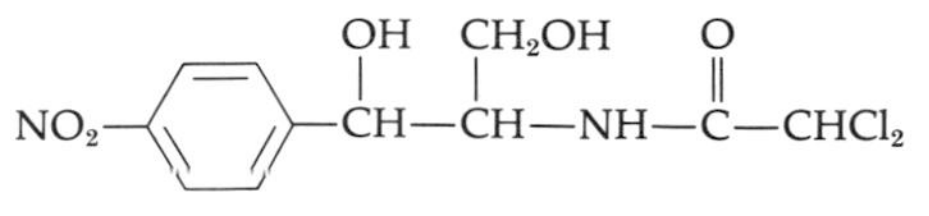

Chloramphenicol

Figure 18–12 Structure of chloramphenicol.

Macrolides

Macrolides are a group of antibiotics especially effective against gram-positive bacteria, a spectrum of activity similar to that of penicillin G. One of the better known macrolides is *erythromycin* (Figure 18–13), and its mode of action is to inhibit protein synthesis. Erythromycin is a bacteriostatic drug that is readily absorbed after oral administration,

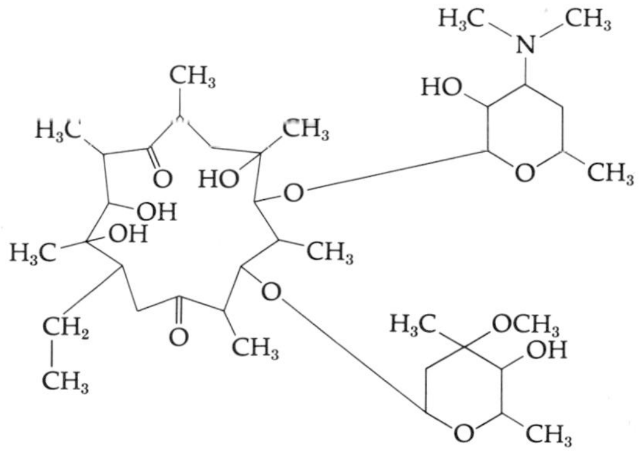

Erythromycin

Figure 18–13 Structure of erythromycin, a representative macrolide.

and it is also one of the least toxic antibiotics. In addition, erythromycin is active against organisms that become resistant to penicillin (streptococcal and staphylococcal infections), and it is often prescribed for individuals who may be allergic to penicillin. Erythromycin is the drug of choice for treating diphtheria and has recently been used in treating Legionnaires' disease.

Lincomycins

Lincomycins have an antimicrobial spectrum similar to that of erythromycin. They also exert their antimicrobial effect by inhibition of protein synthesis and are bacteriostatic. One commonly used drug in this group is *lincomycin.* It is only slightly toxic and is used as a penicillin alternative to treat

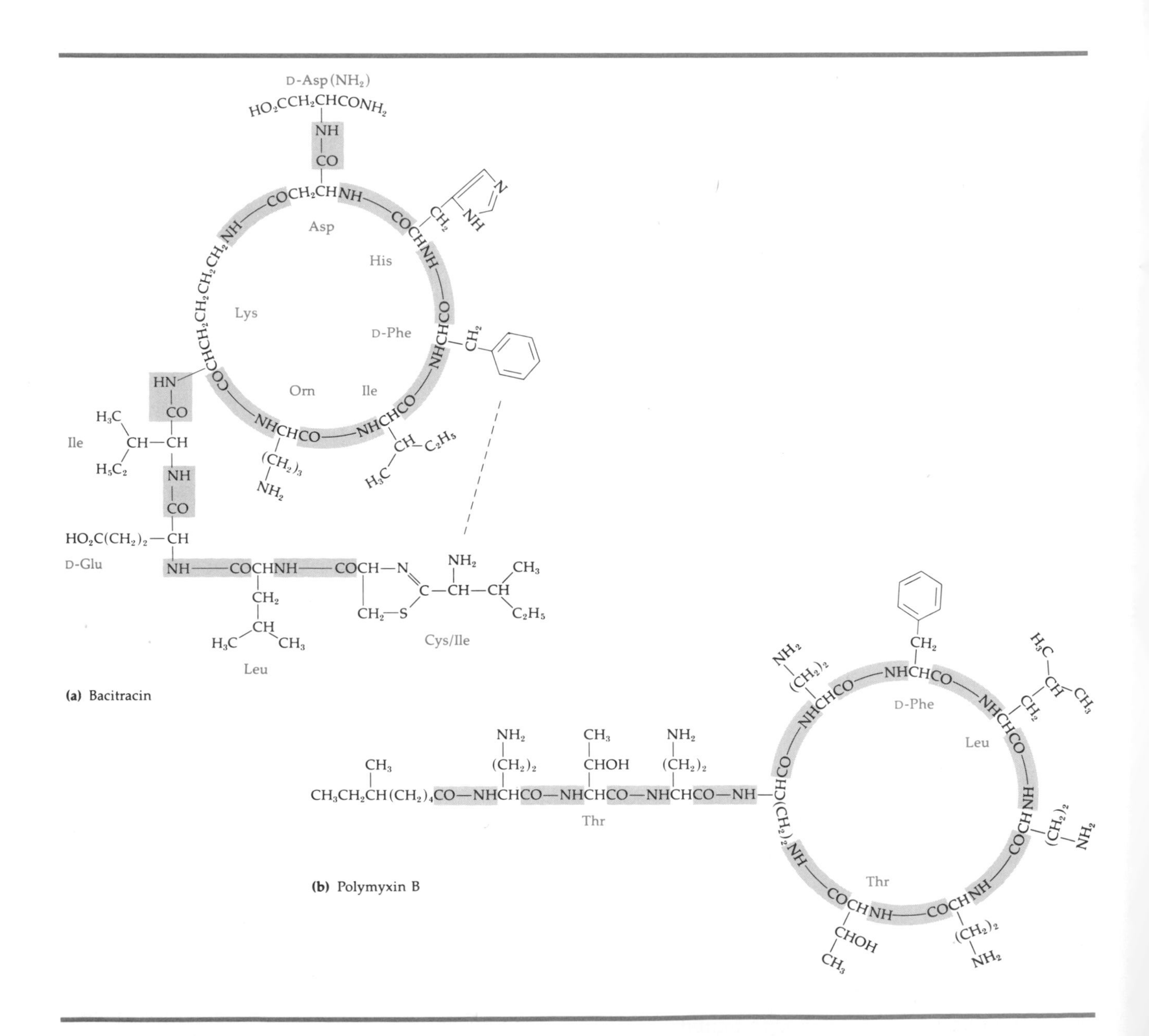

Figure 18–14 Structure of representative polypeptide antibiotics, showing the amino acid subunits (names in color) and peptide bonds (shading). **(a)** Bacitracin. **(b)** Polymycin B.

streptococcal and staphylococcal infections. It is administered orally or by injection.

Polypeptides

A number of antibiotics are **polypeptides,** chains of amino acids linked by peptide bonds (see Chapter 2). Two well-known examples are *bacitracin* and *polymyxin B* (Figure 18–14). Both drugs are poorly absorbed after oral administration.

Bacitracin is bactericidal mostly against gram-positive bacteria, such as staphylococci and streptococci, although it is also used against some gram-negative microorganisms, like *Neisseria* and *Hemophilus influenzae.* Its mode of action is to inhibit cell wall synthesis. Because of its toxic effect on the kidneys, bacitracin is given internally only when the microorganisms being treated are resistant to other drugs and when its administration and effects can be monitored very carefully. It is typically used as a topical application for treating superficial infections.

Polymyxin B is one of a group of five different strains of polymyxins designated as A, B, C, D, and E. It is a bactericidal used effectively against gram-negative bacteria, and its mode of action is to injure microbial plasma membranes. For this reason, it is particularly effective against *Pseudomonas* infections. It is also used to reduce bacterial populations prior to intestinal surgery and to treat superficial infections. As a group, polymyxins cause kidney and brain damage if their use is prolonged.

Rifamycins

Rifamycins are a group of bactericidal antibiotics especially effective against gram-positive bacteria, although they have also been used to combat infections caused by gram-negative bacteria, chlamydias, and poxviruses. Their mode of action is the specific inhibition of the main enzyme needed for bacterial RNA synthesis. A commonly used rifamycin is the antibiotic *rifampin.* It is interesting to note that taking rifampin may cause urine, feces, saliva, tears, and sweat to become orange-red. Although toxicity for rifampin is low, it often leads to the development of drug-resistant strains when used for tuberculosis therapy and is therefore used in combination with other drugs, such as INH and ethambutol.

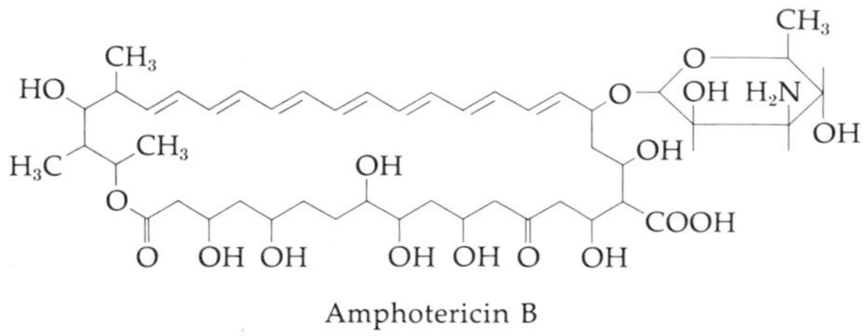

Amphotericin B

Figure 18–15 Structure of a representative polyene antibiotic, amphotericin B.

Polyenes

Two of the more commonly used **polyene** antibiotics are *nystatin* and *amphotericin B* (Figure 18–15). Both drugs are fungicidal and exert their action by damaging the fungal plasma membranes by combining with the sterols of those membranes. This is another good example of selective toxicity. Since bacterial plasma membranes (except for mycoplasmas) do not contain sterols, the drugs affect fungi and not bacteria. Nystatin is used to treat local infections of the vagina and skin. It is often given with tetracyclines for *Candida* infections of the digestive tract. Amphotericin B is used extensively to treat histoplasmosis, coccidioidomycosis, and blastomycosis. One disadvantage of amphotericin B is its toxicity to the kidneys.

Griseofulvin

Griseofulvin is a fungistatic drug whose mode of action seems to be interference with the function of the mitotic spindle in eucaryotic cells. Its selective action against certain fungi may be due to the ability of those fungi to take up and concentrate the drug. Its primary use is to control fungal infections of the skin, such as ringworm, and it has the advantage of low toxicity. It is administered orally.

Antiviral Drugs

Antiviral drugs are few in number and limited in their application because of their high toxicity. Because a virus uses the host cell's metabolic machinery, it is difficult to damage the virus without damaging the host as well. It is therefore particularly difficult to develop nontoxic drugs for systemic use. Several of the first few antiviral drugs approved for use have been designated for topical application to the eye or skin.

One example of an antiviral drug in clinical use is **idoxuridine,** a pyrimidine analog that inhibits DNA synthesis and has proved very effective against an eye infection caused by the herpes simplex virus. Another antiviral drug is **vidarabine,** a purine nucleoside obtained from *Streptomyces* culture. Like idoxuridine, it inhibits viral DNA synthesis and is used topically to treat herpes simplex eye infections. Vidarabine has recently been approved for intravenous use in the treatment of other herpes infections, such as herpes encephalitis.

A promising recent development in antiviral drug therapy is the experimental use of a drug called **2-deoxy-D-glucose** to treat human genital herpes infections. This type of infection accounts for about 13% of all venereal disease in the United States. The disease is of particular importance because the virus that causes it may infect the fetus at the time of delivery. Of even more concern is the evidence that the virus may be implicated in cervical cancer. 2-Deoxy-D-glucose has low toxicity and diffuses rapidly into most body tissues. It apparently acts by inhibiting the synthesis of viral envelopes during viral multiplication.

Amantadine (adamantanamine) is a synthetic drug of unusual structure (Figure 18–16) that seems to prevent the release of nucleic acid from certain viruses into the host cell. It is used in tablet form to prevent infection by influenza A in high-risk individuals, and it also may have some therapeutic effect early in the disease.

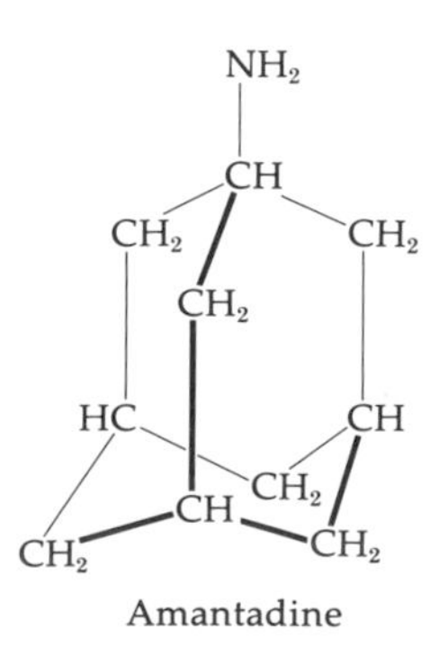

Figure 18–16 Structure of amantadine, a synthetic drug effective in preventing influenza A. It consists of three six-carbon rings joined in an unusual way, and an amino group.

A novel approach to viral chemotherapy that may prove fruitful in the future centers around **interferon,** a type of protein secreted by certain host cells in response to viral infections (see Chapter 15). Interferon inhibits cell infection by all kinds of viruses. Large-scale production of interferon by animal cell culture or by recombinant DNA techniques in bacteria will eventually make enough interferon available for large clinical studies to determine its effectiveness as a chemotherapeutic agent against viruses (and cancer, too). Meanwhile, scientists are also searching for drugs that will stimulate interferon production in the host animal.

Antiprotozoan and Antihelminthic Drugs

For hundreds of years, two tropical plants, cinchona and ipecacuanha, provided the only drugs to treat parasitic infections. The cinchona tree is the source of **quinine,** and ipecacuanha contains **emetine.** But recent work with synthetic chemotherapeutic agents has led to an ever-increasing number of alternative drugs. Unfortunately, protozoans and helminths have eucaryotic cells, which are structurally and functionally similar to human cells. Therefore, chemical agents used to treat these infections often have severe side effects on the human host. Eggs and encysted stages of parasites are difficult to eradicate. A patient may be cured of the actively growing parasite and still harbor dormant stages that will begin to grow at a later time. Preventive measures such as vector control and sanitary living conditions are safer and more effective disease control procedures than chemotherapy.

A synthetic derivative of quinine called **chloroquine** was developed in 1934 and is the drug of choice for treatment of malaria. Chloroquine interferes with DNA replication and transcription and acts as a frameshift mutagen.

Organic arsenicals (Carbarsone) may contain up to 29% arsenic and are used to treat amoebiasis. Synthetic and natural emetine are also used to treat amoebic infections. **Metronidazole** inhibits certain oxidases and is used against *Trichomonas vaginalis.*

Niclosamide inhibits oxidative phosphorylation in treatment of tapeworm infections. **Pyrivinium,** for pinworms, prevents carbohydrate absorption by the parasite. **Antimony compounds** are used to treat helminthic infections, especially schistosomiasis. Antimony interferes with glycolysis. **Metrifonate** is a new drug available for research purposes. It acts as a cholinesterase inhibitor in *Schistosoma* and has a lower toxicity than antimony.

A summary of the major antimicrobial drugs described in this chapter is presented in Table 18–4.

Table 18–4 Summary of Commonly Used Antimicrobial Drugs

Antimicrobial Drug	Effect on Microorganisms	Mode of Action	Clinical Use/Toxicity
Synthetic Drugs:			
Sulfonamides	Bacteriostatic	Competitive inhibition	*Neisseria meningitidis* meningitis and some urinary tract infections
Isoniazid (INH)	Bacteriostatic	Competitive inhibition	With streptomycin and rifampin to treat tuberculosis
Para-aminosalicylic acid (PAS)	Bacteriostatic	Competitive inhibition	With streptomycin to treat tuberculosis
Ethambutol	Bacteriostatic	Competitive inhibition	With INH to treat tuberculosis (less toxic than PAS)
Antibiotics:			
Penicillin, natural (Penicillins F, G, X, K, O, V)	Bactericidal	Inhibition of cell wall synthesis	Gram-positive bacteria
Penicillin, semisynthetic (Ampicillin, methicillin, oxacillin, and others)	Bactericidal	Inhibition of cell wall synthesis	Broad spectrum; resistant to penicillinase and stomach acid
Aminoglycosides	Bactericidal	Inhibition of protein synthesis	
Streptomycin			Gram-negative bacteria
Neomycin			Topical ointment
Kanamycin			*Staphylococcus*
Gentamicin			Gram-negative bacteria
Cephalosporins	Bactericidal	Inhibition of cell wall synthesis	Penicillin-resistant *Staphylococcus*
Tetracyclines	Bacteriostatic	Inhibition of protein synthesis	Very broad spectrum; some toxic side effects include tooth discoloration and liver and kidney damage
Chloramphenicol	Bacteriostatic	Inhibition of protein synthesis	*Salmonella;* aplastic anemia occasional side effect
Macrolides (Erythromycin and others)	Bacteriostatic	Inhibition of protein synthesis	Diphtheria; Legionnaires' disease; various infections in individuals allergic to penicillin
Lincomycins	Bacteriostatic	Inhibition of protein synthesis	Spectrum similar to erythromycin; used as a penicillin alternative
Polypeptides			
Bacitracin	Bactericidal	Inhibition of cell wall synthesis	Gram-positive bacteria primarily; toxic to kidneys; usually used as topical ointment
Polymyxin B	Bactericidal	Injury to plasma membranes	Gram-negative bacteria, including *Pseudomonas;* toxic to brain and kidneys
Rifamycins (Rifampin)	Bactericidal	Inhibition of RNA synthesis	Gram-positive bacteria, some gram-negatives, chlamydias, and poxviruses; often used with INH and ethambutol to minimize drug resistance

(continued on next page)

Table 18–4, continued

Antimicrobial Drug	Effect on Microorganisms	Mode of Action	Clinical Use/Toxicity
Antibiotics, continued:			
Polyenes (Nystatin, amphotericin B, and others)	Fungicidal	Injury to plasma membranes	*Candida* infections of skin and vagina (nystatin); histoplasmosis, coccidioidomycosis, and blastomycosis (amphotericin B)
Griseofulvin	Fungistatic	Inhibition of mitotic spindle function	Fungal infections of skin, primarily ringworm
Antiviral Drugs:			
Idoxuridine	Antiviral	Inhibition of DNA synthesis	Herpes eye infections
Vidarabine	Antiviral	Inhibition of viral multiplication	Herpes eye infections, chickenpox, viral encephalitis
2-Deoxy-D-glucose	Antiviral	Inhibition of viral envelope synthesis (?)	Genital herpes infections (experimental use)
Amantadine	Antiviral	Inhibition of nucleic acid release from viruses into host cells	Influenza A prophylaxis and treatment
Interferon	Antiviral	Induction of other antiviral proteins that block viral protein synthesis	All viruses
Antiprotozoan and Antihelminthic Drugs:			
Quinine and chloroquine	Antiprotozoan	Inhibition of DNA and RNA synthesis	Malaria
Organic arsenicals	Antiprotozoan	Interference with enzyme functions	Amoebic infections
Emetine	Antiprotozoan	Treatment of symptoms	Amoebic infections
Metronidazole	Antiprotozoan	Inhibition of certain oxidases	*Trichomonas* infections
Niclosamide	Antihelminthic	Inhibition of oxidative phosphorylation	Tapeworm infections
Pyrivinium	Antihelminthic	Inhibition of carbohydrate absorption	Pinworm infections
Antimony compounds	Antihelminthic	Inhibition of glycolysis	Schistosomiasis and other helminthic infections

TESTS FOR MICROBIAL SUSCEPTIBILITY TO CHEMOTHERAPEUTIC AGENTS

We will now turn to the methods used to determine the effectiveness of a chemotherapeutic agent in the treatment of infectious diseases. Different microbial species and strains have different degrees of susceptibility to different chemotherapeutic agents. Moreover, the susceptibility of a microorganism may change with time, even during therapy with a specific chemotherapeutic agent. For these reasons, a physician must know the sensitivities of the disease-causing microorganism before a particular treatment can be started. Several tests may be employed that will tell the physician which chemotherapeutic agent is most likely to combat a specific pathogen.

Tube Dilution Test

In the **tube dilution test** for determining antibiotic activity, a series of culture tubes is prepared as follows (Figure 18–17). To each tube is added a liquid medium and a different concentration of a chemotherapeutic agent. Then the tubes are inoculated with test organisms. After 16 to 20 hours of incubation at 35°C, the tubes are examined for turbidity. Lack of turbidity indicates lack of microbial growth. Tubes showing no turbidity are noted, and the minimum inhibitory concentration (MIC) is

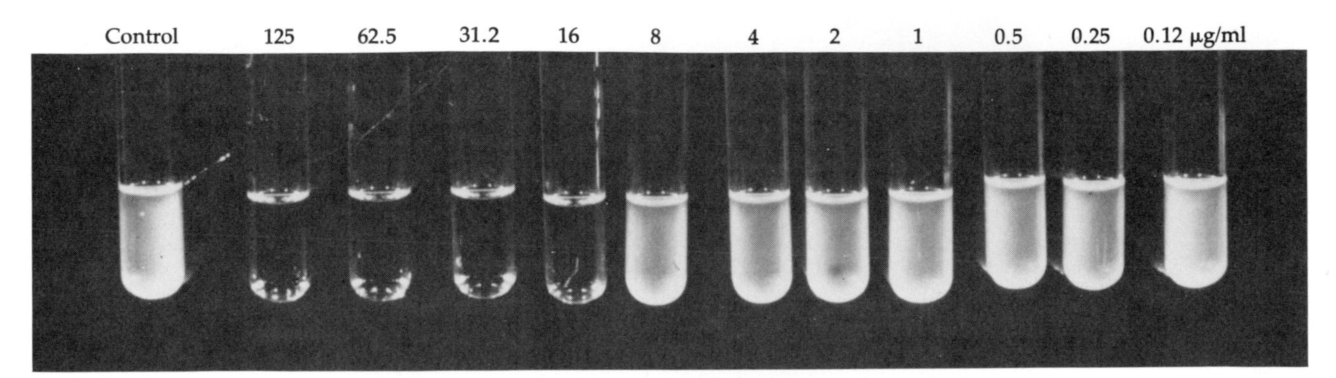

Figure 18–17 Tube dilution test for determining antibiotic activity. This series of tubes shows a test for the sensitivity of *Staphylococcus aureus* to the antibiotic ampicillin. A liquid medium and a different concentration of ampicillin was added to each tube, except no ampicillin was added to the control tube. The tubes were then inoculated with *S. aureus* and incubated. Turbidity (cloudiness) in a test tube indicates microbial growth. The tube containing the lowest concentration of the chemotherapeutic agent capable of preventing growth of the test microorganism is said to contain the MIC (minimum inhibitory concentration). This test shows that the minimum concentration of ampicillin needed to inhibit the growth of *S. aureus* is 16 μg/ml, since growth occurred in tubes with lower concentrations of the antibiotic.

thus determined. The **minimum inhibitory concentration (MIC)** is the lowest concentration of chemotherapeutic agent capable of preventing growth of the test microorganism.

How does one know if the inhibition of growth is permanent, that is, if the bacteria have actually been killed? If the cultures that show no growth of the microorganisms are then subcultured into broth tubes that contain no antibiotic, the lowest concentration of the chemotherapeutic agent that results in no growth of the subcultures can be noted. This is referred to as the **minimum bactericidal concentation (MBC)** of the chemotherapeutic agent. Based on the MIC and MBC, a distinction can be made between a bactericidal and bacteriostatic chemotherapeutic agent or between bactericidal and bacteriostatic concentrations of the same agent. If the concentration of the chemotherapeutic agent required to kill the test microorganisms is only two to four times as much as the inhibitory concentration, then the chemotherapeutic agent is regarded as bactericidal. If, instead, the concentration required for killing is many times higher than the inhibitory concentration, the drug is termed bacteriostatic. Before using the drug for combating the microorganism in a human host, it is also important to determine what serum concentrations of the drug can be attained in clinical use.

Agar Diffusion Method

In the **agar diffusion method,** also known as the **Kirby–Bauer test,** a Petri plate containing an agar medium is inoculated ("seeded") uniformly over the entire surface with a standardized amount of a test organism. The microorganisms can be mixed with a small amount of melted agar, which is then poured as an overlay onto the agar medium in the Petri plate. Next, filter paper disks impregnated with known concentrations of chemotherapeutic agents are placed on the solidified agar surface (Figure 18–18). During incubation, the chemotherapeutic agents diffuse from the disks into the agar. As the agent diffuses farther from the disk, its concentration progressively decreases. If the chemotherapeutic agent is effective, a zone of inhibition is observed in the area immediately around the disk. Beyond the zone of inhibition, growth is present, and the concentration of the drug at the edge of that zone represents the MIC. The diameter of the zone of inhibition may be measured with a ruler. Since the size of the zone of inhibition is affected by the diffusion rate of the chemotherapeutic agent, a wider zone does not always indicate greater antimicrobial activity. Other factors that affect the size of the zone of inhibition are the depth and type of medium, incubation conditions, and

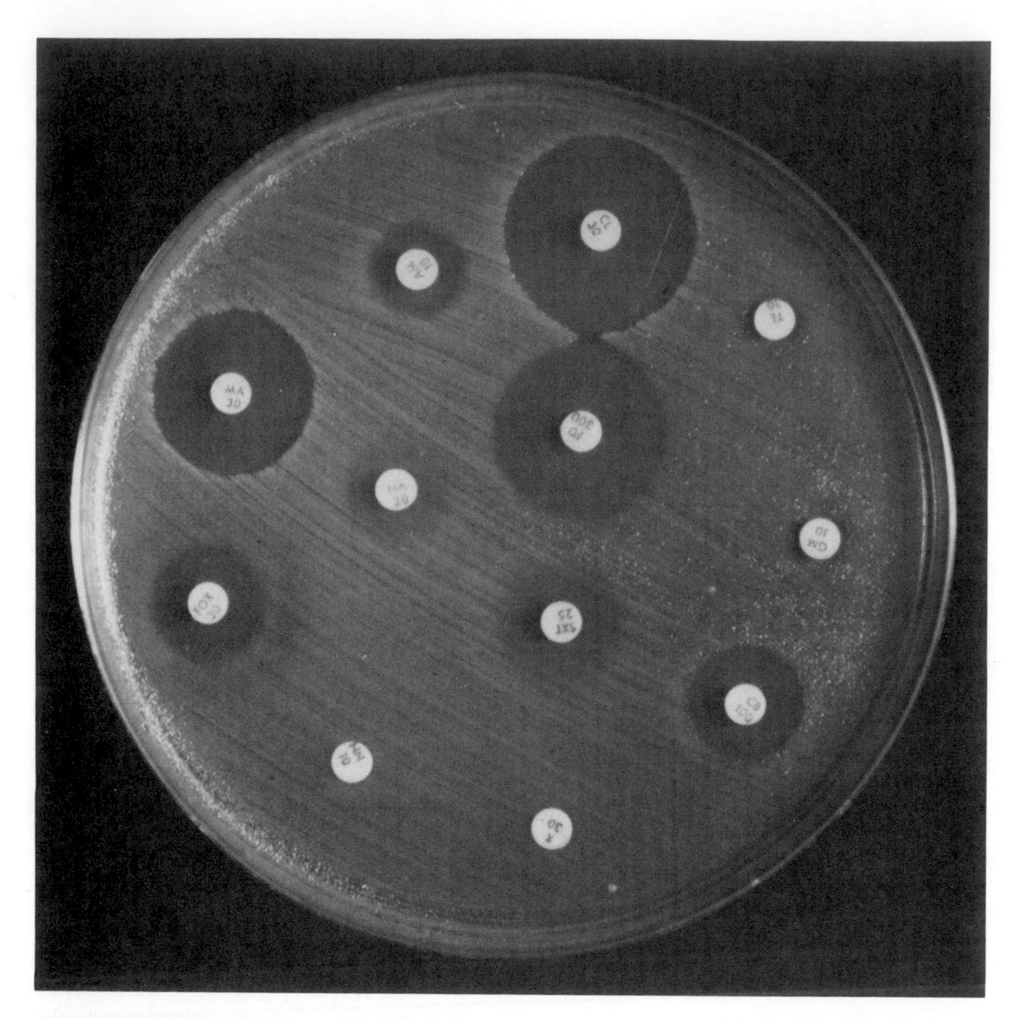

Figure 18–18 Agar diffusion method for determining the activity of chemotherapeutic agents. Each disk contains a different agent, which diffuses into the surrounding agar. Clear zones where bacterial growth is inhibited occur around disks with agents that are effective against the microorganism being tested.

the initial concentration of the chemotherapeutic agent. Therefore, these tests are always performed under conditions standardized for these factors.

Automated Tests

Recently, **automated tests** of microbial sensitivity to drugs have been introduced. These tests measure the inhibitory effect of drugs in a liquid medium by using light scattering to determine bacterial growth. Such tests are widely employed because of their speed and simplicity, yielding results within a few hours. Automated tests are being used increasingly by laboratories called upon to perform large numbers of tests.

DRUG RESISTANCE

A constant threat to the continued usefulness of our arsenal of antibiotics is **drug resistance.** For a generation, diseases such as streptococcal pneumonia and gonorrhea have been controlled by penicillin, although often in increasing doses. But the pathogens of both these diseases have recently developed drug-resistant strains, that is, organisms that resist the antimicrobial effects of the drug (see box). This resistance may arise by a number of mechanisms, and it may be expressed in several ways. In some cases, it is due to the ability of a particular microorganism to destroy the antibiotic. For example, as noted earlier, some staphylococci produce the enzyme penicillinase, which destroys natural penicillins. In other cases, the structure sensitive to the antibiotic may be lacking in the resistant microorganism. Bacterial L forms are resistant to penicillins because they do not synthesize cell walls, the structures sensitive to penicillins. Such changes may be temporary, resulting from environmental changes, or permanent, resulting from mutations in the microbial DNA. It is, of course, the latter type that causes most concern, since in the presence of the drug, the drug-resistant mutants will freely multiply and outgrow the sensitive microorganisms.

An even more common and more serious threat to chemotherapy is hereditary drug resistance carried by *plasmids,* extrachromosomal genetic elements. Some plasmids, including those called *resistance (R) factors,* can be transferred between bacterial cells in the same population and between different but closely related bacterial populations (see Figures 8–20a and 8–24). R factors often con-

Centers for Disease Control

MMWR

Morbidity and Mortality Weekly Report

Penicillin-Resistant Gonococci

Strains of gonococci resistant to all forms of penicillin have emerged—penicillinase-producing *Neisseria gonorrhoeae* (PPNG). From March 1976, when the first PPNG isolate was identified, through April 1980, a total of 1022 cases of PPNG in the United States were reported to the Centers for Disease Control (CDC) (see graph). Although the incidence of PPNG remains low, 186 cases were reported in the period from January to April 1980, reflecting a 66% increase compared to the same period in 1979.

(continued on next page)

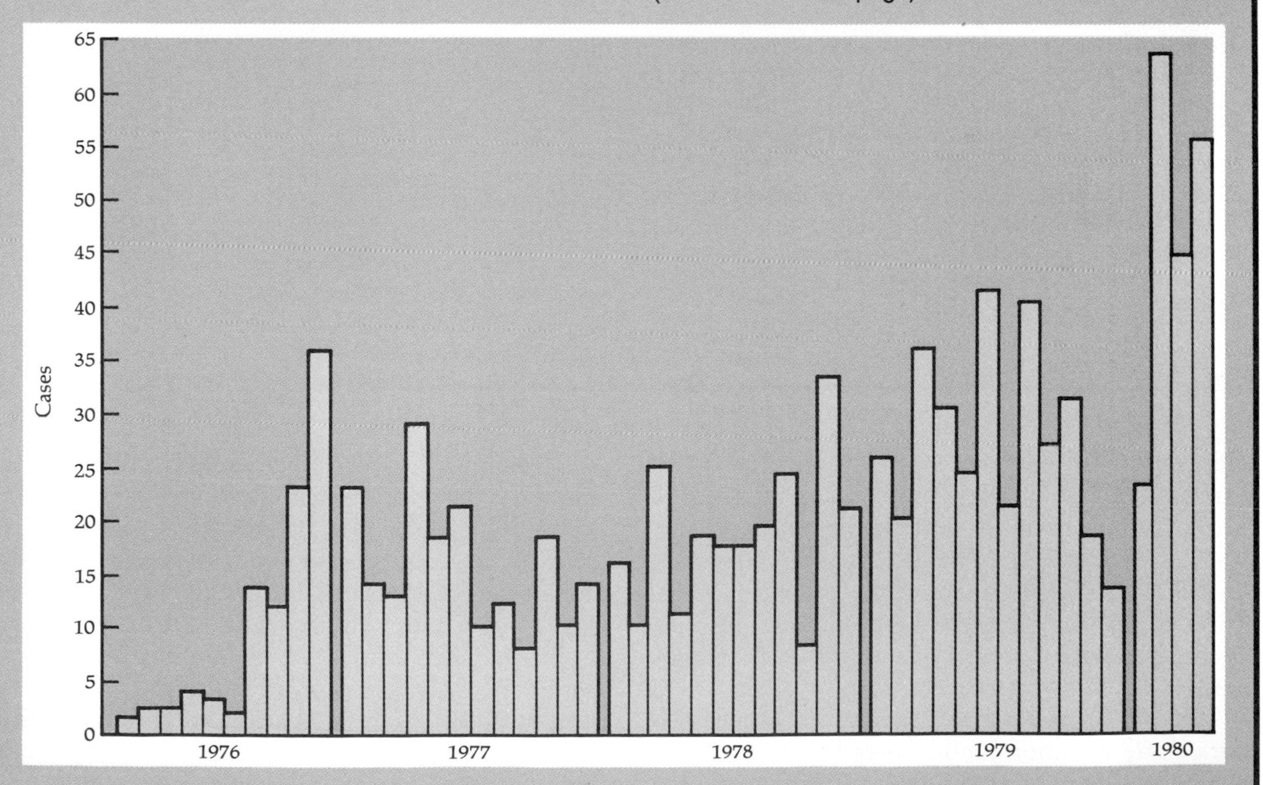

Cases of PPNG by month in the United States, March 1976–April 1980 (includes Commonwealth of Northern Mariana Islands, Federated States of Micronesia, Guam, Puerto Rico, and Virgin Islands). (Source of graph: *MMWR* 5/30/80, p. 243.)

Importation from Other Countries

A significant proportion of PPNG cases have been related to importation of infection from Southeast Asia. Military personnel or their dependents have been responsible for a majority of the imported cases. Twenty-seven countries in Europe, Asia, Africa, Oceania, and North America have reported cases of PPNG to the World Health Organization. The organism accounts for about 30% of all recent gonococcal isolates in the Philippines and 16% in the Republic of Singapore. (Among the factors thought to contribute to the high prevalence of PPNG in the Philippines and Singapore is the preventive use of oral penicillins, especially by prostitutes.) Singapore and the Philippines are the only two countries with proven high prevalence of PPNG, but epidemiologic assessment of PPNG cases imported into Europe indicates that the organisms are also prevalent in West Africa.

Endemic Focus in the United States

In the latter part of 1979 and the early months of 1980, studies suggested an increasing occurrence of PPNG cases within the United States. In the period from August 1 to October 17, 1980, 149 cases of PPNG infection were reported in Los Angeles County, California, representing a sharp increase compared to cases reported earlier. Unlike PPNG cases in most other areas of the United States, only 6% of the Los Angeles County cases in 1980 could be traced to infection acquired outside of the United States. Sustained disease transmission has occurred among county residents; as many as six persons have been consecutively infected in a single chain of transmission. The establishment of an endemic focus of PPNG infections in Los Angeles and recent increases in cases reported from other metropolitan areas, such as New York City, emphasize the need for continuing surveillance for these cases.

Delays in identifying early cases of persistent infection contribute to PPNG transmission. All patients treated for gonorrhea should have a post-treatment culture taken three to seven days after treatment. All positive isolates should be tested for penicillinase production. Spectinomycin is the antibiotic of choice for all patients who continue to be infected. In addition, the Centers for Disease Control specifically recommend spectinomycin for the initial treatment of uncomplicated anogenital gonorrhea in patients who have recently returned from countries with a high prevalence of PPNG infections. Initial treatment with spectinomycin should also be considered for cases in areas where more than 5% of gonococcal isolates are penicillin-resistant.

Spectinomycin-Resistant PPNG

In May 1981, a PPNG strain resistant to spectinomycin was reported from Travis Air Force Base, California. The patient had acquired the infection in the Philippines. Although four spectinomycin-resistant isolates of *N. gonorrhoeae* have been reported previously, this is the first to be penicillinase-producing also. The CDC now recommends that all isolates found to be PPNG also be tested for spectinomycin resistance. Patients who have uncomplicated anogenital infections caused by spectinomycin-resistant PPNG should be treated with cefoxitin and probenecid, or with SMX/TMP.

Sources: *MMWR Annual Summary 1979; MMWR 28:*85 (1979); *MMWR 29:*241, 381, 541 (1980); *MMWR 30:*221 (1981).

tain genes for resistance to several chemically unrelated antibiotics. The mechanism by which R factors bring about drug resistance is based on genes in the R factors that code for specific enzymes. These enzymes degrade the antibiotic or inhibit its action. This mechanism has accounted for the spread of resistant strains of staphylococci in hospitals.

Strains of bacteria resistant to antibiotics are particularly common among people who work in hospitals, where the antibiotics are in constant use. *Staphylococcus aureus,* which is carried in the nasal passages and which readily develops antibiotic resistance, is a common opportunistic pathogen found in hospitals where it infects surgical wounds.

To minimize the chance of survival of resistant mutants, drugs should be used discriminately. For example, patients should always finish the prescribed dosage of antibiotic, even though the apparent symptoms may have disappeared. Another way to reduce the development of resistant strains is through the administration of two or more drugs simultaneously. If a strain is resistant to one of the drugs, the other may destroy it. The probability that the same organism will acquire resistance to both drugs is very small—unless genes for resistance to both are carried on R factors.

It sometimes happens that the chemotherapeutic effect of two drugs given simultaneously is greater than the effect from either alone. This phenomenon is referred to as the **synergistic effect.** For example, in the treatment of bacterial endocarditis, penicillin and streptomycin produce a synergistic effect that neither drug alone can elicit. Perhaps damage to the bacterial cell wall by penicillin makes it easier for streptomycin to enter. Other combinations of drugs may be **antagonistic.** For example, the simultaneous use of penicillin and tetracycline is often *less* effective than either drug used alone. The bacteriostatic drug tetracycline may interfere with the action of penicillin, which requires growth for its action. Combinations of antimicrobial drugs should only be used: (1) to prevent or minimize the emergence of resistant strains, (2) to take advantage of the synergistic effect, (3) to provide optimal therapy in life-threatening situations before a diagnosis can be established with certainty, and (4) to lessen the toxicity of individual drugs by reducing the dosage of each in combination.

Drug resistance is not the only problem with antimicrobial therapy. Antibiotics that kill microorganisms will usually have no effect on toxins that may already have been released into the host's body fluids. Antimicrobial drugs are also affected by their inability to diffuse to certain areas of the body, the presence of blood clots, pH, and abscesses that harbor microorganisms.

Antibiotics are clearly one of the great triumphs of medical science, although they are not without their problems and disadvantages. Our present concern is with the misuse of antibiotics, which can make the emergence of resistant strains of pathogens more likely. The need for more antiviral drugs is also a pressing need. As research uncovers more aspects of the internal workings of cells, chemicals effective against viral synthesis will probably make their appearance. There is considerable similarity between the requirements for cancer therapy and the requirements for treating viral disease. Perhaps research for one will benefit the other.

STUDY OUTLINE

INTRODUCTION (p. 459)

1. A chemotherapeutic agent is a chemical that combats disease in the body.
2. An antimicrobial drug is a chemical that destroys disease-causing microorganisms with minimal damage to host tissues.
3. Antimicrobial drugs may be synthetic drugs (prepared in the laboratory) or antibiotics (produced by bacteria or fungi).

HISTORICAL DEVELOPMENT (pp. 459–461)

1. The first chemotherapeutic agent, discovered by Paul Ehrlich, was salvarsan, used to treat syphilis.
2. Sulfa drugs came into prominence in the late 1930s.
3. Alexander Fleming discovered the first antibiotic, penicillin, in 1929; its first clinical trials were done in 1940.

CRITERIA FOR ANTIMICROBIAL DRUGS (pp. 461–462)

1. Antimicrobial agents should have selective toxicity for microorganisms.
2. They should not produce hypersensitivity in most patients.
3. They should be stable during storage, be soluble in body fluids, and be retained in the body long enough to be effective.

4. Ideally, they should not induce drug resistance.
5. They should not cause excessive harm to normal flora.
6. They should have broad-spectrum activity.

ACTION OF ANTIMICROBIAL DRUGS (pp. 462–466)

1. General action is either by directly killing microorganisms (i.e., bactericidal) or by inhibition of growth (i.e., bacteriostatic).
2. Some agents, such as penicillin, inhibit cell wall synthesis in bacteria.
3. Other agents, such as chloramphenicol, erythromycin, tetracyclines, and streptomycin, inhibit protein synthesis by acting on 70S ribosomes.
4. Agents such as polymyxin B cause injury to plasma membranes.
5. Rifamycin, nalidixic acid, and 5-fluorocytosine inhibit nucleic acid synthesis.
6. Agents such as sulfanilamide, PAS, and nitrofurans act as antimetabolites.

SURVEY OF COMMONLY USED ANTIMICROBIAL DRUGS (pp. 466–476)

Synthetic Antimicrobial Drugs (pp. 466–468)

1. Sulfonamides are bacteriostatic; they are used to treat bacterial meningitis and urinary tract infections; they can cause allergies and drug resistance.
2. Isoniazid, PAS, and ethambutol are very effective against *M. tuberculosis;* they may be used in combination.
3. Other synthetic chemotherapeutic agents are nitrofurans, nalidixic acid, 5-fluorocytosine, and trimethoprim.

Antibiotics (pp. 468–473)

1. Penicillin is bactericidal and has low toxicity; several semisynthetic penicillins are available.
2. Aminoglycosides include streptomycin, neomycin, kanamycin, and gentamicin; all inhibit protein synthesis.
3. Cephalosporins inhibit cell wall synthesis and are used against penicillin-resistant strains and in treating people allergic to penicillin.
4. Tetracyclines are bacteriostatic toward many bacteria, including rickettsias and chlamydias; they have the broadest spectrum of antibacterial activity.
5. Chloramphenicol is bacteriostatic; a side effect of prolonged use is aplastic anemia.
6. Macrolides, such as erythromycin, are bacteriostatic; erythromycin is used to treat Legionnaires' disease.
7. Lincomycins inhibit protein synthesis and are used to treat streptococcal and staphylococcal infections.

8. Polypeptides, such as bacitracin and polymyxin B, inhibit cell wall synthesis; they are used to treat some gram-negative bacterial infections.
9. Rifamycins are bactericidal and are especially effective against gram-positive bacteria.
10. Polyenes, such as nystatin and amphotericin B, are used to treat fungal infections.
11. Griseofulvin interferes with eucaryotic cell division and is used primarily to treat skin infections caused by fungi.

Antiviral Drugs (pp. 473–474)

1. Antiviral drugs are limited because of toxicity.
2. Currently used antiviral drugs include idoxuridine, vidarabine, 2-deoxy-D-glucose, and amantadine.

Antiprotozoan and Antihelminthic Drugs (pp. 474–475)

1. Chloroquine, organic arsenicals, emetine, and metronidazole are used to treat protozoan infections.
2. Antihelminthic drugs include niclosamide, pyrivinium, and antimony; metrifonate is available for investigational purposes.

TESTS FOR MICROBIAL SUSCEPTIBILITY TO CHEMOTHERAPEUTIC AGENTS (pp. 476–478)

1. These tests are used to determine the degree of susceptibility of different microorganisms to chemotherapeutic agents.
2. They help to determine which chemotherapeutic agent is most likely to combat a specific pathogen.

Tube Dilution Test (pp. 476–477)

1. In this test, the microorganism is grown in tubes of liquid media containing different concentrations of a chemotherapeutic agent.
2. The minimum inhibitory concentration (MIC) is the lowest concentration of chemotherapeutic agent capable of preventing microbial growth.
3. The lowest concentration of chemotherapeutic agent that kills bacteria is called the minimum bactericidal concentration (MBC).

Agar Diffusion Method (pp. 477–478)

1. In this test, also known as the Kirby–Bauer test, a bacterial culture is inoculated on an agar medium, and filter paper disks impregnated with chemotherapeutic agents are overlayed on the culture.
2. After incubation, the absence of microbial growth around a disk is called a zone of inhibition.

3. The concentration of chemotherapeutic agent at the edge of the zone of inhibition is the MIC.

Automated Tests (p. 478)

1. These tests determine the amount of microbial growth in a liquid medium containing a chemotherapeutic agent by measuring the amount of light that passes through a culture.
2. Such tests are rapid and simple to perform.

DRUG RESISTANCE (pp. 478–481)

1. Drug resistance refers to the ability of a microorganism to resist the antimicrobial effects of a chemotherapeutic agent.
2. Hereditary drug resistance is carried by plasmids called resistance (R) factors.
3. Resistance can be minimized by the discriminate use of drugs.
4. Combinations of drugs may be used to minimize the development of resistant strains; to employ a synergistic effect; to provide therapy prior to diagnosis; to use small concentrations of each drug to lessen toxicity.

STUDY QUESTIONS

REVIEW

1. Define a chemotherapeutic agent. Distinguish between a synthetic chemotherapeutic agent and an antibiotic.
2. Ehrlich discovered the first __________. Fleming discovered ________; it was the first ____________.
3. List and explain five criteria used to identify an effective antimicrobial agent.
4. Fill in the following table.

Antimicrobial Agent	Synthetic or Antibiotic	Method of Action	Principal Use
Sulfonamides			
Isoniazid			
PAS			
Ethambutol			
Penicillin, natural			
Penicillin, semisyn-thetic			

(continued on next page)

Antimicrobial Agent	Synthetic or Antibiotic	Method of Action	Principal Use
Aminoglycosides			
Cephalosporins			
Tetracyclines			
Chloramphenicol			
Macrolides			
Lincomycin			
Polypeptides			
Polyenes			
Griseofulvin			

5. Identify three methods of action of antiviral drugs. Give an example of a currently used antiviral drug for each method of action.
6. What similar problems are encountered with antiviral, antifungal, antiprotozoan, and antihelminthic drugs?
7. Describe the tube dilution test for microbial susceptibility. What information can you obtain from this test?
8. Describe the agar diffusion test for microbial susceptibility. What information can you obtain from this test?
9. Define drug resistance. How is it produced? What measures can be taken to minimize drug resistance?
10. List the advantages of using two chemotherapeutic agents simultaneously to treat a disease. What problem can be encountered using two drugs?

CHALLENGE

1. Why are antiviral drugs like idoxuridine effective if host cells also contain DNA?
2. The following data were obtained from an agar diffusion test.

Antibiotic	Zone of Inhibition
A	15 mm
B	0 mm
C	7 mm
D	15 mm

(a) Which antibiotic was most effective against the bacteria being tested?
(b) Which antibiotic would you recommend for treatment of a disease caused by this bacterium?
(c) Was antibiotic *A* bactericidal or bacteriostatic? How can you tell?

3. The following results were obtained from a tube dilution test for microbial susceptibility.

Tube #	Antibiotic Concentration	Growth	Growth in Subculture
1	200 μg	−	−
2	100 μg	−	−
3	50 μg	−	+
4	25 μg	+	+

(a) The MIC for this antibiotic is ____________.
(b) The MBC for this antibiotic is ____________.

FURTHER READING

Actor, P. (Convener). 1975. Laboratory evaluation of antimicrobial agents. In *Developments in Industrial Microbiology.* 16:161–214. Washington, D.C.: American Institute of Biological Sciences. Current status of new antibacterial, antifungal, and antiprotozoan drugs.

Baldry, P. 1976. *The Battle Against Bacteria.* Cambridge, England: Cambridge University Press. The history of the fight against diseases and the development of antibacterial drugs.

Ball, A. P., J. A. Gray, and J. M. Murdoch. 1978. *Antibacterial Drugs Today.* Baltimore, MD: University Park Press. Comprehensive coverage of methods of action, side effects, toxicity, and drug interaction of antibacterial drugs.

Kobayashi, G. S. and G. Medoff. 1977. Antifungal agents: recent developments. *Annual Review of Microbiology* 31:291–308. Methods of action and uses for antifungal drugs.

Whitley, R. J. and C. A. Alford. 1978. Developmental aspects of selected antiviral chemotherapeutic agents. *Annual Review of Microbiology* 32:285–300. Inhibition of viral entry, protein synthesis, and nucleic acid replication to treat local and systemic infections.

Wishnow, R. M. and J. L. Steinfeld. 1978. The conquest of the major infectious diseases in the United States: a bicentennial retrospect. *Annual Review of Microbiology* 30:427–450. A concise summary of control of diseases including tuberculosis, cholera, malaria, and yellow fever.

PART

IV

MICROORGANISMS AND HUMAN DISEASE

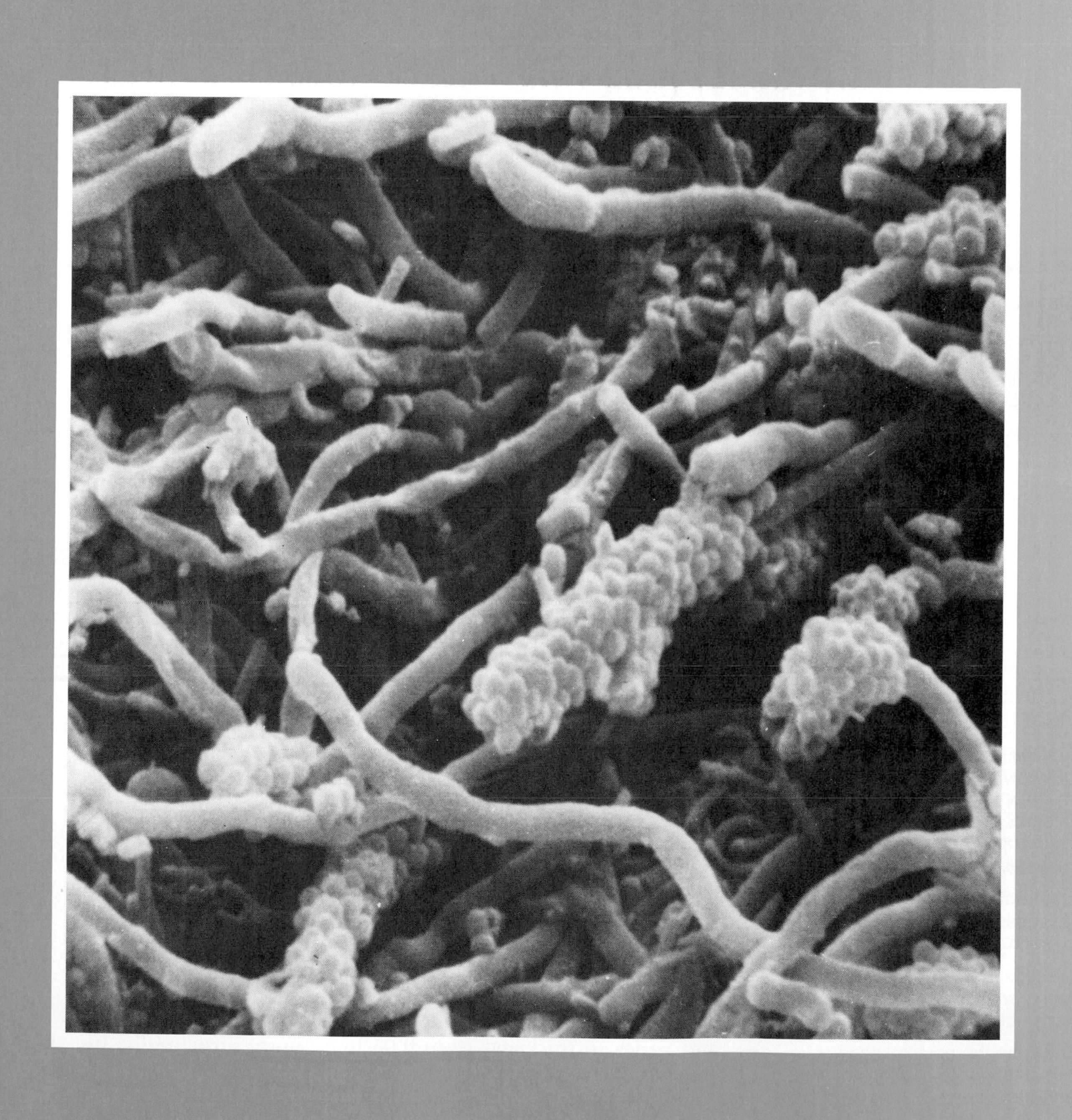

19

Microbial Diseases of the Skin and Eyes

Objectives

After completing this chapter you should be able to

- Describe the structure of the skin and the ways in which pathogens can invade the skin.
- Provide examples of normal skin flora, giving their locations and ecological roles.
- Differentiate between staphylococci and streptococci and list skin infections caused by each.
- List the etiologic agent, method of transmission, and clinical symptoms of the following skin infections: acne; warts; smallpox; chickenpox; measles; German measles; cold sores.
- Differentiate between the types of mycoses and provide an example of each.
- Discuss the role of bacteria, fungi, and viruses in conjunctivitis.
- Describe the epidemiology of neonatal gonorrheal ophthalmia and trachoma.

In Chapter 15, we saw that the human body possesses a number of defenses that contribute to *nonspecific resistance,* that is, resistance against many types of pathogens. These defenses include the skin and mucous membranes, antimicrobial substances (interferon, complement, and properdin), phagocytosis, the inflammatory response, and fever.

The skin, which covers and protects the body, constitutes the first line of defense against pathogens. Basically, the skin is an inhospitable place for most microorganisms, since the secretions of the skin are acidic and most areas of the body are low in moisture. Moreover, much of the body is exposed to radiation, which discourages microbial life. Some parts of the body, however, such as the axilla (armpit), have enough moisture to support microbial growth, and the excretions there tend to contain more organic matter than elsewhere on the body. Thus, the region of the axilla can support relatively large bacterial populations, while regions like the scalp support rather small numbers of microorganisms. The skin is a physical as well as ecological barrier, and it is almost impossible for pathogens to penetrate it, although some may enter through openings not readily apparent to the human eye.

STRUCTURE AND FUNCTION OF THE SKIN

The skin of an average adult occupies a surface area of about 1.9 m^2 and varies in thickness from 0.05 to 3.0 mm. Structurally, the skin consists of two principal parts: the epidermis and the dermis (Figure

19–1a). The *epidermis* is the outer, thinner portion composed of several layers of epithelial cells. The outermost layer of the epidermis, which is in contact with the external environment, consists of dead cells that, like hair and nails, contain a waterproofing protein called *keratin*. This layer, when unbroken, serves as an effective physical barrier against light, heat, microorganisms, and many chemicals.

The *dermis* is the inner, thicker portion of skin that is composed mainly of connective tissue. It contains numerous blood vessels, lymph vessels, nerves, hair follicles, and sweat and oil glands (Figure 19–1a and b). A hair follicle is a small tube in which a hair grows. Sweat glands originate in the dermis and convey their secretion, called perspiration, through ducts that terminate on the surface of the skin as sweat pores. Oil glands contain ducts that feed sebum into the hair follicles. The anatomic structure of hair follicles, sweat gland ducts, and oil gland ducts and their proximity to the surface of the skin provide potential passageways for microorganisms to enter the skin and penetrate deeper tissues.

Perspiration is a mixture of water, salt (mostly NaCl), urea, amino acids, glucose, and lactic acid, plus small amounts of other compounds, such as the enzyme lysozyme. As such, it provides moisture and some nutrients for microbial growth. However, the salt creates a hypertonic environment that dehydrates microorganisms, and lysozyme is capable of breaking down cell walls of certain bacteria. *Sebum* is a mixture of lipids (unsaturated fatty acids), proteins, and salts that prevents hairs from drying out. The fatty acids inhibit the growth of certain pathogens. Sebum passes upward through the hair follicles and spreads over the surface of the skin, where it prevents excessive evaporation of moisture from the skin. Like perspiration, sebum is nutritive for certain microorganisms, most notably the corynebacteria. When oil glands of the face become enlarged because of accumulated sebum, blackheads develop. And since sebum supports the growth of some microorganisms, pimples or boils may then develop. The color of blackheads is due to oxidized oil, not dirt.

In the lining of body cavities such as the mouth, nasal passages, urinary tract, genital tract, and gastrointestinal tract, the outer protective barrier changes structure, becoming essentially a layer of specialized epithelial cells. For example, some cells are ciliated, while others secrete mucus from glands beneath this external layer of epithelial cells—hence the name *mucous membrane (mucosa)*. In the respiratory system the mucous layer traps particles, including microorganisms, and the ciliary movement sweeps them upward out of the body (see Figure 15–3b). (Coughing up mucus demonstrates the functioning of the cilia.) Mucous membranes are frequently acidic, and this tends to limit their microbial populations. The eyes, meanwhile, are mechanically washed by tears, and the enzyme lysozyme contained in the tears destroys the cell walls of gram-positive bacteria and some gram-negative bacteria.

NORMAL FLORA OF THE SKIN

Although the skin is generally inhospitable for most microorganisms, certain factors tend to support the growth of some microbes, establishing them as part of the normal flora. For example, some anaerobic bacteria of the genus *Propionibacterium* live in hair follicles, where their growth is supported by oily secretions from oil glands. These same bacteria can produce an acid (propionic acid) that helps maintain the pH of the skin between 3 and 5 and has a bacteriostatic effect on many other microorganisms, preventing their colonization. On more superficial skin surfaces, certain aerobic bacteria, in the presence of oxygen, can metabolize fatty acids. As a result, these bacteria contribute to the normal level of the skin's fatty acids, which in turn helps maintain a pH that is unfavorable for more harmful microbial growth. In addition, the normal flora contains certain microorganisms that secrete antimicrobial substances and resist salt and drying, giving them an advantage over potential pathogens that attempt to colonize the skin.

Even though the skin is considered a single site, its normal flora varies in different regions. For example, the flora of the face reflects that of the oropharynx (portion of the throat behind the mouth), whereas the flora of the anal region is influenced by microorganisms of the lower gastrointestinal tract. Although vigorous scrubbing of the skin with soap and water or disinfectants will rid it temporarily of most surface bacteria, microorganisms present in hair follicles and sweat glands will soon reestablish the normal floral population.

Scanning electron micrographs show that bacteria tend to clump on the skin in small groups.

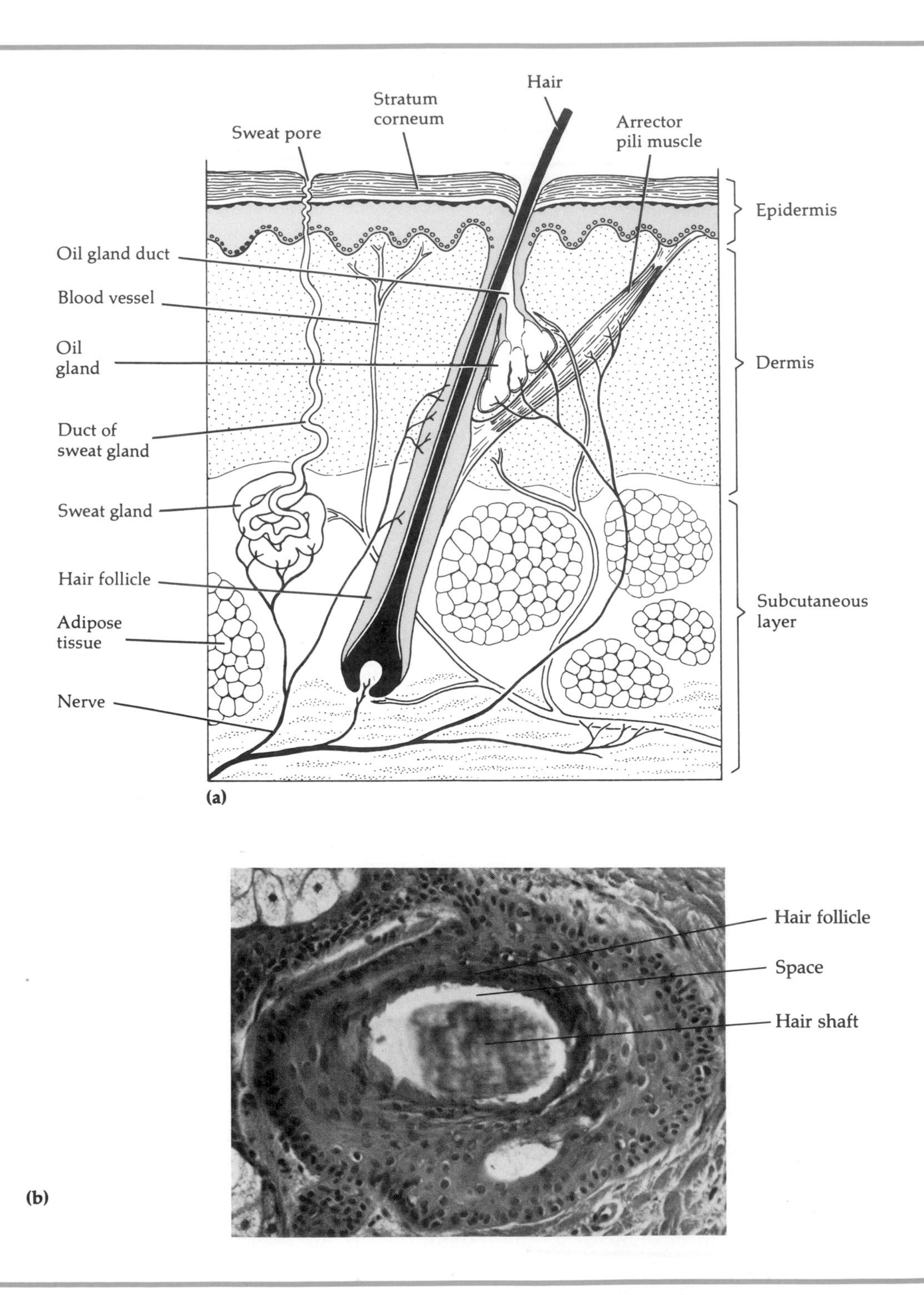

Figure 19–1 Structure of the skin. **(a)** Diagram illustrating the principal parts of the skin. Note the potential passageways between the hair follicle and hair shaft through which microbes may penetrate deeper tissues. Microbes may also enter the skin through sweat pores. **(b)** Cross section through a hair follicle in a human scalp (approx. ×400). Note the space around the hair shaft.

Many are fairly large, spherical, gram-positive bacteria of the genera *Staphylococcus* (Figure 19–2) and *Micrococcus*. Both genera are capable of producing antimicrobial substances that prevent colonization of the skin by pathogens and help maintain the balance of skin flora. Also part of the skin's normal flora are gram-positive pleomorphic rods referred to as *diphtheroids*. Some diphtheroids, such as *Propionibacterium acnes*, are typically anaerobic and inhabit hair follicles. Others, like *Corynebacterium xerosis*, are aerobic and inhabit the surface of the skin. A yeast belonging to the genus *Pityrosporum* is capable of growth on oily skin secretions and is a frequent inhabitant of the skin.

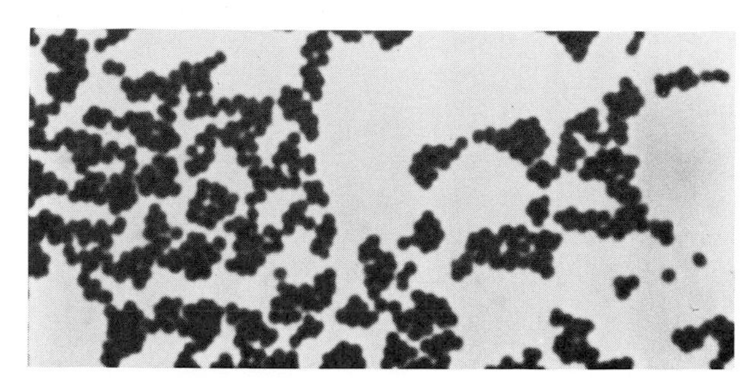

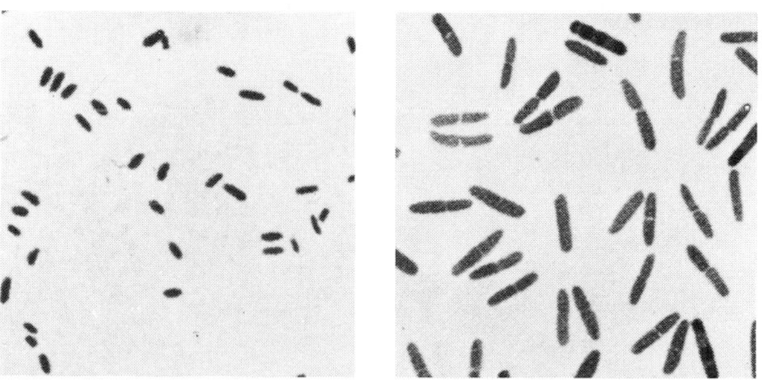

Figure 19–2 Normal bacterial flora on the skin. Staphylococci *(top)*, *Propionibacterium acnes (lower left)*, and *Corynebacterium xerosis (lower right)* are all commonly found and do not normally cause disease.

BACTERIAL DISEASES OF THE SKIN

Two genera of bacteria, *Staphylococcus* and *Streptococcus* (commonly referred to as staphylococci and streptococci), are frequent causes of skin-related diseases and merit special discussion. We will also discuss these bacteria in later chapters as pathogens in relation to other organs and conditions. Superficial staphylococcal and streptococcal infections of the skin are very common. The bacteria frequently come into contact with the skin and are fairly well adapted to the physiological conditions in this area of the body. Both genera also produce invasive enzymes and damaging toxins.

Staphylococcal Skin Infections

Staphylococci are spherical, gram-positive bacteria about 0.5 to 1.5 μm in diameter. They tend to form irregular, grapelike clusters of cells because they divide at random points about their circumference and the daughter cells do not completely separate.

Staphylococci are facultative anaerobes. They can ferment glucose and other sugars with the production of acid. In the presence of oxygen they use aerobic respiration. You may recall from Chapter 5 that hydrogen peroxide may form in the electron transport system and that hydrogen peroxide is lethal to cells. Staphylococci produce the enzyme **catalase,** which breaks down this hydrogen peroxide into water and oxygen. A rapid detection procedure for the presence of catalase (and therefore the presence of staphylococci) is to drop hydrogen peroxide on a colony. The presence of bubbles indicates the release of oxygen as a result of the reaction.

Staphylococcus aureus is the most pathogenic of the staphylococci (see Figure 10–11a). Typically, it forms golden yellow colonies on agar. Practically all pathogenic strains of *S. aureus* produce **coagulase,** an enzyme that coagulates (clots) the fibrin in blood. The fibrin clot may protect the microorganisms from phagocytosis and isolate them from other defenses of the host. A positive coagulase test is the most important test in identifying species of *Staphylococcus* (Figure 19–3). The coagulase test is used to distinguish *S. aureus* from other species of *Staphylococcus*. *S. aureus* also produces a number of toxins, several of which may damage tissues. Some toxins, called *enterotoxins,* affect the gastrointestinal tract, an aspect that will be discussed in detail in later chapters. Other toxins, called *leukocidins,* destroy phagocytic leukocytes, aid the staphylococcal cells in invading body tissues, and are responsible for the formation of pus and the naming of *S. aureus* as a *pyogenic* (pus-producing) coccus.

G+ cocci

H_2O_2 + catalase → O_2 + H_2O

Figure 19–3 Coagulase test for *Staphylococcus aureus*. If the staphylococcus secretes coagulase, the fibrin in plasma clots, indicating that the strain is pathogenic *(left)*. This is a positive coagulase test. If no coagulase is secreted, there is no clotting of the fibrin in plasma and the result is a negative coagulase test *(right)*.

At any given time, about 30% of the human population carries potentially pathogenic, coagulase-positive *S. aureus*. It is primarily an inhabitant of the nasal passages, from which it is a transient inhabitant of the skin. It is relatively well adapted to life on the skin because it is resistant to environmental stresses such as radiation, drying, and high osmotic pressures (high salt concentration).

Staphylococcus aureus is a very common problem in the hospital environment. It is present on the skin of patients, hospital personnel, and hospital visitors and is constantly exposed to antibiotics to which it rapidly develops resistance. Later we will see how, as opportunists, these organisms are responsible for many serious infections of surgical wounds and other artificial breaches of the skin. However, as noted earlier, a natural weakness in the skin barrier is represented by the hair follicles that pass through the epidermal layer. Infections of the hair follicles often occur as **pimples.** When the follicle of an eyelash is infected, it is called a **sty.** A more serious hair follicle infection is the **furuncle (boil),** which is a type of **abscess,** a localized region of pus surrounded by inflamed tissue. When the body fails to wall off the bacterial infection, and neighboring tissue is progressively invaded, the extensive damage is called a **carbuncle.** This is a hard, round, deep inflammation of tissue under the skin that causes an abscess. At this stage, the patient usually exhibits the symptoms of generalized illness with fever. Whenever the organisms are able to gain access to the bloodstream and increase rapidly in numbers, the condition is termed **septicemia.** In the days before effective antibiotics, the skin and mucous membranes were the most important barriers to death from septicemia.

Because of the prevalence of drug-resistant staphylococci, especially in hospitals, it is essential to determine drug sensitivity of the infecting organism. Many staphylococci produce penicillinase, which degrades natural penicillins. Until the drug sensitivity is determined, patients are usually treated with methicillin or oxacillin (semisynthetic penicillin derivatives to which penicillinase-producing strains are susceptible) or cephalosporin. Since established abscesses do not respond to antibiotic therapy because the bloodstream doesn't reach them, they should be drained.

Impetigo is a highly contagious skin infection almost always caused by *Staphylococcus aureus*. (*Streptococcus pyogenes* may also be involved.) It is a superficial skin infection characterized by isolated *pustules* (small, round elevations containing pus) that become crusted and rupture (Figure 19–4). It is a disease most commonly observed in children under about ten years of age, and its occurrence is often related to poor sanitary conditions. It is apparently spread largely by contact, but may also originate in some minor abrasion or premature removal of a wound scab. **Pemphigus neonatorum,** or impetigo of the newborn, is a particularly troublesome problem in hospital nurseries. To prevent outbreaks, hexachlorophene-containing skin lotions are commonly used.

In impetigo infections, ruptured pustules release fluids that can infect adjacent areas, spreading the infection. Also, there is always the risk of infection of the underlying tissue and eventual entry into the bloodstream. In some cases of impetigo, the outer skin layers peel off in extensive sheets, apparently a reaction to the staphylococcal toxins. This is known as the **scalded skin syndrome.**

Impetigo may be treated with penicillin, erythromycin, cephalothin, lincomycin, and vancomycin.

Streptococcal Skin Infections

Like staphylococci, **streptococci** are gram-positive, spherical bacteria. But unlike the staphylococci, the cells grow in chains. Prior to division, the individual cocci elongate and divide to form pairs. When the dividing pairs do not separate, chaining occurs (see Figure 10–11b). The bridges between the individual cocci consist of cell wall material that has not cleaved. Streptococci are called facultative anaerobes by *Bergey's Manual.* However, because their metabolism is strictly fermentative (they cannot use oxygen), they may better be regarded as aerotolerant anaerobes. They do not produce catalase. Although many nonpathogenic streptococci are commonly found as inhabitants of the mouth, gastrointestinal tract, and upper respiratory system, some streptococci are responsible for important skin infections.

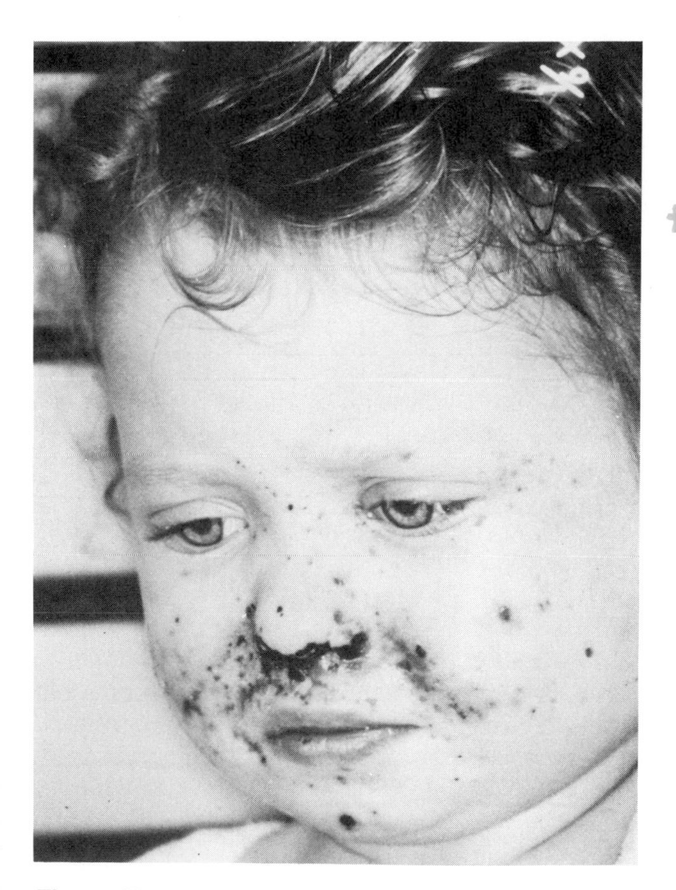

Figure 19–4 Impetigo caused by staphylococci and streptococci on the face of a three-year-old child. Note the characteristic pustules.

As streptococci grow, they secrete several toxins and enzymes into the medium. Among these are **hemolysins,** and depending on the type of red blood cell destruction caused by the hemolysins, streptococci can be divided into alpha-hemolytic, beta-hemolytic, and gamma-hemolytic streptococci (see Chapter 10 and Color Plate IIA and IIB). In addition, beta-hemolytic streptococci are further differentiated into a number of immunologic groups, designated A through O, based on different carbohydrates (antigens) in their cell walls.

From the standpoint of human disease, group A beta-hemolytic streptococci are the most important. The most common species of this group is *Streptococcus pyogenes.* Group A beta-hemolytic streptococci can be further subdivided into over 55 immunologic types on the basis of a protein, called the *M protein,* distributed on the surface of the pili (Figure 19–5). The M protein contributes to pathogenicity by its antiphagocytic properties.

In addition to hemolysins and M proteins, the pathogenicity of group A beta-hemolytic streptococci is related to a number of other substances elaborated by the cells. These include *erythrogenic toxin* (responsible for the scarlet fever rash), *deoxyribonucleases* (enzymes that degrade DNA), *NADase* (an enzyme that breaks down NAD), *streptokinases* (enzymes that dissolve blood clots), *hyaluronidase* (an enzyme that dissolves hyaluronic acid, the cement substance of connective tissue), and *leukocidins* (enzymes that kill white blood cells).

Beta-hemolytic streptococci are among the most susceptible of all pathogens to antimicrobial drugs. Penicillin is the drug of choice, although erythromycin is also frequently used. Tetracyclines are no longer recommended because many streptococcal strains are now resistant. Group A beta-hemolytic streptococci cause a wide variety of

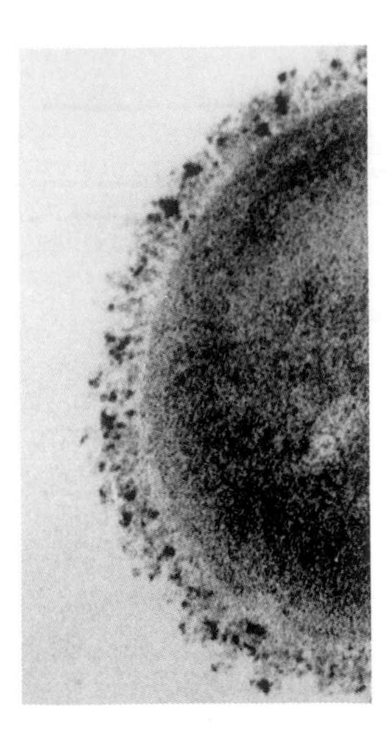
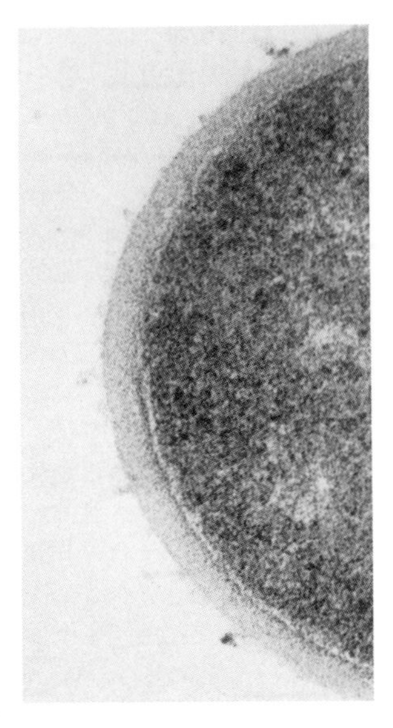

Figure 19–5 Electron micrographs showing portions of group A beta-hemolytic streptococci (approx. ×90,000). The cell on the left is from a strain that has M protein on its pili. The cell on the right lacks M protein.

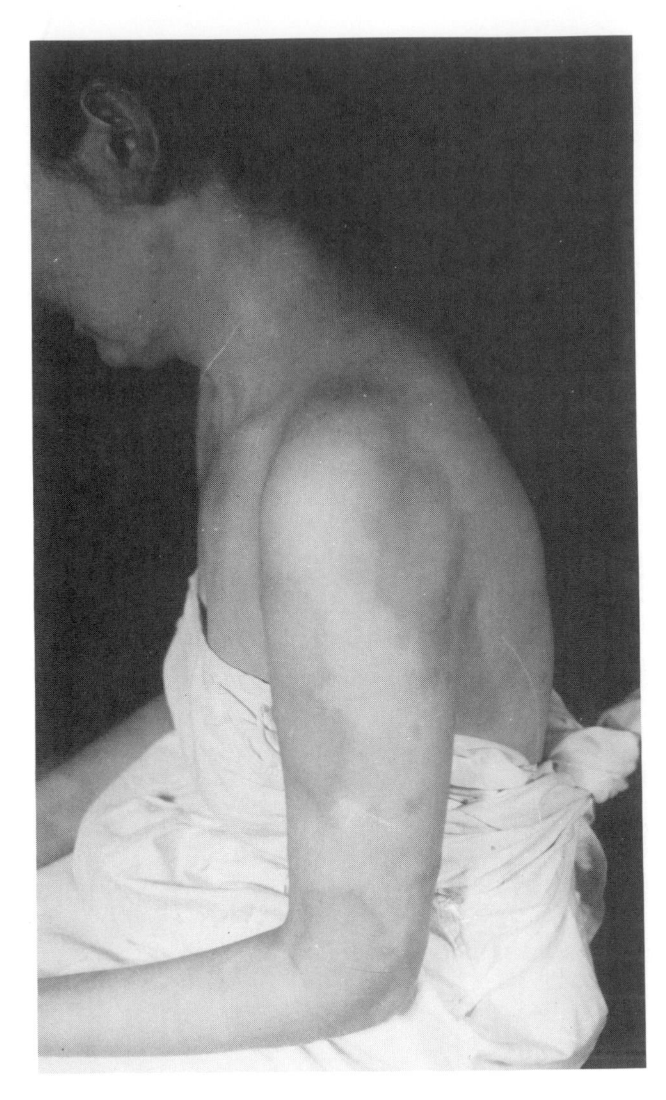

Figure 19–6 Erysipelas caused by group A beta-hemolytic streptococci. The dark patches of skin on the arm and shoulder are areas that have been reddened by streptococcal toxins.

diseases, a number of which will be discussed in later chapters. One species, *Streptococcus pyogenes*, may be implicated in some cases of impetigo.

Streptococcus pyogenes is the most common cause of **erysipelas,** a disease in which the skin erupts into reddish patches that enlarge with a thickening and swelling at the margin (Figure 19–6). The reddening is due to toxins produced by the streptococci as they progressively invade new areas. Usually, the skin outbreak is preceded by a streptococcal infection elsewhere in the body, such as a streptococcal sore throat. It is not even certain if the infection of the skin is from an external invasion or if it is reached by some systemic route. Erysipelas is most likely to occur in the very young and the very old. The drugs of choice for the treatment of erysipelas are penicillin and erythromycin.

Infections by Pseudomonads

Pseudomonads have the ability to cause opportunistic skin infections. They are gram-negative rods that are nonfermentative, oxidase-positive, and typically strictly aerobic. Among the more prominent pseudomonads is *Pseudomonas aeruginosa*, a species found primarily in soil, water, and on plants, and one that is able to proliferate on minimal organic material. Because of its resistance to many antibiotics, it often becomes dominant as a commensal when more susceptible bacteria are suppressed. After the introduction of broad-spectrum antibiotics, *P. aeruginosa* became a principal agent of nosocomial (hospital-acquired) infections (see Chapter 20).

P. aeruginosa is an intestinal tract resident in about 10% of healthy individuals. It is also found from time to time in moist areas of the human skin, such as the groin and axilla, and in saliva. It can multiply in almost any moist environment containing even trace amounts of organic matter, including eye drops, weak antiseptic solutions, soaps, cosmetics, sinks, fuels, humidifiers, and swimming pools. Interestingly enough, *P. aeruginosa* is highly resistant to chlorine, and whirlpool baths and sauna-type bathing pools often become highly contaminated with pseudomonads. This has led to a number of outbreaks of superficial skin infections in persons using such facilities. Competition swimmers are often troubled with **otitis externa,** or swimmer's ear, a pseudomonad infection of the portion of the ear external to the eardrum.

P. aeruginosa produces several exotoxins that account for much of its pathogenicity. It also has an endotoxin. Except for superficial skin infections and otitis externa, infection by *P. aeruginosa* is rare in healthy individuals. However, it often causes respiratory infections in hosts already compromised by immunologic deficiencies (natural or drug-induced) or chronic pulmonary disease. (Respiratory infections are discussed in Chapter 23.) *P. aeruginosa* is also a very common and serious opportunist in burn patients, particularly those with second- and third-degree burns. Infection may produce blue-green pus caused by the bacterial pigment *pyocyanin.* Of major concern in many hospitals is the ease with which *P. aeruginosa* may be carried into hospitals with flowers or plants sent by well-wishers. Since the organism may be transmitted from flowers to burn patients or individuals with fresh surgical incisions, many hospitals do not permit such patients to receive flowers.

Polymyxins and gentamicin are among the few drugs that are effective against *P. aeruginosa.*

Acne

Acne is probably the most common skin disease of humans. More than 65% of all teenagers have the problem to one degree or another. Although its occurrence decreases after the teen years, the scarring from severe cases often remains.

Acne begins when the sebum produced by the oil glands finds its channels to the skin surface blocked. The sebum accumulation leads to the formation of the familiar whiteheads, and later blackheads. In some persons the accumulation of sebum ruptures the hair follicle lining. Bacteria, including *Staphylococcus* species and *Propionibacterium acnes,* a diphtheroid commonly found on the skin, become involved at this point. *P. acnes* is able to metabolize the sebum, forming free fatty acids that cause an inflammatory response by the body. It is this inflammation that leads to the scarring of acne. Picking or scratching the lesions—or even the pressure of tight collars or other clothing in contact with lesions—increases the chances of scar formation. Makeup frequently aggravates the condition, but diet—including consumption of chocolate—has been demonstrated to have no significant effect on the incidence of the disease.

Topical applications of preparations containing benzoyl peroxide are often useful, and antibiotics such as tetracycline have also been used. In any event, severe cases should be treated by a dermatologist rather than with household remedies.

VIRAL DISEASES OF THE SKIN

Warts

Papovaviruses can cause skin cells to proliferate into uncontrolled growths, called **warts.** Warts are generally benign (noncancerous), and regress spontaneously. It is possible to spread warts by contact-transfer of the viruses, and sexual contact has been implicated in the transmission of warts on the genitals. After infection, there is an incubation period of several weeks before appearance of the wart. Warts may be removed by liquid nitrogen cryotherapy, cantharidin, salicylic acid, trichloroacetic acid, dinitrochlorobenzene, bleomycin sulfate, and surgical excision.

Smallpox (Variola)

It is estimated that during the Middle Ages, 80% of the population of Europe could expect to contract **smallpox** during their lives. Many would die, and those who recovered from the disease carried disfiguring scars. It was even more devastating when peoples such as the American Indian, with no previous population exposure, came into contact with the disease.

Smallpox is caused by a poxvirus known as the smallpox (variola) virus. There are two basic forms of this disease: **variola major,** with a mortality rate

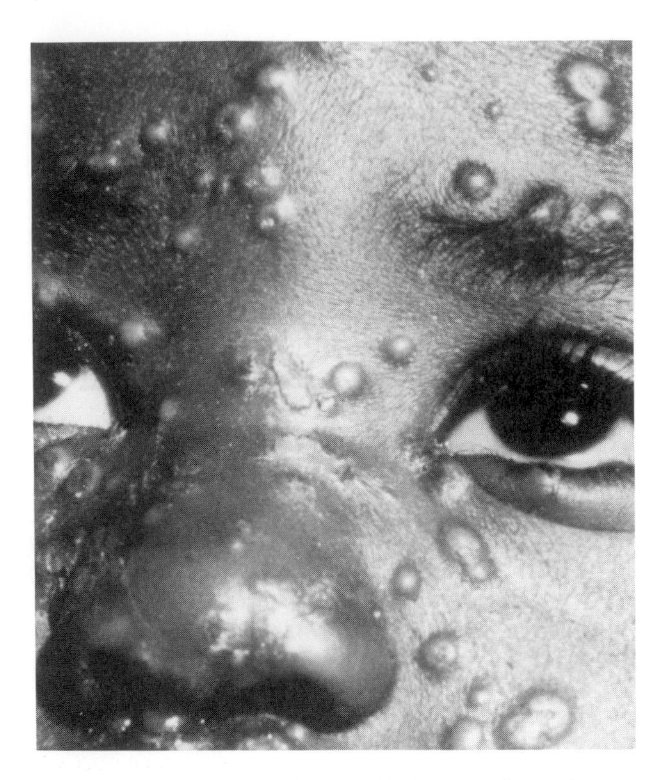

Figure 19–7 Lesions characteristic of smallpox. In some instances, the lesions are separated from each other. In other instances, the lesions run together.

of 20% or higher, and **variola minor,** with a mortality rate of less than 1%. If there is any significant difference between the viruses causing the different forms of the disease, it cannot be detected by laboratory tests, and recovery from one form of the disease leads to effective immunity against the other.

The transmission of smallpox, and the progression of the disease, roughly parallels that of other viral skin diseases. The viruses are transmitted by the respiratory route. They infect many internal organs before eventual movement through the bloodstream **(viremia)** leads to infection of the skin and the more recognizable symptoms. Symptoms include fever, malaise, headache, severe backache, and, occasionally, abdominal pain. The growth of the virus in the epidermal layers of the skin causes the rash (Figure 19–7). The rash begins as small red spots *(macular)* that develop into small raised spots *(papular)* and then fluid-filled blisters *(vesicular)*. Finally, the blisters form *pustules,* small elevations containing pus. Similar lesions are found in the mucous membranes of the oral cavity. At any given time, the lesions in all areas of the body are at the same stage of development.

It has long been observed that persons who recover from smallpox are immune to further infection. In Chapters 1 and 16, we discussed the early attempts to artificially induce such immunity and the successful use of vaccination by Edward Jenner in the eighteenth century.

Smallpox was the first disease deliberately eradicated from the human population, perhaps because there are no animal host reservoirs for the disease. It is believed that the last victim recovered from a natural case of variola minor in 1977 in Somalia, Africa. The last outbreak in the United States was initiated by a man who succumbed in New York City in 1947, probably after acquiring the disease in Mexico. The result was a mass inoculation of several million persons. The eradication of smallpox was accomplished by a concerted effort coordinated by the World Health Organization. In nations such as India, rewards were given to persons reporting cases of smallpox, and all contacts were then vaccinated.

At the present time, the smallpox virus collections that exist in laboratories are the most likely source of new infections. And this is not merely a hypothetical concern, as there have already been several such episodes. One death from a laboratory-associated infection occurred in England in 1978. The virus was apparently carried from the laboratory, through the ventilation system, and into a telephone booth on another floor of the building, where the victim, a medical photographer, was infected. In another episode, this one in Germany, the virus escaped from the laboratory where it was being used and caused the disease in a part of the hospital considerably removed from the laboratory.

Routine vaccination for smallpox was discontinued in the United States in 1971, a number of years before the recent worldwide eradication of the disease. Several persons, usually infants, had died each year as a side effect of the vaccine. (The live variola minor virus given as a vaccination multiplies, producing an ulcerative lesion at the site of vaccination, and from there can be spread throughout the body.) This had been an acceptable risk so long as smallpox represented a credible threat, particularly by importation by air travelers. But as the disease became less and less common in the world, the risk could no longer be justified.

Chickenpox (Varicella)

Chickenpox is a relatively mild childhood disease. After gonorrhea, it is the second most common reportable infectious disease in the United States. Chickenpox is acquired by infection of the respiratory system, and the infection localizes in skin cells after a period of about two weeks. The infected skin is vesicular for 3 to 4 days. The vesicles fill with pus, rupture, and form a scab before healing (Figure 19–8a). At any given time, vesicles may be in a variety of developmental stages. The vesicular rash can also appear in the mouth and throat. When chickenpox occurs in adults, which is not frequent because of the high incidence in childhood, it is a more severe disease with significant mortality rates.

The herpesvirus causing chickenpox (herpes zoster virus) appears to remain latent in nerve cells following recovery from the childhood infection. Later in life some trauma, such as a serious illness or even psychological stress, appears to be able to activate the virus into renewed growth. The result is a related disease, **herpes zoster,** or **shingles.** Vesicles similar to chickenpox occur but are localized in distinctive areas (Figure 19–8b). Most typically, there is a girdlelike distribution about the waist, although facial shingles and infections of the upper chest also occur. The infection is usually limited to one side of the body at a time because of the unilateral distribution of the nerve branches. The infection follows the distribution of the affected cutaneous sensory nerves (see Figure 21–1). Occasionally, such nerve infections can result in nerve damage that impairs vision or even results in cases of paralysis. Shingles is simply a different expression of the same virus that causes chickenpox, expressed differently because the patient now has partial immunity. Exposure of children to cases of shingles has been known to lead to cases of chickenpox.

As with many viral systemic diseases, there is always some danger of involvement of the central nervous system (brain and spinal cord). In the United States, approximately 100 persons each year die of complications of chickenpox, usually from encephalitis. However, this is a very small percentage of the total reported number of cases.

Measles (Rubeola)

Measles is an extremely contagious disease spread by the respiratory route. Since a person with measles is infectious before symptoms appear, quarantine is not an effective measure of prevention. The causative agent is a paramyxovirus called the measles virus, resembling the viruses causing influenza or mumps. It is also closely related to the virus causing distemper in dogs. In actuality,

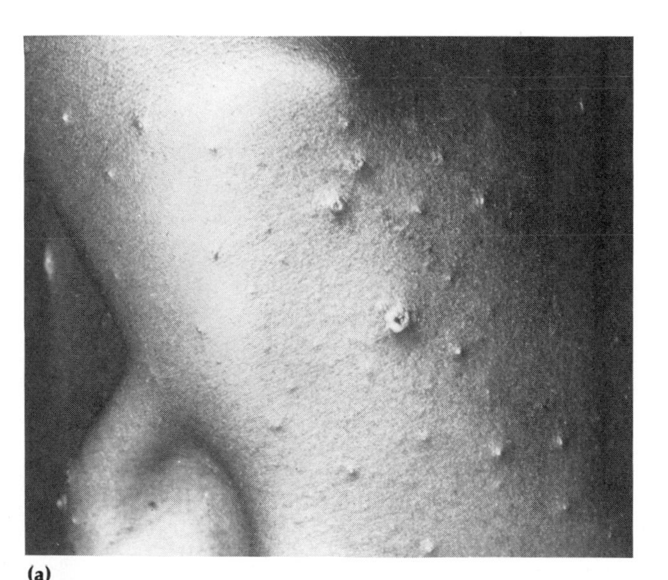

(a)

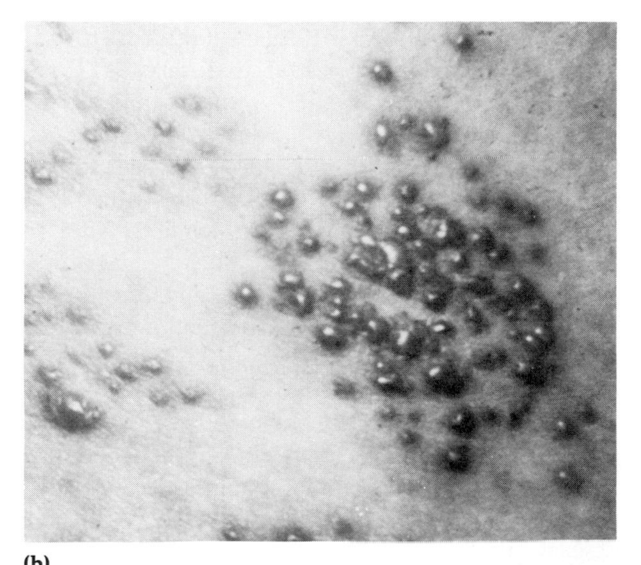

(b)

Figure 19–8 Typical lesions associated with **(a)** chickenpox and **(b)** shingles.

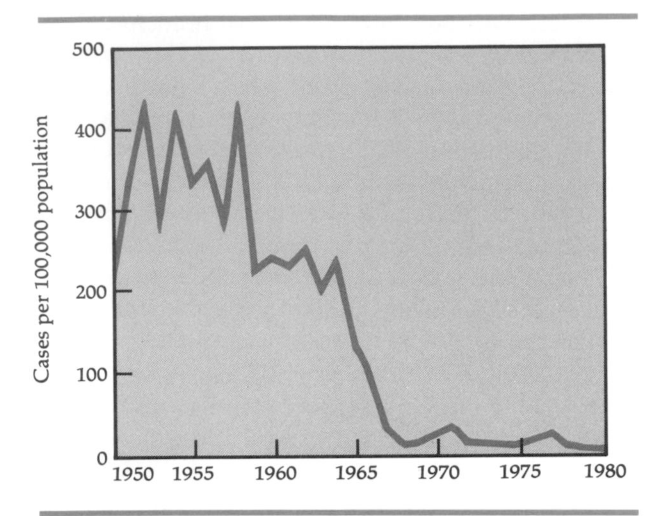

Figure 19–9 Reported measles (rubeola) cases, by year, in the United States, 1950–1980.

humans are the only reservoir in most parts of the world, although monkeys are susceptible. Therefore, measles can potentially be eradicated, much as smallpox was (Figure 19–9).

Measles is an extremely dangerous disease, particularly for the very young and the elderly. It is frequently complicated by middle ear infection or pneumonia caused by the virus itself or by secondary, bacterial infection. Encephalitis occurs in approximately one in every thousand measles victims, often leaving survivors with permanent brain damage. As many as one case in a thousand also may be fatal, usually because of encephalitis or respiratory problems. The virulence of the virus seems to vary with different epidemic outbreaks. Complications such as encephalitis appear about a week after the rash appears.

The sequence of the disease is similar to that observed in smallpox and chickenpox. Infection is initiated in the upper respiratory system. After an incubation period of 10 to 12 days there are symptoms resembling a common cold—sore throat, headache, and cough. Shortly thereafter, a papular rash appears on the skin (Figure 19–10). Lesions of the oral cavity include the diagnostically useful *Koplik spots* (tiny red patches with central white specks) on the oral mucosa opposite the molars.

Measles viruses may remain latent in the recovered patient, as with a number of other viral diseases that give a very solid immunity. The virus may affect only a few cells at a time. In some persons this may lead to an autoimmune reaction. In fact, one school of thought holds that the measles–encephalitis symptoms are part of an autoimmune reaction. Such latent infections have been suspected as the cause of several autoimmune-type diseases, such as **subacute sclerosing panencephalitis (SSPE).** This disease occurs in patients, usually children or young adults, who have had a previous history of measles. There is a sudden, rapid, and fatal degeneration of the nervous system.

An effective vaccine for measles, based on an attenuated live virus, is now recommended for all young children. Immunity seems to be permanent and effectiveness is approximately 95%. Early experience with the vaccine showed that children who received it in the first year of life had a relatively low immunity, and many of these children had to be revaccinated. Now the vaccine is not administered until the child is at least 15 months of age. Frequently, the vaccine is of a trivalent type, consisting of measles, mumps, and rubella (MMR).

Measles has a tendency to die out in isolated populations or small towns because of a lack of newly infectious cases and susceptible persons. When reintroduced after many years, the virus may have more severe effects than normal. And if introduced into groups with no previous exposure, measles can have a very high mortality rate.

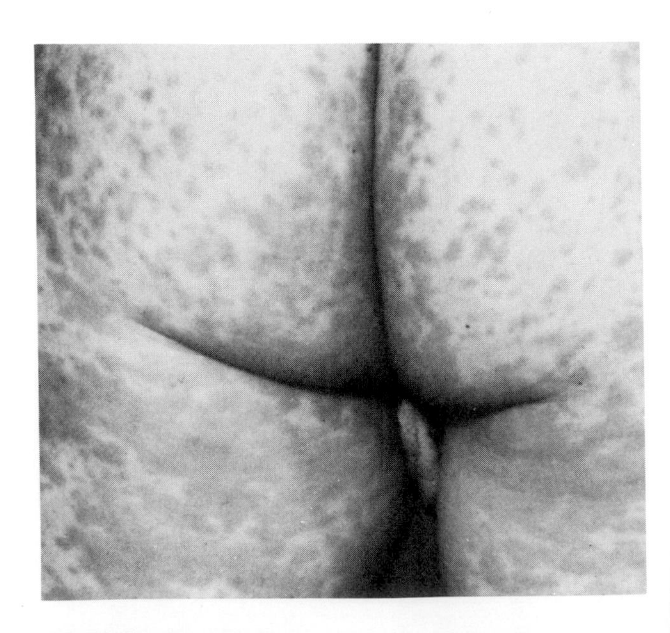

Figure 19–10 Papular rash typical of measles (rubeola).

German Measles (Rubella)

Rubella is caused by a togavirus (rubella virus). It is a much milder disease than rubeola and often goes undetected (a subclinical case). A macular rash accompanied by a light fever are the usual symptoms (Figure 19–11). Transmission is airborne, and an incubation time of two to three weeks is the norm. The seriousness of this disease was not really appreciated until 1941, when the association was made between serious birth defects and maternal infection during the first trimester (three months) of pregnancy—the now familiar **congenital rubella syndrome.** There is about a 35% incidence of serious fetal damage when the disease occurs at this time, often resulting in a stillborn child or in deafness, eye cataracts, heart defects, or mental retardation. Fifteen percent of babies with congenital rubella syndrome die in their first year.

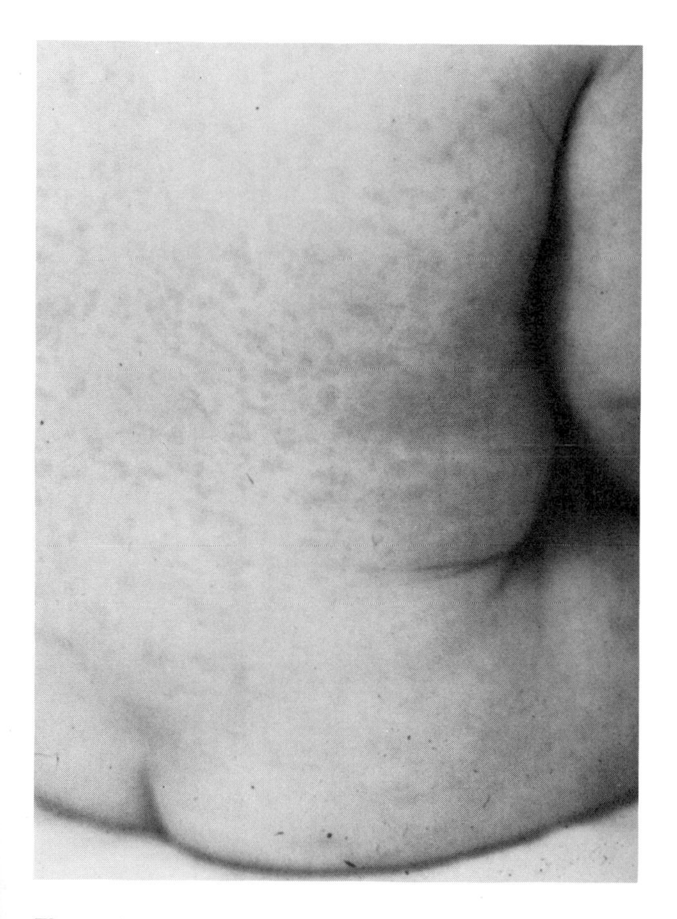

Figure 19–11 Macular rash characteristic of German measles (rubella).

The last major epidemic of rubella in the United States was during 1964 to 1965. About 20,000 severely damaged children from this epidemic are still alive today. The immune status of women of childbearing age is therefore the main concern in the prevention of this disease.

Recovery from clinical or subclinical cases of German measles appears to give a firm immunity, and about ten years ago, a rubella vaccine became available. Follow-up studies indicate that in the last ten years, antibody levels have declined little among those children who were the first to receive the vaccine at an age greater than one year. However, the question of permanent immunity from vaccination will not be answered for a number of years. Nevertheless, vaccinating children prevents them from spreading the disease, even if immunity is not permanent. It may prove necessary to revaccinate women as they enter their childbearing years. There have been a number of cases in which women who did not know they were pregnant received the vaccine. The evidence from these cases suggests that the vaccine itself is not damaging to the fetus. The vaccine is still not recommended for pregnant women, however. Immunoglobulin for passive immunization against rubella is available and is sometimes recommended for exposed persons living in the same household as a pregnant woman.

Cold Sores (Herpes Simplex)

The most striking characteristic of the herpes simplex virus, a herpesvirus, is its ability to remain latent for long periods of time without expressing pathogenicity. During periods of latency, the virus is probably infecting a very few cells at a time and propagating itself slowly. The initial infection usually occurs in infancy. Frequently this is subclinical, but perhaps as many as 15% of the cases develop lesions, usually in the oral mucous membrane (Figure 19–12). These infections subside but recur in the form of what are known as **cold sores** or **fever blisters.** Appearance of these lesions is usually associated with some trauma; excessive exposure to UV radiation from the sun is a frequent cause, as are the hormonal changes associated with menstrual periods or even emotional upset. These infections are due to *herpes simplex type 1 virus,*

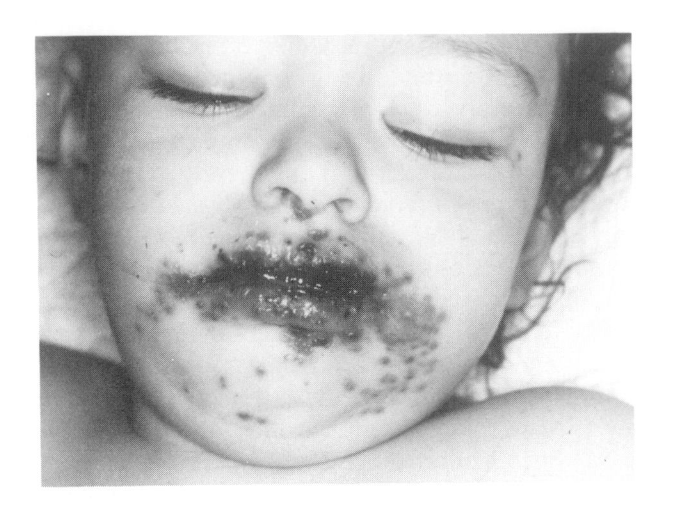

Figure 19–12 Cold sore blisters caused by herpes simplex virus.

which is transmitted primarily by oral or respiratory routes. Another infection by a very similar virus, *herpes simplex type 2 virus* (differentiated from type 1 antigenically and by its effect on cells in tissue cultures) is transmitted primarily by sexual contact. We will discuss this in a later chapter in our discussion of diseases of the urinary and genital systems. There is not yet any effective treatment available for cold sores or genital herpes infections.

FUNGAL DISEASES OF THE SKIN

The skin is most susceptible to microorganisms able to resist high osmotic pressure and low moisture. It is not surprising, therefore, that fungi cause a number of skin disorders. Any fungal infection of the body is referred to as a **mycosis.**

Scabies—An Itchy Infection by a Parasitic Arthropod

Scabies is a disease of human skin caused by the mite *Sarcoptes scabiei,* a parasitic arthropod. The disease is common among school children and is also found in adults. Sometimes it occurs as a nosocomial infection in hospital personnel treating patients with symptoms described as "pruritic dermatitis."

The fingers, wrist, and elbows are the most frequent sites of infection. The mites burrow into the skin and fill the tunnel with their eggs and feces. The eggs hatch and new mites mature, mate, and lay more eggs—perpetuating the life cycle. The symptoms of scabies are the result of hypersensitivity reactions to the mites. Symptoms first occur two to six weeks after the initial infection. The main symptom is itching, especially when the skin is warm (for example, when in bed at night). Red, raised lesions ("erythematous papules") develop, which may become infected with bacteria through scratching. More advanced, chronic cases may result in generalized eczema.

Diagnosis is made by examination of the skin with a 10× hand lens. Burrows can sometimes be seen, and the mites can be picked out with a needle for microscopic examination. Alternatively, scrapings from lesions may be examined microscopically for mites. Scabies is treated by topical application of gamma benzene hexachloride (Kwell). Clothing, bedding, and other personal objects that may contain mites must be thoroughly cleaned. Scabies is transmitted by direct contact with infected persons or with fomites carrying female mites.

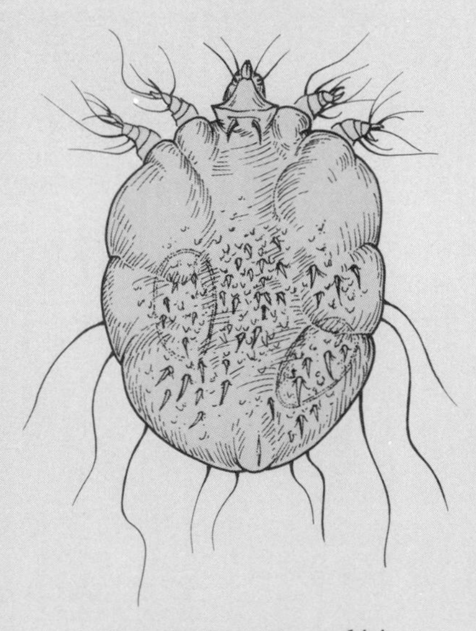

Female *Sarcoptes scabiei.*

Superficial Mycoses

Superficial mycoses are limited to the outermost layers of the skin or hair. **Tinea nigra** is an almost asymptomatic superficial mycosis caused by *Cladosporium werneckii.* The infection causes brown or black blotches on the palms of the hands. The patient is usually more affected by the appearance of the dark spots than by the infection itself. The infection can be treated successfully with topical application of 3% sulfur and 2% salicylic acid or tincture of iodine. Griseofulvin, an antifungal antibiotic that is taken orally, is effective against some superficial fungal infections.

Cutaneous Mycoses

A very common type of cutaneous mycosis or *dermatomycosis* is **ringworm,** or **tinea.** Fungi associated with ringworm include members of the genera *Microsporum* (see Color Plate IVA), *Trichophyton,* and *Epidermophyton.* The name ringworm arose from the ancient Greek belief that the infections, which tend to expand in a circular form, were caused by a worm. The Romans incorrectly associated ringworm with lice, and the term tinea is derived from the Latin for "insect larvae."

Tinea capitis, or ringworm of the scalp, is fairly common among elementary school children and may result in patches of baldness (Figure 19–13a). It is usually transmitted by contact with fomites. Also, dogs and cats are frequently infected with fungi that cause ringworm in children. Ringworm of the groin, or jock itch, is known as **tinea cruris** and ringworm of the feet, or athlete's foot, is known as **tinea pedis** (Figure 19–13b).

The nails of the feet or hands can also be infected by fungal diseases, as indeed can almost any portion of the body. The affinity seems to be for the keratin-containing epidermis. Hair and nails in particular have a high keratin content. However,

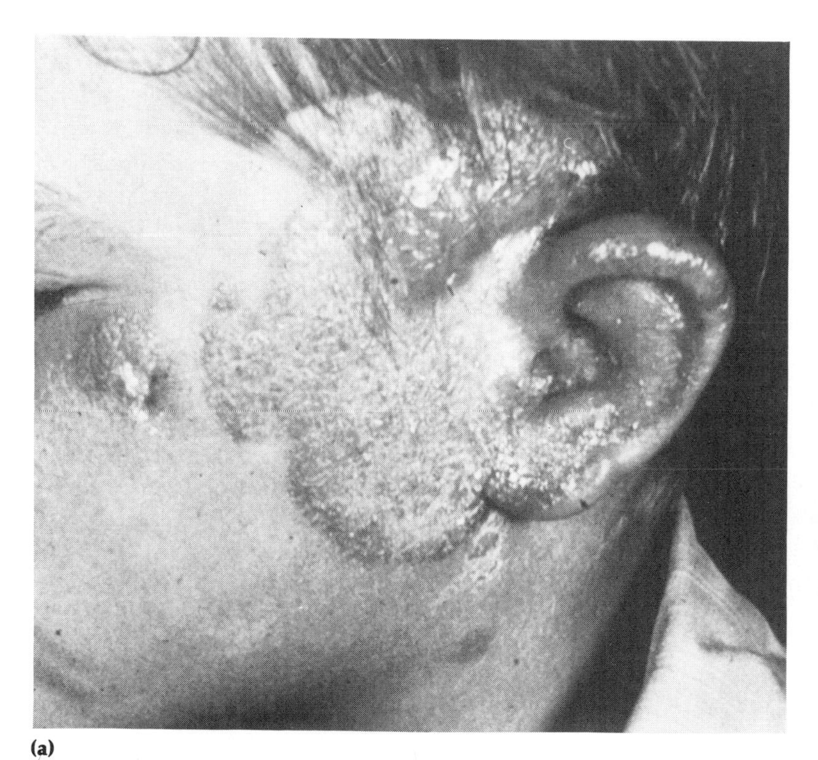

(a)

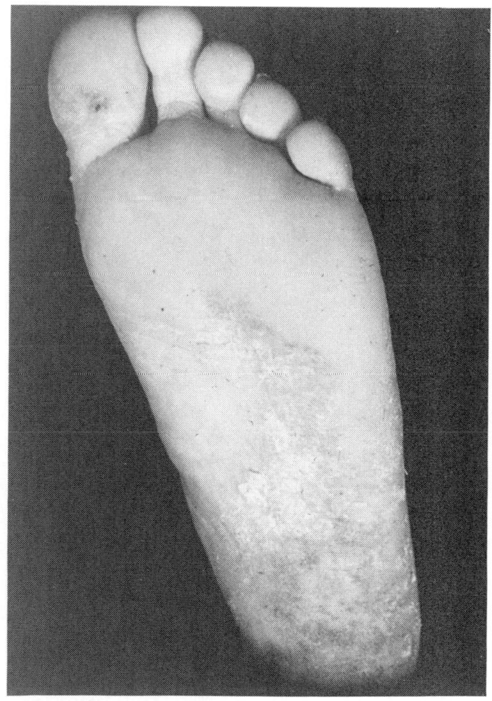

(b)

Figure 19–13 Dermatomycoses. **(a)** A severe case of ringworm on the side of a child's head *(tinea capitis).* **(b)** Ringworm of the foot, or athlete's foot *(tinea pedis).*

these fungi do not seem to be able to invade the lower tissue and organs of the body, even when deliberately injected.

Treatment of fungal skin diseases is usually limited to topical antifungal chemicals of the imidazole and polyene groups (miconazole and amphotericin B). The antibiotic griseofulvin is used particularly in nail infections that cannot be reached topically. It has to be taken for long periods (weeks or months), so it is not used if a topical treatment is effective.

Subcutaneous Mycoses

Subcutaneous mycoses are caused by fungi that are inoculated into a wound. The fungus grows and forms a nodule just under the skin. The nodule may then ulcerate and spread via the lymphatic system to form a series of subcutaneous nodules along the lymphatic vessels. In one type of subcutaneous mycosis, **chromomycosis,** the lesions are pigmented ("chromo") red or violet. This mycosis is caused by several black molds, including *Phialophora verruscosa* and *Fonsecaea pedrosoi.* Flucytosine and bendazole, given orally, can be effective in treatment. Another subcutaneous mycosis is **maduromycosis,** and it is caused by the fungus *Allescheria boydii.* Maduromycosis may be referred to as a **mycetoma** or fungal tumor. The site of infection exhibits slow swelling and eventual draining from many fistulas. Polyenes, which have not yet been used extensively, may be of value, since the causative agent is susceptible in vitro. Surgical drainage is also important.

Candidiasis

Candida albicans is a yeastlike fungus (Figure 19–14a and b and Color Plate IVC), that commonly grows on mucous membranes in the genitourinary tract and on the oral mucosa. *Candida albicans* is a very common cause of **vaginitis** (see Chapter 25) and is also common in the newborn, where it appears as an overgrowth of the oral cavity and is referred to as **thrush** (Figure 19–14c). It is also found in the elderly and those debilitated by diseases such as diabetes and cancer. Becuase it is not affected by antibacterial drugs, it sometimes overgrows tissue when the normal bacterial flora is suppressed by antibiotics. If **candidiasis,** as infections caused by this fungus are called, becomes systemic, as may happen in immunosuppressed individuals, it can result in a fulminating disease and death.

Nystatin is used topically for *Candida* infections. Flucytosine, taken orally in combination with amphotericin B, is also used to treat candidiasis. A summary of systemic mycoses is presented in Table 11–3. A summary of diseases associated with the skin is presented in Table 19–1.

INFECTIONS OF THE EYE

The epithelial cells covering the eye may be considered, in one sense, as a continuation of the skin or mucosa. Microorganisms associated with the eye are usually from the skin and upper respiratory tract. The most common organisms associated with the eye are *Staphylococcus epidermidis, S. aureus,* and diphtheroids.

Bacterial Infections

A number of bacteria may cause eye infections, largely by infection of the conjunctiva. The role of cosmetics should be mentioned here. Microorganisms can not only survive in eye cosmetics, but can multiply there as well. The cosmetics may be sterile prior to being opened, but an initial application can inoculate normal flora into the cosmetics, and subsequent applications can serve to inoculate microorganisms into the eye.

Conjunctivitis refers to an inflammation of the conjunctiva, the mucous membrane that lines the eyelid and covers the outer surface of the eyeball. One form of conjunctivitis, which occurs primarily in children, is called **contagious conjunctivitis** or **pinkeye.** Contagious conjunctivitis is caused by *Hemophilus aegyptius.*

Neonatal gonorrheal ophthalmia (inflammation of the eye in the newborn) is estimated to be responsible for 10% of all cases of blindness. Infection of the newborn occurs during passage through the birth canal. The causative agent, *Neisseria gonorrhoeae,* is transmitted from a mother with gonorrhea. All newborn infants have their eyes washed with 1% silver nitrate, tetracycline, or erythromycin. This procedure prevents the growth of *Neisseria*

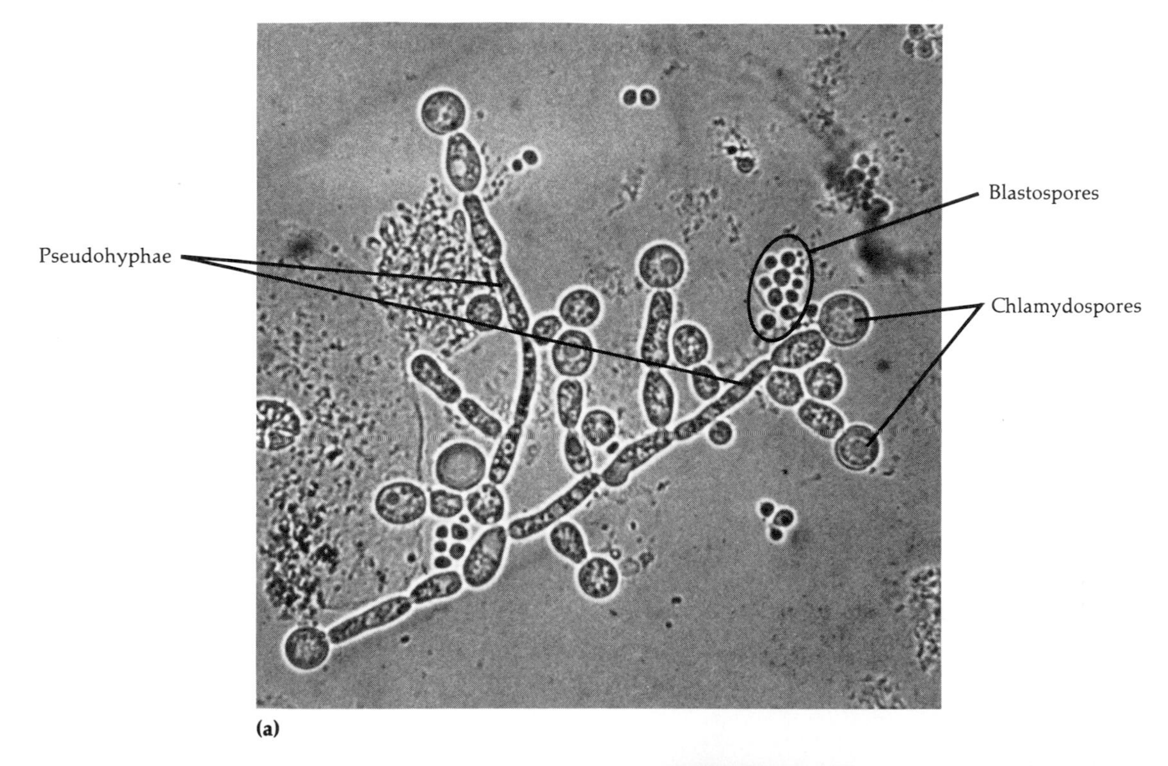

(a)

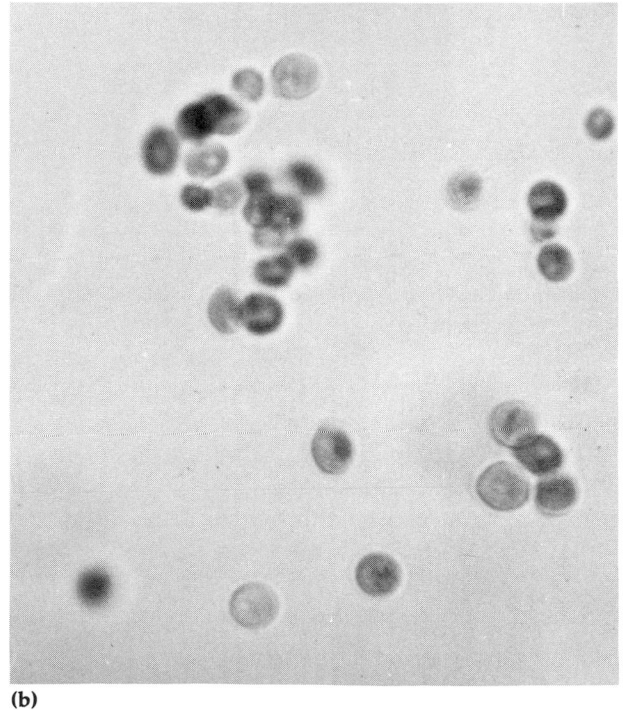
(b)

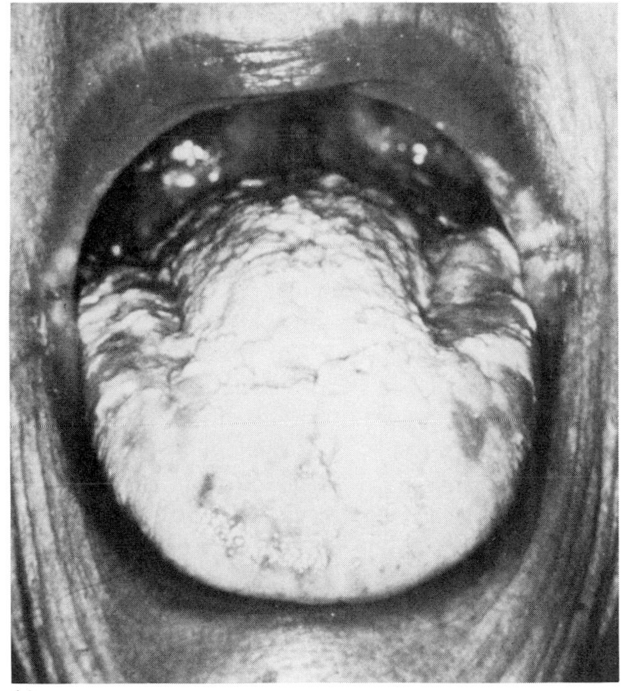
(c)

Figure 19–14 Candidiasis. **(a)** Photomicrograph of *Candida albicans*, an opportunistic fungal pathogen that causes candidiasis. Note the spherical chlamydospores, the smaller blastospores, and the pseudohyphae. **(b)** The yeast form of *Candida albicans* (×2460). *Candida albicans* tends to form pseudohyphae in the body but usually grows in the yeast form in cultures. **(c)** Thrush, or oral candidiasis.

Table 19–1 Summary of Diseases Associated with the Skin

Disease	Causative Agent	Treatment
Bacterial Diseases:		
Impetigo	*Staphylococcus aureus;* occasionally, *Streptococcus pyogenes*	Penicillin, vancomycin, cephalothin, erythromycin, lincomycin
Erysipelas	*Streptococcus pyogenes*	Penicillin, erythromycin
Otitis externa and burn infections	*Pseudomonas aeruginosa*	Polymyxins, gentamicin
Acne	*Propionibacterium acnes* and *Staphylococcus* species	Benzoyl peroxide, tetracyclines
Viral Diseases:		
Warts	Papovavirus	May be removed by liquid nitrogen cryotherapy, cantharidin, salicylic acid, dinitrochlorobenzene, bleomycin sulfate, surgical excision
Smallpox (variola)	Poxvirus (smallpox virus)	None
Chickenpox (varicella)	Herpesvirus (herpes zoster virus)	None
Measles (rubeola)	Paramyxovirus (measles virus)	None
German measles (rubella)	Togavirus (rubella virus)	None
Cold sores (herpes simplex)	Herpesvirus (herpes simplex virus)	None
Fungal Diseases:		
Tinea nigra	*Cladosporium werneckii*	3% sulfur and 2% salicylic acid, tincture of iodine, griseofulvin
Ringworm (tinea)	*Microsporum, Trichophyton, Epidermophyton* species	Griseofulvin, miconazole, amphotericin B
Chromomycosis	*Phialophora verruscosa, Fonsecaea pedrosoi*	Flucytosine and bendazole
Maduromycosis	*Allescheria boydii*	Polyenes might be useful and surgical drainage
Candidiasis	*Candida albicans*	Nystatin, flucytosine and amphotericin B in combination

in the event that the mother's infection was undiagnosed or untreated. Penicillin is used for treatment of neonatal gonorrheal ophthalmia.

Trachoma is the greatest single cause of blindness in the world today. Trachoma is an infection of the epithelial cells of the eye caused by *Chlamydia trachomatis,* a species of bacteria that grows only as obligate intracellular parasites (see Figure 10–15b). Trachoma causes scar tissue to form on the cornea. Millions of people in the world now have the disease and many millions have been blinded by the infection. It is transmitted largely by contact from objects such as towels or by fingers carrying the organism from one eye to another. Flies are also suspected of carrying the causative agent. In some parts of the world almost all children become infected early in life. Opportunistic bacterial infections frequently contribute to the pathogenicity of the disease. Immunity of a partial nature is generated by recovery. Antibiotics, such as sulfonamide and tetracycline, are useful in treatment, particularly in the early stages. However, the control of the disease lies more in sanitation and health education than in treatment.

A disease that is somewhat similar, but much milder, is frequently found in the more developed countries. **Inclusion conjunctivitis,** which is caused by the same organism as trachoma, differs from trachoma in that it does not involve the cornea but is an infection of the conjunctiva. This disease is apparently acquired by newborn infants as they pass through the birth canal. But it also appears to spread in unchlorinated waters such as swimming pools (swimming pool conjunctivitis).

Viral Infections

Epidemic keratoconjunctivitis first occurred in the mainland United States during 1941 to 1942, prob-

Table 19–2 Summary of Diseases Associated with the Eyes

Disease	Causative Agent	Treatment
Bacterial Diseases:		
Contagious conjunctivitis (pinkeye)	*Hemophilus aegyptii*	Sulfonamides
Neonatal gonorrheal ophthalmia	*Neisseria gonorrhoeae*	Silver nitrate, tetracycline, or erythromycin for prevention; penicillin for treatment
Trachoma	*Chlamydia trachomatis*	Sulfonamide, tetracycline
Inclusion conjunctivitis	*Chlamydia trachomatis*	Sulfonamide, tetracycline
Viral Diseases:		
Epidemic keratoconjunctivitis	Adenovirus	None
Herpetic keratitis	Herpesvirus (herpes simplex type 1)	Idoxuridine, vidarabine

ably imported from Hawaii. Epidemic keratoconjunctivitis is caused by an adenovirus, and the infection occurs after trauma to the eye. Airborne particles can cause sufficient damage to allow the virus to grow.

Herpetic keratitis is usually caused by herpes simplex type 1 virus. This is a localized infection of the cornea that may recur with an epidemiology similar to cold sores. It is characterized by inflammation and corneal ulcers that may be quite deep. Swelling of regional lymph nodes accompanies the infection, and damage to the cornea can lead to blindness. Invasion of the central nervous system resulting in encephalitis occasionally occurs in adults.

Idoxuridine, an analog of the DNA component thymidine, can be used in the treatment of herpetic keratitis when applied topically. Vidarabine (adenine arabinoside or ara-A) is also effective against herpetic keratitis and has also been approved for use against herpes-caused encephalitis. These herpes infections represent, then, one of the few instances in which we have effective antiviral chemotherapeutic agents.

A summary of diseases associated with the eyes is presented in Table 19–2.

Now that you have a general understanding of the structure of the skin and its relationship to pathogens, we will turn to diseases associated with wounds and bites, as well as nosocomial infections.

STUDY OUTLINE

INTRODUCTION (p. 488)

1. The skin is a physical and chemical barrier against microorganisms.
2. Moist areas of the skin (axilla) support larger populations of bacteria than dry areas (scalp).

STRUCTURE AND FUNCTION OF THE SKIN (pp. 488–489)

1. The outer portion of the skin, called the epidermis, contains keratin.
2. The inner portion of the skin, or dermis, contains hair follicles, sweat ducts, and oil glands that open to the surface.
3. Sebum and perspiration are secretions of the skin that can remove organisms from openings in the dermis.
4. Sebum and perspiration provide nutrients for some microorganisms.

5. Microorganisms can gain access to the body through the openings in the dermis.
6. Mucous membranes line the body cavities. Mucus traps microorganisms and facilitates their removal from the body.
7. The eye is mechanically washed by tears that contain lysozyme.

NORMAL FLORA OF THE SKIN (pp. 489–491)

1. Members of the genus *Propionibacterium* metabolize oil from the oil glands and colonize hair follicles.
2. Microorganisms that live on human skin are resistant to high salt concentrations and desiccation.
3. Antimicrobial substances secreted by normal flora can inhibit the growth of potential pathogens.
4. The normal flora of the face is determined by the oropharyngeal organisms, and the normal flora of the anal region is influenced by organisms in the lower gastrointestinal tract.

BACTERIAL DISEASES OF THE SKIN (pp. 491–495)

Staphylococcal Skin Infections (pp. 491–493)

1. Staphylococci are gram-positive bacteria, often in grapelike clusters. They are facultative anaerobes, catalase-positive.
2. Practically all pathogenic strains of *S. aureus* produce coagulase.
3. Pathogenic *S. aureus* are pyogenic and may produce enterotoxins and leukocidins.
4. About 30% of the human population are carriers of pathogenic *S. aureus*.
5. Localized infections (pimples, abscesses, and carbuncles) result from *S. aureus* entering openings in the skin.
6. *S. epidermidis* and *Micrococcus* species are related to *S. aureus* and are normally nonpathogenic.
7. Many strains of *S. aureus* produce penicillinase.
8. Impetigo is a highly contagious superficial skin infection caused by *S. aureus*.

Streptococcal Skin Infections (pp. 493–494)

1. Streptococci are gram-positive cocci, often in chains. They are strictly fermentative, catalase-negative.
2. Streptococci are classified according to their hemolytic enzymes and cell wall antigens.
3. Group A beta-hemolytic streptococci (including *S. pyogenes*) are the most important human pathogens.

4. Group A beta-hemolytic streptococci produce a number of virulence factors, including erythrogenic toxin, deoxyribonuclease, NADase, streptokinases, and hyaluronidase.
5. Streptococci are susceptible to penicillin.
6. Erysipelas is a skin infection characterized by reddish patches and caused by *S. pyogenes*.

Infections by Pseudomonads (pp. 494–495)

1. Pseudomonads are gram-negative rods. They are strict aerobes that are oxidase-positive.
2. *Pseudomonas aeruginosa* is an opportunistic pathogen found primarily in soil, water, and on plants. About 10% of the human population carry *P. aeruginosa* in their intestinal tracts.
3. *P. aeruginosa* produces an endotoxin and several exotoxins.
4. Diseases caused by *P. aeruginosa* include otitis externa and post-burn infections.
5. Infections have a characteristic blue-green pus caused by a water-soluble pigment, pyocyanin.
6. Polymyxin and gentamicin are useful in treating *P. aeruginosa* infections.

Acne (p. 495)

1. *Propionibacterium acnes* can metabolize sebum that has become trapped in hair follicles.
2. Metabolic end products (fatty acids) cause an inflammatory response known as acne.
3. Treatment with benzoyl peroxide and tetracycline is somewhat effective.

VIRAL DISEASES OF THE SKIN (pp. 495–500)

Warts (p. 495)

1. Papovavirus causes skin cells to proliferate, resulting in a benign growth called a wart.
2. Warts are spread by direct contact.
3. Warts may regress spontaneously or be removed chemically or physically.

Smallpox (Variola) (pp. 495–496)

1. Variola virus causes two types of skin infections: variola major and variola minor.
2. Smallpox is transmitted by the respiratory route, and the virus is moved to the skin via the bloodstream.

3. At any given time, the vesicular or pustular lesions of the skin and mucous membranes will be in the same stage of development.
4. The only host for smallpox is humans. WHO has announced the eradication of this disease.
5. Routine vaccination with the live virus is no longer used.

Chickenpox (Varicella) (p. 497)

1. Chickenpox is caused by herpes zoster virus.
2. The virus is transmitted by the respiratory route and localized in skin cells, causing a vesicular rash.
3. After recovery from chickenpox, the virus can remain latent in nerve cells. Subsequent activation of the virus results in shingles.
4. Shingles is characterized by a vesicular rash along the affected cutaneous sensory nerves.
5. Complications of chickenpox include encephalitis.

Measles (Rubeola) (pp. 497–498)

1. Measles is caused by measles virus (paramyxovirus) and transmitted by the respiratory route.
2. After incubation in the upper respiratory tract, papular lesions appear on the skin and Koplik spots on the oral mucosa.
3. Complications of measles include middle ear infections, pneumonia, encephalitis, and secondary bacterial infections.
4. Measles virus may remain latent following infection and cause autoimmune reactions such as subacute sclerosing panencephalitis.
5. Vaccination with attenuated live virus provides effective long-term immunity.

German Measles (Rubella) (p. 499)

1. The rubella virus (togavirus) is transmitted by the respiratory route.
2. A macular rash and light fever may occur in an infected individual, or the disease may be asymptomatic.
3. Congenital rubella syndrome may affect a fetus when a woman contracts rubella during the first trimester of pregnancy.
4. Congenital rubella syndrome includes stillbirth, deafness, eye cataracts, heart defects, and mental retardation.
5. Vaccination with live rubella virus provides immunity of unknown duration.
6. Passive immunization with immunoglobulin is available for exposed persons who are in close contact with a pregnant woman.

Cold Sores (Herpes Simplex) (pp. 499–500)

1. Herpes simplex type 1 virus is transmitted primarily by oral and respiratory routes.
2. Infection of skin and mucosal cells results in cold sores.
3. The virus remains latent in nerve cells, and cold sores can recur when the virus is activated.

FUNGAL DISEASES OF THE SKIN (pp. 500–502)

Superficial Mycoses (p. 501)

1. Tinea nigra is characterized by discolored blotches on the palms.
2. It is treated with topical antiseptics.

Cutaneous Mycoses (pp. 501–502)

1. *Microsporum*, *Trichophyton*, and *Epidermophyton* can cause dermatomycoses called ringworm, or tinea.
2. These fungi grow on keratin-containing epidermis, such as hair and nails.
3. Ringworm and athlete's foot are usually treated with topical application of antifungal chemicals.

Subcutaneous Mycoses (p. 502)

1. Chromomycosis and mycetoma result from fungi that are inoculated into a wound.
2. The fungi grow and produce subcutaneous nodules along the lymphatic vessels.
3. Flucytosine and bendazole are used to treat these infections.

Candidiasis (p. 502)

1. *Candida albicans* causes infections of mucous membranes and is a common cause of vaginitis and thrush (oral mucosa).
2. *C. albicans* is an opportunistic pathogen that may proliferate when normal bacterial flora are suppressed.
3. Topical or systemic antifungal chemicals may be used to treat candidiasis.

INFECTIONS OF THE EYE (pp. 502–505)

1. *Staphylococcus epidermidis*, *S. aureus*, and diphtheroids are normal flora of the eye.
2. These microorganisms are usually from the skin and upper respiratory tract.

Bacterial Infections (pp. 502–504)

1. Conjunctivitis is caused by a number of bacteria.
2. Contagious conjunctivitis, or pinkeye, is caused by *Hemophilus aegyptius* and occurs in children.
3. Neonatal gonorrheal ophthalmia is caused by the transmission of *Neisseria gonorrhoeae* from an infected mother to a newborn during passage through the birth canal.
4. All newborn infants are treated with 1% silver nitrate, tetracycline, or erythromycin to prevent the growth of *Neisseria*.
5. Trachoma is caused by *Chlamydia trachomatis* and causes scar tissue to form on the cornea.
6. Trachoma is transmitted by fingers, fomites, and perhaps flies.
7. Inclusion conjunctivitis is an infection of the conjunctiva caused by *Chlamydia trachomatis*. It is transmitted to infants during birth, and also transmitted in unchlorinated swimming water.

Viral Infections (pp. 504–505)

1. Epidemic keratoconjunctivitis follows trauma to the eye and is caused by an adenovirus.
2. Herpetic keratitis causes corneal ulcers. The etiology is herpes simplex type 1 that occasionally invades the central nervous system, resulting in encephalitis.
3. Idoxuridine and vidarabine are effective against herpes keratitis.

STUDY QUESTIONS

REVIEW

1. Discuss the usual mode of entry of bacteria into the skin. Compare bacterial skin infections with those caused by fungi and viruses with respect to method of entry.
2. Complete the following table of epidemiology.

Disease	Etiological Agent	Clinical Symptoms	Method of Transmission
Acne			
Pimples			
Warts			
Smallpox			
Chickenpox			
Measles			
German measles			
Cold sores			

3. How do these infections differ: tinea nigra, mycetoma, athlete's foot? In what ways are they similar?
4. Select a bacterial and viral eye infection and discuss the epidemiology of each.
5. A laboratory test used to determine the identity of *Staphylococcus aureus* is its growth on mannitol salt agar. The medium contains 7.5% sodium chloride (NaCl). Why is it considered a selective medium for *S. aureus?*
6. Why are people immunized against rubella, since the symptoms of the disease are mild or even inapparent?
7. Explain the relationship between shingles and chickenpox.
8. Why are the eyes of all newborn infants washed with an antiseptic or antibiotic?
9. What is the leading cause of blindness in the world?
10. An opportunistic yeast that causes skin infections is ______________.
11. Identify the following diseases based on the symptoms.

Symptoms	Disease
Koplik spots	
Pustular rash	
Vesicular rash	
Macular rash	
Recurrent "blisters" on oral mucosa	
Corneal ulcer and swelling of lymph nodes	
Pustular rash, appearance same over entire body	

12. What complications can occur from herpes simplex type 1 infections?

CHALLENGE

1. You have isolated an organism from what appears to be impetigo. The organism is gram-positive cocci in singles, pairs, and small groups. What test would help you quickly determine whether your isolate is *Staphylococcus* or *Streptococcus?* What is the result of this test if the organism is *Staphylococcus?*
2. A hospitalized patient recovering from surgery develops an infection that has blue-green pus and a grapelike odor. What is the probable etiology? How might the patient have acquired this infection?
3. Is it necessary to treat a patient for warts? Explain briefly.

FURTHER READING

di Mayorca, G. (chairperson). 1978. Human papovaviruses. In *Microbiology 1978*, pp. 419–442. Washington, D.C.: American Society for Microbiology (Proceedings). Biology and oncogenicity of papovaviruses.

Henderson, D. A. October 1976. The eradication of smallpox. *Scientific American* 235:25–33. A summary of the incidence of smallpox and investigation of the last endemic infection.

Marples, M. J. January 1969. Life on the human skin. *Scientific American* 220:108–115. A study of the microflora of the skin and the relationship between bacterial populations and hair, moisture, and temperature.

Peterson, A. F. (convener). 1976. The microbiology of the eye. In *Developments in Industrial Microbiology* 17:13–47. Washington, D.C.: American Institute of Biological Science. A survey of normal eye flora and transient flora and diseases associated with contact lenses and cosmetics.

Tachibana, D. K. 1976. Microbiology of the foot. *Annual Review of Microbiology* 30:351–375. An interesting report on the normal flora of the foot, foot infections, and foot odor.

FURTHER READING FOR PART IV

The following medical microbiology textbooks include principles of microbiology with an emphasis on bacterial pathogens. Some texts also include viral, fungal, and parasitic diseases.

Davis, B. D., R. Dulbecco, H. N. Eisen, and H. S. Ginsberg. 1980. *Microbiology*, 3rd ed. New York: Harper and Row.

Freeman, B. A. 1979. *Burrow's Textbook of Microbiology*, 21st ed. Philadelphia, PA: W. B. Saunders.

Hoeprich, P. D. (ed.). 1977. *Infectious Diseases*. New York: Harper and Row.

Joklik, W. K., H. P. Willett, and D. B. Amos (eds.). 1980. *Zinsser Microbiology*, 17th ed. New York: Appleton-Century-Crofts.

Mandell, G. L., R. G. Douglas, and J. E. Bennett. 1979. *Principles and Practice of Infectious Disease*, 2 vols. New York: Wiley.

20

Nosocomial Infections and Diseases Associated with Wounds and Bites

Objectives

After completing this chapter you should be able to

- Discuss the importance of nosocomial infections.
- Write a general rule for the prevention of nosocomial infections.
- Discuss the causative agents, symptoms, and method of transmission of tetanus and gas gangrene.
- List four bacterial infections that result from animal bites.
- Identify the best control measure(s) for mosquito-borne diseases.
- Identify the causative agent, vector, symptoms, and treatment for malaria, yellow fever, dengue, arthropod-borne encephalitis, and relapsing fever.
- Compare and contrast three rickettsial diseases with respect to causative agent, vector, reservoir, and treatment.
- Discuss the epidemiology of plague.

When the skin is broken by wounds, surgery, or the bites of animals, the body must rely on its internal defenses to combat infection. Recall from Chapter 15 that some of these internal defenses are nonspecific and include antimicrobial substances, phagocytosis, the inflammatory response, and fever. Other internal defenses, as outlined in Chapter 16, represent specific defenses against pathogens and include antibodies. Failure of the internal defenses encourages the continued growth of invading microorganisms.

NOSOCOMIAL INFECTIONS

The word **nosocomial** is derived from the Greek word for "hospital," and refers to hospital-acquired infections. The Centers for Disease Control estimate that 5 to 10% of all hospital patients acquire some type of nosocomial infection, with surgical patients particularly susceptible. After certain types of operations, such as surgery on the large intestine and amputations, the infection rate approaches 30%. Many other hospital procedures, such as catheterization, intravenous feeding, respiratory aids, and injections, also bypass the normal outer defenses of the body and therefore represent sources of infection (see box).

Normal flora from the human body presents a particular danger to hospital patients. Such patients often receive skin punctures and deeper punctures (into a vein, for instance), which can introduce microorganisms into their bodies. Tubs used to bathe patients must be disinfected between uses so that bacteria from the last patient will not contaminate the next. And respirators and humidifiers afford a

Centers for Disease Control
MMWR
Morbidity and Mortality Weekly Report

Endotoxic Reactions Associated with the Reuse of Cardiac Catheters

During a two-week period in July 1978, three cases of suspected endotoxic reactions (fever, chills, and hypotension) occurred in patients undergoing cardiac catheterization at a Massachusetts hospital. An investigation revealed that reusable intravascular catheters, although sterile, were contaminated with endotoxin and that this contamination was related to procedures employed to clean and disinfect the catheters.

At the time the endotoxic reactions occurred, used catheters were rinsed with hospital distilled water, wiped to remove clotted blood, and then soaked in Detergicide, a quaternary ammonium compound. Next, the bore of the catheter was flushed continuously for 2 hours with distilled water, followed by a second flush with 1 liter of commercial pyrogen-free, sterile, distilled water. Finally, the catheters were wrapped and gas-sterilized. The administration set for the delivery of the pyrogen-free fluid had not been changed regularly.

The hospital-supplied distilled water contained greater than 0.2 nanograms (ng) of endotoxin per ml even when samples were taken directly from the storage tanks. Cultures of samples taken from the cardiac catheterization laboratory's distilled water tap and from the research laboratories contained up to 310 colonies of *Pseudomonas cepacia* per ml, while storage tank samples were sterile.

Five previously used, but subsequently cleaned and sterilized, catheters contained levels of endotoxin ranging from 0.3 ng to 7.4 ng per catheter, as measured by the Limulus Amebocyte assay. (A level of 0.05 ng is considered pyrogenic by the laboratory performing the assay.) All catheters were sterile when cultured. New, sterile catheters were free of endotoxin. After only one use and cleaning, catheter flushes yielded excessive endotoxin.

This investigation resulted in the recommendation that disposable catheters be used. When this was not possible, it was recommended that pyrogen-free, sterile, distilled water be used in the cleaning procedure, and that both the outer and inner surfaces of the catheter be flushed. Detergicide was freshly prepared daily, using pyrogen-free, sterile, distilled water. After the catheters were cleaned and packaged, they were stored at 4°C until sterilized. In the five-month period since this outbreak, there have been no further suspected endotoxin reactions.

Editorial Note: The precise mechanism by which catheters used for cardiac catheterization were contaminated with endotoxin was not proven, but this investigation suggests that contamination was introduced during the cleaning procedure. Hospital supplies of distilled water, used in cleaning, were contaminated. Furthermore, the detergent-disinfectant preparation that was used in the preliminary cleaning, an aqueous quaternary ammonium formulation, has been shown to be Ineffective against *P. cepacia* and may even permit the selective growth of this and other microorganisms. Presumably, viable microorganisms were introduced during cleaning from the distilled water, the Detergicide failed to kill the contaminants or permitted their growth, and the final sterilization killed the contaminants but allowed high levels of residual endotoxin to persist. Although a variety of detergent-disinfectants may be used safely for environmental sanitation of floors, surfaces, and the like, these agents must be used with utmost caution with critical medical devices that come in contact with any normally sterile body tissue. An effective sterilizing procedure, such as ethylene oxide, will kill microorganisms but will not remove endotoxin.

This report further highlights the hazards that may result from reuse of some medical devices. Some, but not all, medical devices can be cleaned and safely sterilized. If practical and economical, disposable devices are preferable to reusable devices that are difficult to sterilize.

Source: *MMWR* 28:25 (1/26/79)

suitable growth environment for some bacteria and a method of transmission (aerosols). These sources of nosocomial infections must be kept scrupulously clean and disinfected, and materials used for bandages and intubation should be sterilized prior to use. Moreover, packaging used to maintain sterility should be removed aseptically.

Burn patients are very susceptible to nosocomial infections because their skin is no longer an effective barrier to microorganisms. A **burn** is tissue damage caused by heat, electricity, radiation, or chemicals. The damaging agent causes protein destruction and cell death in affected tissues. Burns are classified as first degree, second degree, or third degree, depending on severity. Criteria include color, presence or absence of sensation, blister formation, and depth of tissue damage. No matter how serious a burn appears on the surface, the systemic effects are far more serious than the local effects. One of the principal systemic effects is bacterial infection.

At one time, gram-positive *Staphylococcus aureus* was the primary cause of nosocomial infections, and antibiotic-resistant strains of the organisms are still a significant factor. Today, however, gram-negative bacteria such as *Escherichia coli* and *Pseudomonas aeruginosa* are usually involved (Table 20–1). Most bacteria causing nosocomial infections are not particularly harmful under ordinary conditions, but are opportunists. As already noted in Chapter 19, *P. aeruginosa* has the ability to cause opportunistic skin infections—especially in surgical and burn patients. *P. aeruginosa*, as well as other gram-negative bacteria, tend to be more difficult to control with antibiotics because of their R factors, which carry genes that determine resistance to antibiotics (Chapter 8).

Table 20–1 Six Bacteria That Account for Over 70% of All Nosocomial Infections

Bacterium	% of Infections
Escherichia coli	21.4
Staphylococcus aureus	11.3
Streptococcus group D	10.8
Pseudomonas species (79% *P. aeruginosa*)	10.1
Proteus species (72% *P. mirabilis*)	8.7
Klebsiella species	8.5

Another group of patients susceptible to nosocomial infections are cancer patients who have had their immune system suppressed by radiation and immunosuppressive drugs. Organ transplant patients receiving immunosuppressive drugs are also susceptible.

Accredited hospitals should have an infection control committee, and most hospitals at least have an infection control nurse or epidemiologist. In this way, problem areas such as antibiotic-resistant strains of bacteria and improper sterilization techniques can be identified. The infection control officer should make periodic examinations of equipment to determine the amount of microbial contamination. Samples should be taken from tubing, catheters, respirator reservoirs, and other equipment. Samples from all fomites must be cultured on differential and selective media.

WOUNDS

A **wound** is a physical injury to the body and may involve a serious rupture of the skin requiring stitches, a simple abrasion, a puncture by a tack or pin, or the bite of an animal. All of these represent penetration of the body's defenses and may result in disease.

Gangrene

If a wound results in interrupted blood supply to a tissue (a condition known as *ischemia*), the wound becomes anaerobic. Ischemia leads to *necrosis*, or death of the tissue. The death of soft tissue due to loss of blood supply is called **gangrene** (Figure 20–1). Substances released from dying and dead cells provide nutrients suitable for many bacteria. Various species of the genus *Clostridium*, which are gram-positive, endospore-forming anaerobes widely found in soil and in the intestinal tracts of humans and domesticated animals, grow readily in such conditions. *Clostridium perfringens* (See Figure 10–12c) is the most common species involved, but other clostridia and an assortment of other bacteria may also be found in such wounds.

Given ischemia and subsequent necrosis, **gas gangrene** can develop, especially in muscle tissue. As the *C. perfringens* microorganisms grow, they

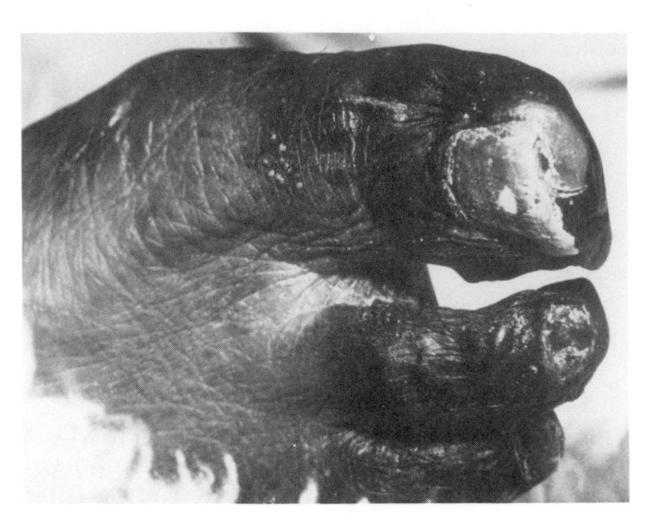

Figure 20–1 Foot of a patient with gangrene, a disease caused by the bacterium *Clostridium perfringens.* Note the black, necrotic tissue on the toes.

ferment carbohydrates in the tissue and produce gases (carbon dioxide and hydrogen), causing the tissue to swell. The microorganisms also produce necrotizing exotoxins and enzymes such as collagenase, proteinase, deoxyribonuclease, and hyaluronidase. Together, these further interfere with blood supply and favor the spread of the infection. As tissue necrosis spreads, there is opportunity for increased microbial growth, hemolytic anemia, and, ultimately, severe toxemia and death.

One complication of improperly performed abortions is the invasion of the uterine wall by *C. perfringens,* which is resident in the genital tract of about 5% of all women. This can lead to gas gangrene and result in a life-threatening invasion of the bloodstream.

Surgical removal of necrotic tissue, called debridement, and amputation are the most common medical treatments for gas gangrene. And instances of gas gangrene in regions such as the abdominal cavity can be treated in *hyperbaric chambers,* where the patient is subjected to an oxygen-rich atmosphere under pressure. The oxygen saturates the infected tissues and prevents the growth of the obligately anaerobic clostridia. Prompt cleaning of serious wounds, accompanied by precautionary antibiotic treatment, are the most effective steps in preventing gas gangrene. Penicillin, clindamycin, and chloramphenicol are antibiotics with good activity against *C. perfringens.*

Tetanus

Another important disease due to clostridial bacteria growing in the damaged tissue of a wound is **tetanus** (Figure 20–2). The causative organism, *Clostridium tetani,* is an obligate anaerobic, endospore-forming, gram-positive rod. It is particularly common in soil contaminated with animal fecal wastes. In the American Civil War, it was observed that the cavalry had a higher rate of tetanus than the infantry, possibly for this reason.

The symptoms of tetanus are due to an extremely potent neurotoxin *(tetanospasmin)* that is released upon the death and lysis of the growing cell (Chapter 14). The bacteria themselves do not spread from the infection site, however; thus tetanus is sometimes called infectious but noncommunicable. There is often no marked inflammation

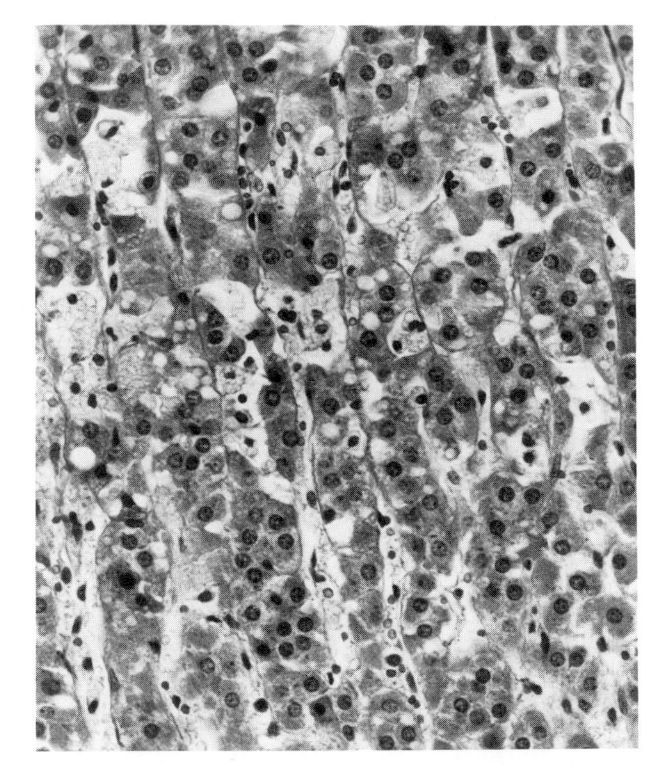

Figure 20–2 Necrotic tissue at the site of infection in a case of tetanus (×130). The light areas are the regions of dead cells, which have pulled away from their neighbors in many cases. Such tissue is low in oxygen concentration, creating an anaerobic situation in which *Clostridium tetani* (not seen at this magnification) may grow.

or observable infection of the wound. Symptoms include spasms, usually near the infected area, followed by contraction of the muscles of the jaw, which prevents opening of the mouth (lockjaw). Gradually other skeletal muscles become involved, including those involved in swallowing. Death results from spasms of the respiratory muscles.

Because, as in gas gangrene, the organism is an obligate anaerobe, the wound must provide anaerobic growth conditions. Improperly cleaned deep puncture wounds and even wounds with little or no bleeding serve admirably. Of the approximately 100 tetanus cases occurring each year in the United States (about a third of which are fatal), many arise from rather trivial injuries.

When a wound is severe enough to be brought to the attention of a physician, the decision is usually made to provide protection against tetanus. Most persons in this country have received DPT immunization, which includes tetanus toxoid to stimulate antibody formation against tetanus toxin. A booster of toxoid may be given when a dangerous wound is received. Even after 20 years or so, the body "remembers" how to manufacture the antibody (the anamnestic response, see Chapter 16). Very rapidly, much more so than on first exposure to the toxoid, a high level of antibody is restored. If a person has not had previous immunity, however, antibodies resulting from an initial toxoid injection given after the injury may appear too late to prevent the disease. In such a case, a temporary immunity, lasting for about a month, can be given by injection of immune globulin containing antibodies to tetanus toxoid. The immune globulin is usually from the serum of humans previously immunized against tetanus. It is also possible to use preformed tetanus antibodies from immunized horses, but this practice increases the chances of hypersensitivity reactions to the serum (serum sickness).

Removal of damaged tissue and use of antibiotics such as penicillin are also useful in therapy. However, once the toxin has attached to the nerves, such therapy is of little use.

In the less developed areas of the world, tetanus affecting the severed umbilical cord of infants is a major cause of death. Worldwide, there are probably several hundred thousand cases of tetanus from all causes each year.

ANIMAL BITES AND SCRATCHES

Animal bites can result in serious wound infections. The animals responsible are usually domestic animals, such as dogs and cats, because of their close contact with humans. Such animals often harbor *Pasteurella multocida,* a nonmotile, gram-negative rod similar to that which causes plague (discussed later in this chapter). *P. multocida* produces acid, but no gas, from glucose, but not from lactose. It also produces hydrogen sulfide and catalase. This bacterium causes **pasteurellosis,** a disease that can take the form of a local tissue infection, a systemic infection, or a respiratory tract infection. Antibiotics, such as penicillin and tetracycline, are usually effective in treatment.

Clostridium species and other anaerobes, such as species of *Bacteroides* and *Fusobacterium,* also may cause infections of deep animal bite wounds. Rat bites typically cause infections by organisms of the genus *Streptobacillus* or *Spirillum,* more commonly in Asiatic countries than in the United States. The disease produced, often called **rat bite fever,** is characterized by recurring fever, arthritis, and infections of the lymph vessels.

Minor cat scratches sometimes result in an ulcerated lesion, perhaps also involving a fever and a generalized illness known as **cat scratch fever.** The probable cause is a member of the genus *Chlamydia.* Although chlamydia have not been isolated from these wounds, patients have antibodies to these bacteria. The cat is apparently unaffected, with no sign of illness. Similar infections can result from wounds by sticks or similar objects. Fatal cases are rare and recovery is spontaneous. Tetracyclines have been shown to be of value in reducing the duration of the disease.

Perhaps the most dangerous disease transmitted by animal bites is **rabies,** a viral disease of the nervous system that will be discussed in Chapter 21.

ARTHROPOD BITES

Arthropods are animals with jointed appendages and include mites and ticks and insects such as mosquitoes, fleas, and lice (see Chapter 11). Certain microorganisms harmless to arthropods may be transmitted to other animals in which they cause disease. Our concern here is to consider several diseases transmitted by arthropod vectors.

Malaria

Malaria is characterized by chills and fever, and often by vomiting and severe headache. These symptoms typically alternate in one- to three-day cycles, with asymptomatic periods in between.

Malaria is found in any part of the world where the protozoan parasite that carries it is present in human hosts and where the mosquito vector, certain species of *Anopheles* (see Figure 11–27b), is found. The disease was once common in the United States, but effective mosquito control and a reduction in the number of human carriers caused the reported cases to drop to under 100 by 1960 (Figure 20–3). In recent years, however, there has been an upward trend in the number of U.S. cases, reflecting a worldwide resurgence of malaria, increased travel to malarial areas, and the increasing number of immigrants from malarial areas. In tropical Asia, Africa, and Central and South America, malaria is still a serious problem. Some cases of malaria are due to blood transfusions or unsterilized syringes used by drug addicts.

The causative organisms of malaria are the spore-forming protozoans (sporozoans) of the genus *Plasmodium*. Four pathogenic species are recognized, each of which causes a distinctive form of the disease. The most dangerous species is probably *Plasmodium falciparum*, a species that is also widespread geographically. Also widely distributed is *Plasmodium vivax*. *Plasmodium malariae* and *Plasmodium ovale* have either a lower infection rate or cause a geographically restricted, relatively milder disease.

When *Plasmodium* first enters a human host, the form of the microorganism called the *sporozoite* enters the liver cells, where it undergoes schizogony (see Figure 11–18). The sporozoites are later released into the bloodstream to infect red blood cells, although some may remain in the liver, only to enter red blood cells months or years later and cause a recurrence of malaria in the patient.

The release of the form called the *merozoite* into the bloodstream causes the paroxysms (recurrences of symptoms) of chills and fever that are the main symptoms of malaria. The fever reaches 40°C and is followed by a sweating stage as the fever subsides. Between paroxysms the patient feels normal. Anemia results from the loss of red blood cells,

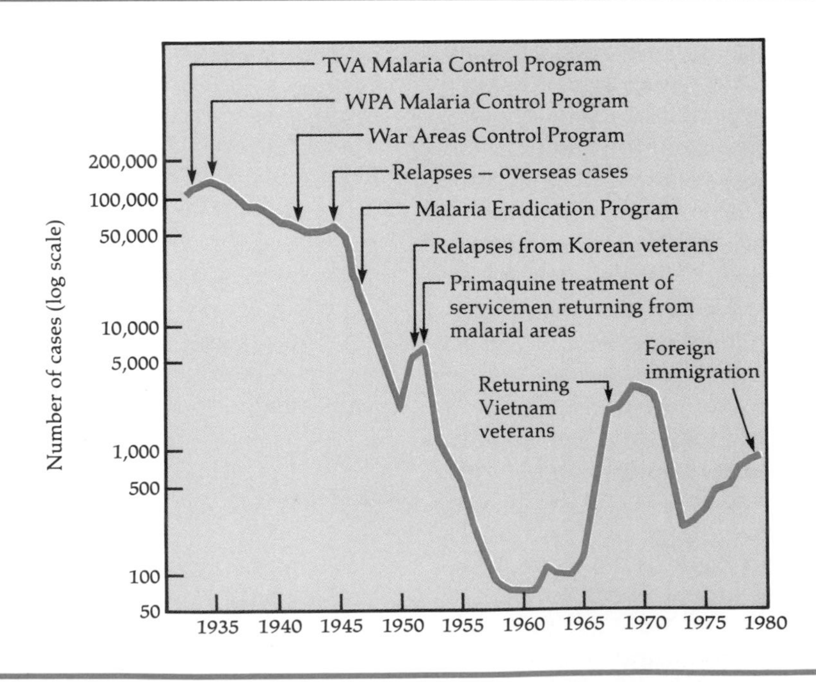

Figure 20–3 Reported cases of malaria in the United States, 1933–1979.

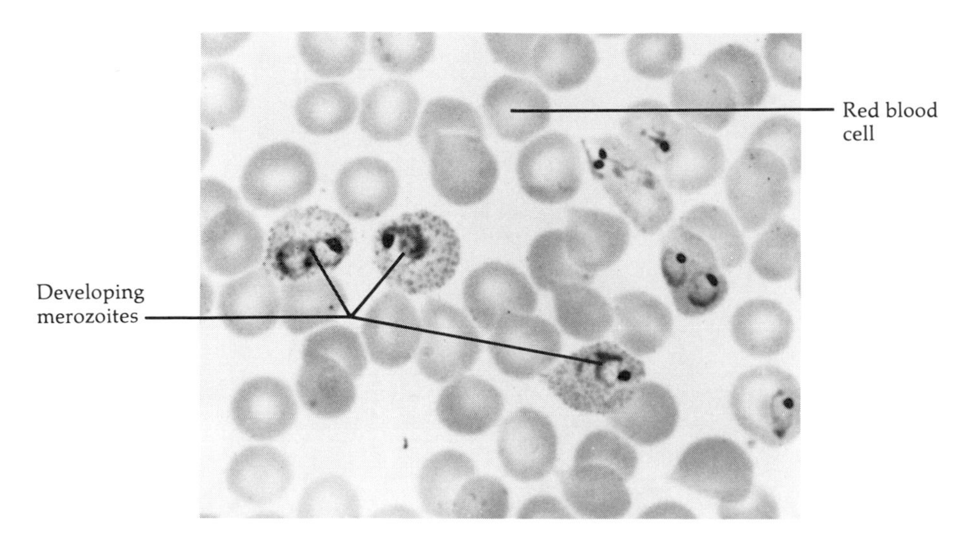

Figure 20–4 Blood smear from a malarial patient. Laboratory diagnosis of malaria is based on observation of *Plasmodium* in red blood cells. Early in its development, the parasite, called a merozoite at this stage, has a ring form.

and hypertrophy of the liver and spleen are added complications.

Laboratory diagnosis of malaria is made by examination of a blood smear (Figure 20–4 and Color Plate IVD). Although there is no effective immunity to malaria, populations in areas where it has long been endemic have generally developed some resistance to it, and individuals may tend to have a less severe disease. In particular, individuals who have the genetic sickle-cell trait or sickle-cell anemia—common in many areas where malaria is endemic—are resistant to malaria (see p. 390). Considerable effort is being expended on the search for an effective vaccine.

Malaria has long had a fairly effective chemotherapeutic treatment with quinine. Quinine and its derivatives (such as primaquine and chloroquine) decrease the number of parasites and minimize symptoms, but do not cure the disease. Travelers from Western countries intending to go to malarial areas should take such drugs for some time before, during, and after their visits. Actually, chemical treatment and prevention of malaria is still difficult, because all of the varied stages of the parasite must be killed or controlled. Moreover, resistance to the more commonly used drugs is now becoming a problem.

African Trypanosomiasis

African trypanosomiasis is a protozoan disease that affects the nervous system. The disease is caused by *Trypanosoma brucei* (see Color Plate IVE), a flagellate that is injected by the bite of a tsetse fly *(Glossina)*. Reservoirs for the disease are domestic cattle and wild antelope that frequent the riverbank habitats of the tsetse fly in Africa. During early stages of the disease, trypanosomes may be found in blood, although they are few in number. During later stages, the trypanosomes move into the cerebrospinal fluid.

Symptoms of the disease include decreasing physical activity and mental acuity. The name **sleeping sickness** is derived from the lethargy of the host. Untreated, death is almost inevitable, as the host enters a coma. Chemotherapeutic agents, such as Bayer 205 (sodium suramin) and pentamidine isethionate, are available to treat the circulatory and central nervous system stages.

Yellow Fever

Arboviruses, or arthropod-borne viruses, are viruses that can reproduce either in arthropods, such as mosquitoes, or in humans. Although a number of

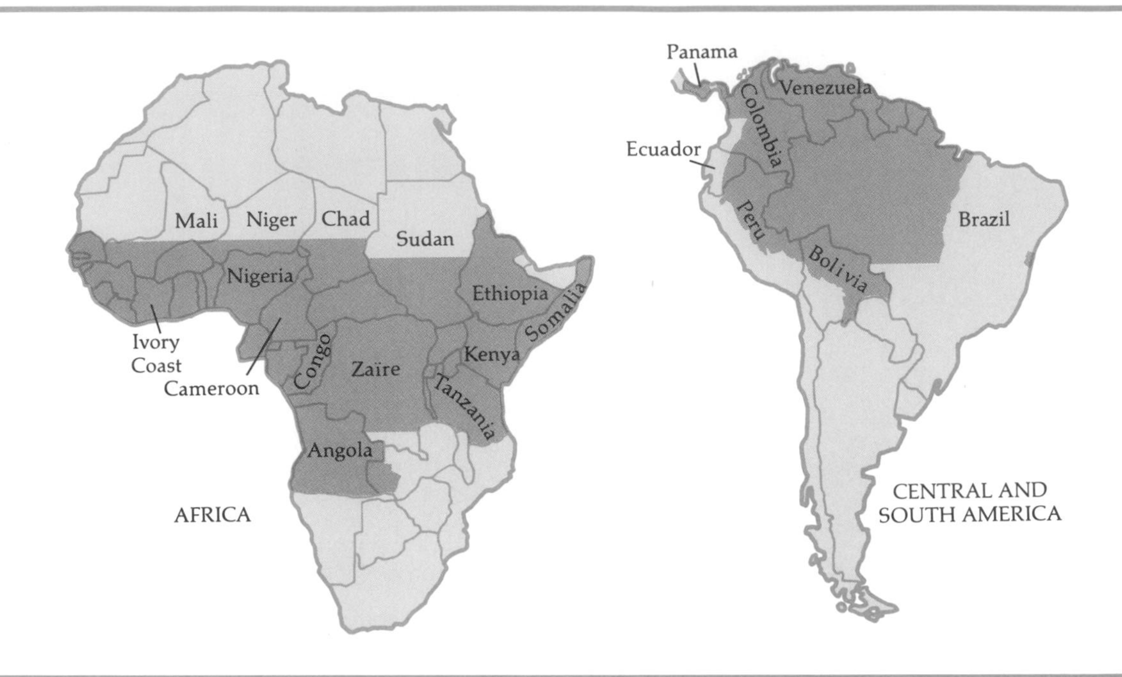

Figure 20–5 Maps showing the yellow fever endemic zones (shown in color).

serious human diseases are caused by such viruses, no disease is evident in the arthropods themselves that serve as vectors. Worldwide, there are probably more than 100 arbovirus-caused diseases. **Yellow fever** is caused by an arbovirus (yellow fever virus) and is historically important because it was the first such virus discovered and the first confirmation that an insect could transmit a virus.

The yellow fever virus is injected into the skin by the mosquito. The virus then spreads to local lymph nodes, where it multiplies, and from the lymph nodes to the liver, spleen, kidney, and heart, where it may persist for days. In the early stages of the disease, the person experiences fever, chills, headache, and backache, followed by nausea and vomiting. This is followed by jaundice, a yellowing of the skin due to the deposition of bile pigments in the skin and mucous membranes, a result of liver damage. (The yellow color of the skin is what gave the disease its name.) In severe cases, the virus produces lesions in the infected organs and hemorrhaging occurs.

Yellow fever is still endemic in many tropical areas such as Central America, tropical South America, and Africa (Figure 20–5). Monkeys constitute a natural reservoir for the virus. Control of the disease depends on localized control of the mosquito vector, usually *Aedes aegypti* (see Figure 11–27a), as was done during the building of the Panama Canal many years ago, or on immunization of the exposed population.

There is no specific treatment for yellow fever. In the early stages of the disease, diagnosis can be made using the serum of patients to neutralize known yellow fever virus strains before inoculating the virus into mice. If the serum contains virus-specific antibodies, the virus is neutralized and the mice do not become ill. The vaccine in use is an attenuated live viral strain and yields a very effective immunity with few side effects.

Dengue

A rather similar, but milder, disease is **dengue,** also a mosquito-borne *(Aedes aegypti)* viral disease endemic in the Caribbean and other tropical climates around the world. It is caused by dengue fever virus, an arbovirus, and is characterized by fever, muscle and joint pain, and rash. The muscle and joint pain experienced by sufferers has led to the name **breakbone fever.** Other than the painful symptoms, classic dengue fever is a relatively mild disease and is rarely fatal. A more serious form of dengue, **dengue hemorrhagic fever,** is characterized by bleeding from the skin, gums, and gastrointestinal tract and sometimes circulatory failure and shock.

Serological tests that are useful in diagnosis include complement fixation, hemagglutination, and neutralization. Control measures are directed at eliminating the *Aedes* mosquitoes. There is no specific treatment that is effective against the virus.

Arthropod-Borne Encephalitis

Encephalitis caused by a mosquito-borne arbovirus (encephalitis virus) is rather common in the United States. A number of different clinical types of this disease have been identified, all of which cause an inflammation of the brain, meninges, and spinal cord. Active cases of these diseases are characterized by chills, headache, and fever. As the disease progresses, mental confusion and coma are observed. There is evidence that subclinical infections are commonplace.

Outbreaks of encephalitis are reported regularly when the mosquito vectors and reservoirs for the virus coincide. The viral reservoir is mainly birds, and the mosquito vector is usually a member of the genus *Culex*.

Horses are frequently infected by such viruses, accounting for terms such as **Eastern equine, Western equine,** and **Venezuelan equine encephalitis (EEE, WEE, and VEE). St. Louis encephalitis (SLE)** virus does not infect horses. **California encephalitis** is another common type. All of these diseases and some Far Eastern types—**Japanese B encephalitis,** in particular—cause a rather high percentage of neurological damage and a mortality rate of about 8%. Figure 20–6 shows the incidence of encephalitis in the United States over a nine-year period. The increase in the summer months is due to the presence of adult mosquitoes during these months.

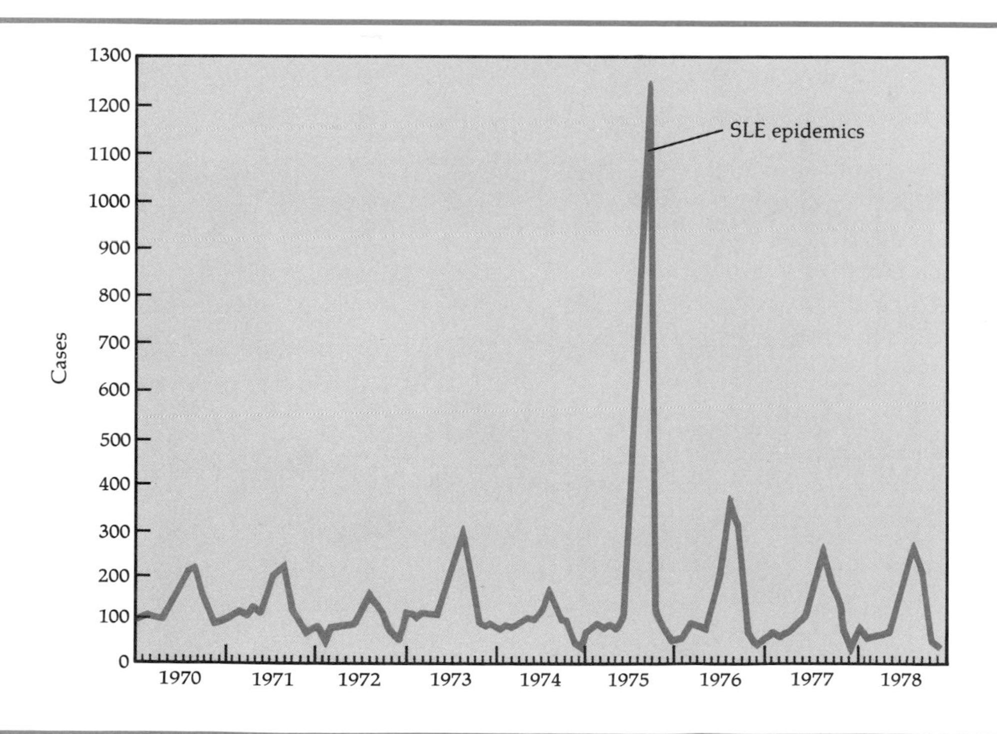

Figure 20–6 Reported cases of encephalitis by month of onset, United States, 1970–1978.

WEE, VEE, SLE, and California encephalitis tend to cause relatively mild diseases in humans, except for occasional complications. Of those prevalent in the United States, EEE is the most severe, with a considerable mortality rate and a high incidence of brain damage, deafness, and similar neurological problems.

Diagnosis of arthropod-borne encephalitis is usually made by serological tests such as complement fixation. The most effective control measure is elimination of the *Culex* mosquitoes. Individuals accidentally exposed may be given immune serum to provide passive immunity, but there is no specific treatment for the disease.

Typhus

Rickettsias, bacteria that are obligate intracellular parasites of eucaryotes, are responsible for a number of diseases spread by arthropod vectors. There are several related rickettsial diseases that differ primarily in their severity and in their arthropod vector. These include epidemic typhus, endemic murine typhus, and the spotted fevers.

Epidemic typhus (louse-borne typhus) is caused by *Rickettsia prowazekii* and carried by the human body louse *Pediculus vestimenti* (see Figure 11–28b). The pathogen is found in the intestinal tract of the louse and is excreted by it. The pathogen is not

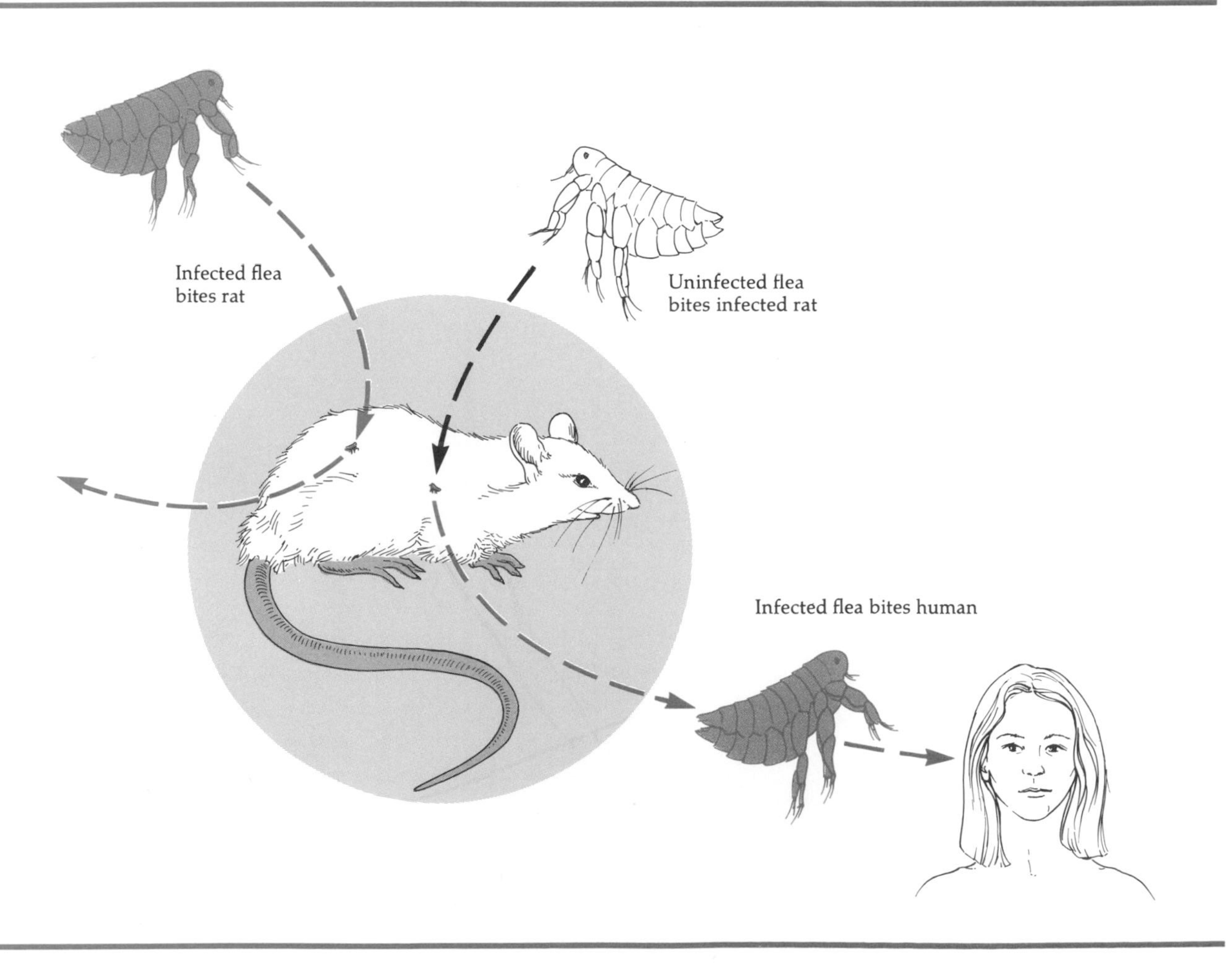

Figure 20–7 The cycle of endemic murine typhus. Normally, the disease is maintained among the rat population, and humans are infected only by accidental contact with the insect vector.

transmitted directly by the bite itself, but rather when the feces of the louse are rubbed into the wound when the bitten host scratches the bite. The disease can flourish only in crowded and unsanitary conditions, when lice can transfer readily from an infected host to a new host. The disease results in a high and prolonged fever for two or more weeks. Stupor and a rash of small red spots caused by subcutaneous hemorrhaging are characteristic, as the rickettsias invade blood vessel linings. Mortality rates can be very high if the disease is untreated.

Tetracycline- and chloramphenicol-type antibiotics are usually effective against epidemic typhus, but the elimination of conditions in which the disease can flourish is more important in its control. Laboratory diagnosis is based on the Weil–Felix test and a complement fixation reaction. Vaccines are available for military populations, which historically have always been highly susceptible to the disease. Recovery from the disease gives a solid immunity and also renders a person immune to the related endemic murine typhus.

Endemic murine typhus occurs sporadically rather than in epidemics. The term *murine* refers to the fact that rodents, such as rats and squirrels, are the common hosts for this type of typhus. Endemic murine typhus is transmitted by rat fleas *(Xenopsylla cheopis)* (see Figure 11–28c), and the pathogen responsible for the disease is *Rickettsia typhi*, a common inhabitant of rats. Humans become involved when the flea finds an opportunity to transfer to a human host (Figure 20–7). The disease is considerably less severe than the epidemic form, with a mortality rate of less than 5%. Except for the reduced severity of the disease, endemic murine typhus is clinically indistinguishable from epidemic typhus.

Tetracycline and chloramphenicol are effective in treating endemic murine typhus, and diagnosis of the disease is the same as for epidemic typhus. Rat control and avoidance of rats are preventative measures for the disease.

Rocky Mountain Spotted Fever

A number of other rickettsial diseases are grouped as **spotted fevers,** the best known being **Rocky Mountain spotted fever,** caused by *Rickettsia rickettsii* (see Color Plate IG). Despite its name (it was first recognized in the Rocky Mountain area), it is most prevalent in the southeastern United States and Appalachia (Figure 20–8). Clinically, the disease tends to resemble the typhus fevers.

Figure 20–8 Reported cases of Rocky Mountain spotted fever in the United States in 1979. The total number of cases reported was 1070.

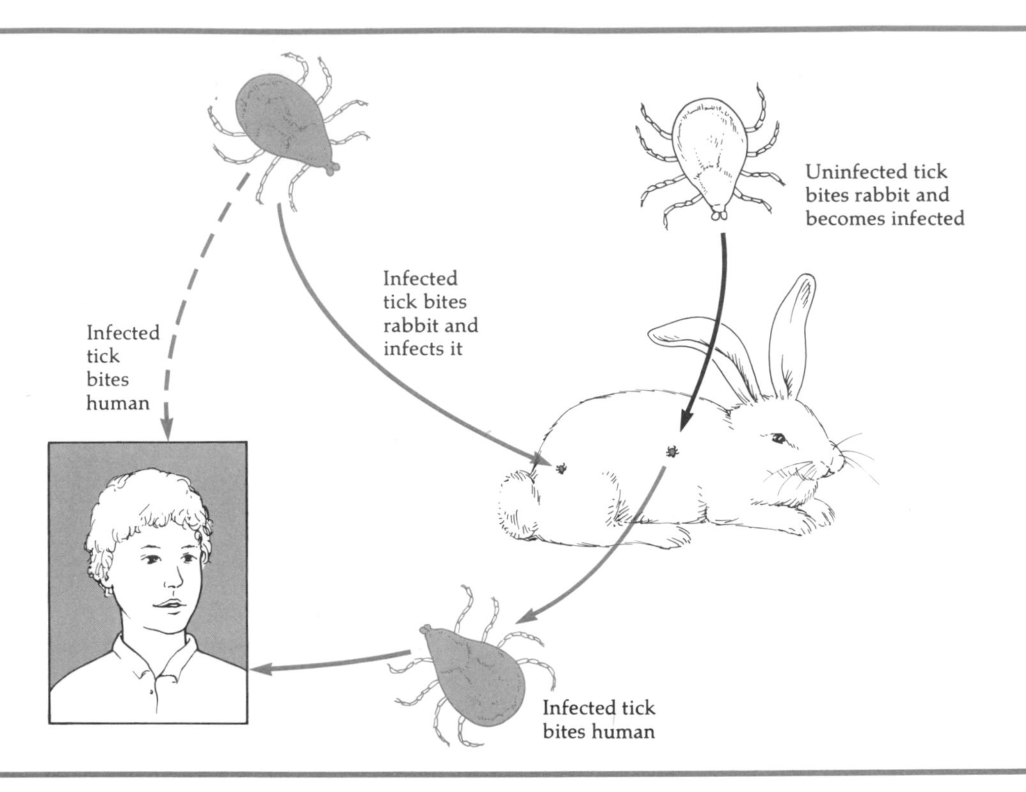

Figure 20–9 The cycle of Rocky Mountain spotted fever. The wild animal is not an essential part of the cycle, since the entire reproductive cycle of the rickettsia can take place in the tick. Details of the tick, *Dermacentor*, are shown in Figure 11–28a.

The rickettsias are parasites of ticks (*Dermacentor* species), the arthropod vectors of the disease (Figure 20–9, and see Figure 11–28a). The rickettsias can pass from generation to generation of tick through tick eggs, a mechanism called *transovarian passage*. In the Eastern states, dog ticks are responsible for most cases of the disease, while in the Rocky Mountain area, wood ticks are mainly responsible. No humans or animals are needed to serve as reservoirs for the reproduction of the rickettsias.

Serological tests, primarily complement fixation, may be used in diagnosis. In addition, guinea pig inoculation results in a characteristic illness that may also be used in diagnosis. Although the mortality rate is currently less than 5%, in the days before antibiotic therapy the rate was much higher. Chloramphenicol and tetracyclines are used effectively to treat the disease.

It should be mentioned that there are a number of other tick-borne rickettsial diseases distributed in various regions of the world.

Plague

Few diseases have had a more dramatic effect on human history than the **plague,** known in the Middle Ages as the Black Death. This term comes from one of the characteristic signs of plague, the blackish areas on the skin caused by hemorrhages. In the fourteenth century, this disease destroyed perhaps one-fourth of the total population of Europe. And near the turn of the twentieth century, as many as 10 million people are reported to have died of plague in India in a 20-year period.

The disease is caused by a gram-negative, rod-shaped bacterium, *Yersinia pestis* (see Color Plate IIIC). One factor in the virulence of the plague bacterium is its ability to survive and proliferate inside phagocytic cells, emerging eventually as increased numbers of highly virulent organisms, rather than being destroyed. Plague is normally a disease of rats, transmitted from one rat to another by the rat flea *(Xenopsylla cheopis)* (see Figure 11–28c). In the United States, particularly the far West and Southwest, the disease is endemic in wild rodents, especially ground squirrels, prairie dogs, and chipmunks **(sylvatic plague).** European rats, introduced into the United States many years ago, are the primary reservoirs of plague. The rat flea, if its host dies, seeks a replacement, which may be another rodent or a human. Plague-infected fleas are hungry for a meal because the growth of the bacteria blocks the digestive tract and the blood the flea ingests is quickly regurgitated. Occasional cases of the disease appear each year in the United States (0 to 20 cases each year) among persons who come into contact with live or recently dead rodents.

In parts of the world where contact with rats is common, the classic form of plague is still prevalent. Bacteria enter the bloodstream of humans from the flea bite and proliferate in the lymph and blood, causing an overwhelming infection. The lymph nodes in the groin and armpit become enlarged and fever develops, part of the body's defensive reactions to the infection. Such swellings are called *buboes,* which accounts for the name **bubonic plague** (Figure 20–10). The mortality rate for untreated bubonic plague is about 50 to 75%. Death from bubonic plague, if it occurs, is usually within less than a week of the appearance of symptoms.

A particularly dangerous condition arises when the bacteria are carried by the blood to the lungs, causing a form of the disease called **pneumonic plague.** The mortality rate for this type of plague is nearly 100%, and the disease is easily spread by airborne droplets, like flu. Even today, this disease cannot be controlled if it is not recognized in time. Pneumonic plague can rarely be controlled after 12 to 15 hours of fever. Great care must be taken to prevent airborne infection of persons in contact with those suffering from pneumonic plague. Persons with pneumonic plague usually die within three days.

Diagnosis is most commonly done by isolation and identification of the bacterium by fluorescent-antibody and phage tests. Animal inoculations are also sometimes used. Serological tests must demonstrate at least a fourfold rise in titer to be diagnostic. Persons exposed to infection can be given prophylactic antibiotic protection. A number of antibiotics, including streptomycin and tetracycline, are effective. Recovery from the disease gives a reliable immunity. No vaccines are available

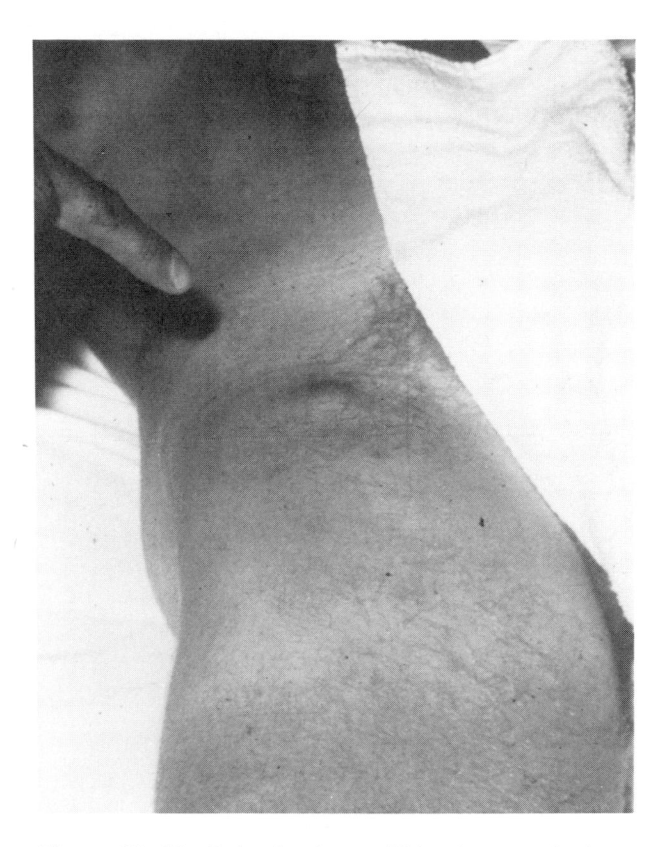

Figure 20–10 Bubonic plague. This photograph shows a bubo in the groin region of the patient. The lymph nodes swell as the body attempts to fight the bacterial infection.

except for laboratory personnel and military personnel likely to be at unusual risk. The long-term effectiveness of the plague vaccine is not high.

Control of the classic rat-based plague is largely due to modern sanitation procedures. Sanitary garbage disposal eliminates a source of food for rats, and ratproofing of buildings denies rats access to hiding places in attics and walls.

Relapsing Fever

All members of the spirochete genus *Borrelia* cause **relapsing fever.** The disease is transmitted by soft ticks *(Ornithodorus)* that feed on rodents. The incidence of relapsing fever increases during the summer months with the increased activity of rodents and arthropods. The disease is characterized by a fever, sometimes in excess of 40.5°C, accompanied by jaundice and rose-colored spots. After 10 days, the fever subsides. Three or four relapses may occur, each shorter and less severe than the initial fever. Diagnosis is made by observation of spirochetes in the patient's blood.

A summary of diseases associated with wounds and bites is presented in Table 20–2.

Table 20–2 Summary of Diseases Associated with Wounds and Bites

Disease	Causative Agent	Mode of Transmission	Treatment
Wounds:			
Gangrene	*Clostridium perfringens*	Contamination of open wound by clostridial endospores	Debridement, amputation, hyperbaric chamber, penicillin, clindamycin, chloramphenicol
Tetanus	*Clostridium tetani*	Contamination of open wound by clostridial endospores	Penicillin before neurotoxin reaches nerves
Animal Bites and Scratches:			
Pasteurellosis	*Pasteurella multocida*	Animal bites from domestic animals such as cats and dogs	Penicillin and tetracycline
Rat bite fever	*Spirillum minor* or *Streptobacillus moniliformis*	Rat bite	Penicillin or tetracycline
Cat scratch fever	*Chlamydia* species (?)	Cat scratch or bite	Tetracycline
Arthropod Bites:			
Malaria	*Plasmodium* species	*Anopheles* mosquito	Quinine and its derivatives
African trypanosomiasis	*Trypanosoma brucei*	*Glossina* species (tsetse fly)	Bayer 205 (sodium suramin) and pentamidine isethionate
Yellow fever	Arbovirus (yellow fever virus)	*Aedes aegypti* mosquito	None
Dengue	Arbovirus (dengue fever virus)	*Aedes aegypti* mosquito	None
Arthropod-borne encephalitis	Arbovirus (encephalitis virus)	*Culex* mosquito	None
Epidemic typhus	*Rickettsia prowazekii*	*Pediculus vestimenti* (louse)	Tetracycline and chloramphenicol
Endemic murine typhus	*Rickettsia typhi*	*Xenopsylla cheopis* (rat flea)	Tetracycline and chloramphenicol
Rocky Mountain spotted fever	*Rickettsia rickettsii*	*Dermacentor andersoni* and other species (tick)	Tetracycline and chloramphenicol
Plague	*Yersinia pestis*	*Xenopsylla cheopis* (rat flea)	Tetracycline and streptomycin
Relapsing fever	*Borrelia* species	*Ornithodorus* species (soft ticks)	Tetracycline

STUDY OUTLINE

NOSOCOMIAL INFECTIONS (pp. 513–515)

1. Nosocomial infections are acquired during the course of stay in a hospital.
2. About 5 to 10% of all hospitalized patients acquire nosocomial infections.
3. Normal flora are often responsible for nosocomial infections when introduced into the body through medical procedures such as surgery and catheterization.
4. Gram-negative bacteria (e.g., *E. coli* or *P. aeruginosa*) are most often involved in nosocomial infections.

5. Patients with large intestinal surgery, burns, or immunosuppressed conditions are most susceptible to nosocomial infections.
6. Hospital infection control personnel are responsible for overseeing proper storage and handling of equipment and supplies.

WOUNDS (pp. 515–517)

1. A wound is a physical injury involving a rupture of the skin.
2. It represents penetration of the body's defenses and may result in disease.

Gangrene (pp. 515–516)

1. The death of soft tissue due to ischemia is called gangrene.
2. Microorganisms may grow on nutrients released from gangrenous cells.
3. Gangrene is especially susceptible to the growth of anaerobic bacteria such as *Clostridium perfringens,* the causative agent of gas gangrene.
4. Gas gangrene may be prevented by prompt cleansing of wounds and antibiotic therapy.
5. *C. perfringens* may invade the uterine wall during improperly performed abortions.

Tetanus (pp. 516–517)

1. Tetanus is due to a localized infection of a wound by *Clostridium tetani.*
2. *C. tetani* produces the neurotoxin tetanospasmin, which causes the symptoms of tetanus: spasms, contraction of muscles controlling the jaw, and death resulting from spasms of respiratory muscles.
3. *C. tetani* is an anaerobe that will grow in unclean wounds and wounds with little bleeding.
4. Acquired immunity results from DPT immunization that includes tetanus toxoid.
5. Following an injury, an immunized person may receive a booster of tetanus toxoid. An unimmunized person may receive human or horse immune globulin.
6. Debridement and antibiotics may be used to control the infection.

ANIMAL BITES AND SCRATCHES (p. 517)

1. Pasteurellosis is caused by *Pasteurella multocida* that may be introduced by the bite of a dog or cat.
2. Anaerobic bacteria such as *Clostridium, Bacteroides,* and *Fusobacterium* may cause infections of deep animal bite wounds.
3. Rat bite fever is caused by *Streptobacillus* or *Spirillum.*

4. Cat scratch fever is probably caused by *Chlamydia*.

ARTHROPOD BITES (pp. 517–526)

1. Arthropods are animals with jointed appendages and include, mites, ticks, and insects.
2. Arthropods may act as vectors for pathogenic microorganisms.

Malaria (pp. 518–519)

1. The symptoms of malaria are chills, fever, vomiting, and headache, occurring in one- to three-day cycles.
2. Malaria is transmitted by *Anopheles* species. The causative agent is any one of four species of *Plasmodium*.
3. Sporozoites reproduce in the liver and release merozoites into the bloodstream.
4. Laboratory diagnosis is based on microscopic observation of merozoites in red blood cells.
5. Quinine and chloroquine are useful in killing the merozoites.

African Trypanosomiasis (p. 519)

1. African trypanosomiasis is a disease caused by the protozoan *Trypanosoma brucei* and transmitted by the bite of the tsetse fly *(Glossina)*.
2. The disease affects the nervous system of the human host, causing lethargy and eventually coma. It is commonly called sleeping sickness.

Yellow Fever (pp. 519–520)

1. Yellow fever is caused by an arbovirus (yellow fever virus). The vector is the mosquito *Aedes aegypti*.
2. The virus multiplies in lymph nodes and is then disseminated to liver, spleen, kidney, and heart.
3. Symptoms include fever, chills, headache, nausea, and jaundice.
4. Diagnosis is based on the presence of viral neutralizing antibodies in the host.
5. No treatment is available. An attenuated, live viral vaccine is available.

Dengue (pp. 520–521)

1. Dengue is caused by an arbovirus (dengue fever virus) and is transmitted by the mosquito *Aedes aegypti*.
2. Symptoms are fever, muscle and joint pain, and rash.

3. Hemorrhagic fever involving bleeding from the skin, gums, and gastrointestinal tract is a more serious form of the disease.
4. Laboratory identification is based on the presence of antibodies.

Arthropod-Borne Encephalitis (pp. 521–522)

1. Symptoms of encephalitis are chills, headache, fever, and eventually coma.
2. Many types of arboviruses (encephalitis viruses) transmitted by *Culex* species mosquitoes cause encephalitis. The reservoir is birds.
3. Horses are frequently infected by EEE, WEE, and VEE viruses.
4. Outbreaks occur primarily during the summer months when adult mosquitoes are present.
5. Diagnosis is based on serological tests.
6. Immune globulin may provide passive immunity, but no treatment is available.
7. Elimination of the vector is the most effective control measure.

Typhus (pp. 522–523)

1. The human body louse *Pediculus vestimenti* transmits *Rickettsia prowazekii* in its feces while biting.
2. Epidemic typhus is prevalent in crowded and unsanitary living conditions that allow the proliferation of lice.
3. The symptoms of typhus are rash, prolonged high fever, and stupor.
4. Endemic murine typhus is a less severe disease.
5. *Rickettsia typhi* is transmitted from rodents to humans by the rat flea.
6. Tetracyclines and chloramphenicol are used to treat typhus.
7. Laboratory diagnosis is based on the Weil–Felix test.

Rocky Mountain Spotted Fever (pp. 523–524)

1. *Rickettsia rickettsii* is a parasite of ticks (*Dermacentor* species) in the southeastern United States, Appalachia, and the Rocky Mountain states.
2. The rickettsia may be transmitted to humans, in whom it causes Rocky Mountain spotted fever.
3. Chloramphenicol and tetracyclines are effective to treat the disease.

Plague (pp. 524–525)

1. Plague is caused by *Yersinia pestis.* The vector is usually the rat flea *(Xenopsylla cheopis).*
2. Reservoirs for plague include European rats and North American rodents.

3. Symptoms of plague include bruises on the skin and enlarged lymph nodes (buboes).
4. The bacteria may enter the lungs and cause pneumonic plague.
5. Laboratory diagnosis is based on isolation and identification of the bacteria.
6. Antibiotics are effective to treat plague, but they must be administered promptly after exposure to the disease.
7. Control of the rat population is effective to decrease the incidence of plague.

Relapsing Fever (pp. 525–526)

1. Relapsing fever is caused by *Borrelia* species and transmitted by soft ticks *(Ornithodorus).*
2. The reservoir for the disease is rodents.
3. Symptoms include fever, jaundice, and rose-colored spots. Symptoms recur three or four times after apparent recovery.
4. Laboratory diagnosis is based on the presence of spirochetes in the patient's blood.

STUDY QUESTIONS

REVIEW

1. Why are nosocomial infections of concern to health care personnel?
2. If you were going to make a sign for a hospital that would help prevent nosocomial infections, what would that sign say?
3. Why is *Clostridium perfringens* likely to grow in gangrenous wounds?
4. Compare and contrast the symptoms of gas gangrene and tetanus, keeping in mind that they are both wound infections.
5. If *Clostridium tetani* is relatively sensitive to penicillin, then why doesn't penicillin cure tetanus?
6. What treatment is used against tetanus under the following conditions?
   (a) Before a person suffers a deep puncture wound
   (b) After a person suffers a deep puncture wound
7. Why is the following description used for wounds that are susceptible to C. *tetani* infection? ". . . improperly cleaned deep puncture wounds . . . ones with little or no bleeding . . ."
8. List four bacterial infections that might result from animal bites.
9. What is the most effective control measure for mosquito-borne diseases?

10. Fill in the following table.

Disease	Causative Agent	Vector	Symptoms	Treatment
Malaria				
Yellow fever				
Dengue				
Arthropod-borne encephalitis				
Relapsing fever				

11. Provide the following information for three rickettsial diseases: name of disease; causative agent; vector; reservoir; treatment.

12. Provide the following information with regard to plague: causative agent; vector; U.S. reservoir; control; treatment; prognosis (probable outcome).

13. Differentiate between bubonic plague and pneumonic plague with regard to transmission and symptoms.

CHALLENGE

1. Most of us have been told that a "rusty nail" causes tetanus. What do you suppose is the origin of this adage?

2. Why is the incidence of arboviral arthropod-borne encephalitis in the United States higher in the summer months? Why do dengue and yellow fever usually occur in tropical areas?

3. Three patients in a large hospital acquired infections of *Pseudomonas cepacia* during their stay. Two patients had septicemia and one had a wound infection. All three patients had been given cryoprecipitate, which is prepared from blood that has been frozen in a standard plastic blood transfer pack like the one shown below. The transfer pack is then placed in a water bath to thaw. Explain the probable origin of the nosocomial infections. What characteristics of *Pseudomonas* would allow it to be involved in this type of infection?

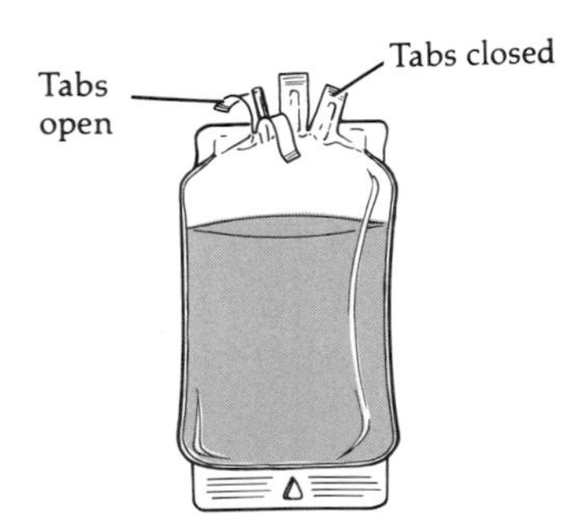

FURTHER READING

Greenberg, B. July 1965. Flies and disease. *Scientific American* 213:92–99. Addresses the question, Do flies actually transmit disease?

Gregg, C. T. 1978. *Plague!* New York: Charles Scribner's Sons. An interesting and factual account of plague in the twentieth century.

Hubbert, W. T., W. F. McCulloch, and P. R. Schurrenberger. 1975. *Diseases Transmitted from Animals to Man.* 6th ed. Springfield, IL: C. C. Thomas. Thorough coverage of bacterial, fungal, parasitic, and viral diseases transmitted to humans by animal bites and vectors.

Kadis, S., T. C. Montie, and S. J. Ajl. March 1969. Plague toxin. *Scientific American* 220:92–100. The method of action of *Yersinia* toxin.

Mandell, G. L., R. G. Douglas, and J. E. Bennett. 1979. Nosocomial infections, pp. 2213–2256. In *Principles and Practice of Infectious Disease,* vol. 2. New York: Wiley. Discussions of specific nosocomial infections, such as pneumonia.

Polk, H. C., and H. H. Stone (eds.). 1977. *Hospital Acquired Infections in Surgery.* Baltimore, MD: University Park Press. Discusses nosocomial infections, including those associated with wounds and burns, and how to prevent them.

See also Further Reading for Part IV at the end of Chapter 19.

21

Microbial Diseases of the Nervous System

Objectives

After completing this chapter you should be able to

- Differentiate between meningitis and encephalitis.
- Discuss the epidemiology of meningococcal meningitis, *H. influenzae* meningitis, and cryptococcosis.
- Define aseptic meningitis.
- Discuss the epidemiology of leprosy and poliomyelitis, including method of transmission, etiology, disease symptoms, and preventative measures.
- List the etiology, methods of transmission, and reservoirs for rabies.
- Describe postexposure and preexposure treatments for rabies.
- List the characteristics of slow virus diseases.
- Provide evidence that slow virus diseases are caused by transmissible agents resembling viruses.

ORGANIZATION OF THE NERVOUS SYSTEM

The human nervous system consists of billions of nerve cells (neurons) and supporting cells (neuroglia) that are organized into two divisions: the central nervous system and the peripheral nervous system (Figure 21–1). The **central nervous system** consists of the brain and spinal cord. It is the control center for the entire body, picking up sensory information from the environment, interpreting the information, and sending out impulses that coordinate body activities. The **peripheral nervous system** consists of all the nerves that branch off from the brain and spinal cord. These nerves are the lines of communication between the central nervous system, the various parts of the body, and the external environment.

The brain is protected from physical injury by skull bones, and the spinal cord is protected by the backbone. Under these bones, both the brain and spinal cord are covered by three continuous membranes called *meninges*. These are the outermost dura mater, the middle arachnoid, and the innermost pia mater (Figure 21–2). Between the pia mater and arachnoid membranes is a space called the *subarachnoid space,* in which *cerebrospinal fluid* circulates. The fluid also circulates in spaces within the brain called *ventricles.* A very interesting feature of the brain is the presence of a *blood–brain barrier.* This barrier consists of capillaries that permit certain substances to pass from the blood into the brain but restrict others. Unlike other capillaries of the body, these capillaries are less permeable and therefore more selective in terms of which materials can pass. Oxygen and glucose, for example,

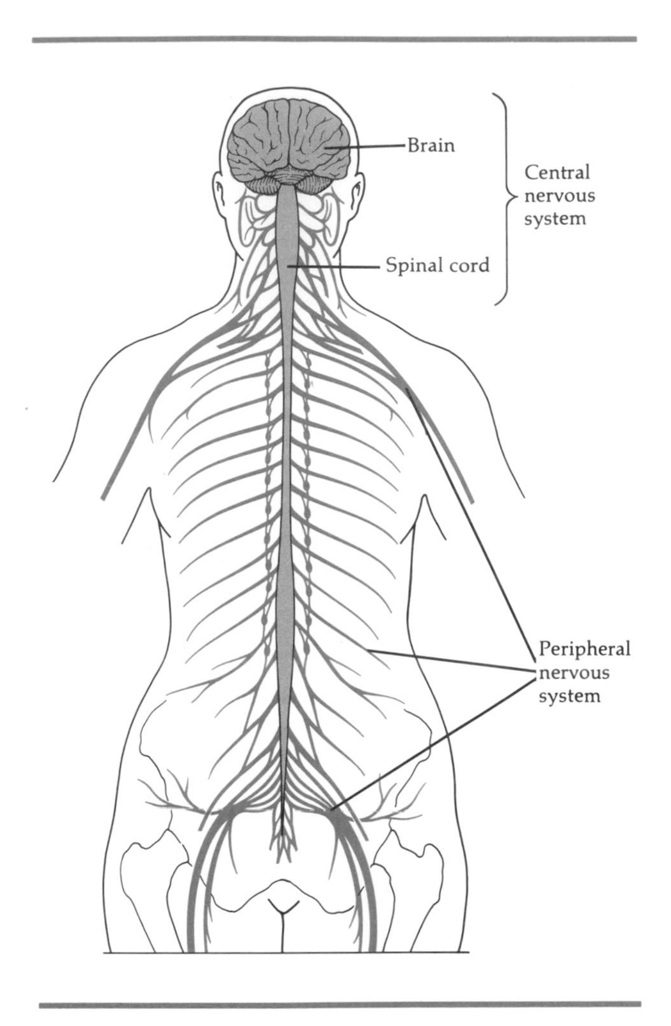

Figure 21–1 Organization of the nervous system.

easily cross the barrier, while microorganisms and most antibiotics normally do not.

Even though the central nervous system has considerable protection, it can still be invaded by microorganisms in several ways. For example, microorganisms can gain access from trauma, such as skull or backbone fracture, or from a medical procedure, such as a lumbar puncture (spinal tap), in which a needle is inserted into the subarachnoid space of the spinal meninges to obtain a sample of fluid for diagnosis. Also, some microorganisms can move along peripheral nerves. But probably the most common route of central nervous system invasion is by the bloodstream and lymphatic system (Chapter 22). Since cerebrospinal fluid communicates with the lymphatic system of the body, invading microorganisms can enter the cerebrospinal fluid itself. Should an infection occur in the meninges, the condition is called **meningitis.** An infection of the brain itself is called **encephalitis.**

If the central nervous system is resistant to invasion by pathogens, it is also resistant to the entrance of many antibiotics and chemotherapeutic agents used in the treatment of such infections. Although an organism causing meningitis may be sensitive to an antibiotic in a test tube, the antibiotic cannot be clinically effective if it is unable to penetrate the blood-brain barrier.

In Chapter 19, the normal flora of the skin was described, and in subsequent chapters, the normal flora of other regions of the body will be discussed. The nervous system, however, like other parts of the body not in contact with the outside environment, is normally sterile; it has no normal flora.

Now that you have a general understanding of the nervous system, we will examine several diseases that affect it.

MENINGITIS

As noted earlier, meningitis is an infection of the meninges. It is an inflammation resulting in swelling and an excess of cerebrospinal fluid. In 1979, 2724 cases of meningococcal infection were reported in the United States. The highest incidence was in children less than four years old. In this section we will discuss several different microorganisms that can cause meningitis.

Meningococcal Meningitis

One form of meningitis, called **meningococcal meningitis** or **cerebrospinal fever**, is caused by the bacterium *Neisseria meningitidis*. The neisseriae are essentially aerobic, nonmotile, gram-negative cocci.

N. meningitidis usually inhabits the portion of the throat behind the nose without causing any symptoms. This carrier state may last from several days to several months and provides the reservoir for the organism. At the same time, the carrier state enhances the immunity of the carrier. If an individual without adequate immunity acquires the microorganism, usually through contact with a healthy carrier, the resulting throat infection can lead to

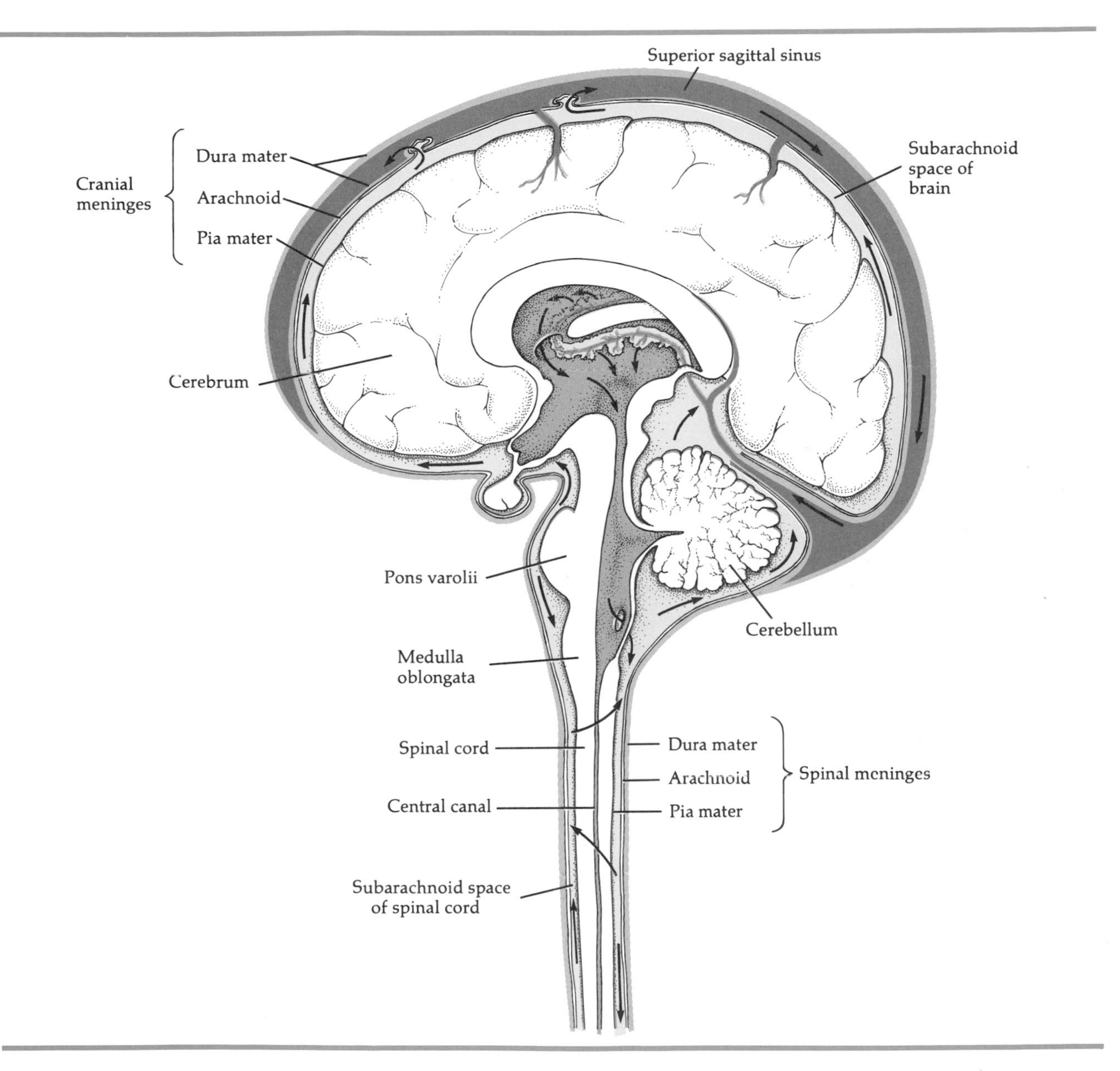

Figure 21–2 Meninges and cerebrospinal fluid (the fluid is indicated by color).

bacteremia followed by meningitis. Most of the bacteria observed in cerebrospinal fluid are found in leukocytes that enter the fluid in response to the infection. The ability of the bacteria to leave the blood to make contact with the meninges also seems to be related to the concentration of microorganisms in the blood and to the length of time they circulate. Most of the symptoms associated with meningitis are thought to be due to an endotoxin produced by the bacteria.

Meningitis begins with symptoms similar to those of a mild cold. This is followed by a sudden fever, severe headache, and pain and stiffness in the neck and back. The pain is associated with swelling of the meninges and excess fluid. Neurological complications, such as convulsions, deafness, blindness, and minor paralysis are not uncommon. Purplish spots may also appear on the skin. The *N. meningitidis* endotoxin not only causes extensive blood vessel damage, but also heightens sensitivity to further exposure to endotoxin. This phenomenon of increased sensitivity is called the *Shwartzman phenomenon*. Death can occur in a few hours in severe cases.

Meningococcal meningitis is a disease primarily of the very young, usually striking children less than four years of age, with the highest incidence in the first year of life. Children are usually born with maternal immunity and only become susceptible as the immunity weakens at about six months of age. The death rate in untreated cases can be as high as 85%. But if antibiotic therapy is begun soon enough, the mortality rate drops to less than 1%. Explosive epidemics among young adults in military training camps also used to be quite common, and just such an outbreak occurred in a school in Houston not too long ago. Apparently, the close quarters promote the transmission from carriers to susceptible individuals. However, a recently developed vaccine consisting of purified capsular polysaccharide has proved effective in protecting military recruits, and this expression of the disease is becoming a thing of the past, although the vaccine is still not in use in the general population.

Diagnosis is based on isolation of the bacteria from blood cultures and lumbar punctures, often found in phagocytes in smears. The spinal fluid may actually be turbid with heavy bacterial growth. Nasal swabs can be streaked onto selective and differential media to detect carriers. *N. meningitidis* ferments glucose and fructose, but not sucrose. It will not grow on agar devoid of blood, and it must be incubated in an environment with 5 to 10% carbon dioxide. This concentration of carbon dioxide can be obtained by the use of a *candle jar* (Figure 21–3). The organism is very susceptible to desiccation and killed when heated at 55°C for 30 minutes. Since *N. meningitidis* is very sensitive to handling once it is removed from the body, care must be taken to ensure its survival until the cultures taken can be used in diagnosis.

N. meningitidis, like many other pathogens, is relatively resistant to phagocytosis. Encapsulated strains are the most likely to prove virulent. At one time, sulfonamide was the drug of choice, since it passes readily from the blood into cerebrospinal fluid. However, resistance to the sulfa drugs has now become common, and other drugs have come into use. Although many antibiotics are unable to pass from the blood into the fluid, a few are capable

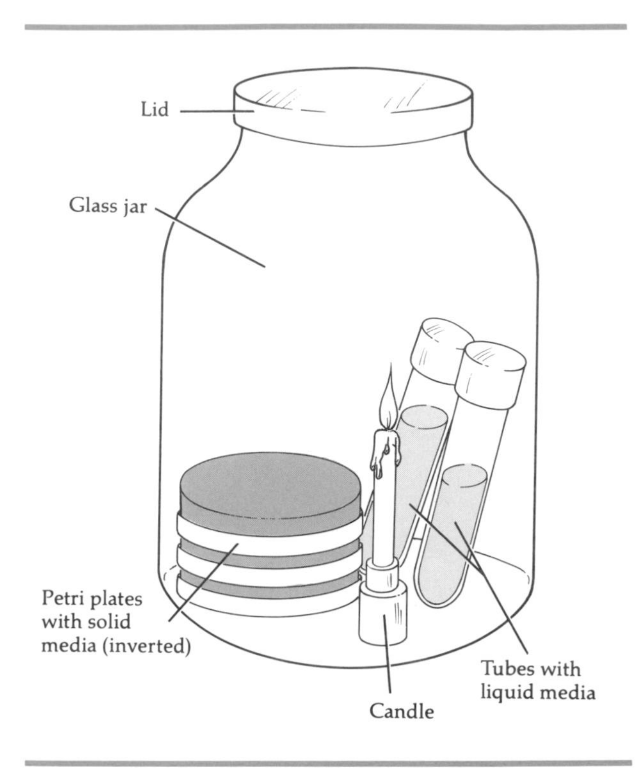

Figure 21–3 Candle jar technique. Plates and tubes inoculated with *Neisseria meningitidis* are placed in a jar with a lighted candle. The cap is placed on the jar. The candle will extinguish when the atmosphere contains approximately 10% carbon dioxide.

of passage when the meninges are seriously infected. Examples include penicillin, chloramphenicol, and rifampin.

Hemophilus influenzae Meningitis

Hemophilus influenzae is a nonmotile, aerobic, nonendospore-forming, gram-negative, pleomorphic bacterium (Figure 21–4). A member of the normal throat flora, it can also cause meningitis. The bacterium, which is encapsulated, is further divided into six types on the basis of its antigenic capsular carbohydrates. Only one type is of real importance in medical microbiology. The virulence of *H. influenzae* is related to its capsule; its endotoxin does not appear to play a major role in pathogenicity, as does that of *N. meningitidis*.

The name *Hemophilus influenzae* is somewhat misleading. The microorganism was erroneously

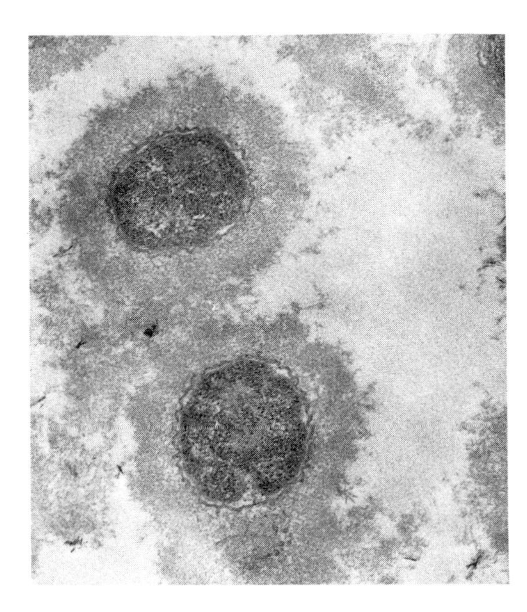

Figure 21–4 Electron micrograph of *Hemophilus influenzae* type b (×36,000). Note the thick layer of capsular material around the cells, which has been accented in this case by treatment with antibody to the type b capsule.

thought to be the causative agent of the influenza pandemics of 1890 and the first world war—thus the term *influenzae*. The etiologic agent of influenza is now known to be a virus. However, *H. influenzae* was probably a secondary invader during those pandemics. The term *Hemophilus* refers to the fact that the microorganism requires factors in blood for growth (*hemo*, meaning "blood"). Actually, *Hemophilus influenzae* grows poorly, if at all, on blood agar. However, it can be cultured on chocolate agar, which is prepared from lysed red blood cells. Lysis releases the X and V factors required for its growth (see Chapter 10).

H. influenzae is by far the most common cause of bacterial meningitis among children under four years of age (Figure 21–5). The incidence of the disease is related to the antibody titer in the blood, whether passively acquired from the mother or actively formed. In children from about 2 to 36 months, antibody levels are minimal. Thereafter, as antibody levels increase, the incidence of the disease drops considerably.

Meningitis caused by *H. influenzae* follows the same basic pattern as meningitis caused by *N. meningitidis*. The carrier state in children is as high as 30 to 50%. Following a viral infection of the respiratory tract, *H. influenzae* gets into the blood from the throat and from there invades the meninges. Untreated, the fatality rate approaches 100%. Another serious disorder produced by *H. influenzae* is acute epiglottitis, which will be discussed in Chapter 23.

Laboratory diagnosis of *H. influenzae* is based on spinal fluid and blood analysis. If a Gram-stained spinal fluid smear shows gram-negative pleomorphic rods, the presence of *H. influenzae* can be assumed. A positive quellung test (see Chapter 4) establishes the diagnosis. Culturing the microorganisms from spinal fluid and blood on blood agar also confirms the diagnosis.

Virtually all persons treated early can be cured. The original drug of choice was ampicillin. However, about 10% of *H. influenzae* strains are resistant to the drug. Therefore, the initial therapy now consists of administration of ampicillin and chloramphenicol in combination. Only a few strains resistant to both ampicillin and chloramphenicol have been found.

The mortality rate of treated meningitis from *H. influenzae* is less than 10%. However, 30% of those who recover have permanent neurological damage. There is no effective immunization.

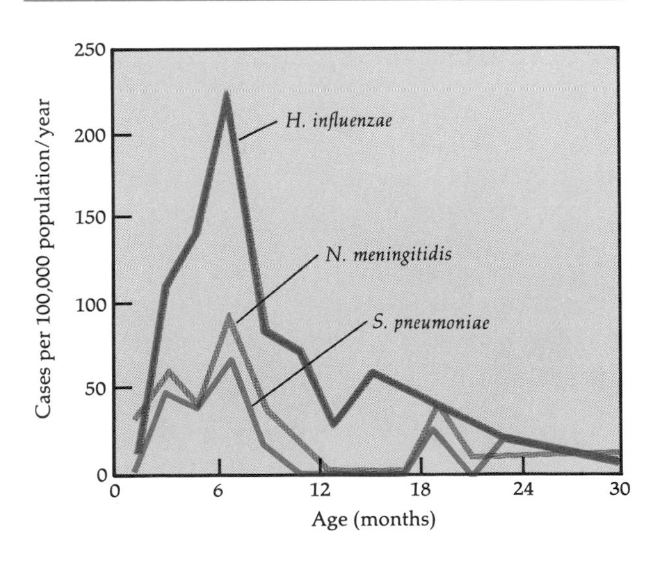

Figure 21–5 Incidence of meningitis caused by *Hemophilus influenzae*, *Neisseria meningitidis*, and *Streptococcus pneumoniae* in children under 30 months of age.

Cryptococcus neoformans Meningitis (Cryptococcosis)

Fungi of the genus *Cryptococcus* are spherical cells resembling yeasts that reproduce by budding and contain polysaccharide capsules, some much thicker than the cells themselves (Figure 21–6). Only one species, *Cryptococcus neoformans,* is pathogenic for humans. The organism is widely distributed in soil, especially soil enriched by pigeon droppings. It is also found in pigeon roosts and nests on window ledges of urban buildings. It is thought to be transmitted by the inhalation of dried infected pigeon droppings.

The disease caused by *C. neoformans* is called **cryptococcosis.** Inhalation of *C. neoformans* initially causes infection of the lungs, and often the disease does not proceed beyond this stage. However, it can spread through the bloodstream to other parts of the body, including the brain and meninges, especially in immunosuppressed individuals. The disease is usually expressed as chronic meningitis.

Laboratory diagnosis of cryptococcosis is by identification of the encapsulated microorganism in an India ink wet mount and by observation of its virulence in mice. Untreated meningitis due to *C. neoformans* is usually fatal. The drug of choice in treatment is amphotericin B.

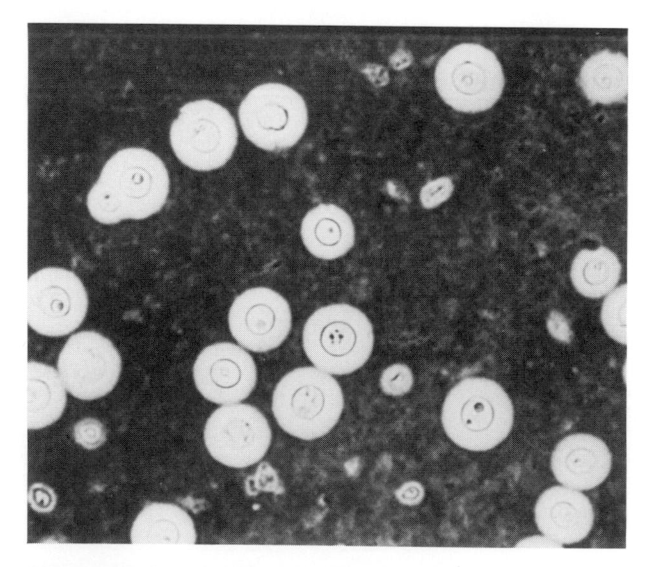

Figure 21–6 *Cryptococcus neoformans.* This is the causative agent of *Cryptococcus neoformans* meningitis (cryptococcosis), a form of chronic meningitis. Note the prominent capsule visible in this India ink preparation.

Other Causes of Meningitis

As many as 50 different microorganisms have been implicated in meningitis. In addition to those already noted, several others are worth mentioning here. **Tuberculosis meningitis,** caused by the bacterium *Mycobacterium tuberculosis,* sometimes occurs as a complication of pulmonary tuberculosis. **Clostridial meningitis** can occur as a result of a penetrating injury in which the bacterium *Clostridium perfringens* gains access to the body. *Flavobacterium meningosepticum,* a bacterium that contaminates hospital equipment and solutions, can cause epidemics of **flavobacterial meningitis** in newborns, especially the premature. The bacterium *Listeria monocytogenes* can cause **listerial meningitis,** especially in individuals with underlying diseases or impaired immunologic systems. **Pneumococcal meningitis** occurs as a secondary infection in individuals with pneumococcal pneumonia, and is caused by *Streptococcus pneumoniae.* If bacteria or fungi cannot be isolated as a cause of meningitis, then a viral cause is suspected and the condition is referred to as **aseptic meningitis.** The mumps virus, for example, is known to cause aseptic meningitis, as are the poliovirus, coxsackieviruses, echoviruses, and others. In general, aseptic meningitis is not as serious as most bacterial forms.

LEPROSY

The nonmotile, acid-fast rod causing **leprosy** is *Mycobacterium leprae.* It is closely related to the tuberculosis pathogen, *Mycobacterium tuberculosis* (see Figure 10–14b). The organism was first isolated and identified about 1870 by G. W. Hansen of Norway, signifying one of the first direct links ever made between a specific bacterium and a disease. Leprosy is sometimes called **Hansen's disease** in an effort to avoid using the dread name of leprosy.

M. leprae is probably the only bacterium that grows primarily in the peripheral nervous system (see Color Plate IC), although it can also grow in skin cells. In general, the organism shows a preference for the outer, cooler portions of the human body. A very slow generation time of about 12 hours has been estimated.

The microorganism has never been grown on artificial media. However, in 1960, *M. leprae* obtained from patients was first grown in the footpads of mice. For the next nine years, this provided the only means of culturing and testing the organism. Then, in 1969, armadillos were experimentally infected with *M. leprae*. But unlike mice, the armadillos acquired the disease. It is now known that armadillos can contract a leprosylike disease in the wild, although they are not considered a source of human infection. Armadillos are used to study the disease and evaluate the effectiveness of chemotherapeutic agents.

There are two main forms of leprosy, although intermediate forms are also recognized. The *tuberculoid (neural) form* is characterized by regions of the skin that have lost sensation and are surrounded by a border of nodules (Figure 21–7a). The **lepromin test,** which is analogous to the tuberculin test for tuberculosis, is positive in most cases of this form of the disease. In this test, lepromin, an extract of lepromatous tissue, is injected into the skin.

In the *lepromatous (progressive) form* of leprosy, skin cells are infected, and disfiguring nodules form all over the body (Figure 21–7b). Mucous membranes of the nose tend to be affected, which leads to the lion-faced appearance associated with this type of leprosy. Deformation of the hand into a clawed form and considerable necrosis of tissue may also occur. The progression of the disease is unpredictable, and remissions may alternate with rapid deterioration. The lepromin test is negative in the lepromatous form of the disease.

Leprosy is spread by transfer of the bacteria in exudates (discharges) from lesions to minor abrasions on the skin. Inanimate objects, such as clothing, may transmit the disease in a similar manner. Leprosy, however, is not very contagious, and transmission usually occurs only between persons in fairly intimate and prolonged contact. The time from infection to the appearance of symptoms is usually measured in years, although children may have a much shorter incubation period. Death is not usually a result of the leprosy itself, but more often is due to complications caused by other bacteria, such as tuberculosis.

Much of the public fear of leprosy can probably be traced to biblical and historical references to the disease. In the Middle Ages, lepers were rigidly excluded from normal European society and sometimes were even given bells to wear so that people could avoid contact with them. This may actually have contributed to the near disappearance of the disease in Europe. But patients with leprosy are no longer kept in isolation. One special treatment center in the United States presently treats about 500 leprosy patients on an ambulatory basis. These outpatients are free to carry out normal daily routines

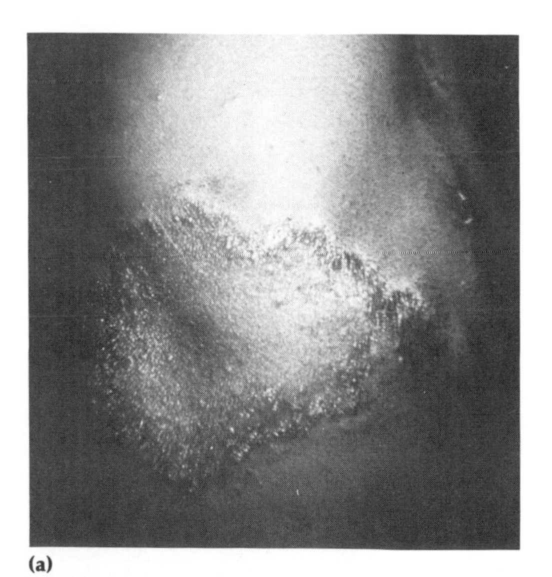
(a)

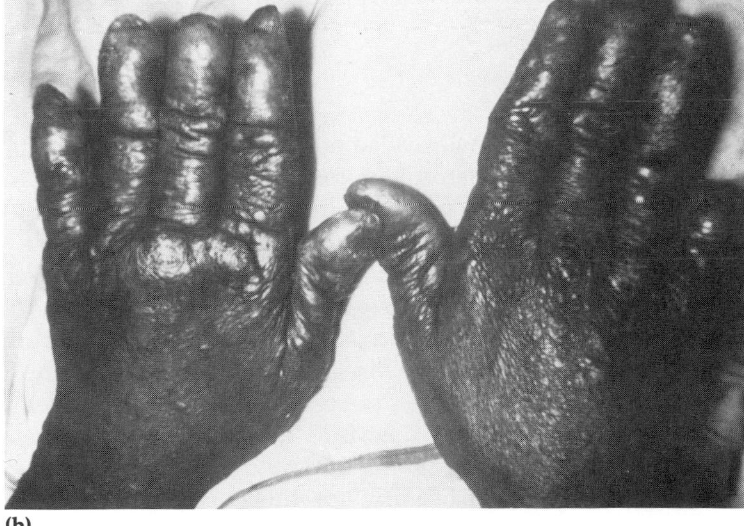
(b)

Figure 21–7 Symptoms of leprosy. **(a)** Tuberculoid leprosy. This lesion below the shoulder on the side of a patient shows a ring of nodules around a relatively clear area in the center. **(b)** Lepromatous leprosy. Tissue necrosis is obvious on the hands of this patient.

and return to the hospital for periodic treatment. Patients can be made noncommunicable within four to five days with sulfone drugs.

The number of leprosy cases in the United States has been gradually increasing, with about 150 reported each year. Estimates place the total number of cases at about 2600. Most patients are treated at The National Leprosy Hospital in Carville, Louisiana. Many of these cases are imported, for the disease is usually found in tropical climates. Millions of persons suffer from leprosy today, most of them in Asia and Africa. Some researchers believe that unsanitary living conditions, rather than climate, are responsible for the prevalence of disease in these areas. Also, some individuals are resistant to infection by *M. leprae*, and there is evidence for genetic predisposition to the infection.

Laboratory diagnosis is by identification of acid-fast rods in scrapings from lesions or fluid from incisions over nonulcerated lesions. (The bacteria may be absent from tuberculoid nodules.) The lepromin test is also useful in some cases. Although the disease is still difficult to cure, prolonged treatment with sulfone drugs (such as dapsone) has been effective in arresting its progress. Rifampin is also a promising drug now under investigation. Children living with leprous parents are vaccinated with bacille Calmette–Guérin (BCG), which results in a positive lepromin test, as well as a positive tuberculin test.

POLIOMYELITIS

Poliomyelitis is informally called **polio** or **infantile paralysis.** The latter name is misleading because paralysis in infants is less common or severe than in adults, probably because of the protection from maternal antibodies. At one time, and this is still the case in parts of the world where sanitation is poor, most cases of polio did occur in infants. However, the paralytic form of the disease was not common, and in fact still is not in poorer countries. In more developed regions of the world, as sanitation improved, there was less chance of developing the disease in infancy, particularly for those in the upper and middle socioeconomic groups. Then, when infection did occur in adolescence or early adulthood, the paralytic form of the disease occurred more frequently.

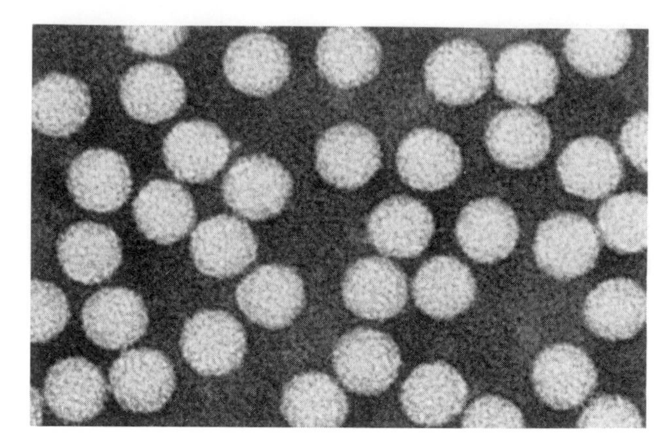

Figure 21–8 Polioviruses ($\times$290,000).

The symptoms of polio are usually no more than a few days of headache, sore throat, fever, and perhaps nausea. Often the disease is never diagnosed at all. Moreover, it can also be confused with aseptic meningitis, with symptoms of pain and stiffness of the back and neck. Relatively few cases are paralytic, probably no more than 1 to 2%.

The *poliovirus* (actually three different serotypes cause the disease) is a picornavirus (Figure 21–8). Humans are the only known natural host for polioviruses, and since polioviruses are more stable than most other viruses, they can remain infectious for relatively long periods in water and food, facilitating their transmission. The primary mode of transmission is ingestion of water contaminated with feces containing the virus. For this reason, polio occurs mainly in the summer months in temperate regions, when people participating in water sports are readily exposed to the virus.

Infection is initiated by ingestion of the virus, and its primary multiplication is in the throat and small intestine (Figure 21–9). This accounts for the sore throat and nausea. Next, the virus invades the tonsils and the lymph nodes of the neck and ileum (terminal portion of small intestine). From the lymph nodes, the virus enters the blood, causing viremia. In most individuals, the viremia is only transient, the infection does not progress past the lymphatic stage, and clinical disease does not result. If the viremia is persistent, the virus penetrates the capillary walls and enters the central nervous system. Once in the central nervous system, the virus displays a high affinity for nerve cells, particularly motor nerve cells in the upper spinal

Centers for Disease Control
MMWR
Morbidity and Mortality Weekly Report

Poliomyelitis Prevention in the United States

Poliovirus vaccines, used widely since 1955, have dramatically reduced the incidence of poliomyelitis in the United States. The annual number of reported cases of paralytic disease declined from more than 18,000 in 1954 to less than 20 in 1973–1978. The risk of poliomyelitis is generally very small in the United States today, but epidemics are certain to occur if the immunity of the population is not maintained by immunizing children beginning in the first year of life.

The proportion of the U.S. population fully immunized against poliomyelitis appears to have declined in recent years. The United States Immunization Survey in 1978 indicated that only 60% of 1- to 4-year-old children had completed primary vaccination against poliomyelitis. Rates for infants and young children in disadvantaged urban and rural areas were even lower. Recent intensive immunization efforts have reversed this downward trend, but clearly there remain many unimmunized (or incompletely immunized) children.

Laboratory surveillance of enteroviruses shows that the circulation of wild polioviruses has diminished markedly. Inapparent infection with wild strains no longer contributes significantly to establishing or maintaining immunity, making universal vaccination of infants and children even more important.

Poliovirus Vaccines

Two types of poliovirus vaccines are currently licensed in the United States: Oral Polio Vaccine (OPV) and Inactivated Polio Vaccine (IPV). Although IPV and OPV are both effective in preventing poliomyelitis, OPV (containing live attenuated viruses) is the vaccine of choice for primary immunization of children in the United States when the benefits and risks for the entire population are considered. OPV is preferred because it induces intestinal immunity, is simple to administer, is well accepted by patients, results in immunization of some contacts of vaccinated persons, and has a record of having essentially eliminated disease associated with wild polioviruses in this country. The choice of OPV as the preferred polio vaccine in the United States has also been made by the Committee on Infectious Diseases of the American Academy of Pediatrics and a special expert committee of the Institute of Medicine, National Academy of Sciences.

Some poliomyelitis experts contend that greater use of IPV in the United States for routine vaccination would provide continued control of naturally occurring poliovirus infections and simultaneously reduce the problem of OPV-associated disease. They argue that there is no substantial evidence that OPV and currently available IPV differ in their ability to protect individuals from disease. They question the public health significance of higher levels of gastrointestinal immunity achieved with OPV. Finally, they question whether the transmission of vaccine virus to close contacts contributes substantially to the level of immunity achieved in the community.

Some countries prevent poliomyelitis successfully with IPV. However, because of many differences between these countries and the United States, particularly with respect to risks of exposure to wild polioviruses and the ability to achieve and maintain very high vaccination rates in the population, their experiences with IPV may not be directly applicable here. Based on current achievements in the United States with other vaccines, it is doubtful that a sufficient number of persons would regularly receive vaccination with IPV to sustain the present level of poliomyelitis protection in the community and to prevent recurrence of outbreaks.

Prospective vaccinees or their parents should be made aware of the polio vaccines available and the reasons why recommendations are made for giving specific vaccines at particular ages and under certain circumstances. Furthermore, the benefits and risks of the vaccines for individuals and the community should be stated so that vaccination is carried out among persons who are fully informed. [Specific recommendations are supplied in the source given below.]

Source: *MMWR* 28:510 (11/2/79)

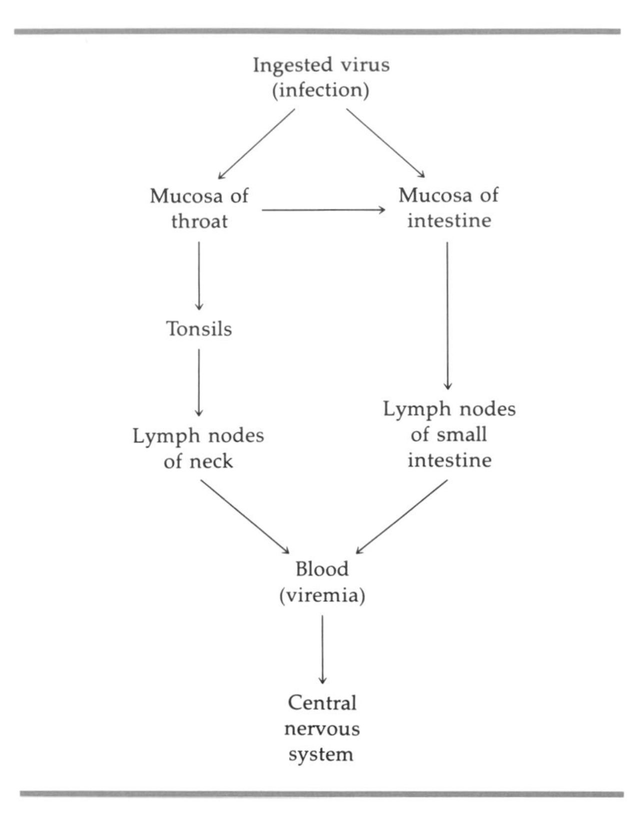

Figure 21–9 Route taken by poliovirus in the body.

cord called anterior horn cells. The virus does not infect the peripheral nerves or the muscles. As the virus multiplies within the cytoplasm of the motor nerve cells, the cells die, resulting in paralysis.

Diagnosis of polio is usually based on isolation of the virus from feces and throat secretions. Cell cultures can be inoculated and cytopathic effects on the cells observed. Serological titrations using viral neutralization tests are also common.

The incidence of polio in the United States has decreased markedly since the availability of the polio vaccines, and now outbreaks usually occur only in segments of the population lacking proper immunization, perhaps because of cultural or religious restrictions. This pattern is found in other developed countries, as well. Elsewhere, the disease usually occurs in infancy and is seldom paralytic.

The development of the first polio vaccine was made possible by the introduction of practical techniques for cell culture, since the virus does not grow in any common laboratory animal. The polio vaccine was the prototype for the vaccines for mumps, measles, and rubella.

Two vaccines are available. The *Salk vaccine,* which was developed in 1954, uses viruses inactivated by treatment with formalin. It requires a series of injections. The effectiveness of this vaccine is high, perhaps 90% against paralytic polio. The antibody levels decline with time and booster shots are needed every few years to maintain full immunity. Nonetheless, several European countries have almost eliminated polio in their populations using only the Salk vaccine.

The *Sabin vaccine,* developed more recently, contains three living, attenuated strains of the virus and is more popular in the United States than the Salk vaccine. It is less expensive to administer, and most people prefer taking a sip of orange-flavored drink containing the virus to having a series of injections. The immunity achieved with the Sabin vaccine resembles that acquired by natural infection. One disadvantage is that on rare occasions—one in several million cases—one of the attenuated strains of the virus (type 3) seems to revert to virulence and cause the disease. A number of these cases have been secondary contacts, rather than the person receiving the vaccine. This illustrates how recipients of the vaccine may also immunize contacts.

Some medical scientists, including Salk himself, have suggested that a return to the Salk vaccine may be desirable despite its disadvantages. As the level of polio in the population declines, the cases that can be linked to the live vaccine become a more and more significant portion of the total cases. There is no reason to think that polio could not be almost totally eliminated by an effectively administered program of immunization in this country. Whatever the advantages of either the Salk or Sabin vaccine, both have been used with remarkable effects (Figure 21–10). See the box for a discussion of poliomyelitis prevention in the United States.

RABIES

Rabies is an acute infectious disease that usually results in fatal encephalitis. The causative agent is the *rabiesvirus,* typically transmitted by the bite of an animal that contains the virus in its saliva. The virus is a rhabdovirus that has a characteristic bullet shape (Figure 21–11) and contains single-stranded RNA.

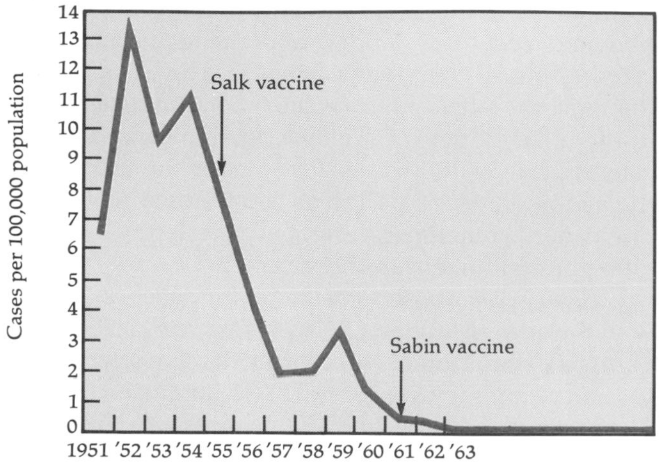

Figure 21–10 Number of cases of poliomyelitis reported in the United States, 1951–1979. Note the decrease after introduction of Salk vaccine (1955) and Sabin vaccine (1961). There were fewer than 0.025 cases per 100,000 population in 1979.

A skin wound or abrasion, usually inflicted by a rabid animal, is the principal portal of entry. It is also possible to contract the disease by inhalation of aerosols of the virus, which may be present in bat caves, for example. It is suspected that aerosol infection may bypass vaccine-based immunity, as indicated by laboratory accidents. The virus may also enter minute skin abrasions in contact with viral-containing fluids.

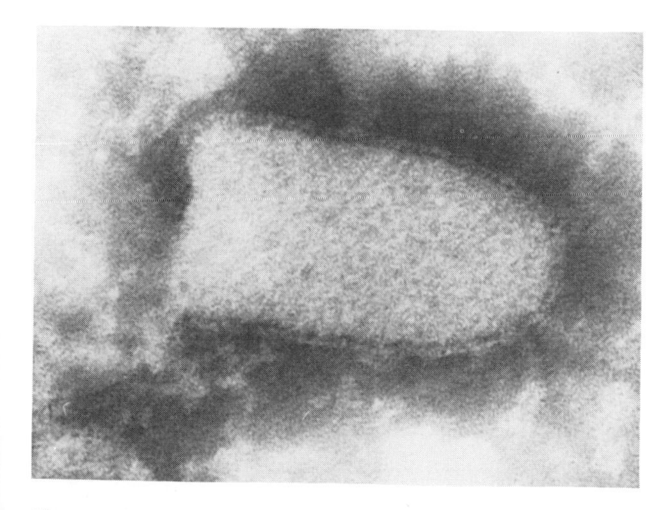

Figure 21–11 A single, bullet-shaped rabiesvirus (×280,300).

Initially, the virus multiplies in skeletal muscle and connective tissue but remains localized for periods ranging from days to months. Then it travels along the peripheral nerves to the central nervous system, where it produces encephalitis. In the relatively few cases of survival, which are due to the intensive, modern supportive care, neurological damage is often severe.

The initial symptoms of rabies include spasms of the muscles of the mouth and pharynx when swallowing liquids. In fact, even the mere sight or thought of water can set off the spasms—thus the common name *hydrophobia* (fear of water). The final stages of the disease result from extensive damage to the nerve cells of the brain and spinal cord.

Dogs with **furious rabies** are at first restless, then highly excited, snapping at anything within reach. When paralysis sets in, the flow of saliva increases as swallowing becomes difficult and nervous control is progressively lost. The disease is almost always fatal within a few days. Some animals suffer from **dumb (paralytic) rabies,** in which there is only minimal excitability. This is especially common in cats. The animal remains relatively quiet and even unaware of its surroundings but may snap irritably if handled.

Laboratory diagnosis of rabies in humans and infected animals is based on several findings. When the patient or animal is alive, a diagnosis can sometimes be confirmed by immunofluorescent studies in which viral antigens are detected in saliva, serum, or cerebrospinal fluid. After death, diagnosis is confirmed by preparing smears from the brain of the infected person or animal. In the past, the diagnostic procedure was to examine the smears for **Negri bodies,** which are masses of viruses or unassembled viral subunits in the cytoplasm of the infected nerve cells (Figure 21–12). Today, however, diagnosis is usually made by a fluorescent-antibody test performed on the smears (see Color Plate IIIA).

Any person bitten by an animal that is positive for rabies must take antirabies treatment. Other indications for antirabies treatment are any unprovoked bite inflicted by a dog, cat, skunk, bat, fox, coyote, bobcat, or raccoon not available for observation and examination. Treatment after a dog or cat bite is determined by the origin of the animal and presence of rabies in the area. (Bites from rodents and rabbits do not require antirabies treatment.)

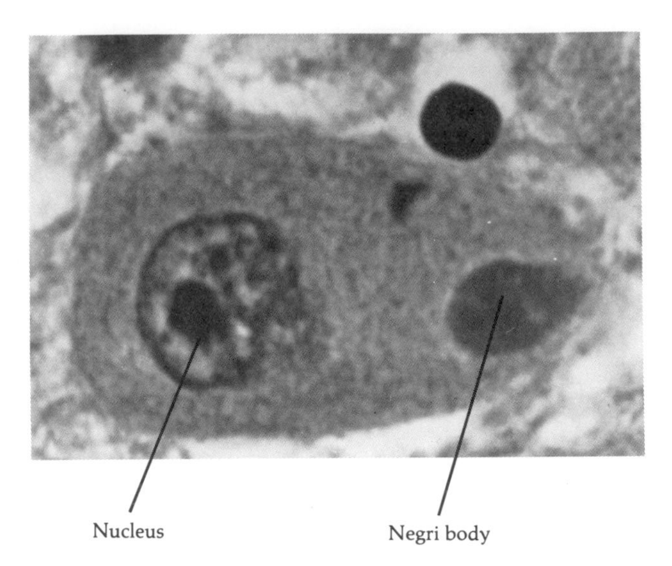

Figure 21–12 A nerve cell of the brain containing an oval Negri body.

Until recent years, the *Pasteur treatment* was standard for the prevention of rabies in exposed individuals (Figure 21–13a). To prepare the Pasteur vaccine, a rabbit was first injected with the rabies-virus, which multiplied in the brain and spinal cord, killing the animal. Then the brain tissue was homogenized in a blender, and the virus inactivated by physical or chemical means. The virus was then incorporated into a vaccine and administered to the infected person. The vaccination procedure consisted of 14 to 21 injections under the skin of the abdomen over a period of two to three weeks. The patient sometimes developed an allergy to the rabbit brain antigens in the vaccine.

Today, postexposure treatment for rabies begins with the administration of a passive immunization. Human **rabies immune globulin (RIG)** is preferred to horse antirabies serum (ARS) because ARS causes serum sickness in many patients (see Chapter 17). Passive immunization is followed by active immunization. Active immunization is achieved with rabiesvirus grown in human diploid cell culture or duck embryo culture. The virus is chemically inactivated and then concentrated to produce the vaccine **human diploid cell vaccine, HDCV,** or **duck embryo vaccine, DEV** (Figure 21–13b). When available, HDCV is the vaccine of choice because it produces a higher antibody titer than DEV. For post-

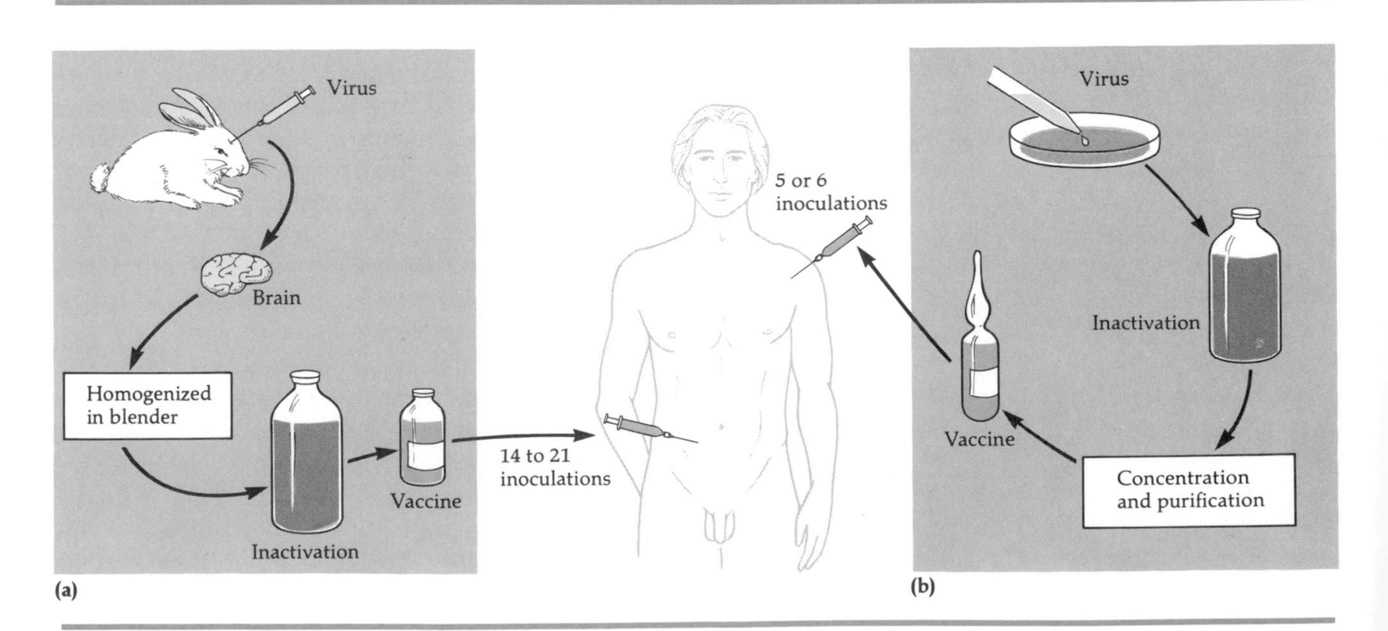

Figure 21–13 Preparation of rabies vaccines. **(a)** Pasteur vaccine. **(b)** Human diploid cell vaccine (HDCV). HDCV is more effective than the Pasteur vaccine in stimulating a high rabies antibody titer. It requires fewer injections and is much less likely to cause allergic reactions.

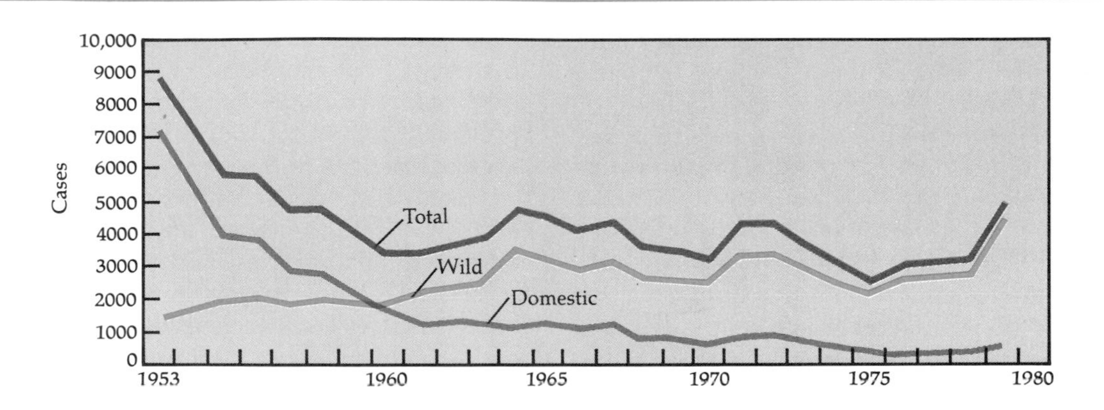

Figure 21–14 Reported cases of rabies in wild and domestic animals in the United States, 1953–1979. The disease has declined in domestic animals, probably because of effective vaccination programs. But note that while fewer cases are now reported in dogs, cats, cattle, and oxen (domestic animals), more cases are reported in skunks, bats, and raccoons (wild animals).

exposure treatments, five intramuscular injections of HDCV are administered over a four-week period, and a sixth dose is recommended two months later. When DEV is used, 23 subcutaneous injections must be given. At the end of the series of injections, the patient's serum is tested for rabies antibodies. If adequate antibody titer is not detected, one or more booster doses are given.

Preexposure, active immunization is required for persons such as veterinarians and laboratory personnel. When HDCV is available for preexposure treatment, three injections are given at weekly intervals. Alternatively, three or four doses of DEV may be given.

Rabies occurs all over the world. Some islands such as Australia, New Zealand, and Hawaii are free of the disease, a condition maintained by rigid quarantine. In North America, rabies is widespread among wildlife (Figure 21–14), with the skunk the wildlife animal most often reported. In 1975, skunks accounted for 46% of animal rabies cases, bats 19%, foxes 10%, and raccoons 10%. Among domestic animals, cattle accounted for 6% of the reported animal cases, dogs 6%, and cats 4%. Worldwide, however, dogs are the most common carrier of rabies. Rabies is seldom, if ever, found in squirrels, rabbits, rats, or mice. In fact, the Centers for Disease Control record only two cases in squirrels. But rabies has long been endemic in vampire bats of South America. And while insect-eating bats in the United States were not known to have rabies until 1953, since that time, bats have been found to be responsible for transmission of rabies to humans and domestic animals in a number of cases. It is not known whether rabies was first introduced into North American bats around 1953 or if it always existed and was simply not detected until then. Rabies is not always fatal in bats, as it is in other animals.

In the United States, there are up to 4000 cases of rabies diagnosed in animals each year, but in recent years, only one to four cases in humans. Worldwide, there are probably about 1000 human cases each year.

SLOW VIRUS DISEASES

There are a number of fatal diseases of the human central nervous system thought to be caused by what are termed **slow viruses**—although no one is absolutely sure of the causative agents. The name refers to the slow progress of these diseases.

A typical slow virus disease is **sheep scrapie.** The infected animal rubs against fencing and walls until raw areas appear on its body. Over a period of several weeks or months, the animal gradually loses motor control and eventually dies. The agent, whatever it is, can be passed to other animals, such as mice, by injection of brain tissue from one animal to the next. The scrapie agent can cause a disease of mink indistinguishable from **transmissible mink encephalopathy,** a disease found in minks fed mutton at mink farms.

Humans suffer from diseases similar to scrapie. For example, there is **Creutzfeldt–Jakob disease,** which appears at a higher than normal rate in certain Middle Eastern Jewish cultures that eat sheep brain. Some tribes in New Guinea have suffered from an apparent slow virus disease called **kuru.** This disease appears most often in women and children, perhaps because one of the funeral rituals involves rubbing the brain of the dead man over the female kin. Cannibalism may also be involved in transmission. Kuru can be transmitted to chimpanzees by injection of bacteria-free filtrates or the brains of kuru victims.

Many years ago, when rabies vaccine virus was grown in rabbit brains and contained a certain number of rabbit brain cells, a disease known as **experimental allergic encephalitis** was transmitted to some vaccine recipients. The disease, which can still be reproduced in laboratory animals, is similar to multiple sclerosis, exhibiting progressive deterioration of the central nervous system with periods of remission. It is possible that diseases such as multiple sclerosis may be due to similar etiologic agents. In multiple sclerosis, the myelin sheath (lipid) covering the nerve cells in the central nervous system is destroyed, interrupting the nerve impulses.

Other possible slow virus diseases are Parkinson's disease, premature senility, and even rheumatoid arthritis. The slow virus diseases may not be attributable directly to the activity of the causative agents, but may be a form of autoimmune disease. Normal immune reactions do not seem to be provoked in the host, nor does the patient exhibit normal reactions to disease, such as fever.

Some authorities divide the slow virus diseases into two groups: those apparently caused by unconventional viruses and those that may be associated with unusual immunologic effects of known viruses such as measles. Measles virus can cause the latent expression of subacute sclerosing panencephalitis, which shows many characteristics of a slow virus disease. Evidence of measles antibodies in the cerebrospinal fluid of patients with multiple sclerosis has raised the possibility that multiple sclerosis is a complication of prior measles.

The unconventional viruses, and some scientists call them "agents" rather than viruses, cannot be inactivated by ultraviolet radiation and are resistant to formalin and heat treatments. It is possible that these agents do not possess nucleic acid at all. And if they do have normal viral nucleic acids, then perhaps they lack the normal viral protein coat, or perhaps the coat is a polysaccharide capsule, if indeed a coat exists at all. Certain plant pathogens, called viroids, are simply naked strands of RNA, and slow "viruses" may resemble them. In any event, the slow virus diseases will be the subject of intense research.

A summary of diseases associated with the nervous system is presented in Table 21–1.

Table 21–1 Summary of Diseases Associated with the Nervous System

Disease	Causative Agent	Mode of Transmission	Treatment
Meningococcal meningitis	*Neisseria meningitidis*	Contact with a healthy carrier via respiratory tract	Penicillin, chloramphenicol, rifampin
Hemophilus influenzae meningitis	*Hemophilus influenzae*	Via respiratory tract in individuals with viral respiratory infections	Ampicillin in combination with chloramphenicol
Cryptococcosis	*Cryptococcus neoformans*	Inhalation of dried infected pigeon droppings	Amphotericin B
Leprosy	*Mycobacterium leprae*	Transfer of exudates from lesions or inanimate objects	Sulfone drugs (dapsone), rifampin
Poliomyelitis	Poliovirus	Ingestion of virus	None
Rabies	Rabiesvirus	Bite of a rabid animal	RIG and HDCV or DEV after exposure

STUDY OUTLINE

ORGANIZATION OF THE NERVOUS SYSTEM (pp. 533–534)

1. The central nervous system consists of the brain protected by the skull bones and the spinal cord protected by the backbone.
2. The peripheral nervous system consists of the nerves that branch from the central nervous system.
3. The central nervous system is covered by three layers of membranes called meninges. Cerebrospinal fluid circulates between the inner and middle meninges and in the ventricles of the brain.
4. The blood–brain barrier normally prevents many substances such as antibiotics from entering the brain.
5. Microorganisms can enter the central nervous system through trauma, along peripheral nerves, and through the bloodstream and lymphatic system.
6. An infection of the meninges is called meningitis. An infection of the brain is called encephalitis.

MENINGITIS (pp. 534–538)

Meningococcal Meningitis (pp. 534–536)

1. *Neisseria meningitidis* causes meningococcal meningitis. This bacterium is found in the throat of healthy carriers.
2. The bacteria probably gain access to the meninges through the bloodstream. The bacteria may be found in leukocytes in cerebrospinal fluid.
3. Early symptoms resemble a cold but progress to pain and stiffness in the neck and back. Neurological complications may develop. Purplish spots appear on the skin.
4. The disease occurs most often in young children. Military recruits are vaccinated with purified capsular polysaccharide to prevent epidemics in training camps.
5. Encapsulated *N. meningitidis* bacteria are resistant to phagocytosis.
6. Diagnosis is based on isolation and identification of the bacteria in blood or cerebrospinal fluid.
7. *N. meningitidis* must be cultured on media containing blood and incubated in an atmosphere containing 5 to 10% carbon dioxide.

Hemophilus influenzae Meningitis (pp. 536–537)

1. *Hemophilus influenzae* is cultured on chocolate agar containing X and V factors.

2. *H. influenzae* is the most common cause of meningitis in children under four years old.
3. The disease often occurs as a secondary infection following a viral infection such as influenza.
4. Diagnosis is based on identification of the organism in spinal fluid by a positive quellung test.

Cryptococcus neoformans Meningitis (Cryptococcosis) (p. 538)

1. *Cryptococcus neoformans* is an encapsulated yeastlike fungus that causes *Cryptococcus neoformans* meningitis.
2. The disease may be contracted by inhalation of dried infected pigeon droppings.
3. The disease begins as a lung infection and then spreads to the brain and meninges.
4. Immunosuppressed individuals are most susceptible to *Cryptococcus neoformans* meningitis.
5. Diagnosis is based on observation of the fungus.

Other Causes of Meningitis (p. 538)

1. Many bacteria including *Mycobacterium tuberculosis, Clostridium perfringens, Flavobacterium,* and *Listeria* can cause meningitis.
2. Aseptic meningitis describes a condition in which bacteria and fungi are not the cause. Viruses are usually suspected.
3. Viruses that cause aseptic meningitis include mumps virus, poliovirus, coxsackievirus, and echovirus.

LEPROSY (pp. 538–540)

1. *Mycobacterium leprae* causes leprosy, or Hansen's disease.
2. *M. leprae* cannot be cultured on artificial media. It is grown in the footpads of mice or in armadillos.
3. The tuberculoid form of the disease is characterized by loss of sensation in the skin surrounded by nodules. The lepromin test is positive.
4. In the lepromatous form, disseminated nodules and tissue necrosis occur. The lepromin test is negative.
5. Leprosy is not highly contagious and is spread by prolonged contact with exudates and fomites.
6. Patients with leprosy are made noncommunicable within four or five days with sulfone drugs and then treated as outpatients.
7. Untreated individuals often die of secondary bacterial complications such as tuberculosis.

8. The disease occurs primarily in the tropics. Many people are treated at the National Leprosy Hospital in Carville, LA.
9. Laboratory diagnosis is based on observation of acid-fast rods in lesions or fluids.
10. Children living with leprous parents are vaccinated with BCG, which results in a positive lepromin test.

POLIOMYELITIS (pp. 540–542)

1. The symptoms of poliomyelitis are usually headache, sore throat, fever, stiffness of the back and neck, and occasionally (1 to 2% of the cases) paralysis.
2. Poliovirus (picornavirus) is found only in humans and is transmitted by ingestion of water contaminated with feces.
3. Poliovirus first invades lymph nodes of the neck and small intestine. Viremia and spinal cord involvement may follow.
4. At present, outbreaks of polio in the United States are uncommon because of the use of vaccines.
5. The Salk vaccination involves injection of formalin-inactivated viruses and boosters every few years. The Sabin vaccine contains three attenuated live strains of poliovirus and is administered orally.
6. Diagnosis is based on isolation of the virus from feces and throat secretions or the presence of viral neutralizing antibodies in the patient's serum.

RABIES (pp. 542–545)

1. Rhabdovirus causes an acute, usually fatal, encephalitis called rabies.
2. The virus enters through the bite of a rabid animal and multiplies in skeletal muscle and connective tissue.
3. Encephalitis occurs when the virus moves along peripheral nerves to the central nervous system.
4. Symptoms of rabies include spasms of mouth and throat muscles followed by extensive brain and spinal cord damage.
5. Laboratory diagnosis may be made by direct immunofluorescent tests of saliva, serum, cerebrospinal fluid, or brain smears.
6. Reservoirs for rabies include skunks, bats, foxes, and raccoons. Domestic cattle, dogs, and cats may get rabies. Rodents and rabbits seldom get rabies.
7. Rabies is fatal in all animals except bats.
8. The Pasteur treatment for rabies involves multiple subcutaneous injections of rabiesvirus grown in rabbit brain tissue.

9. Current postexposure treatment includes administration of human rabies immune globulin (RIG) followed by multiple intramuscular injections of human diploid cell culture virus (HDCV) or duck embryo culture virus (DEV).
10. Preexposure immunization consists of injections of DEV or HDCV.
11. Rabies may be contracted by inhalation of aerosols or invasion through minute skin abrasions.

SLOW VIRUS DISEASES (pp. 545–546)

1. Slow virus diseases are of uncertain etiology. The diseases progress slowly in the host.
2. Sheep scrapie and transmissible mink encephalopathy are examples of slow virus diseases that are transferrable from one animal to another.
3. Creutzfeldt–Jakob disease and kuru are human diseases similar to scrapie that occur in isolated groups of people who eat brains.
4. Multiple sclerosis and Parkinson's disease may be slow virus diseases.
5. Slow virus diseases may result from immunologic complications of conventional viral diseases (i.e., subacute sclerosing panencephalitis following measles).
6. Slow virus diseases may be caused by "agents" that resemble viruses but possibly lack nucleic acid or protein coats. They may resemble viroids, naked pieces of infectious RNA.

STUDY QUESTIONS

REVIEW

1. Define the following terms: meningitis; encephalitis.
2. Fill in the following table.

Causative Agent	Disease	Susceptible Population	Method of Transmission	Treatment
N. meningitidis				
H. influenzae				
C. neoformans				

3. Briefly explain the derivation of the name *Hemophilus influenzae*.
4. To what does the term *aseptic meningitis* refer? Look up *aseptic* in the glossary. Why is this word used in a description of meningitis?
5. Provide the following information on leprosy: etiology, method of

transmission, symptoms, treatment, prevention, and susceptible population.

6. Provide the following information on poliomyelitis: etiology, method of transmission, symptoms, prevention.
7. Compare and contrast the Salk and Sabin vaccines with respect to composition, advantages, and disadvantages.
8. Provide the etiology, method of transmission, reservoirs, and symptoms for rabies.
9. Outline the procedures for treating rabies after exposure. Outline the procedures for preventing rabies prior to exposure. What is the reason for the differences in the procedures?
10. Describe the symptoms and laboratory test results that would lead to a diagnosis of slow virus disease.
11. Provide evidence that slow virus diseases are caused by transmissible agents that resemble viruses.
12. Why are meningitis and encephalitis generally difficult to treat?

CHALLENGE

1. A BCG vaccination will result in positive lepromin and tuberculin tests. What is the relationship between leprosy and tuberculosis?
2. Why aren't Salk and Sabin vaccines considered treatments for poliomyelitis?
3. Compare the Pasteur treatment with current treatments for rabies. What are the advantages of HDCV over DEV and the Pasteur treatment?

FURTHER READING

Dead or Alive? September 1977. *Scientific American* 237:96–100. A discussion of the decline in vaccinations against polio in the United States.

Dean, G. July 1970. The multiple sclerosis problem. *Scientific American* 223:40–46. A discussion of the theory that a latent virus is the cause of multiple sclerosis.

Fucillo, D. A., J. E. Kurent, and J. L. Sever. 1974. Slow virus diseases. *Annual Review of Microbiology* 28:231–264. A summary of information on viral diseases caused by conventional and unconventional agents.

Kaplan, M. M., and H. Koprowski. January 1980. Rabies. *Scientific American* 242:120–134. An overview of rabies and a comparison of available vaccines.

Spector, D. H., and D. Baltimore. May 1975. The molecular biology of poliovirus. *Scientific American* 232:24–31. Information on viral multiplication is obtained from laboratory-grown poliovirus.

See also Further Reading for Part IV at the end of Chapter 19.

22

Microbial Diseases of the Cardiovascular and Lymphatic Systems

Objectives

After completing this chapter you should be able to

- List the symptoms of septicemia.
- Explain the importance of infections that develop into septicemia.
- Discuss the epidemiology of puerperal sepsis, bacterial endocarditis, and myocarditis.
- Discuss the cause, treatment, and prevention of rheumatic fever.
- Describe the epidemiology of tularemia, brucellosis, anthrax, and listeriosis.
- List the causative agents and methods of transmission of infectious mononucleosis and serum hepatitis.
- Describe the causative agent, method of transmission, and effect on world health of toxoplasmosis, American trypanosomiasis, and schistosomiasis.

The **cardiovascular system** consists of the heart, blood, and blood vessels. The **lymphatic system** consists of the lymph, lymph vessels, lymph nodes, and lymphoid organs (tonsils, appendix, spleen, and thymus gland). Since the function of both systems is to circulate various substances throughout the body, they may also serve as vehicles for the spread of infection.

STRUCTURE AND FUNCTION OF THE CARDIOVASCULAR SYSTEM

The center of the cardiovascular system is the *heart* (Figure 22–1). The function of the heart is to keep the blood circulating through the body so it can deliver certain substances to tissue cells and remove other substances from them. The heart is composed of four chambers, two on the right side and two on the left. The upper chambers, called *atria,* receive blood from different parts of the body, while the lower chambers, called *ventricles,* pump blood to various parts of the body. Valves between the chambers prevent the backward flow of blood.

Deoxygenated blood, which contains more carbon dioxide than oxygen, arrives at the right atrium from all parts of the body. From here, it passes into the right ventricle and is then pumped through the pulmonary arteries into the lungs. Within the lungs, carbon dioxide is exchanged for oxygen and the blood is now oxygenated, that is, it contains more oxygen than carbon dioxide. The oxygenated blood returns through pulmonary veins to the left atrium.

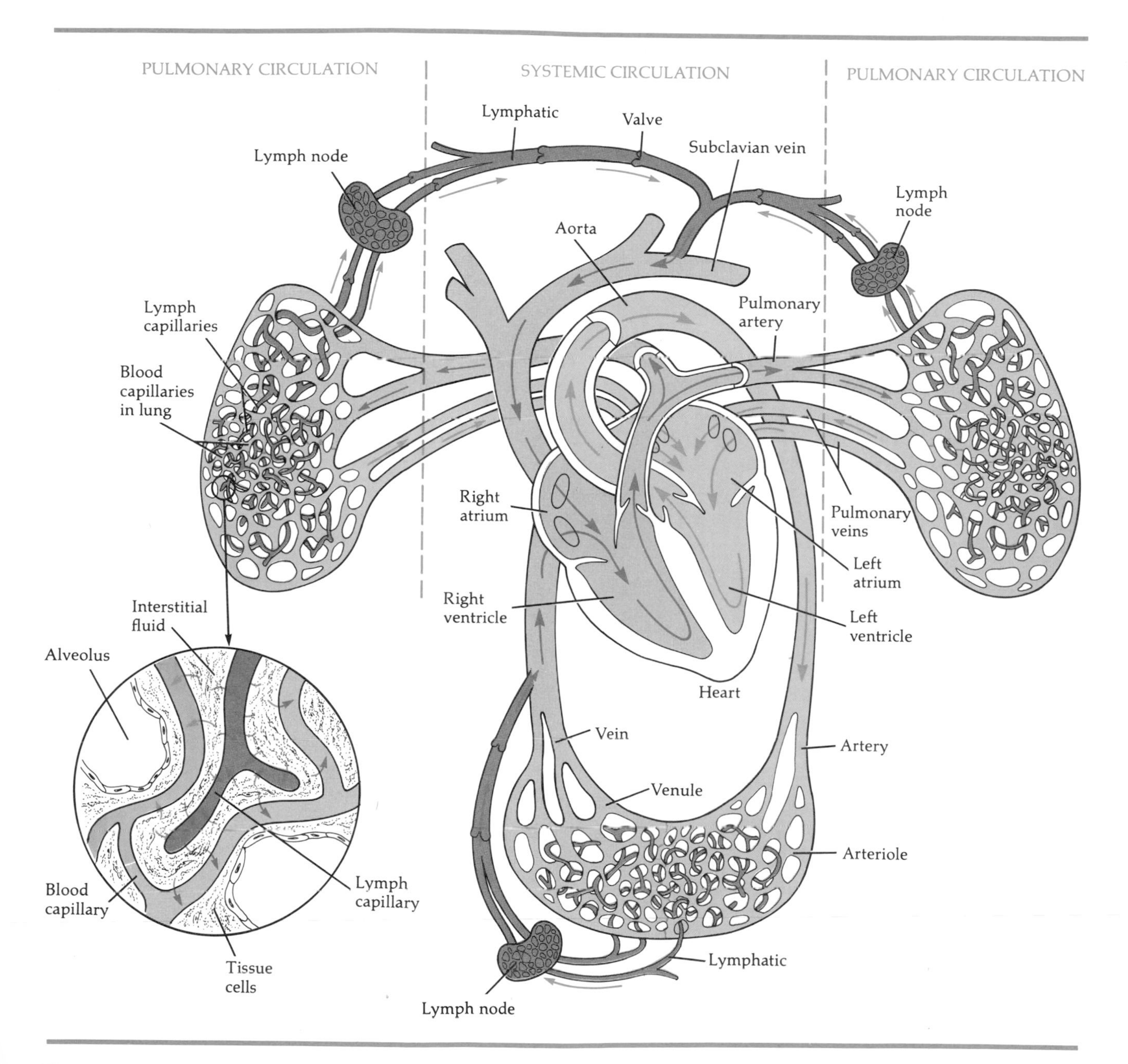

Figure 22–1 The cardiovascular and lymphatic systems.

It then passes into the left ventricle and from here is pumped through the aorta to all parts of the body. As the blood circulates to all tissue cells, the cells give up carbon dioxide and take in oxygen. This produces deoxygenated blood, which returns to the right atrium to start another cycle.

Blood vessels are the tubes that carry circulating blood throughout the body. An *artery* is a blood vessel that carries blood away from the heart. Branching off the aorta are numerous arteries that deliver oxygenated blood to different tissues of the body. When they reach their destinations, the arteries branch into smaller vessels called *arterioles*, which in turn branch into even smaller vessels called *capillaries*. Capillary walls are only one cell thick, and it is through these walls that blood and tissue cells exchange materials. The capillaries that take the deoxygenated blood away from the tissue cells converge to form small veins called *venules*. Venules unite to form *veins,* and veins return the blood to the right atrium.

The *blood* itself is a mixture of a liquid called plasma and cells. It is the plasma that really transports dissolved nutrients to body cells and removes wastes from the cells. The cells are known as red blood cells, white blood cells, and platelets. Red blood cells, or erythrocytes, carry oxygen and also carbon dioxide (although most of the carbon dioxide in blood is dissolved in the plasma). White blood cells, or leukocytes, play several roles in defending the body against infection, as discussed in Part III. Some white cells, in particular the neutrophils, are phagocytes. And the B cells and T cells, two types of lymphocytes, play a key role in immunity. Platelets, or thrombocytes, function in blood clotting.

STRUCTURE AND FUNCTION OF THE LYMPHATIC SYSTEM

As part of the overall pattern of circulation, some plasma filters out of the blood capillaries and into spaces between tissue cells called *interstitial spaces.* The filtered fluid around and between tissue cells is called *interstitial fluid.* Also surrounding tissue cells are microscopic lymphatic vessels called *lymph capillaries,* which are larger and more permeable than blood capillaries. As the interstitial fluid moves around the tissue cells, it is picked up by the lymph capillaries and is now referred to as *lymph* (Figure 22–1). Since lymph capillaries are so permeable, they readily pick up microorganisms or their products. From lymph capillaries, lymph is transported into larger lymph vessels called *lymphatics,* which contain valves that keep the lymph moving toward the heart. Eventually, all the lymph is returned to the blood just before the blood enters the right atrium. This pattern of circulation returns proteins and fluid that have filtered from the plasma back to the blood.

At various points along the lymphatic system are oval or bean-shaped structures called *lymph nodes,* through which lymph flows. Within the lymph nodes are cells capable of phagocytosis (fixed macrophages) and antibody production (B cells), which help to clear the lymph of microorganisms. At times, the number of microorganisms circulating through nodes is so great that the nodes themselves become infected, causing the nodes to become enlarged and tender.

The lymphatic system also consists of *lymphoid organs.* These include the tonsils, appendix, spleen, and thymus gland. We have already considered the role of the thymus gland in processing T cells in Chapter 16.

BACTERIAL DISEASES OF THE CARDIOVASCULAR AND LYMPHATIC SYSTEMS

Septicemia

Although blood is normally sterile, the presence of even moderate numbers of microorganisms is not usually harmful. However, if the defenses of the blood and lymphatic system fail, the microorganisms may proliferate in the blood without control, a condition called **septicemia** or **blood poisoning.**

Clinically, septicemia is characterized by fever and a decrease in blood pressure. The decrease in blood pressure may further result in ischemia and shock. Another symptom of septicemia is the appearance of **lymphangitis,** inflamed lymph vessels shown as red streaks under the skin running along the arm or leg from the infection site. Sometimes the streaks end at a lymph node, where the lymphocytes attempt to prevent the further passage of the invading microorganisms.

The organisms most frequently associated with septicemia are gram-negative rods, although a few gram-positive bacteria and fungi are also implicated. Among the gram-negative rods are *Escherichia coli, Serratia marcescens, Proteus mirabilis, Enterobacter aerogenes, Pseudomonas aeruginosa,* and *Bacteroides* species. These bacteria enter the blood from a focus of infection in the body. As you may recall, the cell wall of many gram-negative bacteria contains endotoxins that are released upon the lysis of the cell. It is the endotoxin that actually causes the symptoms. Once released, the endotoxin damages blood vessels, resulting in the low blood pressure and subsequent shock. In some cases, antibiotics may aggravate the condition by causing the lysis of large numbers of bacteria, which in turn release more of the damaging endotoxins. Many of these gram-negative rods are of nosocomial origin, introduced into the bloodstream by medical procedures that bypass the normal barriers between the environment and the blood. At one time, however, it was probably more common for infections of the bloodstream to be caused by gram-positive bacteria such as staphylococci, which are common skin inhabitants.

Puerperal Sepsis

A nosocomial infection that frequently leads to septicemia is **puerperal sepsis,** also called **puerperal fever** or **childbirth fever.** This begins as an infection of the uterus as a result of childbirth or abortion. *Streptococcus pyogenes,* a group A beta-hemolytic streptococcus, is the most frequent cause, although other organisms may cause infections of this type. Recall from Chapter 19 that the virulence of group A beta-hemolytic streptococci is due to an M protein on their pili that has antiphagocytic properties.

Puerperal sepsis progresses from an infection of the uterus to an infection of the abdominal cavity (peritonitis), and in many cases to septicemia. At one Paris hospital between 1861 and 1864, of the 9886 women who gave birth, 1226 (12%) died of such infections. At that time, the death rate was frequently twice that. These deaths were largely unnecessary, since some 20 years before, Dr. Oliver Wendell Holmes, in America, and an Austrian physician, Ignaz Semmelweiss, had clearly demonstrated that the disease was transmitted by the hands and instruments of the attending midwives or physicians and that disinfection of hands and instruments could prevent such transmission. Yet Louis Pasteur, in 1879, still thought it necessary to lecture physicians as to the cause of the disease.

Antibiotics and modern hygienic practices have now made puerperal sepsis an uncommon complication of childbirth. Penicillin and erythromycin are effective against *S. pyogenes* in the treatment of the disease. Infections related to improperly performed abortions—not strictly puerperal sepsis—are often of mixed bacterial types, many of them caused by anaerobic bacteria of the *Bacteroides* or *Clostridium* genera. Chloramphenicol and penicillin are effective in these cases.

Bacterial Endocarditis

The wall of the heart consists of three layers. The outer layer, called the *pericardium,* is a sac that encloses the heart. The middle layer, known as the *myocardium,* is the thickest layer and consists of cardiac muscle tissue. The inner layer is a lining of epithelium called the *endocardium.* This layer not only lines the heart muscle itself, but also covers the valves in the heart. An inflammation of the endocardium is called **endocarditis.**

One type of bacterial endocarditis is called **subacute bacterial endocarditis.** It is characterized by fever, anemia, general weakness, and heart murmur. It is usually caused by alpha-hemolytic streptococci, although at times beta-hemolytic streptococci or staphylococci *(Staphylococcus epidermidis)* may be involved. About 5 to 10% of the cases are due to enterococci. The condition probably arises from a focus of infection elsewhere in the body, such as infections of the teeth or tonsils. Microorganisms released by tooth extractions or tonsillectomies enter the blood and find their way to the heart. Normally, such bacteria would be quickly cleared from the blood by the defensive mechanisms of the body. But in individuals with abnormal heart valves, either due to congenital heart defects or due to diseases such as rheumatic fever or syphilis, the bacteria lodge in the preexisting lesions. Within the lesions, the bacteria multiply and become entrapped in blood clots that protect them from phagocytes and antibodies. As multiplication progresses and the clot gets larger, pieces of the clot break off and may occlude blood vessels or lodge in the kidneys. In time, the function of the heart valves is impaired. Left untreated, subacute bacterial endocarditis is invariably fatal.

Another type of bacterial endocarditis is called **acute bacterial endocarditis** and is caused by *Staphylococcus aureus* or *Streptococcus pneumoniae* (Figure 22–2). These organisms find their way from the initial site of infection to normal or abnormal heart valves, producing rapid destruction of heart valves that is frequently fatal in days or weeks. Strains of *Staphylococcus* that produce penicillinase are usually sensitive to semisynthetic penicillin derivatives such as methicillin or oxacillin. *Streptococcus pneumoniae* responds well to penicillin. *Streptococcus* may also cause **pericarditis,** inflammation of the sac around the heart (pericardium).

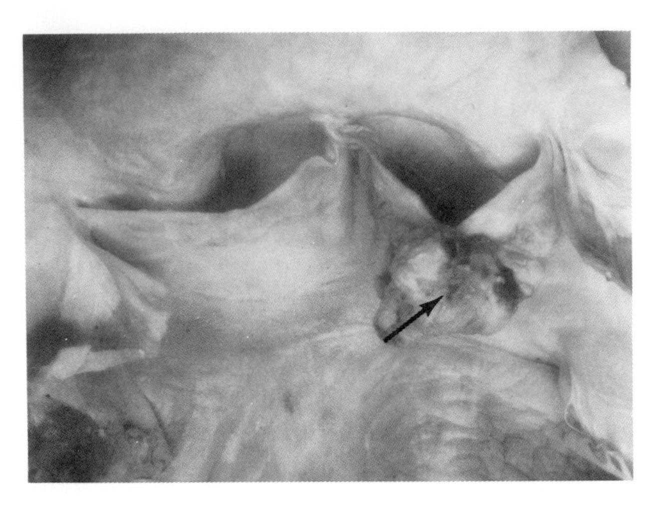

Figure 22–2 Acute bacterial endocarditis. The arrow points to an infected, ulcerated lesion on a heart valve.

Laboratory diagnosis of bacterial endocarditis consists of isolating the bacteria from a blood culture. Penicillin is the most commonly used drug in treatment. It is well to use it prophylactically in susceptible persons, such as those suffering from rheumatic fever or undergoing tooth extractions or tonsillectomies. Erythromycin is also effective in many cases.

Rheumatic Fever

Group A beta-hemolytic streptococcal infections such as those caused by *Streptococcus pyogenes,* sometimes lead to **rheumatic fever,** which is generally considered an autoimmune type of complication (see Chapter 17). The disease is usually expressed as an arthritis, particularly in older persons. It frequently also takes the form of an inflammation of the heart, causing damage to the valves. Other symptoms include fever and malaise. Although recovery from rheumatic fever usually occurs with no permanent damage to the joints, there may be permanent heart damage.

The disease is usually precipitated by a streptococcal sore throat. One to five weeks later, evidence of heart abnormalities appear, by which time the original infection has probably disappeared. Reinfection with streptococci, for example another sore throat, causes renewed attacks and further damage to the heart. Perhaps as many as 3% of children with untreated beta-streptococcal infections contract rheumatic fever. However, many of these cases are essentially subclinical.

The exact mechanism by which streptococci produce rheumatic fever is still obscure, although some data point to an immunologic reaction that remains localized in the heart and joints. For example, infection with group A beta-hemolytic streptococci may result in the deposition of streptococcal antigens such as M proteins in the joints and heart. When the body produces antibodies against the antigens, the antigen–antibody reaction might cause the damage. It is also possible that a streptococcus antigen cross-reacts with heart muscle, inducing the formation of antibodies that subsequently damage the heart. Persons with symptoms of rheumatic fever damage have, in fact, relatively high titers of antibody against streptococcal antigens. And further evidence that rheumatic fever might be the result of an immunologic reaction is seen in the ability to induce arthritic pain in patients by injection of sterile filtrates of streptococcal antigens. And finally, postmortem studies reveal large deposits of antibody and the C3 component of complement in damaged heart tissue.

Laboratory diagnosis of beta-hemolytic streptococci consists of obtaining a culture from the patient, usually a throat swab, and streaking it on a blood agar plate. If the colonies are surrounded by clear areas of hemolysis, the microorganisms are Gram-stained. The appearance of gram-positive cocci occurring in chains is a positive diagnosis. However, streptococci may no longer be present at the time of diagnosis.

Cases of rheumatic fever have been declining in this country because of early treatment of streptococcal infections. In 1969, more than 3000 cases were reported; in 1978, only 850. Penicillin is administered to rheumatic fever patients as a prophylactic measure. The penicillin will not alleviate the rheumatic fever symptoms (which are treated with antiinflammatory drugs), but it will prevent subsequent streptococcal infections that may cause a recurrence of rheumatic fever.

Tularemia

Tularemia is a disease of the lymph nodes caused by a small, gram-negative, facultatively anaerobic, pleomorphic, rod-shaped bacterium, *Francisella*

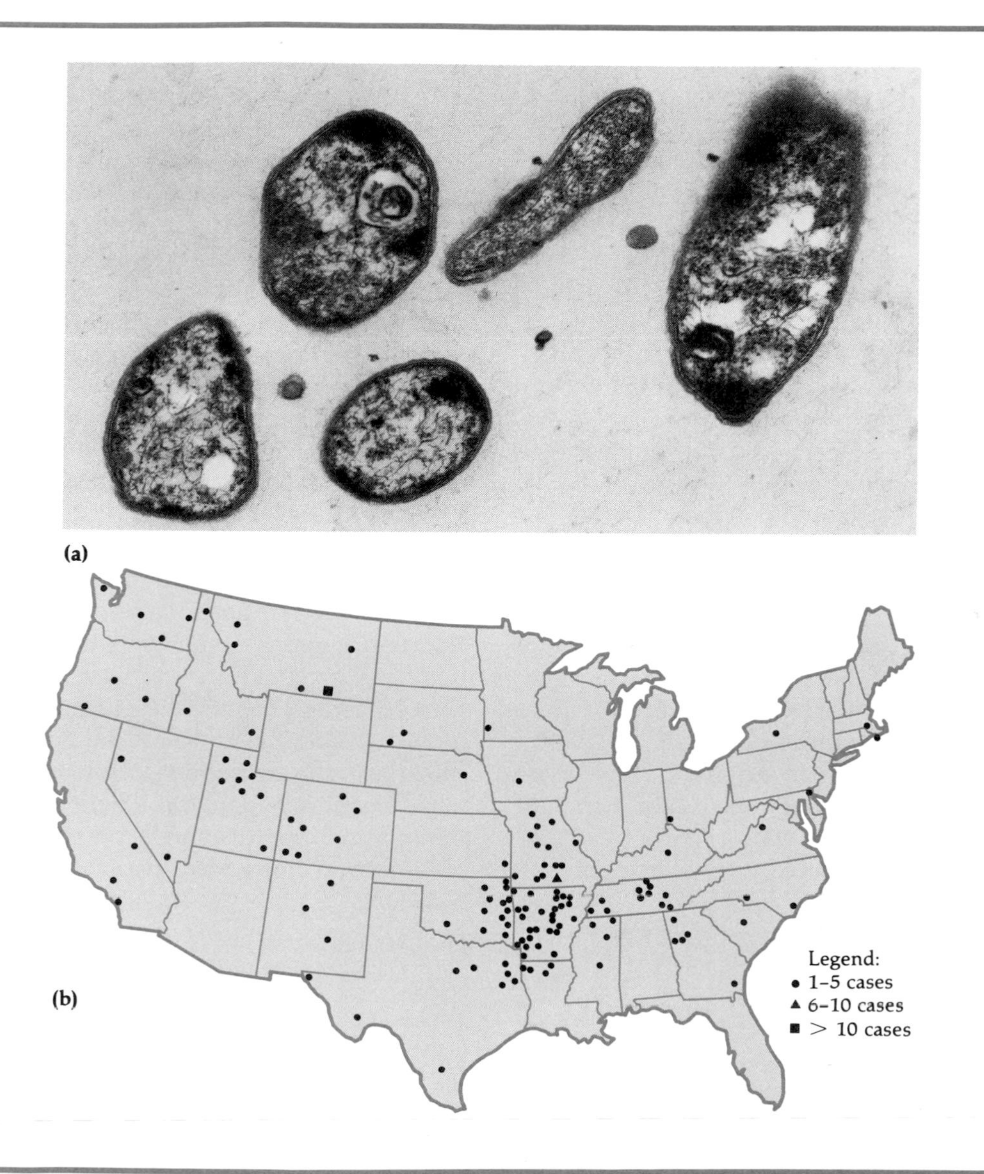

Figure 22–3 **(a)** Electron micrograph of several cells of *Francisella tularensis* (×30,000). **(b)** The geographic distribution of cases of tularemia reported in the United States for 1979. A total of 196 cases were reported.

tularensis. The microorganism was named for Tulare County, California, where it was originally studied (Figure 22–3).

As the disease develops, local inflammation and later a small ulcer appear at the site of infection, usually on the hands. About a week after infection, the lymph nodes enlarge, many containing pockets filled with pus. If the disease is not contained by the lymphatic system, the microorganisms may produce septicemia, pneumonia, and abscesses throughout the body.

Tularemia was first observed in ground squirrels in Tulare County in 1911. Humans most frequently acquire the infection through minor skin abrasions or by rubbing the eyes after handling small wild mammals while hunting. Probably 90% of the cases in this country are contracted from rabbits, with an estimated 1% of all U.S. rabbits infected. Microorganisms in undercooked meat from infected rabbits may also enter through the gastrointestinal tract. Moreover, the disease may be spread by the bites of arthropods such as deer flies (see Figure 11–28d), ticks, or rabbit lice, particularly in the western states, although small-animal contacts are still a more likely cause in the central and eastern states. And tularemia can also be contracted by contact with water that has been contaminated by infected rodents, and even by bites from animals that have fed on dead, diseased animals. Finally, laboratory workers have also been infected by inhaling infectious droplets.

F. tularensis is difficult to grow from serum samples, even on highly specialized media, because of their growth requirement for the amino acid cysteine in amounts greater than those usually present in nutrient media. Colonies may not appear for more than a week, although most cultures will grow within 96 hours. Although no well-defined toxins have been identified to account for its virulence, *F. tularensis* survives for long periods within body cells and phagocytic cells. In fact, resistance to phagocytosis may also account for the low infective dose required. Although naturally acquired immunity is usually permanent, recurrences have been reported. An attenuated live vaccine is available for high-risk laboratory workers.

Serological confirmation of a diagnosis of tularemia is routinely and rapidly accomplished by slide agglutination from isolated bacteria. The antisera are available commercially. Fluorescent-labeled antibody and a delayed hypersensitivity skin test are also used for laboratory identification.

Streptomycin is the antibiotic of choice, but prolonged administration is necessary to avoid relapses. Tetracyclines are also effective, but relapses are less likely with streptomycin. The intracellular location of the microorganism is a problem in chemotherapy.

Brucellosis (Undulant Fever)

The bacterium causing **brucellosis,** like that of tularemia, favors intracellular growth, moves through the lymphatic system, and may travel to many organs via the bloodstream. The microorganisms involved are very small, gram-negative, aerobic rods. They are difficult to grow on artificial media, requiring not only the addition of carbon dioxide to the atmosphere of the incubator, but also specially enriched media. There are four main species: *Brucella abortus,* most commonly found in cattle and transmitted by unpasteurized milk or milk products (Figure 22–4); *Brucella melitensis,* trans-

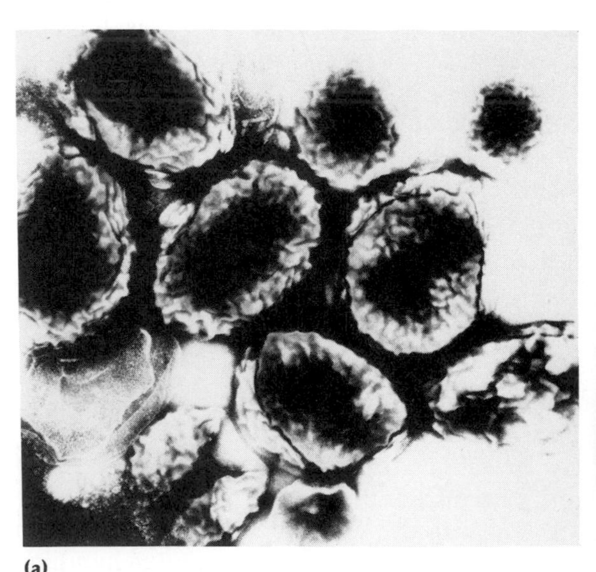
(a)

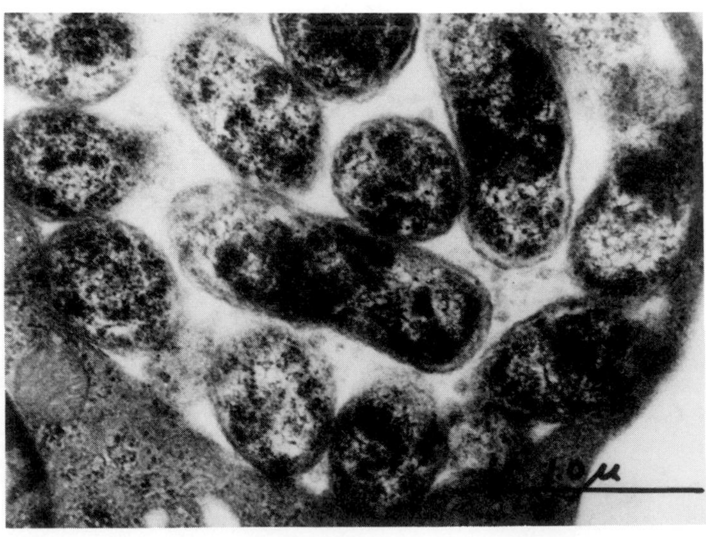

(b)

Figure 22–4 **(a)** *Brucella abortus*, the species transmitted by cattle (×34,200). **(b)** Phagocyte infected with *B. abortus*. Many brucellae are in the large vacuole (×31,000).

mitted by goats; *Brucella suis*, transmitted by swine; and *Brucella canis*, transmitted by dogs, especially beagles. All species are pathogenic to humans and, for that matter, other mammals. *B. suis* and *B. melitensis* infections are more severe than *B. abortus* infections, with a fatality rate of 2 or 3%. About 200 cases of brucellosis are reported each year in the United States. Most of these cases result from contact with swine carcasses.

The organisms apparently enter the body by passing through minute abrasions either in the skin or mucous membrane of the mouth, throat, or intestinal tract. Once in the body, the microorganisms are ingested by fixed macrophages, in which they multiply, and are carried via lymphatics to the lymph nodes. From here, the microorganisms may involve the liver, spleen, or bone marrow. The growth of the microorganisms inside the phagocytes partly accounts for their virulence and resistance to antibiotic therapy and antibodies. Any immunity achieved is not reliable. A vaccine is used for cattle but not for humans.

Symptoms of brucellosis include a general malaise, weakness, and aching with an insidious onset. The fever, particularly with *B. melitensis* infections, typically spikes to about 40°C (104°F) each evening, thus the term **undulant fever,** another name for brucellosis. Brucellosis is often a chronic disease that may last for many years. Periods of latency with few symptoms may alternate with renewed episodes of chills and fever.

Diagnosis is sometimes difficult and is most accurate when the organism can be isolated. Blood, as well as cerebrospinal fluid, urine, and even bone marrow, may be tested. The bacteria in smears of tissue specimens are seen as small coccobacillary organisms growing intracellularly, and special staining or fluorescent-antibody methods will differentiate them from similar intracellular bacteria. The serological tests for human infections are not always reliable. However, there is a simple agglutination test that can be used on serum or milk that identifies brucellosis-positive cattle.

Most strains of *Brucella* are difficult to treat with antibiotics. Streptomycin alone does not reach intracellular bacteria surviving in phagocytes. Tetracycline alone, or in combination with streptomycin in severe cases, is usually effective. However, relapse commonly occurs unless therapy is prolonged for at least three to four weeks.

Anthrax

In 1877, Robert Koch isolated *Bacillus anthracis*, the bacterium causing **anthrax** in animals (see Color Plate IF). His careful work to demonstrate that the rod-shaped, endospore-forming bacterium swarming in the blood of the dead animal was indeed responsible for anthrax eventually resulted in what we now know as Koch's postulates (Chapter 13). These postulates are the criteria that ideally must be met to prove the microbial etiology of a disease.

Animal herders in Europe had long associated certain fields with anthrax because of the high incidence of the disease in animals that grazed there. The disease is primarily one of grazing animals such as cattle or sheep and is still endemic in some parts of the United States. The *B. anthracis* endospores are ingested with the grass on which the endospores rest. The bacillus is a large, aerobic, gram-positive microorganism that is apparently able to grow slowly in certain soil types with specific moisture conditions. The endospores have survived in soil tests for 60 years. Grazing animals ingest the endospores, which pass through the stomach into the intestinal tract. There the endospores germinate, pass through the intestinal mucosa, and invade the blood, causing a fatal, fulminating septicemia. The organism is encapsulated, which probably improves its resistance to phagocytosis. The toxins produced also inhibit phagocytosis. It is also possible for animals to be infected more directly by infection of small cuts or abrasions on the mouth. Human gastrointestinal anthrax has been reported in the Soviet Union, but never in the United States.

There are only a handful of human anthrax cases in the United States each year (Figure 22–5). Persons at risk are those handling animals that may be infected and persons handling hides, wool, and other products from certain foreign countries. Goat hair from the Middle East has been a repeated source of contamination, as have handicrafts containing animal hides. Anthrax in humans differs from that observed in animals. Contact with materials containing anthrax endospores can cause a malignant pustule at the site where the organism enters the skin through a cut or abrasion (Figure 22–6). This pustular infection is sometimes kept localized by the defenses of the body, but there is always danger of septicemia. Probably the most dangerous form of anthrax is pulmonary anthrax.

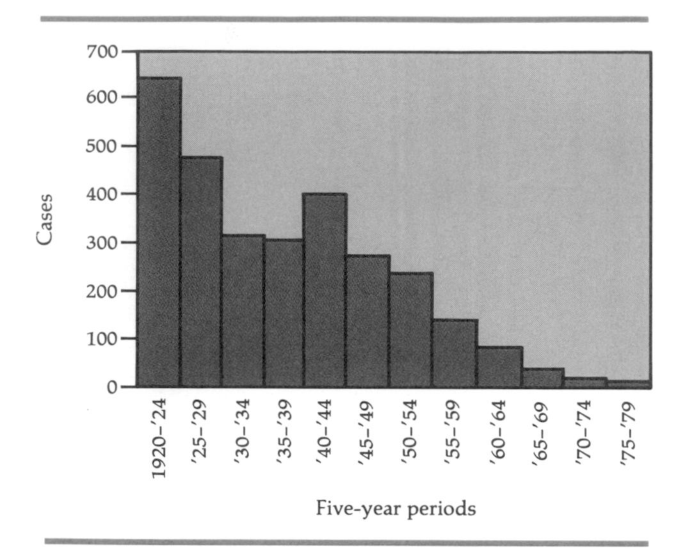

Figure 22–5 Reported cases of human anthrax in the United States, by five-year periods, 1920–1979.

This occurs when the endospores are inhaled, resulting in **woolsorters disease,** a dangerous form of pneumonia. It begins abruptly with high fever, difficulty in breathing, and chest pain. This disease eventually results in septicemia, and the death rate is high.

Animals surviving naturally acquired anthrax are resistant to reinfection. Second attacks in humans are also extremely rare. Vaccines composed of killed bacilli produce no significant immunity, and the best vaccine for humans appears to be a preparation of the protective antigen of the lethal toxin recovered from culture filtrates. Frequent boosters are necessary. Although a reliable vaccine for the protection of animals is available, the sporadic occurrence of bovine anthrax fails to provide ranchers with enough incentive to vaccinate their livestock routinely.

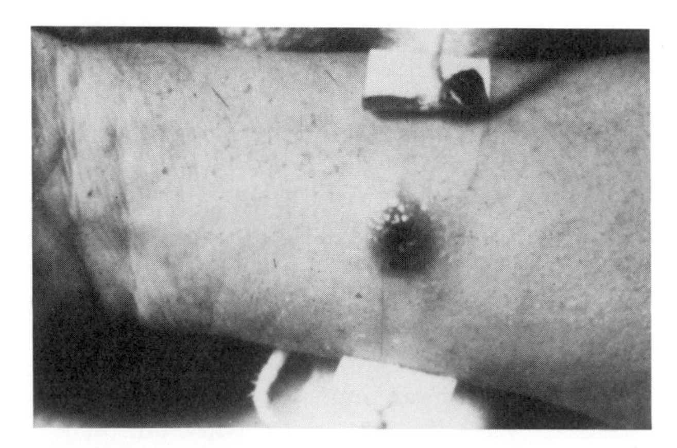

Figure 22–6 A cutaneous pustule of anthrax on a human arm.

Diagnosis of anthrax is best achieved by isolation of the bacterium. There are a number of morphological and biological tests that may be used for precise identification. Animal pathogenicity testing can also be performed, but serological tests are not in common use.

Penicillin is the drug of choice in treatment. Tetracycline, erythromycin, and chloramphenicol are also effective. However, once septicemia is well advanced, antibiotic therapy may prove useless, probably because of the presence of exotoxins that are not affected by the death of the bacteria.

Listeriosis

Listeria monocytogenes is an intracellular parasite associated with a wide variety of mammals, fish, and birds. **Listeriosis** can be contracted by inhalation, ingestion of contaminated raw milk, or direct contact with infected animals. Meningitis, encephalitis, and monocytosis (an increase in the number of monocytes in the blood) are the most common symptoms of human listeriosis. Bacteremia spreads the microorganisms throughout the body, causing granulomatous, necrotic lesions. Untreated infections have a mortality rate of 70%. The bacterium can also be transferred from mother to newborn at birth. Neonatal infections develop within one to four weeks after birth as meningitis, and such infections have a high mortality rate.

Prevention of listeriosis includes pasteurization of milk. An infected pregnant woman should receive prompt treatment.

VIRAL DISEASES OF THE CARDIOVASCULAR AND LYMPHATIC SYSTEMS

Myocarditis

As noted earlier, the heart muscle, also called the myocardium, is sometimes subject to an inflammatory disease called **myocarditis.** Although usually the result of a viral infection, it may also be caused by bacteria, fungi, or protozoans. The most frequently encountered viruses belong to the *coxsackievirus (enterovirus) group*. The virus apparently reaches the heart via the blood or lymph after infecting the respiratory or gastrointestinal tract. In the heart, the infecting virus damages the myocardium, leaving scar tissue. Sometimes pericarditis occurs as a result of myocarditis. **Myocarditis of the newborn,** also caused by a coxsackievirus, is often fatal.

Infectious Mononucleosis

Infectious mononucleosis ("mono") is caused by a herpesvirus, the *Epstein–Barr (EB) virus* (Figure 22–7). Infectious mononucleosis is an acute disease, primarily affecting the lymph nodes. It is characterized by enlarged and tender nodes, enlarged spleen, fever, sore throat, headache, nausea, and general weakness. The virus enters the body in oral secretions, multiplies in lymphatic tissue, and infects white blood cells. Infectious mononucleosis results in a proliferation of atypical lymphocytes (mononuclear white blood cells, thus the origin of its name). In a sense, this resembles the proliferation of leukocytes found in leukemia, and infectious mononucleosis has in fact been described as a self-limiting cancer. Fortunately, it is rarely fatal, although there has been one report of a family with an apparent genetic anomaly in which infectious mononucleosis became cancerous.

The disease is transmitted by direct or indirect contact, such as kissing and drinking from the same bottle. It has been called the kissing disease, and its association with kissing may indicate that a fairly large inoculum is necessary for transmission. It does not seem to spread in households, so aerosol transmission is unlikely. The normal incubation period seems to be from two to six weeks.

The peak incidence of the disease occurs in persons of about 15 to 25 years of age. College populations, particularly those from the upper socioeconomic strata, have a higher incidence of the disease. About 50% of college students have no immunity and about 15% of these can expect to come down with the disease. Lower socioeconomic groups tend to acquire the disease in early childhood, where they obtain a subclinical case and immunity. In some parts of the world, 90% of the children over four years of age have antibodies to

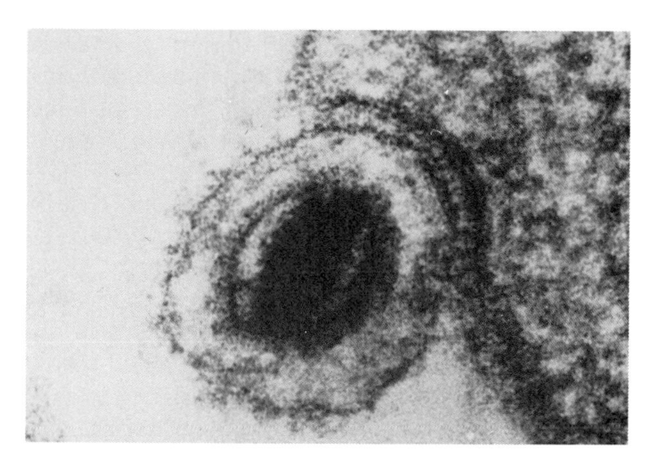

Figure 22–7 An enveloped, polyhedral Epstein–Barr (EB) virus caught in the act of infecting a human cell grown in culture. This particular virus particle was isolated from cells of Burkitt's lymphoma.

the disease. Probably 80% of the general population of the United States eventually acquires the disease in one form or another.

Diagnosis is usually made by a serological test for heterophil antibodies (see Chapter 16). These antibodies are not specific for the EB virus but cause the nonspecific agglutination of sheep red blood cells. They are developed to high titers and are quite persistent. The antigen that causes the appearance of these antibodies is unknown. Heterophil antibodies are also formed in response to other conditions, and there are simple procedures to differentiate these from the antibodies formed in response to infectious mononucleosis.

It is also possible to identify specific antibodies against the EB virus by immunofluorescent techniques, and it is desirable to use such a test for confirmation of the diagnosis. The antibodies against the virus provide a good immunity to subsequent attacks. Viruses that persist as a latent infection after the apparent infection subsides may help to maintain the persistence of antibodies. The persisting virus is often harbored and shed in the oral secretions as a persistent latent infection.

Burkitt's Lymphoma

As mentioned in Chapter 12, the EB virus is the same organism that has been linked to two human cancers: **Burkitt's lymphoma** and **nasopharyngeal**

cancer. There is a considerable amount of evidence linking the EB virus to these human malignancies. For example, in cell cultures, EB virus will infect only human lymphoid cells, transforming some of the cells into EB virus-carrying cells that have malignant characteristics. There is no doubt that EB virus is potentially oncogenic. When inoculated into monkeys it produces fatal lymphomas.

The low frequency of Burkitt's lymphoma and nasopharyngeal cancer, despite the wide distribution of the virus, can perhaps be explained by the existence of viral strains of varying oncogenic potential. Or it may be explained on the basis of environmental or genetic factors. But whatever the reason, most people infected with EB virus do not develop lymphomas, and many Burkitt's lymphomas do not contain demonstrable EB virus.

Serum Hepatitis: Hepatitis B and Non-A Non-B Hepatitis

Hepatitis is an inflammation of the liver that may be caused by allergic reactions, toxic chemicals, and viruses. Symptoms include loss of appetite, malaise, nausea, diarrhea, fever, chills, and jaundice. There are two major forms of viral hepatitis. While both of them affect the liver as the primary organ, one of them, **hepatitis A,** or **infectious hepatitis,** is spread mainly by the oral-intestinal route. It is caused by the *hepatitis A virus (HAV),* which will be discussed in Chapter 24. The other form, **hepatitis B** or **serum hepatitis,** is associated with the blood and will be discussed in this chapter. It is caused by a different virus, the *hepatitis B virus (HBV),* or another hepatitis virus now rather clumsily called the *non-A non-B (NANB) hepatitis virus.* (Alternative names used include *hepatitis C* and *hepatitis* B_2.) It is now apparent that there is more than one NANB virus. NANB hepatitis resembles hepatitis B in both the way the disease progresses and its epidemiology.

The hepatitis B virus is the best studied of the serum hepatitis viruses. Although HBV has never been grown in cell culture, the serum from patients with hepatitis B contains three distinct structures that possess hepatitis B surface antigen, HB_sAg (Figure 22–8). The most numerous are the *spherical particles,* measuring 22 nm in diameter. They appear to consist only of HB_sAg. Also present are larger particles, called *Dane particles,* the least common form. These particles consist of an outer coat containing HB_sAg and an inner core antigen called HB_cAg that contains double-stranded DNA and DNA polymerase. The Dane particles measure 42 nm in diameter and are assumed to represent intact

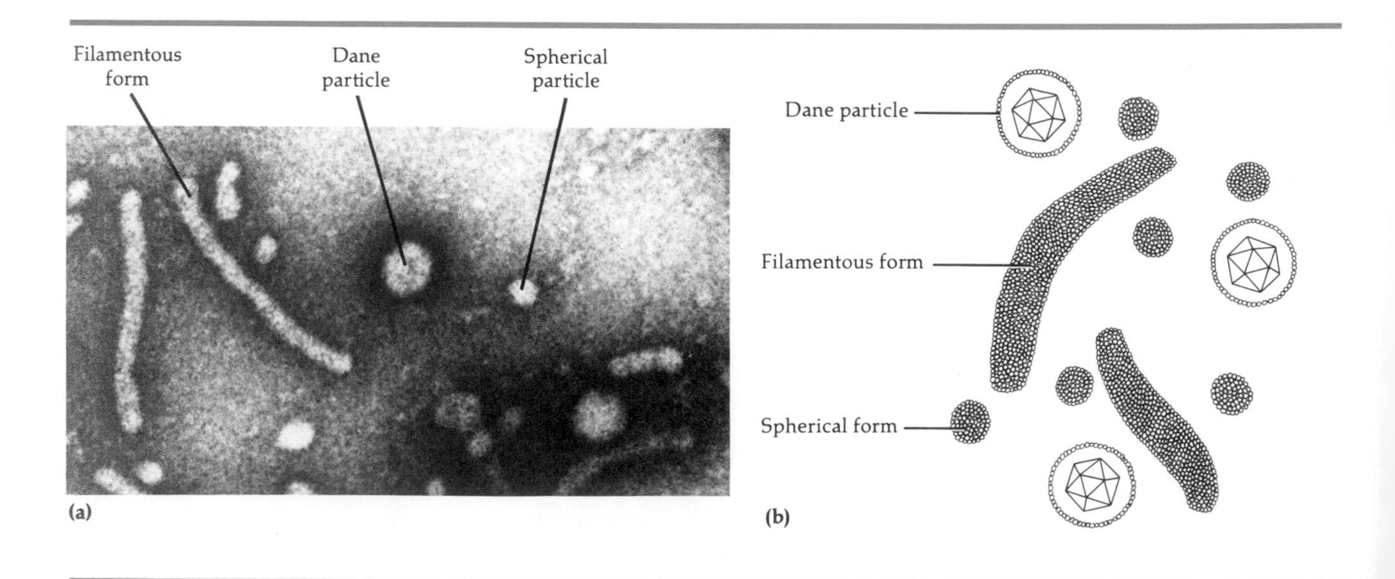

Figure 22–8 (a) Electron micrograph showing the three distinct types of hepatitis B particles discussed in the text. **(b)** Diagrammatic representation.

HB virions. Overproduction of the surface coat of the Dane particle apparently results in the spherical particles. The third structure present in patients with serum hepatitis is the *filamentous form.* It is 22 nm in diameter and ranges in length from 50 nm to greater than 230 nm. Filamentous forms are quite numerous and appear to consist only of HB_sAg. It is assumed that filamentous forms might be aggregates of spherical particles.

The normal route of transmission of hepatitis B is by any transfer of virus-carrying blood from one individual to another. Blood transfusions and contaminated equipment such as syringes are common modes. Persons in daily contact with blood, such as doctors, nurses, dentists, and medical technologists, have a considerably higher incidence of this disease than does the population at large. Doctors, for example, have five times, and dentists two or three times, the normal rate of infection. Recent evidence indicates that hepatitis B can be transmitted by any secretion of fluid from the body, such as saliva, sweat, breast milk, and semen. Spread of the disease by sexual transmission is also common. And occasionally, mosquitoes and oral ingestion may spread the disease. The incubation period is about 60 to 180 days, which makes tracing the origin of an outbreak difficult.

A serious problem with this disease is the difficulty of detecting the virus in blood intended for transfusion. The virus may be present in a person's blood for years after the symptoms of the disease have subsided. Blood donated for transfusion cannot be sterilized to remove the virus without damaging the blood. There are a number of methods for detecting HB_sAg, the major surface antigen, but none of them totally eliminates the possibility that the virus may be present. In view of this, transfusions are only given in cases of genuine need, where the risk is justified.

The mortality rate from hepatitis B is low, usually less than 1% for the population at large. However, it is often higher in certain outbreaks and in certain groups, particularly the elderly.

The inability of researchers to successfully cultivate the virus in cell culture has hampered development of a vaccine. At the present time, however, a vaccine is being prepared by harvesting the virus from human blood. The blood of a hepatitis patient contains the hepatitis B virus and HB_sAg. It was noted that when the mixture was heated, it was not infectious but could trigger antibody production. The vaccine is given in two doses spaced one month apart followed by a booster six months later. Estimates are that the vaccine will be available for use by high-risk personnel by 1982 (see box). No effective therapy for the disease exists.

An increasing number of cases of hepatitis B has been reported in this country. In 1969, there were 5909 cases; in 1978, 15,016. One estimate by the Centers for Disease Control is that the disease is much underreported and that the actual number of cases may be as high as 150,000 each year. Carrier states occur after recovery in at least 10% of the cases. Subclinical infections, particularly in children, are common.

Recently it has become apparent that the major type of posttransfusion hepatitis in the United States is not hepatitis B but rather non-A non-B hepatitis. One estimate is that 90% of transfusion hepatitis cases are of the NANB type. Although detection methods for screening blood for hepatitis B antigens have succeeded in lowering the rate of hepatitis B infections, they have not yet identified specific viral antigens for NANB hepatitis.

PROTOZOAN AND HELMINTHIC DISEASES OF THE CARDIOVASCULAR AND LYMPHATIC SYSTEMS

Toxoplasmosis

Toxoplasmosis is a disease of blood and lymph vessels caused by the protozoan *Toxoplasma gondii,* a small, crescent-shaped organism (Figure 22–9a). *T. gondii* is a sporozoan, like the malarial parasite.

Cats seem to be an essential part of the life cycle of *T. gondii* (Figure 22–9b). Random tests on urban cats have shown that a large number are infected with the organism, which causes no apparent illness in the cat. The organism undergoes its only sexual phase in the intestinal tract of the cat, with oocysts shed in the cat's feces, thereby contaminating the food or water ingested by other animals. The protozoan reproduces asexually in these new hosts. A human eating the uncooked meat of such an animal, or a cat eating an infected mouse, will become infected with the parasite.

The two main modes of infection for humans appear to be by eating undercooked meats and by inhaling the dried feces of cats while cleaning a litter box. One recent outbreak of toxoplasmosis resulted when 34 persons frequented a riding stable with a large resident population of cats. Some surveys have shown that approximately 30% of the population carries antibodies for this organism, an indication of a high rate of subclinical, unrecognized infections.

Microbiology in the News

A Vaccine for Hepatitis

October 1980—A bout of hepatitis can be just plain misery. The ailment can produce high fever, nausea and fatigue, and sometimes damages the liver so severely the victim dies. The most serious form of hepatitis is type B: as many as 200 million people are carriers of the viral infection, which is spread through close personal contact. But hepatitis B—once known as serum hepatitis—may soon join smallpox and polio as a scourge of the past. Last week researchers announced that an experimental vaccine offers virtually complete protection against the devastating illness.

In a major clinical trial, reports epidemiologist Dr. Wolf Szmuness of the New York Blood Center, the new vaccine proved 92% effective in immunizing a high risk group of 1083 male volunteers. All of them were homosexuals. Because gay men often change sexual partners frequently, thus increasing the chance of exposure to the virus, their risk of catching hepatitis B is more than ten times higher than that of the general population. They usually transmit the virus through infected semen or saliva. Other common sources of hepatitis B are contaminated blood and hypodermic needles.

Ingenious Solutions

Developing the new vaccine posed a unique challenge. Most antiviral vaccines are prepared by growing large quantities of the virus in tissue culture. Heat is then used to make the virus harmless without destroying its ability to stimulate immunity. But the hepatitis B virus does not grow in tissue culture. In 1971 Dr. Saul Krugman of New York University found an ingenious solution: the harvesting of the virus from human blood.

Krugman based his work on an earlier finding by Dr. Baruch Blumberg of Philadelphia's Fox Chase Cancer Center. In examining the blood of an Australian aborigine, Blumberg discovered a substance that was clearly related to the B virus. It turned out to be an antigen for hepatitis B, capable of triggering antibodies that fight the disease. Krugman took samples of blood containing the hepatitis virus and the "Australia" antigen and heated them. This mixture, he observed, was not infectious and offered protection against the illness. "Once it was possible to collect large amounts of antigen," said Krugman, "a vaccine became feasible."

Two Doses

Under the direction of Dr. Maurice R. Hilleman, the new vaccine has been perfected by the Merck Institute for Therapeutic Research in West Point, Pennsylvania. It is given in two doses spaced one month apart, followed by a booster shot six months later. In the *New England Journal of Medicine*, Szmuness reports that only eleven of the 549 vaccinated volunteers got hepatitis. In a control group injected with a placebo, 70 out of 534 came down with the disease. Szmuness believes that the cases among vaccinated individuals occurred before they acquired full immunity.

The vaccine must be tested for at least another year, but it could become available in 1982. Initially, experts will probably recommend vaccinations only for people at high risk. Surgeons, dentists, and nurses, for example, are particularly susceptible to hepatitis B because they frequently come in contact with blood. The vaccine may confer still another benefit. Since researchers suspect that hepatitis B is linked to a form of liver cancer called hepatoma, the vaccine may provide protection against this disease as well.

Source: Jean Seligmann, *Newsweek*, 10/13/80, p. 132.

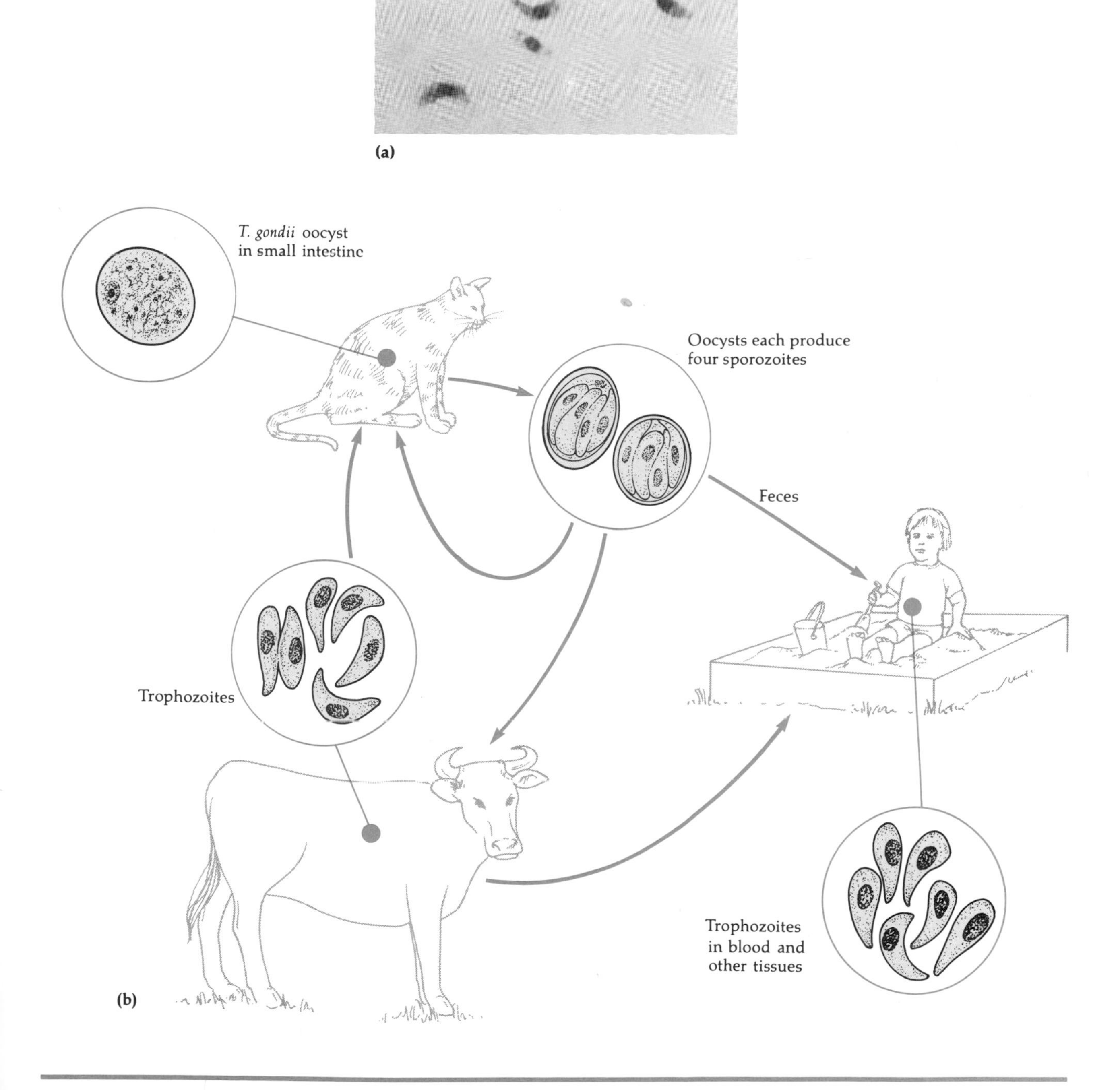

Figure 22–9 (a) Crescent-shaped trophozoites of *Toxoplasma gondii*, the causative agent of toxoplasmosis. **(b)** Life cycle of *T. gondii*. The domestic cat is the definitive host in which the sporozoan reproduces sexually to produce oocysts, which are passed out in feces. Four sporozoites form within each oocyst. The sporozoites are then ingested by one of the intermediate hosts. They mature into trophozoites in blood and other tissues.

The disease is a rather undefined mild illness, and the primary danger is in the congenital infection of the fetus. Thus, pregnant women should avoid cleaning cat litter boxes or having any unnecessary contact with cats, and eating raw meats is also inadvisable. The results on the fetus are drastic, including convulsions, severe brain damage, blindness, and death. The mother is probably unaware of the disease, which is transmitted across the placenta.

T. gondii can be isolated and grown in cell cultures. The preferred serological tests make use of a fluorescent antibody method or a form of hemagglutination testing. The latest recommendation is to use the two tests in conjunction. A rising titer in a pregnant woman may be an indication for a therapeutic abortion.

Treatment of toxoplasmosis is with pyrimethamine in combination with either trisulfapyrimidines or sulfadiazine.

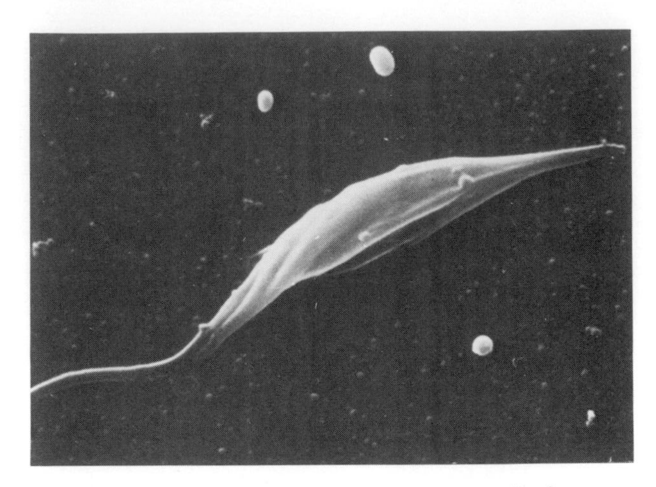

Figure 22–10 Scanning electron micrograph of *Trypanosoma cruzi*, the protozoan that causes American trypanosomiasis (Chagas' disease) (×5600).

American Trypanosomiasis (Chagas' Disease)

An example of a protozoan disease of the cardiovascular system is **American trypanosomiasis (Chagas' disease).** The causative agent is *Trypanosoma cruzi*, a flagellated protozoan (Figures 22–10 and 11–17). The disease occurs in the southern portions of the United States through Mexico, Central America, and into South America to Argentina. Although only a few cases have been reported in the United States, the incidence of the disease is 40 to 50% in some rural areas of South America.

The reservoir for *T. cruzi* includes a wide variety of wild animals, including rodents, opossums, and armadillos. The arthropod vector is the reduviid bug, often called the "kissing bug" (see Figure 11–28e) because of its tendency to bite individuals near the lips. The trypanosomes grow in the gut of the bug, and the bug defecates while feeding. Maturing trypanosomes are passed in the bug's feces, deposited on the skin of humans or animals that the insect bites, and rubbed into the bite wound or other skin abrasions by scratching, or into the eye by rubbing. At the site of inoculation, the trypanosomes reproduce and pass through various stages of their life cycle. A swollen lesion develops at the inoculation site (often near the eye). The organisms infect regional lymph nodes and are carried by the blood to other organs of the body. From the blood, the organisms invade the liver and macrophages of the spleen and more frequently the heart. Although central nervous system involvement in adults is rare, it is not uncommon in children. Trypanosomes are transferred from the body to a reduviid bug when the bug bites an infected animal or person.

Laboratory diagnosis of American trypanosomiasis is by observing the trypanosomes in blood, growing the microorganisms in blood cultures, or using fluorescent-labeled antibody on blood smears. Treatment of the disease is very difficult when chronic, progressive stages have been reached. However, an experimental drug called Lampit may be effective in clearing the blood of trypanosomes in early stages of the disease.

Schistosomiasis

Schistosomiasis is a disease caused by a flatworm parasite, a fluke. The disease is not contracted in the United States, but over 400,000 immigrants to the United States have the disease, and millions are affected in Asia, Africa, South America, and the Caribbean. Schistosomiasis is a major world health problem. Species of the genus *Schistosoma* causing the disease vary with geographical location, but the characteristics of the disease are similar.

Waters become contaminated with ova excreted with human wastes (Figure 22–11). A motile, ciliated form of *Schistosoma* called a *larva* is released from the ova, and at this point it enters certain species of snails. The lack of a suitable snail host is a primary reason why schistosomiasis cannot be transmitted in the United States. Eventually, the pathogen emerges from the snail in an infective form called the *cercaria*. These cercariae have forked swimming tails, and when they contact the skin of a person wading or swimming in the water, they discard the tail and enzymatically penetrate the

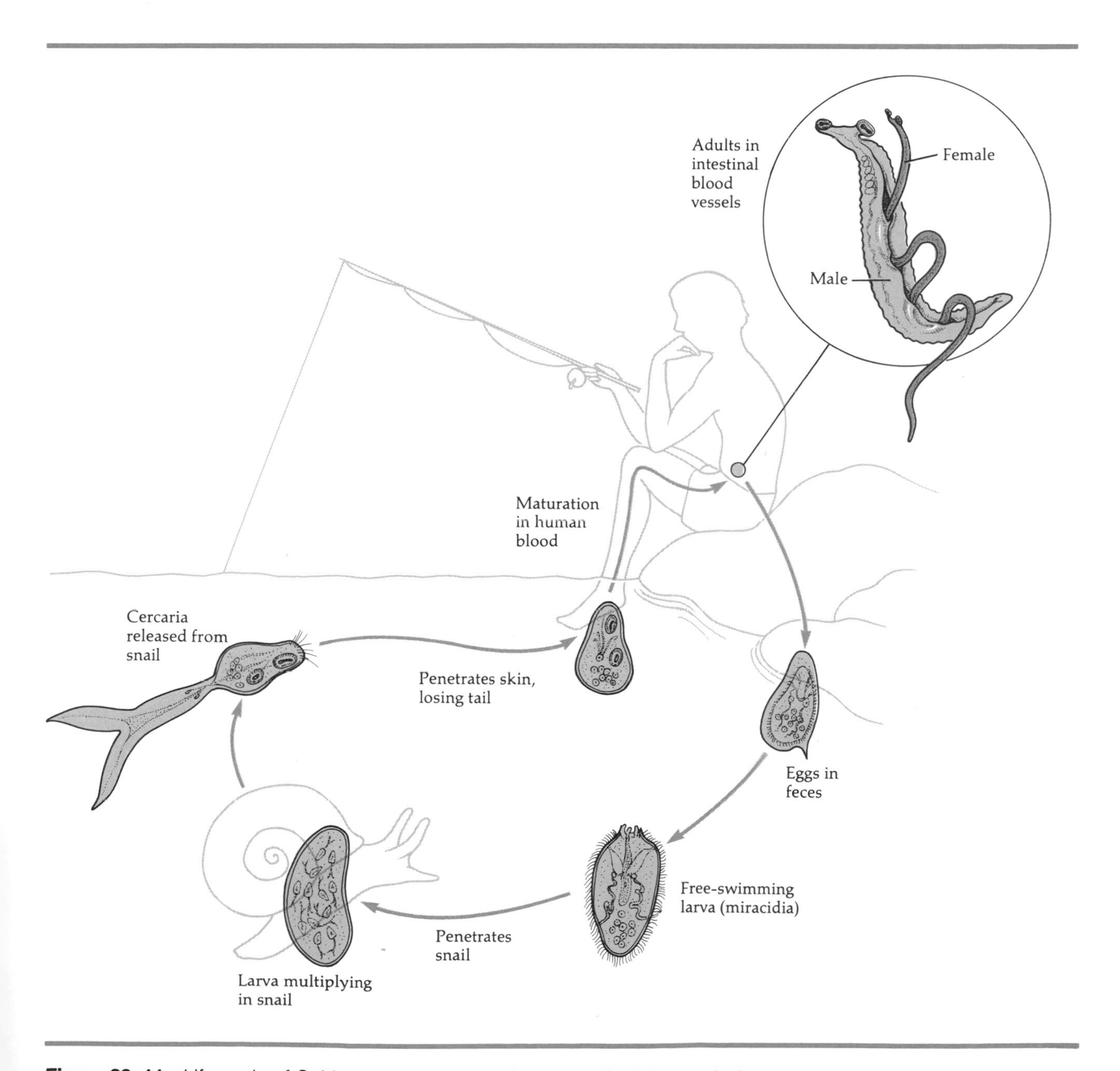

Figure 22–11 Life cycle of *Schistosoma*, the causative agent of schistosomiasis.

Table 22–1 Summary of Diseases Associated with the Cardiovascular and Lymphatic Systems

Disease	Causative Agent	Mode of Transmission	Treatment
Bacterial Diseases of the Cardiovascular and Lymphatic Systems:			
Puerperal sepsis and infections related to abortions	Primarily *Streptococcus pyogenes; Clostridium* and *Bacteroides* species often cause postabortion infections	Unsanitary conditions in childbirth and abortions	Penicillin and erythromycin for *S. pyogenes;* chloramphenicol and penicillin for *Bacteroides* and *Clostridium* species
Endocarditis			
Subacute bacterial	Many organisms, especially alpha-hemolytic streptococci	Bacteremia localizing in the heart	Varies with agent; penicillin and erythromycin for alpha-hemolytic streptococci
Acute bacterial	*Staphylococcus aureus, Streptococcus pneumoniae*	Bacteremia localizing in the heart	Semisynthetic penicillin derivatives such as methicillin or oxacillin for *S. aureus* and penicillin for *S. pneumoniae*
Pericarditis	*Streptococcus pneumoniae*	Bacteremia localizing in the heart	Penicillin
Rheumatic fever	Group A beta-hemolytic streptococci	Inhalation leading to streptococcal sore throat	Penicillin (for sore throat)
Tularemia	*Francisella tularensis*	Animal reservoir (rabbits); skin abrasions, ingestion, inhalation, bites	Streptomycin, tetracycline
Brucellosis	*Brucella* species	Animal reservoir (cows); ingestion in milk, direct contact with skin abrasions	Streptomycin, tetracycline
Anthrax	*Bacillus anthracis*	Reservoir is soil or animals; skin abrasions, inhalation or ingestion of heavy spore concentrations	Penicillin, tetracycline, erythromycin, chloramphenicol
Listeriosis	*Listeria monocytogenes*	Animal reservoir; ingestion	Tetracycline or penicillin
Viral Diseases of the Cardiovascular and Lymphatic Systems:			
Myocarditis	Several agents, especially coxsackievirus (enterovirus)	Coxsackievirus: Inhalation and ingestion	None
Pericarditis	Same as above	Complication of viral myocarditis	None
Infectious mononucleosis	Epstein–Barr virus	Oral secretions, kissing	None
Hepatitis B (serum hepatitis)	Hepatitis B virus	Predominantly parenteral (transfusions, syringes, etc.); also by close contact	None
Protozoan and Helminthic Diseases of the Cardiovascular and Lymphatic Systems:			
Toxoplasmosis	*Toxoplasma gondii*	Animal reservoir: cats; inhalation and ingestion	Pyrimethamine in combination with either trisulfapyrimidines or sulfadiazine
American trypanosomiasis	*Trypanosoma cruzi*	Bite of reduviid bug	Lampit in early stages
Schistosomiasis	*Schistosoma* species	Contaminated water; cercariae enter the body through skin	Stibophen, niridazole, hycanthone

skin. They are then carried by the bloodstream to the veins of the liver or urinary bladder. The larvae mature into an adult form in which the slender female lives in a cleft of the body of the male. The union produces a supply of new ova, some of which cause local tissue damage from defensive body reactions. Others enter the water to continue the cycle.

The damage caused by the disease is often due to liver destruction, but sometimes other organs such as the lungs or urinary system become involved. Abscesses and ulcers are formed in response to the infestation, and allergic reactions by the body cause blockage of various ducts and blood vessels. Swimmers in lakes in the northern United States are sometimes troubled by **swimmer's itch.** This is a cutaneous allergic reaction to cercariae similar to those of human schistosomiasis penetrating their skin. However, those parasites mature only in wildfowl and not in humans, so the infection does not progress beyond penetration of the skin and a local inflammatory response.

Laboratory diagnosis consists of microscopic identification of the flukes or their ova in fecal and urine specimens, intradermal tests, and serological tests such as complement fixation and precipitin tests.

Treatment is difficult. Several drugs, such as stibophen, niridazole, and hycanthone, are moderately effective, particularly in the early stages of the disease, but none is completely satisfactory. Sanitation and elimination of the host snail are the best forms of control.

A summary of diseases associated with the cardiovascular and lymphatic systems is presented in Table 22–1.

STUDY OUTLINE

INTRODUCTION (p. 552)

1. The heart, blood, and blood vessels make up the cardiovascular system.
2. Lymph, lymph vessels, lymph nodes, and lymphoid organs constitute the lymphatic system.

STRUCTURE AND FUNCTION OF THE CARDIOVASCULAR SYSTEM (pp. 552–554)

1. The heart circulates substances to and from tissue cells.
2. Deoxygenated blood enters the right atrium from all parts of the body. It passes to the right ventricle and to the lungs.
3. Oxygenated blood returns to the left atrium. It passes to the left ventricle and to all tissue cells.
4. Arteries and arterioles transport blood away from the heart. Veins and venules bring blood to the heart.
5. Exchange of materials between blood and tissue cells occurs at capillaries. Capillaries connect arterioles to venules.
6. Blood is a mixture of plasma and cells.
7. Red blood cells carry oxygen. White blood cells are involved in defending the body against infection. Platelets are involved in blood clotting. Plasma transports dissolved substances.

STRUCTURE AND FUNCTION OF THE LYMPHATIC SYSTEM (p. 554)

1. Fluid that filters out of capillaries into spaces between tissue cells is interstitial fluid.
2. Interstitial fluid enters lymph capillaries and is called lymph.
3. Lymphatics return lymph fluid to the blood.
4. Lymph nodes contain fixed macrophages and B cells.
5. Tonsils, appendix, spleen, and thymus gland are lymphoid organs.

BACTERIAL DISEASES OF THE CARDIOVASCULAR AND LYMPHATIC SYSTEMS (pp. 554–560)

Septicemia (pp. 554–555)

1. The growth of microorganisms in blood is called septicemia.
2. Symptoms include fever and decreased blood pressure and lymphangitis (inflamed lymph vessels). Septicemia may lead to ischemia and shock.
3. Septicemia usually results from a focus of infection in the body.
4. Gram-negative rods are usually implicated. Endotoxin causes the symptoms.

Puerperal Sepsis (p. 555)

1. Puerperal sepsis begins as a uterine infection following childbirth or abortion; it may progress to peritonitis or septicemia.
2. *Streptococcus pyogenes* is the most frequent cause.
3. Dr. Oliver Wendell Holmes and Ignaz Semmelweiss demonstrated that puerperal sepsis was transmitted by the hands and instruments of midwives and physicians.
4. Puerperal sepsis is now uncommon because of modern hygiene and antibiotics.
5. Anaerobic bacteria, including *Bacteroides* and *Clostridium,* may cause infections from improperly performed abortions.

Bacterial Endocarditis (pp. 555–556)

1. The outer layer of the heart is the pericardium. The inner layer is the endocardium.
2. Subacute bacterial endocarditis is usually caused by alpha-hemolytic streptococci, although other gram-positive cocci may be involved.
3. The infection arises from a focus of infection such as a tooth extraction.
4. Preexisting heart abnormalities are predisposing factors.
5. Symptoms include fever, anemia, and heart murmur.

6. Acute bacterial endocarditis is usually caused by *Staphylococcus aureus* or *Streptococcus pneumoniae*.
7. The bacteria produce rapid destruction of heart valves.
8. *Streptococcus* may also cause pericarditis.
9. Laboratory diagnosis is based on isolation and identification of the bacteria from blood.

Rheumatic Fever (p. 556)

1. Rheumatic fever is an autoimmune complication of group A beta-hemolytic streptococcal infections.
2. Rheumatic fever may be expressed as arthritis or inflammation of the heart, resulting in permanent heart damage.
3. Antibodies against *S. pyogenes* react with streptococcal antigens deposited in joints or heart valves or cross-react with the heart muscle.
4. Rheumatic fever may be a sequel to a streptococcal infection such as streptococcal sore throat. Streptococci may not be present at the time of rheumatic fever.
5. The incidence of rheumatic fever in the United States has declined because of prompt treatment of streptococcal infections.
6. Rheumatic fever is treated with antiinflammatory drugs.

Tularemia (pp. 556–558)

1. Tularemia is caused by *Francisella tularensis*. The reservoir is small wild mammals, especially rabbits.
2. Tularemia is contracted by handling diseased carcasses, eating undercooked meat or diseased animals, and by the bite of certain vectors (i.e., deer flies).
3. Symptoms may include ulceration at the site of entry followed by septicemia and pneumonia.
4. *F. tularensis* is difficult to culture in vitro. In vivo it is resistant to phagocytosis.
5. Laboratory diagnosis is based on a slide agglutination test on isolated bacteria.

Brucellosis (Undulant Fever) (pp. 558–559)

1. Brucellosis is caused by *Brucella abortus*, *B. melitensis*, *B. suis*, and *B. canis*.
2. Symptoms include malaise and fever that spikes each evening (undulant fever).
3. The bacteria enter through minute breaks in the mucosa of the digestive

system, reproduce in macrophages, and spread via lymphatics to liver, spleen, or bone marrow.

4. A vaccine for cattle is available.
5. Diagnosis is based on isolation of the bacteria.

Anthrax (pp. 559–560)

1. *Bacillus anthracis* causes anthrax. Endospores can survive in soil up to 60 years.
2. Grazing animals acquire an infection after ingesting the endospores.
3. Humans contract anthrax by handling hides from infected animals. The bacteria enter through cuts in the skin or through the respiratory tract.
4. Entry through the skin results in a malignant pustule that may progress to septicemia. Entry through the respiratory tract may result in pneumonia.
5. A vaccine for cattle is available.
6. Diagnosis is based on isolation and identification of the bacteria.

Listeriosis (p. 560)

1. *Listeria monocytogenes* causes meningitis, encephalitis, and monocytosis.
2. It is contracted by contact with infected animals and ingestion of contaminated milk.
3. Pasteurization of milk will kill *Listeria.*

VIRAL DISEASES OF THE CARDIOVASCULAR AND LYMPHATIC SYSTEMS (pp. 560–563)

Myocarditis (pp. 560–561)

1. Inflammation of the heart muscle (myocardium) is called myocarditis.
2. Coxsackievirus (enterovirus) is the most common cause. It reaches the heart from a respiratory or gastrointestinal infection.

Infectious Mononucleosis (p. 561)

1. Infectious mononucleosis is caused by the Epstein–Barr (EB) virus (herpesvirus).
2. The virus multiplies in lymphatic tissue and causes the proliferation of atypical lymphocytes.
3. The disease is transmitted by ingestion of saliva from infected individuals.
4. Diagnosis is made by a serological test for heterophil antibodies and indirect fluorescent-antibody technique.

Burkitt's Lymphoma (pp. 561–562)

1. EB virus is implicated in Burkitt's lymphoma and nasopharyngeal cancer.
2. EB virus is potentially oncogenic.

Serum Hepatitis: Hepatitis B and Non-A Non-B Hepatitis (pp. 562–564)

1. Inflammation of the liver is called hepatitis. Symptoms include loss of appetite, malaise, fever, and jaundice.
2. Viral causes of hepatitis are hepatitis A virus (infectious hepatitis), hepatitis B virus (HBV), and non-A non-B hepatitis virus (NANB). HBV and NANB are referred to as serum hepatitis.
3. The virus has never been cultured. Viral antigens can be detected in the serum of infected patients.
4. Spherical particles (HB_sAg), Dane particles (HB_sAg and HB_cAg) and filamentous forms (HB_sAg) are found in patients with hepatitis B.
5. HBV is transmitted by blood transfusions and contaminated syringes.
6. Carrier states occur after recovery, and subclinical infections are common.
7. Ninety percent of transfusion hepatitis is estimated to be NANB.
8. Screening methods for NANB are not yet available.

PROTOZOAN AND HELMINTHIC DISEASES OF THE CARDIOVASCULAR AND LYMPHATIC SYSTEMS (pp. 563–569)

Toxoplasmosis (pp. 563–566)

1. Toxoplasmosis is caused by the sporozoan *Toxoplasma gondii.*
2. *T. gondii* undergoes sexual reproduction in the intestinal tract of domestic cats, and oocysts are eliminated in cat feces.
3. Animals that ingest contaminated food or water may contract toxoplasmosis.
4. Humans contract the infection by ingesting undercooked meat from an infected animal or inhaling dried cat feces.
5. Subclinical infections are probably common because the disease symptoms are rather mild.
6. Congenital infections may occur. Symptoms include convulsions, brain damage, blindness, and death.
7. A rising antibody titer in a pregnant woman may indicate the need for a therapeutic abortion.

American Trypanosomiasis (Chagas' Disease) (p. 566)

1. *Trypanosoma cruzi* causes Chagas' disease. The reservoir includes many wild animals. The vector is the "kissing bug."

2. The infection begins in the lymph nodes and is disseminated via the bloodstream to other organs.
3. Observation of the trypanosomes in blood confirms diagnosis.

Schistosomiasis (pp. 566–569)

1. Species of the blood fluke *Schistosoma* cause schistosomiasis.
2. Eggs eliminated with feces hatch into larvae that infect the intermediate host. Free-swimming cercariae are released and penetrate the skin of a human.
3. The adult flukes live in the veins of the liver or urinary bladder in humans.
4. Adult flukes reproduce, and eggs are excreted or remain in the host.
5. Symptoms are due to the host's defense to eggs that remain in the body.
6. Swimmer's itch is a cutaneous allergic reaction to cercariae that penetrate the skin. The definitive host for this fluke is wildfowl.
7. Observation of eggs or flukes in feces, skin tests, or indirect serological tests may be used for diagnosis.

STUDY QUESTIONS

REVIEW

1. What are the symptoms of septicemia?
2. How can septicemia result from a single focus of infection such as an abscess?
3. Differentiate between endocarditis, myocarditis, and pericarditis. How are these infections contracted?
4. Complete the following table.

Disease	Frequent Causative Agent	Predisposing Condition(s)
Puerperal sepsis		
Subacute bacterial endocarditis		
Acute bacterial endocarditis		
Myocarditis		

5. Describe the probable cause of rheumatic fever. How is rheumatic fever treated? How is it prevented?

6. Fill in this table.

Disease	Causative Agent	Method of Transmission	Reservoir	Symptoms	Prevention
Tularemia					
Brucellosis					
Anthrax					
Listeriosis					

7. Plot the temperature of a patient with brucellosis for a one-week period.
8. Differentiate between HB and NANB hepatitis viruses.
9. List the causative agents and methods of transmission of infectious mononucleosis and serum hepatitis.
10. Cite evidence implicating EB virus in Burkitt's lymphoma. Offer an explanation with regard to the large number of people who are infected with EB virus who do not get Burkitt's lymphoma or nasopharyngeal cancer.
11. What is the probable result of untreated cutaneous anthrax?
12. List the causative agent, method of transmission, and reservoir for schistosomiasis, toxoplasmosis, and American trypanosomiasis. Which disease are you most likely to get in the United States? Where are the other diseases endemic?

CHALLENGE

1. How is blood to be used for transfusions tested for HBV? For NANB?
2. Indirect fluorescent-antibody tests on the serum of three pregnant women provided the following information.

Patient	Antibody Titer		
	Day 1	Day 5	Day 12
Patient *A*	1:256	1:256	1:256
Patient *B*	1:256	1:512	1:1024
Patient *C*	0	0	0

   Which of these women may have toxoplasmosis? What advice might be given to each woman with regard to toxoplasmosis?

3. A 19-year-old man went deer hunting. While on the trail, he found a partially dismembered dead rabbit. The hunter picked up the front paws for good luck charms, which he gave to another hunter in the party. The rabbit had been handled with bare hands that were bruised and scratched from the hunter's work as an automobile mechanic. Festering sores on his hands, legs, and knees were noted two days later. What infectious disease do you suspect the hunter has? How would you proceed to prove it?

FURTHER READING

Freimer, E. H., and M. McCarty. December 1965. Rheumatic fever. *Scientific American* 213:66–74. Discusses the link between streptococcal infection and rheumatic disease.

Henle, W., G. Henle, and E. T. Lennette. July 1979. The Epstein–Barr virus. *Scientific American* 241:48–59. A summary of information on one of the most common viruses infecting humans.

Melnick, J. L., G. R. Dreesman, and F. B. Hollinger. July 1977. Viral hepatitis. *Scientific American* 237:44–62. Epidemiology and immunology of hepatitis virus.

See also Further Reading for Part IV at the end of Chapter 19.

23

Microbial Diseases of the Respiratory System

Objectives

After completing this chapter you should be able to

- Describe how microorganisms are prevented from entering the respiratory system.
- List the methods of transmission of respiratory system infections.
- Characterize the normal flora of the upper and lower respiratory systems.
- Describe the causative agent, symptoms, preferred treatment, and laboratory identification tests for ten bacterial diseases and three viral diseases of the respiratory system.
- Compare these skin tests: Dick test; Schick test; tuberculin skin test; histoplasmin test.
- Describe vaccines that are available to prevent respiratory system infections.
- Discuss several fungal, protozoan, and helminthic diseases of the respiratory system.

Since the upper respiratory system is in direct contact with the air we breathe—air contaminated with microorganisms and endospores—it is a major portal of entry for pathogens. In fact, respiratory system infections are the most common type of infection—and also among the most damaging. Pathogens that enter the body via the respiratory system can sometimes cause infections in other parts of the body, as in the case of measles, mumps, and rubella. Another very serious problem with respiratory infections is the ease with which they are spread, both by direct contact, such as droplets emitted during sneezing, coughing, or talking, and by fomites. Transmission by droplets is especially common with highly communicable respiratory infections such as pneumonic plague and influenza.

STRUCTURE AND FUNCTION OF THE RESPIRATORY SYSTEM

It is convenient to think of the respiratory system as being composed of two divisions: the upper respiratory system and the lower respiratory system. The **upper respiratory system** consists of the *nose* and *throat* and structures associated with them, including the middle ear and the eustachian tubes (Figure 23–1). Emptying into the nose are ducts from the sinuses and the nasolacrimal ducts from the lacrimal apparatus (see Figure 15–2). Emptying into the upper portion of the throat are the eustachian tubes from the middle ear.

The upper respiratory system has several anatomical defenses against airborne pathogens. Coarse hairs in the nose filter

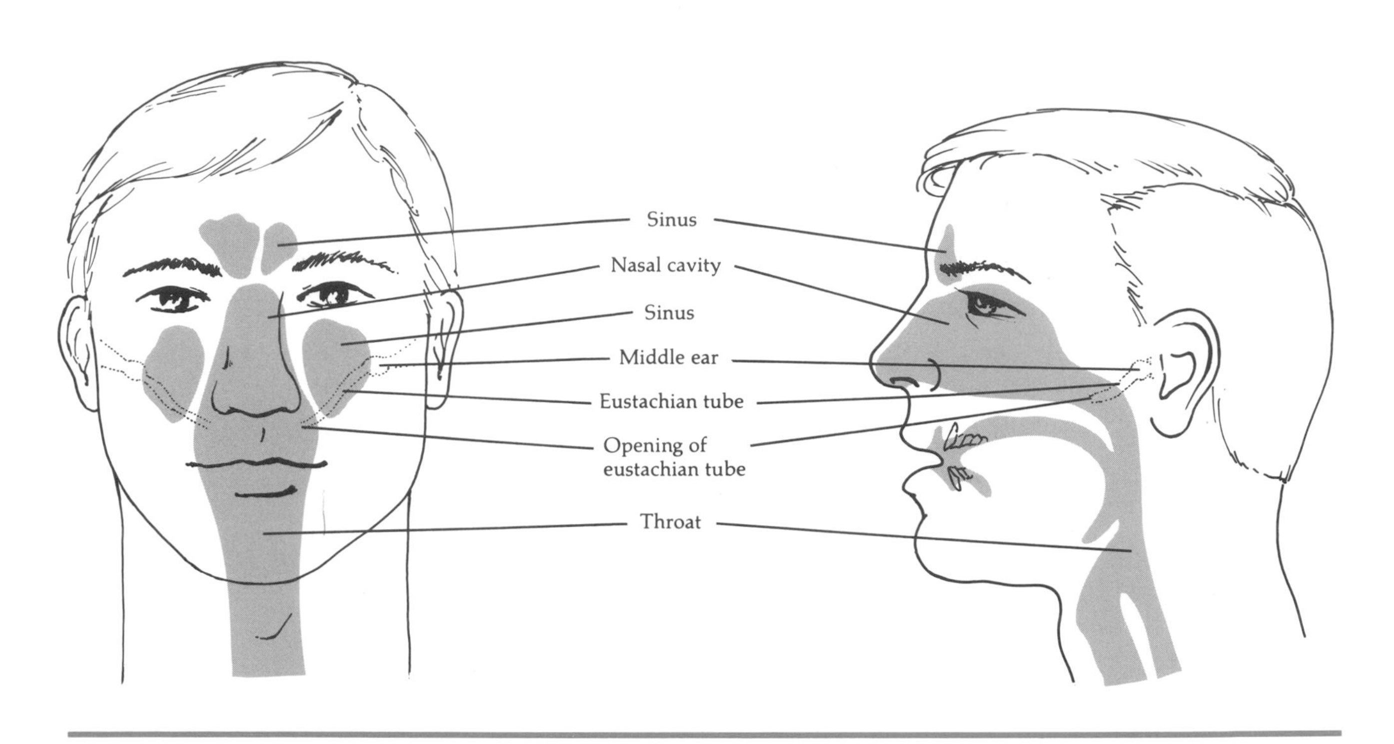

Figure 23–1 Structures of the upper respiratory system.

large dust particles from the air. And the nose is lined with a mucous membrane that contains numerous mucus-secreting cells and cilia. The upper portion of the throat also contains a ciliated mucous membrane. The mucus moistens inhaled air and traps dust and microorganisms, especially particles larger than 4 to 5 μm. The cilia assist by moving dust and microorganisms toward the mouth for elimination. At the junction of the nose and throat, masses of lymphoid tissue, the tonsils and adenoids, provide immunity to certain infections. Occasionally, however, these tissues become infected and help spread infection to the ears via the eustachian tubes. Since the nose and throat are connected to the sinuses, nasolacrimal apparatus, and middle ear, it is not uncommon for infections to spread from one region to another.

The **lower respiratory system** consists of the *larynx* (voice box), *trachea* (windpipe), *bronchial tubes*, and *alveoli* (Figure 23–2). Alveoli are the air sacs that make up the lung tissue, and it is here that oxygen and carbon dioxide are exchanged between the lungs and blood. The double-layered membrane around the lungs is called the *pleura*. A ciliated mucous membrane lines the lower respiratory system down to the smaller bronchial tubes, helping to prevent microorganisms from reaching the lungs. Particles trapped in the larynx, trachea, and larger bronchial tubes are moved up toward the throat by a ciliary action referred to as the ciliary escalator (see Figure 15–3). The ciliary escalator keeps the trapped particles moving toward the throat. Coughing, sneezing, and clearing the throat speed up the escalator, whereas smoking and air pollution inhibit it. If microorganisms actually reach the lungs, then certain phagocytic cells—the *alveolar macrophages*, or *dust cells*—usually locate, ingest, and destroy most of them. Also, IgA antibodies are found in secretions such as respiratory mucus, saliva, and tears. These antibodies help protect mucosal surfaces of the respiratory system from many pathogens. Thus, the body has several mechanisms for minimizing airborne infections. But if all these mechanisms fail, then the microorganism wins the host–parasite competition and a respiratory disease results.

NORMAL FLORA OF THE RESPIRATORY SYSTEM

The normal flora of the mouth will be considered in Chapter 24 as part of the gastrointestinal tract. The normal flora of the nasal cavity includes diphtheroids; staphylococci, including *S. aureus;* micrococci; and *Bacillus* species. All are gram-positive. The portion of the throat near the nose may contain streptococci, including *Streptococcus pneumoniae,* as well as the gram-negative organisms *Hemophilus influenzae* and *Neisseria meningitidis.* Despite the presence of potentially pathogenic microorganisms as part of the normal flora of the upper respiratory

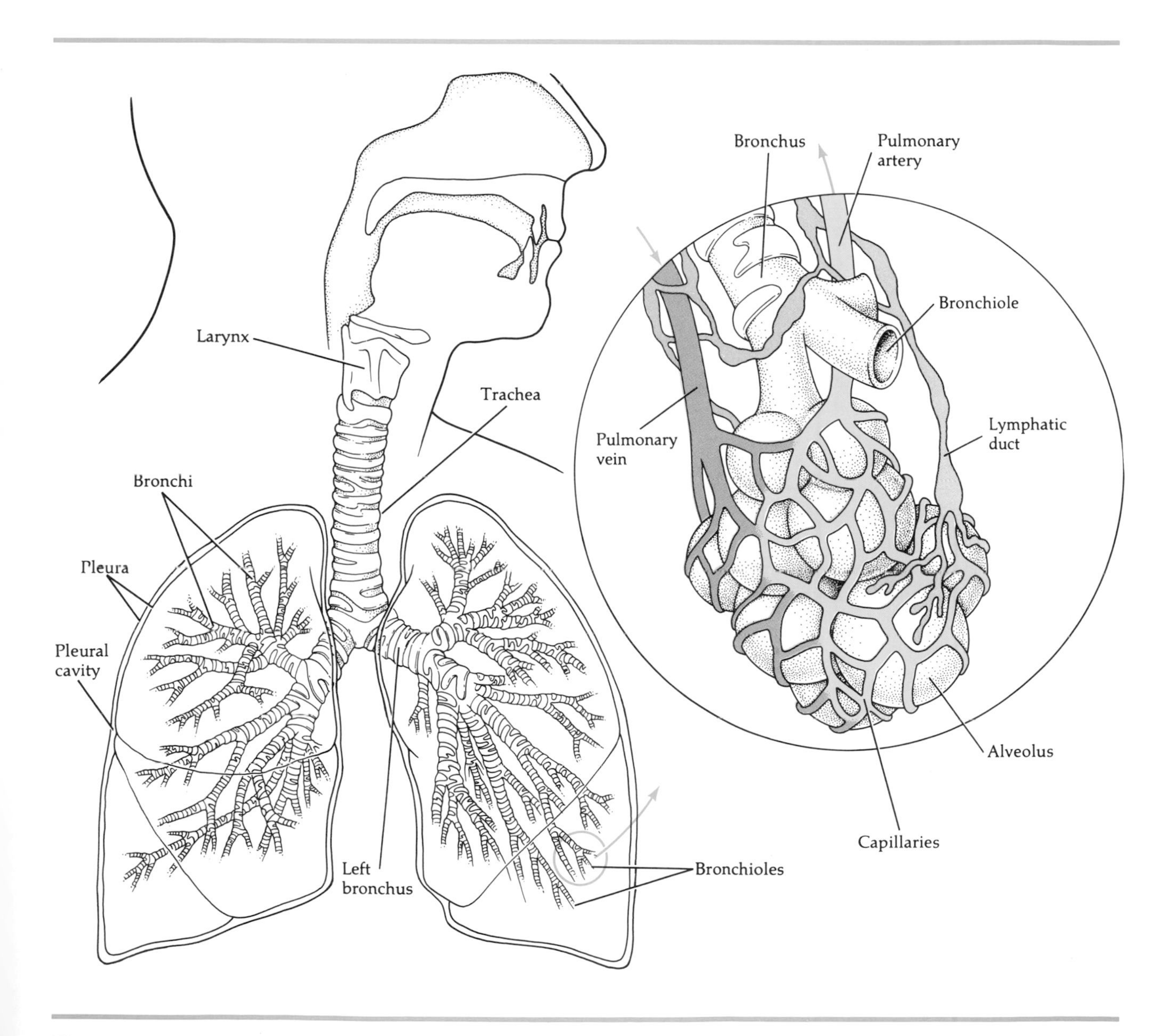

Figure 23–2 Structures of the lower respiratory system.

system, epidemics are rare and morbidity is low because of microbial antagonism. Certain microorganisms of the normal flora suppress the growth of other microorganisms through competition for nutrients and production of inhibitory substances. For example, alpha-hemolytic streptococci limit the growth of pneumococci in the throat. And while the trachea may contain a few bacteria, the lower respiratory tract is usually sterile because of the normally efficient functioning of the ciliary escalator in the bronchial tubes.

BACTERIAL DISEASES OF THE UPPER RESPIRATORY SYSTEM

Streptococcal Sore Throat ("Strep" Throat)

Streptococcal sore throat ("strep" throat) is an upper respiratory infection caused by beta-hemolytic group A streptococci. They are gram-positive bacteria. The most important causative agent is *Streptococcus pyogenes,* the same bacterium responsible for erysipelas, acute bacterial endocarditis, rheumatic fever, and glomerulonephritis.

Streptococcal sore throat is characterized by inflammation of the mucous membrane of the throat (pharyngitis) and fever. Frequently, the tonsils become inflamed (tonsilitis) and the lymph nodes in the neck become enlarged and tender. Another frequent complication is infection of the middle ear, called otitis media (to be discussed shortly).

Laboratory diagnosis consists of streaking throat swabs of bacteria from the inflamed mucous membrane onto blood agar to determine the hemolysins produced (see Color Plate IIB). Since most group A strains of *S. pyogenes* are considerably more sensitive to the antibiotic bacitracin, an agar plate test using paper disks impregnated with the antibiotic is useful if serological tests are not available.

All beta-hemolytic group A streptococci are sensitive to penicillin, the drug of choice. Erythromycin is also effective. Unfortunately, many strains are now resistant to tetracyclines. Treatment of streptococcal sore throat is important in the prevention of rheumatic fever (Chapter 22).

Resistance to streptococcal diseases is type-specific. Since there are more than 55 serological types of group A streptococci, a person who has recovered from infection by one type is not necessarily immune to infection by another type. Immunity to streptococcal infections in general does not usually occur.

Strep throat is most commonly transmitted by respiratory secretions, but epidemics spread by unpasteurized milk were once frequent.

Scarlet Fever

When the *S. pyogenes* strain causing streptococcal sore throat produces an *erythrogenic* (reddening) *toxin,* the infection is called **scarlet fever.** The strains producing this toxin have been lysogenized by a bacteriophage. As you may remember, this means that the genetic information of a bacterial virus has become incorporated into the chromosome of the bacterium, altering its characteristics. The toxin causes a pink-red skin rash and a high fever, probably a generalized cutaneous hypersensitivity reaction of the skin to the circulating toxin. The tongue has a spotted, strawberrylike appearance, and then, as it loses its upper membrane, becomes very red and enlarged. As the disease runs its course, the affected skin frequently peels off as if sunburned. Complications such as deafness may occur.

Scarlet fever seems to increase and decline in severity and frequency over time and in different geographical locations, but it has generally been on the decline in recent years. It is a communicable disease spread mainly by inhalation of infective droplets or dust contaminated by an infected person. Penicillin is effective in treating the original streptococcal infection. Immunity is developed to the toxin of the type of streptococcus that was contracted, but there are many types of streptococci, and immunity to each type is highly specific. Therefore, recovery from scarlet fever does not ensure immunity to streptococcus infections in general. Immunity to scarlet fever can be measured by the *Dick test,* in which erythrogenic toxin is injected into the skin. Local reddening, reaching a maximum at about 24 hours, indicates that the individual is not immune to scarlet fever.

Diphtheria

Another bacterial infection of the upper respiratory system is **diphtheria,** which at one time was a major cause of death in children. The disease begins as a sore throat and fever followed by general malaise and swelling of the neck. The pathogenesis of diphtheria is better understood than that of most other infectious diseases. The organism responsible is *Corynebacterium diphtheriae,* which is a gram-positive, non-endospore-forming, somewhat pleomorphic rod. The dividing cells are often observed to fold together to form V- and Y-shaped figures resembling Chinese letters (Figure 23–3).

C. diphtheriae is often found in the throats of symptomless carriers, having adapted to life in a generally immunized population. The bacterium is well adapted to airborne transmission and is very resistant to drying. Once spread to the upper respiratory system of a susceptible individual, usually by respiratory transmission, the bacteria become established and rapidly increase in number. Although they do not invade tissues, these bacteria do synthesize a powerful exotoxin that is carried by a lysogenic bacteriophage (Chapter 12). The toxin circulates in the bloodstream, interfering with vital processes in many body cells, especially protein synthesis. Only 0.01 mg of this highly virulent toxin is enough to kill a 91 kg (200 pound) person.

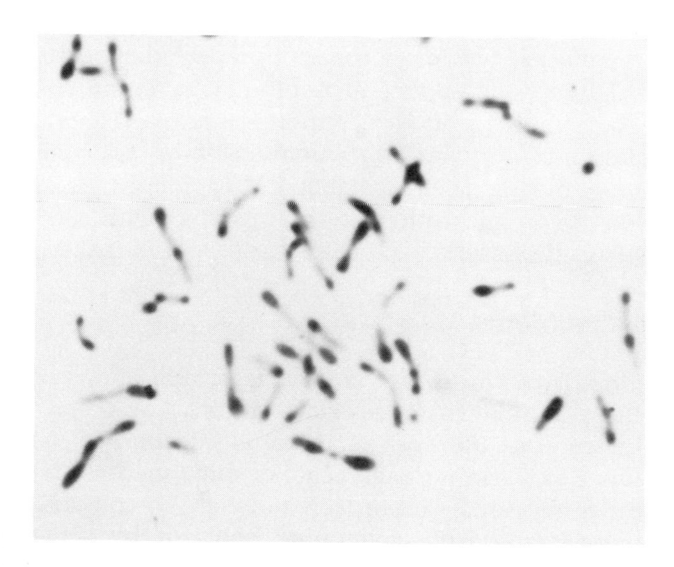

Figure 23–3 The organism causing diphtheria, *Corynebacterium diphtheriae.*

Thus, if antitoxin therapy is to be effective in treatment, it must be administered before the toxin enters the tissue cells. When organs such as the heart or kidneys are affected by the toxin, the disease can be rapidly fatal. In other cases, nerve involvement may occur, resulting in partial paralysis. A characteristic of diphtheria is the formation in the throat of a tough, grayish membrane containing fibrin and dead human and bacterial cells (*diphthera,* meaning "leather"). This membrane, which forms in response to the infection, can totally block the passage of air to the lungs.

Laboratory diagnosis can usually be made by isolating the organisms from the primary lesion. Typical colonies are formed when grown on special media (grayish colonies on a Loeffler slant, black colonies on tellurite agar). Once isolated, the microorganisms are tested for toxigenicity by the guinea pig virulence test or the gel diffusion test. In the *guinea pig test,* the microorganisms are injected into the skin on one side of a shaved guinea pig. After four hours, the guinea pig is injected with antitoxin. Thirty minutes later, microorganisms are injected on the opposite side. If the microorganisms are toxigenic, a characteristic lesion will form between 48 and 72 hours at the site injected before the antitoxin was administered. In the *gel diffusion test,* a strip of filter paper containing diphtheria antitoxin is placed on agar containing antitoxin-free calf serum. Then the microorganisms are streaked at a right angle to the filter paper. If the microorganisms are toxigenic, a visible line of antigen–antibody precipitate will form (see Figure 16–16).

As noted earlier, antitoxin therapy must be administered without delay, for once the diphtheria toxin enters susceptible cells, it is no longer neutralized by antitoxin. Even though antibiotics such as penicillin, tetracyclines, and erythromycin are used to control the growth of the bacteria, they do not neutralize the diphtheria toxin. Thus, they should only be used in conjunction with antitoxin in the treatment of diphtheria.

Part of the normal immunization program in the United States is the *DPT vaccine,* which is effective against diphtheria, pertussis (whooping cough), and tetanus. The "D" stands for diphtheria toxoid, an inactivated toxin that causes the formation of antibodies against the diphtheria toxin. Immunity

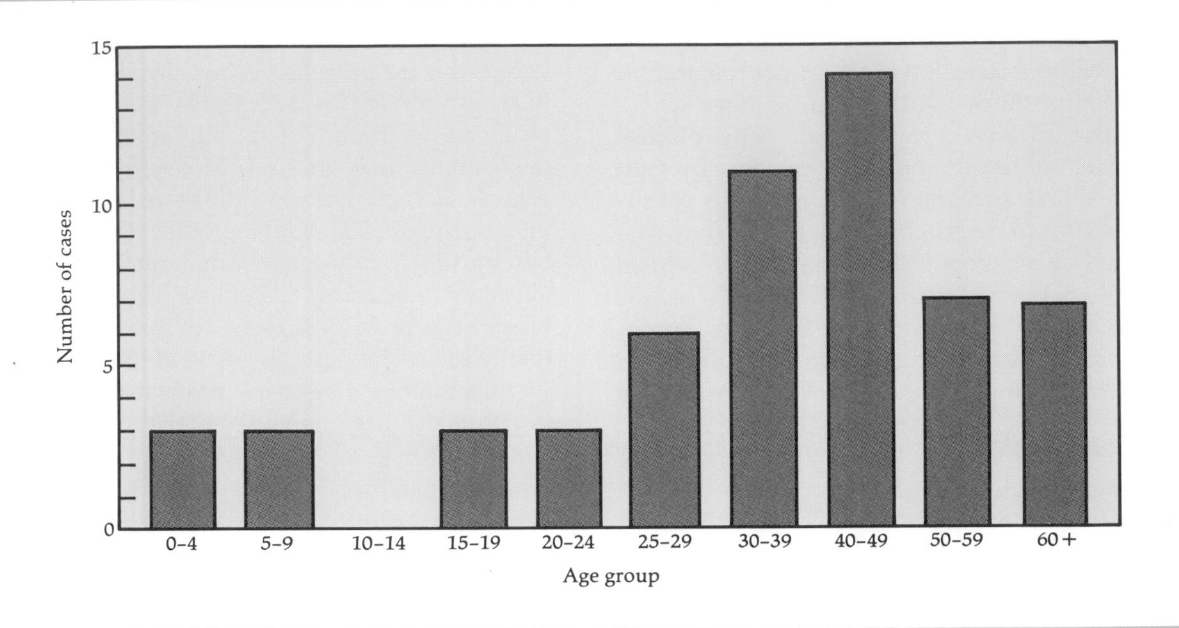

Figure 23–4 Age distribution of diphtheria cases in the United States in 1979. Note that most cases now occur in the adult population. Many of these are cases of cutaneous diphtheria.

to diphtheria can be measured by the *Schick test,* in which a standardized amount of toxin is injected into the skin. The absence of a skin reaction shows that circulating antibodies are sufficient to neutralize the toxin, indicating that the subject is immune.

The number of diphtheria cases reported in the United States each year is presently less than 100, but the death rate in 1976 and 1977, for example, was about 5%. In young children the disease occurs mainly in groups that, for one reason or another, have not been immunized (Figure 23–4).

In the past, diphtheria was spread mainly to healthy carriers by droplet infection. With the advent of immunization, however, the carrier rate has decreased dramatically. Since humans are the only important reservoir for the diphtheria microorganism, total eradication is theoretically possible, but only if universal immunization becomes a reality.

Cutaneous Diphtheria

Diphtheria is also expressed as what is called **cutaneous diphtheria,** in which slow-healing skin ulcerations are formed. In this form of the disease, there is minimal systemic circulation of the toxin. Cutaneous diphtheria is fairly common in tropical countries. In the United States, this disease is responsible for most of the reported cases of diphtheria in persons over 30 (Figure 23–4). In these older groups, immunity is weakened. When the disease was more common, repeated subclinical infections served to reinforce the immunity. Many adults also lack immunity because routine immunization was either unavailable or less common during their childhood. Respiratory cases have been known to arise from contact with cutaneous diphtheria.

Otitis Media

One of the more uncomfortable complications of the common cold (described shortly) or of any infection of the nose or throat is infection of the inner ear, leading to earache, or **otitis media.** The inner ear can also be infected directly by contaminated water from swimming pools or by some severe trauma such as injury to the eardrum and certain skull fractures. The infecting microorganisms generate the formation of pus, which builds up pressure against the eardrum, causing it to become inflamed and painful. The condition is more common in early childhood, probably because the eustachian tube connecting the middle ear to the throat is smaller then and more easily blocked by infection. Enlarged adenoids may be a contributing factor for children with an unusually high frequency of the infection.

A number of bacteria may be involved. The bacteria most commonly isolated are the gram-positive *Staphylococcus aureus*, alpha-hemolytic *Streptococcus pneumoniae*, and assorted beta-hemolytic streptococci, and the gram-negative *Hemophilus influenzae*. Antibiotics such as penicillin and erythromycin are useful in therapy.

VIRAL DISEASE OF THE UPPER RESPIRATORY SYSTEM

Common Cold (Coryza)

There are a number of different viruses involved in the etiology of the **common cold (coryza).** Most of them are generally classified as *coronaviruses* (see Figure 12–9a) or *rhinoviruses*. These viruses are enveloped helical viruses containing single-stranded RNA. Their surface is covered by distinct projections.

There are probably more than 90 serotypes of these viruses; so although immunity of a sort may develop against one viral strain, it is ineffective against others. Immunity is based on IgA antibodies to single serotypes and has a reasonably high short-term effectiveness. Isolated populations may develop a group immunity, with colds disappearing until a new set of viruses is introduced. There is some indication that older persons tend to get fewer and milder colds, perhaps because antibodies to many serotypes have accumulated with time. Although a vaccine can be developed against almost any single viral serotype, so many serotypes exist that a single vaccine against the common cold seems unlikely. However, recent findings indicate that certain serotypes appear with higher frequency and persistence in certain geographical areas. The possibility of developing a vaccine for such an area may prove to be practical.

The symptoms of the common cold are familiar to all of us. They include sneezing, excessive nasal secretion, and congestion. The infection can easily spread from the throat to the sinus cavities, the lower respiratory system, and the middle ear, leading to complications of laryngitis and otitis media. The uncomplicated common cold usually is not accompanied by fever. The virus is spread to new hosts by the discharge of virus-laden mucus during coughing or sneezing or by contact with the mucus directly or on fomites. These fluids also promote survival of the organisms by protecting them from environmental stress.

The incidence of colds averages about two per person per year, making them by far the most common infectious disease. Colds definitely increase in frequency with cold weather. The viruses prefer a temperature slightly below that of normal body temperature—such as might be found in the upper respiratory system, which is open to the outside environment. However, colds will die out in isolated populations even in extremely cold climates as immunity develops in the population. And colds are still found in warm climates where the temperatures vary little. No one knows exactly why colds seem to increase with colder weather in temperate zones. It is not known if the closer indoor contact promotes epidemic-type transmission or if there are actually physiological changes that make one more susceptible.

Since colds are caused by viruses, antibiotics are of no use in treatment. The common cold will usually run its course to recovery in about a week's time. Recovery time is not lessened by taking over-the-counter drugs, although such drugs may lessen the severity of certain signs and symptoms.

BACTERIAL DISEASES OF THE LOWER RESPIRATORY SYSTEM

Whooping Cough (Pertussis)

Infection by the bacterium *Bordetella pertussis* results in **whooping cough (pertussis).** *B. pertussis* is a small, nonmotile, gram-negative coccobacillus that forms a capsule when virulent. It is an obligate aerobe.

Whooping cough, which is primarily a childhood disease, can be quite severe. The initial stage, which is called the *catarrhal stage*, resembles a common cold. Prolonged sieges of coughing characterize the second, or *paroxysmal stage*. The bacterium

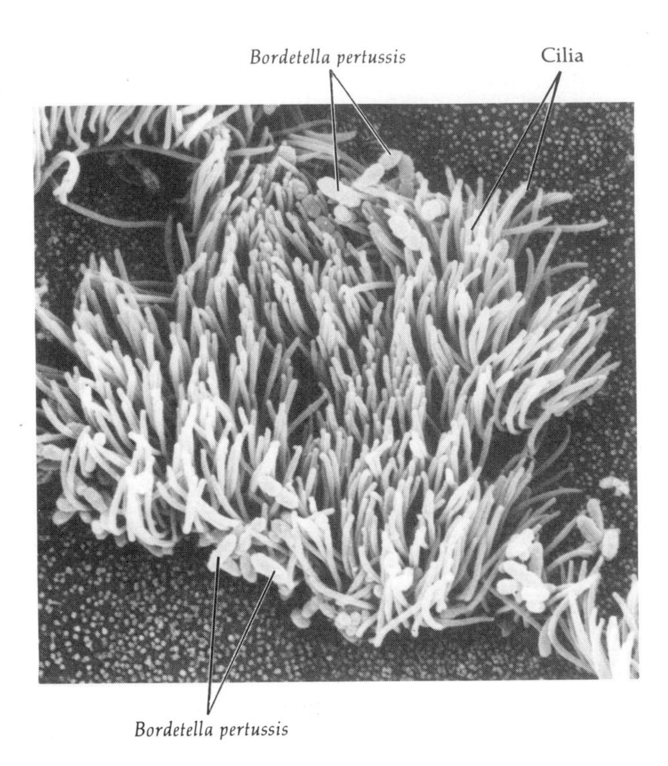

Figure 23–5 Scanning electron micrograph of the whooping cough pathogen, *Bordetella pertussis*, growing among the cilia of respiratory epithelial cells (×3000).

grows in dense masses in the trachea and larger bronchi and causes the production of thick mucus that impedes ciliary action (Figure 23–5). (As noted earlier, the trachea and larger bronchi are usually kept clear of mucus by ciliary action.) The infection leads to desperate attempts to cough up these mucus accumulations. Gasping for air between coughs causes a whoop sound, the reason for the name of the disease. Coughing episodes occur several times a day over an interval of one to six weeks. The *convalescence stage*, the third stage, may last for months. The disease is of unusually long duration for a respiratory infection.

B. pertussis produces an endotoxin as part of its cell wall that resembles that of other pathogenic gram-negative bacteria. It also produces an exotoxin in the cytoplasm that is apparently released along with the endotoxin when the cell autolyses upon death. The precise role of the toxins in pathogenesis has yet to be determined. The organism does not invade tissues. Very high white blood cell counts (sometimes six to eight times normal levels) are frequently seen in pertussis.

The organism is cultured from a throat swab inserted through the nose on a thin wire and held in the throat while the patient coughs. *B. pertussis* forms characteristic small, glistening, white colonies on Bordet–Gengou medium. A fluorescent-antibody test or agglutination by antiserum is used to confirm identification. Mild cases of whooping cough require no specific treatment. Severe cases, especially in infants, are usually treated with erythromycin. Tetracyclines and chloramphenicol may also be used. Although the clinical response to antibiotics may not result in rapid improvement, the drugs do render the patient noninfectious. If secondary pneumonia develops as a complication, antibiotic therapy must be continued.

After recovery, immunity is good. A vaccine prepared from heat-inactivated whole bacteria (the "P" in DPT vaccine stands for *pertussis*) is a part of the regular immunization schedule for children. Vaccination is usually recommended at about two months of age, when the titer of maternal antibodies in the baby's circulation is dropping and the child's own antibodies have not yet replaced them.

There are presently about 2000 cases of whooping cough each year in the United States, with deaths usually fewer than 10 per year. Infants are less capable of coping with the effort of coughing to maintain an airway and have a relatively high mortality rate.

Tuberculosis

Tuberculosis is an infectious disease caused by the bacterium *Mycobacterium tuberculosis*. The microorganism is a slightly bent or curved slender rod (see Figure 10–14b) and an obligate aerobe. The rods are slow growing, sometimes forming filaments, and have a tendency to grow in clumps. The appearance of their growth on the surface of liquid media is moldlike, which suggested the genus name *Mycobacterium* (from the Greek mýkēs, meaning "fungus") (see Color Plate IID). These bacteria are relatively resistant to normal staining procedures. When stained by the Ziehl-Neelson technique of steaming in carbolfuchsin dye, they cannot be decolorized with a mixture of acid and alcohol and are therefore classified as *acid-fast*. This characteristic reflects the unusual composition of the cell wall, which contains large amounts of lipid materials. These lipids may also be responsible for the resistance of the mycobacteria to environmental stresses such as drying. In fact, these bacteria can survive for weeks in dried sputum and are also very resistant to sunlight and to chemical antimicrobials used as antiseptics and disinfectants. Therefore, particular care must be taken in handling possibly infected material.

Tuberculosis is characterized by the formation of nodules in the infected tissue. Infection of the lungs is most common and is usually acquired by inhalation of the pathogen. The disease can also be transmitted via the gastrointestinal tract or by direct contact with infected material. Once the bacteria reach the lungs, the body responds to their presence by walling off the organisms, forming a small nodule called a *tubercle* in two to four weeks. Bacteria can survive in tubercles indefinitely. The tubercle, in time, may undergo necrosis, changing to cheeselike consistency called a *caseous lesion*. Caseous lesions heal by scar formation and calcification and are then referred to as *Ghon complexes*, which may be recognized in chest X rays for the remainder of the individual's life (Figure 23–6).

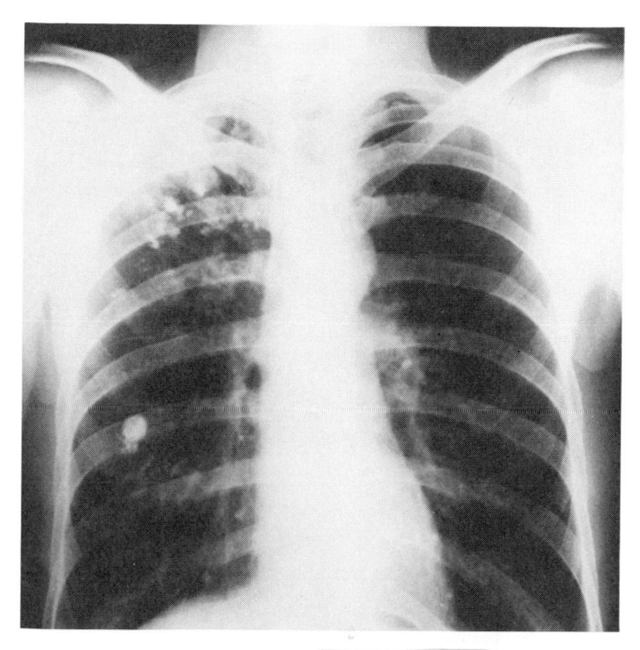

Figure 23–6 X ray showing Ghon complexes in a case of tuberculosis. The lower area of the right lung (left side of photo) has a large complex, and numerous smaller lesions are visible higher up. The left lung (right side of photo) is clear.

The protective mechanisms that isolate the pathogen from surrounding tissue are sometimes defeated. The caseous lesion may rupture and spread into a blood or lymphatic vessel or into the airway. Bacteria are carried by the blood and the lymphatic system to new foci of infection in the lungs or in other organs. This condition is called **miliary tuberculosis,** the name derived from the numerous millet seed-sized tubercles formed in these tissues. This condition leads to a progressive disease characterized by loss of weight, coughing, and general loss of vigor. (Before Koch's discovery of the causative agent, tuberculosis was commonly called *consumption*.) Hypersensitivity reactions may be involved, promoting the progression of the disease. If a cavity opens into an airway, the mycobacteria spread to other parts of the lung and are carried out of the patient by coughing.

Among persons infected with *M. tuberculosis*, only a small number develop active cases of the disease. Variations in host resistance are related to genetic differences and the presence of other illness and probably physiological and environmental factors such as malnutrition, overcrowding, and stress. Tuberculosis is a good example of the importance of ecological balance between host and parasite in infectious disease, since infection with *M. tuberculosis* is clearly a necessary but not a sufficient cause of clinical disease.

The bacteria entering the lungs may be ingested by wandering macrophages. Unfortunately, these bacteria resist lysis by the macrophages, and eventually the macrophages die and release the pathogen, which has generally increased in numbers. This intracellular growth, which shields the bacteria from drugs, plus the very slow reproductive rate of the organisms (many antibiotics work well only on growing cells) make it difficult to treat the microorganisms with chemotherapy.

The first effective antibiotic in the treatment of tuberculosis was streptomycin, although rifampin is currently preferred. Probably the most commonly used drug today is the synthetic chemical isoniazid (INH). A combination of INH and rifampin is particularly effective against phagocytized bacteria. The sulfa derivative para-aminosalicylic acid (PAS) is effective, and ethambutol has come into use recently. Antibiotics and synthetics are usually given in combination to minimize the development of resistance. Such resistance becomes more likely in diseases such as tuberculosis that

Figure 23–7 The geographical occurrence of tuberculosis in the United States, 1976–1978. The map shows reported cases per 100,000 population by county. The darker the area in the figure, the greater the incidence of tuberculosis.

require prolonged chemotherapy, giving time for spontaneous mutants to arise that are resistant to the drug.

Persons infected with tuberculosis develop sensitized T cells against the bacterium, the basis for the *tuberculin skin test*. In this test, a purified protein derivative (PPD) of the tuberculosis bacterium is injected cutaneously. If a person has been exposed to tuberculosis, sensitized T cells react with these proteins, causing a delayed hypersensitive reaction in about 48 hours. This reaction appears as an induration (hardening) and reddening of the area about the injection site. Probably the most accurate tuberculin test is the *Mantoux test*, in which dilutions of 0.1 ml of antigen are injected and the reacting area of the skin is measured. However, a number of similar tests are also in common use. A positive tuberculin test in the very young is a probable indication of an active case of tuberculosis. In older persons it may only indicate a previous exposure to the disease and immunity to it, but not an active case. Nonetheless, it is an indication that further examination is needed, such as X rays for lung lesions, as well as attempts to isolate the bacterium.

Confirmation of tuberculosis by isolation of the bacterium is complicated by the very slow growth of the pathogen. Eight weeks may be needed for the growth of an isolate, with at least a week required for the appearance of an exceedingly minute colony. Special media, such as the Lowenstein–Jensen medium, with a high lipid content in the form of glycerol or eggs is recommended.

Another species, *Mycobacterium bovis*, is a pathogen mainly of cattle. It is the cause of **bovine tuberculosis**, which is transmitted to humans by contaminated milk or food. Bovine tuberculosis seldom spreads from human to human. But before the days of pasteurized milk and the development of control methods such as tuberculin testing of cattle herds, this disease was a common form of tuberculosis in humans. *M. bovis* infections cause tuberculosis that primarily affects the bones or lymphatic system. At one time, a common manifestation of this type of tuberculosis was deformation of the bone structure, resulting in hunchbacks.

The *BCG vaccine* is a live culture of *M. bovis* that has been made avirulent by long cultivation on artificial media. (The name stands for "the bacillus of Calmette and Guerin," the persons who originally isolated the strain.) The vaccine is fairly effective in prevention of tuberculosis and has been in use since the 1920s. In the United States it is used among high-risk groups only, since there is some question about the efficiency and safety of the vaccine. Apparently, the vaccine enhances cell-mediated immunity, and persons who have received the vaccine show a positive reaction to tuberculin skin tests.

There are still about 30,000 new cases of tuberculosis and 3000 deaths reported each year in the United States. In recent years, the death rate has been decreasing faster than the case rate. Patients are no longer isolated once chemotherapy has rendered their disease noncontagious. Tuberculosis in the United States is most prevalent among the poor and among Eskimos and American Indians (Figure 23–7), probably because of environmental factors, although genetic factors no doubt play a role as well. Tuberculosis cases in the developing countries of the world are estimated to number in the tens of millions.

Pneumococcal Pneumonia

The term **pneumonia** is generally applied to any pulmonary infection. The classic form, representing about 75% of reported pneumonias, is caused by the pneumococcus *Streptococcus pneumoniae* and is referred to as **pneumococcal pneumonia.** Other microorganisms that cause pneumonia are beta-hemolytic streptococci and other streptococci, *Staphylococcus aureus, Klebsiella pneumoniae, Hemophilus influenzae, Mycobacterium tuberculosis, Legionella pneumophila, Mycoplasma pneumoniae,* and certain viruses. It should be noted that a considerable number of bacteria with normally low virulence have been known to cause pneumonias. Most of these bacteria are opportunistic and include *Proteus, Serratia,* and *Pseudomonas* species, and even *Escherichia coli.*

Streptococcus pneumoniae is a gram-positive, catalase-negative, ovoid bacterium. Because it usually forms cell pairs, it was formerly called *Diplococcus pneumoniae.* The organism produces a heavy capsule that makes the pathogen resistant to phagocytosis (Figure 23–8a). These capsules are also the basis of serological differentiation of pneumococci into some 83 serotypes. Before antibiotic therapy

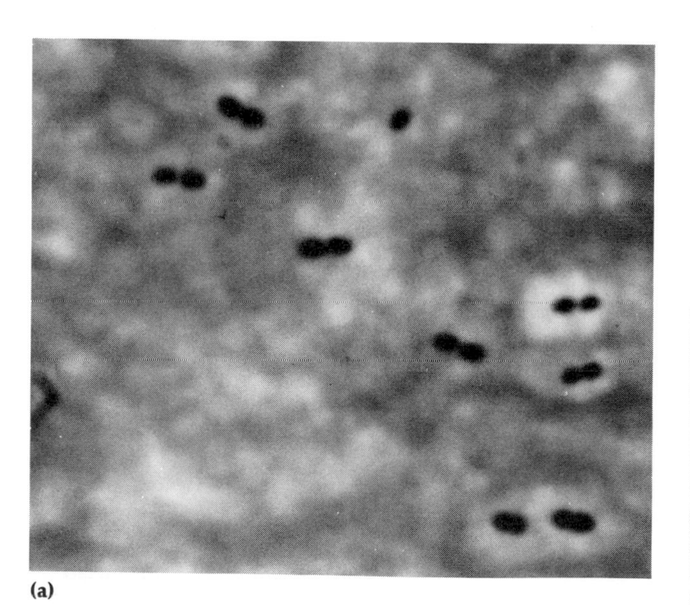

(a)

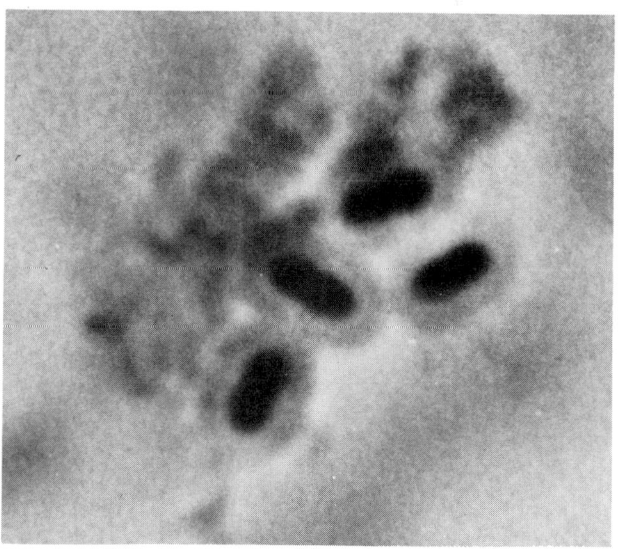

(b)

Figure 23–8 The pneumococcus *Streptococcus pneumoniae.* **(a)** Note the heavy capsule, accounting for the virulence of the organism, and the paired arrangement (×6800). **(b)** Quellung reaction after exposure to pneumococcus antiserum (×12,000).

became so effective, antisera directed at these capsular antigens were used in the therapy of the disease.

Pneumococcal pneumonia involves both the bronchi and the alveoli. Symptoms are fever, hard breathing, and chest pain. The lungs have a reddish appearance because of dilated blood vessels. Alveoli fill with some red blood cells, polymorphonuclear lymphocytes (PMNs), and edema fluid in response to the infection. The sputum is often rust colored from blood coughed up from the lungs. Pneumococci may invade the bloodstream, the pleural cavity surrounding the lung, and occasionally the meninges. No bacterial toxin has been clearly related to pathogenicity.

An initial diagnosis can be made from a chest X ray, since the fluid in the lungs produces recognizable shadows on the X ray. Confirmation of the diagnosis can be made by isolation of the pneumococci from the throat, sputum, and other fluids. Pneumococci can be distinguished from other gram-positive alpha-hemolytic streptococci by their inhibition of growth near a disk of optochin (ethylhydrocuprein hydrochloride) or by the bile solubility of broth cultures. They may be serologically typed with antisera to the capsular polysaccharide by the quellung reaction (from the German *quellung,* meaning "swelling"). In this reaction, the bacteria are observed under a microscope, and when they react with a positive antiserum, there is an

Centers for Disease Control

MMWR

Morbidity and Mortality Weekly Report

Pneumococcal Polysaccharide Vaccine

Despite the use of antibiotic therapy, morbidity and mortality from pneumococcal disease remain public health problems. Pneumococcal pneumonia, which probably affects as many as 400,000 to 500,000 people annually in the United States, has a fatality rate of 5 to 10%, and may be particularly serious for some segments of the population, such as the elderly. Furthermore, the emergence of pneumococcal strains that are resistant to antibiotics heralds additional problems.

Polyvalent polysaccharide vaccine against disease caused by *Streptococcus pneumoniae* (pneumococcus) was licensed in the United States in 1977. This 14-valent polysaccharide vaccine contains purified capsular material of pneumococci extracted separately from those types of organisms to be combined in the final vaccine. Each dose of the vaccine contains 50 μg of each polysaccharide. The 14 particular types of pneumococci in the vaccine available in the United States—American types 1, 2, 3, 4, 6, 8, 9, 12, 14, 19, 23, 25, 51, and 56—cause at least 80% of all bacteremic pneumococcal disease seen in this country.

The majority of adults respond to vaccine with a severalfold rise in antibody measured by radioimmunoassay. Immunity is provided only against the pneumococcal types in the vaccine, although, theoretically, there might be some degree of cross-protection among immunologically similar types.

Nasopharyngeal acquisition of the pneumococcal types included in the vaccine appears to be reduced by immunization. Furthermore, there has been no evidence among the immunized of any increase in diseases caused by other gram-positive or gram-negative microorganisms. Elevated antibody levels appear to persist for at least two years after immunization. In experience to date, the vaccine has proved to be safe; side effects, although frequent, are not severe.

The vaccine is recommended for special populations and individuals, and *not* for mass immunization of healthy people. Certain closed groups, such as those in residential schools and nursing homes, who are at enhanced risk of pneumococcal disease may benefit from immunization. Individuals over two years of age with dysfunction of the spleen (due to sickle cell disease or other causes) should benefit from immunization, as should those with diabetes mellitus and impairment of cardiorespiratory, hepatic, and renal systems. Such individuals are at increased risk of pneumococcal disease. Because the risk of and fatality rate from pneumococcal disease increase with increasing age, the benefits of vaccination should increase with increasing age.

Source: *MMWR* 27:25 (1/27/78).

apparent swelling of the capsule (Figure 23–8b).

It should be noted that there are many healthy carriers of the pneumococcus. Virulence seems based mainly on a lowering of resistance because of stress, and many illnesses of the elderly terminate in pneumococcal pneumonia. Viral infections of the respiratory system are a frequent precursor of pneumonia.

Recurrence of pneumonia is not uncommon, but a different serological type is usually responsible. Before chemotherapy was available, the fatality rate was as high as 25%. This has now been lowered to less than 1% for younger patients treated early in the course of the disease. Penicillin is the drug of choice, but a few strains of penicillin-resistant pneumococci have been reported. A vaccine called Pneumovax has been developed from the purified capsular material of the 14 types of pneumococci that cause at least 80% of the pneumococcal pneumonia in the United States. It is aimed at the most susceptible group, the elderly, and is also used for other debilitated individuals (see box).

Klebsiella Pneumonia

Klebsiella pneumoniae, the causative agent of ***Klebsiella* pneumonia,** is a member of the enteric group of bacteria, but is not uncommon in the throat and mouth of healthy persons. The organism is an encapsulated gram-negative rod, and its virulence is strongly related to the presence of the capsule. It causes a type of pneumonia most commonly found in persons who are chronically debilitated, with malnutrition often a contributing factor. Probably 1–3% of bacterial pneumonias are of this type. Male alcoholics past 40 years of age are probably the most susceptible. Symptoms resemble those of pneumococcal pneumonia, but a prime difference is the formation of lung abscesses and permanent lung damage with *Klebsiella* pneumonia.

The fatality rate in untreated cases can be very high, perhaps exceeding 85%. Early treatment is vital, but considering the most susceptible population, is not usually received. *K. pneumoniae,* like most gram-negatives, is not sensitive to penicillin. Therefore, the normal penicillin treatment for pneumococcal pneumonia would be harmful, because it would suppress the normal gram-positive flora of the throat that affords some protection against invading pathogens. It is essential to make an identification quickly, since delay allows unimpeded progress of the disease, which is very dangerous. For *K. pneumoniae* pneumonia, the antibiotics of choice are cephalosporins or gentamicin, but drug resistance is common.

Mycoplasmal Pneumonia (Primary Atypical Pneumonia)

Typical pneumonia has a bacterial origin. If a bacterial agent is not isolated, the pneumonia is considered atypical, and viral agents are usually suspected. However, there could be another source of infection—mycoplasmas—which are bacteria that do not form cell walls. The mycoplasmas do not grow under the normal conditions used to recover most bacterial pathogens. Because of this, pneumonias caused by mycoplasmas are often confused with viral pneumonias.

The bacterium *Mycoplasma pneumoniae* is the causative agent of **mycoplasmal pneumonia (primary atypical pneumonia).** This type of pneumonia was first discovered when such atypical infections responded to tetracyclines, an indication of a nonviral agent. The disease is endemic in the population and a fairly common cause of pneumonia in young adults and children. Most infections with mycoplasmas result in upper respiratory infections, and relatively few develop into pneumonia. The mortality rate is less than 1%.

Isolates from throat swabs and sputum are grown on a medium containing horse serum and yeast extract, forming distinctive colonies of a "fried egg" appearance. The colonies are so small that they usually have to be observed with a hand lens or microscope. *Mycoplasma pneumoniae* is beta-hemolytic for guinea pig red blood cells, a trait that differentiates it from other mycoplasmas. Conclusive identification is made by fluorescent-antibody methods. Microscopically, the mycoplasmas are highly varied in appearance because of the lack of cell wall (see Figure 10–16). Their flexibility allows them to pass through filters with pores as small as 0.2 μm in diameter, which is small enough to hold back most other bacteria.

Diagnosis based on the recovery of organisms may not be useful in treatment, because as much as two weeks are required for the slow-growing organisms to develop. A complement fixation test, using antigens from *M. pneumoniae* cells, can be used to test for a significantly rising titer of circulating antibodies, although this procedure also takes two to three weeks. Early treatment with tetracycline and erythromycin shortens the illness.

Legionnaires' Disease

A type of pneumonia identified only recently is **Legionnaires' disease,** or **legionellosis,** which has already been discussed in Chapters 1 and 13. As you may recall, this disease first received public attention in 1976 when a series of deaths occurred among members of the American Legion who had attended a meeting in Philadelphia. A total of 182 persons, 29 of whom died, contracted a pulmonary disease, apparently at this meeting. Because no obvious bacterial cause could be found for what came to be known as Legionnaires' disease, the deaths were attributed to viral pneumonia. Closer investigation over a period of time, mostly using techniques directed at locating a suspected rickettsial agent, eventually identified the organism. It was a hitherto unknown bacterium, a gram-negative rod, now known as *Legionella pneumophila* (see Figure 1–1a and Color Plate IIIB).

The disease is characterized by a high temperature, cough, and general symptoms of pneumonia. No person-to-person transmission seems to be involved. Recent work has shown that the bacterium can be isolated with some frequency from natural waters. Of particular interest is the fact that the organism can grow well in some air-conditioning cooling tower waters. This would help explain some of the epidemics associated with hotels and urban business districts and would account for the increased prevalence of the disease in the summer. Although this disease had been previously unidentified, it seems now to have been fairly commonplace. In 1978, 761 cases were confirmed, and in 1979, 578 cases. Fluorescent-antibody tests have been developed to identify the disease agent, which is readily treated by erythromycin or rifampin as the drugs of choice.

Psittacosis (Ornithosis)

The term **psittacosis (ornithosis)** was derived from association of the disease with psittacine birds such as parakeets and parrots. It was later found that the disease can also be contracted from many other birds, such as pigeons, chickens, ducks, and turkeys. Therefore, the more general term ornithosis has come into use.

The causative organism is *Chlamydia psittaci,* a gram-negative, obligately intracellular bacterium (see Figure 10–15b). One way chlamydias differ from the rickettsias is in forming tiny **elementary bodies** as one part of their life cycle. Elementary bodies are resistant to environmental stress. This allows their airborne transmission, rather than requiring an infective bite, which transfers the infective agent directly from one host to another as do most rickettsias. Elementary bodies also specifically enhance their own phagocytosis by host cells, starting the infective cycle.

Psittacosis is a form of pneumonia, usually with fever, headache, and chills. Subclinical infections are very common, and stress appears to enhance susceptibility to the disease. Disorientation and even delirium indicates that the nervous system may be involved in some cases.

The disease is not usually transmitted from one human to another, but is spread by contact with the droppings and other exudates of fowl. One of the most common modes is by inhalation of dried particles from droppings. The birds themselves usually have symptoms of diarrhea, ruffled feathers, respiratory illness, and a generally droopy appearance. The parakeets and other birds sold commercially in pet stores are usually, but not always, free of the disease. However, pet store employees and persons involved in the commercial rearing of turkeys are at greater risk of the disease.

Diagnosis is made by isolation of the bacterium in embryonated eggs or mice or by cell culture. The organism can then be identified by its specific antigens by a fluorescent-antibody staining technique. If organisms are not isolated, a complement fixation test to detect rising serum antibody titer in the patient can be used. No vaccine is available, but tetracyclines are effective antibiotics in treating humans and animals. Immunity after recovery is only moderately effective.

Most years, fewer than 100 cases and very few deaths are reported in the United States. The main danger is in late diagnosis. At one time, before antibiotic therapy was available, mortality was about 20%.

Q Fever

In Australia in the mid-1930s, a disease was reported that was characterized by a fever lasting one or two weeks, chills, chest pain, severe headache, and other evidence of a pneumonia-type infection. The disease was rarely fatal. In the absence of an obvious cause, the affliction was labeled **Q** (for query) **fever**—much as one might say X fever. The causative agent was subsequently identified as the obligately parasitic, intracellular bacterium *Coxiella burnetii*, a rickettsia (Figure 23–9). Most rickettsias are not resistant enough to survive airborne transmission, but this organism is an exception.

The organism is a parasite of a number of arthropods, cattle ticks being the one most commonly involved in human transmission, since they spread the disease among dairy herds. Q fever is transmitted among animals by direct bite, although arthropod bites are rarely, if ever, involved in transmission to humans. In humans, the disease is most commonly associated with dairy operation and dairy products. The organisms are shed in the feces, milk, and urine of infected cattle. Ingestion of unpasteurized milk and inhalation of aerosols generated in dairy barns are responsible for most human infections. Many dairy workers have acquired at least subclinical infections. The pasteurization temperature of milk, which was originally aimed at elimination of tuberculosis organisms, was raised slightly to ensure the killing of *C. burnetii*. Workers in meat- and hide-processing plants are also at risk.

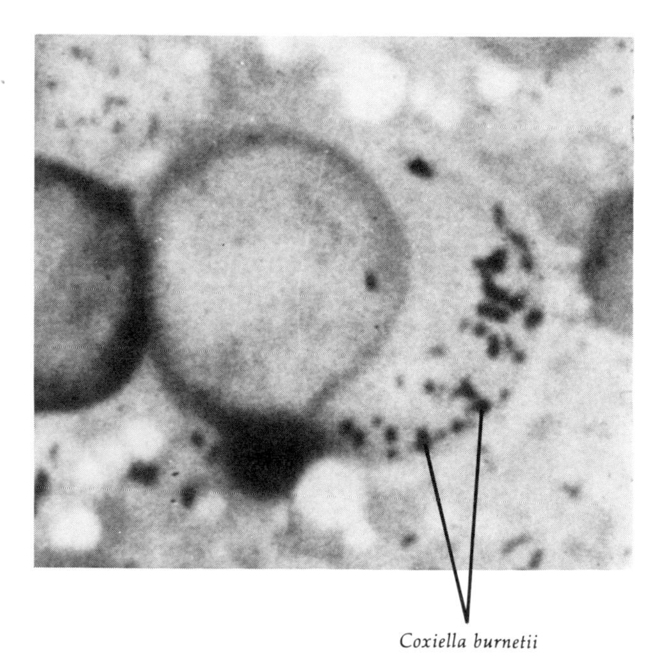

Figure 23–9 The rickettsial organism causing Q fever, *Coxiella burnetii*. Observe that the rickettsias are inside the cells.

Most cases of Q fever in the United States are reported from the western states. The disease is endemic to California, Arizona, Oregon, and Washington. A vaccine for laboratory workers and similar high-risk personnel is available. Tetracyclines are very effective in treatment.

Diagnosis may be made by identification of the organism by isolation and growth in egg embryos or in cell culture. Handling of the organism by laboratory personnel testing for *Coxiella*-specific antibodies in the patient's serum can be avoided by using serological tests, mostly complement fixation and agglutination types.

Acute Epiglottitis

Acute epiglottitis is a serious disorder caused by *Hemophilus influenzae*. The disease usually starts in the respiratory system. It is characterized by the spread of infection from the throat to the epiglottis, a cartilage lid over the voice box. The epiglottis becomes swollen, cherry red, and edematous. The resulting obstruction in the voice box necessitates immediate hospitalization for a possible tracheostomy. The onset of the disease is sudden, often ending fatally within 24 hours.

VIRAL DISEASES OF THE LOWER RESPIRATORY SYSTEM

Viral Pneumonia

Viral pneumonia can occur as a complication of influenza and even chickenpox. A number of enteric and other viruses have been shown to cause viral pneumonia, but viruses are isolated and identified in less than 1% of pneumonia-type infections, since few laboratories are equipped to properly test clinical samples for viruses. Nevertheless, in about 15% of pneumonia cases no cause is determined, and many of these cases may be, and are even assumed to be, of viral etiology if mycoplasmal pneumonia has been ruled out.

Influenza ("Flu")

Next to the common cold, the developed countries of the world are probably more aware of **influenza ("flu")** than any other disease. The disease is characterized by chills, fever, headache, and general muscular aches. Recovery normally occurs in a few days, and coldlike symptoms appear as fever subsides. Incidentally, diarrhea is not a normal symptom of the disease, and the intestinal discomforts attributed to "stomach flu" are probably due to some other cause.

The *influenza virus* consists of eight distinct RNA fragments of differing lengths enclosed by an inner layer of protein and an outer lipid bilayer. Embedded in the lipid bilayer are numerous projections that characterize the virus (Figure 23–10). There are two types of projections: *hemagglutinin (H) spikes* and *neuraminidase (N) spikes.* The H spikes, of which there are about 500 on each virus,

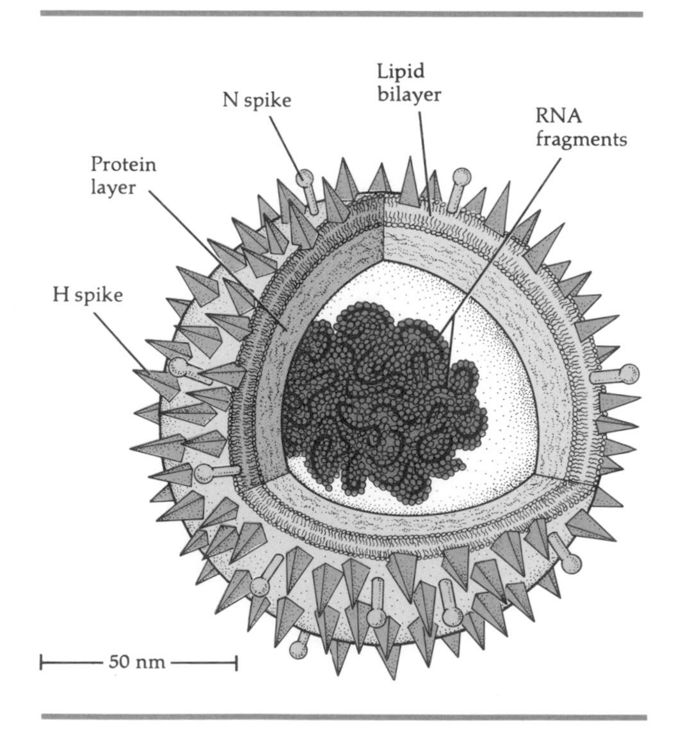

Figure 23–10 Detailed structure of the influenza virus.

allow the virus to recognize and attach to body cells before infection. Antibodies against the influenza virus are directed mainly at these spikes. The term *hemagglutinin* refers to the agglutination (clumping) of red blood cells that occurs when the viruses are mixed with them. This reaction is important in serological tests such as the hemagglutination-inhibition test used to identify influenza viruses. The N spikes, of which there are about 100 per virus, differ from the H spikes in appearance and function. Apparently, they help the virus to separate from the infected cell as it makes its exit following intracellular reproduction. N spikes also stimulate formation of antibodies, but these are of less importance in resistance to the disease than those produced in response to the H spikes.

Viral strains are identified by variations in their H and N antigens. The different forms of the antigens are assigned numbers—for example, H_0, H_1, H_2, H_3, and N_1, N_2. Each number reflects a substantial alteration in the protein makeup of the spike. The first antigenic type was isolated in 1933 and named H_0N_1. The Asiatic flu pandemic of 1957 was of type H_2N_2, and the Hong Kong virus of 1968 was designated H_3N_2. The influenza viruses are also classified into major groups by the antigens of their protein coats. These groups are A, B, and (rarely) C. The A-type viruses are responsible for pandemics. Variations in the H and N antigens occur among the A-type viruses. Historically, such changes, termed *antigenic shifts,* occur every 10 or 12 years and evade most of the immunity developed in the human population. The B-type virus also circulates, but is usually responsible for more geographically limited and milder infections. These antigenic shifts are probably due to a major genetic recombination. Since influenza viral RNA occurs in several pieces, recombination is likely in cases where an organism is infected by more than one strain. Recombination between the RNA of animal viral strains (found in swine, horses, and birds, for example) and the RNA of human strains may be involved.

Between episodes of such major antigenic shifts there are minor year-to-year variations in the antigenic makeup. The virus, for example, may still be designated as H_3N_2, but viral strains reflecting minor antigenic changes within the antigenic group arise. These are sometimes assigned names related to the locality where they were first identified, for example, Victoria or Russia. These minor variations are called *antigenic drifts*. They usually reflect a mutation in which perhaps only a single amino acid in the protein makeup of the H or N spike has been altered. Such minor, one-step mutations are probably a response to selective pressure by antibodies (usually IgA in the mucous membranes) that neutralize all viruses except for the new mutations. Such mutations can be expected in about one in each million multiplications of the virus.

The usual result of antigenic drift is that a vaccine effective against H_3, for example, will be less effective against H_3 isolates circulating 10 years after the event. There will have been enough drift in that time to largely evade the antibodies stimulated originally by the earlier strain. In 1977, the Russian strain of H_1N_1 circulated in the United States. It essentially affected only persons under the age of about 25, apparently because it closely resembled a strain that had circulated in the 1950s. This early reappearance is somewhat out of the normal character of the disease and has led to speculation that an escape from a laboratory collection might have been involved in the later outbreak.

Development of a vaccine for influenza is not practical. Although it is not difficult to make a vaccine for a given strain of the virus, the problem is to identify a new strain of a circulating virus and develop and distribute the new vaccine for it in time. Unless the strain appears to be unusually virulent, this is also considered to be of questionable value. The usual practice is to limit vaccine administration to the elderly, to hospital personnel, and to similar high-risk groups. The vaccines are often *multivalent*, that is, directed at several strains in circulation at the time. At present, influenza viruses for manufacturing vaccines are grown in egg embryo cultures. One new approach to the problem of antigenic variation may be to utilize genetic engineering techniques. A laboratory bacterium can theoretically be made to produce, very quickly, a large number of antigens to the influenza virus when genes for this purpose have been inserted into the bacterial DNA.

Almost annual epidemics of the "flu" spread rapidly through large populations. The disease is so readily transmissible that epidemics are quickly propagated through susceptible populations. The reason the populations are susceptible is that the virus changes character with time and evades the immune defenses. The death rate from the disease is not high, usually less than 1%, and these are mainly among the very young and very old. However, so many persons are infected in a major epidemic that the total number of deaths is often high. Very often the cause of death is not the influenza virus, but secondary bacterial infections. The bacterium *Hemophilus influenzae* was named under the mistaken belief that it was the principal pathogen causing influenza rather than a secondary invader. *Staphylococcus aureus* and *Streptococcus pneumoniae* are other prominent secondary bacterial invaders.

In any discussion of influenza, the great pandemic of 1918–19 must be mentioned. More than 20 million people died worldwide. No one is sure why it was so unusually lethal. Usually, the very young and very old are the principal victims, but in 1918–19, young adults had the highest mortality. Considerable effort has been expended in examining the question of this unusual pattern of virulence. Recent exhumation of bodies of victims of the epidemic buried in the permafrost above the Arctic Circle failed to recover viable isolates of the virus. It is possible that the 1918 viral strain became endemic in swine in the United States.

In 1976, a recruit at Fort Dix, New Jersey, died of influenza caused by a swine flu strain with many of the characteristics of the 1918 pandemic strain. This precipitated a national campaign of preventive inoculations for swine influenza. Many feel this was ill advised, particularly since there proved to be no widespread transmission from the initial focus of infection. Also, the incidence of *Guillain–Barré syndrome (GBS)* was five to six times higher in persons who received flu vaccine that season (1976–77) than in unvaccinated persons. (GBS is characterized by a paralysis that is usually self-limited and reversible, although approximately 5% of the cases are fatal.) The reasons for the association between flu vaccination and GBS that year are still a mystery, and the Centers for Disease Control found no association between flu vaccination and GBS for either the 1978–79 or 1979–80 flu season.

The questions raised by the 1976–77 flu vaccination program highlight the trade-offs that must be considered in instituting a large-scale inoculation program against a new strain of flu virus. In order to be effective, the program must be started as soon as possible, but this does not allow enough time for thorough testing of the new vaccine.

While little can be done to alleviate a viral disease such as influenza except to treat symptoms, the bacterial complications are amenable to treatment with antibiotics. A pandemic such as in 1918–19 would be unlikely to have the same mortality rate today because many of these deaths were probably due to secondary bacterial infections.

An antiviral drug, *amantadine,* has been found to significantly reduce the symptoms of influenza if administered promptly. In prophylactic use, it apparently reduces the rate of infection and illness by perhaps as much as 70%. The mode of action is to inhibit the uncoating of type A viruses in the cell (it has no effect on type B viruses). The inhibition of uncoating interferes with the reproductive cycle of the virus.

FUNGAL DISEASES OF THE LOWER RESPIRATORY SYSTEM

Histoplasmosis

Histoplasmosis superficially resembles tuberculosis. In fact, it first became recognized as a disease in the United States when X ray surveys showed lung lesions in many persons who were tuberculin-test negative. Although the lungs are the most likely initial organs to be infected, the organisms cause lesions in almost all organs of the body. Normally, symptoms of the disease are rather ill defined and mostly subclinical, passing for minor respiratory infections. In a few cases, histoplasmosis becomes progressive and is a severe, generalized disease. This occurs in only a small number of infections, perhaps less than 0.1%.

The causative organism, *Histoplasma capsulatum,* is a dimorphic fungus, that is, it has a yeastlike morphology in tissue growth, and it forms a filamentous mycelium carrying reproductive conidia in soil or artificial media (Figure 23–11). In the body, the yeastlike form is found intracellularly in phagocytic cells.

While the disease is rather widespread throughout the world, it has a limited geographical range in the United States (Figure 23–12). For example, of 771 cases reported in 1978, 617 of these were from the states of Missouri and Indiana. In general, the disease is found in the states adjoining the Mississippi and Ohio rivers. More than 75% of the population in some of these states have antibodies against the infection. In other states, Maine, for example, a positive test is a rather rare event. Approximately 50 deaths are reported each year from histoplasmosis. Considering that the total numbers of infected persons may run into the millions, the death rate is low.

Humans acquire the disease from airborne conidia produced under certain moisture and pH conditions, particularly where droppings from birds have accumulated. Birds themselves do not have the disease, but their droppings provide nutrients, particularly a source of nitrogen, for the fungus.

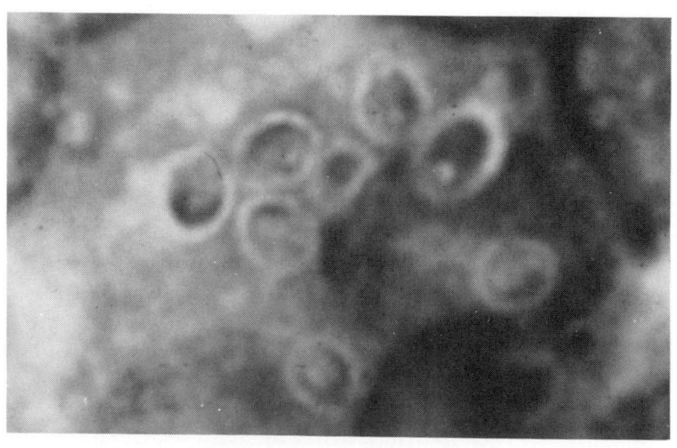
(a)

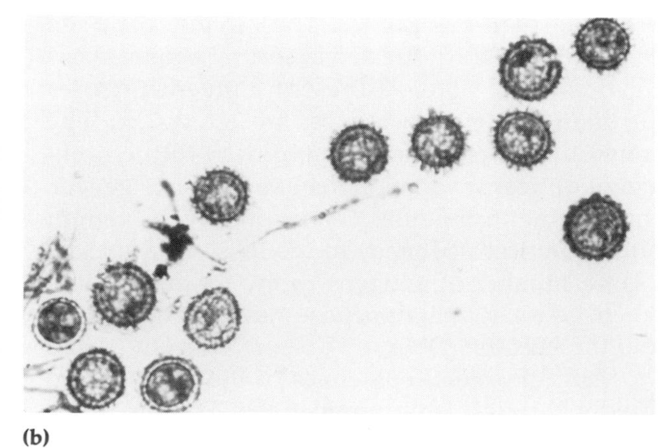
(b)

Figure 23–11 *Histoplasma capsulatum*, the dimorphic fungus that causes histoplasmosis. **(a)** Yeastlike form typical of growth in tissue. **(b)** Spores of the filamentous, spore-forming phase found in soil.

Figure 23–12 The dark area indicates the geographical distribution of histoplasmosis in the United States.

The *histoplasmin test*, a skin test similar to the tuberculin test, is useful in screening. However, there is some cross reaction with immune states caused by other fungal diseases.

Clinical signs and history, histoplasmin reaction, complement fixation tests, and isolation of the organism are important in proper diagnosis. The organism is found in the spleen and bone marrow, as well as in phagocytic cells. Currently, the most effective chemotherapy is with amphotericin B or ketoconazole and other orally active imidazoles. Less toxic than the polyenes like amphotericin B, the imidazoles have recently received FDA approval to be prescribed for systemic mycoses in the United States.

Coccidioidomycosis

Another fungal pulmonary disease, also rather restricted geographically, is **coccidioidomycosis.** The causative organism is *Coccidioides immitis*, a dimorphic fungus. The spores are found in dry, highly alkaline soils of the American Southwest and in similar regions of South America and northern Mexico. In tissue, the organism forms a thick-walled body filled with spores (Figure 23–13a). In soil, it forms filaments that reproduce by formation of arthrospores (Figure 23–13b). The arthrospores are carried by the wind to transmit the infection. Arthrospores are often so abundant that simply driving through an endemic area, particularly during a dust storm, can result in infection.

The symptoms of coccidioidomycosis include chest pain and perhaps fever, coughing, and loss of weight. Most infections are inapparent, and almost all cases resolve themselves by recovery in a few weeks, resulting in a sound immunity. However, in less than 1% of the infections, there is a progressive disease resembling tuberculosis that disseminates throughout the body. The resemblance to tuberculosis is so close that isolation of the organism is necessary to properly diagnose coccidioidomycosis. A tuberculinlike skin test is used in screening for cases. In highly endemic areas, more than half of those tested are found to be positive reactors. Farm workers are more likely than others to be infected, and predisposing factors such as fatigue and poor nutrition can lead to more severe disease. Amphotericin B is used to treat the disease.

Approximately 50 to 100 deaths each year are reported from this disease in the United States.

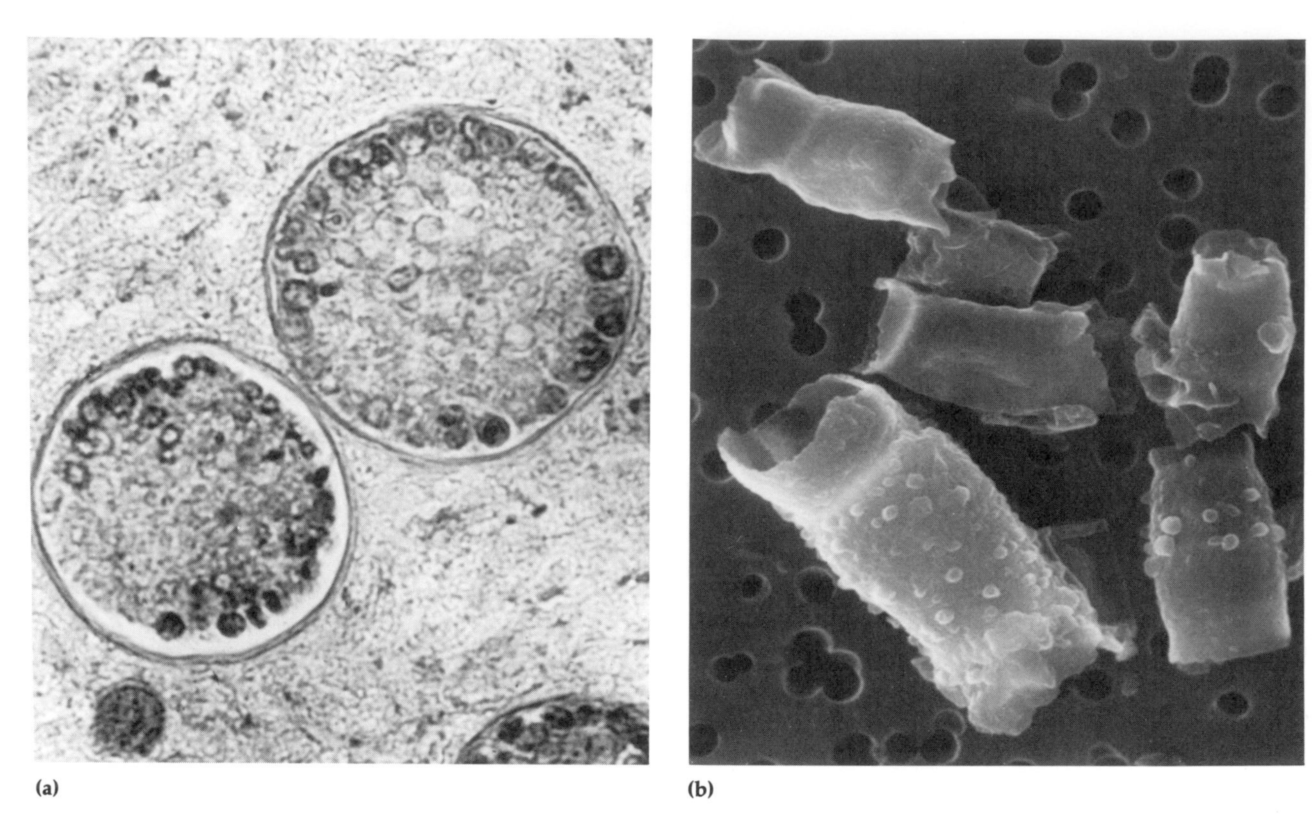

Figure 23–13 *Coccidioides immitis*. **(a)** Spore-filled spherical bodies in tissue. **(b)** Scanning electron micrograph showing infectious arthrospores from the soil that have been isolated on a Nuclepore filter ($\times 4500$).

Most cases are reported from southern California and Arizona desert regions, with only a scattering of cases reported from other states (Figure 23–14).

Blastomycosis (North American Blastomycosis)

Blastomycosis is usually called **North American blastomycosis** to differentiate it from a similar South American blastomycosis. It is caused by the fungus *Blastomyces dermatitidis*, a dimorphic fungus found most often in the Mississippi valley, where it probably grows in soil. Approximately 30 to 60 deaths are reported each year, although most infections are asymptomatic. The infection begins in the lungs and may spread rapidly. Cutaneous ulcers commonly appear and there is extensive abscess formation and tissue destruction. The organism can be isolated from pus and biopsy sections. Amphotericin B is usually an effective treatment.

Other Fungi Involved in Respiratory Disease

There are many other opportunistic fungi that may cause respiratory disease, particularly in immunosuppressed hosts or if there is exposure to massive numbers of spores. **Aspergillosis** is caused by the spores of *Aspergillus fumigatus* (see Color Plate IVB) and other species of *Aspergillus*, which are fairly widespread in decaying vegetation. Compost piles are ideal sites for growth, and farmers and gardeners are the ones most likely to be exposed to infective doses of such spores. Similar pulmonary infections occur with exposure of certain individuals to spores of mold genera such as *Rhizopus* or *Mucor*. Such diseases, particularly invasive infections of pulmonary aspergillosis, can be very dangerous. Predisposing factors include an impaired immune system, cancer, and diabetes. As with most systemic fungal infections, there is only a lim-

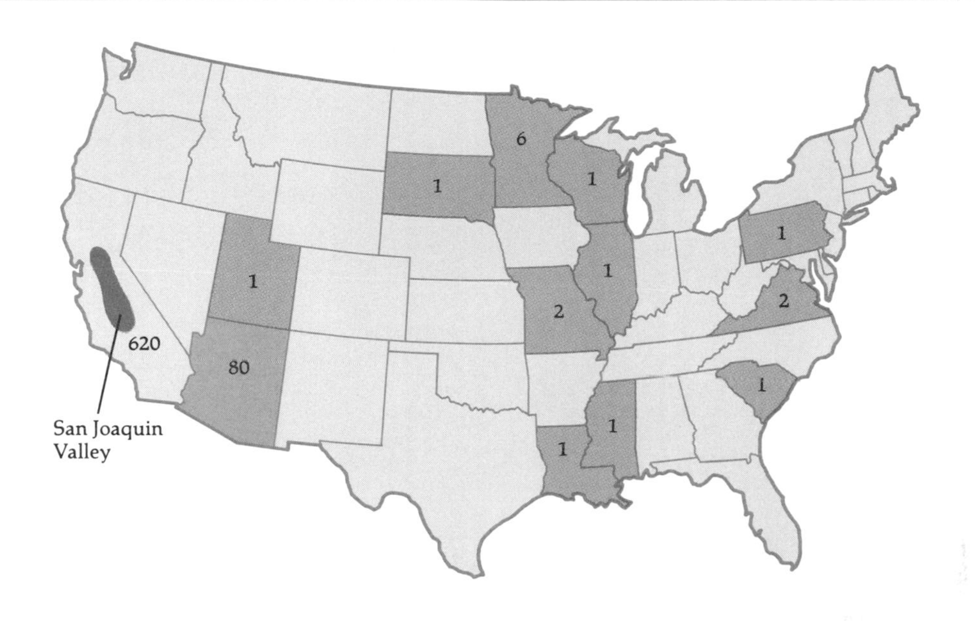

Figure 23–14 Cases of coccidioidomycosis in the United States, 1979. Because most of the cases occur in the San Joaquin Valley in California, this disease is also called San Joaquin Valley fever.

ited arsenal of antibiotics available; amphotericin B has proved the most useful.

PROTOZOAN AND HELMINTHIC DISEASES OF THE LOWER RESPIRATORY SYSTEM

Pneumocystis pneumonia is caused by the sporozoan *Pneumocystis carinii.* The disease occurs throughout the world and may be endemic in hospitals. The life cycle of this protozoan is not well known. However, outbreaks in hospitals suggest that the parasite may be transmitted by direct contact between humans. Patients receiving immunosuppressive drugs or who have leukemia are especially susceptible to infection. *Pneumocystis* causes the alveoli to become filled with a frothy exudate. Diagnosis is based on the recovery of *Pneumocystis* from the respiratory system. Untreated infections are usually fatal. Pentamidine is the drug of choice.

Humans are occasional intermediate hosts for the tapeworm *Echinococcus granulosus* (see Figure 11–22). **Hydatidosis** can remain asymptomatic until the cyst reaches a diameter of 10 cm or more. The cyst may rupture under pressure, and daughter cysts will then form. Infections are acquired by direct contact with dogs. The infections are often acquired in childhood but remain symptomless until adulthood. Allergic reactions to the cyst may occur. Other symptoms include a cough and the spitting of blood. The disease occurs in areas where dogs are used to herd sheep.

Paragonimiasis is caused by the lung fluke *Paragonimus westermanni* (see Figure 11–19), acquired from eating undercooked crayfish. Adult worms are found in fibrous cysts in humans. Symptoms include the spitting of blood, and the disease may resemble tuberculosis. Diagnosis is based on microscopic identification of adult flukes in surgical specimens or ova in sputum. A skin test and immunologic tests are also used in diagnosis. The disease is most common in the Far East but has been reported in the United States. Chloroquine has been used to treat infections.

See Table 23–1 for a summary of diseases associated with the respiratory system.

Table 23–1 Summary of Diseases Associated with the Respiratory System

Disease	Causative Agent	Mode of Transmission	Treatment
Bacterial Diseases of the Upper Respiratory System:			
Streptococcal sore throat ("strep" throat)	Streptococci, especially *Streptococcus pyogenes*	Respiratory secretions	Penicillin and erythromycin
Scarlet fever	Erythrogenic toxin-producing strains of *Streptococcus pyogenes*	Respiratory secretions	Penicillin and erythromycin
Diphtheria	*Corynebacterium diphtheriae*	Respiratory secretions, healthy carriers	Antitoxin and penicillin, tetracyclines, erythromycin
Cutaneous diphtheria	*Corynebacterium diphtheriae*	Respiratory secretions, healthy carriers, direct contact	Antitoxin and penicillin, tetracyclines, erythromycin
Otitis media	Several agents, especially *Staphylococcus aureus, Streptococcus pneumoniae,* beta-hemolytic streptococci, and *Hemophilus influenzae*	Complication of sore throat	Penicillin or erythromycin
Viral Disease of the Upper Respiratory System:			
Common cold	Coronaviruses (rhinoviruses)	Respiratory secretions	None
Bacterial Diseases of the Lower Respiratory System:			
Whooping cough	*Bordetella pertussis*	Respiratory secretions	Erythromycin, tetracyclines, chloramphenicol in severe cases; none in mild cases
Tuberculosis	*Mycobacterium tuberculosis*	Respiratory secretions; infrequently by food, especially milk	Isoniazid and rifampin
Pneumococcal pneumonia	*Streptococcus pneumoniae*	Healthy carriers; primarily a disease following viral respiratory infection or other stress	Penicillin
Klebsiella pneumonia	*Klebsiella pneumoniae*	Primarily a disease in debilitated hosts, e.g., alcoholics	Cephalosporins or gentamicin
Mycoplasmal pneumonia	*Mycoplasma pneumoniae*	Respiratory secretions probably	Tetracycline, erythromycin
Legionnaires' disease	*Legionella pneumophila*	Unknown; thought to be aerosols from contaminated air-conditioning cooling tower water	Erythromycin, rifampin
Psittacosis (Ornithosis)	*Chlamydia psittaci*	Animal reservoir: aerosols of dried droppings and other exudates of birds; person-to-person transmission is rare	Tetracyclines
Q fever	*Coxiella burnetii*	Animal reservoir: aerosols in dairy barns and similar places; unpasteurized milk	Tetracyclines
Acute epiglottitis	*Hemophilus influenzae*	Complication of sore throat	Ampicillin in combination with chloramphenicol started immediately
Other bacterial pneumonias	*Proteus, Serratia,* and *Pseudomonas* species; *Escherichia coli,* and other bacteria	Bacteria are opportunistic	Varies with agent

Table 23–1, continued

Disease	Causative Agent	Mode of Transmission	Treatment
Viral Diseases of the Lower Respiratory System:			
Viral pneumonia	Several viruses	Aerosols; complication of other viral diseases	None
Influenza	Influenza viral strains; many serotypes	Respiratory secretions	Amantadine
Fungal Diseases of the Lower Respiratory System:			
Histoplasmosis	*Histoplasma capsulatum*	Animal reservoir: aerosols of dried bird droppings; person-to-person transmission is rare	Amphotericin B, ketoconazole
Coccidioidomycosis	*Coccidioides immitis*	Soil organism: aerosols of dust	Amphotericin B
Blastomycosis	*Blastomyces dermatitidis*	Soil organism: aerosols of dust	Amphotericin B
Other fungal pneumonias	Species of *Aspergillus, Rhizopus, Mucor,* and other genera	Aerosols of dust containing opportunistic fungi	Amphotericin B
Protozoan and Helminthic Diseases of the Lower Respiratory System:			
Pneumocystis pneumonia	*Pneumocystis carinii*	Direct contact (?)	Pentamidine
Hydatidosis	*Echinococcus granulosis*	Direct contact with dogs	Surgical removal of cysts
Paragonimiasis	*Paragonimus westermanni*	Ingestion of crayfish	Chloroquine

STUDY OUTLINE

INTRODUCTION (p. 577)

1. Infections of the upper respiratory system are the most commonly encountered type of infection.
2. Pathogens that enter the respiratory system can infect other parts of the body.
3. Respiratory infections are transmitted by direct contact, droplets, and fomites.

STRUCTURE AND FUNCTION OF THE RESPIRATORY SYSTEM (pp. 577–579)

1. The upper respiratory system consists of the nose, throat, and associated structures such as the middle ear and eustachian tubes.
2. Coarse hairs in the nose filter large particles from the air.
3. The ciliated mucous membrane of the nose and throat traps airborne particles and removes them from the body.
4. Lymphoid tissue, tonsils, and adenoids provide immunity to certain infections.

5. The lower respiratory system consists of the larynx, trachea, bronchial tubes, and alveoli.
6. The ciliary escalator of the lower respiratory system helps prevent microorganisms from reaching the lungs.
7. Microorganisms in the lungs may be phagocytized by alveolar macrophages.
8. Respiratory mucus contains IgA antibodies.

NORMAL FLORA OF THE RESPIRATORY SYSTEM (pp. 579–580)

1. Normal flora of the nasal cavity are usually gram-positive bacteria including diphtheroids, staphylococci, micrococci, and *Bacillus.*
2. The throat may contain *Streptococcus pneumoniae, Hemophilus influenzae,* and *Neisseria meningitidis.*
3. The lower respiratory system is usually sterile because of the action of the ciliary escalator.

BACTERIAL DISEASES OF THE UPPER RESPIRATORY SYSTEM (pp. 580–583)

Streptococcal Sore Throat ("Strep" Throat) (p. 580)

1. This infection is caused by beta-hemolytic group A streptococci, the group to which *Streptococcus pyogenes* belongs.
2. Symptoms of this infection are inflammation of the mucous membrane and fever; tonsillitis and otitis media may also occur.
3. Production of beta hemolysins and inhibition by bacitracin are used to identify this group of streptococci.
4. Immunity to streptococcal infections does not usually develop, in part because there are many serological types.
5. Streptococcal sore throat is usually transmitted by droplets, but has been associated with unpasteurized milk.

Scarlet Fever (p. 580)

1. Streptococcal sore throat caused by an erythrogenic toxin producing *S. pyogenes* results in scarlet fever.
2. *S. pyogenes* produces erythrogenic toxin when in lysogeny with a phage.
3. Symptoms include a pink rash, high fever, and a red, enlarged tongue.
4. Immunity to scarlet fever is measured by the Dick test, a skin test with erythrogenic toxin.

Diphtheria (pp. 581–582)

1. Diphtheria is caused by exotoxin-producing *Corynebacterium diphtheriae.*
2. Exotoxin is produced when the bacteria are lysogenized by a phage.

3. The exotoxin inhibits protein synthesis, which may result in heart, kidney, or nerve damage.
4. A membrane containing fibrin and dead human and bacterial cells forms in the throat and may block air passage.
5. Laboratory diagnosis is based on isolation of the bacteria and appearance of growth on differential media.
6. Toxigenicity is determined by the guinea pig test or gel diffusion test.
7. Antitoxin must be administered to neutralize the toxin, and antibiotics can stop growth of the bacteria.
8. Routine immunization in the United States includes diphtheria toxoid in the DPT vaccine.
9. The Schick test is a skin test used to determine immunity to diphtheria.

Cutaneous Diphtheria (p. 582)

1. Slow-healing skin ulcerations are characteristic of cutaneous diphtheria.
2. This disease is probably due to minimal dissemination of the exotoxin in the bloodstream.

Otitis Media (pp. 582–583)

1. Earache, or otitis media, occurs as a complication of nose and throat infections or directly through inoculation from an external source.
2. Pus accumulation causes pressure on the eardrum.
3. Bacterial causes include *Staphylococcus aureus*, *Streptococcus pneumoniae*, beta-hemolytic streptococci, and *Hemophilus influenzae*.

VIRAL DISEASE OF THE UPPER RESPIRATORY SYSTEM (p. 583)

Common Cold (Coryza) (p. 583)

1. The common cold is usually caused by coronaviruses (rhinoviruses).
2. These viruses contain single-stranded RNA and are helical and enveloped.
3. There are probably more than 90 serotypes.
4. Symptoms include sneezing, nasal secretions, and congestion.
5. Sinus infections, lower respiratory tract infections, laryngitis, and otitis media may occur as complications of a cold.
6. Mucus discharged during sneezing protects the virus from environmental stress before contact with a new host.
7. The viruses prefer temperatures slightly cooler than body temperature.
8. The incidence of colds increases during cold weather because of greater indoor contact or physiological changes.

BACTERIAL DISEASES OF THE LOWER RESPIRATORY SYSTEM (pp. 583–591)

Whooping Cough (Pertussis) (pp. 583–584)

1. Whooping cough is caused by *Bordetella pertussis.*
2. The initial stage of whooping cough resembles a cold and is called the catarrhal stage.
3. The proliferation of bacteria blocks the trachea and bronchi, causing deep coughs characteristic of the paroxysmal (second) stage.
4. The convalescence (third) stage may last for months.
5. Laboratory diagnosis is based on isolation of the bacteria on Bordet–Gengou medium followed by serological tests.
6. Regular immunization for children includes dead *B. pertussis* cells as part of the DPT vaccine.

Tuberculosis (pp. 584–587)

1. Tuberculosis is caused by *Mycobacterium tuberculosis.*
2. Large amounts of lipids in the cell wall account for the acid-fast characteristic as well as resistance to drying and disinfectants.
3. Lesions formed by *M. tuberculosis* are called tubercles; necrosis results in a caseous lesion that may calcify and appear in an X ray as a Ghon complex.
4. New foci of infection may develop when a caseous lesion ruptures and releases bacteria into blood or lymph vessels; this is called miliary tuberculosis.
5. Miliary tuberculosis is characterized by weight loss, coughing, and loss of vigor.
6. *M. tuberculosis* may be ingested by wandering macrophages; the bacteria reproduce in the macrophages and are liberated when the macrophages die.
7. The tuberculin skin test is used to determine whether a person has been exposed to tuberculosis.
8. A positive tuberculin skin test may indicate an active case of tuberculosis or prior exposure and immunity to the disease.
9. Laboratory diagnosis is based on isolation of the bacteria and requires up to eight weeks incubation.
10. *Mycobacterium bovis* causes bovine tuberculosis and can be transmitted to humans by unpasteurized milk.
11. *M. bovis* infections usually affect the bones or lymphatic system.
12. BCG vaccine for tuberculosis consists of a live, avirulent culture of *M. bovis.*

Pneumococcal Pneumonia (pp. 587–589)

1. Pneumococcal pneumonia is caused by encapsulated *Streptococcus pneumonia*.
2. Symptoms of this disease are fever, difficult breathing, chest pain, and rust-colored sputum.
3. Alveoli fill with red blood cells and edema fluid; the appearance of the lungs in an X ray can be used for diagnosis.
4. The bacteria can be identified by the production of alpha hemolysins, inhibition by optochin, and bile solubility.
5. A vaccine called Pneumovax consists of purified capsular material from 14 serotypes of *S. pneumonia*.

Klebsiella Pneumonia (p. 589)

1. *Klebsiella pneumoniae* causes *Klebsiella* pneumonia.
2. *Klebsiella* pneumonia results in lung abscesses and permanent lung damage.

Mycoplasmal Pneumonia (Primary Atypical Pneumonia) (pp. 589–590)

1. *Mycoplasma pneumoniae* causes primary atypical pneumonia, or mycoplasmal pneumonia.
2. *M. pneumoniae* produces small "fried egg" colonies after two weeks' incubation on enriched media containing horse serum and yeast extract.
3. A complement fixation test may be used to diagnose the disease based on a rising antibody titer.

Legionnaires' Disease (p. 590)

1. This disease is caused by the gram-negative rod *Legionella pneumophila*.
2. The bacterium can grow in water such as air-conditioning reservoirs and then be disseminated in the air.
3. This pneumonia does not appear to be transmitted from person to person.
4. Fluorescent-antibody tests are used for laboratory diagnosis.

Psittacosis (Ornithosis) (p. 590)

1. *Chlamydia psittaci* is transmitted by contact with contaminated droppings and exudates of fowl.
2. Elementary bodies allow the bacteria to survive outside a host.
3. The bacteria are isolated in embryonated eggs, mice, or cell culture; identification is based on fluorescent-antibody staining.
4. Commercial bird handlers are most susceptible to this disease.

Q Fever (p. 591)

1. Obligately parasitic, intracellular *Coxiella burnetii* causes Q fever.
2. The disease is usually transmitted to humans through unpasteurized milk or inhalation of aerosols in dairy barns.
3. Laboratory diagnosis is made by culturing the bacteria in embryonated eggs or cell culture.

Acute Epiglottitis (p. 591)

1. *Hemophilus influenzae* causes acute epiglottitis, which may result in obstruction of the voice box.
2. Immediate treatment involves a tracheostomy.

VIRAL DISEASES OF THE LOWER RESPIRATORY SYSTEM (pp. 591–594)

Viral Pneumonia (pp. 591–592)

1. Approximately 15% of pneumonia cases are thought to be due to viral etiology.
2. A number of viruses may cause pneumonia as a complication of infections such as influenza.
3. The etiologies are not usually identified in a clinical laboratory because of the difficulty in isolating and identifying viruses.

Influenza ("Flu") (pp. 592–594)

1. Influenza is caused by influenza virus and is characterized by chills, fever, headache, and general muscular aches.
2. Hemagglutinin (H) and neuraminidase (N) spikes project from the outer lipid bilayer of the virus.
3. Viral strains are identified by antigenic differences in the H and N spikes; they are also divided by antigenic differences in their protein coats (A, B, and C).
4. Antigenic shifts alter the antigenic nature of the H and N spikes, making natural immunity and vaccination of questionable value.
5. Deaths during an influenza epidemic are usually due to secondary bacterial infections.
6. Multivalent vaccines are available for the elderly and other high-risk groups.

FUNGAL DISEASES OF THE LOWER RESPIRATORY SYSTEM (pp. 594–597)

Histoplasmosis (pp. 594–595)

1. *Histoplasma capsulatum* causes a subclinical respiratory infection that only occasionally progresses to a severe, generalized disease.

2. The disease is acquired by inhalation of airborne conidia.
3. A skin test, the histoplasmin test, is used to determine actual cases and prior exposure to the disease.

Coccidioidomycosis (pp. 595–596)

1. Inhalation of the airborne arthrospores of *Coccidioides immitis* can result in coccidioidomycosis.
2. Most cases are subclinical, but predisposing factors such as fatigue and poor nutrition can lead to a progressive disease resembling tuberculosis.

Blastomycosis (North American Blastomycosis) (p. 596)

1. *Blastomyces dermatitidis* is the causative agent of blastomycosis.
2. The infection begins in the lungs and may spread to cause extensive abscesses.

Other Fungi Involved in Respiratory Disease (pp. 596–597)

1. Opportunistic fungi can cause respiratory disease in immunosuppressed hosts, especially when large numbers of spores are inhaled.
2. Among the fungi are *Aspergillus, Rhizopus,* and *Mucor.*

PROTOZOAN AND HELMINTHIC DISEASES OF THE LOWER RESPIRATORY SYSTEM (p. 597)

1. Immunosuppressed patients or patients with certain forms of cancer are susceptible to an endemic hospital sporozoan, *Pneumocystis carinii.*
2. Humans infected with the tapeworm *Echinococcus granulosus* may have hydatid cysts in their lungs.
3. The lung fluke *Paragonimus westermanni* is acquired from eating undercooked crayfish in endemic areas.

STUDY QUESTIONS

REVIEW

1. Describe how microorganisms are prevented from entering the upper respiratory system. How are they prevented from causing infections in the lower respiratory system?
2. Respiratory diseases are usually transmitted by ______________. List other methods by which they can be transmitted.
3. How does the normal flora of the respiratory system illustrate microbial antagonism?

4. Complete the following table.

Disease	Causative Agent	Symptoms	Treatment
Streptococcal sore throat			
Scarlet fever			
Diphtheria			
Whooping cough			
Tuberculosis			
Pneumococcal pneumonia			
Klebsiella pneumonia			
Legionnaires' disease			
Psittacosis			
Q fever			
Acute epiglottitis			

5. How is otitis media contracted? What causes otitis media? Why was otitis media included in a chapter on diseases of the respiratory system?

6. Provide the causative agent, symptoms, and treatment for three viral diseases of the respiratory system. Separate the diseases according to whether they infect the upper or lower respiratory system.

7. Compare and contrast primary atypical pneumonia and viral pneumonia.

8. A patient has been diagnosed as having pneumonia. Is this sufficient information to begin treatment with antimicrobial agents? Briefly discuss why or why not.

9. List the causative agent, method of transmission, and endemic area for the following fungal diseases: histoplasmosis; coccidioidomycosis; blastomycosis.

10. Under what conditions can the saprophytes *Aspergillus* and *Rhizopus* cause infections?

11. Why is the DPT vaccine routinely given to children in the United States? Of what does it consist?

12. Briefly describe the procedures and positive results of the following skin tests, indicating what a positive test reveals: Dick test; Schick test; tuberculin test; histoplasmin test.

13. Match the bacteria in question 4 to the following laboratory test results.

Gram-positive cocci
- beta-hemolytic, bacitracin inhibition ____________
- alpha-hemolytic, optochin inhibition ____________

Gram-positive rods
- not acid-fast ____________
- acid-fast ____________

Gram-negative rods
facultative anaerobes
require X and V factors ______________
X and V not required ______________
aerobes
rods ______________
coccobacillus ______________

14. Complete the following table.

Disease	Causative Agent	Stage of Parasite in Humans	Method of Transmission
Pneumocystis pneumonia			
Hydatidosis			
Paragonimiasis			

CHALLENGE

1. Differentiate between the following.
   (a) *S. pyogenes* causing "strep" throat and *S. pyogenes* causing scarlet fever.
   (b) Diphtheroids and *C. diphtheriae.*
2. Why is vaccination against influenza felt to be of questionable value?
3. Discuss reasons for the increased incidence of colds during cold weather.
4. Provide reasons for each of the following.
   (a) Penicillin should not be used indiscriminately.
   (b) A combination of antimicrobial drugs is used to treat tuberculosis.

FURTHER READING

Al-Doory, Y. 1975. *The Epidemiology of Human Mycotic Diseases.* Springfield, IL: C. C. Thomas. Includes histoplasmosis, coccidioidomycosis, and blastomycosis.

Fraser, D. W. and J. E. McDade. October 1979. Legionellosis. *Scientific American* 241:82–99. A summary of the story of the mysterious epidemic that led to the discovery of a new bacterium.

Kaplan, M. M. and R. G. Webster. December 1977. The epidemiology of influenza. *Scientific American* 237:88–106. The relationship between antigenic shift in the virus and influenza pandemics.

Wilkinson, H. W. 1978. Group B streptococcal infection in humans. *Annual Review of Microbiology* 32:41–57. An analysis of a relatively new group of human pathogens.

See also Further Reading for Part IV at the end of Chapter 19.

24

Microbial Diseases of the Digestive System

Objectives

After completing this chapter you should be able to

- Describe the antimicrobial features of the digestive system.
- Describe the events that lead to the formation of dental caries.
- Provide the etiologic agent, suspect foods, symptoms, and treatment for each of the following: botulism; staphylococcal food poisoning; salmonellosis; typhoid fever; bacillary dysentery (shigellosis); Asiatic cholera; gastroenteritis.
- List the etiologic agent, method of transmission, site of infection, and symptoms for mumps, cytomegalovirus inclusion disease, and infectious hepatitis.
- Compare and contrast giardiasis, balantidiasis, and amoebic dysentery.
- Discuss how most infections of the gastrointestinal system can be prevented.

The major sites of invasion of the body are the skin and the mucous membranes of the respiratory and digestive systems. The most common routes by which pathogens exit the body are also the respiratory and digestive systems. Many pathogens are excreted with feces as a result of digestive system diseases or diseases of other organs that are carried by the blood to the digestive system. Although various respiratory diseases are the leading cause of illness in the United States, diseases of the digestive system rank second.

Most diseases of the digestive system result from the ingestion of food and water that contain microorganisms and their toxins. Good sanitation practices by food handlers, including proper handwashing and fly control, prevent contamination of food. And modern methods of sewage treatment and disinfection of drinking water help break the fecal–oral cycle of disease. Preservation of foods by heat and refrigeration also minimizes contamination by disease microorganisms.

STRUCTURE AND FUNCTION OF THE DIGESTIVE SYSTEM

The **digestive system** may be conveniently divided into two principal groups of organs (Figure 24–1). The first group is referred to as the *gastrointestinal (GI) tract* or *alimentary canal,* essentially a tubelike structure that includes the mouth, pharynx (throat), esophagus (food

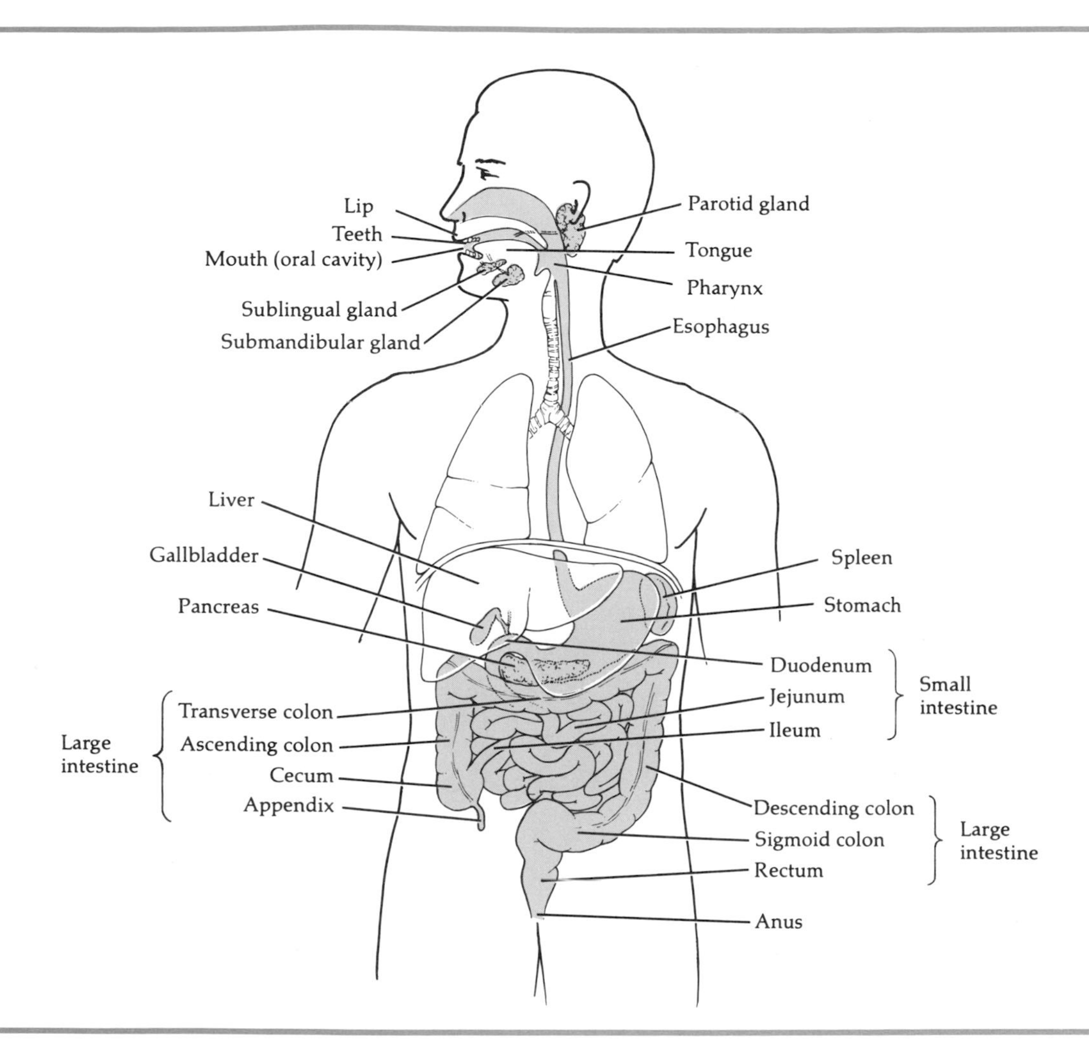

Figure 24–1 Anatomy of the human digestive system.

tube), stomach, small intestine, and large intestine. The second group of organs, the *accessory structures*, consists of the teeth, tongue, salivary glands, liver, gallbladder, and pancreas. Except for the teeth and tongue, the accessory structures lie outside the tract and produce secretions released into the tract by ducts.

The purpose of the digestive system is to digest foods, that is, to break them down into small molecules that can be taken up and used by body cells. The combined action of enzymes in saliva, stomach enzymes, pancreatic enzymes, intestinal enzymes, and bile from the gallbladder converts ingested nutrients into the end products of digestion. Thus, carbohydrates are reduced to simple sugars, fats to fatty acids and glycerol, and proteins to amino acids. These end products of digestion then pass from the small intestine into the blood or lymph, a process called *absorption*, for distribution to body cells.

By the time food leaves the small intestine, digestion and absorption are almost complete. As the food moves through the large intestine, water, vitamins, and nutrients are absorbed from it and microbial cells are added to it. The resulting undigested solids, called *feces*, are stored in the rectum until eliminated from the body through the anus, a process called *defecation*.

ANTIMICROBIAL FEATURES OF THE DIGESTIVE SYSTEM

Despite its intimate contact with food, the digestive system has several mechanisms to prevent infection. The digestive system, like the skin and respiratory system, is in contact with the external environment. And, like the skin and respiratory system, it has certain adaptations to protect it from microorganisms. First of all, saliva contains lysozyme, mucins, and IgA antibodies. Lysozyme exerts antimicrobial activity by bringing about the lysis of gram-positive bacteria. Both mucins and IgA antibodies can coat bacteria and prevent their attachment to the surfaces of the teeth, gums, and mucous membrane of the mouth. Second, some of the stomach cells produce hydrochloric acid (HCl), which destroys many types of microorganisms. Third, the small and large intestines contain patches of lymphoid tissue in their walls that provide immunity against certain microorganisms.

Finally, the normal flora of the intestines inhibits the growth of certain other microorganisms. Like normal flora of other parts of the body, the microorganisms normally found in the intestines use a number of strategies for outgrowing pathogens and other invaders. Some members of the normal flora secrete acid by-products of metabolism, thus creating an inhospitable, acidic environment for pathogens. Others combat invaders by more efficiently using the available nutrients. Some microorganisms of the flora also have shorter generation times, crowding out any newcomers.

Elaboration of highly specific growth inhibitors may also assume a role in maintaining the flora and checking the growth of pathogens. For example, certain strains of *E. coli* can produce a lethal inhibitory substance called *colicin* that will kill several strains of *E. coli* but has no effect on other strains of the same species. There are 22 different kinds of colicin labeled A to V. Similar substances are produced by bacteria other than enterics. Strains of *Pseudomonas aeruginosa* produce *pyocins*, and strains of *Bacillus megaterium* produce *megacins*. Together, colicins, pyocins, and megacins are collectively known as *bacteriocins*. The formation of a given bacteriocin is due to a corresponding plasmid called a *bacteriocinogen*. Bacteriocins are proteins that attach to specific outer membrane receptors of certain microorganisms. Some bacteriocins act by altering membrane permeability, while others inhibit protein synthesis after entering the cell. It is worth noting that if the delicate balance of normal flora is changed by antibiotic therapy, normally harmless microorganisms may become opportunistic pathogens.

NORMAL FLORA OF THE DIGESTIVE SYSTEM

Bacteria heavily populate most of the digestive system. The mouth may contain millions of bacteria in each milliliter of saliva, with various bacteria distributed in specific areas (Figure 24–2). For example, streptococci, which utilize sugars to form lactic acid, are found in different regions of the mouth.

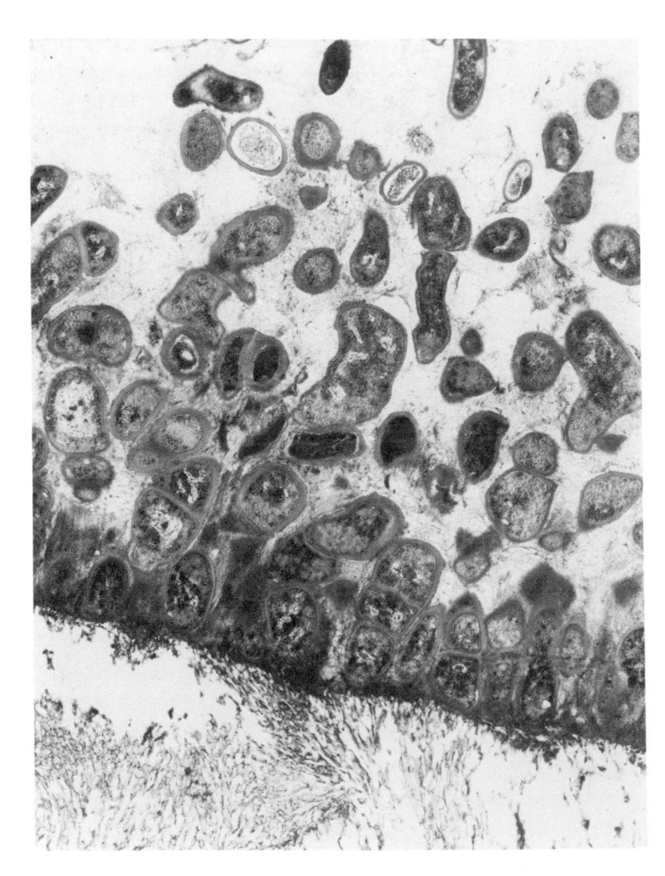

Figure 24–2 Normal flora of the mouth ($\times$11,000). This remarkable transmission electron micrograph shows a section through the enamel of a tooth and the microorganisms on its surface; no tooth decay or other oral disease was present in conjunction with this plaque (see Figure 24–5).

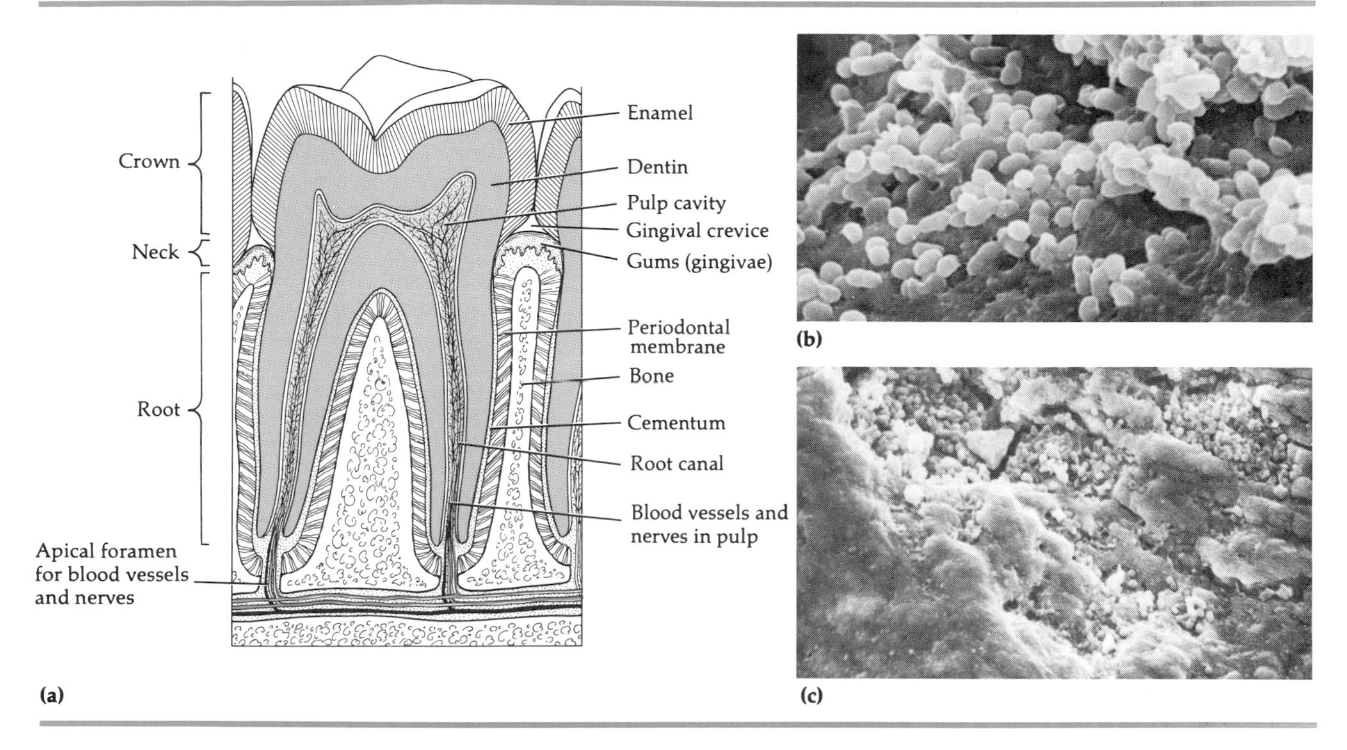

Figure 24–3 Dental caries. **(a)** Structure of a normal tooth. **(b)** Spherical cells of *Streptococcus mutans* adhering to tooth enamel (×3700). **(c)** Scanning electron micrograph showing carious lesion of tooth enamel with numerous cells of *Streptococcus mutans* in the eroded area (×1300).

Streptococcus salavarius colonizes the surface of the tongue, *S. mitis* primarily inhabits the mucosa of the cheek, and *S. sanguis* colonizes the teeth. Strict anaerobes, such as *Bacteroides* and *Fusobacterium,* and spirochetes inhabit the spaces where the gums and teeth merge.

The stomach and small intestine have relatively few microorganisms, a result of the hydrochloric acid produced by the stomach and the rapid movement of food through the small intestine. By contrast, microbial populations in the large intestine are enormous, exceeding 100 billion bacteria per gram of feces. (Up to 40% of fecal mass is microbial cell material.) The large intestinal population is mostly anaerobes of the genera *Lactobacillus* and *Bacteroides* and facultative anaerobes such as *E. coli, Enterobacter, Citrobacter,* and *Proteus.* Most of these bacteria assist in the enzymatic breakdown of foods, and some of them synthesize useful vitamins, such as niacin, vitamins B_1, B_2, B_6, and B_{12}, folic acid, biotin, and vitamin K.

BACTERIAL DISEASES OF THE DIGESTIVE SYSTEM

In the discussion of bacterial diseases of the digestive system that follows we will divide the diseases into those associated with the mouth and those associated with the lower digestive system.

Diseases of the Mouth

Dental caries (tooth decay)

Dental caries (tooth decay) involve a gradual softening of the enamel and dentin of a tooth (Figure 24–3). If the condition remains untreated, various bacteria may invade the pulp, causing death (necrosis) of the pulp and abscess of the bone surrounding the tooth. (Such a tooth must be treated by root canal therapy.) You may be surprised to learn that dental caries were not common until about the seventeenth century. In human remains from older times, only about 10% or less of teeth contain caries. The introduction of table sugar, or sucrose, into the diet is highly correlated with our present level of caries in the Western world. Oral bacteria convert sucrose and other carbohydrates into acid, which in turn attacks the tooth enamel.

Bacteria that convert sugar into lactic acid are commonplace in the oral cavity, but most of them are not important in tooth decay. For caries formation, there must be a localization of the acid production. If a single bacterial species can be said to be cariogenic (causing caries), that species is *Streptococcus mutans,* a gram-positive coccus (Figure 24–3b and c). The organisms (some feel it is a group of related organisms) become established in the mouth at the time the teeth erupt from the gums. They adhere to the teeth, especially in crevices and contact points between teeth, and not to the epithelial surfaces of the mouth.

The ability of *S. mutans* to adhere to the teeth is due to their production of a sticky polysaccharide, *dextran,* from sucrose. An enzyme on the surface of the bacterium catalyzes this synthesis: the sucrose is hydrolyzed into its component monosaccharides, glucose and fructose, and the energy released in the hydrolysis is used for the polymerization of the glucose to form dextran (Figure 24–4). The dextran forms a capsule around the bacterium and causes it to stick to the tooth. The masses of bacterial cells, dextran, and other polymers and debris adhering to teeth constitute **dental plaque** (Figure 24–5).

S. mutans ferments the fructose released from sucrose, as well as other sugars, to produce lactic acid and other acids. The acid environment also provides a suitable medium for other bacteria, such as lactobacilli, *Veillonella,* and *Actinomyces.* It is the localization of the acid production by plaque that initiates dental caries.

Studies with germfree (gnotobiotic) animals have shown that both sucrose and bacteria are necessary for tooth decay. Other carbohydrates, such as glucose or starch, are not converted to dextran and hence do not promote plaque formation or cause many caries. "Sugarless" chewing gums and candies contain sweet-tasting carbohydrates such as mannitol or sorbitol that are not metabolized by cariogenic bacteria to form dextran.

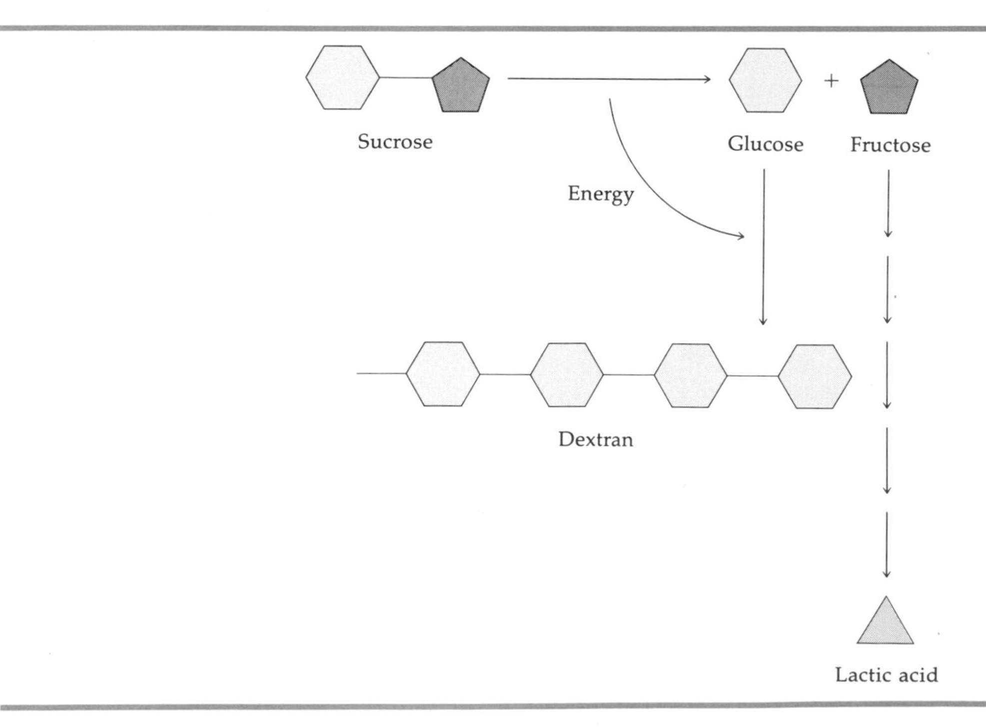

Figure 24–4 Conversion of sucrose to dextran and lactic acid by *Streptococcus mutans.* The sticky dextran holds the bacteria against tooth surfaces, and the lactic acid breaks down tooth enamel, leading to the formation of caries.

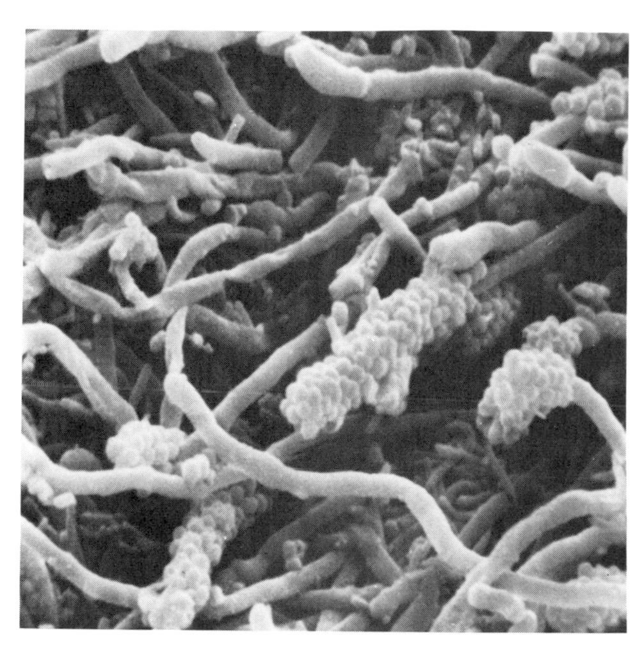

Figure 24–5 Dental plaque. These microorganisms have accumulated after three days without brushing.

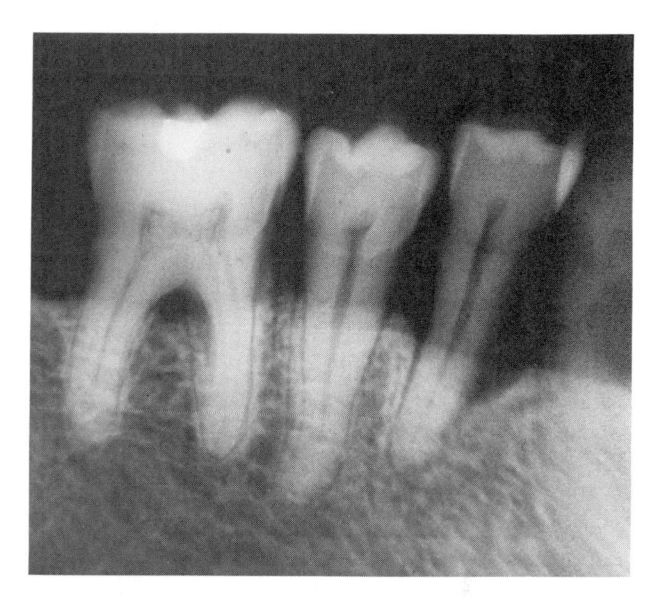

Figure 24–6 Radiograph of a severe case of periodontal disease. The bone has degenerated away from the upper portions of the two left teeth and away from all but the tip of the right tooth.

No drugs are now available to prevent tooth decay. A vaccine against caries, based on IgA antibodies in the saliva, is theoretically possible, and animal tests have given encouraging results. However, at present the main defense strategies in caries prevention are minimal ingestion of sucrose; brushing, flossing, and professional cleaning to remove plaque; and the use of fluorides to provide resistance of the teeth against acidic action.

Periodontal disease

Even persons who avoid tooth decay may, in later years, lose their teeth to **periodontal disease.** Periodontal disease is a collective term for a number of conditions characterized by inflammation and degeneration of the gums, supporting bone, periodontal ligament, and cementum (Figure 24–6). The disease is probably initiated by bacterial growth in the gingival crevice, although the exact mechanism is unclear. In any event, plaque deposited in the gingival crevice causes an initiating inflammatory reaction. Species of *Actinomyces, Nocardia,* and *Corynebacterium* (among other genera) are involved in this plaque formation. This can eventually lead to formation of abscesses, bone destruction, and general tissue necrosis.

Some periodontal disease can be controlled by frequent removal of plaque by brushing, flossing, and professional cleaning. Other forms of the disease may require antibiotics such as penicillin and tetracycline.

Diseases of the Lower Digestive System

In this section, we will consider diseases of the lower digestive system that are produced from two different mechanisms. In the first mechanism, called *bacterial intoxication,* the disease results from the consumption of food or drink contaminated by microbial growth. Because large numbers of microorganisms and their toxins are present initially, symptoms occur quickly, usually between 1 and 48 hours after ingestion. In the second mechanism, called *bacterial infection,* the disease results from microbial growth in the tissues of the host.

Botulism

One form of bacterial intoxication is **botulism.** It is caused by *Clostridium botulinum,* an obligately anaerobic, endospore-forming, gram-positive rod that is found in soil and in the sediments of many freshwater bodies. The microorganism produces an extremely potent exotoxin, a neurotoxin that blocks the transmission of the nerve impulses across synapses. The toxin is highly specific for the synaptic end of the nerve, where it blocks the release of acetylcholine, a chemical necessary for nerve activity. Persons suffering from botulism undergo a progressively flaccid paralysis for 1 to 10 days that may culminate in death from respiratory and cardiac failure. Nausea, but no fever, may precede the neurological symptoms. The initial neurological symptoms vary, but nearly all sufferers have double or blurred vision. Other symptoms include difficulty in swallowing and general weakness. Incubation time varies, but symptoms typically appear within a day or two.

Botulism was first described as a clinical disease in the early 1800s, when it was known as the sausage disease (*botulus*, meaning "sausage"). Blood sausage was made by filling a pig intestine with blood and ground meats, tying shut all the openings, boiling it for a short time, and smoking it over a wood smoke fire. The sausage was then stored at room temperature. Its preparation involved most of the requirements for an outbreak of botulism: an attempt at preservation that kills competing bacteria but allows survival of more heat-stable botulinum endospores, anaerobic conditions, and an incubation period for toxin production. Most botulism results from attempts at preservation that fail to eliminate the *C. botulinum* endospore. Foodborne botulism is not a common disease: only 65 cases in the United States were reported to the Centers for Disease Control in 1978 and only 12 in 1979.

There are several serological types of the toxin, produced by different strains of the pathogen, which differ considerably in their virulence and other factors. *Type A* toxin is probably the most virulent. There have been deaths from type A toxin in which the food was only tasted without swallowing. It is even possible to absorb lethal doses through skin breaks while handling laboratory samples. Without treatment, there is a 60 to 70% mortality rate. The endospores are also the most heat-resistant of all botulinum strains. In the United States, type A is found mainly in California, Washington, Colorado, Oregon, and New Mexico. Some eastern states have never had a type A outbreak. The type A organism is usually proteolytic (the breakdown of proteins by clostridia releases amines with unpleasant odors), but obvious spoilage odor is not always apparent in low-protein foods such as corn or beans.

Type B toxin is responsible for most European outbreaks of botulism and is the most common type in the eastern United States. The mortality rate, without treatment, is perhaps 25%. Type B botulism organisms occur in both proteolytic and nonproteolytic strains. Nonproteolytic strains are less likely to give warning by obvious spoilage.

Type E toxin is produced by botulism organisms that are often found in marine or lake sediments. Therefore, outbreaks often involve fish or other seafood and are particularly common in the Pacific Northwest, Alaska, and the Great Lakes area. The endospore of type E botulism is less heat-resistant and is usually destroyed by simple boiling. Types A and B will survive much higher temperatures. Type E is nonproteolytic, which minimizes chances of detecting spoilage in high-protein foods such as fish. It is also probably the only important food-borne pathogen capable of producing a toxin at refrigerator temperatures. Type E also requires less strictly anaerobic conditions for growth.

Eskimos and Coastal Indians in the Alaska area probably have the highest rate of botulism (mostly type E) in the world. The problem arises from attempts to preserve food without using precious fuel for heating or cooking. Cooking the food is important because all botulinal toxin is very heat labile and will be destroyed by most ordinary cooking methods that bring the food to a boiling temperature. The difficulty in getting prompt treatment for botulism among isolated ethnic groups is reflected in the mortality rate of 40% observed in the 1950 to 1973 period for type E.

Botulinal toxin is not formed in acid foods below pH 4.7. Foods such as tomatoes, therefore, may safely be preserved without the use of a pressure cooker. A concern that newer strains of tomatoes did not contain enough acid for reliable preservation by boiling does not seem to have been confirmed—but some recipes suggest addition of more acid. There have been cases of botulism in acidic foods that normally would not have supported the growth of the botulism organisms. Most of these episodes are related to mold growth that metabolized enough acid to allow the initiation of growth of the botulism organisms.

Sausage is rarely a cause of botulism today, largely because of the addition of nitrites to sausages and bacon. It was discovered that nitrites prevent botulinum growth following germination of the endospores. There is considerable research aimed at minimizing or eliminating this use of nitrites because nitrites are involved in the formation of carcinogenic (cancer-causing) nitrosamines.

Botulism can also occur as a result of the pathogen growing in wounds in a manner similar to clostridia causing tetanus or gas gangrene. Occasional episodes of **wound botulism** occur in this manner.

Because the botulism organisms do not seem to be able to compete successfully with the normal intestinal flora, growth and production of toxin is not generally an important factor in botulism in adults. However, in recent years, the ability of the botulism organisms to grow in the intestinal tract of infants and cause botulism has been confirmed. In 1977 and 1978, there were, respectively, 43 and 36 cases of **infant botulism** recognized. In 1979, there were 25 such cases reported. Infants have many opportunities to ingest soil and other materials contaminated with the endospores of the organism, but 30% of recent cases have been associated with honey. Endospores of *C. botulinum* are recovered with some frequency from honey, and the recommendation is to not feed honey to infants under one year of age (see box). There is no problem with older children or adults with a normally established intestinal flora.

The treatment of botulism relies heavily on supportive care. Antibiotics are of almost no use because the toxin is preformed. Antitoxins aimed at neutralization of toxins are available and are usually administered as a trivalent ABE type. The antitoxin will not affect the toxin already attached to the nerve ends and is probably most effective with type E, and less with types A and B. The toxin is quite firmly bound and recovery is slow. Extended

Centers for Disease Control

MMWR

Morbidity and Mortality Weekly Report

Honey Exposure and Infant Botulism—California

Of the 43 documented cases of infant botulism reported from California in 1977, 13 have had a history of ingestion of honey before the onset of constipation, the first symptom of most cases. Of foods fed to babies who developed infant botulism in California, only honey was found to contain *Clostridium botulinum* organisms. No honey specimens containing *C. botulinum* organisms contained preformed botulinal toxin. In three California cases, *C. botulinum* was isolated from honey fed the affected infants; in each case the infant had type B illness, and the honey sample contained type B organisms. In a fourth California case, no honey was available for culture; however, a jar of honey of the identical brand and size as that consumed by the patient, purchased at the market where the family shopped, contained type A botulism organisms. This case was type A botulism. Of over 60 honey specimens tested in California, about 13% have contained *C. botulinum.* This finding has been confirmed independently by four laboratories elsewhere in the United States. In two other states, *C. botulinum* type B was isolated from honey fed to two type B cases.

Since honey ingestion occurred in less than one-third of all California cases of infant botulism, development of infant botulism involves additional risk factors. However, since honey is not an essential food for infants, the California Department of Health concurs with the recent recommendation of the Sioux Honey Association that honey not be fed to infants under one year of age.

Editorial note: Infant botulism, a disease apparently resulting from intraintestinal toxin production by *Clostridium botulinum*, was first recognized as a distinct clinical entity in late 1976. Since then, cases have been identified with increasing frequency throughout the United States.

C. botulinum endospores are present in soil and on the surface of many vegetables. When vegetables are canned commercially, they are subjected to sufficient heat (≥123°C or ≥253.4°F) and pressure to destroy the botulism endospores. The repeated finding of botulinal organisms of the same type in infants with botulism and a history of honey ingestion, and the honey itself, suggests that honey may have been the source of infection for these and perhaps other infants. Although ingestion of honey was recorded in only a third of the cases and was therefore not the only risk factor, it appears prudent that honey not be recommended as a food for infants. The safety of honey as a food for older children and adults remains unquestioned.

Source: *MMWR* 27:249 (7/21/78).

respiratory assistance may be needed, and some neurological impairment may persist for months.

Botulism is diagnosed by inoculation of mice with samples from patient serum, stools, or vomitus specimens, which contain the toxin. Identification of the toxin in food can also be done by mouse inoculation. Different sets of mice are immunized with type A, B, or E antitoxins. All the mice are inoculated with the test toxin, and if, for example, those protected with type A antitoxin are the only survivors, the toxin is type A.

Staphylococcal food poisoning (staphylococcal enterotoxicosis)

A much more common form of bacterial intoxication food poisoning is caused by an exotoxin called an *enterotoxin* (*entero* refers to the intestine), produced by *Staphylococcus aureus.* This organism, we have seen, is a common inhabitant of the nasal passages, from which it often contaminates the hands. It is a gram-positive, catalase-positive coccus growing in grapelike clusters (see Figure 10–11a and Color Plate IA). *S. aureus* is a very common cause of opportunistic infections in humans, and certain strains produce a plasmid-coded enterotoxin that causes **staphylococcal food poisoning,** or **staphylococcal enterotoxicosis.** The disease is characterized by nausea, vomiting, and diarrhea one to six hours after ingesting contaminated food. The symptoms usually disappear in 24 hours. The disease has been identified in 25% of the outbreaks of food poisoning of known etiology in recent years. Although the mortality rate is almost nil, no immunity is developed.

S. aureus actually produces several toxins that damage tissues or increase the organism's virulence. The enterotoxins causing food poisoning are classified as serological types A (which is responsible for most cases) through D. Production of this toxin is usually correlated with production of an enzyme that coagulates blood plasma. Such bacteria are described as *coagulase-positive* (see Chapter 10). There is no direct pathogenic effect attributable to the enzyme, but it is useful in tentatively identifying likely virulent types. Coagulase-positive strains are considered potentially pathogenic.

Staphylococci are comparatively resistant to environmental stresses, as we discussed in a previous chapter. Vegetative cells tolerate 60°C for half an hour, which is a fairly high resistance to heat. They are also resistant to drying, radiation, and high osmotic pressures, which helps them survive on skin surfaces and grow in foods in which high osmotic pressure or salts inhibit growth of competitors. The enterotoxin itself is very heat-resistant for a polypeptide and will withstand boiling temperatures for as long as 30 minutes.

Handling food with fingers contaminated with staphylococci will inoculate foods. Food handlers may also have staphylococcal lesions, such as boils on their hands or arms. Food inoculated in this manner, if left at room temperature for a few hours, will have toxin formed in it, and the toxin will not be destroyed by normal cooking temperatures.

Certain foods have an unusually high incidence of association with staphylococcal food poisoning. Custard and cream pies, for example, have a high osmotic pressure because of their sugar content, have been heated enough to eliminate most competing organisms, and are readily inoculated with *S. aureus* by handling. Another example is ham, which contains many salts as preservatives or flavoring. These curing agents inhibit competing organisms more than staphylococci. Poultry products are too often handled and allowed to stand at room temperatures—but in fact, almost every food has been incriminated in staphylococcal food poisoning outbreaks at one time or another. On the other hand, foods such as hamburger are rarely associated with any type of food poisoning because staphylococci are not competitive with the normal flora in such foods, and also because they are cooked immediately before consumption. Picnic foods that are prepared early in the day, are not kept chilled, and are not eaten until later are a potential source of staphylococcal food poisoning.

Staphylococcal food poisoning is usually diagnosed by the symptoms, particularly the short incubation time characteristic of an intoxication. If the food has not been reheated and the bacteria killed, the organism can be recovered and grown. To produce enough toxin, a bacterial population of about a million per gram of food is required. *S. aureus* isolates can be tested by *phage typing,* which is useful in tracing the source of the contamination (Chapter 9). Because the bacteria will grow well at 10% sodium chloride, this is often used in selective media for their identification. Pathogenic staphylococci usually ferment mannitol, produce blood hemolysins, coagulase, and form yellow colonies. They cause no obvious spoilage when growing in foods. There is no simple procedure for identification of the toxin in foods. It requires a complex extraction and serological testing against antisera, procedures that are not commercially available.

Because contamination of foods cannot be avoided completely, the most reliable method of preventing staphylococcal food poisoning is adequate refrigeration during storage to prevent toxin formation.

Since considerable water and electrolytes are lost during episodes of vomiting and diarrhea, treatment for the disease consists of replacing the lost water and electrolytes.

Salmonellosis (*Salmonella* gastroenteritis)

As noted earlier, in bacterial infections the disease results from microbial growth in body tissues rather than the ingestion of food or drink already contaminated by microbial growth. Compared with bacterial intoxications, bacterial infections, such as salmonellosis, are usually characterized by longer incubation periods (from 12 hours to 2 weeks), needed for the growth of the organism in the tissues of the host, and by some fever indicative of the host's response to the infection.

The *Salmonella* bacteria (named for a Dr. Salmon, nothing to do with the fish) are gram-negative, non-endospore-forming, usually motile rods that ferment glucose, producing acid and gas. Their normal habitat is the intestinal tracts of humans and many animals. All salmonellae are considered pathogenic to one degree or another. The relatively mild infection that is known as **salmonellosis,** *Salmonella* **gastroenteritis,** or *Salmonella* **food infection,** is caused by organisms that require an infective dose usually in the millions. The nomenclature of the *Salmonella* organisms is confusing, with several schemes in existence. In addition to a few recognized species, there are hundreds of serotypes, either named or with combinations of numbers and letters referring to antigens of the capsules, flagella, or the cell itself (somatic).

The disease has an incubation time of about 12 to 36 hours. As the infecting bacteria eventually die, they autolyse, and their cellular endotoxins (see Chapter 14) are released. There is usually a moderate fever accompanied by nausea, abdominal pain and cramps, and diarrhea. The mortality rate is, overall, very low, probably less than 1%. However, the death rate is much higher in infants and among the very old, usually due to septicemia. There is a good deal of individual variation in response to the infection. The severity and incubation time may depend on the number of *Salmonella* ingested. Normally, recovery will be complete in a few days, but a number of recovered persons can be expected to become carriers.

Salmonellosis is probably greatly underreported. The 20,000 to 30,000 cases reported each year are probably only 1 to 10% of those that actually occur.

Meat products are particularly susceptible to contamination by *Salmonella,* and if mishandled, growth to infective levels occurs rather quickly. The sources of the bacteria are the intestinal tracts of many animals, and meats can be contaminated readily in processing plants. Poultry, eggs, and egg products are often involved in salmonellosis. The organisms are generally destroyed by normal cooking that heats the food to an internal temperature of about 68°C (145°F). However, foods can be contaminated and mishandled after cooking.

Prevention of salmonellosis depends on good sanitation practices to avoid contamination and proper refrigeration to avoid increases in bacterial numbers. Bear in mind that contaminated food may also contaminate a surface such as a cutting board with *Salmonella.* The original food prepared on the board may be cooked, but a subsequent food prepared on the board, such as a salad, may not be.

Diagnosis generally depends on isolation of the organisms from leftover foods or patient stools. Antibiotic therapy is not usually useful in salmonellosis. Treatment consists of replacing water and electrolytes lost through diarrhea.

Typhoid fever

A few species of *Salmonella* are much more virulent than others and produce an intestinal disease in which they cross the intestinal wall and become invasive. The most virulent species, *Salmonella typhi,* causes the bacterial infection **typhoid fever.** The incubation period is normally about two weeks, much longer than salmonellosis. Diarrhea is usually absent, but fever and malaise lasting for two or three weeks are typical. In severe cases there may be perforation of the intestinal wall. The organism becomes disseminated in the body and can be isolated from the blood, urine, and feces. Before the days of proper sewage disposal, water treatment, and food sanitation, typhoid was an extremely common disease.

A substantial number of recovered patients become carriers, harboring the pathogen in the gallbladder, and continue to shed the pathogen for several months. A certain number of such carriers will shed the organism indefinitely. Most of us are familiar with the term "Typhoid Mary." This was Mary Mallon, who worked as a cook in New York state in the early part of the century and was responsible for several outbreaks of typhoid and three deaths. Her case became well known through the attempts of the state to restrain her from working at her chosen trade. In 1979, there were 71 typhoid carriers listed with the Centers for Disease Control. In the same year, there were 528 new cases of typhoid in the United States. More than half of the cases in recent years have been acquired during foreign travel. Many of those who acquire typhoid in the United States are migrant workers and others using unsanitary facilities. Normally, there are fewer than 10 deaths each year.

Despite its known toxicity, chloramphenicol has often been the drug of choice for treating typhoid fever. However, drug sensitivity testing is required because of the appearance of resistant strains. Ampicillin and trimethoprim-sulfamethoxazole are usually effective alternatives. Immunization for typhoid is not normally done except for high-risk laboratory and military personnel.

Bacillary dysentery (shigellosis)

Bacillary dysentery (shigellosis) is a severe form of diarrhea that is characterized by the appearance of mucus and blood in the stools. Symptoms include abdominal cramps and fever. Bacillary dysentery (*dys+enteron*, meaning "sick gut") is a bacterial infection caused by a group of gram-negative rods of the genus *Shigella*. These intestinal bacteria of humans or higher primates are not as invasive as the salmonellae, and *Shigella* infections are usually limited to the large intestine.

There are four species of pathogenic *Shigella: S. sonnei, S. dysenteriae, S. flexneri*, and *S. boydii*. *S. sonnei*, the most common in the United States, causes a relatively mild dysentery. At the other extreme, *S. dysenteriae* infection results in a severe dysentery and prostration. Ulcerations are formed in the intestinal mucosa, which eventually heal with the formation of scar tissue. Although some strains of *S. dysenteriae* produce an endotoxin (neurotoxin), its role in pathogenesis is still not clear. It is known, however, that toxin production alone is not sufficient to cause virulence; invasiveness seems to be the more important factor. Based on tests with human volunteers, only a few hundred cells will suffice to cause shigellosis.

Diagnosis is usually based on recovery of the organisms from rectal swabs. Differentiation from amoebic (protozoan) dysentery, which is discussed later, is usually made from examination of the feces. Leukocytes are more likely to be present in high numbers in bacillary dysentery.

In recent years, the number of cases of bacillary dysentery reported in the United States has been about 15,000 to 20,000, with 20 to 35 (around 0.2%) deaths. However, the death rate in certain tropical areas may be much higher, perhaps 20%. The disease is probably more common than reported. Some cases of so-called tourist diarrhea may be milder forms of bacillary dysentery. Bacillary dysentery is a particular problem with institutionalized patients and is also common on Indian reservations. Some immunity seems to result from recovery, but no satisfactory vaccine has yet been developed.

In severe cases of bacillary dysentery, antibiotic therapy and fluid and electrolyte replacement are indicated. At present, ampicillin is the drug of choice, but some resistance has appeared. Chloramphenicol and trimethoprim-sulfamethoxazole are frequent alternatives. In view of the high rate of antibiotic resistance, a vaccine would be of value among high-risk groups.

Asiatic cholera

During the 1800s, the bacterial infection called **Asiatic cholera** crossed Europe and North America in repeated epidemics. Today, it is endemic in Asia, particularly India, with only occasional outbreaks in Western countries. These outbreaks are due to temporary lapses in sanitation, and the number of cases is limited. In the United States the disease is a rarity, although 11 cases occurred in Louisiana in

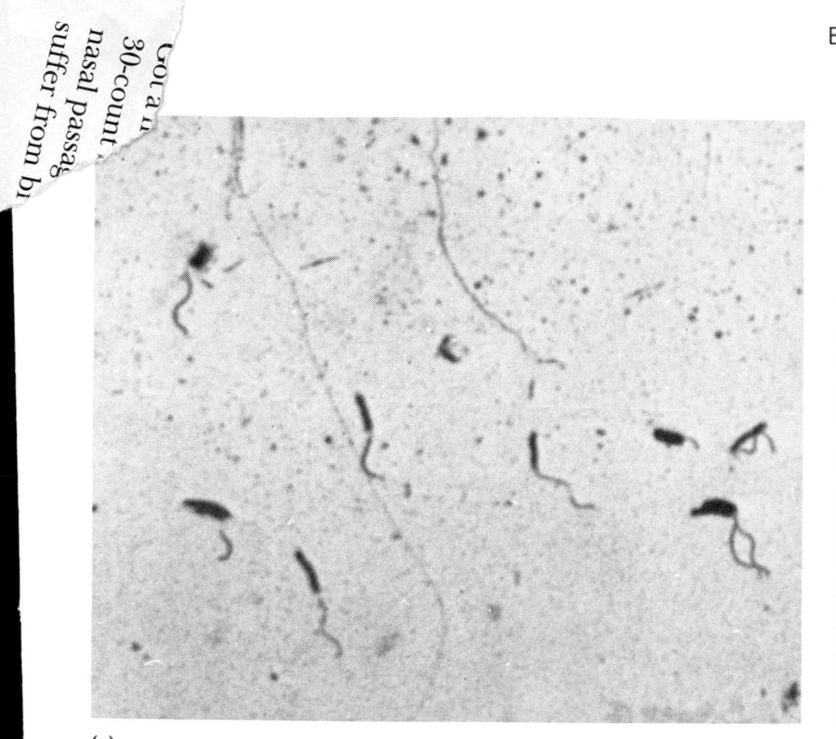

(a)

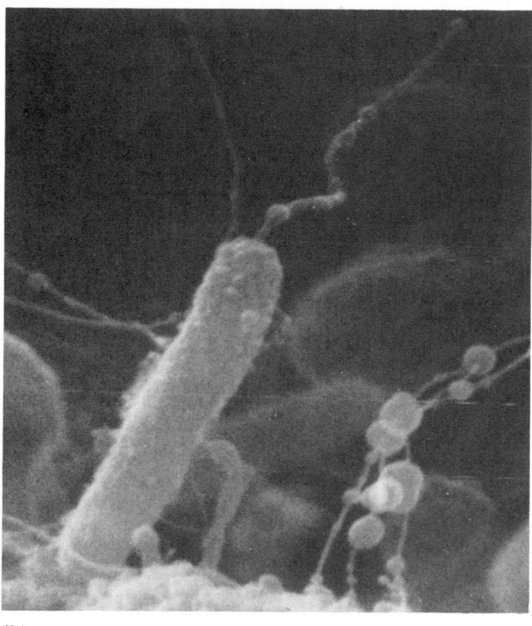

(b)

Figure 24–7 *Vibrio cholerae*, the causative agent of Asiatic cholera. The bacteria appear to vibrate when viewed microscopically in live cultures, the source of the genus name. **(a)** Light micrograph of cells of *V. cholerae* that have been treated with a flagella stain. **(b)** Scanning electron micrograph of a single cell of *V. cholerae* (×30,000).

1978. These were traced to the eating of crabs reared in marshes in that area. The causative organism, *Vibrio cholerae*, is a gram-negative rod, slightly curved, with a single flagellum (Figure 24–7). The disease, as with other enteric bacteria, is spread by the fecal-oral route, frequently from contaminated water. The incubation period is usually less than three days.

The organisms remain in the intestinal tract and are not invasive. They produce an exotoxin, called *enterotoxin*, which causes an extraordinary permeability of the intestinal wall. Body fluids and mineral electrolytes move through the wall and are excreted from the body. While semisolid stools may appear early in the disease, the excretions soon take on the typical appearance of "rice water stools." This appearance is due to the masses of intestinal mucus, epithelial cells, and bacteria. The sudden loss of these fluids and electrolytes (12 liters of fluid can easily be lost in one day) causes shock, collapse, and often death. As a result of the loss of fluid, the blood becomes so viscous that vital organs are unable to function properly. Violent vomiting is also a characteristic of the disease. The severity of the disease varies considerably, usually because of different strains of the pathogen. The number of subclinical cases may be several times those that result in recognized disease.

Isolation of the bacteria can be made readily from the feces and is aided by the ability of the organism to grow in media that are alkaline enough to suppress the growth of many other organisms.

The most reliable control methods are based on principles of sanitation, particularly of water supplies. A vaccine is available, but is only moderately effective, and then only for a few months. Recovery from the disease results in a much more effective immunity based on the antigenic activity of both the cells and the enterotoxin. However, because of antigenic differences among bacterial strains, the same person may have cholera more than once. Most victims in endemic areas are children.

Chemotherapy (tetracycline or chloramphenicol) is usually employed but is not particularly effective. Treatment is essentially based on replacement of the lost fluids and electrolytes. Untreated cases of cholera may have a 50% mortality rate, whereas the rate can be as low as 1% with proper supportive care.

Vibrio parahaemolyticus gastroenteritis

Vibrio parahaemolyticus was first recognized as a pathogen in Japan in 1950. It is morphologically similar to *V. cholerae,* but differs in that it is halophilic, requiring 2 to 7% sodium chloride for growth. It is the most common cause of **gastroenteritis** in Japan, resulting in thousands of reported cases annually. It is also a common cause of gastroenteritis in many nations of southeast Asia. And the organism is present in coastal waters of the continental United States and Hawaii. Crustaceans such as shrimp and crabs have been associated with several outbreaks caused by *V. parahaemolyticus* in the United States in recent years. Symptoms include a burning sensation in the stomach and abdominal pain. Vomiting and watery stools not unlike those of Asiatic cholera are also characteristic. The organism has also been known to cause cutaneous infections by cuts from contact with contaminated clams and oysters. Incubation time is normally less than 24 hours. Recovery usually follows in a few days and the fatality rate is low. Antibiotics such as tetracycline, chloramphenicol, and penicillin are used only in severe cases.

Since *V. parahaemolyticus* has a requirement for sodium, usually supplied as sodium chloride, isolating media containing 2 to 7% sodium chloride are used in diagnosis of the disease.

Escherichia coli gastroenteritis

Probably the most familiar bacterium to microbiologists is *Escherichia coli,* one of the most numerous microorganisms in the human intestinal tract. *E. coli* is normally harmless, but certain strains acquire the ability to cause gastroenteritis by one of three mechanisms. Some produce an enterotoxin that is plasmid coded. These are referred to as *enterotoxigenic E. coli (ETEC).* Others invade the epithelial lining of the large intestine and are called *enteroinvasive E. coli (EIEC).* Still others have a distinct but not understood mechanism and are known as *enteropathogenic E. coli.* They are the primary cause of **epidemic diarrhea** in nurseries, and probably the major cause of **traveler's diarrhea** acquired in foreign countries. In practice, it is difficult to differentiate between isolates of pathogenic and nonpathogenic *E. coli.* In adults , the disease is usually self-limiting, and chemotherapy is not attempted. In any event, antibiotic resistance is widespread in pathogenic strains. Traveler's diarrhea is treated by replacement of water and electrolytes. Infant diarrhea (epidemic diarrhea in nurseries) has been successfully treated with antibiotics. However, nurseries must maintain constant surveillance to administer antibiotics that are still effective.

Recently, an antibiotic called doxycycline has been used to treat traveler's diarrhea. The drug is long-lasting enough that a single dose each day will protect against diarrhea. The protection seems to last about a week after the drug is stopped. Unlike most antidiarrhea drugs, which are excreted through the kidneys, doxycycline is excreted through the large intestine. Unfortunately, the drug cannot be taken by children under the age of nine because it can permanently stain the teeth, or by pregnant women because of damage to the fetus.

Gastroenteritis caused by other gram-negative bacteria

In addition to the gram-negative organisms discussed so far, a number of others are known to cause gastrointestinal disorders in humans, and more are being discovered all the time. *Yersinia enterocolitica* and *Campylobacter fetus* (see Figure 4–2h) are two gram-negative rods whose importance as human enteric pathogens has only become apparent in recent years. They cause diarrhea and other gastrointestinal symptoms that vary in severity. Poultry (alive or dressed), dogs, raw milk, and contaminated water have all been implicated as sources of human infection by these bacteria.

Clostridium perfringens gastroenteritis

Probably one of the more common, if unrecognized, forms of food poisoning in the United States is caused by *Clostridium perfringens,* a large gram-positive, endospore-forming, obligately anaerobic rod (see Figure 10–12c). This is the same organism responsible for human gas gangrene. It produces a wide array of toxins that damage tissues and cause gastrointestinal disturbances.

Most outbreaks are associated with meats or stews containing meats. The organism has a nutritional requirement for amino acids that is met by such foods, and the meats are usually cooked, which lowers the oxygen level enough for clostridial growth. The endospores survive most heating, and the generation time of the vegetative bacterium is less than 20 minutes under ideal conditions. Large populations may therefore build up rapidly when foods are being held for serving, or when inadequate refrigeration leads to too-slow cooling.

The organism grows in the intestinal tract and produces an exotoxin that causes the typical symptoms of abdominal pain and diarrhea. Most cases are probably mild and self-limiting and thus never clinically diagnosed. The symptoms usually appear about 8 to 12 hours after ingestion. The toxin alters the permeability of the intestinal wall, and the resulting loss of water and electrolytes produces the diarrheal symptoms.

Diagnosis is usually accomplished by isolation and identification of the organism in stool samples. Treatment consists of replacing lost water and electrolytes.

VIRAL DISEASES OF THE DIGESTIVE SYSTEM

Mumps

The *parotid glands* are located just below and in front of the ear (see Figure 24–1). Although any of the salivary glands may become infected as a complication of nasopharyngeal infections, the parotid glands are a particular target of the *mumps virus.* The virus itself is a *paramyxovirus,* the same group to which the measles virus belongs. Morphologically, the mumps virus is an enveloped helical virus containing single-stranded, fragmented RNA (Figure 24–8).

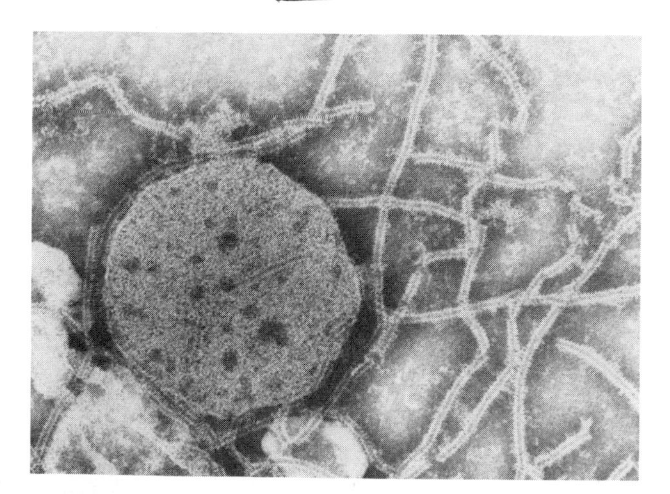

Figure 24–8 Transmission electron micrograph of a mumps virus; a single virus particle and released helical nucleocapsids are shown (×45,600).

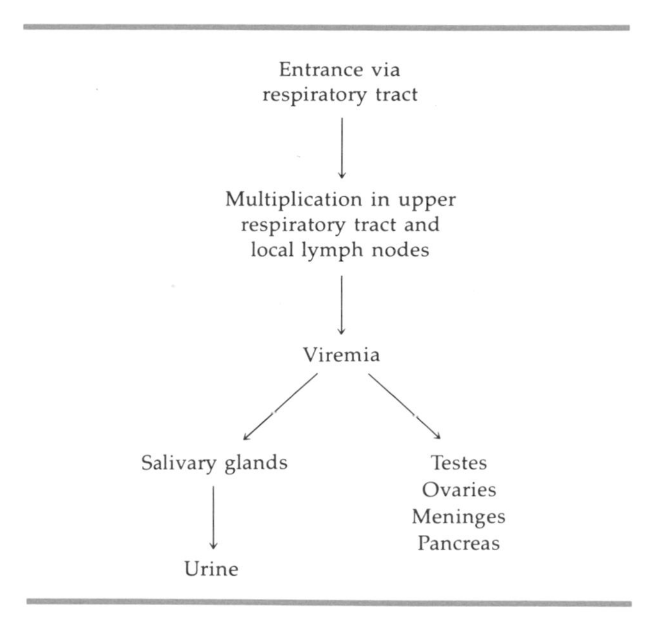

Figure 24–9 Pathogenesis of mumps.

Mumps typically begins with painful swelling of one or both parotid glands 16 to 18 days after exposure. The virus is transmitted in saliva and respiratory secretions, and its portal of entry is the respiratory tract. Once the viruses multiply in the respiratory tract and local lymph nodes in the neck, they reach the salivary glands via the blood (Figure 24–9). Viremia (viruses in the blood) begins several days before the onset of mumps and before the virus appears in saliva. The virus is present in the blood and saliva for 3 to 5 days after the onset of the disease, and the presence of the virus in the urine is common 10 days or so after the onset of the disease. Mumps is characterized by inflammation and swelling of the parotid glands, fever, and extreme pain during swallowing. About 4 to 7 days later, inflammation of the testes may occur (orchitis). This happens in about 20 to 35% of males past puberty. Sterility is a possible consequence, but rarely occurs. Other complications of mumps include meningitis, inflammation of the ovaries, and pancreatitis. Mumps is less common in children than chickenpox or measles because the disease is not as highly infectious, and so many children escape infection.

An effective vaccine is available, often administered as part of the trivalent measles–mumps–rubella (MMR) vaccine. The number of cases of mumps has dropped sharply since the introduction of the vaccine in 1968. For example, in 1971 there were 125,000 cases of mumps reported, whereas in 1978 there were a little under 17,000 cases. Second attacks are rare, and cases involving only one parotid gland, or, for that matter, subclinical cases, are as effective in conferring immunity as bilateral mumps.

Serological diagnosis is not usually necessary. But if laboratory confirmation of diagnosis based on symptoms is desired, the virus can be isolated by embryonated egg or cell culture techniques. The virus can be identified by hemagglutination-inhibition tests.

Cytomegalovirus (CMV) Inclusion Disease

The term *cytomegalovirus (CMV)* is applied to herpesviruses that produce a cellular response characterized by cytomegaly (large size) and the presence of intranuclear inclusion bodies. CMV is classified with herpesviruses because of its morphology (Figure 24–10a), chemical composition, and characteristic intranuclear inclusion body present in infected cells (Figure 24–10b). CMV is responsible for **cytomegalovirus (CMV) inclusion disease,** a severe, often fatal disease of newborns that usually affects the salivary glands, kidneys, lungs, and liver.

CMV commonly produces latent infections that may be activated by pregnancy, blood transfusions, or immunosuppression for transplanted organs. A mother with a latent infection may trans-

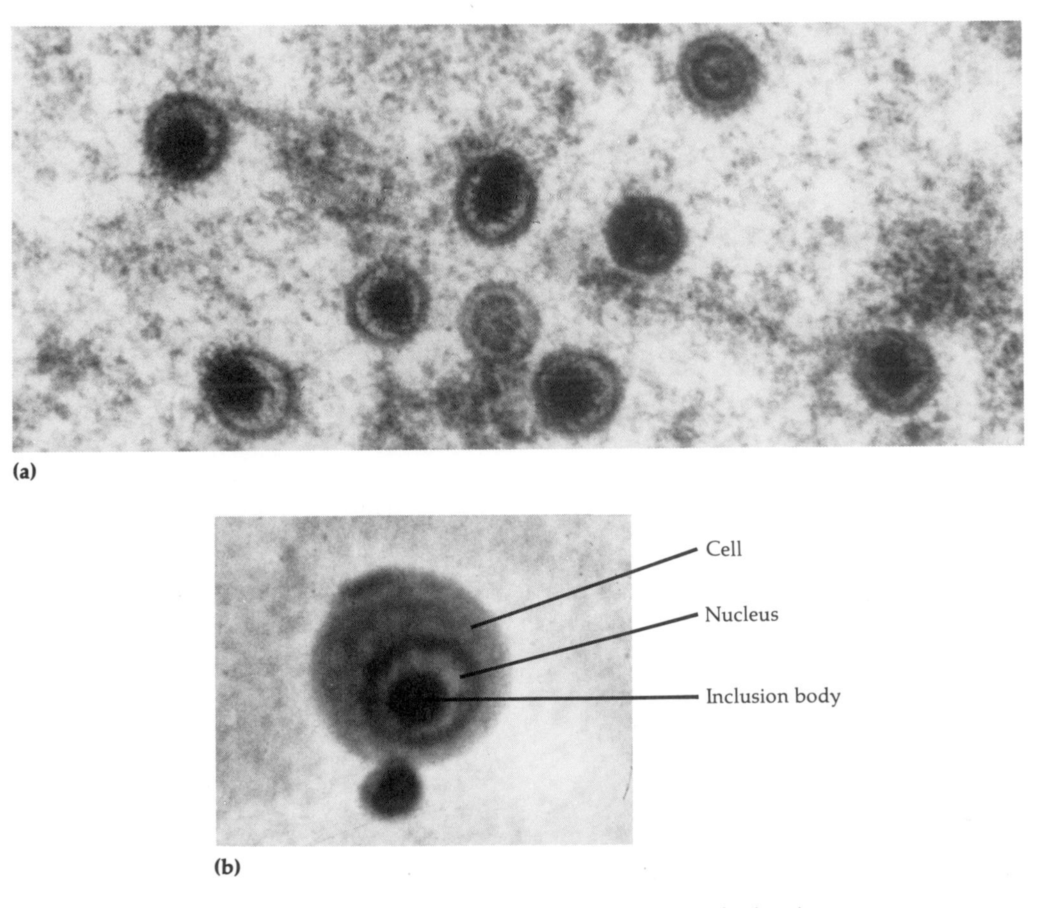

Figure 24–10 Cytomegalovirus (CMV). **(a)** Transmission electron micrograph showing individual particles in the cytoplasm of an infected cell (approx. ×110,000). **(b)** Light micrograph of a single cell infected with cytomegalovirus, showing the dark, intranuclear inclusion body.

mit CMV to the fetus via placental transfer or by excreting it into the genital tract at the time of birth. The virus may also become latent in the infant. When disease occurs in an infant up to 4 months of age, a rare occurrence, symptoms include liver and kidney malfunction, pneumonia, neurological complications, and eventual death. The cells of the various organs affected by CMV greatly increase in size, and the nucleus enlarges and contains a distinctive inclusion body (Figure 24–10b). When CMV infection occurs in older patients, it is associated with cancer patients, recipients of organ transplants receiving immunosuppressive therapy, and open heart surgery patients who receive large volumes of blood.

Laboratory diagnosis consists of the demonstration of intranuclear inclusion bodies and cytomegalic cells in specimens, isolation of CMV in tissue culture, and serological tests such as immunofluorescence, hemagglutination, and complement fixation.

Despite the relative rarity of active CMV inclusion disease, infection seems to be worldwide and common. From 10 to 18% of all stillborns show the characteristic CMV lesions, and antibodies appear in the blood of 53% of the population in the United States between 18 to 25 years of age and 81% of the population above 35 years of age. A vaccine for CMV inclusion disease is being studied.

Hepatitis A (Infectious Hepatitis)

The *hepatitis A virus (HAV)* is the etiological agent of a very important viral disease of the liver called **hepatitis A (infectious hepatitis).** Morphologically, the HAV resembles picornaviruses. It contains single-stranded RNA and lacks an envelope (Figure 24–11). HAV, hepatitis B virus (HBV), and non-A non-B hepatitis (NANB) virus all affect the liver but differ in their routes of transmission. Whereas HBV and NANB virus are primarily transmitted by contaminated needles (parenteral route), HAV is primarily transmitted by the fecal-oral route. However, both HAV and HBV viruses have been known to have been transmitted by either route. (Hepatitis B and NANB hepatitis were discussed in Chapter 22.)

After its typical entrance via the oral route, HAV multiplies in the epithelial lining of the intestinal tract. Viremia eventually occurs, and the virus spreads to the liver, kidney, and spleen. The virus is shed in the feces and can also be detected in the blood and urine. The danger of spreading HAV from an infected person is greatest when viruses are shed in the feces. Contamination of food or drink by the feces is favored by the resistance of HAV to disinfectants such as chlorine at ordinary concentrations in water. Moreover, shellfish that live in contaminated waters, such as oysters, are also a source of infection.

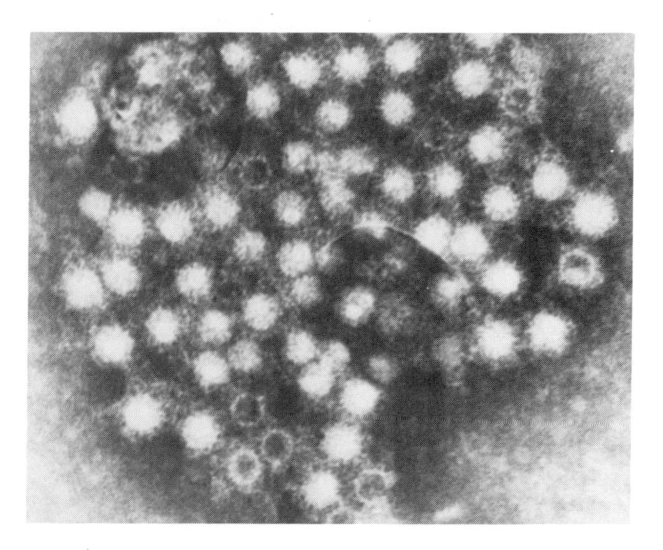

Figure 24–11 Hepatitis A viruses (× 163,000).

The initial symptoms of hepatitis A are anorexia (loss of appetite), general malaise, nausea, diarrhea, abdominal discomfort, fever, and chills. These symptoms may last from two days to three weeks. Eventually, jaundice, with typical yellowing of the skin due to liver infection, appears. At this point, the liver becomes tender and enlarged.

The mortality is low (less than 1%), and infections, particularly in children, are often unrecognized. Something like 45 to 75% of tested adults are found to have serum antibodies, yet few of these have ever been aware of a diagnosed illness. In 1978, there were approximately 30,000 reported cases of hepatitis A. The actual incidence has been estimated at half a million. Hepatitis A is generally a milder disease than hepatitis B. Most cases resolve in a month or six weeks, and recovery yields good immunity.

The hepatitis A virus or antibodies can be identified by serological tests, including complement fixation, immune adherence, and radioimmunoassay. No specific treatment for the disease exists. However, the incidence of hepatitis A may be decreased by passive immunization; that is, persons exposed to the disease can be given immune globulin from many individuals to avoid developing the disease. The incubation period of one to six weeks for hepatitis A is much shorter than that for hepatitis B, but it is still long enough to make epidemiological studies for the source of the infections difficult.

MYCOTOXINS

Some fungi produce toxins called **mycotoxins,** which cause blood diseases, nervous system disorders, kidney damage, and liver damage. Some mycotoxins are associated with poisonous mushrooms, such as *Amanita,* or *Claviceps purpurea* (see Figure 11–11). *Claviceps* causes *ergot poisoning,* which is caused by the ingestion of rye contaminated with the mycotoxin. A mycotoxin of current interest is called *aflatoxin.* It is produced by the fungus *Aspergillus flavus,* a common mold. Aflatoxin is highly toxic and can cause serious damage to livestock when their feed is contaminated with *A. flavus.* Although the risk to humans is unknown, there is strong circumstantial evidence that aflatoxin may contribute to cirrhosis of the liver and cancer of the liver in parts of the world, such as India and Africa, where food is subject to aflatoxin contamination.

PROTOZOAN DISEASES OF THE DIGESTIVE SYSTEM

Giardiasis

Giardia lamblia, a flagellated protozoan (Figure 24–12) that is able to attach firmly to the intestinal wall, is the cause of a prolonged diarrheal disease in humans called **giardiasis.** The disease, which sometimes persists for weeks, is also characterized by malaise, nausea, weakness, weight loss, and abdominal cramps. There are frequent outbreaks of giardiasis in the United States, and approximately 13,000 cases were reported in 1978. A substantial number of the population, perhaps as many as 20%, are carriers. These carriers are mostly asymptomatic.

The organism is also shed by a number of wild mammals, and the disease is not an uncommon infection of backpackers drinking from wilderness waters. Most outbreaks in the United States are transmitted by contaminated water supplies. And

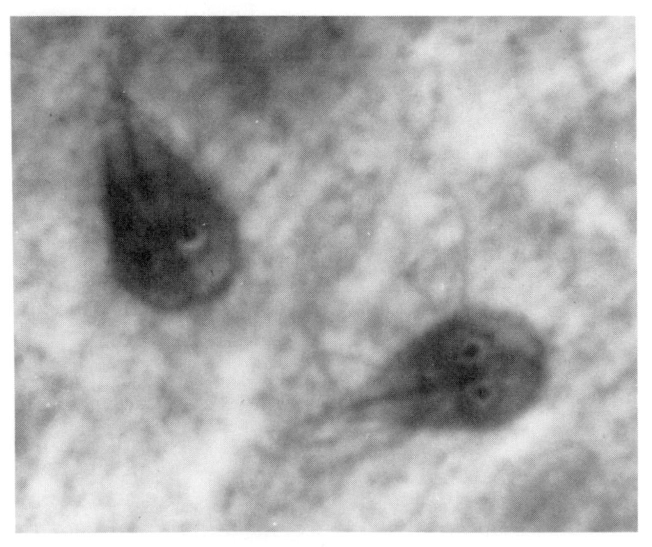

(a)

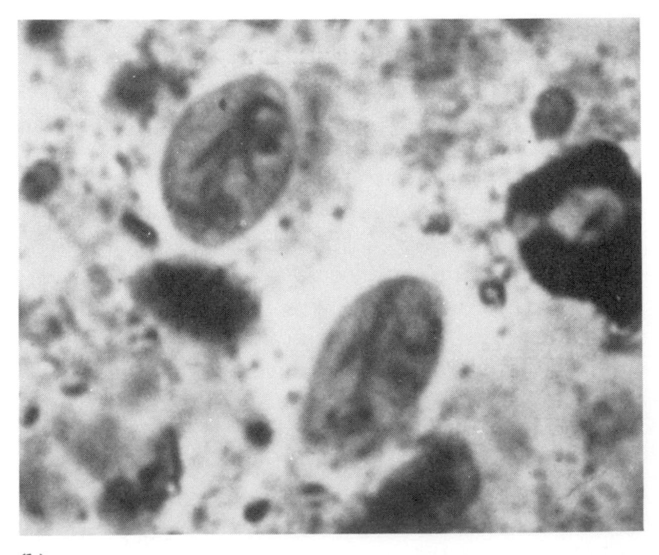

(b)

Figure 24–12 *Giardia lamblia,* a flagellated protozoan causing giardiasis. **(a)** Vegetative form (trophozoite). **(b)** Cyst form. See also Figure 11–15.

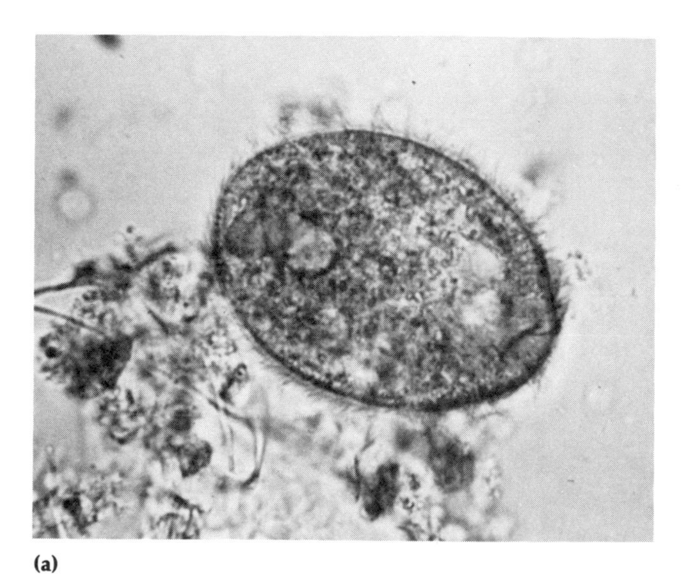
(a)

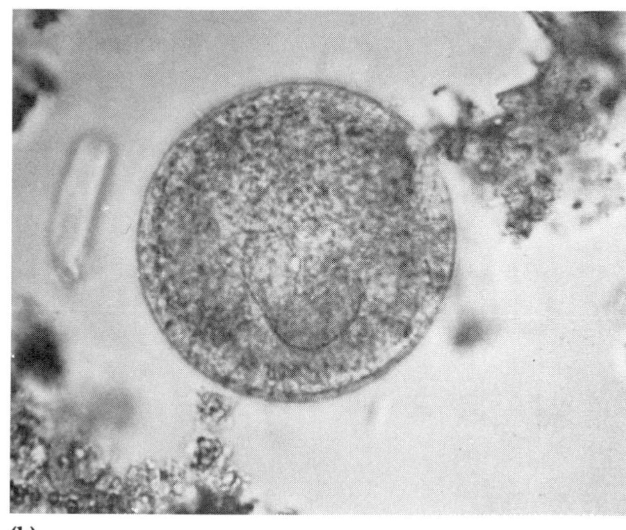
(b)

Figure 24–13 *Balantidium coli.* **(a)** Vegetative form (ciliated trophozoite). **(b)** Spherical cyst.

because the cyst stage of the protozoan is insensitive to chlorine, filtration of water supplies is usually necessary to eliminate the cysts from water.

Diagnosis of giardiasis is made by identification of the protozoan in the feces. Treatment by metronidazole (Flagyl) and quinacrine hydrochloride (Atabrine) have been effective.

Balantidiasis (Balantidial Dysentery)

Balantidium coli is a ciliated protozoan that causes **balantidiasis,** or **balantidial dysentery.** The organism is the only ciliate known to be pathogenic for humans and is the largest intestinal protozoan of humans. Like *Giardia lamblia, B. coli* exists in both a vegetative form (trophozoite) and a cyst form (Figure 24–13). *B. coli* lives in the large intestine, where it can, in rare cases, invade the epithelial lining, causing ulceration and fatal dysentery. Typically, the disease is mild, consisting of abdominal pain, nausea, vomiting, diarrhea, and weight loss.

Humans acquire *B. coli* by ingesting cysts from food and water contaminated by feces containing cysts. Following ingestion, the cysts descend to the colon, where the cyst walls dissolve and the released vegetative cells feed on bacteria, fecal debris, and the host's cells. As feces pass through the colon and are dehydrated, encystment occurs. Some cysts form after the feces are discharged. Subsequent food and water contamination initiates another cycle.

Laboratory diagnosis of balantidiasis is based on demonstrating vegetative cells in feces. Treatment consists of administering chlortetracycline or oxytetracycline, followed by diiodohydroxyquin, if necessary.

Amoebic Dysentery (Amoebiasis)

Amoebic dysentery (amoebiasis) is found worldwide, spread mostly by food or water contaminated by the protozoan amoeba *Entamoeba histolytica*. The cysts of the amoeba are ingested in contaminated food or water (Figure 24–14). Although stomach acid (HCl) can destroy vegetative cells, it has no effect on the cysts. In the intestinal tract, the cyst wall is digested away, releasing the vegetative forms. The vegetative forms then multiply in the epithelial cells of the large intestinal wall. A severe dysentery results, and the feces characteristically contain blood and mucus. The vegetative forms feed on red blood cells and cause tissue destruction in the gastrointestinal tract. Severe infections may result if perforation of the intestinal wall occurs.

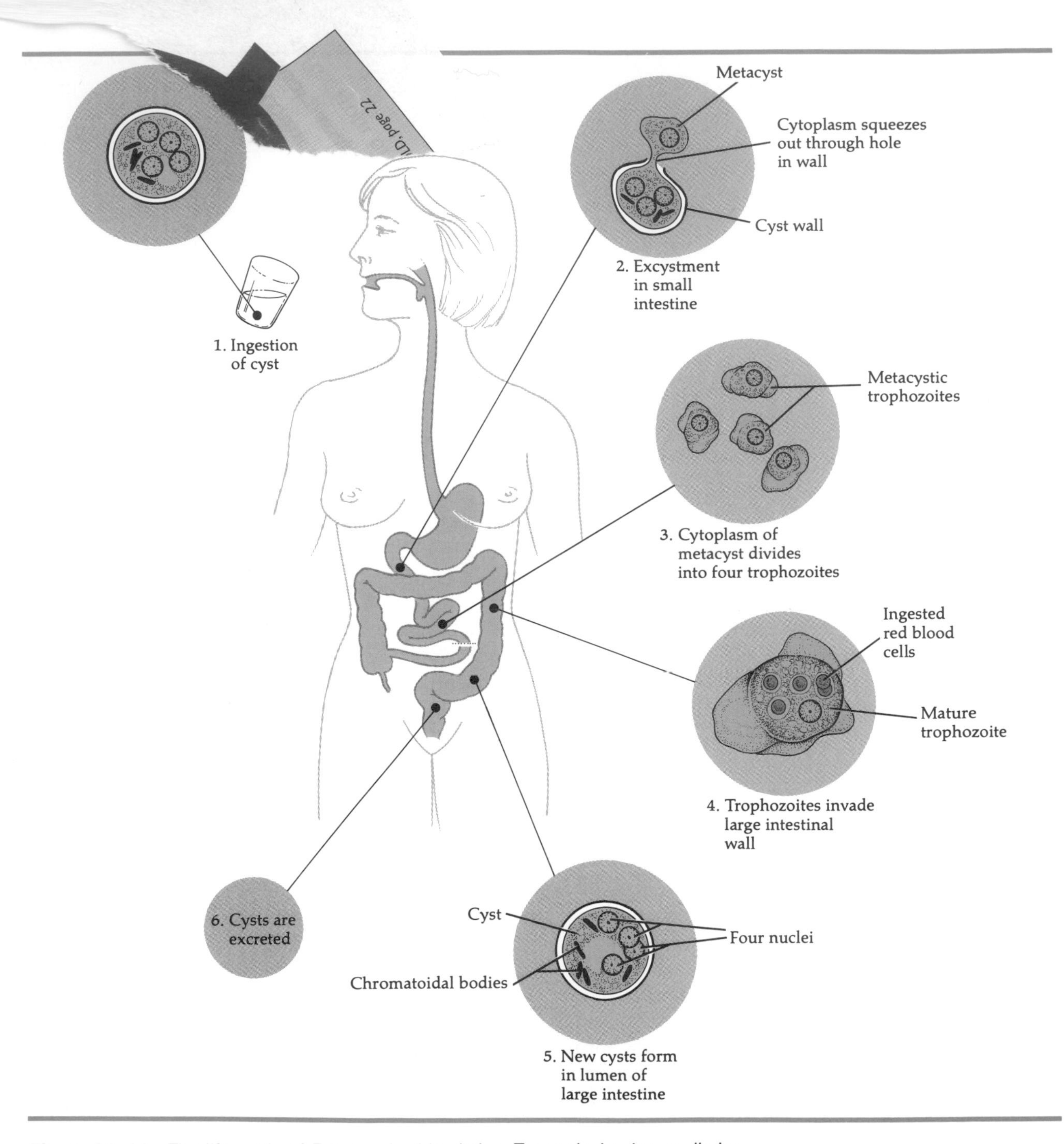

Figure 24–14 The life cycle of *Entamoeba histolytica*. Transmission is usually by ingestion of cysts. The trophozoite lives in the intestinal epithelial layer, feeding on red blood cells. New cysts are formed by trophozoites in the lumen of the large intestine.

Abscesses may have to be treated surgically, and invasion of other organs, particularly the liver, is not uncommon. In recent years, more than 3000 cases per year have been reported in the United States, with a mortality rate approaching 1%.

Diagnosis is largely dependent upon recovery and identification of the organisms (red blood cells observed within the trophozoite stage of an amoeba are considered an indication of *E. histolytica*). There are several serological tests that can also be used for diagnosis, including immunodiffusion and fluorescent-antibody tests. Metronidazole is the drug of choice, except for pregnant women, where it is contraindicated.

HELMINTHIC DISEASES OF THE DIGESTIVE SYSTEM

Tapeworm Infestation

Most **tapeworm infestations** of humans occur from eating undercooked beef, pork, or fish, intermediate hosts for the tapeworm. When forms of the tapeworm called cysticerci, encysted in muscles of the intermediate host, are ingested by a human, they develop into adult tapeworms. The adults attach to the intestinal wall of the human host and shed eggs with the host's feces. In areas where human excrement contaminates animal pastures and food, the life cycle continues (Figure 24–15).

Usually the symptoms of an infestation are so mild that the host is unaware of the parasite. But anemia can result in severe cases, since some tapeworms can absorb vitamin B_{12}, a substance required for red blood cell formation. Tapeworms rarely invade tissue.

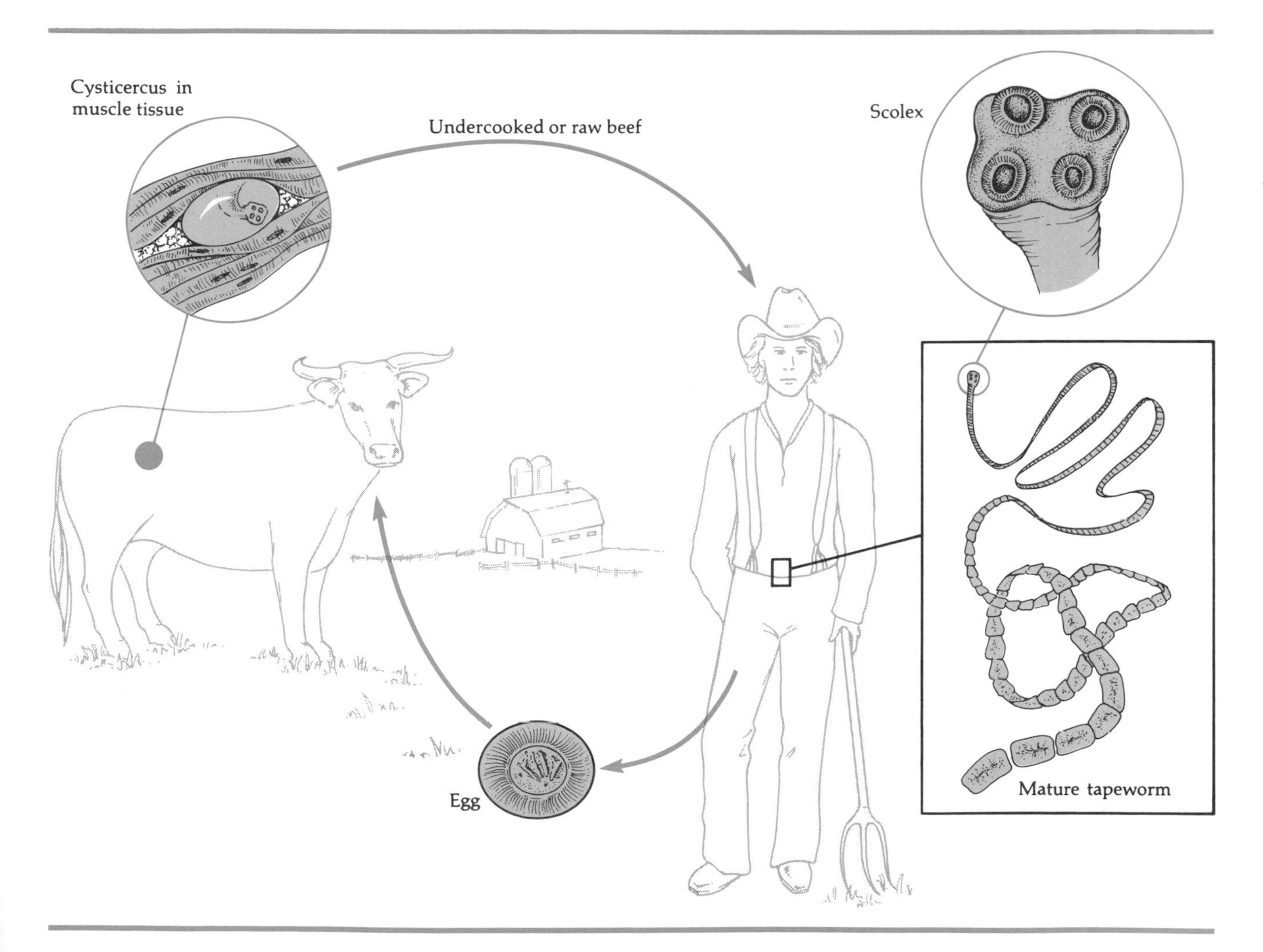

Figure 24–15 Life cycle of the beef tapeworm, *Taenia saginata.* The adult tapeworm lives in the intestine of the human, the definitive host. Tapeworm proglottids and eggs are eliminated with feces and ingested by grazing steers, which serve as the intermediate host. The tapeworm eggs hatch, and cysticerci form in the steers' muscles, to be consumed later by humans. The pork tapeworm, *Taenia solium* (see Color Plate IVF and IVG), has a similar life cycle.

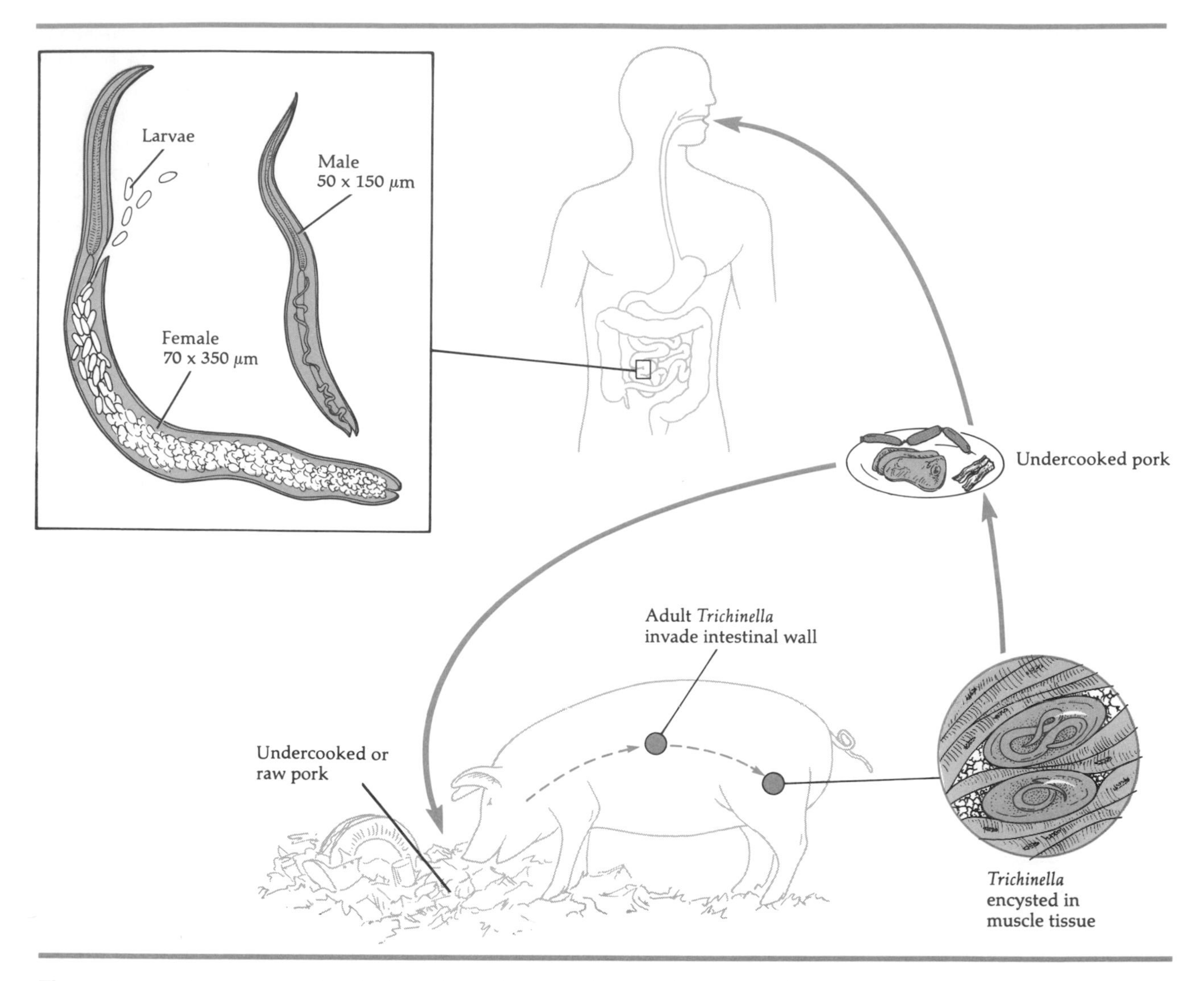

Figure 24–16 Life cycle of *Trichinella spiralis*, the causative agent of trichinosis.

Taenia saginata, the beef tapeworm, may reach a length of 6 m or more. The pork tapeworm, *Taenia solium*, is normally from 2 to 7 m in length. *Diphyllobothrium latum*, the fish tapeworm, is found in pike, trout, perch, and salmon. Fully developed, they may be from 3 to 6 m in length.

Today, less than 0.1% of the cattle slaughtered in the United States are infected with tapeworm, and only a few swine are infested. At one time, however, tapeworm infestations were very common. Laboratory diagnosis of human infection consists of identifying the tapeworm in feces. The drugs of choice for eliminating tapeworms are quinacrine hydrochloride (Atabrine), niclosamide, and paromomycin.

Trichinosis

Most infestations with the small roundworm *Trichinella spiralis*, called **trichinosis,** are insignificant. The larvae, in encysted form, are located in muscles of the host. Routine autopsies of human diaphragm muscles in 1970 showed that about 4% of those tested carried this parasite. Severe cases of trichinosis can be fatal—perhaps in a few days. The severity of the disease is generally in proportion to the amount of infestation. Ingestion of undercooked pork is the most common vehicle of infection, but the flesh of other animals that feed on garbage, bears, for example, are also the source of outbreaks.

Any ground meat may also contain the parasite because of contamination from machinery previously used to grind contaminated meats. Eating raw sausage or hamburger is a poor habit. One case is on record of trichinosis acquired by chewing fingernails after handling infected pork. Prolonged freezing of meats containing *Trichinella* tends to eliminate them but should not be considered a substitute for thorough cooking.

The organism *T. spiralis* is encysted in the muscles of intermediate hosts such as pigs in the form of a short worm, 1 to 2 mm in length (Figure 24–16). When ingested by humans with the flesh of an infected animal, the cyst wall is removed by digestive action in the intestine, where it matures into the adult form. The adult worms spend only about a week in the intestinal mucosa and produce larvae that invade tissue. Eventually, the encysted larvae localize in muscle (common sites include the diaphragm and eye muscles), where they are barely visible to the naked eye in biopsied specimens.

Symptoms of trichinosis include fever, swelling about the eyes, and gastrointestinal upset. Small hemorrhages under fingernails are often observed. Biopsy specimens, as well as a number of serological tests, can be used in diagnosis. Treatment consists of the administration of thiabendazole to kill intestinal worms and corticosteroids to reduce inflammation.

In recent years, the number of cases reported annually has varied from 67 to 252. Deaths are rare, and in most years none occur.

See Table 24–1 for a summary of diseases associated with the mouth and gastrointestinal system.

Table 24–1 Summary of Diseases Associated with the Mouth and Gastrointestinal System

Disease	Causative Agent	Mode of Transmission	Prevention or Treatment
Bacterial Diseases of the Mouth:			
Dental caries	*Streptococcus mutans*, lactobacilli, *Veillonella*, *Actinomyces*	Bacteria use sucrose to form plaque; bacteria produce acids	Restrict ingestion of sucrose; brushing, flossing, and professional cleaning to remove plaque; fluoridation
Periodontal disease	*Actinomyces*, *Nocardia*, *Corynebacterium*	Plaque initiates an inflammatory response	Same as above plus antibiotics such as penicillin and tetracycline
Bacterial Diseases of the Lower Digestive System:			
Botulism	*Clostridium botulinum*	Ingestion of exotoxin in food, usually improperly preserved	Antitoxin
Staphylococcal food poisoning (enterotoxicosis)	*Staphylococcus aureus*	Ingestion of exotoxin in food, usually improperly refrigerated	Replace lost water and electrolytes
Salmonellosis	*Salmonella* species	Ingestion of contaminated food and drink	Replace lost water and electrolytes
Typhoid fever	*Salmonella typhi*	Ingestion of contaminated food and drink	Ampicillin and trimethoprim-sulfamethoxazole, chloramphenicol
Bacillary dysentery (shigellosis)	*Shigella* species	Ingestion of contaminated food and drink	Replace lost water and electrolytes; ampicillin, chloramphenicol, or trimethoprim-sulfamethoxazole
Asiatic cholera	*Vibrio cholerae*	Ingestion of contaminated food and drink	Replace lost water and electrolytes
Vibrio parahaemolyticus gastroenteritis	*Vibrio parahaemolyticus*	Ingestion of contaminated shellfish	Tetracycline, chloramphenicol, or penicillin in severe cases
Escherichia coli gastroenteritis	Enterotoxigenic, enteroinvasive, and enteropathogenic strains of *E. coli*	Ingestion of contaminated food and drink	None usually, but doxycycline has been used in some cases for traveler's diarrhea
Clostridium perfringens gastroenteritis	*Clostridium perfringens*	Ingestion of contaminated food and drink	Replace lost water and electrolytes

(continued on next page)

Table 24–1, continued

Disease	Causative Agent	Mode of Transmission	Prevention or Treatment
Viral Diseases of the Digestive System:			
Mumps	Mumps virus	Saliva and respiratory secretions	None, but a vaccine is available (MMR)
Cytomegalovirus (CMV) inclusion disease	Cytomegalovirus (CMV)	Placental transfer and multiple blood transfusion	None
Hepatitis A	Hepatitis A virus (HAV)	Ingestion of contaminated food and drink	None, but passive immunization is available
Protozoan Diseases of the Digestive System:			
Giardiasis	*Giardia lamblia*	Ingestion of contaminated water	Metronidazole and quinacrine hydrochloride
Balantidiasis (balantidial dysentery)	*Balantidium coli*	Ingestion of contaminated food and water	Antibiotics (chlortetracycline, oxytetracycline) and diiodohydroxyquin
Amoebic dysentery (Amoebiasis)	*Entamoeba histolytica*	Ingestion of contaminated food and water	Metronidazole
Helminthic Diseases of the Digestive System:			
Tapeworm infestation	*Taenia saginata* (beef tapeworm), *T. solium* (pork tapeworm), *Diphyllobothrium latum* (fish tapeworm)	Ingestion of contaminated food and water	Quinacrine hydrochloride, niclosamide, and paromomycin
Trichinosis	*Trichinella spiralis*	Ingestion of contaminated food, especially improperly cooked pork	Thiabendazole and corticosteroids

STUDY OUTLINE

INTRODUCTION (p. 608)

1. Diseases of the digestive system are the second most common illnesses in the United States.
2. Diseases of the digestive system usually result from ingestion of microorganisms and their toxins in food and water.
3. The fecal-oral cycle of transmission can be broken by proper disposal of sewage and proper preparation and storage of foods.

STRUCTURE AND FUNCTION OF THE DIGESTIVE SYSTEM (pp. 608–609)

1. The gastrointestinal (GI) tract, or alimentary canal, consists of the mouth, pharynx, esophagus, stomach, small intestine, and large intestine.
2. The teeth, tongue, salivary glands, liver, gallbladder, and pancreas are accessory structures.
3. In the GI tract, with mechanical and chemical help from the accessory

structures, carbohydrates are broken down to simple sugars, fats to fatty acids and glycerol, and proteins to amino acids.

4. Feces, resulting from digestion, are stored in the rectum and eliminated by defecation.

ANTIMICROBIAL FEATURES OF THE DIGESTIVE SYSTEM (p. 610)

1. Saliva contains lysozyme, mucins, and IgA antibodies.
2. Hydrochloric acid in the stomach destroys many microorganisms.
3. Lymphoid tissue in the small and large intestines provides immunity against some microorganisms.
4. Normal flora of the intestines inhibits the growth of some microorganisms through the production of acids and rapid growth.
5. Some of the normal flora produce bacteriocins that bind to outer membrane receptors of susceptible bacteria and kill them.

NORMAL FLORA OF THE DIGESTIVE SYSTEM (pp. 610–611)

1. *S. salivarius* is found on the tongue, *S. mitis* on the cheek mucosa, and *S. sanguis* on the teeth.
2. *Bacteroides, Fusobacterium,* and spirochetes are found in anaerobic niches between the teeth and gums.
3. The stomach and small intestine have few resident microorganisms.
4. The large intestine is the habitat for *Lactobacillus, Bacteroides, E. coli, Enterobacter, Citrobacter,* and *Proteus.*
5. Bacteria in the large intestine assist in degradation of food and synthesize useful vitamins.
6. Up to 40% of fecal mass is microbial cells.

BACTERIAL DISEASES OF THE DIGESTIVE SYSTEM (pp. 611–621)

Diseases of the Mouth (pp. 611–613)

1. Dental caries begin with erosion of tooth enamel and dentin, which exposes the pulp to bacterial infection.
2. *S. mutans,* found in the mouth, utilizes sucrose to form dextran from glucose and lactic acid from fructose.
3. The sticky dextran capsule allows bacteria to adhere to teeth, forming dental plaque.
4. Acid produced during carbohydrate fermentation destroys tooth enamel at the site of the plaque.
5. Carbohydrates such as glucose, starch, mannitol, and sorbitol do not lead to the production of dextran by cariogenic bacteria and do not promote tooth decay.

6. Caries are prevented by restricting ingestion of sucrose and by physical removal of plaque; a vaccine against *S. mutans* is theoretically possible.
7. Inflammation of the gums, bone destruction, and necrosis of tissues around the teeth may be caused by *Actinomyces, Nocardia,* and *Corynebacterium;* this is called periodontal disease.

Diseases of the Lower Digestive System (pp. 613–621)

1. Bacterial intoxications result from ingestion of preformed bacterial toxins.
2. Onset of symptoms occurs between 1 and 48 hours after ingestion of the toxin.
3. Antibiotics are of little use because symptoms are due to an exotoxin.

Botulism (pp. 613–616)

1. Botulism is caused by an exotoxin produced by *C. botulinum* growing in foods.
2. The toxin is a neurotoxin that inhibits transmission of nerve impulses, resulting in respiratory and cardiac failure.
3. Serological types of botulinum toxin vary in virulence, with type A being the most virulent.
4. Blurred vision and nausea occur in 1 to 2 days, progressively flaccid paralysis follows for 1 to 10 days. There is as high as 60 to 70% mortality in untreated cases.
5. *C. botulinum* will not grow in acidic foods or in an aerobic environment.
6. Endospores are killed by using a pressure cooker for canning. Addition of nitrites to foods inhibits outgrowth after endospore germination.
7. The toxin is heat labile and destroyed by boiling (100°C) for 5 minutes.
8. Wound botulism occurs when *C. botulinum* grows in anaerobic wounds.
9. Infant botulism results from the growth of *C. botulinum* in an infant's intestines.
10. For diagnosis, mice protected with antitoxin are inoculated with toxin from the patient or foods.

Staphylococcal food poisoning (staphylococcal enterotoxicosis) (pp. 616–617)

1. Staphylococcal food poisoning is due to ingesting enterotoxin produced in improperly stored foods.
2. *S. aureus* is inoculated into foods during preparation. The bacteria grow and produce enterotoxin in food stored at room temperature.
3. The polypeptide exotoxin is not denatured by boiling for 30 minutes.
4. Foods with high osmotic pressure and those not cooked immediately before consumption are most often the source of staphylococcal enterotoxicosis.

5. Nausea, vomiting, and diarrhea occur 1 to 6 hours after eating, and the symptoms last approximately 24 hours.
6. Laboratory identification of *S. aureus* isolated from foods can confirm diagnosis.

Salmonellosis (*Salmonella* gastroenteritis) (p. 617)

1. Bacterial infections result from bacterial growth in the GI tract.
2. Incubation times range from 12 hours to 2 weeks to allow sufficient growth for bacterial cells and their products to produce symptoms.
3. Antibiotic therapy is useful to treat bacterial infections.
4. Adequate sewage treatment and good personal hygiene will help prevent these infections.
5. Salmonellosis, or *Salmonella* gastroenteritis, is caused by *Salmonella* species endotoxin.
6. Symptoms include nausea, abdominal pain, and diarrhea and occur 12 to 36 hours after ingestion of large numbers of *Salmonella*. Septicemia may occur in infants and the elderly.
7. Mortality is less than 1% and recovery may result in a carrier state.
8. Heating food to 68°C will usually kill *Salmonella*.
9. Laboratory diagnosis is based on isolation and identification of *Salmonella* from feces.

Typhoid fever (pp. 617–618)

1. A few species of *Salmonella* cause typhoid fever.
2. Fever and malaise occur after a two-week incubation. Symptoms last two to three weeks.
3. *Salmonella* is harbored in the gallbladder of carriers.

Bacillary dysentery (shigellosis) (p. 618)

1. Bacillary dysentery is caused by four species of *Shigella*.
2. Symptoms include blood and mucus in stools, abdominal cramps, and fever. Infections of *S. dysenteriae* result in ulceration of the intestinal mucosa.
3. Isolation and identification of the bacteria from rectal swabs is used for diagnosis.

Asiatic cholera (pp. 618–619)

1. *V. cholerae* produces an exotoxin that alters membrane permeability of the intestinal mucosa, resulting in vomiting, diarrhea, and loss of body fluids.

2. The incubation period is approximately three days. The symptoms last for a few days. Untreated cholera may have a 50% mortality rate.

Vibrio parahaemolyticus gastroenteritis (p. 620)

1. Gastroenteritis may be caused by the halophile *V. parahaemolyticus.*
2. Onset of symptoms begins within 24 hours after ingestion of contaminated foods. Recovery occurs within a few days.
3. The disease is contracted by ingesting contaminated crustaceans or handling contaminated mollusks.

Escherichia coli gastroenteritis (p. 620)

1. Gastroenteritis may be caused by enterotoxigenic, enteroinvasive, or enteropathogenic strains of *E. coli.*
2. The disease occurs as epidemic diarrhea in nurseries and as traveler's diarrhea.
3. The disease is usually self-limiting and does not require chemotherapy.

Gastroenteritis caused by other gram-negative bacteria (p. 620)

1. *Yersinia enterocolitica* and *Campylobacter fetus* cause gastroenteritis.
2. They may be transmitted by poultry, dogs, raw milk, and contaminated water.

Clostridium perfringens gastroenteritis (pp. 620–621)

1. A self-limiting gastroenteritis is caused by *C. perfringens.*
2. Endospores survive heating and germinate when food (usually meats) are stored at room temperature.
3. Exotoxin produced when the bacteria grow in the intestines is responsible for the symptoms.
4. Diagnosis is based on isolation and identification of the bacteria in stool samples.

VIRAL DISEASES OF THE DIGESTIVE SYSTEM (pp. 621–624)

Mumps (pp. 621–622)

1. Mumps virus (paramyxovirus) enters and exits the body through the respiratory tract.
2. About 16 to 18 days after exposure, the virus causes inflammation of the parotid glands, fever, and pain during swallowing. About 4 to 7 days later, orchitis may occur.
3. After onset of the symptoms, the virus is found in the blood, saliva, and urine.

4. Vaccination is available in measles–mumps–rubella (MMR) vaccine.
5. Diagnosis is based on symptoms or hemagglutination-inhibition tests of viruses cultured in embryonated eggs or cell culture.

Cytomegalovirus (CMV) Inclusion Disease (pp. 622–623)

1. CMV (herpesvirus) causes intranuclear inclusion bodies and cytomegaly of host cells.
2. CMV may be latent in a host and activated by pregnancy, blood transfusions, or immunosuppression drugs.
3. The virus may cross the placenta and cause cytomegalovirus inclusion disease in the newborn.
4. Malfunction of the salivary glands, kidneys, lungs, and liver may result from this disease.
5. Diagnosis is based on appearance of the host cells and serological tests.

Hepatitis A (Infectious Hepatitis) (pp. 623–624)

1. Hepatitis A virus causes infectious hepatitis.
2. HAV is ingested in contaminated food or water, grows in the cells of the intestinal mucosa, and spreads to the liver, kidney, and spleen in the blood.
3. The virus is eliminated with feces.
4. Symptoms include anorexia, malaise, nausea, diarrhea, fever, chills, and jaundice.
5. Recovery is complete in four to six weeks.
6. Diagnosis is based on serological tests for the virus or antibodies.

MYCOTOXINS (p. 624)

1. Mycotoxins are toxins produced by some fungi.
2. They affect the blood, nervous system, kidney, or liver.

PROTOZOAN DISEASES OF THE DIGESTIVE SYSTEM (pp. 624–627)

Giardiasis (pp. 624–625)

1. *G. lamblia* grows in the intestines of humans and wild animals and is transmitted in contaminated water.
2. Symptoms of giardiasis are malaise, nausea, weakness, and abdominal cramps that persist for weeks.
3. Diagnosis is based on identification of the protozoan in feces.

Balantidiasis (Balantidial Dysentery) (p. 625)

1. *B. coli* causes balantidial dysentery when growing in the large intestine.
2. Infections are acquired by ingesting cysts in contaminated food and water.
3. *B. coli* may cause ulceration of the intestinal wall and fatal dysentery.
4. Diagnosis is based on observation of trophozoites in feces.

Amoebic Dysentery (Amoebiasis) (pp. 625–627)

1. Amoebic dysentery is caused by *E. histolytica* growing in the large intestine.
2. The amoeba feeds on red blood cells and GI tract tissues. Severe infections result in abscesses.
3. Diagnosis is confirmed by observation of trophozoites in feces.

HELMINTHIC DISEASES OF THE DIGESTIVE SYSTEM (pp. 627–629)

Tapeworm Infestation (pp. 627–628)

1. Tapeworms are contracted by eating undercooked beef, pork, or fish containing encysted larvae (cysticerci).
2. The scolex attaches to the intestinal mucosa of humans (the definitive host) and matures into an adult tapeworm.
3. Eggs are shed in the feces and must be ingested by an intermediate host.
4. Adult tapeworms may be undiagnosed in a human. Severe infestations can result in anemia, as the parasite competes with the host for vitamin B_{12}.

Trichinosis (pp. 628–629)

1. *T. spiralis* larvae encyst in muscles of humans, swine, and other mammals to cause trichinosis.
2. The roundworm is contracted by ingesting undercooked meat containing larvae.
3. Adults mature in the intestine and lay eggs. The new larvae migrate to invade muscles.
4. Symptoms include fever, swelling around the eyes, and gastrointestinal upset.

STUDY QUESTIONS

REVIEW

1. List at least five antimicrobial features of the digestive system and describe their activity.

2. What are bacteriocins? How are they produced and what do they do?
3. Provide examples of representative normal flora, if any, in these parts of the GI tract: mouth; stomach; small intestine; large intestine; rectum.
4. List the general symptoms of gastroenteritis. Since there are many etiologies, on what is the laboratory diagnosis usually based?
5. Differentiate between bacterial intoxication and bacterial infection with regard to prerequisite conditions, etiologic agents, onset, duration of symptoms, and treatment.
6. Complete the following table.

Disease	Etiologic Agent	Suspect Foods	Symptoms	Treatment
Botulism				
Staphylococcal food poisoning				
Salmonellosis				
Bacillary dysentery				
Asiatic cholera				
Gastroenteritis				
Traveler's diarrhea				

7. Differentiate between salmonellosis and typhoid fever.
8. You probably listed *E. coli* in answer to questions 2 and 6. Explain why this one bacterial species could be both beneficial and harmful.
9. Provide the information asked for in this table.

Disease	Etiologic Agent	Method of Transmission	Site of Infection	Symptoms
Mumps				
CMV inclusion disease				
Infectious hepatitis				

10. Define mycotoxin.
11. Explain how the following diseases differ and how they are similar: giardiasis; balantidiasis; amoebic dysentery.
12. Differentiate between amoebic dysentery and bacillary dysentery.
13. Diagram the life cycle for a human tapeworm.
14. Diagram the life cycle for trichinosis, including humans in the cycle.
15. How can bacterial and protozoan infections of the GI tract be prevented?

CHALLENGE

1. Look at your diagrams for questions 13 and 14. Indicate places in the life cycles that could be easily broken to prevent these diseases.
2. Why is a human infection of trichinosis considered a "dead end" for the parasite?
3. Complete the following table.

Disease	Conditions Necessary for Microbial Growth	Basis for Diagnosis	Prevention
Botulism			
Staphylococcal food poisoning			
Salmonellosis			

4. Differentiate between infectious hepatitis and serum hepatitis.

FURTHER READING

Centers for Disease Control. *Salmonella Surveillance Report.* Published throughout each year. Tabulation of incidence and reports on current investigations.

Centers for Disease Control. *Shigella Surveillance Report.* Published throughout each year. Tabulation of data and background on recent food-borne or water-borne outbreaks.

Cooke, M. E. 1974. *Escherichia coli and Man.* London: Churchill Livingstone. Chapters on *E. coli* as normal flora, as a pathogen, and its antibiotic resistance.

Faust, E. C., P. F. Russel, and R. C. Jung. 1970. *Craig and Faust's Clinical Parasitology,* 8th ed. Philadelphia, PA: Lea and Febiger. Comprehensive coverage of protozoan and helminthic infections for medical personnel.

Savage, D. C. 1977. Microbial ecology of the gastrointestinal tract. *Annual Review of Microbiology* 31:107–133. An examination of the ecological niches occupied by gastrointestinal flora.

Smibert, R. M. 1978. The genus *Campylobacter. Annual Review of Microbiology* 32:673–709. A detailed study of the biology of *Campylobacter* and its relation to humans.

See also Further Reading for Part IV at the end of Chapter 19.

25

Microbial Diseases of the Urinary and Genital Systems

Objectives

After completing this chapter you should be able to

- List the normal flora of the urinary and genital systems and their habitats.
- Describe methods of transmission for urinary and genital system infections.
- List microorganisms that may cause cystitis and pyelonephritis and give the predisposing factors for these diseases.
- Describe the cause and treatment of glomerulonephritis.
- List the etiologic agent, symptoms, method of diagnosis, and treatment for each of the following: leptospirosis; gonorrhea; syphilis; NGU; LGV; chancroid; granuloma inguinale; candidiasis; trichomoniasis.
- Discuss the epidemiology of genital herpes.
- List genital diseases that can cause congenital and neonatal infections and explain how these infections can be prevented.

The **urinary system** consists of a group of organs that regulate the chemical composition of the blood and excrete waste products of metabolism. The **genital,** or **reproductive, system** consists of a series of organs that produce gametes for propagation of the species and, in the female, support and nourish the developing embryo. Since the urinary and genital systems are closely related anatomically, they will be discussed in the same chapter. In fact, some diseases that affect one system also affect the other, especially in the male. Microbial diseases of both systems tend to cause only minor discomfort at first. But if left untreated, these diseases may spread beyond their original locations and cause serious complications.

Both systems open to the external environment and thus have a portal of entry for microorganisms that can cause disease. Normal flora of these and other body systems also can cause opportunistic infections of the urinary and genital systems.

STRUCTURE AND FUNCTION OF THE URINARY SYSTEM

The **urinary system** consists of two *kidneys,* two *ureters,* a single *urinary bladder,* and a single *urethra* (Figure 25–1). The kidneys contain microscopic functional units called *nephrons.* As blood circulates through the kidneys, the nephrons control the concentration and volume of blood by removing and adding selected amounts of water and solutes and excreting wastes. The wastes,

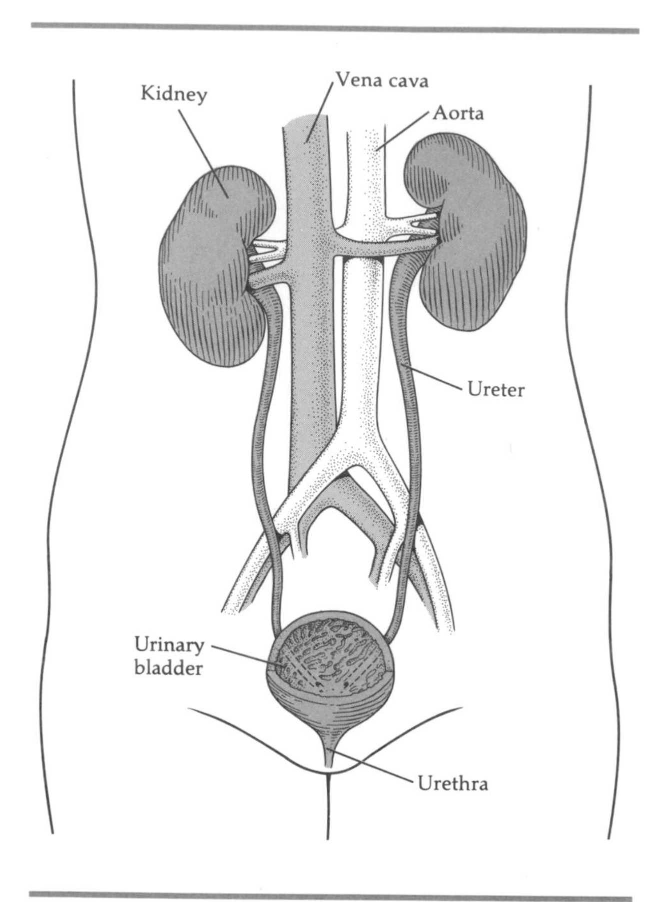

Figure 25–1 Organs of the urinary system.

which include urea, uric acid, creatinine, and various salts, together with water, are referred to as *urine*. The urine passes down the ureters into the urinary bladder, where it is stored prior to elimination from the body. Elimination occurs through the urethra. In the female, the urethra conveys only urine to the exterior. In the male, the urethra is a common tube for both urine and seminal fluid.

The urinary tract has certain characteristics that help prevent infection. At the point where the ureters enter the urinary bladder, there are valves that normally prevent the backflow of urine to the kidneys. This same mechanism also helps to shield the kidneys from lower urinary tract infections. In addition, the acidity of normal urine has some antimicrobial properties. And the flushing action of urine to the exterior prevents microorganisms from setting up foci of infection. Finally, it appears that antibody-forming cells cluster in regions of the urinary system where infection does occur.

STRUCTURE AND FUNCTION OF THE GENITAL SYSTEM

The **female reproductive system** consists of two *ovaries*, two *uterine (fallopian) tubes*, the *uterus*, the *vagina*, and *external genitalia* (Figure 25–2). The ovaries produce female sex hormones and ova (eggs). When an ovum is released in the process called ovulation, it enters a uterine tube, and if viable sperm are present, fertilization occurs. The fertilized ovum (zygote) descends the uterine tube and enters the uterus, where it implants in the inner wall. Here it remains while it develops into an embryo and later a fetus. At the time of birth, the fetus is expelled from the uterus through the vagina. The vagina also serves as a copulatory canal. The external genitalia consist of the clitoris, labia, and glands that produce a lubricating secretion during copulation.

The **male reproductive system** consists of two *testes*, a system of *ducts*, *accessory glands*, and the *penis* (Figure 25–3). The testes produce male sex hormones and sperm. The newly produced sperm are moved into the epididymis, where they are stored until ejaculation. To gain exit from the body, the sperm cells pass through a series of ducts (epididymis, ductus (vas) deferens, ejaculatory duct, and urethra). During ejaculation, contractions of the ductus (vas) deferens and ejaculatory duct propel the sperm toward the urethra. Along the route, the seminal vesicles, prostate gland, and bulbourethral glands secrete an alkaline fluid into seminal fluid. This alkaline fluid buffers the acidic environment of the vagina, an important mechanism, since sperm cells are killed in an acidic environment. On ejaculation, the seminal fluid leaves the body through the urethra. A valve system prevents the seminal fluid from entering the bladder.

NORMAL FLORA OF THE URINARY AND GENITAL SYSTEMS

Normal urine in the urinary bladder and the organs of the upper urinary tract are sterile. The urethra, however, does contain a normal resident flora that includes *Streptococcus*, *Bacteroides*, *Mycobacterium*, *Neisseria*, and a few enterobacteria, and urine becomes contaminated with skin flora during passage.

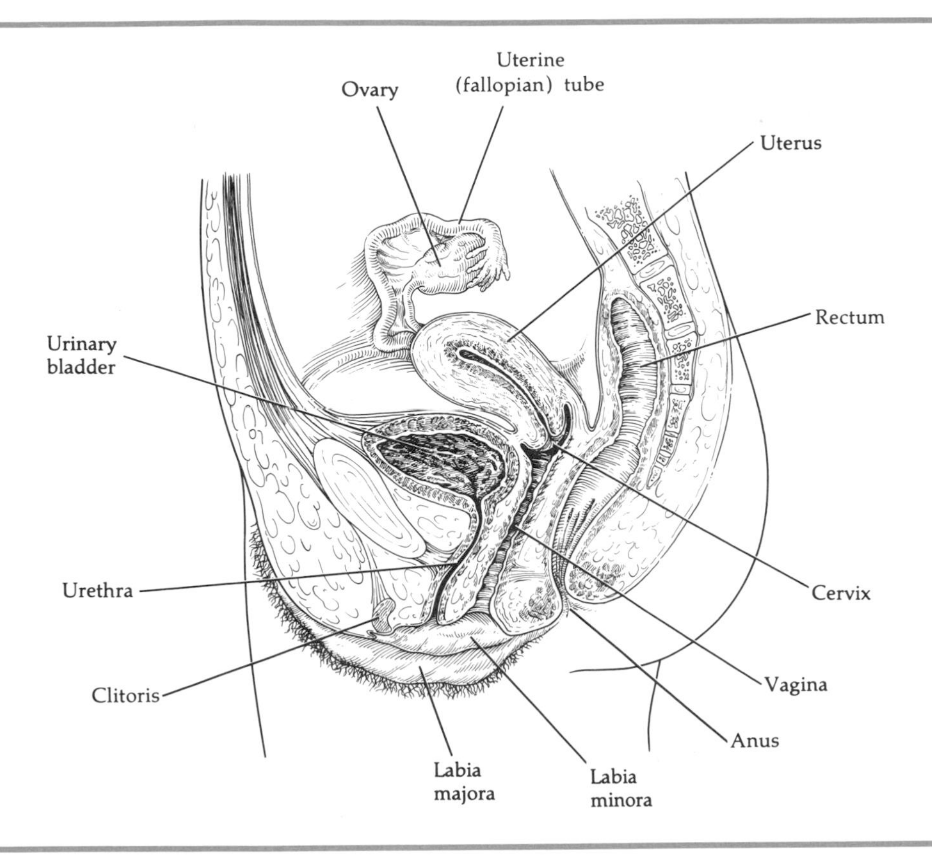

Figure 25–2 Female organs of reproduction.

In the female genital system, the normal flora of the vagina is greatly influenced by sex hormones. For example, within a few weeks after birth, the female infant's vagina is populated by lactobacilli. Under the influence of estrogens, which were transferred from maternal to fetal blood, glycogen accumulates in the lining cells of the vagina. Lactobacilli convert the glycogen to lactic acid, and the pH of the vagina becomes acidic. As the effects of the estrogens diminish several weeks after birth, other bacteria—including corynebacteria and a variety of cocci and bacilli—become established and dominate the flora, and the pH becomes more neutral until puberty. At puberty, estrogen levels increase, lactobacilli again dominate, and the vagina again becomes acidic. During the reproductive years, small numbers of other bacteria and yeasts become part of the flora. When the female reaches menopause, estrogen levels again decrease, the flora returns to that of childhood, and the pH again becomes neutral.

BACTERIAL DISEASES OF THE URINARY SYSTEM

Many infections of the urinary system appear to be opportunistic and related to a number of predisposing factors. These factors include nervous system disorders, toxemia associated with pregnancy, diabetes mellitus, and obstructions to the flow of urine, such as tumors and kidney stones. Because the anus, from which feces are excreted, is located close to the urethra, it is not unusual, especially in females, for the urinary tract to become contaminated with intestinal bacteria. These are usually *Escherichia coli* and other gram-negative enteric bacteria, although pseudomonads, *Proteus* species, fecal streptococci, and staphylococci are also commonly found in such infections. *Pseudomonas* infections, as elsewhere, are unusually troublesome to treat. The fungus *Candida albicans* is also an opportunistic agent of urinary tract infections. Moreover, a number of sexually acquired diseases may cause

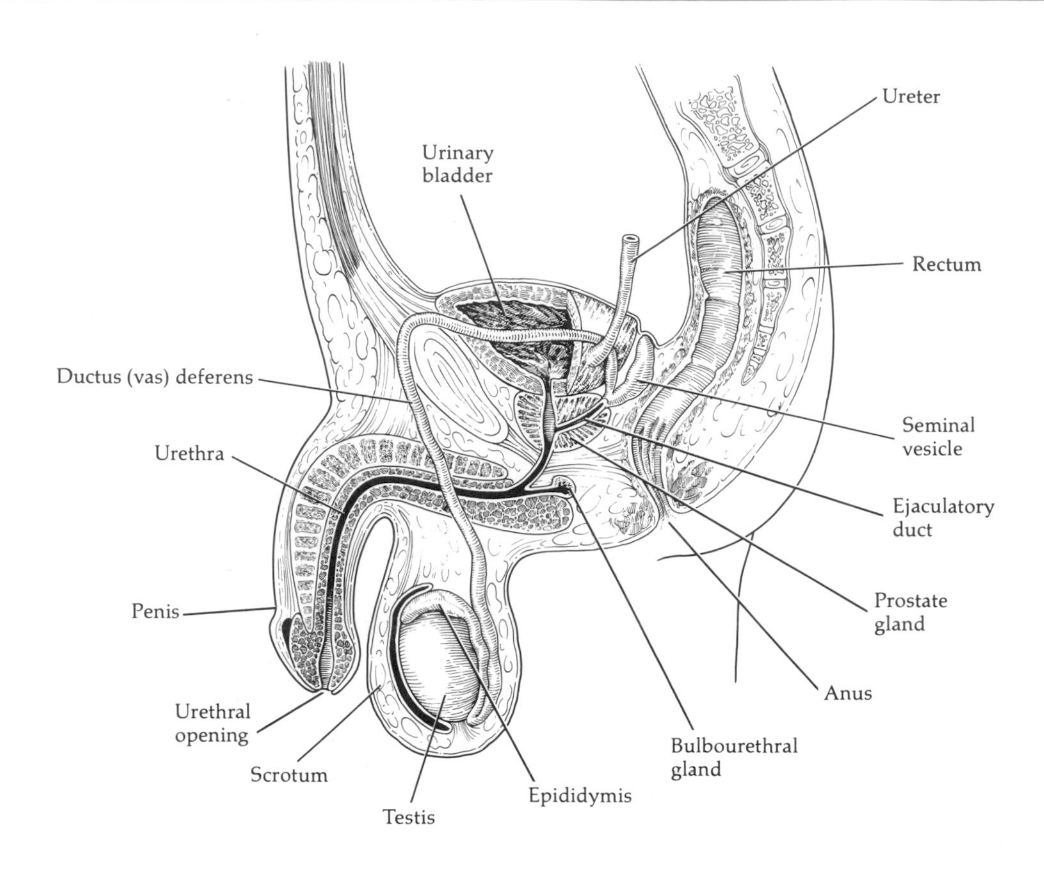

Figure 25–3 Male organs of reproduction.

inflammation in the urinary tract.

Infections usually cause inflammation of the affected tissue. Inflammation of the urethra is called *urethritis;* of the urinary bladder, *cystitis;* and of the ureters, *ureteritis*. The most significant danger from lower urinary tract infections is that they may affect the kidneys (pyelonephritis) and impair their function. The kidneys are also sometimes affected by systemic bacterial diseases such as typhoid fever or leptospirosis, and as a result, the pathogens causing these diseases may be found in the excreted urine. Treatment of diseases of the urinary tract depends on diagnostic isolation of the causative organism and determination of its antibiotic sensitivity. Normal urine contains fewer than 10,000 bacteria per milliliter. When more than 100,000 bacteria per milliliter are found, it usually indicates an infection.

Many infections of the urinary tract are of nosocomial origin. (In fact, about 35% of all nosocomial infections occur in the urinary tract.) Operations on the urinary bladder or prostate gland and catheterization for draining the urinary bladder of urine are procedures that introduce bacteria into the bladder and ureters. *E. coli* causes more than half of the nosocomial infections of the urinary tract, although fecal streptococci, *Proteus*, *Klebsiella*, and *Pseudomonas* also commonly cause such infections.

Cystitis

Cystitis is an inflammation of the urinary bladder and is very common, especially among females. The female urethra has many microorganisms in the area around its opening, and it is shorter than the male urethra, so that microorganisms can traverse it more readily between voidings. Careless personal hygiene or sexual intercourse may facilitate such transfer. Contributing factors in females include gastrointestinal system infections and preexisting infections of the vagina, uterus, or urethra. In males, cystitis may be associated with infections of the gastrointestinal system, kidneys, or urethra. Bacteria frequently associated with cystitis in both sexes are the gram-negative rods *Escherichia coli*, *Proteus vulgaris*, and *Pseudomonas aeruginosa*.

Treatment of cystitis depends on the bacterium responsible and involves administration of sulfonamides, chloramphenicol, kanamycin, penicillin G, or polymyxin.

Pyelonephritis

Pyelonephritis is an inflammation of one or both kidneys involving the nephrons and the renal pelvis (the opening into the ureter). The disease is generally a complication of infection elsewhere in the body. In females it is often a complication of lower urinary tract infections. The causative agent in about 75% of the cases is *Escherichia coli*. Other bacteria associated with pyelonephritis are *Enterobacter aerogenes*, *Proteus* species, *Pseudomonas aeruginosa*, *Streptococcus pyogenes*, and staphylococci. Should pyelonephritis become chronic, scar tissue forms in the kidneys and their function is severely impaired. Depending on the etiological agent, a number of antibiotics can be used to treat pyelonephritis, including chloramphenicol, gentamicin, kanamycin, polymyxin, methicillin, penicillin, vancomycin, and tetracycline.

Leptospirosis

Leptospirosis is a disease characterized by chills, fever, headache, and muscular pain. Jaundice may result from localization of the pathogen in the liver and kidneys. The human mortality rate is less than 10%, with kidney failure the most common cause of death. Recovery results in solid immunity. There are about 100 cases per year in the United States.

The causative organism is the spirochete *Leptospira interrogans* (Figure 25–4). Leptospiras have a characteristic cell shape—an exceedingly fine spiral wound so tightly that it is barely distinguishable under the darkfield microscope. *L. interrogans*, like other spirochetes, is hard to see under a normal light microscope and is therefore rarely Gram stained. It is an obligate aerobe that can be readily grown in a variety of artificial media supplemented with rabbit serum. Leptospirosis is a disease of wild mammals, and the microorganism is excreted in the urine of infected animals. Humans and animals in contact with water contaminated by such urine are most likely to contract the disease, which can be transmitted through mucous membranes. Domestic dogs are commonly immunized against the disease.

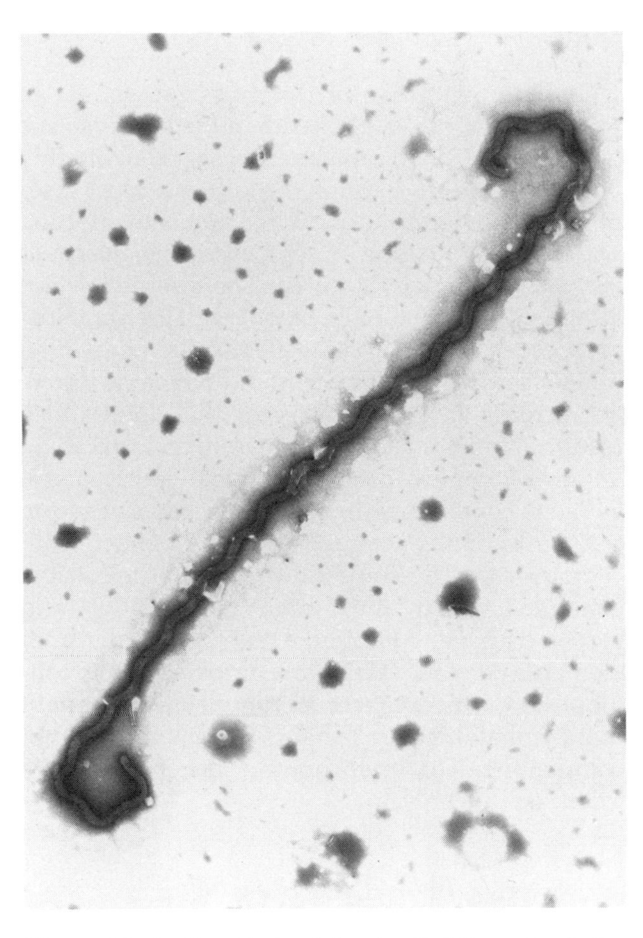

Figure 25–4 Negatively stained cell of the spirochete *Leptospira interrogans*, the causative agent of leptospirosis (×12,000). The "hooked" ends are commonly seen in preparations of this organism, and remnants of the threadlike axial filament are barely visible on the hooks.

Diagnosis by serology is complicated by the immunologic overlap between the various pathogenic serotypes of *L. interrogans*. Serum antibodies usually appear during the second week of illness and reach a maximum titer during the third or fourth week. However, they are not easily identified because of the numerous antigenic types encountered. Agglutination tests, however, can be done with suspensions of killed leptospiras, pooled to contain the most common antigenic strains. The treatment of leptospirosis is generally unsatisfactory, partly because the disease is frequently recognized late. The drug of choice appears to be penicillin.

Glomerulonephritis

Glomerulonephritis, or **Bright's disease,** is an inflammation of the glomeruli in the kidneys. The glomeruli are blood capillaries that assist in filtering blood as it passes through the kidneys. Most cases of glomerulonephritis occur as a sequel to infection with beta-hemolytic streptococci *(Streptococcus pyogenes)*, especially type 12. The disease is characterized by fever, high blood pressure, and the presence of protein and red blood cells in the urine. The presence of the red cells and protein is due to an increase in permeability of the glomeruli, a result of inflammation. Glomerulonephritis is an immune-complex disease. Soluble streptococcal antigens combine with specific antibodies to form antigen–antibody complexes that interact with complement. The complexes are deposited in the glomeruli, where they cause inflammation and kidney damage. Although a few patients die from glomerulonephritis and some develop chronic conditions, most people recover completely. Treatment of the initial infection consists of administration of chloramphenicol, erythromycin, lincomycin, penicillin, or vancomycin.

BACTERIAL DISEASES OF THE GENITAL SYSTEM

Most diseases of the genital system are transmitted by sexual activity and are therefore called **sexually transmitted diseases (STDs).** These diseases are also called **venereal diseases (VDs),** an older term derived from the name Venus, the Roman goddess of love. Most of these diseases can be readily cured with antibiotics if treated early and can largely be prevented by the use of condoms. Nevertheless, STDs are a major U.S. public health problem.

Gonorrhea

By far the most common reportable communicable disease in the United States is **gonorrhea,** an STD caused by the gram-negative diplococcus *Neisseria gonorrhoeae* (see Figure 10–9a). An ancient disease, gonorrhea was described and given its present name by the Greek physician Galen in A.D. 150. At the present time, the number of cases seems to have reached a plateau after a steady and steep increase for many years (Figure 25–5). About 1 million cases in the United States are reported to the Centers for Disease Control each year, and the true number of cases is probably 3 to 4 million. Over 60% of the cases are in the 15 to 24 age group.

To be infective, the gonococcus must attach, by means of pili, to the mucosal cells of the epithelial wall. One experimental vaccine is aimed at these pili to prevent attachment. The organism does not infect the layered squamous cells characteristic of the external skin, but invades the spaces separating mucosal cells. Mucosal cells are found in the oral-pharyngeal area, male and female genitalia, external genitalia of prepubertal females, and the eyes, joints, and rectum. The invasion sets up an inflammation, and when leukocytes move into the inflamed area, the characteristic pus formation results.

Males become aware of a gonorrheal infection by painful urination and discharge of pus-containing material from the urethra (Figure 25–6). About 80% of infected males show these obvious symptoms after an incubation period of only a few days, and most others in less than a week. In untreated cases, recovery may eventually occur without complication, but when complications do occur, they can be serious. In some cases, there is scarring of the urethra, partially blocking it. And sterility may result when the testes become infected or when the tube carrying sperm from the testes, the ductus (vas) deferens, becomes blocked by scar tissue.

In females, the disease is more insidious. In the early stages, very few women are aware of the infection. Later, however, there may be abdominal

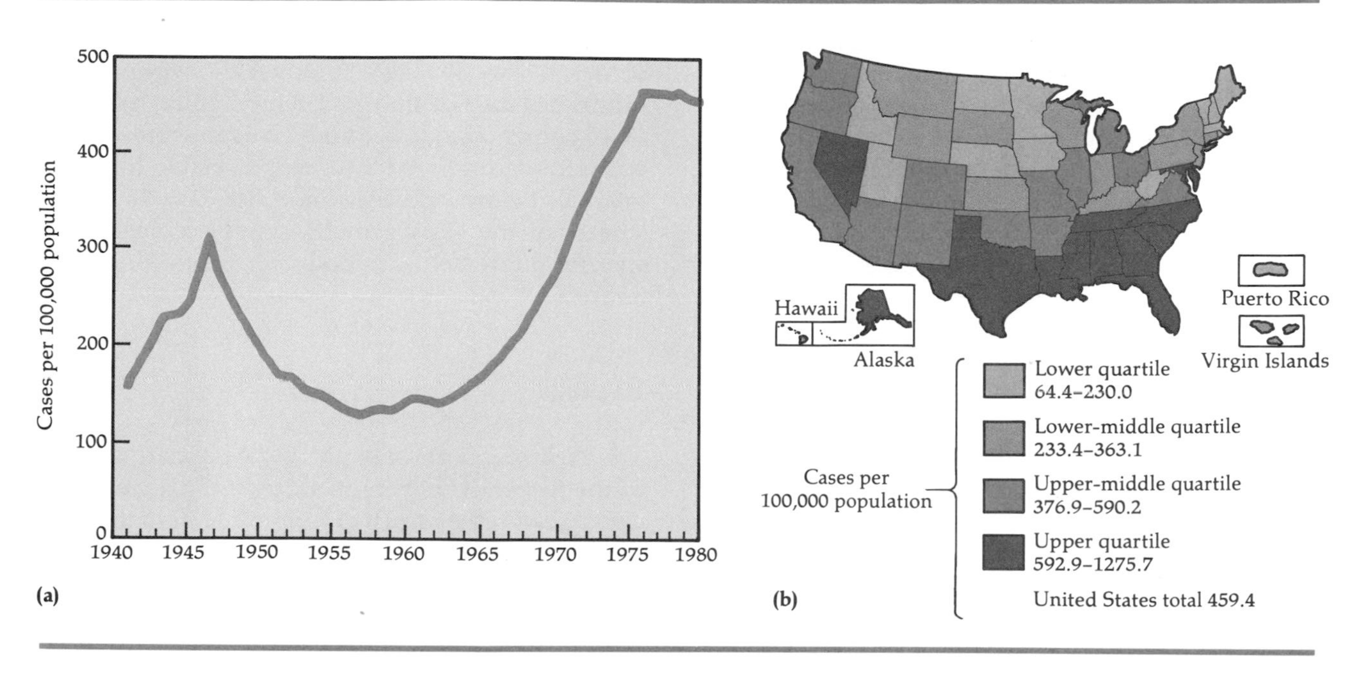

Figure 25–5 Incidence and distribution of gonorrhea. **(a)** Incidence of gonorrhea in the United States, 1941–1979. **(b)** Geographical frequency of cases reported in 1979.

pain as the infection spreads to the uterus and causes a chronic infection of the uterine tubes called *salpingitis*. This is a fairly common occurrence if the disease is not treated in the early, mild, and often undiagnosed stage. The results of gonorrheal infection of the uterine tubes are often serious. Passage of ova from the ovaries to the uterus can be blocked by scar tissue, and only about 20% of such cases of sterility can be reversed surgically. This can also lead to pregnancy taking place in the uterine tubes, rather than in the uterus, a life-threatening condition called *ectopic pregnancy*. Another complication of gonorrhea in females is a condition called *pelvic inflammatory disease (PID)*. PID is a collective term for any extensive bacterial infection of the pelvic organs, especially the uterus, uterine tubes, and ovaries. PID is most commonly caused by gonorrhea.

In both sexes, untreated gonorrhea may become a serious, systemic infection. Complications of gonorrhea may involve the joints, heart **(gonorrheal endocarditis)**, meninges **(gonorrheal meningitis)**, eyes, pharynx, or other parts of the body. **Gonorrheal arthritis**, which is caused by the growth of the gonococcus in fluids in joints, occurs in about 1% of gonorrhea cases. Joints commonly affected include the wrist, knee, and ankle.

Gonorrheal eye infections occur most often in newborns. If the mother is infected, then the eyes of the infant can become infected as it passes through the birth canal. This condition, **ophthalmia neonatorum**, can result in blindness. Because

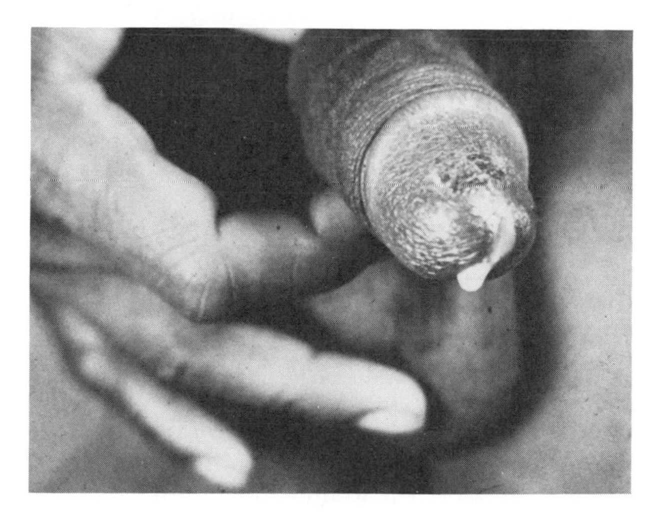

Figure 25–6 Pus-containing discharge from the urethra of a male with an acute case of gonorrhea.

of the seriousness of this condition, and the difficulty of being certain that the mother is free of gonorrhea, silver nitrate in dilute solution, tetracycline, or erythromycin is placed in the eyes of all newborn infants. This is required by law in most states. Gonorrheal infections can also be transferred by hand contact from infected sites to the eyes of adults.

Gonorrheal infections can be acquired at any point of sexual contact; pharyngeal and anal gonorrhea are not uncommon. The symptoms of **pharyngeal gonorrhea** often resemble those of the usual septic sore throat, and **rectal gonorrhea** can be rather painful.

Increased sexual activity with multiple partners and the fact that the disease in the female may go unrecognized have contributed considerably to the increased incidence of gonorrhea and other venereal diseases. The widespread use of oral contraceptives has also contributed to the increase. Oral contraceptives tend to increase the moisture content and raise the pH of the vagina, at the same time exposing more susceptible mucosal cells. All these factors predispose the female to infection.

There is no immunity to reinfection as a result of recovery from gonorrhea. Penicillin has been an effective treatment for gonorrhea over the years, although the dosages have had to be much increased because of the appearance of penicillin-resistant bacteria. Penicillin resistance is usually due to the presence of a gene for penicillinase (an enzyme that degrades penicillin). The penicillinase gene may be carried by the bacterial chromosome or by a plasmid (see Chapter 8). Gonococci having such plasmids first appeared in 1976. The plasmid is transmissible among gonococci and may have been derived from *Hemophilus influenzae* or other gram-negative bacteria. Spectinomycin is the favored alternative antibiotic for strains carrying penicillinase plasmids (see pp. 479–480). For complications such as endocarditis and meningitis, the antibiotic therapy is usually more extended.

Diagnosis of acute gonorrhea is made by identifying the organism in the infectious discharges of both men and women. The organisms are observed as gram-negative cocci in pairs, flattened at the point of contact. They are mostly found in phagocytic leukocytes (Figure 25–7 and Color plate IE). In chronic gonorrhea, examination of Gram-stained material is less helpful, and cultural or fluorescent-antibody techniques are more reliable. Cultivation of the nutritionally fastidious bacterium requires an atmosphere containing carbon dioxide. The gonococcus is very sensitive to adverse environmental influences (desiccation and temperature) and survives poorly outside the body. It even requires special transporting media to keep it viable for short intervals before cultivating it. After 12 to 16 hours, smears of the early growth may be stained with specific fluorescent antibody for rapid identification.

Syphilis

The earliest reports of **syphilis** date back to the end of the fifteenth century in Europe. This was coincidental with the return of Columbus from the New World, and there has always been some speculation that syphilis may have been introduced to Europe by his men. Although there is no way to know for sure, it appears that the disease described in earlier times was rather more severe than at present. Whether this is because of a diminished virulence of the bacterium or because of an increased resistance in the population is also unknown.

In contrast to the precipitous increase in the number of gonorrhea cases, the number of syphilis cases in the United States in the 1960s and '70s has

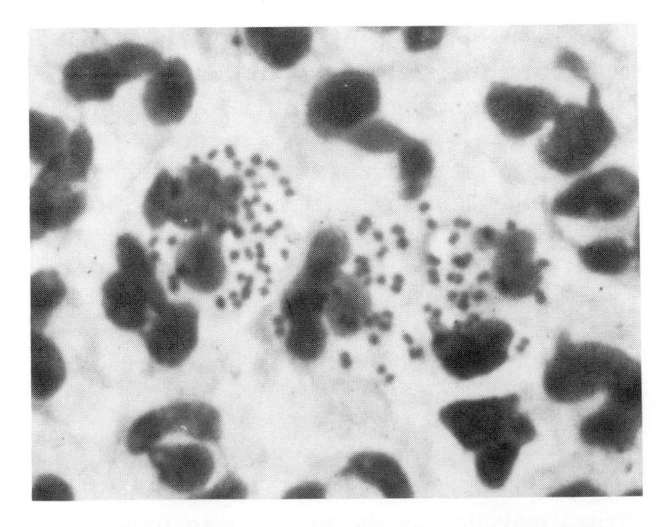

Figure 25–7 A smear from a patient with gonorrhea. Gram-negative diplococci in phagocytic leukocytes is a probable indication of gonorrhea, although it should be confirmed by isolation and identification of the bacterium.

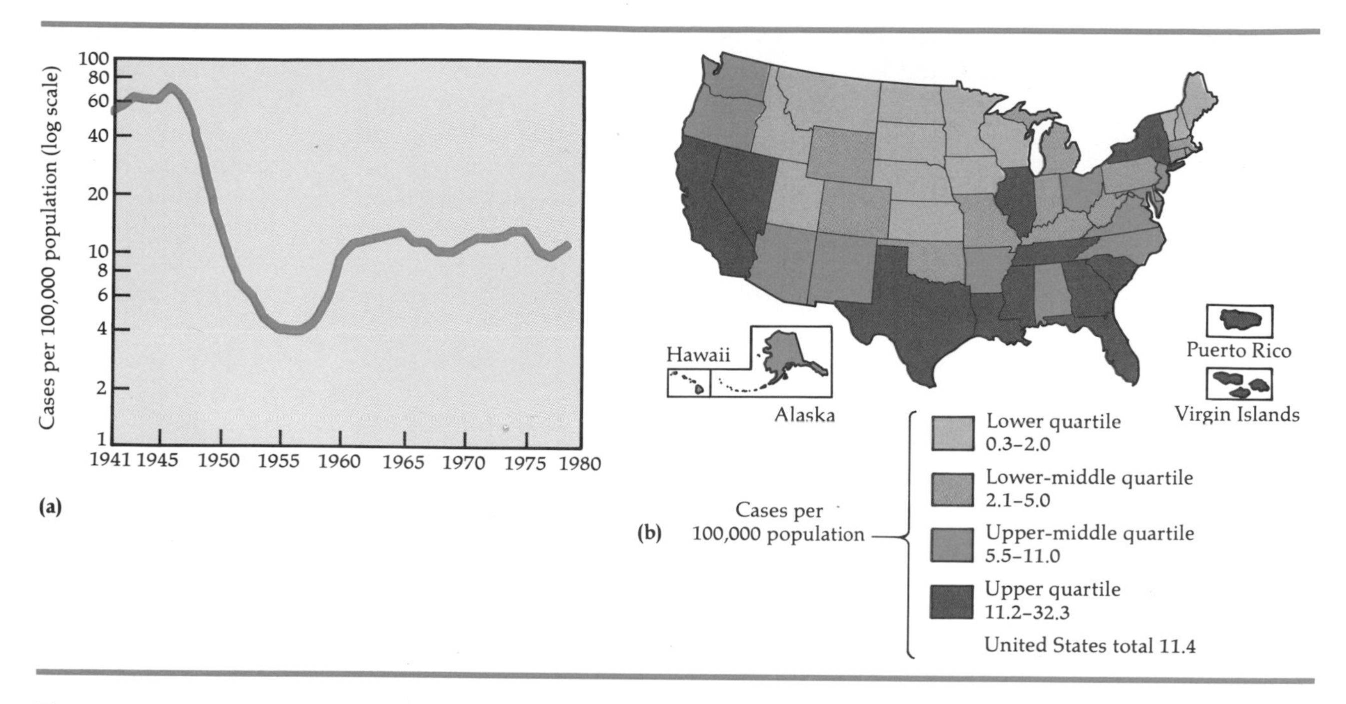

Figure 25–8 Incidence and distribution of syphilis. **(a)** Number of cases in the United States, 1941–1979. **(b)** Geographical frequency of cases reported in 1979.

remained fairly stable (Figure 25–8). About 25,000 cases were reported in 1979. This relative stability of incidence compared to gonorrhea is remarkable, since the epidemiology of the two diseases is quite similar and concurrent infections by both diseases are not uncommon. An important difference, however, is that syphilis has a longer incubation time, which facilitates the tracing of cases in time for effective treatment before the new contacts also become infective.

The causative organism is a gram-negative spirochete, *Treponema pallidum*. The spirochete is a thin, tightly coiled helix no more than 15 μm in length (see Figure 10–2). Since the narrowness of the *Treponema* (0.09 to 0.5 μm) makes them hard to see with an ordinary light microscope, living cultures taken from fluid exudates of syphilitic lesions are viewed with a darkfield microscope, the thin spirochetes appearing lighted against a black background. The number of bacteria in such fluids is so high that there is usually no difficulty in finding them. The pathogenic strains of the spirochetes have never been successfully cultured in vitro. Instead, they are propagated in rabbits.

Syphilis is transmitted by sexual contact of all kinds, with syphilitic infections of the genitals or other body parts. The organism is apparently able to penetrate the intact mucous membranes, but skin is penetrated only through minor breaks or abrasions. The highest incidence is in the 20 to 39 age group, which is the most sexually active. The incidence of syphilis among male homosexuals is about five times that of the general population.

The incubation period averages about three weeks, but can in fact range from two weeks to several months. The disease progresses through several recognized stages.

In the **primary stage** of the disease, the initial symptom is a small, hard-based **chancre,** or sore, which usually appears at the site of infection (Figure 25–9a). The chancre is painless, and a serous exudate is formed in the center. This fluid is highly infectious, and examination with the darkfield microscope shows many spirochetes. In a few weeks, this lesion disappears, possibly a result of developing local immunity. None of this causes any undue distress. In fact, many women are entirely unaware of the formation of the chancre, which is

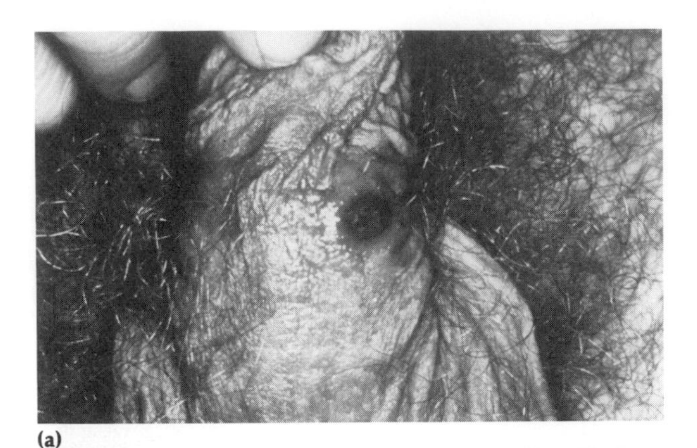

(a)

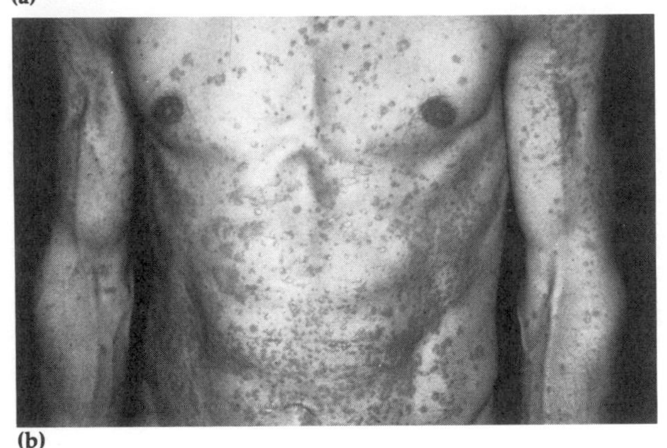

(b)

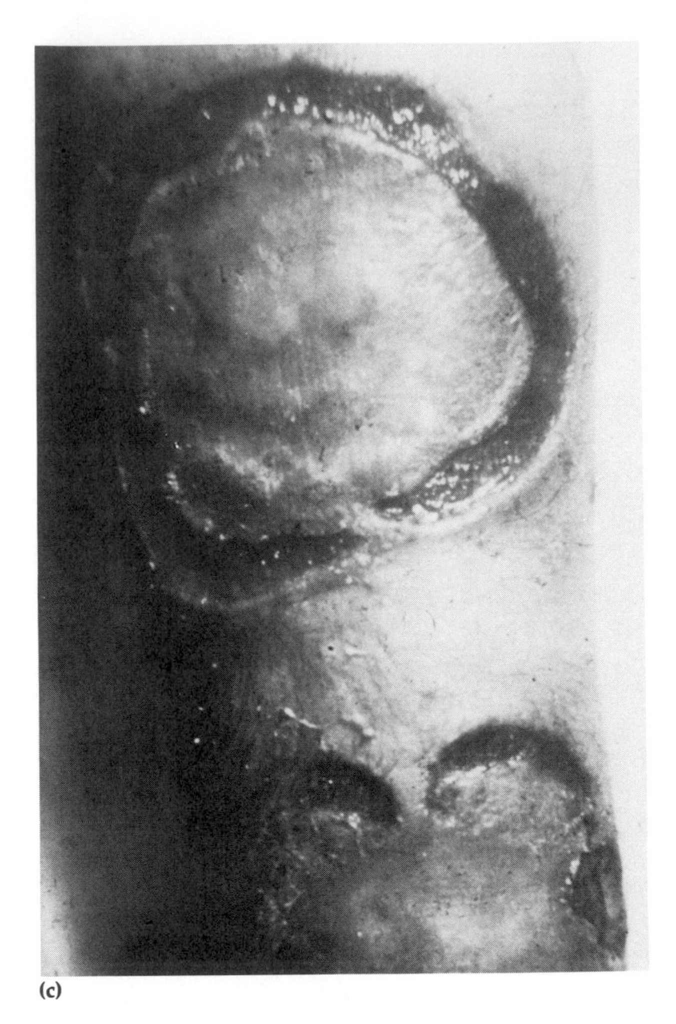

(c)

Figure 25–9 Characteristic lesions associated with syphilis at various stages of development. **(a)** Chancre of primary stage on a male. **(b)** Skin rash of secondary stage. **(c)** Gummas of tertiary stage on the back of an arm.

commonly on the cervix of the uterus. And in males, the chancre sometimes forms in the urethra and is not visible. Serological diagnostic tests are positive in about 80% of patients in the primary stage. During this stage, bacteria enter the bloodstream and lymphatic system and become widely distributed in the body.

Several weeks after the primary stage (the time varies), the disease enters the **secondary stage,** characterized mainly by skin rashes of varying appearance (Figure 25–9b). Other symptoms often observed are the loss of patches of hair, malaise, and mild fever. The rash is widely distributed on the skin and is also found in the mucous membranes of the mouth, throat, and cervix. The lesions of the rash at this stage also contain many spirochetes and are very infectious. Dentists or other medical workers coming into contact with fluid from these lesions can easily become infected by entrance of the spirochete through minute breaks in the skin. Such nonsexual transmission is always possible, but because the organisms do not survive long on environmental surfaces, they are very unlikely to be transmitted by such things as toilet seats. Serological tests for syphilis become almost uniformly positive at this stage.

The symptoms of secondary syphilis usually subside after a few weeks, and the disease enters a **latent period.** After about two years, the disease is not normally infectious, except for transmission from mother to fetus. The majority of cases do not progress beyond the secondary stage, even without treatment, perhaps because of developing immunity. Some resistance to reinfection is observed at this period.

Because the symptoms of primary and secondary syphilis are not disabling, it is not uncommon for persons to enter the latent period without having received medical attention. In less than half of these cases, the disease reappears as a **tertiary stage.** This stage occurs only after an interval of many years—usually at least 10. It has been suggested that most of the symptoms of tertiary syphilis are due to hyperimmune reactions of the body to surviving spirochetes. Many of the lesions are not unlike those formed by the body in response to infections by *Mycobacterium tuberculosis.* These lesions, called *gummas,* are rubbery masses of tissue that appear in many organs and sometimes on the external skin (Figure 25–9c). Although many of these lesions are not particularly harmful, some can cause extensive tissue damage, such as perforation of the palate (roof of the mouth), which interferes with speech, or deafness or blindness from lesions in the central nervous system. Few, if any, organisms are found in the lesions of the tertiary stage, and they are not considered very infectious. Today, cases of syphilis allowed to progress to this stage have become much less common.

One of the most distressing and dangerous forms of syphilis, called **congenital syphilis,** is transmitted across the placenta to the unborn fetus. Damage to mental development and other neurological symptoms are among the more serious consequences. This type of infection is most likely to occur in pregnancies during the latent stage of syphilis. Pregnancy in the primary or secondary stage is likely to produce a stillbirth.

In its primary and secondary stages, syphilis can often be diagnosed by darkfield examination of *T. pallidum* in fluid from lesions. During all stages of syphilis, serological tests based on detection of antibodies of more than one type are of great value. Lipid materials formed by the body in response to the spirochete infection apparently stimulate the appearance of *reagin-type antibodies.* These antibodies are not directed at the spirochete itself but are still indicative of a syphilitic infection. The *VDRL (Venereal Disease Research Laboratory) slide test* and a number of other similar rapid screening tests for syphilis are based on detection of these antibodies. The antigen used in such slide tests is not that of the treponemal organisms. Rather, it is an extract of beef heart (cardiolipin), which seems to contain lipids similar to those formed by the body that stimulated the reagin production.

Reagin tests will increase or decrease in titer as the disease progresses or regresses. This makes them useful in following the success of treatment, because changes in disease symptoms are difficult to interpret and of little use for this purpose. Most slide tests of this type are likely to result in a percentage of false positive reactions for syphilis. Therefore, any positive reactions should be confirmed by a more exacting test based on spirochete-type antigens. One such test is the *fluorescent treponemal antibody absorption (FTA-ABS) test.* This is an indirect immunofluorescence test in which a preparation of an avirulent, cultivated strain of *T. pallidum* is allowed to react with a sample of a patient's serum on a microscope slide (see Figure 16–23b). Antibodies in the serum will combine with the spirochetes. In the next step, excess serum is washed from the slide and fluorescent dye-tagged antibodies against human gamma globulin (anti-HGG) are added. The anti-HGG antibodies will combine with any human antibodies already on the spirochetes. The slide is again washed and examined with a special fluorescent-type microscope. The spirochetes are detected by their fluorescent glow when struck by ultraviolet light.

Penicillin continues to be effective in the treatment of syphilis, and no significant antibiotic resistance has appeared. Penicillin is used in a long-acting formulation that remains effective in the body for about two weeks, after which the dosage is usually repeated. One reason for this prolonged exposure is that the spirochete is slow growing and penicillin is effective only against growing, not dormant, organisms. In penicillin-sensitive persons, a number of other antibiotics such as erythromycin and the tetracyclines have also proved effective. Antibiotic therapy aimed at gonorrhea and other infections is not likely to eliminate syphilis as well, because the therapy is usually administered over too short a term. Penicillin therapy is especially effective during the primary stage. But antibiotics are generally not effective in tertiary syphilis, probably because the organisms are usually not present.

Nongonococcal Urethritis (NGU)

The term **nongonococcal urethritis (NGU),** also known as **nonspecific urethritis (NSU),** can refer to any inflammation of the urethra not caused by *Neisseria gonorrhoeae.* Symptoms include pain during urination and a watery discharge.

Although nonmicrobial factors such as trauma (passage of a catheter) or chemical agents (alcohol and certain chemotherapeutic agents) can cause this condition, probably at least 40% of the cases of NGU are acquired sexually. In fact, NGU may be the most common sexually transmitted disease in the United States today. Although it is not a notifiable disease and exact data are lacking, the Centers for Disease Control estimate that 4 to 9 million Americans have NGU. Since the symptoms are often mild in males, and females are usually asymptomatic, many cases of infection go untreated. Physicians often treat NGU as a male urological disease rather than as an STD. Complications are not common, but can be serious. Males may develop inflammation of the epididymis. And in females, inflammation may cause sterility by blocking the cervix or uterine tubes. As in gonorrhea, the bacteria can be passed from mother to infant during birth, infecting the eyes.

NGU may be caused by chlamydias. Chlamydias are very small, gram-negative bacteria and are obligate intracellular parasites (see Figure 10–15b). Since techniques for culturing chlamydias are not readily available, diagnosis is usually made by examining a smear or culture of the male discharge for *Neisseria gonorrhoeae*. If *N. gonorrhoeae* is not present, the disease is assumed to be NGU. Chlamydial infections can be treated successfully with tetracycline. Other bacteria that may also be implicated in NGU include *Mycoplasma hominis* and hemolytic staphylococci and streptococci. Infections by these microorganisms may require treatment by other antibiotics. NGU resulting from infections by the yeast *Candida albicans* or the protozoan *Trichomonas vaginalis* will be discussed later in this chapter.

Lymphogranuloma Venereum (LGV)

There are a number of STDs that are more common to the tropical areas of the world. For example, *Chlamydia trachomatis*, the cause of trachoma (see Chapter 19) and perhaps a major cause of NGU, is responsible for **lymphogranuloma venereum (LGV),** a disease found in much of the tropical or near-tropical world. In the United States, most cases occur in the southeastern states, with 250 cases reported in 1979.

After a latent period of 7 to 12 days, a small lesion appears at the site of infection, usually on the genitals. The lesion ruptures and heals without scarring. From one week to two months later, the microorganisms invade the lymphatic system, causing the regional lymph nodes to become enlarged and tender. Suppuration may also occur. The inflammation of the lymph nodes results in scarring that may occasionally obstruct the lymph vessels, leading to edema of the genital skin and massive enlargement of the external genitalia in males and rectal narrowing in females. In females, rectal narrowing results from involvement of the lymph nodes in the rectal region.

For diagnosis, pus may be aspirated from infected lymph nodes. When properly stained with an iodine preparation, the chlamydias may be seen as inclusions in infected cells. The isolated organisms can also be grown in cell culture or in embryonated eggs. Tetracycline and cycloserine antibiotics, as well as the sulfonamides, are effective in treatment. Enlarged nodes may be slow to subside, even after successful antibiotic therapy. Several serological tests are available, but all are troubled with frequent false positive reactions.

Chancroid (Soft Chancre) and Granuloma Inguinale

The STDs known as chancroid and granuloma inguinale are also rather more common in tropical areas. In the United States, there are about 800 cases of chancroid reported per year and fewer than 100 cases of granuloma inguinale. However, these diseases are seen with some frequency in sailors and overseas military personnel.

In **chancroid (soft chancre),** a swollen, painful ulcer forms on the genitalia that also involves an infection of the adjacent lymph nodes. About a week elapses between infection and appearance of the symptoms. Infected lymph nodes in the groin area may even break through and discharge pus to the surface. These ulcers are highly infective, as are the other genital lesions that occur. Lesions may also occur on such diverse areas as the tongue and lips. The causative organism is *Hemophilus ducreyi*, a small gram-negative rod that can be isolated from exudates of lesions. Sulfonamides, tetracyclines, and a number of other drugs are effective in treatment.

Granuloma inguinale is characterized by an initial chancre that ulcerates on or about the genitals. It may easily be confused with chancroid and primary syphilis. Without treatment, however, the ulcer spreads, and large areas of genital tissue and other tissues may be destroyed in time. The disease is most common in India, the Caribbean, and Southeast Asia. Epidemiology of the disease is not completely understood, but it is assumed that the disease is sexually transmitted, although not highly communicable. Male homosexuals are the most common sufferers of granuloma inguinale, although it also represents about 2 to 3% of the venereal disease found in newly inducted black military recruits in the United States. There is probably a great deal of individual resistance to the disease.

The causative organism is *Calymmatobacterium granulomatis*, a small, gram-negative, encapsulated bacillus, very similar in many respects to *Klebsiella pneumoniae.* In smears of scrapings from lesions, the encapsulated *C. granulomatis* can often be seen with large mononuclear cells called *Donovan bodies* (Figure 25–10). The bacteria can be grown in embryonated eggs and laboratory media. Diagnosis depends on demonstrating the presence of Donovan bodies in infected lesions and tissues. Gentamicin and chloramphenicol are the drugs of choice, although tetracyclines and streptomycin are also effective.

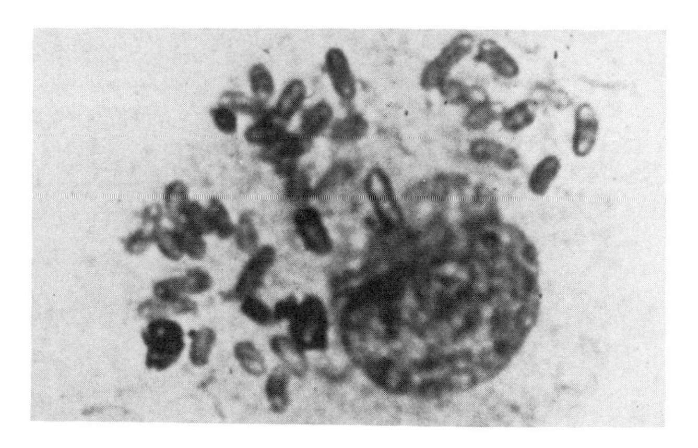

Figure 25–10 Numerous rodlike Donovan bodies are seen in the cytoplasm of this cell from a case of granuloma inguinale. (The darkly-stained nucleus is clear but the plasma membrane around the cell is not.)

VIRAL DISEASE OF THE GENITAL SYSTEM

Genital Herpes

Much more troublesome to the U.S. population than granuloma inguinale is an STD caused by the *Herpes simplex virus.* The Herpes simplex virus (see Figures 12–6b and 12–8a) occurs as type 1 and type 2. Herpes simplex type 1 is the virus primarily responsible for the common cold sore or fever blister (see Chapter 19). Herpes simplex type 2, and sometimes type 1, causes a disease of sexual transmission that affects the genitalia, called **genital herpes.** The lesions appear after an incubation period of about a week, causing a burning sensation. After this, vesicles appear. Urination may be painful in both sexes, and the patient finds it quite uncomfortable to walk and is even irritated by clothing. The vesicles contain fluid that is infectious. Usually, the vesicles heal in a couple of weeks.

The incidence of genital herpes has increased so much over the past decade that it is now a very common sexually transmitted disease in the United States. Females are less likely to be aware that they have the disease. This may be serious because there is danger that a newborn infant may become infected at birth. Such *neonatal herpes* may be a mild asymptomatic infection or it may become rapidly fatal. And infant survivors may be left with serious damage to the central nervous system. Therefore, for females known to have active genital herpes infections, delivery by Caesarian section may be advisable. Occasionally, the infection takes place earlier, in the uterus, also causing birth defects.

As in the case of cold sores, herpesviruses infecting the genitalia may enter a latent state in nerve cells. Various traumas and hormonal changes may cause the reappearance of the lesions. Some persons are more susceptible than others to frequent recurrences of genital herpes lesions.

Genital herpes infection seems to be associated with a higher than normal rate of cervical cancer. It is also associated with a higher rate of miscarriages. Although herpes infections of the eye and certain other herpes infections have been successfully treated with antiviral drugs, there is no proven, effective treatment for genital herpes. The

Microbiology in the News

Drug Shows Promise Against Herpes

July 1981—The herpesviruses have until recently been one of the body's most persistent antagonists. Nearly everyone harbors one of the five types of herpes in latent form, as the intractable residue of ailments ranging from cold sores to chicken pox, from mononucleosis to venereal disease. The viruses lodge in a variety of cells and when the body is weakened, they replicate and wreak their own peculiar havoc. Victims of herpes can experience painful blistering of the skin, lung damage leading to pneumonia, liver infections, blood clots, or encephalitis.

Cancer or transplant patients taking drugs that suppress natural immunity are at particularly high risk of herpes infection, which is painful for most and fatal in a few. Until recently, no effective treatment for herpes existed, but a promising new drug may appear on the market soon. Recently clinical trials of acyclovir, manufactured by the Burroughs Wellcome Co., indicate that it can eliminate active forms of herpesvirus for at least the duration of immunosuppressive therapy.

Results of several trials were reported in the 27 June *Lancet* and the 9 July *New England Journal of Medicine.* A trial conducted by Charles Mitchell and others at the University of Minnesota involved 11 recipients of heart or liver transplants and 13 victims of cancer or aplastic anemia, each of whom had an active form of herpes simplex virus (of the types associated with cold sores and venereal disease). Treatment with acyclovir significantly shortened the duration of pain and virus replication. Similar results were obtained in a study by Sunwen Chou and others at Stanford University, involving ten heart transplant patients, and in a study by Rein Saral and others at Johns Hopkins University, involving 20 recipients of bone marrow transplants, all of whom had active herpes simplex infections.

These are the first results of clinical trials with acyclovir to be reported and additional studies must be completed before it can be licensed. The drug, whose chemical name is acycloguanosine, was discovered at Burroughs Wellcome in 1974 during a search for analogs of guanine—a building block of DNA—that would selectively stop DNA replication. There is no indication that the drug will eradicate the dormant (or nonreplicating) form of herpes, and indeed, in the three reported studies, the virus recurred in most patients after treatment with acyclovir ceased.

Still, the number of persons who would benefit from it is thought to number in the hundreds of thousands. About 70 million people annually get cold sores; about 20 million have contracted the herpes venereal disease, with 5 million new cases each year. The Food and Drug Administration may allow a topical ointment preparation of acyclovir on the market by the end of the year; injectable, oral, and ophthalmic preparations are also being considered.

At this early stage of its testing, one of the drug's major advantages appears to be its lack of toxicity. Vidarabine, an existing drug effective against several herpesviruses, has been licensed only for use in herpes simplex encephalitis cases, because of side effects and animal tests indicating it is a mutagen, carcinogen, and teratogen. No similar problems have yet appeared with acyclovir.

Source: R. Jeffrey Smith, *Science* 213:524, July 31, 1981.

chemical 2-deoxy-D-glucose applied topically showed promise in a recent study, but further testing will be necessary to establish its effectiveness. For an update on drug treatment for herpes, see the box. The virus may be isolated in cell culture, where it produces cytopathic effects that allow it to be differentiated from type 1 viruses. This differentiation may also be made with serological tests.

FUNGAL DISEASE OF THE GENITAL SYSTEM

Candidiasis

Candida albicans is a yeastlike fungus that commonly grows on mucous membranes of the mouth, intestinal tract, and genitourinary tract (see Figure 19–14a and Color Plate IVC). As already discussed in Chapter 19, *C. albicans* is the cause of thrush (oral candidiasis). It is also responsible for occasional cases of NGU in males and a common infection of the mucous membrane of the vagina called **vulvovaginal candidiasis.** The lesions of vulvovaginal candidiasis resemble those of thrush but produce more irritation, severe itching, and a thick, yellow, cheesy discharge. *C. albicans* is an opportunistic pathogen. Predisposing conditions to the disease include pregnancy, diabetes, certain tumors, and intensive treatment with broad-spectrum antibacterial drugs or immunosuppressive drugs.

Vulvovaginal candidiasis is diagnosed by microscopic identification of the fungus in scrapings of lesions and by isolation of the fungus in culture. Treatment consists of the application of nystatin to accessible lesions and the administration of amphotericin B to treat severe systemic infections.

PROTOZOAN DISEASE OF THE GENITAL SYSTEM

Trichomoniasis

The protozoan *Trichomonas vaginalis* is a fairly normal inhabitant of the vagina in females and of the urethra in many males (Figures 25–11 and 11–16). If the usual acidity of the vagina is disturbed, the protozoan may overgrow the normal microbial population of the genital mucosa, causing **trichomoniasis**. (Males rarely have any symptoms as a result of the presence of the organism.) In response to the protozoan infection, the body accumulates leukocytes at the infection site, causing a purulent discharge that is profuse, yellowish or light cream in color, and characterized by a disagreeable odor.

Diagnosis is easily made by identification of the organisms by microscopic examination of the discharge. Male carriers may show the organism in semen or urine. Treatment is by oral metronidazole, which readily clears the infection. The organism can also be isolated and grown on laboratory media.

The major microbial diseases of the urinary and genital systems are summarized in Table 25–1.

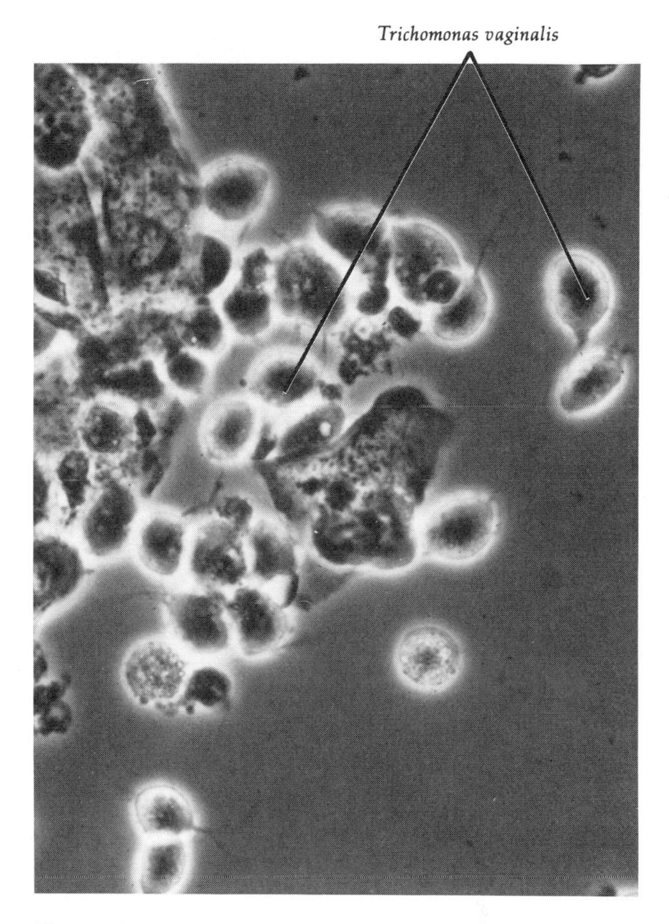

Figure 25–11 *Trichomonas vaginalis*, the protozoan causing trichomoniasis.

Table 25–1 Summary of Diseases Associated with the Urinary and Genital Systems

Disease	Causative Agent	Mode of Transmission	Treatment
Bacterial Diseases of the Urinary System:			
Cystitis (bladder infection)	*E. coli, Proteus vulgaris, Pseudomonas aeruginosa,* and others	Opportunistic infections	Sulfonamides, chloramphenicol, kanamycin, penicillin G, or polymyxin
Pyelonephritis (kidney infection)	*E. coli* is most frequently implicated; other agents include *Enterobacter aerogenes, Proteus* species, *Pseudomonas aeruginosa, Streptococcus pyogenes,* and staphylococci	From systemic bacterial infections or infections of the lower urinary tract	Depending on agent, antibiotics include chloramphenicol, gentamicin, kanamycin, polymyxin, methicillin, penicillin, vancomycin, and tetracycline
Leptospirosis (kidney infection)	*Leptospira interrogans*	Direct contact with infected animals, urine of infected animals, or contaminated water	Penicillin
Glomerulonephritis (kidney infection)	Certain strains of *Streptococcus pyogenes*	Sequel to infection by certain strains of *S. pyogenes* in another part of the body	Chloramphenicol, erythromycin, lincomycin, penicillin, or vancomycin for the initial infection
Bacterial Diseases of the Genital System:			
Gonorrhea	*Neisseria gonorrhoeae*	Direct contact, especially sexual contact	Penicillin for nonresistant strains; spectinomycin for resistant strains
Syphilis	*Treponema pallidum*	Direct contact, especially sexual contact	Penicillin, erythromycin, tetracycline
Nongonococcal urethritis (NGU)	Chlamydias or other bacteria, including *Mycoplasma hominis,* and hemolytic staphylococci and streptococci	Sexual contact or opportunistic infections	Tetracycline, erythromycin, or other antibiotics
Lymphogranuloma venereum (LVG)	*Chlamydia trachomatis*	Direct contact, especially sexual contact	Tetracycline, cycloserine antibiotics, and sulfonamides
Chancroid (soft chancre)	*Hemophilus ducreyi*	Direct contact, especially sexual contact	Tetracycline and sulfonamides
Granuloma inguinale	*Calymmatobacterium granulomatis*	Probably sexual contact	Gentamicin, chloramphenicol, tetracycline, streptomycin
Viral Disease of the Genital System:			
Genital herpes	Herpes simplex virus type 2, and occasionally type 1	Direct contact, especially sexual contact	No effective treatment
Fungal Disease of the Genital System:			
Vulvovaginal candidiasis (occasionally urethritis)	*Candida albicans*	Opportunistic pathogen, may be transmitted by sexual contact	Nystatin and amphotericin B
Protozoan Disease of the Genital System:			
Trichomoniasis (usually vaginal infection)	*Trichomonas vaginalis*	Opportunistic infection, may be transmitted by sexual contact or contact with fomites	Metronidazole

STUDY OUTLINE

INTRODUCTION (p. 639)

1. The urinary system regulates the chemical composition of the blood and excretes waste products of metabolism.
2. The genital system produces gametes for reproduction and, in the female, also supports the growing embryo.
3. The urinary and genital systems are closely related anatomically, and both open to the external environment.
4. Microbial disease of these systems may result from infection from an outside source or opportunistic infection by normal body flora.

STRUCTURE AND FUNCTION OF THE URINARY SYSTEM (pp. 639–640)

1. The nephrons of the kidneys remove urea, uric acid, creatinine, salts, and water from the blood and form urine.
2. Urine leaves the kidneys through ureters to the urinary bladder and is eliminated through the urethra.
3. Valves prevent urine from flowing back to the urinary bladder and kidneys.
4. The flushing action of urine and the acidity of normal urine have some antimicrobial value.

STRUCTURE AND FUNCTION OF THE GENITAL SYSTEM (p. 640)

1. The female genital system consists of two ovaries, two uterine (fallopian) tubes, the uterus, the vagina, and external genitalia.
2. The vagina functions as the birth canal and copulatory canal.
3. The male genital system consists of two testes, ducts, accessory glands, and the penis.
4. Seminal fluid leaves the male body through the urethra.

NORMAL FLORA OF THE URINARY AND GENITAL SYSTEMS (pp. 640–641)

1. *Streptococcus*, *Bacteroides*, *Mycobacterium*, *Neisseria*, and enterobacteria are resident flora of the urethra.
2. The urinary bladder and upper urinary tract are sterile under normal conditions.
3. Lactobacilli dominate vaginal flora during reproductive years.

BACTERIAL DISEASES OF THE URINARY SYSTEM (pp. 641–644)

1. Opportunistic gram-negative bacteria from the intestines often cause urinary tract infections, especially in females.

2. Predisposing factors of urinary tract infections include nervous system disorders, toxemia, diabetes mellitus, and obstructions to the flow of urine.
3. Urethritis, cystitis, and ureteritis describe inflammation of tissues of the lower urinary tract.
4. Pyelonephritis may result from lower urinary tract infections or systemic bacterial infections.
5. Diagnosis and treatment of urinary tract infections depends on the isolation and antibiotic sensitivity testing of the etiological agents.
6. More than 100,000 bacteria per milliliter of urine indicates an infection.
7. About 35% of all nosocomial infections occur in the urinary system. *E. coli* causes more than half of these infections.

Cystitis (pp. 642–643)

1. Cystitis, inflammation of the urinary bladder, is common in females.
2. Microorganisms at the opening of the urethra and along the length of the urethra, careless personal hygiene, and sexual intercourse contribute to the high incidence of cystitis in females.

Pyelonephritis (p. 643)

1. Inflammation of the kidneys, or pyelonephritis, is usually a complication of lower urinary tract infections.
2. About 75% of pyelonephritis cases are caused by *E. coli.*

Leptospirosis (pp. 643–644)

1. The spirochete *L. interrogans* causes leptospirosis.
2. Leptospirosis is characterized by chills, fever, headache, and jaundice.
3. Serological identification is difficult because of many serotypes of *L. interrogans.*

Glomerulonephritis (p. 644)

1. Glomerulonephritis, or Bright's disease, is an immune-complex disease occurring as a sequel to a beta-hemolytic streptococcal infection.
2. Antigen–antibody complexes are deposited in the glomeruli, where they cause inflammation and kidney damage.

BACTERIAL DISEASES OF THE GENITAL SYSTEM (pp. 644–651)

1. Most diseases of the genital system are sexually transmitted diseases (STDs).

2. Most STDs can be prevented by the use of condoms and treated with antibiotics.

Gonorrhea (pp. 644–646)

1. *N. gonorrhoeae* causes gonorrhea.
2. Gonorrhea is the most common reportable communicable disease in the United States.
3. *N. gonorrhoeae* attaches to mucosal cells of the oral-pharyngeal area, genitalia, eyes, joints, and rectum by means of pili.
4. Symptoms in males are painful urination and pus discharge. Blockage of the urethra and sterility are complications of untreated cases.
5. Females may be asymptomatic until the infection spreads to the uterus and uterine tubes. Blockage of the uterine tubes and pelvic inflammatory disease are complications of untreated cases.
6. Gonorrheal endocarditis, gonorrheal meningitis, and gonorrheal arthritis are complications that can affect both sexes if gonorrheal infections are untreated.
7. Ophthalmia neonatorum is an eye infection acquired by infants during passage through the birth canal of an infected mother.
8. Penicillin has been used effectively to treat gonorrhea. In 1976, penicillinase-producing strains of *N. gonorrhoeae* appeared.
9. Direct fluorescent-antibody tests can be used to confirm identification of gram-negative diplococci isolated from patients.

Syphilis (pp. 646–649)

1. Syphilis is caused by *T. pallidum*, a spirochete that has not been cultured in vitro. Laboratory cultures are grown in rabbits.
2. *T. pallidum* is transmitted by direct contact and can invade intact mucous membranes or penetrate through breaks in the skin.
3. The primary lesion is a small, hard-based chancre at the site of infection. The bacteria invade the blood and lymphatic system and the chancre spontaneously heals.
4. The appearance of a widely disseminated rash on the skin and mucous membranes marks the secondary stage. Spirochetes are present in the lesions of the rash.
5. The patient enters a latent period after spontaneous healing of the secondary lesions.
6. At least 10 years after the secondary lesion, tertiary lesions called gummas may appear on many organs.
7. Congenital syphilis, resulting from *T. pallidum* crossing the placenta during the latent period, may result in neurological damage in the newborn.

8. *T. pallidum* is identifiable by darkfield microscopy of fluid from primary and secondary lesions.
9. Many serological tests, such as VDRL and FTA-ABS, are used to detect the presence of antibodies against *T. pallidum* during any stage of the disease.

Nongonococcal Urethritis (NGU) (pp. 649–650)

1. NGU, or nonspecific urethritis, is an inflammation of the urethra not caused by *N. gonorrhoeae.*
2. About 40% of NGU cases are caused by *Chlamydia* and are usually transmitted sexually.
3. Symptoms of NGU are often mild or lacking.
4. Congenital eye infections may occur.
5. *Mycoplasma hominis,* hemolytic staphylococci, and hemolytic streptococci have been implicated in NGU.

Lymphogranuloma Venereum (LGV) (p. 650)

1. *C. trachomatis* causes LGV, primarily a disease of tropical and subtropical regions.
2. The initial lesion appears on the genitalia and heals without scarring.
3. The bacteria are spread in the lymph system and cause enlargement of lymph nodes, obstruction of lymph vessels, and edema of the genital skin.
4. The bacteria are isolated and identified from pus taken from infected lymph nodes.

Chancroid (Soft Chancre) and Granuloma Inguinale (pp. 650–651)

1. Chancroid is a swollen, painful ulcer on the mucous membranes of the genitalia or mouth caused by *H. ducreyi.*
2. *C. granulomatis* causes granuloma inguinale.
3. Granuloma inguinale begins as an ulcer on the genitalia that ulcerates and spreads to other tissues.
4. Diagnosis is based on observation of mononuclear cells called Donovan bodies.

VIRAL DISEASE OF THE GENITAL SYSTEM (pp. 651–652)

Genital Herpes (pp. 651–652)

1. *Herpes simplex* type 2 causes an STD called genital herpes.
2. Symptoms of the infection are painful urination, irritation, and fluid-filled vesicles.

3. Neonatal herpes may be contracted during birth. It may be asymptomatic or it may result in neurological damage or infant fatalities.
4. The virus may enter a latent stage in nerve cells. Vesicles reappear following trauma and hormonal changes.
5. The disease is not contagious during the latent periods.
6. Genital herpes is associated with cervical cancer and miscarriages.

FUNGAL DISEASE OF THE GENITAL SYSTEM (p. 653)

Candidiasis (p. 653)

1. *C. albicans* causes NGU in males and vulvovaginal candidiasis in females.
2. Vulvovaginal candidiasis is characterized by lesions that produce itching and irritation.
3. Predisposing factors for candidiasis are pregnancy, diabetes, tumors, and broad-spectrum antibacterial chemotherapy.

PROTOZOAN DISEASE OF THE GENITAL SYSTEM (p. 653)

Trichomoniasis (p. 653)

1. *T. vaginalis* causes trichomoniasis when the pH of the vagina increases.
2. Diagnosis is based on observation of the protozoan in purulent discharges from the site of infection.

STUDY QUESTIONS

REVIEW

1. List the normal flora of the urinary system and show their habitats in Figure 25–1.
2. List the normal flora of the genital system and show their habitats in Figures 25–2 and 25–3.
3. How are urinary tract infections transmitted?
4. Explain why *E. coli* is frequently implicated in cystitis in females. List some predisposing factors for cystitis.
5. Name one organism that causes pyelonephritis. What are the portals of entry for organisms that cause pyelonephritis?

6. Complete the following table.

Disease	Causative Agent	Symptoms	Method of Diagnosis	Treatment
Leptospirosis				
Gonorrhea				
Syphilis				
NGU				
LGV				
Chancroid				
Granuloma inguinale				

7. Name one fungus and one protozoan that can cause genital system infections. What symptoms would lead you to suspect these infections?

8. Describe the symptoms of genital herpes. What is the etiological agent? When is this infection least likely to be transmitted?

9. What is glomerulonephritis? How is it transmitted?

10. List the genital infections that cause congenital and neonatal infections. How can transmission to a fetus or newborn be prevented?

CHALLENGE

1. Why can frequent douching be a predisposing factor to vulvovaginal candidiasis or trichomoniasis?

2. Leptospirosis is a kidney infection of humans and other animals. How is this disease transmitted? What types of activities would increase one's exposure to this disease?

3. Matching.

___Candidiasis	(a) Spirochetes that can be cultured in vitro
___Chancroid	(b) Gram-negative diplococci
___Gonorrhea	(c) Spirochetes that are cultured in vivo
___Granuloma inguinale	(d) Gram-negative intracellular parasites
___Leptospirosis	(e) Gram-negative rods
___LGV	(f) Donovan bodies
___Syphilis	(g) Pseudohyphae
___Trichomoniasis	(h) Flagellate

4. The tropical skin disease called yaws is transmitted by direct contact. The causative agent is *Treponema pertenue* and it is indistinguishable from *T. pallidum*. The appearance of syphilis in Europe coincides with the first importation of slaves. How might *T. pertenue* have evolved into *T. pallidum* in the temperate climate of Europe?

FURTHER READING

Rafferty, K. A. October 1973. Herpes viruses and cancer. *Scientific American* 229:26–33. Evidence that herpesviruses cause cancer in humans.

Rapp, F. 1978. Herpesviruses, venereal disease, and cancer. *American Scientist* 66:670–674. Differentiates between the association of a virus in cancer and the cause of the cancer.

Rosebury, T. 1971. *Microbes and Morals.* New York: Viking Press. A history of venereal diseases from Fracastor through modern antibiotics, including laboratory diagnosis, treatment, and the prospects for vaccines.

Schachter, J. Chlamydiae. *Annual Review of Microbiology* 34:285–309. A good review of the clinical features and epidemiology of chlamydial infections.

See also Further Reading for Part IV at the end of Chapter 19.

PART

V

MICROBIOLOGY, THE ENVIRONMENT, AND HUMAN AFFAIRS

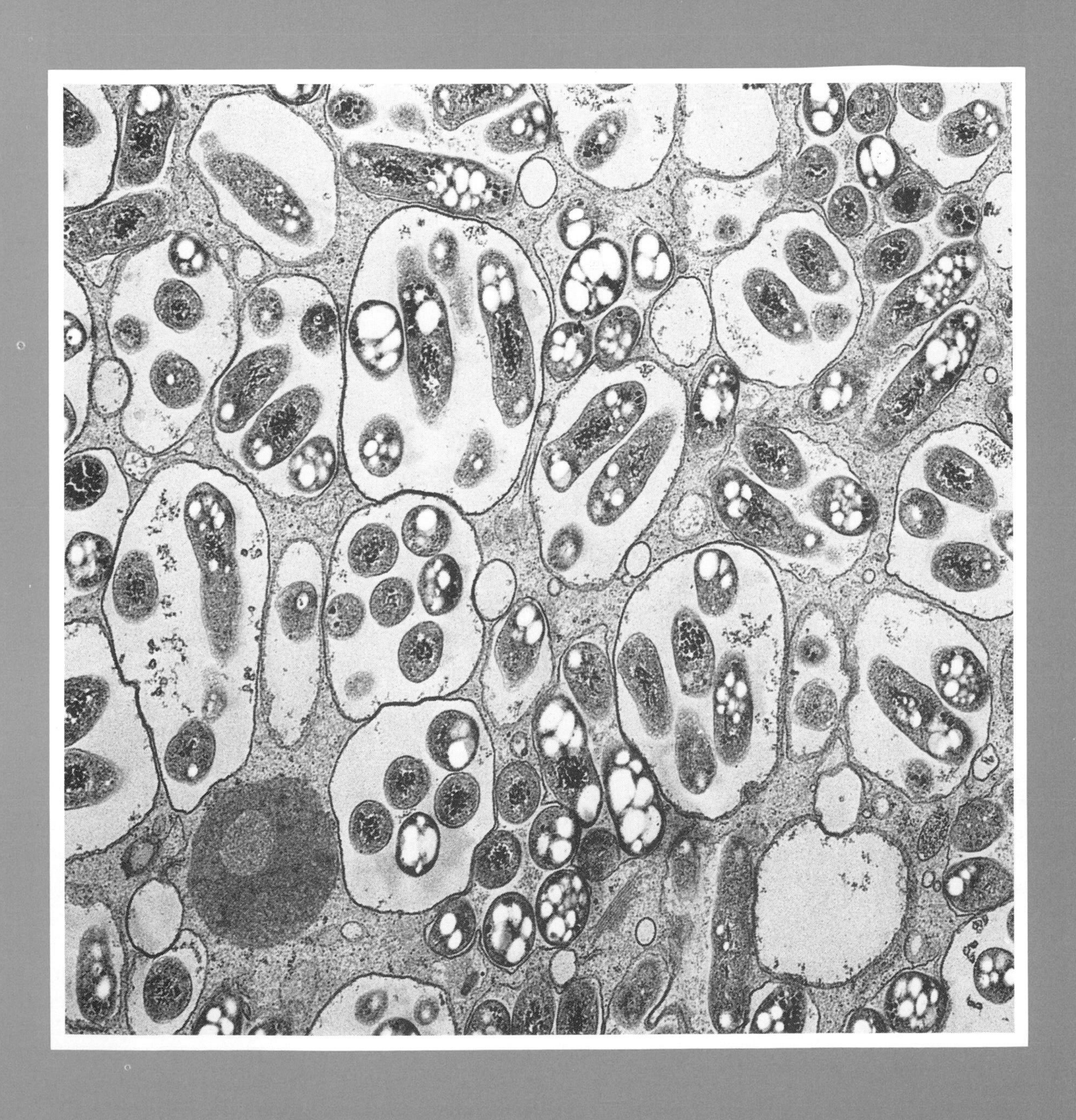

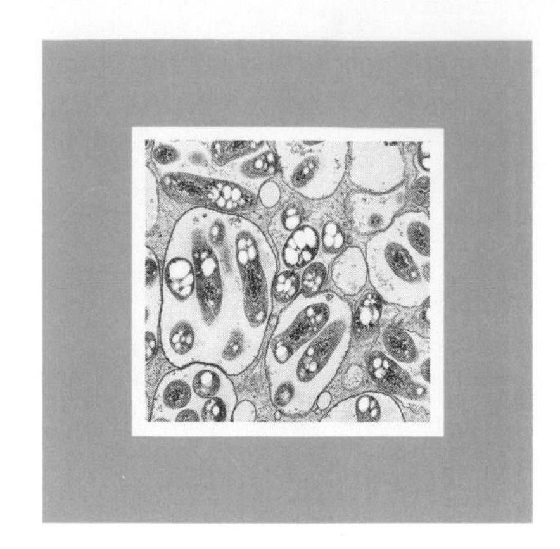

26

Soil and Water Microbiology

Objectives

After completing this chapter you should be able to

- Explain how the components of soil affect soil microflora.
- Outline the carbon and nitrogen cycles and explain the role of microorganisms in these cycles.
- Describe the freshwater and seawater habitats of microorganisms.
- Explain how water is tested for bacteriologic quality.
- Compare primary, secondary, and tertiary sewage treatment.
- List some of the biochemical activities that take place in an anaerobic sludge digester.
- Define each of the following terms: BOD; septic tank; oxidation pond; activated sludge; trickling filter.
- Discuss the causes and effects of eutrophication.

Many people immediately associate bacteria with disease or food spoilage, thinking of the microorganisms as things to be avoided or controlled, as things that can hurt them. Actually, microorganisms that are harmful represent only a very small fraction of the total. In fact, bacteria and other microorganisms are intimately involved in the very maintenance of life on earth and perform many essential functions. In this chapter, we will consider the importance of microorganisms as they relate to soil and water.

SOIL MICROBIOLOGY AND CYCLES OF THE ELEMENTS

THE COMPONENTS OF SOIL

Soil is a relatively complex mixture of solid inorganic matter (rocks and minerals), water, air, and living organisms and the products of their decay, in which numerous physical and chemical changes are taking place. The uppermost layer of soil, called the topsoil, is the most important as far as living organisms are concerned because of their direct contact with this layer. Topsoils vary considerably in physical texture, chemical composition, origin, depth, and fertility.

Rock Particles and Minerals

Most soil consists of a mixture of rock and mineral fragments that have been formed by the weathering of preexisting rock. The weathering processes include precipitation, wind, and temperature fluctuations (especially involving the freezing of water), leading to physical and chemical breakdown. As a result of weathering, elements such as silicon, aluminum, and iron are added to the soil. Calcium, magnesium, potassium, sodium, phosphorus, and small amounts of other elements are also present in the mineral component of soil.

Soil Water

Soil also contains water, and the amount present depends on the amount of precipitation as well as on the climate and drainage. The water is found between the soil particles and adheres to the soil particles. Through the solvating property of water, various inorganic and organic constituents of soil are dissolved in the soil water and made available to the living inhabitants of the soil.

Soil Gases

The gases in the soil are essentially the same as those found above the soil, and they include carbon dioxide, oxygen, and nitrogen. However, the relative proportions of soil gases may be different from those of atmospheric gases because of the biological processes occurring in soil. For example, as a result of respiration, soil contains a higher proportion of carbon dioxide and a lower proportion of oxygen. Soil gases are found mainly between soil particles or dissolved in the soil water.

Organic Matter

All organic matter in soil is derived from the remains of microorganisms, plants, and animals, their waste products, and the biochemical activities of various microorganisms. The organic matter in soil consists of carbohydrates, proteins, lipids, and other materials, and much of it is ultimately decomposed to inorganic substances such as ammonia, water, carbon dioxide, and various nitrate, phosphate, and calcium compounds. A great portion of the organic matter of soils is of plant origin, principally from dead roots, wood and bark, and fallen leaves. A second source of organic matter is from the vast numbers of bacteria, fungi, algae, protozoans, small animals, and viruses, which may total billions per gram of fertile soil. The breakdown of organic substances through the action of these soil organisms produces and maintains a continuous supply of inorganic substances that plants and other organisms require for growth. In most agriculturally important soils, organic matter comprises from 2 to 10%. Swamps and bogs, by contrast, contain a higher content of organic matter, up to 95% in some peat bog soils. The waterlogged soil of swamps and bogs is an anaerobic environment, and the microbial decomposition of organic matter occurs only slowly.

A considerable part of the organic matter in soil occurs as **humus,** a dark-colored material composed chiefly of organic materials that are relatively resistant to decay. In this regard, humus is partially decomposed organic matter.

The addition of organic matter, either completely or partially decomposed, is essential to continued soil fertility. Moreover, because of their spongy nature, organic materials loosen the soil, preventing the formation of heavy crusts and increasing the proportion of pore spaces in the soil, which in turn increases aeration and water retention.

Organisms

Fertile soil contains a great many animals, ranging from microscopic forms, including numerous nematodes, to larger forms, including insects, millipedes, centipedes, spiders, slugs, snails, earthworms, mice, moles, gophers, and reptiles. Most of these animals are beneficial in that they promote some mechanical movement of the soil, thereby keeping the soil loose and open. In addition, all soil organisms contribute to the organic matter of soils as a result of their waste products and eventual death.

Soil also contains the root systems of higher plants and enormous numbers of microorganisms.

Without the presence of microorganisms, especially bacteria, the soil would soon become unfit to support life. Bacteria affect the soil in many ways. Most decompose organic matter into simple products. In these reactions, nutrients are made available for reuse. Other soil bacteria are associated with the transformation of nitrogen and sulfur compounds, so that usable supplies of nitrogen and sulfur are continually maintained.

The soil microflora

The soil constitutes one of the main reservoirs of microbial life. A good agricultural soil the size of a football field usually contains a microbial population approaching the weight of a cow eating grass on that field. But the metabolic capabilities of these vast numbers of microorganisms is probably about 100,000 times that of the grazing cow. However, measurement of the carbon dioxide evolved from soil and other evidence indicate that these organisms are existing in a condition of near starvation at low reproductive rates. When usable nutrients are added to soil, the microbial populations and their activity rapidly increase until the nutrients are depleted, and then they return to lower levels.

The most numerous organisms in soil are bacteria (Table 26–1). A typical garden soil will have millions of bacteria in each gram. The population is highest in the top few centimeters of the soil, declining rapidly with depth. The populations are usually estimated by plate counts on nutrient media, and the actual numbers are probably greatly underestimated. No single nutrient medium or growth condition can possibly meet all the myriad nutrient or other requirements of soil microorganisms.

Actinomycetes, although bacteria, are usually considered separately in enumerating soil populations. Found in soil in large numbers, they produce a substance called *geosmin*, which gives fresh soil its characteristic musty odor. The numbers reported may reflect only the formation of asexual spores and fragmentation of the mycelium of these filamentous organisms. The actual biomass (total mass of living organisms in a given volume) of the actinomycetes is probably about that of the conventional bacteria. Interest in these bacteria was greatly stimulated by the discovery that they include genera, particularly *Streptomyces*, that produce valuable antibiotics.

Fungi are found in soil in much fewer numbers than bacteria and actinomycetes. Because many of the counted fungal colonies arise from the germination on media of asexual spores, the true relationship between the count and the actual fungal population is questionable. Estimates of the biomass of fungi indicate that it probably equals that of the bacteria and actinomycetes combined. This is because the dimensions of the fungal mycelium are many times greater than that of bacterial cells. Molds greatly outnumber yeasts in soil.

Algae and cyanobacteria sometimes form visible **blooms** (abundant growths) on the surface of moist soils but are also found in dry desert soils. As might be expected of photosynthetic organisms, they are located mainly on the surface layer, where sunlight, water, and carbon dioxide are most abundant. Nonetheless, significant numbers of algae and cyanobacteria are found more than 50 cm below the surface. The environmental contribution of these microorganisms is significant in only special cases. For example, fixation of atmospheric nitrogen (discussed shortly) by some species of cyanobacteria in grasslands, tundra regions, and after rainfall in deserts is significant in soil fertility.

Pathogens in soil

Human pathogens, adapted to life as parasites, find the soil an alien, hostile environment. Even

Table 26–1 Distribution of Microorganisms in Numbers Per Gram of Typical Garden Soil at Various Depths

Depth (cm)	Bacteria	Actinomycetes	Fungi	Algae
3–8	9,750,000	2,080,000	119,000	25,000
20–25	2,179,000	245,000	50,000	5,000
35–40	570,000	49,000	14,000	500
65–75	11,000	5,000	6,000	100
135–145	1,400	—	3,000	—

Source: Adapted from M. Alexander, *Introduction to Soil Microbiology*, 2nd ed., Wiley, New York, 1977.

relatively resistant enteric pathogens such as *Salmonella* species, when introduced into soil, have been observed to survive for only a period of a few weeks or months. Most human pathogens that can survive in soil are endospore-forming bacteria. Endospores of *Bacillus anthracis,* which causes anthrax in animals, can survive in certain soils for decades, finally germinating when ingested by grazing animals. Considerable care is needed in disposing of the body of an animal infected with anthrax to avoid seeding the soil with the endospores from the dead animal. *Clostridium tetani* (tetanus), *Clostridium botulinum* (botulism), and *Clostridium perfringens* (gas gangrene) are also examples of endospore-forming pathogens whose normal habitat is the soil. From soil they are introduced into foods or wounds, where they grow and elaborate toxins.

Pathogens of plants are much more likely to be normal soil inhabitants. Most plant pathogens are fungi, mainly because fungi are capable of growth in the low moisture typical of plant surfaces. (Bacterial diseases of plants are uncommon.) Many of the rusts, smuts, blights, and wilts affecting plants are caused by fungi that pass part of their life cycle in soil.

Some of the microorganisms found in soil are insect pathogens, potentially useful for pest control. *Bacillus thuringiensis,* for example, is a soil bacterium pathogenic to the larvae of many insects and is now used as an aid in their control. The ingested endospore germinates, and the growing bacillus forms a toxic protein crystal that eventually kills the insect. There are a number of other promising insect pathogens, found in soil, including viruses and fungi, that are under investigation as biological pesticides.

MICROORGANISMS AND BIOGEOCHEMICAL CYCLES

Perhaps the most important role of soil microorganisms is their participation in **biogeochemical cycles,** that is, the recycling of certain chemical elements so that they can be used over and over again. Among these elements are carbon, nitrogen, sulfur, and phosphorus. Were it not for the activities of microorganisms in the various biogeochemical cycles, essential elements would become depleted and all life would cease. The first biogeochemical cycle we will consider is the carbon cycle.

Carbon Cycle

As you already know, all organic compounds contain carbon. Most of the inorganic carbon used to synthesize organic compounds comes from the carbon dioxide (CO_2) in the atmosphere (Figure 26–1). Some carbon is also dissolved in water.

The first step in the carbon cycle involves the utilization of carbon dioxide in photosynthesis by photoautotrophs such as cyanobacteria, green plants, algae, and green and purple sulfur bacteria. Carbon dioxide is incorporated into organic compounds by photoautotrophs. In the next step in the cycle, chemoheterotrophs consume the organic compounds—animals eat photoautotrophs, especially green plants, as well as other animals. Thus, the organic compounds are digested and resynthesized. In this way, the original carbon atoms of carbon dioxide are transferred from organism to organism.

Some of the organic molecules are used by chemoheterotrophs, including animals, to satisfy their energy requirements. In the process of res-

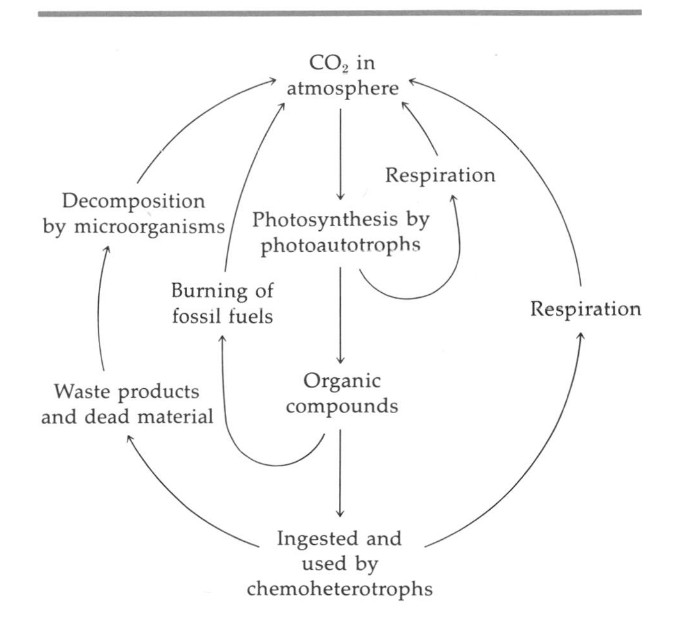

Figure 26–1 Carbon cycle.

piration, carbon dioxide is released into the atmosphere. This carbon dioxide becomes immediately available to start the cycle over again. However, much of the carbon remains with the organisms until they excrete wastes or die. When the organisms die, the organic compounds are deposited in the soil and decomposed by microorganisms, principally bacteria and fungi. As a result of decomposition, organic compounds are degraded and carbon dioxide is returned to the atmosphere. Although the carbon dioxide of the atmosphere represents only about 0.03% of the atmospheric gases, it is essential for the synthesis of new living matter.

Carbon is stored in rocks such as limestone ($CaCO_3$) and dissolved in oceans (as CO_3^- ions), and it is also stored in organic forms such as coal and petroleum. Burning such fossil fuels releases carbon dioxide into the atmosphere.

Nitrogen Cycle

Nitrogen is needed by all organisms for the synthesis of protein, nucleic acid, and other nitrogen-containing compounds. Molecular nitrogen (N_2) comprises almost 80% of the earth's atmosphere; the atmosphere over every acre of fertile soil contains more than 30,000 tons of nitrogen. Despite its abundance, however, no eucaryote is able to make direct use of nitrogen, a molecular gas. Instead, the nitrogen must be fixed (combined) with other elements such as oxygen and hydrogen. The resulting compounds, which include nitrate ion (NO_3^-) and ammonium ion (NH_4^+), are then used by other organisms. The chemical and physical forces operating in the soil, water, and air, together with the activities of specific microorganisms, are important factors in converting nitrogen to usable forms.

Practically all of the nitrogen in the soil exists in organic molecules, primarily as proteins. When an organism dies, the process of decomposition results in the hydrolytic breakdown of proteins into amino acids. The amino groups of amino acids are removed in a process called **ammonification,** so called because ammonia (NH_3) is formed (Figure 26–2). Ammonification is brought about by aerobic and anaerobic bacteria and fungi and may be represented as follows:

$$\text{Proteins from dead cells} \xrightarrow[\text{}]{\text{Microbial decomposition}} \text{Amino acids}$$

$$\text{Amino acids} \xrightarrow{\text{Microbial ammonification}} \text{Ammonia } (NH_3)$$

Microbial growth releases extracellular proteolytic enzymes that accomplish this simplification of chemicals. The fate of the ammonia thus produced depends on soil conditions. Since ammonia is a gas, it may rapidly disappear from dry soil. But in moist soil, it becomes solubilized in water, and ammonium ions (NH_4^+) are formed.

$$NH_3 + H_2O \rightarrow NH_4OH \rightarrow NH_4^+ + OH^-$$

Ammonium ions are used by bacteria and plants for amino acid synthesis.

The next sequence of reactions in the nitrogen cycle involves the oxidation of the ammonium ion to nitrate, a process called **nitrification.** Living in the soil are two genera of bacteria, *Nitrosomonas* and *Nitrobacter,* that are capable of oxidizing ammonia to nitrate in two successive stages. In the first stage, *Nitrosomonas* oxidizes ammonia to nitrites:

$$\underset{\text{Ammonium ion}}{NH_4^+} \xrightarrow{\textit{Nitrosomonas}} \underset{\text{Nitrites}}{NO_2^-}$$

In the second stage, *Nitrobacter* oxidizes nitrites to nitrates:

$$\underset{\text{Nitrites}}{NO_2^-} \xrightarrow{\textit{Nitrobacter}} \underset{\text{Nitrates}}{NO_3^-}$$

Nitrate is the form of nitrogen most commonly used by plants, which use it primarily for protein synthesis.

At various points in the cycle, atmospheric nitrogen is either added or removed. The loss of nitrogen from the cycle involves a process called **denitrification,** the conversion of nitrates to nitrogen gas, and may be represented as follows:

$$\underset{\text{Nitrates}}{NO_3^-} \rightarrow \underset{\text{Nitrites}}{NO_2^-} \rightarrow \underset{\text{Nitrous oxide}}{N_2O} \rightarrow \underset{\text{Atmospheric nitrogen gas}}{N_2}$$

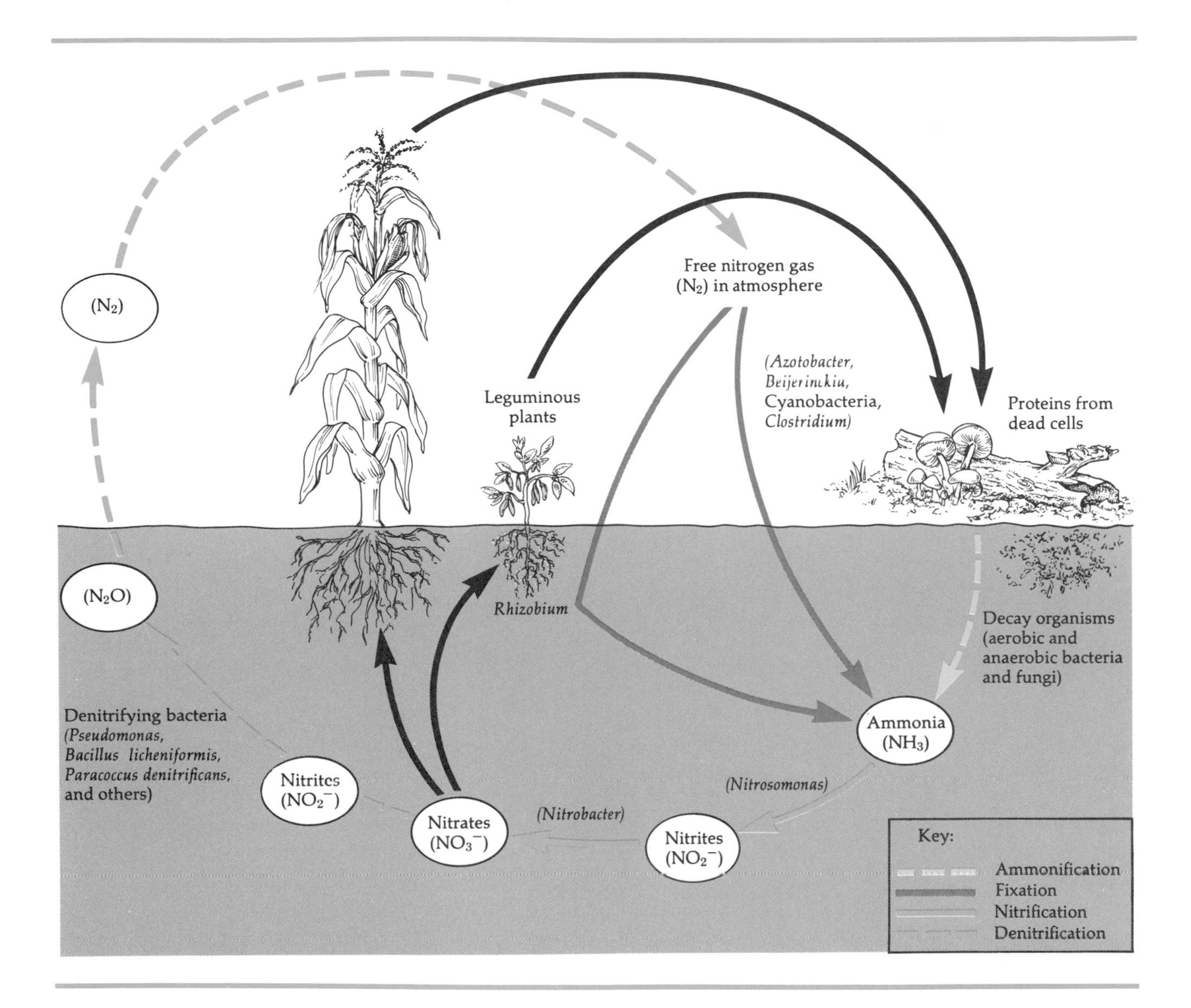

Figure 26–2 Nitrogen cycle.

Pseudomonas species appear to be the most important group of bacteria involved in denitrification in soils. A number of other genera, including *Paracoccus, Thiobacillus,* and *Bacillus,* also have species capable of carrying out denitrification reactions. Denitrifying bacteria are aerobic, but under anaerobic conditions, they can use nitrate in place of oxygen as a final electron acceptor. (This process is called anaerobic respiration; see Chapter 5.) Thus, denitrification is often active in waterlogged soils depleted of oxygen. Since denitrifying bacteria deliver nitrogen to the atmosphere at the expense of removing nitrates from the soil, this is an unfavorable process from the standpoint of soil fertility.

The final phase of the nitrogen cycle deals with the conversion of nitrogen into ammonia, a process called **nitrogen fixation.** Only certain species of bacteria and cyanobacteria are capable of this process. The nitrogenase enzyme responsible for nitrogen fixation probably arose early in the history of the planet, before the atmosphere contained oxygen (the enzyme requires anaerobic conditions) and before nitrogen-containing compounds were available from decaying organic matter. Nitrogen fixation is brought about by two types of organisms: nonsymbiotic and symbiotic.

Nonsymbiotic (free-living) *bacteria* are found in particularly high concentrations in the *rhizosphere,* the region where the soil and roots make contact, especially in grasslands. Among the nonsymbiotic bacteria that can fix nitrogen are some aerobic species such as *Azotobacter.* These organisms apparently shield the nitrogenase enzyme from oxygen by, among other things, a very high rate of oxygen utilization that paradoxically minimizes the diffusion of oxygen into the cell where the enzyme is located. Another nonsymbiotic obligate aerobe that fixes nitrogen is *Beijerinckia.* Some anaerobic bacteria, such as certain species of *Clostridium,* also fix nitrogen. Cyanobacteria usually carry their nitrogenase enzymes in specialized cells called *heterocysts* that neatly provide anaerobic conditions for

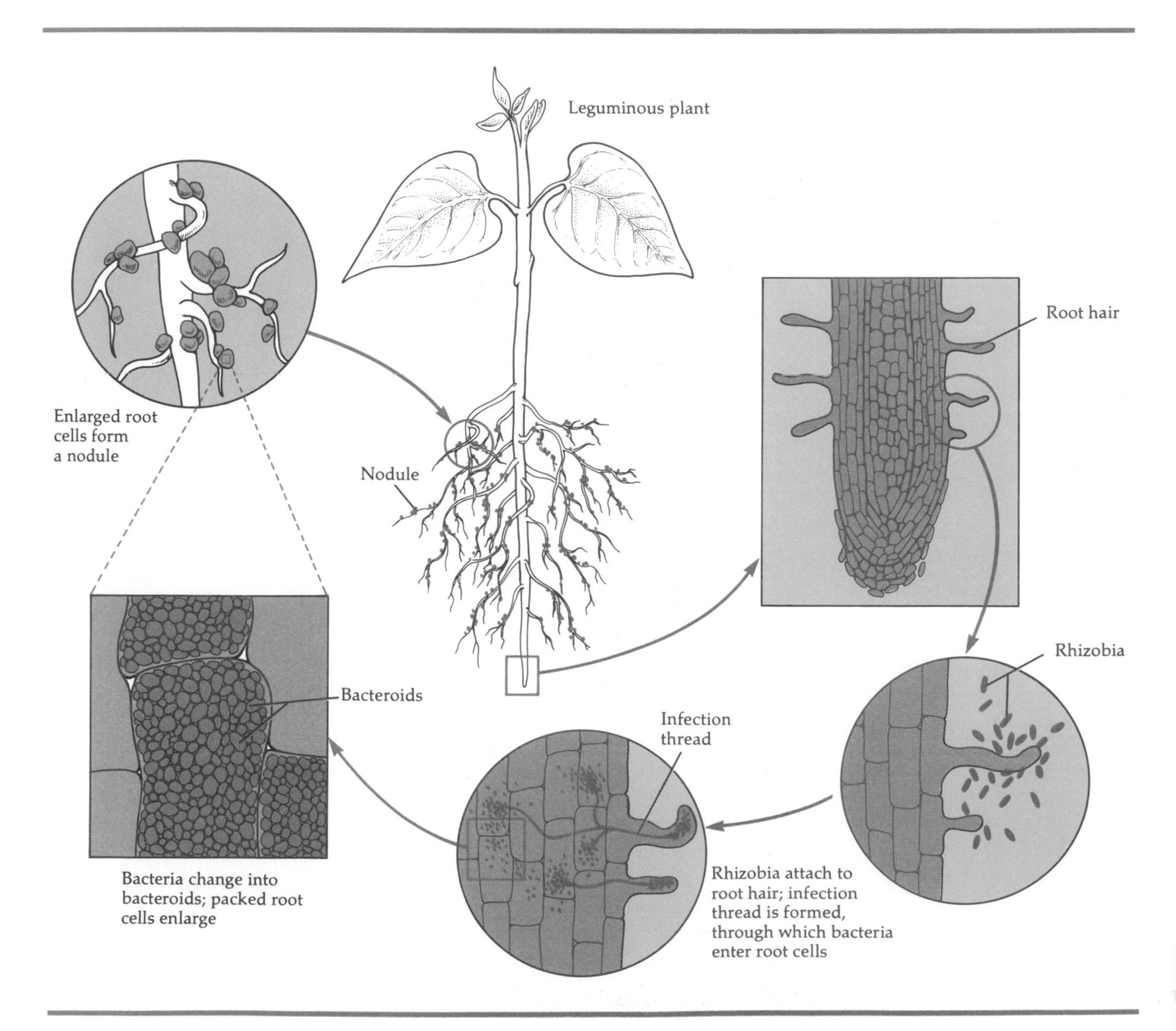

Figure 26–3 Stages in the formation of a root nodule.

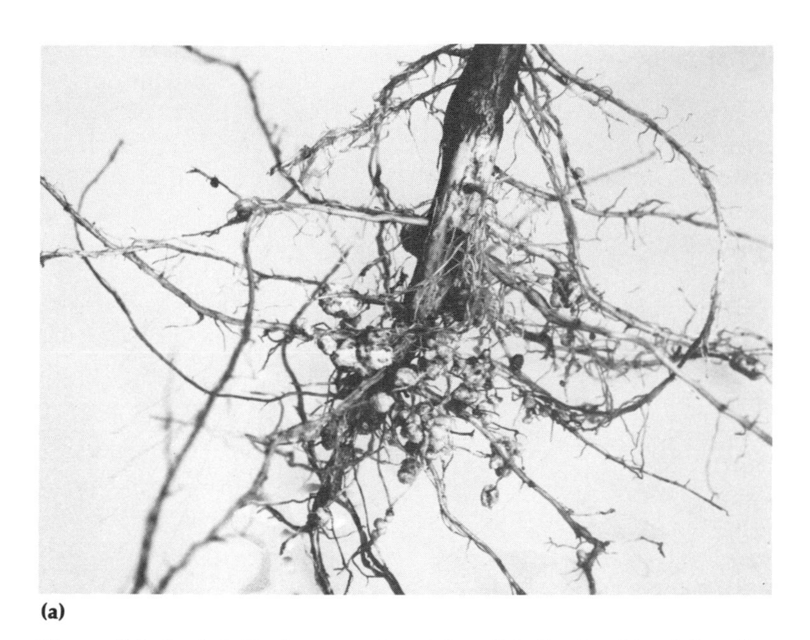

(a)

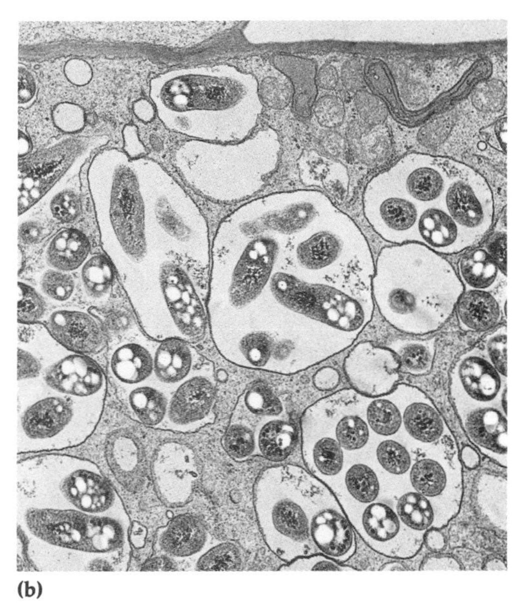

(b)

Figure 26–4 **(a)** Soybean root nodules. **(b)** Bacteroids of the nitrogen-fixing bacterium *Rhizobium japonicum* in vacuoles in an infected cell of a soybean root nodule (×12,000).

fixation. These microorganisms are especially important in flooded soils, such as rice paddy fields. A dominant obligate anaerobic nitrogen-fixing microorganism is the bacterium *Clostridium pasteurianum*. Other nonsymbiotic nitrogen-fixing bacteria include certain species of the facultatively anaerobic *Klebsiella, Enterobacter,* and *Bacillus* and the photoautotrophic *Rhodospirillum* and *Chlorobium*.

Most of the nonsymbiotic nitrogen-fixing organisms are capable of fixing large amounts of nitrogen under laboratory conditions. But in the soil, there is usually a shortage of usable carbohydrates, which are needed for energy to reduce nitrogen to ammonia for incorporation into protein. Nevertheless, the bacteria make important contributions to the nitrogen economy of areas such as grasslands, forests, and the arctic tundra.

Symbiotic (mutualistic) nitrogen-fixing bacteria serve an even more important role in helping plant growth and crop production. In a symbiotic relationship, two organisms of different species live together, each benefiting from the relationship. Such a relationship is illustrated by members of the genus *Rhizobium* (symbiotic nitrogen-fixing bacteria) with the roots of leguminous plants such as soybeans, beans, peas, peanuts, alfalfa, and clover. These agriculturally important plants represent only a few of the thousands of known leguminous species, many of which are bushy plants or small trees found in poor soils in many parts of the world.

The rhizobia bacteria are specific for a particular leguminous species. They attach to the root of the host legume, usually at a root hair (Figure 26–3). An indentation is formed in the root hair in response to the bacterial infection, leading to formation of an *infection thread* that passes down the root hair into the root itself. The bacteria follow this infection thread and enter the cells in the root. Inside these cells, the bacteria alter in morphology, changing into larger forms called *bacteroids* that eventually pack the plant cell. The root cells are stimulated by this infection to form a tumorlike *nodule* of bacteroid-packed cells (Figure 26–4). Nitrogen is then fixed by a symbiosis between the plant and the bacteria. The plant furnishes anaerobic conditions and growth nutrients for the bacteria, and the bacteria fix nitrogen to be incorporated into plant protein.

Millions of tons of such nitrogen are fixed each year in the world. There are similar examples of symbiotic nitrogen fixation in nonleguminous plants such as alder trees. These trees are among the first to appear in forests after fires or glaciation. The alder tree is symbiotically infected with an actinomycete *(Frankia)* and forms nitrogen-fixing root nodules. About 100 pounds per acre of nitrogen can be fixed each year by the growth of alder trees, a valuable addition to the forest economy.

An important contribution to the nitrogen economy of forests is made by **lichens**—a symbiosis between a fungus and an alga or cyanobacterium. When one symbiont is a nitrogen-fixing cyanobacterium, nitrogen is fixed that eventually enriches the forest soil. These cyanobacteria can also fix significant amounts of nitrogen on desert soils directly following infrequent rains, and on the surface of arctic tundra soils. In the Orient, rice paddies can accumulate heavy blooms of such nitrogen-fixing organisms. The cyanobacteria also form a symbiosis with a small floating fern plant, *Azolla*, which grows thickly in rice paddy waters. So much nitrogen is fixed by these organisms that other nitrogenous fertilizers are often unnecessary for rice cultivation.

Other Biogeochemical Cycles

Microorganisms are also important in cycles involving other elements, such as sulfur. In addition, there are important microbial transformations of potassium, iron, manganese, mercury, selenium, zinc, and other minerals. The various chemical reactions in these cycles are often essential in making the minerals available to plants in soluble form for their metabolism.

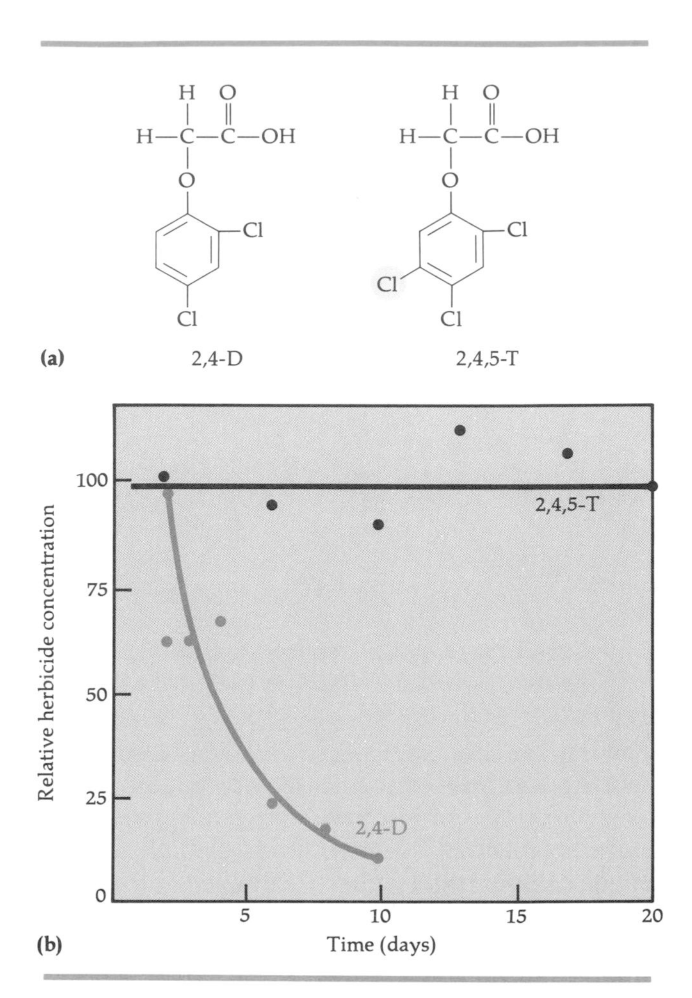

Figure 26–5 Slight structural differences affect biodegradability. **(a)** Comparison of the structures of 2,4-D and 2,4,5-T. **(b)** Comparison of the rates of microbial decomposition of 2,4-D and 2,4,5-T.

DEGRADATION OF PESTICIDES AND OTHER SYNTHETIC CHEMICALS

We seem to take it for granted that soil microorganisms will degrade materials entering the soil. Natural organic matter such as falling leaves or animal residues are, in fact, readily degraded. However, in this industrial age, there are many chemicals, such as agricultural pesticides and plastics, that enter the soil in large amounts—chemicals that do not occur in nature. Many of these synthetic chemicals are highly resistant to degradation by microbial attack. A well-known example is the insecticide DDT. When first introduced, the property of *recalcitrance* (resistance to degradation) was considered quite beneficial because one application remained effective in the soil for an extended time. But it was soon found that because of their fat solubility, such chemicals tended to accumulate and concentrate in certain parts of the food chain. Eagles and other predatory birds began accumulating DDT in their tissues from contamination of their food supply and suffered impaired reproductive ability. (For example, eggs had soft shells and failed to hatch.) Not all synthetic chemicals are as recalcitrant as DDT. Some are made up of chemical bonds and subunits that are subject to attack by bacterial enzymes. But small differences in chemical structure can make large differences in biodegradability. The classic example is that of two herbicides—2,4-D, the common chemical used to kill lawn weeds, and 2,4,5-T, which is used to kill shrubs. The addition of a single chlorine atom to the structure of 2,4-D is sufficient to extend its life in soil from a few days to an indefinite period (Figure 26–5).

AQUATIC MICROBIOLOGY AND SEWAGE TREATMENT

Aquatic microbiology refers to the study of microorganisms and their activities in natural waters such as lakes, ponds, streams, rivers, estuaries, and the sea. Here we will consider the microbial content of freshwater and seawater.

High nutrient levels in water are generally reflected by high microbial numbers. Water contaminated by inflows from sewage systems or biodegradable industrial organic wastes is also relatively high in bacterial counts. Similarly, ocean estuaries (fed by rivers) have higher nutrient levels and therefore higher microbial counts than other shoreline waters. In water, particularly water with low nutrient concentrations, microorganisms have a tendency to grow on stationary surfaces and on particulate matter, rather than being randomly suspended in the water. In this way, the microorganism has contact with more nutrients than if it were freely floating with the current. Many bacteria whose main habitat is water have appendages, such as holdfasts, that attach to various surfaces. One example is *Caulobacter* (see Figure 10–17d). Some bacteria also have gas vesicles that they can fill and empty to adjust buoyancy.

FRESHWATER MICROBIAL FLORA

A typical lake or pond can be used to represent the various zones and the kinds of microbial flora in a body of freshwater (Figure 26–6). The **littoral zone** is the region along the shore that has considerable rooted vegetation and where light penetrates to the bottom. The **limnetic zone** consists of the surface of the open water area away from the shore. The **profundal zone** is the deeper water under the limnetic zone. And the **benthic zone** is the sediment at the bottom.

Microbial populations of freshwater bodies tend to be differentiated mainly by the availability of oxygen and light. Light is in many ways the more important of the two because the photosynthetic algae are the main source of organic matter, and hence energy, for the lake. These organisms are the primary producers of a lake that supports a population of bacteria, protozoans, fish, and other aquatic life. They are located in the limnetic zone. In areas of the limnetic zone where sufficient oxygen is available, pseudomonads and species of *Cytophaga, Caulobacter,* and *Hyphomicrobium* can be found. Oxygen does not diffuse into water very well, as any aquarium owner knows. Microorganisms growing on nutrients in stagnant water quickly use up the dissolved oxygen in the water, leading to the death of the fish and odors associated with anaerobic activity (from hydrogen sulfide and organic acids, for example). Wave action in shallow layers, or water movement as in rivers, tends to increase oxygen levels and aid in the growth of aerobic populations of bacteria. This improves the quality of the water and aids in degradation of polluting nutrients.

Deeper waters of the profundal zone have low oxygen concentrations and also less light. Algal growth near the surface often filters the light, and it is not unusual for photosynthetic microorganisms deeper down to utilize wavelengths of light differing from that of surface layer photosynthesizers. The purple and green sulfur bacteria are found in deeper waters such as this. These are anaerobic photosynthetic organisms that metabolize hydrogen sulfides to sulfur and sulfates in the bottom sediments of the benthic zone. The sediment also includes bacteria such as *Desulfovibrio,* which use sulfate (SO_4^{2-}) as an electron acceptor, reducing it to hydrogen sulfide (H_2S) (see box), which is responsible for the rotten-egg odor of many lake muds. Methane-producing bacteria are also part of these anaerobic, benthic populations. In swamps, marshes, or bottom sediments, they produce methane gas. *Clostridium* species are common in bottom sediments and may include botulism organisms, especially those causing outbreaks of botulism in water fowl.

In freshwater systems, then, the primary producers are the photosynthetic algae and cyanobacteria. These are eventually consumed by other aquatic life and are degraded to more elemental nutrients by bacteria of the limnetic and benthic zones of the lake.

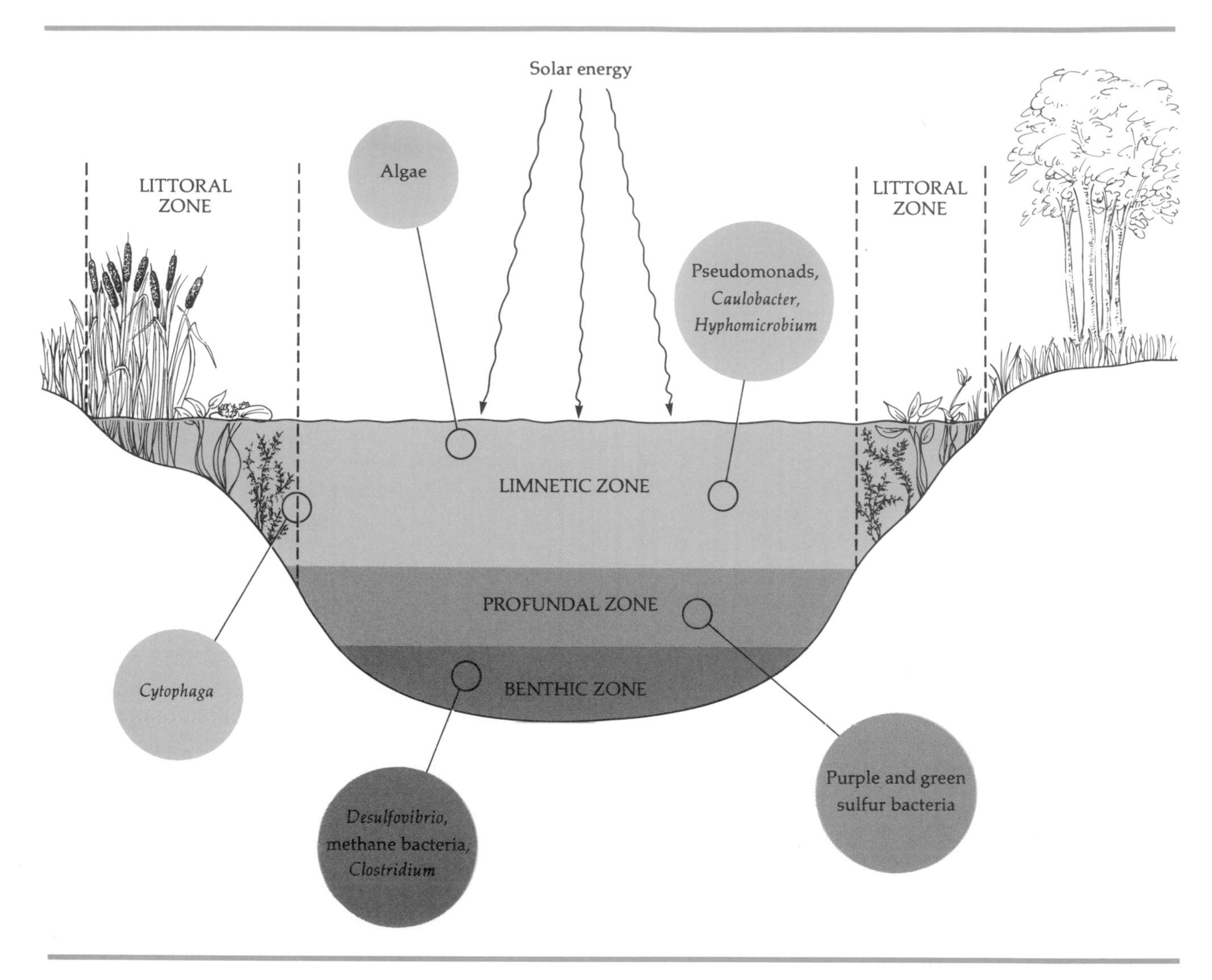

Figure 26–6 The zones in a lake or pond, and some representative microorganisms.

SEAWATER MICROBIAL FLORA

The open ocean is relatively high in osmotic pressure and low in nutrients. The pH also tends to be higher than the optimum for most microorganisms. Populations of bacteria in such waters tend to be much lower, therefore, than in estuaries and most small freshwater bodies, which are fed by rivers and streams and have higher microbial counts. Much of the microscopic life of the ocean is composed of photosynthetic diatoms and other algae. These are largely independent of preformed organic nutrient sources, using energy from photosynthesis and using atmospheric carbon dioxide for carbon. These organisms constitute the marine **phytoplankton** community, the basis of the oceanic food chain. Oceanic bacteria benefit from the eventual death and decomposition of phytoplankton and also attach to their living bodies. Protozoans in turn feed on bacteria and the smaller phytoplankton. Krill, shrimplike crustaceans, feed on the phytoplankton and in turn are an important part of the food supply of larger sea life. Many fish and whales are also able to feed directly on the phytoplankton.

Microbial luminescence is an interesting minor aspect of deep-sea life. Certain algae produce light flashes when agitated by wave action or a boat wake. Many bacteria are also luminescent, and some have established symbiotic relationships with benthic-dwelling fish. These fish sometimes use the glow of their residing bacteria as an aid in attracting and capturing prey in the complete darkness of the ocean depths. Bioluminescent organisms have an enzyme called luciferase that picks up electrons from flavoproteins in the electron transport chain and then emits some of the electron's energy as a photon of light.

EFFECTS OF POLLUTION

Water that moves down below the ground surface undergoes a filtering action that tends to remove microorganisms. For this reason, water from springs and deep wells is generally of good quality.

Transmission of Infectious Diseases

Contamination of water supplies by pathogenic microorganisms is an important factor in the spread of many diseases. Sometimes this takes the form of

The Ecological Niche of *Desulfovibrio*

Bacteria of the genus *Desulfovibrio* look ordinary enough: they are curved rods with polar flagella. And they are found in a variety of habitats, including freshwater, seawater, and water-logged soils. Their particular niche in the ecosystem, however, is an interesting one. *Desulfovibrio* bacteria are anaerobic chemoheterotrophs. For carbon sources, they use compounds such as lactic acid and pyruvic acid that are the end products of fermentation from other anaerobic microorganisms. *Desulfovibrio* oxidize these compounds to acetic acid and carbon dioxide. For energy, they carry out anaerobic respiration, using sulfate (or other sulfur compounds) as their electron acceptor. Sulfate is the end product of protein composition by other anaerobes. The sulfate (SO_4^{2-}) is reduced to hydrogen sulfide (H_2S).

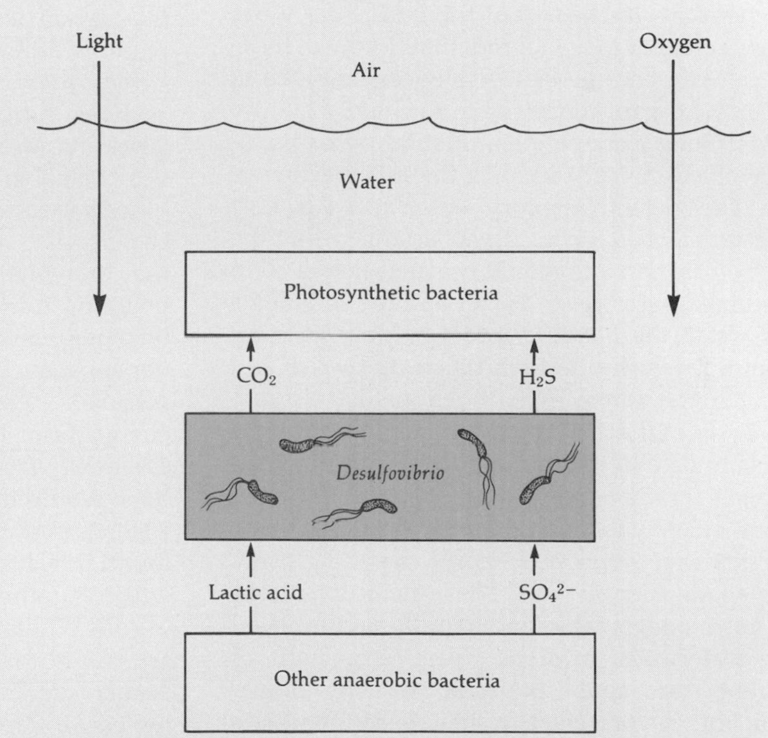

The diagram shows the position of *Desulfovibrio* relative to other bacteria in a body of water. Growth of other anaerobic bacteria provides lactic acid and sulfate for *Desulfovibrio*. *Desulfovibrio*, in turn, provides carbon dioxide and hydrogen sulfide photosynthetic bacteria in higher layers of the water to use.

ingestion of bacterial pathogens, as in typhoid fever or Asiatic cholera. Protozoan diseases, such as amoebic dysentery or giardial dysentery, are spread by cysts carried in water. Relatively large parasites, such as the free-swimming trematode cercaria, the causative agent of schistosomiasis, are also dependent upon water for their spread to human hosts (see Chapters 11 and 22). In developed parts of the world, such types of contamination are minimized by education of the population and by sophisticated water treatment systems.

Chemical Pollution

Chemical contamination of water is a more difficult problem. Great amounts of industrial and agricultural chemicals leached from the land enter water in forms resistant to biodegradation. Many of these chemicals can become biologically concentrated in some of the organisms in the food chain.

A striking example of industrial water pollution involved mercury used in the manufacture of paper. The metallic mercury was allowed to flow into waterways as waste. It was assumed that the mercury was inert and would remain segregated in the sediments. However, bacteria in the sediments incorporated the mercury into a soluble chemical compound, which was then taken up by fish and shellfish in the waters. When such seafood is a substantial part of the diet, the mercury concentrations can accumulate in the human body with devastating effects on the nervous system.

Immediately following World War II, synthetic detergents were developed, rapidly replacing many of the soaps then in use. These new detergents were not biodegradable and rapidly accumulated in the waterways. In some rivers, large rafts of detergent suds could be seen traveling downstream; and in some cities, a small head of bubbles might appear on the surface of a glass of water. These detergents were replaced in 1964 by new biodegradable formulations. These, however, brought new problems. Substantial amounts of phosphates were added to many of these detergents to improve cleaning effectiveness. Unfortunately, phosphates pass nearly intact through most sewage treatment systems and can cause **eutrophication** of lakes. Eutrophication, meaning "well nourished," refers to the situation in which the addition of large amounts of nutrients to water results in massive growth of algae and subsequent oxygen loss from the water. In the short run, these algae produce oxygen. However, the algae eventually die and are degraded by bacteria, a process that uses up the oxygen in the water. Undigested remnants settle to the bottom and hasten the filling of the lake.

Eutrophication can also result from the addition of raw sewage, agricultural runoff, or industrial wastes to a lake (Figure 26–7). However, eutrophication is not always a result of chemical pollution, but may occur as a natural process. Algae and cyanobacteria are able to grow by using photosynthesis for energy and atmospheric carbon dioxide dissolved in water to construct their organic compounds. Therefore, the major component of the carbon structures of algae and cyanobacteria requires little or no addition of nutrients to water. Given light and relatively small amounts of nutrients such as nitrogen and phosphorus compounds, lakes and streams can undergo sudden blooms of heavy algal growth.

As mentioned earlier, many species of cyanobacteria also have the capacity to fix nitrogen from the atmosphere. Thus, even in waters in which the polluting nitrogen is minimal, it is possible to get blooms of these cyanobacteria. This is where the phosphates from detergents become involved. Because these procaryotic organisms get their energy from light and their carbon and nitrogen essentially from the atmosphere, they find most of their requirements for growth even in very clean waters. About the only major growth nutrient not found in water in sufficient amounts is phosphate, which is rather insoluble and tends to be retained in soil. When a few parts per million of phosphate are added to a lake, the result can be a bloom of cyanobacteria. Eventually, these organisms die and the bacteria metabolizing their remains use up the oxygen in the water in a manner similar to that resulting from the addition of excess nutrients such as sewage. But even small amounts of phosphate lead to these results, and for this reason, phosphate-containing detergents are illegal in many localities.

Coal-mining wastes, particularly in the eastern United States, are very high in sulfur content, mostly iron sulfides (FeS_2). In the process of obtaining energy from the oxidation of the ferrous ion

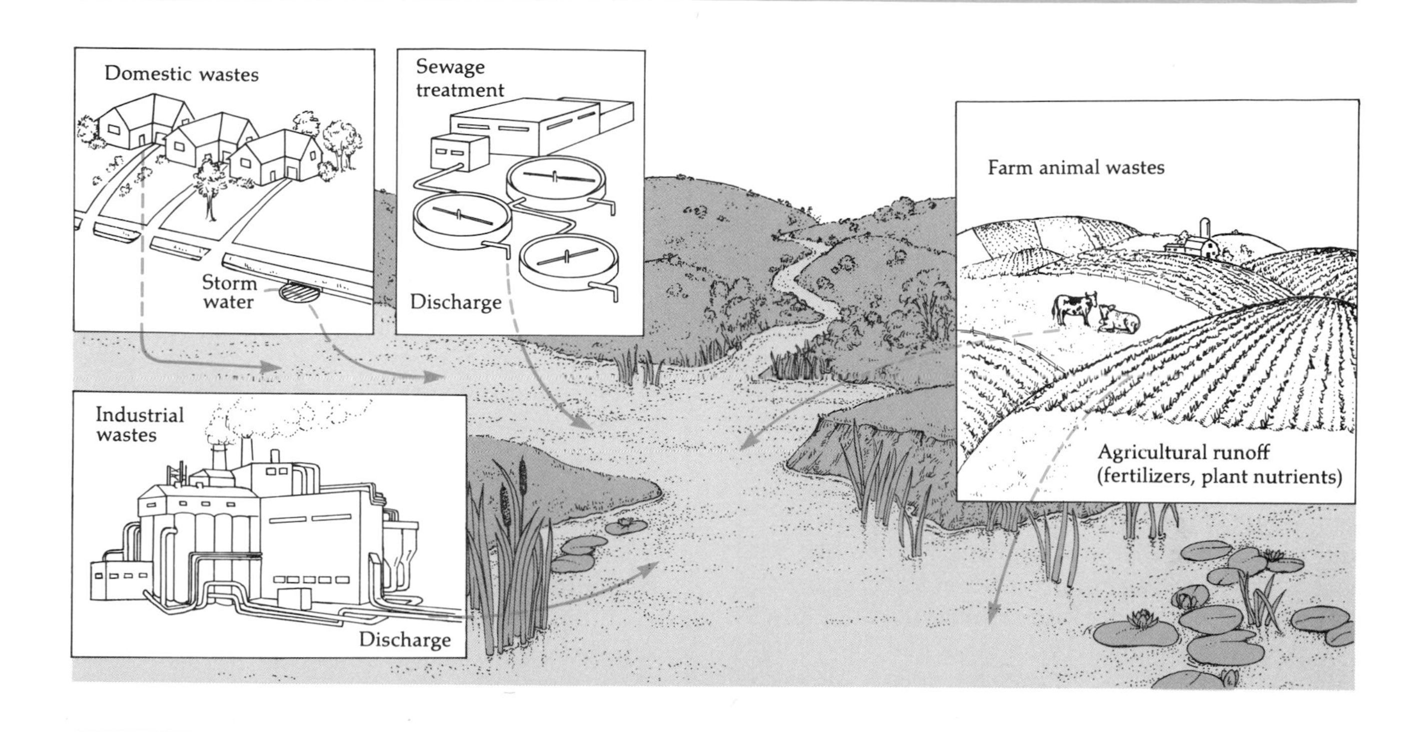

Figure 26–7 Sources of nutrients, mainly phosphorus and nitrogen compounds, that contribute to eutrophication.

(Fe^{2+}), bacteria such as *Thiobacillus ferrooxidans* convert the sulfide into sulfates. The sulfates enter streams as sulfuric acid, lowering the pH and damaging aquatic life. The low pH also promotes the formation of insoluble iron hydroxides, causing a yellow precipitate often seen clouding such polluted waters. On the other hand, *T. ferrooxidans* may some day aid in the recovery of otherwise unprofitable grades of uranium and other ores. Solutions containing ferric ion (Fe^{3+}) can be washed through deposits of insoluble uranium compounds, oxidizing the uranium so that it forms soluble compounds. In this process, Fe^{3+} is reduced to Fe^{2+}. The Fe^{2+} can be reoxidized by *T. ferrooxidans* and the leaching process continued.

TESTS FOR WATER PURITY

Historically, most of our concern about water purity has been related to the transmission of disease. Tests to determine the safety of water (many of these tests are also applicable to foods) have therefore been developed.

It is not practical to look only for pathogens in water supplies. For one thing, if we were to find the pathogen causing typhoid or cholera in the water supply system, it would already be too late to prevent an outbreak of the disease. Moreover, such pathogens would probably be present only in relatively small numbers and might easily be missed by sampling.

The tests for water safety in use today are aimed at the detection of certain **indicator organisms.** There are several criteria for the selection of an indicator organism. The most important is that it be consistently present in human intestinal wastes in substantial numbers so that its detection is a good indication that human wastes are entering the water. The indicator organisms should also survive in the water at least as well as the more likely pathogenic organisms in the water. The indicator organisms must also be detectable by simple tests that can be carried out by persons with relatively little technical training, for example, water treatment plant operators in small communities.

In the United States, the usual indicator organisms are the *coliform* bacteria. Coliforms are defined as aerobic or facultative anaerobic, gram-negative, non-endospore-forming, rod-shaped bacteria that ferment lactose with acid and gas formation within 48 hours at 35°C. Because some coliforms are not solely enteric bacteria but are more commonly found in plant and soil samples, many standards for food and water specify the determination of *fecal coliforms*. The predominant fecal coliform is *Escherichia coli,* which constitutes a large proportion of the human intestinal population. There are specialized tests to distinguish between fecal coliforms and nonfecal coliforms. It is important to note that coliforms are not themselves pathogenic under normal conditions, although they can cause opportunistic urinary bladder infections such as in hospital patients with catheters.

The United States Environmental Protection Agency Drinking Water Standards specify the minimum number of water samples to be examined each month and the maximum number of coliform organisms permitted in each 100 ml of water.

For detection and enumeration of coliforms, selective and differential media are required and specialized methods are used. Two basic methods are recognized. The first is the **multiple-tube fermentation technique.** In this technique, coliforms are detected in three stages: presumptive, confirmed, and completed (Figure 26–8a). In the *presumptive test,* dilutions from the water sample are added to tubes of lactose broth medium with a pH indicator and incubated for 24 to 48 hours. Fermentation of lactose to acid and gas (trapped in an inverted small vial) is considered a positive reaction. In a common version of the presumptive test, 15 tubes of lactose broth are inoculated with samples of the water being tested, with 5 tubes receiving water samples of 10 ml each, 5 tubes samples of 1 ml each, and 5 tubes samples of 0.1 ml each. If no acid or gas forms in any of the tubes, the water is usually considered satisfactory. Statistically, this would be considered an indication of less than 2.2 coliforms per 100 ml of water. (Most Probable Number tables are consulted; see Chapter 6 and Appendix D.)

If any of the tubes in the presumptive test are positive, additional tests are performed, since organisms other than coliforms are also capable of lactose fermentation and gas production. In the *confirmed test,* samples are streaked from the positive presumptive tubes at the highest dilutions onto plates of differential media, eosin methylene blue (EMB) agar that contains lactose. Since coliforms produce acid from lactose, and the eosin and methylene blue dyes are absorbed under acid conditions, the coliforms form dark-centered colonies with or without metallic sheens. These colonies indicate a positive confirmed test. In the *completed test,* separate colonies are selected from the confirmed test and inoculated in lactose broth and on a nutrient agar slant for 24 hours at 35°C. If acid and gas are produced in lactose broth, and the isolated microorganism is a gram-negative non-endospore-forming rod, it indicates a positive completed test.

Coliforms may also be detected by the **membrane filter method.** The water sample of 100 ml is drawn through a paper-thin membrane filter with pores of about 0.45 μm, which retains the bacteria on the membrane surface. The filter membrane is then placed on a pad of suitable nutrient medium. Coliform bacteria growing on the nutrient medium soaking through the membrane form colonies with a distinctive appearance. If only one colony is detected per 100 ml of water, the water is usually considered of satisfactory quality for drinking. For uses of water other than drinking, the number of coliforms allowable in a water sample varies with the intended human use.

WATER TREATMENT

When water is obtained from uncontaminated reservoirs fed by clear mountain streams or from deep wells, it requires minimal treatment to make it safe to drink. Many cities, however, obtain their water from badly polluted sources such as rivers that receive municipal and industrial wastes, as well as leaking nuclear cooling water wastes, upstream. In order to allow as much particulate suspended matter as possible to settle out, very turbid (cloudy) water is allowed to stand in a holding reservoir for a period of time (Figure 26–9). The water then undergoes **flocculation treatment,** that is, removal of colloidal materials such as clay, which would remain in suspension indefinitely. Among the

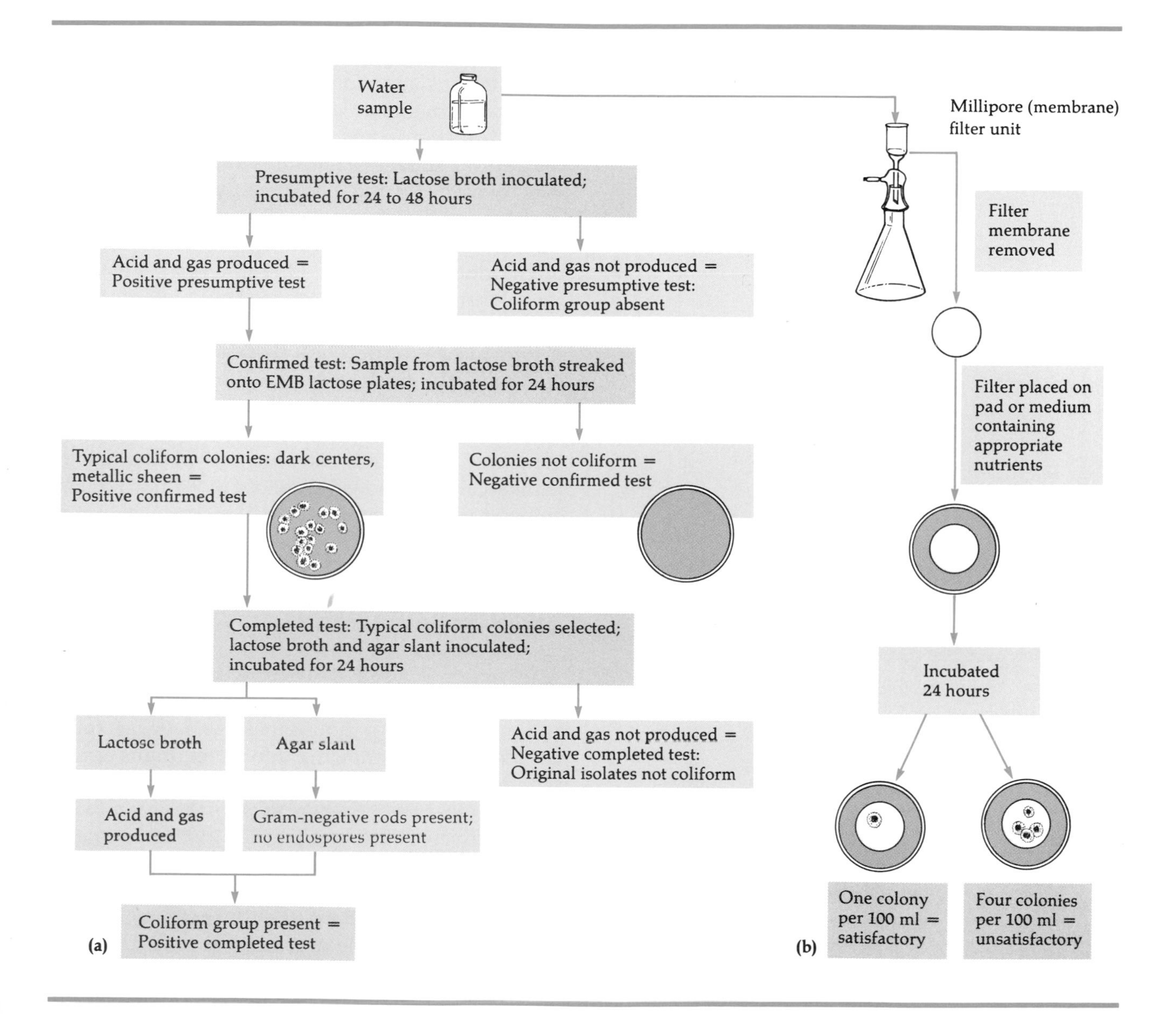

Figure 26–8 Analysis of drinking water for coliforms. **(a)** Multiple-tube fermentation technique. **(b)** Membrane filter method.

more common flocculant chemicals used is aluminum potassium sulfate (alum). This chemical forms a floc, which slowly settles out, carrying colloidal material entrapped in it to the bottom. Large numbers of viruses and bacteria are also removed by this treatment. You might be interested to know that alum was used to clear muddy river water during the first half of the nineteenth century in the military forts of the American West, long before the concept of a germ theory of disease. Similarly, in the early 1800s, many cities in Germany filtered muddy river water through sand beds to clarify it. There were observations at the time that people who used such filtered water did not have the same incidence of cholera during outbreaks.

After flocculation treatment, water is passed through beds of sand to accomplish **sand filtration.** This treatment removes about 99% of the bacteria and viruses still remaining from flocculation. Some protozoan cysts, such as those of *Giardia lamblia,*

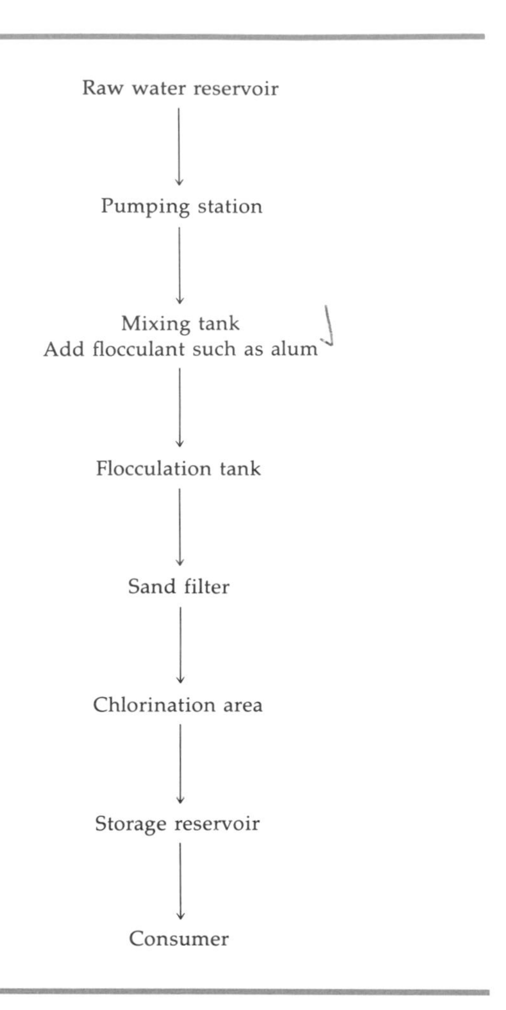

Figure 26–9 Steps involved in water treatment in a typical municipal water purification plant.

appear to be removed from water only by such a filtration treatment. The microorganisms are trapped in the sand beds, mostly by surface absorption. Bacteria do not penetrate the torturous routing of the sand beds, even though the openings may be larger than the organisms that are filtered out. These sand filters are periodically backflushed to clear them of accumulations. Future water systems may replace or supplement sand filtration with filters of activated charcoal (carbon). Charcoal has the advantage of removing some dissolved organic chemical pollutants in solution as well as particulate matter.

Before entering the municipal distribution system, the water is *chlorinated.* Chlorination is very effective in killing pathogenic bacteria but less effective in the destruction of protozoan cysts and viruses. Because organic matter neutralizes chlorine, constant attention by the plant operators is needed to maintain effective levels of chlorine. There has been some concern that chlorine itself may be a health hazard, reacting with organic contaminants of the water to form carcinogenic compounds. For the moment, this possibility is considered minor compared to the proven usefulness of chlorination of water. One potential substitute for chlorination might be ozone treatment of water. Ozone (O_3) is a highly reactive form of oxygen that is formed by electric spark discharges and ultraviolet light. The fresh odor of air following an electrical storm or around an ultraviolet light bulb is due to ozone. Ozone for water treatment would be generated electrically at the site of treatment. Unlike chlorine, however, ozone does not remain behind to provide continued protection in the water system.

SEWAGE TREATMENT

After water is used it becomes **sewage.** Sewage includes all the dishwashing, handwashing, or bath water from a household, as well as toilet wastes. Rainwater flowing into street drains enters the sewage system in some cities, as do certain industrial wastes. Sewage is mostly water with comparatively little particulate matter, perhaps only about 0.03%. Even so, in large cities, this solid portion of sewage can amount to more than 1000 metric tons of solid material daily that must be disposed of. Much of the organic matter load of sewage is dissolved in liquid form.

Until environmental awareness intensified in recent years, a surprising number of large cities in this country had only rudimentary sewage treatment systems—or no system at all. Raw sewage, untreated or nearly so, was simply discharged into rivers or oceans. A flowing, well-aerated stream is capable of considerable self-purification. Therefore, until increases in populations and their wastes began to exceed this capability, this casual treatment of municipal wastes caused little complaint. Most situations of this kind in the United States are now being corrected. However, at this time, 90% of the sewage of the 125 largest cities on the Mediterranean Sea release all sewage into this body of water with no treatment at all. This is surprising, considering that many of these cities are supposed to be part of the well-developed world.

Primary Treatment

The usual first step in sewage treatment is called **primary treatment** (Figure 26–10a). In this essentially physical process, incoming sewage receives preliminary treatment: large floating materials are screened out; the sewage is allowed to flow through settling chambers to remove sand and similar gritty material; skimmers remove floating oil and grease; and floating debris is shredded and ground. Following this, the sewage passes through sedimentation tanks, where solid matter is allowed to settle out. (The design of primary settling tanks varies.) Sewage solids collecting on the bottom are called **sludge;** in this instance, it is called primary sludge. From 40 to 60% of suspended solids are removed from sewage by this settling treatment, and flocculating chemicals are sometimes added to increase the removal of solids. The design of primary settling tanks varies. Biological activity is not particularly important in primary treatment, although some digestion of sludge and dissolved organic matter can occur during longer holding times. The sludge is removed on a continuous or intermittent basis, and the effluent (the liquid flowing out) then undergoes secondary treatment.

Biochemical Oxygen Demand

Primary treatment removes approximately 25 to 35% of the **biochemical oxygen demand (BOD)** of the sewage. BOD is an important concept in sewage treatment and in the general ecology of waste treatment. It is a measure of the biologically degradable organic matter in water and is determined by measuring the amount of oxygen required by bacteria to metabolize it. The classic method is to use special bottles with an airtight stopper (Figure 26–11). The bottle is filled with the test water or dilutions of the test water. The water is initially aerated to provide a relatively high level of dissolved oxygen and, if necessary, seeded with bacteria. The bottles are incubated for five days at 20°C, and the decline in dissolved oxygen levels in the water is determined by a chemical or electronic testing method. The more oxygen used up as the bacteria utilize the

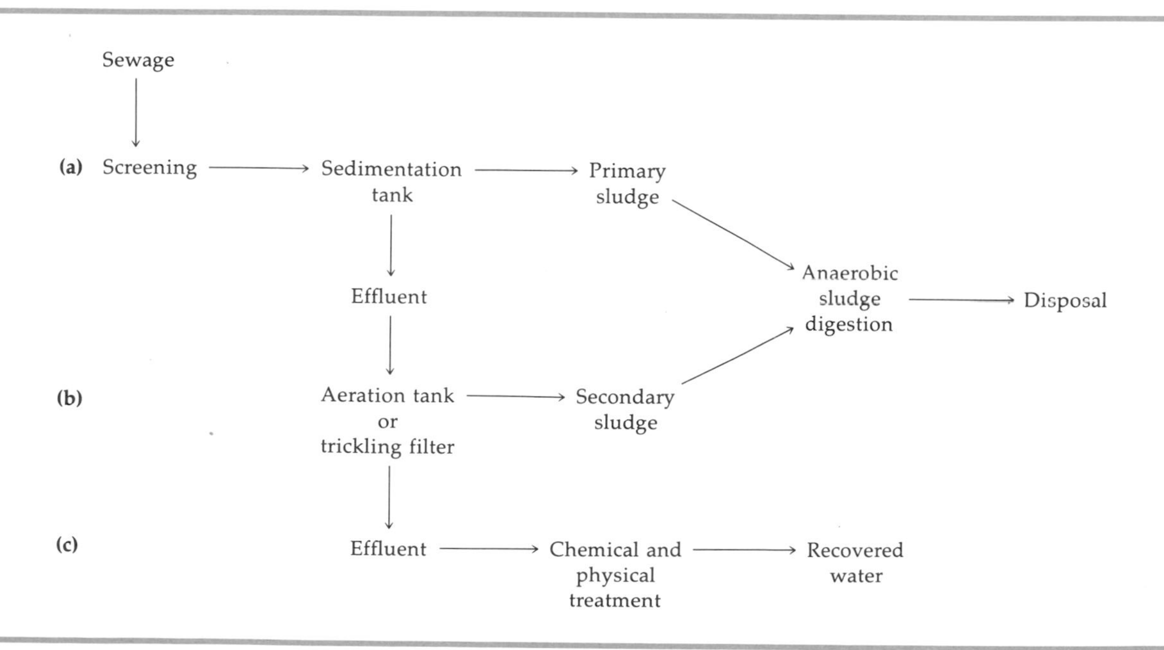

Figure 26–10 Steps involved in waste treatment. **(a)** Primary treatment. **(b)** Secondary treatment. **(c)** Tertiary treatment.

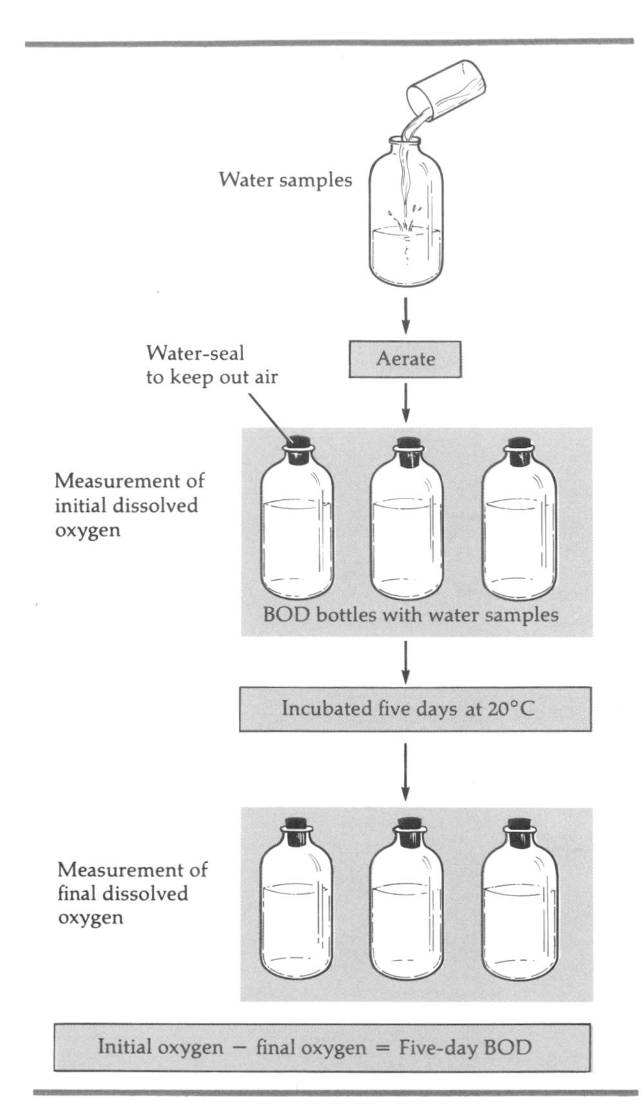

Figure 26–11 Biochemical oxygen demand (BOD).

organic matter in the sample, the greater the BOD—which is usually expressed in terms of milligrams per liter of oxygen used. Normal dissolved oxygen solubility in water is about 10 mg/l. Typical BOD values may be 200 mg/l. Therefore, oxygen can be depleted rapidly from sewage within a short period of time if bacteria are present.

Secondary Treatment

After primary treatment, the greater part of the BOD remaining in the sewage is in the form of dissolved organic matter. **Secondary treatment,** which is primarily biological, is designed to remove most of this organic matter and reduce the BOD. Essentially, this process involves high aeration to encourage the growth of aerobic bacteria and other microorganisms that oxidize the dissolved organic matter to carbon dioxide and water. Two basic methods of secondary treatment in common use are activated sludge systems and trickling filters.

In the **activated sludge system,** air is added to the effluent from primary treatment in aeration tanks (Figure 26–10b). The sludge present in the effluent contains large numbers of metabolizing bacteria, together with yeasts, molds, and protozoans. These organisms grow and rapidly oxidize organic matter, including the organic materials in the secondary sludge. After several hours, the liquid passes to a tank where the sludge settles to the bottom in a manner similar to the nonsoluble solids in primary treatment (Figure 26–12). Part of the sludge is removed for disposal; about 20% of the sludge is recycled to the activated sludge tanks as a "starter" for the next sewage batch. The effluent is sent on for final treatment. Occasionally, when aeration is stopped, the sludge will float rather than settle out, a phenomenon called *bulking*. When this happens, the organic matter in the floc now flows out with the discharged effluent, often causing serious local pollution problems. A considerable amount of research has been expended on the causes of bulking and its possible prevention. It is apparently caused by the growth of filamentous bacteria of various types, although the sheathed bacterium *Sphaerotilus natans* is often mentioned as the primary offender. Activated sludge systems are quite efficient, removing 75 to 95% of the BOD from sewage.

Trickling filters are the other commonly used method of secondary treatment. In this method, the sewage is sprayed over a bed of rocks or plastic (Figure 26–13). The bed is about 6 feet deep, and the rocks forming the bed are about 2 to 4 inches in diameter. Activity of the system depends on the growth of a slime layer on the surface of the rocks. Actually, there is no "filtering" action involved as such. The slimy, gelatinous film of aerobic microorganisms forming on the rocks' surface is in many ways functionally similar to the organisms found in activated sludge systems. Because air circulates

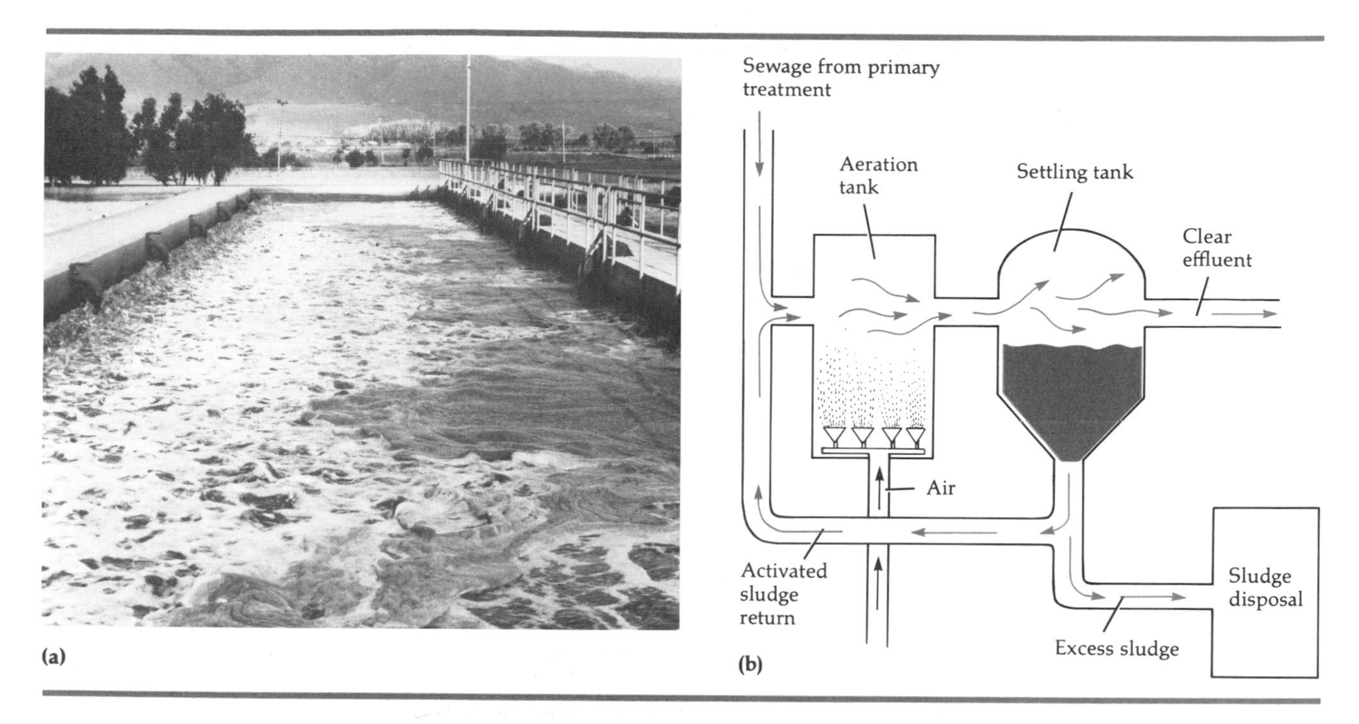

Figure 26–12 Activated sludge system of secondary treatment. **(a)** An aeration tank at the San Jose–Santa Clara Water Pollution Control Plant in San Jose, California. At this facility, sewage passes through four such tanks during secondary treatment. **(b)** Diagram of an activated sludge system.

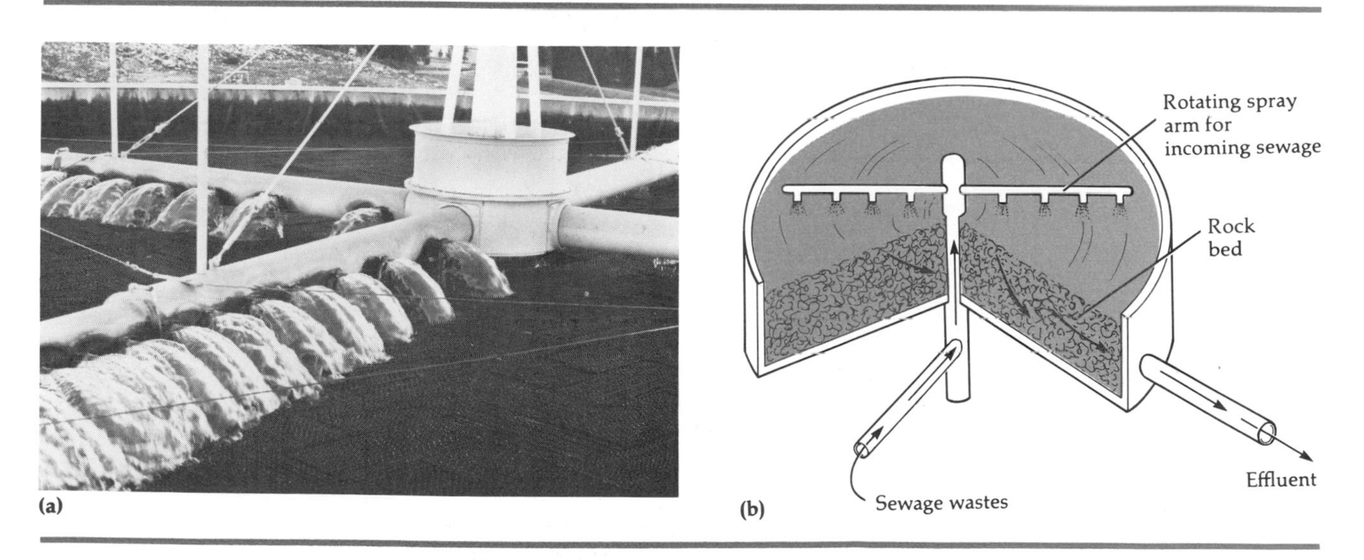

Figure 26–13 Trickling filter of secondary treatment. **(a)** A modern trickling filter at the Sunnyvale, California, Water Pollution Control Plant. This type of tank, which is more than 18 feet high, is filled with a plastic honeycomb. Microorganisms grow on the enormous surface area of the plastic filling and act on liquid flowing down through the maze of channels. **(b)** Diagram of a rock bed trickling filter.

throughout the rock bed, these aerobic microorganisms in the slime layer are able to oxidize into carbon dioxide and water much of the organic matter trickling over the surfaces. Trickling filters are less efficient in removing BOD than are activated sludge systems, removing 80 to 85% of the BOD. On the other hand, they are usually less troublesome to operate and less subject to problems from overloads or toxic sewage.

Sludge Digestion

Sludge accumulates in primary sedimentation tanks (primary sludge) and further accumulates following activated sludge and trickling filter secondary treatments. These sludges are often pumped to **anaerobic sludge digesters** for further treatment (Figure 26–14). The process of sludge digestion is carried out in large tanks from which oxygen is almost completely excluded. In secondary treatment, emphasis is placed on maintaining aerobic conditions in order to achieve conversion of organic matter to carbon dioxide, water, and solids that can settle out. By comparison, anaerobic fermentations are metabolically inefficient, and the microorganisms leave large amounts of only partially digested organic materials in the form of fatty acids, alcohols, and similar products that retain much of their original BOD. An anaerobic sludge digester is designed to encourage the growth of anaerobic bacteria, especially methane-producing bacteria, that decrease organic solids by degrading them to soluble substances and gases, primarily methane (60 to 70%) and carbon dioxide (20 to 30%). The gases are routinely used as a fuel for heating the digester and are frequently also used in the operation of equipment in the plant.

The first stage of activity in an anaerobic sludge digester is essentially a simple fermentation and results in the formation of organic acids and considerable amounts of carbon dioxide. Microorganisms metabolize these organic acids and other fermentation products, forming considerable hydrogen and carbon dioxide gases. The presence of hydrogen and carbon dioxide, as well as the necessary temperature and pH, encourages the growth of methane-producing bacteria, ancient forms of

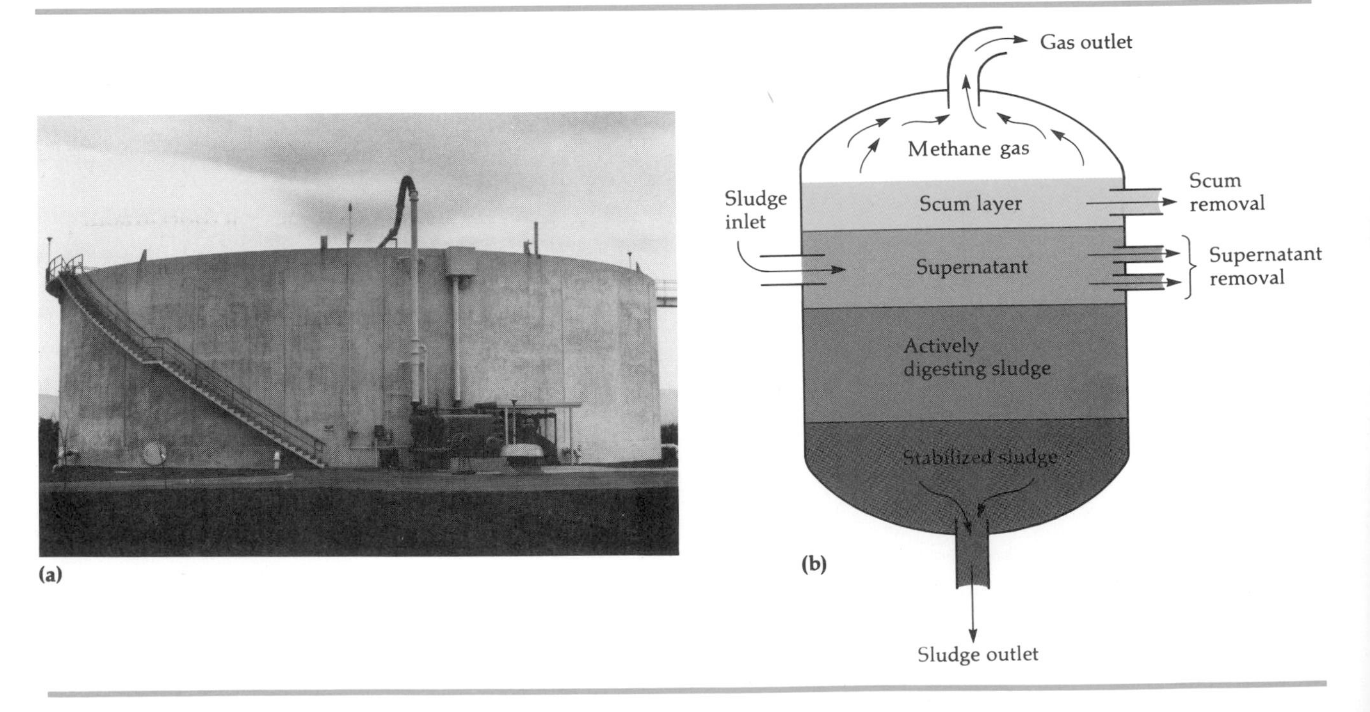

Figure 26–14 Sludge digestion. **(a)** An anaerobic sludge digester at the San Jose–Santa Clara Water Pollution Control Plant. Methane from the sludge digesters is used to generate electricity for the plant. **(b)** Diagram of a sludge digester.

anaerobic bacteria that derive energy from reducing carbon dioxide with hydrogen to form methane.

Formation of methane and carbon dioxide is a relatively innocuous end product of sludge, comparable to the carbon dioxide and water from aerobic treatment. However, there are still considerable amounts of undigested sludge remaining, although it is relatively stable and inert. This sludge is pumped to shallow drying beds or filters for reducing the volume and then carried away for disposal. Some seacoast plants barge sludge out to sea; it is also used for landfill and as a soil conditioner. Sludge has about a fifth of the value of normal commercial lawn fertilizers but has desirable soil-conditioning qualities much like soil humus.

Septic Tanks

Homes and businesses not connected to municipal sewage systems often use a **septic tank** (Figure 26–15), a device that is similar in principle to primary treatment. Sewage enters a holding tank where suspended solids settle out. The sludge in the tank must be pumped out periodically and disposed of. The effluent flows through a system of perforated piping into a leaching (soil drainage) field. The effluent entering the soil is decomposed by soil microorganisms. These systems work well when not overloaded and the drainage system is properly sized to the load and soil type. Heavy clay soils require more extensive drainage systems because of poor soil permeability. Sandy soils may cause chemical or bacterial pollution of nearby water supplies.

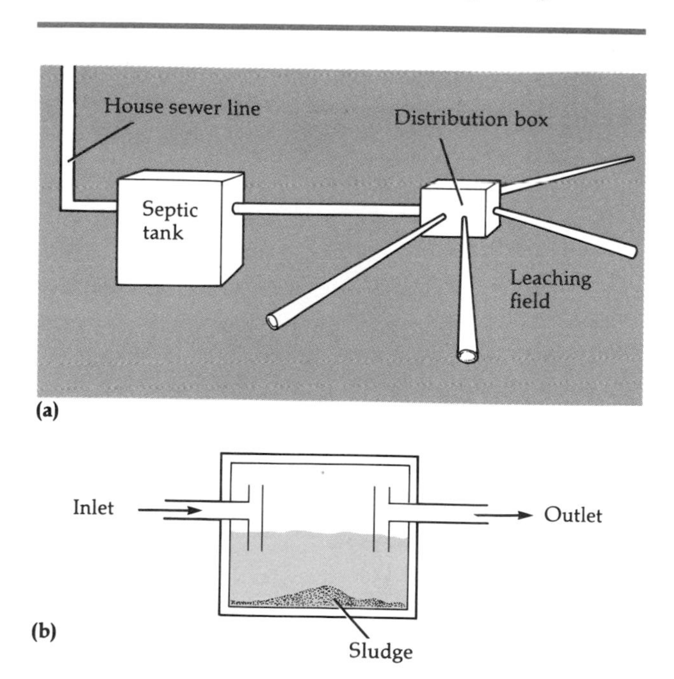

Figure 26–15 Septic tank system. **(a)** Overall plan. **(b)** Diagram of the inside of a septic tank.

Oxidation Ponds

Many small communities and many industries use **oxidation ponds,** also called **lagoons** or **stabilization ponds,** for water treatment. These are inexpensive to build and operate but require considerable land. Designs vary, but most incorporate a two-stage system. The first stage is analogous to primary treatment and is deep and anaerobic. Sludge settles out in this stage. Effluent is then pumped into an adjoining pond or system of ponds that are shallow enough to be aerated by wave action, corresponding to secondary treatment. Because it is difficult to maintain aerobic conditions for bacterial growth in ponds with so much organic matter, the growth of algae is encouraged to produce oxygen. Bacterial action in decomposing the organic matter in the wastes generates carbon dioxide. This encourages the growth of algae, which uses carbon dioxide in photosynthetic metabolism, producing oxygen, which in turn encourages activity of aerobic microorganisms in the sewage. Considerable amounts of organic matter in the form of algae accumulate, but this is not a problem because the oxidation pond, unlike a lake, already has a heavy nutrient load.

Some small sewage-producing operations, such as isolated campgrounds or highway rest stop areas, use an **oxidation ditch** for sewage treatment. In this method, a small oval channel in the form of a race track is filled with sewage water. A paddle wheel similar to that on a Mississippi steamboat propels the water in a sort of self-contained flowing stream.

Tertiary Treatment

As we have seen, primary and secondary treatments of sewage do not remove all of the biologically degradable organic matter. When added to a flowing stream in amounts that are not excessive, this is not a serious problem. Eventually, however, the pressures of increased populations may change this picture, and additional treatments may be required. Even now, primary and secondary treatments are inadequate in certain situations, such as when the effluent is discharged into small streams or recreational lakes. Thus, some communities have tertiary treatment plants. Lake Tahoe in the Sierra Nevada mountains, surrounded by extensive development, is the site of one of the best known tertiary systems designed to deal with such problems.

Secondary treatment plant effluent still contains some residual BOD. It also contains about 50% of the original nitrogen and 70% of the original phosphorus—which can have a great impact on a lake's ecosystem. Tertiary treatment is designed to remove essentially all of the BOD, nitrogen, and phosphorus. Tertiary treatment depends less on biological treatment than on physical and chemical treatments (Figure 26–10c). Nitrogen is converted to ammonia and evaporated into the air in stripping towers. Some systems encourage the formation of volatile nitrogen gas by denitrifying bacteria. Phosphorus is precipitated out by combination with chemicals such as lime, alum, and ferric chloride. Filters of fine sands and activated charcoal remove small particulate matter and dissolved chemicals. Finally, chlorine is added to the purified water to kill or inhibit any remaining microorganisms and to oxidize any remaining odor-producing substances.

Tertiary treatment provides water that is suitable for drinking, but the process is extremely costly. Secondary treatment is less costly, but the water contains many nutrients that cause water pollution. Much work is being done at present to design secondary treatment plants in which the effluent can be used for irrigation. This would eliminate a source of water pollution, provide nutrients for plant growth, and reduce the demand on already scarce water supplies. The soil would act as a trickling filter to remove chemicals and microorganisms before the water reaches ground and surface water supplies. Even now, wastewater with coliform counts below 2.2/100 ml are being used to irrigate food crops, orchards, and pastures; and water with coliform counts below 23/100 ml are being used to irrigate landscaping and recreational areas.

STUDY OUTLINE

SOIL MICROBIOLOGY AND CYCLES OF THE ELEMENTS (pp. 664–673)

THE COMPONENTS OF SOIL (pp. 664–667)

1. Soil consists of solid inorganic matter, water, air, and living organisms and products of their decay.
2. Rock and mineral fragments result from weathering of preexisting rock.
3. Organic and inorganic materials are dissolved in soil water.
4. Soil gases are the same as atmospheric gases. However, the proportions vary due to biological processes in the soil.
5. Organic matter in the soil called humus comes from plants, microorganisms and animals, their waste products, and the biochemical activities of microorganisms.
6. Organic matter provides nutrients for growth and affects the soil water and gases.
7. Microorganisms in the soil decompose organic matter and transform nitrogen- and sulfur-containing compounds into usable forms.

8. Bacteria are the most numerous organisms in the soil.
9. The characteristic musty odor of soil is due to the geosmin produced by actinomycetes.
10. Soil is not a reservoir for human pathogens except for some endospore-forming bacteria.
11. Insect and plant pathogens are found in the soil.

MICROORGANISMS AND BIOGEOCHEMICAL CYCLES (pp. 667–672)

1. In biogeochemical cycles, certain chemical elements are recycled.
2. Microorganisms are essential for biogeochemical cycles.

Carbon Cycle (pp. 667–668)

1. CO_2 is fixed into organic compounds by photoautotrophs.
2. These organic compounds provide nutrients for chemoheterotrophs.
3. Chemoheterotrophs release CO_2 used by photoautotrophs.

Nitrogen Cycle (pp. 668–672)

1. Microorganisms decompose proteins from dead cells and release amino acids.
2. Ammonia is liberated by microbial ammonification of the amino acids.
3. Ammonia is oxidized to nitrates by nitrifying bacteria.
4. Denitrifying bacteria reduce nitrates to molecular nitrogen (N_2).
5. N_2 is converted into ammonia by nitrogen-fixing bacteria.
6. Nitrogen-fixing bacteria include free-living genera such as *Azotobacter*; cyanobacteria; and the symbiotic bacteria *Rhizobium* and *Frankia*.
7. Ammonium and nitrate are used by bacteria and plants to synthesize amino acids.

Other Biogeochemical Cycles (p. 672)

1. Microorganisms are also involved in the transformation of other elements, including sulfur, potassium, iron, manganese, mercury, and zinc.
2. These reactions make minerals available to plants in soluble form for their metabolism.

DEGRADATION OF PESTICIDES AND OTHER SYNTHETIC CHEMICALS (pp. 672–673)

1. Many synthetic chemicals such as pesticides and plastics are recalcitrant (resistant to degradation).
2. Recalcitrance is based on the nature of the chemical bonding.

AQUATIC MICROBIOLOGY AND SEWAGE TREATMENT (pp. 673–686)

1. The study of microorganisms and their activities in natural waters is called aquatic microbiology.
2. Natural waters include lakes, ponds, streams, rivers, estuaries, and the sea.
3. The concentration of bacteria in water is proportional to the amount of organic material in the water.
4. Most aquatic bacteria tend to grow on surfaces rather than in a free-floating state.

FRESHWATER MICROBIAL FLORA (pp. 673–674)

1. Numbers and location of microbial flora depend on the availability of oxygen and light.
2. Photosynthetic algae are the primary producers of a lake. They are found in the limnetic zone.
3. Pseudomonads, *Cytophaga*, *Caulobacter*, and *Hyphomicrobium* are found in the limnetic zone, where oxygen is present.
4. Microbial growth in stagnant water uses available oxygen and can result in the death of fish and cause odors.
5. The amount of dissolved oxygen is increased by wave action.
6. Purple and green sulfur bacteria are found in the benthic zone, where there is light and H_2S, but no oxygen.
7. *Desulfovibrio* reduces SO_4 to H_2S in benthic mud.
8. Methane-producing bacteria are also found in the benthic zone.

SEAWATER MICROBIAL FLORA (pp. 674–675)

1. The open ocean is not favorable for most microorganisms because of its high osmotic pressure, low nutrients, and high pH.
2. Phytoplankton, consisting mainly of diatoms, are the primary producers of the open ocean.
3. Some algae and bacteria are bioluminescent. They possess the enzyme luciferase, which can emit light.

EFFECTS OF POLLUTION (pp. 675–677)

1. Microorganisms are filtered from water that percolates into ground water supplies.
2. Some pathogenic microorganisms are transmitted to humans in water supplies.

3. Recalcitrant chemical pollutants may be concentrated in animals in an aquatic food chain.
4. Mercury is metabolized by certain bacteria into a soluble compound that is concentrated in animals.
5. Nutrients such as phosphates cause eutrophication of aquatic ecosystems.
6. Eutrophication means "well-nourished." It is the result of the addition of pollutants or natural nutrients.
7. Coal-mining wastes contain sulfides that are converted into sulfuric acid by *Thiobacillus ferrooxidans*.
8. The metabolic activities of *T. ferrooxidans* can be used to recover uranium ore.

TESTS FOR WATER PURITY (pp. 677–678)

1. Tests for the bacteriologic quality of water are based on the presence of indicator organisms.
2. The most common indicator organisms are coliforms.
3. Coliforms are aerobic or facultative anaerobic, gram-negative, non-endospore-forming rods that ferment lactose with the production of acid and gas within 48 hours at 35°C.
4. Fecal coliforms, predominately *E. coli*, are used to indicate the presence of human feces.
5. The multiple-tube fermentation technique and membrane filter method are used to detect the presence of coliforms.
6. The most probable number (MPN) of coliforms per 100 ml of water is determined statistically from the results of the multiple-tube fermentation test and directly from the membrane filter method.
7. The allowable number of coliforms in water varies with the intended human use of the water.

WATER TREATMENT (pp. 678–680)

1. Drinking water is held to allow suspended matter to settle.
2. Flocculation treatment uses a chemical such as alum to coalesce and then settle colloidal material.

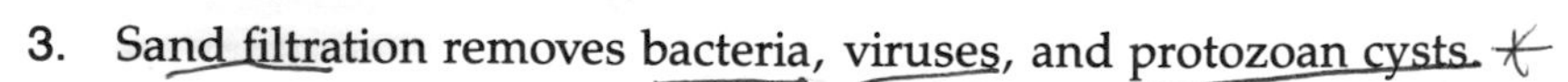

3. Sand filtration removes bacteria, viruses, and protozoan cysts.
4. Finally, drinking water is disinfected with chlorine to kill remaining pathogenic bacteria.

SEWAGE TREATMENT (pp. 680–686)

1. Domestic wastewater is called sewage.
2. It includes dishwashing, handwashing, bath water, and toilet wastes.

Primary Treatment (p. 681)

1. Primary sewage treatment is the removal of solid matter called sludge.
2. Biological activity is not very important in primary treatment.

Biochemical Oxygen Demand (pp. 681–682)

1. Primary treatment removes approximately 25 to 35% of the biochemical oxygen demand (BOD) of the sewage.
2. BOD is a measure of the biologically degradable organic matter in water.
3. It is determined by the amount of oxygen bacteria require to degrade the organic matter.

Secondary Treatment (pp. 682–684)

1. Secondary treatment is the biological degradation of organic matter in sewage after primary treatment.
2. Activated sludge and trickling filters are methods of secondary treatment.
3. Microorganisms degrade the organic matter aerobically.
4. Secondary treatment removes most of the BOD, about 50% of the nitrogen, and 30% of the phosphorus.

Sludge Digestion (pp. 684–685)

1. Sludge is placed in an anaerobic sludge digester, where bacteria degrade organic matter and produce simpler organic compounds, methane, and CO_2.
2. The methane produced in the digester is used to heat the digester and to operate other equipment.
3. Excess sludge is periodically removed from the digester, dried, and disposed of at sea, as landfill, or as soil conditioner.

Septic Tanks (p. 685)

1. Septic tanks may be used in rural areas to provide primary treatment of sewage.
2. They require a large leaching field for the effluent.

Oxidation Ponds (p. 685)

1. Small communities may use oxidation ponds for secondary treatment.
2. These require a large area to build an artificial lake.

Tertiary Treatment (pp. 685–686)

1. Tertiary treatment employs physical filtration and chemical precipitation to remove all of the BOD, nitrogen, and phosphorus from water.

2. Tertiary treatment provides drinkable water, whereas secondary treatment provides water usable for irrigation.

STUDY QUESTIONS

REVIEW

1. Write a one-sentence description of each of the following soil constituents: solid inorganic matter; water; gases; organic matter.
2. The precursor to coal is peat found in bogs. Why does peat accumulate in bogs?
3. The metabolic activities of microorganisms often produce acids. Why would microbial growth increase the solid inorganic matter in soil?
4. Compare and contrast humus and recalcitrant chemicals.
5. Diagram the carbon cycle in the presence and absence of oxygen. Name at least one microorganism that is involved at each step.
6. Fill in the following table with the information provided below.

Process	Chemical Reactions	Microorganisms
Ammonification		
Nitrification		
Denitrification		
Nitrogen fixation		

Choices:

$NO_3^- \rightarrow N_2$
$N_2 \rightarrow NH_3$
$-NH_2 \rightarrow NH_3$
$NH_3 \rightarrow NO_2^-$
$NO_2^- \rightarrow NO_3^-$

Bacillus
Rhizobium
Nitrosomonas
Azotobacter
Nitrobacter
Proteolytic bacteria

7. Compare and contrast the physical conditions of the ocean with those of freshwater.
8. Matching.

___ $CO_2 + H_2S \xrightarrow{\text{light}} C_6H_{12}O_6 + SO$ (a) Methane-producing bacteria
___ $SO_4 + H^+ \rightarrow H_2S$ (b) *Desulfovibrio*
___ $CO_2 + H_2 \rightarrow CH_4$ (c) Photosynthetic bacteria

9. Indicate which reactions in question 8 require atmospheric oxygen (O_2).
10. Outline the treatment process for drinking water.
11. What is the purpose of a coliform count on water?
12. If coliforms are not normally pathogenic, why are they used as indicator organisms for bacteriologic quality?

13. The following processes are employed in wastewater treatment. Match the type of treatment with the processes. Each choice can be used once, more than once, or not at all.

Processes:
___Leaching field
___Removal of solids
___Biological degradation
___Activated sludge
___Chemical precipitation of phosphorus
___Trickling filter
___Results in drinking water
___Effluent can be used for irrigation
___Produces methane

Types of treatment:
(a) Primary
(b) Secondary
(c) Tertiary

14. Define BOD.
15. Why is activated sludge a more efficient means of removing BOD than a sludge digester?
16. Why are septic tanks and oxidation ponds not feasible for large municipalities?
17. Explain the effect of dumping untreated sewage into a lake on the eutrophication of the lake. The effect of sewage that has primary treatment? The effect of sewage that has secondary treatment?

CHALLENGE

1. A sewage treatment plant in Pacifica, CA, is a primary treatment plant with a secondary digester. Outline the flow of water and solids through this plant. What is the final quality of the water with respect to BOD, nitrogen, and phosphorus?
2. Here are the formulas of two detergents that have been manufactured:

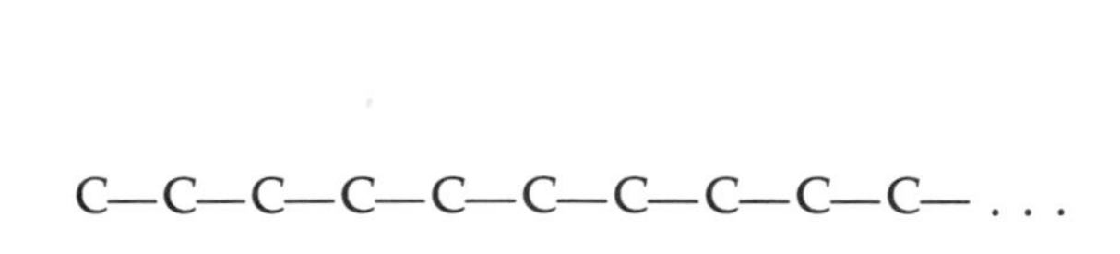

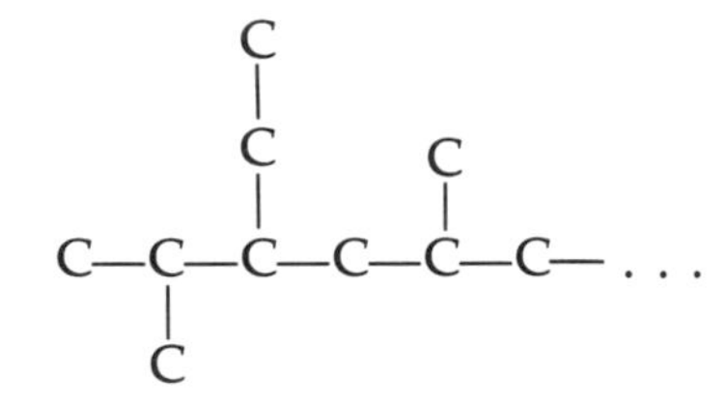

Which of these would be recalcitrant, and which would be readily degraded by microorganisms? (*Hint:* Refer to the degradation of fatty acids in Chapter 5.)

3. What is the MPN of a water sample with these multiple-tube fermentation test results: 10 ml portions: 4 positive for acid and gas; 1 ml portions: 3 positive for acid and gas; 0.1 ml portions: 2 positive. (See MPN table, Appendix D.)

FURTHER READING

Alexander, M. 1977. *Introduction to Soil Microbiology.* New York: Wiley. A text on microbial ecology and the role of microorganisms in the transformation of carbon, nitrogen, sulfur, phosphorus, and iron.

Brock, T. D. 1978. *Thermophilic Microorganisms and Life at High Temperatures.* New York: Springer-Verlag. The microorganisms, their metabolism, and growth requirements of the unique environment of the hot springs in Yellowstone National Park.

Gutnick, D. L. and E. Rosenberg. 1977. Oil tankers and pollution: a microbiological problem. *Annual Review of Microbiology* 31:378–396. The role of microorganisms in the removal of petroleum from an ecosystem.

Imhoff, K., W. J. Muller, and D. K. B. Thistlewayte. 1971. *Disposal of Sewage and Other Water-Borne Wastes.* Ann Arbor, MI: Ann Arbor Science. An excellent reference on wastewater treatment plant designs, methods of disposal, and biological treatment of sewage.

Jannasch, H. W. and W. O. Wirsen. June 1977. Microbial life in the deep sea. *Scientific American* 236:42–52. Sea floor experiments to determine the effects of high barometric pressure and low temperature on microorganisms.

Nickerson, W. J. (convener). 1975. Biodegradation of refractory molecules. *Developments in Industrial Microbiology* 16:67–118. Washington, D.C.: American Institute of Biological Science. Papers on the bacterial degradation of pesticides, rubber, petroleum, and insoluble natural materials.

Sieburth, J. M. 1979. *Sea Microbes.* New York: Oxford University Press. Microorganisms and their adaptation for living in ocean waters.

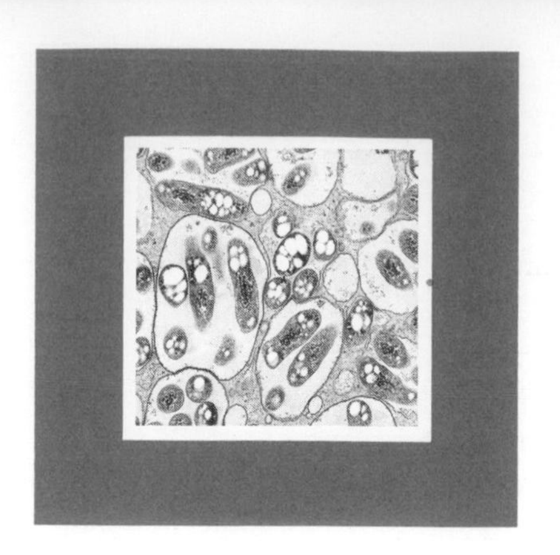

27

Food and Industrial Microbiology

Objectives

After completing this chapter you should be able to

- Provide a brief history of the development of food preservation.
- Explain why sterilization of canned foods is important.
- Describe thermophilic anaerobic spoilage, flat sour spoilage, and spoilage by mesophilic bacteria.
- Describe low-temperature preservation.
- Define pasteurization and explain why dairy products are pasteurized.
- Provide examples of chemical food preservatives and explain why they are used.
- Outline at least four examples of the beneficial activities of microorganisms in food production.
- Describe the role of microorganisms in the production of single-cell protein (SCP), alternative energy sources, and chemotherapeutic agents.

We will now turn our attention from soil and aquatic microbiology to food and industrial microbiology. In this chapter, we will examine some of the principles relating to food spoilage and preservation, foodborne infections and food poisoning, and food production by microorganisms. We will also take a look at some important aspects of industrial microbiology and the impact of genetic engineering on industrial microbiology.

FOOD SPOILAGE AND PRESERVATION

Modern civilization and the support of large populations could not exist were it not for effective methods of preserving food. In fact, civilization arose only after agriculture permitted a year-round stable food supply in a single site and people were able to give up the nomadic hunting cultures. Many of the methods of food preservation used today were probably discovered by chance in centuries past. Primitive people observed that dried meat or salted fish resisted decay. And the nomads no doubt observed that animal milk that soured resisted further decomposition and was still palatable. Moreover, if the curd of the soured milk was pressed to remove moisture and allowed to ripen (in effect, cheesemaking), it was even more effectively preserved and also sometimes developed many desirable characteristics. And farmers learned that if grains were kept dry, they did not become moldy.

All such phenomena are readily understood by anyone familiar with the physical methods used to control the growth of microorganisms. Bacteria and fungi require a minimum amount of available moisture for growth. Drying food or adding salt or sugar lowers the available moisture and prevents spoilage. And the acidity found in the natural fermentation of milk or vegetable juices, such as in sauerkraut, also prevents the growth of many spoilage bacteria. These methods of preservation are still in use today. However, a tour of any supermarket will demonstrate that heat sterilization, pasteurization, refrigeration, and freezing are now much more popular methods for the control of food spoilage. These modern methods provide a preserved food much nearer its natural state and palatability than preserved food available to people in older times.

Preservation by heat (canning) requires considerable sophistication, and refrigeration of foods is an even more technical and recent development. In some parts of the world, ice was used for food preservation and could be kept for warm seasons in specially insulated structures, but it was difficult to transport and its use for much of the world was only a curiosity.

Preservation by heat had its origin in the early 1800s, before microorganisms were known to cause human disease and food spoilage. The technology of canning was promoted by military necessity of the day. When armies were small professional organizations, they could be supported off the land, supplemented by portable rations of dried grains, cheese, and wine. But with the rise of large armies of citizen soldiers, a development of the French revolution, this was no longer sufficient. Thus, the French government offered a prize to anyone who could devise a method of preserving food, particularly meat. The prize was won by a confectioner, Appert, who sometime around 1810 showed how food could be preserved by sealing it in tightly stoppered containers and boiling it for specified periods of time. The key to his method was the detailed tables of times required for different foods and container sizes. As we know today, there are many endospore-forming bacteria that will survive hours of such boiling. Failures were of manageable proportions, however, and usually attributed to faulty sealing.

The concept of heat preservation was followed quickly by the invention of the metal can (Figure 27–1). Temperatures above the boiling point of water were introduced into processing and were obtained by oil or salt baths. The use of steam under pressure in closed containers, as done today, was introduced only late in the eighteenth century, when reliable pressure controls and safety valves were developed. By then, of course, a better theoretical basis was also available.

Actually, it is not difficult to preserve foods by heating a properly sealed container. Commercially,

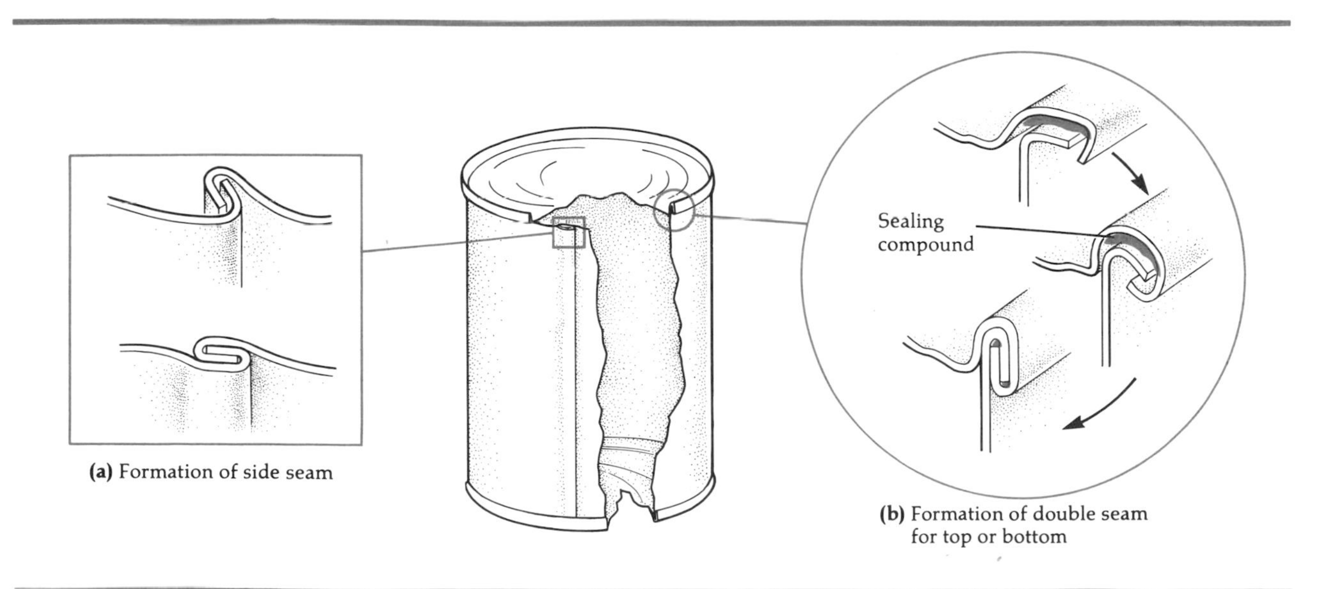

Figure 27–1 Sealing of a metal can. **(a)** Side seam. **(b)** Top and bottom seam.

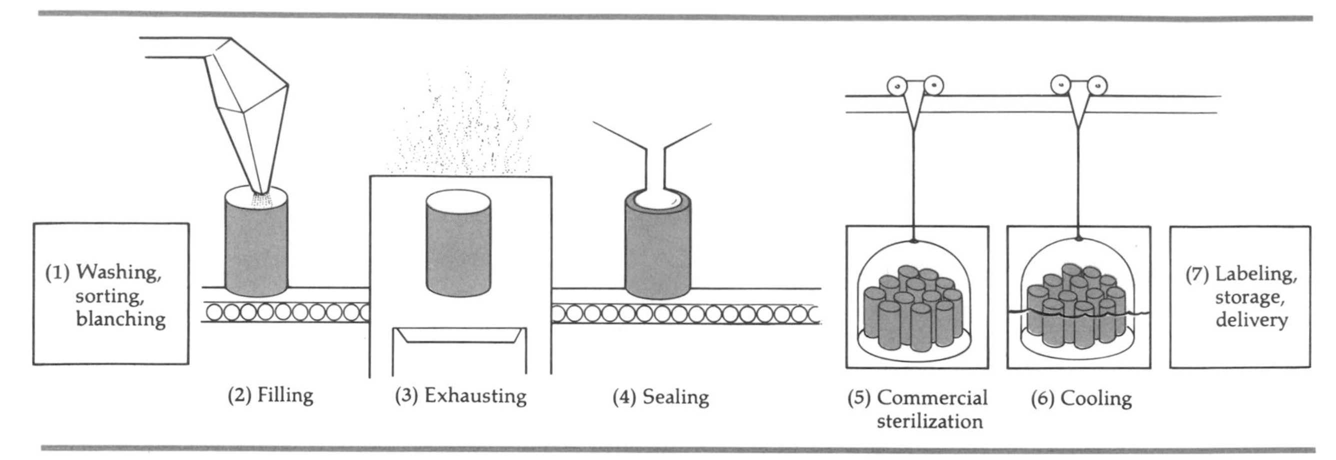

Figure 27–2 Industrial canning. (1) In blanching, the product is treated with hot water or live steam to soften the product so that the can can be filled properly, to eliminate air from the product, and to destroy enzymes that might alter flavor, texture, or color of the product. (2) After the cans are cleaned, they are filled with the product, as full as possible, to eliminate dead space. (3) In exhausting, air is removed to reduce corrosion and spoilage. (4) Sealing is accomplished by machine and is designed to make tight seals (see Figure 27–1). After sealing, the cans are marked with code numbers indicating product, grade, and packing date in case of recall. (5) Commercial sterilization is designed to destroy *C. botulinum*. (6) Cooling of the cans follows sterilization. (7) The cans are then labeled, stored, and delivered.

the problem is to do it with the minimum amount of heat needed to kill spoilage organisms and dangerous microorganisms, such as the endospore-forming botulism bacterium. Heating degrades the quality of food and much research was needed to provide the exact degree of heat treatment to minimize these quality changes.

Canned goods today undergo what is called **commercial sterilization** (Figure 27–2). The minimum preservation process is aimed at destroying the endospore-forming pathogen *Clostridium botulinum*. If this is accomplished, then any other significant spoilage or pathogenic bacterium will also be destroyed. To ensure this, the 12D treatment is applied, by which a theoretical population of botulism bacteria would be decreased by 12 logarithmic cycles. In other words, if there were 10^{12} (1,000,000,000,000) botulism organisms in a can, there would be only one survivor. Since this is a most unlikely population, this is considered quite safe. Certain thermophilic endospore-forming bacteria have endospores that are more resistant to heat treatment than *C. botulinum*. However, these bacteria are obligate thermophiles and will not grow at normal room temperatures, generally remaining dormant until about 45°C. Therefore, they are not a problem under normal storage temperatures.

CANNED FOOD SPOILAGE

If canned foods are incubated at high temperatures, such as in a truck in the hot sun or next to a steam radiator, the thermophilic anaerobic bacteria that often survive commercial sterilization (not as rigorous as true sterilization) can germinate and grow. This **thermophilic anaerobic spoilage** is a fairly common cause of spoilage in low-acid canned foods. The can is usually swollen from gas production and the contents have a lowered pH and a sour odor (Figure 27–3). A number of thermophilic species of *Clostridium* can cause this type of spoilage. When thermophilic spoilage occurs and the can is not swollen by gas production, it is termed **flat sour spoilage.** This type of spoilage is caused by thermophilic organisms such as *Bacillus stearothermophilus*, which is found in the starch or sugars used in food preparation. Many industries have standards for the numbers of such thermophilic organisms permitted in raw materials. Both types of canned goods spoilage depend upon storage at higher than normal temperatures, permitting the growth of bacteria whose endospores are not destroyed by normal processing.

Canned food spoilage by mesophilic bacteria is due to can leakage or underprocessing. Nor-

Figure 27–3 Severe can swelling due to gas production, as in thermophilic anaerobic spoilage.

mally, the bacteria are killed by proper processing. The presence of non-endospore-formers strongly suggests can leakage, while underprocessing is more likely to result in spoilage by endospore-formers. Contamination by can leakage often occurs during the cooling down of cans after heat processing. The hot cans are sprayed with cooling water or drawn on wheeled carts through a trough filled with water. As the can cools, a vacuum is formed inside and water can be sucked past the heat-softened sealant in the crimped lid. Contaminating bacteria in the cooling water are drawn with the water into the can. Spoilage due to underprocessing and can leakage is likely to result in odors of putrefaction, at least in high-protein foods, and occurs at normal storage temperatures. In such types of spoilage, there is always the potential of botulism bacteria being present.

A summary of the types of canned food spoilage is presented in Table 27–1.

Home canning is an important use of heating for food preservation. Because of the possibility of botulism food poisoning from improper canning methods, persons involved in home canning should obtain reliable directions and follow them exactly.

Table 27–1 Types of Canned Food Spoilage

	Indications of Spoilage	
Type of Spoilage	Appearance of Can	Contents of Can
Low- and Medium-Acid Foods (pH above 4.5):		
Flat sour *(Bacillus stearothermophilus)*	Possible loss of vacuum on storage	Appearance not usually altered; pH markedly lowered; sour; may have slightly abnormal odor; sometimes cloudy liquid
Thermophilic anaerobic *(Clostridium thermosaccharolyticum)*	Can swells, may burst	Fermented, sour, cheesy, or butyric odor
Sulfide spoilage *(Desulfotomaculum nigrificans)*	Can flat	Usually blackened, "rotten egg" odor
Putrefactive anaerobic *(Cl. sporogenes)*	Can swells, may burst	May be partially digested; pH slightly above normal; typical putrid odor
Aerobic endospore-formers *(Bacillus* species)	Usually no swelling, except in cured meats when nitrate and sugar are present	Coagulated evaporated milk, black beets
High-acid foods (pH below 4.5):		
Flat sour *(B. coagulans)*	Can flat, little change in vacuum	Slight pH change; off odor and flavor
Butyric anaerobic *(Cl. butyricum)*	Can swells, may burst	Fermented, butyric odor
Non-endospore-formers (mostly lactic acid bacteria)	Can swells, usually bursts, but swelling may be arrested	Acid odor
Yeasts	Can swells, may burst	Fermented; yeasty odor
Molds	Can flat	Surface growth; musty odor

Source: Data from the National Canners Association, 1950 6th Street, Berkeley, CA 94710.

Some acidic foods, such as tomatoes or preserved fruits, are preserved by heating at temperatures of 100°C or below. As a rule, the only spoilage organisms of consequence in acidic foods are molds, yeasts, and occasionally species of acid-tolerant non-endospore-forming bacteria. These organisms are the only ones capable of growth at the pH of these foods and are easily killed by temperatures less than 100°C. One problem with such acidic foods can be the *sclerotia* of certain species of molds. Sclerotia are specialized resistant bodies that can survive temperatures of 80°C for a few minutes.

LOW-TEMPERATURE PRESERVATION

Low temperatures slow down the reproduction time of microorganisms. Even so, some molds and bacteria can grow at a significant rate at temperatures below the freezing point of water, 0°C. Food typically does not freeze solid at temperatures several degrees below this point. Growth in foods a few degrees below freezing is common, and even growth at −18°C has been reported. A properly set refrigerator maintains a temperature of 0° to 7°C. Many microorganisms will grow slowly at these temperatures and will alter the taste and appearance of foods stored for too long a time. Pathogenic bacteria, with a few exceptions (such as the clostridia that cause type E botulism), will not grow at correctly set refrigerator temperatures. Freezing will not kill bacteria outright in significant numbers, but frozen bacterial populations are dormant and decline slowly in numbers with time. Some parasites such as the roundworms causing trichinosis are killed by several days' freezing. Some important temperatures associated with microorganisms and food spoilage are shown in Figure 27–4.

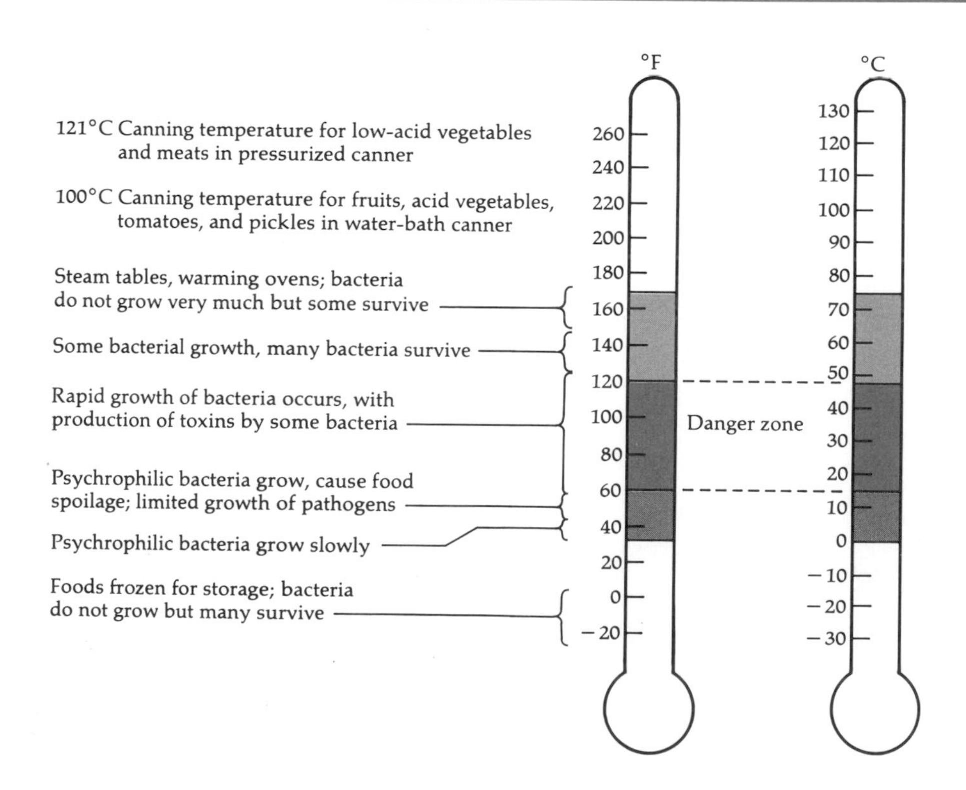

Figure 27–4 Some important temperatures associated with microorganisms and food spoilage.

RADIATION AND FOOD PRESERVATION

There has been considerable research, particularly for military applications, into the use of ionizing radiation for food preservation. The European fishing industry has also done considerable work with ionizing-type radiation in attempts to lower counts of bacteria in newly caught fish. Lower counts improve the keeping quality during shipment in cold storage on ships. Pasteurization-type applications may eventually be practical, but actual sterilization by radiation probably causes too much alteration in the taste of most foods for commercial use. Microwaves, such as used for cooking, kill bacteria only by their heat, and they have little or no direct radiation effect.

PASTEURIZATION

Louis Pasteur began his career as a microbiologist when he was commissioned in 1871 by the French brewing industry to investigate the causes of spoilage in beer and wine. He concluded that spoilage was due to the growth of microorganisms and he worked out a method of eliminating them with mild heating. Too much heat would alter the beverage characteristics to an unacceptable degree, but it was found unnecessary to kill all the microorganisms. The principle of the selective killing of microorganisms, now called **pasteurization,** was later applied to other foods, milk in particular (see Chapter 7). Other products, such as ice cream, yogurt, and beer, all have their own pasteurization times and temperatures, which often differ considerably.

CHEMICAL PRESERVATIVES

Chemical preservatives are frequently added to foods to retard spoilage. Among the more common additives used are sodium benzoate, sorbic acid, and calcium propionate. These chemicals are simple organic acids, or salts of organic acids, which the body readily metabolizes and which are generally judged to be safe in foods. Sorbic acid and sodium benzoate prevent mold growth in certain acidic foods such as cheese and soft drinks. Such foods, usually with a pH of 5.5 or less, are most susceptible to mold-type spoilage. Calcium propionate is an effective fungistat used in bread. It prevents the growth of surface molds and the *Bacillus* species of bacteria that causes ropy bread. These organic acids do not work by any pH effect in inhibiting mold growth, but interfere with mold metabolism or the integrity of the cellular membrane.

Sodium nitrate and sodium nitrite are found in many meat products such as ham, bacon, wieners, and sausage. The active ingredient is sodium nitrite. The sodium nitrate is a reservoir that certain bacteria in the meats convert to sodium nitrite. These bacteria use nitrate as a substitute for oxygen under anaerobic conditions, much as in denitrification in soil. The nitrite has two main functions: to preserve the pleasing red color of the meat by reacting with blood components in the meat, and to prevent the germination and growth of any botulism endospores that might be present. There has been some concern that the reaction of nitrites with amino acids can form certain carcinogenic products known as **nitrosamines,** and the amount of nitrites added to foods has generally been reduced lately for this reason. However, the established value of the nitrites in preventing botulism has prevented its total elimination, and since nitrosamines are formed in the body from other sources, the added risk posed by limited use of nitrates and nitrites in meats may not be as high as once thought.

FOOD-BORNE INFECTIONS AND MICROBIAL INTOXICATIONS

We have seen in Chapter 24 that illness resulting from microbial growth in food is associated with two principal mechanisms. In the mechanism known as food-borne infection, the contaminating microorganism may infect the person who ingests contaminated food. As the pathogen grows in the host, the damaging toxins are produced. Diseases caused by this mechanism include gastroenteritis, typhoid fever, and dysentery. In the other mechanism, called microbial intoxication, the toxin is formed by microbial growth in the food and then ingested with the food. Diseases associated with this mechanism include botulism, staphylococcal food poisoning, and mycotoxicoses (intoxications caused by fungal toxins such as ergot or aflatoxin). Several microbial intoxication diseases are described in Table 27–2.

Table 27–2 Food-Borne Microbial Intoxications

Disease	Foods Involved	Prevention	Clinical Features	Duration
Aflatoxin	Moldy grains	Avoid contaminated grains	Low doses may induce liver cancer; high doses cause general liver damage; no human cases	
Bacillus cereus intoxication	Custard, cereal, starchy foods	Refrigeration of foods	Cramps, diarrhea, nausea, vomiting	Onset: 8–16 hours Duration: Less than 1 day
Botulism	Canned foods	Proper canning procedures; boiling food prior to consumption	Nausea, vomiting, headache, vertigo, respiratory paralysis	Onset: 2 hours–6 days Duration: weeks
Ergotism	Moldy grains	Avoid contaminated grains	Burning abdominal pain, hallucinations	Onset: 1–2 hours Duration: months
Methyl mercury poisoning	Freshwater or ocean fish	Stop dumping mercury into waters	Blurred vision, numbness, apathy, coma	Onset: 1 week Duration: may be chronic
Mushroom poisoning	*Amanita* species	Don't eat poisonous mushrooms	Vomiting, hepatic necrosis, neurotoxic effects	Onset: Less than 1 day Duration: Less than 10 days
Paralytic shellfish poisoning	Bivalve mollusks during red tide (dinoflagellate blooms)	Avoid eating mollusks during red tide	Tingling, rash, fever, respiratory paralysis	Onset: Less than 1 hour Duration: Less than 12 hours
Scombroid poisoning	Histaminelike substance produced by *Proteus* growing on ocean fish	Refrigeration of fish	Headache, cramps, hives, shock (rare)	Onset: minutes–hour
Staphylococcal intoxication	High osmotic pressure foods not cooked before eating	Refrigeration of foods	Nausea, vomiting, diarrhea	Onset: Less than 1 day Duration: Less than 3 days

Dairy products, which are often consumed without cooking, are particularly likely to transmit food-related diseases. Standards for sanitation in the dairy industry are therefore very stringent. Because most dairy milk is drawn mechanically and promptly enters cooled holding tanks, most of the bacteria in milk today are gram-negative psychrophiles. Grade A pasteurized cultured products, such as buttermilk and cultured sour cream, have high counts of lactic acid bacteria as a natural result of their method of manufacture. Therefore, the standards require only that they not contain more than 10 coliforms per milliliter. Pasteurized Grade A milk is required to have a standard plate count of less than 20,000 bacteria per milliliter and not more than 10 coliforms per milliliter. Grade A dry milk products are required to have a standard plate count of less than 30,000 bacteria per gram or a coliform count of less than 90 per gram.

Unpasteurized (raw) dairy products are available commercially in about half the states across the country. Three states have certifying boards that certify raw milk from dairies that have met required standards for cleanliness, including inspection of dairy herds for disease. Some people prefer raw milk because of the lack of additives or because they feel it has not been altered chemically. Pasteurization may denature enzymes found in milk. Certified raw milk has a maximum standard for bacterial counts of 10,000 per milliliter and 10 coliforms per milliliter. The microbial standards guaranteed by certification do not prevent transmission of disease in raw milk; in fact, certified milk regularly causes outbreaks of salmonellosis.

ROLE OF MICROORGANISMS IN FOOD PRODUCTION

Cheese

The United States leads the world in the manufacture of cheese, producing more than 1.5 million

Figure 27–5 During cheesmaking, curd is poured into cheesecloth or some other porous material to permit the drainage of whey.

tons each year. Although there are many types of cheeses, all require the formation of a **curd,** which can then be separated from the main liquid fraction, or **whey** (Figure 27–5). Except for a few unripened cheeses, such as Ricotta or cottage cheese, the curd undergoes a microbial ripening process. The curd is made up of the milk protein, **casein,** and is usually formed by the action of an enzyme, **rennin,** aided by acidic conditions provided by inoculation with certain lactic acid-producing bacteria. These lactic acid bacteria also provide the characteristic flavors and aromas of fermented dairy products.

Cheeses are generally classified by their hardness. The more moisture lost from the curd and the harder it is compressed, the harder the cheese. Romano and Parmesan cheeses, for example, are classified as very hard cheeses; Cheddar and Swiss are hard cheeses. Limburger, blue, or Roquefort cheeses are classified as semisoft; and Camembert is an example of a soft cheese. The hard Cheddar or Swiss cheeses are ripened by lactic acid bacteria growing anaerobically in the interior. (A *Propionibacterium* species of bacteria in Swiss cheese produces carbon dioxide that forms the holes.) The longer the incubation time, the more the acidity and the sharper the taste of the cheese. Such hard interior-ripened cheeses can be quite large. Semisoft cheeses, such as Limburger, are ripened by bacteria and other contaminating organisms growing on the surface. Blue and Roquefort cheeses are ripened by *Penicillium* molds inoculated into the cheese. The texture of the cheese is loose enough that adequate oxygen can reach the aerobic molds. The growth of the *Penicillium* molds is visible as blue-green clumps in the cheese. Camembert cheese is ripened in small packets so that the *Penicillium* mold growing aerobically on the surface will still have its enzymes diffuse into the cheese for ripening.

Other Dairy Products

Butter is made by churning cream until the fat phase forms globules of butter separated from the liquid buttermilk fraction. Lactic acid bacteria are allowed to grow in cream to provide flavor from the *diacetyls* that give these products the typical buttery flavor and aroma until the desired acidity is reached. Actually, today, buttermilk is not usually a by-product of buttermaking, but is made by deliberately inoculating skim milk with bacteria that form lactic acid and the diacetyls. The inoculum is allowed to grow for 12 or more hours before cooling and packaging. Sour cream is made by inoculating cream with organisms similar to those used to make buttermilk.

A wide variety of slightly acidic dairy products, probably a heritage of a nomadic past, are found around the world. Many of them are part of the daily diet in the Balkans, Eastern Europe, and Russia. One such product is yogurt, which is also popular in the United States. Commercially, yogurt is made from low-fat milk by first evaporating much of the water in a vacuum pan. This is followed by inoculation with a species of lactic acid-producing *Streptococcus* that grows at elevated temperatures. Incubation at about 45°C for several hours gives the acidity to the product, and the addition of stabilizers aids in the formation of the thick texture. A second bacterial inoculum provides the characteristic flavor of the yogurt. A proper balance between the flavor-producing and acid-producing organisms is the secret of good-quality yogurt. Kefir and kumiss are popular beverages in Eastern Europe. The usual lactic acid-producing bacteria are supplemented with a lactose-fermenting yeast to give these drinks an alcoholic content of 1 or 2%.

Nondairy Fermentations

Microorganisms are also used in baking. The sugars in bread dough are fermented to alcohol and carbon dioxide by yeasts, just as in making alcoholic beverages (discussed in the next section). The carbon dioxide makes the typical bubbles of leavened bread, while the alcohol evaporates during baking. Some types of bread such as rye or sourdough also involve the growth of lactic acid bacteria to give the typical tart flavoring.

Fermentation is also used in producing foods such as sauerkraut, pickles, and olives. In Asia, soy sauce is produced in extremely large amounts. The soy sauce process makes use of molds for the formation of starch-degrading enzymes to produce fermentable sugars. This principle is used in making other Asian fermented foods, including sake, or Japanese rice wine. In soy sauce production, molds such as *Aspergillus oryzae* are grown on wheat bran and then allowed to act, along with lactic acid-type bacteria, on cooked soybean and crushed wheat mixtures. After this process has produced fermentable carbohydrates, a prolonged fermentation results in soy sauce.

Alcoholic Beverages

Microorganisms are involved in the production of almost all alcoholic beverages. Beer and ale are made by the fermentation of grains by yeasts, as are a number of other alcoholic beverages (Table 27–3). Because yeasts are unable to use starch directly, it is necessary to convert the starch from the grain to glucose and maltose, which the yeasts can ferment into ethanol and carbon dioxide. This conversion, called **malting,** involves the sprouting of starch-containing grains such as malting barley, which are then dried and ground. This product, called **malt,** contains starch-degrading enzymes to convert cereal starches into fermentable carbohydrates that can be utilized by yeasts. Distilled spirits such as whiskey, vodka, or rum are made by yeast fermentation of carbohydrates from such sources as cereal grains, potatoes, and molasses. The alcohol is then distilled to make a concentrated alcoholic beverage.

Table 27–3 Production of Alcoholic Beverages by Yeasts

Beverage	Yeast	Method of Preparation	Function of Yeast
Beer	*Saccharomyces cerevisiae* or *S. carlsbergensis*	Barley malt and starch adjuncts mixed with warm water; after enzymatic starch conversion, wort is filtered, then boiled with hops, and finally fermented with yeast	Converts sugar into alcohol and carbon dioxide; produces changes in proteins and other minor constituents that modify flavor
Rum	*S. cerevisiae* or other yeasts	Blackstrap molasses containing 12 to 14% fermentable sugar; ammonium sulfate and occasionally phosphates may be added as nutrients; distilled after fermentation	Sugar converted to alcohol, which is then removed by distillation
Scotch	*S. cerevisiae* (generally a top yeast)	Grain mash cooked with peated malt and fermented; distillate aged in oak casks at least three years; then blended with grain whiskey	Produces alcohol and congeneric substances (acids, esters, various alcohols), which, with the peated malt, give characteristic Scotch flavor
Bourbon	*S. cerevisiae*	Grain mash consisting of corn (at least 51%); generally with rye cooked with malt and fermented; matured in charred oak barrels	Same as for Scotch whiskey, but flavor is characteristic of bourbon whiskey
Wine	*S. ellipsoideus*, various strains	Grapes with sugar concentration up to 22%; allowed to ferment with special strain of yeast, or with yeast naturally present on the grape; primary fermentation succeeded by a period of storage for maturation	Converts sugar into alcohol and also produces changes in minor constituents that modify flavor and bouquet; amount of alcohol varies according to type of wine

Source: W. S. Spector (ed), *Handbook of Biological Data*, Saunders, Philadelphia, PA, 1956.

Grape testing and picking → Crushing → Sulfite addition to prevent spolage → Fermentation → Pressing → Clarification and aging → Filtration → Bottling

Figure 27–6 Basic steps employed in making red wine. In making white wine, the pressing step comes before fermentation.

Wines are typically made from grapes, which contain sugars that can be used directly by yeasts for fermentation, without the malting needed to make a substrate for yeasts in beers and whiskies (Figure 27–6). Grapes usually need no additional sugars, but wines from other fruits may be supplemented with sugars to ensure enough alcohol production.

Lactic acid bacteria are also important in making wine from certain grapes that are particularly acidic because of the presence of malic acid. These bacteria convert malic acid to the less acidic lactic acid in a process called **malolactic fermentation.** The result is a less acidic, better-tasting wine than would otherwise be produced.

Winemakers who allowed wine to be exposed to air found that it became sour from growth of acetic acid-forming bacteria growing aerobically on the alcohol. The result was bad wine, or vinegar. This process is now used deliberately to make vinegar. Ethanol is first made by yeast fermentation anaerobically. The ethanol is then aerobically oxidized to acetic acid by acetic acid-producing bacteria *(Acetobacter).* The basic process of vinegar production is shown in Figure 27–7.

SINGLE-CELL PROTEIN (SCP) PRODUCTION

In the future, as human populations expand and arable land does not, microorganisms may become much more important to the food requirements of the world. Protein is in particularly short supply, and microorganisms can double their weight in a few hours, and often in less than an hour, when they are provided with suitable substrates. Many of these substrates can be converted from unusable forms into **single-cell protein (SCP).**

A factory of microorganisms provided with the proper growing conditions can produce hundreds of thousands of metric tons of proteinaceous material yearly on very little land. The substrates may be unusable by humans directly—for example, cellulose, methanol, or petroleum hydrocarbons. Petroleum yields the most food for its weight because it is almost totally reduced, but because it is needed as a fuel, its costs are beginning to make

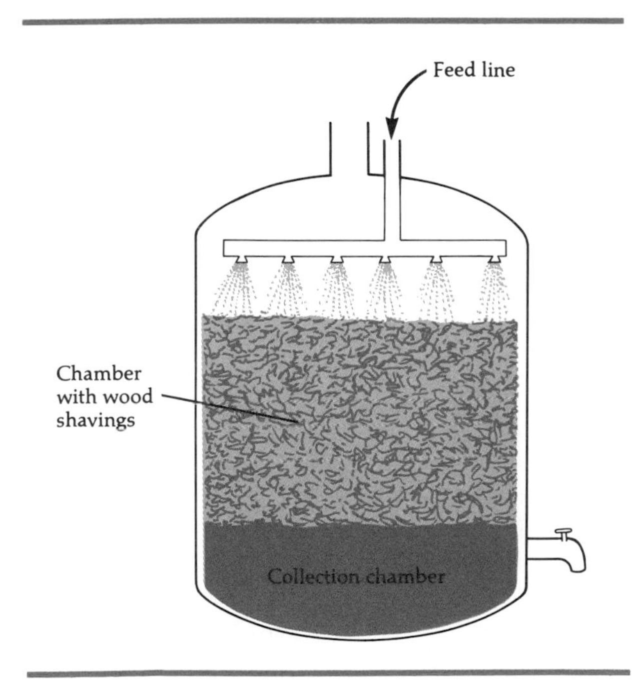

Figure 27–7 Vinegar production. In this particular vinegar generator, ethanol trickles through wood shavings that are coated by *Acetobacter.* As a result, the bacteria oxidize the ethanol to acetic acid, which is removed periodically.

it an unlikely source. Many wastes from meat packing, paper production, agriculture, and forestry might also serve. Photosynthetic organisms are of special interest because of their inexhaustible energy source. However, algae and photosynthetic bacteria tend to make relatively unpalatable end products, except for the alga *Spirulina,* which has been used in dried form from the shores of a lake in Chad, Africa, for centuries. *Spirulina* was also eaten by ancient Aztecs in Mexico. Experiments on eating *Spirulina* have been conducted in the United States, France, and Mexico, and as a result, the United States and Mexico permit the sale of *Spirulina.*

Among all the microorganisms from which single-cell protein can be made, bacteria are the most attractive because they have short generation times compared to yeasts and molds. Cost-effective processes have been experimentally developed that can produce about 5 tons of SCP per day. And compared to yeasts, they can use a greater variety of substrates. Single-cell protein would probably not be consumed directly by humans for several reasons, palatability included. At some loss in efficiency, it would probably be fed to animals first. Among potential problems in direct ingestion are the high nucleic acid contents of the food organisms, which could exceed the body's capacity to metabolize them and precipitate ailments such as gout.

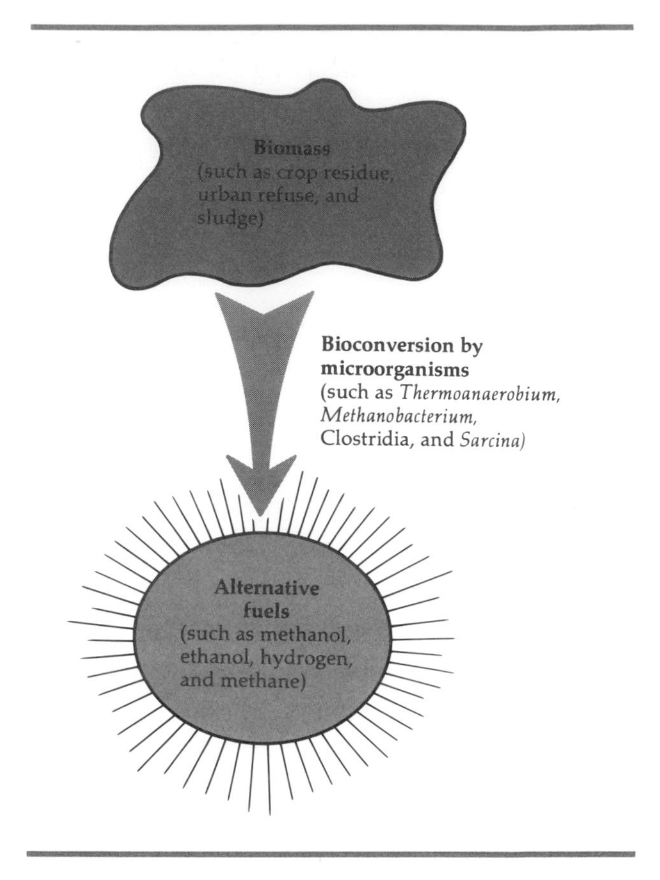

Figure 27–8 Bioconversion.

ALTERNATIVE ENERGY SOURCES FROM MICROORGANISMS

Each year, the United States produces many hundreds of metric tons of organic waste from crops, forests, and municipalities. This waste is called **biomass,** and so far it has been disposed of by burning or placing in a dump. Conversion of biomass into alternative fuel sources, called **bioconversion,** would help meet our energy needs as well as help dispose of much solid waste (Figure 27–8). The bioconversion of organic materials into alternative fuels by microorganisms is a developing industry. The processes are similar to those in a sewage treatment plant digester. Biomass provides a fermentable carbohydrate for microorganisms that produce methanol, ethanol, hydrogen, and methane. Gasohol (90% gasoline + 10% ethanol) is presently being sold commercially.

In another approach to alternative fuels, scientists have teamed up two microorganisms—an alga and a bacterium. The alga converts sunlight and carbon dioxide into organic compounds which the bacterium processes into unsaturated fatty acids. These fatty acids can then be used as a fuel (see box).

INDUSTRIAL MICROBIOLOGY AND GENETIC ENGINEERING

For years, a number of industries have used conventional microbial genetics to breed better microbial strains for their purposes. Until recently, however, the gene pool available to scientists was limited to what nature has provided. But now the gene pool has been greatly expanded by gene manipulation. This entire new field of industrial microbiology is based on genetic engineering.

In Chapter 8, we examined the principles involved in genetic engineering. Recall that the DNA of one organism can be enzymatically fragmented, and isolated genetic units can be incorporated into easily handled, fast-growing bacteria, using recombinant DNA techniques. These genes can even come from higher animals and humans. There has been some concern that potentially harmful bacterial strains might be released into the environment with serious results, but proper controls have made this unlikely. In any event, there is little doubt that the genetic engineering techniques now developed will be rapidly improved and will have a tremendous impact on human civilization and the earth as a whole. Conventional industrial microbiology and the techniques of genetic engineering are vital industries in the production of medicines, hormones, food products, and chemicals, to name a few.

The field of medicine should benefit from the increased production of useful drugs such as interferon or insulin, which are now laboriously and expensively extracted from sources such as transfused blood or slaughtered animals. Vaccines can

Microbiology in the News

Chlorella: Algae to Fuel the World?

Chlorella may turn out to be the biggest name in gasoline since Standard Oil. This tiny, one-celled green alga looks like it has the chance to become a giant competitor in the alternative fuel's race. Isolated from blooms that appear on fresh water lakes in the spring and autumn, *Chlorella* grows quickly and will spread three to five kilograms per square meter overnight. In Japan, *Chlorella* has been used as protein and vitamin supplements in foods like ice cream, yogurt, and bread, but it may soon be filling tanks instead of stomachs.

Recently, two Canadian scientists from the University of Toronto reported preliminary results from a microbial fuel system they devised which uses *Chlorella* and a modified bacterium. *Chlorella,* although not exactly in the driver's seat, plays the principal role of energy supplier. The scientists use this tiny alga to take up carbon dioxide and churn out raw energy in the form of organic compounds. Then these organic goodies are gobbled up by *Arthrobacter* AK 19, a bacterium isolated from the soils around natural gas wells.

AK 19 takes up oil better than a sponge absorbs water. It can accumulate over 85% of its body weight as fat. Under the electron microscope, AK 19 looks like a transparent plastic bag almost bursting with round oil droplets. Better yet, by the time *Chlorella* has relinquished its organic morsels to the greedy gulps of AK 19, these oily materials have had their carbon chains reorganized and they have become unsaturated fatty acids. So, AK 19 winds up storing organic ooze that is not too different from crude oil.

How much oil can *Chlorella* and AK 19 produce? The scientists calculate that although they have to use a lot of light (200 watts per square meter), for each hectare of growing microbes it is possible to produce the equivalent of 50,000 barrels of crude oil a year.

Comparing this microbial oil to gasohol made from the ethanol of maize or sugar cane, the microbial oil requires no land to cultivate its crop, so it is not produced at the sacrifice of agricultural soil that could be used to grow food for people or livestock. And although both gasohol and microbial oil need carbon dioxide and sunshine, some algal species are so photosynthetically efficient that they can produce oil crops every two days, whereas maize and sugar can require months of cultivation.

It is possible that microbial oil will not only compete with petroleum fuels, but it may also supply the food industry with a substitute to animal fats and vegetable oils. However, the Toronto team thinks that the microbial fuel may best be put to use as the liquid energy required for space travel.

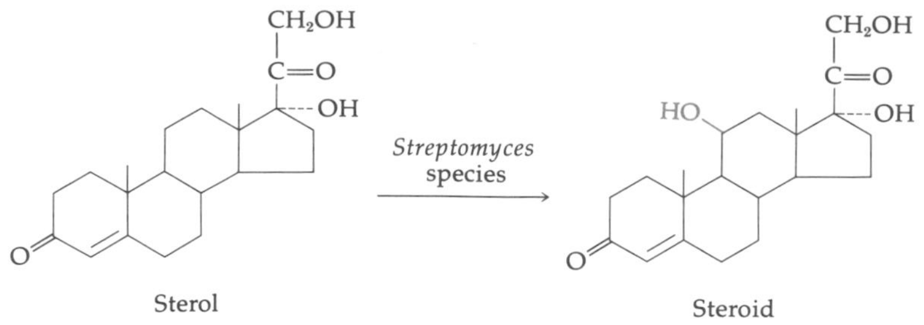

CH_2OH / $C{=}O$ / OH — *Streptomyces* species → HO / CH_2OH / $C{=}O$ / OH

Sterol — Steroid

Figure 27–9 Conversion of a precursor compound, such as a sterol, into a steroid by *Streptomyces*.

Table 27–4 Some Microbial Enzymes Produced Commercially

Enzyme	Microorganism	Use of Enzyme
Amylase	*Aspergillus niger* *Aspergillus oryzae* *Bacillus subtilus*	Baking: flour supplement Brewing: mashing Food: precooked foods, syrup Pharmaceuticals: digestive aids Starch: cold-water laundry Textile: desizing agent
Cellulase	*Aspergillus niger*	Food: liquid coffee concentrate
Dextransucrase	*Leuconostic mesenteroides*	Pharmaceuticals: dextran
Glucose oxidase	*A. niger*	Food: glucose removal from egg solids Pharmaceuticals: test papers
Invertase	*Saccharomyces cerevisiae*	Candy: prevents granulation in soft center Food: artificial honey
Lactase	*Saccharomyces fragilis*	Dairy: prevents crystallization of lactose in ice cream and concentrated milk
Lipase	*A. niger*	Dairy: flavor production in cheese
Pectinase	*A. niger*	Wine and juice: clarification
Penicillinase	*B. subtilis*	Medicine: diagnostic agent
Protease	*A. oryzae*	Brewing: beer stabilizer Baking: breadmaking Food: meat tenderizer Pharmaceutical: digestive aid Textile: desizing
Streptodornase	*Streptococcus pyogenes*	Medicine: reagent, wound debridement

Source: J. R. Porter, Microbiology and the food and energy crisis, *ASM News* 40(11):822, November 1974.

be made safer and more effective by synthesis of the essential protein antigenic fraction of the antigen rather than through use of the entire killed or weakened microorganisms. Antibiotics are an important microbiological industrial product, produced in surprisingly large tonnages in the United States, with nearly half of the antibiotics fed to animals to increase meat yields. The search for new and modified antibiotics is ever-increasing as worldwide microbial resistance to antibiotics continues to rise alarmingly.

Steroids (Figure 27–9) are a very important group of chemicals that include substances such as cortisone, an antiinflammatory, and estrogens and progesterone, both used in birth control pills. Even though it is difficult to recover steroids from animal sources, they can be synthesized by microorganisms from sterols or related compounds derived from animals and plants. Certain strains of molds and bacteria (particularly streptomycetes) can convert chemical groups on sterols and similar molecules to produce steroids.

Microorganisms now yield many metric tons of important food products such as citric acid, lactic acid, enzymes, and amino acids. Certain amino acids such as methionine and lysine tend to be deficient in some cereal crops. Therefore, lysine produced by microorganisms is used as a nutritional supplement. Glutamic acid is widely used to make the flavor enhancer, monosodium glutamate. Both lysine and glutamic acid can be produced in large amounts by genetically manipulated bacteria. Citric acid constitutes an estimated 60% of all food additives produced in the United States. It is widely used to provide tartness to food and as an antioxidant. It is produced by a mold, *Aspergillus niger*, using molasses as a substrate.

Enzymes are widely used in industry. For example, amylases are used in the production of syrups from corn starch, in making paper sizing (a coating for smoothness), and in the production of glucose from starch. And probably half of the bread baked in this country is made with proteases to adjust the amounts of glutens (protein) in wheat to make improved or uniform baked goods. Other proteolytic enzymes are used as meat tenderizers. Several enzymes produced commercially by microorganisms are listed in Table 27–4.

As petroleum products decrease in availability and increase in price, the classic microbial fermentations once used industrially to produce such products as butanol, acetone, and the like may be revived. Currently, petroleum derivatives are used as the starting material for the synthesis of these chemicals.

Agriculture could also benefit from improved species of the nitrogen-fixing bacterium *Rhizobium*. Some day it might be possible to engineer nitrogen-fixing bacteria to make a symbiosis with roots of cereal crops much like the useful and familiar legume–*Rhizobium* symbiosis.

STUDY OUTLINE

FOOD SPOILAGE AND PRESERVATION (pp. 695–696)

1. The earliest methods of preserving foods were drying, the addition of salt or sugar, and fermentation.
2. In 1810, Appert devised a canning process that involved keeping air out of sealed cans and heating food in the cans.
3. Later in the nineteenth century, steam under pressure (pressure cooking) was used in canning.
4. Today, commercial sterilization is heating canned foods to the minimum temperature necessary to destroy *C. botulinum* endospores while minimizing alteration of the food.
5. The commercial sterilization process uses sufficient heat to reduce a population of *C. botulinum* by 12 logarithmic cycles (12D treatment).
6. Endospores of thermophiles can survive commercial sterilization.

CANNED FOOD SPOILAGE (pp. 696–698)

1. Canned foods stored above 45°C may undergo thermophilic anaerobic spoilage.

2. Thermophilic anaerobic spoilage may be accompanied by gas production; if no gas is formed, it is called flat sour spoilage.
3. Spoilage by mesophilic bacteria may be due to improper heating procedures or leakage.
4. Acidic foods may be preserved by heating to 100°C because microorganisms that survive are not capable of growth in a low pH.

LOW-TEMPERATURE PRESERVATION (p. 698)

1. Many microorganisms, although few pathogens, will grow slowly at refrigeration temperatures (0 to 7°C).
2. Freezing may not kill bacteria, but it will prevent their growth.

RADIATION AND FOOD PRESERVATION (p. 699)

1. Research is being done to determine the utility of ionizing radiation to sterilize foods.
2. Microwaves kill bacteria by heating the food.

PASTEURIZATION (p. 699)

1. Pasteur determined that spoilage of food was done to only certain microorganisms and that it was not necessary to kill all microorganisms to prevent spoilage.
2. Pasteurization is the application of heat to selectively kill microorganisms without greatly altering the quality of the food.

CHEMICAL PRESERVATIVES (p. 699)

1. Sodium benzoate, sorbic acid, and calcium propionate are added to foods to inhibit the growth of molds.
2. These chemicals are metabolized by humans.
3. Sodium nitrate and sodium nitrite are added to meats to perserve the red color and to prevent *C. botulinum* endospore germination.
4. Nitrates are converted to nitrites by some bacteria using anaerobic respiration; nitrites are the active ingredient.

FOOD-BORNE INFECTIONS AND MICROBIAL INTOXICATIONS (pp. 699–700)

1. A food-borne infection involves the growth of a pathogen in the host.
2. Gastroenteritis, typhoid fever, and dysentery are food-borne infections.
3. Microbial intoxication results from ingestion of a toxin produced by microbial growth in food.
4. Botulism and staphylococcal food poisoning are examples of microbial intoxications.

5. Pasteurized dairy milk may contain gram-negative psychrophiles.
6. Pasteurized cultured milk products (i.e., buttermilk and sour cream) contain large numbers of lactic acid bacteria.
7. Public health service standards for Grade A pasteurized milk are a standard plate count of less than 20,000 bacteria per milliliter and less than 10 coliforms per milliliter.
8. Cultured milk products must have less than 10 coliforms per milliliter.
9. Dry milk products must have a standard plate count of less than 30,000 bacteria per gram and a coliform count of less than 90 per gram.
10. The species of bacteria in unpasteurized (raw) milk are not controlled, and raw milk may transmit disease.

ROLE OF MICROORGANISMS IN FOOD PRODUCTION (pp. 700–703)

Cheese (pp. 700–701)

1. The milk protein casein curdles due to the action of acid produced by lactic acid bacteria and the enzyme rennin.
2. Cheese is the curd separated from the liquid portion of milk, called whey.
3. Hard cheeses are produced by lactic acid bacteria growing in the interior of the curd.
4. The growth of microorganisms in cheeses is called ripening.
5. Semisoft cheeses are ripened by bacteria growing on the surface.
6. Soft cheeses are ripened by *Penicillium* growing on the surface.

Other Dairy Products (p. 701)

1. Old-fashioned buttermilk was produced by lactic acid bacteria growing during the buttermaking processing.
2. Commercial buttermilk is made by letting lactic acid bacteria grow in skim milk for 12 hours.
3. Sour cream, yogurt, kefir, and kumiss are produced by the growth of lactobacilli, streptococci, and/or yeast in low-fat milk.

Nondairy Fermentations (pp. 701–702)

1. Sugars in bread dough are fermented by yeast to ethyl alcohol and CO_2; the CO_2 causes the bread to rise.
2. Sauerkraut, pickles, olives, and soy sauce are the result of microbial fermentations.

Alcoholic Beverages (pp. 702–703)

1. Carbohydrates obtained from grains, potatoes, or molasses are fermented by yeast to produce ethanol in the production of beer, ale, and distilled spirits.

2. The sugars in fruits such as grapes are fermented by yeast to produce wines.
3. *Acetobacter* oxidizes ethanol in wine to acetic acid (vinegar).

SINGLE-CELL PROTEIN (SCP) PRODUCTION (pp. 703–704)

1. Microorganisms can grow and produce proteinaceous cell material from otherwise unusable substrates such as cellulose.
2. The microbial cells can be made into a source of food called single-cell protein.
3. *Spirulina* is cultured and used as SCP.

ALTERNATIVE ENERGY SOURCES FROM MICROORGANISMS (p. 704)

1. Organic waste, called biomass, can be converted by microorganisms into alternative fuels, a process called bioconversion.
2. Fuels produced by microbial fermentation are methane, methanol, ethanol, and hydrogen.

INDUSTRIAL MICROBIOLOGY AND GENETIC ENGINEERING (pp. 704–707)

1. Inexpensive production of hormones, vaccines, and antibiotics may be made possible by genetic engineering.
2. Microorganisms produce lactic acid, citric acid, enzymes, and amino acids that are used in food manufacture and other industrial processes.

STUDY QUESTIONS

REVIEW

1. Why are attempts made to preserve foods?
2. List five methods used to preserve foods.
3. Matching.

___Sorbic acid	(a) Sterilization
___5°C	(b) Pasteurization
___72°C for 15 seconds	(c) Low-temperature preservation
___121°C for 15 minutes	(d) Chemical preservation
___12D treatment	

4. Define pasteurization.
5. Pasteurization does not kill all microorganisms, so food can still spoil. Why then are dairy products pasteurized?
6. Provide one advantage and one disadvantage of the addition of nitrites to foods.

7. Outline the production of cheese, comparing the production of hard and soft cheeses.
8. Outline the production of wine.
9. Provide at least two advantages and two disadvantages of SCP.
10. How can microorganisms provide energy sources? What metabolic processes can result in fuels?
11. Appert's invention was the forerunner of hermetically sealed cans. Why do hermetically sealed cans prevent food spoilage?
12. Why is a can of blackberries preserved by heating to 100°C instead of at least 116°C characteristic of commercial sterilization?
13. Under what conditions would you expect to find thermophiles spoiling foods? Psychrophiles?
14. Discuss the importance of industrial microbiology.

CHALLENGE

1. What bacteria seem to be most frequently used in the production of food? Can you guess why?
2. Why do the following work to preserve food?
   (a) Fermentation
   (b) Salting
   (c) Drying.
3. Explain the processes involved in the production of sourdough bread.

FURTHER READING

Amerine, M. A. August 1964. Wine. *Scientific American* 211:46–56. The biochemistry of yeast fermentation, grape strains, and aging.

Industrial microbiology. September 1981. *Scientific American* 245. The September issue is devoted entirely to the topic of industrial microbiology and genetic engineering.

Jay, J. M. 1978. *Modern Food Microbiology.* 2nd ed. New York: Van Nostrand. Contains chapters on the necessary conditions for microbial growth in food, methods of preserving food, types of food spoilage, and food-borne diseases.

Kang, K. S. (convener). 1978. Microbial products in foods. In *Developments in Industrial Microbiology* 19:71–131. Washington, D.C.: American Institute of Biological Science. Current technology and applications for the production of food from agricultural wastes and microbial products in food, including single-cell protein.

Zeikus, J. G. 1980. Chemical and fuel production by anaerobic bacteria. *Annual Review of Microbiology* 34:423–464. A summary of the current technology and practicality of producing hydrocarbon fuels and organic solvents from agricultural and municipal wastes.

APPENDIX A

Classification of Bacteria According to *Bergey's Manual of Determinative Bacteriology*, 8th edition

KINGDOM PROCARYOTAE

DIVISION I. The Cyanobacteria

DIVISION II. The Bacteria

PART 1.
Phototrophic Bacteria
Order I. *Rhodospirillales*
Family I. *Rhodospirillaceae*
Genus I. *Rhodospirillum*
Genus II. *Rhodopseudomonas*
Genus III. *Rhodomicrobium*
Family II. *Chromatiaceae*
Genus I. *Chromatium*
Genus II. *Thiocystis*
Genus III. *Thiosarcina*
Genus IV. *Thiospirillum*
Genus V. *Thiocapsa*
Genus VI. *Lamprocystis*
Genus VII. *Thiodictyon*
Genus VIII. *Thiopedia*
Genus IX. *Amoebobacter*
Genus X. *Ectothiorhodospira*
Family III. *Chlorobiaceae*
Genus I. *Chlorobium*
Genus II. *Prosthecochloris*
Genus III. *Chloropseudomonas*
Genus IV. *Pelodictyon*
Genus V. *Clathrochloris*

PART 2.
The Gliding Bacteria
Order I. *Myxobacterales*
Family I. *Myxococcaceae*
Genus I. *Myxococcus*
Family II. *Archangiaceae*
Genus I. *Archangium*
Family III. *Cystobacteraceae*
Genus I. *Cystobacter*
Genus II. *Melittangium*
Genus III. *Stigmatella*
Family IV. *Polyangiaceae*
Genus I. *Polyangium*
Genus II. *Nannocystis*
Genus III. *Chondromyces*
Order II. *Cytophagales*
Family I. *Cytophagaceae*
Genus I. *Cytophaga*
Genus II. *Flexibacter*
Genus III. *Herpetosiphon*
Genus IV. *Flexithrix*
Genus V. *Saprospira*
Genus VI. *Sporocytophaga*
Family II. *Beggiatoaceae*
Genus I. *Beggiatoa*
Genus II. *Vitreoscilla*
Genus III. *Thioploca*
Family III. *Simonsiellaceae*
Genus I. *Simonsiella*
Genus II. *Alysiella*
Family IV. *Leucotrichaceae*
Genus I. *Leucothrix*
Genus II. *Thiothrix*
Families and Genera of Uncertain Affiliation:
Genus *Toxothrix*
Family *Achromatiaceae*
Genus *Achromatium*
Family *Pelonemataceae*
Genus *Pelonema*
Genus *Achroonema*
Genus *Peloploca*
Genus *Desmanthos*

PART 3.
The Sheathed Bacteria
Genus *Sphaerotilus*
Genus *Leptothrix*
Genus *Streptothrix*
Genus *Lieskeella*
Genus *Phragmidiothrix*
Genus *Crenothrix*
Genus *Clonothrix*

PART 4.
Budding and/or Appendaged Bacteria
Genus *Hyphomicrobium*
Genus *Hyphomonas*
Genus *Pedomicrobium*
Genus *Caulobacter*
Genus *Asticcacaulis*
Genus *Ancalomicrobium*
Genus *Prosthecomicrobium*
Genus *Thiodendron*

Genus *Pasteuria*
Genus *Blastobacter*
Genus *Seliberia*
Genus *Gallionella*
Genus *Nevskia*
Genus *Planctomyces*
Genus *Metallogenium*
Genus *Caulococcus*
Genus *Kusnezovia*

PART 5.
The Spirochetes
Order I. *Spirochaetales*
Family I. *Spirochaetaceae*
Genus I. *Spirochaeta*
Genus II. *Cristispira*
Genus III. *Treponema*
Genus IV. *Borrelia*
Genus V. *Leptospira*

PART 6.
Spiral and Curved Bacteria
Family I. *Spirillaceae*
Genus I. *Spirillum*
Genus II. *Campylobacter*
Genera of Uncertain Affiliation:
Genus *Bdellovibrio*
Genus *Microcyclus*
Genus *Pelosigma*
Genus *Brachyarcus*

PART 7.
Gram-Negative Aerobic Rods and Cocci
Family I. *Pseudomonadaceae*
Genus I. *Pseudomonas*
Genus II. *Xanthomonas*
Genus III. *Zoogloea*
Genus IV. *Gluconobacter*
Family II. *Azotobacteraceae*
Genus I. *Azotobacter*
Genus II. *Azomonas*
Genus III. *Beijerinckia*
Genus IV. *Derxia*
Family III. *Rhizobiaceae*
Genus I. *Rhizobium*
Genus II. *Agrobacterium*
Family IV. *Methylomonadaceae*
Genus I. *Methylomonas*
Genus II. *Methylococcus*
Family V. *Halobacteriaceae*
Genus I. *Halobacterium*
Genus II. *Halococcus*
Family VI. *Legionellaceae*
Genus I. *Legionella*
Genera of Uncertain Affiliation:
Genus *Alcaligenes*
Genus *Acetobacter*
Genus *Brucella*
Genus *Bordetella*
Genus *Francisella*
Genus *Thermus*

PART 8.
Gram-Negative Facultatively Anaerobic Rods
Family I. *Enterobacteriaceae*
Genus I. *Escherichia*
Genus II. *Edwardsiella*
Genus III. *Citrobacter*
Genus IV. *Salmonella*
Genus V. *Shigella*
Genus VI. *Klebsiella*
Genus VII. *Enterobacter*
Genus VIII. *Hafnia*
Genus IX. *Serratia*
Genus X. *Proteus*
Genus XI. *Yersinia*
Genus XII. *Erwinia*
Family II. *Vibrionaceae*
Genus I. *Vibrio*
Genus II. *Aeromonas*
Genus III. *Plesiomonas*
Genus IV. *Photobacterium*
Genus V. *Lucibacterium*
Genera of Uncertain Affiliation:
Genus *Zymomonas*
Genus *Chromobacterium*
Genus *Flavobacterium*
Genus *Haemophilus* (*H. vaginalis*)
Genus *Pasteurella*
Genus *Actinobacillus*
Genus *Cardiobacterium*
Genus *Streptobacillus*
Genus *Calymmatobacterium*
Parasites of *Paramecium*

PART 9.
Gram-Negative Anaerobic Bacteria
Family I. *Bacteroidaceae*
Genus I. *Bacteroides*
Genus II. *Fusobacterium*
Genus III. *Leptotrichia*
Genera of Uncertain Affiliation:
Genus *Desulfovibrio*
Genus *Butyrivibrio*
Genus *Succinivibrio*
Genus *Succinimonas*
Genus *Lachnospira*
Genus *Selenomonas*

PART 10.
Gram-Negative Cocci and Coccobacilli
Family I. *Neisseriaceae*
Genus I. *Neisseria*
Genus II. *Branhamella*
Genus III. *Moraxella*
Genus IV. *Acinetobacter*
Genera of Uncertain Affiliation:
Genus *Paracoccus*
Genus *Lampropedia*

PART 11.
Gram-Negative Anaerobic Cocci
Family I. *Veillonellaceae*
Genus I. *Veillonella*
Genus II. *Acidaminococcus*
Genus III. *Megasphaera*

PART 12.
Gram-Negative, Chemolithotrophic Bacteria
a. Organisms oxidizing ammonia or nitrite
Family I. *Nitrobacteraceae*
Genus I. *Nitrobacter*
Genus II. *Nitrospina*
Genus III. *Nitrococcus*
Genus IV. *Nitrosomonas*
Genus V. *Nitrosospira*
Genus VI. *Nitrosococcus*
Genus VII. *Nitrosolobus*
b. Organisms metabolizing sulfur
Genus *Thiobacillus*
Genus *Sulfolobus*
Genus *Thiobacterium*
Genus *Macromonas*
Genus *Thiovulum*
Genus *Thiospira*
c. Organisms depositing iron or manganese oxides
Family I. *Siderocapsaceae*
Genus I. *Siderocapsa*
Genus II. *Naumanniella*
Genus III. *Ochrobium*
Genus IV. *Siderococcus*

PART 13.
Methane-Producing Bacteria
Family I. *Methanobacteriaceae*
Genus I. *Methanobacterium*

Genus II. *Methanosarcina*
Genus III. *Methanococcus*

PART 14.
Gram-Positive Cocci

a. Aerobic and/or facultatively anaerobic
Family I. *Micrococcaceae*
Genus I. *Micrococcus*
Genus II. *Staphylococcus*
Genus III. *Planococcus*
Family II. *Streptococcaceae*
Genus I. *Streptococcus*
Genus II. *Leuconostoc*
Genus III. *Pediococcus*
Genus IV. *Aerococcus*
Genus V. *Gemella*
b. Anaerobic
Family III. *Peptococcaceae*
Genus I. *Peptococcus*
Genus II. *Peptostreptococcus*
Genus III. *Ruminococcus*
Genus IV. *Sarcina*

PART 15.
Endospore-Forming Rods and Cocci

Family I. *Bacillaceae*
Genus I. *Bacillus*
Genus II. *Sporolactobacillus*
Genus III. *Clostridium*
Genus IV. *Desulfotomaculum*
Genus V. *Sporosarcina*
Genus of Uncertain Affiliation:
Genus *Oscillospira*

PART 16.
Gram-Positive, Asporogenous Rod-Shaped Bacteria

Family I. *Lactobacillaceae*
Genus I. *Lactobacillus*
Genera of Uncertain Affiliation:
Genus *Listeria*
Genus *Erysipelothrix*
Genus *Caryophanon*

PART 17.
Actinomycetes and Related Organisms

Coryneform Group of Bacteria
Genus I. *Corynebacterium*
a. Human and Animal Parasites and Pathogens
b. Plant Pathogenic Corynebacteria
c. Non-pathogenic Corynebacteria
Genus II. *Arthrobacter*
Genera incertae sedis
Brevibacterium
Microbacterium
Genus III. *Cellulomonas*
Genus IV. *Kurthia*
Family I. *Propionibacteriaceae*
Genus I. *Propionibacterium*
Genus II. *Eubacterium*
Order I. *Actinomycetales*
Family I. *Actinomycetaceae*
Genus I. *Actinomyces*
Genus II. *Arachnia*
Genus III. *Bifidobacterium*
Genus IV. *Bacterionema*
Genus V. *Rothia*
Family II. *Mycobacteriaceae*
Genus I. *Mycobacterium*
Family III. *Frankiaceae*
Genus I. *Frankia*
Family IV. *Actinoplanaceae*
Genus I. *Actinoplanes*
Genus II. *Spirillospora*
Genus III. *Streptosporangium*
Genus IV. *Amorphosporangium*
Genus V. *Ampullariella*
Genus VI. *Pilimelia*
Genus VII. *Planomonospora*
Genus VIII. *Planobispora*
Genus IX. *Dactylosporangium*
Genus X. *Kitasatoa*
Family V. *Dermatophilaceae*
Genus I. *Dermatophilus*
Genus II. *Geodermatophilus*
Family VI. *Nocardiaceae*
Genus I. *Nocardia*
Genus II. *Pseudonocardia*
Family VII. *Streptomycetaceae*
Genus I. *Streptomyces*
Genus II. *Streptoverticillium*
Genus III. *Sporichthya*
Genus IV. *Microellobosporia*
Family VIII. *Micromonosporaceae*
Genus I. *Micromonospora*
Genus II. *Thermoactinomyces*
Genus III. *Actinobifida*
Genus IV. *Thermomonospora*
Genus V. *Microbispora*
Genus VI. *Micropolyspora*

PART 18.
The Rickettsias

Order I. *Rickettsiales*
Family I. *Rickettsiaceae*
Tribe I. *Rickettsieae*
Genus I. *Rickettsia*
Genus II. *Rochalimaea*
Genus III. *Coxiella*
Tribe II. *Ehrlichieae*
Genus IV. *Ehrlichia*
Genus V. *Cowdria*
Genus VI. *Neorickettsia*
Tribe III. *Wolbachieae*
Genus VII. *Wolbachia*
Genus VIII. *Symbiotes*
Genus IX. *Blattabacterium*
Genus X. *Rickettsiella*
Family II. *Bartonellaceae*
Genus I. *Bartonella*
Genus II. *Grahamella*
Family III. *Anaplasmataceae*
Genus I. *Anaplasma*
Genus II. *Paranaplasma*
Genus III. *Aegyptionella*
Genus IV. *Haemobartonella*
Genus V. *Eperythrozoon*
Order II. *Chlamydiales*
Family I. *Chlamydiaceae*
Genus I. *Chlamydia*

PART 19.
The Mycoplasmas

Class I. *Mollicutes*
Order I. *Mycoplasmatales*
Family I. *Mycoplasmataceae*
Genus I. *Mycoplasma*
Family II. *Acholeplasmataceae*
Genus I. *Acholeplasma*
Genera of Uncertain Affiliation:
Genus *Thermoplasma*
Genus *Spiroplasma*
Mycoplasma-like Bodies in Plants

APPENDIX B

Pronunciation of Scientific Names

RULES OF PRONUNCIATION

Vowels. Pronounce all the vowels in scientific names.

Diphthongs. Two vowels written together are a diphthong. Pronounce diphthongs as one vowel.

Accent. The accented syllable is either the next to last or third to last syllable.

The accent is on the next to last syllable:

1. When the name contains only two syllables, for example, péstis.
2. When the next to last syllable is a diphthong, for example, *Amoéba*.
3. When the vowel of the next to last syllable is long, for example, *Treponéma*.

The vowel in the next to last syllable is long in words ending in these suffixes:

Suffix	Example
-ales	orders such as Eubacteriáles
-ina	*Sarcína*
-anus, -anum	*pasteuriánum*
-uta	*diminúta*

The vowel in the next to last syllable is short in words ending in these suffixes:

Suffix	Example
-atus, -atum	*caudátum*
-ella	*Salmonélla*

The accent is on the third to last syllable in family names. Families end in -aceae and the final -ae is pronounced as ē as in Micrococcáceae.

Consonants. When *c* or *g* is followed by ae, e, oe, i, or y it has a soft sound. When *c* or *g* is followed by a, o, oi, or u it has a hard sound. When a double *c* is followed by e, i, or y it is pronounced as ks (i.e., cocci).

PRONUNCIATION OF ORGANISMS IN THIS TEXT

Pronunciation key:

a	hat	ē	see	o	hot	th	thin
ā	age	ė	term	ō	go	u	cup
ã	care	g	go	ô	order	u̇	put
ä	father	i	sit	oi	oil	ü	rule
ch	child	ī	ice	ou	out	ū	use
e	let	ng	long	sh	she	zh	seizure

Acetobacter xylinum ä-sē-tō-bak´tėr zī´lin-um

Actinomyces israelii ak-tin-ō-mī´sēs is-rā´lē-ī

Aedes aegypti ä-e´dēz ē-jip´tē

Aeromonas ā-rō-mō´nas

Agaricus augustus ä-gãr´i-kus au´gus-tus

Ajellomyces ä-jel-lō-mī´sēs

Allescheria boydii al-lesh-er´ē-ä boi´dē-ī

Amanita phalloides am-an-ī´ta fal-loi´dēz

Anabaena an-ä-bē´nä

Anopheles an-of´el-ēz

Aquaspirillum bengal a-kwä-spī-ril´lum ben´gal

Arthroderma är-thrō-dėr´mä

Ascaris lumbricoides as´kar-is lum-bri-koi´dēz

Aspergillus flavus a-spėr-jil´lus flā´vus
A. fumigatus fü-mi-gä´tus
A. niger nī´jėr
A. oryzae ô-rī´zē
A. tamarii ta-mãr´ē-ī

Azotobacter vinelandii ä-zō-tō-bak´tėr vin-lan´dē-ī

Bacillus anthracis bä-sil´lus an-thrā´sis
B. brevis bre´vis
B. cereus se´rē-us
B. licheniformis lī-ken-i-fôr´mis
B. polymyxa po-lē-miks´ä
B. sphaericus sfe´ri-cus
B. stearothermophilus ste-rō-thėr-mä´fil-us
B. subtilis su´til-us
B. thermoacidurans thėr-mō-as-id´ür-ans
B. thuringiensis thu̇r-in-jē-en´sis

Bacteroides fragilis bak-tė-roi´dēz fra´jil-is
B. hypermegas hī-pėr-meg´äs

Balantidium coli bal-an-tid´ē-um kō´lē

Beggiatoa bej-jē-ä-tō´ä

Beijerinckia bī-jė-rink´ē-ä

Blastomyces dermatitidis blas-tō-mī´sēs dėr-mä-tit´i-dis

Bordetella pertussis bôr-de-tel´lä pėr-tus´sis

Borrelia bôr-rel´ē-ä

Brucella abortus brü-sel´lä ä-bôr´tus
B. canis kā´nis
B. melitensis me-li-ten´sis
B. suis sü´is

Calvulina cristata kal-vū-lī´nä kris´tä-tä

Calymmatobacterium granulomatis kal-im-mä-tō-bak-ti´rē-um gran-ū-lō´mä-tis

Campylobacter fetus kam-pī-lō-bak´tėr fē´tus

Candida albicans kan´did-ä al´bi-kans

Carpenteles kär-pen´tel-ēz

Caulobacter bacteroides kô-lō-bak´tėr bak-tėr-oi´dēz

Cephalosporium sef-ä-lō-spô´rē-um

Chlamydia psittaci kla-mi´dē-ä sit´tä-sē
C. trachomatis trä-kō´mä-tis

Chlorella klô-rel´lä

Chlorobium klô-rō´bē-um

Chroococcus turgidus krō-ō-kok´kus tėr´gi-dus

Chrysops krī´sops

Cladosporium werneckii kla-dō-spô´rē-um wėr-ne´kē-ī

Claviceps purpurea kla´vi-seps pu̇r-pu̇-rē´ä

Clonorchis sinensis klo-nôr´kis si-nen´sis

Clostridium acetobutylicum klôs-tri´dē-um a-sē-tō-bū-ti´li-kum
C. bifermentens bī-fėr-men´tans
C. botulinum bo-tū-lī´num
C. butyricum bü-ti´ri-cum
C. nigrificans nī-gri´fi-cans
C. pasteurianum pas-tyėr-ē-ā´num
C. perfringens pėr-frin´jens
C. sporogenes spô-rä´jen-ēz
C. subterminale sub-tėr-mi-na´lē
C. tetani te´tan-ē
C. thermosaccharolyticum thėr-mō-sak-kär-ō-li´ti-cum

Coccidioides immitis kok-sid-ē-oi´dēz im´mi-tis

Corynebacterium diphtheriae kô-rī-nē-bak-ti´rē-um dif-thi´rē-ä
C. xerosis ze-rō´sis

Coxiella burnetii käks-ē-el´lä bėr-ne´tē-ī

Cryptococcus neoformans kryp-tō-kok´kus nē-ō-fôr´manz

Culex kū´leks

Cytophaga sī-täf´äg-ä

Dermacentor andersoni dėr-mä-sen´tôr an-dėr-sō´nī

Desulfovibrio dē-sul-fo-vib´rē-ō

Didinium nasutum dī-di´nē-um nä-sūt´um

Diphyllobothrium latum dī-fil-lo-bo´thrē-um lā´tum

Echinococcus granulosus ē-kīn-ō-kok´kus gra-nū-lō´sus

Ectothiorhodospira mobilis ek-tō-thī-ō-rō-dō-spī´rä mō´bil-is

Emmonsiella em-mon-sē-el´lä

Endothia parasitica en-dō´thē-ä par-ä-si´ti-kä

Entamoeba histolytica en-tä-mē´bä his-tō-li´ti-kä

Enterobacter aerogenes en-te-rō-bak´tėr ā-rä´jen-ēz
E. cloacae klō-ā´kä

Enterobius vermicularis en-te-rō´bē-us ver-mi-kū-lar´is

Epidermophyton ep-i-dėr-mō-fī´ton

Erwinia herbicola ėr-wi´nē-ä hėr-bik´ō-lä

Erysipelothrix e-ri-si-pe-lō´thriks

Escherichia coli esh-ėr-i´kē-ä kō´lē

Euglena ū-glē´nä

Eurotium emericella yėr-ō´shum em-ėr-ē-sel´lä

Filobasidiella fi-lō-ba-si-dē-el´lä

Flavobacterium meningosepticum flā-vō-bak-ti´rē-um me-nin-jō-sep´ti-kum

Fonsecaea pedrosoi fon-se´kē-ä pe-drō´sō-ī

Francisella tularensis fran-sis-el´lä tü-lä-ren´sis

Frankia frank´ē-ä

Fusobacterium fü-sō-bak-ti´rē-um

Giardia lamblia jē-är´dē-ä lam´lē-ä

Gloecapsa glō´cap-sä

Glossina gläs-sē´nä

Hemophilus aegyptius hē-mä´fi-lus ē-jip´ti-us
H. ducreyi dü-krā´ē
H. haemoglobinophilus hē-mō-glä-bi-nä´fi-lus
H. influenzae in-flü-en´zē
H. parainfluenzae pa-ra-in-flü-en´zē

Histoplasma capsulatum hiss-tō-plaz´mä cap-su-lä´tum

Homo sapiens hō´mō sā´pē-ens

Hyphomicrobium hī-fō-mī-krō´bē-um

Klebsiella pneumoniae kleb-sē-el´lä nü-mō´nē-ī

Lactobacillus acidophilus lak-tō-bä-sil´lus a-si-dä´fi-lus
L. bulgaricus bul-ga´ri-kus
L. casei kā´sē-ī

Legionella pneumophila lē-jä-nel´lä nü-mō´fi-lä

Leptospira interrogans lep-tō-spī´rä in-tėr´rä-ganz

Leuconostoc mesenteroides lü-kō-nos´tok mes-en-ter-oi´dēz

Listeria monocytogenes lis-te´rē-ä mo-nō-sī-tô´je-nēz

Lycoperdon perlatum lī-kō-pėr´don pėr-lā´tum

Micrococcus luteus mī-krō-kok´kus lū´tē-us

Micromonospora purpureae mī-krō-mo-nä´spô-rä pur-pu-rē´ī

Microsporum mī-krō-spô´rum

Moraxella lacunata mô-raks-el´lä la-kü-nä´tä

Mucor mū´kôr

Mycobacterium bovis mī-kō-bak-ti´rē-um bō´vis
M. leprae lep´rē
M. smegmatis smeg-ma´tis
M. tuberculosis tü-ber-kū-lō´sis

Mycoplasma hominis mī-kō-plaz´mä ho´min-nis
M. pneumoniae nu-mō´nē-ī

Naegleria fowleri nī-gli´rē-ä fou´lėr-ī

Nannizia nan-nī´zē-ä

Necator americanus ne-kā´tôr ä-me-ri-ca´nus

Neisseria gonorrhoeae nī-se´rē-ä go-nôr-rē´ä
N. meningitidis me-nin-ji´ti-dis

Nitrobacter nī-trō-bak´tėr

Nitrosomonas nī-trō-sō-mō´näs

Nocardia asteroides nō-kär´dē-ä as-tėr-oi´dēz
N. lurida lėr´id-ä

Oocystis ō-ō-sis´tis

Ornithodorus ôr-nith-ō´dô-rus

Paracoccidioides brasiliensis par-ä-kok-sid´ē-oi-dēz bra-sil-ē-en´sis

Paracoccus denitrificans pa-rä-kok´kus dē-nī-tri´fi-kanz

Paragonimus westermanni par-ä-gōn´e-mus we-ster-man´nī

Paramecium multimicronucleatum par-ä-mē´sē-um multē-mī-krō-nü-klē´ä-tum

Pasteurella multocida pas-tyėr-el´lä mul-tō´ci-dä

Pediculus vestimenti ped-ik´ū-lus ves-ti-men´tē

Pediococcus pe-dē-ō-kok´kus

Penicillium griseofulvum pe-ni-sil´lē-um gri-sē-ō-fůl´vum
P. marneffei mar-nif´fē-ī
P. notatum nō-tä´tum

Petriellidium pet-rē-el-li´dē-um

Phialophora verrucosa fē-ä-lo´fô-rä ver-rü-kō´sä

Phormidium luridum fôr-mi´dē-um lė-rid´um

Phytophthora infestans fī-tof´thô-rä in´fes-tans

Pitryosporum pit-rē-ō-spô´rum

Plasmodium falciparum plaz-mō´dē-um fal-sip´är-um
P. malariae mä-lā´rē-e
P. ovale ō-vä´lē
P. vivax vī´vaks

Plesiomonas shigelloides ple-sē-ō-mō´nas shi-gel-loi´dēz

Pneumocystis carinii nü-mō-sis´tis kär-i´nē-ī

Propionibacterium acnes prō-pē-on-ē-bak-ti´rē-um ak´nēz

P. freudenreichii frü-den-rik´ē-ī

Proteus mirabilis prō´tē-us mi-ra´bi-lis
P. vulgaris vul-ga´ris

Pseudomonas aeruginosa sū-dō-mō´nas ā-rü-ji-nō´sä
P. cepacia se-pā´sē-ä
P. diminuta di-mi-nü´tä

Quercus kwer´kus
Rhizobium rī-zō´bē-um
Rhizopus arrhizus rī-zō´pùs är-rī´zus
R. nigricans nī´gri-kans
Rhodospirillum rubrum rō-dō-spī-ril´lum rüb´rum
Rickettsia prowazekii ri-ket´sē-ä prou-wä-ze´kē-ī
R. rickettsii ri-ket´sē-ī
R. typhi tī´fē
Saccharomyces carlsbergensis sak-ä-rō-mī´sēs kärls-bėr´gen-sis
S. cerevisiae se-ri-vis´ē-ī
S. ellipsoideus ē-lip-soi-dē´us
S. fragilis fra´ji-lis
Salmonella cholerae-suis sal-mōn-el´lä kol-ėr-ä-sü´is
S. enteritidis en-tėr-it´id-is
S. paratyphi pa-rä-tī´fē
S. schottmulleri shot-mül´lėr-ī
S. typhi tī´fē
S. typhimurium tī-fi-mùr´ē-um
Sarcina maxima sär-sī´nä maks´ē-mä
Sarcoptes scabiei sär-kop´tēs skā-bē´ī
Sartorya sär-tô´rē-ä
Schistosoma japonicum skis-tō-sō´mä ja-pon´i-cum
Serratia marcescens ser-rä´tē-ä mär-ses´sens
Shigella boydii shi-gel´lä boi´dē-ī
S. dysenteriae dis-en-te´rē-ī
S. flexneri fleks´nėr-ī
S. sonnei sōn´nē-ī
Sphaerotilus natans sfe-rä´ti-lus nā´tans
Spirillum minor spī-ril´lum mī´nôr
Spirulina spī-rü-lī´nä
Sporosarcina ureae spô-rō-sär-sī´nä yė´rē-ī
Sporothrix schenkii spô-rō´thriks shen´kē-ī
Staphylococcus aureus staf-i-lō-kok´kus ô´rē-us
S. epidermidis e-pi-dėr´mi-dis
Stigmatella aurantica stig-mä´tel-lä ô-ran-tī´cä
Streptobacillus moniliformis strep-tō-bä-sil´lus mō-ni-li-fôr´mis
Streptococcus faecalis strep-tō-kok´kus fē-kāl´is
S. mitis mī´tis
S. mutans mū´tans
S. pneumoniae nü-mō´nē-ī
S. pyogenes pī-äj´en-ēz
S. salivarius sa-li-vā´rē-us
S. sanguis san´gwis
Streptomyces aureofaciens strep-tō-mī´sēs ô-rē-ō-fa´si-ens
S. erythraeus ā-rith´rē-us
S. fradiae frā´dē-ī
S. griseus gri-sē´us
S. kanamyceticus kan-ä-mī-sē´ti-kus
S. nodosus nō-dō´sus
S. noursei nôr´sē-ī
S. parvullus pär-vū´lus
S. rimosus ri-mō´sus
S. venezuelae ve-ne-zü-e´lē
Taenia saginata te´nē-ä sa-ji-nä´tä
T. solium sō´lē-um
Talaromyces ta-lä-rō-mī´sēs
Thiobacillus ferrooxidans thī-ō-bä-sil´lus fer-rō-oks´i-danz
Toxoplasma gondii toks-ō-plaz´mä gon´dē-ī
Trachelomonas trä-kel-ō-mōn´as
Treponema pallidum tre-pō-nē´mä pal´li-dum
T. pertenue pėr-ten´ū-ē
Triatoma trī-ä-tō´mä
Trichinella spiralis tri-kin-el´lä spī-ra´lis
Trichomonas vaginalis trik-ō-mōn´as va-jin-al´is
Trichophyton trik-ō-fī´ton
Trypanosoma brucei tri-pan-ō-sō´mä brü-sē´ī
T. cruzi kruz´ē
T. gambiense gam-bē-ens´
Veillonella vī-lo-nel´lä
Vibrio cholerae vib´rē-ō kol´ėr-ī
V. parahaemolyticus pa-rä-hē-mō-li´ti-kus
Xenopsylla cheopis ze-nop-sil´lä chē-ō´pis
Yersinia enterocolitica yėr-sin´ē-ä en-tėr-ō-kōl-it´ic-ä
Y. pestis pes´tis

APPENDIX C

Word Roots Used in Microbiology

a-, an- absence, lack. Examples: abiotic, in the absence of life; anaerobic, in the absence of air.

-able able to, capable of. Example: viable, ability to live or exist.

aer- air. Examples: aerobic, in the presence of air; aerate, to add air.

albo- white. Example: *Streptomyces albus*, produces white colonies.

amoeb- change. Example: ameboid, movement involving changing shapes.

amphi- around. Example: amphitrichous, tufts of flagella at both ends of a cell.

amyl- starch. Example: amylase, enzyme that degrades starch.

ana- up. Example: anabolism, building up.

ant-, anti- opposed to, preventing. Example: antimicrobial, a substance that prevents microbial growth.

asco- bag. Example: ascus, a bag-like structure holding spores.

aur- gold. Example: *Staphylococcus aureus*, gold pigmented colonies.

aut-, auto- self. Example: autotroph, self feeder.

bacillo- a little stick. Example: bacillus, rod-shaped.

basid- base, pedestal. Example: basidium, a cell that bears spores.

bio- life. Example: biology, the study of life and living organisms.

blast- bud. Example: blastospore, spores formed by budding.

bovi- cattle. Example: *Mycobacterium bovis*, a bacterium found in cattle.

brevi- short. Example: *Lactobacillus brevis*, a bacterium with short cells.

butyr- butter. Example: butyric acid, formed in butter, responsible for rancid odor.

campylo- curved. Example: *Campylobacter*, curved rod.

carcin- cancer. Example: carcinogen, a cancer-causing agent.

-caryo, -karyo a nut. Example: eucaryote, a cell with a membrane-enclosed nucleus.

caseo- cheese. Example: caseous, cheese-like.

caul- a stalk. Example: *Caulobacter*, appendaged or stalked bacteria.

chlamydo- covering. Example: chlamydospores, spores formed inside hypha.

chloro- green. Example: chlorophyll, green pigmented molecule.

chrom- color. Examples: choromosome, readily stained structure; metachromatic, intracellular colored granules.

chryso- golden. Example: *Streptomyces chryseus*, golden colonies.

-cide killing. Example: bactericide, an agent that kills bacteria.

cili- eye-lash. Example: cilia, a hairlike organelle.

cleisto- closed. Example: cleistothecium, completely closed ascus.

co-, con- together. Example: concentric, common center, together in the center.

cocci- a berry. Example: coccus, a spherical-shaped cell.

coeno- shared. Example: coenocyte, cell with many nuclei not separated by septa.

col- colo- colon. Examples: colon, large intestine; *Escherichia coli*, bacterium found in large intestine.

conidio- dust. Example: conidia, spores developed at end of aerial hypha, never enclosed.

-cul small form. Example: particle, a small part.

cyano- blue. Example: cyanobacteria, blue-green pigmented organisms.

cyst- bladder. Example: cystitis, inflammation of the urinary bladder.

cyt- cell. Example: cytology, the study of cells.

de- undoing, reversal, loss, removal. Example: deactivation, becoming inactive.

di-, diplo- twice, double. Example: diplococci, pairs of cocci.

dia- through, between. Example: diaphragm, the wall through or between two areas.

dys- difficult, faulty, painful. Example: dysfunction, disturbed function.

ec-, ex-, ecto- out, outside, away from. Example: excrete, to remove materials from the body.

en-, em- in, inside. Example: encysted, enclosed in a cyst.

entero- intestine. Example: *Enterobacter*, bacterium found in the intestine.

epi- upon, over. Example: epidemic, disease over all the people.

erythro- red. Example: erythma, redness of the skin.

eu- well, proper. Example: eucaryote, a proper cell.

exo- outside, outer layer. Example: exogenous, from outside the body.

extra- outside, beyond. Example: extracellular, outside the cells of an organism.

flagell- a whip. Example: flagellum, a projection from a cell; in eucaryotic cells, it pulls cells in a whiplike fashion.

flav- yellow. Example: *Flavobacterium*, cells produce yellow pigment.

fruct- fruit. Example: fructose, fruit sugar.

-fy to make. Example: magnify, to make larger.

galacto- milk. Example: galactose, monosaccharide from milk sugar.

gamet- to marry. Example: gamete, reproductive cell.

gastr- stomach. Example: gastritis, inflammation of the stomach.

gel- to stiffen. Example: gel, solidified colloid.

-gen an agent that initiates. Example: pathogen, any agent that produces disease.

-genesis formation. Example: pathogenesis, production of disease.

germ, germin- bud. Example: germ, part of an organism capable of developing.

-gony reproduction. Example: schizogony, multiple fission producing many new cells.

halo- salt. Example: halophyll, an organism that can live in high salt concentrations.

haplo- one, single. Example: haploid, half the number of chromosomes or one set.

hema-, hemato-, hemo- blood. Example: *Hemophilus*, bacterium that requires nutrients from red blood cells.

hepat- liver. Example: hepatitis, inflammation of the liver.

herpes creeping. Example: herpes, or shingles, lesions appear to creep along the skin.

hetero- different, other. Example: heterotroph, obtains organic nutrients from other organisms; other feeder.

hist- tissue. Example: histology, the study of tissues.

hom-, homo- same. Example: homofermenter, an organism that produces only lactic acid from fermentation of a carbohydrate.

hydr-, hydro- water. Example: dehydration, loss of body water.

hyper- excess. Example: hypertonic, having a greater osmotic pressure in comparison to another.

hypo- below, deficient. Example: hypotonic, having a lesser osmotic pressure in comparison to another.

im- not, in. Example: impermeable, not permitting passage.

inter- between. Example: intercellular, between the cells.

intra- within, inside. Example: intracellular, inside the cell.

io- violet. Example: iodine, a chemical element that produces a violet vapor.

iso- equal, same. Example: isotonic, having the same osmotic pressure when compared to another.

-itis inflammation of. Example: colitis, inflammation of the large intestine.

kin- movement. Example: streptokinase, an enzyme that lyses or moves fibrin.

lacti- milk. Example: lactose, the sugar in milk.

leuko- whiteness. Example: leukocyte, white blood cell.

lip-, lipo- fat, lipid. Example: lipase, an enzyme that breaks down fats.

-logy the study of. Example: pathology, the study of changes in structure and function brought on by disease.

lopho- tuft. Example: lophotrichous, having a group of flagella on one side of a cell.

luc-, luci- light. Example: luciferin, substance in certain organisms that emits light when acted upon by the enzyme luciferase.

lute-, luteo- yellow. Example: *Micrococcus luteus*, yellow colonies.

-lysis loosening, to break down. Example: hydrolysis, chemical decomposition of a compound into other compounds as a result of taking up water.

macro- largeness. Example: macromolecules, large molecules.

meningo- membrane. Example: meningitis, inflammation of the membranes of the brain.

meso- middle. Example: mesophile, an organism whose optimum temperature is in the middle range.

meta- beyond, between, transition. Example: metabolism, chemical changes occuring within a living organism.

micro- smallness. Example: microscope, instrument used to make small objects appear larger.

-mnesia memory. Examples: amnesia, loss of memory; anamnesia, return of memory.

-monas a unit. Example: *Methylomonas*, a unit (bacterium) that utilizes methane as its carbon source.

mono- singleness. Example: monotrichous, having one flagellum.

morpho- form. Example: morphology, the study of form and structure of organisms.

multi- many. Example: multinuclear, having several nuclei.

mur- wall. Example: murein, component of bacterial cell walls.

mus-, muri- mouse. Example: murine typhus, a form of typhus endemic in mice.

mut- to change. Example: mutation, a sudden change in characteristics.

myco-, -mycetoma, -myces a fungus. Example: *Saccharomyces*, sugar fungus, a genus of yeast.

myxo- slime, mucus. Example: Myxobacteriales, an order of slime-producing bacteria.

necro- a corpse. Example: necrosis, cell death or death of a portion of tissue.

nigr- black. Example: *Aspergillus niger*, fungus that produces black conidia.

ob- before, against. Example: obstruction, impeding or blocking up.

oculo- eye. Example: monocular, pertaining to one eye.

-oecium, -ecium a house. Examples: perithecium, ascus with an opening that encloses spores; ecology, the study of the relationships between organisms and between an organism and its environment (household).

-oid like, resembling. Example: coccoid, resembling a coccus.

-oma tumor. Example: lymphoma, a tumor of the lymphatic tissues.

-ont being, existing. Example: schizont, a cell existing as a result of schizogony.

ortho- straight, direct. Example: orthomyxovirus, a virus with a straight, tubular capsid.

-osis, -sis condition of. Examples: lysis, the condition of loosening; symbiosis, the condition of living together.

pan- all, universal. Example: pandemic, an epidemic affecting a large region.

para- beside, near. Example: parasite, an organism that "feeds beside" another.

peri- around. Example: peritrichous, projections from all sides.

phaeo- brown. Example: Phaeophyta, brown algae.

phago- eat. Example: phagocyte, a cell that engulfs and digests particles or cells.

philo-, -phil liking, preferring. Example: thermophile, an organism that prefers high temperatures.

-phore bears, carries. Example: conidiophore, a hypha that bears conidia.

-phyll leaf. Example: chlorophyll, the green pigment in leaves.

-phyte plant. Example: saprophyte, a plant that obtains nutrients from decomposing organic matter.

pil- a hair. Example: pilus, hairlike projection from a cell.

plano- wandering, roaming. Example: plankton, organisms drifting or wandering in water.

plast- formed. Example: plastid, formed body within a cell.

-pnoea breathing. Example: dyspnea, difficulty in breathing.

pod- foot. Example: pseudopod, footlike structure.

poly- many. Example: polymorphism, many forms.

post- after, behind. Example: posterior, places behind (a specific) part.

pre-, pro- before, ahead of. Examples: procaryote, cell with the first nucleus evolutionarily; pregnant, before birth.

pseudo- false. Example: pseudopod, false foot.

psychro- cold. Example: psychrophile, an organism that grows best at cold temperatures.

-ptera wing. Example: Diptera, order of true flies, insects with two wings.

pyo- pus. Example: pyogenic, pus-forming.

rhabdo- stick, rod. Example: Rhabdovirus, an elongated, bullet-shaped virus.

rhin- nose. Example: rhinitis, inflammation of mucous membranes in the nose.

rhizo- root. Examples: *Rhizobium*, bacterium that grows in plant roots; mycorrhiza, mutualism between a fungus and the roots of a plant.

rhodo- red. Example: *Rhodospirillum*, red-pigmented spiral-shaped bacterium.

rod- gnaw. Example: rodents, class of mammals with gnawing teeth.

rubri- red. Example: *Clostridium rubrum*, red-pigmented colonies.

rumin- throat. Example: *Ruminococcus*, bacterium associated with a rumen.

saccharo- sugar. Example: disaccharide, a sugar consisting of two simple sugars.

sapr- rotten. Example: *Saprolegnia*, fungus that lives on dead animals.

sarco- flesh. Example: sarcoma, a tumor of muscle or connective tissues.

schizo- split. Example: schizomycetes, organisms that reproduce by splitting, an early name for bacteria.

scolec- worm. Example: scolex, the head of a tapeworm.

-scope, -scopic watcher. Example: microscope, an instrument used to watch small things.

semi- half. Example: semicircular, having the form of half a circle.

sept- rotting. Example: aseptic, free from bacteria that could cause decomposition.

septo- partition. Example: septum, a cross-wall in a fungal hypha.

serr- notched. Example: serrate, with a notched edge.

sidero- iron. Example: *Siderococcus*, a bacterium capable of oxidizing iron.

siphon- tube. Example: Siphonaptera, order of fleas, insects with tubular mouths.

soma- body. Example: somatic cells, cells of the body other than gametes.

speci- particular thing. Examples: species, the smallest group of organisms with similar properties; specify, to indicate exactly.

spiro- coil. Example: spirochete, bacterium with a coiled cell.

sporo- spore. Example: sporangium, a structure that holds spores.

staphylo- grape-like cluster. Example: *Staphylococcus*, bacterium that forms clusters of cells.

-stasis arrest, fixation. Example: bacteriostasis, cessation of bacterial growth.

strepto- twisted. Example: *Streptococcus*, bacterium that forms twisted chains of cells.

sub- beneath, under. Example: subcutaneous, just under the skin.

super- above, upon. Example: superior, quality or state of being above others.

sym-, syn- together, with. Examples: synapse, the region of communication between two neurons; synthesis, putting together.

-taxi to touch. Example: chemotaxis, response to the presence (touch) of chemicals.

taxis- orderly arrangement. Example: taxonomy, the science dealing with arranging organisms into groups.

thallo- plant body. Example: thallus, an entire macroscopic fungus.

therm- heat. Example: thermometer, an instrument used to measure heat.

thio- sulfur. Example: *Thiobacillus*, a bacterium capable of oxidizing sulfur-containing compounds.

-tome, -tomy to cut. Example: appendectomy, surgical removal of the appendix.

-tone, -tonic strength. Example: hypotonic, having less strength (osmotic pressure).

tox- poison. Example: antitoxic, effective against poison.

trans- across, through. Example: transport, movement of substances.

tri- three. Example: trimester, three-month period.

trich- a hair. Example: peritrichous, hairlike projections from cells.

-trope turning. Example: geotropic, turning towards the earth (pull of gravity).

-troph food, nourishment. Example: trophic, pertaining to nutrition.

-ty condition of, state. Example: immunity, condition of being resistant to disease or infection.

undul- wavy. Example: undulating, rising and falling, presenting a wavy appearance.

uni- one. Example: unicellular, pertaining to one cell.

vaccin- cow. Example: vaccination, injection of a vaccine (originally pertained to cows).

vacu- empty. Example: vacuoles, an intracellular space that appears empty.

vesic- bladder. Example: vesicle, a bubble.

vitr- glass. Example: in vitro, in culture media in a glass (or plastic) container.

-vorous eat. Example: carnivore, an animal that eats other animals.

xantho- yellow. Example: *Xanthomonas*, produces yellow colonies.

xeno- strange. Example: axenic, sterile, free of strange organisms.

xero- dry. Example: xerophyte, plant that tolerates dry conditions.

xylo- wood. Example: xylose, a sugar obtained from wood.

zoo- animal. Example: zoology, the study of animals.

zygo- yoke, joining. Example: zygospore, spore formed from the fusion of two cells.

-zyme ferment. Example: enzyme, protein in living cells that catalyzes chemical reactions.

APPENDIX D

Most Probable Numbers (MPN) Table

MPN Index for Various Combinations of Positive and Negative Results When Five 10-ml Portions, Five 1-ml Portions, and Five 0.1-ml Portions Are Used (see pp. 162, 164)

No. of Tubes Giving Positive Reaction out of 5 of 10 ml Each	5 of 1 ml Each	5 of 0.1 ml Each	MPN Index per 100 ml
0	0	0	<2
0	0	1	2
0	1	0	2
0	2	0	4
1	0	0	2
1	0	1	4
1	1	0	4
1	1	1	6
1	2	0	6
2	0	0	5
2	0	1	7
2	1	0	7
2	1	1	9
2	2	0	9
2	3	0	12
3	0	0	8
3	0	1	11
3	1	0	11
3	1	1	14
3	2	0	14
3	2	1	17
3	3	0	17
4	0	0	13
4	0	1	17
4	1	0	17
4	1	1	21
4	1	2	26
4	2	0	22
4	2	1	26
4	3	0	27
4	3	1	33
4	4	0	34
5	0	0	23
5	0	1	31
5	0	2	43
5	1	0	33
5	1	1	46
5	1	2	63
5	2	0	49
5	2	1	70
5	2	2	94
5	3	0	79
5	3	1	110
5	3	2	140
5	3	3	180
5	4	0	130
5	4	1	170
5	4	2	220
5	4	3	280
5	4	4	350
5	5	0	240
5	5	1	350
5	5	2	540
5	5	3	920
5	5	4	1600
5	5	5	≧2400

Source: *Standard Methods for the Examination of Water and Wastewater*, 13th ed., American Public Health Association, New York, 1971.

APPENDIX E

Methods for Taking Clinical Samples

To diagnose a disease, it is often necessary to obtain a sample of material that may contain the disease-causing organism. Samples must be taken aseptically. The sample container should be labeled with the patient's name, room number (if hospitalized), date, time, and medications being taken. Samples must be transported to the laboratory immediately for culture. Delay in transport may result in growth of some organisms and their toxic products may kill other organisms. Pathogens tend to be fastidious and die without their optimum environmental conditions.

In the laboratory, samples from infected tissues are cultured on differential and selective media in an attempt to isolate and identify any pathogens or organisms that are not normally found in association with that tissue.

Wound or Abscess Culture

1. Cleanse the area with a sterile swab moistened in sterile saline.
2. Disinfect the area with 70% ethyl alcohol or iodine solution.
3. If the abscess has not ruptured spontaneously, a physician will open it with a sterile scalpel.
4. Wipe the first pus away.
5. Touch a sterile swab to the pus taking care not to contaminate the surrounding tissue.
6. Replace the swab in its container and properly label the container.

Ear Culture

1. Clean the skin and auditory canal with 1% tincture of iodine.
2. Touch the infected area with a sterile cotton swab.
3. Replace the swab in its container.

Eye Culture

This procedure is usually performed by an opthalmologist.

1. Anesthetize the eye with topical application of a sterile anesthetic solution.
2. Wash the eye with sterile saline solution.
3. Collect material from the infected area with a sterile cotton swab. Return the swab to its container.

Blood Culture

1. Close room windows to avoid contamination.
2. Clean skin around selected vein with 2% tincture of iodine on a cotton swab.
3. Remove dried iodine with gauze moistened with 80% isopropyl alcohol.
4. Draw a few milliliters of venous blood.
5. Aseptically bandage puncture.

Urine Culture

1. Provide the patient with a sterile container.
2. Instruct the patient to collect a mid-stream sample. This is obtained by voiding a small volume from the

bladder before collection. This washes away extraneous skin flora.

3. A urine sample may be stored under refrigeration (4–6°C) for up to 24 hours.

Fecal Culture

For bacteriological examination, only a small sample is needed. This may be obtained by inserting a sterile swab into the rectum or feces. The swab is then placed in a tube of sterile enrichment broth for transport to the laboratory. For examination for parasites, a small sample may be taken from a morning stool. The sample is placed in a preservative (polyvinyl alcohol, buffered glycerol, saline, or formalin) for microscopic examination for eggs and adult parasites.

Sputum Culture

1. A morning sample is best as microorganisms will have accumulated while the patient is sleeping.
2. Patient should rinse his/her mouth thoroughly to remove food and normal flora.
3. Patient should cough deeply from the lungs and expectorate into a sterile glass wide-mouth jar.
4. Care should be taken to avoid contamination of personnel.
5. In cases such as tuberculosis where there is little sputum, stomach aspiration may be necessary.
6. Infants and children tend to swallow sputum. A fecal sample may be of some value in these cases.

Glossary

abscess A localized accumulation of pus.

acetyl coenzyme A (acetyl-CoA) A substance composed of an acetyl group and a carrier molecule called coenzyme A (CoA); provides the means for pyruvic acid to enter the Krebs cycle.

acetyl group

$$CH_3-\overset{\overset{\displaystyle O}{\|}}{C}-$$

acid A substance that dissociates into one or more hydrogen ions and one or more negative ions.

acid dyes A salt in which the color is in the negative ion; used for negative staining.

acid-fast stain A differential stain used to identify bacteria that are not decolorized by acid-alcohol.

acquired immunity The ability, obtained during the life of the individual, to produce specific antibodies.

actinomycetes Bacteria included in Part 17, order Actinomycetales in *Bergey's Manual*.

activated sludge Aerobic digestion used in secondary sewage treatment.

activation energy The minimum collision energy required for a chemical reaction to occur.

actively acquired immunity Production of antibodies by an individual in response to an antigen.

active transport Net movement of a substance across a membrane against a concentration gradient; requires energy.

acute disease A disease in which symptoms develop rapidly but last for only a short time.

adenine A purine nucleic acid base that pairs with thymine in DNA and uracil in RNA.

adenocarcinoma Cancer of glandular tissue.

adenosine diphosphate (ADP) The substance formed when ATP is split and energy released.

adenosine diphosphoglucose (ADPG) Glucose activated by ATP; precursor for glycogen synthesis.

adenosine triphosphate (ATP) An important intracellular energy source.

adjuvant A substance that improves the effectiveness of an antigen.

aerial mycelium A mycelium composed of fungal hyphae that project above the surface of the growth medium and produce asexual spores.

aerobic Requiring oxygen (O_2) to grow.

aerobic respiration Respiration in which the final electron acceptor in the electron transport chain is oxygen (O_2).

aerotolerant anaerobe An organism that does not use oxygen (O_2) but is not affected by its presence.

agar A complex polysaccharide derived from a marine alga and used as a solidifying agent in culture media.

agglutination A joining together or clumping of cells.

agglutinin An antibody that causes an agglutination reaction.

agranulocyte A leukocyte without granules in the cytoplasm; includes lymphocytes and monocytes.

aldehyde An organic molecule with the functional group

$$-\underset{\underset{\displaystyle O}{\|}}{C}-H$$

algae A group of photosynthetic eucaryotes; some are included in the kingdom Protista and some in the kingdom Plantae.

alkaline Having more OH^- ions than H^+ ions; pH is greater than 7.

allergen An antigen that evokes a hypersensitivity response.

allergy See *hypersensitivity*.

allosteric inhibition The mechanism whereby an inhibitor binds to the allosteric site of an enzyme, preventing the enzyme from acting on the substrate.

allosteric site The site on an enzyme at which an inhibitor binds.

allosteric transition The process in which an enzyme's activity is changed because of binding on the allosteric site.

alpha-fetoprotein (AFP) An antigen associated with liver and embryonal carcinomas.

alum Aluminum sulfate.

Ames test A procedure using bacteria to identify potential carcinogens.

amino acid An organic acid containing an amino group and a carboxyl group.

aminoglycoside An antibiotic consisting of amino sugars and an aminocyclitol ring; for example, streptomycin.

amino group —NH_2

ammonification Removal of amino groups from amino acids to form ammonia.

amoeba An organism belonging to the Protozoa phylum that moves by means of pseudopods.

amphibolic pathway A pathway that is anabolic and catabolic.

amphitrichous Having tufts of flagella at both ends of a cell.

anabolism All synthesis reactions in a living organism.

anaerobic Not requiring oxygen (O_2) for growth.

anaerobic respiration Respiration in which the final electron acceptor in the electron transport chain is an inorganic molecule other than oxygen (O_2); for example, a nitrate or sulfate.

anaerobic sludge digester Anaerobic digestion used in secondary sewage treatment.

anal pore A site in certain protozoans for elimination of waste.

anamnestic response A rapid rise in antibody titer following exposure to an antigen after the primary response to that antigen.

anaphylaxis A hypersensitivity reaction involving IgE antibodies, mast cells, and basophils.

anemia A decreased number of red blood cells or decreased percentage of hemoglobin in the blood.

angstrom (Å) A unit of measurement equal to 10^{-10} m, 10^{-4} μm, and 10^{-1}nm.

animal virus A virus that multiplies in animal tissues.

anion An ion with a negative charge.

anorexia Loss of appetite.

antibiotic An antimicrobial agent produced naturally by a bacterium or fungus.

antibody A protein produced by the body in response to an antigen and capable of combining specifically with that antigen.

antibody titer The amount of antibody in serum.

anticodon The three nucleotides by which a transfer RNA recognizes an RNA codon.

antigen Any substance that, when introduced into the body, causes antibody formation.

antigenic determinant site A specific region on the surface of an antigen against which antibodies are formed.

antigenic drift Minor variations in the antigenic makeup of a virus that occur with time.

antigen modulation The process whereby tumor cells shed their specific antigens, thus avoiding the host's immune response.

antihuman gamma globulin Antibodies that react specifically with human antibodies.

antimetabolite Any substance that interferes with metabolism by competitive inhibition of an enzyme.

antimicrobial agent A chemical that destroys pathogens without damaging body tissues.

antiseptic A chemical for disinfection of the skin, mucous membranes, or other living tissues.

antiserum A solution containing antibodies.

antitoxin A specific antibody produced by the body in response to a bacterial exotoxin or its toxoid.

apoenzyme The protein portion of an enzyme, which requires activation by a coenzyme.

A protein staphylococcal cell wall protein that prevents phagocytosis.

arthrospore An asexual fungal spore formed by fragmentation of a septate hypha.

artificially acquired active immunity The production of antibodies by the body in response to a vaccination.

artificially acquired passive immunity The transfer of humoral antibodies formed by one individual to a susceptible individual, accomplished by injection of antiserum.

ascospore A sexual fungal spore produced in an ascus, formed by the Ascomycetes.

ascus A saclike structure containing ascospores.

asepsis The absence of contamination by unwanted organisms.

asexual reproduction Reproduction without opposite mating strains.

asthma An allergic response characterized by bronchial spasms and difficult breathing.

atom The smallest unit of matter that can enter into a chemical reaction.

atomic number The number of protons in the nucleus of an atom.

atomic weight The total number of protons and neutrons in the nucleus of an atom.

atopy The spontaneous development of an allergy.

attenuation Lessening of virulence of a microorganism.

autoclave Equipment for sterilization by steam under pressure, usually operated at 15 psi and 121°C.

autoimmunity An immunologic response against a person's own tissue antigens.

autotroph An organism that uses carbon dioxide (CO_2) as its principal carbon source.

auxotroph A mutant microorganism with a nutritional requirement not possessed by the parent.

axial filament The structure for motility found in spirochetes.

bacillus Any rod-shaped bacterium; when written as a genus, refers to rod-shaped, endospore-forming, facultative anaerobic, gram-positive bacteria.

bacteremia A condition of bacteria in the blood.

bacteria All living organisms with procaryotic cells.

bacterial growth curve A graph indicating the growth of a bacterial culture over time.

bactericidal Capable of killing bacteria.

bacteriochlorophyll The light-absorbing pigment found in green sulfur and purple sulfur bacteria.

bacteriocinogens Plasmids containing genes for the synthesis of bacteriocins.

bacteriocins Toxic proteins produced by bacteria that kill other bacteria.

bacteriophage A virus that multiplies in bacterial cells.

bacteriostatic Capable of inhibiting bacterial growth.

bacteroid Enlarged *Rhizobium* cells found in root nodules.

basal body A structure that anchors flagella to the cell wall and plasma membrane.

base A substance that accepts hydrogen ions and is capable of uniting with water to form an acid.

base analog A chemical that is structurally similar to the normal nitrogenous bases in nucleic acids but with altered base-pairing properties.

base pairs The arrangement of nitrogenous bases in nucleic acids based on hydrogen bonding; in DNA, base pairs are A–T and G–C; in RNA, base pairs are A–U and G–C.

base substitution The replacement of a single base in DNA by another base, causing a mutation.

basic dye A salt in which the color is in the positive ion; used for bacterial stains.

basidiospore A sexual fungal spore produced in a basidium, characteristic of the Basidiomycetes.

basidium A pedestal that produces basidiospores; found in basidiomycetes.

basophil A granulocyte that readily takes up basic dye.

B cells Stem cells that are processed in the spleen and bone marrow and differentiate into plasma cells that secrete antibodies.

BCG vaccine A live, attenuated strain of *Mycobacterium bovis* used to provide immunity to tuberculosis.

benign tumor A noncancerous tumor.

benthic zone The sediment at the bottom of a body of water.

Bergey's Manual The taxonomic reference on bacteria.

beta-oxidation The removal of two carbon units of a long chain of fatty acid to form acetyl-CoA.

binary fission Bacterial reproduction by division into two daughter cells.

binomial nomenclature The system of having two names (genus and species) for each organism.

biochemical oxygen demand (BOD) A measure of the biologically degradable organic matter in water.

biochemical pathway A sequence of enzymatically catalyzed reactions occurring in a cell.

biochemistry The science of chemical processes in living organisms.

bioconversion Changes in organic matter brought about by the growth of microorganisms.

biogenesis The concept that living cells can only arise from preexisting cells.

biogeochemical cycles The recycling of chemical elements by microorganisms for use by other organisms.

biological transmission Transmission of a pathogen from one host to another when the pathogen reproduces in the vector.

bioluminescence The emission of light from the electron transport chain of certain living organisms.

biomass Organic matter produced by living organisms and measured by weight.

blastospore An asexual fungal spore produced by budding from the parent cell.

blocking antibody An antibody that reacts with an allergen to prevent a hypersensitivity reaction.

blood–brain barrier Cell membranes that allow some substances to pass from the blood to the brain but restrict others.

blooms (algal) Abundant growth of microscopic algae, producing visible colonies in nature.

booster dose The administration of antigens to elicit an anamnestic response.

brightfield microscope A microscope that uses visible light for illumination; the specimens are viewed against a white background.

broad-spectrum antimicrobial agent A chemical that has antimicrobial activity against many infectious microorganisms.

Brownian movement The movement of particles including microorganisms in a suspension due to bombardment by the moving molecules in the suspension.

bubo An enlarged lymph node caused by inflammation.

budding asexual reproduction beginning as a protuberance from the parent cell that grows to become a daughter cell; also, release of an enveloped virus through the plasma membrane of an animal cell.

buffer A substance that tends to stabilize the pH of a solution.

burst size The number of newly synthesized bacteriophage particles released from a single cell.

burst time The time required from bacteriophage adsorption to release.

cancer A malignant, invasive cellular tumor that has the capability of spreading throughout the body or body parts.

capsid The protein coat of a virus that surrounds the nucleic acid.

capsomere A protein subunit of a capsid.

capsule An outer, viscous covering on some bacteria composed of a polysaccharide or polypeptide.

carbohydrates Organic compounds composed of carbon, hydrogen, and oxygen, with the hydrogen and oxygen present in a 2:1 ratio; includes starches, sugars, and cellulose.

carbon cycle The series of processes that converts carbon dioxide (CO_2) to organic substances and back to carbon dioxide in nature.

carbon skeleton The basic chain ring of carbon atoms in a molecule; —C—C—C—, for example.

carboxyl group —COOH.

carbuncle An inflammation of the skin and subcutaneous tissue due to dissemination of a furuncle.

carcinoembryonic antigen (CEA) An antigen found in the normal fetus and malignant colon.

carcinogen Any cancer-producing substance.

cardiolipin A beef heart extract used in the venereal disease research laboratory (VDRL) slide test to detect antibodies against syphilis.

carrier An individual who harbors a pathogen but exhibits no signs of illness.

casein Milk protein.

catabolism All decomposition reactions in a living organism.

catalase An enzyme that catalyzes the breakdown of hydrogen peroxide to water and oxygen.

catalyst A substance that affects the rate of a chemical reaction, usually increasing the rate, but isn't changed in the reaction.

catheter A narrow, hollow tube that can be inserted into a body cavity for withdrawal of fluids.

cation A positively charged ion.

cell The basic microscopic unit of structure and function of all living organisms.

cell culture Animal cells grown in vitro.

cell theory The principle that all living things are composed of cells.

cellular immunity An immune response that involves the binding and elimination of antigens by T cell lymphocytes.

cellulose A polysaccharide that is the main component of plant cell walls.

centrioles Paired, cylindrical structures found in the centrosome of eucaryotic cells.

centrosome A dense area of cytoplasm near the nucleus of eucaryotic cells; involved in mitosis.

cephalosporin An antibiotic produced by the fungus *Cephalosporium* that inhibits the synthesis of gram-positive bacterial cell walls.

cercaria A free-swimming larva of trematodes.

chancre A hard sore, the center of which ulcerates.

chemical bond Attractive force between atoms forming a molecule.

chemical element A fundamental substance composed of atoms that have the same atomic number and behave the same way chemically.

chemical energy The energy of a chemical reaction.

chemically defined medium A culture medium in which the exact chemical composition is known.

chemical reaction The process of making or breaking bonds between atoms.

chemistry The science of the interactions of atoms and molecules.

chemoautotroph An organism that uses an inorganic chemical as an energy source, and carbon dioxide (CO_2) as a carbon source.

chemoheterotroph An organism that uses organic molecules as a source of carbon and energy.

chemostat An apparatus to keep a culture in log phase indefinitely.

chemotaxis A response to the presence of a chemical.

chemotherapy Treatment of a disease with chemical substances.

chemotroph An organism that uses oxidation–reduction reactions as its primary energy source.

chitin A glucosamine polysaccharide that is the main component of fungal cell walls and arthropod skeletons.

chlamydospore An asexual fungal spore formed within a hypha.

chlorophyll *a* The light-absorbing pigment in cyanobacteria, algae, and plants.

chloroplast The organelle that performs photosynthesis in photoautotrophic eucaryotes.

chromatophore An infolding in the plasma membrane where bacteriochlorophyll is located in photoautotrophic bacteria.

chromosome The structure that carries hereditary information.

chronic disease An illness that develops slowly and is likely to continue or recur for long periods of time.

cilia Relatively short cellular projections that move in a wavelike manner.

ciliate A member of the protozoan phylum Ciliata that uses cilia for locomotion.

cisternae Stacked elements of the Golgi complex.

class A taxonomic ranking between phylum and order.

clone A population of cells that are identical to the parent cell.

coagulase A bacterial enzyme that causes blood plasma to clot.

coccobacillus A bacterium that is an oval shaped rod.

coccus A spherical or ovoid bacterium.

codon A group of three nucleotides in DNA or mRNA that specifies the insertion of an amino acid into a protein.

coenocytic hyphae Fungal filaments that are not divided into uninucleate cell-like units because they lack septa.

coenzyme A nonprotein substance that is associated with and that activates an enzyme.

coenzyme A A coenzyme that functions in decarboxylation.

cofactor The nonprotein component of an enzyme.

colicins Bacteriocins produced by *Escherichia coli.*

coliforms Aerobic or facultatively anaerobic, gram-negative, non-spore-forming, rod-shaped bacteria that ferment lactose with acid and gas formation within 48 hours at 35°C.

collagen The main structural protein of muscles.

collagenase An enzyme that degrades collagen.

collision theory The principle that chemical reactions are due to energy gained as particles collide.

colony A clone of bacterial cells on a solid medium that is visible to the naked eye.

colony-forming units Visible units counted in a plate count, which may be formed from a group of cells rather than from one cell.

combined vaccine A vaccine that contains several immunizing agents; for example, DPT.

commensalism A system of interaction in which two organisms live in association and one is benefited while the other is neither benefited nor harmed.

commercial sterilization A process of treating canned goods aimed at destroying the endospores of *Clostridium botulinum.*

communicable disease Any disease that can be spread from one host to another.

competence The physiological state in which a recipient cell can take up and incorporate a large piece of donor DNA.

competitive inhibition The process by which a chemical competes with the normal substrate for the active site of an enzyme.

complement (C) A group of 11 serum proteins involved in phagocytosis and lysis of bacteria.

complement fixation The process in which complement combines with an antigen–antibody complex.

complete antigen An antigen with reactivity and immunogenicity.

complete digestive system A digestive system with a mouth and an anus.

completed test The final test for detection of coliforms in the multiple-tube fermentation test.

complex medium A culture medium in which the exact chemical composition is not known.

complex virus A virus with a complicated structure, such as a bacteriophage.

compound A substance composed of two or more different chemical elements.

compound light microscope An instrument with two sets of lenses that uses visible light as the source of illumination.

condensation reaction A chemical reaction in which a molecule of water is released.

condenser A lens system located below the microscope stage that directs light rays through the specimen.

confirmed test The second stage of the multiple-tube fermentation test, used to identify coliforms on solid, differential media.

congenital disease A disease present at birth as a result of some condition that occurred in utero.

conidiophore An aerial hypha bearing conidiospores.

conidiospore An asexual spore produced in a chain from a conidiophore.

conjugation The transfer of genetic material from one cell to another involving cell-to-cell contact.

conjugative plasmid A plasmid with genes for carrying out conjugation.

constitutive enzyme An enzyme that is produced regardless of how much substrate is present.

contact inhibition The cessation of animal cell movement and division due to contact.

contagious disease A disease that is easily spread from one person to another.

continuous cell line Animal cells that can be maintained through an indefinite number of generations in vitro.

convalescent period The period of recovery from a disease.

corepressor The molecule (end product) that brings about repression of a repressible enzyme.

corticosteroids Steroid hormones released by the adrenal gland, derivatives of which are used to treat inflammatory diseases.

counterstain A stain used to give contrast in a differential stain.

coupled reaction Two chemical reactions that must occur simultaneously.

covalent bond A chemical bond in which the electrons of one atom are shared with another atom.

cresols A mixture of isomers from petroleum.

crisis The phase of a fever characterized by vasodilation and sweating.

cristae Foldings of the inner membrane of a mitochondrion.

cross-over Process by which a portion of a chromosome is exchanged with a portion of another chromosome.

cross-reactivity Principle by which an antibody that binds a given antigen will also bind other molecules with similar antigenic sites.

culture Microorganisms that grow and multiply in a container of culture medium.

culture medium The nutrient material prepared for growth of microorganisms in a laboratory.

curd The solid part of milk that separates from the liquid when the milk is fermented.

cutaneous mycosis A fungal infection of the epidermis, nails, and hair.

cuticle Nonliving outer covering of helminths.

cyanobacteria Members of the kingdom Monera, formerly called blue-green algae.

cyst A sac with a distinct wall containing fluid or other material; also, a protective capsule of some protozoans.

cysticercus Encysted tapeworm larva.

cystitis Inflammation of the urinary bladder.

cytochrome oxidase An enzyme that oxidizes cytochrome *c*.

cytochromes Enzymes that function as electron carriers in respiration and photosynthesis.

cytocidal Resulting in cell death.

cytomegaly Enlarged cells.

cytopathic effects (CPE) Tissue deterioration caused by viruses.

cytoplasm In a procaryote, everything inside the plasma membrane; in a eucaryote, everything inside the plasma membrane and external to the nucleus.

cytoplasmic streaming The flowing of cytoplasm in a eucaryotic cell.

cytosine A pyrimidine nucleic acid base that pairs with guanine.

cytostome The mouthlike opening in some protozoans.

cytotrophic Binding to cells; for example, IgE antibodies bind to target cells.

D- Prefix describing a stereoisomer.

Dane particle A structure in the serum of a patient with hepatitis B having hepatitis B virus (HBV) surface antigens.

darkfield microscope A microscope that has a device to scatter light from the illuminator so that the specimen appears white against a black background.

deamination Removal of an amino group.

debridement Surgical removal of necrotic tissue.

decarboxylation Removal of carbon dioxide (CO_2) from an amino acid.

decimal reduction time The time (in minutes) required to kill 90% of a bacterial population at a given temperature.

decolorization The process of removing a stain.

decomposition reaction A chemical reaction in which bonds are broken to produce smaller parts from a large molecule.

definitive host An organism that harbors the adult, sexually mature form of a helminthic parasite.

degenerative disease Loss of function due to wearing down of parts.

dehydration The removal of water.

dehydrogenation The loss of hydrogen atoms from a substrate.

delayed hypersensitivity Cell-mediated hypersensitivity.

denaturation A change in the molecular structure of a protein.

denitrification The reduction of nitrates to nitrites or nitrogen gas.

dental plaque A combination of bacterial cells, dextran, and debris adhering to the teeth.

deoxyribonucleic acid (DNA) The nucleic acid of genetic material.

deoxyribose A five-carbon sugar contained in DNA nucleotides.

dermatophyte A fungus that causes a cutaneous mycosis.

dermis The inner portion of the skin.

desensitization The prevention of allergic inflammatory responses.

desiccation The removal of water.

detergents Any substance that reduces the surface tension of water.

dextran A polymer of glucose.

diacetyl $CH_3COCOCH_3$ produced from carbohydrate fermentation.

diagnosis Identification of a disease.

diapedesis The process by which leukocytes (PMNs) move out of blood vessels.

Dick test A skin test to determine immunity to scarlet fever.

differential count The number of each kind of leukocyte in a sample of 100 leukocytes.

differential interference contrast (DIC) microscope A microscope that provides a three-dimensional image.

differential medium A solid culture medium that makes it easier to distinguish colonies of the desired organism.

differential stain A stain that distinguishes objects on the basis of reactions to the staining procedure.

diffusion The net movement of molecules or ions from an area of higher concentration to an area of lower concentration.

digestion The process of breaking down substances physically and chemically.

digestive vacuole An organelle in which substrates are broken down enzymatically.

dimorphism The property of having two growth forms.

dioecious Referring to organisms in which organs of different sexes are located in different individuals.

diplobacilli Rods that divide and remain attached in pairs.

diplococci Cocci that divide and remain attached in pairs.

diploid A cell or organism with two sets of chromosomes.

direct contact A method of spreading infection from one host to another through some kind of close association of the hosts.

direct count Enumeration of cells by observation through a microscope.

direct F.A. test A fluorescent-antibody test to detect the presence of an antigen.

disaccharide A sugar consisting of two monosaccharides.

disease Any change from a state of health.

disinfectant Any substance used on inanimate objects to kill or inhibit the growth of microorganisms.

dissimilation plasmids Plasmids containing genes coding for the production of enzymes that catalyze the catabolism of certain unusual sugars and hydrocarbons.

dissociation Transformation of a compound into positive and negative ions in solution.

disulfide bond Two atoms of sulfur held together by a covalent bond (S—S).

donor cell A cell that gives DNA to a recipient cell in recombination.

Donovan bodies Inclusion bodies in large mononuclear cells infected with *Calymmatobacterium granulomatis*.

DPT vaccine A combined vaccine used to provide active immunity, containing diphtheria and tetanus toxoids and killed *Bordetella pertussis* cells.

droplet infection The transmission of infection by small liquid droplets carrying microorganisms.

dysentery A disease characterized by frequent, watery stools.

eclipse period The time during viral multiplication when complete, infective virions are not present.

edema An abnormal accumulation of fluid in body parts or tissues, causing swelling.

electrical energy Energy from the flow of electrons.

electron A negatively charged particle in motion around the nucleus of an atom.

electronic configuration The arrangement of electrons in shells or energy levels in an atom.

electron microscope A microscope that uses a flow of electrons instead of light to produce an image.

electron transport system A series of compounds that transfers electrons from one compound to another, generating ATP by oxidative phosphorylation.

electrophoresis The separation of substances (for example, serum proteins) by their rate of movement through an electric field.

Elek test An immunodiffusion test to detect toxin-producing strains of *Corynebacterium diphtheriae*.

elementary body An infectious form of *Chlamydia*.

Embden–Meyerhof pathway See *glycolysis*.

emulsify To mix two liquids that do not dissolve in each other.

encephalitis Inflammation of the brain.

encystment Formation of a cyst.

endemic disease A disease that is constantly present in a certain population.

endocarditis Inflammation of the lining of the heart.

endocytosis The process of moving material into a eucaryotic cell.

endogenous Originating from within an organism.

endoplasmic reticulum A membrane network in eucaryotic cells connecting the plasma membrane with the nuclear membrane.

endospore A resting structure formed inside some bacteria.

endotoxin Part of the outer portion of the cell wall of most gram-negative bacteria; a lipopolysaccharide.

end-product repression Mechanism whereby an end product of a pathway inhibits the synthesis of an enzyme required for formation of that end product.

energy The capacity to do work.

enrichment culture A culture medium used for preliminary isolation that favors the growth of a particular microorganism.

enterics The common name for bacteria in the family Enterobacteriaceae.

enterotoxin A staphylococcal exotoxin that causes diarrhea.

Entner–Doudoroff pathway An alternate pathway for the oxidation of glucose to pyruvic acid.

envelope An outer covering surrounding the capsid of some viruses.

enzyme A protein that catalyzes chemical reactions in a living organism.

enzyme induction The process by which a substance can cause the synthesis of an enzyme.

enzyme repression The process by which a substance can stop the synthesis of an enzyme.

enzyme–substrate complex A temporary union of an enzyme and its substrate.

eosinophil A granulocyte whose granuoles take up the stain eosin.

epidemic disease A disease acquired by many people in a given area in a short time.

epidemiology The science dealing with when and where diseases occur and how they are transmitted.

epidermis The outer portion of the skin.

equilibrium The point of even distribution.

ergot A substance produced in sclerotia by the fungus *Claviceps purpurea* that causes contraction of arteries and uterine muscle.

erythema Redness of the skin.

erythrocyte Red blood cell.

erythrogenic toxin A substance produced by some streptococci that causes erythema.

ester linkage The bonding between two organic molecules (R) as $R{-}\overset{\overset{\displaystyle O}{\|}}{C}{-}O{-}R$.

ethambutol An antimicrobial agent that interferes with the synthesis of RNA.

etiology The study of the cause of a disease.

eucaryote A cell with DNA enclosed within a distinct membrane-bounded nucleus.

eutrophication The addition of organic matter and subsequent removal of oxygen from a body of water.

exchange reaction A chemical reaction that has both synthesis and decomposition components.

exocytosis The process of exporting material from a eucaryotic cell.

exogenous Originating from a source outside the body.

exotoxins Protein toxins released from bacterial cells into the surrounding medium.

extracellular enzyme An enzyme released from a cell to break down large molecules.

extreme halophile An organism that requires a high salt concentration for growth.

facilitated diffusion The transfer of a substance across a plasma membrane from an area of higher concentration to an area of lower concentration mediated by carrier proteins (permeases).

facultative anaerobe An organism that can grow with or without oxygen (O_2).

facultative halophile An organism capable of growth in, but not requiring, 1 to 2% salt.

family A taxonomic group between order and genus.

fat An organic compound consisting of glycerol and fatty acids.

fatty acids Long hydrocarbon chains ending in a carboxyl group.

fecal coliforms Coliform organisms found in the human intestine, capable of fermenting lactose at 44.5°C.

feedback inhibition Inhibition of an enzyme in a particular pathway by the accumulation of end product from the pathway.

fermentation The enzymatic degradation of carbohydrates in which the final electron acceptor is an organic molecule, ATP is synthesized by substrate-level phosphorylation, and oxygen (O_2) is not required.

fever An abnormally high body temperature.

F factor Fertility factor; a plasmid found in the donor cell in bacterial conjugation.

filamentous form A structure in the serum of a patient with hepatitis B having hepatitis B virus (HBV) surface antigens.

filtration Passage of a liquid or gas through a screen-like material.

fimbria *See* pilus.

fixed macrophage Macrophage that is located in a certain organ or tissue, for example, in liver, lungs, spleen, or lymph nodes.

fixing (in slide preparation) The process of attaching the specimen to the slide.

flagella Thin, whiplike appendages that arise from one or more locations on the surface of a cell and are used for cellular locomotion.

flagellate A member of the protozoan phylum Mastigophora that uses flagella for locomotion.

flaming The process of sterilizing an inoculating loop by holding it in an open flame.

flat sour Thermophilic spoilage of canned goods not accompanied by gas production.

flavoproteins Enzymes that function as electron carriers in respiration.

flocculation Removal of colloidal material by addition of a chemical that causes the colloidal particles to coalesce.

flora The microbial population of an area, such as of human skin.

fluid mosaic model A way of describing the dynamic arrangement of phospholipids and proteins comprising the plasma membrane.

fluke A flatworm belonging to the class Trematoda.

fluorescence The ability to give off light of one color when exposed to light of another color.

fluorescent-antibody technique A diagnostic tool using antibodies labeled with fluorochromes and viewed through a fluorescent microscope.

fluorescent microscope A microscope that uses an ultraviolet light source to illuminate specimens that will fluoresce.

fluorochromes Dyes used to stain bacteria that fluoresce when illuminated with ultraviolet light.

focal infection A systemic infection that began as an infection in one place.

fomite A nonliving object that can spread infection.

formalin A 37% aqueous solution of formaldehyde that may contain 15% methanol.

fractional sterilization The use of free-flowing steam not under pressure.

frameshift mutation A mutation due to the addition or deletion of one or more bases in DNA.

FTA–ABS test An indirect fluorescent-antibody test used to detect syphilis.

functional groups Arrangement of elements in organic molecules that are responsible for most of the chemical properties of those molecules.

fungi Organisms that belong to the kingdom Fungi; eucaryotic chemoheterotrophs.

furuncle An infection of a hair follicle.

gamete A male or female reproductive cell.

gametocyte A male or female *Plasmodium* cell.

gamma globulin The serum fraction containing immunoglobulins (antibodies).

gangrene Tissue death due to loss of blood supply.

gastroenteritis Inflammation of the stomach and intestine.

gene A segment of DNA or a sequence of nucleotides in DNA that codes for a functional product.

generation time The time required for a cell or population to double in number.

genetic engineering Manufacturing and manipulating genetic material in vitro.

genetics The science of heredity.

genotype The genetic makeup of an organism.

genus The first name of the scientific name (binomial).

geosmin An alcohol produced by actinomycetes that has an earthy odor.

germ Part of an organism capable of developing.

germicidal Capable of killing microorganisms.

germicidal lamp An ultraviolet light (wavelength = 260 nm) capable of killing bacteria.

germination The process of starting to grow from a spore.

germ theory The principle that microorganisms cause disease.

glomerulonephritis Inflammation of the glomeruli of the kidneys, but not a result of a kidney infection.

glucan A polysaccharide component of yeast cell walls.

glycerol An alcohol; $C_3H_5(OH)_3$.

glycogen A polysaccharide stored by some cells.

glycolysis The main pathway for the oxidation of glucose to pyruvic acid.

glycoprotein Carbohydrate and protein complex.

Golgi complex An organelle involved in the secretion of certain proteins.

graft-versus-host (GVH) disease A condition that occurs when a transplanted tissue has an immune response to the tissue recipient.

Gram stain A differential stain that divides bacteria into two groups, gram-positive and gram-negative.

granulocyte A leukocyte with granules in the cytoplasm; includes neutrophils, basophils, and eosinophils.

griseofulvin A fungistatic antibiotic.

guanine A purine nucleic acid base that pairs with cytosine.

Guillain–Barré syndrome A neurological syndrome, of unknown etiology, marked by muscular weakness and paralysis.

gumma A rubbery mass of tissue characteristic of tertiary syphilis and tuberculosis.

halogen One of the following elements: fluorine, chlorine, bromine, iodine, or astadine.

hanging-drop preparation A slide prepared so that microorganisms can be viewed live in a liquid suspension.

H antigen A flagellar antigen of enterics.

haploid A cell or organism with one of each type of chromosome (eucaryotic).

hapten An antigen that has reactivity and not immunogenicity.

heavy metals Certain elements with specific gravity greater than 4 that are used as antimicrobial agents; for example, silver (Ag), copper (Cu), and mercury (Hg).

helminth A parasitic roundworm or flatworm.

helper T cells Cells that interact with an antigen before B cells interact with the antigen.

hemagglutination Clumping of red blood cells.

hemagglutination-inhibition Process whereby an antibody inhibits viral hemagglutination.

hemoflagellate A parasitic flagellate found in the circulatory system of its host.

hemolysins Enzymes that lyse red blood cells.

heparin A substance that prevents clotting.

hermaphroditic Having both male and female reproductive capacities.

heterocyst A large empty cell in certain cyanobacteria.

heterophilic antibody An antibody that reacts with distantly related antigens.

heterotroph An organism that requires an organic carbon source.

hexachlorophene A chlorinated phenol used as an antiseptic.

hexose monophosphate shunt *See* pentose phosphate pathway.

hexylresorcinol A benzene derivative used as an antihelminthic.

Hfr A bacterial cell in which the F factor has become integrated into the chromosome.

H (hemagglutination) spikes Projections from the outer lipid bilayer of influenza virus.

High-temperature short-time (HTST) pasteurization 72°C for 15 seconds.

histamine A substance released by tissue cells that causes vasodilation.

histiocyte *See* fixed macrophage.

histocompatibility antigens Antigens on the surface of human cells.

histones Proteins associated with DNA in eucaryotic chromosomes.

histoplasmin test A skin test to detect antibodies against *Histoplasma*.

HLA complex *See* major histocompatibility complex (MHC).

holoenzyme An enzyme consisting of an apoenzyme and a cofactor.

host An organism infected by a pathogen.

hot-air sterilization The use of an oven at 170°C for approximately 2 hours.

humoral immunity Immunity produced by antibodies dissolved in body fluids, mediated by B cells.

humus Organic matter in soil.

hyaluronic acid A mucopolysaccharide that holds together certain cells of the body.

hydrogen bond A bond between a hydrogen atom covalently bonded to oxygen or nitrogen and another covalently bonded oxygen or nitrogen.

hydrogen ion (H^+) A proton.

hydrolysis The breakdown of water.

hydroxyl group OH^-.

hyperbaric chamber An apparatus to hold gases at pressures greater than one atmosphere.

hypersensitivity Altered, enhanced immune reactions leading to pathologic changes.

hypertonic Describing a solution that has a higher concentration of solutes than an isotonic solution.

hyphae Long filaments of cells in fungi or actinomycetes.

hypothermic factor A bacterial substance that affects the host's body temperature.

hypotonic Describing a solution that has a lower concentration of solutes than an isotonic solution.

iatrogenic disease A disease caused by health professionals while administering health care.

icosahedron A polyhedron with 20 triangular faces and 12 corners.

ID_{50} The bacterial concentration required to produce a demonstrable infection in 50% of the test host population.

idiopathic disease A disease of undetermined cause.

IgA The class of antibodies found in secretions.

IgD Antibodies found on B cells.

IgE The class of antibodies involved in hypersensitivities.

IgG The most abundant antibodies found in serum.

IgM The first antibodies to appear after exposure to an antigen.

illuminator A light source.

imidazoles Antimetabolites that inhibit the action of histamine.

immediate hypersensitivity Allergic reactions involving humoral antibodies.

immune adherence The attachment of the phagocyte to an antigen.

immune-complex disease A condition in which antibodies are formed against the self.

immunity The body's defense against a particular microorganism.

immunization A process that produces immunity.

immunoassay Detection of an antigen or antibody by serological methods.

immunodiffusion test A test consisting of precipitation reactions carried out in an agar-gel medium.

immunoelectrophoresis Identification of proteins by electrophoretic separation and then serological testing.

immunofluorescence Procedures using the fluorescent-antibody technique.

immunogenicity The ability to stimulate antibody production.

immunoglobulin (Ig) An antibody.

immunologic competence The ability to perform specific immune reactions.

immunologic disease A disease caused by the immune system attacking the self (autoimmunity) or overreacting (hypersensitivity).

immunologic enhancement The binding of antibodies to tumor antigens.

immunologic paralysis Immunologic tolerance requiring constant exposure to high concentrations of an antigen.

immunologic surveillance The body's immune response to cancer.

immunologic tolerance A state of specific unresponsiveness to an antigen following initial exposure to that antigen.

immunosuppression Inhibition of the immune response.

immunotherapy Treatment to cause the production of immune responses.

IMViC Biochemical tests used to identify enterics, including indole, methyl red, Voges–Proskauer, and citrate.

inactivated viral vaccine A vaccine consisting of virions that have been treated with formalin or ultraviolet radiation.

inapparent infection *See* subclinical infection.

incidence The fraction of the population that contracts a disease during a particular length of time.

inclusion Material inside a cell.

incompatible transfusion Transfer of red blood cells to an individual with isoantibodies to the red blood cell antigen.

incomplete digestive system A digestive system with one opening (mouth) for intake of food and elimination of waste.

incubation period The time interval between the actual infection and first appearance of any signs or symptoms of disease.

indirect contact Transmission of pathogens by agents such as food and water.

indirect F.A. test A fluorescent-antibody test to detect the presence of specific antibodies.

individual immunity Resistance of an individual to diseases that affect other individuals of the same species.

inducer A substrate that brings about an increased amount of an enzyme.

inert Inactive.

infection Growth of microorganisms in the body.

infection thread An invagination in a root hair that allows *Rhizobium* to infect the root.

infectious disease A disease caused by pathogens.

inflammation A host response to tissue damage characterized by reddening, pain, heat, and swelling.

inherited disease A disease passed from parent to child through gametes.

initial body An intracellular stage of *Chlamydia.*

inoculate To introduce microorganisms into a culture medium or host.

inorganic compounds Small molecules not usually containing carbon.

interferon An antiviral protein produced by certain animal cells in response to a viral infection.

intermediate host An organism that harbors the larval stage of a helminth.

intoxication Poisoning.

invasiveness The ability of microorganisms to establish residence in a host.

in vitro "In glass"; not a living organism.

in vivo Within a living organism.

iodophor A complex of iodine and a detergent.

ion A negatively or positively charged atom or group of atoms.

ionic bond A chemical bond formed when atoms gain or lose an electron in the outer energy levels.

ionization Separation of a molecule into groups of atoms with electrical charges.

ionizing radiation High-energy radiation that causes ionization; for example, X rays and gamma rays.

isoantibody An antibody present in some individuals that reacts with a specific isoantigen.

isoantigen An antigen present in some individuals; for example, blood group antigens.

isomers Two molecules with the same chemical formula but different structures.

isoniazid (INH) A bacteriostatic agent used to treat tuberculosis.

isotonic Referring to a solution in which osmotic pressure is equal across a membrane.

isotope A form of a chemical element in which the number of neutrons in the nucleus is different from the other forms of that element.

jaundice Yellowing of the skin due to deposition of bile pigments in the skin and mucous membranes.

K antigen Capsular antigen of enterics.

keratin A protein found in epidermis, hair, and nails.

killer T cells Cells that destroy antigens.

kinases Bacterial enzymes that break down fibrin (blood clots).

kinetic energy The energy of motion.

kingdom The highest category in the taxonomic hierarchy of classification.

kinins Substances released from tissue cells that cause vasodilation.

Kirby–Bauer test An agar-diffusion test to determine microbial susceptibility to chemotherapeutic agents.

Koch's postulates Criteria used to determine the causative agent of infectious diseases.

Krebs cycle A pathway that converts two-carbon compounds to carbon dioxide (CO_2), transferring electrons to NAD^+ and other carriers.

L- Prefix describing a stereoisomer; L-amino acids are more commonly found in proteins.

lag phase The time interval in a bacterial growth curve with no growth.

larva The sexually immature stage of a helminth or arthropod.

latent viral infection A condition in which a virus remains in the host without producing disease for long periods of time.

LD_{50} The lethal dose for 50% of the inoculated hosts within a given period of time.

lepromin test A skin test to determine the presence of antibodies to *Mycobacterium leprae*.

leukemias Cancers characterized by abnormally high numbers of leukocytes.

leukocidins Substances produced by some bacteria that can destroy neutrophils.

leukocyte White blood cell.

leukocytic pyrogen A protein produced by PMNs that causes an increase in body temperature.

leukocytosis-promoting factor A substance released by inflamed tissues that increases the production of granulocytes.

leukopenia A condition in which the number of leukocytes is less than normal.

L form A mutant bacterium with a defective cell wall.

lichen A symbiosis between a fungus and an alga or cyanobacterium.

ligase An enzyme that joins together pieces of DNA.

limnetic zone The surface zone of a body of water away from the shore.

lincomycin An antibiotic that inhibits protein synthesis.

lipase An exoenzyme that breaks down fats into their component fatty acids and glycerol.

lipid A molecule composed of glycerol and fatty acids; fat.

lipopolysaccharide A molecule consisting of a lipid and a polysaccharide, forming the outer layer of gram-negative cell walls.

lipoprotein A molecule consisting of a lipid and protein.

liposome A fatty globule that may be used to administer chemotherapeutic agents.

littoral zone The region along the shore of an inland body of water where there is considerable vegetation and where light penetrates to the bottom.

local infection An infection in which pathogens are limited to a small area of the body.

localized reaction An anaphylaxis-type reaction such as hayfever, asthma, and hives.

logarithmic decline phase Period of bacterial death.

logarithmic phase Period of bacterial growth or logarithmic increase in cell number.

lophotrichous Having two or more flagella at one end of a cell.

lymphangitis Inflammation of the lymph vessels.

lymphocyte A granulocyte.

lymphokines Polypeptides released by T cells.

lymphoma Cancer of lymphoid tissue.

lyophilization Freeze-drying; freezing a substance and evaporating the ice in a vacuum.

lysis Disruption of the plasma membrane.

lysogeny A state in which phage DNA is incorporated into the host cell without lysis.

lysosome An organelle containing digestive enzymes.

lysozyme An enzyme in eucaryotic secretions capable of lysing gram-positive cell walls.

lytic cycle A sequence for replication of phages that results in host cell lysis.

macrolides Antibiotics that inhibit protein synthesis; for example, erythromycin.

macromolecules Large organic molecules.

macrophage A phagocytic cell; an enlarged monocyte.

macular rash Small red spots appearing on the skin.

major histocompatibility complex (MHC) The genes that code for the histocompatibility antigens; also known as human leukocyte antigens (HLA).

malaise A feeling of general discomfort.

malignant Cancerous.

mannan A polysaccharide component of yeast cell walls.

margination The process by which PMNs stick to the lining of blood vessels.

mast cell Type of cell found throughout the body that contains histamine and other substances that stimulate vasodilation.

maximum growth temperature The highest temperature at which a species can grow.

mechanical energy The energy involved in movement.

mechanical transmission The process by which arthropods transmit infections by carrying pathogens on their feet and other body parts.

megacin A bacteriocin produced by *Bacillus megaterium.*

meiosis The process that leads to the formation of haploid gametes in a diploid organism.

membrane filter A screen-like material with pores small enough to retain microorganisms.

meningitis Inflammation of the meninges covering the central nervous system.

merozoite A trophozoite of *Plasmodium* found in red blood cells.

mesophile An organism that grows between 25° and 40°C.

mesosome An irregular fold in the plasma membrane of a procaryotic cell.

messenger RNA (mRNA) The type of RNA molecule that directs the incorporation of amino acids into proteins.

metabolic disease A disease resulting from an abnormality in the biochemistry of bodily functions.

metabolism The sum of all the chemical reactions that occur in a living cell.

metacercaria The encysted stage of a fluke in its final intermediate host.

metachromatic granule The intracellular volutin stored by some bacteria.

metastasis The spread of cancer from a primary tumor to other parts of the body.

microaerophile An organism that grows best in an environment with less oxygen (O_2) than is found in air.

micrometer (μm) A unit of measure equal to 10^{-6} m.

microorganism A living organism too small to be seen with the naked eye; includes bacteria, fungi, protozoans, microscopic algae, and viruses.

microphage A granulocytic phagocyte.

microtubule The structure of the proteins comprising eucaryotic flagella and cilia.

minimal bactericidal concentration (MBC) The lowest concentration of a chemotherapeutic agent that will result in no growth.

minimal inhibitory concentration (MIC) The lowest concentration of a chemotherapeutic agent that will prevent growth of the test microorganism.

minimum growth temperature The lowest temperature at which a species will grow.

miracidium The free-swimming, ciliated larva of a fluke that hatches from the egg.

missense mutation A mutation that results in substitution of an amino acid in a protein.

mitochondria Organelles containing the respiratory ATP-synthesizing enzymes.

mitosis The division of the cell nucleus, often followed by division of the cytoplasm of the cell.

mixed culture A culture containing more than one kind of microorganism.

MMWR A weekly publication of the Centers for Disease Control containing data on notifiable diseases and topics of special interest.

molds Fungi that form mycelia and appear as cottony tufts.

molecular biology The science dealing with proteins of living organisms.

molecular weight The sum of the atomic weights of all atoms making up a molecule.

molecule A combination of atoms forming a specific chemical compound.

Monera The kingdom to which all procaryotic organisms belong.

monoclonal antibodies Specific antibodies produced by in vitro clones of B cells hybridized with cancerous cells.

monolayer A single layer of cells due to the cessation of cell division by contact inhibition.

monomers The units that combine to form polymers.

monotrichous Having a single flagellum.

morbidity The incidence of a specific notifiable disease.

mordant A substance added to a staining solution to make it stain more intensely.

morphology The external appearance.

mortality The deaths from a specific notifiable disease.

most probable number (MPN) A statistical determination of the number of coliforms per 100 ml of water or food.

motility The ability of an organism to move by itself.

M protein A heat- and acid-resistant protein of streptococcal cell walls.

multiple-tube fermentation test A method of detecting the presence of coliforms.

murein *See* peptidoglycan.

mutagen An agent in the environment that brings about mutations.

mutation Any change in the base sequence of DNA.

mutation rate The probability of a gene mutating each time a cell divides.

mutualism A symbiosis in which both organisms are benefited.

mycelium A mass of long filaments of cells that branch and intertwine, typically found in molds.

mycetoma A chronic infection caused by certain fungi and *Nocardia* characterized by a tumor-like appearance.

mycology The science dealing with fungi.

mycosis A fungal infection.

mycotoxin A toxin produced by a fungus.

myeloperoxidase A lysosomal enzyme that causes chlorine atoms to bind to bacteria and viruses.

myocarditis Inflammation of the heart muscle.

naked virus A virus without an envelope.

nanometer (nm) A unit of measurement equal to 10^{-9} m, 10^{-3} μm, and 10 Å.

native immunity Resistance to disease that is present at birth.

naturally acquired active immunity Antibody production in response to an infectious disease.

naturally acquired passive immunity The natural transfer of humoral antibodies; transplacental transfer.

necrosis Tissue death.

negative stain A procedure that results in colorless bacteria against a stained background.

neoplastic disease A disease resulting in new growth of cells; tumors.

neutralizing antibody An antibody that inactivates a bacterial exotoxin or virus.

neutron An uncharged particle in the nucleus of an atom.

nicotinamide adenine dinucleotide (NAD) A coenzyme that functions in the removal and transfer of H^+ and electrons from substrate molecules.

nicotinamide adenine dinucleotide phosphate (NADP) A coenzyme similar to NAD.

nitrification The oxidation of nitrogen from ammonia to nitrites and nitrates.

nitrogen cycle The series of processes that converts nitrogen (N_2) to organic substances and back to nitrogen in nature.

nitrogen fixation The conversion of nitrogen (N_2) into ammonia.

nitrosamine A carcinogen formed by the combination of nitrite and amino acids; nitroso-: —N=O.

N (neuraminidase) spikes Projections from the outer lipid bilayer of influenza virus.

nomenclature The system of naming things.

noncommunicable disease A disease that is not transmitted from one person to another.

nonimmune phagocytosis The process by which a phagocyte engulfs a microorganism after trapping it against a surface.

nonionizing radiation Radiation that does not cause ionization; for example, ultraviolet radiation.

nonsense codon A special terminator codon that does not code for any amino acid.

nonsense mutation A base substitution in DNA that results in a nonsense codon.

nonspecific resistance Host defenses that tend to afford protection from any kind of pathogen.

normal flora Microorganisms that colonize an animal without causing disease.

nosocomial infection An infection that develops during the course of a hospital stay and was not present at the time the patient was admitted.

notifiable disease A disease that physicians must report to the public health service.

nuclear envelope The double membrane that separates the nucleus from the cytoplasm in a eucaryotic cell.

nucleic acid A macromolecule consisting of nucleotides; for example, RNA and DNA.

nucleic acid hybridization The process of combining single complementary strands of DNA.

nucleoid The region in a bacterial cell containing the chromosome.

nucleoli Areas in a eucaryotic nucleus where rRNA is synthesized.

nucleoplasm The gellike fluid within the nuclear envelope.

nucleoprotein A macromolecule consisting of protein and nucleic acid.

nucleotide A compound consisting of a purine or pyrimidine base, a five-carbon sugar, and a phosphate.

nucleus The part of a eucaryotic cell that contains the genetic material; also, the part of an atom consisting of the protons and neutrons.

numerical taxonomy A method of comparing organisms on the basis of many characteristics.

nutrient broth (agar) A complex medium made of meat extracts that may contain agar.

nutritional deficiency disease A disease caused by lack of nutrients.

O antigen A cell antigen of enterics.

objective lens In a compound light microscope, the lens closest to the specimen.

obligate anaerobe An organism that is unable to use oxygen.

ocular lens In a compound light microscope, the lens closest to the viewer.

oligodynamic action The ability of small amounts of a heavy metal compound to exert antimicrobial activity.

oncogene A gene that can bring about malignant transformation.

oncogenic virus A virus that is capable of producing tumors.

operator The region of DNA adjacent to structural genes that controls their transcription.

operon The operator site and structural genes it controls.

opportunistic pathogen An organism that does not ordinarily cause a disease but can become pathogenic under certain circumstances.

opsonic index The ratio of the phagocytic index between immune serum and normal serum.

opsonization The enhancement of phagocytosis by coating microorganisms with certain serum proteins (opsonins).

optical density A measure of the amount of light that is absorbed by (or does not pass through) a culture.

optimum growth temperature The temperature at which a species grows best.

oral groove On some protozoans, the site at which nutrients are taken in.

order A taxonomic classification between class and family.

organelles Membrane-bounded structures within eucaryotic cells.

organic compounds Molecules that contain carbon.

osmosis The net movement of solvent molecules across a selectively permeable membrane from an area of higher concentration to an area of lower concentration.

osmotic pressure The force with which a solvent moves from a solution of lower solute concentration to a solution of higher solute concentration.

Ouchterlony test An immunodiffusion test.

oxidation The removal of electrons from a molecule or the addition of oxygen to a molecule.

oxidation pond A method of secondary sewage treatment.

oxidation–reduction (redox) reaction A coupled reaction in which one substance is oxidized and one is reduced.

oxidative phosphorylation The synthesis of ATP coupled with electron transport.

pandemic disease An epidemic that occurs worldwide.

papular rash A skin rash characterized by raised spots.

para-aminosalicylic acid (PAS) A competitive inhibitor used to treat tuberculosis.

parasite An organism that derives nutrients from a living host.

parasitism A symbiosis in which one organism (the parasite) exploits another (the host) without providing any benefit in return.

parenteral route Deposition directly into tissues beneath the skin and mucous membranes.

passive agglutination test A procedure to detect antibodies that react with soluble antigens by first attaching the antigens to insoluble particles.

passively acquired immunity Immunity acquired when antibodies produced by another source are transferred to the individual who needs them.

pasteurization The process of mild heating to kill particular spoilage organisms or pathogens.

pathogenic Disease-causing.

pathology The study of disease.

Paul–Bunnell test A hemagglutination test used to diagnose infectious mononucleosis.

pellicle The flexible covering of some protozoans.

penicillins A group of antibiotics produced either by *Penicillium* (natural penicillins) or by adding side chains to the beta-lactam ring (semisynthetic penicillins).

pentaglycine bridge A five-glycine chain component of certain bacterial cell walls.

pentose phosphate pathway A metabolic pathway that can occur simultaneously with glycolysis to produce pentoses and $NADH_2$ without ATP production.

peptide A chain of two (di-), three (tri-) or more (poly-) amino acids.

peptide bond A bond joining the amino group of one amino acid to the carboxyl group of a second amino acid with the loss of a water molecule.

peptidoglycan The structural molecule of bacterial cell walls consisting of the molecules N-acetylglucosamine, N-acetylmuramic acid, pentaglycine, and peptide side chains.

peptone Short chains of amino acids produced by the action of acids or enzymes on proteins.

peritrichous Having flagella distributed over the entire cell.

permease A carrier protein in the plasma membrane.

petrochemical An organic compound derived from petroleum.

pH The symbol for hydrogen ion concentration; a measure of the relative acidity of a solution.

phage *See* bacteriophage.

phage typing A method of identifying bacteria using specific strains of bacteriophages.

phagocyte A cell capable of engulfing and digesting particles that are harmful to the body.

phagocytic index The average number of bacteria ingested per leukocyte.

phagocytosis The ingestion of solids by cells.

phagolysosome A digestive vacuole.

phase-contrast microscope A compound light microscope that allows examination of structures inside cells through the use of a special condenser.

phenol C_6H_5OH; carbolic acid.

phenol coefficient A standard of comparison for the effectiveness of disinfectants; the disinfecting action of a chemical is compared to phenol for the same length of time on the same organism under identical conditions.

phenotype The external manifestations of the genetic makeup of an organism.

phosphate group A portion of a phosphoric acid molecule attached to some other molecule;

```
        O
        ‖
  —O—P—O—
        |
        OH
```

phospholipid A complex lipid composed of glycerol, two fatty acids, and a phosphate group.

phosphorylation The addition of a phosphate group to an organic molecule.

photoautotroph An organism that uses light for its energy source and carbon dioxide (CO_2) as its carbon source.

photoheterotroph An organism that uses light for its energy source and an organic carbon source.

photophosphorylation The production of ATP by photosynthesis.

photoreactivation The process of repairing ultraviolet damage in the presence of visible light.

photosynthesis The light-driven synthesis of carbohydrate from carbon dioxide (CO_2).

phototroph An organism that uses light as its primary energy source.

phylogeny The evolutionary history of a group of organisms.

phylum A taxonomic classification between kingdom and class.

pilus An appendage on a bacterial cell for attachment and conjugation.

pinocytosis The engulfing of liquid by a cell.

plaque A clearing in a confluent growth of bacteria due to lysis by phages.

plaque-forming units Visible plaques counted, perhaps due to more than one phage.

plasma The liquid portion of blood in which the formed elements are suspended.

plasma cells Cells produced by B lymphocytes and which manufacture specific antibodies.

plasma membrane The selectively permeable membrane enclosing the cytoplasm of a cell; the outer layer in animal cells, internal to the cell wall in other organisms.

plasmid A small cyclic DNA molecule in bacteria in addition to the chromosome.

plate count A method of determining the number of bacteria in a sample by counting the number of colony-forming units on a solid culture medium.

platelet-activating factor An anaphylaxis mediator that causes the lysis of platelets.

pleomorphic Having many shapes.

pneumonia Inflammation of the lungs.

polar molecule A molecule with an unequal distribution of charges.

poly-β-hydroxybutyric acid A lipid storage material unique to bacteria.

polyene An antimicrobial agent that alters sterols in eucaryotic plasma membranes and contains more than four carbon atoms and at least two double bonds.

polyhedron A many-sided solid.

polykaryocyte A multinucleated giant cell.

polymer A molecule consisting of a sequence of similar units or monomers.

polymerase An enzyme that synthesizes specific polymers.

polymorphonuclear (PMN) leukocyte A neutrophil.

polymyxins A group of antibiotics that causes disintegration of phospholipids.

polyribosome An mRNA strand with several ribosomes attached to it.

portal of entry The avenue by which a pathogen gains access to the body.

portal of exit The route by which a pathogen leaves the body.

positive (direct) selection A procedure for picking out mutant cells.

potential energy Energy that is stored.

pour plate method A method of inoculating a solid nutrient medium by mixing bacteria in the melted medium and pouring the medium into a Petri plate to solidify.

precipitation reaction A reaction between multivalent soluble antigens and multivalent antibodies to form aggregates.

precipitin An antibody that causes a precipitation reaction.

predisposing factor Anything that makes the body more susceptible to a disease or alters the course of a disease.

presumptive test The first step in the detection of coliforms; used to determine MPN values.

prevalence The fraction of a population having a specific disease at a given time.

primary cell line Human tissue cells that grow only a few generations in vitro.

primary infection An acute infection that causes the initial illness.

primary response The antibody production to the first contact with an antigen.

primary treatment Physical removal of solid matter from wastewater.

procaryote A cell whose genetic material is not enclosed in a nuclear membrane.

prodromal period The time following incubation when the first symptoms of illness appear.

product The substance formed in a chemical reaction.

profundal zone The deeper water under the limnetic zone in an inland body of water.

proglottid A body segment of a tapeworm containing male and female organs.

promoter site The starting point on DNA for transcription of RNA by RNA polymerase.

properdin system Three serum proteins that function with complement to kill bacteria.

prophage Phage DNA inserted into the host cell's DNA.

prostaglandins Hormonelike substances that are synthesized in many tissues and circulate in the blood, and the exact function of which is unknown.

prosthetic group A coenzyme that is bound tightly to its apoenzyme.

protease An exoenzyme that breaks down proteins into their component amino acids.

protein A sequence of amino acids that forms a helix and then folds into a globular structure.

proteolytic An enzyme capable of hydrolyzing a protein.

Protista The kingdom to which protozoans belong; unicellular, eucaryotic organisms.

proton A positively charged particle in the nucleus of an atom.

protoplast A gram-positive bacterium without a cell wall.

protovirus theory The concept that viral genes become integrated in host DNA, creating new gene sequences that may result in malignancies.

protozoa A unicellular eucaryotic organism belonging to the kingdom Protista.

provirus Viral DNA that is integrated into the host cell's DNA.

provirus theory The concept that RNA tumor viruses are transmitted from host to host.

pseudohypha A short chain of cells that results from the lack of separation of daughter cells after budding.

pseudopods Extensions of a cell that aid in locomotion and feeding.

psychogenic disease A condition in which emotional factors contribute to disease.

psychrophile An organism that grows best at 15°C and does not grow above 20°C.

psychrotroph An organism that is capable of growth at 4°C and above 20°C.

pure culture A culture with only one kind of microorganism.

purines The class of nucleic acid bases that includes adenine and guanine.

pus An accumulation of dead phagocytes, dead bacterial cells, and fluid.

pustules Small elevations containing pus.

pyelonephritis Inflammation of the nephrons and renal pelvis due to bacterial infection.

pyocin A bacteriocin produced by *Pseudomonas pyocyaneus*.

pyrimidines The class of nucleic acid bases that includes uracil, thymine, and cytosine.

pyrogen A substance that causes a rise in body temperature.

quaternary ammonium compound A cationic detergent with four organic groups attached to a central nitrogen atom, used as a disinfectant.

quellung reaction Swelling of a bacterial capsule in the presence of a specific antibody.

quinine An antimalarial drug derived from the cinchona tree and effective against sporozoites in red blood cells.

quinones Low molecular weight, nonprotein carriers in the electron transport chain.

radiant energy Energy traveling through space; radiation.

radioimmunoassay A method of measuring the amount of a compound using radioactive antigens.

reactants Substances that are combined in a chemical reaction.

reactivity The ability of an antigen to combine with an antibody.

reagin A substance made in response to a treponemal infection characterized by its ability to combine with lipids; reagin–lipid complex will fix complement.

recalcitrance Being resistant to degradation.

recipient cell A cell that receives DNA from a donor cell in recombination.

recombinant DNA A DNA molecule produced by recombination.

recombinant DNA techniques *See* genetic engineering.

recombination The process of joining pieces of DNA from different sources.

redia A trematode larval stage which may reproduce asexually one or two times before developing into a cercaria.

reducing medium A culture medium containing ingredients that will remove dissolved oxygen from the medium to allow the growth of anaerobes.

reduction The addition of electrons to a molecule; the gain of hydrogen atoms.

refractive index The relative velocity with which light passes through a substance.

regulator gene The gene that codes for a repressor protein.

renin A proteolytic enzyme obtained from a calf's stomach.

replica plating A method of inoculating a number of plates of solid culture media from an original plate of medium to obtain patterns of colonies exactly like the original plate.

replication fork The point where DNA separates and new strands of DNA will be synthesized.

replicative form A double-stranded RNA molecule produced during the multiplication of certain RNA viruses.

repressor A protein that binds to the operator site to prevent transcription.

reservoir of infection A continual source of infection.

resistance The ability to ward off diseases through nonspecific and specific defenses.

resistance (R) factor A group of bacterial plasmids carrying genes that determine resistance to antibiotics.

resistance transfer factor (RTF) A group of genes for replication and conjugation on the R factor.

resolution The ability to distinguish fine detail with a magnifying instrument.

respiration An ATP-generation process in which chemical compounds are oxidized and the final electron acceptor is usually an inorganic molecule; also, the process by which living organisms produce carbon dioxide (CO_2).

restriction enzyme An enzyme that cuts DNA.

reticuloendothelial system (RES) A system of fixed macrophages located in the spleen, liver, lymph nodes, and bone marrow.

reverse transcriptase RNA-dependent DNA polymerase; an enzyme that synthesizes a complementary DNA from an RNA template.

reversible reaction A chemical reaction in which the end products can readily revert to the original molecules.

R genes Genes that code for enzymes that inactivate certain drugs carried on the R factor.

rhizosphere The region in soil where the soil and roots make contact.

ribonucleic acid (RNA) The class of nucleic acids that comprises messenger RNA, ribosomal RNA, and transfer RNA.

ribose A five-carbon sugar that is part of ribonucleotide molecules and RNA.

ribosomal RNA The RNA molecules that form the ribosomes.

ribosomes The site of protein synthesis in a cell, composed of RNA and protein.

rifamycin An antibiotic that inhibits bacterial RNA synthesis.

ring precipitin test A precipitation test performed in a capillary tube.

RNA-dependent RNA polymerase An enzyme that synthesizes a complementary RNA from an RNA template.

root nodule A tumor like growth on the roots of certain plants containing a symbiotic nitrogen-fixing bacterium.

roundworms Animals belonging to the phylum Aschelminthes.

Sabin vaccine A preparation containing three attenuated strains of polio virus administered orally.

Salk vaccine A preparation of a formalin-inactivated polio virus that is injected.

salt A substance that dissolves in water to cations and anions, neither of which is H^+ or OH^-.

saphrophyte An organism that obtains its nutrients from dead organic matter.

sarcina A group of eight bacteria that remain in a packet after dividing.

sarcoma A cancer of fleshy, nonepithelial tissue or connective tissue.

satellite phenomenon The growth of *Hemophilus* around other bacterial colonies which supply X and V factors on a solid culture medium.

saturation The condition in which the active site on an enzyme is occupied by the substrate or product at all times.

scanning electron microscope An electron microscope that provides three-dimensional views of the specimen magnified about 10,000 times.

Schaeffer–Fulton stain An endospore stain that uses malachite green to stain the endospores and safranin as a counterstain.

Schick test A skin test to detect the presence of antibodies to diphtheria.

schizogony The process of multiple fission, where one organism divides to produce many daughter cells.

sclerotia The reddish hardened ovaries of a grain that are filled with mycelia of the fungus *Claviceps purpurea.*

scolex The head of a tapeworm, containing suckers and possibly hooks.

secondary infection An infection caused by an opportunistic pathogen after a primary infection has weakened the host's defenses.

secondary treatment Biological degradation of the organic matter in wastewater, following primary treatment.

secretion The production and release of fluid containing a variety of substances from a cell.

secretory granule A vesicle containing protein produced from the Golgi complex.

selective medium A culture medium designed to suppress the growth of unwanted bacteria and encourage the growth of desired microorganisms.

selective permeability The property of a plasma membrane to allow certain molecules and ions to move through the membrane while restricting others.

selective toxicity The property of some antimicrobial agents to be toxic for a microorganism and nontoxic for the host.

semiconservative replication The process of DNA replication in which each double-stranded molecule of DNA contains one original strand and one new strand.

sense codon A codon that codes for an amino acid.

sepsis The presence of unwanted bacteria.

septate hypha A hypha consisting of uninucleate cell-like units.

septicemia A condition characterized by the multiplication of bacteria in the blood.

septic tank A tank, built into the ground, in which wastewater is treated by primary treatment.

septum A crosswall dividing two parts.

serial dilution The process of diluting a sample several times.

serology The branch of immunology concerned with the study of antigen–antibody reactions in vitro.

serotonin A mediator of anaphylaxis that causes smooth muscle constriction and vasoconstriction.

serum The liquid remaining after blood plasma is clotted, and which contains immunoglobulins.

serum sickness A hypersensitivity reaction due to the formation of immune complexes following exposure to an antigen.

sewage Domestic wastewater.

sex pilus A pilus used for the transfer of genetic material during bacterial conjugation.

sexual dimorphism The distinctly different appearance of adult male and female organisms.

sexual reproduction Reproduction that requires two opposite mating strains, usually designated male and female.

signs Changes due to a disease that a physician can observe and measure.

simple stain A method of staining microorganisms with a single basic dye.

single-cell protein (SCP) A food substitute consisting of microbial cells.

skin test The intradermal injection of an antigen or antibody to determine susceptibility to an antigen.

slide agglutination test A method of identifying an antigen by combining it with a specific antibody in a slide.

slime layer *See* capsule.

slow-reacting substance of anaphylaxis (SRS-A) A chemical released by target cells after being bound by IgE antibodies.

slow virus infection A disease process that occurs gradually over a long period of time, caused by a virus.

sludge Solid matter obtained from sewage.

smear A thin film of material on a slide.

soap A surface-active agent made from animal fats and lye (NaOH).

soil A mixture of solid inorganic matter, water, air, organic matter, and living organisms.

solubility The ability to be dissolved, usually in water.

solute A substance dissolved in another substance.

somatic Relating to the cell itself.

species The most specific level in the taxonomic hierarchy; the second name in a scientific binomial.

species immunity The resistance of an animal species to diseases of another species.

specific resistance *See* immunity.

spherical particle A particle in the serum of a patient with hepatitis B having hepatitis B virus (HBV) surface antigens.

spheroplast A gram-negative bacterium lacking a complete cell wall.

spike A carbohydrate–protein complex that projects from the surface of certain enveloped viruses.

spirillum A spiral or corkscrew-shaped bacterium.

spirochete A corkscrew-shaped bacterium with an axial filament.

spontaneous generation The idea that life could arise spontaneously from nonliving matter.

spontaneous mutation A mutation that occurs without a mutagen.

sporadic disease A disease that occurs occasionally in a population.

sporangiospore An asexual fungal spore formed within a sporangium.

sporangium A sac containing one or more spores.

spore A reproduction structure formed by fungi and actinomycetes.

sporogenesis The process of spore and endospore formation.

sporozoite A trophozoite of *Plasmodium* found in mosquitoes, infective for humans.

stability The condition of not deteriorating with time.

staphylococcus Broad sheet of cells.

stationary phase The period in a bacterial growth curve when the number of cells dividing equals the number dying.

stereoisomers Two molecules consisting of the same atoms, arranged in the same manner but differing in their relative positions; mirror images.

sterile Free of microorganisms.

steroids A specific group of chemical substances, including cholesterol and hormones.

sterol A lipid–alcohol found in the plasma membranes of fungi and *Mycoplasma*.

strain A group of cells all derived from a single cell.

streak plate method A method of inoculating a single culture medium by spreading microorganisms over the surface of the medium.

streptobacilli Rods that remain attached in chains after cell division.

streptococci Cocci that remain attached in chains after cell division.

structural gene A gene that codes for an enzyme.

sty An infection of an eyelash follicle.

subacute disease A disease with symptoms between acute and chronic.

subclinical infection An infection that does not cause a noticeable illness.

subcutaneous mycosis A fungal infection of tissue beneath the skin.

substrate Any compound with which an enzyme reacts.

substrate-level phosphorylation The synthesis of ATP by direct transfer of a high-energy phosphate group from an intermediate metabolic compound to ADP.

sulfa drugs Any synthetic chemotherapeutic agent containing sulfur and nitrogen; *see* sulfonamides.

sulfonamides Bacteriostatic compounds that interfere with folic acid synthesis by competitive inhibition.

superficial mycosis A fungal infection localized in surface epidermal cells and along hair shafts.

suppressor T cells Cells that inhibit an immune response.

surface-active agent (surfactant) Any compound that decreases the tension between molecules lying on the surface of a liquid.

surface tension The property of a liquid that makes its surface contract to the minimal area.

susceptibility The lack of resistance to a disease.

sylvatic Belonging in the woods.

symbiosis The living together of two different organisms.

symptom A change in body function that is felt by the patient due to a disease.

syndrome A specific group of signs or symptoms accompanying a particular disease.

synergistic effect The principle whereby the effectiveness of two drugs used simultaneously is greater than either drug used alone.

synthesis reaction A chemical reaction in which two or more atoms combine to form a new, larger molecule.

synthetic chemotherapeutic agent An antimicrobial agent that is prepared in a laboratory.

synthetic drug A chemotherapeutic agent that is prepared from chemicals in a laboratory.

systemic (generalized) infection An infection throughout the body.

systemic mycosis A fungal infection in deep tissues.

systemic reaction An anaphylaxis-type reaction producing shock and asphyxia.

T antigen An antigen in the nucleus of a tumor cell.

tapeworm A flatworm belonging to the class Cestoda.

target cell A basophil or mast cell to which IgE antibodies bind.

taxon A taxonomic category.

taxonomy The science of classification.

T cells Stem cells processed in the thymus gland responsible for cellular immunity.

teichoic acid A polysaccharide found in gram-positive cell walls.

temperature phage A bacteriophage existing in lysogeny with a host cell.

terminator site The site on DNA at which transcription ends.

tertiary treatment Physical and chemical treatment of wastewater to remove all BOD, nitrogen, and phosphorus, following secondary treatment.

tetracyclines Broad-spectrum antibiotics that interfere with protein synthesis.

tetrad A group of four cocci.

tetrahedron A four-sided solid structure.

T-even bacteriophage A complex virus with double-stranded DNA that infects bacteria; for example, T2, T4, T6.

thallus The entire vegetative structure or body of a fungus or lichen.

thermal death point (TDP) The temperature required to kill all the bacteria in a liquid culture in 10 minutes at pH 7.

thermal death time (TDT) The length of time required to kill all bacteria in a liquid culture at a given temperature.

thermophile An organism whose optimum growth temperature is between 50 and 60°C.

thermophilic anaerobic spoilage Spoilage of canned foods due to the growth of thermophilic bacteria.

thymine A pyrimidine nucleic acid base in DNA that pairs with adenine.

tincture An alcoholic or aqueous solution.

tinea A cutaneous fungal infection; ringworm.

tissue culture *See* cell culture.

total magnification Magnification of a specimen determined by multiplying the ocular lens magnification by the objective lens magnification.

toxemia Symptoms due to toxins in the blood.

toxigenicity The capacity of a microorganism to produce a toxin.

toxin Any poisonous substance produced by a microorganism.

toxoid An inactivated toxin.

trace element A chemical element required in small amounts for growth.

transamination The transfer of an amino group from an amino acid to an organic acid.

transcription The process of synthesizing RNA from a DNA template.

transduction The process of transferring a piece of DNA from one cell to another by bacteriophage.

transfer factor A protein released by killer T cells that sensitizes other lymphocytes.

transfer RNA The class of molecules that brings the amino acids to the site where they are incorporated into proteins.

transformation The process in which genes are transferred from one bacterium to another as "naked" DNA in solution; also, the changing of a normal cell into a cancerous cell.

transient flora Microorganisms that are present on an animal for a short time without causing a disease.

translation The use of RNA as a template in the synthesis of protein.

transmission electron microscope An electron microscope that provides high magnifications of thin sections of a specimen.

transplantation The replacement of injured or diseased organs and tissues with natural ones.

transverse fission *See* binary fission.

transverse septum A crosswall that separates genetic material into two daughter cells in binary fission.

trickling filter A method of secondary sewage treatment.

trophozoite The vegetative form of a protozoan.

tube agglutination test A method for determining antibody titer using serial dilutions of serum mixed with antigen.

tuberculin test A skin test used to detect the presence of antibodies to *Mycobacterium tuberculosis.*

tumor Excessive tissue due to uncontrolled cell growth.

tumor-specific transplantation antigen (TSTA) A viral antigen on the surface of a transformed cell.

turbidity The cloudiness of a suspension.

turnover number The number of substrate molecules metabolized per enzyme molecule per second.

12D treatment A sterilization process that results in a decrease of the bacterial population by 12 logarithmic cycles.

UDP-N-acetyl glucosamine (UDPNAG) A compound necessary for the biosynthesis of peptidoglycan.

ultrastructure Fine detail not seen with a compound light microscope.

ultraviolet (UV) radiation Radiation from 10 to 390 nm.

uncoating The separation of viral nucleic acid from its protein coat.

undulating membrane A highly modified flagellum on some protozoans.

universal donor An individual with blood type O, which lacks isoantigens.

universal recipient An individual with blood type AB, which lacks isoantibodies.

uracil A pyrimidine nucleic acid base in RNA that pairs with adenine.

vaccination The process of conferring immunity using a vaccine.

vaccine A preparation of killed, inactivated, or attenuated microorganisms or toxoids to induce artificially acquired active immunity.

vacuole An intracellular inclusion, in eucaryotic cells, surrounded by a plasma membrane containing raw food; in procaryotic cells, surrounded by proteinaceous membrane containing gas.

valence The combining capacity of an atom or molecule.

vasodilation Dilation or enlargement of blood vessels.

VDRL test A rapid screening test to detect the presence of antibodies against *Treponema pallidum.*

vector An arthropod that carries disease-causing organisms from one host to another.

vegetative cells Cells involved with obtaining nutrients, as opposed to reproduction or resting.

venereal disease A sexually transmitted disease.

vesicle The expanded, terminal area of the Golgi complex; also, a fluid-filled blister.

V factor NAD or NADP.

vibrio A curved or comma-shaped bacterium.

viral hemagglutination The ability of certain viruses to cause agglutination of red blood cells.

viremia The presence of viruses in the blood.

virion A fully developed complete viral particle.

viroid An infectious piece of "naked" RNA.

viropexis The process by which an animal cell takes in a complete virus; phagocytosis.

virulence The degree of pathogenicity of a microorganism.

virus A submicroscopic, parasitic filterable agent consisting of a nucleic acid surrounded by a protein coat.

visible light Radiation from 400 to 700 nm, which the human eye can see.

Vollmer patch test A method of determining the allergen causing a skin hypersensitivity reaction.

volutin Stored phosphate in a procaryotic cell.

wandering macrophage A macrophage that leaves the blood and migrates to infected tissue.

Wassermann test A complement fixation test used to diagnose syphilis.

Weil–Felix test A serological test used to diagnose certain rickettsial diseases.

wheal An area of edema of the skin resulting from a skin test.

whey The fluid portion of milk that separates from the curd.

Widal test An agglutination test used to detect typhoid fever.

wide-spectrum antimicrobial agent *See* broad-spectrum antimicrobial agent.

wound A physical injury to the body involving a rupture of the skin.

X factor A precursor necessary to synthesize cytochromes.

yeast A unicellular fungus belonging to the class Ascomycetes.

Ziehl–Neelsen stain An acid-fast staining method.

zone of inhibition The area of no bacterial growth around an antimicrobial agent in the agar diffusion test.

zoonosis A disease that occurs primarily in wild and domestic animals but can be transmitted to humans.

zygospore A sexual fungal spore characteristic of the Zygomycetes.

zygote A fertilized ovum produced by the fusion of two gametes.

zymogen An enzyme storage inclusion in eucaryotic cells.

Acknowledgments

Note: CDC = Centers for Disease Control, Public Health Service, U.S. Department of Health and Human Services, Atlanta, Georgia.
BPS = Biological Photo Service.

Chapter 1 Figures: 1–1: CDC. 1–2: BPS. 1–6: A. Fleming. *British Journal of Experimental Pathology* 10:226–236 (1929). 1–7a: Z. Skobe, Forsyth Dental Center/BPS. 1–7b: BPS. 1–7c: H. S. Wessenberg and G. A. Antipa. *J. Protozool.* 17:250–270 (1970). 1–7d: J. Robert Waaland, University of Washington/BPS. 1–7e: Bernard Roizman, University of Chicago/BPS. 1–10: Z. Skobe, Forsyth Dental Center/BPS.

Chapter 2 Figures: 2-6 and 2–13: R. E. Dickerson and I. Geis, *Chemistry, Matter, and the Universe.* Menlo Park, CA: Benjamin/Cummings Publishing Co., 1976. Copyright © 1976 R. E. Dickerson and I. Geis.

Chapter 3 Figures: 3–1a: Unitron Instruments, Inc., Plainview, NY. 3–2: J. Robert Waaland, University of Washington/BPS. 3–3b: CDC. 3–4a: Richard Rodewald, University of Virginia/BPS. 3–4b: E. Golub, Purdue University/BPS. 3–8a: John Mayhew, Milwaukee, Wisconsin. 3–8b and c: CDC.

Chapter 4 Figures: 4–1a and b: Z. Skobe, Forsyth Dental Center/BPS. 4–1c: J. B. Baseman, University of North Carolina School of Medicine. Appeared in *Infect. Immun.* 17:174–186 (1977). 4-2a: CDC. 4–2b: T. J. Beveridge, University of Guelph/BPS. 4–2c and d: Leon J. Le Beau, University of Illinois Hospital at the Medical Center, Chicago/BPS. 4–2e: CDC. 4–2f: G. T. Cole, University of Texas at Austin/BPS. 4–2g: T. J. Beveridge, University of Guelph/BPS. 4–2h: Stanley C. Holt, University of Massachusetts/BPS. 4–4 and 4–5c upper left: CDC. 4–5c lower left: BPS. 4–5c middle: T. J. Beveridge, University of Guelph/BPS. 4–5c right: L. E. Simon, Rutgers University, New Brunswick, NJ. 4–6: CDC. 4–7: Stanley C. Holt, University of Massachusetts/BPS. 4–10: M. E. Bayer and T. W. Starkey. *Virology* 49:236–256 (1972). 4–11c: D. Branton, Harvard University. 4–12: G. Cohen-Bazire, Pasteur Institute. 4–13: T. J. Beveridge, University of Guelph/BPS. 4–17: L. E. Simon, Rutgers University, New Brunswick, NJ. 4–18b: Stanley C. Holt, University of Massachusetts. 4–18c left: S. Schroeder, MIT. 4–18c middle and right: CDC. 4–20a: E. B. Small, University of Maryland, courtesy of G. A. Antipa, San Francisco State University. 4–20b: H. S. Wessenberg and G. A. Antipa. *J. Protozool.* 17:250–270 (1970). 4–21b: C. L. Sanders, Batelle-Pacific Northwest Laboratories/BPS. 4–22a: D. Branton, Harvard University. 4–23a and 4–24a: G. E. Palade, Yale University Medical School. 4–25a: Keith R. Porter, University of Colorado. 4–25b: C. K. Levy, *Elements of Biology.* Reading, MA: Addison-Wesley Publishing Co., 1978. 4–26: G. E. Palade, Yale University Medical School. 4–27: M. C. Ledbetter and K. R. Porter, *Introduction to the Fine Structure of Plant Cells.* New York: Springer-Verlag, 1970.

Chapter 6 Figures: 6–5: Leon J. Le Beau, University of Illinois Hospital at the Medical Center, Chicago/BPS. 6–6: Richard Humbert, Stanford University/BPS. 6–7a, b, and d: CDC. 6–7c: R. J. Hawley, Georgetown University School of Dentistry. 6–9: CDC. 6–10b: L. E. Simon, Rutgers University, New Brunswick, NJ.

Chapter 7 Figures: 7–2b, 7–4, and 7–7: Richard Humbert, Stanford University/BPS. 7–9: P. B. Price, "Skin Antisepsis," in J. H. Brewer (ed.), *Lectures on Sterilization.* Durham, NC: Duke University, 1957.

Chapter 8 Figures: 8–2a: Jack D. Griffith, University of North Carolina, Chapel Hill. 8–2b: G. F. Bahr, M.D., Armed Forces Institute of Pathology. 8–4a: J. Cairns, Imperial Cancer Research Fund Laboratory, Mill Hill, London. 8–5b: O. L. Miller, Jr. and B. R. Beatty, Oak Ridge National Laboratory. 8–23a: Rod Welch, Stanford University School of Medicine/BPS.

Chapter 9 Figures: 9–2: J. William Schopf, University of California, Los Angeles/BPS. 9–4: Leon J. Le Beau, University of Illinois Hospital at the Medical Center, Chicago/BPS. 9–5: B. D. Davis, et al., *Microbiology*, 3rd ed. Hagerstown, MD: Harper & Row, 1980. 9–6: After P. H. A. Sneath, "The Construction of Taxonomic Groups," in *Microbial Classification*, G. C. Ainsworth and P. H. A. Sneath (eds.). New York: Cambridge University Press, 1962.

Chapter 10 Figures: 10–1a: J. C. Burnham, Medical College of Ohio. 10–1b and c: J. Robert Waaland, University of Washington/BPS. 10–2: CDC. 10–3: T. J. Beveridge, University of Guelph/BPS. 10–4: CDC. 10–5a and b: Stanley C. Holt, University of Massachusetts/BPS. 10–5c: CDC. 10–6: G. T. Cole, University of Texas at Austin/BPS. 10–7: Leon J. Le Beau, University of Illinois Hospital at the Medical Center, Chicago/BPS. 10–8 and 10–9: CDC. 10–10: Z. Skobe, Forsyth Dental Center/BPS. 10–11a: Leon J. Le Beau, University of Illinois at the Medical Center, Chicago/BPS. 10–11b and 10–12a and b: CDC. 10–12c: Leon J. Le Beau, University of Illinois Hospital at the Medical Center, Chicago/BPS. 10–13: T. J. Beveridge, University of Guelph/BPS. 10–14 a, b, d, and e: CDC. 10–14c: R. J. Hawley, Georgetown University School of Dentistry. 10–15a: Willy Burgdorfer, Rocky Mountain Laboratory, Hamilton, MT. 10–15b: Randall C. Cutlip, National Animal Disease Center, Ames, IA. 10–16: Michael G. Gabridge, University of Illinois at Urbana–Champaign. 10–17a: Stanley C. Holt, University of Massachusetts/BPS. 10–17b: Karen Stephens, Indiana University; *FEMS Microbiology Letters* 9:189–192. 10–17c and d: T. J. Beveridge, University of Guelph/BPS. Box photos: R. Malcolm Brown, Jr., University of North Carolina, Chapel Hill.

Chapter 11 Figures: 11–2: CDC. 11–3: C. Robinow, University of Western Ontario/BPS. 11–4b: CDC. 11–7: C. Emmons, Phoenix, AZ. 11–8: CDC. 11–9a: Richard Humbert, Stanford University/BPS. 11–9b and c: J. Robert Waaland, University of Washington/BPS. 11–10: Armed Forces Institute of Pathology Photograph, Neg. No. 75-7779-2. 11–14 and 11–27: CDC.

Chapter 12 Figures: 12–3b: Jack D. Griffith, University of North Carolina, Chapel Hill. 12–4b: Alyne K. Harrison, Viral Pathology Branch, CDC. 12–5b: Frederick A. Murphy, Viral Pathology Branch, CDC. 12–6b: Alyne K. Harrison, Viral Pathology Branch, CDC. 12–7a: Frederick A. Murphy, Viral Pathology Branch, CDC. 12–7c: Robley C. Williams, University of California, Berkeley, and Harold W. Fisher, University of Rhode Island. 12–8a: Bernard Roizman, University of Chicago. 12–8b: C. Garon and J. Rose, National Institute of Allergy and Infectious Diseases. 12–9a and b: Frederick A. Murphy, Viral Pathology Branch, CDC. 12–9c: G. H. Smith, National Cancer Institute. 12–10: Richard Humbert, Stanford University/BPS. 12–12: Gail Wertz, School of Medicine, University of North Carolina, Chapel Hill. 12–15: C. Morgan, H. M. Rose, and B. Mednis. *Journal of Virology* 2:507–516 (1968). 12–19b: D. T. Brown, et al. *Journal of Virology* 10:524–536 (1972). 12–20: T. O. Diener, U. S. Dept. of Agriculture; T. Koller and J. S. Sogo, Swiss Federal Institute of Technology, Zurich. 12–21a: Leon J. Le Beau, University of Illinois Hospital at the Medical Center, Chicago/BPS. 12–21b: S. Dales and S. L. Wilton, University of Western Ontario.

Chapter 13 Figures: 13–1: Dwayne C. Savage, University of Illinois at Urbana–Champaign. 13–3: M. W. Jennison and the Dept. of Biology, Syracuse University.

Chapter 14 Figures: 14–1: Dwayne C. Savage, University of Illinois at Urbana–Champaign. 14–2: Stanley C. Holt, University of Massachusetts/BPS. 14–3: Frederick A. Murphy, Viral Pathology Branch, CDC. 14–4: John P. Bader, National Cancer Institute, National Institute of Health, Bethesda, MD.

Chapter 15 Figures: 15–1: L. Winograd, Stanford University School of Medicine/BPS. 15–3a: K. E. Muse, Duke University Medical Center/BPS. 15–4: Richard Rodewald, University of Virginia/BPS. 15–6f: J. G. Hadley, Battelle-Pacific Northwest Laboratories/BPS. 15–10a: Schreiber, et al. *J. Exp. Med.* 149:870–882 (1979).

Chapter 16 Figures: 16–10: Richard Rodewald, University of Virginia/BPS. 16–16 and 16–17: Leon J. Le Beau, University of Illinois Hospital at the Medical Center, Chicago/BPS. 16–18: BPS. 16–19: Richard Humbert, Stanford University/BPS. 16–23c: CDC.

Chapter 17 Figures: 17–1a: A. L. Sullivan, et al. *J. Exp. Med.* 134:1408 (1971). 17–2: Leon J. Le Beau, University of Illinois Hospital at the Medical Center, Chicago/BPS. 17–7: A. Calin, Stanford University School of Medicine. 17–10a and c: Andrejs Liepins, Sloan-Kettering Institute for Cancer Research.

Chapter 18 Figures: 18–1: A. Fleming. *British Journal of Experimental Pathology* 10:226–236 (1929). 18–2: M. E. Bayer, Fox Chase Cancer Center, Philadelphia. 18–17 and 18–18: Leon J. Le Beau, University of Illinois Hospital at the Medical Center, Chicago/BPS.

Chapter 19 Figures: 19–1b: L. Winograd, Stanford University School of Medicine/BPS. 19–2 bottom: CDC. 19–2 top, 19–3, and 19–4: Leon J. Le Beau, University of

Illinois Hospital at the Medical Center, Chicago/BPS. 19–5: P. Patrick Cleary, University of Minnesota School of Medicine/BPS. 19–6: Armed Forces Institute of Pathology Photograph, Neg. No. 58-6180-6. 19–7 and 19–8a: World Health Organization, Geneva, Switzerland. 19–8b: CDC. 19–9: Centers for Disease Control. Annual summary 1979: reported morbidity and mortality in the United States. *Morbidity and Mortality Weekly Report* 1980; 28(54). 19–10 and 19–11: CDC. 19–12: P. Weary, University of Virginia School of Medicine. 19–13 and 19–14a: CDC. 19–14b: Leon J. Le Beau, University of Illinois Hospital at the Medical Center, Chicago/BPS. 19–14c: Goodman, American Society for Microbiology Slide Collection.

Chapter 20 Figures: 20–1: Armed Forces Institute of Pathology Photograph, slide number 79-18280-2. 20–2: Armed Forces Institute of Pathology Photograph, Neg. No. N-98656. 20–3, 20–6, and 20–8: Centers for Disease Control. Annual summary 1979: reported morbidity and mortality in the United States. *Morbidity and Mortality Weekly Report* 1980; 28(54). 20–4: CDC. 20–5: Centers for Disease Control. *Health Information for International Travel.* 20–10: CDC.

Chapter 21 Figures: 21–4: F. L. A. Buckmire, The Medical College of Wisconsin. 21–5: D. W. Fraser, et al., *J. Infect. Dis.*, 127:271, 1973. 21–6: Goodman, American Society for Microbiology Slide Collection. 21–7: CDC. 21–8: Bruce A. Phillips, University of Pittsburgh Medical School. 21–10 and 21–14: Centers for Disease Control. Annual summary 1979: reported morbidity and mortality in the United States. *Morbidity and Mortality Weekly Report* 1980; 28(54). 21–11: Alyne K. Harrison, Viral Pathology Branch, CDC. 21–12: Frederick A. Murphy, Viral Pathology Branch, CDC.

Chapter 22 Figures: 22–2: Armed Forces Institute of Pathology Photograph, Neg. No. N-64686. 22–3a: U.S. Army Medical Research Institute of Infectious Diseases, Ft. Detrick, MD. 22–3b and 22–5: Centers for Disease Control. Annual summary 1979: reported morbidity and mortality in the United States. *Morbidity and Mortality Weekly Report* 1980; 28(54). 22–4a: J. J. Cardamone, Jr., University of Pittsburgh Medical School. 22–4b: Marilyn J. Tufte, University of Wisconsin—Platteville. 22–6: CDC. 22–7 and 22–8a: Jack D. Griffith, University of North Carolina, Chapel Hill. 22–9a: CDC. 22–10: Stephen G. Baum, Albert Einstein College of Medicine/BPS.

Chapter 23 Figures: 23–3: CDC. 23–4: Centers for Disease Control. Annual summary 1979: reported morbidity and mortality in the United States. *Morbidity and Mortality Weekly Report* 1980; 28(54). 23–5: K. E. Muse, Duke University Medical Center/BPS. 23–6: R. B. Morrison, M.D., Austin, TX. 23–7: Centers for Disease Control. Annual summary 1979: reported morbidity and mortality in the United States. *Morbidity and Mortality Weekly Report* 1980; 28(54). 23–8a: Leon J. Le Beau, University of Illinois Hospital at the Medical Center, Chicago/BPS. 23–8b: Glen L. Goodhart, V. A. Hospital, Philadelphia, PA. 23–9: CDC. 23–11a: Glenn D. Roberts, Mayo Clinic, Rochester, MN. 23–11b: J. E. Steadham, Texas Dept. of Health Resources, Austin. 23–13a: CDC. 23–13b: G. T. Cole, University of Texas at Austin/BPS.

Chapter 24 Figures: 24–2: M. A. Listgarten, University of Pennsylvania/BPS. 24–3b and c and 24–5: Z. Skobe, Forsyth Dental Center/BPS. 24–6: M. A. Listgarten, University of Pennsylvania/BPS. 24–7a: CDC. 24–7b: G. T. Cole, University of Texas at Austin/BPS. 24–8: Frederick A. Murphy, Viral Pathology Branch, CDC. 24–10a: Jack D. Griffith, University of North Carolina, Chapel Hill. 24–10b: CDC. 24–11: Albert Z. Kapikian, National Institute of Health, Bethesda, MD. 24–12a: CDC. 24–12b: Armstrong, American Society for Microbiology Slide Collection. 24–13: CDC.

Chapter 25 Figures: 25–4: David Bromley, West Virginia University Medical Center. 25–5 and 25–8: Centers for Disease Control. Annual summary 1979: reported morbidity and mortality in the United States. *Morbidity and Mortality Weekly Report* 1980; 28(54). 25–6, 25–7, 25–9, 25–10, and 25–11: CDC.

Chapter 26 Figures: 26–4a: R. Toja, Charles F. Kettering Laboratory. 26–4b: E. H. Newcomb and S. R. Tandon, University of Wisconsin/BPS. 26–5b: J. S. Whiteside and M. Alexander. *Weeds* 8:204 (1960). 26–8a: T. D. Brock and K. M. Brock, *Basic Microbiology with Applications,* 2nd ed. Englewood Cliffs, NJ: Prentice-Hall, 1978. 26–12a, 26–13a, and 26–14a: Carl W. May/BPS.

Chapter 27 Figures: 27–3: Carl W. May/BPS. 27–4: T. D. Brock and K. M. Brock, *Basic Microbiology with Applications,* 2nd ed. Englewood Cliffs, NJ: Prentice-Hall, 1978. 27–5: George deGennaro Studios, in D. M. Townsend, *Cheese Cookery.* Tucson, AZ: H.P. Books, 1980.

Color Plates: IA, IB, ID, and IIE: Leon J. Le Beau, University of Illinois Hospital at the Medical Center, Chicago/BPS. IIA and IIB: L. M. Pope and D. R. Grote, University of Texas, Austin/BPS. IIIA: Richard W. Emmons, Viral and Rickettsial Disease Laboratory, Dept of Health Services, State of California. All others: CDC.

Index

Note: Italic type refers to figures or tables.